ROUTLEDGE HANDBOOK OF CHINESE MEDICINE

The *Routledge Handbook of Chinese Medicine* is an extensive, interdisciplinary guide to the nature of traditional medicine and healing in the Chinese cultural region, and its plural epistemologies. Established experts and the next generation of scholars interpret the ways in which Chinese medicine has been understood and portrayed from the beginning of the empire (third century BCE) to the globalisation of Chinese products and practices in the present day, taking in subjects from ancient medical writings to therapeutic movement, to talismans for healing and traditional medicines that have inspired global solutions to contemporary epidemics. The volume is divided into seven parts:

- *Longue Durée* and Formation of Institutions and Traditions
- Sickness and Healing
- Food and Sex
- Spiritual and Orthodox Religious Practices
- The World of Sinographic Medicine
- Wider Diasporas
- Negotiating Modernity

This handbook therefore introduces the broad range of ideas and techniques that comprise pre-modern medicine in China, and the historiographical and ethnographic approaches that have illuminated them. It will prove a useful resource to students and scholars of Chinese studies, and the history of medicine and anthropology. It will also be of interest to practitioners, patients and specialists wishing to refresh their knowledge with the latest developments in the field.

Vivienne Lo 羅維前 is Professor of Chinese History at University College London. She has published widely on the ancient and medieval history of medicine in China and in diaspora. Her research interests include medical manuscripts, medical imagery and the history of nutrition.

Michael Stanley-Baker 徐源 is Assistant Professor in History at the School of Humanities, and of Medical Humanities at the Lee Kong Chian School of Medicine, at Nanyang Technological University, Singapore. An historian of Chinese medicine and religion, particularly Daoism, he works on the early imperial period as well as contemporary Sinophone communities. Currently completing a monograph on medicine and religion as related genres of practice in China, he also produces digital humanities tools and datasets to study the migration of medicine across spatio-temporal, intellectual and linguistic boundaries.

Dolly Yang 楊德秀 is a postdoctoral research associate at the Institute of History and Philology, Academia Sinica, Taiwan. She received a PhD in 2018 from University College London for her investigation into the institutionalisation of therapeutic exercise in Sui China (581–618 CE). She has a particular interest in examining the use of non-drug-based therapy in early medieval China, allied to a passion for translating and analysing ancient Chinese medical and self-cultivation texts.

ROUTLEDGE HANDBOOK OF CHINESE MEDICINE

*Edited by Vivienne Lo and Michael Stanley-Baker,
with Dolly Yang*

Routledge
Taylor & Francis Group
LONDON AND NEW YORK

First published 2022
by Routledge
2 Park Square, Milton Park, Abingdon, Oxon OX14 4RN

and by Routledge
605 Third Avenue, New York, NY 10158

Routledge is an imprint of the Taylor & Francis Group, an informa business

© 2022 selection and editorial matter, Vivienne Lo and Michael Stanley-Baker; individual chapters, the contributors

The right of Vivienne Lo and Michael Stanley-Baker to be identified as the authors of the editorial material, and of the authors for their individual chapters, has been asserted in accordance with sections 77 and 78 of the Copyright, Designs and Patents Act 1988.

The Open Access version of this book, available at www.taylorfrancis.com, has been made available under a Creative Commons Attribution-Non Commercial-No Derivatives 4.0 license.

Trademark notice: Product or corporate names may be trademarks or registered trademarks, and are used only for identification and explanation without intent to infringe.

British Library Cataloguing-in-Publication Data
A catalogue record for this book is available from the British Library

Library of Congress Cataloging-in-Publication Data
Names: Lo, Vivienne editor. | Stanley-Baker, Michael, 1971– editor. | Yang, Dolly, editor.
Title: Routledge handbook of Chinese medicine / edited by Vivienne Lo and Michael Stanley-Baker with Dolly Yang.
Identifiers: LCCN 2021030176 | ISBN 9780415830645 (hardback) | ISBN 9781032149776 (paperback) | ISBN 9781135008970 (adobe pdf) | ISBN 9781135008963 (epub) | ISBN 9781135008956 (mobi)
Subjects: LCSH: Medicine, Chinese.
Classification: LCC R601 .R68 2022 | DDC 610.951—dc23
LC record available at https://lccn.loc.gov/2021030176

ISBN: 978-0-415-83064-5 (hbk)
ISBN: 978-1-032-14977-6 (pbk)
ISBN: 978-0-203-74026-2 (ebk)

DOI: 10.4324/9780203740262

Typeset in Bembo
by codeMantra

Printed in the United Kingdom
by Henry Ling Limited

To Jennifer,
Geoffrey and Rowan
With all my love
Michael (Papa)

To all my teachers, friends and family
human and otherwise
the spirits, the demons, the dogs and the horses
Vivienne

In memory of
Charles 'Chip' Chace (1958–2018)
whose dedication and enthusiasm for the study of Chinese medicine,
and great generosity in sharing his knowledge, have been a model for
generations of scholars and practitioners

and

Ma Kanwen 馬堪溫 (1927–2016)
whose life and contributions to the history of medicine have been an
inspiration to us all

CONTENTS

List of figures   *xiii*
List of tables   *xvii*
List of contributors   *xix*
Acknowledgements   *xxvi*
Conventions and abbreviations   *xxvii*

    An introduction   1
    Michael Stanley-Baker and Vivienne Lo

PART 1
***Longue Durée* and formation of institutions and traditions**   **11**

1 *Yin, yang,* and five agents (*wuxing*) in the *Basic Questions* and early Han (202 BCE–220 CE) medical manuscripts   13
   Chen Yun-Ju

2 *Qi* 氣: a means for cohering natural knowledge   23
   Michael Stanley-Baker

   Appendix: Categories of *qi* in the *Inner Canon*   39
   Jiang Shan

3 Re-envisioning Chinese medicine: the view from archaeology   51
   Vivienne Lo and Gu Man

4 The importance of numerology, part 1: state ritual and medicine   72
   Deborah Woolf

5	The importance of numerology, part 2: medicine: an overview of the applications of numbers in *Huangdi neijing* *Deborah Woolf*	91
6	Therapeutic exercise in the medical practice of Sui China (581–618 CE) *Dolly Yang*	109
7	The canonicity of the *Yellow Emperor's Inner Classic:* Han through Song *Stephen Boyanton*	120
8	Pre-standardised pharmacology: Han through Song *Asaf Goldschmidt*	133
9	Developments in Chinese medicine from the Song through the Qing *Charles Chace*	146

PART 2
Sickness and healing — **161**

10	Ancient pulse taking, complexions and the rise of tongue diagnosis in modern China *Oliver Loi-Koe*	163
11	Case records *Yi'an* 醫案 *Nancy Holroyde-Downing*	181
12	Acupuncture illustrations *Huang Longxiang and Wang Fang*	189
13	Anatomy and surgery *Li Jianmin* 李建民 *Translated by Michael Stanley-Baker and Vivienne Lo*	206
14	History of disease: pre-Han to Qing *Di Lu*	217
15	Pre-modern madness *Hsiu-fen Chen*	230
16	Late imperial epidemiology, part 1: from retrospective diagnosis to epidemics as diagnostic lens for other ends, 1870s to 1970s *Marta Hanson*	245
17	Late imperial epidemiology, part 2: new material and conceptual methods, 1980s to 2010s *Marta Hanson*	263

18 Folk medicine of the Qing and Republican periods: a review of
    therapies in Unschuld's Berlin manuscripts   282
    Nalini Kirk

PART 3
Food and sex   **301**

19 What not to eat – how not to treat: medical prohibitions   303
    Vivienne Lo and Luis F-B Junqueira

20 Chinese traditional medicine and diet   320
    Vivienne Lo

21 Food and dietary medicine in Chinese herbal literature
    and beyond   328
    Paul D. Buell

22 The sexual body techniques of early and medieval China –
    underlying emic theories and basic methods of a non-reproductive
    sexual scenario for non-same-sex partners   337
    Rodo Pfister

23 Sexing the Chinese medical body: pre-modern Chinese medicine
    through the lens of gender   356
    Wang Yishan

24 Gynecology and obstetrics from antiquity to the early
    twenty-first century   368
    Yi-Li Wu

25 The question of sex and modernity in China, part 1: from *xing* to
    sexual cultivation   381
    L. A. Rocha

26 The question of sex and modernity in China, part 2: from new
    ageism to sexual happiness   389
    L. A. Rocha

PART 4
Spiritual and orthodox religious practices   **399**

27 Daoism and medicine   401
    Michael Stanley-Baker

28	Buddhist medicine: overview of concepts, practices, texts, and translations *Pierce Salguero*	417
29	Time in Chinese alchemy *Fabrizio Pregadio*	427
30	Daoist sexual practices for health and immortality for women *Elena Valussi*	444
31	Numinous herbs: stars, spirits and medicinal plants in Late Imperial China *Luis Fernando Bernardi Junqueira*	456

PART 5
The world of Sinographic medicine: a diversity of interlinked traditions — **473**

32	Transmission of Persian medicine into China across the ages *Chen Ming* 陳明 *Translated by Michael Stanley-Baker*	475
33	Vietnam in the pre-modern period *Leslie de Vries*	493
34	History and characteristics of Korean Medicine *Yeonseok Kang* *Translated by Jaehyun Kim*	503
35	Chinese-style medicine in Japan *Katja Triplett*	513
36	A brief history of Chinese medicine in Singapore *Yan Yang* 杨妍	524
37	Minority medicine *Lili Lai* 賴立里 *and Yan Zhen* 甄艳	537

PART 6
Wider diasporas — **549**

| 38 | Early modern reception in Europe: translations and transmissions
Eric Marié | 551 |

39	The emergence of the practice of acupuncture on the medical landscape of France and Italy in the twentieth century *Lucia Candelise*	564
40	Entangled worlds: traditional Chinese medicine in the United States *Mei Zhan*	576
41	The migration of acupuncture through the *Imperium Hispanicum*: case studies from Cuba, Guatemala, and the Philippines *Paul Kadetz*	585
42	Long and winding roads: the transfer of Chinese medical practices to African contexts *Paul Kadetz*	599
43	Translating Chinese medicine in the West: language, culture, and practice *Sonya Pritzker*	613

PART 7
Negotiating modernity — **623**

44	The Declaration of Alma Ata: the global adoption of a 'Maoist' model for universal healthcare *Paul Kadetz*	625
45	Communist medicine: the emergence of TCM and barefoot doctors, leading to contemporary medical markets *Xiaoping Fang*	638
46	Contested medicines in twentieth-century China *Nicole Elizabeth Barnes*	649
47	Public health in twentieth-century China *Tina Phillips Johnson*	659
48	Encounters with Linnaeus? Modernisation of pharmacopoeia through Bernard Read and Zhao Yuhuang up to the present *Lena Springer*	669
49	*Yangsheng* in the twenty-first century: embodiment, belief and collusion *David Dear*	687

50 Liquorice and Chinese herbal medicine: an epistemological
   challenge   707
   Anthony Butler

51 Decontextualised Chinese medicines: their uses as health
   foods and medicines in the 'global North'   721
   Michael Heinrich, Ka Yui Kum and Ruyu Yao

Index   743

FIGURES

2.1	气 in *Shuowen jiezi*	24
2.2	*Qi* graphs with fire component, 380 BCE–100 CE	25
2.3	*Qi* graphs with grain component, Warring States to 168 BCE	25
3.1	*Ci shu* 刺數 (Principles of Piercing)	52
3.2	Mawangdui chart *Daoyin tu* 導引圖 (Chart of Guiding and Pulling)	56
3.3	The Laoguanshan 老官山 Lacquer Figurine	59
3.4	The Shuangbaoshan 雙包山 Lacquer Figurine	61
4.1	The Four Emblematic animals, *siling* 四靈, from the author's private photo collection	77
4.2	*Xishuipo* 西水坡 tomb (M45) in Henan, showing clamshell figures representing a dragon in the East and tiger in the West, with a ladle in the North (Sun and Kistemaker 1997: 116, fig 6.2)	78
4.3	*Fuxi* and *Nüwa* from Wuliang shrine, chamber 2. Ink rubbing of carved stone relief, Eastern Han (Chavannes 1909:1, plate 60, no. 123)	79
4.4	The Seasons and the Two Phases of lesser and greater *yin* and *yang* – representation by the author	80
4.5	*Liuren* astrolabe (with Northern ladle in the centre), from Shuanggudui 雙古堆 tomb, sealed 165 BCE. After *Kaogu* 1978.5: 340. Drawing by Li Xiating. Copied with permission from Harper, 1999: 840	82
5.1	The Day Court (or cord-hook) diagram – representation by the author	96
5.2	Earth between each season: as Four Gates which radiate out as diagonals from the centre of a two-dimensional representation of a Six Dynasties (220–589) liuren 六壬 shi divination board. Reproduced from Steavu, 2017: 200, on the basis of Abe no Seimei to Onmyōdō ten 安倍晴明と陰陽道展 [Abe no Seimei and the Way of *Yin* and *Yang*: The Exhibition], 53	98
5.3	Bronze Man acupuncture figurine, front view, Chinese woodcut. Wellcome Collection (Ming dynasty): 4.0 International (CC BY 4.0)	99
5.4	Shuangbaoshan 雙包山 figurine, showing channels (Lo 2007: 419)	100
5.5	Five Agents with Sovereign Fire and Minister Fire: modern representation by the author	102

Figures

10.1	Pulse taking: A treatise on the pulse written by Wang Shuhe, fourth century CE. Illustration shows taking the pulse. Wellcome collection: 4.0 International (CC BY 4.0)	168
10.2	European pulse taking. 1682 edition of *Specimen medicinae* by Andreas Cleyer. Courtesy of the New York Academy of Medicine Library	169
10.3	From a 1529 edition of Xue Ji's *Xueshi yi'an* 薛氏醫案, showing a pure red tongue and prescribing *Touding qingshen san* 透頂清神散 (Extremely Penetrating Clear-the-Spirit Powder)	172
10.4	The Foetus and placenta. Parrenin [1715?], v. XV, p. 254. Image permission of BnF	175
12.1	Sun Simiao *Mingtang* Triptych (reproduction). All images from Huang Longxiang (2003)	191
12.2	Diagrams of Acu-moxa Bronzes of the facsimile of stone inscriptions of *Tongren shuxue zhenjiu tujing* 銅人腧穴針灸圖經 (Illustrated Manual of Acupoints of the Bronze Figure) preserved in the Imperial Household Agency (Kunaichō 宮内庁) of Japan. All images from Huang Longxiang (2003)	193
12.3	Hua Shou 滑壽, 1341, diagrams of the acu-moxa points of the fourteen channels in *An Elucidation*. Edition of the tenth year of Kansei era, Japan (1798) printed in *Zhenjiu yixue dianji daxi* 針灸醫學典籍大系 (Great Compilation of Acupuncture and Moxibustion Medical Dictionary). All images from Huang Longxiang (2003)	194
12.4	Japan Ancient Medicine Information Center, 1978, the *Great Compilation of Acupuncture and Moxibustion*, Diagrams of Ten Channels in *Chan Jing* 產經 (Classic of Childbirth); printed in *Zhenjiu yixue dianji daxi* 針灸醫學典籍大系 (Great Compilation of Acupuncture and Moxibustion Medical Dictionary). All images from Huang Longxiang (2003)	197
12.5	Yang Jie, 1113, *Huanzhong tu* 環中圖 (Diagrams of the Circulatory Course of the Channels), as cited in *Mananpō* 萬安方 (*Myriad Relief Prescriptions*), edition of the second year of the Enkyo era of the Kamakura period (1309), Japan, collected in the Cabinet Library of the National Archives of Japan. All images from Huang Longxiang (2003)	198
12.6	Zhu Gong 朱肱, 1118，Illustrations of channels in *Chongjiazheng huorenshu* 重校正活人書 (Revised Lifesaving Book), Song dynasty edition, preserved in Seikado Library, Japan. All images from Huang Longxiang (2003)	199
12.7	Wu Qian 吳謙, Qing, 1742, 'Diagrams of the Eight Extra Channels' in *Yizhong jinjian: Cijiu xinfa yaojue* 醫宗金鑒・刺灸心法要訣 (Golden Mirror of the Medical Tradition: Essential Teachings on Acupuncture and Moxibustion). All images from Huang Longxiang (2003)	200
18.1	*Jingzheng huoshi* 驚症火式 (Fire Approach to Pathoconditions of Fright, MS 8036), p. 15. Staatsbibliothek zu Berlin – PK	289
18.2	*Michuan dieda bofang* 秘傳跌打鈸方 (Secretly Transmitted Recipes [Treating Injuries from] Knocks and Falls, [Strung Together like] Cymbals, MS 8111), p. 7. Staatsbibliothek zu Berlin – PK	290
23.1	Different levels of *yin-yang* pairs of Essence and Blood. Reproduction from Charlotte Furth (1999) *A Flourishing yin*, p. 49. Credit: University of California Press	359

Figures

27.1	Parasites from *Taishang chu sanshi jiuchong baosheng jing* 太上除三尸九蟲保生經 (Most High Scripture on Conserving Life and Expelling the Three Corpses and Nine Worms) DZ 871 9b15a, Five Dynasties (907–960 CE)	409
27.2	Talisman from *Taishang chu sanshi jiuchong baosheng jing* 太上除三尸九蟲保生經 (Most High Scripture on Conserving Life and Expelling the Three Corpses and Nine Corpse-worms) DZ 871 17a, Five Dynasties (907–960 CE)	411
29.1	The three stages of the alchemical process represented by means of trigrams of the *Book of Changes*	430
29.2	The refining of Essence, Breath, and Spirit and the corresponding trigrams of the *Book of Changes*. Li Daochun (fl. 1288–92), *The Harmony of the Centre*, 2.6a–b	430
29.3	The twelve stages of the Lesser Celestial Circuit (*xiao zhoutian* 小周天)	435
29.4	The Waterwheel (*heche*). *Chart of the Inner Warp* (*Neijing tu*), detail	436
29.5	The River Chariot (*heche*). Xiao Tingzhi, *The Great Achievement of the Golden Elixir: An Anthology* (*Jindan dacheng ji*), 9.3b	437
31.1	Example of twelve 'secret characters' (on the left). *Miben fuzhou quanshu* 秘本符咒全書 (Compendium of Secret Talismans), Shanghai: Jingzhi tushuguan, 1912–1949, n.p	462
31.2	Example of six 'elaborated talismans' (on the right). *Miben fuzhou quanshu*, n.p	463
31.3	Altar for the Heavenly Physician. The sixteenth-century statue at the centre is Shennong, the God of Medicine and Agriculture. He is surrounded by talismans, medicinal jars and ritual objects, with herbs displayed in front of him for blessing. Photo by author 14 November 2019, private altar, Shanghai	464
31.4	Nineteenth-century medicinal jar inscribed with talismans. Such jars are filled with medicinal herbs and then placed in the altar for the Heavenly Physician to bestow upon them His blessings. This ritual, still pervasive among Chinese folk healers today, is believed to maximise the healing power of herbs. Photo by author 14 November 2019, private altar, Shanghai	464
31.5	'Divine Tree that Heals Disease'. This drawing depicts an ancient divine tree and its two girl-spirits (on the left) in Hanling 韓嶺 village, Ningbo, in the late nineteenth century. When falling sick, villagers would pray and burn incense to the tree. Its wondrous healing powers led locals to erect a special temple for its worshipping. *Dianshizhai huabao* 點石齋畫報 (Illustrated Lithographer), vol. 4 upper, (1884–1889): 1	465
36.1	Number of TCM Practitioners on the Register, 2004–2015	527
38.1	Pulse diagram, *Specimen Medicinae Sinicae,* p. 20. Courtesy of the New York Academy of Medicine Library	556
38.2	Conflation of acupoints and pulse points, Specimen Medicinae Sinicae, p. 68. Courtesy of the New York Academy of Medicine Library	557
48.1	Bernard Read and Liu Ju-Ch'iang's Flora Sinensis (1927). Courtesy of Needham Research Institute	677
48.2	'Catalogue of the Li Shih-Chen Exhibition' (1954) History of Medicine Museum, Shanghai. Courtesy of Needham Research Institute	678
49.1	'A decision. A life for living. The Butterfly Allure Gentleman's *Yangsheng* SPA centre'. Poster. Author's own photograph	692

50.1 Glycyrrhiza glabra Wellcome Collection: Glycyrrhiza glabra (Liquorice or Licorice). Credit: Rowan McOnegal. Attribution-NonCommercial 4.0 International (CC BY-NC 4.0) 709
50.2 Chemical structure of glycyrrhizin. The left part of the molecule is hydrophilic while that on the right is hydrophobic (Drawn by Hazel Nicholson) 710
50.3 Chemical structure of glycyrrhetinic acid, a hydrolytic product of glycyrrhizin metabolism (Drawn by Hazel Nicholson) 711
50.4 Cross section of a micelle. A drug may be contained within the micelle (Drawn by Hazel Nicholson) 716
51.1 Artemisinin from *Artemisia annua* 724

TABLES

1.1	*Yin-Yang* Correspondence in Mawangdui Manuscripts the *Discussion of the Culminant Way in Under-heaven* and the *Cheng*	16
1.2	*Yin-Yang* of the Body in Section Four of the *Basic Questions*	17
1.3	Correspondences between Imperial Bureaucratic Structures and Twelve Viscera	18
1.4	Selected Correspondence between Five Agents and Phenomena in *Suwen* 5	19
1.5	Mutual Generation or Overcoming among Five Agents in *Suwen* 25	19
1.6	The Affinities of the Five Agents to *Yin-Yang* Channels in *Suwen* 22	19
2.1	*Qi* typologies organised by number	43
3.1	Categories of Remedies and Arts from the *Han Imperial Catalogue* as Compared with Related Manuscripts Excavated from the Mawangdui Tomb	57
4.1	The Sexagenary Cycle by the author	76
5.1	Channels and Climates	94
5.2	Six Channels, Climates, Five Agents as Represented in Wang Bing's *Wuyun liuqi* Chapters of 762 CE	102
6.1	Medical teaching staff at the Imperial Medical Academy during the reigns of Wendi and Yangdi	111
6.2	*Yangsheng* texts in the medical section of *Suishu jingjizhu* 隋書經籍志 (The Catalogue of the Imperial Library of the Sui)	112
6.3	*Yangsheng* texts quoted in *Zhubing yuanhou lun*	116
7.1	Titles of *Inner Classic* Texts Mentioned in Other Texts	121
7.2	Publications of the Song Bureau for Editing Medical Texts (*Jiaozheng yishu ju* 校正醫書局)	126
8.1	Important *Materia Medica* Collections and Number of Drugs Included	141
8.2	Important Formularies and Number of Formulæ Included	142
19.1	Survey Based on the Database *Beijing Airusheng shuzihua jishu yanjiu zhongxin* (2016)	312
19.2	Survey Based on the Database *Beijing Airusheng Zhongyi dianhai*, 1st edition, 2016	313
20.1	Five Agent Correspondences with Flavours, Seasons, Grains and Meat	323
21.1	Song dynasty (960–1279) Official Herbals	331
22.1	The Thirty Rubrics of Scroll 28 *On the Chamber-Intern* in *Ishinpō*	339
29.1	The *yueti najia* (Matching Stems of the Moon) device	432

29.2	The twelve 'sovereign hexagrams' (*bigua*) and their relation to other duodenary series: earthly branches (*dizhi*), bells and pitch-pipes (*zhonglü*), months of the year, and 'double hours' (*shi*)	433
36.1	TCM Associations in Singapore	529
37.1	Policies and Projects Issued by the Government of China	538
37.2	State Administration Departments Mentioned in the Article	539

CONTRIBUTORS

Nicole Elizabeth Barnes, Assistant Professor of History at Duke University, is the author of 'Serving the People: Chen Zhiqian and the Sichuan Provincial Health Administration, 1939–1945', in *China and the Globalization of Biomedicine*, edited by David Luesink, William Schneider and Zhang Daqing (2019). She is currently revising a manuscript entitled *Protecting the National Body: Gender and Public Health in Wartime Sichuan, 1937–1945*.

Stephen Boyanton received his PhD in East Asian history from Columbia University, with a dissertation on 'The *Treatise on Cold Damage* and the Formation of Literati Medicine: Social, Epidemiological, and Medical Change in China, 1000–1400'. He currently works as an independent scholar and translator in Chengdu, Sichuan, China.

Paul D. Buell, PhD, Sinologist, Mongolist and Turkologist with a strong interest in the history of medicine and food. Co-author of *Crossroads of Cuisine* (2020) on the role of Central Eurasia in food, and medical exchanges. Lead author of forthcoming multi-volume work on Arabic Medicine in China.

Anthony Butler is a graduate in chemistry from King's College, University of London. After some years in the USA at Cornell University he returned to Britain to a post at the University of Sussex and then at the University of St Andrews. He taught there in both the Chemistry Department and the Medical School. He is a Fellow of the Royal College of Physicians of Edinburgh. Since retirement he has concentrated on writing about the history of medicine and struggling with the Chinese language.

Lucia Candelise is a medical anthropologist and historian of medicine. She is a senior fellow at the Institut des Sciences Sociales (ISS), Lausanne University and at the CEPED Institut Recherche Développement, Université de Paris. She works on how Chinese medicine is being received in Europe and Africa and on medical knowledge as an immaterial cultural heritage.

Charles "Chip" Chace graduated from the New England School of Acupuncture in 1984 where he also began studying the Chinese language. He maintained a longstanding interest in the medical literature of China and translated and authored a wide variety of books and articles on pre-modern approaches to acupuncture and Chinese medicine, as well as translations of 'The Yellow Emperor's Systematic Classic of Acupuncture and Moxibustion' (*Huangdi zhenjiu jiayi jing* 黃帝針灸甲乙經), and Li Shizhen's 'Exposition on the Eight Extraordinary Vessels' (*Qijing bamai kao* 奇經八脈考). He practised Chinese medicine in Boulder, Colorado, for over thirty years, and was a long-standing member of the faculty of the Seattle Institute of Oriental Medicine. He passed away in 2018 before this volume came to print. He is fondly remembered for his kindness and generosity, both as a clinician and as a mentor.

Chen Hsiu-fen 陳秀芬 is a Professor of History at the National Chengchi University, Taiwan. She is widely published in English and Chinese on the topics of madness, emotions and health promotion in late imperial China. She also studies medical images, material cultures of drugs and religious healing in Ming-Qing China. Chen's first Chinese monograph, *Yangsheng yu xiushen: Wanming wenren de shenti shuxie yu shesheng jishu* 養生與修身: 晚明文人的身體書寫與攝生技術 (Nourishing Life and Cultivating the Body: Writing the Literati's Body and Techniques for Preserving Health in the Late Ming), was honoured by the National Science Council (Taiwan) as one of the best humanities publications of 2010.

Chen Ming 陳明 is a Professor in the Department of South Asian Studies at Beijing University (PKU). He researches the history of cultural communication between China and central and south Asia in the medieval period mainly, but not exclusively, in subjects related to medicine. He has published six books in Chinese on the history of Sino-Indian medical cultural exchange as well as studies on the medical manuscripts along the Silk Road, such as *Zhongu yiliao yu wailai wenhua* 中古醫療與外來文化 (Foreign Medicine and Culture in Medieval China, 2013), *Silu yiming* 絲路醫明 (Medical Culture along the Silk Road, 2017), and *Dunhuang de yiliao yu shehui* 敦煌的醫療與社會 (Medicine and Social Life in Dunhuang, 2018).

Chen Yun-Ju 陳韻如 is an historian of Middle Imperial China, currently an Assistant Research Fellow at the Institute of History and Philology, Academia Sinica. Her research focuses on the emergence of authorities in knowledge and knowledge practices in pre-modern China, especially with regard to medicine between the ninth and thirteenth centuries.

Leslie de Vries is a Lecturer in East Asian Studies at the University of Kent. He received his PhD in Oriental Languages and Cultures from Ghent University. Previously, he was also a Research Fellow of the 'Beyond Traditions' project at the University of Westminster. His research focuses on the history of medicine and religion in China, Vietnam and Japan.

David Dear is a film maker, photographer and script writer. He spent over a decade living in Beijing studying popular medico-religious practices and martial arts and their interconnections and relationships with other aspects of Chinese culture, both high and low.

Xiaoping Fang is an Associate Professor of Chinese History at Monash University, Australia. His current research interests focus on the history of medicine, health and disease in twentieth-century China. He is the author of *Barefoot Doctors and Western Medicine in China* (2012) and *China and the Cholera Pandemic: Restructuring Society under Mao* (2021).

Contributors

Asaf Goldschmidt is an historian, who received his PhD from the University of Pennsylvania and is currently a Professor of East Asian Studies at Tel Aviv University. His main research interests are the history of Chinese medicine and science, the Song dynasty, and public health, drug trade and consumption, and medical practice. His current projects focus on the history of the Imperial Pharmacy in China and on medical case records.

Gu Man 顧漫 is Director of the Department of Chinese Medical Literature, China Institute for the History of Medicine and Medical Literature, China Academy of Chinese Medical Sciences. He was a Postdoctoral Fellow in the Department of Chinese Language and Literature, Fudan University (2008–2010). His major research concerns the collation and digitisation of ancient Chinese medical manuscripts written on bamboo slips or silk. He is widely published on the concepts and techniques of acupuncture and the construction of classical Chinese medicine as revealed in the Tian Hui Laoguanshan manuscripts.

Marta Hanson is Associate Professor of the History of Chinese Science and Medicine at the Department of the History of Medicine, The Johns Hopkins University. She has published widely on the history of epidemics, disease, and public health in China; disease maps in East Asia; Chinese arts of memory; the healer's body in Chinese medicine; and Sino-European cross-cultural medical history.

Michael Heinrich is Professor of Ethnopharmacology and Medicinal Plant Research (Pharmacognosy) and was previously the head of the research cluster 'Biodiversity and Medicines' at the UCL School of Pharmacy. His main research focuses on anti-inflammatory agents, the safety of herbal medicines and modern ethnopharmacological approaches, including value chains of medicinal plants (especially from Asian countries).

Nancy Holroyde-Downing is an acupuncturist and herbalist with an MA in the History of Medicine from SOAS and a PhD in History from UCL. She has maintained a practice in Chinese medicine for over 30 years, coupled with teaching and mentoring, and has contributed to publications on the history of medicine in China.

Huang Longxiang 黃龍祥, Chief Researcher at the China Academy of Chinese Medical Sciences, has been engaged in research into the theory of traditional Chinese medicine for over thirty years. Major publications include *Evidence-based Surface Anatomy for Acupuncture: acupuncture integrated with surface anatomy and imaging* (2011), and *Zhongguo zhenjiu xueshushi dagang* 中國針灸學術史大綱 *The Historical Development of Acupuncture* (2002).

Jiang Shan is a post-doctoral researcher from the School of Medical Humanities, Peking University. Her research interests include the premodern theory of acu-moxa and the history of Sino-Japanese medical and cultural exchange. She received her PhD in the China Academy of Chinese Medical Sciences, with a dissertation titled 'A Study of *Qi* in Classical Acupuncture-Moxibustion Theory'. Part of the work of Jiang's *qi* research was published in the book 'Needling and *Qi*: Discovering *Qi* in Ancient Acumoxa Classics' (2018).

Tina Phillips Johnson is Professor of History at Saint Vincent College in Latrobe, Pennsylvania, and a Research Associate at the Asian Studies Center, University of Pittsburgh. Her current research focuses on women's health in 20th-century China. She also holds a Master of Public Health degree and works on contemporary public health initiatives in rural China, most recently infant nutrition and rural medical waste management.

Luis Fernando Bernardi Junqueira 林友樂 earned his master's degree in Chinese History at Fudan University and is currently a PhD candidate in History at University College London. Sponsored by the Wellcome Trust, his research explores the impact of transnational spiritualism and psychical research on China's healthcare market during the first half of the twentieth century.

Paul Kadetz is a Senior Lecturer in the Institute for Global Health and Development at Queen Margaret University in Edinburgh, and a Professor of Practice in the Shiley-Marcos School of Engineering at The University of San Deigo. He is also an Honorary Fellow in the China Centre for Health and Humanity, UCL. Paul completed his doctorate in Development Studies at the University of Oxford and works at the intersections of global health, international development, and critical medical anthropology. Recent publications include *The Handbook of Welfare in China* (2017), *Creating Katrina, Rebuilding Resilience* (2018), and the forthcoming *Encyclopedia of Health Humanities* (joint editor-in-chief).

Yeonseok Kang holds a PhD from the College of Korean Medicine. He is an Associate Professor of Medical History at the Wonkwang University, Iksan, and Director of Planning for the Institute of Korean Medicine Education and Evaluation in Seoul. His research on the history of Korean medicine is continuing and his publications include a study on *Hyangyak jipsungbang* 鄉藥集成方 (Compendium of Recipes Ready-made with Local Korean Herbs) and the use of Hyangyak's medicine in Korea and China during the 15th Century.

Nalini Kirk is a sinologist and a practitioner of Chinese medicine. She has worked as a researcher at the Institute of Chinese Life Sciences in Berlin, where she co-authored the *Dictionary of the Bencao gangmu, Vol. 3 – Persons and Literary Sources*. She was a fellow of the excellence cluster TOPOI and is currently working on her PhD at Freie Universität Berlin.

Ka Yui Kum specialises in Traditional Chinese Medicine (TCM) and nuclear magnetic resonance (NMR) metabolomics for herbal quality assessment and received his doctorate from the UCL School of Pharmacy in 2020. He is also a successful public engagement organiser with funding from the Royal Society of Biology and UCL.

Lili Lai 賴立里 is Associate Professor of Anthropology at the School of Health Humanities, Peking University, with research interests in the body, everyday life, and medical practices. Her monograph, Hygiene, Sociality, and Culture in Contemporary Rural China, was published in 2016. Now she is working on a book Gathering Medicines: Nation and Knowledge in China's Mountainous South (with Professor Judith Farquhar, University of Chicago).

Li Jianmin 李建民 is a Research Fellow at the Institute of History and Philology, Academia Sinica, Taiwan. He also teaches at several universities in Taiwan. His specialty is the history of Chinese medicine and of health and healing in ancient China. He is the author of numerous books and articles on medicine and culture in ancient China, and recently *The Vulnerable Surgeons and the Fading of Surgery in the History of Sinitic Medicine* (2018).

Vivienne Lo 羅維前 is Professor of Chinese History at University College London. She has published widely on the ancient and medieval history of medicine in China and in diaspora. Her research interests include medical manuscripts, medical imagery and the history of nutrition.

Oliver Loi-Koe is an Emergency Doctor working at a London hospital. He aims to combine medical specialities with general practice, as well as Chinese medicine. He has an MA from University College London. His research interests are the origins and development of Chinese medicine and he is training in acupuncture.

Di Lu 蘆笛 is currently a Zvi Yavetz Fellow at Tel Aviv University. He received his PhD from University College London in 2017. His research mainly focuses on Sino-European exchange of scientific and medical knowledge in the eighteenth through early twentieth centuries. Currently he is writing a book-length transnational history of the caterpillar fungus.

Eric Marié is a member of the Faculty of Medicine, University of Montpellier. He has dedicated himself to the study of Chinese medicine for more than 30 years, as a physician, an historian and a sinologist. Doctor of Chinese medicine (China) and PhD in history of civilisations (France), he has taught various aspects of Chinese medicine in several universities in Europe and in China. Author of ten books, in particular *Grand Formulaire de Médicine Chinoise* (1991), *Précis de Médecine Chinoise* (2008) and *Le Diagnostic par les Pouls en Chine et en Europe* (2011).

Rodo Pfister conducts studies into the history of medicine, meditation, consciousness, emotion and sexuality, both in ancient and medieval China. At present he is working on the *Modular Sourcebook of Early and Medieval Chinese Medicine – History of the Body, Consciousness and Healing*, and on bubonic plague history in early medieval China.

Fabrizio Pregadio earned his PhD in Civilizations of East Asia at the Ca' Foscari University of Venice and has taught at universities in Italy, Germany, the United States, and Canada. His research interests are Daoist thought and religion, alchemy, views of the human being, and self-cultivation traditions. He is the editor of the *Encyclopedia of Taoism* (2008), and his publications include studies and translations of several alchemical classics. He is currently working on a monograph on the Daoist master Liu Yiming 劉一明 (1734–1821).

Sonya Pritzker is Assistant Professor in the Department of Medicine, David Geffen School of Medicine, UCLA, where she is affiliated with the Division of General Internal Medicine and the UCLA Center for East-West Medicine. Her research focuses on the global clinical translation of Chinese medicine in both text and practice.

Leon Antonio Rocha works at the Bargaining, Organising, Campaigns and Education Department, University and College Union (UCU). He received his PhD in history and philosophy of science and medicine from University of Cambridge, and has taught at Yale University, University of Liverpool, and University of Lincoln.

Pierce Salguero is a transdisciplinary humanities scholar interested in the role of Buddhism in the crosscultural exchange of medical ideas. He is Associate Professor of Asian History and Religious Studies at Penn State University's Abington College. He recently published *Buddhism and Medicine: An Anthology* (2 vols.) with Columbia University Press.

Contributors

Lena Springer gained her PhD in Sinology at the University of Vienna. She researches the transmission of medical heritage. As a research fellow of Sichuan University, she investigates multi-ethnic folk medicines in China's West. In a database team at Charité Medical University, Berlin, she is designing the translation, scientific identification, and interdisciplinary accessibility of Chinese historical pharma-recipes. She has recently joined a project at King's College London researching scientists' and shamans' efforts at reforestation in Southwest China and Siberia. Springer publishes on ethnographic archivers and medical-history-writers in China, spatial and social migration to Europe and the anthropology of science.

Michael Stanley-Baker 徐源 is Assistant Professor in History at the School of Humanities, and of Medical Humanities at the Lee Kong Chian School of Medicine, at Nanyang Technological University, Singapore. An historian of Chinese medicine and religion, particularly Daoism, he works on the early imperial period as well as contemporary Sinophone communities. Currently completing a monograph on medicine and religion as related genres of practice in China, he also produces digital humanities tools and datasets to study the migration of medicine across spatio-temporal, intellectual and linguistic boundaries. He is the editor of *Situating Medicine and Religion In Asia: Methodological Insights and Innovations* forthcoming next year.

Katja Triplett was Professor of the Study of Religions at Göttingen and is currently based at Leipzig University. Her doctorate in the study of religions is from Marburg University, where she also studied Japanese linguistics and cultural anthropology. Among her recent publications is *Buddhism and Medicine in Japan* (2019).

Elena Valussi is a Senior Lecturer in the History Department at Loyola University, Chicago. She has published several articles on the intersection of gender, religion and body practices in Late Imperial China and modern Daoism. Her more recent work focuses on Daoist intellectual history, printing, gender and religion in the late Qing and Republican periods. She is currently co-editing a book on spirit-writing techniques in Chinese History, and she is the Vice-President of the Society for the Study of Chinese Religions.

Wang Fang 王芳 is an Associate Researcher at the China Academy of Chinese Medical Science Acupuncture Institute. She has been engaged in the international communication of traditional Chinese medicine for over twenty years. Her major publications include *International Acupuncture Textbooks* (2009).

Yishan Wang 王颐姗 is currently a doctoral candidate in Anthropology at the School of Anthropology and Museum Ethnography, University of Oxford. She undertook a Master's degree in Gender Studies at University College London before going to Oxford to study the MPhil in Medical Anthropology. Her research concentrates on deconstructing the discursive construction of sexual normality in the context of Chinese medicine.

Deborah Woolf is an historian and researcher on Chinese Medicine. As a practising acupuncturist, she has over twenty years' experience in teaching Chinese Medicine and Philosophy. She has an MA in Chinese Health and Humanities and is currently a PhD candidate in history at University College London. She is particularly interested in divination, hemerology, iatromancy and the more esoteric practices embedded in Chinese medical thought.

Yi-Li Wu is Associate Professor of Women's and Gender Studies and of History at the University of Michigan, Ann Arbor. Her research on the history of Chinese medicine focuses on the multiple intersections of society, culture, gender, and the body in the sixteenth to nineteenth centuries. Her publications include *Reproducing Women: Medicine, metaphor, and childbirth in late imperial China* (University of California Press, 2010) and articles on medical illustration, breast cancer, forensic medicine, bone setting, Sino-Korean medicine, and Chinese views on Western anatomical science.

Dolly Yang 楊德秀 is a postdoctoral research associate at the Institute of History and Philology, Academia Sinica, Taiwan. She received a PhD in 2018 from University College London for her investigation into the institutionalisation of therapeutic exercise in Sui China (581–618 CE). She has a particular interest in examining the use of non-drug-based therapy in early medieval China, allied to a passion for translating and analysing ancient Chinese medical and self-cultivation texts.

Yan Yang 杨妍 is a research fellow at the National University of Singapore. She obtained her PhD in Chinese Studies (Specialisation in History) from the National University of Singapore. Her research interests are in the history of Chinese medicine and the history of the Overseas Chinese. She has published articles on the history of Chinese medicine in Singapore.

Ruyu Yao is a postdoctoral researcher at the Institute of Medicinal Plant Development, Chinese Academy of Medical Sciences, Beijing, China. In 2018 he obtained his Ph.D. degree at the University of Zurich, Switzerland, with a study on the traditional uses, quality assessment and value chain analysis of *goji*.

Mei Zhan is Associate Professor of Anthropology at the University of California, Irvine. She is the author of *Other-Worldly: making Chinese medicine through transnational frames* (2009). Her current ethnographic project examines how the invention of a new 'classical medicine' in entrepreneurial China strives to 'bring medicine back to life'.

Yan Zhen 甄艳 is Professor of the History of Minority Medicine in China, especially the history of Tibetan medicine, at the Department of Minority Medicine, China Institute for History of Medicine and Medical Literature, affiliated with the China Academy of Chinese Medical Sciences. She has published on Tibetan medicine, with three academic books on Tibetan *materia medica*, health cultivation and a Tibetan medicine bibliography.

ACKNOWLEDGEMENTS

It is impossible to thank all those who have contributed in small and large ways to a project nearly a decade in the making. In addition to the writers whose excellent chapters are featured in this volume, many invisible hands have made this project possible. Dolly Yang and Penelope Barrett not only brought their extensive knowledge of Chinese medicine to bear, but also their patient and persistent attention to detail in preparing the style guidelines and editing the manuscript. Sally Stewart generously shared her extensive editorial experience refining finer points of language and punctuation. Emily Pickthall, Georgina Bishop and Stephanie Rogers at Routledge have carefully shepherded this project through its long gestation. Apart from our home universities, Department III of the Max Planck Institute for the History of Science, Peking University Health Science Center and Domaine Le Puget have all provided a home for this project allowing important meetings and consultations to take place, with many of those who feature in this volume. Nanyang Technological University (Ref no.: SUG M4082222.100) and UCL History have generously contributed to the production cost of this project. We are particularly indebted to the Wellcome Trust for supporting the research of Professor Lo and for providing substantial funds towards making this whole volume Open Access (Wellcome Trust References: 201616/Z/16/Z and 217661/Z/19/Z), and to all the authors who contributed as well. In addition, we are grateful for the guidance, friendship and intellectual companionship of Judith Farquhar, Michael Loewe, Donald Harper, Ma Jixing, Timothy Barrett, Andrew Wear, Zhou Xun, Liu Changhua, Wang Shumin, Geoffrey Samuel, Angelika Messner, Shigehisa Kuriyama, Dagmar Schäfer, Francesca Bray, Marc Kalinowski, Roel Sterckx, Catherine Despeux, Guo Liping, Stewart Allen, Sonja Brentjes, Helen Verran, Elisabeth Hsu, Asaf Goldschmidt, Teriyuki Kubo, Pierce Salguero, Elaine Leong and many others. Both of the editors give their heartfelt thanks to their long-suffering families. Jennifer Cash, in particular, has been a mainstay and constant companion throughout; better than dancing backwards in high heels, Jennifer read multiple chapter drafts, raised their young family while Michael travelled for editorial rounds, all while budgeting the family accounts, publishing her own book and creating her own company.

CONVENTIONS AND ABBREVIATIONS

The history of Anglophone contact with Chinese medicine has produced many styles of translation, which we have sought to represent by not imposing standardised translations of text titles or terminology.

All translated terms and titles are accompanied by pinyin and Chinese characters for the term, and readers can use these to cross reference them.

AN INTRODUCTION

Michael Stanley-Baker and Vivienne Lo

It is a delight and a cause for celebration for us to finally introduce this edition of the Routledge Handbook of Chinese Medicine. It represents a mammoth effort on which both of us have laboured in equal measure, spanning almost a decade and multiple life and professional changes. It has also drawn together a community of, mostly young, researchers who are actively shaping the history and anthropology of the field – and it is a huge field with both diachronic and synchronic depth and breadth, with rich and ever-growing quantities of ancient medical manuscripts on subjects that still have contemporary academic and practical relevance. There are chapters on the origins of the basic principles of *yin* and *yang* and the Five Agents, on early Chinese anatomy and physiology, on herbs, food, sex and self-cultivation and the ways in which ancient ideas and practices have been enhanced, reinvented and negotiated over two millennia.

The range of the project has grown as we embraced all the topics which increasingly became essential to do justice to the rubric 'Chinese Medicine'. At the outset, we had to debate the concept of a Handbook. This is not an instructional manual. Far from it. Albeit that both editors have been practitioners of Chinese medicine at some point in their careers and are concerned with the *know-how* of how we know the body, there is little here that will have direct relevance in the clinic. Our authors are concerned with the nature of medicine and healing in the Chinese cultural region, its plural epistemologies and the way it has been portrayed from the beginning of the empire in the third century BCE to the globalisation of Chinese products and practices in the present day. In what way, then, is this collection a Handbook?

Contributors to the Routledge Handbook Series aim to provide cutting-edge overviews of current scholarship in the Humanities and Social Sciences with an authoritative guide to theory and method, the key sub-disciplines and current major debates. We have therefore selected topics and designed the style and arrangement of the chapters so that they can be a primary port of call for all those readers seeking an introduction to the broad range of ideas and techniques that have made up pre-modern and traditional medicine in China, and the historiographical and ethnographic approaches that have illuminated them. The book is intended for entry-level researchers, interested and engaged general readers, practitioners and patients, but also specialists in Chinese medicine who feel the need to refresh their knowledge and understanding with the latest developments in the field.

DOI: 10.4324/9780203740262-1

We should first attempt to define the limits of Chinese medicine. This is a famously slippery subject. What do the geopolitical 'Chinese', and eclectic notion of 'medicine' mean as separate terms and then in combination? There was no term in Chinese for 'Chinese medicine' (*zhongyi* 中醫) in China before the arrival of an identifiable and different style of medicine originating elsewhere, i.e. the European medicine that appeared with the Jesuits and surgeons of the seventeenth and eighteenth centuries, forcing a recognition of difference and a local crisis of identity. Until the arrival of this powerful and foreign medicine in China, there were only the terms for 'medicine' of one kind or another, without any imperial or national designation. As China had been the self-styled centre of the world since 221 BCE and had produced copious writings on the body, often collected in the imperial library and/or state-sponsored, there was no need to designate where the medical orthodoxy came from!

Long before the Jesuits' arrival in China, a medicine from the Chinese cultural region had, in fact, been identified by other and distant language communities. At the turn of the fourteenth century, at the end of one of the so-called Silk Routes, far from Chinese borderlands, physicians, translators and scholars gathered at the court of Rashid al-Din, the Judeo-Muslim scholar and Vizier. They had travelled to Mongolian Ilkhanid Persia from the Hindu Kush and the slopes of the Himalayas, from China, Arabia and Europe. Rashid al-Din sponsored translations of medical works from around the world, producing a monumental collectanea of medical knowledge, including Chinese works. Compiled in Persian, and finalised in 1313, the *Tansuqnama-i ilkhan dar funun-I'ulum-i khatayi* (The Treasure Book of the Ilkhan on Chinese Science and Techniques) contains some of the earliest extant evidence of the classical treatises of Chinese medicine. As a distant and exotic unknown, Chinese science and medicine could be named and explored. In this great intellectual melting pot, Chinese medical practitioners were thought very skilled in pulse diagnosis and there were earnest attempts to characterise the distinctive aspects of this foreign medicine from far away.

Longue Durée and formation of institutions and traditions

Historians have come a long way from representing the classics of Chinese medicine as a revealed truth of pre-history, delivered by the Yellow Emperor in conversation with his ministers and advisers. Much of the classical literature was indeed framed in debates with his ministers and advisers which acted as a rhetorical device for authors who aimed not so much at introducing contesting viewpoints, but at standardising information. They promoted a view of the body as a microcosm of the state and the state as a microcosm of the cosmos, where imperial power was naturalised as a mandate from Heaven offered to the incumbent ruler and extended to the very depths of every human being. Joseph Needham described this view as 'organismic': it aligned the physiology and functions of the body with state priorities of political unity and unification and aimed to remove blockages to the flow of imperial power and to place at the centre of that authority a divine ruler, or the heart, interpolating the wise ministers and their executive functions into the organs and bowels and their activities.

There is little doubt that what we now regard as Chinese medicine crystallised in the centuries around the turn of the first millennium of our common era in a period that spanned the late Warring States, and the Qin and Han empires. The early impulse to collate and standardise the disparate writings and beliefs of ritualists, doctors and diviners of the late Zhou period within a coherent whole was certainly a feature of the larger imperial drive for standardisation (of the calendar, writing, weights and measures, roads, etc.). Writers on medicine attributed the traditional sources of medical knowledge to icons of a legendary past – sages, cultural heroes and mythical rulers. To the Red Emperor or Divine Husbandman, Shennong 神農 was attributed the tradition of trying and testing drugs and food. He was the legendary patron of the earliest *materia medica* (Chapter 8) and the pharmacological traditions that were more thoroughly systematised after the Song, but still invoked his name (Chapter 9). The most celebrated legendary patron of scholarly medicine, however, was the Yellow Emperor (Huangdi 黃帝; also translated as 'Yellow Thearch'). A founding culture hero, he is said to have developed many of the inventions that bring together universalising, cosmological concepts with ideas of law, punishment, calendrics, divination practices and medicine (Chapter 7).

New archaeological sources have completely transformed our understanding of the sociocultural contexts within which this new formalised imperial medicine came into being. The last half-century of research has seen the excavation of manuscripts from the late Warring States and Han tombs (ca. third to first centuries BCE), which continue to be unearthed as China digs up ancient cemeteries in a sustained fever of urban construction (Chapter 3). The texts, mostly inscribed on silk and bamboo in numerous editions, testify to a wide range of bodily treatments and remedy and recipe traditions that co-existed in everyday practice at least for the nobility of ancient China. They provide the earliest evidence of many longstanding medical traditions and body cultivation practices: remedy collections, of acupuncture, sexual cultivation, breathing meditations and therapeutic exercise. Some of the practices did not survive as orthodox medicine but many, such as therapeutic exercise, massage and acupuncture, were adopted as fields of study in later imperial medical institutions (Chapter 6).

The excavated texts also reveal myriad and anonymous scholarly voices whose work is compiled in brief excerpts in the classical Yellow Emperor corpus. They also testify to many aspects of everyday life that led to the innovations of the medicine of *yin*, *yang*, the Five Agents and *qi* – observing and recording the seasons and the animal world, circadian rhythms, the rituals that governed everyday life and the pervasive importance of the cultures of numerology (Chapters 4 and 5).

Why does a medicine that is so culturally situated in the homologies of Chinese empire remain relevant today? The terms *yin*, *yang*, the Five Agents and *qi* (Chapters 1 and 2) form the linguistic glue that sustains our imagination of China's long, coherent tradition of medicine. But to what degree does the Chinese medicine provided on the streets of Beijing, Taipei, Tokyo, London and San Francisco reflect the medicine of the imperial courts of the ancient world or medieval Daoist or Buddhist healing practices? There is much yet to be gained by applying analytic frameworks already well developed in the domains of social and cultural critiques to these open questions. While within the constraints of a single volume we have not been able to pay as much attention as we would have liked to environmental, climatic or economic changes, to the histories that focus on the deep structuring effect of external patterns and events, much is said by the contributors about internal structures of medicine and healing, the enduring Chinese mentalities, social norms and taboos. While there is a rough chronological ordering of the chapters throughout this Handbook and a pinpointing of sudden change and the causes of those change, we pay as much attention to recurring patterns in conceptions of health and illness, to the plural epistemologies of practice and their incremental transformation over roughly the two millennia since the unification of China in 221 BCE, and thus to the *longue durée* of medical practice in China.

Sickness and healing

Certain types of medical knowledge were derived, from the Han period onwards, by observing the tissues, bones and viscera of the physical body (Chapter 13). As in ancient Greece and other cultures with longstanding written medical traditions, it is abundantly clear to us that deep surgery could not be therapeutically effective – except under the extreme conditions of, for example, war and childbirth. In the treatment of living bodies, it was more practical and effective to imagine physiology and anatomy as homologous with the order of the external worlds. This was demonstrated, for instance, in the *yin* and *yang* channels portrayed on models and diagrams illustrating acupuncture practices (Chapter 12).

The salience of medical discourse in imperial China is clear: it portrayed a cosmos wherein the healthy body was confluent with natural laws, observed as cycles and phases of Heaven and Earth, dissonance with which produced illness (Chapter 14). Physicians who could perceive and identify these dissonances could diagnose and predict disorders at a more fundamental level using pulse, palpation and facial diagnosis, and thereby treat their patients, including for emotional disorders and their behavioural manifestations (Chapters 10 and 15).

In addition to these cosmological patterns of disease, symptoms of epidemic disease were, from the Han period onwards, classified as 'cold damage'. This syndrome framed feverish diseases and other infectious conditions in the language of cycles of *yin* and *yang* mapped on to body physiology. Similarly, we can see the roots of a genre of medical case histories that captured changing and plural approaches to diagnosis and treatment (Chapter 11). The plural natures of the Chinese medical traditions are most vividly illustrated in surviving manuscript collections which have not been shaped by subsequent editors, and perhaps reflect more faithfully the heterogeneity of medicine from below, of local healers, of women and of itinerate doctors (Chapter 18). Case histories were a genre that grew exponentially in late imperial times (late fourteenth to twentieth centuries), a period which also saw increasing numbers of epidemics, the history of which is a controversial subject that has inspired subtle historiographical reflections in this Handbook (Chapters 16 and 17).

Food and sex

Historiographical reflections are also very much a feature of the section on food and sex (Chapters 25 and 26) which traces the emergence of a gendered body, in both medical and sexual scenarios (Chapters 23 and 24). Writing the sexual body was surely the earliest context within which the sensory aspects of *yin* and *yang* and *qi* first developed (Chapter 22). The anatomy and function of the sexual organs were the subject of intense scrutiny, but more significantly *yin* and *yang* and *qi* formed a terminological code applied to the sensory realms of the inner body before these concepts became the building blocks of a medical orthodoxy. Heat, passion, pain and the total physical dimensions of emotion and well- or ill-being all came to be described in these terms.

Pleasure in China, be it of sexual, culinary or spiritual design, was also the source of many prohibitions concerned with restraint and particular practices (Chapters 19–21). Daoist sexual teachings on health and immortality were a significant legacy of these traditions and trained and disciplined women's desires (Chapter 30).

Practices of the spirits and religious orthodoxies

Much of this book is concerned with what we might now consider religious healing and the innovations in treatments inspired by the religious institutions that formed estates within the state. Their healing practices were not professionally or imperially endorsed as medicine, despite continuities of physiology and cosmology that carried over into native religious traditions: Buddhists and Daoists were prohibited from competing with physicians from the Tang period onwards. Communications with deities and divine entities, enlisting their support to empower treatments and to intervene in the aetiology of a sickness, have been commonplace in medicine in China (Chapter 31). Divination and the cultures of calculation that evoke the authority of the spirits were at the heart of the construction of early Chinese medicine and structured the time-honoured practice of selecting auspicious days for undertaking medical treatment. Manipulation of time was also a key technique in the practice of Chinese alchemy affecting both the compounding of material substances in pursuit of the elixir of life and the inner cultivation that refined the essences of the body (Chapter 29).

Divination features strongly in Daoism where healing regularly involved ritual practices to rid the body of demonic infestation, but is this Daoist medicine as it is currently purveyed on the internet? While a broad variety of therapeutic and cultivation repertoires were assimilated into and developed within Daoist traditions, the moniker of 'Daoist medicine' itself is modern. Abroad, Daoist medicine was a creation of the European and American countercultures during the latter half of the last century. The term came to be used widely to refer to the whole of classical Chinese medicine and self-cultivation cultures as they became meaningful to communities interested in mind-body-spirit healing. Daoism has, therefore, been conflated with Chinese medicine by those interested in locating holism and spirituality in Asia. At home in China, Daoist medicine served as a politically sanctioned label to dissociate from the more volatile *qigong*, sheltering a broad array of regional and orally transmitted practices under the epistemic umbrella of scripturally authorised practice (Chapter 27).

Disentangling the history of religions and healing in China forces us to embrace the diversity of the interlinked traditions. Given the enormous impact of Indian medicine on the Chinese medical landscape and on local religious forms, that came through the introduction of Buddhism (Chapter 28) – monastic provision for the sick, the (quasi-illicit) compounding of drugs by monks and nuns and power of the charismatic healers and deities such as the Medicine Buddha – one has to interrogate the assumption of a monolithic and discrete culturally specific imperial, Daoist or national 'Chinese medicine'.

The worlds of Sinographic medicine: a diversity of interlinked traditions

If you look at the plethora of cultural influences and healing modalities covered in this *Handbook*, you will get some sense of how diverse and interconnected healing in China has been despite, and also paradoxically because of, its grand and eclectic scholarly literature which embraces and monopolises diversity (Chapter 37). The history of medicine and healing in the region must be contextualised within a broader frame that is not limited to the political or linguistic boundaries considered today to be 'Chinese'. Rather, it extends with migrations, adaptations and transformations into nearby regions. Some of these continuities have been mediated by geographic proximity, by political history or by trade, while other regions shared common forms of written language, while using linguistically distinct vernaculars.

The commonalities of writing in Korea, Japan, pre-colonial Vietnam and, more recently, Singapore have allowed for more direct sharing, negotiation and localisation of medical theory (Chapter 33–36). The overland exchange of material culture with Mongolia, Tibet, India and all the regions and cultures along the routes to Persia and beyond (Chen Ming Chapter 32) illustrates the tensions between the historical imaginations of Asia writ large, and a narrower cultural ecumene of 'East Asia'.

These commonalities invite consideration of how the exchange of knowledge of medical materials, bodily and universal cosmography, as well as forms of textuality are reconciled within a common Sinographic sphere, or 'Graphbund', a notion which adapts the

Euro-origin term of Sprachbund. Freeing the history of medicine from its Chinese borders, and considering the forms of its regional porosity, allows for much more nuanced narratives and comparison. Looking westwards to the medieval records of multi-graph and multi-lingual exchange recovered from the library cave at the Buddhist shrines in Dunhuang (modern Gansu, northwest China), we find a rich vein of materials, which enable us to study the notions of periphery and centre. Recent studies of these sources, which lie mainly in the British Library and the Bibliothèque nationale de France, show the extensive circulation of official medical texts produced in the successive Chinese capitals, while also attesting to dynamic local medical cultures and materials, as well as inter-regional exchange, that are all but omitted in official canons and transmitted works.

Wider diasporas

Chinese medicine has been shaped most radically by its multi-faceted encounters in the processes of globalisation from the first major diaspora during the Mongolian Empire to the transmissions and translations of early modern Europe and then the multi-faceted encounters with biomedicine in the twentieth century and beyond. We find it intersecting with new debates about sphygmology in eighteenth-century France, in divisions between Northern and Southern elites in Vietnam, shoring up ethnic communities, with knowledge and the ability to organise in first colonial and then technocratic Singapore, as a vehicle for Chinese soft diplomacy and development in communist African states or as a cultural and economic riposte to the inequities of profit-driven biomedicine in post-Spanish colonies across the world (Chapter 38–42 and 46).

In China itself, the arrival of a new medical knowledge and methods from overseas took place during the time when the empire was weak and being attacked by imperial powers from America, Europe and Japan. While the Qing dynasty lost its increasingly frail grip on central power, new medical techniques were brought in, primarily by Christian missionaries. The first Treaty of Tianjin (1858) granted foreigners immunity from Chinese law as well as unrestricted movement, affecting the relationship between religion and medicine in China in new ways, which nevertheless echoed earlier eras when priests and monks exercised medical authority. Rights under the treaty allowed foreigners to acquire property as well as the power to reside beyond the confines of treaty ports for the first time. Ecclesiastical institutions saw their opportunity, and in the 1890s set up missionary clinics in larger towns and urban areas across the country. It became widely known that converts could be won over by free medical care where preaching had failed. The earliest wave of European-style hospitals and medical schools was due to the efforts of religious institutions, a fact which echoes the medieval introduction of Indian medicine via Buddhist hospitals and rural care in the community. As in the Tang dynasty, this kind of missionary medicine appealed particularly to the rural and urban poor with no social resources. By contrast, however, the secondary political and religious agendas of these institutions led the wealthy, who could afford to pay for expert medical care, to be suspicious of foreign missionary medicine and to defer to traditional medicine.

European anatomical texts were available in China from the early eighteenth century via Jesuit translation but, without verifiable treatment methods to complement them, were considered little more than curious intellectual exotica. The middle of the nineteenth century saw a dramatic change, with the introduction of spectacular new foreign techniques, mainly in anaesthesia and surgery. Nevertheless, despite their dramatic appeal (including the removal of cataracts, tumours, stones and cysts), the nature of these treatments aligned

European surgeons with humbler practitioners of artisanal medicine, rather than with literate scholars. Even before the time of medical records, many forms of minor and superficial surgery had been performed routinely in China, such as lancing abscesses, bloodletting, removing projectiles, suturing wounds, hernia repair, haemorrhoid surgery and castration, as well as acupuncture. By the nineteenth century, a few foreign miracle drugs, in particular quinine and chloroform, were regularly included within Chinese *materia medica* repertoires. The prevention of smallpox also formed an arena wherein foreign and indigenous technologies were negotiated.

Negotiating modernity

Later in the Qing (Manchu) administration, the cause of technological 'Westernisation' was conflated with 'modernisation' and championed by Chinese officials who advocated the use of foreign military methods and tools. Conservative in outlook, they formed a hard core of the Self Strengthening reform movement (1861–95), their slogan for which was *Zhongti xiyong* 中體西用 'Chinese Learning for our foundation, Western learning for practical application'. A minor modernisation programme ensued during which the Tianjin Medical School was established in 1881 as the first national institute where physicians could train in 'Western medicine'.

The vital need for reform was underscored by the debacle of the country's defeat in the Sino-Japanese War (1894–95) and the subsequent fiasco of the Boxer Uprising, which intensified the grip by imperialist powers on the failing Qing state. Chinese sentiment was divided: conservative court officials thoroughly rejected institutional modernisation, while the reformers lobbied for thoroughgoing Westernisation as the sole solution. Numerous intellectuals went overseas to study natural science or medicine, most commonly in Japan, which had instigated a thorough reform programme in the wake of the Meiji Restoration (1868). Overseas medical training continued to be a dominant motif through the early twentieth century in the biographies of major revolutionary writers and reformist politicians, including Dr Sun Yat-sen (1866–1925).

The early twentieth century saw increased foreign activity in China. After the fall of the last imperial house in 1911, the Republican government (1911–49) was more open to trade and diplomacy, but invested little in health. *Health* diplomacy provided one way through which the colonial powers could establish a foothold in China. A century of warfare, rebellion and plague had taken its toll and the second Sino-Japanese War (1937–45) and Chinese Civil War (1945–49) were looming. Foreign public health initiatives such as the League of Nations-Health Organization (LON-HO) and the Rockefeller Foundation's International Health Division sponsored programmes aiming to improve the health of the population through modern scientific nutrition and hygiene in line with the teachings of the Johns Hopkins University School of Public Health (Chapter 47). Modernisers issued a legal challenge to medical practitioners still working in traditional ways, but this had the opposite of the intended effect by galvanising opposition to a new level of professionalisation aligned with the state, and thus a new 'Chinese' medicine was born.

Following the Communist takeover of China, there was a concerted official effort to validate Chinese medicine as compatible with medical modernity. The profession of medical history gained a new salience as it offered a form of legitimacy to Chinese medicine. A precedent for this can be found in histories of Chinese medicine and physicians throughout the imperial period, particularly hagiographies of eminent medical figures. What was significant in the post-1949 accounts was how the past was explicitly used (and in some ways re-created and re-configured) to serve the present in this new post-'liberation' discipline. Where many other instances of 'modernisation' saw the discarding of the 'pre-scientific' past in efforts to construct 'modern' scientific medical systems, China's revolutionary government sought to incorporate and scientise its medical past. This led to political tension, particularly during the Cultural Revolution (1966–76), and confrontations with traditional knowledge, typified by a ruthless re-inscription of tradition and a push to purge Chinese society of superstition and religion.

The years 1953 and 1954 saw the first institutions to teach a newly modern Chinese medicine in Shanghai, Chengdu, Guangzhou, Beijing and Nanjing. The very future of traditional medicine was, however, threatened by the strident ideological debates that foreshadowed the radical social and economic campaigns of the Great Leap Forward (1958–61). During this time, great efforts were devoted to modernisation, standardisation and institution-building for traditional medicine, in order to serve the people. It was at this time that the new term Traditional Chinese Medicine or TCM came into being, coined by Ma Kanwen 馬堪溫 (1927–2016), marking its regional and temporal genesis as it began to migrate abroad.

During the Cultural Revolution (1966–76), there was an extraordinary experiment that embodied Mao Zedong's (1893–1976) spirit of continuing revolution and deploying the masses for the health of the nation, the Barefoot Doctors Campaign, named for the farming people who went barefoot in the paddy fields. Over a million people, mainly farmers, village healers and young graduates, were given a short training in anatomy, bacteriology, maternal and infant care, together with the basics of public health as well as the Chinese *materia medica* and acupuncture, after which they were sent out into the countryside. The effect is very difficult to assess, but many believe that this was one of the few positive elements of the period, and in general a success. As a propaganda campaign, it certainly had a powerful legacy in global health when the 1978 Alma Ata Declaration embraced its principles of popular participation in low-tech health for the masses and the model remains an inspiration for development work (Chapters 44 and 45).

Traditional medicine was, in Mao Zedong's famous adage, 'a national treasure' to be exploited for its benefits to the masses. Inventing a tradition for our times, twentieth-century Chinese leaders were never totally modern, as they retained the vision of a glorious tradition with holistic aspirations – now a post-modern imagination of bodily resonance, a holistic medicine no longer struck through with imperial homologies, but still with a commitment to following the harmony of Heaven and Earth by embodying the rhythm of the seasons *for the universal good of the people.*

But their brave new socialist vision of medicine was also one of material substance and could simultaneously embrace modern reductionist approaches to phytochemistry (Chapter 48). It was fit enough to survive the encounter with early twentieth-century pharmacology and the marriage of old and new would lead ultimately to extracting a substance's bioactive metabolites to make a modern medicine which carried with it the power of tradition (Ibid.).

Despite the new material modernity, the cultivation of *qi* and the spiritual dimensions of ancient China survive vividly in the global cultures of the martial arts and therapeutic movement, and also in meditation regimes. They survive in the work of contemporary Chinese sexologists looking for holistic concepts of sexual well-being (Chapter 26). They have in addition been instrumental in carving out communities for practitioners within which individuals can find common interests in the care of the body, whether for the cultivation of socialist bodies, or in the more individualistic and family oriented domains of the age of reform (Chapter 49).

It is useful to study the translation and analysis of remedies and recipe books in order to learn more about the transmission of knowledge. The process of translating tangible details and material practices brings the modern translator up against similar problems of identification and interpretation of substances and methods that have bedevilled merchants, scholars, pundits and simple end-users of old (Chapter 43).

Ironically, it is this very propensity for change, and an inherent ability to adapt to contemporary situations as well as absorb influences from abroad, that has ensured the durability of this quasi-imperial medicine. To some degree, it has been the polysemic nature of the terminology of Chinese medicine that has ensured a continuity of practice as terms like *qi* segue seamlessly into the imagination of an 'energy' coursing through our bodies as if we can plug ourselves into the national grid. Similarly, we celebrate the 'treasure house' of Chinese medicine and its empirical use of thousands of plants as medicine for providing us with the latest wonder drug to treat malaria, artemisinin – even though that substance, disease concepts, explanations for efficacy and manner of delivery are all changed beyond recognition.

The imagination of a coherent and pervasive Chinese tradition, easily identified, is therefore illusory, ephemeral. Those who have tried to capture it, as a counterpoint to a Western tradition, inevitably run the risk of addressing the subject in crude and unsustainable ways. Given the number of peoples who have lived in, and moved through, China, the origins of Chinese medicine are certainly not entirely Chinese, and historians therefore must embrace an equivalent diversity of healers and healing practices.

The end of Chinese medicine?

Returning to the question posed at the beginning of this introduction and, indeed, implicit in this entire Handbook: what are the limits of Chinese medicine or TCM? At what stage is Chinese medicine no longer Chinese medicine? This is not just a semantic riddle, but an important academic question that rests on contextualised definitions where one necessarily has to state the realms of the discourse in order to make sense of it (Chapter 51). To turn this question on its head, while there have not been many new 'natural' materials added to the Chinese *materia medica* after the Qing dynasty (since 1911), we must wonder whether we can discover new Chinese medicines and therefore develop the tradition. Undoubtedly over the millennia, many substances, such as frankincense, cloves and American ginseng, have been introduced into the Chinese tradition and attributed new properties. Recently, for example, Chinese medical practitioners have begun to employ maca, which originated from South America, for new medicinal purposes. It does not yet have an official monograph in Chinese pharmacopoeia, but the community of Chinese practitioners commonly recognise it as strengthening the yang of kidney based on TCM theory. Working outside of China, Chinese settlers also borrow from the indigenous healing traditions they encounter and draw them into the Chinese recipes and traditions. For example, Peranakan Chinese settlers in the Malay Peninsula use local plants, yet describe them in Hokkien and other southern Chinese dialects.

Many Chinese people and the Beijing government itself are, rightfully, very proud and excited about the discovery of artemisinin and its origin in Chinese medicine; in this historically and politically charged process, this innovation has been construed as a great national achievement for Chinese science and TCM. The product was developed according to a historical book description and it is therefore representative of an ethnopharmacologically driven drug development. Although it was originally extracted with ether, which has never been used in Chinese medicine, it was found that the ancient technique of wringing the juice extracts was actually more effective. Also, in this process, other compounds, which have anti-malaria properties, are extracted. However, artemesinin can simply be regarded as a medicine with an interesting chemistry and the pharmacological properties used in treating the modern disease, malaria, have been proved via modern scientific strategies. How we got to test it in the first place may be no more relevant than the use of aspirin being grounded in centuries of pre-modern European and Islamic healers employing the willow tree for fever and other illnesses.

Modern science has very few strategies for testing multi-ingredient remedies and traditional herbal remedies, such as goji, ginseng and gingko, as well as those like Padma, are marketed because of their profile as TCM preparations. They are all, however, used outside of their original contexts so have but a tenuous connection with pre-modern medicine. Thus, when single herbs or substances that originate in the Chinese *materia medica* are not used within either TCM or pre-modern Chinese conceptual and practical frameworks, they are evidently not Chinese medicine in any traditional way. One must also distinguish between new diagnostic practices where there is some integration of Chinese medical theory such as the use of electronic devices to stimulate acupuncture 'channels' and the use of techniques such as ether to extract bioactive metabolites. The latter was not invented in a Chinese medical context nor is it a practice common to medical practitioners working with TCM. Fruits and herbs that were in the ancient *materia medica* become phytomedicines with usages supported by modern scientific evidence and little or no reference to traditional theory except perhaps as a marketing exercise to evoke the romance of ancient authorities. The modern uses of *cordyceps* or goji in healthcare in what has been called 'the global North' (Chapter 51) have transformed these substances into some sort of a power food or food supplement which are, therefore, no longer traditional medicines. Moreover, once the active ingredient has been derived, or the essence extracted, and a substance is packaged and mass delivered as an injection, as an extract or as a pill, with indications not listed in the pre-modern texts, or in modern TCM practice, can it still be thought of as Chinese or traditional, despite its name and history?

Finally, are there ways to bridge the epistemological gaps between the modern and traditional (Chapter 50) or does Chinese medicine have its own kind of modernity that defies these tidy binaries? This Handbook could never give an exhaustive account of Chinese medicine. The more we cover the more we realise has been omitted. In the end, the answers to these and many questions we have not yet considered are for future researchers, who we hope will benefit in framing their work from the studies in this Handbook.

PART 1

Longue Durée and formation of institutions and traditions

1
YIN, YANG, AND FIVE AGENTS (*WUXING*) IN THE *BASIC QUESTIONS* AND EARLY HAN (202 BCE–220 CE) MEDICAL MANUSCRIPTS

Chen Yun-Ju

Few concepts lie so close to the core of medical thinking and healing in ancient and Imperial China (221 BCE–1911 CE) as *yin-yang* 陰陽 and five agents (also frequently translated as five elements, or five phases, these all render the Chinese term *wuxing* 五行, which are identified as Metal *jin* 金; Water *shui* 水; Wood *mu* 木; Fire *huo* 火, and Earth *tu* 土). In short, the terminology and concepts of *yin-yang* and five agents in Han times gradually came to serve to understand and explain almost all phenomena and their transformations in the universe, including the structures and functions of the human body, the occurrence and development of disorders, and principles of treatments. This section is going to describe the concepts of *yin-yang* and five agents of Han China (202 BCE–220 CE) in the realm which we recognise nowadays as medicine, by drawing examples from excavated medical manuscripts as well as the received medical classic, *Basic Question* (*Suwen* 素問). This work, along with three other works, thought to have been compiled later, constitutes the *Inner Canon of Yellow Emperor* (*Huangdi neijing* 黃帝內經, hereafter, the *Inner Canon*) – the foundational medical classical doctrine for learned Chinese medicine to date (Chapter 7 in this volume).[1]

Yin-yang and five agents (with examples drawn from excavated early Han manuscripts)

The first dictionary in Imperial China, *Explaining Graphs and Analyzing Characters* (*Shouwen jiezi* 說文解字), compiled by the Eastern Han scholar Xu Shen 許慎 in 100 CE, defines *yang* as height (*gao* 高) and brightness (*ming* 明) and *yin* as shadow (*an* 暗), the southern side of rivers, and the northern side of mountains. In earlier times, the *Book of Songs* (*Shijing* 詩經), which may be dated between c. 1000 and c. 800 BCE, used *yin* and *yang* characters to refer to cloudy and sunny weather and presumably to shady and sunny parts of a hill (Kuang 1992: 8–10).

Medical manuscripts preserved in elite families' tombs of early Han China disclose more information about the relationships between the concepts of *yin, yang,* and the human body. Among the most informative manuscripts are Zhangjiashan tomb 247 (found in present-day

Jiangling county, Hubei province; earliest date of burial in 186 BCE – latest date of burial in 156 BCE, excavated in 1983–84) and Mawangdui tomb 3 (found in present-day Changsha city, Hunan province; buried in 168 BCE, excavated in 1973). Roughly speaking, the relationships between *yin*, *yang*, and the body in the Zhangjiashan and Mawangdui manuscripts can be divided into at least six forms, which I will introduce in sequence in the following paragraphs.

The first relationship is that titles of some channels (or vessels, *mai* or *mo* 脈) of the body bear the term *yin* or *yang*. One of the Zhangjiashan manuscripts, known as the *Channel Book* (*Maishu* or *Moshu* 脈書), states that there are twelve *yang* channels and ten *yin* channels in total.[2] It also mentions nine more channels bearing the term *yin* or *yang* in their titles, such as channels of minor *yang* (*shaoyang* 少陽); *yang* brightness (*yangming* 陽明); major *yin* (*taiyin* 泰陰), and so forth (*Jiangling Zhangjiashan hanjian zhengli xiaozu* 1989: 72–3). Notably, the *yin* and *yang* channels described in the *Channel Book* make no reference to *qi* (see Introduction). Two Mawangdui manuscripts known as *Cauterization Canon of the Eleven Channels of the Foot and Forearm* (*Zubi shiyi mai jiujing* 足臂十一脈灸經) and *Cauterization Canon of the Eleven Yin and Yang Channels* (*Yinyang shiyi mai jiujing* 陰陽十一脈灸經) provide the earliest textual evidence of *yin* and *yang* channel titles totalling eleven channels. These included two major *yang* channels, two *yang* brightness channels, two minor *yang* channels, two major *yin* channels, two minor *yin* channels, and one dull *yin* channel (Harper 1998: 192–212). Another Mawangdui manuscript, *Death Signs of the Yin and Yang Channels* (*Yinyang mai sihou* 陰陽脈死候), uses similar names for the channels. It moreover mentions the case that the disturbance of *yin* channels would lead to death (Harper 1998: 81–90).

The second form of the association of *yin* and *yang* with *qi* can be found in one Zhangjiashan manuscript, known by the modern title, *Pulling Book* (*Yinshu* 引書). This gives rise to the second form by which *yin* and *yang* were linked to the body. It states:

> The reason why nobility get illness is that they do not harmonize their joys and passions. If they are joyful then the *yang qi* is in excess. If they are angry then the *yin qi* is in excess. On accounts of this, if those that follow the Way are joyful then they quickly exhale [warm breath] (*xu* 呴), and if they become angry they increasingly puff out [moist breath] (*chui* 吹), in order to harmonize it. They breathe in the quintessential *qi* of heaven and earth to make *yin* substantial, hence they will be able to avoid illness.
>
> (Translation is based on Lo 2001a: 26–27)

貴人之所以得病者，以其喜怒之不和也。喜則陽氣多，怒則(陰)氣多，是以道者喜則急呴(呴)，怒則劇炊(吹)，以和之。吸天地之精氣，實其(陰)，故能毋病。[3]

This account, alongside other language in the Zhangjiashan manuscripts, expresses how *yin*, *yang*, and *qi* first came to relate to each other in the healing arts (Lo 2001a; Lo and Li 2010: 380).

The third form in which *yin-yang* relates to the body is the association of *yin* and *yang* with the location of injuries and disorders. After introducing channels, the *Channel Book* categorises location of injuries and disorders according to *yin* and *yang*. It states:

> In all cases the three *yang* are the *qi* of heaven. Of their illness only if there is breaking of the bones and splitting of the [skin] will death occur. In all cases the three *yin* are the *qi* of earth. They decay the viscera and rot the bowels which is where the killing resides.
>
> (Translation is based on Lo 2000: 26)

Yin, Yang and five agents

凡三陽，天氣殹，其病唯折骨列裂□死。凡三陰，地氣殹，死脈殹，腐臧爛腸而主殺。

(M3.1, 3.2)[4]

The location of three *yang* injuries concerns the outside body, while that of three *yin* disorders relates to the inner body (Lo 2000: 26).

The fourth form is that concepts of *yin* and *yang* were used to select auspicious or inauspicious locations. One Mawangdui text, titled in modern times *Book of the Generation of the Fetus* (*Taichan shu* 胎產書, indexed in Harper 1998 as MSV), suggests locations for burying the afterbirth which can bring positive influences on a child throughout his lifetime (Li 1994; Harper 1998: 367, 374).

The fifth form is the notion that *yin* serves as a primary element of human life. Although many passages in both the Zhangjiashan and Mawangdui manuscripts mention cultivation of the body, they seldom refer to *yin* and *yang*. When referring to *yin-yang* in the bodily cultivation, they mostly identify *yin* (not *yang*) as the primary element of life to be preserved or cultivated (Harper 1998: 122). For instance, the *Pulling Book* says that 'those who are good at regulating their *qi* and solidifying their *yin* will benefit their body (能善節其氣而實其 [陰]，則利其身矣)' (Ibid.). One Mawangdui manuscript, with the modern title *Discussion of the Culminant Way in Under-heaven* (*Tianxia zhidao tan* 天下至道談, MSVII.B), describes arts of 'eight benefits' (*bayi* 八益), a set of regulated techniques through which men bring women to orgasm and in eight steps. According to the *Discussion*, if men fail to conduct these arts and to circumvent 'seven detriments' (*qisun* 七損) because of inadequate sexual practice, he would lose half of his lifetime supply of *yin qi* by his fortieth year, and would not be able to use his *yin qi* in his seventieth year; by contrast, men should enjoy increasingly strengthened *yin qi*, keenness of the ears, brightness of the eyes, and lightness of the body, if they practise the eight benefits and refrain from the seven detriments.

The sixth form, a rather indirect manner by which the body was conceptualised through *yin* and *yang*, is the correspondence between *yin*, *yang*, and male and female sexual activities. The last passage of the *Discussion of the Culminant Way* mentions the correspondence (Table 1.1) before saying that 'this is called the calculation of *yin* and *yang*, and the intrinsic pattern of female and male'. The correspondence between *yin* and *yang* is further elaborated in a text titled *Cheng* 稱 (*Weighing Factors*) in the Mawangdui manuscript (MS.B). In both of these manuscripts, *yin* and *yang* serve to categorise objects and phenomena in contrast to one another.

As for the five agents, they are absent from the excavated Han manuscripts which have been published at the time of going to press in early 2020 (Editors: The *Tianhui* 天回 manuscripts, also known for the Laoguanshan 老官山 district in Chengdu where they were uncovered, and which are still in press, reference the five organs and five colours (Chapter 3 in this volume)).[5] The term only appears in modern titles of certain excavated Han manuscripts. The description closest to the concept of five agents with which we are familiar (see below) can only be found in the *Book of the Generation of the Fetus*. This book accounts the development of the foetus from the fourth through the eighth months as follows: in the fourth month, Water was received and then blood was 'formed' (*cheng* 成); in the fifth month, Fire was received and then *qi* was formed; in the sixth month, Metal was received and then sinews (*jin* 筋) were formed; in the seventh month, Wood was received and then bone was formed; in the eighth month, Earth was received and then 'skin' (*fuge* 膚革) was formed. The foregoing account of Water, Fire, Metal, Wood, and Earth that appear in the controlling sequence and each of them engenders different bodily parts is reminiscent of the five agents (Harper 1998: 79). However, the book makes no reference to the very term *wuxing* itself.

The range of foregoing associations between *yin*, *yang*, and the body in early Han manuscripts differs conspicuously from the range of associations in the *Basic Questions*. To be very

Table 1.1 Yin-Yang Correspondence in Mawangdui Manuscripts the *Discussion of the Culminant Way in Under-heaven* and the *Cheng*

Yin	Yang	Sources
Internal	External	The *Discussion of the Culminant Way in Under-heaven* (Translation is based on Harper 1998: 438)
Category female (*cipin* 雌牝)	Category male (*xiong mu* 雄牡)	
Rubbed inside (*li* 裡)	Rubbed outside (*biao* 表)	
Earth	Heaven	*Cheng* (Translation is based on Lo 2000: 27)
Autumn	Spring	
Night	Day	
Insignificant states	Important states	
Not having things to do	Having things to do	
Contracting	Stretching	
Minister	Ruler	
Below	Above	
Women	Men	
Child	Father	
Younger brother	Elder brother	
Younger	Older	
Base	Noble	
Staying poor	Achieving worldly success	
Mourning the dead	Taking a wife and having children	
Being organised by others	Organising others	
(The body as a) Host	(Disorders conceived as a spirit- entity like a) Guest	
Conscripts	Commander/Soldiers	
Silence	Speech	
Receiving	Giving	

brief, the sections in the *Basic Questions* were amended and re-arranged repeatedly from at least the Han dynasty onwards, and the *Basic Questions* should not be regarded as a single book but as a collection of short essays from various medical lineages (Chapter 7 in this volume). Within the extant version of the *Basic Questions*, sections sixty-six to seventy-one and seventy-four are highly likely to have been interpolated by a Tang official Wang Bing 王冰 in 762. These seven sections will not be discussed in the following description.

The following paragraphs will draw some passages about the concepts of *yin-yang* and five agents from the *Basic Questions* as an example to illustrate the implications of these concepts in medical works apart from excavated manuscripts. This section ends by introducing a scholarly discussion on the political and cultural milieu of Western Han China in which the above-mentioned concepts of *yin-yang* and five agents were embedded.

Yin-yang and five agents with examples drawn from the *Basic Questions*

The concepts of *yin-yang* and five agents in the *Basic Questions* do not appear in any one single comprehensive section, but spread out in various sections that often differ from one another. Many passages in the *Basic Questions* adopt different ways of categorising phenomena into five groups without mentioning the term five agents at all, such as the five flavours (*wuwei* 五味), or the five viscera. The inconsistencies between these numerous accounts make it difficult to provide a systematic and exhaustive description of *yin-yang* and five agents as they appear in the *Basic Questions*. The aim of this sub-section then is to describe the more conspicuous examples in which *yin*, *yang*, and five agents are used to organise the understanding of the body and its processes.

Yin and *Yang*

One of the most noticeable differences between the application of *yin-yang* concepts in excavated early Han medical manuscripts and in the *Basic Questions* is the fact that the latter assigns *yin* and *yang* to a much wider range of objects and phenomena. For instance, *yin* and *yang are* not only used to distinguish the six channels types (e.g. *Suwen* 22; see Table 1.6 for the list of *yin-yang* channel titles). They also organise relationships between body parts and inform the division of five viscera (*wuzang* 五臟) as *yin* and six bowels (*liufu* 六腑) as *yang* (*Suwen* 4; see Table 1.2).

Emphasis is put on the systematic functions of the five viscera and six bowels, rather than their anatomical structures. For instance, the kidneys were considered to govern the reproductive function and to store 'essence' (*jing* 精), the vital fluid closest to *qi*, which manifests as semen in men, but which circulates throughout the body providing nourishment and generative energy. The liver stored 'blood' (*xue* 血), which also differs from the biomedical understanding of blood but was understood as a general type of bodily fluid containing life energy and particularly relating to reproduction (such as menses and human milk).[6]

Authors of the *Basic Questions* moreover explained the systematic functions of the 'twelve viscera' (*shier zang* 十二臟) via imperial bureaucratic titles and structures. Each of the five organs was assigned a domain of visceral functions in the way officials were charged with certain bureaucratic duties (*Suwen* 8, Table 1.3). The Han understanding of the human body in the light of bureaucratic systems gives expression to one important feature of medical thinking in the *Basic Questions*: that is, the correspondence between governance, the body, and the cosmos (Sivin 1995).

Discussions about *yin-yang* aspects of body parts in the *Basic Questions* often appear in the context of explaining the occurrence of disorders; treatments; and the maintenance of health. For instance, after listing the *yin-yang* quality of body parts, section four of the *Basic Questions* claims that disorders in winter and spring lie in *yin*; disorders in summer and autumn lie in *yang*. The harmony of *yin* and *yang* inside the body is by no means merely a matter of the body enclosed under the skin but of correspondence with the configuration of *yin* and *yang* forces in the universe. For instance, section five in the *Basic Questions* says

Table 1.2 Yin-Yang of the Body in Section Four of the *Basic Questions*

Yin	*Yang*
Internal	External
the Abdomen	the Back
the Five viscera (*wuzang* 五臟): heart (*xin* 心); liver (*gan* 肝); spleen (*pi* 脾); lung (*fei* 肺) and kidney (*shen* 腎)	the Six bowels (*liufu* 六腑): gallbladder (*dan* 膽); stomach (*wei* 胃); large intestine (*dachang* 大腸); small intestine (*xiaochang* 小腸); bladder (*pangguang* 膀胱) and triple burner (*sanjiao* 三焦)

Table 1.3 Correspondences between Imperial Bureaucratic Structures and Twelve Viscera

Viscera	Bureaucratic Title	Bureaucratic Function
Heart	Ruler	Spirit brilliance
Lung	Chancellor and mentor	Order and moderation
Liver	General	Planning and deliberation
Gallbladder	Rectifier	Decisions and judgement
Danzhong 膻中	Minister and envoy	Joy and happiness
Spleen and Stomach	Official for grain storage	Five flavours
Large intestine	Official transmitter along the Way	Changes and transformations
Small intestine	Official recipient of what has been perfected	The transformation of things
Kidneys	Official operator with force	Technical skills and expertise
Triple burners	Official operator of channels	The path of water originates in them
Bladder	Official regional rectifier	Body liquids are stored in it

Translation based on Unschuld *et al.* (2011: 155–8).

that 'Heaven is insufficient in the northwest, thus the northwest is *yin* and human ears and eyes on the right are not as good as those on the left' (translation based on Hanson 2011: 33). Descriptions of *yin* and *yang* channels often appear in the context of choosing channels to be treated for certain disorders, especially treatments by needling (*ci* 刺) (e.g. *Suwen* 24 and 32).

Some sections in the *Basic Questions* discuss *yin* and *yang* as the origins of myriad things in the universe, and claim that following the pattern of *yin* and *yang* is the way to maintain health. Section two of the *Basic Questions* states that *yin*, *yang*, and the four seasons were the origins of myriad things; if only a sage is able to follow *yin*, *yang*, and four seasons, then he will not suffer disorders.[7] In a dialogue between the Yellow Emperor and his minister Qibo 岐伯, Qibo says that a wise man who is able to conduct eight benefits (*bayi* 八益) will then enjoy a strong body when he was old. This is the way that a sage 'ruled his body and himself' (*zhishen* 治身) (*Suwen* 5). The notion that a man who was shrewd enough to follow normal transformations of phenomena in the cosmos was valued as wise as a sage reveals the synthesis of the cosmos-state-body, in which correspondences between the body and the cosmos also extended to the emperor's fitness for governance (For the further discussion of nurturing the body, see Chapter 6 in this volume).[8]

Five agents

The *Basic Question* bears witness to a wider scope of phenomena that had been systematically corresponded to five agents, which is exemplified in section five (Table 1.4).

Furthermore, each of the five agents was considered to act either as an adversary overcoming another agent, or as a promoter generating a third one. However, the received *Basic Questions* leaves us merely partial cycles of the mutual generation and overcoming among each of the five agents, which is evident in its section twenty-five (Table 1.5).

The affinities of the five agents to organs could further serve as a theoretical ground for treatments. An example can be found in Qibo's explanation to ways of following five seasons and five agents to treat disorders (*Suwen* 22). One of his explanations reads as follows: 'The Kidneys are dominant (*zhu* 主) [in] winter. The minor *yin* and major *yang* channels of the foot dominate the treatments' (see Table 1.6 for the list of the affinities of the five agents to *yin-yang* channels in *Suwen* 22).

Table 1.4 Selected Correspondence between Five Agents and Phenomena in *Suwen* 5

Five Agents	Direction	Quality	Flavour	Viscera	Body Part
Wood	East	Wind	Sourness	Liver	Sinews
Fire	South	Heat	Bitterness	Heart	Blood
Earth	Centre	Dampness	Sweetness	Spleen	Fleshes
Metal	West	Dryness	Pungency	Lung	Skin and hair
Water	North	Coldness	Savoury	Kidney	Bone marrow

Table 1.5 Mutual Generation or Overcoming among Five Agents in *Suwen* 25

Sequence of Generation	Wood (generates) →Fire → Earth → Metal → Water →Wood
Sequence of Overcoming	Earth (overcomes)→Water→Fire→Metal → Wood→Earth

Table 1.6 The Affinities of the Five Agents to *Yin-Yang* Channels in *Suwen* 22

Viscera	Season	Channel Title
Liver	Spring	Dull *yin* and minor *yang* channels of the foot (*zu jueyin shaoyang* 足厥陰少陽)
Heart	Summer	Minor *yin* and major *yang* channels of the hand (*shou shaoyin taiyang* 手少陰太陽)
Spleen	Long summer (*changxia* 長夏)	Major *yin* and *yang* brightness channels of the foot (*zu taiyin yangming* 足太陰陽明)
Lung	Autumn	Major *yin* and *yang* brightness of the hand (*shou taiyin yangming* 手太陰陽明)
Kidney	Winter	Minor *yin* and major *yang* channels of the foot (*zu shaoyin taiyang* 足少陰太陽)

The bodily affinities of *yin/yang* and five agents should be regarded as homologies rather than symbolic metaphors. In other words, the affinities of *yin/yang* and five agents were not understood as analogies or allusions, but as a common fundamental quality or energy shared between the bodily parts and their agents.

Medicine of systematic correspondence in Han China: a new style of thought

The systematic correspondence of *yin-yang* and five agents under discussion here, though inconsistent and scattered across received Han medical sources, reflects a new style of thought. In the Shang dynasty (ca. 1600–1046 BCE), disorders were often attributed to ancestors and demons; accordingly, typical treatments were to pacify ancestors or expel demons. In contrast, in the foregoing medical manuscripts and classics, the occurrence of disorders was attributed to disharmony between *yin* and *yang* in the internal body, to the insufficiency of *yin* power of the body, and to changes of environment (such as cycles of the four seasons). Therapies that treated the body internally without recourse to invisible, conscious agents came to the fore, such as drug therapy and acupuncture.[9] These new theories of pathogenesis and corresponding treatments are also evident in the eminent early Western Han physician Chunyu Yi's 淳于意 (b. 216 or 206 BCE) memoirs. In his memoir, Chunyu used the terms *yin* and *yang* extensively to explain the occurrence and development of his patients' disorders (Hsu 2010). The stark contrast between earlier disease theories and Han aetiology and treatments is referred to by scholars as a 'new style of thought' (Unschuld 2003: 319).

This new style of thought in medicine was embedded in a long-term development of cosmological thought that could be traced back at least to the Warring States period (475–221 BCE) and which reached mature synthesis of universal phenomena via numerical categories in the Western Han.[10] Nevertheless, a consensus among historians has not yet been reached on a specific time when *yin-yang* and five agents unambiguously referred to *qi*. It is now thought that the *Spring and Autumn of Master Lü* (*Lüshi chunqiu* 呂氏春秋), a work composed by Lü Buwei 呂不韋 (d. 235 BCE), Chancellor of the State of Qin, had pivotal importance as the earliest literary source to link *qi* to *yin-yang* and five agents extensively (Lloyd and Sivin 2002: 264). Another historian argues that before the late Western Han, it is difficult to find solid evidence of the connection between *yin-yang*, five agents, and *qi* (Nylan 2010: 399). In the Western Han, the link between *yin*, *yang*, and *qi* appears much more often than the association of five agents with *qi*. This more frequent appearance of linking *qi* with *yin-yang* is exemplified in the *Master Huainan* (*Huainanzi* 淮南子) compiled under the patronage of the King of Huainan Liu An 劉安 (?179–122 BCE) and in a few writings in the *Abundant Dew on the Spring and Autumn Annals* (*Chunqiu fanlu* 春秋繁露) reasonably attributed to Dong Zhongshu 董仲舒 (?179–?104 BCE). Until the first century, the *Canon of Supreme Mystery* (*Taixuan jing* 太玄經), composed by Yang Xiong 揚雄 (53 BCE–18 CE) and completed in 4 CE, marked the fullest integration of *qi*, *yin-yang*, and five agents (Lloyd and Sivin 2002: 264–5).

Historical reasons for the prominence of synthesising *qi* with *yin-yang* and five agents in Han times are still a question awaiting further research. One scholarly suggestion is that this cosmological synthesis made the newly established Han imperial bureaucratic system appear to be a natural feature of the cosmos, and thus a pertinent way of governing reunited Chinese territory. This synthesis resulted less from common underlying mentalities but from the needs of the Han government to justify its imperial bureaucratic institutions (Sivin 1995). Another suggestion is that the Western Han thinkers, such as Dong Zhongshu proposed the notion that since *yin* and *yang* consisted of both heaven and human, the latter could affect the natural world if they understood the principles of *yin-yang* cosmos and conducted self-cultivation. This notion served as a critique of the emperor's attempt to influence the natural world and gain immortality via offering sacrifices to spirits (Puett 2002: 287–315).

Concluding remarks

To sum up, this section has described how a variety of systematic correspondence between *yin*, *yang*, five agents, and the body, which appeared in manuscripts and in received texts during Han times, discloses a pluralism of healing arts in early imperial China. The pronounced systematic correspondence manifests itself in the schemes of *yin-yang* and five agents in the *Basic Questions*. In contrast, Zhangjiashan manuscripts mentioned little about *qi* and five agents (Lo 2000, 2001a). Mawangdui manuscripts also barely refer to circulation of *qi* and systematic correspondence in terms of five agents, but list a considerable amount of non-theoretical recipes and healing methods (Harper 1998: 90–8). Chunyu Yi's memoir similarly talked much about *yin*, *yang*, and *qi*, rather than about the five agents when explaining the occurrence and treatments of disorders. The development towards synthesising *qi* and *yin-yang* with five agents, to a greater or lesser degree, together shows a new style of thought; that is, the attribution of the occurrence of disorders became disharmony of *yin* and *yang* in the internal body or environmental patterns, rather than demons or ancestors in Shang times.

The rising prominence of systematic correspondence in Han healing arts paralleled a broader trend among Han thinkers who attempted to use the correspondence to draw heaven; earth; body; and state together conceptually. One possible reason why some Han thinkers advocated the heaven-body-state synthesis is that this concept would help them naturalise the bureaucratic system of Han state. Another possible reason is that Han officials attempted to regulate the emperor's power. The heaven-body-state synthesis became more elaborated and applied to wider scopes in healing arts in later periods, which will be mentioned in other sections.

Notes

1 For further discussion about the development of the concepts of *yin-yang* and five agents from its earliest concrete appearance during the late Warring States period (ca. 400 BCE) to its prominence in the Han dynasty (ca. 200 CE), see Harper (1999: 860–6); Lloyd and Sivin (2002: 193–203, 253–71); and Nylan (2010).
2 Many of the manuscripts are untitled. In these cases, modern archaeologists and editors have given them names, referred to here as their 'modern titles'.
3 *Jiangling Zhangjiashan hanjian zhengli xiaozu* 1990: 86.
4 Ibid. 1989: 73.
5 Recent scholarship on philosophy (Queen 1996), politics (Loewe 2004: 496–7), and medicine (Hsu 2010: 23) has observed that by the second century BCE, ideas of five agents were less elaborated than the concepts of *yin* and *yang*, which were much more prominent.
6 For discussion of the difference between *xue* 血 (blood) in medicine of imperial China and blood in biomedicine, see Furth (1986).
7 Puett (2002) emphasises that in early China, the concept that the cosmos is a self-generating and spontaneous system arose not only against dominant sacrificial practices held by states as attempts to win the support of ancestors and gods, but also against self-divination claims that humans potentially owned divine power. For the concepts of sage's self-cultivation in early Imperial China, see Csikszentmihalyi (2004: 141–60).
8 Besides arts of self-cultivation, possible therapeutic implications of *yin-yang* and five agents in the *Inner Canon* contain narratives which depict the proficient physician as capable of determining the occurrence and development of disorders through the theoretical framework of *yin-yang* or five agents, such as through changes of *yin* and *yang* forces in the body and through five agents seasonal patterns or irregularities. Though the *Inner Canon* does not flesh out details of pharmacotherapy and acupuncture, it mentions a general principle of restoring *yin* and *yang* forces to harmony by replenishing (*bu* 補) the deficient force or dispersing (*xie* 洩/泄) the surplus one.
9 To be sure, the older ideas still coexisted with new concepts of *yin*, *yang*, and *qi* to varying degree in texts of healing arts after the Han dynasty; for example, see Stanley-Baker's article (2014) on the fourth-century *Declarations of the Perfected* (*Zhen'gao* 真誥) and Lo's analysis (2001b) of the *Yellow Emperor's Toad Canon* (*Huangdi hama jing* 黃帝蛤蟆經), whose compilation dates to the early Tang dynasty (618–907 CE).
10 For possible ways in which the *Inner Canon* serves as valuable textual evidence of this development, see Unschuld (2003: 325–49). Nylan (2010) argues that the fullest expression of correlative thinking is during the late first century BCE.

Bibliography

Pre-modern sources

Suwen 素問 (Basic Questions), reference to *Huangdi neijing Suwen* 黃帝內經素問 (The Yellow Emperor's Inner Canon, Basic Questions) Zhou to Han, 3rd c. BCE to 3rd c. CE, anon., Guo Aichun 郭靄春 (ed.) (1992) *Huangdi neijing Suwen jiaozhu* 黃帝內經素問校注 (The Yellow Emperor's Inner Canon, Basic Questions, with Annotations), Beijing: Renming weisheng chubanshe.

Modern sources

Csikszentmihalyi, M. (2004) *Material Virtue: ethics and the body in early China*, Leiden: Brill.
Furth, C. (1986) 'Blood, body and gender: medical images of the female condition in China 1600–1850', *Chinese Science*, 7: 43–66.
Hanson, M. (2011) *Speaking of Epidemics in Chinese Medicine: disease and the geographic imagination in late imperial China*, London: Routledge.
Harper, D. (1998) *Early Chinese Medical Literature: the Mawangdui medical manuscripts*, New York: Columbia University Press.
—— (1999) 'Warring states: natural philosophy and occult thought', in M. Loewe and E.L. Shaughnessy (eds) *The Cambridge History of Ancient China: from the origins of civilization to 221 BC*, Cambridge: Cambridge University Press, pp. 813–84.
Hsu, E. (ed.) (2001) *Innovation in Chinese Medicine*, Cambridge: Cambridge University Press.
—— (2010) *Pulse Diagnosis in Early Chinese Medicine: the telling touch*, Cambridge: Cambridge University Press.
Jiangling Zhangjiashan hanjian zhengli xiaozu 江陵張家山漢簡整理小組 (1989) 'Jiangling Zhangjiashan hanjian maishu shiwen 江陵張家山漢簡脈書釋文' (Interpretation of the book of the *Mai* on Han bamboo slips from Zhangjiashan, Jiangling), *Wenwu*, 7: 72–4.
—— (1990) 'Jiangling Zhangjiashan hanjian yinshu shiwen 江陵張家山漢簡引書釋文' (Interpretation of the book of pulling on Han bamboo slips from Zhangjiashan, Jiangling), *Wenwu*, 10: 82–6.
Kuang Zhiren 鄺芷人 (1992) *Yinyang wuxing jiqi tixi* 陰陽五行及其體系 ('*Yin-Yang* and the Five Agents' and Its System), Taipei: Wenjin chubanshe.
Li Jianmin 李建民 (1994) 'Mawangdui Han mu boshu Yucang maibao tu jianzheng 馬王堆漢墓帛書「禹藏埋胞圖」箋証' (Textual research on the silk writing entitled 'diagram for burying afterbirths from Mawangdui'), *Zhongyang yanjiu yuan lishi yuyan yanjiusuo jikan* 中央研究院歷史語言研究所集刊, 65.4: 725–832.
Lloyd, G.R.E. and Sivin, N. (2002) *The Way and the Word: science and medicine in early China and Greece*, New Haven, CT: Yale University Press.
Lo, V. (2000) 'Crossing the *neiguan* 內關 "Inner Pass": a *nei/wai* 內外 "inner/outer" distinction in early Chinese medicine', *East Asian Science, Technology, and Medicine*, 17: 15–65.
—— (2001a) 'The influence of nurturing life culture on the development of Western Han acumoxa therapy', in E. Hsu (ed.) *Innovation in Chinese Medicine*, Cambridge: Cambridge University Press, pp. 19–50.
—— (2001b) 'Huangdi hama jing (Yellow Emperor's Toad Canon)', *Asia Major*, 14: 61–99.
Lo, V. and Li, Jianmin (2010) 'Manuscripts, received texts, and the healing arts', in M. Nylan and M. Loewe (eds) *China's Early Empires: a re-appraisal*, Cambridge: Cambridge University Press, pp. 367–97.
Loewe, M. (2004) *The Men Who Governed Han China: companion to a biographical dictionary of the Qin, Former Han and Xin period*, Brill: Leiden.
Nylan, M. (2010) '*Yin-yang*, five phases, and *qi*', in M. Nylan and M. Loewe (eds) *China's Early Empires: a re-appraisal*, Cambridge: Cambridge University Press, pp. 398–414.
Nylan, M. and Loewe, M. (eds) (2010) *China's Early Empires: a re-appraisal*, Cambridge: Cambridge University Press.
Puett, M. (2002) *To Become a God: cosmology, sacrifice, and self-divinization in early China*, Cambridge, MA: Harvard University Asia Center for the Harvard-Yenching Institute, Harvard University Press.
Queen, S. (1996) *From Chronicle to Canon: the hermeneutics of the Spring and Autumn according to Tong Chung-shu*, Cambridge: Cambridge University Press.
Sivin, N. (1995) 'State, cosmos, and body in the last three centuries B.C.', *Harvard Journal of Asiatic Studies*, 55.1: 5–37.
Stanley-Baker, M. (2014) 'Drugs, destiny, and disease in medieval China: situating knowledge in context', *Daoism: Religion, History and Society*, 6: 113–56.
Unschuld, P.U. (2003) *Huang di nei jing Suwen: nature, knowledge, imagery in an ancient Chinese medical text*, Berkeley: University of California Press.
Unschuld, P.U., Tessenow, H. and Zheng Jinsheng (2011) *Huang di nei jing su wen: an annotated translation of Huang Di's Inner Classic – Basic Questions*, Berkeley: University of California Press.

2
Qi 氣

A means for cohering natural knowledge[1]

Michael Stanley-Baker

Pneuma, ki, prana, air, breath, vital energy, energy, magnetism – these terms and more have variously been employed to translate the term *qi* over the long history of the Western reception of the world.[2] It is so frequently used in English language contexts that it has now even entered some English language dictionaries. Many Anglophone representations of *qi* have, however, portrayed it as a singular, incomprehensible and universal force that borders on the quasi-mystical. This paper points out how such portrayals short-change one of the most fundamental topics in the history of science in China. *Qi* does not simply point towards a mystical unity between the material and unmanifest worlds, but plays a variety of roles in Chinese medical practice, as well as in other pre-modern sciences. Recent anthropological studies argue that *qi* functions as a linguistic code, providing social and intellectual coherence among practitioners of Chinese medicine and drawing together multiple ways of knowing the world.

Qi, standardised nowadays as 氣 or 气, and less commonly found as 吃, 噢, 炁, 氖, and 餼, is normally referred to as *the* fundamental universal substance in Chinese cosmology and physiology. It has played a fundamental role in the fields of philosophy, self-cultivation and ritual practice, as well as in the natural sciences such as medicine, alchemy and astronomy. It is related to the breath, to the vitality of the body, to primordial cosmic substance, to food, to the stars and more. Modern dictionaries give up to thirty-five different definitions of the single character alone. The definitions mostly come in compound terms, pointing to different areas of the vast semantic field such as *qixiang* 氣象 (lit. *qi*-image or weather pattern), *qisu* 氣俗 (fashions and customs) or the modern *qigong* 氣功 (physical exercises which manipulate *qi* in the body – see below).[3] Compound forms beginning with the term 氣 number over 230, a figure which does not begin to address the much wider variety of compounds where modifiers prefix the term, such as *shenqi* 蜃氣 (mirage), *yongqi* 勇氣 (heroic *qi*, or courage), *guqi* 穀氣 (grain *qi*, or nourishment from food) or *tianqi* 天氣 (heavenly or natural *qi*) – which in modern Chinese refers to weather – but in middle and classical Chinese can include air, destiny, seasonal patterns or even daily time markers. *Qi* has played an important role in practices as diverse as military strategy, literary writing, calligraphy, painting, music and the art of conversation.

While many of these uses rely on similar conceptions of *qi*, and cluster together related material practices, many others use the term in ways that bear little or no relation to one another. This is worth bearing in mind when considering how *qi* has functioned in the history of Chinese science. The traditional and secondary literature has emphasised the concept of *qi* as a unitary, singular substrate of the material universe in all its diversity, whereas in practice, it is frequently divided into typologies, and articulates a wide variety of refined observations of material transformation. It is thus worth paying attention not just to 'what *qi* is', but also to how the term functions to impute conceptual and material continuity – a coherence which draws different technical disciplines into a common conversation. Readers not interested in early Chinese language may want to skip past the sections on grammar and writing, to the conceptual history which follows.

The grammar of *qi*

Some language philosophers argue that the grammar of *qi* may point towards fundamental ways of thinking about the world in early China. *Qi* is a 'mass noun', that is, a general type of 'stuff' that cannot be counted as individual things, such as carrots, but is uncountable like water. Mass nouns cannot be measured on their own as units, but are distinguished by types, or by specific measures: a cup of water, a pot of wine, a basket of food. Early Chinese mass noun phrases, such as 'three wines' (*sanjiu* 三酒), refer to three types of wine rather than three units. Thus, 'six *qi*' refers to different types of *qi* (Harbsmeier et al. 1998: 312–21).[4] *Qi* nevertheless has quantity – physicians diagnose disease in terms of 'excess' (*shi* 實) or 'deficient' (*xu* 虛) *qi*. Graham (1989, 402) argues that this way of thinking about *qi* points to a fundamental difference between Chinese thought and Western philosophy, because it counts types of *qi* as subsets of a single whole, rather than as discrete particulars which can be added up. While such general claims unavoidably tend towards essentialism, they are nevertheless provocative ideas for thinking about a term which became 'the backbone of Han cosmology and physical thought' (Sivin 2000: 123–4). A full grammatical study of *qi* remains to be undertaken, but many agree that it could provide useful insights.

Imaging and sounding qi

Figure 2.1 气 in *Shuowen jiezi*

The early forms of the character can tell us about the conceptual roots and phonetics behind the term, and demonstrate that *qi* is not a static concept (Figure 2.1).[5] Different meanings have emerged over time, and scholars are not quite agreed on the order in which they emerged. The longstanding traditional argument by Xu Shen 許慎 (30–124 CE) was that the earliest and original meaning of the graph referred to 'vapours', that is, clouds, mist and fog, and that from this the meaning later extended to the breath and the stuff of the cosmos.[6] Although this story was developed quite late, in the first century CE, it has been influential ever since. Modern palaeographers note, however, that older forms of this graph dating back to the second millennium BCE (三, 气 and 气) were used to write a quite different word, pronounced 'xjət 乞', which meant 'seek, arrive, to reach'.[7] The graph was only used as a sound-loan (*jiajie* 假借) to represent the term for 'mists and vapours' (pronounced $k^h jei^C$), starting from around 400 BCE.

This phonetic reconstruction of the pronunciation bears a striking similarity to words which mean 'anger', or 'the forced expulsion of air along with physical effort' in

Qi 氣

other related languages, namely Austroasiatic (kʰis, kissa, kᴴɛs) and Sino-Tibetan (khʊs) (Schuessler 2007: 60, 423). This corresponds to the Chinese use of *qi* as 'to become angry', which is still in use today. It is therefore not certain that the earliest root meaning of the word was 'vapours', and its early roots may come from more physio-emotional domains (Figure 2.2).[8]

In fact, the earliest extant graph describing *qi* as the movement of breath in the body also uses the fire component. Graph A below, from an early inscription on body cultivation dating to roughly 380 BCE, combines the 'vapours' component with the fire component (火 or 灬). The following graphs are arranged roughly chronologically, ending in the first century CE.[9] Graphs B and C also represented *qi*, but combine fire with the sound components 既 and 旡 (kjeiᶜ), and no vapours. The latter form (D) became adopted throughout the imperial period to refer to special forms of embodied *qi* (Stanley-Baker 2019b; Zhu Yueli 1982) (Figure 2.3).

During roughly the same period, we also find *qi* written with the components for grain 米 and food 食 (E-G). At this time, the graph 氣 represented a homophone (early Chinese xjət, modern Chinese *xi*), meaning 'to provision guests', 'grain' and 'sacrifice' (Schuessler 2007: 423). Over time, the graph 氣 became used as a sound-loan for the term *qi*, and these usages reversed (Qiu Xigui 1995: 260), so that by the early Han dynasty, it was fairly established to write *qi* as we find it in the Mawangdui medical manuscripts (Graph H). We can observe the relations between the graphs, as E closely parallels B above, exchanging 米 for 火. E and F use the sound component 既 (kjeiᶜ), together with the grain component 米 or food component 食. G replaces the sound component 既 with 'vapour' 气, and inserts grain 米.

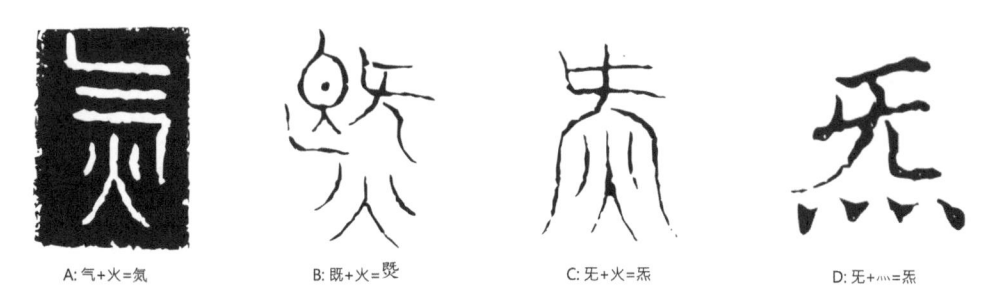

A: 气+火=氡   B: 既+火=燹   C: 旡+火=炁   D: 旡+灬=炁

Figure 2.2 *Qi* graphs with fire component, 380 BCE–100 CE

E: 既+米=槩   F: 既+食   G: 食+气+米=餼   H: 气+米=氣

Figure 2.3 *Qi* graphs with grain component, Warring States to 168 BCE

25

By the early Han dynasty, the iconic form in Graph H became the standard way to represent the multi-valent concept *qi*, while 餼 and its variant 氣 began to represent the word 'provisioning' or 'sacrifice'.[10]

The early phonology and palaeography do not support an origin myth of the concept (and graph) of *qi* emerging from a primordiality of mists and fog, and later becoming adapted to mean the breath. We see at least three different primary meaning components in play – mists, fire and grain – being used to represent the spoken word. From this perspective, Xu Shen's unilinear narrative reads like the Daoist cosmogenesis story retold in a new guise: where the Dao gives birth to the one, which then divides into two, then three, then gives birth to the 10,000 things (*Daode jing* 42; Lau 1963), or the *Zhuangzi*'s notion of primordial chaos (*hundun* 混沌) that is the basis of all things. Spoken and written languages have different and sometimes separate rules and logics, which morph existing forms and sounds to transmit new meanings. Spoken language is, as a rule, prior to the written – and it is evocative to imagine the sounds and sights of heavy breathing after exertion as 'k^h-jei^C', not unlike that of being in a rage, taking visual form on a cold morning or Tibetan mountainside as misty breath, similar to steam produced by a fire below. But the reality of how the term *qi* was derived lies more in the visual slippage between grain and fire, and the phonetic slippage between kjei^C, xjət and k^{h}jei^C, and the creative linguistic inventions of people exchanging ideas over time. The question of Xu Shen's origin story thus deserves to be considered in a new light. While we can cast doubt on his linguistic arguments, the historiographical question should not be whether Xu Shen's origin story was right or wrong, but what conditions in this period made it intuitive for intellectuals to resort to Daoistic cosmographic arguments when explaining natural phenomena like *qi*. A more thorough examination of this question than space allows for here would surely offer some provocative insights.

Emerging concepts of qi

Early sources show a developing understanding of *qi* over time, and indicate that its cosmological dimensions came into play about the year 400 BCE.[11]

Technical literature saw a growing interrelationship between the notion of *qi*, the five agents and *yin-yang*, but these were not present early on. While by the Han dynasty *yin-yang* and the five agents describe phases or states of *qi*, expressed most fully in the medical classics, these relationships were absent in all but very late Warring States literature. Prior to this, *qi* was not considered to be the fundamental unifying 'stuff' of the universe. *Yin* and *yang* were regarded as something akin to shadow and light, but did not achieve the status of universal cosmological dualism in the way they did in the Han period. The following passage is attributed to the sixth-century BCE Physician He (醫和), but probably recorded in the fifth- to fourth-century BCE narratives that make up the *Zuozhuan* 左傳:

> Heaven has six vapours (*qi* 氣). They descend to generate the five tastes; radiate to make the five colours; are called forth to make the five sounds; and in excess produce the six illnesses. The six vapours are: yin and yang, wind and rain, dark and bright. They divide to make the four seasons; form a sequence to make the five nodes; and make calamity when they exceed.[12]

(Zuozhuan 左傳 *41)*

As Harper argues, we see here *yin* and *yang* not as cosmic poles, but simply as shady or sunny aspects, which generate coolness or heat. The six *qi* form weather and seasonal patterns which affect the body, causing illness when in excess. *Qi* here is causal to the body, but it is not yet that which circulates within the body.

By the fourth century BCE, *qi* took on a cosmological dimension. The passage below, attributed as fourth century BCE, describes the formation of the cosmos from a state of primordial chaos, as one might envision the settling of a bowl of muddy water, the light contents drifting upwards and the heavier settling below:

> When *qi*, shapes and matter, were still indistinguishably blended together, that state is called chaos (*hundun* 混沌). All things were mixed in it, and had not yet been separated from one another…The purer and lighter (*qingqing* 清輕) drifted upwards, making the heavens; the grosser and heavier (*zhuozhong* 濁重) settled downwards, forming the earth. The *qi* that rushed together combined to make humans.[13]
>
> (*Chongxu zhide zhen jing*, 1.2a)

Because our access to this text is through a second-century CE or later reconstruction, what we see above may be later ideas that have crept in (Graham 1986; Kohn 2008b). This basic idea was reiterated in the *Huainanzi* 淮南子 (139 BCE), and belies a Han dynasty penchant for emphasising the parallel relationship between Heaven and Earth, with humanity in the middle (trans. Major *et al.* 2010: Section 3.1). Nevertheless, the notion of *qi* as an emergent product of the separation of the cosmic primordium is akin to ideas in the *Daode jing* and *Zhuangzi*, and should be understood as emerging during the fourth to third centuries (Graham 1989: 101).

Qi as a medium for self-cultivation

A number of early texts discuss how to cultivate the body, and one's own self, and demonstrate different relationships with *qi*-practice. The earliest surviving document to describe circulating *qi* in the body is the source of graph A above, a text carved into a twelve-sided ornamental jade block called the *Xingqi ming* 行氣銘 (ca. 380 BCE).[14] It describes a practice where the breath is retained in the lower regions of the body, collects there, rises up and circulates back down again, in this way ensuring life. It is clearly cognate with other self-cultivation literature of the period, a topic that was discussed in multiple contemporary texts, such as the *Zhuangzi* 莊子 and *Guanzi* 管子[15]:

> To circulate the breath (*xing qi*), breathe deeply so there is great volume. When the volume is great, the breath will expand. When it expands, it will move downwards. When it has reached the lower level, fix it in place. When it is in place, hold it steady. Once it is steady, it will become like a sprouting plant. Once it sprouts, it will grow. As it grows, it will retrace its path. When retracing its path, it will reach the Heaven area. The Heaven impulse forces its way downward. Whoever acts accordingly will live, whoever acts contrariwise will die.[16]
>
> (*Xingqi ming* 行氣銘)

It emerged at a time when the term began to be debated in philosophical texts which have been passed down in the received tradition. These works began to discuss physical exercises and meditative practice in relation to spiritual and cosmological reflection. Some texts exhibit mixed attitudes towards *qi*-practice, distinguishing between its roles in bodily versus spiritual cultivation. Chapter 15 of the *Zhuangzi* famously criticised 'huffing and puffing' exercises that resembled the 'ambling of bears' and 'stretching of birds'. It argued that these attempts to attain longevity were beneath the Daoist sage who accepted the order of things.[17] Nevertheless, the work considers *qi* as playing a fundamental role in life itself, and *qi*-practice an important role in spiritual cultivation:

> Man's life is a coming-together of breath. If it comes together, there is life; if it scatters, there is death…
> The True Man breathes with his heels; the mass of men breathe with their throats.[18]
>
> *(Zhuangzi 22,6)*

The *Zhuangzi*'s practice of the True Man, then, attended to the presence of *qi* throughout the body, but did not seek to manipulate it for longevity.

Other texts from this period did not suggest a conflict between cultivating health and longevity, and spiritual practice. The earliest chapters in the *Guanzi* 管子 are considered to be fourth century BCE but edited by a second-century BCE community. These chapters advocated the alignment of the body, the four limbs, the *qi* and the heart-mind (Roth 1999: 109 ff.):

> When the four limbs are aligned
> And the blood and vital breath (*qi*) are tranquil;
> Unify your awareness, concentrate your mind,
> Then your eyes and ears will not be overstimulated
> And even the far-off will seem close at hand.[19]
>
> *(Guanzi 49)*

Such alignment bestowed clear health benefits: supple skin, clear and acute sight and hearing, supple muscles and strong bones.[20] Such maintenance also included proper diet: overeating would impair the *qi* and cause the body to deteriorate, eating too little caused the bones to wither and the blood to congeal:

> Overfilling yourself with food will impair your vital energy (*qi*),
> And cause your body to deteriorate.
> Over-restricting your consumption causes the bones to wither
> And the blood to congeal.
> The mean between overfilling and over-restricting:
> This is called 'Harmonious Completion'.
> It is where the vital essence (*jing*) lodges
> And knowledge is generated.[21]
>
> *(Guanzi 49)*

We see in these passages early vestiges of spiritual practice in concert with *yangsheng* 養生 regimes of exercise and diet, and also a more direct expression for the rationale of bodily cultivation as a source of moral bearing.

While Confucius does not mention *qi*, Mengzi 孟子 (Mencius) (372–289 BCE? 385–302/03 BCE?), considered the next great Confucian philosopher after the master himself, advocated the embodied cultivation of 'vast, flooding *qi*' (*haoran zhi qi* 浩然之氣), which, when cultivated properly, would 'fill the space between Heaven and Earth'.[22] For Mencius, his flood-like *qi* was cultivated because he 'Knows [how to] speak'. It is through ethical thoughts and behaviour that the heart-mind does not disturb the *qi*:

The will commands the *qi*. The *qi* fills the body. The *qi* stops when the will takes over. Therefore, it is said, 'Take hold of your will and do not disorder your *qi*'. …If [vast, flooding *qi*] is nourished with integrity and is not harmed, it will fill the space between Heaven and Earth. This *qi* unites with Rightness (*yi* 義) and the Dao; without them, the *qi* will starve. This *qi* is born from accumulated Rightness and not an occasional show of Rightness. Action that is below the standard set in one's heart-mind (*xin* 心) will starve the *qi*…. The heart-mind must not forget the *qi*, but growth must not be forced.[23]

(*Mengzi* 3)

We see here *qi* taking a central role in self-cultivation. Mencius is at pains to qualify experiences of interiority, distinguishing between thought and the experience of *qi*. There is a phenomenology of *qi* at play here: for Mencius, it is important to define these experiences, qualify their relationships and establish a hierarchy of importance. *Qi* follows the will; thus, it is important to cultivate the mind (the seat of the will), and the *qi* will nourish naturally. It is his rightness of speech that allows him to cultivate vast, flooding *qi*. Notably, like the *Zhuangzi*, he does not advocate 'forcing' the growth of *qi*, which he compares to the pulling up of crops in order to help them grow faster, only to result in their dying.

The political cosmos: qi and the five agents

In the third century BCE, we find that *qi* plays a critical role in conceptualisations of the relationship between the state and the cosmos. It is traditionally said that the major catalyst in the *yin-yang* and five agents theory was the work of technical master Zou Yan 鄒衍 (305 BCE–240 BCE). While this cannot be proved, as all of his works are now lost, the earliest text which articulates the five agents as a cycle of different qualities or kinds of *qi* comes from the same period. It is, however, a political work. The *Annals of Lü Bowei* (*Lüshi chunqiu* 呂氏春秋) is an encyclopaedic work compiled for the benefit of the first emperor of China, purportedly the summation of knowledge of thousands of technical consultants (*fangshi* 方士) hired by the Qin court. In this treatise, produced in order to aid in state management, natural observation becomes so state-oriented that natural forces are construed as the mechanisms behind the state's success or failure. The following passage describes the five agents in terms of cyclical changes in political rule, as natural as the changing seasons, and reflections of the changing will of Heaven:

Whenever a true king is about to rise, Heaven invariably sends omens to the people below first. In the time of the Yellow Emperor, Heaven first made large earthworms and mole crickets appear. The Yellow Emperor said, 'The *qi* of earth is getting strong', and so he took yellow as his colour and earth as pattern for his activities. In the time of Yu, Heaven first made grass and evergreens appear. Yu said, 'The *qi* of wood is getting strong', and so he took deep blue-green as his colour and wood as his pattern for his activities.[24]

(*Lüshi chunqiu* 13.2.1)

This formulation was a significant shift in thinking about the right to rule, or Heavenly Mandate (*tianming* 天命), a notion established by the Zhou dynasty. That simpler formulation claimed that Heaven's assent was granted on the basis of the new monarch's virtuous conduct, in contrast to the previous ruler's corruption. The *Zuozhuan* formulation, however, normalises dynastic change as part of the controlling cycle of the five agents.[25] We see the right to rule transitioning from an argument about morality and a personal relationship with Heaven, to one about natural forces, visible by observable cyclical signs, and communicated through the medium of *qi*. Morality is no longer located within a body that cultivates the accumulation of *qi*, but rather is reflected in the entire universe, the dynamism and change of which is managed through a notion of stable systems of cyclical change.

Through proper observation of these cycles, one can determine the ascending *qi*, and by adhering to that achieve success. This shift also reflects a changing notion of Heaven (*tian* 天), which during the early Zhou dynasty referred to an anthropomorphic deity or pantheon (scholars are divided on this), but in later philosophical and technical works came to refer to 'Nature', or the forces of the natural world more generally. This relationship between the state, *qi* and nature formed much of what we might call the 'political science' of the time – that is, a science of observations of omens and portents in order to determine the relationships between the state's actions and the natural world – one which was inclusive of moral and ethical behaviours or, rather, did not differentiate between material and social processes. These are strongly reflected in texts of the early Western Han dynasty, such as the *Luxuriant Dew of the Spring and Autumn Annals* (*Chunqiu fanlu* 春秋繁露) and the *Huainanzi*, which assembled technical knowledge and portent divination as part of the necessary toolkits for monarchs to maintain the well-being of the unified empire.

Qi in early medical writing

Excavated texts and figurines from the third and second centuries BCE describe the gradual emergence of *qi*-based medicine, and show the role of physical, bodily sensations in determining the function and nature of *qi* and how to use it. These works include the *Vessel Book* (*Maishu* 脈書) and the *Pulling Book* (*Yinshu* 引書) (Lo trans. 2014) – both discovered in tombs dated 186 BCE in Zhangjiashan in Hubei – as well as others found in Mawangdui in Changsha (168 BCE), in Shuangbaoshan 雙包山 outside Mianyang in South-west China (ca. 118 BCE), and those recently discovered in Laoguanshan, also known as Tianhui in Chengdu (ca. 157–141 BCE). These texts and objects reveal how, prior to the formation of the received medical classics, medical knowledge gradually became more systematic, reaching its full form in the *Yellow Emperor's Inner Canon* (*Huangdi neijing* 黃帝內經, ca. first century CE), which defined Chinese medicine for the next 2,000 years (Sivin 1993). While these earlier works contain some correlations between the five agents, vessel locations, acupoints (or 'cavities' *xue* 穴), and *yin-yang* theory, they do not reach anything like the theoretical integration of the *Inner Canon*. What we do see is evidence of an increasingly refined language to codify bodily experience.

Qi was considered to have different material qualities in different texts. It was depicted with the metaphor of fluid flowing in waterways in the *Inner Canon,* while the contemporary *Toad Canon* (*Hama jing* 蛤蟆經) describes it as an energetic 'ball' moving according to the lunar cycle around the body. Lo (2001a) points out that the latter text is the first instance of a 'circulation' of *qi*, an imagination grounded in astro-calendrics and mathematical modelling of repeated *qi* movements around the body, whereas the 'flows' of *qi* in earlier Han literature and the *Inner Canon* were grounded in the phenomenological experience of self-cultivation practice.

Early figurines engraved with channels of *qi* have been discovered in Mianyang and Tianhui; the latter is punctuated by starry dots corresponding to acupuncture 'cavities' (*xue* 穴) described in the later *Inner Canon* (Chengdu Municipal Institute 2015; Zhou Qi *et al.* forthcoming). These indicate the early emergence of theories about the network of channels (*jingluo* 經絡) terms which refer to the warp threads of a loom and a skein of silk. The presence of small models of weaving looms in the Tianhui tomb has encouraged recent scholarship to pay attention to the material origins of this metaphor, which may reflect the influence of textile production on cosmological thought and modes of literary composition

(Zürn 2020). Texts, written on bamboo slips, were literally 'woven together' with thread. This metaphor has further implications for rethinking the role of *ji* 機 (mechanism), a term referring both to the mechanism of a crossbow trigger and to the weaving loom (Lo forthcoming). The term plays an important role in early Daoist philosophy as a source of life and dynamism as well as in the use of acupuncture cavities in the *Inner Classic*.

Qi is also described in water-like metaphors of flow and circulation in these early texts, which portray generalised flows of *qi* through the limbs towards particular regions, rather than the anatomically specific channels of imperial medicine. This came to the fore at a time when the central state was concerned with maintaining a fluid infrastructure, internal control and stable borders. The trope of describing *qi* in terms of water draws on an older metaphor that was widespread throughout early texts (Allan 1997), but perhaps most directly encapsulated in the *Dao de jing* 道德經:

> The Highest good is like water. Because water excels in benefiting the myriad creatures without contending with them and settles where none would like to be, it comes close to the way.[26]

(Dao de jing 8)

But it was not only material metaphors which shaped the way *qi* was conceived. This early literature articulates a diagnostic sensibility that is attuned to internal experiences of the body, which emerged from the body-sensations germane to a *qi*-cultivation practice (Lo 1998, 2001b). These experiences include 'heat, burning pain, breathing difficulties, gas in the alimentary canal and excessive emotion' (Lo 1998: 237).

The *Pulling Book* (*Yinshu* 引書) prescribes a variety of stretching exercises, categorised either in terms of the animals they imitate, or in the diseases they are designed to treat, with specific therapeutic ends. They thus appear more aligned with the 'huffing and puffing' exercises critiqued in the *Zhuangzi*. The descriptions of *qi* are tactile – where it states that 'Squeezing the toes benefits the *qi* of the feet', the suggestion is that *qi* is experienced from warmth, or tensile strength (Ibid.). Numbness is described as a deficiency of *qi*, suggested by the metaphor of 'rushing water', perhaps evocative of what we call 'pins and needles':

> When suffering with there being less *qi* in the two hands, both the arms cannot be raised equally and the tips of the fingers, like rushing water, tend to numbness. Pretend that the two elbows are bound to the sides, and vigorously swing them. In the morning, middle of the day and middle of the night. Do it altogether one thousand times. Stop after ten days.[27]

(Yinshu 2)

The dynamic movement of the limbs and their experience of sensation are associated with the presence or absence of *qi*. Thus, we see that *qi* in medical writing is much more than the breath, but comes to stand for a range of qualitatively distinct interior experiences with direct therapeutic implications.

These internal sensations should be read primarily as patient-centred or first-person viewpoints, rather than the detached observation of physicians. They thus constitute a different perspective on embodied *qi* than later channel theory, which required trained knowledge of body topology and palpation performed by someone other than the patient.

Isolated references describe acupuncture, here defined as inserting a needle in order to stimulate the flow of *qi*. *Maishu* 6 contains the earliest description of acupuncture, defined as using a needle to stimulate *qi* flow. Here, *qi* flows downwards like water and, where it flows upwards, it is seen as pathogenic, a sign of excess, as of a vessel filled up and overflowing. This situation is remedied by cautery, but in cases where the *qi* is unstable, needling can be applied:

> The channels are valued by the sages. As for *qi* it benefits the lower and harms the upper; follows heat and distances coolness. So, the sages cool the head and warm the feet. Those who treat illness take the surplus and supplement the insufficiency. So if *qi* goes up, not down, then when you see the channel that has over-reached itself, apply one cauterisation where it meets the articulation. When the illness is intense then apply another cauterisation at a place two *cun* above the articulation. When the *qi* rises at one moment and falls in the next pierce it with a stone lancet at the back of the knee and the elbow.[28]
> *(Maishu 6)*

This passage does not use named, memorised acupuncture points, but refers simply to the inside of the elbows or knees. Neither does it name any specific channels, or their corresponding inner organs, nor does five agents theory come either into the diagnosis or into the theorisation of the function of the acupuncture point. The needling simply stimulates the generalised flow of *qi*, whether upwards or downwards, which can be sensed in the patient's experience of heat or cold.

Qi thus formed the medium for expressing and experiencing graded tactile experiences of pleasure, pain and passion. It was associated with burning pain, which travels through the body accompanied by fever. *Qi* rising with heat to the upper body, with an accumulation of uncomfortable feelings, and opposite sensations in the limbs, is almost always pathological. Other signs associated with rising *qi* include breathing difficulties and problems of the alimentary canal – both of which ingest *qi* – such as a hot mouth, split tongue, dry throat, choking on food, pain in the throat, exhaustion and coughing. The presence of excessive *qi* in the alimentary canal was recognised through burping and flatulence, whereas absence of *qi* in the lesser *yin* channel can be indicated by wheezing, dimness of the eyes, a sense of suspension around the heart, sallowness in the complexion, anger, a feeling of alarm and of being trapped, lack of appetite and coughing of blood (*Maishu* 72–3, Lo 1998: 233–5).

Qi also extended to the emotions: anger and fear, associated with interoception of the heart and greater chest area. As such, disturbed emotions experienced there ran counter to the therapeutic, calming meditations on the same region we see in the earlier cultivation texts such as the *Guanzi*. Violent anger not released (through battle) will transform into an internal abscess. Excessive and extreme emotions of the nobility – whether anger or joy – lead to bodily harm. *Qi* is also associated with pleasant signs – the arousal of *qi* is a sign of increasing strength and vigour, *qi* welling up in the body signifies sexual excitement and the release of sexual tension is described in these texts with the same language as the illumination of the spirit (*shenming* 神明), a fully embodied, ecstatic state of penetrating insight and clairaudience (*Yinshu* 83, 86; Lo 1998: 236–8).

The *yangsheng* and medical literature of the late Warring States and early Han dynasty thus evinced an increasingly refined sense of internal states of the body. It drew on metaphors and spiritual implications present in the received philosophical and political literature, but focussed much more on direct, self-perceived experience of the body. The adaptation of these internal experiences, derived from exercise and cultivation practices, was a process of gradual medicalisation as these became pathological signs and symptoms, which physicians could draw on during patient interviews to determine the course of a disease.

Qi 氣

Qi in imperial medicine: the Yellow Emperor's Inner Canon

The imperial medical tradition begins with the *Yellow Emperor's Inner Canon*, composed in the first to second century CE, which has been the foundation of Chinese medical theory ever since. In this work, we find a much more thorough conceptualisation of the relationship between *qi*, *yin-yang* and the five agents. With this comes a radically more sophisticated descriptive terminology for *qi*, and a more complex understanding of physiological processes. It was listed first in the catalogue of medical writings in the Han dynasty imperial bibliography, indicating its prominent status, and it has been regarded as the seminal authority on medical theory from then on.

Here, we find the fullest expression of a synthetic theory of *yin-yang* and five agents (Chapter 1 in this volume) in concert with *qi*. These principles are not laid out in a systematic way, but are compiled from many different shorter passages from different earlier sources, many of which have since been excavated (Keegan 1988). Over time, systematic summaries of these theories have been compiled and put together for later generations of students (Chapter 7 in this volume).

The terms for different *qi* from this period take part in a language of orthodoxy and uprightness.[29] **True *qi* (*zhenqi* 真氣)** indicates the basic, normal *qi* which flows through the channels, which is formed after a process of digestion, absorption and transformation into the bodily system. This is sometimes interchangeable with **upright** or ***orthodox qi* (*zhengqi* 正氣)**, which refers to normative *qi* of a healthy body, and is distinguished from invading pathogenic, ***deviant qi* (*xieqi* 邪氣)**. The latter was identified in classical medicine with meteorological factors, such as wind, damp, dry, cold, heat and summer heat. The term 'deviant' (*xie* 邪) was a touchstone for boundary-marking and debate with other contemporaries, who maintained that illness was the result of ghosts and demons permeating the body with their sepulchral influences – thus entailing ritual instead of medical treatment (Li Jianmin 2009).

Essential *qi* (*jingqi* 精氣) plays a fundamental role as an elementary and subtle substance. Essence (*jing*) refers to semen when external to the body, but inside the body this subtle fluid is considered to flow throughout the form as the most dynamic and vital of the material substances, second only to *qi*. Essence, *qi* and ***spirit* (*shen* 神)** are considered to be on an ascending scale of subtlety and power.

Original *qi* (*yuanqi* 元氣 or primary *qi*) is inherited from one's parents, stored in the kidneys and provides the basic motive force of life in the body. It is the primary impetus for growth, as well as the functions of the viscera. In later Daoist works, it comes to refer to ***pre-natal* (*xiantian* 先天) *qi***, which was variously the allotment from Heaven, and also that inherited from one's parents.

The topology of the imperial landscape, criss-crossed with riverways and trade routes, is thought to have influenced the emerging forms of the **network of channels (*jingluo* 經絡)**. Interchangeable terms include the ***jingmai* 經脈** and **mailuo 脈絡**, where *mai* refers to pathways along the body. Although described in earlier texts and figurines, their role in illnesses and therapy only became fully theorised in detail in the *Yellow Emperor's Inner Canon*. The flow of *qi* can be stimulated by acupuncture at specific points along these channels – if it is insufficient or excessive, stagnant or too quick, torpid or clear. These channels contain and communicate *qi* throughout the body, much like the system of rivers and canals which conveyed vital supplies across the Han state.

Military and infrastructure metaphors play a role in the conceptualisation of how the channels transport '**supply *qi***' (*ying qi* 營氣, also translated as '**nutritive** or **camp *qi***') and '**defensive *qi***' (*wei qi* 衛氣, also **guard *qi***). The former brings nourishment to the limbs and the surface of the body, and the latter protects the exterior of the body from pathogenic invasion. These images reflected the political conditions of the Han state, which was constantly in danger of invasion along its northern borders, and would use warriors in border outposts who radiated outwards from the fort to repel invaders, while supported by a supply chain which brought nourishment and resources to these defenders.

The language of political orthodoxy was also influential. Conditions where the movement of *qi* was not healthy for the body, and ran against the normal directions of flow, were described as **rebellious (*ni* 逆), chaotic (*luan* 亂)** and reckless, and in need of regulation and **control (*zhi* 治)**, the latter was the classical term for cure, and also for establishing order, often through violent means. The political tenor of these metaphors reflects the historical conditions of the Han state in which these new technological developments emerged, and the close intertwining of its *qi*-based cosmology with the needs of a unified, central state to maintain strict order.

It was not specified whether the channels communicated blood, *qi* or both – as *qi* was considered to flow with, or in, the blood – but the **movement of the channel** or **pulse (*dongmai* 動脈)** was indicative of the state of the internal organs. The channels could be palpated at numerous locations – along the side of the neck, along the radial bone at the wrist at three positions leading up the arm away from the wrist crease, on the dorsal surface of the foot and elsewhere. A fairly sophisticated typology emerged which distinguished qualities of movement – marking speed, rhythm, depth, volume and textures like stringy or greasy or rough. These diagnostics became the direct object of attention in the Eastern Han work, the *Pulse Classic Maijing* 脈經 (trans. Yang 1997), which lays out sensory qualities that have been influential ever since.

Beyond these important types of *qi*, early medical writers organised their observations of natural processes within numerological frameworks to categorise different kinds of *qi* (Chapters 4, 5 and 19 in this volume). Jiang Shan's (2017) doctoral work on the concept of *qi* in the classical corpus identifies a number of typologies of *qi*, and performs a thorough analysis of the different kinds of meanings attributed to the single character *qi*. She also goes on to identify exhaustive lists of typologies of *qi* that are organised by numerologically significant numbers: one, two, four, five, six, nine and twenty-seven. These are both summarised in the Appendix.

These lists show, on the one hand, that *qi* was distinguished on the basis of where it came from, what produced it, where it circulated and what it was associated with – be it blood internal to the body, or natural forces external to the body. On the other hand, they also demonstrate the role of numerology in the organisation and structuring of observational data. The use of numerologically significant numbers to organise and typologise *qi* was not simply born out of a desire to construct a uniform vision of the natural world and the body in it, or to identify a microcosmic reflection of the outer world in the interior of the body. It also acted as a mnemonic, using convenient, memorable numbers around which to structure observations (and sensations) of the material (and inner) world. Some numbers organised multiple clusters of *qi*-types: the five agents, for example, were used to organise different and largely unrelated typologies: calendrical and cosmological progressions, diagnostics of facial complexion, the five external vapours and the *qi* of the five internal organs.

The marrying of these multiple kinds of observation with the same numerology makes a tacit claim that these typologies are related, but without specifying exactly how or why – simply that they are. This is not the same as a claim to causal relation. These organisational typologies allow for the cataloguing and aggregation of observational data, when the mechanisms behind what is observed have not yet been made clear – allowing them to be clarified later, but also without requiring that they are ever clarified at all. Jiang Shan (Ibid.) argues that 'the method of numerological logic started to be merged into deductive medical theory', as medical thinkers began to explore whether these explanatory models were useful descriptions of the world.

As classical medical theories became more widely accepted in the imperial period, Daoist and other communities developed practices coherent with these ideas, but which extended beyond curative treatment. Campany (2002: 18–21) refers to these as being conceived within the 'pneumatic idiom'.[30] These practice repertoires included the collection, storing and circulating of *qi* in the body with exercises such as 'guiding and pulling' (*daoyin* 導引) pathogenic *qi* or illness out of the body, circulating *qi* (*xingqi* 行氣) within the body and sexual cultivation through 'merging *qi*' (*heqi* 合氣). The loss of semen in sexual regimes was considered to deplete one's original *qi*, which is stored in the kidneys, linked to one's physical inheritance from one's parents, and one's store of life in this body. Thus, semen retention was considered an important longevity practice. Dietetic schemas and fasting regimes linked different foods to different levels of moral or intellectual cultivation, asserting that a diet of *qi* made one divine or spirit-like. Meditative practices aspired to 'foetal breathing' (*taixi* 胎息) through the abdomen and/or the skin, and medical and cultivation exercises advocated the holding or stopping of breath altogether (*biqi* 閉氣, *buxi* 不息). Such interoceptive practices, which could have visual qualities through synaesthesia, were but a step away from Daoist visualisation of body gods – divinities within the body which kept it functioning like a managed internal bureaucracy, or which communicated with divine powers in the Heavens above (Stanley-Baker 2012, 2019a). The later inner alchemy tradition extrapolated from physical alchemy to internal bodily transformation with a focus on the graded transmutation of bodily essence (*jing* 精) into *qi* into spirit (*shen* 神), which culminated in an immortal (but not fully physical) body of the practitioner (Chapters 29 and 30 in this volume). The mutability of *qi* thus led to a porosity between curing and salvation, between healing, longevity and immortality, and led to many religious sects incorporating medical elements within their repertoires (Chapter 27 in this volume).

Qi took on a fundamental role in the neo-Confucian philosophy of Zhu Xi 朱熹 (1130–1200), as the basic material substance of the universe, and counterpart to *li* 理 (principle), the guiding internal order of the material world. Reflections on the interrelationship of *qi* and *li* became a discursive space for considering dynamic change in the natural world and developing the natural sciences generally, for example, in the writings of Song Yingxing 宋應星 (1587–1666?) (Schäfer 2011: 50–89; Cullen 1990). Neo-Confucian thought was very influential on the major medical theorists of the Song, Yuan and Ming dynasties, and played out in their diagnostic theories, notions of circulation and transformation, and the ways they understood pre-natal (that which is innate and inherited) and post-natal *qi* (that derived from the natural world once one is born) (Chapter 9 in this volume; Meng Qingyun 2002).

Modern forms

Over the nearly 200 years since the arrival of Western medicine in China, *qi* has been co-opted into a broad set of conversations about modernity, identity, subjectivity and valid forms of knowledge. With the ever-increasing influence of Western medicine from the nineteenth century onwards, Chinese medicine and its fundamental notions have undergone conceptual transformation, as well as outright rejection. In an attempt to converge Western physics and Chinese science, the medical elect Tang Zonghai 唐宗海 (1851–1908) adapted notions of the then-innovative steam engine to produce the new medical concept of *qi*-transformation (*qihua* 氣化) (Lei 2012). The logician Yan Fu 嚴復 (1854–1921) argued in 1909 that *qi*, among other traditional concepts like heart-mind (*xin* 心), Heaven (*tian* 天) and Way (*dao* 道), were logically incoherent. Literary authors and scholars such as Chen Duxiu 陳獨秀 (1879–1942) and Lu Xun 魯迅 (1881–1936) denied the validity of Chinese medicine altogether (Unschuld and Andrews 2018: 99–105). 'Outdated' notions like *qi* were criticised by the Chinese medical practitioner Zhu Lian 朱璉 (1909–78), whose textbook *New Acupuncture* (*Xin zhenjiue xue* 新針灸學) argued that *qi* had no basis in reality, and excluded channel theory entirely from its description of over 360 acupuncture points (Taylor 2005: 20–7). Zhu's rejection of channel theory was not eventually accepted but, since the integration of Chinese medicine with modern healthcare in the 1950s, medical textbooks have largely de-emphasised cosmological speculation about *qi*.

A survey of three popular textbooks in China and the US (Maciocia 1994; Kaptchuk 1983; Chen Xinnong 1987) is instructive. Chen's widely used *Chinese Acupuncture and Moxibustion*, which is the basis for the US licensure exams, re-defines *qi* in terms of physiological 'function', reducing the large universal claims of cosmology to observable processes limited to the body:

> Qi is too rarefied to be seen and its existence is manifested in the functions of the *zang-fu* organs. All vital activities of the human body are explained by changes and movement of *qi*.
>
> *(Chen Xinnong 1987: 46)*

Quantum physics-like terms appear in Ted Kaptchuk's *Web That Has No Weaver*, a widely celebrated introductory textbook, which describes *qi* as an oscillation between energy and matter (Kaptchuk 1983: 35).

These works reveal roughly the same typologies, which were probably established as part of pedagogical norms after the rationalisation of Traditional Chinese Medicine (TCM) in the 1950s. They divide up their presentation of *qi* into its functions, types and pathologies. Little or no attention is paid to other aspects of *qi* in terms of its historical uses in divination, rationalisation of natural processes, astronomy, expressive arts, nor to means of circulating the *qi* by *qigong* or other internal cultivation. Some translation choices indicate varying relations to traditional thought. The rendering of *zongqi* 宗氣 as 'pectoral *qi*' by Chen interprets the term entirely in terms of its function and location, avoiding the Confucian inflections of the renderings by Kaptchuk and Macioccia as 'ancestral *qi*'. All three texts discuss original *qi* as a substance inherited from the parents during the pre-natal stage, making no allusions to the Daoist notion of the cosmic primordium.

Five agents theory is presented early on in these textbooks as a general theory without reference to the cosmological framing of classical works, and quickly passed over in favour of discussion of the organs (*zangfu* 臟腑). These are treated as material objects and processes, 'materialist entities within a body-enveloped-by-skin' (Hsu 2007: 98). Discussion of 'Organ *qi*' in Chen is subsumed under the section of general pathologies, and not treated as a distinct category of *qi*, whereas Kaptchuk and Maciocia emphasises *zangfu qi* as spatially local to, and characteristic of, the *zangfu* organs. This may well be due to the fact that it is more intuitive in Chinese than in English to assume that *qi* takes on the qualities of the time and space in which it is located, and the categorical boundaries between some *qi* are porous to the fluidity of *qi* itself.

While reformed Chinese medicine (commonly referred to as TCM) seems to have set aside more traditional cosmological frames of reference, this is not true of all forms of *qi*-practice.

Qi continues to function as a fully operational term in religious and self-cultivation contexts, perhaps most conspicuously in *qigong*, which was the subject of wide popular fascination in the late eighties and early nineties in mainland China when it was known as *qigong re* 氣功熱 (*qigong* fever). Those studying this moment have been at pains to wrestle with how the different kinds of *qigong* they encountered were accommodated or not within broader social, political and epistemological frames of modernity. Scholars like David Palmer (2007) and Nancy Chen (2003) have described how the communities formed through shared bodily practice established alternate political spaces for private and political expressions that may have run counter to norms expected in the socialist state. Elisabeth Hsu (1999) closely reads the traditional literature and contrasts it with one *qigong* healer's teachings and practice, framing these within anthropological studies of ritual healing and ethnomedicine. She argues that *qigong* positions the physical body within a phenomenologically sensible relationship to the outside, palpable world, mediated by the universal continuum of *qi*, constituted as a form of the 'body ecologic'. Thomas Ots (1994) early on took a phenomenological position, focussing on the interoceptive experience of the body as a vehicle for personal self-exploration and means for psychological self-transformation. The interiority of *qigong* is also the core of two studies by the philosopher Yasuo Yuasa (1987, 1993), who argues that the phenomenology of *qi*-practice is the basis of a distinctively East Asian mode of embodiment. This phenomenological approach has been influential on East Asian scholars such as Cai Biming 蔡璧名 (1997; 2011) whose historical studies enquire into the relationship between *qi* and East Asian modes of knowing the world and self.

The epistemological itch: what to do with qi?

Taken collectively, the above-mentioned studies of *qigong* all touch on a common anxiety or disconcertment. As Verran (2014) argues, disconcertment points to deeper, tacit misalignments between basic ways how people constitute and deal with the world around them. Scratching this epistemological itch, unpacking the reasons for the discomfort, can help reveal the contours of the disagreements, and perhaps allow for their better coordination. These attempts to scratch the epistemological itch presented by *qi*-practice, that is, to explore how people deal with *qi* in a modern world which denies the epistemological grounds of *qi*-practice, proceed through emic explanation (by uncovering the internal rationales and logics of *qi*-practice), etic interpretation (situating those practices within alternative theoretical models) or justification (intellectual and socio-political legitimation of these and subordinate practices as culturally valid). This itch is felt in myriad ways, whether through political crackdowns on *qigong*, intellectual movements to criticise Chinese medicine within China (*piping zhongyi yundong* 批評中醫運動) and outside (quackwatch.org), or through insurers' denial of coverage for *qi*-based practice. More subtle forms of enlightenment parochialism construct *qi*-practice into a Eurocentric genealogy as 'mind-body medicine', beginning with Mesmerian parlour tricks, and addressing *qi* only as a late 1960s newcomer, entirely ignoring its longstanding cultural, intellectual and medical history within China (Harrington 2008). The epistemological itch does not arise from such acts of boundary-marking alone, but rather from the fact that, despite them, intelligent, educated, modern people continue to engage in *qi*-practice. How does a concept similar to the ancient Greek *pneuma* and the early modern 'animal magnetism' abandoned in Western culture continue to hold wide currency in East Asian thought, language and habitual practice, even as these societies have fully adopted a scientific practice and worldview? And why do more Westerners increasingly adopt *qi*-based practices, whether sitting meditation, moving *qigong* or TCM?

The contexts are many, and so are the different lives of *qi*. In the Sinophone world, the term *qi* is widely distributed across many domains of life simply through language alone. Words containing the character refer to steam, the breath, flavour, atmospheric conditions, weather, the feeling in a room, anger, tone of voice, the weight of calligraphy brushstrokes, the scent of flowers and even scientific terms for oxygen, gas and others. Even as the world of scientific knowledge denies the reality of *qi*, scientific terms written in Chinese depend on the term as a linguistic component of meteorology, gases and electricity. The appearance of the term in so many diverse domains and concepts inherits a tacit, linguistic assent to its universality.

Yan Fu's rejection of *qi* as logically incoherent neglects that the function of *qi* is fundamentally coher*ing*: it expresses a world that is coherent, intuitively drawing together many discrete areas and asserting their mutual relevance and the potential for humans to interact with them in comparable ways (Stanley-Baker 2019b). The tangible atmosphere in a concert hall as the last strings go quiet, the texture of a conversation, or the aroma of food, constitutes an external field in which one finds oneself, along with others. At the same time, they are internally sensed experiences that are related to emotions, desire, discomfort, bloatedness, physical pain or the pleasure of relaxation. Whether or not users of these words assent to a materiality of *qi*, it is indelibly present in their lived linguistic worlds, and for that reason will always remain an intuitive concept.

Clearly, for many, *qi* is palpable, convincing and produces results. It is not difficult to experience, and requires no religious belief framework, just the patience and willingness to try simple, physical exercises such as *zhanzhuang* 站樁 (standing like a post and holding the two palms opposite one another). Whether the *qigong* practitioner's sensation of something present between the hands, of body relaxation, of the dull ache of pressure at an acupuncture point or the travelling pain which radiates out in a line during needling, *qi* can be sensed as practitioners describe it.

But it is another thing to assert that these sensations indicate the presence of a common *thing* called *qi* (Farquhar 1994: 34–5). Such claims participate in larger socio-epistemic, or cosmopolitical debates, which resist the notion that one normative form of rationality should hold sway (Stengers 2010; Farquhar 2002). The sometime dramatic successes of *qi*-practice, inexplicable to the language and praxis of science, give them the air of the 'miraculous'. Such performances draw attention to *qigong* as a spectacle, but in the same stroke relegate it as epistemically secondary and subaltern to rational science (Zhan Mei 2009: 91–118). This can be particularly egregious in the case of the grandiose claims of some *qi*-healers and masters, and the mutability of *qi* as a material substance affords no clear boundary-markers between valid medical claims and fantasy (Chapter 49 in this volume).

The search among Europeans and Americans for 'authentic' or 'pure' Eastern tradition, teachers and lifeways through their forms of *qi*-practice is well-documented as part of larger 'romanticist critiques of modernity' (Scheid 2002: 43), new Eastern religiosity (Unschuld 2009: 202–3) or an enlightened form of orientalism (Phan 2017). The distinctively Western psychotherapeutic *imaginaire* of Chinese medicine privileges 'authentic expressions of self' over consideration of the historical, philosophical and technical applications of the term *qi* (Pritzker 2014: 46–9,122–32). While these frameworks are argued by some to be 'authentic', they have also produced one of the stumbling blocks for better integration of Chinese medicine with modern globalised medicine (Wegmüller 2015). The appeal to a liberationist holism associated with *qi* also ignores the political history both of *qi* and of 'holism' itself. Historically, holism has been a functional term in fascist and Maoist authoritarian regimes (Scheid 2016). The emergence of *qi* itself as fundamental cosmic stuff, intertwined with *yin-yang* and five agents, was situated within the political cosmology of the Qin and Han states as an assertion of an 'all-embracing interdependence' bound up in new forms of sanctioned violence (Lewis 1990: 218; Hsu 1999).

The very slipperiness of *qi* is thus both its undoing and also its power. Its infinite permeability and constant transformation and movement make it impossible to capture in defined, limited and repeatable laboratory experiments that are the stuff of rational science, despite repeated attempts by researchers in the Mainland and on Taiwan. Yet, scholars who have paid attention to '*qi*-talk', that is, practitioner discourse about *qi* and the role it plays in the work they do, also reveal important functions of this discourse. They draw our attention to what the language of *qi* actually does. It is widely acknowledged that to remove *qi* from Chinese medical theory is to reduce it to minor anatomical exercises, related to trigger points, and the practice of 'dry needling' or intra-muscular stimulation. Notions of *qi* are fundamental to the epistemology of Chinese medicine, and afford it the conceptual framework to draw together generations of accumulated natural observations – whether or not their causal relations are coherent to modern science. The language of *qi* that connects, for example, migraine headaches with dryness in the liver, leading to its inability to capture liver *qi*, which rises up to the head and causes the headache, facilitates the use of acupuncture on the gallbladder and liver channels on the lower leg and foot which alleviate the headache.

Scheid (2013) points out that *qi* and other Sinophone terms function to create a translingual community of practice, as these terms and concepts are shared in common across Chinese, Japanese, Korean, English and other languages – however, they may variously interpret them. Ho (2006; 2015) and Pritzker (2014) describe how *qi*-talk enables a discourse within which bodyworkers can articulate their tactile sensations and assessments of their patients. While this has a social function, designating insiders and outsiders depending on their fluency with *qi* language, it is also a language for various styles of practice, wherein professionals compare their treatment rationales and tactile sensibilities.

This tactile focus accords with Lo's insight that the early formation of classical medicine arose from the phenomenological experience and observations of pain, pleasure and sensation (Lo 2001b; 1999; 1998). Herein the slipperiness of *qi*-talk is not a generalised vagueness about cosmic unity, but rather a finely tuned discourse, a subtle differentiation of nuanced shifts and changes that can be observed in others or experienced in oneself. Its permeability to the self and the other is resistant to rigid quantification, and can also give expression to the lived bodily subject or *leib* of phenomenologists like Ots (1994) and Yasuo (1987; 1993). The tacit, subjective, masterful sense of touch, to which *qi*-talk gives expression, excludes objective forms of knowing that have historically emerged from a particularly visual mode (Daston and Galison 2010).

Regardless of popular mischaracterisations of *qi* as mysticism, the language of *qi* is a storehouse of long-held cultural memories, which reflects changing values over time. Whether or not objective science argues that *qi*-talk is like that of the blind men who failed to identify the elephant, the kind of skilled knowledge that it takes to stroke the creature's trunk and identify its textures and moods is articulated in the language of *qi*.

Appendix: Categories of *qi* in the *Inner Canon*, by Jiang Shan 姜姍

This Appendix is designed to complement Stanley-Baker's chapter on *qi* as a mass noun. As he points out, mass nouns are distinguished by types, or by specific measures. Below, I set out some of the different types of *qi* that can be found in the *Yellow Emperor's Inner*

Canon (and below translated from Unschuld 2016), focussing on the connotations of the single character *qi*.

1. An abbreviation for a variety of different types of *qi*

The fact is: When man exhales once, his vessels move twice, and the passage of his *qi* covers a distance of 3 inches. When he inhales once, his vessels move twice again, and the passage of his *qi* covers a distance of 3 inches.

Lingshu 15 (Unschuld 2016: 15.239)

This passage comes from a chapter on the *ying* 營氣 'supply *qi*' and *wei qi* 衛氣 'defensive *qi*'. *Qi* here refers to both types depending on the context, and thus functions linguistically as a pronoun.

2. Formless elements of the body, as opposed to the physical form

Bo Gao replied: Wind and cold harm the physical appearance. Grief, fear, fury and rage harm the *qi*. When the *qi* harm a long-term depot, they cause a disease in that long-term depot. When cold harms the physical appearance, the physical appearance will show a reaction. When wind harms the sinews and vessels, the sinews and vessels will show a reaction.

Lingshu 6 (Ibid.: 6.127)

In this passage, we see the physical form being contrasted with the interior *qi*. Note that the former is visible externally to others, whereas the latter is sensed internally to self.

3. The functioning of any body parts

If there are 50 movements without a single intermittence, all the five long-term depots are supplied with *qi*. If within 40 movements there is one intermittence, one long-term depot is without *qi*. If within 30 movements there is one intermittence, two long-term depots are without *qi*. If within 20 movements there is one intermittence, three long-term depots are without *qi*.

Lingshu 5 (Ibid.: 5.116)

In this passage which discusses the movement of supply *qi* throughout the body, *qi* refers to the function of the internal viscera. In other contexts, it can refer to the function of other organs, or the limbs.

4. *Qi* in the channels

Qi Bo replied: It is simply impossible that there is a place not passed through by the *qi*. This is like the flow of water, like the movement of sun and moon – they will never stop.

Lingshu 18 (Ibid.: 18.248)

This might be the most common type of *qi* in the *Inner Canon*. Here, the author borrows the water metaphor to indicate that its flow is unending. Note also that elsewhere in the *Yellow Emperor's Inner Canon*, the description of *qi* moving like the sun and moon provides us with the earliest extant reference to regular cycles and the concept of *qi* circulating around the body.

5. Characteristic or tendency

Huang Di asked Qi Bo: Now, the *qi* of the four seasons, they all differ in their physical appearance. The emergence of each disease is linked to a certain location where it develops. As for the way of cauterisation and piercing, what specifications exist?

Lingshu 19 (Ibid.: 19.259)

Qi here refers to a tendency, feature or aspect of, in this case, the four seasons. This kind of use was common in ancient texts. It refers to a summation of the total characteristics of a specific type of thing.

6. A counterpart of blood

Whether the long-term depots are firm or brittle, and whether the short-term repositories are large or small, how much grain [they have received], and of what length the vessels are, whether the blood is clear or turbid, and whether the *qi* are many or few, whether the twelve conduits transmit much blood and little *qi*, or little blood and much *qi*, and whether overall they contain much blood and much *qi* or little blood and little *qi*, all this can be quantified.

Lingshu 12 (Ibid.: 12.212)

Blood (*xue* 血) was considered the *yin* portion of a larger, generalised *qi*, against which a narrower concept of *qi* was considered the *yang* component (Sivin 1987: 51–2). The pairing of blood and *qi* dates back at least to the works attributed to Confucius in the early fifth century BCE, for whom *xieqi* 邪氣 was a compound referring to an individual's vitality in general (Ibid.: 46). This pairing became more formally theorised in the *Yellow Emperor's Inner Canon*, as we see here. Blood was considered substantial and steady, against *qi* which was dynamic and clear. Where blood housed and formed the root of the *qi* that was the force that moved blood along through the vessels, as well as keeping it contained within the vessels.

7. An overall term for the various *qi* that flowed through the body

Qi Bo: The major differentiation of the *qi* [is as follows]: The clear [*qi*] ascend and flow into the lung. The turbid ones descend and move into the stomach.

Lingshu 40 (Ibid.: 40.388)

The image here pictures a variety of *qi* flowing throughout the body, in dynamic relation to one another. Yet, the first use of the term above generally refers to all of the *qi* and it is therefore a mass noun in the sense used by Stanley-Baker at the beginning of this chapter.

8. Internal sensations in the body

In the case of a sudden loss of voice, with the *qi* [breathing section] hardened, [for therapy] one chooses the *fu tu* [opening] and removes blood from the basis of the tongue.

Lingshu 21 (Ibid.: 21.270)

Qi could refer to interoception of illness sensations in the body as here, where hardness of the *qi* refers to a substantial sense of something plugged or blocked in the throat. Other similar descriptions might indicate feelings of blockage, knots or accumulation in other regions of the body.

9. Response to needling

It happens that [a patient's] spirit is excited, and the *qi* moves even before a needle has been applied. It happens that the *qi* and the needles confront each other. It happens that the needle was withdrawn, but the *qi* flow by themselves. It happens that the effect is noticeable only after several piercings. It happens that the deployment of the needle causes the *qi* to move contrary to the norms. It happens that despite repeated piercing the disease increases in its severity.

Lingshu 67 (Ibid.: 67.581)

This passage is a typical description of bodily responses to piercing, and furthermore shows the close relationship between *qi* and acumoxa therapy. It is worth pointing out that these reactions are not necessarily local to the site of piercing, but might be observable throughout the rest of the body. It relates directly to needling and is mostly noticed by the therapist; thus, it is differentiated from 4 or 8.

10. Pathogenesis and pathomechanism

When there are evil [qi] in the lung and the heart, the qi remain in the two elbows.
When there are evil [qi] in the liver, the qi flow into the two armpits.
When there are evil [qi] in the spleen, the qi remain in the two thighs.
When there are evil [qi] in the kidneys, the qi remain in the hollows of the
two knees.

Lingshu 71 (Ibid:71.645–46)

While *qi* refers to subtle materials in the natural world, when these invade the body to cause disease, the term *qi* refers to the means by which pathogenic influences spread throughout the body.

11. Substances in nature

The upright 'proper *qi*' (*zhengqi* 正氣) are proper wind coming out of a certain cardinal direction. They are not [the so-called] 'depletion wind'.

Linghsu 75 (Ibid.: 75.655)

When *qi* in the natural world, such as wind, heat, and dryness, remain outside the body, they are not pathogenic, and thus ordered and 'proper'.[31]

12. Intestinal gas

Food ingested will be thrown up again. The stomach duct aches. The abdomen is swollen. [Patients] tend to moan. After defecation and [when intestinal] *qi* [have passed], a joyous feeling sets in, as if something had been shed. The body and all its limbs feel heavy.

Lingshu 10 (Ibid.: 10.179)

Qi here literally refers to gas in the intestines. This may be the most material of *qi* in the *Inner Classic*.

Although these forms of *qi* are closely related to one another, they can be distinguished by the fact that nine out of these twelve forms are 'ideological', that is, grounded in a conceptual theory about the makeup of the body or the natural order. Only three, described in 8, 9 and 12 above, I maintain, are direct, phenomenological observations of own's body or the body of others (patients in particular) (Table 2.1).

Table 2.1 Qi typologies organised by number

Number of qi	Treatise	Connotation
1	I have been informed that every person has essence, *qi, jin* liquids, *ye* liquids, blood and vessels. I was of the opinion that all of them constitute one identical *qi*. Now they are distinguished by six names. I have no idea why this is so. *Lingshu* 30 (Ibid.: 30.351)	Singular overarching category of bodily materials.
2	However, when the *qi* recede in the lower regions, while the camp and guard [*qi*] remain where they are, when cold *qi* mount upwards, and the true and evil [*qi*] attack each other, that is, when two types of *qi* strike at each other, then such a union ends in a swelling. *Lingshu* 35 (Ibid.: 35.379)	True *qi* internal to the body, versus *qi* invading the body from outside. The former is good, the latter is pathogenic or deviant.
	When [the patient] sweats profusely and is soggy, this is so because he has met with very [much] dampness. The yang *qi* is diminished, while the yin *qi* abounds. The two *qi* affect each other. Hence, sweat leaves [the body] and [the patient is] soggy. *Suwen* 43 (Unschuld and Tessenow 2011: 43.651)	*Qi* distinguished by yin and yang.
3	[Triple Burner](*san-jiao* 三焦), as the bypass of [original] *yuan qi*, guards the passing of three *qi* throughout the five *zang* and six *fu*. *Nanjing* 66 (Unschuld 1986: 66.561)	*Qi* which flows through, and is processed by, the upper, medium and lower parts of the body.
4	Comprehensive Discourse on Regulating the Spirit [in Accordance with] the *Qi* of the Four [Seasons]. *Suwen* 2 (Unschuld and Tessenow 2011: 2.45)	The *qi* of the four seasons.
5	The five *qi* take their positions one after another; each of them has [a *qi*] that it dominates. The changes between [periods of] abundance and depletion, this is their regularity. *Suwen* 9 (Ibid.: 9.170)	Five *yun* 運 or cosmic progressions, i.e. five seasons.
	Hence, when the five *qi* enter the nose, they are stored by [heart and] lung. When [heart and] lung have a disease, the nose is not free as a result. *Suwen* 11 (Ibid.: 11.208)	External influences of nature which enter the body and accumulate there.
	I have been informed: in piercing there are five administrative organs serving as five observation points to inspect the five *qi*. The five *qi*, in turn, are the emissaries of the five long-term depots; they are assistants associated with the five seasons. *Lingshu* 37 (Unschuld 2016: 37.387)	The five colours as seen in the facial complexion which are used for diagnosing the state of the corresponding internal organs
	In case [the patient] was formerly wealthy and later became poor, this is called lost essence. The five *qi* stay for long; one suffers from something having merged. *Suwen* 77 (Unschuld and Tessenow 2011: 77.667)	The essential *qi* of the five *zang* organs.

Table 2.1 (Continued)

Number of qi	Treatise	Connotation
6	[A duration of] five days is named *hou*; [a duration of] three *hou* [terms] is named *qi*; [a duration of] six *qi* [terms] is named a season; [a duration of] four seasons is named year. *Suwen* 9 (Ibid.: 9.169)	A definite temporal marker, a period of fifteen days, which, when added together six times make 90 days, or a quarter of a year.
	I have been informed that every person has essence, *qi*, *jin* liquids, *ye* liquids, blood and vessels. I was of the opinion that all of them constitute one identical *qi*. Now they are distinguished by six names. I have no idea why this is so. *Lingshu* 30 (Unschuld 2016: 30.351)	The sum of six different bodily substances in which construct the body.
9	I know that the hundred diseases are generated by the *qi*. When one is angry, then the *qi* rises. When one is joyous, then the *qi* relaxes. When one is sad, then the *qi* dissipates. When one is afraid, then the *qi* moves down. In case of cold the *qi* collects; in case of heat, the *qi* flows out. When one is frightened, then the *qi* is in disorder. When one is exhausted, then the *qi* is wasted. When one is pensive, then the *qi* lumps together. These nine *qi* are not identical. Which diseases generate [these states]? *Suwen* 43 (Unschuld and Tessenow 2011: 43.651)	Nine different pathogenic states or movements of *qi*, which become so due to emotions or external influences.
27	Of conduit vessels there are twelve. Of network vessels there are fifteen. Altogether there are 27 *qi*, including above and below. Where they exit, these are the wells. Where they move in swift currents, these are the creeks. Where they flow, these are the transport [openings]. Where they permit passage, these are the streams. Where they enter, these are the confluences. The passage of all 27 *qi* touches the transport [openings] of the five [long-term depots]. The joints where [structures] intersect, they constitute 365 meeting points. *Lingshu* 35 and 12 (Unschuld 2016: 35.379)	*Qi* in the twelve main channels and fifteen collaterals.

Notes

1 Funding for the research, writing and publication of this paper was generously provided by the Max Planck Institute for the History of Science Dept. III, under the project 'Charting Interior and Exterior Worlds'; the Kolleg Forschungsgruppe 'Multiple Secularities' at the University of Leipzig, and Nanyang Technological University, under the project 'Situating Medicine, Religion and *Materia Medica* in China and Beyond'.

  I would also like to thank friends who have helped review drafts of this paper or offered helpful comments on particular issues: Ash Henson, Terry Kleeman, Pierce Salguero, Misha Tadd, Rodo Pfister and Yan Liu. Any errors or misrepresentations are entirely my own.

2 A number of scholarly, book-length works on *qi* make for important references in the historiography of *qi* scholarship. Kuroda Genji's (1977) *Ki no kenkyū* posthumously publishes a collection of his papers. Onozawa *et al.*'s (1978) *Ki no shisō* 気の思想 collects chapters on topics across the imperial period, ranging from early paleography through early imperial medicine and other technical arts, medieval Daoist and Buddhist cultivation, and spending the lion's share of ink on Neo-Confucian thought. The general arc of this narrative can also be seen in Kubny's (2002) German language dissertation and monograph. A conference volume on *qi* from the Institute of Ethnology

(Minzu xueyuan 2000), Academia Sinica, Taipei collects cultural studies ranging from paleography, analysis of early excavated literature to modern lab studies, as well as numerous descriptions of contemporary *qi* experiences and cultivation practice. Sakade Yoshinobu's (2007) collection of papers on early medieval medicine and Daoist cultivation practices presents *qi* in the title, tacitly asserting *qi* as a focal medium, but does not articulate a consistent case through the work. In general, Sakade's work over his career has dealt with *qi*-practice. The phenomenologist Yasuo Yuasa (1993, and see below), grounded in Western phenomenology and Kyoto school philosophy, also asserted *qi* as a phenomenological medium through which the East Asian body is constituted. Cai Biming 蔡璧名 (1997) uses close philological reading of early and pre-imperial medical and cultivation works to unpack a similar reading of Chinese embodiment, an approach she later (2011) explicitly argues for as a phenomenology.

3   See Hanyu da cidian chubanshe (2007).
4   Harbsmeier *et al.* (1998) and Graham (1989: 402) disagree over whether *qi* should be considered a mass noun, which is enumerated by container terms such as 'cup of water', or whether it is a generic noun referring to a generalised category of thing, divisible into different types, as in 'the six domestic animals'. Nevertheless, both agree that six *qi* refer to types, not units, of *qi*.

   Both writers critique with an even bolder hypothesis put forward by Chad Hansen (1983 [2020]); 1992), who argues that all Chinese nouns are mass nouns. While this hypothesis has since been discredited, nevertheless it has inspired the above reflection about the role of mass nouns in Chinese thought.
5   I would particularly like to thank Ash Henson for going multiple rounds on this with me. Any mistakes and errors remain my own.
6   *Shuowen jiezi* 說文解字, 2.221 on 气; see Figure 2.1.
7   Yu Xingwu 于省吾 and Yao Xiaosui 姚孝遂 (1996: 3371–9); Ji Xusheng 季旭昇 (2014: 58). Huang Dekuan 黃德寬 (2007: 3218) argues that this form was used to refer to various time markers and stages of completion (until 迄, already/thus 既, completely 訖). Ji Xusheng 季旭昇 (2014: 58) finds Xu Shen's tale unlikely but, unable to provide a better alternative, agrees that his narrative will have to suffice for the present.Graphs from He Linyi 何琳儀 (1998: 1199). Old Chinese (OC) pronunciation, spoken between 1250 BCE and the Han dynasty (202 BCE), from Schuessler (2007: 423).
8   Thanks to Donald Harper for a lengthy conversation on this point and about dating concepts such as 'stuff in nature' and 'breath'. Notably by the time of the later Warring States and Han, the meanings of 'cosmic stuff' and 'vapors' were well-established. See Harper (1999) for in-depth discussion of sources.
9   Graphs in Figures 2.2 and 2.3 are from Xiaoxue Tang. In graph A, also see Li Ling 李零 (2006: 270); He Linyi 何琳儀 (1998: 1200). In graph H (from the Mawangdui bamboo slips), see also Chen Jian'gong 陳建貢 and Xu Min 徐敏 (1991: 475).
10 Li Xueqin 李學勤 and Zhao Ping'an 趙平安 (2013: 648); see entry by Zhang Yujin 張玉金.
11 These 'received' sources coming to us through the written tradition have been passed on till today through being copied and recopied over time, but may therefore introduce copyist and editorial changes that may subtly shift interpretation and nuance to suit their later eras. Scholars thus differentiate them from 'excavated' sources found in tombs and caves – such editions were produced much closer to the time of original composition, and therefore contain fewer changes.
12 *Zuozhuan* 左傳, 41 (Zhao 1) 708b–9a; Harper 1999: 862.
13 *Chongxu zhide zhen jing* 沖虛至德眞經 1.2a, translation informed by Needham and Wang (1956).
14 Once possessed by the collector Li Mugong 李木公 from Hefei, the jade knob is currently housed in the Tianjin Museum. Needham and Wang (1956: 143) date the piece to the mid-sixth century, following Wilhelm (1948). Guo Moruo 郭沫若 (1972) puts it at 380 BCE, while Chen Banghuai 陳邦懷 (1982: 344, n. 3) argues for a late Warring States dating.
15 The *Zhuangzi* is a multi-authored work thought to contain chapters by the fifth- to fourth-century sage Zhuang Zhou 莊周, his students and later editors. The *Guanzi* is traditionally attributed to Guanzhong 管仲 (eighth to seventh century BCE), but is mostly considered to have been composed in the fourth century BCE.

   On the context of these texts in the Warring States discourses on health and spiritual cultivation, see Stanley-Baker (2019a: 11–12); Rickett (1998: 19); Harper (1998: 125–26; Roth (1999: 161–3) and Kohn (2008a: 14–15). For an overview of these arguments, see Yang (2018: 118–22).
16 Rickett (1998: 19) with minor modifications.

17 *Zhuangzi jishi* 莊子集釋, 15.535–7. For a translation, see Watson (2013: 15.119).
18 *Zhuangzi jishi* 莊子集釋, 22.6; Watson (2013: 43.177).
19 *Guanzi Jin zhu jin yi* 管子今註今譯, 49.778–9; Roth (1999: 82, 90); Rickett (1998: 51, 53).
20 *Guanzi Jin zhu jin yi* 管子今註今譯, 49.778; Roth (Ibid.: 76); Rickett (Ibid.: 48–9).
21 *Guanzi Jin zhu jin yi* 管子今註今譯, 49,779; Roth (Ibid.: 90); Rickett (Ibid.: 53).
22 *Mengzi*, 3.54b–55a; Rainey (1998: 91).
23 Ibid.
24 *Lüshi chunqiu jiao shi* 呂氏春秋校釋 13/2.1; translation from Nylan (2010: 399–400). See also Knoblock and Riegel (2000: 13.2.282–3).
25 Five-agent theory incorporates two cycles of interaction. In the generating (*sheng* 生) cycle, each agent generates another in a cycle of mutual support. In the controlling or conquering (*ke* 克), different phases mutually suppress, control or conquer one another. This pattern of stimulus and control produces an overall stability and constancy.
26 *Dao de jing* 道德經 8; Lau (1963: 12).
27 *Yinshu* 引書 2; Lo (1888: 348)
28 *Maishu* 脈書 6; Lo (1998: 340)
29 The following section draws from an unpublished synopsis of Jiang Shan 姜姍 (2017), provided by the author. Details of her findings are summarised in the Appendix: Categories of *qi* in the Inner Canon.
30 This phrase translates the term *qi* with the Greek *pneuma*, a common rendering among some Sinologists. While it is widely adopted among scholars of religion, others (Libbrecht 1990) have argued for the cultural differences between *qi* and *pneuma*.
31 Editor's note: On the development of the concept of pathogenic *qi* from the earlier notions of directional winds, see Kuriyama and Barlow (1994).

Bibliography

Pre-modern sources

Chongxu zhide zhen jing 沖虛至德眞經 (Exegesis of the Authentic Scripture on the Ultimate Virtue of Unfathomable Emptiness) att. Liezi 列子 (c. 450–375 BCE), DZ 668.
Guanzi jinzhu jinyi 管子今註今譯 (Modern Annotated and Translated Master Guan), Guan Zhong 管仲 (c. 725–645 BCE), in Li Mian 李勉 (717–788 CE) (ed.) (2013), Taipei: Taiwan shangwu yinshuguan.
Lingshu 靈樞 (Numinous Pivot), Anon., in *Huangdi neijing lingshu yijie* 黃帝內經靈樞譯解, in Yang Weijie 楊維傑 (ed.) (1984), *Zhongguo yiyao congshu* 中國醫藥叢書 (Chinese Medicine Series), Taipei: Tailian guofeng chubanshe.
Lüshi chunqiu jiao shi 呂氏春秋校釋 (Collation and Annotation of the Master Lü's Spring and Autumn Annals) c. 239 BCE, Lü Buwei 呂不韋 (292–235 BCE), in Chen Qiyou 陳奇猷 (ed.) (1984), Shanghai: Xuelin chubanshe.
Mengzi 孟子, Mencius (ca 372–289 BCE), in Ruan Yuan 阮元 (ed.) (1815/1976) *Mengzi zhushu fujiao kanji shisan jing jiaoshu: fujiao kanji* (*Nanchang fuxue kanben*) 孟子注疏附校勘記 十三經注疏：附校勘記 (南昌府學刊本), Taipei: Yiwen yinshu guan.
Shuowen jiezi 說文解字 (Explaining Depictions of Reality and Analyzing Graphs of Words) 121 CE, Xu Shen 許慎 (58–147 CE), in Duan Yucai 段玉裁 (1735–1815) (ed.) (1815/1981), Shanghai: Shanghai guji.
Zhuangzi jishi 莊子集釋 (Collected Explanations of Zhuangzi) late 4th century BCE, Zhuang Zhou 莊周 (c. 369–286 BCE), in Guo Qingfan 郭慶藩 and Wang Xiaoyu 王孝魚 (eds) (1995), Beijing: Zhonghua shuju.
Zuozhuan 左傳 (Zuo's Commentary) c. late 4th century BCE, in Ruan Yuan 阮元 (ed.) (1976/1815) *Shisan jing jiaoshu: fujiao kanji* (*Nanchang fuxue kanben*) 十三經注疏：附校勘記(南昌府學刊本), Taipei: Yiwen yinshu guan.

Modern sources

Allan, S. (1997) *The Way of Water and Sprouts of Virtue*, Albany: State University of New York Press.
Cai Biming 蔡璧名 (1997) *Shenti yu ziran: yi Huangdi neijing suwen wei zhongxin lun gudai sixiang chuangtong zhong de shenti guan* 身體與自然：以《黃帝內經素問》為中心論古代思想傳統中的身體觀 (Body and Nature: A Discussion of Embodied Perspectives in Traditional Ancient Thought,

Based on the *Yellow Emperor's Inner Classic: Basic Questions*), [Taipei]: Guoli Taiwan daxue chuban weiyuanhui.

——— (2011) 'Jibing changyu yu zhijue xianxiang: Shanghan lun zhong de 'fan' zheng de shenti gan' 疾病場域與知覺現象:《傷寒論》中「煩」證的身體感 (Sensory perception in the field of disease: embodied sensation of 'vexation' in the treatise on cold damage), *Taida zhongwen xuebao*, 34: 1–54.

Campany, R.F. (2002) *To Live as Long as Heaven and Earth: a translation and study of Ge Hong's traditions of divine transcendents*, Berkeley: University of California Press.

Chen Banghuai 陈邦怀 (1982) 'Zhanguo xingqi ming kaoshi' 战国行气铭考释 (A study and explanation of the warring states inscription on circulating *qi*), *Guwenzi yanjiu* 古文字研究, 7: 187–92.

Chen Jian'gong 陳建貢 and Xu Min 徐敏 (1991) *Jiandu boshu zidian* 簡牘帛書字典 (A Dictionary of Bamboo Slips and Silk Manuscripts), Shanghai: Shanghai shuhua chubanshe.

Chen, N.N. (2003) *Breathing Spaces: qigong, psychiatry, and healing in China*, New York: Columbia University Press.

Chen Xinnong (ed.) (1987) *Chinese Acupuncture and Moxibustion*. Beijing: Foreign Languages Press.

Chengdu Municipal Institute of Cultural Relics and Archaeology and Jingzhou Conservation Center (2015) 'The Laoguanshan cemetery of the Han Dynasty in Tianhui Town, Chengdu City', *Chinese Archaeology*, 15.1: 61–72.

Cullen, C. (1990) 'The science/technology interface in seventeenth-century China: Song Yingxing on *qi* and the *wu xing*', *Bulletin of the School of Oriental and African Studies*, 53.2: 295–318.

Daston, L. and Galison, P. (2010) *Objectivity*, New York: Zone Books.

Farquhar, J. (1994) *Knowing Practice : the clinical encounter of Chinese medicine*, Boulder: Westview Press.

——— (2002) *Appetites: food and sex in post-socialist China*, Durham, NC: Duke University Press.

Graham, A.C. (1986) *Studies in Chinese Philosophy and Philosophical Literature*, New York: State University of New York Press.

——— (1989) *Disputers of the Tao: philosophical argument in ancient China*, La Salle: Open Court.

Guo Moruo 郭沫若 (1972) 'Gudai wenzi zhi bianzheng de fazhan' 古代文字辯證的發展 (The dialectics of the development of ancient writing), *Kaogu xuebao*, 29: 2–13.

Hansen, C. (1983 [2020]) *Language and Logic in Ancient China*, Socorro, NM: Advanced Reasoning Forum.

——— (1992) *A Daoist Theory of Chinese Thought*, New York: Oxford University Press.

Hanyu da cidian chubanshe 漢語大詞典出版社 (2007) *Hanyu da cidian* 漢語大詞典 (Great Dictionary of the Chinese Language), Hong Kong: Shangwu yinshuguan youxian gongsi.

Harbsmeier, C., Robinson, K. and Needham, J. (1998) *Science and Civilisation in China. Vol. 7, Part I, Language and Logic*, Cambridge: Cambridge University Press.

Harper, D. (1998) *Early Chinese Medical Literature: the Mawangdui medical manuscripts*, London; New York: Kegan Paul.

——— (1999) 'Warring states: natural philosophy and occult thought', in M. Loewe and E.L. Shaughnessy (eds) *The Cambridge History of Ancient China: from the origins of civilization to 221 BC*, Cambridge: Cambridge University Press, pp. 813–84.

Harrington, A. (2008) *The Cure Within: a history of mind–body medicine*, New York: W.W. Norton.

He Linyi 何琳儀 (1998) *Zhanguo guwen zidian: Zhanguo wenzi shengxi* 戰國古文字典: 戰國文字聲系 (Warring States Paleographic Dictionary: Warring States phonological systems), Beijing: Zhonghua shuju.

Ho, E.Y. (2006) 'Behold the power of *Qi*: the importance of *qi* in the discourse of acupuncture', *Research on Language and Social Interaction*, 39: 411–40.

——— (2015) '*Qi* (Chinese)', in K. Tracy, C. Ilie and T. Sandel (eds) *The International Encyclopedia of Language and Social Interaction*, Hoboken, NJ: Wiley–Blackwell, pp: 1–5.

Hsu, E. (1999) *The Transmission of Chinese Medical Knowledge*, Cambridge: Cambridge University Press.

——— (2007) 'The biological in the cultural: the five agents and the body ecologic in Chinese medicine', in P. David and U. Stanley (eds) *Holistic Anthropology: emergence and convergence*, New York; Oxford: Berghahn Books, pp. 91–126.

Huang Dekuan 黃德寬 (2007) *Gu wenzi puxi shuzheng* 古文字譜系疏證 (Explanatory Verification of the Genealogies of Ancient Characters), Beijing: Shangwu yinshuguan.

Ji Xusheng 季旭昇 (2014) *Shuowen xinzheng* 說文新證 (New Verifications of the *Shuowen*), Taipei: Yiwen.

Jiang Shan 姜姍 (2017) *Jingdian zhenjiu lilun zhi qi yanjiu* 经典针灸理论之气研究 (*Qi* in Classical Acupuncture Theory), PhD thesis, China Academy of Chinese Medical Sciences.

Kaptchuk, T.J. (1983) *The Web That Has No Weaver: understanding Chinese medicine*, New York: Congdon and Weed.

Keegan, D.J. (1988) *The Huang–ti nei–ching: the structure of the compilation, the significance of the structure*, PhD thesis, University of California.

Knoblock, J. and Riegel, J. (2000) *The Annals of Lü Buwei*, Stanford, CA: Stanford University Press.

Kohn, L. (2008a) *Chinese Healing Exercises: the tradition of Daoyin*, Honolulu: University of Hawai'i Press.

——— (2008b) 'Liezi 列子', in F. Pregadio (ed.) *The Encyclopedia of Taoism*, London: Routledge, pp. 654–56.

Kuriyama, S. (1994) 'The imagination of winds and the development of the Chinese conception of the body', in A. Zito and T. Barlow (eds) *Body, subject and power in China*, Chicago, IL: University of Chicago Press, pp. 23–41.

Kuroda Genji 黒田源次 (1977) *Ki no kenkyū* 気の研究 (Researches on *ki*), Tokyo: Tōkyō bijutsu.

Kubny, M. (2002) *Qi – Lebenskraftkonzepte in China: Definitionen, Theorien und Grundlagen*, PhD thesis, Heidelberg University.

Lau, D.C. (1963) *Dao de jing*, Harmondsworth; New York: Penguin Books.

Lei, S.H.-l. (2012) 'Qi-transformation and the steam engine: the incorporation of western anatomy and re-conceptualisation of the body in nineteenth-century Chinese medicine', *Asian Medicine*, 7.2: 319–57.

Lewis, M.E. (1990) *Sanctioned Violence in Early China*, Albany: State University of New York Press.

Li Jianmin 李建民 (2009) 'They shall expel demons: etiology, the medical canon and the transformation of medical techniques before the Tang', in J. Lagerwey and M. Kalinowski (eds.) *Early Chinese Religion*, Vol. 2, Leiden; Boston, MA: Brill, pp. 1103–50.

Li Ling 李零 (2006) *Zhongguo fang shu zheng kao* 中国方术正考 (Primary Investigation of Chinese Technical and Numerological Arts), Beijing: Zhonghua shuju.

Li Xueqin 李学勤 and Zhao Ping'an 赵平安 (eds) (2013) *Ziyuan* 字源 (Etymology of Chinese Characters), Tianjin: Tianjin guji chubanshe.

Libbrecht, U. (1990) 'Prāṇa = pneuma = ch'i?', in W.L. Idema and E. Zürcher (eds) *Thought and Law in Qin and Han China: studies dedicated to Anthony Hulsewé on the occasion of his eightieth birthday*, Leiden; New York: Brill, pp. 42–62.

Lo, V. (1998) *The Influence of 'Yangsheng' Culture on Early Chinese Medicine*, PhD thesis, School of Oriental and African Studies.

——— (1999) 'Tracking the pain: *jue* and the formation of a theory of circulating *qi* through the channels', *Sudhoffs Archiv*, 83.2: 191–211.

——— (2001a) '*Huangdi Hama jing* (Yellow Emperor's Toad Canon)', *Asia Major*, XIV.2: 61–99.

——— (2001b) 'The influences of nurturing life culture on the development of Western Han acumoxa therapy', in E. Hsu (ed.) *Innovation in Chinese Medicine*, Cambridge: Cambridge University Press, pp. 19–50.

——— (2014) *How To Do the Gibbon Walk: a translation of the Pulling Book* (ca. 186 BCE), Cambridge: Needham Research Institute Working Papers 3.

——— (forthcoming) 'Looms of life: weaving a new medicine in early China,' *Asian Medicine*.

Maciocia, G. (1994) *The Practice of Chinese Medicine: the treatment of diseases with acupuncture and Chinese herbs*, New York: Churchill Livingstone.

Major, J.S., Queen, S.A., Roth, H.D., Meyer, A.S., Puett, M. and Murray, J. (2010) *The Huainanzi: a guide to the theory and practice of government in early Han China*, New York: Columbia University Press.

Meng Qingyun 孟庆云 (2002) 'Song Ming lixue dui zhongyixue lilun de yingxiang' 宋明理学对中医学理论的影响 (Influence of neo-confucianism in the Song and Ming dynasties on the theory of traditional Chinese medicine), *Zhonghua yishi zazhi*, 32: 131–4.

Needham, J. and Wang, L. (1956) *Science and Civilisation in China, Vol. 2, History of Scientific Thought*, Cambridge: Cambridge University Press.

Nylan, M. (2010) 'Yin-yang, five phases, and *qi*', in M. Nylan and M. Loewe (eds.) *China's Early Empires: a re-appraisal*, Cambridge: Cambridge University Press, pp. 398–414.

Onozawa Seiichi 小野沢精一, Fukunaga Mitsuji 福永光司 and Yamanoi Yū 山井湧 (1978) *Ki no shisō: Chūgoku ni okeru jinenkan to nigen-kan no tenkai* 気の思想: 中国における自然観と人間観の展開 (Thoughts on Qi: The Emergent Distinction of the Human-centric Perspective and the Nature-Centric Perspective in China), Tōkyō: Tōkyō Daigaku Shuppankai.

Ots, T. (1994) 'The silenced body–the expressive Leib: on the dialectic of mind and life in Chinese cathartic healing', in T.J. Csordas (ed.) *Embodiment and Experience: the existential ground of culture and self*, New York: Cambridge University Press, pp. 116–36.
Palmer, D.A. (2007) *Qigong Fever: body, science, and utopia in China*, New York: Columbia University Press.
Phan, T. (2017) *American Chinese Medicine*, PhD thesis, University College London.
Pritzker, S. (2014) *Living Translation: language and the search for resonance in U.S. Chinese medicine*, New York; Oxford: Berghahn Books.
Qiu Xigui 裘錫圭 (1995) *Wenzi xue gaiyao* 文字學概要 (Compendium of Paleography), Taipei: Wanjuan lou.
Rainey, L. (1998) 'Mencius and his vast, overflowing *qi* (*haoran zhi qi*)', *Monumenta Serica*, 46: 91–104.
Rickett, W.A. (1998) *Guanzi: political, economic, and philosophical essays from early China: a study and translation*, Vol. 2, Princeton, NJ: Princeton University Press.
Roth, H.D. (1999) *Original Tao: inward training and the foundations of Taoist mysticism*, New York: Columbia University Press.
Sakade Yoshinobu (2007) *Taoism, Medicine and Qi in China and Japan*, Osaka: Kansai University Press.
Schäfer, D. (2011) *The Crafting of the 10,000 Things: knowledge and technology in seventeenth–century China*, Chicago, IL: University of Chicago Press.
Scheid, V. (2002) *Chinese Medicine in Contemporary China: plurality and synthesis*, Durham, NC: Duke University Press.
——— (2013) 'Constraint 鬱 as a window on approaches to emotion–related disorders in East Asian medicine', *Culture, Medicine, and Psychiatry*, 37.1: 2–7.
——— (2016) 'Holism, Chinese medicine and systems ideologies: rewriting the past to imagine the future', in S. Atkinson, J. Macnaughton, J. Richards, A. Whitehead and A. Woods (eds) *The Edinburgh Companion to the Critical Medical Humanities*, Edinburgh: Edinburgh University Press, pp. 66–86.
Schuessler, A. (2007) *ABC Etymological Dictionary of Old Chinese*, Manoa: University of Hawai'i Press.
Sivin, N. (1987) *Traditional Medicine in Contemporary China: a partial translation of revised outline of Chinese medicine (1972): with an introductory study on change in present day and early medicine*, Ann Arbor: Center for Chinese Studies, University of Michigan.
——— (1993) 'Huang ti nei ching 黃帝內經', in M. Loewe (ed.) *Early Chinese Texts: a bibliographical guide*, Berkeley: Society for the Study of Early China, pp. 196–215.
——— (2000) 'Book review: Christoph Harbsmeier, science and civilisation in China, vol. 7, rhe social background, part 1: language and logic in traditional China', *East Asian Science Technology and Medicine*, 17: 121–33.
Stanley-Baker, J. 徐小虎 (1994) 'The physiology and anatomy of Chinese calligraphy', conference paper presented at *The Nature of Chinese Painting*, Sept. 12–14, Norwich: Sainsbury Centre.
Stanley–Baker, M. (2012) 'Palpable access to the divine: Daoist medieval massage, visualisation and internal sensation', *Asian Medicine*, 7.1: 101–27.
——— (2019a) 'Health and philosophy in pre– and early imperial China', in P. Adamson (ed.) *Health: a history*, Oxford; New York: Oxford University Press, pp. 7–42.
——— (2019b) 'Qi', in M. Meulenbeld (ed.) *Critical Terms for Religious Studies: an international conference*, Hong Kong: Hong Kong Polytechnic University, pp. 174–88.
Stengers, I. (2010) *Cosmopolitics*, trans. R. Bononno, Vol. 1, Minneapolis: University of Minnesota Press.
Taylor, K. (2005) *Chinese Medicine in Early Communist China, 1945–63: a medicine of revolution*, London: RoutledgeCurzon.
Unschuld, P.U. (1986) *Nan Jing: the classic of difficult issues*, Berkeley: University of California Press.
——— (2009) *What Is Medicine? Western and Eastern approaches to healing*, Berkeley: University of California Press.
——— (2016) *Huang di nei jing ling shu: the ancient classic on needle therapy*, Berkeley: University of California Press.
Unschuld, P.U. and Andrews, B. (2018) *Traditional Chinese Medicine: heritage and adaptation*, New York: Columbia University Press.
Unschuld, P.U. and Tessenow, H. (2011) *Huang di nei jing su wen: an annotated translation of Huang Di's inner classic – basic questions*, *2 Volumes*, Berkeley: University of California Press.

Verran, H. (2014) 'Working with those who think otherwise', *Common Knowledge*, 20.3: 527–39.

Watson, B. (2013) *The Complete Works of Zhuang Zi*, New York: Columbia University Press.

Wegmüller, K. (2015) 'The problem with *qi*: vitalism, science and the soul of traditional Chinese medicine', *Journal of Chinese Medicine*, 108: 49–55.

Wilhelm, H. (1948) 'Eine Chou–Inschrift über Atemtechnik', *Monumenta Serica*, 13: 385.

Xiaoxue Tang (2020) https://xiaoxue.iis.sinica.edu.tw/, accessed 28/7/2020.

Yan Fu 嚴復 (1909/1931) *Mingxue qianshuo* 名學淺說 (Logic Primer), Shanghai: Shangwu yinshu guan.

Yang, D. (2018) *Prescribing 'Guiding and Pulling': the institutionalisation of therapeutic exercise in Sui China (581–618 CE)*, PhD thesis, University College London.

Yang, S.Z. (1997) *The Pulse Classic: a translation of the Mai Jing*, Boulder, CO: Blue Poppy Press.

Yasuo Yuasa 湯浅泰雄 (1987) *The Body: toward an Eastern mind–body theory*, trans. T.P. Kasulis, Albany: State University of New York Press.

—— (1993) *The Body, Self–Cultivation, and Ki–Energy (Ki, shugyō, shintai)*, Albany: State University of New York Press.

Yu Xingwu 于省吾 and Yao Xiaosui 姚孝遂 (1996) *Jiagu wenzi gulin* 甲骨文字詁林 (A Treasury of Commentaries on the Oracle–bone Characters), Beijing: Zhonghua shuju.

Zhan, M. (2009) *Other–Worldly: making Chinese medicine through transnational frames*, Durham, NC: Duke University Press.

Zhongyang Yanjiuyuan Minzuxue Yanjiusuo 中央研究院民族學研究所 (ed. (2000) *Qi de wenhua yanjiu – Wenhua, qi yu chuangtong yixue xueshu yantao hui lunwen ji* 氣的文化研究 – 文化、氣與傳統醫學學術研討會論文集 (Cultural Researches on *Qi*: Conference Volume on Culture, *Qi* and Traditional Chinese Medical Arts), Taipei: Zhongyang yanjiuyuan minzuxue yanjiusuo.

Zhou Qi, Liu Changhua, Gu Man, Luo Qiong and Liu Yang (forthcoming) 'Research on the lacquered channel figurine excavated from a Han tomb in Tian Hui, Sichuan', *Asian Medicine*.

Zhu Yueli 朱越利 (1982) 'Qi/qi er zi yitong bian' 炁气二字异同辨 (Discussion of commonalities and differences between two characters for *qi*), *Shijie zongjiao*, 1: 50–58.

Zürn, T.B. (2020) 'The Han imaginaire of writing as weaving: intertextuality and the Huainanzi's self–fashioning as an embodiment of the Way', *Journal of Asian Studies*, 79.2: 367–402.

3

RE-ENVISIONING CHINESE MEDICINE

The view from archaeology[1]

Vivienne Lo and Gu Man

If you read almost any textbook about Chinese medicine, you will find that the identity and authority of traditional Chinese medicine remain firmly rooted in references to classical writings. There are thousands of pre-modern Chinese medical texts, but the classical works most often cited are those named after the legendary culture bringers of Chinese civilisation: the Yellow Emperor or Thearch, *Huangdi* 黃帝, patron of warfare and putative author of the *Huangdi neijing* 黃帝內經 'Yellow Emperor's Inner Canon', a title that refers to several medical compilations; and the Divine Farmer, *Shennong* 神農, who is said to have invented agriculture and led people away from lives of hunting and gathering, and is also credited with writing the classical *bencao* 本草 (*materia medica* or remedy books). Both medical corpuses arguably date to late Han (202 BCE–220 CE) (Chapter 7 in this volume; Unschuld et al. 2011; Unschuld 2003). This was a time when treatises concerned with law, medicine, philosophy, ritual, military strategy and governance (and admixtures thereof) were individually copied on to scrolls made of silk, or wooden or bamboo slips tied together with silk thread; they were then collated into numerous compilations, some editions of which survive to the present day in printed form (Figure 3.1).

Recensions of the medical treatises attributed to the Yellow Emperor and the Divine Farmer were carved in woodblock by imperial order, and printed in the Song period (960–1279 CE), a millennium or so later; from this time onwards, their content stabilised. However, what was once represented as immutable knowledge, medical truths revealed in conversations between these legendary ancient sages and equally fabulous interlocutors, now appears to have been anonymous and multi-authored – the result was a knowledge that shifted by degrees and that was flexible enough to adapt to changing cultures and fashions. That is, the key concepts of *Yin*, *Yang*, the Five Agents and *Qi*, described by Chen Yun-ju in Chapter 1, have proved so malleable that they can even be successfully interpreted by those seeking to find a place for traditional medicine in the modern Chinese world (Lei 2015).

Modern traditions cling tight to the legends of medical antiquity, obscuring historical research into exactly when, where and how classical medicine originally took shape. Fortunately, late twentieth- and early twenty-first-century archaeology has recovered many of the original silk and bamboo manuscripts (Ma 1992; Harper 1998; Lo 2014). The manuscripts have made a marked contribution to clarifying many aspects of the history of healing and

Figure 3.1   *Ci shu* 刺數 (Principles of Piercing, facsimile)

medicine in the ancient world. They help us to date the classical texts; they permit new analyses of the nature of medicine and healing in the ancient world and its transmission. They situate knowledge geographically and socially in ways that are not possible, for example, with the sources available for the history of ancient Greek medicine where ancient texts only survive through much later, and often extensively edited, editions. From the increasing number of early manuscripts excavated and retrieved in China, we now know much more about the importance of healing in early Chinese scholarship, and the range of healing techniques used in the late Warring States and early empires.

Such was the importance of medical manuscripts that editions were placed in tombs of the late Warring States and early empires as part of the all-important provisions and conspicuous display that served to enhance the status of the family and sustain the deceased in the after-life. The texts themselves were, therefore, valuable objects which attracted a deep reverence as a kind of scripture: their purported antiquity and lack of reference to actual authorship conferred an unchallenged authority on the knowledge contained therein. Mere possession of a manuscript could enhance one's personal power and influence. Practitioners in possession of these texts acquired great prestige, something all the more important in a world where most physicians were itinerant and needed ways to establish their reputations quickly to their changing clientele. Not only were there benefits for the scholars and physicians who wrote, compiled and used the texts, and the scribes who copied them, but also the noblemen who sponsored and collected the texts.

Thanks to an ever-increasing number of archaeological digs, in ancient sites often revealed as an unintentional consequence of the building boom that has accompanied China's rapid urban development, a large number of medical manuscripts and artefacts have been recovered. With them, we are also now able to date more or less accurately when classical medical ideas began to take shape, and the debates that contributed to that process. Medical manuscripts tend to tell very different stories from the Song dynasty canonical works preserved in print. They are easier to situate and reveal more diverse forms of healing, ones distinctively local or religious. We can, therefore, now fill in great gaps in the social and cultural histories of the medical innovation that occurred during the 400 or so years of Qin and Han rule, and to begin to analyse why the latter, relatively stable, ruling house with its investment in administrative and bureaucratic unification should have provided the context for concerted attempts to standardise medicine.

It seems that medical practitioners of the Western Han period initially acquired knowledge through oral tradition: teachers would pass on their knowledge to their selected disciples by word of mouth. Committing medical knowledge to writing was also a respected and well-established practice during the Han period, encouraged by the imperial court, which collected technical writings. They could reinforce their claim to authority by handing down their knowledge, and eventually the texts themselves, to their pupils in a process shrouded in esoteric ritual. This marked the texts as precious and their transmission as exclusive and underscored the bond between teacher and pupil.

Although it recounts an episode between figures we know to be mythical, the Yellow Emperor and Lei Gong 雷公 (Lord Thunder), an extract from the *Lingshu* 靈樞 (Numinous Pivot) recension of the *Yellow Emperor's Inner Canon* gives an insight into the kind of ritual that may have accompanied the transmission of high-prestige texts from master to disciple during Western Han:

> The Yellow Emperor then entered the purification chamber with him. They cut their forearms and smeared the blood on their mouths. The Yellow Emperor chanted this incantation: 'Today is the True *Yang*. Smearing the blood, I transmit the remedies. May he who dares to turn his back on these words himself bear calamities'. Lord Thunder bowed twice and said: As a young child, I receive them,' The Yellow Emperor then with the left hand gripped his hand and with his right gave him the texts, saying: 'Pay heed! Pay heed!
>
> (Huangdi neijing lingshu, 48 'Jinfu'; translation modified from Harper 1998: 63)

Remedies, medical theory of the channels and a wide range of self-cultivation and the healing arts were transmitted in this way in manuscript form, before some of the same manuscripts, or copies thereof, were committed to the earth as mortuary goods. We are, therefore, aware of multiple copies of some of the same texts circulating in the course and hinterland of the Yangzi valley, with important variations which, when compared to the later printed version, permit analysis of individual interpolation and innovation as part of community, rather than exclusively individual, endeavour. It is to a summary of these manuscripts, and the radical effect they have had on the history of medicine, that this chapter is dedicated.

The manuscripts and artefacts

We have listed manuscripts and artefacts concerned with the body and its care by chronological order of the original tomb or site, as far as that dating can be established, and provide references for the excavation reports and transcripts in the bibliography for this chapter. The quantity and quality of our evidence are naturally as varied as the circumstances of the finds themselves, but the growing number of relevant sites and discoveries indicated allows us to come to some preliminary hypotheses about the flow of medical knowledge in the early empires, the contexts for medical innovation, the communities involved and their cultural influences. The following list grew out of a skeletal version prepared as an appendix for Lo and Li (2010: 391–7).

Shuihudi 睡虎地 (Hubei, Yunmeng 雲夢), tomb 11: *terminus ad quem* 217 BCE.

Twelve tombs were accidentally discovered at the Shuihudi site in 1975–76 by irrigation workers. They date to between the late Warring States and Qin period. Altogether 1,155 complete slips and eighty fragments were discovered in the inner coffin of the tomb owner of tomb 11, who has been identified as Xi 喜, an official of the Qin period. Xi had held positions as a scribe, a clerk and a prison officer. The tomb was a rectangular vertical pit tomb with two compartments, dividing coffin and burial goods. This style was common in Chu, but some of the burial customs were characteristic of Qin (Harper and Kalinowski 2017: 25). Four of the texts bore their own titles: *Yushu* 語書, *Xiaolü* 效律, *Fengzhenshi* 封診式 and *Rishu* 日書 and represent an overall content that primarily represents legal cases, but also state chronicles and communications with officials. The legal documents also include some cases concerned with forensic medicine, and the rights and treatment of those with contagious disease such as leprosy (Hulsewé 1985).

The 166 bamboo slips of *Rishu jiazhong* 日書甲種 (Daybook A) and 257 slips of *Rishu yizhong* 乙種 (Daybook B) contain rich evidence of Yin, Yang and the Five Agents structuring ideas about everyday practice. They bear testimony to the influence of hemerology, the selection of auspicious days and avoidance of inauspicious days in protecting one's health, and the healing arts. Understanding the cycles and development of the human body is also a feature of these texts and they are fascinating for the illustrations and charts they include. See, for example, the methods for predicting the fortune of a child according to the Branch sign attributed to the birth date. Daybooks are also a valuable repository of information about the role of spirits and demons in causing illness, and ritual and magical cures. Daybooks have also been discovered at four other tombs at Fangmatan 放馬灘 tomb 1 in Gansu, Tianshui 天水, Kongjiapo 孔家坡 tomb 8, Jiudian 九店 tomb 56 and Zhoujiatai 周家臺 all in Hubei.

Liye 里耶 (Liye town, Longshan county 龍山縣, Hunan), well 1: *terminus ad quem* 222–208 BCE

The site of a Warring States–Qin city was discovered within the area of Liye Township, on the banks of the waters of Youshui 酉水, on the main branch of the Yuanshui 沅水. The excavation, sponsored by the Institute of Archaeology of Hunan, began in April 2002, and recovered two ancient wells.

Well No. I was discovered three metres below the ground surface. Excavations of the deposits within the well, which consisted of eighteen layers and twenty-eight sub-layers, yielded large quantity of relics, including 36,000 inscribed bamboo slips. The slips are mostly official documents of the Qin dynasty from Qianling county 遷陵縣, Dongting prefecture 洞庭郡. They include government decrees, files and communications, registries of names and goods, etc.

One set of slips records the patient consultations, diagnosis and treatments, of a physician by the name of Jing 靜, also known as Qianling Yi 遷陵醫. Jing used moxibustion, acupuncture, hot ironing (*weifa* 熨法) and prescribed pharmacological remedies. Some of the bamboo slips record Qin Shihuang's 秦始皇 decree to search for the drugs of immortality (*xianyao* 仙藥) and record local responses to the decree. Medical recipes from the bamboo slips published are mostly fragmented, but they are numbered systematically. Some are similar to those of *52 bingfang* 五十二病方 at Mawangdui and the Western Han medical manuscripts on bamboo slips collected by Peking University.

Zhoujiatai 周家臺 (Hubei, Shashi 沙市, Guanju 關沮 district), tomb 30: *terminus ad quem* c. 209–206 BCE

The archaeologists excavated the Eastern Han Tomb 26 at Xiaojiacaochang 蕭家草場 in November 1992, and the Qin Tomb 30 at Zhoujiatai 周家臺 in Hubei in June 1993. The occupant of Zhoujiatai tomb was probably a minor official responsible for tax collection, deceased between the ages of thirty and forty. Three hundred and eighty-seven fragments of bamboo slips found in the tomb at Zhoujiatai were divided into three groups, respectively, devoted to the calendar, to medical recipes and to a Day Book. Of the three manuscript finds, the calendar, which included all of the *ganzhi* 干支 days (Chapter 4 in this volume) of the 34th year of the reign period of the first Emperor, Qinshi huangdi 秦始皇帝 (213 BCE), has been deemed the most valuable. The wooden slips are the calendar of the first year of the reign of the second emperor, Qin ershi 秦二世. The third group includes seventy-three bamboo fragments containing medical recipes, *zhuyoushu* 祝由術 (treating diseases by prayer), *zeji* 擇吉 (selecting auspicious days), divination and farming activities. These medical remedies are very similar to those found at Liye 里耶 and Mawangdui.

Zhangjiashan 張家山 (Hubei, Jiangling 江陵 county), tomb 247: *terminus ad quem* c. 187–180 BCE

Tomb 247 at the Jiangling, Zhangjiashan site in Hubei was excavated in late 1983 and early 1984. The occupant was probably a minor official, interred between the regency of dowager empress Lü 呂后 of the Western Han dynasty (188–180 BCE) and the first year of the reign of the emperor Wendi 文帝 (179 BCE). This dating makes it roughly contemporary with the Mawangdui tomb described below. The tomb was a rectangular vertical pit tomb with two rooms, not grand in scale and, apart from the manuscripts, was unremarkable. A register in the tomb unusually listed the manuscripts among the burial goods. Along with 178 bamboo slips with writings on medicine, there were documents relating to judicial matters, administration, military strategy and mathematical calculations. The medical texts, unusually with their own ancient titles written on the back of the first bamboo slip, included *Maishu* 脈書 (Writings on the Channels; sixty-five bamboo slips) and *Yinshu* 引書 (Writings on 'Pulling'; 113 bamboo slips). The former is similar in content to the Mawangdui texts, given the modern titles *Yinyang 11 mai jiujing* 陰陽十一脈灸經, *Maifa* 脈法 and *Yinyang mai sihou* 陰陽脈死候, discussed below, but is preserved in a more complete state than any of these. It mainly describes the course of superficial anatomical channels which relate to the *jingluo* 經絡 acupuncture channels and collateral pathways as standardised in classical medicine. Without this, potentially anachronistic, lens, the channels described are a landscape of the body as it was shaped by the seams and valleys between muscles and bone, to the pulses joined up: a somatography (in its sense of a descriptive science of the body), which not only approximates to the routes common in later acupuncture theory, but also to what we now understand as the course of the veins and arteries, and to a sensory experience of the nervous system. Uniquely, the description of the channels is preceded by a list of illnesses arranged in a hierarchy from head to toe (with exclusively female illnesses congregated around the breasts and genitals). The latter text, *Yinshu*, is the earliest extant manual of therapeutic exercise, known as *daoyin* 導引 (guiding and pulling) (trans. Lo 2014). *Daoyin* of this period primarily prescribes stretching exercises for 'pulling' pain and treating ailments which in part mimic the movements of animals. It bears a distinct relation to the Mawangdui chart *Daoyin tu* 導引圖, with which it shares, with important variations, some of the exercise names, techniques and purposes (Figure 3.2).

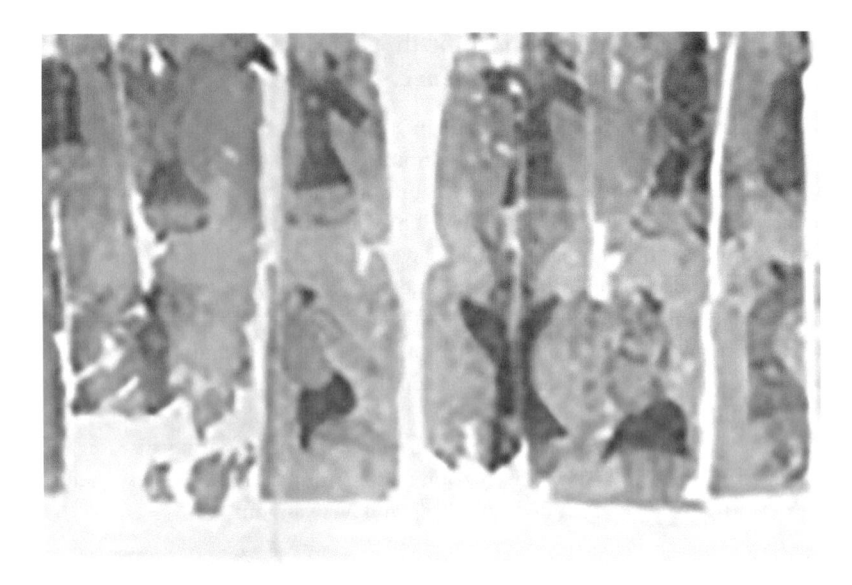

Figure 3.2 Mawangdui chart *Daoyin tu* 導引圖 (Chart of Guiding and Pulling)

Shuanggudui 雙古堆 (Fuyang 阜陽, Anhui), tomb no. 1: tomb closed during the reign of Wendi 文帝 (r. 179–157 BCE)

In 1977, 130 of the bamboo fragments discovered in the Han tomb of the second-generation Marquis of Ruyin 汝陰侯 Xiahou Zao 夏侯竈 at Shuanggudui were found to contain at least ten lost texts. The text known as *Wanwu* 萬物 (The Myriad Creatures) records over seventy drugs, and thirty identifiable illness categories. There are also techniques that scholars believe are associated with the immortality cults, such as a method for making oneself weightless.

Mawangdui 馬王堆 (Hunan, Changsha), tomb 3: *terminus ad quem* 168 BCE

Excavated in 1973, the vertical pit tomb contained a large cache of texts mainly written on silk, but also bamboo and wood. The tomb occupant died as a young adult of about thirty years and is presumed to be the son of Li Cang 利蒼, Marquis of Dai 軑侯, chancellor of the Kingdom of Changsha 長沙國 and his wife who are buried in tombs 2 and 1, respectively. The thirty or so manuscripts that were buried in their son's tomb are broadly divisible into works of philosophy, astronomy, divination, government affairs and military strategy. There was also a substantial corpus of medical literature, some that resembles classical medical theory, but much of which has not otherwise been transmitted. According to the team of scholars involved in collating and editing this material, there are fifteen separate medically related texts inscribed on seven manuscripts. Five of the seven medical manuscripts are written on silk, while the remaining two are on bamboo and bamboo and wooden slips, respectively (Harper 1998: 22–30). The content is taken to tally broadly with the categories set out in the *Fangjilüe* 方技略 (Remedies and Arts) section of *Hanzhi* 漢志 (the Bibliographic Treatise of the History of the [Western] Han, which recorded copies of manuscripts for the imperial library) and the modern titles given to the Mawangdui medical texts reflect this analogy (*Hanshu*, 30; Chen 1983). Sadly, we have few texts which would represent the immortality

Table 3.1 Categories of Remedies and Arts from the *Han Imperial Catalogue* as Compared with Related Manuscripts Excavated from the Mawangdui Tomb

Han Imperial Catalogue: Record of Remedies and Arts	Mawangdui Texts (Modern Editorial Titles)
Yijing 醫經 (Medical canons) 7 Titles, 216 volumes	*Zubi 11 mai jiujing* 足臂十一脈灸經 (Moxibustion Canon of the 11 Channels of the Arms and Legs) *Yinyang 11 mai jiujing* 陰陽十一脈灸經 (Moxibustion Canon of Yin and Yang and the 11 Channels, Recensions A and B) *Maifa* 脈法 (Pulse Techniques) *Yinyang maisihou* 陰陽脈死候 (Prognosis of Death on the Yin and Yang Channels)
Jingfang 經方 (Canonical remedies) 11 titles, 274 volumes	*52 bingfang* 五十二病方 (52 Medical Remedies)
Fangzhong 房中 (Arts of the bedchamber) 8 titles, 186 volumes	*Yangsheng fang* 養生方 (Remedies for Nourishing Life) *Zaliao fang* 雜療方 (Miscellaneous Therapeutic Remedies) *Taichan shu* 胎產書 (Writings on Gestation) *Shiwen* 十問 (Ten Questions) *He yinyang* 合陰陽 (Uniting *Yin* and *Yang*) *Za jinfang* 雜禁方 (Various Prohibited Remedies) *Tianxia zhidao tan* 天下至道談 (On the Culminant Way under Heaven)
Shenxian 神仙 (Immortal beings) 10 titles, 205 volumes	*Quegu shi qi* 去穀食氣 (Abstaining from Grain and Eating Qi) *Daoyin tu* 導引圖 (Chart of Guiding and Pulling)

category. There is only one lengthy medical recipe and remedy text, but it is a substantial work, and there are many other such collections, notably at Wuwei and in recent discoveries such as those Laoguanshan. There are two collections of recipes which are more concerned with knowledge of *fangzhong* 房中 'arts of the bedchamber'. The proportion of medical texts to other texts in the Mawangdui tomb is not so far removed from the ratio of the number of texts in *Fangji* (35), a category which embraced a wide range of medical remedies, body cultivation and healing arts, to the number in the other categories, which include the six arts, the arts of calculation, the military and of philosophy (*Hanshu*, 30; Chen 1983: 1–3) (Table 3.1).

Laoguanshan 老官山 (Sichuan, Chengdu, Tianhui 天回鎮 town), tomb 3: *terminus ad quem* c. 157–141 BCE

Between July 2012 and August 2013, four earthen shaft pit tombs aligned north south with wooden chambers were excavated by Chengdu Municipal Institute of Cultural Relics and Archaeology in a cemetery of the Western Han dynasty located among the Tumen community of Tianhui Town, Jinniu 金牛區 District, Chengdu (Xie *et al.* 2014: 59–70). Despite the fact that all the tombs had been looted, 951 bamboo slips were discovered in the North II and South II compartments of the bottom chamber of tomb M3. Most of the texts they contain are of medical interest, although there are twenty pieces of bamboo slips among them, which are fragments of a legal document.

On the basis of their subjects, it has been suggested that the occupant was a physician or medical official. Six texts of the M3-121 medical manuscripts from the North II compartment concern medical theory, the causes and symptoms of disease, the nature of the *mai* 脈,

the channels and vessels of the acupuncture body, diagnosis, acupuncture and possibly moxibustion therapy. They have been given the following titles by the modern editors:

Maishu-shangjing 脈書上經 (Channel Writings – the upper);
Maishu-xiajing 脈書下經 (Channel Writings – the lower);
Zhi liushi bing heqi tan fa 治六十病和劑湯法 (Methods for Mixing Decoctions for Treating the Sixty Illnesses);
Ci shu 刺數 (Principles of Piercing) (see Figure 3.1);
Nishun wuse maizang yan jingshen 逆順五色脈藏驗精神 (Contrary and Consistent [Indications] of the Five Complexions in the Channels and *Zang*-organs in Assessing the Spirit).
Ba li 犮理 (Principles of *Ba* Therapy). It has been suggested that 'Ba' therapy relates to moxibustion treatment.

In *Maishu-shangjing*, the introductory phrase 'Bixi yue 敝昔曰 (Bixi said)' seems to call on the authority of the medical scriptures passed on by Canggong 倉公 (Chunyu Yi 淳于意, b. 215 BCE, Qi Kingdom, fl. 176 BCE) and the semi-legendary character Bian Que (trad. dates c. sixth to fifth century BCE). Chunyu Yi held the honorary title of 'Granary Chief' of the Kingdom of Qi and his biography is included in *Shiji* (105: 2785–820). The biography records that Chunyu Yi was given copies of Bian Que 扁鵲 and the Yellow Emperor's *Mai shu* 脈書, the sexual cultivation text *Jieyinyang* 接陰陽 (Receiving *Yin* and *Yang*), *Shishen* 石神 (Stone Spirit), *Yaolun* 藥論 (Treatise on Drugs), *Wuse zhen bing* 五色診病 (Diagnosing Illness from the Five Colours), *Zhiren sisheng* 知人死生 (Knowing a person's prognosis of life or death) and *Kuiduo yin yang wai bian* 揆度陰陽外變 (Observe and Estimate External Transformations of *Yin* and *Yang*), by which one might know a person's prognosis of life or death. The dates that the texts were copied seem to fall within the reign periods of Empress Lü 呂后 and Emperor Wen 文帝 of the Western Han dynasty.

The M3–139 medical manuscripts from the South II compartment included a text known as the *Liaoma shu* 療馬書 (Writings on Treating Horses) and *Jingmai* 經脈 (Standard Channels). The dates that the texts were copied seem to be earlier than that of the M3–121 medical manuscripts.

Altogether the Laoguanshan manuscripts mirror many of the themes in manuscripts from the Zhangjiashan and Mawangdui tomb sites. The different versions of *Maishu*, for example, are the same type of texts in that the themes, character choices and arrangement of the passages can be compared, paying attention to what is identical and where there are variations, and much can be learnt from cross referencing. Dating to a decade or two later, the Laoguanshan texts also appear to represent a later stage of theoretical elaboration.

Among the 620 grave goods, mostly of lacquered wooden and pottery wares but also bronze and iron, a black lacquered wooden figurine similar to that discovered twenty years earlier at the Shuangbaoshan tomb site in modern Mianyang, described below, was also inscribed in red with some twenty-two or more channels. It is, however, distinguished from the earlier discovery by its tiny size, which at 14.9 cm tall is about half the height, another set of white lines and a web of what appear to be at least one hundred points, some of them with familiar acupoint names scratched into the lacquer. The diffuse porous wood that the figure was carved from was from a broad-leaved tree. The points are layered with some of random size having been scratched simultaneously with the white lines, and other round and concave spots which were carved before the lacquer undercoat. These can be compared to descriptions in some of the received texts (Liang *et al.* 2015: 91–3; Huang 2017: 131–44) (Figure 3.3).

Figure 3.3 The Laoguanshan 老官山 Lacquer Figurine

Palaeoanthropological research into four human skulls (one male, three females) that were also discovered in the Laoguanshan tomb group suggests that the tomb occupants had many similarities to modern South Asians and differences from modern Mongolian groups and other ancient peoples (Yuan *et al.* 2018).

The Western Han medical manuscripts on bamboo slips collected by Peking University (tomb or other site excavations unknown): *terminus ad quem* c. 141–87 BCE

There are 3,346 bamboo slips dating to the middle Western Han period in the collection of Peking University, on which there are nearly twenty texts. The slips themselves are well preserved in various calligraphic styles, and beautifully finished with shallow sloping cuts across the reverse side which facilitate restoring the slip sequences. Among them, 711 bamboo slips are medical texts, of which 516 are complete and 185 are broken. There are no titles for these medical texts, which are mainly medical prescriptions.

It has been estimated that Peking University Medical Manuscripts were copied during the reign period of Emperor Wudi 武帝 of the Western Han dynasty, between the dating of the Mawangdui tomb and the Wuwei medical manuscripts and roughly contemporary with the Laoguanshan tomb texts. Compared with other unearthed medical prescriptions, preliminary analysis suggests that the Peking University Medical Manuscripts have been edited more carefully and are both more complex and practically oriented (Li and Yang 2011: 88–9).

Nanyue 南越 (Guangzhou, Xianggang Hill 象崗山): Yuanshuo 元朔 and Yuanshou 元狩 era of Wudi (c. 128–117 BCE)

When it was excavated in 1983, the Nanyue 南越 royal tomb of King Zhao Mei 趙眜 (incorrectly recorded as Zhao Hu 趙胡 in *Shiji* and *Hanshu*) yielded several kilograms of the 'five-coloured medical stones' (*wuse yaoshi* 五色藥石): cinnabar, lead, amethyst, sulphur and malachite. The tomb also contained various pharmaceutical implements, such as copper and iron pestles and copper mortars. These finds suggest that the rulers of the Nanyue kingdom, like the Han aristocracy, were engaged in the pursuit of longevity and immortality.

Shuangbaoshan 雙包山 (Sichuan, Mianyang 綿陽): 118 BCE

A lacquered wooden figurine was unearthed from a shallow tomb northwest of Chengdu in 1993. From the size and mortuary goods, which included one hundred large lacquerware horses and twenty chariots, we can tell that the tomb owner was of high rank and probably connected with the military. The figurine is naked and stands erect. Despite slightly elongated arms and legs and unclear gender, it is notable for its anatomical accuracy. Unlike grave figures – which are typically dressed in the attire of officials, entertainers or servants with handsome faces and hands respectfully hidden inside wide-sleeved gowns, their hair in topknots – the demeanour and posture of the figurine are designed to facilitate inspection of the physical body. It stands 28.1 cm tall and has nineteen lines drawn in red lacquer mostly running between the head and the hands and feet (Figure 3.4). The arrangement of the channels is different from that described in the texts of *Maishu* found at Mawangdui and Zhangjiashan, and also from the system of channels in the *Jingmai* 經脈 section of *Lingshu* 靈樞 (*Huangdi neijing*: The Numinous Pivot). It can be compared to the tiny figurine more recently recovered from the Laoguanshan site described above, which has many more channels and captions indicating organs and acupuncture points. The inconsistency of detail between the manuscripts, the two figurines and the medical canons preserved in print indicates that various theories of, and beliefs about, the channels coexisted from the Late Zhou to the middle of the Western Han period.

Mancheng 滿城 (Zhongshan 中山, Mancheng County, Hebei): 113 BCE

The Western Han tomb of King Jing 靖王 (Liu Sheng 劉勝) of the Kingdom of Zhongshan 中山 was excavated in 1968. It was found to contain a number of artefacts relating to the healing arts and physical cultivation. They include nine needles (four gold and five silver) often thought to be for acupuncture, but more probably mainly for sewing together the jade suit in which Prince Jing was interred, an implement engraved with the characters *yi gong* 醫工 (medical professional), a bronze drug spoon and ornate case, a flask for medicinal wine, and a bronze phallus and small stone spheres associated with female sexual practices.

Juyan 居延 (Han Zhangye 張掖 commandery, Juyan and Jianshuijinguan 肩水金關 areas): (c. first century BCE to third century CE)

The Juyan Han bamboo slips refer to the administrative records kept by the officials (*wei* 尉 'commandant') in charge. 10,000 bamboo slips were discovered in 1930–31 by the Swedish researcher Folk Bergman in the Ejina 額濟納 river basin of Gansu, but the majority – over 19,400 – were discovered between 1973 and 1984 by the Juyan Archaeological Team. Unlike

Figure 3.4 The Shuangbaoshan 雙包山 Lacquer Figurine

most of the other finds recorded here that were deliberately placed in tombs, these military sites *accidentally* preserved records used in daily life, sometimes in rubbish tips; they are therefore invaluable for what they reveal about day-to-day military medical practice.

The Juyan slips date, at the earliest, to the first year of the *Zhaodi* 昭帝 reign period (86–81 BCE) of the Western Han, while the latest date could be from Western Jin (317–420 CE), the fourth year of the *Taikang* 太康 reign period of Jin Wudi 晉武帝 (284 CE) (Xie 2005).

Scattered through these records, there are case histories, treatments, evidence of medical infrastructures, the deployment of medical officials, and disease and mortality among officers and soldiers in the border region. They provide a fascinating window onto the state of medicine in the late mid-Western Han period.

Marquis of Haihun's tomb 海昏侯漢墓 **(Purple Gold City** 紫金城**, Nanchang** 南昌**, Jiangxi), tomb 1:** *terminus ad quem* 59 BCE, the date when the probable tomb occupant died.

Excavated in March 2011 at a very well-preserved burial site for Western Han marquises, the tomb has been identified as that of Liu He 劉賀 (d. 59 BCE), first of the Marquises of Haihun, and is located on Guodun Hill in Datangping Township, Xinjian District, in modern Nanchang. The burial mound was constructed in the shape of a truncated pyramid, the grave built in a *jia* 甲-shaped design with a rectangular wooden coffin chamber. The wooden tomb was partitioned into the main chamber, the entrance passage and an outside surrounding corridor, which was large enough to act as a passageway and the storage area for the mortuary goods. The artefacts unearthed from the tomb included gold, bronze and jade objects, lacquered, wooden and pottery wares, textiles, bamboo manuscripts and wooden tablets.

Over 5,200 bamboo slips and wooden tablets were recovered from four lacquered cases in the tomb. They included official documents which were exchanges between the Marquis of Haihun and his consort and the emperor and empress dowager, and imperial edicts and memorials from central government officials. Other writings related to and including texts on prayer, rites and ceremonies, social etiquette, historical annals, philosophy and governance which we know well from the received printed literature: *Shijing* 詩經 (Book of Poetry), *Liji* 禮記 (Book of Rites), *Lunyu* 論語 (Analects of Confucius), *Chunqiu* 春秋 (the Spring-and-Autumn Annals), *Xiaojing* 孝經 (Classic of Filial Piety), the *Zhuzi* 諸子 (various expository writers) and *Shifu* 詩賦 (poetry and rhapsodies). There were also less well-known technical writings related to *shushu* 數術 (geomancy, astrology, divination and mathematics) and *fangji* 方技 (remedies and arts), as well as the ancient Chinese *liubo* board game for two gamers 六博棋. Two hundred slips relate to care of the body with content similar to those indicated in the *Fangji* treatise of the *Hanshu*, described above: they combine information on the arts of the bedchamber and health preservation in a single work, and medical recipes. There are clear intersections with the Mawangdui literature described above. One text referring to the *shidao* 十道 (Ten ways) appears to be based on the *badao* 八道 (Eight ways) described in the Mawangdui sexual cultivation text known as *Tianxia zhidao tan* 天下至道談 (On the Culminant Way under Heaven). It adds the two categories *xuzhi* 虛之 (make it empty) and *shizhi* 實之 (make it full) (Chapter 22 in this volume). Just like works in the *Yellow Emperor's Inner Canon* the text are expressed in the form of a series of questions and answers with the Yellow Emperor. In this case, as in the Mawangdui text, the Yellow Emperor is talking to Rongcheng 容成 (one of the Huangdi's legendary teachers). There is also a remedy text which preserves recipes for eliminating *gu chong* 蠱蟲 (poisonous worms).

Duhuang 敦煌 (Gansu 甘肅, Shule River 疏勒河 areas and Majuanwan 馬圈灣): (c. 65 BCE–21 CE)

The Dunhuang Han bamboo slips refer to the bamboo and wood slips unearthed at the fire beacon towers of the Han dynasty in the Shule River Basin of the Hexi Corridor 河西走廊 in Gansu Province since the early twentieth century. They were named after they were first discovered in the area of Dunhuang County of the Han dynasty. The earliest materials were first discovered by M. A. Stein in the 1900s and published by E. Chavannes in London in 1912. Later, based on the photos provided by E. Chavannes, Luo Zhenyu 羅振玉 and Wang Guowei 王國維 compiled and published the famous work *Liusha Zhuijian* 流沙墜簡 (Bamboo Slips Left in the Flowing Desert) in 1914. In 1979, the Gansu Provincial Cultural Relics and Archaeology Team and the Dunhuang County Cultural Center discovered a fire beacon tower in Majuanwan, which was missing from Stein's investigation. 1,127 bamboo slips were unearthed. The Dunhuang Han bamboo slips have the same nature to Juyan Han bamboo slips, and also spanned a long period. Among them, there are some fragmented medical remedies. Recent research reveals that the content of these bamboo slips should be part of a medical book. A man named Anguo 安國 whose name appears in a medical remedy appears to refer to Han Anguo 韓安國 recorded in *Shiji*, who was appointed Grandee Secretary 御史大夫 during the reign of Emperor Wudi 漢武帝 of the Western Han dynasty (Hirose Kunio 2018, 1: 47–51).

Hantanpo 旱灘坡 (outside Gansu, Wuwei 武威 city) unspecified tomb: first century CE, from Guangwu 光武 Jianwu 建武 19 to Zhangdi 章帝, 43–76 CE (Chen Zhi 1974/1988)

Discovered by farmers repairing an irrigation system in 1972, the *tudong danshi mu* 土洞單室墓 (earthern single room) cave tomb yielded two sets of wooden slips totalling ninety-two altogether. Details of the original site of the tomb have proved hard to establish but the tomb occupant is thought to have been a senior doctor. Seventy-eight documents filled a hemp bag that had been placed at the head of the tomb. The texts are mostly concerned with the healing arts and record the names of over 100 drugs, including details of the preparations, dosages and methods of ingestion (trans. Yang and Brown 2017: 241–301). The material on acupuncture and moxibustion includes details of the duration of needling, acu-moxa points and taboos and prohibitions as to when to avoid treatment. Before the discovery of the Laoguanshan manuscripts and figurine, there was no pristine, extant evidence dating to the Western Han of named acupoints, so the references in the Hantanpo, Wuwei manuscripts were the earliest. In addition, there are some fragmented medical manuscripts as recorded in the introduction to Zhang Lei (2018), Zhou Zuliang (2014) and Wang Qixian (2015).

Preliminary conclusions and future research

A brief survey of the contents of these manuscripts leads to the conclusion that it is very difficult to differentiate neatly what constituted a classical 'medicine' from other ways of healing and cultivating the body, a fact that dogs all historical scholarship which attempts to define the early Chinese healing arts according to a modern sensibility about what constitutes medicine, without specifying the realm of discourse within which the analysis is to operate.

The physical grouping of the manuscripts in the tombs often provides an indication of the contemporary classification of knowledge of the healing arts. For instance, the Mawangdui manuscripts recording the earliest extant theories of physiology were buried together with treatises on exercise, on breathing and sexual techniques, on herbs, on skin-deep surgery and on remedy collections full of magical and ritual procedures. Recent research into the kind of literature categorised, together with remedy collections and standard works on medicine, has, therefore, begun to build a deeper and richer view of the healing arts and medical innovation in Chinese society.

The geographic distribution of the manuscripts arguably permits both a spatial and temporal analysis of the origins and transmission of classical Chinese medical ideas. The variety of medical theories shown in the excavated manuscripts, for example, demonstrate such a degree of intellectual ferment in major cultural centres of the early empires that the burden of proof now falls on those who still hold that classical medicine came to fruition in pre-imperial times. If, then, classical Chinese medicine was fundamentally a medicine that came to maturity along with the inauguration of empire, then where in the empire did it originate?

It has proved attractive to many scholars to imagine a cultural centre of Chinese medical knowledge, together with the pre-existing histories of the origins of Chinese philosophy, that cite the authority and centrality of the legendary Jixia 稷下 academy of Qi, in modern Shandong. Certainly, the biographies of Canggong and Bian Que in Sima Qian's *Shiji* (listed as the first of the twenty-four standard histories) are centred on Qi and the northeast, with the former appointed state Granary Chief and the semi-legendary Bian Que cast as an itinerate physician circumambulating the state in search of work. Working against the centrality of Qi to this story is the growing evidence of a southern and southwestern tradition centred on the sources and course of the Yangzi River, the location of most of the tomb manuscripts cited above. Of course, it may be that the anonymous authors of texts that cite the authority of Bian Que were also from Qi, and that the mortuary copies are evidence of the south-southwesterly flow of knowledge and an hierarchical landscape of knowledge.

Yet, by the time of the Qin and Han empires, there had been many centuries of cultural communication within the geographic entity that was to become imperial China, especially along the network of rivers, and latterly the canals. Over the course of the Han period, we can also see a practical medicine, in the form of collections of remedies and recipes, travelling westwards along the Hexi corridor in modern Gansu, along with the expansion of military settlements towards the frontiers. There were also key centres of scholarly activity in the southwest at Chengdu, not to speak of the imperial centre at Chang'an. The old kingdoms of Chu 楚 and Shu 蜀, where there were distinctive technologies related to water and weaving, also seem to have left their mark on the language and culture of medicine (Lo and Gu *forthcoming*; Lo 2020). It, therefore, seems more balanced to assume that the medical writing that came to be enshrined in printing a millennium or more from the dating of these manuscripts had multiple origins. Thus, while classical Chinese physiology imagined a new, integrated, medical body that was consistent with the politically unified body of China, and its physiological functions famously reflected those of the centralised administration of state, the manner of its conception, and indeed all the techniques and medicines it embraced, was as diverse and plural as China has always been.

The Mawangdui manuscripts contain the earliest extant set of *Yin–Yang* correspondences, arranged as two lists (Chapter 1 in this volume). Today, these correspondences are known mainly through their use in food, medicine and the martial arts. But since they were recorded in second-century BCE ritual and philosophy, rather than exclusively medical or nutritional contexts (and therefore not described above), this underlines the importance of proper contextualisation beyond the immediate remit of this chapter. On the one hand, the correspondences juxtapose oppositions, such as hot and cold, up and down, male and female, pairs which seem to be the kind of binaries that are valid cross-culturally and easily lend themselves to universal and allopathic medical distinctions. On the other, the social and political pairs such as host–guest; big, important states–small, weaker states; ruler–minister make the treatise iconic of its own time, a late Warring States to early imperial production when the political environment was transitioning from two centuries of fierce combat towards achieving a fragile unity (Chapter 1 in this volume).

There is much to be learnt from the broader context within which medicine was emerging in writing during the early empires. Those legal case histories concerned with medical matters discovered at the Shuihudi site (*terminus ad quem* 217 BCE), for example, should be analysed together with the medical case histories of Chunyu yi (fl. 176 BCE), which are significant both legally and medically (Furth 2007: 125–51; trans. Hsu 2010). The manuscript finds have also brought into focus the intricately linked worlds of diviner and physician and their shared numerological cultures (Li 2000). Predicting the future, fortune telling and determining the future course of illness could all involve written calendrical and astronomical/astrological computations, a belief in the regularity of interventions in the lives of human beings by spirits and spirit ancestors; and a knowledge of related spells and exorcism to try to control those interventions (Li 1993; Harper 2001: 99–120; Lo 2001b: 61–100; Cook 2006; Raphals 2013; Chapters 4 and 5 in this volume). At the centre of these calendrical cultures were the *rishu* 日書, the Day Books which, like the Assyrian cuneiform hemerological texts, tend to assume that 'each day of the year was assigned to a specific deity or patron saint, in whose honour special ceremonies and services had to be performed' (Chace as quoted in

Harper 2017: 2). It follows that certain days were more or less auspicious for specific kinds of activities. The first of the Day Books to be recovered in China was from Shuihudi tomb 11; it bore the title *Rishu* on the back of the first slip; subsequently, five more tombs have revealed untitled texts that are structured similarly. Until the recovery of these manuscripts, very little was known about the nature and use of Day Books since they were evidently not of sufficient prestige to be either listed in the bibliographical treatises of the imperial library or preserved and printed in their original form.

Common characteristics include the organisation of prescriptions and prohibitions topically, around rubrics for everyday activities such as agriculture, travel, marriage or childbirth; or according to a particular hemerological system; and they were never organised according to the calendrical sequence (Harper and Kalinowski 2017: 6–7; Chapters 4 and 5 in this volume). Harper has described the texts as 'miscellanies' that copy multiple texts to the same manuscript in ways that were locally significant, mixing hemerological with non-hemerological texts such as rituals, incantations and magic. Day Books were used across the social spectrum and are popular, in its sense of pervasive, throughout the late Warring States and early imperial world. While they do not appear as a separate rubric in the Han bibliographic treatises, their use is perhaps implicated in other types of literature such as that can be found in the standard histories. The quality of these texts therefore varies extensively. Hemerological cultures survive to this day in China in the form of popular almanacs that tell their readers about how to select good and bad days for both everyday events and important life occasions. What once were guides to avoiding inauspicious days or drawing on auspicious forces based on calendrical cultures shared throughout society, and an integral part of state orthodoxy, have therefore proved extraordinarily tenacious, particularly in the prescriptions for popular almanac and dietary cultures still used in everyday life today (Chapter 19 in this volume).

Concerted attempts by the Republican and then the Communist governments in China have been made to supress such 'superstitious' elements in Chinese medicine in favour of scientisation and modernisation. The manuscript finds force us to embrace a history of Chinese medicine full of religious, ritual and magical practices; yet, ironically, it is just this history that has been the context for extraordinary medical discoveries, such as the early and medieval Chinese use of nitrates under the tongue for being struck by evil, treating heart pain and prolonging life, thousands of years before the discovery of the use of nitrates under the tongue for hypertension and other vascular events (Webb et al. *New England Journal of Medicine* 2008 as quoted in Butler 2009: 184). The extent to which those discoveries were empirical in nature, discovered through observation and sensory experience, and/or driven by arcane theoretical considerations remains moot. There is no doubt that healthcare beliefs and practices that were originally grounded in belief in the possibility of lengthening one's life and even achieving immortality remain meaningful today, albeit re-cast for a modern audience: movement and exercise traditions, deep breathing, dietary regimes, the burning of incense, and *fengshui*, the geomantic siting of buildings, objects and graveyards, all remain meaningful and take their place in our modern world.

Much has been made of the 'empirical' nature of early Chinese remedy literature as compared to the theoretically driven classical acupuncture theory that fills the classical Yellow Emperor corpus. Historians of medicine tend to mean by empirical that herbs and medical substances were tried and tested, albeit within the norms of early China, and not framed within the emergent medical discourses of *Yin, Yang* and the Five Agents. After all, to assign Yin and Yang qualities and flavours associated with the Five Agents to hundreds, and ultimately many thousands, of substances and herbal remedies was a massive project, much more difficult than theorising the twelve channels of acupuncture theory. Several hundred years later, during the first centuries CE, we can see evidence of much greater theoretical elaboration, such as we find post the *Shennong* (Divine Farmer) *materia medica*, with the beginning of monographs of the substances used medically, and the classification of all according to the Five Agent rubrics and associated flavours and potencies.

Instead, the illnesses and remedies recorded in the Han manuscripts barely refer to ideas of causation beyond identifying inauspicious relations with the spirit world and spirit ancestors; the manuscripts do not expound any theory about the physiology or aetiology of disease. There are eclectic treatments for every kind of illness from haemorrhoids to forms of madness, which embrace ritual incantation, spitting magic, recipes, including herbs and minerals and household substances commonly used in traditional medicine today, and those that are not such as certain types of excrement, dirt and filth that would be classified in the history of medicine as the *dreckapotheke* (Harper 1998: 148–86; Lo 2002: 99–128). A shared range of healing and restorative techniques are demonstrated in the literature and substances recovered from Mawangdui, Zhangjiashan and the Shuangudui tombs, as well as the tombs of the Marquis of Haihun and King Zhao Mei. Drug taking, techniques of meditation, the arts of the bedchamber, movement, breathing to lighten the spirit and ensure the body's endurance and longevity were pervasive styles of self-cultivation and alchemical practice among the aristocracy and educated elite of early China. Strengthening and potentising the body through these unique practices of self-cultivation was also the earliest context within which we see the concepts of *Yin, Yang* and *Qi* developing as physiological substrates of the inner body (Lo 2001a: 19–50). The most obvious distinction between early Chinese self-cultivation and medical literature is that the former recorded how it felt to be more or less well, rather than the dispassionate clinical observations of other people's diseases. Sensations of pleasure, pain and enhanced qualities of the emotion and spirit common in this literature came to be critical in the formation of classical medical concepts such as Yin and Yang, thereby mitigating any radical disintegration of the human being into mind, body and spirit in later medical thought (Chapters 6 and 49 in this volume; Lo 2001a).

From the remedy and recipe literature excavated from these ancient tombs, it has been possible to identify large numbers of drugs and complex multi-ingredient remedies that were already in use in Early China. Some of these are unique to the ancient world and others, like the complex prescription for the Five *Long* 癃, a urinary disorder, are still in use (although one can neither assume the recipes are the same nor the ingredients go under the same name) (Barrett and Lo 2015: 86–96). We do not yet have any excavated remedy collections attributed to Shennong, the Divine Farmer. Many of the remedy collections are unattributed. However, more recently, the name of the legendary Bian Que has been associated with the remedy collections among the Peking University manuscripts and those excavated from the Laoguanshan site (personal communication Li Ling); we see the beginnings of the tailoring of the remedy traditions, and standard collections calling on the authority of the mythical founders of medicine.

The material remains of foods and drugs are also commonly found in ancient Chinese tombs. Many of the drugs dating to the Eastern Han period (25–220 CE), in particular, had ritual purposes and were thought to have magical and apotropaic efficacy: gold, cinnabar, *realgar*, *azuritum*, *arsenopyritum* and *codonopsis* root, to name but a few, were employed to lighten the spirit and expel demons and other harmful forces that could possess a victim. Drug taking for longevity and immortality was widespread among the Han aristocracy, many of whom were engaged in the practice. The first emperor Qinshi Huangdi 秦始皇帝 and Han emperor Wudi 武帝 are famous for giving greater privileges to shamans than their court officials (Ngo Van Xuyet 1976: 17–18).

Shiji 史記, the 'Records of the Historian', depicts the first emperor as wary of challenges to his power and deeply fearful of harm from evil spirits (*Shiji*, 6.248, 263). He and his consorts were particularly fond of the skills of men from Yan 燕 and Qi 齊 who, it is said, transmitted the arts of Zou Yan 騶衍 and whose school was purported to be the institutional origin of the *yin-yang* philosophies. Some advised him on elixirs of immortality, others on the power of the *wuxing* (Five Agents) and its relationship to political legitimacy. Likewise, the authors of *Hanshu* 漢書 portray Wudi 武帝 (reigned 141–87 BCE) as easy prey to the machinations of magicians and other such 'charlatans' who made capital out of his obsession with immortality (Ngo Van Xuyet 1976: 20). Many of his advisors on immortality, alchemy and spirit world were given great privileges (*Hanshu*, 97.3952–53; *Shiji*, 28.1385).

The archaeological record testifies that the immortality practices were widespread, with the royal tomb of King Zhao Mei 趙眜 yielding several kilograms of the 'five-coloured medical stones' (*wuse yaoshi* 五色藥石): cinnabar, lead, amethyst, sulphur and malachite (Guangzhou Xianggang Hanmu fajuedui 1984). His tomb at Nanyue 南越, which dates to the Yuanshuo 元朔 and Yuanshou 元狩 era of Wudi (c. 130–120 BCE), was far in the south in modern Guangzhou, at Xianggang 象崗山 Hill. The tomb also contained various pharmaceutical implements, such as copper and iron pestles and copper mortars.

All of these manuscript finds and artefacts, excavated in the late twentieth and twenty-first centuries, open up new horizons for young scholars interested in the history and archaeology of medicine and we hope that this brief survey will be of some assistance to them. We can only begin to imagine the potential of the manuscripts for reorienting many academic fields. Barely touched upon in this brief introduction, for example, but the subjects of some of the following chapters are the extensive sources for the history of gender and sexuality (Chapters 22–24 in this volume). There are new sources for the study of animal medicine, and its intersection with human medicine, for those looking for new Chinese precedents for a body with different boundaries, for studies of cross-cultural alternatives to understanding the Anthropocene or developing a historical phenomenology in new histories of the senses, for bio-prospecting for new medicines and, equally, for understanding the role of magic and ritual in medicine, both in its own right, but also as rich contexts for the history of innovation.

Note

1 This work was supported by the Wellcome Trust [Reference: 201616/Z/16/Z].

Bibliography

Excavation reports and studies in Chinese

Beijing daxue chutu wenxian yanjiusuo 北京大學出土文獻研究所 (2011) 'Beijing daxue cang xihan zhushu gaishuo' 北京大學藏西漢竹書概說 (A general survey of the Western Han bamboo slips collected by Peking University), *Wenwu*, 6: 49–56.

Chen Zhi 陳直 ([1974], 1988) 'Wuwei Hantanpo chutu yiyaofang huikao' 武威旱灘坡漢墓出土醫藥方彙考 (Collected research on the Han dynasty Wuwei Hantanpo tomb excavated medical prescriptions), in Chen Zhi (ed). *Wenshi kaogu luncong* 文史考古論叢, Tianjin: Tianjin guji chubanshe, pp. 300–6.

Dong Yuan 董源 (1995) 'Wanwu zhong bufen zhiwu mingcheng gujin kao'《萬物》中部分植物名稱古今考 (Textual research on the names of medicinal herbs in *Wanwu*), *Zhongguo keji shiliao*, 4: 77–83.

Gansusheng bowuguan 甘肅省博物館, Wuwei xian wenhuaguan 武威縣文化館 (1973) 'Wuwei Hantanpo Hanmu fajue jianbao' 武威旱灘坡漢墓發掘簡報 (Report on the Wuwei Hantanpo Han tomb excavations), *Wenwu*, 12: 18–22.

Guangzhou Xianggang Hanmu fajuedui 廣州象崗漢墓發掘隊 (1984) 'Xihan Nanyuewang mu fajue chubu baogao' 西漢南越王墓發掘初步報告 (Preliminary report on the Western Han tomb of the King of Nanyue), *Kaogu*, 3: 222–30.

He Zhiguo 何志國 (1994) 'Woguo zuizao de renti jingmai qidiao' 我國最早的人體經脈漆雕 (China's earliest lacquered human figurine with channels), *Zhongguo wenwubao*, 17th April.

Hirose Kunio 廣瀬薰雄 (2018) 'Dunhuang Hanjianzhong suojian Han Anguo shou yi yiyao fang de gushi' 敦煌漢簡中所見韓安國受賜醫藥方的故事 (The story of Han Anguo accepting the prescription on the bamboo slips of the Han dynasty in Dunhuang), *Zhongyiyao Wenhua*, 1: 47–51.

Hubei Xiaogan diqu dierqi yigong yinong wenwu kaogu xunlian ban 湖北孝感地區第二期亦工亦農文物考古訓練班 (1976) 'Hubei Yunmeng Shuihudi shiyi zuo Qin mu fajue jianbao' 湖北雲夢睡虎地是一座秦墓發掘報告 (A short excavation report on eleven Qin tombs at Shuihudi in Yunmeng, Hubei), *Wenwu*, 9: 51–61.

Hunansheng wenwu kaogu yanjiusuo 湖南省文物考古研究所, Xiangxi tujiazu miaozu zizhizhou wenwuchu 湘西土家族苗族自治州文物處, Longshanxian Wenwu Guanlichu 龍山縣文物管理所 (2003) 'Hunan Longshan Liye Zhan'guo – Qindai gucheng yihao jing fajue jianbao' 湖南龍山里耶戰國 – 秦代古城一號井發掘簡報 (Excavation report of well no. 1 of the ancient Warring States – Qin town Liye of Longshan), *Wenwu*, 1: 4–35.

Huang Longxiang 黃龍祥 (2017) 'Laoguanshan chutu xihan zhenjiu muren kao' 老官山出土西漢針灸木人考 (Study on the excavated wooden carved acupuncture statue of the Western Han dynasty in Laoguanshan), *Zhonghua yishi zazhi*, 3: 131–44.

Jiangxisheng wenwu kaogu yanjiuyuan 江西省文物考古研究院, Beijing daxue chutu wenxian yanjiusuo 北京大學出土文獻研究所 and Jingzhou wenwu baohu zhongxin 荊州文物保護中心 (2018) 'Jiangxi Nanchang xihan haihunhou Liuhe mu chutu jiandu' 江西南昌西漢海昏侯劉賀墓出土簡牘 (The bamboo slips and wooden tablets unearthed from Marquis of Haihun's tomb of the Western Han dynasty in Nanchang, Jiangxi), *Wenwu*, 11: 87–96.

Li Jiahao 李家浩 and Yang Zesheng 楊澤生 (2011) 'Beijing daxue cang Handai yijian jianjie' 北京大學藏漢代醫簡簡介 (Introduction to the Han dynasty bamboo manuscripts held at Peking University), *Wenwu*, 6: 88–9.

Liang Fanrong 梁繁榮, Zeng Fang 曾芳, Zhou Xinglan 周兴兰, Xie Tao 谢涛, Lu Yinke 卢引科, Wang Yi 王毅 and Jiang Zhanghua 江章华 (2015) 'Chengdu Laoguanshan chutu jingxue xiuqi renxiang chutan' 成都老官山出土經穴髹漆人像初探 (Preliminary study on a lacquer figure with meridian points marked of the western Han dynasty unearthed in Laoguanshan, Chengdu), *Zhongguo zhenjiu*, 1: 91–3.

Liu Changhua 柳長華, Gu Man 顾漫, Zhou Qi 周琦, Liu Yang 刘阳 and Luo Qiong 罗琼 (2017) 'Sichuan Chengdu Tianhui Hanmu yijian de mingming yu xueshu yuanliu kao' 四川成都天回漢墓醫簡的命名與學術源流考 (On the naming and academic origin of the medical bamboo slips unearthed from the Han tombs at Tianhui town in Chengdu, Sichuan), *Wenwu*, 12: 58–69.

Ma Jixing 馬繼興 (1996) 'Shuangbaoshan Hanmu chutu de zhenjiu jingmai qimu renxing' 雙包山漢墓出土的針灸經脈漆木人形 (A wooden figurine painted with *jingmai* network for acupuncture unearthed from a Han tomb at Shuangbaoshan), *Wenwu*, 4: 55–65.

Museum of the Zhouliangyuqiao 周梁玉橋 Site, Jingzhou, Hubei (1999) 'Guanju Qinhan mu qingli jianbao' 關沮秦漢墓清理簡報 (Excavation of the Qin and Han tombs at Ganju, Jingzhou, Hubei), *Wenwu*, 6: 26–47.

Wang Qixian 王奇賢 and Zhang Xiancheng 張顯成 (2015) 'Chutu sanjian sheyi jiandu yanjiu zongshu' 出土散見涉醫簡牘研究綜述 (A summary of research into the excavated records on bamboo and wood concerned with medicine found at various sites), *Guji zhengli yanjiu xuekan*, 6: 179–85.

Wenhuabu guwenxian yanjiushi 文化部古文獻研究室 and Anhui Fuyang diqu bowuguan Fuyang Hanjian zhenglizu 安徽阜陽地區博物館阜陽漢簡整理組 (1988) 'Fuyang Hanjian' 阜陽漢簡 (Han bamboo manuscripts from Fuyang), *Wenwu*, 4: 36–47, 54.

Wu Rongzheng 吳榮政 (2013) 'Liye Qinjian wenshu dang'an chutan' 里耶秦簡文書檔案初探 (A preliminary study of the Qin bamboo manuscripts and documents from Liye), *Xiangtan daxue xuebao*, 6: 141–6.

Xie Tao 謝濤, Wu Jiabi 武家璧, Suo Dehao 索德浩 and Liu Xiangyu 劉祥宇 (2014) 'Chengdushi Tianhuizhen Laoguanshan Hanmu' 成都市天回鎮老官山漢墓 (The Laoguanshan Han tombs at Tianhui town, Chengdu city), *Kaogu*, 7: 59–70.

Yang Jun 楊軍 and Xu Changqing 徐長青 (2016) 'Nanchangshi Haihunhou Hanmu' 南昌市海昏侯漢墓 (The Han tomb of the Marquis of Haihun, Nanchang city), *Kaogu*, 7: 45–62.

Yuan Haibing 原海兵, Xie Tao 謝濤 and He Kunyu 何錕宇 (2018) 'Chendushi Tianhuizhe Laoguanshan chutu lugu de guancha yu celiang' 成都市天回鎮老官山漢墓出土顱骨的觀察與測量 (Observation and measurement on the human skull from the Laoguanshan cemetery, Chengdu city), *Bianjiang kaogu yanjiu*, 1: 261–86.

Zhang Defang 張德芳 (2013) *Dunhuang Majuanwan Hanjian jishi* 敦煌馬圈湾漢簡集釋 (Collected Notes on the Han Bamboo Slips from Dunhuang Maquanwan), Lanzhou, Gansu: Gansu wenhua chubanshe.

Zhang Lei 張雷 (2018) *Qinhan jiandu yifang jizhu* 秦漢簡牘醫方集注 (Collected and Annotated Qin and Han Bamboo and Wood Manuscripts on Prescriptions), Beijing: Zhonghua shuju.

Zhongguo shehui kexueyuan kaogu yanjiusuo 中國社會科學院考古研究所 and Hebeisheng wenwu guanlichu 河北省文物管理處 (1980) *Mancheng Hanmu fajue baogao* 滿城漢墓發掘報告 (Excavation Report of Han Tombs in Mancheng), Beijing: Wenwu chubanshe.

Zhongguo zhongyi kexueyuan Zhongguo yishi wenxian yanjiusuo 中國中醫科學院中國醫史文獻研究所, Chengdu wenwu kaogu yanjiusuo 成都文物考古研究所 and Jingzhou wenwu baohu zhongxin 荊州文物保護中心 (2017) 'Sichuan Chengdu Tianhui Hanmu yijian zhengli jianbao' 四川成都天回漢墓醫簡整理簡報 (A briefing report on the medical bamboo slips from the Han tombs of Tianhui in Chengdu, Sichuan), *Wenwu*, 12: 48–57.

Zhongyiyan (Ma Jixing) 鐘依研 (馬繼興) (1972) 'Xihan Liusheng mu chutu de yiliao qiju' 西漢劉勝墓出土的醫療器具 (Medical appliances unearthed in the tomb of Liu Sheng in the Western Han Dynasty), *Kaogu*, 3: 49–53.

Zhou Zuliang 周祖亮 and Fang Yilin 方懿林 (2014) *Jianbo yiyao wenxian jiaoshi* 簡帛醫藥文獻校釋 (Correction and Explanation of Medical Literature on Bamboo and Silk), Beijing: Xueyuan chubanshe.

List of transcripts

Gansu jiandu bowuguan 甘肅簡牘博物館, Gansu sheng wenwu kaogu yanjiusuo 甘肅省文物考古研究所, Gansu sheng bowuguan 甘肅省博物館, Zhongguo wenhua yichan yanjiuyuan guwenxian yanjiushi 中國文化遺產研究院古文獻研究室, Zhongguo shehui kexueyuan jianbo yanjiu zhongxin 中國社會科學院簡帛研究中心 (2011) *Jianshuijinguan Hanjian* 肩水金關漢簡 (The Han Bamboo Slips from the Jinguan Site in Jianshui County), Vol. I, Shanghai: Zhongxi shuju.

Gansusheng bowuguan 甘肅省博物館 and Wuwei xian wenhuaguan 武威縣文化館 (1975) *Wuwei Handai Yijian* 武威漢代醫簡 (The Han Medical Bamboo Slips from Wuwei), Beijing: Wenwu chubanshe.

Gansusheng wenwu kaogu yanjiusuo 甘肅省文物考古研究所 (1991) *Dunhuang Hanjian* 敦煌漢簡 (The Han Bamboo Slips from Dunhuang), Beijing: Zhonghua shuju.

Hunan sheng wenwu kaogu yanjiusuo 湖南省文物考古研究所 (2012) *Liye Qinjian* 里耶秦簡 (The Qin Bamboo Slips from Liye), Vol. I, Beijing: Wenwu chubanshe.

—— (2017) *Liye Qinjian* 里耶秦簡 (The Qin Bamboo Slips from Liye), Vol. II, Beijing: Wenwu chubanshe.

Ma Jixing 馬繼興 (1992) *Mawangdui guyishu kaoshi* 馬王堆古醫書考釋 (Textual Research and Interpretation of Ancient Medical Texts of Mawangdui), Changsha: Hunan kexue jishu chubanshe.

—— (2015) *Zhongguo chutu guyishu kaoshi yu yanjiu* 中國出土古醫書考釋與研究 (Textual Research and Interpretation of China's Excavated Ancient Medical Writings), Shanghai: Shanghai kexue jishu chubanshe.

Mawangdui boshu zhengli xiaozu (1985) *Mawangdui Hanmu Boshu* 馬王堆漢墓帛書 (Silk Manuscripts from the Mawangdui Han Tombs), Vol. IV, Beijing: Wenwu chubanshe.

Museum of the Zhou Liangyuqiao 周梁玉橋 Site, Jingzhou, Hubei (ed.) (2001) *Guanju Qinhan mu jiandu* 關沮秦漢墓簡牘 (The Bamboo Slips and Wooden Tablets from the Guanju Qin and Han Tombs), Beijing: Zhonghua shuju.

Qiu Xigui 裘錫圭 (ed.) (2014) *Changsha Mawangdui Hanmu jianbo jicheng* 長沙馬王堆漢墓簡帛集成 (The Silk and Bamboo Manuscripts from the Mawangdui Han Tombs at Changsha), Beijing: Zhonghua shuju.

Shuihudi Qinmu zhujian zhengli xiaozu 睡虎地秦墓竹簡整理小組 (ed.) (1990) *Shuihudi Qinmu zhujian* 睡虎地秦墓竹簡 (The Bamboo Slips from the Shuihudi Qin Tomb), Beijing: Wenwu chubanshe.

Zhangjiashan 247 Hao Hanmu zhujian zhengli xiaozu (2006) *Zhangjiashan Hanmu zhujian (247 hao mu)* 張家山漢墓竹簡（二四七號墓）(The Han Bamboo Slips from Zhangjiashan Han Tomb no. 247), Beijing: Wenwu chubanshe.

Zhongguo shehui kexueyuan kaogu yanjiusuo (1980) *Juyan Hanjian* 居延漢簡: *jia and yi bian* (甲乙編) (The Han Slips from Junyan), Beijing: Zhonghua shuju.

Pre-modern sources

Hanshu 漢書 (Records of the Han Dynasty) 111 CE, Ban Gu 班固 *et al.* (1962), Beijing: Zhonghua shuju.
Hanshu yiwenzhi huibian 漢書藝文志注釋彙編 (Annotated Compilation of the Bibliographic Treatise of the Han), Chen Guoqing 陳國慶 (ed.) (1983), Beijing: Zhonghua shuju.
Huangdi neijing 黃帝內經 (Yellow Emperor's Inner Canon), 762 CE, Wang Bing 王冰 (ed.) (1994), Beijing: Zhongyi guji chubanshe.
Shiji (Records of the Grand Historian) 91 BCE, Sima Qian 司馬遷 (145–86 BCE) (1969), Hong Kong: Zhonghua shuju.

Modern sources

Barrett, P., Flower, A. and Lo, V. (2015) 'What's past is prologue: Chinese medicine and the treatment of recurrent urinary tract infections', *Journal of Ethnopharmacology*, 167: 86–96.
Butler, A. (2009) 'Saltpetre in early Chinese medicine', *Asian Medicine: Tradition and Modernity*, 5.1: 173–65.
Cook, C.A. (2006) *Death in Ancient China: the tale of one man's journey*, Leiden: Brill.
Furth, C., Zeitlin, J. and Hsiung, Ping-chen (2007) *Thinking with Cases: specialist knowledge in Chinese cultural history*, Honolulu: University of Hawai'i Press.
Harper, D. (1998) *Early Chinese Medical Literature: the Mawangdui medical manuscripts*, London; New York: Kegan Paul International.
——— (2001) 'Iatromancy, diagnosis and prognosis in Early China', in E. Hsu (ed.) *Chinese Medicine and the Question of Innovation; Festschrift in Commemoration of Lu Gweidjen*, Cambridge: Cambridge University Press, pp. 121–66.
Harper, D. and Kalinowski, M. (eds) (2017) *Books of Fate and Popular Culture in Early China: the daybook manuscripts of the Warring States, Qin, and Han*, Leiden; Boston, MA: Brill.
Hsu, E. (2010) *Pulse Diagnosis in Early Chinese Medicine: the telling touch*, Cambridge: Cambridge University Press.
Hulsewé, A.F.P. (1985) *Remnants of Ch'in Law: an annotated translation of the Ch'in legal and administration rules of the 3rd century B.C. discovered in Yün-meng prefecture, Hu-pei province, in 1975*, Leiden: Brill.
Lei, Sean Hsiang-Lin (2015) *Neither Donkey Nor Horse: medicine in the struggle over China's modernity*, Studies of the Weatherhead East Asian Institute, Chicago, IL: University of Chicago Press.
Li Jianmin 李建民 (2000) *Sisheng zhi yu: Zhou-Qin-Han maixue zhi yuanliu* 死生之域：周秦漢脈學之源流 (The Boundary between Life and Death: Sources of Vessel Theory in the Zhou, Qin and Han dynasties), Taipei: Academia Sinica.
Li Ling 李零 (1993) *Zhongguo fang shu kao* 中國方術考 (Research into China's Technical Literature), Beijing: Zhongguo renmin.
Lo, V. (2001a) 'The influence of nurturing life culture on the development of western Acumoxa therapy', in E. Hsu (ed.) *Chinese Medicine and the Question of Innovation, Festschrift in Commemoration of Lu Gweidjen*, Cambridge: Cambridge University Press, pp. 19–50.
——— (2001b) 'Huangdi Hama jing (Yellow Emperor's Toad Canon)', *Asia Major*, 14.2: 61–99.
——— (2002) 'Spirit of stone: technical considerations in the treatment of the jade body', *Bulletin of SOAS*, 65.1: 99–128.
——— (2014) *How To Do the Gibbon Walk: a translation of the Pulling Book (ca. 186 BCE)*, Needham Research Institute Working Papers 3, Cambridge: Needham Research Institute (e-book: http://www.nri.org.uk/yinshu.pdf).
——— (2022) 'Looms of life: weaving and medicine in early China', *Asian Medicine*, forthcoming.
Lo, V. and Li Jianmin (2010) 'Manuscripts, received texts and the healing arts', in M. Nylan and M. Loewe (eds) *China's Early Empires*, Cambridge: Cambridge University Press, pp. 367–97.
Lo, V. and Gu Man (forthcoming) 'Water as homology in the construction of classical Chinese medicine', in C. A. Cook, S. Blader and C. Foster (eds) *Thinking about Early China with Sarah Allan*, Albany: SUNY.

Ma Jixing 馬繼興 (1992), *Mawangdui guyishu kaoshi* 馬王堆古醫書考釋, Hunan: Hunan kexue jishu chubanshe.

Ngo Van Xuyet (1976) *Divination Magie et Politique dans la Chine Ancienne*, Paris: Presses Universitaires de France.

Raphals, L. (2013) *Divination and Prediction in Early China and Ancient Greece*, Cambridge: Cambridge University Press.

Unschuld, P.U. (2003) *Huang Di Nei Jing Su Wen: nature, knowledge, imagery in an ancient Chinese medical text, with an appendix, the Doctrine of the five periods and six Qi in the Huang Di Nei Jing Su Wen*, Berkeley: University of California Press.

Unschuld, P.U., Tessenow, H. and Zheng, J. (2011) *Huang Di Nei Jing Su Wen: an annotated translation of Huang Di's inner classic – basic questions*, Vol. 1, Berkeley: University of California Press.

Xie Guihua (2005) 'Han bamboo and wooden medical records discovered in military sites from the north-western frontier regions', in V. Lo and C. Cullen (eds), *Medieval Chinese Medicine*, London; New York: RoutledgeCurzon, pp. 104–32.

Yong Yang and Brown, M. (2017) 'The Wuwei medical manuscripts: a brief introduction and translation', *Early China*, 40: 241–301.

4

THE IMPORTANCE OF NUMEROLOGY, PART 1

State ritual and medicine

Deborah Woolf

Numerology (the science of numbers) was central to philosophy, and to the structuring of everyday life in early and medieval China; hence, it was also central to the formation of Chinese medicine at that time. It was said that, 'Heaven and Earth have their regular ways, and men like these for their pattern, imitating the brilliance of Heaven, and matching the forms of Earth'[1] (*Zuozhuan*, Duke Zhao 25.1457).[2] This chapter argues that numerology is also the key to unlocking the history of Chinese medicine since its inception: innovations and adaptations to the classical tradition in Chinese medicine have been modulated and stabilised by numbers. This historical factor has been overlooked in recent views of Chinese medicine, as the proponents of TCM ('Traditional' Chinese Medicine) of modern China embraced selected elements of modern science to align themselves with what has been perceived as a dominant and 'Western' mode of thinking that has rejected numerology as superstitious (Introduction and Chapter 45 in this volume; Lei 2014).

Keightley (1990) pointed to various cultural elements, such as hierarchical distinctions, massive mobilisation of labour, rituals and boundaries, an ethic of service and emulation, deep faith and a general optimism, which established a core 'Chinese' identity that can be traced continuously from Shang (1600–1046 BCE) to Han (202 BCE–220 CE) times, in an expanding geographic area. Surprisingly, Keightley omitted numerology as one of the key cultural practices that provided continuity to Chinese identity for over three and half millennia.

> Knowing the structure of numbers results in a better comprehension of the world.
> *(Robinet 2011: 48)*

Numbers have been used within many cultures to represent the unfolding of structure and form: in architecture, music, physics and medicine (Critchlow 1969; Crump 1990; Verran 2010). Their organising principles provide a continuous thread, seamlessly establishing calendrical and ritual continuities in many aspects of everyday life. Despite the enormous diversities of peoples and cultures that have co-existed within the territory that we associate with China, its changing socio-political structures and periods of administrative unity and fragmentation, numbers and their complex web of relationships have underpinned change and provide an important constant: a constant that has paradoxically allowed for transformation and innovation, both through the transferability of numerical systems across different domains of practice, and in the way numbers provide a grounding element of familiarity in new theories about the world.

Numerology and divination

Changes in the socio-political landscape of ancient China, from diviner kings, to kingly states and then imperial rule, were underpinned by numbers. Numbers were used in astronomy and to describe time, in terms of its rhythms and cycles, and therefore in the traditions of calendrical calculations. From Shang times, the calendar was an essential part of establishing power: calibrating the movements of the heavens was a way to control the actions of the people. Every dynasty established their own calendar at the start of their rule: thus, the calendar was used to define and adjust times of taxation, rituals and celebrations to the specific requirements of each new regime, aligning the people with the rhythms of the heavens.

> Derived from the pattern of celestial cycles, calendars embody the order of the natural world (*li* 理), which in turn holds the key to number (*shu* 數).
>
> *(Ho 1985: 153)*

By the Spring and Autumn period (771–476 BCE), and particularly during the Warring States (475–221 BCE), a culture of 'numbers and techniques' (*shushu* 數術, hereafter *shushu*) had produced a range of numerical systems (Granet 1934: 127–248; Harper 1979; Kalinowski 1991, 1996; Li 2006b).

Shushu, in its relationship to natural philosophy and occult arts (Harper 1998: 46), was listed as one of the six categories of the bibliographic treatise of the *Hanshu* 漢書, the 'Treatise on Literature' (*Yiwenzhi* 藝文志).[3] The *shushu* category, which focused on the use of numbers and calculation, was a broad category, including texts on astro-calendrical sciences: calendrical computations, mapping constellations and divination related to stars and climate. It also included divination by means of *yin-yang* and five agents (*wuxing* 五行), the rubrics of Wood, Fire, Earth, Metal and Water (hereafter *wuxing*) and their sets of correspondences. Even in the Han dynasty, they continued to use the earliest known types of divination, turtle[4] and milfoil (Li 2006b: 42–51).[5] The *Yiwenzhi* also included divination by dreams, omens, signs portents, hemerology and physiognomy (Kalinowski 2005:110). The astro-calendrical section was the largest section in the *shushu* category, and was compiled by the Grand Astrologer (*Taishi ling* 太史令), who controlled the activities of state astrologers and calendar officials, in an attempt to subject multifarious astro-calendrical practices to state control.[6]

Shushu texts, through calculations and techniques, gave the theoretical basis for the more practical area of medicine exemplified in the *Yiwenzhi* category of 'Recipes and Methods' (*fangji* 方技) (Li 2006b: 17). *Shushu* and *fangji* categories together comprised around one-third of the total texts listed in the *Yiwenzhi* (Harper 1998: 52; Li 2006a: 73–78). Many manuscripts excavated from tombs from the late Warring States onwards include *shushu* and *fangji* texts, which also demonstrate the pervasive importance of these uniquely Chinese traditions throughout social strata (Liu 1994; Li 2000, 2006a, 2006b). This treasure trove of divinatory texts, manuscripts on silk and bamboo, excavated from ancient and medieval tombs (such as Jiudian 九店 Fangmatan 放馬灘, Shuihudi 睡虎地, Mawangdui 馬王堆 and Kongjiapo 孔家坡, dating between ca. 300 and 142 BCE) has broadened our understanding of the underlying beliefs and uses of divination, *shushu*, in everyday life, by local communities (Harper and Kalinowski 2017: 92).

Daybooks, *rishu* 日書, discovered in many of these tombs were, as their name suggests, collected miscellanies of information giving auspicious and inauspicious activities for each day (hemerology), keyed to calendrical divination (Li 2006b: 102–71). The influence of *shushu* culture, shown by the quantity and diversity of excavated texts, was widespread, throughout society. These texts reveal a less controlled historical source than that of the imperial standard histories, arguably providing a more authentic source than the early and medieval received texts, which were collated, edited and commented upon repeatedly up to the Song dynasty editions (960–1279).

Among the excavated texts, medical texts were often present, along with daybooks and other technical literature. The divinatory texts were grounded on a common base of *shushu* knowledge, including the calendar, *wuxing* and astrological ideas (Harper and Kalinowski 2017:193). The medical texts have shown that the acupuncture traditions did not arise from one single lineage, but coalesced from a melting pot of ideas and practices, out of which later classical medical texts emerged (Li 2000b). Excavated medical texts were, however, already using terms such as *qi*, *yin-yang* and *wuxing*, revealing the common roots and inter-relatedness of *shushu* and medicine: both drew on the same numerological roots which maintained continuity within this manuscript culture (Harper and Kalinowski 2017: 192). Iatromancy, the use of divination for medical purposes (physician as diviner), was used in the earliest recorded divinations: in Shang plastromancy, the king first divined to find out if the illness was from the ancestors, then exactly from which ancestor and what would be the appropriate sacrifice.[7] Cook (2006) describes the many iatromantic interventions made to diagnose and treat a travelling government official of Chu 楚 buried in 316 BCE. The art of prognosis and diagnosis, essential to medical practice (a physician needed to know if a person was too ill to be successfully treated, or they may lose their life by risking treatment and being held responsible for their patient's death), also drew from the numerological and divinatory practices that worked together to predict the future (Liu 1994: 116, 377–8, 436, 447).[8]

Shushu culture was an enduring part of Chinese thought, which provided the underlying framework for correlative schemes, cosmological thought and calendrical computations and practices up to and including the Qing Dynasty (1644–1911) (Ibid.: 365–94; 431–6; Kalinowski 2004: 234–6; Smith, 2012; Li 2006a). Numbers, essential to *shushu* culture, provided the continuity for many of the changes and transformations in medical thought. In this chapter, we will trace this process in relation to the key numbers which came to structure the medical traditions.

Huangdi neijing Suwen

The *Huangdi neijing* 黃帝內經 (Yellow Emperor's Inner Canon, hereafter *Neijing*), combining the *Suwen* 素問 (Simple Questions) and *Lingshu* 靈樞 (Numinous Pivot) recensions, is a compendium of medical knowledge, with a long and complex textual history.[9] The former is more concerned with acupuncture theories and dietary practices (Porkert 1974; Yamada 1979; Lu and Needham 1980; Sivin 1987, 1993; Keegan 1988). It is also concerned with the interconnectedness between humans and the cosmos (Li 2000b: 166), and therefore is infused with the numerical cultures of *shushu*. The latter, a later text, is focused on the practice of acupuncture. Historians variously date the two compilations between 100 BCE and 100 CE (Chapter 7 in this volume).

The *Neijing* has been analysed and discussed in relation to its origins, dating, content and structure (Ibid.). Keegan (1988: 13–17) asserts that there is no original Han text of the *Neijing*; rather, it is a compilation of small excerpts from a variety of earlier works. Many of the terms used and theories discussed are similar to those seen in texts which date to the late Warring States, and early imperial periods, such as the *Lüshi Chunqiu* 呂氏春秋 (Master Lü's Spring and Autumn Annals), a compendium of early Chinese thought,[10] and the *Huainanzi* 淮南子 (Masters from Huainan), a collection of philosophical, geographic and cosmological essays with a strong Daoist basis.[11] From the early compilations of medical texts up to the Tang period (618–907), texts carrying the name *Huangdi* in their title evolved incrementally, absorbing new material and repeatedly reforming into other, slightly different texts (Ibid.: 49–66). These various separate master/disciple teachings were the result of many pedagogical encounters involving a master discussing and explaining the texts orally. The question and answer format used in these styles of knowledge transmission is reflected in the structure of the compilation where *Huangdi* 黃帝 (Yellow Emperor, a mythical culture bringer)[12] and a variety of other mythical figures, his advisors and teachers, possibly represent different medical traditions (Kong 2010: 411). The Chinese dialogue form emphasised learning and consensus, unlike the Greek tradition which used dialogue for disputation and competitive public debates. Within the general framework of agreement, a broad range of different viewpoints could be developed and discussed (Lloyd and Sivin 2002: 247). Dialogue was not designed to stimulate debate but to unify diverse opinions: it allowed room for inconsistencies and seemingly irreconcilable versions, without the need to prove a single proposition (Chapter 2 in this volume; Nylan 1997: 135–6). Greater ambiguity allowed for denser association of various topics, which was more suited to the complexities of the human body and medical practice (Lloyd and Sivin 2002: 248).

Commentaries and evolution of the *Huangdi neijing Suwen* recension

By the end of the Liang dynasty (502–557), Quan Yuanqi 全元起 (of whom little is known) compiled the first major commentary of the *Suwen*.[13] In 751, Wang Bing 王冰 took Quan's commentated version as his master copy, corrected, rearranged and added new text. The peace and stability of early Tang encouraged innovation and Wang Bing brought to his recension of the *Suwen* many changes, with an emphasis on self-cultivation, and the importance of aligning oneself with the movements of the cosmos (Unschuld 2003: 41). Wang's finished version crystallised in 762 and comprised eighty-one chapters: he had added seven chapters, which he claimed were secret knowledge from earlier times, received as a master-disciple transmission (Ibid.: 46). These seven comprehensive discourses (*Suwen* 66–74)[14] on *Wuyun liuqi* 五運六氣 (Five Circuits and Six *Qi*) amounted to one-third of the received version of the *Suwen*. No-one knows where these extra chapters came from but they may have been influenced by Wang's early Daoist studies.[15] The so-called Wang Bing chapters of the *Suwen* were based on an already well-established tradition of *shushu* and *fangji* (Harper 1988: 52, 71; Li 2006b). Gao Baoheng 高保衡 (fl. 1050–65) et al., eleventh-century editors in charge of the Imperial Medical Editorial Office's 1067 edition of the *Suwen*,[16] suggested that these interpolated chapters were from the 'Comprehensive Discourses on *Yin-yang*' (*Yinyang dalun* 陰陽大論), a text which is now lost.[17]

A short history of numerology in Chinese thought

The Tang period *Wuyun liuqi* chapters of the *Suwen* represent a first peak of the *shushu* traditions as they applied to medicine and, as we shall see, these changes in the tradition were based on a combination of ancient numerical systems: namely the sexagenary (or sixty-unit) calendrical cycle, the dualistic *yin-yang*, four, five and six (Li 2006a, 2006b).[18] Below is a brief overview of the chronological appearance of these key numbers. I have selected these numerical bases, over other significant numerical traditions, for their particular relevance to the formation of the *Wuyun liuqi* tradition in classical Chinese medicine. For this reason, equally important numerical indexes in Chinese medical culture and divination, such as the *ba gua* 八卦 (Eight Hexagrams),[19] have been omitted.[20]

Shang: the number four and the sexagenary cycle

Shang rulers who established the earliest known civilisation on the Yellow River plain, left written characters on bones, bronzes, jade, stone and ceramics. The first records of divination[21] were found on ox-bones and turtle shells. The process of divination comprised the boring of holes in the bone or turtle shell plastron, then inserting hot pokers into the holes, proclaiming the question out loud and interpreting the resulting cracks (Keightley 1988) (Li 2006a: 218–32; 2006b: 42–7). Both the divination and its interpretation were recorded on the bone or turtle shell (Keightley 2000: viii–ix).

Shang time was enumerated according to the cycle of the ten Heavenly Stems (*tiangan* 天干, hereafter Stems) and twelve Earthly Branches (*dizhi* 地支, hereafter Branches) to form a sixty-unit cycle. The Stems and Branches were odd or even: odd Stems were paired with odd Branches; even Stems paired with even Branches. When combined, they formed a continuous and recurring cycle of sixty Stem-Branch binomes (the sexagenary cycle, n1–n60), rather than a possible 120 combinations (Table 4.1).

This recurring cycle of sixty (hours, days, months, years) was used to record important dates, such as the dates of divinations on the oracle bones. The Ten Stems were also used retrospectively to name the Shang kings: the calendar of ancestor worship of the Shang kings structured daily life by the end of Shang times (Allan 1991; Keightley 2000: 47–9; Smith 2010).

Shang rulers divided their territory by Four, the four lands (*situ* 四土); land beyond their central domain was categorised into four directions (*sifang* 四方) (Keightley 2000: 66, 69). Allan (1991: 101–11) suggests that the shape of the Shang lands was symbolised by the turtle carapace itself: a model of the Shang cosmos, with domed sky held above the Earth by four

Table 4.1 The Sexagenary Cycle by the author

n1 - 10	s1-b1	s2-b2	s3-b3	s4-b4	s5-b5	s6-b6	s7-b7	s8-b8	s9-b9	s10-b10
	甲子	乙丑	丙寅	丁卯	戊辰	己巳	庚午	辛未	壬申	癸酉
n11-20	s1-b11	s2-b12	s3-b1	s4-b2	s5-b3	s6-b4	s7-b5	s8-b6	s9-b7	s10-b8
	甲戌	乙亥	丙子	丁丑	戊寅	己卯	庚辰	辛巳	壬午	癸未
n21-30	s1-b9	s2-b10	s3-b11	s4-b12	s5-b1	s6-b2	s7-b3	s8-b4	s9-b5	s10-b6
	甲申	乙酉	丙戌	丁亥	戊子	己丑	庚寅	辛卯	壬辰	癸巳
n31-40	s1-b7	s2-b8	s3-b9	s4-b10	s5-b11	s6-b12	s7-b1	s8-b2	s9-b3	s10-b4
	甲午	乙未	丙申	丁酉	戊戌	己亥	庚子	辛丑	壬寅	癸卯
n41-50	s1-b5	s2-b6	s3-b7	s4-b8	s5-b9	s6-b10	s7-b11	s8-b12	s9-b1	s10-b2
	甲辰	乙巳	丙午	丁未	戊申	己酉	庚戌	辛亥	壬子	癸丑
n51-60	s1-b3	s2-b4	s3-b5	s4-b6	s5-b7	s6-b8	s7-b9	s8-b10	s9-b11	s10-b12
	甲寅	乙卯	丙辰	丁巳	戊午	己未	庚申	辛酉	壬戌	癸亥

The importance of numerology, part 1

pillars in the four ordinal corners (northeast, southeast, northwest, southwest). Four was also symbolised by the *ya* 亞 design found on many Shang inscriptions (Ibid.: 89–91).

The Shang sky was also seen as composed of Four quarters. The rotation of the constellation of the Big Dipper, or Northern Ladle, *beidou* 北斗, around a central/still point (the pole star) over the course of the year, defined the relationship between the directions and the seasons. *Beidou* was in the south in summer, north in winter, east in spring and west in autumn.[22] Each of the four quarters had a constellation, which could be seen in its appropriate direction and season, the Four images, *sixiang* 四象 (Figure 4.1).

By the Han dynasty, these came to be regularly symbolised by four emblematic animals, divine creatures (*siling* 四靈), representing the Four cardinal directions, thought to aid the transition from the material to the spirit world (Wong 2006: 119): the Green Dragon (*canglong* 蒼龍) in the east; Vermillion Bird (*zhuque* 朱雀) in the south; White Tiger (*baihu* 白虎) in the west and Black Warrior (*xuanwu* 玄武) or Dragon and Tortoise in the north (Sun and Kistemaker 1997: 114–8; Loewe 2005: 42). The *Tianguan shu* 天官書 (Treatise on Celestial Offices) of the *Shiji*[23] describes the asterisms of each direction and their effects on state administration and society.[24]

Although these symbolic relationships were consolidated during Han times (202 BCE–220 CE), studies of early astronomy indicate that these symbols may have played a part in Shang thought (Pankenier 2013:39), and excavations such as this Neolithic tomb (4000 BCE) excavated in Puyang, Henan province in 1987–88, may point to even earlier traces (Figure 4.2).

Figure 4.1 The Four Emblematic animals, *siling* 四靈, from the author's private photo collection

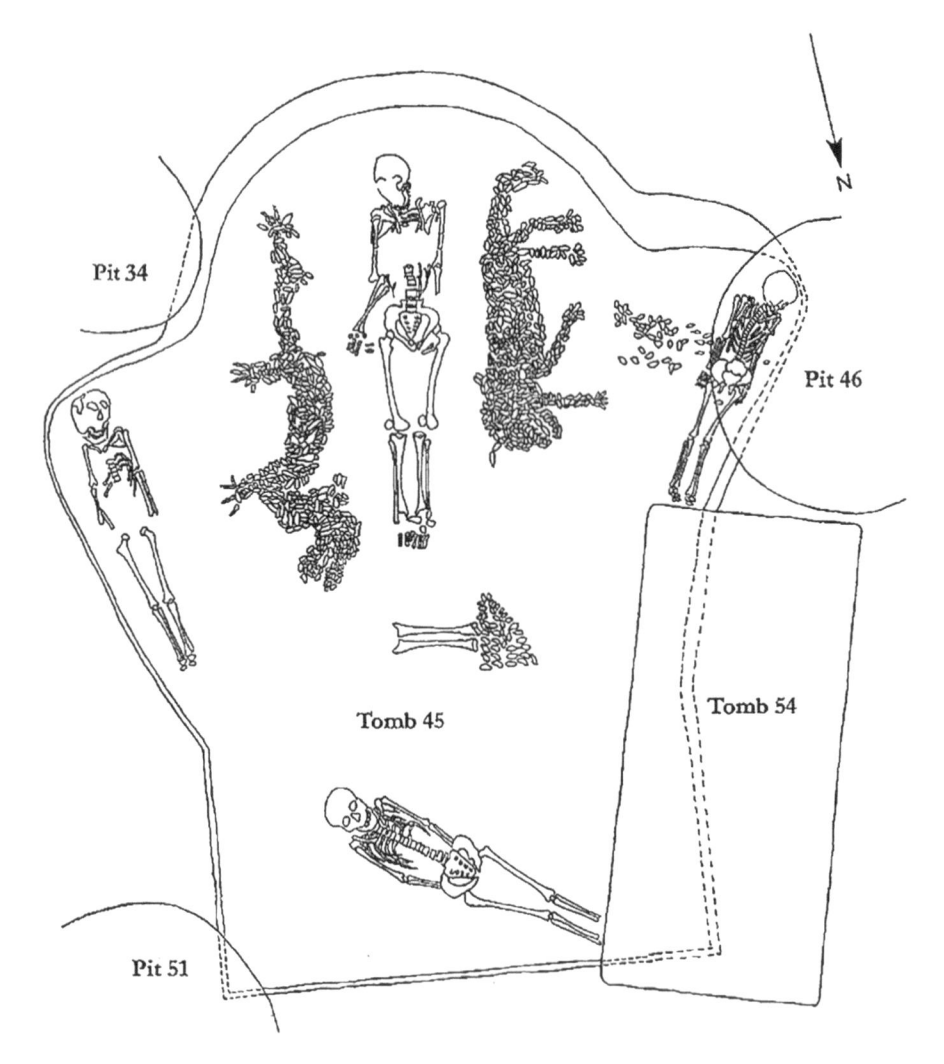

Figure 4.2 Xishuipo 西水坡 tomb (M45) in Henan, showing clamshell figures representing a dragon in the East and tiger in the West, with a ladle in the North (Sun and Kistemaker 1997: 116, fig 6.2)

The centre was the place of power, where Shang kings were in constant contact with the heavens through their divination. Four was therefore important in early Chinese spatial design, but in medicine we will see that its significance was also concerned with the way the four seasons related to regimes of health, and the aetiology of disease, particularly in relation to the way unseasonal factors led to epidemic disease.

Numerology of the Warring States (475–222 BCE): the number two

By the time of the Zhou collapse (ca. 256 BCE), there were numerous astro-calendrical divinatory systems (Li 2006b), such as the *Zidanku* 子彈庫 silk manuscript, a hemerological text used for affairs of the state and the military (Ibid.: 141).[25] A set of bamboo slips from

Yinqueshan 銀雀山 tomb also relate *yin-yang* and *wuxing* to the years, seasons and months, to influence the activities of the ruler, government and the military (Harper and Kalinowski 2017: 80–1).[26] Warring States bureaucracy had, in theory, been arranged according to a structure based on cosmology and numerology: offices of administration and ritual relied on the calendar and divination (Lewis 1999: 42–6). The calendar became the means to organise the collection of taxes necessary to support rituals and sacrifices (Sterckx 2007: 840). Philosophical debate concerning cosmology and the origins of human beings was framed according to the duality of Heaven (*tian* 天) and Earth (*di* 地), represented in the material world by *yang* 陽 and *yin* 陰, together with *qi* 氣, which powered their continuous interaction.

Heaven and Earth, seen as the origin of all creatures (including humans) were personified as the *yin-yang* couple of Fuxi 伏羲 and Nüwa 女媧. Fuxi was the first of the Three Sovereigns (*sanhuang* 三皇), mythological rulers associated with the origins of Chinese culture. He was credited with inventing the trigrams of the *ba gua*, with kingship, writing and numbers. Nüwa in contrast tamed the flood, repaired the pillars holding up the sky and created human beings (*Huainanzi*, 6.479–80).[27] Between them, they brought order to an unruly world (Lewis 2006). They were associated with various *yin-yang* pairs, such as the sun and moon, time and space, or the round Heaven and the square Earth of early Chinese cosmological design. They were often depicted intertwined, half-snake/half-human: to enable the transition between the worlds of form and spirit, life and death.

> Heaven and Earth are the father and mother of the ten thousand beings. When they are united they achieve bodily form; when they are scattered they achieve a new beginning.
> (*Zhuangzi*, 5.19)

Yang (Heaven) and *yin* (Earth) often mirrored each other: Fuxi held a set square, *ju* 榘 – a *yang* (male) symbol to regulate the four directions of the square *yin* (Earth). Nüwa held a compass, *gui* 規 – a *yin* (female) instrument to 'encompass' and observe the circular *yang* (Heaven) (Girardot 1983: 202–7) (Figure 4.3).

Figure 4.3 Fuxi and Nüwa from Wuliang shrine, chamber 2. Ink rubbing of carved stone relief, Eastern Han (Chavannes 1909:1, plate 60, no. 123)

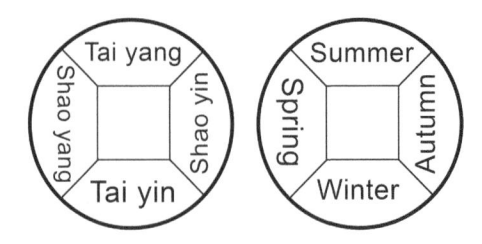

Figure 4.4 The Seasons and the Two Phases of lesser and greater *yin* and *yang* – representation by the author

The rhythm and interplay of Heaven and Earth, and their abstraction as *yin* and *yang*, was symbolised by the flow of time through the four seasons. For example in the *Chunqiu fanlu* 春秋繁露 (Luxuriant Dew of the Spring and Autumn Annals)[28]

> Hence in spring choose *shaoyang*, in summer choose *taiyang*, in autumn choose *shaoyin*, in winter choose *taiyin*.
>
> (*Chunqiu fanlu yizheng*, 24:218)

The four-fold system of young/growing (*shao* 少 'lesser') and mature/old (*tai* 太 'greater'), where spring and summer were designated as two phases of lesser and greater *yang*, and autumn and winter were designated as phases of lesser and greater *yin*, is conceptualised in Figure 4.4.

Numbers and imperial ritual: Qin (221–206 BCE) and Western Han (202 BCE–9 CE): the number five

By the time of the unification of China by the Qin (221 BCE), the system of *wuxing* (five agents) correspondences had started to become more prevalent in government administration, court structure and thinking: it began to appear in ritual, medical and calendar texts.

This is seen in the *Lüshi Chunqiu*, a rich and complex compendium on the philosophy of government, compiled by a group of scholars in 239 BCE, under the patronage of Lü Buwei 呂不韋 (251–235 BCE).[29] Lü, a merchant from Wei 衛, became the chancellor of Qin 秦, and regent to the King of Qin's son, Zheng 政 (259–210 BCE), personal name Ying 嬴. Lü assembled large numbers of scholars to make Qin an intellectual hub to advise Zheng (who became the first Emperor of China, Qin Shi Huangdi 秦始皇帝 in 221 BCE) in the art of rulership. Lü hoped to create a coherent philosophy for the new empire, in which the ruler coordinated the affairs of Heaven, Earth and humans, keeping everything in harmony. To this end, the structure of the *Lüshi chunqiu* related to the cosmological ideas underpinning the philosophy: the Almanacs (*ji* 紀) represented Heaven, twelve Almanacs, each comprising five chapters, making sixty chapters in all (representing a full sexagenary cycle of the Heavenly Stems and Earthly Branches). The first chapter of each Almanac gave correspondences for each of the twelve solar months of the year: constellations, Stems, colours, numbers, pitch pipe tones, tastes and smells. The five chapters in each Almanac represented each of the five agents, *wuxing*, thought to control each season's workings (notably the fifth, which correlated with Earth, was not given the same weighting as the other four agents, which we will turn to presently). The Examinations (*lan* 覽) represented Earth, with eight sections of eight chapters, mirroring the sixty-four hexagrams of the *Yijing*. The six discourses (*lun* 論), containing six chapters each, corresponded to humans and their affairs (Knoblock and Riegel 2000: 32–46). We will see later that six is a number that relates to the physical form of humans.

The five agents had matured from early divinatory practices, into the conceptual basis for all activities (Kalinowski 1991). These rubrics were beginning to dominate ritual, divination, calendrical arts and medicine by the third century BCE (Lo 2013: 60). The plasticity of *wuxing* and *yin-yang* calculations allowed inconsistencies among diverse and sometimes contradictory systems, to achieve a standardised, coherent and correlative cosmology (Wang 2000: 117). Five first came to prominence as the five agents or processes, often translated as five elements or five phases (Lo 2013: 64). The *locus classicus* for these terms is the five types of conduct in the *Hongfan* 洪範 ('Great Plan' chapter of the *Shujing* 書經 *Book of Documents*).[30] The 'Great Plan' was an exposition of the patterns of proper relationships and government, traditionally given to the mythical Emperor Yu 禹 by Heaven. Late Warring States debates among Confucian, or *ruist*, communities about the embodiment of morality and ethics as five types of action (also *wuxing* 五行) have recently come to light with the excavation of a *Wuxing jing* 五行經 and a commentary on this work (Csikszentmihalyi 2004). This takes a different direction than other materialistic discourses which represented the five materials of water, fire, wood, metal and soil (*wucai*五材), as five basic resources for any human livelihood (fire and water are the basis of making food; wood and metal are the basis of building dwellings; soil is the basis of agriculture), whereas technical masters whose ideas later structured medical knowledge identified these as five kinds of material process (sinking, rising, bending, yielding, planting) (Graham 1989: 326; Rosemont 1991: 278). Zou Yan 鄒衍 (305–240 BCE), a philosopher of the legendary Jixia Academy, traditionally credited with systematising *yin-yang* and *wuxing* theory, put forward the idea that dynasties rose and fell according to the 'mutual conquest' (*xiangke* 相剋 or *xiangsheng* 相勝) cycle of the five agents, as described in the *Lüshi chunqiu*.[31] This cycle was used to legitimise the process of regime change: Water overcame Fire; Fire overcame Metal; Metal overcame Wood; Wood overcame Earth and Earth overcame Water. Five agents cycles thus structured the timing and quality of imperial rituals (Knoblock and Riegel 2000: 283), which, in turn, raised the status of practices previously associated with five agents: mainly diviners, astrologers and physicians (Li 2006a: 69).

During the Western Han (202 BCE–9 CE), the importance of astro-calendrical practices was deepened and reinforced; this was reflected in debates on calendar reforms (Kalinowski 2005:110). The sacrifice to Heaven by the Emperor at the Temple of Heaven (*Tiantan* 天壇) was performed at the capital, Chang'an 長安 in 31 BCE, according to astro-calendrical practices. A Bright Hall (*Mingtang* 明堂) (Henderson 1984: 67–74) was erected at the capital and completed around 5 CE (Kalinowski 2005: 111).[32] The 'Bright Hall' was a designed according to astro-calendrical principles: in theory, the Emperor moved from room to room according to the 'Monthly Ordinances' (*yueling* 月令), a politico-ritual calendar of religious and administrative activities of the ideal state.[33] At the end of the Han when the liturgy of the 'Bright Hall' was adopted by the Emperor were the 'Monthly Ordinances' progressively adopted into the state calendar (Kalinowski 2004: 233; Li 2006a: 70).

Philosophical ideas and *shushu* traditions relating to cosmology which had developed from the second half of the fourth century BCE onwards (Kalinowski 2004: 233) were established as the ideological foundations of cosmology and correlative thinking by the end of the Eastern Han (Kalinowski 2005: 111). The rubrics found in the *Lüshi chunqiu* and *Huainanzi*, although based on earlier texts such as the *Zuozhuan* 左傳 (Zuo Commentary), were now further systematised and applied across many more domains (Kalinowski 2004: 234).[34]

The *shi* 式 or *shipan* 式盤 (mantic astrolabe, cosmic or diviner's board; Li 2006a: 74, 84; Li 2006b: 91) seen in various guises, on Han bronze mirrors (TLV mirrors) (Cammann 1948;

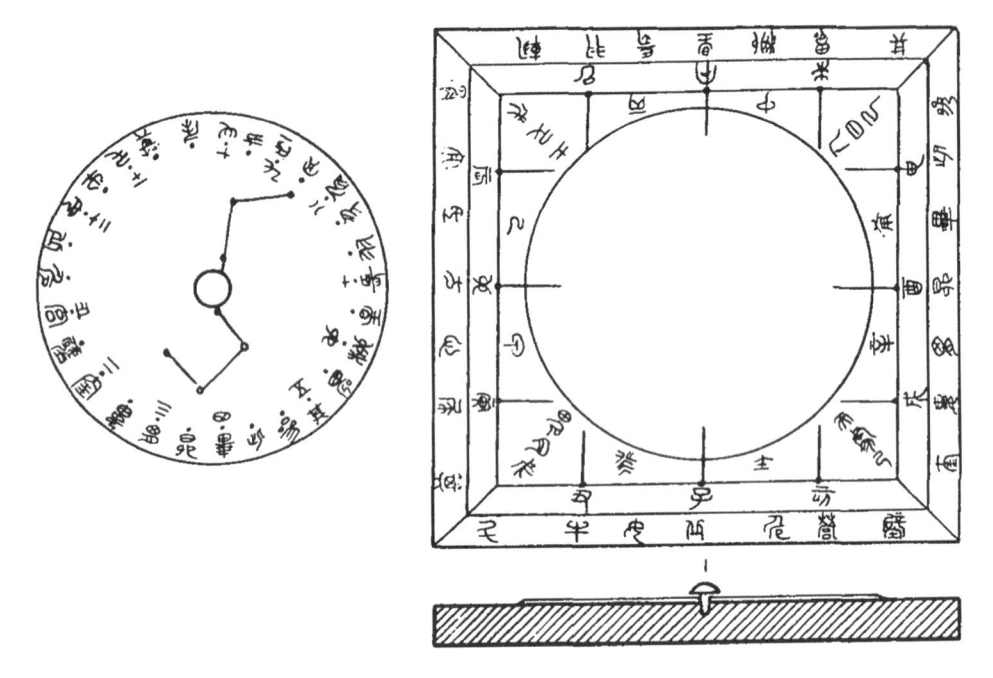

Figure 4.5 *Liuren* astrolabe (with Northern ladle in the centre), from Shuanggudui 雙古堆 tomb, sealed 165 BCE. After *Kaogu* 1978.5: 340. Drawing by Li Xiating. Copied with permission from Harper, 1999: 840

Loewe 1979: 60–85; Li 2006b: 138) and game boards (*liubo* 六博) (Tseng 2002, 2004; Li 2006b: 134–8),[35] as well as mantic devices, was a space-time reference frame, placing the directions, Stems, Branches and other symbols to model the cosmos (Harper and Kalinowski 2017: 153–4; Li 2006b: 69–85, 134–8) (Figure 4.5).

This symbolic framework, no longer directly related to astronomy, allowed adaptation to different Heavenly and Earthly circumstances, forming the basis for all subsequent divinatory practices (Harper and Kalinowski 2017: 160).

The *liuren* 六壬 ('Six *ren* 壬 days' astrolabe),[36] comprising a fixed, square, Earth plate and a rotating, round, Heaven plate (Li 2006b: 88–9), was one of the earliest applications of this framework. Rotating the Heaven plate imitated the movement of the Northern Ladle *beidou* over the course of the year. The movement between the Heaven and Earth boards allowed a dialogue between the deities above and the diviner (Ho 2003: 127–8). The Ladle's handle was used to determine the orientation of the board, with relation to the heavens and time of divination, and to point to a segment of the Earth plate (Raphals 2013: 343). This system was used to analyse complex situations such as for weather forecasts and military matters, as well as more mundane, everyday issues.[37]

Eastern Han (25–220): the number six

While the Western Han was a time of unification, administrative reform, military expansion and intense philosophical debate, by the time of the Eastern Han, there were many geographical and political areas of disintegration. State treasuries and granaries were depleted.

The army could not manage to defend the now vast borders from the people the Western Han rulers had defeated with their expansionist policies. The court had become estranged from the rest of the country. The Emperors of the Western Han had relied on contracts with kinsmen, followers and the great land-owning families for the maintenance of their power. The Eastern Han Emperors ruled over a much more fragmented group of wealthy clans, many of whom came to be dominated by affinal relationships, while the court itself was under the power of the eunuchs. The Yellow River changed its course from north to south, causing homelessness and unrest among the peasants: and there were many migrations south. The Eastern Han empire disintegrated from within, due to in-fighting between the elite families and eunuchs, while the large landowners and clans, with their own armies, effectively became independent states. Various peasant and religious rebellions saw an end to the Han empire in 220 CE, after strong generals from powerful clans had increasingly taken over culminating around 189 CE, and giving rise to the warring factions of the subsequent Three Kingdoms period (Twitchett and Loewe 1986).

During this period of general disintegration, the ritual and philosophical foundations of Han power, as seen in texts such as the *Chunqiu fanlu* 春秋繁露 (Luxuriant Dew of the Spring and Autumn Annals) were systematically reviewed and consolidated in the Han conference, held in the White Tiger Hall in the year 58 CE, an imperially led attempt to systematise the various pre-existing numerical traditions.[38]

For some three hundred years, six had been presented along with five, symbolising the interaction between Heaven and Earth, as the six potential causes of disease.

> Heaven and Earth produce the six *qi*, and make use of the five agents [...] when they are in excess, there is obscurity and disorder, and people lose their [proper] nature.
> (*Zuozhuan*, Duke Zhao 25.1457)

The rubrics of five and six were assigned to Heaven and Earth: Heaven could be associated with the climates (as the movements of the heavens gave seasons and climates on Earth); then Earth was associated with changes in physical structure, plants, tastes and bodies.

> Heaven has six *qi*, Earth gives birth to five tastes: divide them to have four seasons, order them to have five nodes, if you go beyond them, then there is calamity.
>
> [As for the six *qi*], *Yin* excess causes cold illness, *yang* excess causes heat illness, wind excess causes illness of the extremities, rain excess causes illness of the abdomen, excess of darkness causes illnesses involving confusion, excess of brightness causes illnesses of heart/mind.
> (*Zuozhuan*, Duke Zhao 1.1222)

Heaven and Earth were always associated with five and six, respectively, but the ways in which they were associated were not always the same. This association thus provided continuity but also allowed change, as their specific relationships to five and six could change.

For example, human beings, living between Heaven and Earth, were seen in medical texts such as the *Neijing Suwen*, as mirroring the macrocosm. Within a human body, the rubrics of five and six were assigned opposite relationships: the innermost, purest, most Heavenly things were counted by five, while the outermost, coarser and more Earthly things were counted by six. There was a continuous dance between five and six: the six climates of Heaven, if unseasonal, directly affected the five innermost organs in humans, the *zang* 臟 (the *yin* storage organs). In contrast, if the five sapors of the Earth (water and grains) were tainted, this affected the six more exterior organs, the *fu* 腑 (*yang* transforming and transporting organs).

> Hence, if there is evil *qi* of Heaven, then the response is harm to humans' five *zang* organs. If there is cold or heat of food and drink then the response is harm to humans' six *fu* organs.
>
> (Suwen, 2.78)[39]

However, even from the Warring States, the *qi* of Heaven (representing the macrocosm) had been enumerated as the six *qi* (*liuqi* 六氣): the climates (wind, cold, dryness, dampness, heat and fire).

> Just like [the one who] follows the normality of Heaven and Earth, and wards off the changes of the six *qi*, by means of travelling in the illimitable.
>
> (Zhuangzi, 1:4)

The *qi* of Earth (the microcosm) manifested as the *wuxing*. This six-fold view of the cosmos was used to represent the climates, and as humans mirrored their environment, these six climates were linked to bodily conditions, which, if imbalanced or unseasonal, became the Six Excesses (*liuyin* 六淫). Heaven had a circulation of *qi* through each season/climate, in turn; Earth had a circulation of *qi* through each agent, in turn.

The two numerical systems of five and six, which had separately begun to structure traditions of medical thought and practice as a feature of early imperial ritual, became increasingly integrated and systematised in medicine in medieval times. They were fundamental to the medical innovations of Wang Bing in the eighth century as he added the *Wuyun liuqi* (Five Circuits and Six *Qi*) chapters of the Yellow Emperor's corpus.

Notes

1. A note on the use of capitals: I have used capitals for any concept that is significantly different in Chinese thought, e.g. Heaven is more than just the sky and is not related to the Christian concept of life after death. I capitalise a Number when I am referring to its numerological significance. Where I refer to the specific Chinese category of one of the *wuxing*, they are capitalised as agents, similarly for the Chinese category of organs, Viscera and Storehouses.
2. All translations are mine.
3. The *Yiwenzhi* is the earliest existing bibliographic catalogue of the Imperial library, found in the *Hanshu* 漢書 (*Book of Han*), Chapter 30, compiled by Ban Gu 班固 (32–92).
4. Holes were bored in turtles' carapaces, then heated to 'read' the meanings of the resulting cracks.
5. The stalks of the milfoil plant were used to perform a complex calculation, resulting in a hexagram, *gua* 卦. A hexagram is a stack of six horizontal *yang* (unbroken) or *yin* (broken) lines (*yao* 爻). There are sixty-four hexagrams in the *Yijing* 易經 (Classic of Change), a collection of prognosticatory texts, credited to the mythical ruler Fuxi, with commentaries by Confucius, but probably compiled over several centuries from the Western Zhou period (ca. 1100–770 BCE), evolving to its present state in the Warring States (475–221 BCE).
6. The Grand Astrologer, a position of chronicler of state affairs since at least Zhou times, was also in charge of observing phenomena in Heaven and Earth (such as unusual events – portents) to align the activities of humans with those of Heaven and Earth through the calendar (Hucker 1985: 482). He was an astronomer, historian and diviner, deciding auspicious days for state rituals (Bo 1985: 261).

7 See plastron set *bingbian* 丙編 12–21, dated to period I, ca. 1240–1181 BCE, in Keightley (1985: 76–90).
8 See Despeux (2005: 176–205) for a comparison of facial diagnosis in divination and medicine.
9 See Sivin (1993: 196–215) for a comprehensive analysis.
10 The *Lüshi Chunqiu* was compiled around 239 BCE by Lü Buwei 呂不韋 (290–235 BCE). See Knoblock and Riegel (2000).
11 The *Huainanzi* was written by the retainers of the Prince of Huainan, Liu An 劉安 (ca. 179–122 BCE) and compiled by Liu An; it was presented to the court of Han Emperor Wu (157–87 BCE) in 139 BCE. See Major *et al.* (2010).
12 *Huangdi* was never represented as a master, but always a disciple of his Daoist ministers and sages (Seidel 1992: 51).
13 A text called *Huangdi Suwen* was listed in the *Suishu* 隋書 (Book of the Sui, 636). The author was identified as Yuan Quanji by the authors of the 'New Tang History', *Xin Tangshu* 新唐書 (compiled in 1060). Quan Yuanqi's *Suwen* commentary was lost by the early twelfth century at the latest. Xu Chunfu 徐春甫 (1520–96, an official of the *Taiyiyuan* 太醫院, *Great Medical Office*) named it the 'Instructions and Explanation of the Suwen' (*Suwen xunjie* 素問訓解) in the mid-sixteenth century, based on the Imperial Medical Editorial Office's 1067 view that it was the first commentary of the *Suwen* (Unschuld 2003: 24).
14 Wang Bing included the names but no text for chapters 72 and 73, which were already missing from Quan's edition: later Song editors added these in, based on texts written in Tang-Song (Unschuld 2003).
15 Wang was also the supposed author (under a Daoist pseudonym) of several other apocryphal texts on calendar and medicine (Despeux 2001: 129).
16 Wang Bing's preface to the *Suwen*, as well as that of Gao Baoheng *et al.* are translated in Veith 1949, although Veith's introduction to Chinese medical concepts and translation of *Suwen* 1–34 is not particularly literal or useful. Its interest lies mainly in being the earliest partial translation of the *Suwen* into English.
17 This text was mentioned by Zhang Ji 張機 (150–219, style name Zhongjing 仲景) in his preface to the 'Treatise on Cold Damage and Miscellaneous Diseases' (*Shanghan zabinglun* 傷寒雜病論) which was the first systematic organisation of herbal remedies to be used when there was an external pathogen causing disease.
18 Despeux (2001, 129) hypothesises that although the Wang Bing chapters were a Tang innovation, such methods are not cited more broadly in the *Ishinpō* and *Taiping Shenghui fang*, and thus the movement should be considered as not developed until the eleventh century.
19 Eight trigrams made from stacks of three *yang* (unbroken) or *yin* (broken) lines (*yao* 爻) lines used in divination.
20 See Yang (1997) for a study of the relationships between the *Yijing* and Chinese Medicine.
21 'Divination' is the foretelling of future events or discovery of what is hidden or obscure by supernatural or magical means.
22 The observation of the position of the Dipper 'handle' on the horizon at dusk over the course of the year was called the Dipper Establishment *doujian* 斗建 method. The earliest written definition of this method is from Yinqueshan 銀雀山 tomb 1 (excavated in 1972, dated ca. 130 BCE). See Harper and Kalinowski 2017: 163.
23 The 'Records of the Grand Historian' (*Shiji* 史記) by Sima Qian 司馬遷 (145–86 BCE) was the first dynastic history, spanning 2,500 years to Han Emperor Wu 漢武帝 (r. 141–87 BCE). It was presented to Wu in 91 BCE, and was the basis for all future dynastic histories.
24 See Pankenier's translation of the *Tianguan shu* in (Pankenier 2013: 459–72). Also, see a comparison of the *Huainanzi*'s 'Tianwen' 天文 with the *Shiji* 'Celestial Offices' chapters in Pankenier 2014: 199–224.
25 Previously known as the *Chu* silk manuscript 楚帛書, excavated in 1942, from ca. 300 BCE.
26 About late second century BCE, excavated in 1972.
27 See Major *et al.* (2010: 224–5) for a description of this perfect world, mended and established by Fuxi and Nüwa.
28 Traditionally ascribed to Dong Zhongshu 董仲舒 (ca. 179–104 BCE). See Loewe 2011 for a detailed analysis.
29 Lü's biography is in *Shiji* 85.

30 The *Shujing* 書經 (Book of Documents, also known as the *Shangshu* 尚書) was traditionally believed to be compiled by Confucius (551–479 BCE). It is a compilation of historical records from mythical culture bringers (written retrospectively, probably before 221 BCE) and early dynasties, comprising edicts and speeches by rulers and other important people. The 'Great Plan' is in the Book of Zhou (*Zhoushu* 周書) which was probably written during the Western Zhou period (1046–771 BCE). See Shaughnessy 1993.

31 From *Lüshi Chunqiu* Book 13 part 2, *Yingtong* 應同 (Resonating with the Identical): describing dynasties, starting from Yellow Emperor as Earth; then Great Yu (*Dayu* 大禹) as Wood; Tang of Shang (Shang Tang 商湯) as Metal and King Wen of Zhou (Zhou Wenwang 周文王) as Fire.

32 The structure and function of the 'Bright Hall' has been the subject of much speculation since Han times, when it was attributed to Zhou times. It was referred to in terms of its cosmological significance in the *Lüshi Chunqiu* and *Huainanzi*. See Li (2006b: 103–4) for the development of Four and the Centre to become the Nine (or Twelve) rooms of the Bright Hall.

33 The 'Monthly Ordinances' give astronomical, Stem, Branch and *wuxing* correlations for each of the twelve months of the year: prescribing appropriate ritual ceremonies and behaviour for the ruler to be in resonance and aligned with the cosmos. The earliest version is found in the first section of each of the first twelve chapters of the *Lüshi Chunqiu* (Almanacs); this is also found, almost identical, in the *Liji* 禮記 (Book of Rites, Chapter 6); and in Chapter 5 of the *Huainanzi* (Major 1993: 217–68).

34 The *Zuozhuan* was attributed to Zuo Qiuming 左丘明 (early sixth century BCE) but was compiled during the Warring States. It was thought to be written as a commentary and parallel version of the 'Spring and Autumn Annals' (*Chunqiu* 春秋), a chronicle of the state of Lu and other states during the Eastern Zhou (771–256 BCE).

35 A game of chance, popular in Han (202BCE–220CE), played with six sticks (acting as dice) and two sets of six black and white game pieces.

36 *Liuren* are six days in the sexagenary cycle with Stem ren 壬: n9 *renshen* 壬申; n19 *renwu* 壬午; n29 *renchen* 壬辰; n39 *renyin* 壬寅; n49 *renzi* 壬子 and n59 *renxu* 壬戌.

37 For the *liuren* method, see Kalinowski (1983); Ho (2003: 113–38); Raphals (2013: 167–70).

38 'The Virtuous Discussions of the White Tiger Hall' (*Baihu tongde lun* 白虎通德論) was compiled on the basis of the Han conference under the supervision of the imperial bibliography Ban Gu. On its effect on medical discourse, see Unschuld (2009).

39 *Suwen* 5, *Yinyang yingxiang dalun* 陰陽應象大論 (Comprehensive Discourse on Phenomena Corresponding to *Yin* and *Yang*); Unschuld et al. (2011: 1,120).

Bibliography

Pre-modern sources

Chunqiu fanlu yizheng 春秋繁露義證 (Luxuriant Dew of the Spring and Autumn Annals–Correct Verification), Dong Zhongshu 董仲舒, in Su Yu 蘇輿 (ed.) (1992), *Xinbian zhuzi jicheng* 新編諸子集成 (New Collection of Ancient Philosophers), Beijing: Zhonghua shuju.

Chunqiu zuozhuan zhu 春秋左傳注 (Zuo Commentary with Annotations), Yang Bojun 楊伯峻 (ed.) (1990), Beijing: Zhonghua shuju.

Hanshu 漢書 (Records of the Han Dynasty) 111, Ban Gu 班固 et al. (1962), Beijing: Zhonghua shuju.

Huainanzi jishi 淮南子集釋 (Masters from Huainan – Collected Explanations) 139 BCE, Liu An 劉安, in He Ning 何寧 (ed.) (1998), *Xinbian zhuzi jicheng* 新編諸子集成 (New Collection of Ancient Philosophers), Beijing: Zhonghua shuju.

Huangdi neijing 黃帝內經 (Yellow Emperor's Inner Canon), 762, Wang Bing 王冰 (ed.) (1994), Beijing: Zhongyi guji chubanshe.

Liji zhengyi 禮記正義 (Book of Rites – Corrected Interpretation), Xuan Zheng玄鄭 (127–200) et al., in Ruan Yuan 阮元 (1764–1849) (ed.) (1990), *Shisanjing zhushu* 十三經注疏 (Thirteen Confucian Classics with Notes and Commentaries), Shanghai: Guji chubanshe.

Lüshi chunqiu jishi 呂氏春秋集釋 (Master Lü's Spring and Autumn Annals with Collected Explanations), 239 BCE, Lü Buwei 呂不韋, in Xu Weiyu 許維遹 and Liang Yunhua 梁雲華 (eds) (2009), *Xinbian zhuzi jicheng* 新編諸子集成 (New Collection of Ancient Philosophers), Beijing: Zhonghua shuju.

Shanghan zabing lun 傷寒雜病論 (Treatise on Cold Damage and Miscellaneous Diseases), Zhang Zhongjing 張仲景 (150–219) (2014), Beijing: Kexue jizhu chubanshe.

Shiji (Records of the Grand Historian), Sima Qian 司馬遷 (145–86 BCE) (1969), Xianggang: Zhonghua shuju.

Shangshu zhengyi 尚書正義 (Book of Documents – Corrected Interpretation), 653, Kong Yingda 孔穎達 (574–648) et al., in Ma Xinmin 馬辛民 (ed.) (2000), *Shisanjing zhushu* 十三經注疏 (Thirteen Confucian Classics with Notes and Commentaries), Beijing: Beijing daxue chubanshe.

Zhuangzi jijie 莊子集解 (Master Zhuang - Collected Explanations), Zhuang Zhou 莊周 (ca. 369–286 BCE) in Wang Xianqian 王先謙 (ed.) (2001), Taipei: Shijie shuju.

Modern sources

Allan, S. (1991) *The Shape of the Turtle: myth, art and cosmos in early China*, New York: State University of New York Press.

Bo Shu-ren 薄樹人 (1985) 'Sima Qian – the great astronomer of ancient China', *Chinese Astronomy and Astrophysics*, 9: 261–7.

Cammann, S. (1948) 'The 'TLV' pattern on cosmic mirrors of the Han dynasty', *Journal of the American Oriental Society*, 68.4: 159–67.

Chavannes, E. (1909) *Mission archéologique dans la Chine septentrionale, Partie 1- Planches*, Paris: Imprimerie Nationale.

Cook, C.A. (2006) *Death in Ancient China: the tale of one man's journey*, Leiden: Brill.

Critchlow, K. (1969) *Order in Space: a design sourcebook*, London: Thames and Hudson.

Crump, T. (1990) *The Anthropology of Numbers*, Cambridge: Cambridge University Press.

Csikszentmihalyi, M. (2004) *Material Virtue*, Leiden: Brill.

Despeux, C. (2001) 'The system of the five circulatory phases and the six seasonal influences (Wuyun Liuqi), a source of innovation in medicine under the Song (960–1279)', in E. Hsu (ed.) *Innovation in Chinese Medicine*, Cambridge: Cambridge University Press, pp. 121–66.

—— (2005) 'From prognosis to diagnosis of illness in Tang China: comparison of the Dunhuang manuscript P.3390 and medical sources', in V. Lo and C. Cullen (eds) *Medieval Chinese Medicine: the Dunhuang medical manuscripts*, Abingdon: Routledge Curzon, pp. 176–205.

Girardot, N.J. (1983) *Myth and Meaning in Early Taoism: the theme of chaos (hun-tun)*, Berkeley and Los Angeles: University of California Press.

Graham, A.C. (1989) *Disputers of the Tao: philosophical argument in ancient China*, Chicago, IL; La Salle: Open Court.

Granet, M. (1934) *La Pensée Chinoise*, Paris: Albin Michel.

Harper, D. (1979) 'The Han cosmic board', *Early China*, 4.1: 1–10.

—— (1988) 'A note on nightmare magic in ancient and medieval China', *Tang Studies*, 6: 69–76.

—— (1998) *Early Chinese Medical Literature: the Mawangdui medical manuscripts*, London and New York: Kegan Paul International.

—— (1999) 'Warring states natural philosophy and occult thought', in M. Loewe and E.L. Shaughnessy (eds) *The Cambridge History of Ancient China*, Cambridge: Cambridge University Press, pp. 813–84.

Harper, D. and Kalinowski, M. (eds) (2017) *Books of Fate and Popular Culture in Early China: the daybook manuscripts of the Warring States, Qin, and Han*, Leiden; Boston, MA: Brill.

Henderson, J.B. (1984) *The Development and Decline of Chinese Cosmology*, New York: Columbia University Press.

Hinrichs, TJ and Barnes, L. (eds) (2013) *Chinese Medicine and Healing: an illustrated history*, Cambridge, MA: Harvard University Press.

Ho, P.Y. (1985) *Li, Qi and Shu: an introduction to science and civilization in China*, Hong Kong: Hong Kong University Press.

—— (2003) *Chinese Mathematical Astrology: reaching out to the stars*, Abingdon: Routledge.

Hsu, E. (ed.) (2001) *Innovation in Chinese Medicine*, Cambridge: Cambridge University Press.

Hucker, C.O. (1985) *A Dictionary of Official Titles in Imperial China*, Stanford, CA: Stanford University Press.

Kalinowski, M. (1983) 'Les instruments astro-calendriques des Han et la méthode liu ren', *Bulletin de l'Ecole Française d'Extrême-Orient*, 72: 304–419.

——— (1991) *Cosmologie et Divination Dans La Chine Ancienne. Le Compendium Des Cinq Agents (Wuxing Dayi, VIe Siècle)*, Paris: Ecole Francaise d'Extreme-Orient.

——— (1996) 'Astrologie calendaire et calcul de position dans la Chine ancienne. Les mutations de l'hémérologie sexagésimale entre le IVe et le IIe siècles avant notre ére', *Extreme-Orient, Extreme-Occident*, 18: 71–113.

——— (2004) 'Technical traditions in ancient China and *shushu* culture in Chinese religion', in J. Lagerwey (ed.) *Religion and Chinese Society: ancient and medieval China*, Vol. 1, Hong Kong: Chinese University Press, pp. 223–48.

——— (2005) 'Mantic texts in their cultural context', in V. Lo and C. Cullen (eds) *Medieval Chinese Medicine: the Dunhuang medical manuscripts*, Abingdon: Routledge Curzon, pp. 135–59.

Keegan, D.J. (1988) *The 'Huang-Ti Nei-Ching': the structure of the compilation; the significance of the structure*, Berkeley: University of California Press.

Keightley, D.N. (1985) *Sources of Shang History: the oracle-bone inscriptions of Bronze Age China*, Berkeley: University of California Press.

——— (1988) 'Shang divination and metaphysics', *Philosophy East and West*, 38.4: 367–97.

——— (1990) 'Early civilisation in China: reflections on how it became Chinese', in P.S. Ropp (ed.) *Heritage of China: contemporary perspectives on Chinese civilisation*, Berkeley: University of California Press, pp. 15–54.

——— (2000) *The Ancestral Landscape: time, space and community in late Shang China*, Berkeley: University of California Press.

Knoblock, J. and Riegel, J. (2000) *The Annals of Lu Buwei: a complete translation and study*, Stanford, CA: Stanford University Press.

Kong, Y.C. (2010) *Huangdi Neijing: a synopsis with commentaries*, Hong Kong: The Chinese University of Hong Kong.

Lagerwey, J. and Kalinowski, M. (eds) (2007) *Early Chinese Religion: part one, Shang through Han*, Leiden: Brill.

Lei Hsiang-lin, S. (2014) *Neither Donkey nor Horse: medicine in the struggle over China's modernity*, Chicago, IL: University of Chicago Press.

Lewis, M.E. (1999) *Writing and Authority in Early China*, Albany: State University of New York Press.

——— (2006) *The Flood Myths of Early China*, Albany: State University of New York Press.

Li Jianmin 李建民 (2000a) *Fangshu yixue lishi* 方術醫學歷史 (History of Divination and Medicine), Taipei: Nantian shuju.

——— (2000b) *Sisheng zhi yu: Zhou-Qin-Han maixue zhi yuanliu* 死生之域周秦漢脈學之源流 (The Boundary between Life and Death: Sources of Vessel Theory in the Zhou, Qin and Han Dynasties), Taipei: Academia Sinica.

Li Ling 李零 (2006a) *Zhongguo fangshu xukao* 中国方术续考 (A Study of the Occult Arts of China – Continued), Beijing: Zhonghua shuju.

——— (2006b) *Zhongguo fangshu zhengkao* 中國方術正考 (A Study of the Occult Arts of China), Beijing: Zhonghua shuju.

Liu Lexian 劉樂賢 (1994) *Shuihudi Qin jian rishu yanjiu* 睡虎地秦簡日書研究 (Shuihudi Qin Bamboo Daybooks Research), Taipei: Wenjin chubanshe.

Lloyd, G.E.R. and Sivin, N. (2002) *The Way and the Word: science and medicine in early China and Greece*, New Haven, CT: Yale University Press.

Lo, V. (2013) 'The Han period', in TJ Hinrichs and L. Barnes (eds) *Chinese Medicine and Healing: an illustrated history*, Cambridge, MA: Harvard University Press, pp. 31–64.

Lo, V. and Cullen, C. (eds) (2005) *Medieval Chinese Medicine: the Dunhuang medical manuscripts*, Abingdon: Routledge Curzon.

Loewe, M. (1979) *Ways to Paradise: the Chinese quest for immortality*, London: George Allen and Unwin.

——— (ed.) (1993) *Early Chinese Texts: a bibliographical guide*, Berkeley: University of California Press.

——— (2005) *Faith, Myth, and Reason in Han China*, Indianapolis: Hackett Publishing.

——— (2011) *Dong Zhongshu, a "Confucian" Heritage and the Chunqiu Fanlu*, Leiden; Boston, MA: Brill.

Lu, G.-D. and Needham, J. (1980) *Celestial Lancets: a history and rationale of acupuncture and moxa*, Cambridge: Cambridge University Press.

Major, J.S. (1993) *Heaven and Earth in Early Han Thought: chapters three, four and five of the Huainanzi*, Albany: State University of New York Press.

Major, J.S., Queen, S.A., Meyer, A.S. and Roth, H. (trans and eds) (2010) *The Huainanzi: a guide to the theory and practice of government in early Han China*, New York: Columbia University Press.

Nylan, M. (1997) 'Han classicists writing in dialogue about their own tradition', *Philosophy East and West*, 47.2: 133–88.

Pankenier, D. (1981) 'Astronomical dates in Shang and Western Zhou', *Early China*, 7: 2–37.

—— (2013) *Astrology and Cosmology in Early China: conforming earth to heaven*, New York: Cambridge University Press.

—— (2014) 'The *Huainanzi's* "heavenly patterns" and the *Shiji's* "treatise on the celestial offices": what's the difference?', in S. Queen and M. Puett (eds) *The Huainanzi and Textual Production in Early China*, Leiden: Brill, pp. 199–224.

Porkert, M. (1974) *The Theoretical Foundations of Chinese Medicine: systems of correspondence*, Cambridge, MA: MIT Press.

Raphals, L. (2013) *Divination and Prediction in Early China and Ancient Greece*, Cambridge: Cambridge University Press.

Robinet, I. (2011) *The World Upside Down: essays on Taoist internal alchemy*, trans. F. Pregadio, Mountain View, CA: Golden Elixir Press.

Ropp, P.S. (ed.) (1990) *Heritage of China: contemporary perspectives on Chinese civilisation*, Berkeley: University of California Press.

Rosemont, H. Jr. (1991) *Chinese Texts and Philosophical Contexts: essays dedicated to Angus C.Graham*, La Salle, PA: Open Court.

Seidel, A. (1992) *La divinisation de Lao Tseu dans le Taoisme des Han*, Paris: École Française d'Extrême-Orient.

Shaughnessy, E.L. (1993) '*Shang shu* 尚書 (*Shu ching* 書經)', in M. Loewe (ed.) *Early Chinese texts: a bibliographical guide*, Berkeley: Society for the Study of Early China, pp. 376–89.

Simpson, J.A. and Weiner, E.S.C. (eds) (1989) *The Oxford English Dictionary*, Oxford: Clarendon Press; Oxford; New York: Oxford University Press

Sivin, N. (1987) *Traditional Medicine in Contemporary China*, Ann Arbor: University of Michigan.

—— (1993) 'Huang Ti Nei Ching', in M. Loewe (ed.) *Early Chinese Texts: a bibliographical guide*, Berkeley: University of California Press, pp. 196–215.

Smith, A. (2010) 'The Chinese sexagenary cycle and the ritual origins of the calendar', in J.M. Steele (ed.) *Calendars and Years II: astronomy and time in the ancient and medieval world*, Oxford: Oxbow Books, pp. 1–37.

Smith, K., Bol, P.K., Adler, J.A. and Wyatt, D.J. (1990) *Sung Dynasty Uses of the I Ching*, Princeton, NJ: Princeton University Press.

Smith, R.J. (2012) *The I Ching: a biography*, Princeton, NJ: Princeton University Press.

Steele, J.M. (ed.) (2010) *Calendars and Years II: astronomy and time in the ancient and medieval world*, Oxford: Oxbow Books.

Sterckx, R. (2007) 'The economics of religion in Warring States and early imperial China', in J. Lagerwey and M. Kalinowski (eds) *Early Chinese Religion: part one, Shang through Han*, Leiden: Brill, pp. 839–80.

Sun, X. and Kistemaker, J. (1997) *The Chinese Sky during the Han: constellating stars and society*, Leiden: Brill.

Tseng, Lillian Lan-ying (2002) 'Divining from the game liubo: an explanation of a Han wooden slip excavated at Yinwan', *China Archaeology and Art Digest*, 4.4: 55–62.

—— (2004) 'Representation and appropriation: rethinking the TLV mirror in Han China', *Early China*, 29: 164–215.

Twitchett, D. and Loewe, M. (eds) (1986) *The Cambridge History of China, Vol. 1: the Ch'in and Han empires, 221 BC–AD 220*, Cambridge: Cambridge University Press.

Unschuld, P.U. (2003) *Huang Di Nei Jing Su Wen: nature, knowledge, imagery in an ancient Chinese medical text, with an appendix, the doctrine of the five periods and six Qi in the Huang Di Nei Jing Su Wen*, Berkeley: University of California Press.

Unschuld, P.U. (2009) 'Yin-Yang theory, the human organism, and the *Bai hu tong*: a need for pairing and explaining', *Asian Medicine*, 5: 19–38.

Unschuld, P.U., Tessenow, H. and Zheng, J. (2011) *Huang Di Nei Jing Su Wen: an annotated translation of Huang Di's inner classic – basic questions*, Vol. 1, Berkeley: University of California Press.

Veith, I. (1949) *Huang Ti Nei Ching Su Wen – The Yellow Emperor's Classic of Internal Medicine*, Berkeley: University of California Press.

Verran, H. (2010) *Science and an African Logic*, Chicago, IL: University of Chicago Press.
Wang, A. (2000) *Cosmology and Political Culture in Early China*, Cambridge: Cambridge University Press.
Wong, Pui Yin Marianne (2006) *The 'Siling' (Four Cardinal Animals) in Han Pictorial Art*, PhD thesis, London: School of Oriental and African Studies.
Yamada, Keiji (1979) 'The formation of the Huang-Ti Nei-Ching', *Acta Asiatica*, 36: 67–89.
Yang Li 杨力 (1997) *Zhouyi yu Zhongyi xue* 周易与中医学 (Book of Changes and Chinese Medicine Studies), Beijing: Beijing kexue jishu chubanshe.
Yang, Xiaoneng (2004) *New Perspectives on China's Past: Chinese archaeology in the twentieth century*, New Haven, CT; London: Yale University Press with the Nelson-Atkins Museum of Art, Kansas City.

5
THE IMPORTANCE OF NUMEROLOGY, PART 2 MEDICINE

An overview of the applications of numbers in *Huangdi neijing*

Deborah Woolf

Numerology and Chinese medicine

The texts of classical Chinese medicine testify to creative tensions between numerical continuities with the ancient world and changing conceptions of the human body as they relate to medical practice. They draw on number-related clues to understanding innovation, in the way they draw on, systematise and pluralise the earlier numerical traditions described in Part I. The *Huangdi neijing* 黃帝內經 (Yellow Emperor's Inner Canon), the *locus classicus* of classical Chinese Medicine, is a selection of different texts and traditions that have accumulated over various centuries. This has resulted in the received text having an eclectic range of number systems, which have been deployed at different times and for different effects. The *Huangdi neijing* is in two parts: the *Suwen* 素問 (Simple Questions) and *Lingshu* 靈樞 (Numinous Pivot). In 751, Wang Bing 王冰 began his revision of the earlier version of the *Suwen*, with an emphasis on self-cultivation and the importance of aligning oneself with the movements of the cosmos (Chapter 7 in this volume; Unschuld 2003: 41). His revision of 762 CE included seven extra chapters (*Suwen* 66–74),[1] specifically focused on *Wuyun liuqi* 五運六氣 (Five Movements and Six *qi*) theory, which he claimed were secret knowledge from earlier times, received as a master-disciple transmission (Ibid.: 46). This new version demonstrates an attempt to order what at first sight might seem like a confusing and fragmentary medieval legacy. Wang Bing's focus, seen uniquely in his *Wuyun liuqi* chapters, reflects a post-Han view of the imperial cosmos, based on the interplay between the numbers Five and Six.[2]

This chapter argues that number sequences provided a structural organisation that was the basis for various arrangements over longer periods of time by introducing innovative arrangements which, nonetheless, seemed to ground these innovations in ancient wisdom. We see in the Wang Bing chapters permutations of all the basic numbers sequences introduced chronologically in Part I, beginning with One, and the identification of *qi* as the unifying factor (Yang 1997: 56–7). Human life, itself an expression of the merging of the duality of Earth/*yin* and Heaven/*yang*, was animated by *qi*, the vitality of life.

陰陽者，天地之道也，萬物之綱紀，變化之父母，生殺之本始

DOI: 10.4324/9780203740262-7

> *Yin* and *yang* are the way of Heaven and Earth, the rule and order of the myriad beings, the mother and father of change and transformation, the beginning and end of life and death.
>
> (Suwen 5, 2.61)[3]

The totality of an ideal human lifespan is discussed in the first chapter of the *Suwen*: life was presented therein as dependent on a finite amount of essence (*jing* 精) allotted at birth and gifted by Heaven. This essence was the basis of the growth and development of the human body according to seven- and eight-year periods: men were thought to follow a cycle of eight years, while women followed a seven-year cycle. Human life was in constant interaction between Heaven and Earth, following underlying numerological cycles. The rubric of Four (or doubled here to make the Eight of the male cycle) had been a basic design following the Four directions and the seasons since Shang times. The rubric of Seven appears here as the female cycle but is less pervasive throughout classical medicine as an organising principle, and perhaps reflects a period before gender difference was clearly and consistently articulated as a feature of medicine (Chapter 23 in this volume).

Thereafter, the numerological organisation of the chapters *Suwen* 2, *Suwen* 4 and *Suwen* 5[4] turns to elaborating correspondences that structure human health and the body to the Four seasons and also to the Five Agents (*wuxing*): these numbers structure direction, colour, taste, organ, orifice, animal, planet, etc., through a new integration of the old rubrics of Four and Five. This was a shift from the early Shang idea of Four directions to a more integrated imperial view of the cosmos based on Five, identified as 'correlative cosmology' in the secondary literature (Henderson 1984; Graham 1986; Wang 2000).

Microcosm and macrocosm

The Five Agents interacted in different ways, following cycles of 'mutual conquest' (*xiangke* 相剋 or *xiangsheng* 相勝) and of 'mutual generation' (*xiangsheng* 相生). The earliest introduction and use of Five Agents as an interpretive structure that we know is evident in the 'conquest' cycle: a cycle that regularised the succession of dynasties and their regime change by giving the conquering rulers a cosmological legitimacy. The advice to (mythical) rulers was to devise rituals associated with their ruling Agent, using the colours associated with their Agent, and interpreting omens accordingly (Loewe 2005: 152–4).

> When an emperor or king is about to arise, Heaven must first manifest good omens to the people below. At the time of the Yellow Emperor, Heaven first caused giant mole crickets and earthworms to appear. The Yellow Emperor announced 'the *qi* of Earth is victorious'. Since the *qi* of Earth was victorious, he valued the yellow colour and used Earth as a rule for his affairs.
>
> (Lüshi chunqiu, 13.2, 284)[5]

The 'Great Plan' (*Hongfan* 洪範) chapter of the *Shujing*[6] lists the Agents following the conquest or control cycle sequence.

> Five Agents: the first is Water; the second is Fire; the third, Wood; the fourth, Metal; and the fifth, Earth.
>
> (Shujing, 12.357)

The conquest cycle was a strong feature of the daybooks[7] or almanacs (*rishu* 日書) and other divinatory texts of the Han (Liu 1994: 434; Li 2000a: 131; Harper and Kalinowski 2017: 169). Reflecting relationships of control and violence, the conquest cycle first structured divinations used for the purposes of government and warfare.

> The virtue (of the ruler) is only from good government, good government nourishes the people. This comes only by studying Water, Fire, Metal, Wood, Earth, and grains.
> (Shujing, 4.106)

In contrast, the *xiangsheng* 相生 (mutual generation) cycle, that was to become so important to imperial medical theory, was virtually absent from the daybooks and this contemporary evidence allows us to imagine a chronology for the numerical sequences (Harper and Kalinowski 2017: 169). The *xiangsheng* cycle, one indicating nourishment or support, like a mother to a son, was expressed in Dong Zhongshu's 董仲舒 (ca. 179–104 BCE) *Chunqiu fanlu* 春秋繁露 (Luxuriant Dew of the Spring and Autumn Annals), a text of the Western Han (202 BCE–9 CE) period, which consolidated Five Agent theory to uphold Dong's theory that humans should follow the cycles of the heavens closely and look to the lessons of the past for guidance, especially the hierarchies of social relationships at the basis of the traditions of Kongzi 孔子 (Confucius). The emergence of the *xiangsheng* cycle is a clear reflection of the work of Han scholars in whose interests it was to promote the new imperial priorities of unification, rather than conflict.

> Wood gives birth to Fire, Fire gives birth to Earth, Earth gives birth to Metal, Metal gives birth to Water, these are father and son.
> (Chunqiu fanlu, 38.315)[8]

The change in focus from *ke*/control to *sheng*/nourishing imitated the 'Great Plan', but followed the order of generation instead of conquest:

> Heaven has Five Agents: the first is Wood, the second is Fire, the third is Earth, the fourth is Metal, the fifth is Water. Wood is the beginning of the Five Agents; Water is the end of the Five Agents. These are the sequence of Heaven's order.
> (Chunqiu fanlu, 42.320)[9]

The *Chunqiu fanlu* also gave the correlations and qualities of the Five Agent cycle of generation, relating each Agent to a direction and season: except Earth, which was placed in the centre, establishing its special relationship to Heaven.

> Therefore Wood resides in the Eastern direction and rules spring *qi*, Fire resides in the Southern direction and rules summer *qi*, Metal resides in the Western direction and rules autumn *qi*, Water resides in the Northern direction and rules winter *qi* [...] Earth resides in the centre, it acts as the enrichment of Heaven. As for Earth, it is Heaven's trusted aide.
> (Chunqiu fanlu, 42.322)[10]

The early chapters of the *Suwen* mention both the conquest and generation cycles, but more emphasis is put on the harmonious cycle of life, the growth and regeneration cycle: spring gives birth to summer, just as Wood and the liver give birth to Fire and the heart.

Table 5.1 Channels and Climates

Channel	*Jueyin*	*Shaoyin*	*Shaoyang*	*Taiyin*	*Yangming*	*Taiyang*
Climate	Wind	Fire	Heat	Dampness	Dryness	Cold

> The East generates Wind; Wind generates Wood; Wood generates sour [flavour]; sour [flavour] generates the liver; the liver generates the sinews; the sinews [representing East/Wind/Wood] generate the heart.
>
> (Suwen, 2.67)[11]

This transition from a cosmological polity numerically modelled on the dynamics of warfare and violent control to one of harmonious central imperial governments reflects the priorities of the new administration in medical terms. *Suwen* 8[12] introduces the twelve organs (*zang* 臟, the *yin* storage organs or 'viscera'; and *fu* 腑, the *yang* transporting organs or 'storehouses') as homologous with the collaborative hierarchy of the Emperor and his court officials. Medicine, in this way, echoed the transition from the final decades of the Warring States to the new forms of government of Qin (221–206 BCE) and Han (202 BCE–220 CE) – shifting focus from the conquest to the generation cycle. This centralised government relied on a strong bureaucratic hierarchy to mediate between the Emperor and his people (Unschuld 2003: 325–6).

> The heart holds the office of sovereign, the radiance of the spirits stems from it. The lungs hold the office of chancellor, regulating rhythms stems from it. The liver holds the office of general, conception of plans stems from it.
>
> (Suwen, 3.93)[13]

The reflection between human beings and their environment was emphasised in *Suwen* 3, *Suwen* 6 and *Suwen* 9, which discuss the notion of Six. Six in the macrocosm were the Six climates or Six *qi* (*liuqi* 六氣 (Chapter 4 in this volume), which in the human form corresponded to the *liujing* 六經, the Six Channels (*jing* 經 hereafter body 'Channels'). In this way, the human body became a microcosm of the environment and each Channel was thought to respond to its associated climate, *qi* (Yang 1997: 64–9) (Table 5.1).

This meteorological body thus began to map the ideal harmony between human anatomy and physiology against changes in the external environment, in terms of the combination of Six Channels and Five Agents (Chapter 2 in this volume). While the combination of Five and Six had been a foundational concept first emphasised in the Warring States text, the *Zuozhuan*,[14] then systematised in the Han medicine of the *Huangdi neijing* – these numbers were then combined in innovative ways in the interpolated chapters of *Wuyun liuqi* theory – written more than half a millennium later than the rest of the *Huangdi neijing* material.

Four and Five in classical medicine

Number systems were neither exclusive nor fixed in practice. Different systems co-existed, and the aligning of plural numerological theories responded to both political expediencies and the practicalities of medicine. This plurality in the use of numbers facilitated continuity, even as different systems generated one set of numerical combinations from another.

In the body, the fundamental significance of Four was represented by the Four limbs, dependent on the centre, the trunk; similarly, the Four lands of the Shang of the ancient world had been defined from their centre, the residence of the Shang king. In ritual texts,

Four easily became the basis of Five: Four with their intersection, the centre. The increasing systematisation in the medicine of the *Huangdi neijing* highlighted a fundamental tension between the priority of Four and Five in early imperial Chinese numerology. This tension was to play out in the Yellow Emperor corpus through the spatio-temporal characteristics of the Five Agents, through the position allotted to the agent of Earth; it first emerged in the discourse of auspicious days in the daybooks (Liu 1994: 435; Li 2006a: 303) and was later systematised in *Suwen* 4 and *Suwen* 5.

The calendar provided the cosmological framework (Harper and Kalinowski 2017: 153) which marked time and space in an integration that was codified in the 'cord-hook diagram' (二繩四鉤圖 *ersheng sigou tu* – literally the 'two cords and four hooks' diagram; Li 2006b: 91–105; Li 2006a: 74, 84, 316; Harper and Kalinowski 2017: 162–9). This diagram is the basis for many divinatory systems and the *locus classicus* for its definition is found in the astronomical chapter of the Han dynasty *Huainanzi* 淮南子 (Masters from Huainan, compiled 139 BCE).[15] The rubric of Four is present in the two 'cords' marking the North-South and East-West axes, which also represents the summer-winter solstices and the spring-autumn equinoxes, respectively (Li 2000b: 131, 163; Li 2006b: 69–85). The Branches of the Shang calendar (Chapter 4 in this volume) in the quotation that follows serve to plot the seasons onto the diagram:

> *Ziwu* [b1 and b7], *maoyou* [b4 and b10] are two cords, *Chouyin* [b2 and b3], *chensi* [b5 and b6], *weishen* [b8 and b9], *xuhai* [b11 and b12] are four hooks.
>
> (Huainanzi, 3.207)[16]

The twelve calendrical Branches, represented by the 'b's on the diagram, were evenly spaced around the outside of the cord-hook diagram (b1–b12), each Branch representing a lunar month. Four of these Branches, (b1, b7) and (b4, b10), plotted the ends of the two cords, which also formed a central cross of the vertical and horizontal axes. These four Branches at the North, South, East and West of the diagram were the lunar months containing the two solstices and two equinoxes, representing the midpoint of each of the Four seasons on the four sides of the diagram (Yang 1997: 59–61). Each season plotted three Branches along one side: the solstices and equinoxes in the four directions and the two Branches placed either side of these. The two Branches either side of the solstices and equinoxes were then connected by a right-angled line to the previous or next seasons linking across the corners of the diagram. These right-angled corners of the diagram were the four 'hooks'.

The Stems (s1–s10) add a further dimension to the Four directions, spatial markers that were arranged around the cord-hook diagram just inside these outer seasonal arrangements (Liu 1994: 434; Yang 1997: 252). The ten Stems, unlike the twelve Branches, had an Earth/central component: two Stems (s5 and s6) were placed in the centre (Liu 1994: 159), next to the intersection of the two cords, emphasising the Fifth dimension of the Centre (see Figure 5.1).

This interweaving of Four and Five in the two cords and four hooks design can therefore be interpreted as defining 'spatio-temporal spaces' (Li 2000b: 162–4). The daybooks, in particular, allow us to see critical points at which the Five Agents were refined and standardised from pre-existing correlative systems (Liu 1994: 283, 346–7, 431–40; Li 2006b: 156–171). By the end of the Warring States (475–220 BCE), the Five Agents had, in this way, been used extensively to construct daybooks and divinatory schemes (Harper and Kalinowski 2017: 171).

Where Five Agents were primarily an indication of powers located spatially and directionally, through this process, they became increasingly associated with the passage of time. The calendar, its seasons and months, was thereby enshrined in the late Warring States ritual

Deborah Woolf

Summer – South - Fire

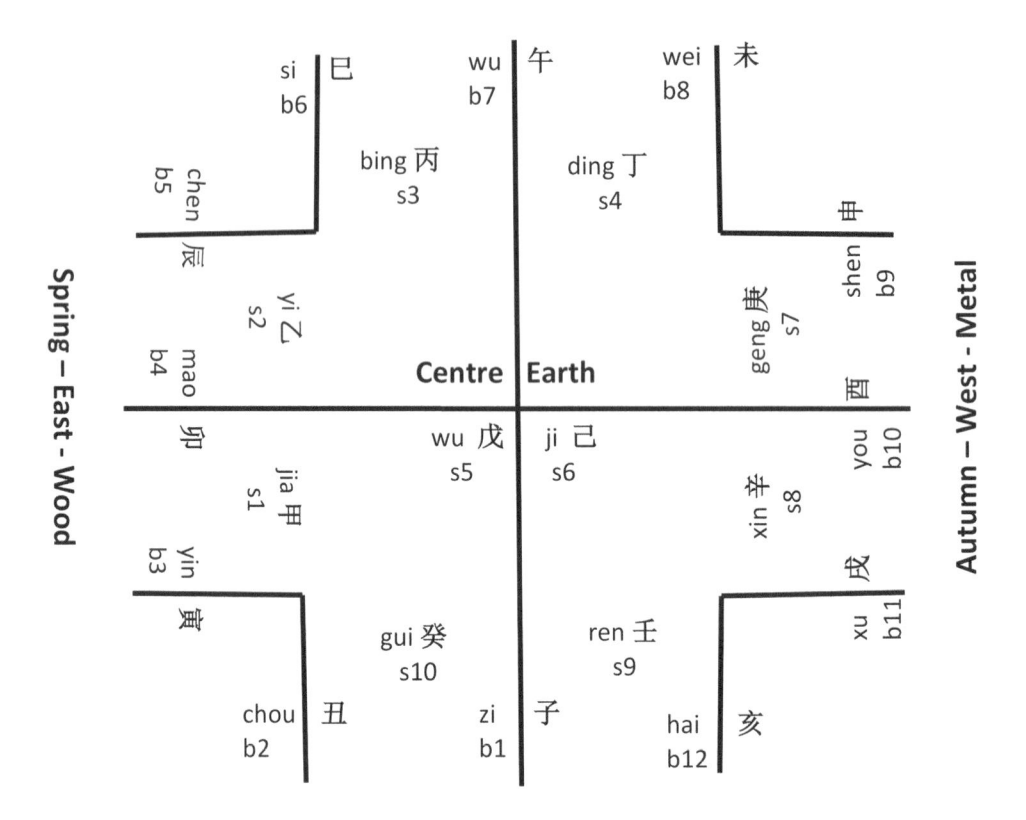

```
        si 巳      wu 午      wei 未
        b6         b7         b8

           bing 丙      ding 丁
           s3           s4
  chen                                shen 甲
  b5                                  b9
      yi                          geng
      s2                          s7
         mao          Centre │ Earth        you
         b4                                 b10
              jia 甲   wu 戊  ji 己    xin
              s1      s5    s6       s8
                 yin                     xu
                 b3                      b11
           gui 癸      ren 壬
           s10        s9

        chou 丑    zi 子     hai 亥
        b2         b1        b12
```

Spring – East - Wood

Autumn – West - Metal

Winter – North - Water

Figure 5.1 The Day Court (or cord-hook) diagram – representation by the author

texts that were not only a feature of everyday life, but also guided the Emperor's behaviour – in this case according to a ritual calendar (the Monthly Ordinances, *yueling* 月令 in the *Lüshi Chunqiu*). It was largely this temporal organisation that provided the basis for the calendar's incorporation into medical practice (Liu 1994: 416; Li 2006a: 62–72).

However, the persistent prevalence of the rubric Four, perhaps more naturally manifest in the cardinal directions and four distinct seasons of North China had to be reconciled with the Five of the Five Agents, which had an overarching political imperative. This was accomplished by dividing the summer season in two, making a fifth season, Late Summer (*changxia* 長夏). The attempt at reconciling potentially conflicting ritual sequences, through adding the Late Summer sequence in time, is first evident in Book Six of the ritual almanacs of the *Lüshi Chunqiu* – an eclectic text, tellingly compiled by Lü Buwei 呂不韋, advisor to King Zheng of Qin the first Emperor of China-to-be (and possibly, according to Sima Qian, his father). The Central Region was, in this text that heralded the rise of the Five Agents, as emblematic of imperial unity, summarily appended to the description of the last month of summer, creating a special fifth season, related to Earth (Knoblock and Riegel 2000: 152–6). In texts such as the *Huainanzi* and the *Chunqiu fanlu* of the mid-Western Han period, the Earth (*tu* 土) was no longer the Centre but just one sequence in the generative cycle of the

Five Agents, just as the ritual centre was regularised as part of an integrated spatial unity of empire itself.[17] Nevertheless, there remained a creative tension between the two rubrics of Four and Five.

The Five Agents, the Four directions and their intersection, the Centre, symbolised all possible movements of life: their effects on the body were described in *Suwen* 4 and *Suwen* 5. Here are the correspondences with the Agent Metal from *Suwen* 5:

> The West generates Dryness; Dryness generates Metal; Metal generates acrid [flavour]; acrid generates the lungs, the lungs generate skin and body hair [...] These in Heaven are Dryness, on Earth are Metal, in the body are skin and body hair, among the *zang* [organs] it is the lung; among the colours it is white; among the tones it is *shang*; among the sounds it is weeping; among the transformations it is coughing; among the orifices it is the nose; among the flavours it is acrid; among the states of mind it is sorrow.
>
> (Suwen, 2.71)[18]

Over the course of the Han dynasty, in texts applying the Five Agents as the framework for all patterns of the cosmos, such as the *Chunqiu fanlu*, while Earth was just one of the Five Agents, it often retained its temporal significance as Late Summer and was emphasised as the most important Agent.

> As for Earth, it is the son of Fire. Among the *wuxing* there is none as valuable as Earth.
> (Chunqiu fanlu, 38.316)[19]

In later discussions, however, immortalised in the *Baihutong delun* 白虎通德論 ('The Virtuous Discussions of the White Tiger Hall',[20] 58 CE – a treatise on politics and philosophy keyed to cosmological ideas such as Five Agents and *yin-yang*), the cosmos was seen as ordered and symmetrical through the alternation of the Four seasons, and thus incompatible with Earth as a discrete fifth phase. Emphasis on the role of the Earth was as an intermediary place of transformation from each season to the next, so it was allotted an eighteen-day period at the end of each of the Four seasons. This mediating role can be seen in early Han astrolabe *shi* instruments for divination. The Earth-related Stems (s5, s6) were no longer placed in the centre, as in the cord-hook diagram, but were in the four corners. These became the 'four gates' *simen* 四門 – see Figure 5.2 – the four diagonals radiating out from the centre (North-East, North-West, South-East and South-West) enabling a smooth transition from one season to the next (Harper and Kalinowski 2017: 163).

The flexibility which the rubrics of numbers Four and Five maintained throughout the early empires was to endow medical theory with an enduring plasticity that was critical to Wang Bing's calculations in his new interpretations of classical medicine in the Tang period (618–907).

Five and Six in medieval medicine

Central to Wang Bing's medieval additions to the classical tradition was a systematisation of the rubrics of Five and Six as embedded in the *Wuyun liuqi* chapters. His attempt to integrate these numerological traditions was not new. It was a Tang systematisation of earlier solutions to numerical disagreements about the total number of organs and channels in the body.

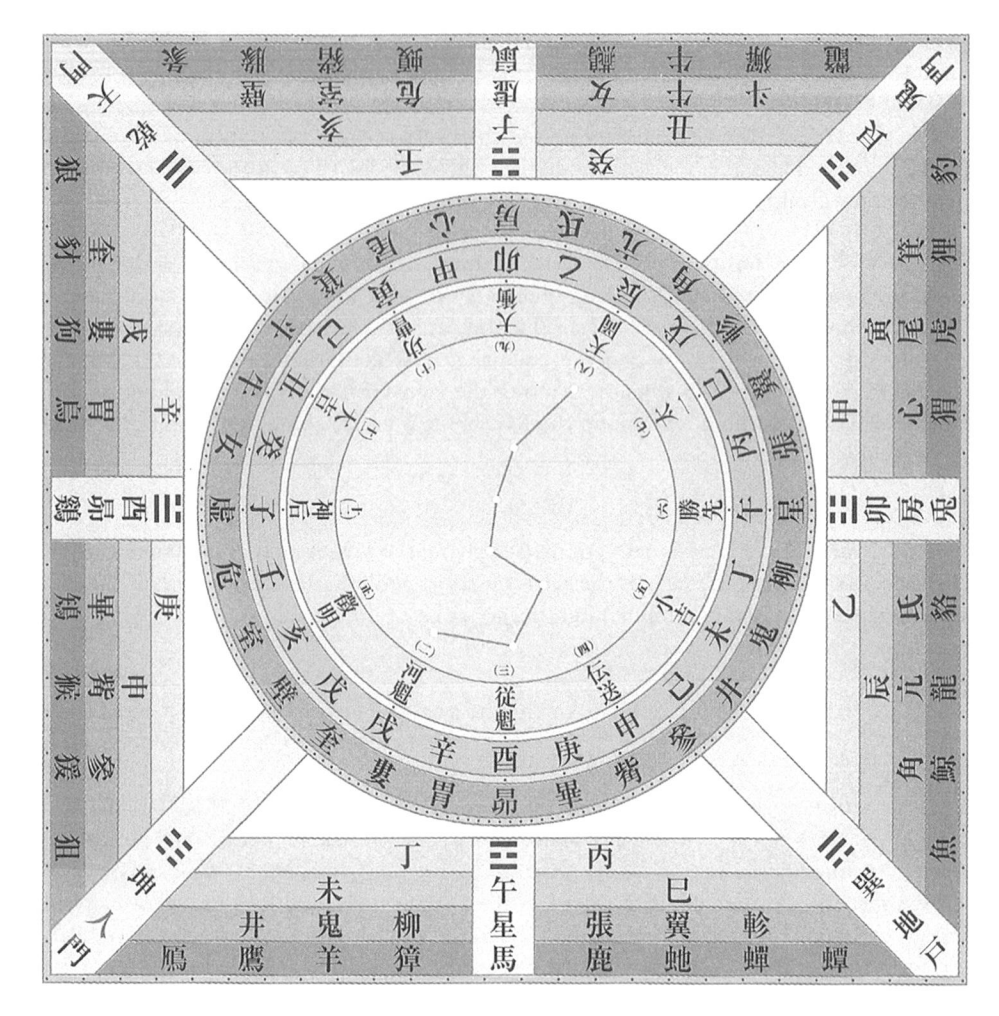

Figure 5.2 Earth between each season: as Four Gates which radiate out as diagonals from the centre of a two-dimensional representation of a Six Dynasties (220–589) liuren 六壬 shi divination board. Reproduced from Steavu, 2017: 200, on the basis of Abe no Seimei to Onmyōdō ten 安倍晴明と陰陽道展 [Abe no Seimei and the Way of Yin and Yang: The Exhibition], 53

Channels

There is evidence of a long running discourse on the priorities of Five and Six in the determining of the number of Channels of the body (vessels *mai* 脈 and Channels *jing* 經).[21] Manuscripts from the two Western Han tombs (Mawangdui 馬王堆 closed in 168 BCE,[22] and Zhangjiashan 張家山 closed in 186 BCE; Li 2000b: 110) reveal numerical ideas about Channels in the human body which exemplify different stages in the rubrics and relationships between Five and Six (Ibid.: 206–9; Figure 5.3).[23]

The twelve Channels of systematised classical medical theory evident in the *Huangdi neijing* developed from various earlier theories, which we can glimpse in the excavated medical texts. These excavated manuscripts listed eleven Channels 'tied' to the feet like threads and at the head like warp threads, each with associated symptoms to be treated by cauterisation techniques.[24] The eleven Channels were differentiated into numerically significant *yin* and *yang* elements (Five *yin* and Six *yang* Channels).[25]

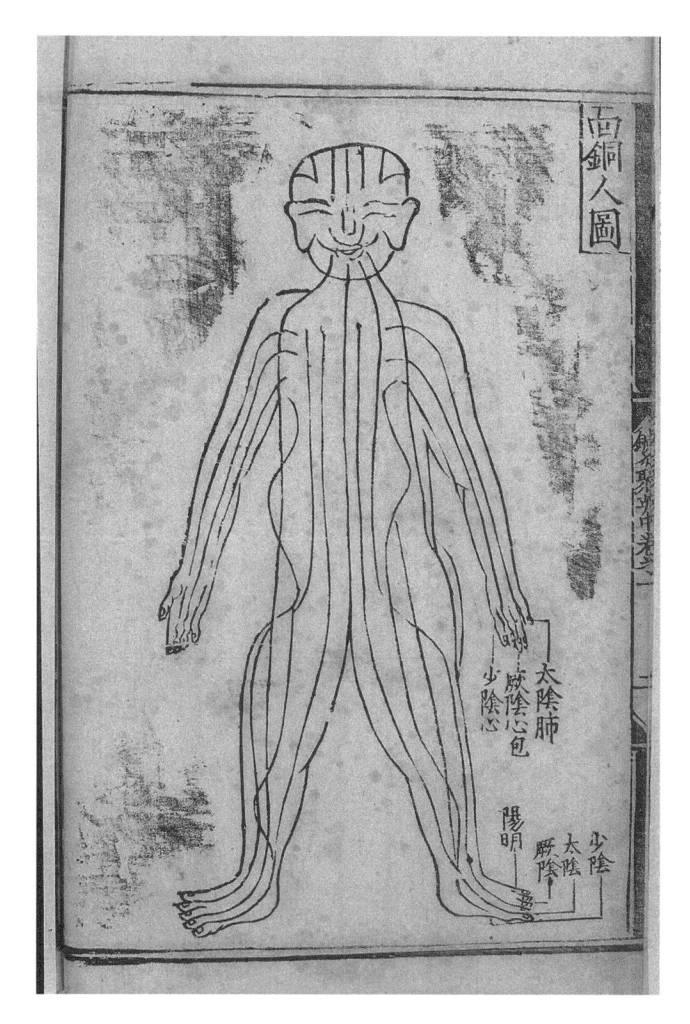

Figure 5.3 Bronze Man acupuncture figurine, front view, Chinese woodcut. Wellcome Collection (Ming dynasty): 4.0 International (CC BY 4.0)

Two wooden lacquered figurines have been found in Han tombs, showing the changing ideas around Channel theory. Both could have been used for various medical purposes, such as self-cultivation, diagnosis or as models of pathology. The Shuangbaoshan 雙包山 figurine excavated in 1993 (latest date 118 BCE) has ten red carved lines, nine lines from the extremities to the head and one line following the spine on the posterior midline, over the head to the nose (Lo 2007: 387–8). These lines appear to be only *yang* Channels (Ibid.: 404; Figure 5.4).[26]

Another figurine surfaced in 2012–13, from the Laoguanshan 老官山 site near Chengdu (tomb M3, closed between 157 and 88 BCE). The Laoguanshan figurine has eleven red lines painted on it and twelve white lines carved on it (Huang 2017). This figurine also has 109 white dots[27] clustered around joints, and bears the names of the five *yin* storage organs (*zang*) near the spine in a sequence that mirrors the conquest cycle of Five Agents. Although the Channels are not identical to those listed in the Mawangdui and Zhangjiashan texts, both figurines appear to be medically relevant, perhaps illustrating stages in Channel theory, before the systematic correspondences of the *Huangdi neijing* (Harper 2014; Zhou 2017).

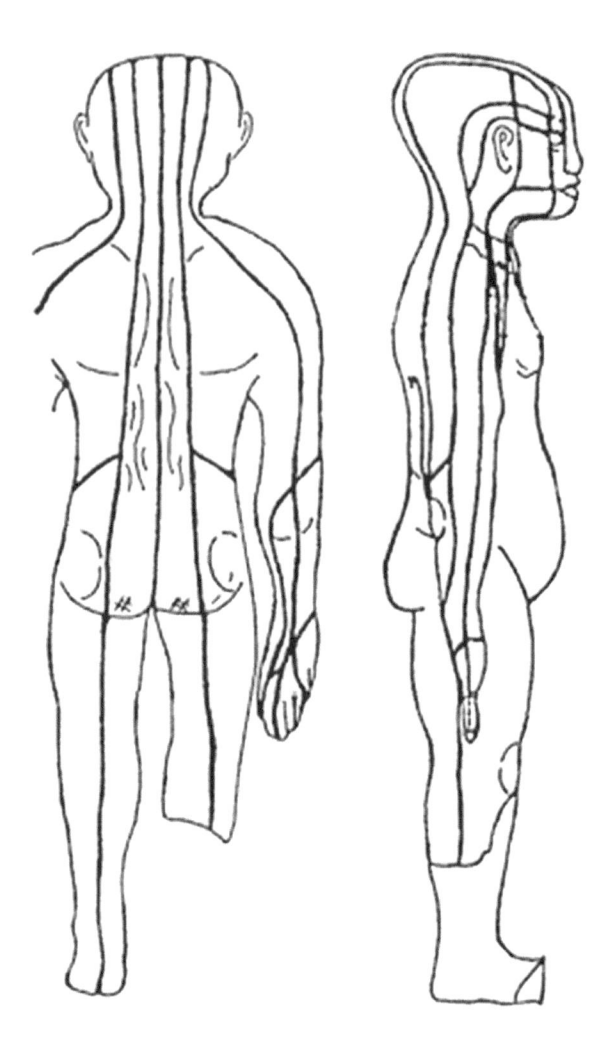

Figure 5.4 Shuangbaoshan 雙包山 figurine, showing channels (Lo 2007: 419)

These examples from tombs show that early Channel theory was evolving, focused on the rubrics of Five and Six. The version of twelve Channels with Six *yang* and Six *yin* as described in the printed texts of the received tradition can be found in *Lingshu* 10.[28] *Lingshu* is also the *locus classicus* of an interconnected circulation, with each Channel linked to one of the internal organs, *zang* or *fu* (Yang 1997: 64; Harper 1998: 87; Li 2000b: 117).

Organs

The *Huangdi neijing* listed a seemingly coherent correspondence system where each Channel was associated with an organ, organised according to Five Agents. The attempts to rationalise Five Agents theory with the six bilateral *yin* and *yang* Channels meant that some innovative compromises had to be made in the enumeration of the organs (Li 2000b: 206–7).

The importance of numerology, part 2

The 'Classic of Difficult Issues' (*Nanjing* 難經 compiled after 100 CE)[29] made the conundrum explicit, and demonstrated the contemporary quandary about reconciling the figures.

> *Zang* are only five, yet *fu* are six, why is this?
> So: the reason *fu* are six, is called *sanjiao*.
>
> *(Nanjing, 3.71)*[30]

The solution to these differing numbers of organs was to add a Triple Burner (*sanjiao* 三焦), an organ attributed a function but no physically visible form. Now there were eleven organs but twelve Channels. How was one to rationalise the five *zang* and *fu* with the Channel system now structured around Six, *yin* and *yang* channels? The *Nanjing* provided a solution again:

> There are twelve Channels, [but] five *zang* and six *fu* only [make] eleven, as for this one [other] Channel, what kind is it?
> So: this one [other] Channel, it is hand *shaoyin* and *xinzhu* [as] separate Channels.
>
> *(Nanjing, 2.52)*[31]

This was the introduction of another non-physical organ: the Heart Protector (*xinbao* 心包), with its associated Channel, Heart Master (*xinzhu* 心主). The heart which we have met representing the emperor in the body was here related to Heavenly, spiritual matters: *xinbao* possessed the more physical functions of the heart. The heart channel, or hand *shaoyin*, was thus associated with spirit (which in Chinese medical parlance refers more to psycho-emotional faculties than to a disincarnate soul) and was not needled on account of the perceived fragility of the heart. Instead, the *xinzhu* channel was used for any physical heart issues and to protect the heart from any external pathogens (*Lingshu*, 20.322).[32]

Additionally, the two organs with a 'name but no form', the Triple Burner and Heart Protector, complemented the two existing Fire organs, the Small Intestine and the Heart, each with their respective channels *taiyang* and *xinzhu*, thus completing the organ-channel relationships between the Five Agents and Six Channels.

Wang Bing made extensive use of the rubrics of Five and Six in *Suwen* 66 to consolidate the links between the Six Channels and the Five Agents.[33] To assimilate the Five Agents to the Six Channels, he divided the Fire Agent into two different types of Fire: Sovereign Fire (*junhuo* 君火) was to be the Fire of the heart, the emperor of the body.[34] Minister Fire (*xianghuo* 相火) was the Fire necessary to maintain physiological functions, and came into play after birth, represented by two organs (Heart Protector, *xinbao* 心包 and Triple Burner, *sanjiao* 三焦) (Figure 5.5).

Each Channel linked internally with the organs and externally with one of the meteorological and climatic factors, the Six *qi* in Heaven (Wind, Cold, Dry, Damp, Heat and Fire) (Sivin 1987: 80–3). The interplay between Five Agents and the Six Channels (through the shared interface of the climates) comprehensively defined the body and its interactions with the Universe, in health and in disease (Table 5.2).

> Cold, Heat, Dry, Damp, Wind, Fire, these are the *yin* and *yang* of Heaven; the three *yin* and three *yang* attend to what is above [Heaven]. Wood, Fire, Earth, Metal and Water, these are the *yin* and *yang* of Earth: birth, growth, change, gathering and storing reflect what is below [Earth].
>
> *(Suwen, 19.524)*[35]

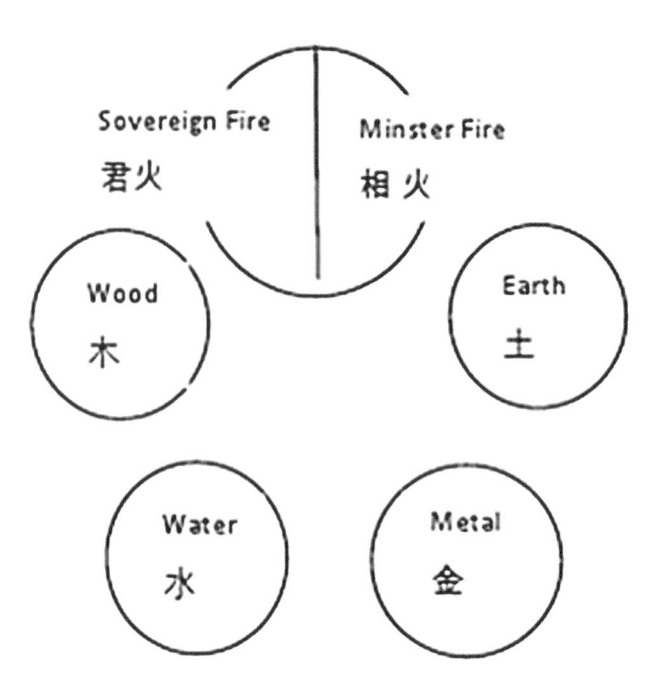

Figure 5.5 Five Agents with Sovereign Fire and Minister Fire: modern representation by the author

Table 5.2 Six Channels, Climates, Five Agents as Represented in Wang Bing's *Wuyun liuqi* Chapters of 762 CE

Channel	*Jueyin*	*Shaoyin*	*Shaoyang*	*Taiyin*	*Yangming*	*Taiyang*
Climate	Wind	Fire	Heat	Dampness	Dryness	Cold
Agent 行	Wood	Sovereign Fire	Minister Fire	Earth	Metal	Water

Wuyun Liuqi 五運六氣 (Five Circuits[36] and Six *qi*): the integration of the numbers Five and Six

Wuyun liuqi was a legacy of the intertwining of divination and medicine in China: incorporating many aspects of the astro-calendrical divinatory systems, based on Five and Six (Porkert 1974: 55–106; Ren 1982; Fang and Xu 1984; Unschuld 1985: 170–1; Liu 1990; Despeux 2001: 121–65; Unschuld 2003: 385–494). Scholars studying the Song dynasty consider this doctrine as one of the more significant intellectual changes (Goldschmidt 2009: 183). These chapters were a complex map of the movements of Heaven and Earth designed to predict seasonal imbalances: if the cosmos moved according to *Wuyun liuqi*, then the body, as a mirror of the macrocosm, would adjust itself accordingly to maintain health, by alignment with the prevailing climates (Goldschmidt 2009: 83). Weather prediction, keyed to the calendar, provided a yardstick to measure whether the seasons and associated climates would come too early, or too late, either of which could be the cause of disease and impel its aetiological processes (Hinrichs 2003: 140). These seven chapters provided a method to establish normative physiological and cosmic cycles in order to diagnose seasonally related illnesses: they offered the hope of prediction, which could lead to the control of devastating anomalies such as epidemics (Unschuld 1985: 141).

The importance of numerology, part 2

The interplay between Five and Six was embodied in the terminology used in the Wang Bing chapters: 'Five Circuits' (*wuyun* 五運) and 'Six *qi*' (*liuqi* 六氣). The Stem and Branch of a year (an ancient calendrical example of the combination of Five and Six)[37] gave the overall climate for the year which would then vary predictably according to Wang Bing's innovations (Yang 1997: 794–7). Each year was then further divided into smaller seasons: a series of five seasons, known as the five host Circuits (*zhuyun* 主運)[38] and a series of six seasons, known as the six host *qi* (*zhuqi* 主氣).[39] The Five host Circuits and the Six Host *qi* followed the expected seasonal climates occurring at set dates: Wind in spring, Cold in winter, etc. To allow for the considerable climatic variation within each season, there were also guests (*ke* 客). These guest Circuits and guest *qi* spanned the same time periods as the host Circuits and *qi* (Yang 1997: 797–800). The relationships between an unchanging host and a variable guest were interpreted through the generating and conquest cycles of Five Agents, introducing further flexibility in the potential to predict climatic variation for the year.[40]

The concept of Five Circuits is specific to the *Wuyun liuqi* chapters, and has some parallels with Five Agents, but refers to a symbolic system keyed specifically to the movements of the Heavenly bodies (sun and moon, constellations and planets), and used uniquely to determine climatic and physiological anomalies. *Suwen* 67 matched the Five Circuits to the Five Agents by repeating exactly the same relationships from *Suwen* 5 where it is stated that, for example:

> The Southern direction generates Heat, Heat generates Fire, Fire generates bitter, bitter generates the heart (*Suwen*, 2.68–70; 19.539–41).[41]

This parallel emphasised their essential similarity, and provided the normative state, since the Circuits from Heaven[42] were harmonised with those of the human body through the resonance of the Agents.[43]

Suwen 69 listed the climates expected in each year of the sexagenary cycle.[44] A particular year would be associated with one of the Five *yang* Great Circuits[45] (active or excess, *taiguo* 太過), or one of the Five *yin* Great Circuits (passive or insufficient, *buji* 不及) and a tendency towards particular pathologies (Unschuld 2003: 410).

> Search for the [season's] arrival [...] when it is not yet [expected] to arrive but arrives, this is called *taiguo* [excessive] when it is [expected] to arrive but does not arrive, this is called *buji* [insufficient].
>
> (*Suwen, 3.103–4*)[46]

If a year had an active Great Circuit, then the *qi* associated with that Circuit was expected to manifest itself violently, giving an excess of the associated climate. During Earth active Great Circuit years, for example, one would have to be vigilant against diseases of Dampness and cold. If it was a year with a passive Great Circuit, then the climate associated with that Circuit was weak. During Earth passive Great Circuit years, one would have to be vigilant against diseases of excess Wind (Ibid.: 409). The active or passive actions of a Great Circuit applied the conquest cycle of the Five Agents to predict the expected climate of the entire year. Further spatio-temporal flexibility was introduced through grouping the Branches in different ways according to their seasonal relationships or according to cycles of climatic change following the cycles of the Six *qi*.

The core Wang Bing chapters, especially *Suwen* chapters 71 and 74, provide exhaustive tables with which a physician could predict the disease profile of a particular year.[47] To do that, he would have to have specialised training to interpret the quality of an individual year and all its shorter (Five and Six) seasons. All these predictions were prefaced on innovative arrangements of the rubrics of Five and Six, as understood in their relations to the *xiangsheng* and *xiangke* cycles of generation and control.

Practical applications of *Wuyun liuqi*

The cyclical flow of Heaven (space: sun, moon, stars) and Earth (time: seasons, day/night) created and maintained the constantly changing climates, the sphere of human existence (Unschuld 2003: 393). *Wuyun liuqi* was an attempt to order what seemed to be in disorder: the occurrence and duration of different climates, following a sexagenary cycle. *Huangdi neijing* and *Wuyun liuqi* theory was based on the premise that climates were the root of disease, and humans could maintain health by adapting to their climatic environment (Yang 1997: 801–2). Climate prediction allowed humans to integrate themselves in their environment, thus following the natural laws of Heaven and Earth to maintain their health (Unschuld 2003: 394).

> Hence as for *yin-yang* and the four seasons, they are the end and beginning of the myriad beings, the root of life and death; if you oppose [them] then there is disaster and harm to life, if you follow then severe illness will not arise, this is called to obtain the way.
> (*Suwen*, 1.39–40)[48]

A map of future climatic variations was important for the prevention of disease (Unschuld 2003: 335) and as advance warning for too much of one particular climate, which could precipitate disease. Climate prediction acted as a standard against which to measure any variations (unseasonal climates), which caused disease especially in people who were unable to adapt to it sufficiently, as they were already depleted.

The overall rule for treatment was to adapt to, or counteract the effect of, the predominant climates calculated according to the specific climatic predictions of that year (*Suwen* 69, 71, 74). If these predicted climates were overwhelming, they could attack the body, so the overwhelming climatic *qi* needed to be calmed and cleared from the body. If the climatic *qi* were weak, the body's *qi* could be strengthened to oppose any pathological influences.

> For each [instance of unruly climate] pacify its *qi*, clear and calm it; then disease *qi* declines and leaves, returning to its place of origin; these are cardinal principles of treatment.
> (*Suwen*, 22.736)[49]

Suwen 74 lists pathological indicators for the whole year and seasons (Circuits and *qi*) of the year, with expected illnesses, ending with some basic rules for herbal medicine, finally exhorting the practitioner to focus on the underlying cause which, we can only assume, could be found by fully applying *Wuyun liuqi* theory.

The culmination of the iatromancy inherent in *Wuyun liuqi* theory may not have had many clinical applications until the Song dynasty (960–1279), three centuries after Wang Bing devised the theories (Despeux 2001: 129), when it came to prominence along with renewed interest in cosmology (Hanson 2008: 344). The physician Pang Anshi 龐安時 (1042–1099) author of 'Discussions on Cold Damage and general disorders' (*Shanghan zongbing lun* 傷寒總病論) incorporated *Wuyun liuqi* into Cold Damage theory (Chapters 4 and 8 in this volume) (Hinrichs 2003: 110). Both *Wuyun liuqi* and Cold Damage theories drew on the Six *qi*, and predicted the source and course of a disease based on unseasonable climates: so together they could be used to predict and treat the epidemics which plagued the dynasty (Hinrichs and Barnes 2013: 114). In 1099, the Director of Studies of the Imperial Medical Service, Liu Wenshu 劉溫舒 (late eleventh century), presented his 'Discussion of the esoterica of the circulatory phases and the seasonal *qi* as formalised in the *Suwen*' (*Suwen rushi yunqi lun'ao* 素問入式運氣論奧)[50] claiming that each season was dominated by certain climatic influences that inevitably caused certain illnesses. Liu explained *Wuyun liuqi* in detail, its application to physiology, pathology, diagnosis, treatment and the integrated use of medicinals: it was the first step in applying *Wuyun liuqi* theory to medicine as a whole (Goldschmidt 2009: 184).

Notes

1. Wang Bing included the names but no text for chapters 72 and 73, which were already missing from Quan's edition: later Song editors added these, based on texts written in Tang-Song (Unschuld 2003).
2. A note on the use of capitals: I have used capitals for any concept that is significantly different in Chinese thought, e.g. Heaven is more than just the sky and is not related to the Christian concept of life after death. I capitalise a Number when I am referring to its numerological significance. Where I refer to the specific Chinese category of one of the *wuxing*, they are capitalised as Agents, similarly for the Chinese category of organs, Viscera and Storehouses.
3. *Suwen* 5, *Yinyang yingxiang dalun* 陰陽應象大論 (Comprehensive Discourse on Phenomena Corresponding to *Yin* and *Yang*); Unschuld and Tessenow (2011: 1, 95).
4. *Suwen* 2, *Siqi diaoshen dalun* 四氣調神大論 (Comprehensive discourse on regulating the spirit according to the *qi* of the Four seasons); Unschuld and Tessenow (2011: 1, 45–57). *Suwen* 4, *Jingui zhenyan lun* 金匱真言論 (Discourse on the true words in the golden cabinet); Unschuld and Tessenow (2011: 1, 83–94). *Suwen* 5, *Yinyang yingxiang da lun* 陰陽應象大論 (Comprehensive Discourse on the Phenomena Corresponding to Yin and Yang); Unschuld and Tessenow (2011: 1, 95–126).
5. 'Master Lü's Spring and Autumn annals' (*Lüshi Chunqiu* 呂氏春秋 239 BCE), Book 13 part 2, *Yingtong* 應同 (Resonating with the Identical); Knoblock and Riegel (2000: 283).
6. The *Shujing*, or *Shangshu* 尚書, is series of historical records compiled between the Western Zhou period (1046–771 BCE) and the Warring States period (475–221 BCE).
7. Daybooks were collected miscellanies of information with information on the auspicious or inauspicious nature of a variety of daily activities, keyed to calendrical divination.
8. *Chunqiu fanlu* 38, *Wuxing dui* 五行對 (Responding to the Five Agents).
9. *Chunqiu fanlu* 42, *Wuxing zhi yi* 五行之義 (Meaning of the Five Agents).
10. Ibid.
11. *Suwen* 5; Unschuld and Tessenow (2011: 1, 106).
12. *Suwen* 8, *Linglan midian lun* 靈蘭秘典論 (Discourse on the secret treatise of the numinous orchid); Unschuld and Tessenow (2011: 1, 155–62).
13. *Suwen* 8; Unschuld and Tessenow (2011: 1, 155–6).
14. *Zuozhuan*, Duke Zhao 1.1222 and 25.1457.
15. *Huainanzi* 3, *Tianwen xun* 天文訓 (The Treatise on the patterns of Heaven); Major (1993: 55–139).
16. Ibid.: 84.
17. See Ibid.: 186–9 for a presentation of various Five Agent cycles, where Earth no longer regulates the Four seasons, but is integrated within the Five Agents, although it is still acknowledged as the starting point, thus emphasising the importance of the Centre.
18. *Suwen* 5, Unschuld and Tessenow (2011: 1, 109).
19. *Chunqiu fanlu* 38, *Wuxing dui* 五行對 (Responding to the Five Agents).
20. Compiled by Ban Gu 班固 (32–92 CE) who also compiled the second official dynastic history, the *Hanshu* 漢書 (Book of the [Former] Han) which was completed ca. 80 CE.
21. *Jing* 經 is a difficult term to translate, since it is used to indicate not only the idea of a vertical warp thread on a loom. *Jing*, combined with *mai* 脈, describes a vessel (sometimes containing blood) along which something flows, in a connected network, like underground streams in the Earth. *Mai* is also the term for 'pulsation', a further link to blood and its impulse from the heart (Harper 1998: 82–3).
22. Mawangdui tomb contained a variety of medical and hygiene texts, including the 'Cauterisation canon of the eleven vessels of the foot and forearm' (*Zubi shiyi mai jiujing* 足臂十一脈灸經) and the 'Cauterisation canon of the eleven *yin* and *yang* vessels' (*Yinyang shiyi mai jiujing* 陰陽十一脈灸經安定) (Ibid.: 22–25).

23 Zhangjiashan tomb had two medical and hygiene manuscripts, the *Vessel Book* (*Maishu* 脈書), listing eleven Channels with related ailments (very similar to the Mawangdui manuscripts); and the *Pulling Book* (*Yinshu* 引書), a more developed version of the hygiene texts from Mawangdui. The prevention of disease was focused on exercises to ward off climatic factors (Wenwu 1989; Engelhardt 2000: 88–89; Li 2006b: 281–295). See Lo (2007) for an analysis of the Mawangdui and Zhangjiashan texts.
24 Channels were vertical and parallel, without intercommunication; health depended on *qi* flowing in a downward direction, so a healthy person had a cold head and warm feet. If an excess of *qi* caused upward movement, cauterisation and exercises were used to move *qi* downwards (Harper 1998: 81, 86).
25 *Yang* Channels, the *qi* of Heaven, caused less fatal ailments. *Yin* Channels, the *qi* of Earth, were called the 'vessels of death', causing fatal disease (Ibid.: 79, 88).
26 See He and Lo (1996) and Lo (2002: 123–5) for an analysis of the Shuangbaoshan figurine.
27 The number 109 is probably not numerically significant, as there may have been many more dots, now erased; or dots of other colours, also no longer visible.
28 *Lingshu*, 5.83–105, *Lingshu* 10, *Jingmai* 經脈 (Channels and Vessels); Unschuld (2016: 175–208).
29 The *Nanjing* comments and discusses the Five Agent systematic correspondence system aiming to reach coherence and consistency.
30 *Nanjing* 38, *Zangfu dushu* 藏府度數 (The Principles of *Zang* and *Fu* Organs); Unschuld (1985: 395).
31 *Nanjing* 25, *Jingluo dashu* 經絡大數 (The Channels and Vessels); (Ibid.: 310).
32 *Lingshu* 71, *Xieke* 邪客 (Evil Guests); (Ibid.: 639–40).
33 *Suwen* 66, *Tianyuanji dalun* 天元紀大論 (Comprehensive Discourse on Arrangements of the Principal [*Qi*] of Heaven); Unschuld and Tessenow (2011: 2, 173–88).
34 *Suwen*, 19.526; (Ibid.: 2, 184).
35 *Suwen* 66; (Ibid.: 2, 181).
36 I have translated *yun* 運 as 'Circuit' to emphasise its regular circulation; it has been translated elsewhere as 'Circulatory Phase' (Despeux 2001) or 'Period' (Unschuld 2003).
37 The Ten Heavenly Stems are Five *yang* and Five *yin*; the Twelve Earthly Branches are Six *yang* and Six *yin*.
38 *Suwen* 66; *Suwen* 67, *Wuyun xing dalun* 五運行大論 (Comprehensive Discourse on the Progression of the Five Circuits); Unschuld and Tessenow (2011: 2, 189–212); *Suwen* 70, *Wuchang zheng dalun* 五常政大論 (Comprehensive Discourse on the Five Regular Policies); (Ibid.: 2, 285–356).
39 *Suwen* 66; *Suwen* 68, *Liuweizhi dalun* 六微旨大論 (Comprehensive Discourse on the Subtle Significance of the Six [*Qi*]); (Ibid.: 2, 213–39).
40 These smaller cycles were like fractals: the larger cycles were mirrored in smaller cycles, so the whole calculation of *Wuyun liuqi* for a particular time period could illustrate very complex patterns of climates.
41 *Suwen* 5; (Ibid.: 1, 107–8). *Suwen* 67; (Ibid.: 2, 205–6).
42 Circuits, *yun* were described in *Suwen* 67 (*Suwen*, 19.531; Ibid.: 2, 192–3) as coloured paths in the sky, with reference to a no longer extant text (Unschuld 2003: 406–7). Circuits were keyed to the lunar mansions (*xiu* 宿) fixed constellations which were used as reference points by ancient Chinese astronomers.
43 Both *yun* and *xing* mean 'to move' and 'to circulate'; however, *yun* also means 'luck, fate or fortune' possibly alluding to circulations in the heavens, whose movements determine the fate of those on Earth.
44 *Suwen* 69, *Qijiao bian dalun* 氣交變大論 (Comprehensive Discourse on the Changes of *Qi* Interaction); Unschuld and Tessenow (2011: 2, 241–84).
45 A Great Cicuit, *dayun* 大運 related to the Circuit of a complete year.
46 *Suwen* 9; (Ibid.: 1, 172).
47 *Suwen* 71, *Liuyuan zhengji dalun* 六元正紀大論 (Comprehensive Discourse on the Policies and Arrangements of the Six Original *Qi*); (Ibid.: 2, 357–533). *Suwen* 74, *Zhizhenyao dalun* 至真要大論 (Comprehensive Discourse on the Essentials of the Perfect Truth); Unschuld and Tessenow 2011: 2, 535–642.
48 *Suwen* 2; (Ibid.: 1, 56).
49 *Suwen* 74; (Ibid.: 2, 587).
50 See Hanson (2008) for more on Liu's book.

Bibliography

Pre-modern sources

Baihutong shuzheng 白虎通疏證 (Comprehensive Discussions in the White Tiger Hall – Commentaries and Verification), 79, Ban Gu 班固 et al. in Chen Li 陳立 (ed.) (1994), *Xinbian zhuzi jicheng* 新編諸子集成 (New Collection of Ancient Philosophers), Beijing: Zhonghua shuju.

Chunqiu fanlu yizheng 春秋繁露義證 (Luxuriant Dew of the Spring and Autumn Annals –Evidence and Analysis), Dong Zhongshu 董仲舒, in Su Yu 蘇輿 (ed.) (1992), *Xinbian zhuzi jicheng* 新編諸子集成 (New Collection of Ancient Philosophers), Beijing: Zhonghua shuju.

Chunqiu zuozhuan zhu 春秋左傳注 (Zuo Commentary with Annotations), Yang Bojun 楊伯峻 (ed.) (1990), Beijing: Zhonghua shuju.

Huainanzi jishi 淮南子集釋 (Masters from Huainan – Collected Explanations) 139 BCE, Liu An 劉安, in He Ning 何寧 (ed.) (1998), *Xinbian zhuzi jicheng* 新編諸子集成 (New Collection of Ancient Philosophers), Beijing: Zhonghua shuju.

Huangdi bashier nanjing 黃帝八十一難經 (Yellow Emperor's Classic of Eighty-One Difficulties), 1st century CE (2007), Beijing: Xueyuan chubanshe.

Huangdi neijing 黃帝內經 (Yellow Emperor's Inner Canon), 762, Wang Bing 王冰 (ed.) (1994), Beijing: Zhongyi guji chubanshe.

Lüshi chunqiu jishi 呂氏春秋集釋 (Master Lü's Spring and Autumn Annals with Collected Explanations) 239 BCE, Lü Buwei 呂不韋, in Xu Weiyu 許維遹 and Liang Yunhua 梁雲華 (eds) (2009), *Xinbian zhuzi jicheng* 新編諸子集成 (New Collection of Ancient Philosophers), Beijing: Zhonghua shuju.

Shanghan zabing lun 傷寒雜病論 (Treatise on Cold Damage and Miscellaneous Diseases), Zhang Zhongjing 張仲景 (150–219) (2014), Beijing: Kexue jizhu chubanshe.

Shangshu zhengyi 尚書正義 (Book of Documents – Corrected Interpretation), 653, Kong Yingda 孔穎達 (574–648) et al., in Ma Xinmin 馬辛民 (ed.) (2000), *Shisanjing zhushu* 十三經注疏 (Thirteen Confucian Classics with Notes and Commentaries), Beijing: Beijing daxue chubanshe.

Zhuangzi jijie 莊子集解 (Master Zhuang – Collected Explanations), Zhuang Zhou 莊周 (ca. 369–286 BCE) in Wang Xianqian 王先謙 (ed.) (2001), Taipei: Shijie shuju.

Modern sources

Despeux, C. (2001) 'The system of the five circulatory phases and the six seasonal influences (Wuyun Liuqi), a source of innovation in medicine under the Song (960–1279)', in E. Hsu (ed.) *Innovation in Chinese Medicine*, Cambridge: Cambridge University Press, pp. 121–166.

Engelhardt, U. (2000) 'Longevity techniques and Chinese medicine', in L. Kohn (ed.) *Daoism Handbook*, Leiden; Boston, MA: Brill, pp. 74–108.

Fang Yaozhong 方药中 and Xu Jiasong 许家松 (1984) *Huangdi neijing suwen yunqi qipian jiangjie* 黃帝内经素问运气七篇讲解 (Explanations of the Seven Great Treatises on Circuits and *qi* in the Yellow Emperor's Inner Canon), Beijing: Renmin weisheng chubanshe.

Goldschmidt, A. (2009) *The Evolution of Chinese Medicine: Song Dynasty, 960–1200*, London; New York: Routledge.

Graham, A.C. (1986) *Yin-Yang and the Nature of Correlative Thinking*, IEAP Occasional Paper and Monograph Series 6, Singapore.

Hanson, M. (2008) 'Hand mnemonics in classical Chinese medicine: texts, earliest images, and arts of memory', *Asia Major*, 21.1: 325–57.

Harper, D. (1998) *Early Chinese Medical Literature: the Mawangdui medical manuscripts*, London; New York: Kegan Paul International.

Harper, D. and Kalinowski, M. (eds) (2017) *Books of Fate and Popular Culture in Early China: the daybook manuscripts of the Warring States, Qin, and Han*, Leiden; Boston, MA: Brill.

He Zhiguo 何志國, and Lo, V. (1996) 'The channels: a preliminary examination of a lacquered figurine from the Western Han period', *Early China*, 21: 81–123.

Henderson, J.B. (1984) *The Development and Decline of Chinese Cosmology*, New York: Columbia University Press.

Hinrichs, TJ (2003) *The Medical Transforming of Governance and Southern Customs in Song Dynasty China (960–1279 C.E.)*, Ph.D. Dissertation, Harvard University.

Hinrichs, TJ and Barnes, L. (eds) (2013) *Chinese Medicine and Healing: an illustrated history*, Cambridge, MA: Harvard University Press.

Huang Longxiang 黄龙祥 (2017) 'Laoguanshan chutu Xi Han zhenjiu muren kao' 老官山出土西汉针灸木人考 (Study of the excavated wooden acupuncture figurine of the Western Han dynasty from Laoguanshan), *Zhonghua yishi zazhi* 中华医史杂志, 47.3: 131–45.

Jiangling Zhangjiashan Hanjian collation group 江陵張家山漢簡整理小組 (1989) 'Jiangling Zhangjiashan Hanjian zhengli xiaozu 江陵張家山漢簡《脈書》釋文' (Notes on the Han bamboo slips 'Mai-shu' from Jiangling county Zhangjiashan), *Wenwu* 文物, 7: 72–4.

Kalinowski, M. (1983) 'Les instruments astro-calendriques ses Han et la méthode Liu Ren', *Bulletin de l'Ecole Française d'Extrême-Orient*, 72: 304–419.

Keegan, D.J. (1988) *The 'Huang-Ti Nei-Ching': the structure of the compilation; the significance of the structure*, Berkeley: University of California.

Knoblock, J. and Riegel, J. (2000) *The Annals of Lu Buwei: a complete translation and study*, Stanford, CA: Stanford University Press.

Li Jianmin 李建民 (2000a) *Fangshu yixue lishi* 方術醫學歷史 (History of Divination and Medicine), Taipei: Nantian shuju.

——— (2000b) *Sisheng zhi yu: Zhou-Qin-Han maixue zhi yuanliu* 死生之域周秦漢脈學之源流 (The Boundary between Life and Death: Sources of Vessel Theory in the Zhou, Qin and Han Dynasties), Taipei: Academia Sinica.

Li Ling 李零 (2006a) *Zhongguo fangshu xukao* 中国方术续考 (A Study of the Occult Arts of China – Continued), Beijing: Zhonghua shuju.

——— (2006b) *Zhongguo fangshu zhengkao* 中國方術正考 (A Study of the Occult Arts of China), Beijing: Zhonghua shuju.

Liu Lexian 劉樂賢 (1994) *Shuihudi Qin jian rishu yanjiu* 睡虎地秦簡日書研究 (Shuihudi Qin Bamboo Daybooks Research), Taipei: Wenjin chubanshe.

Lo, V. (2002) 'Spirit of stone: technical considerations in the treatment of the jade body', *Bulletin of the School of Oriental and African Studies*, 65.1: 99–128.

——— (2007) 'Imagining practice: sense and sensuality in early Chinese medical illustrations', in F. Bray, V. Dorofeeva-Lichtmann and G. Metailie (eds) *Graphics and Text in the Production of Technical Knowledge*, Leiden: Brill, pp. 383–424.

Loewe, M. (2005) *Faith, Myth, and Reason in Han China*, Indianapolis: Hackett Publishing.

Major, J.S. (1993) *Heaven and Earth in Early Han Thought: chapters three, four and five of the Huainanzi*, Albany: State University of New York Press.

Porkert, M. (1974) *The Theoretical Foundations of Chinese Medicine: systems of correspondence*, Cambridge, MA: MIT Press.

Ren Yingqiu 任应秋 (1982) *Yunqi xueshuo* 运气学说 (The Theory of Circuits and Qi), Shanghai: Shanghai kexue jishu chubanshe.

Sivin, N. (1987) *Traditional Medicine in Contemporary China*, Ann Arbor: University of Michigan.

Steavu, D. (2017) 'The allegorical cosmos: the *Shi* 式 board in medieval Taoist and Buddhist sources', in M. Lackner (ed.) *Coping with the Future: theories and practices of divination in East Asia*, Leiden: Brill, pp. 196–232.

Unschuld, P.U. (1985) *Medicine in China: a history of ideas*, Berkeley; Los Angeles: University of California Press.

——— (2003) *Huang Di Nei Jing Su Wen Nature, Knowledge, Imagery in an Ancient Chinese Medical Text: with an appendix: the doctrine of the five periods and six Qi in the Huang Di Nei Jing Su Wen*, Berkeley: University of California Press.

——— (2016) *Huang Di Nei Jing Ling Shu The Ancient Classic on Needle Therapy: the complete Chinese text with an annotated English translation*, Oakland: University of California Press.

Unschuld, P.U. and Tessenow, H. (2011) *Huang Di Nei Jing Su Wen: annotated translation of Huang Di's inner classic – basic questions*, Berkeley: University of California Press.

Wang, Aihe (2000) *Cosmology and Political Culture in Early China*, Cambridge: Cambridge University Press.

Yang Li 杨力 (1997) *Zhouyi yu Zhongyi xue* 周易与中医学 (Book of Changes and Chinese Medicine Studies), Beijing: Beijing kexue jishu chubanshe.

Zhou Qi 周琦 (2017) 'The lacquered figurine from the Tianhui Han tomb', conference paper delivered at *Looms of Life: weaving, medicine and knowledge production in early China*, London: University College London.

6
THERAPEUTIC EXERCISE IN THE MEDICAL PRACTICE OF SUI CHINA (581–618 CE)

Dolly Yang

In an age when there are increasing concerns about reliance on drug-based therapies, and self-care is the public health advice and economics of choice, it is fascinating to realise that government initiatives to promote therapeutic exercise already existed in seventh-century China. This chapter investigates the seventh-century institutional and religious dimensions of a tradition of exercise known as *daoyin* 導引, as a contribution to the history of state-sponsored medical care. The term *daoyin*, its interpretation and its significance in practice have changed in different contexts. By the Sui period (581–618), a wide range of social groups, including the aristocratic elite, physicians, Daoist practitioners and Buddhist monks, were actively engaged in practising, developing and transmitting *daoyin* therapeutic exercise. In comparison with previous administrations, a disproportionately large number of *daoyin* experts were employed at the Imperial Medical Academy during the reign of the second Sui emperor, Yangdi 隋煬帝 (r. 604–618), and it is this phenomenon that is at the heart of my investigation.

Daoyin literally means 'guiding and pulling' (Kohn 2008: 11–12). A definition of *daoyin* can be found in the seventh-century medical text *Zhubing yuanhou lun* 諸病源候論 (Treatise on the Origins and Symptoms of Medical Disorders), which describes the action of 'pulling out the deviant *qi* hidden in the ageing body and, as a result of this pulling, the deviant *qi* is drawn out, giving rise to the name *daoyin*' (27: 766). As a therapeutic exercise, *daoyin* works on the *qi*, the vital force in all living beings (Chapter 2 in this volume) to create a harmonious condition whereby the body of the practitioner is aligned and adjusted to its environment, the seasons, the climate and the larger workings of the cosmos. *Daoyin* involved stretching and contracting the body and limbs, self-massage and controlled breathing and focused intention to 'pull' deviant *qi* out of the body,[1] thereby restoring or enhancing harmony and good health.

Daoyin exercises, with their ancient roots in the tradition of spirit mediums, known as *wu* 巫 (Despeux 1989: 237–40) became prevalent among the elite of the Warring States, Qin and Han dynasties (475 BCE–220 CE) (Harper 1998; Lo 1998, 2007, 2014). A silk manuscript illustrating 44 *daoyin*-style exercises was excavated from a Han tomb, tomb 3 (closed in 168 BCE) at the Mawangdui burial site. Some of the figures were depicted with accompanying captions such as 'pulling deafness', 'pulling knee pain' and 'pulling the warm ailment, evidence that specific *daoyin*-style exercises were already being recommended for treating particular conditions in the second century BCE (Harper 1998: 310–6; Chapter 3 in this volume). A text excavated from another Han tomb, tomb 247 (closed in 186 BCE) at Zhangjiashan 張家山 in Hubei 湖北 province, is unusual in having the title *Yinshu* 引書 (The Book of Pulling) on the back of the first of the bamboo slips. This excavated manuscript is the earliest known systematised description of therapeutic exercise in China, and possibly anywhere in the world, offering a comprehensive step-by-step guide to bodily movement as part of a seasonal health regime for the educated Han social elite. There follows a couple of examples of the kind of exercise found in *Yinshu*:

> Strip No. 11: Massage the lower leg with the foot, thirty times on the *Yin* aspect and thirty times on the *Yang* aspect, alternating. Extend the two feet out straight thirty times. This is called Pulling the *Yang* Muscles and Sinews.
>
> Strip No. 15: Wild Duck Bathing. Interlock the hands behind the back and shake the head. Swivelling and Extending. With the hands interlocked, raise the hands, shaking them behind.
>
> *(Lo 2014: 18, 20)*

The *daoyin* texts of this period form part of the literature of the masters of formulæ (*fangshi* 方士), who were connected with the cult of immortality and transcendence (*xian* 仙) and were specialists in natural philosophy and occult knowledge.[2] Their aristocratic patrons, who had the time and the means to engage in the pursuit of longevity and immortality practices, were 'participants in the culture of secrecy and privileged knowledge' that perpetuated the tradition (Harper 1998: 57).

During the Six Dynasties (220–581 CE), various groups of people took up *daoyin* exercises, as part of the self-cultivation practices known as *yangsheng* 養生. The term *yangsheng* has meant different things in different historical contexts to different people. However, by the time Zhang Zhan 張湛 (fl. 350–400 CE), an aristocrat of northern descent, compiled a text entitled *Yangsheng yaoji* 養生要集 (Compendium of Essentials for Nourishing Life) citing many bodily practices from various sources, the association of *yangsheng* with bodily self-cultivation practices would have been well established. These practices included breathing exercises, *daoyin*, diet, meditation and visualisation; sexual hygiene and other preventative health practices (Chapter 49 in this volume). Many of the *yangsheng* practitioners were members of genteel families with access to medical and *yangsheng* knowledge, who practised *daoyin* for their own physical health and well-being, as well as passing on such knowledge to family members. Daoist religious observances were intertwined with *yangsheng* practices, including *daoyin*, and Buddhists also adopted and adapted indigenous *yangsheng* practices as part of their own spiritual cultivation (Yang 2018). When the Sui government came into power, *daoyin* exercises, already widely popular among circles of literate elites and religious sects, became an important component of state-sponsored medicine.

The Sui dynasty, though remarkably short-lived, was one of the most vigorous dynasties in Chinese history. By 589, Wendi 文帝 (541–604 CE), the founder of the Sui, had conquered much of what we now know as China. This unification under a single authority followed nearly four centuries of political fragmentation after the fall of the Han dynasty in 220 CE. The two emperors of the Sui, Wendi (r. 581–604 CE) and Yangdi 煬帝 (r. 604–618 CE), initiated major construction projects and introduced comprehensive institutional reforms whose purpose was to unify China under a single centralised government. A new capital, *Daxingcheng* 大興城 (modern-day Xian 西安), was built, the eastern capital Luoyang 洛陽 was renovated and a series of canals were constructed linking north and south China.

An important manifestation of these Sui reforms, aimed at centralisation and unification, was the establishment of the Imperial Medical Office (*Taiyishu* 太醫署) and with it a state-sponsored medical education system (Chapters 7 and 8 in this volume). Details of the *Taiyishu* and its antecedents are recorded mainly in the Book of Sui (*Suishu* 隋書, 28: 775), the official history of the Sui dynasty, and the Six Statutes of Tang (*Tang liudian* 唐六典, 14: 410), the state records of the Tang dynasty (618–907 CE).

State medical education had first been established in 443 CE during the Liu Song dynasty (420–479 CE) in the south. During the Northern Wei dynasty (386–534 CE), there were two, relatively low-ranking, teaching posts – a Medical Erudite (*Taiyi boshi* 太醫博士) and a Medical Teaching Assistant (*Taiyi zhujiao* 太醫助教). The growing centralisation of the Sui dynasty can be seen in the evolution of state-sponsored medical education, which was hugely expanded. During Yangdi's reign, the staffing level at the medical education institution had risen to include two Medical Erudites with two Teaching Assistants, twenty *Anmo* Erudites (*Anmo boshi* 按摩博士) with 120 *Anmo* Masters (*Anmo shi* 按摩師) and an Erudite of Incantation and Talismans (*Zhojin boshi* 咒禁博士) (Table 6.1) (Yang 2018: 203–18). It is evident that staff specialising in *anmo* constituted by far the largest grouping in the medical education institution at Yangdi's court. According to the *Tang liudian* (14: 411), Erudites and Masters of *anmo* taught students 'methods of breathing and *daoyin* exercises in order to get rid of eight types of illness: Diseases caused by wind; cold; summer heat; damp; hunger; over-eating; over-exertion and over-indulgence'. Various texts suggest that it was the second Sui emperor, Yangdi, who was responsible for the employment of a large number of *daoyin* experts with state funds (Ibid.: 222–34).

The term *anmo*, often arguably translated as massage, was used interchangeably with *daoyin* during the Sui and Tang periods. For example, two sets of unmistakably *daoyin* exercises, entitled 'The Indian Massage' (*Tianzhuguo anmo* 天竺國按摩) and 'The Massage Technique of Laozi' (*Laozi anmo fa* 老子按摩法), are described as *anmo* in the seventh-century medical text 'Essential Formulas for Urgent Needs Worth a Thousand in Gold' (*Beiji Qianjin yaofang* 備急千金要方, completed in 652 CE by Sun Simiao (581–682 CE)).

In 583 CE, as part of the process of unification and centralisation, Wendi decreed that private collectors who lent their books for copying should be recompensed. As a result, many texts from disparate sources were brought together, augmenting the collection in the Imperial Library and benefitting the work of further compilation (*Suishu* 49: 1297). The Catalogue of the Imperial Library of the Sui (*Suishu jingji zhi* 隋書經籍志) testifies to the fact that the number of medical texts had increased significantly since the Han period (202 BCE–220 CE). While the catalogue of the Imperial Library of the Han (*Hanshu yiwenzhi* 漢書藝文志)

Table 6.1 Medical teaching staff at the Imperial Medical Academy during the reigns of Wendi and Yangdi

Job Titles	No. of Medical Teaching Staff during the Reign of Wendi (581–604)	No. of Medical Teaching Staff during the Reign of Yangdi (604–618)
Medical Erudite	2	2
Medical Teaching Assistant	2	0
Anmo Erudite	2	20
Anmo Master	0	120
Erudite of Incantation and Talismans	2	1

Table 6.2 Yangsheng texts in the medical section of *Suishu jingjizhu* 隋書經籍志 (The Catalogue of the Imperial Library of the Sui)

1.	*Daoyin tu* 導引圖 (*Daoyin* Chart)
2.	*Diwang yangsheng yaofang* 帝王養生要方 (Essential Methods of Nourishing Life for the Emperors) by Xiao Ji 蕭吉
3.	*Longshu pusa yangxing fang* 龍樹菩薩養性方 (Methods of Cultivating Innate Nature by Longshu Bodhisattva)
4.	*Pengzu yangxing jing* 彭祖養性經 (Scripture of Pengzu's Cultivation of Innate Nature)
5.	*Yangshen jing* 養身經 (Scripture of Nourishing the Body)
6.	*Yangsheng shu* 養生術 (Techniques for Nourishing Life) by Zhai Ping 翟平
7.	*Yangsheng yaoji* 養生要集 (Essential Compendium for Nourishing Life) by Zhang Zhan 張湛
8.	*Yangsheng yaoshu* 養生要術 (Essential Techniques for Nourishing Life)
9.	*Yangsheng zhu* 養生注 (Commentary of Nourishing Life)
10.	*Yangsheng zhuan* 養生傳 (Biographies of Nourishing Life)
11.	*Yinqi tu* 引氣圖 (Chart of Pulling the Qi)

records thirty-six medical texts in 868 *juan*, the Sui catalogue lists 256 medical texts in 4,510 *juan* (*Hanshu* 30: 1176–81; *Suishu* 34: 1040–50). Of these 256, at least eleven (Table 6.2), though no longer extant, refer directly to *yangsheng* self-cultivation practices, as indicated by their titles. These lost works are likely to have contained writings and illustrations relating to *daoyin*. For example, there are two charts depicting *daoyin* exercises, and a text assigned to Pengzu 彭祖, a legendary figure who was reputed to have lived for 800 years as a result of his practice of *daoyin* and sexual techniques (*Liexian zhuan*, trans. Kaltenmark 1953: 82–4).

One medical text in the *Suishu jingjizhi*, the aforementioned *Zhubing yuanhou lun*, has a rich collection of *daoyin* exercises. We know this because the text in its entirety has survived to the present day. The text gives descriptions of 1,739 diseases detailing under seventy-one headings their aetiology and symptoms.[3] According to the preface of the Northern Song edition, it was produced by the court physician (*taiyi* 太醫) Chao Yuanfang 巢元方 (fl. c. 605–616 CE) and his colleagues at the decree of the second Sui emperor, Yangdi (Ding 1991, 19–20).[4] Focusing mainly on the description of diseases, the text gives very little information about drugs or other therapies, but does include quotations from *yangsheng* self-cultivation texts, and from two in particular – 'Recipes for Nourishing Life' (*Yangsheng fang* 養生方) and 'Recipes for Nourishing Life and the *Daoyin* Method' (*Yangsheng fang daoyin fa* 養生方導引法). The former offers advice on lifestyle such as diet, sleep and personal hygiene. The latter gives *daoyin* instructions for the cure of various diseases. Almost 200 different *daoyin* exercises are recommended for treating diseases in this state-sponsored medical text (Ding 1993: 2).

So, not only did the department of *daoyin* exercises form the biggest department in the newly established state-sponsored medical school, but the work of that department was also supported by a comprehensive, newly compiled, medical text. The unprecedented and elevated status given to *daoyin* in the Imperial Medical Academy at the Sui court was effectively due to the endorsement and patronage of the second Sui emperor, Yangdi (Yang 2018).

While we can confidently assert that Yangdi's vision of medical care emphasised the role of *daoyin*, it is important to note that Wendi was the one who first established the three specialised medical departments. This suggests that the two Sui emperors would have been aware of the importance of *daoyin* in medical and *yangsheng* practices in both secular and religious spheres. On the secular level, the aristocratic elite, who were important carriers of cultural knowledge, appear to have engaged in self-cultivation practices for their own various needs. Some, spurning official employment, engaged themselves in bodily cultivation as part of a 'hermetic' lifestyle; others, who became officials, used *daoyin* exercises in order to stay healthy and to be able to enjoy worldly power and pleasure (Kohn 2008, 62–8).

The two emperors were keen supporters of Daoism and Buddhism. Upon founding the Sui dynasty, Wendi established the Daxingshan Monastery 大興善寺 as the religious centre for Buddhism, and the Xuandu Abbey 玄都觀 for Daoism, in the new capital Daxingcheng. In the Eastern Capital of Luoyang, Yangdi established four 'Places of Dao' (*daochang* 道場), two of which were Buddhist Monasteries. The other two were Daoist Abbeys (Xiong 2006). The religious observances of both Buddhism and Daoism were intertwined with many *yangsheng* practices, including *daoyin* exercises (Kohn 2006, 2008; Salguero 2012). The religious receptivity of the two emperors would have brought them into contact with various self-cultivation practices intrinsic to these religious traditions. Indeed, in one of the three teaching departments in the Imperial Medical Academy, which was devoted to incantation and talismans, both Buddhist and Daoist rituals were taught for medical purposes (*Tang liudian* 14: 411).

Four groups of people – aristocrats, Daoists, Buddhists and physicians – were conversant with *daoyin* exercises, and each group played an important role in bringing *daoyin* to the Sui court.

Aristocrats

Between 304 and 316 CE, a series of uprisings instigated by northern nomadic tribes, namely Xiongnu 匈奴, Xianbei 鮮卑, Jie 羯, Di 氐 and Qiang 羌, captured the Western Jin (265–316 CE) capital of Luoyang, and established a series of states and kingdoms in the north known as the Sixteen Kingdoms (316–387 CE). The ensuing chaos and devastation prompted mass migration (Lewis 2009). Those who had the means to escape, many of whom were from prominent families, fled south, bringing with them the refinement of the northern cultures. These aristocratic families were important transmitters of cultural knowledge. The southern shift also saw a decentralisation, with wealthy families no longer entirely dependent on the court for their livelihoods or authority. They maintained their power and position by immersing themselves in classical learning, which helped them not only to refine the skills necessary to hold high office but also to establish new authoritative traditions independently (Ibid.).

Some *yangsheng* self-cultivation texts were written and read by aristocrats who had settled in the Jiangnan 江南 region, south of the Yangtze River. For example, Zhang Zhan 張湛 (early fourth century), a southern aristocrat and official, produced a compilation of *yangsheng* practices from a number of *yangsheng* texts in 'Essential Compendium on Nourishing Life' (*Yangsheng yaoji* 養生要集). These include a set of *daoyin* exercises to be performed in the morning and evening:

> 'Before rising at dawn, first clack your teeth together twice seven times, close your eyes and clench your fists, rinse the mouth completely with saliva, and swallow three times. Try to retain the *qi* by holding your breath for as long as you can. Then, exhale steadily and softly. Do this three times. Then sit up, and do wild-wolf and owl neck-twists. Rock to the right and the left while holding your breath for as long as possible. Repeat this three times. Then rise and get down from your bed. Clench your fists while holding your breath and stamp your feet three times. Then, raise one arm up and the other one down while holding your breath for as long as possible. Do this three times. Then, lock your hands together behind your neck and twist to the left and right while holding your breath. Repeat this three times. Then, stretch out both legs; lock your hands together and move them forwards and backwards as much as you can. Do this three times. You should always do this at dawn and dusk. The more you can do it, the better.'[5]

Aristocrats also passed on *yangsheng* knowledge within families. One such was Yan Zhitui 顏之推 (531–591 CE), a scholar and government official from a prominent southern family who was interested in self-cultivation practices. His 'Family Instructions of the Yan Clan' (*Yanshi jiaxun* 顏氏家訓), written for his sons, has a whole chapter entitled '*yangsheng*', within which he refers to the effectiveness of a specific *daoyin* exercise, i.e. 'clacking the teeth' 300 times as a cure for toothache (*yangsheng* 5.15: 327). For this more secular group of aristocrats, *daoyin* exercises were valued for their health benefits and were practised to ensure health and well-being rather than to attain immortality and transcendence.

Daoists

Daoism, in the form of political, institutional and religious organisations, came into being during the second century CE with the collapse of the Han dynasty in 220 CE. Several religious movements, such as the Way of the Celestial Masters (*Tianshi dao* 天師道) and the Way of the Great Peace (*Taiping dao* 太平道), emerged at that time, later acquiring a more coherent collective identity to be known as Daoism (Raz 2012). Key Daoist sects adopted various *xian* techniques and *yangsheng* practices. The influential southern school of Highest Clarity (*Shangqing* 上清) tempered the religious beliefs and self-cultivation practices of northern émigrés, such as the Celestial Masters, producing a syncretic fusion with local ecstatic traditions, alchemical practices of longevity and immortality, and Buddhism (Robinet 2008). This school rose to prominence in the fifth century under the guidance of Tao Hongjing 陶弘景 (456–546 CE), an eminent scholar with a good knowledge of *daoyin* exercises, who is said to have practised them himself (*Liangshu* 51: 742; *Nanshi* 76: 1897). Yangdi was connected with the lineage of the Highest Clarity school, in that he became a nominal disciple of Wang Yuanzhi 王遠知 (528–635 CE), who was himself a disciple of Tao Hongjing (*Jiutangshu* 192: 5125). Renowned for his occultist practices, Wang Yuanzhi was an influential Daoist master, who passed on the self-cultivation techniques he had learned from Tao Hongjing to Pan Shizheng 潘師正 (585–682 CE), the eleventh Highest Clarity patriarch, who, in turn, taught them to Sima Chengzhen 司馬承禎 (647–735 CE). Sima Chengzhen was the author of 'Treatise on the Essential Meaning of Absorbing Qi' (*Fuqi jingyi lun* 服氣精義論), a key text from the Tang period on physical self-cultivation, which shows that *daoyin* exercises were practised and adapted from the very beginning of the Highest Clarity school to well into the Sui and Tang periods (Engelhardt 1987, 1989; Kohn 1987).

The *Shangqing* practitioners performed their daily regimes in order to maintain health, cure illness and prepare them for the ultimate practice, which was the recitation of the Great Cavern Scripture (Stanley-Baker 2013: 113–7). An example of their exercises can be found in 'Declaration of the Perfected' (*Zhengao* 真誥, DZ1016), a fifth-century Daoist text compiled by Tao Hongjing:

> When you get up in the morning, sit up straight, breathing evenly. First, lock your hands together and place them behind your neck. Next, lift your face and look up, pressing your hands firmly against your neck while raising your head upwards. Do this three or four times and then stop. This harmonises essence, increases blood flow, and prevents wind *qi* from entering. Practised over a long period of time, it will keep you free from disease and death.
>
> (*Zhengao* 9)

Buddhists

Buddhism was first introduced into China from India at the beginning of the Eastern Han dynasty (25–220 CE) by Buddhist monks and traders along the Silk Road. Many Buddhist texts with distinctively new ideas and practices were translated into Chinese and developed to suit Chinese practitioners. Rather than a faithful transmission of original practices, these translations were adapted to pre-existing local religious cultures. Such new ideas and practices also influenced medical knowledge during the Sui and Tang periods and were absorbed into the *daoyin* repertoire. Among Buddhist meditation techniques, the *Ānāpānasati*, the mindfulness of breathing technique (Zacchetti 2010), required practitioners to concentrate on inhalation and exhalation by counting the breaths, a technique resembling breathing exercises familiar to practitioners of *yangsheng*.[6]

Both Wendi and Yangdi had a close association with Buddhism. Wendi spent his formative years in the care of a Buddhist nun, and later proclaimed himself to be a Chakravartin (wheel-turning monarch), a benevolent ruler in both sacred and secular realms. During Yangdi's reign, more than 2,000 monks, specialising in occult arts, were housed in the monastery in the Eastern Capital of Luoyang (Xiong 2006: 161). Yangdi was drawn to *Tiantai* Buddhism, founded at Tiantai Mountain in Zhejiang province in southern China. In 591, under the instruction of its founder Zhiyi 智顗 (538–597 CE), he took part in a ritual known as the Bodhisattvasila (the observance of the Bodhisattva's precepts). Zhiyi was knowledgeable about many indigenous self-cultivation practices. In his writings on curing illness, he recommended two main techniques: *Śamatha* (*zhi* 止) and *Vipaśyanā* (*guan* 觀). Techniques of *Śamatha* included focusing attention on the location of an illness, or on the *dantian* (an inch below the navel) or on the soles of the feet. *Vipaśyanā* prescribed, among other practices, a set of breathing exercises with six different sounds for treating five internal organs (Salguero 2012, 2017).

Sun Simiao 孫思邈 (581–682 CE), a seventh-century physician who was influenced by Buddhist thought and practice, recorded a set of 18 *daoyin* exercises referred to as 'the massage technique from India and the Brahmans' method' in his work 'Essential Formulas for Urgent Needs Worth a Thousand in Gold' (*Beiji Qianjin yaofang* 備急千金要方, 27). Some of these exercises, while grounded in local Chinese *yangsheng* techniques and language, nevertheless required the practitioners to sit in the lotus position, a position introduced from India by Buddhists. A further set of 12 *daoyin* exercises entitled 'The *Daoyin* Method of the Brahmans' can be found in 'Collection of Texts for Conserving Health' (*Shesheng zuanlu* 攝生纂錄, DZ578) in the Daoist Canon (*Daozang* 道藏) of 1445 (Schipper and Verellen 2004: 356).

Physicians

The Sui administration was the first to introduce a civil service examination system based on a meritocratic recruitment process. It was designed to replace the recruitment practices of the Han dynasty, which had become highly elitist during the period of the Six Dynasties (Xiong 2006: 123–6). While the earlier system had been nominally meritocratic, testing was rudimentary and entry requirements demanded recommendations from powerful people that were hard to obtain. But in the Sui system, anyone, regardless of their social background, who could pass the oral and written examinations held in various prefectures, would be able to enter officialdom. These factors, the standardisation of means of entry, independence from patronage and the monitoring of standards to ensure skill, would suggest that the new system attracted highly skilled physicians to the imperial court. Although there is

Table 6.3 Yangsheng texts quoted in *Zhubing yuanhou lun*

Title	No. of Quotations
Yangsheng fang 養生方 (*Yangsheng* Recipes)	105
Yangsheng fang daoyinfa 養生方導引法 (*Yangsheng* Recipes: *Daoyin* Methods)	198
Yangsheng fang: zhengao 養生方真誥 (*Yangsheng* Recipes: Declarations of the Perfected) - a Daoist text	1
Yangsheng jinji 養生禁忌 (*Yangsheng* Prohibitions)	2
Yangshengjing yaoji 養生經要集 (Essential Compendium on *Yangsheng*)	1

no concrete evidence of a specific medical component within the civil service examination system during the Sui, there is clear evidence of a medical examination from the Tang period (*Tang huiyao* 82: 1525).

Most learned physicians would have been familiar with the use of *daoyin* exercises, which had been part of self-cultivation and therapeutic treatment since the time of the Warring States, Qin and Han periods (481 BCE–220 CE). The 'Inner Canon of the Yellow Emperor' (*Huangdi neijing* 黄帝内經 ca. first century CE), the oldest received classic of medical theory, mentions *daoyin* as one of seven therapies for treating illness, the others being massage, moxibustion, hot compresses, acupuncture, heat therapy and decoctions (*Lingshu* 42). Illnesses such as atrophy, reversal of *qi*, and hot and cold diseases were to be cured with *daoyin* exercise and self-massage. A chronic breathing condition could be cured with drugs only in combination with *daoyin* exercise (*Suwen* 47).

Political division prior to the Sui's unification was in part responsible for the development of different medical practices in the north and south. According to Fan Kawai (2007: 47–73), the medical tradition of the south was, in many ways, more developed than that of the north. Many physicians from the south became government officials in the north, particularly during the reign of Yangdi, including Chao Yuanfang 巢元方, Xu Zhicang 許智藏, Xu Cheng 許澄, Xu Yinzong 許胤宗, Wu Jingda 吳景達 and members of the Jiang family from Yixing 義興蔣氏.[7] Yangdi, who was sympathetic to southern culture, spent a lot of time in the south, and even married the daughter of Emperor Ming of the southern Western Liang dynasty (542–585 CE).

Fan Kawai (2007: 52–64) also places strong emphasis on the significant Daoist influence on the southern medical tradition, suggesting that the inclusion of *daoyin* exercises in *Zhubing yuanhou lun* was an example of the incorporation of Daoist practices into an official medical text. However, a close examination of *Zhubing yuanhou lun* reveals a significant majority of quotations from *yangsheng* texts and relatively few from Daoist sources (Table 6.3).[8] It is more accurate to suggest that there was an intention to incorporate *yangsheng* practices into a central medical blueprint. The fact that a large number of *yangsheng* self-cultivation texts formed part of the medical literature in the Imperial Library of the Sui court reflects their importance in the medical practice of this period.

To conclude, during the Sui period, people from diverse groups, such as physicians, literate elites, Daoists and Buddhists, were conversant with, and played an important role in, the development of *daoyin*, particularly those who resided in the Sui's capitals or who had personal contact with the Sui emperors. Wendi and Yangdi would have been aware of both the medical and religious applications of *daoyin*. As part of their efforts to unify China under a single, centralised, government, Wendi and Yangdi initiated medical reforms that were unprecedented. Three specialised medical departments, including the department of *anmo* established by Wendi, began the institutionalisation of *daoyin* as an important part of state medical practice. Wendi's medical reform reflected medical knowledge and practices between the Han and the Sui periods, while Yangdi's innovative medical approach, involving employing large numbers of *daoyin* practitioners to teach at the state-sponsored Imperial Medical Academy, created an official medical system with therapeutic exercise as its principal therapy. Yangdi's erudition, his fascination for occultism, his personal connections with the contemporary religious leaders Zhi Yi and Wang Yuanzhi, as well as his love of southern culture, which led to the employment of many southern physicians at the imperial court, would all have contributed to his decision to promote *daoyin* as the primary medical treatment. The legacy of his vision is embodied and preserved in *Zhubing yuanhou lun*.

However, the Sui dynasty lasted for only thirty-eight years. During the Tang period that followed, the number of *daoyin* practitioners employed at the imperial medical education institution was significantly reduced (*Tang Liudian* 14: 409–411). By the time of the Song dynasty (960–1279 CE), *daoyin* was no longer part of orthodox medical training. Nevertheless, *daoyin* remained a widespread practice. In subsequent centuries, with the advancement of printing technology, an increasing number of *daoyin*-related texts were printed with illustrations, making them even more accessible and 'user-friendly' (Despeux 2005: 12–14). Such texts proliferated during the Ming (1368–1644) and Qing (1644–1912) dynasties. Fused with all kinds of Buddhist, Daoist and martial arts practices, *daoyin* has continued to evolve into bodily practices such as *taiji* and *qigong*.

Notes

1. *Xie* 邪, translated here as 'deviant', is a complex concept which can be understood in many different ways. Other common translations of the term in a medical context include 'heteropathy', 'pathogen', 'perversity', 'wayward' and 'evil'. For discussion on *xie*, see Lo and Schroer (2005).
2. In *Hanshu yiwenzhi* 漢書藝文志, the Catalogue of the Imperial Library of the Han, two *daoyin*-related texts can be found listed in the 'Spirit Transcendence' (*shenxian* 神仙) category under the 'Recipes and Techniques' (*fangji* 方技) section: The Yellow Emperor's Miscellaneous Walking and Pulling Exercises (*Huangdi zazi buyin* 黃帝雜子步引) and The Massage Techniques of the Yellow Emperor and Qibou (*Huangdi Qibou anmo* 黃帝岐伯按摩). *Hanshu* 30: 1779–80.
3. Evidence concerning the number of diseases and categories varies. The figure of 1739 diseases under seventy-one categories is based on Ding Guangdi and Ni Hexian (1991).
4. We know very little about Chao Yuanfang, there being no biographical details about him in any official records. Two other possible authors of *Zhubing yuanhou lun*, Wu Jingxian 吳景賢 and Wuxian 吳賢, are also identified in some historical bibliographies prior to the Song period.
5. Although *Yangsheng yaoji* is no longer extant, quotations from it and references to its contents can be found in various other texts. The quotation of this particular *daoyin* exercises can be found in *Yangxing yanming lu* 養性延命錄 (Records of Cultivating Nature and Extending Life, DZ 838), a synoptic text compiled during the early Tang period (618–907 CE).
6. An Shigao 安世高 (fl. c. 148–180 CE), the earliest known translator of Indian Buddhist texts into Chinese, made the first translation of *Ānāpānasati* instructions in the second century CE. See *Da anban shouyi jing* 大安般守意經 (The Great Ānāpānasmṛti Sūtra), Taishō Tripiṭaka 602.
7. See also Shengming yiliao shi yanjiu shi (2015) 'New Perspectives on Chinese History – A separate volume on the History of Medicine' (*Zhongguo shi xinlun – yiliao shi fen ce* 中國式新論 – 醫療史分冊), Taipei: Academia Sinica, pp. 162–9.
8. Apart from *Zhengao* (Declarations of the Perfected) which can be found in the Daoist Canon, all *yangsheng* texts cited in *Zhubing yuanhou lun* are no longer extant. The text '*Yangsheng* Recipes: *Daoyin* Methods', source of nearly all *daoyin* exercises in *Zhubing yuanhou lun*, appears neither in the Sui catalogue nor in any other catalogue, which raises the question of whether such a text existed at all. It is conceivable that medical officials, having consulted an array of *daoyin*-related texts in the Imperial Library, collected them all under the heading of *Yangsheng fang daoyin fa* (Recipes for Nourishing Life and the Daoyin Method) (Yang 2018: 291–8).

Bibliography

Pre-modern sources

Beiji qianjin yaofang 備急千金要方 (Essential Formulas for Urgent Needs Worth a Thousand in Gold) 652, Sun Simiao 孫思邈 (1982), Beijing: Renmin weisheng chubanshe.
Da anban shouyi jing 大安般守意經 (The Great Ānāpānasmṛti Sūtra) Taishō Tripiṭaka 602.
Fuqi jingyi lun 服氣精義論 (Treatise on the Essential Meaning of Absorbing Qi) DZ 830.
Hanshu 漢書 (Book of Han) 111, Ban Gu 班固 *et al.* (1962), Beijing: Zhonghua shuju.
Huangdi neijing suwen 黃帝內經素問 (Inner Canon of the Yellow Emperor: *Plain Questions*) Wang Bing 王冰 (fl. 762.) (ed.) (1995), Zhongyi yanjiuyuan Ming edition repr., Beijing: Zhongyi guji chubanshe.
Huangdi neijing lingshu 黃帝內經靈樞 (Inner Canon of the Yellow Emperor: *Numinous Pivot*), Wang Bing 王冰 (fl. 762.) (ed.) (1995), Zhongyi yanjiuyuan Ming edition repr. Beijing: Zhongyi guji chubanshe.
Ishinpō 醫心方 (Remedies at the Heart of Medicine), Tanba no Yasunori 丹波康賴 (912–95), (1993), Beijing: Renmin weisheng chubanshe.
Liangshu 梁書 (Book of Liang) 635, Yao Silian 姚思廉 (557–637) (1973), Beijing: Zhonghua shuju.
Shiji suoyin 史記索隱 (Seeking the Obscure in the Records of the Grand Historians), Sima Zhen 司馬真 (679–732) (1991), Beijing: Zhonghua shuju.
Suishu 隋書 (Book of Sui), Wei Zheng 魏徵 (580–643) (1973), Beijing: Zhonghua shuju.
Tang huiyao 唐會要 (Institutional History of the Tang Dynasty) (1955), Wang Pu 王溥 (922–982), Beijing: Zhonghua shuju.
Tang liudian 唐六典 (Six Statutes of Tang), Li Lingfu 李林甫 (683–753) *et al.* (1992), Beijing: Zhonghua shuju.
Ui'bang'ryuchui 醫方類聚 (Classified Collection of Medical Remedies) 1477, Kim Ye-mong 김예몽 金禮蒙 (1981), Beijing: Renmin weisheng chubanshe.
Waitai miyao 外台祕要 (Secret Essentials of the Outer Terrace) 752, Wang Tao 王燾 (670–755) (1955), Beijing: Renmin weisheng chubanshe.
Yanshi jiaxun 顏氏家訓 (Family Instructions of the Yan Clan) 6th century, Yan Zhitui 顏之推 in (eds) Zhuang Huiming 莊輝明 and Zhang Yihe 章義和 (2006) *Yanshi jiaxun yizhu* 顏氏家訓譯注 (Family Instructions of the Yan Clan: Translation and Annotations), Shanghai: Shanghai guji chubanshe.
Yangxing yanming lu 養性延命錄 (Records of Cultivating Nature and Extending Life) DZ 838.
Zhengao 真誥 (Declaration of the Perfected) DZ 1016.
Zhengtong Daozang 正統道藏 (The Daoist Canon of the Zhengtong Reign) 1444–1445, Bai Yunji 白雲霽 and Qiu Changchun 丘長春 (eds) (1985), Taipei: Xinwenfeng chuban gongsi.
Zhubing yuanhou lun 諸病源候論 (Treatise on the Origins and Symptoms of Medical Disorders) 610, Chao Yuanfang 巢元方 *et al.*, in Ding Guangdi 丁光迪 and Ni Hexian 倪和憲 (eds) (1991) *Zhubing yuanhou lun jiaozhu* 諸病源候論校注 (Proofread and Comment on Treatise on the Origins and Symptoms of Medical Disorders), Beijing: Renmin weisheng chubanshe.

Modern sources

Alter, J.S. (ed.) (2005) *Asian Medicine and Globalization*, Philadelphia: University of Pennsylvania Press.
Bray, F., Dorofeeva-Lictmann, V. and Métailié, G. (eds) (2007) *Graphics and Text in the Production of Technical Knowledge in China: the warp and the weft*, Leiden: Brill.
Despeux, C. (1989) 'Gymnastics: the ancient tradition', in L. Kohn (ed.) *Taoist Meditation and Longevity Techniques*, Ann Arbor: Center for Chinese Studies, University of Michigan, pp. 223–61.
——— (2005) 'Visual representations of the body in Chinese medical and Daoist texts from the Song to the Qing period (tenth to nineteenth century)', *Asian Medicine*, 1: 10–52.
——— (2006) 'The six healing breaths', in L. Kohn (ed.) *Daoist Body Cultivation: traditional models and contemporary practices*, Magdalena, NM: Three Pines Press, pp. 37–67.
Ding Guangdi 丁光迪 (1993) *Zhubing yuanhou lun yangsheng fang daoyin fa yanjiu* 諸病源候論養生方導引法研究 (A Study on *Yangsheng* Recipes and *Daoyin* Methods in Treatise on the Origins and Symptoms of Medical Disorders), Beijing: Renmin weisheng chubanshe.
Ding Guangdi 丁光迪 and Ni Hexian 倪和憲 (eds) (1991) *Zhubing yuanhou lun jiaozhu* 諸病源候論校注 (Proofread and Comment on Treatise on the Origins and Symptoms of Medical Disorders), Beijing: Renmin weisheng chubanshe.

Engelhardt, U. (1987) *Die klassische Tradition der Qi-Übungen (Qigong): eine Darstellung anhand des Tang-zeitlichen Textes Fuqi jingyi lun von Sima Chengzhen*, Wiesbaden: Franz Steiner.

—— (1989) 'Qi for life: longevity in the Tang', in L. Kohn (ed.) *Taoist Meditation and Longevity Techniques*, Ann Arbor: Center for Chinese Studies, University of Michigan, pp. 263–94.

Fan Kawai 范家偉 (2004) *Liuchao Sui Tang yixue zhi chuancheng yu zhenghe* 六朝隋唐醫學之傳承與整合 (Transmission and Integration of Medicine from the Six Dynasties to the Tang), Hong Kong: Chinese University Press.

—— (2007) *Dayi jingcheng: Tangdai guojia, xinyang yu yixue* 大醫精誠：唐代國家、信仰與醫學 (State, Belief, and Medicine in the Tang), Taipei: Dongda Publishing House.

Harper, D. (1998) *Early Chinese Medical Literature: the Mawangdui medical manuscripts*, London: Kegan Paul International.

Kaltenmark, M. (1953) *Le Lie-sien tchouan*, Pékin: Centre d'Etudes Sinologues.

Kohn, L. (1987) *Seven Steps to the Tao: Sima Chengzhen's Zuowanglun*, Nettetal: Steyler Verlag.

—— (ed.) (1989) *Taoist Meditation and Longevity Techniques*, Ann Arbor: Center for Chinese Studies, University of Michigan.

—— (ed.) (2006) *Daoist Body Cultivation: traditional models and contemporary practices*, Magdalena, NM: Three Pines Press.

—— (2006) 'Yoga and Daoyin', in L. Kohn (ed.) *Daoist Body Cultivation: traditional models and contemporary practices*, Magdalena, NM: Three Pines Press, pp. 123–150.

—— (2008) *Chinese Healing Exercises: the tradition of Daoyin*, Honolulu: University of Hawai'i Press.

Lewis, M.E. (2009) *China between Empires: the Northern and Southern dynasties*, Cambridge, MA: Belknap Press of Harvard University Press.

Lo, V. (1998) *The Influence of Yangsheng Culture on Early Chinese Medical Theory*, PhD thesis, University of London.

—— (2007) 'Imagining practice: sense and sensuality in early Chinese medical illustration', in F. Bray, V. Dorofeeva-Lictmann and G. Métailié (eds) *Graphics and Text in the Production of Technical Knowledge in China: the warp and the weft*, Leiden: Brill, pp. 383–423.

—— (2014) *How To Do the Gibbon Walk: a translation of the Pulling Book (ca. 186 BCE)*, Cambridge: Needham Research Institute.

Lo, V. and Schroer, S. (2005) 'Deviant airs in "traditional" Chinese medicine', in J. Alter (ed.) *Asian Medicine and Globalization*, Philadelphia: University of Pennsylvania Press, pp. 45–66.

Robinet, I. (2008) 'Shangqing', in F. Pregadio (ed.) *The Encyclopedia of Taoism*, London: Routledge, pp. 858–66.

Raz, G. (2012) *The Emergence of Daoism: creation of tradition*, Abingdon, Oxon: Routledge.

Pregadio, F. (ed.) (2008) *The Encyclopedia of Taoism*, London: Routledge.

Salguero, C.P. (2012) '"Treating illness": translation of a chapter from a medieval Chinese Buddhist meditation manual by Zhiyi (538–597)', *Asian Medicine*, 7.2: 461–473.

—— (2017) 'Healing with meditation: "Treating Illness" from Zhiyi's shorter treatise on samatha and vipasyana', in C.P. Salguero (ed.) *Buddhism and Medicine: an anthology of premodern sources*, New York; Chichester, West Sussex: Columbia University Press, pp. 382–9.

Schipper, K.M. and Verellen, F. (eds) (2004) *The Taoist Canon: a historical companion to the Daozang*, Chicago, IL: University of Chicago Press.

Shengming yiliao shi yanjiu shi 生命醫療史研究室 (Research Group of the History of Medicine and Healing) (ed.) (2015) *Zhongguo shi xinlun: Yiliao shi fence* 中國史新論. 醫療史分冊 (A New Chinese History: Volume on History of Medicine), Taipei: Academia Sinica.

Stanley-Baker, M. (2013) *Daoists and Doctors: the role of medicine in six dynasties Shangqing Daoism*, PhD thesis, University College London.

Sterckx, R. (2002) *The Animal and the Daemon in Early China*, Albany: State University of New York Press.

Xiong, V.C. (2006) *Emperor Yang of the Sui Dynasty His Life, Times, and Legacy*, Albany: State University of New York Press.

Wang Shumin 王淑民 and Barrett, P. (2006) 'Profile of a *daoyin* tradition: the "five animal mimes"', *Asian Medicine*, 2.2: 225–53.

Yang, D. (2018) *Prescribing 'Guiding and Pulling': the institutionalisation of therapeutic exercise in Sui China (581–618 CE)*, PhD thesis, University College London.

Zacchetti, S. (2010) 'A "new" early Chinese Buddhist commentary: the nature of the *Da anban shouyi jing* T 602 reconsidered', *Journal of the International Association of Buddhist Studies*, 31.1/2: 421–84.

7

THE CANONICITY OF THE *YELLOW EMPEROR'S INNER CLASSIC*

Han through Song

Stephen Boyanton

Introduction

Chinese medicine today is frequently said to rest on the foundation of a group of texts known as the *Yellow Emperor's Inner Classic* (*Huangdi neijing* 黃帝內經, hereafter the *Inner Classic*, *Neijing* 內經). This point of view is found in medical texts as far back as the Song (960–1279) and is even cited in modern scholarly discussions of Chinese medicine as well. But is it accurate?

The *Inner Classic* is the oldest text in the received tradition of East Asian medicine and has been cited since the Song to explain and justify medical practices. There are, however, many ways in which a text can be authoritative or canonical, and although we now know a great deal about the formation of the *Inner Classic*, we still know very little about how it acquired its current status at the centre of the Chinese medical canon.

The very concept of canonicity is somewhat problematic. There is no clear Chinese equivalent to the term. The notion of a classic text (*jing* 經) comes closest; however, throughout China's history, a variety of texts and groups of texts have been selected as authoritative in various fields, and many of these 'canonical' texts were never given the label 'classic' (an excellent survey of these is found in De Weerdt 1999: 91–4). Even in English, the term 'canon' is far from precise. It can refer to closed canons, such as the Bible and open canons, such as 'the canon' of English literature. No single model of textual authority can be applied to all of these cases. However, rather than lament the imprecision of this term, I wish to make its vagueness a virtue. A more relaxed analytical gaze may enable us to better recognise the various types of textual authority the *Inner Classic* has possessed throughout its long history. Therefore, in this chapter, I do not take the meaning of canonicity as a given but rather as a topic for investigation.

The earliest mention of a *Yellow Emperor's Inner Classic* is found in the bibliographic section (*Yiwen zhi* 藝文志) of the *Hanshu* 漢書 (History of the Han) 6.30, alongside a *Yellow Emperor's Outer Classic* 黃帝外經 (*Huangdi waijing*) and *Inner* and *Outer Classics* by the legendary physician Bian Que (扁鵲, sixth to fifth century BCE) and a Mr White (*Baishi* 白氏), who is otherwise unknown (*Hanshu*, 30.1776). None of the other texts in this list is still extant. Surviving literature mentions six titles associated with the *Yellow Emperor's Inner*

Table 7.1 Titles of *Inner Classic* Texts Mentioned in Other Texts

---

Inquiry into the Fundamental (*Suwen* 素問)
Numinous Pivot (*Lingshu* 靈樞)ᵃ
Systematic Classic of Acumoxa (*Zhenjiu jiayi jing* 針灸甲乙經)
Grand Fundamental (*Taisu* 太素)
Needling Classic (*Zhenjing* 針經)ᵃ
Nine Fascicles (*Jiujuan* 九卷)ᵃ

ᵃ Some scholars believe that these are variant titles for one text. *Numinous Pivot* and *Needling Classic* are most likely separate texts (Keegan 1988: 37–43). The identity of the *Nine Fascicles* is unclear.

Classic (Table 7.1).[1] A great deal of ink has been spilled arguing over which of these was the original text mentioned in the *Hanshu*. A combination of archaeological discoveries and careful investigation of extant texts, however, has led scholars to reject this question entirely and radically rethink our understanding of this group of texts.

The formation of the *Inner Classic* corpus

The various texts titled *Yellow Emperor's Inner Classic* share not only a title, but many passages as well. These shared passages vary little from one *Inner Classic* text to another and are generally quite short, in modern terms no more than a few paragraphs in length. Although their wording varies little from one text to another, they are located in radically different positions within each text – different chapters, different positions within a chapter, etc. (Keegan 1988: 64–6). The explanation of these passages began to become clear after the archaeological discovery of similar passages in excavated texts. In 1973, archaeologists working at a site called Mawangdui (馬王堆) in Hunan province discovered a collection of texts buried along with the occupant of a tomb (Chapter 3 in this volume). Several medical texts were included among them. Since this tomb was sealed in 168 BCE, these texts provide an unparalleled look into the medical world of the early Han. Four of the medical texts deal specifically with the channels (*jing* 經) used in acupuncture and moxibustion, and some of their passages are clearly parallel to parts of the *Numinous Pivot* (*Lingshu* 靈樞) (Harper 1997).

In his seminal dissertation, David Keegan analysed these parallels and demonstrated that our understanding of the *Inner Classic* was hampered by the anachronistic imposition onto it of our concept of a book. Keegan argues that in its earliest stages, the *Inner Classic* is best thought of as a large collection of very small primary texts – the shared passages noted above – which were rearranged in various orders to compose the chapters and books that we know as the *Yellow Emperor's Inner Classic*. The *Inner Classic* is therefore not a single book, and the various extant books bearing that title are actually compilations produced by selection and rearrangement of the primary texts. The primary texts were thus conserved with few changes across all compilations, but how they were put together varied greatly. The question of which extant compilation is the 'original' *Yellow Emperor's Inner Classic* is therefore moot. None of them is; there never was any such creature (Keegan 1988: 110–13).

Having unravelled the structure of the *Inner Classic*, Keegan went further. Drawing on Nathan Sivin's discussion of the transmission of medical learning in the Han and medieval periods, he posited a social mechanism by which the primary texts were compiled in such different orders. Physicians learned medicine by apprenticeship to a master. Initially, learning consisted of observing the master, receiving oral instructions, and being guided in the memorisation of texts. When the master decided the student was ready, he would transmit a text to the student.

The student would copy it and the master would impart an oral explanation. The texts thus transmitted were not necessarily whole books, but rather short texts like the primary texts found in the *Inner Classic*. Students copied texts in the order in which they received them and might receive the same text in slightly different forms from different teachers. The resulting compilation of primary texts was therefore unique to that student. The various *Inner Classic* compilations we have today were produced by editing this sort of *ad hoc* compilation, and the order of primary texts in each compilation therefore differs (Keegan 1988: 219–47; Sivin 1995a).

Keegan's analysis also enabled a more accurate dating of the *Inner Classic* corpus by making it clear that the individual primary texts were composed at different times and predate the earliest compilation of an *Inner Classic* text. Comparison of the Mawangdui channel texts with parallel primary texts in the *Numinous Pivot* reveals that the *Numinous Pivot* texts are substantially more developed (Harper 1997: 86–90). Given that the Mawangdui texts cannot date from later than 168 BCE and that *Records of the Historian* (*Shiji* 史記) – the earliest portions of which were completed c. 100 BCE – makes no mention of the *Inner Classic*, we can safely take 100 BCE as the earliest possible date for its formation. If we take the date of the *Hanshu* as its *terminus ante quem* – the *Numinous Pivot*'s passages must date from between 100 BCE and 92 CE. This evidence has led most scholars to conclude that the primary texts of the *Inner Classic* corpus must have reached something approaching their current state in the first century BCE, though individual texts may have earlier or later dates (Unschuld 2003: 1–3).

The extant compilations of the *Inner Classic* differ considerably in their structure and textual history. These differences have been discussed at length by Nathan Sivin (1993) and Paul Unschuld (2003) and will not be repeated here except as they bear on the question of the *Inner Classic* corpus's textual authority.

The Han dynasty (202 BCE–220 CE)

The relative importance of the *Inner Classic* before the late Han is difficult to reconstruct. We can gain some sense of its relationship with the broader medical context, however, by examining the list of medical books possessed by the Granary Master (*Canggong* 倉公) Chunyu Yi 淳于意 (fl. mid-second century BCE), as recorded in his biography in *Records of the Historian*. Although his biography does not mention the *Inner Classic*, eight of the ten books he received from his teachers are also mentioned in extant *Inner Classic* compilations. The Granary Master presents his possession of these texts as proof of his medical skill and, in the case records included within the biography, is depicted as relying on these texts to justify his treatments.

We must, however, use this source with care. Miranda Brown has pointed out that this biography, along with most of the other sources used in reconstructing early medical history in China, was written by non-physicians for purposes unrelated to medical practice. Furthermore, the sources used in constructing this biography have been shown to be heterogeneous, some of them clearly not the work of a historical Chunyu Yi. Brown argues that the practice of justifying medical treatment by textual reference is not seen elsewhere in the early medical literature. Such justification was, however, part of the record-keeping practices of government officials. Recording and treating illness were among the duties of officials, and the case records in this biography were most likely derived from such records. This makes it impossible to say with certainty that the use of textual justification was common among doctors in this period (Brown 2015: 63–86). Nevertheless, the fact that medical texts are mentioned at all and that many of the same texts are also referred to in the *Inner Classic* does support Keegan's thesis that there was a group of medical texts circulating in the early Han that were seen – by some at least – as authoritative.

In addition to revealing their provenance, the use of a governmental model in the Granary Master's case records may also tell us something about the kind of authority these medical texts – and the *Inner Classic* – claimed. The presence of terms and metaphors related to government has long been recognised in both the *Inner Classic* and medical literature more broadly. By the end of the Warring States, there was widespread acceptance of a homology between the cosmos at large, the state, and the human body (Sivin 1995b). It is possible that medical practitioners adopted a governmental style of justification as part of their effort to bolster their authority. Just as officials could cite law codes in deciding legal cases, doctors could cite medical texts in deciding on diagnosis and treatment. The choice of the Yellow Emperor as an interlocutor in the *Inner Classic* may also reflect this governmental sensibility. The Yellow Emperor was believed to have taught humans the art of government in the distant past and is considered by some scholars (Yates 1997: 10–16) to have been a key figure in the Huang-Lao (黃老) movement, which was purportedly influential in the early Han and made explicit links between ruling the state and caring for the body. Yamada Keiji first put forward the 'working hypothesis' that this choice might reflect a connection between the *Inner Classic* and the Huang-Lao movement (Yamada Keiji 1979: 89). Subsequent research has tended to support this hypothesis and has shown that the *Inner Classic* relies heavily on a cluster of concepts – the Way (*Dao* 道), compliance (*shun* 順), and rebellion (*ni* 逆) – that was fundamental to much of Huang-Lao thought (Peerenboom 1993: 51–3, 64–6; Boyanton 2006). In medical texts like the *Inner Classic*, *shun* and *ni* can refer to compliance with or variance from the natural pattern of things, or to going with or against the normal flow of bodily fluids – such as *qi* along the channels. In Huang-Lao texts, the same terms reflect positions in relation to authority (i.e. of the state).

By the late Han, we have more evidence for the perceived authority of the *Inner Classic*. Its mention in the *Hanshu* bibliography attests to its significance, but the mention of other medical texts side by side with it indicates that it was not the only esteemed medical classic. Likewise, although the titles *Inquiry into the Fundamental* (*Suwen* 素問) and *Nine Fascicles* (*Jiujuan* 九卷) appear in the preface to the *Treatise on Cold Damage* (*Shanghan lun* 傷寒論, c. 206 CE), they are accompanied by a number of other texts (*Shanghan lun*, preface.305).[2] In the late Han, the *Inner Classic* appears to have been one of a large number of important medical texts, highly valued, but lacking its later pre-eminence.

In addition to considering the bibliographic evidence and the intellectual context within which the *Inner Classic* made its earliest claims to canonical status, it is also important to bear in mind its physical aspect as a book. During the Han, texts were generally written on bamboo slips that were then tied together and rolled up. These bamboo slip 'books' were difficult to produce, very heavy, and prone to falling apart if the ties holding them together were not maintained (Nylan 2000: 244). Given the technical limitations of book production, it is unlikely that physicians could produce and carry about large texts like the extant compilations of the *Inner Classic*, and the production of standardised texts would have been all but impossible. The type of authority the *Inner Classic* could claim was limited by the type of book into which it could be formed.

The period of division, Sui, and Tang (220–907)

We have abundant evidence for the *Inner Classic*'s importance throughout the period of division following the fall of the Han and up through the Tang. It is clear from the bibliographic literature and the surviving compilations that it was not only circulating, but also circulating in several compilations and various editions of these compilations.

The earliest surviving compilation of the *Inner Classic* corpus is the *Systematic Classic of Acumoxa* (*Zhenjiu jiayi jing* 針灸甲乙經), or to use its full title, the *Yellow Emperor's Systematic Classic of Acumoxa in Three Parts* (*Huangdi sanbu zhenjiu jiayi jing* 黃帝三部針灸甲乙經). Compiled by Huangfu Mi 皇甫謐 (215–282) between 256 and his death, it is of particular historical importance. Huangfu Mi became an imperial physician (*taiyi* 太醫) under the Jin (265–420) but was already an accomplished and respected scholar and author in other fields. According to his preface, he became interested in medicine after suffering an illness that was treated unsuccessfully (*Huangdi sanbu Zhenjiu jiayijing*, preface, 16–21).[3] The stature of the *Inner Classic* at this time is to some degree attested by the fact that disappointment in the quality of contemporary doctors would lead Huangfu Mi to examine it. More enlightening, however, is how the preface justifies the authority of the *Inner Classic*. It attributes the *Inner Classic* to the Yellow Emperor's investigation of medicine and roots its authority in the fact the he '… inwardly investigated the five viscera and the six bowels; outwardly, surveyed the channels and network vessels, the *qi*, blood, and the complexion; consulted heaven and earth and examined people and things; took the nature allotted by heaven as the basis; and exhaustively studied the marvellous and thoroughly inquired into transformation' (*Huangdi sanbu Zhenjiu jiayijing*, preface, 16). For Huangfu Mi, the authority of the *Inner Classic* lay in its transmission of the knowledge of the sages, a wisdom beyond the ken of contemporary humans.

When he investigated the *Inner Classic*, however, he found it in disarray. The texts were not 'in order' and contained many 'repetitions' and 'misplacements'. He therefore rearranged the texts in a logical order and removed repetitions and unnecessary verbiage to produce a new text (*Huangdi sanbu Zhenjiu jiayijing*, preface, 20). Huangfu Mi's statement has often been taken as an indication that the *Inner Classic* was on the verge of being lost (Unschuld 2003: 22) – and he intended it that way – but given the nature of textual transmission among physicians, described above, it is far more likely that he was simply looking at a doctor's text with the eyes of a literatus accustomed to a completely different book culture. Huangfu Mi approached the *Inner Classic* expecting a 'book' in which all of the contents had fixed locations and logical relationships. The repetitions and variations inevitably produced by physicians' textual practices therefore struck him as textual errors, and he set out to do what any good scholar would under the circumstances: produce a reliable edition.[4]

The next two known compilations of the *Inner Classic* unfortunately provide us with little information about its claims to canonical status save the fact that they were produced by and circulated among the social elite. The Quan Yuanqi 全元起 (fl. sixth century) edition of *Inquiry into the Fundamental* (*Suwen* 素問) has been lost since the Song – fragments quoted in other texts are all that survive – and the preface is missing from the rediscovered copy of *Grand Fundamental* (*Taisu* 太素), compiled by Yang Shangshan 楊上善 (fl. early Tang). The third, however, reaffirms Huangfu Mi's view of the *Inner Classic*'s origins and authority. In the preface, dated 762, to his edition of *Inquiry into the Fundamental*, Wang Bing 王冰 (fl. eighth century) affirms that the *Inner Classic* is the work of the Yellow Emperor and that without such sagely advice one cannot hope to 'be released from the grievous bondage [of illness], complete and guide the true *qi*, enable the common people to attain a long life, or aid the frail and weak to obtain peace'.

Like Huangfu Mi, Wang Bing complains in his preface that the *Inquiry into the Fundamental* was in poor condition when he started. He took a nine-fascicle edition of Quan Yuanqi's edition as his base text but found it necessary to edit it extensively. He removed redundant chapters, corrected passages he deemed corrupt, altered the wording of the dialogues to maintain proper etiquette between ruler and minister, divided chapters he felt combined two originally separate discourses, and added chapters to replace those that had been lost

(*Chongguang buzhu Suwen*, preface, 5–6). Since Quan Yuanqi's edition is no longer available, we cannot be certain what type of text Wang Bing was examining, but it seems likely that once again we are witnessing the reaction of an elite scholar to the very different writing and reading habits of non-elite physicians.

In spite of the sagely pedigree claimed for it, throughout this period, the *Inner Classic* remained one among many esteemed medical texts. In Chen Yanzhi's 陳延之 (c. late fourth to early fifth centuries) *Short Essay on Classical Formulae* (*Jingfang xiaopin* 經方小品), a *Methods of the Yellow Fundamental* (*Huangsu fang* 黃素方) is mentioned in a list of important medical texts; however, even if we assume that this text is related to *Inquiry into the Fundamental*, it is only one text in a list of sixteen (Yan Shiyun and Li Qizhong, 2009: 785). Likewise, in Sun Simiao's 孫思邈 (d. 682) famous list of texts a doctor should study, *Inquiry into the Fundamental*, the *Systematic Classic of Acumoxa*, and the *Needle Classic* are listed first, but they are followed by twelve other medical texts and then by a variety of non-medical texts and methods of divination (*Qianjin yaofang*, 1.1–2).

By the end of the Tang, although the position of the *Inner Classic* vis-à-vis other medical texts was largely unchanged, the nature of the *Inner Classic* as a physical text had changed greatly. What had been a corpus of texts extant only in *ad hoc* compilations produced by and for physicians now also existed in at least four compilations produced by and structured according to the norms of the literati elite. These compilations fixed the form of the text in ways that decreased its mutability and made it more easily shared. The shift from writing on bamboo slips to writing on paper also vastly increased its portability. They altered both the nature of its claims to authority – by limiting that authority to particular books as opposed to the *Inner Classic* corpus as a whole – and the scope of its claims – by allowing larger numbers of people to share the same edition of the text. These developments continued and reached unprecedented levels during the following Song dynasty.

The Song dynasty (960–1279)

The Song was a period of intense transformation in many aspects of Chinese society, economy, government, and thought. The changes most relevant to this discussion are the development of printing, an unprecedented involvement by the government in medical affairs, and the changing status of medical practice as an occupation (Hinrichs 2013; Boyanton 2015: 76–115; Hymes 2015).

Although the technology of woodblock printing was invented during the Tang, it was only in the Song that it became the basis of a large publishing industry. The ability to rapidly produce large numbers of books drove the cost of books down, making literacy and book ownership affordable to a larger segment of the population than ever before. The Song book market generated a wide range of books, from expensive editions that are considered among the finest examples of printing ever produced to cheap editions of poor quality. Books were published on a wide variety of topics, including medical books of many types (Hymes 2015: 542–68).

In addition to increasing the availability of books, printing allowed the circulation of texts in identical editions. Copying texts by hand tended to introduce errors, and when working with manuscripts, it was a standard practice to compare several copies and produce a new edition for one's own use (Hymes 2015: 562). This process meant that for all but the most common texts, any two copies were unlikely to be identical. Mass printing created the possibility of readers widely separated in space or time working from truly identical editions.

The Song government capitalised on this possibility in one of its most innovative interventions in medicine: large-scale, government-sponsored medical publishing. Before the

Song, it was rare for the court to commission medical texts. A total of five medical books were produced for the court prior to the Song. From its inception, the Song broke with this precedent, and the state continued to produce and distribute medical books throughout the Northern Song (960–1127). The height of Song government medical publishing occurred between 1057 and 1069, when the government established the Bureau for Editing Medical Texts (*Jiaozheng yishu ju* 校正醫書局). The court was concerned that the most significant medical texts were in limited circulation in editions that were often erroneous. The Bureau was tasked with correcting this problem by editing and publishing accurate editions of these books. It was the first and almost the only time a Chinese court used medical publication as a means of benefiting the people by improving medical practice (Hinrichs 2013: 102–8).

Over the course of twelve years, the Bureau produced fifteen books (Table 7.2), the last eight of which rapidly became among the most widely cited medical texts in Chinese history.

The influence of these texts appears to have been due to multiple factors that are not easily disentangled. Medical texts produced by the government certainly carried a type of prestige that other medical texts lacked. It is difficult to discern, however, whether their influence derived from this prestige or from their ready availability. The government saw to it that its texts were widely distributed, and private reprints spread them even further (Goldschmidt 2008: 87–93; Miyashita 1967: 135–8).

If we cannot fully explain the success of the Bureau's publications, the prefaces its editors wrote allow us to reconstruct their purpose and reasoning to a far greater extent than is possible for earlier editors of the *Inner Classic*. An analysis of these prefaces reveals an arbitrariness in the process of canonisation that differs strikingly from the aura of inevitability and independent value that surrounds canonical texts.

Table 7.2 Publications of the Song Bureau for Editing Medical Texts (*Jiaozheng yishu ju* 校正醫書局)

Title	Year Published
Yellow Emperor's Inner Classic: Grand Fundamental (*Huangdi neijing taisu* 黃帝內經太素)	c. 1057[a]
Classic of the Numinous Pivot (*Lingshu jing* 靈樞經)	c. 1057[a]
Systematic Classic of Acu-moxa (*Zhenjiu jiayi jing* 針灸甲乙經)	c. 1057[a]
Jiayou Materia Medica (*Jiayou bencao* 嘉祐本草)	1062
Illustrated Classic of Materia Medica (*Bencao tujing* 本草圖經)	1063
Treatise on Cold Damage (*Shanghan lun* 傷寒論)	1065
Essentials of the Golden Coffer (*Jingui yaolüe* 金匱要略)	1066
Classic of the Golden Coffer and Jade Case (*Jingui yuhan jing* 金匱玉函經)	1066
Essential Formulae worth a Thousand Gold (*Qianjin yaofang* 千金要方)	1066
Further Formulae worth a Thousand Gold (*Qianjin yifang* 千金翼方)	1066[b]
Yellow Emperor's Inner Classic: Questions on the Fundamental (*Huangdi neijing suwen* 黃帝內經素問)	1067
Systematic Classic of Acu-moxa (*Zhenjiu jiayi jing* 針灸甲乙經)	1069
Classic of the Pulse (*Maijing* 脈經)	1069
Secret Essentials of the Outer Terrace (*Waitai miyao* 外臺秘要)	1069

[a] For unknown reasons, these texts were never published. No copies are extant.
[b] The Song preface to *Qianjin yifang* is not dated. It is usually assumed that it was edited at the same time as *Qianjin yaofang*.

Although its members changed over the course of its history, a core group of six high-status, well-educated men was responsible for producing the last eight texts. With two exceptions – who participated in the editing of only two of these eight books – the members of this core group were all connected to one another through their association with a man named Gao Ruona 高若訥 (997–1055). He was the father of one editor, the father-in-law of another, and taught medicine to a third, whose brother, in turn, was the fourth. Gao Ruona himself passed the civil service examination in 1024 and held several high positions in the Song government but also nourished an interest in medicine and wrote a text on the Han-dynasty medical text *Treatise on Cold Damage* (Goldschmidt 2008: 79–83). In addition to their connection with Gao Ruona, the core group of editors all shared an active interest in medicine. The two brothers were practising physicians, and their father, like Gao Ruona, had written books on the *Treatise on Cold Damage*.

The original imperial mandate that created the Bureau appears to have included a list of texts to be edited and published. The editors' memorial on submitting the *Jiayou Materia Medica* (*Jiayou bencao* 嘉祐本草) to the throne states that they were entrusted with editing:

> ... the *Divine Farmer's Materia Medica* (*Shennong bencao* 神農本草), the *Numinous Pivot*, the *Grand Fundamental*, the *Systematic Classic* [*of Acumoxa*], and *Inquiry into the Fundamental*, as well as the formularies *Expansive Aid* (*Guangji* 廣濟), [*Formulae worth*] *a Thousand Gold* (*Qianjin* 千金), and *Secret Essentials of the Outer Terrace* (*Waitai miyao* 外臺秘要) ...
>
> (quoted in Okanishi 2010: 1027)

Conspicuously absent from this list are the works of Zhang Ji 张机 (stylename Zhang Zhongjing; 150–219) – author of the *Treatise on Cold Damage*, three of whose books the Bureau published under the direction of these four editors – and the *Classic of the Pulse*, which they published in 1069. Furthermore, they never published the formulary *Expansive Aid*, which was included in the list. There is no evidence that the imperial mandate was ever revised. The final decision about what to edit and publish was left in the hands of the editors, and they chose to publish texts they personally felt were important. The authority of imperial publication was thus conferred on texts that were chosen neither by imperial decree nor because they were universally acknowledged as worthy of canonisation, but merely because a small group of men was able to bend the imperial apparatus to their own purposes.

Among the texts they chose to anoint with imperial blessing were two compilations of the *Inner Classic* corpus: *Inquiry into the Fundamental* and the *Systematic Classic of Acumoxa*. Unlike previous editors, they do not lament finding the texts in a deplorable condition, though editing was still necessary, particularly for the *Systematic Classic of Acumoxa* (*Chongguang buzhu Suwen*, preface, 4–5; *Zhenjiu jiayi jing*, preface, 12–13). The Song publication likely saved these compilations of the *Inner Classic* from being lost. Both *Grand Fundamental* and the *Needle Classic* were extant at the time the Bureau was working, but were lost shortly afterwards. Ironically, however, the success of the Bureau's publications may have contributed to this loss. As the new, authoritative editions became widely available in cheap, printed editions, there was little incentive to copy and preserve manuscripts of other editions. Printing and canonisation thus acted as a two-edged sword, ensuring the survival of the chosen texts and hastening the demise of those not chosen.

The final eight texts published by the Bureau both survived and thrived. They remain the standard editions of these texts even today. An exception to the hegemony of the Bureau's texts, however, is *Numinous Pivot*. As noted in Table 7.2, the *Numinous Pivot* produced by the Bureau in 1057 was not published. In their preface to *Inquiry into the Fundamental*, the Bureau's editors lament the poor quality of the *Numinous Pivot* available to them, and this may account for their failure to produce an edition of it. The extant editions of the *Numinous Pivot* derive from a private edition submitted to the court in 1155 by Shi Song 史崧 (fl. twelfth century), which he produced by comparing an edition of the *Numinous Pivot* held by his family to other texts with parallel passages (Sivin 1993: 203). Once again, we see an element of arbitrariness in the process of canonisation. Although the Song government had desired to publish an edition of the *Numinous Pivot*, the edition which finally received its blessing and became canonical was produced by a single individual with no direct connection to the court.

The Song editors of the *Inner Classic* continued to justify its canonicity through reference to the Yellow Emperor. It would be a mistake, however, to assume that the meaning of this claim was the same as it was during the Han or even the Tang. Even in the earliest periods, the myth of the Yellow Emperor was polysemous. Charles LeBlanc has identified twenty different themes in the stories surrounding the figure of the Yellow Emperor in the ancient sources. He was the founding ancestor of Chinese civilisation, an icon of good governance, a devotee of esoteric practices of sexual self-cultivation, and an immortal transcendent, among many other roles (LeBlanc 1985). Nor was that the end of his transformations: the meanings associated with him continued to grow and develop throughout the imperial period and into modern times (Jochim 1990; Yates 1997: 17–19; Harper 1999; Marsili 2003; Leibold 2006; Shinno 2007). The Song editors' support for the claim that the *Inner Classic* conveyed the wisdom of the Yellow Emperor is particularly significant since the Song witnessed serious challenges to that claim (Brown 2015: 146) alongside a widespread conviction that the ideal method of governance was to found in the 'Way of the sages' (*shengren zhi dao* 聖人之道) (Bol 1994: 126–31, 156–71, 181). The stakes were high. Successfully claiming the Yellow Emperor as the founder of medicine would provide a strong basis for assertions that it was a proper calling for the literati elite. Failure to establish a suitable pedigree could result in it being seen as a profession whose respectability was dubious at best (Boyanton 2015: 90–7, 130–4).

The Song editions of the *Inner Classic* contributed to a revolution in its status. Prior to the Song, the practice of medicine as an occupation was considered beneath the dignity of the elite, literati stratum of society. During the Song, this changed. The Bureau's publications were part of an effort by the imperial government to encourage more elite men to become physicians, an effort which the Bureau's members and an increasing number of literati supported (Boyanton 2015: 78–82; Goldschmidt 2008: 42–57; Hinrichs 2013: 99–117). Although the transformation of medicine into an elite occupation was still incomplete during the Song, increasing numbers of literati became physicians, bringing with them their assumptions about learning, reading, and writing. Previously, medical authors did not so much cite other medical texts as excerpt from them. The medical compendia of the Tang, for example, are compilations of treatment methods from a wide range of earlier texts. Whole sections are excerpted, but they are not used to justify the compiler's assertions. They are simply presented as possible approaches to treatment. Starting in the Song, however, medical authors cited texts as support for their own arguments. They drew on a far smaller range of texts – in particular the publications of the Bureau – but they did so much more frequently. The *Inner Classic* was no longer one of a large group of important medical texts; it was now one of a smaller group of essential medical texts, reference to which was almost mandatory. The canon had narrowed, resulting in a more intense engagement with the texts that composed it.

Conclusion: canonicity as a historical process

In this chapter, I have shown that the nature of the authority claimed for the *Inner Classic* and received varied greatly over the roughly 1,200 years discussed here. These differences resulted primarily from changes in the technology of book production, the types of authority available to supporters of the *Inner Classic*, and the varying assumptions they made about the nature and use of books. The last two are inextricably tied to the changing social status of medicine as an occupation. My findings, therefore, largely parallel Richard Ohmann's first three conclusions in his classic study of literary canon formation, namely:

> (1) A canon – a shared understanding of what literature is worth preserving – takes shape through a troubled historical process. (2) It emerges through specific institutions and practices, not in some historically invariant way. (3) These institutions are likely to have a rather well-defined class base.
>
> *(Ohmann 1983: 219)*

His fourth conclusion – that the ideology of the ruling class is reproduced in canon formation by a subordinate class acting in its perceived self-interest – is, I think, also applicable to the canonisation of the *Inner Classic*. However, limitations of space make it impossible to demonstrate this point here.

Taken as a case study of canonisation, the example of the *Inner Classic* offers several further insights. First, the canonicity of long-lived canonical texts is not static. Changes in social conditions, economic structures, cultural values, and technology all impact the claims that can be made for a text and the types of authority it can be granted. If a text remains important over a long period of time, these changes will inevitably be reflected in the way its canonicity is understood and even in the structure and physical form of the text itself. Second, the history surveyed here suggests that such changes will tend to be obscured by assertions that canonical texts possess eternal value and therefore unchanging form. Third, although canonisation is a restrictive phenomenon that operates by means of exclusion, it serves, somewhat counterintuitively, to enlarge the field of discourse by focusing discussion on a limited number of shared texts. This capacity of canonisation becomes particularly powerful after the advent of printing and the development of a thriving book market.

This chapter concludes with the Song dynasty because the medical canon formed by the Bureau for Editing Medical Texts has remained the core of the Chinese medical canon from that time up to today. This does not, however, indicate that either the canon or the canonicity of the *Inner Classic* has been static since that time. Research has shown that during the last thousand years, other texts have risen to and fallen from prominence and the relative importance of the texts within the canon has fluctuated greatly (Chapter 9 in this volume; Hanson 2003; Leung 2003; Simonis 2015; Vigouroux 2015). The process of asserting, disputing, and renegotiating the *Inner Classic*'s canonicity continues to this day. Far from indicating that its canonical status is in jeopardy, this engagement with the text is precisely the proof of the *Inner Classic*'s continuing significance for East Asian medicine. Canonicity is only fixed when the texts involved are dead; the canonicity of a living text is always an ongoing process.

Notes

1 The extant text, *Classic of Difficulties* (*Nanjing* 難經), is usually described as a commentary on the *Inner Classic*. However, there is no clear relationship between it and any extant *Inner Classic* text, and its history is very unclear (Keegan 1988: 31–3). For that reason, I have excluded it from consideration in this chapter.
2 Adding to the difficulty of dating, this part of the preface is written as commentary in small characters in the disputed *Kōhei* edition of the *Treatise on Cold Damage* (see Qian Chaochen 1993: 674–5).
 Editor's note: *Suwen* is often translated as 'Simple Questions', which readers can find elsewhere in this volume.
3 Miranda Brown points out that the textual history of the *Systematic Classic of Acumoxa* casts some doubt on the provenance of this preface; however, she concludes that the evidence available does not, so far, justify rejecting Huangfu Mi's authorship or disregarding the preface entirely (Brown 2015: 144–5).
4 The fact that this was seen as a necessary step in any serious study of a text in manuscript form is attested by Chao Yuezhi's (晁說之, 1059–1129) complaint that the ubiquity of printed texts had produced a generation of uncritical readers who didn't verify a text's accuracy by comparing it to other copies (Hymes 2015: 562).

Bibliography

Pre-modern sources

Beiji qianjin yaofang 備急千金要方 (Essential Formulae worth a Thousand Gold in Emergencies) 652, 孫思邈, Li Jingrong 李景榮 *et al.* (eds.) (1998) *Beiji qianjin yaofang jiaoshi* 備急千金要方校釋 (Formulae Worth a Thousand Gold in Emergencies with Annotations), Beijing: Renmin weisheng chubanshe.

Chongguang buzhu Huangdi neijing Suwen 重廣補註黃帝內經素問 (Yellow Emperor's Inner Classic: Inquiry into the Fundamental, Expanded with Additional Commentary) 1066, (*Gu Congde ben* 顧從德本 1550), Chen Yongguo 陳勇國 (ed.) (1989), Taipei: Tianzi chubanshe.

Hanshu 漢書 (History of the Han) 92, Ban Gu 班固 (1987), Beijing: Zhonghua shuju.

Huangdi neijing Taisu 黃帝內經太素 (Yellow Emperor's Inner Classic: Grand Fundamental) 7th century, Sago, Masami 左合昌美 (ed.) (2010) *Kōtei daikei taiso shin kōsei* 黃帝內經太素新新校正 (Yellow Emperor's Inner Classic: Grand Fundamental, Newly Annotated), 2nd edition, Tokyo: Japanese Inner Classic Study Society.

Huangdi sanbu zhenjiu jiayi jing 黃帝三部針灸甲乙經 (Yellow Emperor's Systematic Classic of Acumoxa in Three Parts) 282, Huangfu Mi 皇甫謐, Xu Guoqian 徐國仟 and Zhang Canjia 張燦玾 (eds) (1996) *Zhenjiu jiayi jing jiaozhu* 針灸甲乙經校注 (The Systematic Classic of Acumoxa, Edited with Commentary), Beijing: Renmin weisheng chubanshe.

Jingfang xiaopin 經方小品 (A Short Essay on Classical Formulae) late 4th–early 5th century, Chen Yanzhi 陳延之, in Yan Shiyun 彥世芸, Li Qizhong 李其忠 (eds) (2009) *Sanguo liang Jin Nanbeichao yixue zongji* 三國兩晉南北朝醫學總集 (Medical Anthology for the Three Kingdoms, Two Jin, and Northern and Southern Dynasties), Beijing: Renmin weisheng chubanshe, pp. 781–850.

Shanghan lun 傷寒論 (Treatise on Cold Damage) c. 206 [1065], Zhang Ji (張機, ca 150–219), Zhang Xinyong 張新勇 (ed.) (2010) *Zhongjing quanshu zhi Shanghan lun, Jingui yaoliie fanglun* 仲景全書之傷寒論、金匱要略方論 (The Complete Works of Zhongjing, Editions of the Treatise on Cold Damage and Essentials of the Golden Coffer), Beijing: Zhongyi guji chubanshe.

Xinkan Huangdi neijing Lingshu 新刊黃帝內經靈樞 (Yellow Emperor's Inner Classic: Numinous Pivot), *Lingshu (zuishanben)* 靈樞 (最善本) (Numinous Pivot [Best Edition]), Unpublished ms, Japanese Inner Classic Study Society.

Modern sources

Bates, D. (ed.) (1995) *Knowledge and the Scholarly Medical Traditions*, New York: Cambridge University Press.

Bol, P. (1994) *'This Culture of Ours': intellectual transitions in T'ang and Sung China*, Stanford, CA: Stanford University Press.

Boyanton, S. (2006) 'Putting the Yellow Emperor in his place', presented at the American Academy of Religion, Annual Conference, San Diego.
—— (2015) *The Treatise on Cold Damage and the Formation of Literati Medicine: social, epidemiological, and medical change in China 1000–1400*, PhD thesis, Columbia University.
Brown, M. (2015) *The Art of Medicine in Early China: the ancient and medieval origins of a modern archive*, New York: Cambridge University Press.
Chaffee, J. and Twitchett, D. (eds) (2015) *The Cambridge History of China, Vol. 5, Part 2: Sung China, 960–1279*, New York: Cambridge University Press
De Weerdt, H. (1999) 'Canon formation and examination culture: the construction of "guwen" and "daoxue" canons', *Journal of Song-Yuan Studies*, 29: 91–134.
Elman, B. (ed.) (2015) *Antiquarianism, Language, and Medical Philology: from early modern to modern Sino-Japanese medical discourses*, Boston, MA: Brill.
Goldschmidt, A. (2008) *The Evolution of Chinese Medicine: Song dynasty, 960–1200*, New York: Routledge.
Hanson, M. (2003) 'The "golden mirror" in the imperial court of the Qianlong emperor, 1739–1742', *Early Science and Medicine*, 8.2: 111–47.
Harper, D. (1997) *Early Chinese Medical Literature*, New York: Routledge.
—— (1999) 'Physicians and diviners: the relation of divination to the medicine of the *Huangdi neijing* (Inner Canon of the Yellow Thearch)', *Extrême-Orient Extrême-Occident*, 21: 91–110.
Hinrichs, TJ (2013) 'The Song and Jin periods', in TJ Hinrichs and L.L. Barnes (eds) *Chinese Medicine and Healing: an illustrated history*, Cambridge, MA: Harvard University Press, pp. 97–128.
Hinrichs, TJ and Barnes, L.L. (eds) (2013) *Chinese Medicine and Healing: an illustrated history*, Cambridge, MA: Harvard University Press.
Hymes, R. (2015) 'Sung society and social change', in J. Chaffee and D. Twitchett (eds) *The Cambridge History of China, Vol. 5, Part 2: Sung China, 960–1279*, New York: Cambridge University Press, pp. 526–664.
Jochim, C. (1990) 'Flowers, fruit, and incense only: elite versus popular in Taiwan's religion of the Yellow Emperor', *Modern China*, 16: 3–38.
Keegan, D.J. (1988) *The 'Huang-Ti Nei-Ching': the structure of the compilation; the significance of the structure*, PhD thesis, University of California, Berkeley.
LeBlanc, C. (1985) 'A re-examination of the myth of Huang-ti', *Journal of Chinese Religions*, 13: 45–63.
Leibold, J. (2006) 'Competing narratives of racial unity in Republican China: from the Yellow Emperor to Peking Man', *Modern China*, 32: 181–220.
Leung, A.K.C. (2003) 'Medical instruction and popularization in Ming-Qing China', *Late Imperial China*, 24: 130–52.
Loewe, M. (ed.) (1993) *Early Chinese Texts: a bibliographical guide*, Society for the Study of Early China and the Institute of East Asian Studies, Berkeley: University of California.
Marsili, F. (2003) 'The myth of *Huangdi*, the ding vases, and the quest for immortality in the *Shiji*: some aspects of Sima Qian's "Laicism"', *Rivista degli Studi Orientalivol*, 77: 135–68.
Miyashita, Saburō 宮下三郎 (1967) 'Sō Gen no iryō 宋元の医療 (Song-Yuan Medicine)', in Yabuuchi, Kiyoshi 藪内清 (ed.), *Sō Gen jidai no kagaku gijutsushi* 宋元時代の科学技術史 (History of Science and Technology in the Song-Yuan Era), Kyōto: Kyōto Daigaku Jinbun Kagaku Kenkyūjo, pp. 123–70.
Nylan, M. (2000) 'Textual authority in Pre-Han and Han', *Early China*, 25: 205–56.
Ohmann, R. (1983) 'The shaping of a canon: U.S. fiction, 1960–1975', *Critical Inquiry*, 10: 199–223.
Okanishi, Tameto 岡西為人 (2010) *Sō izen iseki kō* 宋以前醫籍考 (Investigation of Medical Texts from the Song and Earlier), Beijing: Xueyuan chubanshe.
Peerenboom, R.P. (1993) *Law and Morality in Ancient China: the silk manuscripts of Huang-Lao*, Albany: State University of New York Press.
Qian Chaochen 錢超尘 (1993) *Shanghan lun wenxian tongkao* 傷寒論文獻通考 (Bibliographic Investigation of the *Treatise on Cold Damage*), Beijing: Xueyuan chubanshe.
Shinno, R. (2007) 'Medical schools and the temples for the three progenitors in Yuan China: a case of cross-cultural interactions', *Harvard Journal of Asiatic Studies*, 67: 89–133.
Simonis, F. (2015) 'Illness, texts, and "schools" in Danxi medicine: a new look at Chinese medical history from 1320 to 1800', in B. Elman (ed.) *Antiquarianism, Language, and Medical Philology: from early modern to modern Sino-Japanese medical discourses*, Boston, MA: Brill, pp. 52–80.

Sivin, N. (1993) '*Huang-ti nei-ching* 黃帝內經', in M. Loewe (ed.) *Early Chinese Texts: a bibliographical guide*, Society for the Study of Early China and the Institute of East Asian Studies, Berkeley: University of California, pp. 196–215.

——— (1995a) 'Text and experience in classical Chinese medicine', in D. Bates (ed.) *Knowledge and the Scholarly Medical Traditions*, New York: Cambridge University Press, pp. 177–205.

——— (1995b) 'State, cosmos, and body in the last three centuries B.C.', *Harvard Journal of Asiatic Studies*, 55: 5–37.

Unschuld, P.U. (2003) *Huang Di nei jing su wen: nature, knowledge, imagery in an ancient Chinese medical text*, Berkeley: University of California Press.

Vigouroux, M. (2015) 'The reception of the circulation channels theory in Japan (1500–1800)', in B. Elman (ed.) *Antiquarianism, Language, and Medical Philology: from early modern to modern Sino-Japanese medical discourses*, Boston, MA: Brill, pp. 105–32.

Yabuuchi, Kiyoshi 藪內清 (ed.) (1967) *Sō Gen jidai no kagaku gijutsushi* 宋元時代の科学技術史 (History of Science and Technology in the Song-Yuan Era), Kyōto: Kyōto Daigaku Jinbun Kagaku Kenkyūjo.

Yamada, Keiji (1979) 'The formation of the *Huang-ti nei-ching*', *Acta Asiatica*, 36: 67–89.

Yan Shiyun 彥世芸 and Li Qizhong 李其忠 (eds) (2009) *Sanguo Liangjin Nanbeichao yixue zongji* 三國兩晉南北朝醫學總集 (Medical Anthology for the Three Kingdoms, Two Jin, and Northern and Southern Dynasties), Beijing: Renmin weisheng chubanshe.

Yates, R. (trans.) (1997) *Five Lost Classics: Tao, Huanglao, and Yin-Yang in Han China*, New York: Ballantine Books.

8
PRE-STANDARDISED PHARMACOLOGY
Han through Song

Asaf Goldschmidt

Throughout history, the most important and widely used method of treatment in Chinese medicine was drug therapy, as evidenced by the amount and variety of extant pharmacological literature. In this chapter, we will delineate the development of this literature from its earliest and eclectic forms to its standardised form that began with the integration of canonical doctrines in the Song dynasty (960–1276). First, we need to understand Chinese doctors' perception of drugs.

In traditional Chinese medicine, there are two distinct terms which fall under the broad Western term of 'drug' – *yao* 藥, hereafter translated as 'drug', and *fang* 方, hereafter 'formula'. Accordingly, literature concerning the broad category of drug therapy in China divides into two predominant literary genres: *materia medica* collections (*bencao* 本草), which discuss simples, and formularies (*fangshu* 方書), which focus on formulæ. A drug is a specific singular medicinal material (originating from plants, animals, or minerals), which can be applied in the treatment of disorders. Drugs, in Chinese medicine, can come in either crude or prepared form. A crude drug (*shengyao* 生藥) – literally 'fresh drug' – is the medicinal material in its natural form. A processed or prepared drug (*shuyao* 熟藥) is a medicinal material that has undergone some preparation, such as frying or steaming, to enhance preservation or alter its medicinal properties.

In general, physicians formulated combinations of drugs usually with complementary effects to achieve the desired clinical results. By doing so, they avoided undesirable side effects usually associated with consuming a single drug. Constructing an effective formula involves more than simply combining drugs of complementary effects to obtain a desired therapeutic result. Physicians established guidelines for achieving optimal efficacy in combining drugs so as to enhance drugs' curative effects and moderate their side effects. The most common guideline is based on four bureaucratic ranks, according to which the ingredients of the formula are categorised. This categorisation reflects the hierarchy of the Imperial Court and was titled accordingly: Monarch (*jun* 君), Minister (*chen* 臣), Assistant (*zuo* 佐), and Envoy (*shi* 使).[1] Formulæ come in various forms, including, for example, decoctions, powders, pills, pellets, ointments, and patches.

Turning to drug therapy literature, the most natural translation of the term *Bencao* into English may be 'basic herbs' or 'fundamental simples'; however, *materia medica* better reflects the essence of these books. *Materia medica* collections often served as reference texts during the education of a physician as they grouped together drugs initially according to their effectiveness and later according to their origin – plants, minerals, and animals. They described the innate characteristics of the various drugs, especially the four thermo-influences or *qi* of drugs (*siqi* 四氣: hot, warm, cold, and cool) and the 'five sapors' (*wuwei* 五味: pungent, sweet, salty, sour, and bitter), and they listed the symptoms each drug alleviates. They often described the ability of the drug to lengthen life, the season and time of day to gather the drug, the desirable part of the plant that serves as a drug, and the preparation methods. The recorded data are essential for understanding the strength and the effects of each drug. In the era before laboratory analysis, the environment (dry vs wet/humid, for example) and the season the plant grew in were the only means to control the concentration of active ingredients in drugs of vegetable origin.[2]

Formularies listed 'proven remedies' or efficacious formulæ accumulated by physicians in their clinical practice. These formularies also often recorded acu-moxa treatments, amulets, incantations, healing rituals, and other practices for treating disorders. Physicians transmitted them to their disciples and at some point the accumulated knowledge was published. Historically, formularies constitute the most important literary genre of clinical Chinese medicine. Formularies grouped together the formulæ according to the major symptoms that they treated. In contrast to *materia medica* collections, which were somewhat encyclopaedic in nature, formularies recorded the decades-long accumulated clinical knowledge of clinical practitioners. Throughout history, physicians recorded their effective formulæ in notes and handed them down to their disciples or family members, rarely expanding their circulation as they were the means of obtaining a livelihood. Unlike *materia medica* collections, formularies did not descend from one constituting classic that was revised over and over again. Rather, formularies were often testaments to the clinical knowledge of a private doctor and did not have the social authority of a canon.

Formative era – the earliest texts – third century BCE to third century CE

It is difficult to determine when drug therapy began in China since we have no documents dating before the third century BCE, but we assume it has been in use from early human civilisation. The earliest surviving texts on drug therapy are not systematic compilations but rather are excavated texts dating to the Han dynasty (202 BCE–220 CE); additionally, one *materia medica* and one formulary have survived to the present from this era. It is probably due to doctors' reluctance to share their proven remedies that we have such scant literature on drug therapy from this period. These texts, at least during this period, were of very little interest to the elite or to government officials who managed the imperial archives.

The earliest textual record of prescriptions dates to the Qin dynasty and was discovered in a well in Liye 里耶, Hunan in 1996. Among the thousands of boards and slips unearthed, at least twelve recorded medical prescriptions and even mentioned an official doctor (Yates 2012). The earliest complete text devoted to medical prescriptions, unearthed from a Han-dynasty tomb originally sealed in 168 BCE in Mawangdui in Hunan Province, is titled *Prescriptions for Fifty-Two Ailments* (*Wushier bingfang* 五十二病方) (Harper 1998). As the name suggests, the text is arranged according to fifty-two categories of ailment. In fact, the text ends with additional prescriptions for several more ailments deviating from the original list written at the beginning of the text, listing in total 283 prescriptions. The diseases predominantly revolve around external injuries such as flesh wounds, warts, haemorrhoids, and snake bites, but there are also other disorders such as injury of child by a demon. The prescriptions are sometimes presented in a sentence or two, and sometimes provide much longer descriptions of symptoms, drug preparation, and therapy. It is important to note that the text often lists prescriptions that include exorcistic incantations or magico-religious operations along with drug formulæ for the treatment of the same disorder. The text cites various methods of treatment, including moxibustion, cupping, and massage, but the application of herbs is the predominant method of treatment. The format of the text is straightforward: a disorder is listed and the treatment is given. No theoretical discussions are found. This leaves us to surmise that either the author assumed that the reader would understand the medical theories underlying the treatments or the treatments themselves were simply pragmatic, with no particular theoretical underpinnings.

The second surviving text recording prescriptions was unearthed from another Han-dynasty tomb originally sealed in the mid-first century CE in Wuwei 武威 in Gansu Province. It is titled *Han Dynasty Medical Bamboo Strips from Wuwei* (*Wuwei Handai yijian* 武威漢代醫簡), later named *Formulary for Curing all Disorders* (*Zhi baibing fang* 治百病方) (Yang and Brown 2017). This text represents a more developed material culture for drug preparations than the *Prescriptions for Fifty-Two Ailments*, listing decoctions, pills, powders, syrups, and many external applications as well. This formulary covers a wider array of disorders than its predecessor, including external injuries, female and children's disorders, and various diseases such as cough, breathing problems, eye and ear problems, and even bleeding, resembling the diversity of later formularies. This formulary's prescriptions are predominantly drug formulæ and the majority of the recorded drugs also appear in the *Bencao jing*, discussed below. The text also discusses the preparation of decoctions, pills, ointments, powders, and so on. It also provides instructions on how and when to administer the prescriptions. There are altogether thirty-six prescriptions in it.

The earliest formulary not from archaeological excavations is the *Treatise on Cold Damage and Miscellaneous Disorders* (*Shanghan zabing lun* 傷寒雜病論), written at the end of the Eastern Han dynasty by an official turned physician, Zhang Zhongjing 張仲景 (150–220 CE).[3] This formulary brought forth a new system of diagnosis based on the six warps (or stages, *liujing* 六經) of disease as well as a well-developed taxonomy of diseases. In addition, each of the formulæ is given a name, and the dosage and method of preparation of the herbs are specifically described. In fact, this book provides by far the greatest single source of formulæ in traditional Chinese medicine. Zhang, who described himself primarily as a collector rather than as a composer of formulæ, was the first person we know of to identify the condition of a patient (the diagnosis) with a particular formula used to treat that condition (Goldschmidt 2009, ch. 3; Brown 2015, ch. 5).

An unknown author or authors wrote the first *materia medica* in Chinese history during the Han dynasty, titled the *Divine Husbandman's Materia Medica* (*Shennong bencao jing* 神農本草經, *Bencao jing* for short). This book, lost not long after its completion, survived due to the fact that a fifth-century Daoist priest, Tao Hongjing 陶弘景, reconstructed it. The *Bencao jing* was the first of a line of encyclopaedic books that listed medicinal materials or drugs in a systematic manner. The full title of the *Bencao jing* includes the name of a legendary emperor, the Divine Husbandman (*Shennong* 神農), who, according to tradition, ruled during the third millennium BCE. He was said to have thrashed all plants and other drug materials to extract their essential qualities, tasted them, and then classified them according to their value as foodstuffs and drugs. According to the same tradition, he poisoned himself and found the antidote in each and every instance.[4]

The *Bencao jing*, now believed to be a compilation dating to the second or first century CE, organised its drugs into three classes. The upper class (*shangpin* 上品), or monarchs, included 120 drugs, which were considered to be non-toxic (or non-curative) and aimed to prolong life. These drugs would not cause harm even if taken in large doses for long periods of time. The middle class (*zhongpin* 中品), or ministers, included another 120 drugs that were designed to prevent diseases and were somewhat toxic (or with curative effects). These drugs could be dangerous depending on dosage and on the other drugs with which they were used. The lower class (*xiapin* 下品), or assistants and envoys, included 125 drugs that were considered to be toxic with side effects, and were specifically used for therapeutic purposes to treat diseases or to produce medicinal effects.[5]

Altogether, the *Bencao jing* listed a total of 365 Chinese medicines of which 252 were of plant origin, 67 from animals, and 46 from minerals. For each drug listed, the canon discussed its thermal influence or *qi* as well as its sapor (see above). It also discussed drug preparation methods, recommended doses, and the correct time to consume the drugs based on its traits. The major treatment strategy offered in the canon is treatment by opposites, for example, heat symptoms or heat disease should be treated by cold or cooling drugs.

In sum, by the end of the third century CE, drug therapy literature, unlike the literature of canonical classical medicine which was under official auspices, was not in wide circulation and was often transmitted within lineages of physicians. It took a major change in both the political and religious scenes to bring drug therapy to the fore.

Second stage – fourth to tenth centuries: the expansion and systematisation of knowledge

After the collapse of the Han dynasty, China splintered into a series of small kingdoms, beginning centuries of political disunity initiating an era known as the Northern and Southern Dynasties (317–589) or as the Six Dynasties period (222–589). North China was dominated by the Sixteen Kingdoms (304–438) ruled by five non-Chinese peoples later to be unified under the Northern Wei dynasty (386–535). The period of disunity and the Sui (581–618) and Tang (618–907) dynasties which followed were an era when drug therapy and its literature flourished and evolved rapidly. Without the early imperial preference for canonical medicine that favoured the cosmology of systematic correspondence and with the ever-growing penetration of Buddhism into China and the expansion of the Daoist religion, this period saw the growing production of practical books discussing drug therapy, including *materia medica* literature and formularies. One of the highlights of this era was the appearance of the 'great' religions in China, namely Buddhism and the Daoist religion. The impact of both on medicine is discussed in depth in Chapters 27 and 28 in this volume, but we have to reiterate the latter's decisive impact on drug therapy, predominantly on *materia medica* literature. During this era, *materia medica* literature expanded significantly, some claim due to Daoists' interest in longevity and their search for the immortality elixir though this is probably only partially, if at all, true.

Materia medica literature advanced dramatically during this era. One of the most important figures in the development of drug therapy literature in Chinese medicine was Tao Hongjing 陶弘景 (456–536). Tao not only reconstructed the original *Bencao jing*, but added to that his original knowledge of *materia medica*, expanding the number of listed drugs from 365 in the original *Bencao jing* to 730 in his work. Additionally, he augmented the information on the herbs' nature, location, and time of harvesting. Tao categorised drugs into seven groups: 'jade and stones', 'herbs and trees', 'insects and animals', 'grains', 'fruits', 'vegetables', and 'miscellaneous', and many later medical practitioners followed his classification system. One of the additional sources he drew on was a collection referred to as *Separate Records of Famous Physicians* (*Mingyi bielu* 名醫別錄). Although both the original and Tao's texts were later lost, they have been reconstructed based on later *bencao* literature. Tao's book entitled [*Divine Husbandman's*] *Materia Medica, with Collected Annotations* (*Bencao jing jizhu* 本草經集注) dominated pharmaceutical literature until the middle of the seventh century CE, when the Tang government sponsored the compilation of an official *materia medica*.

In 659, a group of over twenty officials and physicians led by Su Jing 蘇敬, working under an imperial directive issued by Tang emperor Gaozong (650–683), compiled a new official *materia medica* that followed and relied on Tao Hongjing's *materia medica*. The *Tang Materia Medica* (*Tang bencao* 唐本草, also known as the *Newly Revised Materia Medica* [*Xinxiu bencao* 新修本草]) included a total of 850 drugs. The Tang government made this book the official standard with regard to drug usage, though we are not sure to what extent this regulation was implemented beyond the Imperial Palace and maybe the capital, since without print technology few physicians are likely to have had access to the book or to have been able to read it. This book had special significance, not only because it was the first officially compiled *materia medica* in China but also because it standardised the knowledge of *materia medica* developed since the first century CE and provided illustrations in addition to the written text. The book, however, has not survived intact to the present, and what we know about it is through scattered quotations in other later works.

Su's text was one of many medical compendia compiled in a period when the trade in natural and medical objects between China and its neighbours flourished. Tang China saw a new market for drugs, medicines, and spices. A burgeoning economy and the fact that the Tang capital of Chang'an became a cosmopolitan urban centre, drawing merchants from all over Asia, greatly expanded the number of drugs used as ingredients in pharmaceutical recipes. Buddhist monks who travelled to China also facilitated the transmission of Indian medicine and Indian pharmaceutical knowledge to China. The adoption of knowledge from India, however, was by and large limited (Salguero 2014, 2017).

Materia medica literature was not the only drug therapy literature to flourish during this period, which also saw a growing production of practical books of therapeutic recipes, or formularies. Two of the most important formularies of Chinese medicine, and the first belonging to the 'great formularies' genre, date to the Tang dynasty.

The earliest formularies took the form of manuals often for urgent use rather than comprehensive cover-all books. A good representative of this early formularies is the most important formulary of the Jin dynasty – Ge Hong's 葛洪 (?283–?343) *Emergency Formulas to Keep Up One's Sleeve* (*Zhouhou beiji fang* 肘後備急方, published between 306 and 333).[6] Ge Hong, who is best known for his writings on alchemy, authored this formulary introducing a rational structured presentation of formulæ. This book details simple and inexpensive, yet effective, formulæ and their application by symptom. Ge Hong thereby provided an organised set of formulæ making the text more useful as a reference for physicians and maybe laypeople. Tao Hongjing, mentioned above, further edited and supplemented Ge Hong's formulary, in *One Hundred and One Supplementary Formulas to Keep up One's Sleeve* (*Buque zhouhou baiyi fang* 補闕肘後百一方).[7] In general during this era, collections of formulæ, targeting either specific disorders or specific groups of society (women, men, or children), became popular. This has probably to do with the new notion of 'proven formulæ' (*yanfang* 驗方) that served as the basis for the majority of these publications.

In addition to shorter formularies designed for laypeople's practical reference, the period saw the production of works that aimed to comprehensively collect and organise therapeutic knowledge, something more useful to the learned physician, serious medical scholar, or official educator. In this category, we find Sun Simiao's 孫思邈 (581–682) *Essential Prescriptions Worth a Thousand Gold, for Urgent Need* (*Beiji qianjin yaofang* 備急千金要方, pub. 659), and Wang Tao's 王燾 (670–755) *Arcane Essentials from the Imperial Library* (*Waitai miyao* 外台秘要, pub. 752).[8] These were the two grand formularies of the Tang dynasty. During the Song dynasty, the government printed these two works, establishing them as part of the canonical medical literature. Together, these works assembled thousands of remedies, from which we can learn much not only of Tang medicine but also of daily life. Wang's collection quoted extensively from Six Dynasties works that are no longer extant, preserving much that otherwise would have been lost. Sun's work, which represents one of the greatest works by a single physician, presented to the reader the complete scope of a physician's clinical practice, including ethical dilemmas and healing practices borrowing from religions (Unschuld 1979: 24–35; Sivin 2017).

Another notable formulary of this era, though it was compiled during the tenth century in Japan, is the *Prescriptions at the Heart of Medicine* (*Ishimpō* 醫心方).[9] A Japanese scholar, Tamba Yasuyori 丹波康賴, compiled this formulary between the years 982 and 984 relying on over 200 earlier Chinese medical books, the majority of which have not survived to the present. Citations of these earlier works are very well documented in this text, making it an ideal source, reflecting the state of prescriptions during the Tang dynasty. It is important to note that many manuscripts from Dunhuang include formulæ; however, none is a comprehensive formulary like those listed above (Lo and Cullen 2005; Despeux 2010).

To sum up, drug literature during this stage contained great amounts of drugs but with relative lack of discrimination probably since a certain level of knowledge was presumed of the readers. Drug therapy literature during this stage focused on the treatment of diseases (*bing* 病) rather than differentiating patterns (*zheng* 證), of which there were hundreds. Consequently, the authors of these texts listed under each disease up to twenty formulæ without any indication as to which of them is most effective in a particular situation. Furthermore, many of the formulæ from this era contained a large number of herbs.

Third stage – tenth to thirteenth centuries: standardising drug therapy and integrating with canonical doctrines

The Song dynasty (960–1276) was an era of major changes in every aspect of life. The change with the most impact on drug therapy was the large-scale implementation of print technology leading to the democratisation of medical knowledge (Goldschmidt 2005, 2009). Another change was the fact that two empires established to the west and to the north of the Song limited its land commerce via the Silk Road. Consequently, the Song expanded and intensified maritime trade with Southeast Asia enabling access to a new drug lore not readily available to previous generations (earlier drugs were predominantly from India and Central Asia) (Lo Jung-pang 2012; Schottenhammer 2015). The Song also exhibited a change in society and in social order leading to a growing interest in medicine among scholar-officials and shifting the focus of the bureaucracy to regions previously ignored by the government. These changes created a new context in which drug therapy underwent major transformations in drug therapy literature, predominantly the standardisation of *materia medica* literature and the incorporation of canonical medical doctrine into the discussion of the characteristics and effects of drugs and formulæ (Goldschmidt 2009).

Up to the Song, drug therapy literature flourished and expanded, however, without incorporating canonical doctrines of physiology and pathology into its discussion. Moreover, records of drugs and formulæ were not standardised and often identical drugs or formulæ were listed under different names or different drugs were designated by the same name leading to much confusion. The origin of the transformations in drug therapy can be traced to the Song government's printing of drug therapy manuals thereafter extending doctors' familiarity with the literature. The Song government's wide-ranging and unprecedented book-collection projects led not only to centralised, standardised, and hybridised medical knowledge, but also to equally unprecedented dissemination of newly forged medical texts through the feverish literary activities of scholar-officials (Goldschmidt 2009). Thus, older texts collected by Song officials initially served as the foundation for authoritative government *materia medica* collections and grand-scale formularies. Later, this same foundation was used for new styles of drug therapy literature.

Our discussion of drug therapy literature during the Song should be divided into two stages: the first consisted of printing and disseminating earlier literature, whereas the second consisted of standardising and then integrating canonical doctrines into the contents of these books. Early during the Song dynasty, the imperial government initiated a few projects for the collection and printing of drug therapy manuals resulting in one edition of *materia medica* and two formularies. The founder of the Song dynasty, Taizu 太祖 (r. 960–976), instructed his officials to compile a new *materia medica* collection. The newly printed *Kaibao Materia Medica* (*Kaibao bencao* 開寶本草, pub. 974) was designed to make *materia medica* literature widely available to a broader audience. This manual expanded the number of recorded drugs, adding 134 records, but made no other modifications. During the early years of the Song, the government published two grand formularies. In the year 981, the Imperial Court issued an order to compile the first government-sponsored formulary. A group of editors headed by Jia Huangzhong (賈黃中, 941–990) worked on this project for five years. In 986, the editors completed a gigantic manuscript entitled the *Divine Doctor's Formulary for Universal Relief* (*Shenyi pujiu fang* 神醫普救方). This enormous work consisted of 1,000 substantial chapters, the table of contents alone comprising ten chapters. This work has not survived to the present. In 982, the second emperor of the Song, Taizong 太宗 (r. 976–997), issued a decree instructing Wang Huaiyin (王懷隱, fl. 978–992), who was a medical official of the Hanlin Academy and the Chief Steward of the Palace Medical Service, to compile a new formulary. After ten long years, in 992, Wang and his associates completed work on what had become one of the largest government-sponsored medical compilation projects to survive to the present, the *Imperial Grace Formulary of the Great Peace and Prosperity Reign Period* (*Taiping shenghui fang* 太平聖惠方, pub. 992). It consisted of 100 chapters and included 16,834 different formulæ, including some from Taizong's personal collection of formulæ.[10] In order to stress this book's importance, Emperor Taizong personally wrote a preface for it.

The second stage of drug literature arose from the realisation that the uncritical reprinting of ancient knowledge was insufficient. This realisation occurred when the Emperor Renzong 仁宗 (r. 1023–63) ordered the publication of yet another revised version of ancient *materia medica*, titled *Jiayou Bencao* (嘉祐本草, pub. 1061), during the 1060s. Facing inconsistencies within the text, the editors of the Bureau for Printing Medical Books decided to rejuvenate and revolutionise *materia medica* literature by initiating an empire-wide survey of drug knowledge. They requested all the prefectures as well as entry ports into China to send them samples of all drugs grown at or imported to that locality along with local knowledge about these drugs. The editors then compared the samples with recorded information on drugs in existing *materia medica* collections and compiled the most up to date *materia medica* manual, the *Illustrated Materia Medica* (*Bencao tujing* 本草圖經, pub. 1062). The *Illustrated Materia Medica*, as is evident from its name, also provided illustrations of all the drugs to aid identification. Early in the twelfth century, the Song government sponsored its last *materia medica* combining previously published information with many southern drugs recorded in a Tang-dynasty manual. This created the greatest *materia medica* collection to that date, entitled the *Classified Materia Medica* (*Zhenglei bencao* 證類本草, pub. 1108, 1116), which recorded 1,748 drugs. This book served as the gold standard for drugs until Li Shizhen 李時珍 (1518–93) compiled his *Systematic Materia Medica* (*Bencao gangmu* 本草綱目) in 1596 (Lu Gwei-djen 1966; Sivin 1973; Métailié 2001, 1990; Nappi 2009; Bian 2020).

In 1117, an official-physician named Kou Zongshi (寇宗奭, fl. early twelfth century) privately compiled his *Dilatations on Materia Medica* (*Bencao yanyi* 本草衍義) in which he began to integrate classical doctrines into the explanation of drugs' effects for the first time ever. This process continued for over a century until the complete integration of these doctrines into drug descriptions in Wang Haogu's (王好古, fl. thirteenth century) *Materia Medica for Decoctions* (*Tangye bencao* 湯液本草, pub. c.1238–48). The accessibility to the ancient literature, both canonical and drug therapies, created recognition that there was incompatibility in the knowledge since the traits of drugs and formulæ did not make use of canonical physiology and pathology. This compelled doctors to build bridges and integrate the two genres into one systematic medical tradition (Unschuld 1977; Goldschmidt 2009; Chapters 4 and 5 in this volume).

Song formularies also changed during the early twelfth century. In 1122, the Song government published a second grand formulary – the *Medical Encyclopaedia: A Sagely Benefection of the Zhenghe Reign Period* (*Zhenghe shengji zonglu* 政和聖濟總錄). This formulary consisted of 200 chapters and recorded over 20,000 formulæ. The biggest innovation of this formulary was the fact that throughout the text classical doctrines are interwoven with the discussion of formulæ. This formulary included information collected from practitioners as well as from contemporary and ancient medical texts. The contents of the book are divided into sixty-six categories of general manifestation types or patterns (*zheng* 證), which does not differ from earlier formularies. But instead of just providing the reader with the specific formula designed to alleviate the symptom with limited elaboration on how and why, the *Medical Encyclopaedia* adds to the discussion of the formula a thorough theoretical explanation serving as a background to better understand the treatment. For example, phrases such as 'Treating deficient heart *qi*' or 'Treating original deficiency in the spleen' occur with much greater frequency than in the earlier Song formulary, the *Imperial Grace Formulary*. In contrast to earlier formularies, the number of infusions in this book is limited, with a sharp increase in other forms of prescriptions such as powders, ointments, pellets, and boluses. This change is probably linked to the establishment of the Imperial Pharmacy, which sold these new forms of pre-prepared prescriptions rather than providing the ingredients for formulæ that the patient had to boil at home in order to make the infusion (Goldschmidt 2008).

The standardisation in formulæ took longer and occurred due to an external trigger. In 1076 the Song government established the Imperial Pharmacy initially to regulate the drug market, and later to benefit the people by supplying a wide range of drugs at a subsidised price. In 1110, the Imperial Pharmacy published a formulary listing all the pre-prepared prescriptions on sale at its branches along with the symptoms they treated. The extant version we have lists 297 formulæ, an almost insignificant number in comparison to the *Medical*

Encyclopaedia (Goldschmidt 2008). At the same time, physicians began looking to reconcile the discordant information recorded in the government-printed formularies with both their clinical know-how, accumulated over generations of medical practice, and the physiological and pathological doctrines delineated in the newly available Han-dynasty canons. These tensions concerning the clinical application of formulæ and textual knowledge, along with the need to demarcate themselves from the ready-to-be-applied over-the-counter prescriptions sold at the Imperial Pharmacies, pushed these physicians to create theoretical bridges to span the gap between medical theory and drug therapy practices.

In sum, by the end of the Song dynasty, Chinese drug therapy was transformed. First, it underwent systematisation and standardisation following the many editorial projects sponsored by the imperial government. Formularies became all-inclusive mammoth compilations that contained virtually all earlier practical medical information on record, given that Song officials were now able to obtain copies of the texts. *Materia medica* collections expanded dramatically, more than doubling the number of recorded drugs. Additionally, their contents were standardised by imperial editors and physicians, producing a unified illustrated *materia medica*. Second, during the twelfth and thirteenth centuries, drug therapy was integrated with classical cosmology, physiology, and pathology. This change was much more profound and took a longer time to complete. By the end of the Song dynasty, we find a standardised medicine that discusses drugs and formulæ using terminology originally belonging to classical canonical medicine (Tables 8.1 and 8.2).

Table 8.1 Important *Materia Medica* Collections and Number of Drugs Included

Title	Year of Compilation	Compiler	Number of Drugs	Increase in Number
Divine Husbandmen's Materia Medica, *Shennong bencao jing* 神農本草經	25–220	Anonymous	365	
[Divine Husbandman's] Materia Medica, with Collected Annotations, *Shennong bencao jing jizhu* 神農本草經集注	530–557	Tao Hongjing 陶弘景	730	365
Tang Materia Medica, *Tang bencao* 唐本草	659	Su Jing 蘇敬	850	120
Kaibao Materia Medica, Kaibao bencao 開寶本草	974	Liu Han 劉翰 and Ma Zhi 馬志	984	134
Jiayou Materia Medica, Jiayou bencao 嘉祐本草	1057–61	Zhang Yuxi 掌禹錫	1083	99
Illustrated Materia Medica, *Bencao tujing* 本草圖經	1058–62	Su Song 蘇頌	1186	103
Classified Materia Medica *Zhenglei bencao* 證類本草	1108 and 1116	Tang Shenwei 唐慎微 (Ai Sheng 艾晟)	1748	562
Dilatations on Materia Medica, Bencao yanyi 本草衍義	Compiled 1106–16 (published in 1119)	Kou Zongshi 寇宗奭	472	

Table 8.2 Important Formularies and Number of Formulæ Included

Title	Year	Compiler	Number of Formulæ
Essential Prescriptions Worth a Thousand [for Urgent Need], [Beiji] qianjin yaofang 備急千金要方	659	Sun Simiao 孫思邈	3,500 232 entries (according to Zhongyao xueshi 中藥學史)
Arcane Essentials from the Imperial Library Waitai miyao 外臺秘要	752	Wang Tao 王燾	Over 6,000 1,048 entries (Ibid.)
Imperial Grace Formulary, Taiping shenghui fang 太平聖惠方	978–992	Wang Huaiyin 王懷隱	16,834
Formulary of the Imperial Pharmacy, Taiping huimin hejiju fang 太平惠民和濟局方	Daguan reign 1107–10	Chen Cheng 陳承 et al.	Originally 297, final version 788
Medical Encyclopaedia, Shengji zonglu 聖濟總錄	1122	Shen Fu 申甫	20,000

Notes

1. On these hierarchies and their role in formulæ, see Scheid et al. (2009). For an English version of materia medica, see Bensky, Clavey, and Stöger (2004). For a broader discussion of plants and botany, see Needham and Lu (1986) and (2015).
2. Editor's Note: Research ongoing today shows demonstrable differences in concentrations of active ingredients depending on the location the plants come from, even within the same species (Chapter 48 in this volume).
3. Although this is the earliest surviving formulary, it is not the ancestor of the line of formularies, fangshu 方書, that we later find during the Tang dynasty.
4. For a detailed discussion of Emperor Shennong, see Henricks (1998). For a comprehensive overview of the history of bencao literature up to and during the Song dynasty, see Unschuld (1986: 11–53 and 53–84), respectively. For a textual history of bencao during the Han dynasty, see Schmidt (2006).
5. See Unschuld (1986:11–53) and Schmidt (2006). For a translation of the book, see Wilms (2016).
6. For an additional discussion, see Stanley-Baker (2021).
7. Tao Hongjing also wrote a formulary entitled the Experientially Proven Recipes (Xiaoyan fang 效驗方). He composed this formulary, which summarised his own practical experience, before editing Ge Hong's formulary.
8. For the biography of Sun Simiao, see Sivin (1968).
9. The only partial and somewhat problematic translation of this work appears in Hsia, Veith, and Geertsma (1986).
10. It is important to note that the number of formulæ in the Divine Doctor's Formulary for Universal Relief is unknown. Although the Imperial Grace Formulary included over 16,000 formulæ in a tenth as many chapters as the former, the lengths of chapters in Chinese books varied considerably.

Bibliography

Pre-modern sources

Materia Medica collections

Bencao jing jizhu 本草經集注 ([Divine Husbandman's] Materia Medica, with Collected Annotations) pr. c. 6th century, Tao Hongjing 陶弘景 (456–536 CE), repr. Bencao jing jizhu: jijiao ben 本草经集注辑校本, Shang Zhijun 尚志钧 and Shang Yuansheng 尚元胜 (eds) (1994), Beijing: Renmin weisheng chubanshe.

Bencao gangmu 本草綱目 (Systematic *Materia Medica*) 1596, Li Shizhen 李時珍 (1518–1593), commentary by Liu Hengru 刘衡如 and Liu Shanyong 刘山永 (1998), 2 vols, Beijing: Huaxia chubanshe. Also in Liu Changhua 柳长华 (ed.) (1990) *Li Shizhen yixue quanshu* 李时珍医学全书, Beijing: Zhongguo zhongyiyao chubanshe.

Chongxiu Zhenghe xinxiu jingshi zhenglei beiyong bencao 重修政和新修經史證類備用本草 (Revised Zhenghe Reign *Materia Medica* for Urgent Use, Classified and Verified from the Classics and Histories) 1249, edited by Zhang Cunhui 张存惠 (1993), Beijing: Huaxia chubanshe. This title is often abbreviated to *Zhenglei bencao*.

Shennong bencao jing 神農本草經 (Divine Husbandman's Canon of *Materia Medica*) Late first or second century CE (lost), Anonymous. For the best reconstruction, see Ma Jixing 馬繼興 (ed.) (1995) *Shennong bencao jing jizhu* 神農本草經輯注, Beijing: Renmin weisheng chubanshe. See also Shang Zhijun 尚志鈞 (ed.) (2008) *Shennong bencao jing jiaozhu* 神農本草經校注, Beijing: Xueyuan chubanshe.

Xinxiu bencao 新修本草 (Revised *Materia Medica*) 658, Su Jing 蘇敬 (650–659), (1981), Anwei: Anwei kexue jishu chubanshe.

Formularies

Beiji qianjin yaofang 備急千金要方 (Essential Recipes Worth a Thousand Gold Pieces, for Urgent Need) c. 650–659, Sun Simiao 孫思邈, in Liu Gengsheng 刘更生 and Zhang Ruixian 张瑞贤 (eds) (1993) *Qianjin fang* 千金方, Beijing: Huaxia chubanshe.

Shengji zonglu 聖濟總錄 (Medical Encyclopaedia: A Sagely Benefaction) 1122, attributed to Zhao Ji 趙佶 (Emperor Huizong), in Wang Zhenguo 王振国 (ed.) (2016) *Shengji zonglu jiaozhu* 圣济总录校注, 2 vols, Shanghai: Shanghai kexue jishu chubanshe.

Taiping sheng hui fang 太平聖惠方 (Imperial Grace Formulary of the Great Peace and Prosperous Reign Period) 978–992, Wang Huaiyin 王懷隱 et al. (2016), 2 vols, Beijing: Renmin weisheng chubanshe.

Waitai miyao 外臺秘要 (Arcane Essentials from the Imperial Library) 752, Wang Tao 王燾 (1982), Beijing: Renmin weisheng chubanshe.

Yixin fang (Japanese *Ishimpō*) 醫心方 (Formulas at the Heart of Medicine) 984, Tamba Yasuyori 丹波康賴 (1996), critical reprint with annotations, Beijing: Huaxia chubanshe.

Modern sources

Secondary sources, Chinese

Fu Weikang 傅维康 (1993) *Zhongyao xueshi* 中药学史 (A History of Drug Therapy), Sichuan: Bashu shushe.

Shang Zhijun 尚志钧, Lin Qianliang 林乾良 and Zheng Jinsheng 郑金生 (1989) *Lidai zhongyao wenxian jinghua* 历代中药文献精华 (Essentials of the Literature of Traditional Chinese Drug Therapy), Beijing: Kexue jishu wenxian chubanshe.

Wei Zixiao 魏子孝 and Nie Lifang 聂莉芳 (1994) *Zhongyi zhongyao shi* 中医中药史 (History of Chinese Traditional Drugs in Traditional Chinese Medicine), Taipei: Wenjin chubanshe.

Secondary sources, Western languages

Bensky, D., Clavey, S. and Stöger, E. (2004) *Chinese Herbal Medicine: materia medica*, 3rd edn, Seattle: Eastland Press.

Bian, He (2020) *Know Your Remedies: pharmacy and culture in early modern China*, Princeton: Princeton University Press.

Brown, M. (2015) *The Art of Medicine in Early China: the ancient and medieval origins of a modern archive*, Cambridge: Cambridge University Press.

Chaffee, J.W. and Twitchett, D. (eds) (2015) *The Cambridge History of China: Volume 5, Sung China, 960–1279, Part 2*, Cambridge: Cambridge University Press.

Despeux, C. (ed.) (2010) *Médecine, religion et société dans la Chine médiévale: étude de manuscrits chinois de Dunhuang et de Turfan*, Paris: Collège de France, Institut des hautes études chinoises.

Gillispie, C. (ed.) (1973) *Dictionary of Scientific Biography*, vol. 8, New York: Scribner.

Goldschmidt, A. (2005) 'The Song discontinuity: rapid innovation in Northern Song dynasty medicine', *Asian Medicine*, 1.1: 53–90.
—— (2008) 'Commercializing medicine or benefiting the people – the first public pharmacy in China', *Science in Context*, 21.3: 311–50.
—— (2009) *The Evolution of Chinese Medicine: Song dynasty, 960–1200*, London: Routledge.
Harper, D. (1998) *Early Chinese Medical Literature: the Mawangdui medical manuscripts*, New York: Kegan Paul International.
Henricks, R.G. (1998) 'Fire and rain: a look at Shen Nung 神農 (the divine farmer) and his ties with Yen Ti 炎帝 (the 'Flaming Emperor' or 'Flaming God')', *Bulletin of the School of Oriental and African Studies*, 61.1: 102–24.
Hsia, E., Veith, I. and Geertsma, R. (1986) *The Essentials of Medicine in Ancient China and Japan: Yasuyori Tamba's Ishimpō*, 2 vols., Leiden: Brill.
Hsu, E. (ed.) (2001) *Innovation in Chinese Medicine*, Cambridge: Cambridge University Press.
Lo, Jung-pang and Elleman, B.A. (2012) *China as a Sea Power 1127–1368: a preliminary survey of the maritime expansion and naval exploits of the Chinese people during the Southern Song and Yuan periods*, Singapore: National University of Singapore Press.
Lo, V. and Cullen, C. (eds) (2005) *Medieval Chinese Medicine: the Dunhuang medical manuscripts*, Needham Research Institute Series, London: RoutledgeCurzon.
Lu Gwei-djen (1966) 'China's greatest naturalist, a brief biography of Li Shih-chen', *Physis* 8.4: 383–92 (also published in *American Journal of Chinese Medicine* 1976, 4.3: 209–18).
Métailié, G. (1990) 'Botanical terminology of Li Shizhen in *Bencao gangmu*', in H.M. Said (ed.) *Essays on Science*, Karachi: Hamdard Foundation Pakistan, pp. 140–53.
—— (2001) 'The *Bencao gangmu* of Li Shizhen: an innovation for natural history?' in E. Hsu (ed.) *Innovation in Chinese Medicine*, Cambridge: Cambridge University Press, pp. 221–61.
Nappi, C. (2009) *The Monkey and the Inkpot: natural history and its transformations in early modern China*, Cambridge, MA: Harvard University Press.
Needham, J., and Lu Gwei-Djen (1986) *Science and Civilisation in China*, Vol. 6, Biology and Biological Technology, Part 1: botany, Cambridge: Cambridge University Press.
—— (2015) *Science and Civilisation in China*, Vol. 6, Biology and Biological Technology, Part 4: traditional botany: an ethnobotanical approach, by Georges Métailié, trans. J. Lloyd, Cambridge: Cambridge University Press.
Said, H.M. (ed.) (1990) *Essays on Science*, Karachi: Hamdard Foundation Pakistan.
Salguero, P. (2014) *Translating Buddhism Medicine in Medieval China*, Philadelphia: University of Pennsylvania Press.
—— (ed.) (2017) *Buddhism and Medicine: an anthology of premodern sources*, New York: Columbia University Press.
Scheid, V., Bensky, D., Ellis, A. and Barolet, R. (2009) *Chinese Herbal Medicine: formulas and strategies*, Seattle, WA: Eastland Press.
Schmidt, F.R.A. (2006) 'The textual history of the materia medica in the Han period: a system - theoretical reconsideration', *T'oung Pao*, 92.4: 293–324.
Schottenhammer, A. (2015) 'China's emergence as a maritime power', in J.W. Chaffee and D. Twitchett (eds) *The Cambridge History of China: volume 5, Sung China, 960–1279, part 2*, Cambridge: Cambridge University Press, pp. 437–525.
Sivin, N. (1968) *Chinese Alchemy: preliminary studies*, Cambridge, MA: Harvard University Press. (Harvard Monographs in the History of Science).
—— (1973) 'Li Shih-chen (1518–1593)', in C. Gillispie (ed.) *Dictionary of Scientific Biography*, vol. 8, New York: Scribner, pp. 390–8.
—— (2017) 'Sun Simiao on medical ethics: "The perfect integrity of the great physician" from prescriptions worth a thousand in gold', in P. Salguero (ed.) *Buddhism and Medicine: an anthology of premodern sources*, New York: Columbia University Press, pp. 538–42.
Stanley-Baker, M. (2021) '*Ge xianweng zhouhou beiji fang* 葛仙翁肘後備急方', in Lai Chi-Tim 黎志添 (ed.) *Daozang jiyao – Tiyao* 道藏輯要 提要 [*Companion to the Essentials of the Daoist Canon*]. Hong Kong: Chinese University of Hong Kong Press, pp. 809–19.
Unschuld, P.U. (1977) 'Traditional Chinese pharmacology: an analysis of its development in the thirteenth century', *Isis*, 68.2: 224–48.

———. (1979) *Medical Ethics in Imperial China: a study in historical anthropology*, Berkeley: University of California Press.

——— (1986) *Medicine in China: a history of pharmaceutics*, Berkeley: University of California Press.

Wilms, S. (2016) *Shén Nóng Běncǎo Jīng – The Divine Farmer's Classic of Materia Medica*, Boulder, CO: Blue Poppy Press.

Yang, Y. and Brown, M. (2017) 'The Wuwei medical manuscripts: a brief introduction and translation', *Early China*, 40: 241–301.

Yates, R.D.S. (2012) 'The Qin slips and boards from well no. 1, Liye, Hunan: a brief introduction to the Qin Qianling county archives', *Early China*, 35/36: 291–329.

9
DEVELOPMENTS IN CHINESE MEDICINE FROM THE SONG THROUGH THE QING

Charles Chace[1]

Introduction

When practitioners of Chinese medicine look to pre-modern literature for insight and inspiration, we find different kinds of information in the writings of each era. In terms of practical application, medical literature can be broadly divided into the writings of the last twelve hundred years, and the entirety of what preceded it. We typically refer to the corpus of these earlier writings to deepen our understanding of the larger concepts of Chinese medicine. We may read the *Yellow Emperor's Inner Classic* (*Huangdi neijing* 黃帝內經, Eastern Han 東漢, 25–225 CE), hereafter *Inner Classic*, for a fresh contextual perspective on a difficult case, but as a rule we do not expect to find a specific treatment protocol in the *Inner Classic* that will solve a particular clinical conundrum.

In the literature from the Song dynasty (960–1279) onwards, however, we find precisely this kind of clinical detail. We see how the core ideas presented in the *Inner Classic* and the other seminal texts of our literature were interpreted and extrapolated upon in very concrete and tangible ways. This provides us with a multiplicity of approaches adapted to address a vast range of circumstances and individual presentations. The *Discussion of Cold Damage* (*Shanghan lun* 傷寒論, c. 220), a seminal treatise on the progression and treatment of febrile disease, by Zhang Ji 張機 (stylename Zhongjing 仲景, 150–219) is an apt illustration of this phenomenon (Chapter 8 in this volume).[2] Although now regarded as a cornerstone of medical literature of the first few centuries of the common era, the *Discussion of Cold Damage* became much more influential in the Song, and virtually every clinical perspective on the text in use today was developed in the Song or later.[3] At least in terms of its clinical relevance today, the *Discussion of Cold Damage* is a Song document. Chinese medical writings of the past thousand years may be read as a multi-faceted articulation of the canonical literature in clinical practice. This chapter will provide an overview of some of the most important milestones in the development of that body of knowledge and the circumstances that produced them.

The Song dynasty

In recent years, the Song dynasty has been recognised in Anglophone scholarship as a pivotal period in the development of Chinese medicine (Hinrichs 2003; Leung 2003a, b;

Scheid 2007; Goldschmidt 2009). By then, the foundations of medical theory were well established, and as Asaf Goldschmidt has observed, it was 'during the Song that all the intellectual habits of what is now thought of as Chinese medicine assumed their mature form' (Goldschmidt 2009: 2). Definitive versions of canonical texts were printed, cosmological and physiological principles were more thoroughly integrated into pharmacology and 'medicine' (*yi* 醫) – defined as the administration of drug therapies and, to a lesser extent, acupuncture – eclipsed other ritualistic and shamanistic forms of healing (Hinrichs 2003). The number of medicinals recorded in the *materia medica* doubled during the Northern Song (960–1127) (Shang Zhijun 2002: 34–57). During the Song, the role of the scholar-physician also emerged as a definitive social feature, and medical practice gained an unprecedented level of legitimacy. A number of factors were responsible for these changes.

Governmental oversight

The Song dynasty witnessed unprecedented government involvement in the regulation and distribution of medicine, which was at least partially a political response to the epidemics sweeping the country during this time. This, and a genuine interest in medicine among imperial rulers, prompted the development of official prescription references such as the *Formulary of the Pharmacy Service for Benefiting the People in an Era of Great Peace* (*Taiping huimin hejiju fang* 太平惠民和劑局方, 1107),[4] which were published and distributed by the court. Such formularies contained fixed prescriptions designed to treat specific disorders, making them accessible to the masses, which had the effect of radically reducing the demand for specialists.

Governmental involvement in medical practice gradually extended to the credentialling of practitioners and the production of texts under the Bureau for Revising Medical Texts (*jiaozheng yishu ju* 校正醫書局) (Liang Jun 1995: 97–8). Its interest in both practitioners and the manner in which medicine was practised ultimately gave rise to governmentally run dispensaries, the first of which was established in 1076.

The flowering of Cold Damage

In terms of clinical practice, the flowering of interest in *Discussion of Cold Damage* was among the most significant theoretical developments of the Song. Again, at least partially in response to the epidemics that plagued this time, the Bureau for Revising Medical Texts published ten primary medical texts. These included an authoritative edition of the *Discussion of Cold Damage*, and two other texts concerning Cold Damage theory, as well as three more texts directly relevant to it, testifying to the influence of this approach to treatment (Goldschmidt 2009: 91–5). During this time, concerns with Cold Damage diseases ultimately overshadowed interests in other medical pursuits. Texts such as Pang Anshi's 龐安時 (stylename Anchang 安常, 1042–99) *General Discussion of the Disease of Cold Damage* (*Shanghan zongbing lun* 傷寒總病論, 1100) and Zhu Gong's 朱肱 (stylename Yizhong 翼中, fl. 1088) *Lifesaving Book that Categorises the Presentations of Cold Damage* (*Shanghan leizheng huoren shu* 傷寒類證活人書, 1108) both systematised Cold Damage theory and made it more easily accessible for laypeople.

Perhaps the most significant Cold Damage text of this period was Cheng Wuji's 成無己 (1066–1156) *Annotation and Explanation of the Discussion of Cold Damage* (*Zhujie shanghan lun* 注解傷寒論, 1144). Cheng explained the course of Cold Damage disorders in the context of the classical doctrines linking the channels and viscera.[5] This innovative synthesis became part of a medical orthodoxy that physicians in subsequent eras would actively attempt to dismantle (Goldschmidt 2009: 169; Scheid 2013).

Scholarly medicine

Prodigious population growth during the Song created a need for jobs among the educated (Hymes 1987; Scheid 2007: 37). Beginning in the Northern Song, governmental endorsement of medicine as a profession helped to legitimise medical practice and produced career opportunities for the gentry (Boyanton 2015: 171–220). Fan Zhongyan's 范仲淹 (stylename Xiwen 希文, 989–1052) admonition that '[if] one cannot become a good minister one [should strive] to become a good physician' is indicative of the new-found status of medicine during this time (Scheid 2007: 41). By the Southern Song (1127–1279), literati were taking up the profession in large numbers and physicians began to be regarded as elite exponents of cultivated ideas (Ibid.: 35).

Book printing facilitated intellectual exchange both among physicians and between physicians and scholars, making medicine an increasingly scholarly pursuit. In this environment, Daoist and Buddhist ideals also became deeply integrated into Confucian thought. An important characteristic of this Neo-Confucianism was the principle of personal cultivation through the investigation of things (*gewu* 格物), and the place of humanity in them, for the betterment of society as a whole.[6] Scholars steeped in this sensibility who also had an interest in medicine came to be referred to as *ruyi* 儒醫 (scholar-physicians). In the Southern Song (1127–1279), the term came to denote not only the degree of a physician's intellectual understanding but also his own level of psychological and spiritual cultivation, and embodied the conviction that both of these qualities were essential to effective medical practice (Ibid.: 39–41).

Jin and Yuan dynasties

The elite ranks of scholarly physicians, or at least those claiming that status, blossomed in the Jin 金 and Yuan 元 dynasties (1115–1368). Their interest in adapting the classical medical literature to contemporary medical concerns was a driving force in the evolution of Chinese medicine. Diagnostic focus shifted away from single pathogenic factors such as wind or cold, and descriptions of specific medical disorders (*bing* 病), and towards a new focus on the dynamics of *qi* transformation and how the breakdown of this transformation expressed itself in distinctive presentations (*zheng* 證; Ibid.: 43).

Scholar-physicians of the Jin-Yuan period further challenged the Song orientation towards established formulæ and globally defined disorders by focusing on local disease processes (Boyanton 2015: 164 ff.). Although government involvement in healthcare fostered greater access to medical information among the general population in the form of treatment manuals, and even previously secret hereditary prescriptions were made public, scholar-physicians directly attacked these formulary approaches as a deeply degraded form of medicine. From their perspective, administering fixed formulæ based on identification of a patient's disease was insufficient. Prescriptions should be carefully tailored to the patient and the patient's relationship to their immediate environment.

Systematisation in diagnostics and pharmacology

The organisation and systematisation of diagnostic parameters that had begun in the Later Han with the *Inner Classic* assumed a more mature form in the Jin-Yuan. The paediatrician Qianyi 錢乙 (stylename Zhongyang 仲陽, 1032–1113) and Kou Zongshi 寇宗奭 (fl. 1111–1117) had already begun to articulate the eight necessities (*bayao* 八要) for effective treatment. Synthesising these parameters, Zhang Yuansu 張元素 (stylename Jiegu 潔古, fl. 1151–1254) systematised pattern presentation of the viscera and receptacles (*zangfu* 臟腑) and articulated relationships between the viscera and receptacles with the channels and networks (*jingluo* 經絡), and the six *qi*. It was Zhang Yuansu who expressed the concepts of root diseases (*benbing* 本病), those pathodynamics that are at the root of a problem, and branch diseases (*biaobing* 標病), the typically symptomatic expression of an illness. He also defined what today is considered eight-principle differentiation (*bagang* 八綱), that any presentation should be described in terms of whether it is *yin* or *yang*, excess or deficient, hot or cold, and whether it originates from inside or outside the body (Buck 2009).

A similar consolidation of concepts occurred in the field of pharmacology where there was an ongoing effort to describe drugs in a more comprehensive and coherent manner. Wang Haogu's 王好古 (stylename Haizang 海藏, thirteenth century) *Materia Medica for Decoctions* (*Tangye bencao* 湯液本草, 1298) was especially significant in its articulation of a pharmacology of systematic correspondence, presenting a system for treating specific conditions based on drug qualities derived from Five Phase (or Five Agent) theory (Unschuld 1986: 111).[7] This trend culminated in Zhu Zhenheng's 朱震亨 (stylename Danxi 丹溪, 1281–1358) identification of multiple primary drug qualities that rationalised the apparent contradictions between theoretical and empirical effects of drugs (Ibid.: 187).

The four freat masters of the Jin-Yuan

Of the many scholar-physicians active during the Jin-Yuan period, four in particular advanced a sequence of related ideas that significantly contributed to the development of medical theory during this time. These were Liu Wansu 劉完素 (stylename Shouzhen 守真 c. 1120–1200), Zhang Congzheng 張從正 (stylename Zihe 子和, 1156–1228), Li Gao 李杲 (stylename Dongyuan 東垣, 1180–1251) and Zhu Zhenheng 朱震亨 (stylename Danxi 丹溪, 1281–1358).

Liu Wanusu

For Liu Wansu, all pathogenic factors, regardless of their origin, eventually manifest as pathogenic fire. Therefore, cool and cold medicinals are indicated to treat nearly all conditions. One of Liu's characteristic strategies was to drain heart fire while nourishing kidney water, an idea that would be developed by subsequent thinkers, most notably Zhu Zhenheng. Framing this core idea was a belief that diagnosis and treatment must take into account environmental and climatic conditions, as well as the patient's constitutional predispositions. This would become a defining perspective of all four masters of the Jin-Yuan and a general principle of the time.

Zhang Congzheng

Like Liu Wansu, Zhang Congzheng's primary focus was on the accumulation of exogenous pathogenic factors and improper food intake, and he downplayed the role of internally generated pathologies due to the emotions or constitutional weakness. Zhang's fundamental approach involved the removal of these pathogens through diaphoresis, emesis and purgation, expanding the definition of these strategies as originally outlined in *Discussion of Cold Damage* and employing a wider range of therapies that included acupuncture, moxibustion, massage and exercise. Even more vigorously averse to tonifying medicinals than Liu Wansu, Zhang Congzheng preferred to employ dietary stratagems rather than pharmaceuticals when supplementation was indicated.

Li Gao

Li founded his approach on the *Inner Classic's* assertion that the spleen and stomach were the basis of the body's *qi*, and a belief that they could be compromised due to improper food intake, overexertion and emotional disorders (Welden 2015). Although Xu Shuwei 許叔微 (stylename Heke 知可, 1079–1154) had already emphasised the importance of the stomach during the Song, Li made this insight the centrepiece of his own thinking while integrating Zhang Yuansu's discrimination of viscera and receptacle syndrome patterns.

Where Liu Wansu and Zhang Congzheng's overall approach to therapy stressed the elimination of excess pathogenic factors, those of Li Gao and Zhu Zhenheng focused more on replenishing metabolic insufficiencies. Later scholars came to believe that this shift in emphasis reflected the deteriorating political circumstances during the latter part of Jin-Yuan period. According to Xu Dachun 徐大椿 (stylename Lingtai 靈胎, 1693–1771):

> When in the final years of the Song dynasty the Chinese territory fell into the hands of the enemy [referring to the northerners], the head of state was weak and his ministers failed to show any strength. At the same time, Zhang Yuansu, Li Gao, and others, in establishing their prescriptions, emphasised supplementation of the central palace.
>
> (*Xu Lingtai yixue quanshu*: 140)[8]

A weak central government was analogous to weak digestive function. Strengthening the central government/digestion was the key to both political and individual health (Weldon 2015).

The spleen stomach current became one of the most influential perspectives in subsequent Chinese medical history. Li Gao's most mature work, his *Discussion of the Spleen and Stomach* (*Piwei lun* 脾胃論, thirteenth century), was very influential on later writers.[9] Wang Haogu, Li Shizhen, Ye Gui (see below) and others all acknowledged the centrality of the spleen and stomach even as they incorporated a broader range of concepts into their medical understanding.[10]

Zhu Zhenheng

The last of the four great masters, Zhu Zhenheng is generally regarded as representing the pinnacle of Jin-Yuan medical thought. His ideas reflect a synthesis of previous medical currents produced by a consummately cultivated philosopher-physician. Charlotte Furth has detailed the influence of Zhu's thought and practice of Zhu Xi's 朱熹 (1130–1200) Neo-Confucian philosophy of 'investigating things and extending knowledge' (*gewu zhizhi* 格物致知),[11] a realist philosophical approach to the external, physical world that sought to understand its basic principles and humanity's role in them (Furth 2006). Zhu Zhenheng was a prominent exemplar of this principle, which helped to propel and legitimise medical innovation from the Song to the end of the imperial period (Unschuld 1985: 195–6). As the most outspoken critic of formula books and the simplified Cold Damage therapeutics of his time, Zhu Zhenheng was an icon of the scholar-physician's intellectual agenda.

Zhu Zhenheng's reading of the three preceding Jin-Yuan currents led him to posit that the flaring of 'ministerial fire' (*xianghuo* 相火), the fiery aspect of *qi* that actually drove the physiological activity of the body, was the central cause of disease (Simonis 2010). Based on his reading of the *Inner Classic*, he understood ministerial fire as an active *yang* that gave rise to human desires and physiological functions. When ministerial fire became pathological, it was due to excessive longings and emotions, particularly of a sexual nature, impairing the *yin* and blood. The nourishment of kidney *yin* was the primary means of controlling the ensuing symptoms of fire. This strategy has its basis in the Five Phase paradigm where water is the primary means of mitigating fire along the control (*ke* 克) cycle. Notwithstanding his emphasis on *yin*-nourishing therapies, Zhu regarded the support of the spleen and stomach as a central component of an overall *yin* nourishing strategy.[12]

Zhu Zhenheng's most important ideas are collected in *Danxi's Essential Methods (Danxi xinfa practitioner of the lowest rank.* 丹溪心法, 1347), and in a companion volume, *Extra Treatises based on Investigation and Inquiry* (*Gezhi yulun* 格致餘論, 1347), the title of which reflects his Neo-Confucian orientation.[13]

The Ming dynasty

By the Ming dynasty (1368–1644), all famous doctors were considered scholar-physicians, though the profession was still so poorly regulated that it was quite easy to make the claim that one was a physician, and so it remained a 'Minor Way' (*xiao dao* 小道) (Scheid 2007:45). Nevertheless, medicine had become a well-established means of social advancement. The family history of Li Shizhen 李時珍 (1518–93), among the most iconic figures in Chinese medical history, is emblematic of such a progression. Li's grandfather was an itinerant bell doctor (*lingyi* 鈴醫), a medical practitioner of the lowest rank. Li's father, Li Yuechi 李月池, passed the central state exams, elevating the family's social status considerably, though he remained a physician who became famous for his success in treating patients during a local epidemic.[14] These improved circumstances afforded Li Shizhen himself access to the world of the literati and court physicians making an especially broad range of medical information available to him.

Consolidation of medical knowledge

Li Shizhen

Li Shizhen 李時珍 (stylename Binhu 瀕湖, 1518–93) is among the most iconic figures in Chinese medical history. Each of his three surviving works constituted a definitive contribution to Chinese medicine, particularly with regard to the systematisation of knowledge within its specific field. Paul Unschuld describes Li's *Comprehensive Materia Medica* (*Bencao gangmu* 本草綱目, 1578) as the 'qualitative and quantitative climax in the development of the *materia medica* literature, expanding the genre from a work on drugs to a comprehensive and detailed encyclopaedia of medicine, pharmaceutics, minerology, metallurgy, botany and zoology' (Unschuld 1986: 163). Carla Nappi articulates Li Shizhen as a naturalist, and the *Comprehensive Materia Medica* as a seminal work of naturalist science (Nappi 2009). This magnum opus reflects a comprehensive familiarity with the medical literature and drug use throughout China at that time. In fact, the primary criticism of the work at the time of its publication was its very size, which made it unwieldy and expensive. The organisation of the book further refined the existing classification systems of the *materia medica* literature. Its title alludes to Zhu Xi's *Comprehensive Mirror to Aid in Government* (*Zizhi tongqian gangmu* 資治通鑑綱目), reflecting Li's Neo-Confucian sensibilities (Unschuld 1986: 151). Li's work is divided into individual monographs called *gang* 綱 (mainstays) and the criteria for the internal arrangement within the monographs are referred to as *mu* 目 (a hierarchical catalogue).[15] Li's works on pulse diagnosis, *Pulse Studies of the Lakeside Master* (*Binhu maixue* 瀕湖脈學, 1564) and on the eight Extraordinary Vessels, *Exposition on the Eight Extraordinary Vessels* (*Qijing bamai kao* 奇經八脈考, 1577), have been similarly influential.[16]

Li's *Pulse Studies of the Lakeside Master* organised and codified pulse diagnosis into the twenty-seven pulses that are the basis for Chinese medical pulse diagnosis today.[17] Finally, his *Exposition on the Eight Extraordinary Vessels* synthesised and extrapolated upon previous ideas concerning the Extraordinary Vessels, particularly with regard to herbal medicine.[18] Li's integration of Daoist inner elixir (*neidan* 内丹) principles of internal cultivation into medical practice in this text is representative of the widespread interest in such practices during the Ming.[19]

Zhang Jiebin

Zhang Jiebin 張介賓 (stylename Jingyue 景岳, fl. 1562–1639) was an erstwhile soldier who developed the use of military terminology in medical treatment. He presented a perspective that was a counterpoint to the premise of Liu Wansu and Zhu Zhenheng that an excess of *yang* was the primary source of illness. For Zhang, *yang* was *qi* and *yin* was form, and since one could never have too much *qi*, one could never have too much *yang*. He is remembered as the foremost developer of the Life Gate (*mingmen* 命門) principle of thought and practice. He understood the Life Gate to be the manifestation of primordial *yin* and *yang* (*yuan yinyang* 元陰陽), residing between the kidneys.[20] As such, he stressed the enrichment of genuine or true *yin* (*zhen yin* 真陰). Although the Life Gate was the repository of former heaven (*xiantian* 先天) *yin* and *yang* endowed before birth, he recommended its replenishment through the latter heaven (or postnatal) essence (*houtian jing* 後天精) provided by the nourishment and digestive processes of the spleen and stomach.[21]

Regardless of whether it was the True *Yin* and True *Yang* (*zhenyang* 真陽) that appeared to be compromised, for Zhang Jiebin the problem was invariably rooted in a debilitation of the true *yin*. He, therefore, prioritised the nourishment of postnatal essence (*jing* 精) and blood (*xue* 血) in his treatment strategies, often through the support of spleen and stomach function. In this, his approach to treatment reflects a synthesis of spleen-, stomach- and *yin*-nourishing theories that flowered during the Jin-Yuan period. Zhang's *Categorisation of the Classic* (*Leijing* 類經, 1624) is among the most coherent, and systematically organised of the extant commentaries on the *Inner Classic*. It and Zhang's other most important writings are anthologised in his *Complete Works of Jingyue* (*Jingyue quanshu* 景岳全書, 1624).

Qing dynasty

The Qing (清) dynasty (1644–1911) was marked by a political and ideological conservatism that spawned the evidential scholarship (*kaozheng* 考證) movement, a re-examination of classical sources based on the principle that they were a more reliable basis for the resolution of the problems of the present than more modern ideas (Unschuld 1985: 193; Scheid 2007: 48). Its proponents argued that these earlier texts had not yet been corrupted by the Daoist and Buddhist principles characteristic of Song Neo-Confucianism. The introduction of Western science during this time almost certainly fuelled the evidential scholarship movement by providing multiple models of enquiry that in many cases had much greater predictive accuracy than the theory-laden structures of the past. In medicine, this played out as a search for the true understanding of classical sources, a quest that was tethered to the idea that wisdom must be uncovered through observation coupled with introspection.

Those involved in evidential scholarship not only directed medical interest away from the sage writings of antiquity, but also focused more on the experience of individual physicians in their local environs. This trend centred on physicians in Jiangnan, particularly in and around Suzhou, fostering the invention of a southern medical tradition that was uniquely suited to the local *qi* (*tuqi* 土氣), which endowed the local inhabitants with a weaker (*rouruo* 柔弱) constitution than their northern counterparts (Hanson 1998).

Once again, new approaches to medical learning fostered a plurality of ideas tied to individual scholars. Yet, because none of these currents had become so predominant as to achieve the force of doctrine, they were ultimately identified as part of the problem (Unschuld 1985: 197). This plurality inspired efforts in the community of writers to redefine medical orthodoxy in texts such as the *Essential Readings from the [Orthodox] Medical Lineage* (*Yizong bidu* 醫宗必讀, 1637), and *Golden Mirror of the Medical Lineage* (*Yizong jinjian* 醫宗金鑑, 1739–42), which strove to identify medicine in terms of the canonical literature.[22] The vast range of individual approaches to learning was nevertheless legitimised by its representation in the state-sponsored anthology of Chinese knowledge, the *Complete Collection of the Four Treasuries* (*Siku quanshu* 四庫全書, 1773). A number of physicians, however, distinguished themselves as synthesisers of these two intellectual trends.

Synthesis

Ye Gui

Ye Gui 葉桂 (stylename Tianshi 天士, 1667–1746) was the most influential physician of the Qing period (1644–1912). In addition to his major contributions to the field of Warm Disease theory and practice, Ye was renowned as a virtuoso synthesiser and integrator of virtually all of the medical theories that preceded him, masterfully adapting a vast array of approaches and techniques to the demands of the moment. Ye himself is credited with writing very little and the clinical orientation of Ye's contribution is evident in the case records attributed to him that were collected by his students.

Ye's case records, particularly his *Case Record Guide to Clinical Presentation* (*Linzheng zhinan yi'an* 臨証指南醫案, 1746), became immensely popular after his death. Despite the fact that he was not of the scholarly class, Ye's own family strove to position him not as an innovator but as an exemplar of the scholar-physicians steeped in the literature of antiquity (Hanson 1997: 248).

Among the most significant of Ye's contributions to Chinese medical practice was his explicit emphasis on treatment strategies (*fa* 法) that could be flexibly adjusted, rather than on fixed formulæ. This privileging of methodology over diagnosis pioneered by Yu Chang 喻昌 (stylename Jiayan 嘉言, 1585–1664) was the logical extension of the scholarly concern with specific disease dynamics presenting in individual patients. Although it required a comprehensive understanding of the full spectrum of Chinese medical theories, it allowed for carefully tailored herbal prescribing. Despite this more generalised influence on medical practice, Ye Gui is best known in the West for his role in the development of Warm Disease theory.

Warm Disease

The role of heat as a cause of infectious disease had been a topic of discussion since the term Warm Disease (*wenbing* 溫病) was first mentioned in the *Inner Classic* as either a range of acute-onset febrile disorders or as a specific type of Cold Damage. Building on Liu Wansu's emphasis on fire as a more predominant pathogenic factor than cold, Wu Youxing 吳有性

(stylename Youke 又可, 1582–1652) posited that febrile epidemics (*wenyi* 溫疫) were due to pestilential *qi* (*liqi* 癘氣) that was hot in nature, establishing the conceptual foundation for Warm Disease therapeutics. From the seventeenth through the nineteenth centuries, Warm Disease theory developed as a methodology that was distinct from Cold Damage therapeutics (Hanson 2013: 204). Although the Warm Disease current is commonly positioned as a response to the failures of Cold Damage therapeutics, the concept of Warm Disease is both historically and conceptually a logical extension of the Cold Damage current, the principles of which are grounded in the *Discussion on Cold Damage* (Scheid 2013).[23] Similarly, Ye's development of new strategies for treating Warm Disease was undeniably innovative, and yet it is clear that his thinking was firmly rooted in the classically oriented sensibilities of the scholar-physician.

The one text that Ye Gui may well have written himself, the *Discourse on Warm Disease* (*Wenre lun* 溫熱論, 1792), systematised the progression of pathogenic heat into the body through four layers: defensive, *qi*, nutritive and blood. Within this theoretical structure, the text primarily concerned itself with the practicalities of differential diagnosis and treatment strategy. It is clearly written as a complement to the *Discussion on Cold Damage* and uses many of its ideas and therapies.[24] A significant portion of Ye's *Case Records Based on Clinical Presentation* is devoted to the treatment of Warm Disease patterns, providing concrete examples of how Ye made use of the therapeutic framework he laid out in his *Discourse*.

Xue Xue

Ye's contemporary rival, the scholar and poet Xue Xue 薛雪 (stylename Shengbai 生白, 1681–1770), posited a complementary approach to the four-layer model of disease progression that would form the other pillar of the theoretical framework of Warm Disease. Xue is credited with most thoroughly developing the Three Burner (*sanjiao* 三焦) model of illness that divided the location of pathogenic factors into upper, middle and lower parts of the body, roughly the thorax, epigastrium and lower abdomen. This model was particularly suited to damp heat (*shire* 濕熱) patterns, which he believed were not only different from cold patterns but also from Warm Diseases caused by heat alone (Liu 2001: 21). Thus, Xue's treatment approach reflects an abiding concern with the relative predominance of dampness or heat as it presents in a given location or warmer (Burner) in the body. Such patterns were then treated with particular attention to the stomach and spleen, which he believed were especially vulnerable to attack by damp heat (Qiu Peiran 1984: 278). Xue published a number of works concerning poetry and medicine. His most significant contribution to the Warm Disease current is his *Folio on Damp Heat Diseases* (*Shirebing pian* 濕熱病篇, 1852) edited by Wang Shixiong 王士雄 (stylename Mengying 孟英, 1808–67).[25]

Wu Tang

The person who brought the disparate streams of Warm Disease theory together was Wu Tang 吳瑭 (stylename Jutong 鞠通, 1758–1836). Building upon and substantially redefining the works of preceding Warm Disease physicians, Wu combined Xue Xue's Three Burner model dividing the torso into three regions, with Ye Gui's four-sector (*sifen* 四分) model identifying the relative depth of a pathogenic factor. In the process, he also linked each of the viscera and receptacles to a particular Burner, producing a finely detailed system for locating the progression of pathogenic heat within the body (Liu 2005: 23). His *Systematic Differentiation of Warm Disease* (*Wenbing tiaobian* 溫病條辨, 1798) has been immensely influential in popularising the theory and treatment of Warm Disease. In many cases, the formulæ contained in this text are clearly extrapolations on the treatment strategies and herb combinations developed by Ye Gui. Wu was especially concerned with the loss of essence and believed that 'the words loss of storage of essence should be looked at flexibly and not explained exclusively an overindulgence in sex. Rather it is the case that all human activity can agitate and disturb the essence' (Liu 2001: 23).[26]

Conclusion

Even as a basis for critique, canonical literature has remained the impetus for the most significant developments in Chinese medical thinking over the last millennium. Attempts to understand such texts have produced a dynamic tension between the comparatively liberal or expansive interpretations exemplified by the four great masters of the Jin-Yuan, and relatively conservative or restrictive readings such as those of the Qing philologists. The creative interplay between these two impulses is such that one could not have occurred without the other. Each interpretation has produced a concrete and pragmatic response to the principles presented in the canonical literature. In this, Chinese medicine's perennial return to the classics may be best understood less as the search for an overarching truth than as an ongoing response to the challenge of developing effective therapies in the face of ever-changing circumstances.

Notes

1 Sadly, Charles 'Chip' Chace passed away while this volume was in production. Beloved by generations of Chinese medicine practitioners, Chip made a lasting impact in his published work, his training of many students, his kind humour and generous personality. It is fitting to share his chapter on the legacy of the Jin-Yuan masters, a legacy in which he himself took part. The editors wish to thank Volker Scheid for overseeing the final revisions.
2 Zhang Ji originally composed a single treatise known as the *Discourse on Cold Damage and Miscellaneous Diseases*. The earliest editor of this document, Wang Shuhe (王叔和, 201–280), subsequently divided the text into two parts, the *Discourse on Cold Damage* addressing infectious disease and the *Essential Prescriptions from the Golden Cabinet* (*Jingui yaolüe* 金匱要略) concerned with internal medicine. They have been considered as two separate texts ever since.
3 The modern clinician Huang Huang 黄煌 is among the most influential advocates of *shanghan*-based prescribing in both Asia and the West today. His approach is likely to be far more comprehensible to later *shanghan* thinkers than to Zhang Ji himself (Huang 2007). For an examination of the evolution of *shanghan* thinking, see Scheid (2013). For an overview of the range of clinical applications to the Cold Damage literature, see Chen and Zhang (1998). The best translation of the *Discussion of Cold Damage* to date is by Craig Mitchell, Feng Ye and Nigel Wiseman (Mitchell, Feng and Wiseman 1999).
4 *Taiping huimin heji ju fang* 太平惠民和劑局方 (*Formulary of the Pharmacy Service for Benefiting the People in the Taiping Era*) 1148, Liu Jingyuan 劉景源 (ed.) (1985), Beijing: Renmin weisheng chubanshe.
5 Ed's note: See Chapter 1 in this volume.
6 For a historical survey of the shifting concept of *gewu*, see Elman (2005: 5–9).
7 Ed: On the Five Phases (or Five Agents) (*wuxing* 五行) (see Introduction and Chapter 1 in this volume).
8 For a translation of Xu Dachun's *Yixue yuanliu lun* 醫學源流論 (Discourse on the Origin and Development of Medicine, 1764), see Unschuld (1990).
9 The *Treatise on the Spleen and Stomach* appeared relatively early in the wave of pre-modern Chinese medical texts translated into English beginning in the 1980s. Its appreciation of dietary and environmental aetiological factors, and its approach to modelling complex disease presentations have made it especially influential in the transmission of Chinese medical ideas into the West. Li's notion of '*yin* fire' (*yinhuo* 陰火) arising from a spleen stomach debility at the centre of a host of

other pathologies has been a popular model of pathophysiology for a variety of modern complex disease presentations, including autoimmune diseases. See Li and Flaws (2004).
10 For Wang Haogu's perspectives on the spleen and stomach, see *Wang Haogu yixue quanshu* 王好古醫學全書 (2004: 352ff.). For Li Shizhen's perspectives on the spleen and stomach, see Wang Xiaoping (2003). For a discussion of Li Gao's approach to gynaecology, see Chace (2001).
11 For an overview of the concept of *gewu zhizhi*, and the debates surrounding it, see Andrew Plaks' entry in the *RoutledgeCurzon Encyclopedia of Confucianism*: '*Gewu zhizhi* (Putting all things into the correct conceptual grid, extending to the utmost one's range of comprehension)' (Yao 2003, vol. 1: 226–7). On the influence of *gewu zhizhi* in pharmacology, see Métailié (2001).
12 For a further discussion of Zhu Zhenheng's intellectual legacy, see Furth (2006). For more on Zhu Zhenheng's approach to diseases of the stomach and spleen, see Ni and Damone (1992).
13 Both texts have been translated into English. See Yang (1993, 1994). For further discussion of scholar-physicians in the Jin-Yuan period, see Chen (1997); Ding *et al.* (1999); Leung (2001, 2003a, 2003b).
14 For an examination of itinerant doctors in Chinese medical history, see Qiu (2008); Unschuld (1979).
15 For an English language translation of the *Comprehensive Materia Medica*, see Luo (2003).
16 For a comprehensive collection of Li Shizhen's works, see *Li Shizhen yixue quanshu* 李时珍医学全书 (1999). For a biographical sketch of Li Shizhen, see Chace and Shima (2010: 11–18); Unschuld (1986: 145–8). For further discussion of Li Shizhen's *Comprehensive Materia Medica*, see Métailié (2001, 2010); Nappi (2009b).
17 For an excellent discussion of Li Shizhen's pulse diagnosis, see Li, Morris, Li and Mondot (2011).
18 For a translation of Li's *Exposition on the Eight Extraordinary Vessels*, see Chace and Shima (2010).
19 Li's organisation of this text, pieced together as it is from earlier writings, is also emblematic of a distinctive style of medical writing that developed during this time. For an examination of the influence of this syncretic style of medical writing on Ming dynasty medicine, see Simonis (2010). On inner alchemy, see Chapters 29 and 30 in this volume.
20 On these physiological terms and their relationship to inner alchemy, see Chapters 29 and 30 in this volume.
21 *Houtian jing* 後天精 is often rendered in English as 'postnatal essence', as it is produced from imbibed substances such as food, concentrated into blood, which is further concentrated into essence. For a detailed discussion of Zhang Jiebin's physiology of *mingmen*, and its importance for understandings of sexuality and reproduction, see Wu (2010, ch. 3).
22 For a discussion of the influence of the *Golden Mirror of Medicine* in late imperial China, see Hanson (2003).
23 For a comprehensive collection of case records reflecting the synthesis of these two currents in clinical practice, see Luo *et al.* (2004).
24 For an English language translation and commentary on the *Discourse on Warm Disease*, see Van Wart and Chace (2006).
25 The text is also known as the *Itemised Differentiation of Damp Heat* (*Shire tiaobian* 濕熱條辨).
26 An English language translation of *Systematic Differentiation of Warm Disease* in digital format is forthcoming. See Boyanton, Cm-db.com (Chinese Medicine Database).

Bibliography

Pre-modern sources

Bencao gangmu 本草綱目 (Comprehensive *Materia Medica*) 1578, Li Shizhen 李時珍, Liu Hengru 刘衡如 and Liu Shanyong 刘山永 (eds) (2013), Beijing: Huaxia chubanshe.
Bencao shiyi 本草拾遺 (Gleanings of *Materia Medica*) 739, Chen Zangqi 陳藏器, Shang Zhijun 尚志钧 (ed.) (2002), Hefei: Anhui kexue jishu chubanshe.
Danxi xinfa 丹溪心法 (Danxi's Essential Methods) 1347, Zhu Zhenheng 朱震亨 (1997), Shenyang: Liaoning kexue jishu chubanshe.
Gezhi yulun 格致餘論 (Extra Treatises based on Investigation and Inquiry), 1347, Zhu Zhenheng 朱震亨 (2018), Beijing: Zhongguo yiyao keji chubanshe.
Huangdi neijing 黃帝内經 (Yellow Emperor's Inner Classic), Eastern Han 東漢, 25–225 CE), Wang Bing 王冰 (anno. and ed. in 762) (2003), Beijing: Zhongyi guji chubanshe.

Jinkui yaolüe 金匱要略 (Essential Prescriptions from the Golden Cabinet), c. 220, Zhang Ji 張機 (stylename Zhongjing 仲景) (2016), Beijing: Zhongguo yiyao keji chubanshe.

Leijing 類經 (Categorisation of the Classic) 1624, Zhang Jingyue 張景岳 (2011), Beijing: Zhongguo yiyao keji chubanshe.

Li Shizhen yixue quanshu 李时珍医学全书 (The Complete Medical Writings of Li Shizhen), Liu Changhua 柳长华 (ed.) (1999), Beijing: Zhongguo zhongyiyao chubanshe.

Ming Qing mingyi quanshu dacheng 明清名醫全書大成 (A Complete Compendium of the Medical Writings of Famous Physicians in the Ming and Qing Dynasties), Hu Guoju 胡國巨 *et al.* (eds) (1999), Beijing: Zhongguo zhongyiyao chubanshe.

Piwei lun 脾胃論 (Discussion of the Spleen and Stomach) 1249, Li Gao 李杲 (2007), Beijing: Zhongguo zhongyiyao chubanshe.

Qijing bamai kao 奇經八脈考 (Exposition on the Eight Extraordinary Vessels) 1577, in Liu Changhua 柳长华 (ed.) (1999), *Li Shizhen yixue quanshu* 李时珍医学全书 (The Complete Medical Writings of Li Shizhen), Beijing: Zhongguo zhongyiyao chubanshe.

Shanghan leizheng huoren shu 傷寒類證活人書 (Lifesaving Book that Categorises the Presentations of Cold Damage) 1108, Zhu Gong's 朱肱 (stylename Yizhong 翼中, fl. 1088) (2012), Beijing: Zhongyi guji chubanshe.

Shanghan lun 傷寒論 (Discussion of Cold Damage) c. 220, Zhang Ji 張機 (stylename Zhongjing 仲景) (2010), Beijing: Kexue jishu wenxian chubanshe.

Shanghan zongbing lun 傷寒總病論 (General Discussion of the Disease of Cold Damage) 1100, Pang Anshi 龐安時 (1989), Beijing: Renmin weisheng chubanshe.

Taiping huimin heji ju fang 太平惠民和劑局方 (Formulary of the Pharmacy Service for Benefiting the People in the Taiping Era) 1148, Liu Jingyuan 劉景源 (ed.) (1985), Beijing: Renmin weisheng chubanshe.

Tangye bencao 湯液本草 (*Materia Medica* for Decoctions), 1298, Wang Haogu 王好古 (2013), Beijing: Zhongguo zhongyiyao chubanshe.

Wang Haogu yixue quanshu 王好古醫學全書 (The Complete Medical Writings of Wang Haogu), Sheng Zengxiu 盛增秀 (ed.) (2004), Beijing: Zhongguo zhongyiyao chubanshe.

Wenbing tiaobian 溫病條辨 (Systematic Differentiation of Warm Disease), 1798, Wu Tang 吳塘 (2013), Beijing: Zhongguo yiyao keji chubanshe.

Wenre lun 溫熱論 (Discourse on Warm Disease), 1792, Ye Gui 葉桂 (2007), Beijing: Renmin weisheng chubanshe.

Xu Lingtai yixue quanshu 徐灵胎医学全书 (The Complete Medical Writings of Xu Lingtai), Liu Yang 刘洋 *et al.* (eds) (1999), Beijing: Zhongguo zhongyiyao chubanshe.

Ye Tianshi yixue quanshu 叶天士医学全书 (The Complete Medical Writings of Ye Tianshi), Huang Yingzhi 黄英志 (ed.) (1999), Beijing: Zhongguo zhongyiyao chubanshe.

Yizong bidu 醫宗必讀 (Essential Readings from the [Orthodox] Medical Lineage), 1637, Li Zhongzi 李中梓 (1998), Beijing: Zhongguo zhongyiyao chubanshe.

Yizong jinjian 醫宗金鑑 (Golden Mirror of the Medical Lineage), 1739–1742, Wu Qian 吳謙 *et al.* (1997), Shenyang: Liaoning kexue jishu chubanshe.

Zhujie shanghan lun 注解傷寒論 (Annotation and Explanation of the Discussion of Cold Damage), 1144, Cheng Wuji's 成無己 (1997), Shenyang: Liaoning kexue jishu chubanshe.

Modern sources

Boyanton, S. (2015) *The Treatise on Cold Damage and the Formation of Literati Medicine: Social, Epidemiological, and Medical Change in China, 1000–1400*, PhD thesis, Columbia University.

——— (trans.) (forthcoming) *Wenbing tiaobian* 溫病條辨 (Systematic Differentiation of Warm Disease) by Wu Tang 吳塘, Chinese Medicine Database, http://cm-db.com, accessed 8/3/2020.

Buck, C. (2009) 'Who invented Bagang?', *Journal of Chinese Medicine*, 91: 12–16.

Chace, C. (2001) 'Li Dong-yuan's gynecology; excerpted from the secret treasury of the orchid chamber (*Lanshi mizang* 蘭室秘藏) (Three Types of Menstrual Block and Failure of Movement)', *Open Gate Acupuncture*, http://www.chinesemedicinedoc.com/for-practitioners/chinese-medicine-articles, accessed 1/7/2013.

Chace, C. and Shima, M. (2010) *Exposition on the Eight Extraordinary Vessels*, Seattle, WA: Eastland Press.

Chen Ming 陈明 and Zhang Yinsheng 张印生 (1998) *Shanghan mingyi an jingxuan* 伤寒名医验案精选 (Selected Efficacious Cases by Famous Cold Damage Physicians), Beijing: Xueyuan chubanshe.

Chen Yuanpeng 陳元朋 (1997) *Liang Song de 'shangyi shiren' yu 'ru yi': jianlun qi zai Jin Yuan de liubian* 兩宋的尚醫士人與儒醫：兼論其在金元的流變 (Elites who Esteemed Medicine and Literati Physicians in the Northern and Southern Song Dynasties: with a Discussion of Their Spread and Transformation During the Jin and Yuan), Taipei: Guoli Taiwan daxue chubanshe.

Ding Guangdi 丁光迪 et al. (1999) *Jin-Yuan yixue pingxi* 金元医学评析 (A Critical Assessment of Jin-Yuan Medicine), Beijing: Renmin weisheng chubanshe.

Elman, B.A. (2005) *On Their Own Terms: science in China, 1550–1900*, Cambridge, MA: Harvard University Press.

Elvin, M. and Vogel, H.U. (eds) (2010) *Concepts of Nature: A Chinese-European cross-cultural perspective*, Leiden: Brill.

Furth, C. (2006) 'The physician as philosopher of the way: Zhu Zhenheng (1282–1358)', *Harvard Journal of Asiatic Studies*, 66.2 (December): 423–59.

Goldschmidt, A.M. (2007) 'Epidemics and medicine during the Northern Song dynasty: the revival of cold damage disorders (*shanghan*)', *T'oung Pao*, 93: 53–109.

—— (2009) *The Evolution of Chinese Medicine: Song Dynasty, 960–1200*, London; New York: Routledge.

Hanson M.E. (1997) *Inventing a Tradition in Chinese Medicine: From universal canon to local medical knowledge in south China, the seventeenth to the nineteenth century*, PhD thesis, University of Pennsylvania.

—— (1998) 'Robust northerners and delicate southerners: the nineteenth-century invention of a southern medical tradition', *Positions: Asia Critique*, 6.3: 515–50.

—— (2003) 'The *Golden Mirror* in the imperial court of the Qianlong emperor', 1739–1742', *Early Science and Medicine*, 8: 111–47.

—— (2013) *Speaking of Epidemics in Chinese Medicine: disease and the Geographic Imagination in Late Imperial China*, New York: Routledge.

Hinrichs, TJ (2003) *The Medical Transforming of Governance and Southern Customs in Song Dynasty China*, PhD thesis, Harvard University.

Hsu, E. (ed.) (2001) *Innovation in Chinese Medicine*, Cambridge: Cambridge University Press.

Huang Huang 黄煌 (2007) *Jingfang de meili* 经方的魅力 (The Allure of Classical Formulas), Beijing: Renmin weisheng chubanshe.

Hymes, R.P. (1987) 'Not quite gentlemen? Doctors in Sung and Yuan', *Chinese Science*, 8: 9–76.

Leung A.K.C. (Liang Qizi 梁其姿) (2001) *Song-Yuan Ming de difang yiliao ziyuan chutan* 宋元明的地方醫療資源初探 (An Initial Exploration into Resources for Medical Treatment in the Song, Yuan and Ming Periods), Beijing: Zhongguo shehui lishi pinglun.

—— (2003a) 'Medical learning from the Song to the Ming', in P.J. Smith and R. von Glahn (eds) *The Song-Yuan-Ming Transition in Chinese History*, Cambridge, MA: Harvard University Asia Center, pp. 374–98.

—— (2003b) 'Medical instruction and popularization in Ming-Qing China', *Late Imperial China*, 24: 130–52.

Li, G. and Flaws, B. (2004) *Li Dong-yuan's Treatise on the Spleen and Stomach: a translation of the Pi wei lun*, Boulder, CO: Blue Poppy Press.

Li, S., Morris, W.R., Li, S. and Mondot, M. (2011) *Li Shi-Zhen's Pulse Studies: An Illustrated Guide* (*Li Shizhen mai xiang tu pu* 李時珍脉象圖譜), Beijing: Renmin weisheng chubanshe.

Liang Jun 梁峻 (1995) *Zhongguo gudai yizheng shilüe* 中国古代医政史略 (A Brief History of Medicine and State Policies in Ancient China), Hohhot: Neimenggu renmin chubanshe.

Liu, G.H. (2005) *Warm Pathogen Diseases: a clinical guide*, Seattle, WA: Eastland Press.

Luo Hegu 罗和古 et al. (eds) (2004) *Shanghan wenbing yian* 伤寒温病医案 (Case Records on Cold Damage and Warm Disease), Beijing: Zhongguo yiyaoke chubanshe.

Luo, X. (trans.) (2003) *Compendium of Materia Medica: Bencao gangmu*, Beijing: Foreign Languages Press.

Métailié, G. (2001) 'The *Bencao gangmu* of Li Shizhen: an innovation in natural history?', in E. Hsu (ed.) *Innovation in Chinese Medicine*, Cambridge: Cambridge University Press, pp. 221–61.

—— (2010) 'Concepts of nature in traditional Chinese *materia medica* and botany (sixteenth to seventeenth century)', in M. Elvin and H.U. Vogel (eds) *Concepts of Nature: a Chinese-European cross-cultural perspective*, Leiden: Brill, pp. 345–67.

Mitchell, C., Feng, Y. and Wiseman, N. (1999) *Shang Han Lun* (On Cold Damage) Brookline, MA: Paradigm Publications.

Morris, W. and Mondot, M. (2011) *Li Shi-Zhen's Pulse Studies: an illustrated guide*, Beijing: Renmin weisheng chubanshe.

Nappi, C.S. (2009a) 'Bolatu's pharmacy: theriac in early modern China', *Early Science and Medicine,* 14: 737–64.

——— (2009b) *The Monkey and the Inkpot: natural history and its transformations in early modern China*, Cambridge, MA: Harvard University Press.

Ni Yitian and Damone, R. (1992) 'Zhu Dan Xi's treatment of diseases of the spleen and stomach', *Journal of Chinese Medicine*, 40: 25–9.

Qian Chaochen 钱超尘 and Wen Changlu 溫長路 (eds) (2003) *Li Shizhen yanjiu jicheng* 李时珍研究集成 (Collected Research on Li Shizhen), Beijing: Zhongyi guji chubanshe.

Qiu, Y. (2008) 'Itinerant doctors in Chinese history', *Journal of Chinese Medicine*, 86: 28–33.

Qiu Peiran, 裘沛然 (1984) *Zhongyi lidai ge jia xueshuo* 中医历代各家学说 (Currents of Medical Theory Associated with Physicians from Past Generations), Shanghai: Shanghai kexue jishu chubanshe.

Scheid, V. (1995) 'The great qi, Zhang Xichun's reflections on the nature, pathology and treatment of the *daqi*', *Journal of Chinese Medicine*, 49: 5–11.

——— (2007) *Currents of Tradition in Chinese Medicine, 1626–2006*, Seattle, WA: Eastland Press.

——— (2013) 'Transmitting Chinese medicine: changing perceptions of body, pathology, and treatment in Late Imperial China', *Asian Medicine*, 8.2: 299–360.

Shang Zhijun 尚志钧 (2002) 'Bencao shiyi yanjiu lunwen 本草拾遗研究论文' (Studies on the Gleanings of *Materia Medica*), in Shang Zhijun (ed.) *Bencao shiyi jishi*《本草拾遗》辑释 (Gleanings of *Materia Medica*), Hefei: Anhui kexue jishu chubanshe, pp. 470–528.

Simonis, F. (2010) *Mad Acts, Mad Speech and Mad People in Late Imperial Chinese Law and Medicine*, PhD thesis, Princeton University.

Smith, P.J. and Von Glahn, R. (eds) (2003) *The Song-Yuan-Ming Transition in Chinese History*, Cambridge, MA: Harvard University Asia Center.

Unschuld, P.U. (1979) *Medical Ethics in Imperial China: a study in historical anthropology*, Berkeley: University of California Press.

——— (1985) *Medicine in China: a history of ideas*, Berkeley: University of California Press.

——— (1986) *Medicine in China: a history of pharmaceutics*, Berkeley: University of California Press.

——— (1990) *Forgotten Traditions of Ancient Chinese Traditions*, Brookline, MA: Paradigm Publishers.

Unschuld, P.U. and Zheng Jinsheng 鄭金生 (2012) *Chinese Traditional Healing: the Berlin collections of manuscript volumes from the 16th through the early 20th century* (3 vols), Leiden: Brill.

Van Wart, D. and Chace, C. (2006) 'Ye Tian-Shi's *wen re lun*, part 1, discourse on warm–heat disease, *Lantern*, 3: 8–15.

Wang Xiaoping 王晓萍 (2003) 'Shishu Li Shizhen tiaozhi piwei zhi fangfa 试述李时珍调治脾胃之方法' (The Development of Li Shizhen's Prescription Method on the Regulation and Treatment of the Stomach and Spleen), in Qian Chaochen 钱超尘 and Wen Changlu 溫長路 (eds) *Li Shizhen yanjiu jicheng* 李时珍研究集成 (Collected Research on Li Shizhen), Beijing: Zhongyi guji chubanshe, pp. 624–6.

Welden, J. (2015) *To Bring Order Out of Chaos: literati medicine of the Jin dynasty (1115–1234)*, PhD thesis, University of Hawai'i.

Wu, Y.-L. (2010) *Reproducing Women: medicine, metaphor, and childbirth in late imperial China*, Berkeley; London: University of California Press.

Yang Shou-zhong (1993) *Danxi zhifa xinyao* 丹溪治要心法 (The Heart & Essence of Dan-xi's Methods of Treatment) by Zhu Danxi 朱丹溪 (1347), Boulder, CO: Blue Poppy Press.

——— (trans.) (1994) *Gezhi yulun* 格致餘論 (Extra Treatises based on Investigation and Inquiry) by Zhu Danxi 朱丹溪 (1347), Boulder, CO: Blue Poppy Press.

Yao Xinzhong 姚新中 (ed.) (2003) *RoutledgeCurzon Encyclopedia of Confucianism*, London: RoutledgeCurzon.

PART 2
Sickness and healing

10
ANCIENT PULSE TAKING, COMPLEXIONS AND THE RISE OF TONGUE DIAGNOSIS IN MODERN CHINA

Oliver Loi-Koe

Introduction

The first step for all medical practitioners when treating any patient is to make a judgement about what exactly their patient is suffering from, and what the future holds for them (Chaney 2016: 1). To reach a diagnosis, the physician's finesse in the diagnostic system is challenged. In Chinese medicine, the skills of diagnosis have deployed most of the commonly enumerated sensory organs: the eyes, the ears, the nose, the tongue and the sense of touch. Four essential diagnostic techniques were already recorded in the 61st dialogue in the *Huangdi bashiyi nanjing* 黃帝八十一難經 (The Yellow Emperor's Canon of Eighty-One Difficult Issues, hereafter *Nanjing* Classic of Difficult Questions, c. second century CE):

a   *wang* 望 (visual inspection, patient complexion)
b   *wen* 聞 (auditory, the timbre of the voice)
c   *wen* 問 (asking, propensity for particular flavours)
d   *qie* 切 (palpation of the pulse) (Unschuld, 1986b: 539).[1]

The traditional categories from the *Nanjing* require explanation. To begin with, the term *wen* 聞 seems to elide what Europeans would consider smelling and hearing; and *wen* 問 refers to interrogating a patient's desire for different flavours, but later comes to refer to the taking of a case history rather than any sensory-based diagnostic technique. *Wang* 望 (visual inspection) has always meant looking for patterns of the *wuse* 五色, the five colours of the complexion. However, what I list above in Chinese diagnosis as the skill of the 'tongue' does not involve tasting the patient, but another style of visual inspection, of that of the patient's tongue itself. In almost two millennia since the *Nanjing* scheme was set out, the relative priorities given to these different diagnostic modalities have not been set in stone, but have been subject to changing emphases and interpretations.

The use of *qie* 切 (palpation), through palpating the pulse – referred to as pulse diagnosis – was highly prized for its diagnostic capabilities in the ancient Chinese world. Testament to the longevity of the skill of pulse diagnosis, it still features as a key component of primary clinical encounters in China. From the late Han dynasty (202 BCE–220 CE), the *Shiji* 史記 (Records of the Historian), compiled by *Sima Qian* 司馬遷 (145–146 BCE), offers early examples of pulse diagnosis as part of a set of twenty-five case histories provided as a hagiography for a physician, Chunyu Yi 淳于意 (216–150 BCE), who had been summoned by the imperial court to account for his work (Holroyde-Downing 2017: 114). The majority of these case histories state: 'when I examined the *mai* (pulse)' (Hsu 2001: 57). They do not, for example, provide the comparable, 'when I inspected the tongue', which became one of the most important methods of diagnosis. From these early case histories and widespread references to the pulse in the medical literature that dates to the Han period, we can infer that Chunyu Yi and other physicians alike placed heavy reliance on the pulse, to the extent that it became customary in clinical practice.

Despite the apparent dominance of pulse diagnosis, bountiful evidence suggests that physicians regarded *wang* 望, visual inspection or visual diagnosis, as another useful diagnostic method; evidence of this appears in the *Suwen* 素問 recension of the *Huangdi neijing* 黃帝內經 (Inner Canon of the Yellow Emperor), the Yellow Emperor corpus of classical medical treatises, which states that the physician who diagnoses using both the pulse and the complexion achieves perfection (Holroyde-Downing 2017; Kuriyama 1999: 10). Visual diagnosis was grounded in the observation of a patient's complexion, inclusive of their facial appearance, body state and skin condition; on this basis, one could diagnose the severity of the illness and come to a prognosis of the underlying condition.

Historically, visual diagnosis was intimately entwined with the predictions of fortune telling and divination: one's face exhibited the coloured emanations of the *qi*, from which diviners could portend one's future and doctors could come to a prognosis of illness; the colours determined the likelihood of good health or auspicious circumstances, depending on their correlation with the Five Agents and to the temporal phases of the seasons (Harper 2001: 112–3). The difference between the skills of doctor and diviner was subtle, but involved the integration of knowing the future with different types of knowledge, such as the correlative cosmology of Yin Yang and the Five Agents and/or a range of divinatory schemes. This is well illustrated in the work of physician Sun Simiao 孫思邈 (581–682), which states:

> Hence, if the complexion appears green-blue like young grasses, death [is imminent]; yellow like hovenia-fruit, death [is imminent]; black like soot, death [is imminent]; red like rotten blood, death [is imminent]; white like withered bones, death [is imminent]; This is how death is visible in the five complexions.
>
> *(Unschuld et al. 2011: 188;* Qianjin yifang, *25.298,* Suwen, *3.10 'Wuzang shengcheng pian* 五藏生成篇', *62)*

Glossier colours like the green of the wings of a mandarin duck, the red of the cockscomb, the yellow of the belly of the crab, white like pork fat or the black of a crow's wing were signs that a patient would be more likely to recover (Despeux 2005: 191). Coloured auras in the face betrayed physiological disharmony, a visible expression of strength or weakness, or indeed life or impending death (Kuriyama 1999: 167).

Furthermore, we see in the *Shiji* that complexion diagnosis was the preferred method of legendary physician, Bian Que 扁鵲 (tr. dates c. sixth to fifth century BCE). Bian Que (the stories about whom tend to portray much later attitudes to medicine than the legendary dates suggest) 'did not depend on taking the pulse for his diagnosis; he observed the complexion, the sounds, described the body and could thus deduce the seat of the illness' (*Shiji*, 105.941; Despeux 2005: 180). In another famous record from the *Hanfei zi* 韓非子 (c. 280–233 BCE), the same legendary physician tells Lord Huan 桓 that he is suffering from a mortal illness and requires treatment (*Hanfei zi jijie*, 7.21.161; Brown 2015: 41–62). Despite this warning, Lord Huan feels altogether well and dismisses the warning. Bian Que persists, even returning to the court several times to insist that the illness is worsening and requires treatment; however, his advice is totally ignored. From a distance, Bian Que recognises that the Lord's illness is terminal, and the Lord subsequently dies. Bian Que flees when he sees that the illness is no longer curable. Indeed, during the Han period (202 BCE–220 CE), prognosis proved a vital element in diagnosis: Bian Que knew that the death of a patient did not bode well for the physician whose own life might be in jeopardy. Importantly, the physician's gaze proved invaluable and by the Han period the Bian Que story had become a trope which represented the genius of Chinese medicine and medical practitioners for recognising and treating the underlying causes and patterns of disease before they manifest as symptoms.

The use of the tongue for medical diagnosis initially only appeared incidentally in scholarly medicine. The classical treatise on the pulse, *Maijing* 脈經 (Pulse Classic), attributed to third-century author Wang Shuhe 王叔和 (180–c. 270 CE), states: 'A sick person with a curled tongue and retracted testicles is bound to die' (Yang 1997: 135). The *Maijing* shows that changes in the tongue were correlated with certain illnesses and prognostic patterns, even serving as an indicator of the severity of the illness of the patient and, in this case, their impending death. However, initially, the tongue was not used as an indicator of specific pathologies related to the organs; arguably, this served as a limiting factor in its diagnostic capability. With this in mind, one may re-visit a set of questions proposed by Holroyde-Downing in 2017: how did the tongue become a well-regarded diagnostic tool, and what factors led to its being used as a diagnostic technique with prognostic capability? In the same framework, we ask: what prompted the primacy of pulse diagnosis to be eroded to such a degree that tongue diagnosis could rise in credibility as a diagnostic technique, and when did that happen? The tongue was not always admired and did not always enjoy its current level of prestige as one of the main 'pillars to diagnosis': its rise to fame spanned many centuries (Farquhar 1994: 90; Holroyde-Downing 2017: 29).

Due to the inherent difficulties of mastering pulse diagnosis – as we shall see below, the rise of tongue diagnosis flourished suddenly under particular epidemic, socioeconomic and political changes. There were also important influences from abroad. Kuriyama (1999: 86) argued that the pulse was the language of life and doctors were its interpreters. For nearly two millennia, this may have been an indisputable claim; however, by the sixteenth century, pulse, diagnosis and its transmission came to be fraught with difficulty, and the tongue was set to become the new language of life.

The difficulty of the pulse

A cumbersome mastery

There is a distinct method of pulse diagnostics articulated in the classical works of Chinese medicine. The customary technique detailed throughout the *Huangdi neijing* emphasises palpating the pulse at the *cunkou* 寸口 ('inch opening', a term for the wrist) (Unschuld *et al.* 2011: 127). The index, middle and ring fingers were placed on the radial artery. The spatial positioning of each finger mirrored the spatial organisation of specific organs: the upper position depicted organs above the diaphragm; the middle, organs between diaphragm and navel; the lower, organs in lower body (Hao 1987: 90). Degrees of pressure on the wrist determined the position at which one could sense the condition of the organ/channel (Kuriyama 1999: 166).

Descriptions of pulse qualities were documented in a text titled *Maijing* attributed to third-century author Wang Shuhe (Hanson and Pomata 2020: 25). This early text dating to around 280 CE lists twenty-four diagnostic states of the pulse: Floating, Hollow, Flooding, Slippery, Rapid, Intermittent, Chord-like, Tense, Sunken, Hidden, Leathery, Faltering, Full, Faint, Rough, Thin, Soft, Weak, Empty, Dispersing, Lazy, Slow, Halting, Moving (Hao 1987).

Arguably, this rich and descriptive vocabulary was redolent with meaning; but its pluralistic, lyrical and 'dense tangled mesh of interrelated, interpenetrating sensations' lacked defining parameters (Kuriyama 1999: 340). For example, how did a floating pulse differ from a flooding pulse and through what qualitative or quantitative measurement could this be defined? *Maijing juan* 5 gives us the pulses that heralded death, with animal imagery telling us about imminent failure of the inner organs. Bian Que states that when you examine the pulse, it is a sign of death if the *qi* is like the gathering of a flock of birds, a chariot with but one horse, the pecking of a sparrow, a swimming shrimp or a hovering fish (Lo 2011: 75). Diagnostic interpretation relied on a level of subjectivity that made the reliability and repeatability notoriously difficult (Kuriyama 1999: 75). Unsurprisingly, the subjectivity that lay inherent in pulse interpretation led to difficulties with accurately passing on this clinical skill. It was a challenge that took years of practice and experience to master and presumably required a one-to-one apprenticeship with a renowned or senior physician (Holroyde-Downing 2017: 31).

The limits of pulse diagnosis in acute medical scenarios

Treating diseases medically required an understanding of the cause of the disease. *Waigan bing* 外感病 were diseases caused by exogenous factors. Common causes include conceptual opposites such as *hanxie* 寒邪 (cold evil) and *rexie* 熱邪 (heat evil) (Kuriyama 1999: 20). Case histories reveal that the use of the pulse to distinguish between hot and cold conditions was fraught with difficulty; yet, a heavy reliance was placed on the physician's skill. In *Gujin yi'an an Shanghan* 古今醫案按·傷寒 (Medical Case Histories, Past and Present: Cold Damage), physician Wang Kentang 王肯堂 (1549–1613) records an acute life-threatening incident where the patient's weak and thin pulse, corresponding to 'cold', did not correlate with his physical signs, which were indicative of 'heat'. Instead of pulse diagnosis providing a solution to an ailment, it had the potential to create confusion and contradiction: should the disease be treated based on the pulse or on the physical signs? Remember, diagnostic mistakes led to wrong treatment methods and disastrous consequences, including possible death (Lei 2002: 360). In acute scenarios, the credibility of the pulse could be thrown into the spotlight. For sure, experienced doctors claimed to be able to make accurate assessments with the pulse and to translate those assessments into meaningful and effective therapy. Acute scenarios demanded quick diagnostic decisions, however, which the pulse might not be able to communicate with certainty.

In contrast to the possible confusion of pulse diagnosis, the visual thermodynamic changes of the tongue, particularly in febrile conditions, offered physicians a quick and much easier visual guide to monitor acute changes in presentation. Fever writes clear and rapidly changing signs on the tongue which are relatively easy to read. Tongue qualities were categorised: the colour, shape, moistness, absence or presence of fissures were qualities that could be easily visualised. Judging the colour of the tongue to be red, blue, yellow or black was inherently easier and quicker than judging the pulse to be 'tight' or 'soggy'. The diagrammatic and theoretical detail provided offered a 'quick and easy' diagnosis for both the amateur and skilled physicians. Reliance on touch for gathering information was replaced by the apparently superior ability of the eye to capture the shape and colour of 'reality'. The tongue became a clear and reliable diagnostic companion during times of urgent medical diagnostics – an invaluable and loyal aid which allowed the physician to adapt treatment accordingly (Holroyde-Downing 2017: 9, 20). There is no wonder, then, that it was during periods of widespread epidemics in south China during the Song, and particularly Ming, periods, that we see evidence of an increase in diagnoses being made on the basis of the tongue, latterly in the form of case histories. Southern physicians of the *wenbing xuepai* 溫病學派 (Warm Disease Current of Learning) developed new treatments for febrile epidemic illness for which the appearance of the tongue provided significant information (Ibid.: 27, 70–1, 162–82).

Socioeconomic influences

During the Ming dynasty (1368–1644), the number of scholars passing imperial examinations rose sharply with no proportional increase in jobs being offered in the civil service. With the advent of printing, publishing had become a more widespread activity and the increasing amount of medical writing made medicine a more respectable and scholarly profession. However, there was a dearth of experienced scholar physicians to teach the new class of amateur physicians (Chao 2009: 25). Thereafter, acquisition of pulse diagnosis expertise was faced with obstacles. There were difficulties for an amateur physician – albeit competent in medical theory – to acquire clinical expertise: for one cannot acquire clinical competence through books alone. Physicians competent in medical theory but with sparse clinical experience faced a problem. These physicians were aware of the theories of how to 'cool interior heat' through applying textual knowledge; however, opportunities to apprentice with a notable physician meant that judging degrees of heat and appropriate levels of treatment in practice could prove impossible. The complexity of the pulse led to difficulties in articulating and disseminating knowledge, which led to the primacy of pulse diagnosis being gradually eroded.

Conflict with European conceptions of the body

The Chinese concept of the pulse could not be universally accepted yet; at the same time, it had a certain fascination and could not be easily dismissed. As early as 1313, a Persian physician, Rashid al-Din, had overseen the translation and interpretation of a variety of texts, including sections of the *Maijue* 脈訣 (Pulse Rhymes) corpus, rhymed verses about the art of the pulse, in a manuscript known as *Tansuqnama-i ilkhan dar funun-i 'ulum-i khatayi* (The Treasure Book of the Ilkhan on Chinese Science and Techniques), dated AH 713 (1313) (Lo and Wang 2017: 291; Berlekamp et al. 2015: 60). During his lifetime, the knowledge of Chinese medicine he promoted fascinated physicians as far as Byzantium.[2] Within just over three centuries, similar material from the *Maijue* was beginning to inspire Latin and then French translations (Chapter 38 in this volume). Chinese sphygmology was thought complex and sophisticated and European physicians marvelled at the 'experience, examined and approved for four thousand years' (Floyer 1707: 355). But explanations in classical medical texts also increasingly befuddled European physicians. Palpating the pulse engaged tracts in the body and between organs that intertwined with philosophical concepts of Yin and Yang, the Five Agents and Qi to discern physical conditions (Kuriyama 1999: 8). And one could not see these tracts. Palpating the pulse using two fingers on the lateral part of the wrist was also considerably different from the Chinese method described above. This pictorial contrast is highlighted in Figures 10.1 and 10.2.

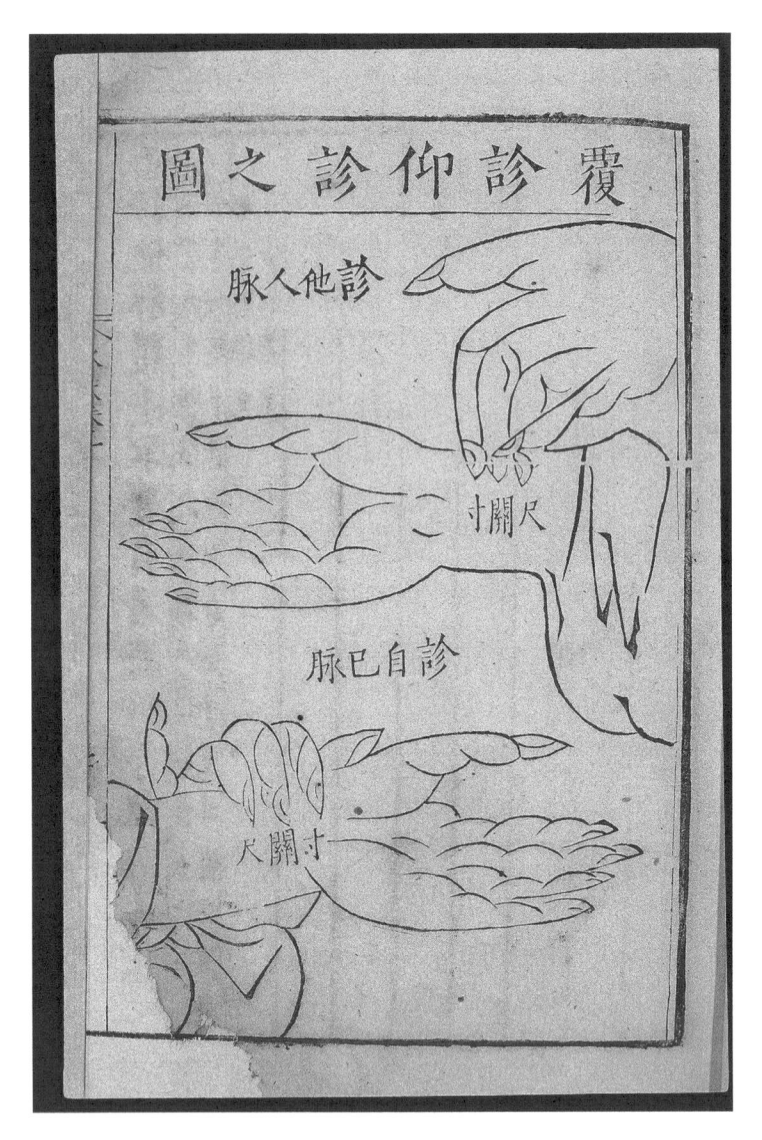

Figure 10.1 Pulse taking: A treatise on the pulse written by Wang Shuhe, fourth century CE. Illustration shows taking the pulse. Wellcome collection: 4.0 International (CC BY 4.0)

Significant differences in the European model lay in the pulse's ancient connection to the heart and Harvey's seventeenth-century theories about blood circulation through the vessels (Asen 2009: 28). European knowledge of the pulse mainly hailed from the Greek doctor Galen (129–200 CE) who wrote extensively on the cause, varieties and functions of the pulse.

For Europeans working in the Galenic tradition, then, there was an inseparable coupling of the pulse and discoveries in anatomy. Furthermore, by the time of the Renaissance (fifteenth to sixteenth centuries), European medical theory was to become even more firmly grounded in anatomy, in the human structure of bones, nerves and vessels as expressed in the iconic sixteenth-century representations of the human form by Da Vinci. Binding the conception of the pulse to anatomical findings, the visible nature of pulsation and the tubes of the arteries themselves encouraged new visualisations (Kuriyama 1999: 33).

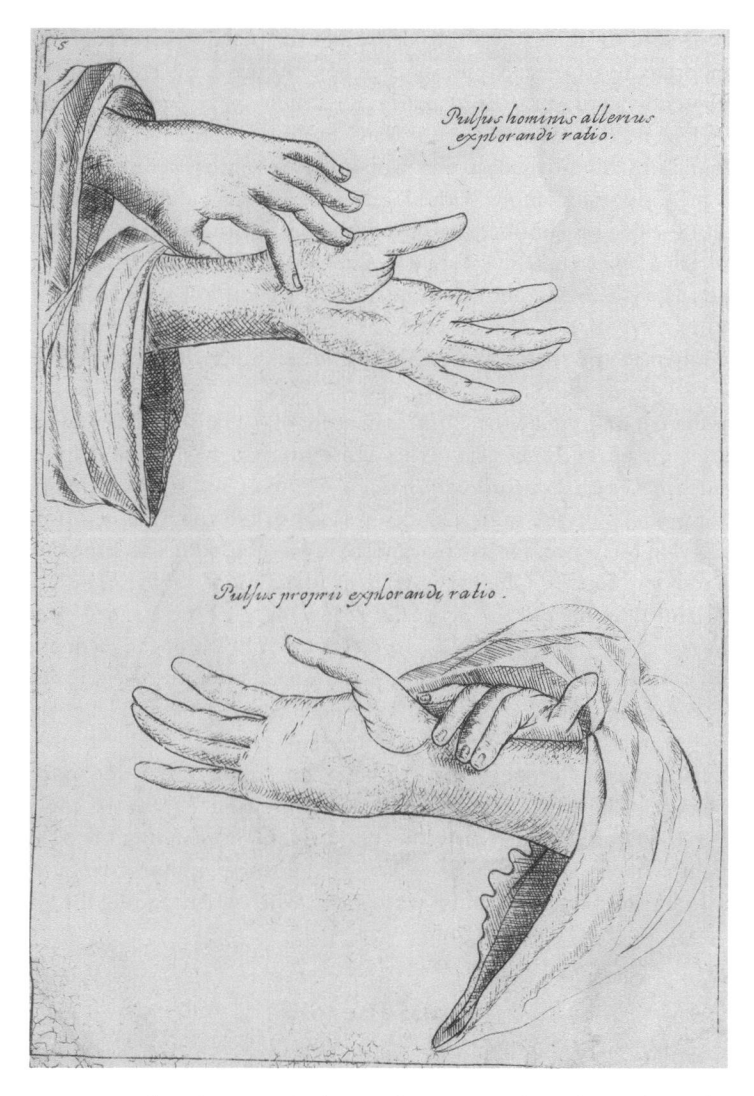

Figure 10.2 European pulse taking. 1682 edition of *Specimen medicinae* by Andreas Cleyer. Courtesy of the New York Academy of Medicine Library

The assumption that anatomy was an exclusive pre-occupation of Europeans, however, has been roundly rebutted in recent years, with scholars such as Catherine Despeux and Li Jianmin contending that the Chinese were just as curious as the Europeans (Despeux 2018: 54; Li Jianmin 2000: 15 and Chapter 13 in this volume). Highlighted as an important example was the execution and dissection of Sun Qing 孫慶 in 16 CE, where organs and vessels were intimately inspected. We also witness the work of Yang Jie 楊介, a physician known to have treated the Emperor Huizong (r. 1101–25), attending and rectifying previous errors about the position of the liver. This subverted theoretical considerations that the liver corresponded to 'spring' and to the 'left' of the five agents and five theories (Despeux 2018: 59). There is plentiful evidence that shows not only consideration for anatomy, but also the pursuit to improve anatomical knowledge. However, anatomical understanding did not, of itself, necessarily lead to innovations in therapy and the practical aspects of pulse diagnosis were more important to Chinese physicians than how the pulse reflected anatomical realities.

By the eighteenth century, the Europeans and Chinese spoke entirely different languages of the pulse (Chapter 38 in this volume). Europeans criticised the lyrical descriptions of the pulse to be found in Chinese medical treatises. Pulses in China were known as the *mai* 脈, a term simultaneously synonymous with blood vessel, channel and pulse depending on the context (Lo 2018: 70). This confused Europeans further. By the nineteenth century, as Latin translations of the *Maijue* circulated widely in Europe, European physicians were to dismiss Chinese pulse theory as free from the 'fundamental anatomical knowledge of the human body' (Kuriyama 1999: 37). It contradicted, displaced and confused European conceptions so that it 'literally made no sense', being unscientific and superstitious according to contemporary European norms (Ibid.).

Europeans did not understand the Chinese concept of the pulse but there were those, such as Charles Ozanam, a French physician who wrote admiringly that the 'Chinese recognise and cure, sometimes with extraordinary success the most recalcitrant illnesses' (Ozanam 1886: 84). Admiration for the pulse's ability to lead to effective Chinese treatments was clearly not followed by respect for its conception; it was praised in the same breath as it was criticised. In response to this, Chinese physicians like Tang Zonghai 唐宗海 (1851–1908), the widely acclaimed founder of the School of Converging Chinese and Western Medicine (*zhongxiyi huitong xuepai* 中西醫匯通學派), argued that whilst European medicine excelled in anatomy, medicine in China had a firmer grasp of the *qi* and thus the way the body functions (Chiang 2015: 35). Many Chinese physicians defended themselves with language and beliefs that were deeply entrenched in traditional Chinese culture.

The tensions between European and Chinese conceptions were not simply about the pulse. Understanding the truth about the body was 'inseparable from the challenge of discovering the truth about people' (Kuriyama 1999: 14). Understanding the pulse was a clue to the truth behind one's identity and one's haptic experiences of illness, recovery and death. In other words, the intimacy of the pulse was a gateway to understanding the 'pulseless soul' (Ibid.: 18).

Visual diagnosis

If Chinese conceptions of the pulse, and pulse diagnosis, had failed to convince nineteenth-century European anatomists because of their very invisibility, Chinese tongue diagnosis was simultaneously on the rise, arguably as a result of its contrasting visuality. As one of the earliest illustrated medical genres to be committed to woodblock print and, therefore, theoretically more easily available than manuscript copies, the tongue diagnosis manuals lent themselves to new forms of teaching and dissemination.

Printing had its early origins in the Tang imperial Buddhist projects under Empress Wu Zetian 武則天 dedicated to copying large quantities of sutras and devotional images for the religious merit of the ruling house (Barrett 2008: 68). With the subsequent Song (960–1279) and Ming dynasties (1368–1644), circumstances in China became increasingly 'favourable to the emergence of printing' (Kobayashi and Sabin 1981; Barrett 2008: 69). An upsurge in printed texts was facilitated by the carving of woodblocks – and through the introduction of ink cake, ink stones, paper and brushes, known collectively as 'four treasures of the scholar's desk' – which allowed mass production of literature (Kobayashi and Sabin 1981: 25). The creation of the *Jiaozheng yishu ju* 校正醫書局 (Bureau for the Revision of Medical Texts, 1057) was significant; it was a government agency responsible for the revision and publication of medical texts which led to an exponential increase in the availability of books, 'printed in just thirty-odd places during the Northern Song, they were published in almost two hundred places in the far smaller territory of the Southern Song' (McDermott 2005: 65). Mass production injected pace and movement into the transmission of medical knowledge – in particular texts pertaining to tongue diagnosis which, by then, could be printed *en masse* with images (Holroyde-Downing 2018: 168).

Alongside these changes, the government bestowed a further degree of formality on medical education with the creation of the *Taiyi ju* 太醫局 (Imperial Medical Service) in 1044 (Goldschmidt 2009: 40). As state exams became simplified and standardised, an examination culture flourished (Hymes 1987: 45; Holroyde-Downing 2017: 52). In timely fashion, economic progression through industrial expansion and commercialisation led to a new class of wealthy people. They could finance the education of their sons, which boosted the chances of a larger proportion of the population passing the imperial examinations and, in turn, gaining acceptance into the civil service and holding government office (Holroyde-Downing 2017: 55). However, an increase in those passing examinations was not met with an increase in civil service posts for them to fill (Hymes 1987: 46; Hinrichs and Barnes 2013: 60). Consequently, many unable to find civil service posts moved laterally and found employment within medicine, 'a dignified calling because through it the gentleman's altruism was made effective. In this respect it nearly equaled high office' (Hymes 1987: 44). Medicine soon blossomed into an increasingly attractive and popular profession which held a respectable social standing (Hymes 1987: 47; Holroyde-Downing 2017: 55).

'See this, do that'

The emergence of a medical text in 1341, during the Yuan dynasty, containing thirty-six drawn representations of the varieties of tongue pathologies revolutionised the application of tongue diagnosis (Holroyde-Downing 2017: 63). Physicians with access to the medical text, *Aoshi shanghan jinjing lu* 敖氏傷寒金鏡錄 (Scholar Ao's Golden Mirror Record of Cold Damage, hereafter *Jinjing Lu*), were shown an illustrated diagnostic guide to these thirty-six tongue images differentiating colour, shape and coating (Ibid.: 64). The tongue drawings were not just mere decorative additions to written theory. They depicted a variety of dynamic changes possible for the tongue and matched these to the corresponding illness. Most significant of all, the text provided a directive to the prescription necessary for treatment. This new, innovative conception integrated drug knowledge with tongue diagnosis: each tongue pathology was keyed to a particular recipe. Changes to the appearance of the tongue reflected a person's illness progression, facilitating diagnosis and treatment merely by looking at the tongue.

One such drawing – documented in Figure 10.3 – notes that a red tongue signals heat within the body and 'you don't need to ask which channel'. Whatever the channel, it was appropriate to use *Touding qingshen san* 透頂清神散 (Extremely Penetrating. Clear the Spirit Powder, *Aoshi shanghan jinjing lu*). A clear and instructive 'see this do that' recommendation foregoes the need to touch the patient, or indeed, use other diagnostic techniques (Holroyde-Downing 2017: 21). In contrast to the lyrical nature of the pulse, the distinct white surface coatings of the tongue could not be mistaken for black surface coatings; likewise, a pale tongue could not be mistaken for a bright one. For the first time, in such a pivotal movement for the tongue as a diagnostic technique, the tongue was codified as a diagnostic system to treat illnesses, at this stage the syndrome known as *shanghan* (cold damage) (Chapter 9 in this volume). This clear and decisive method would have been attractive and

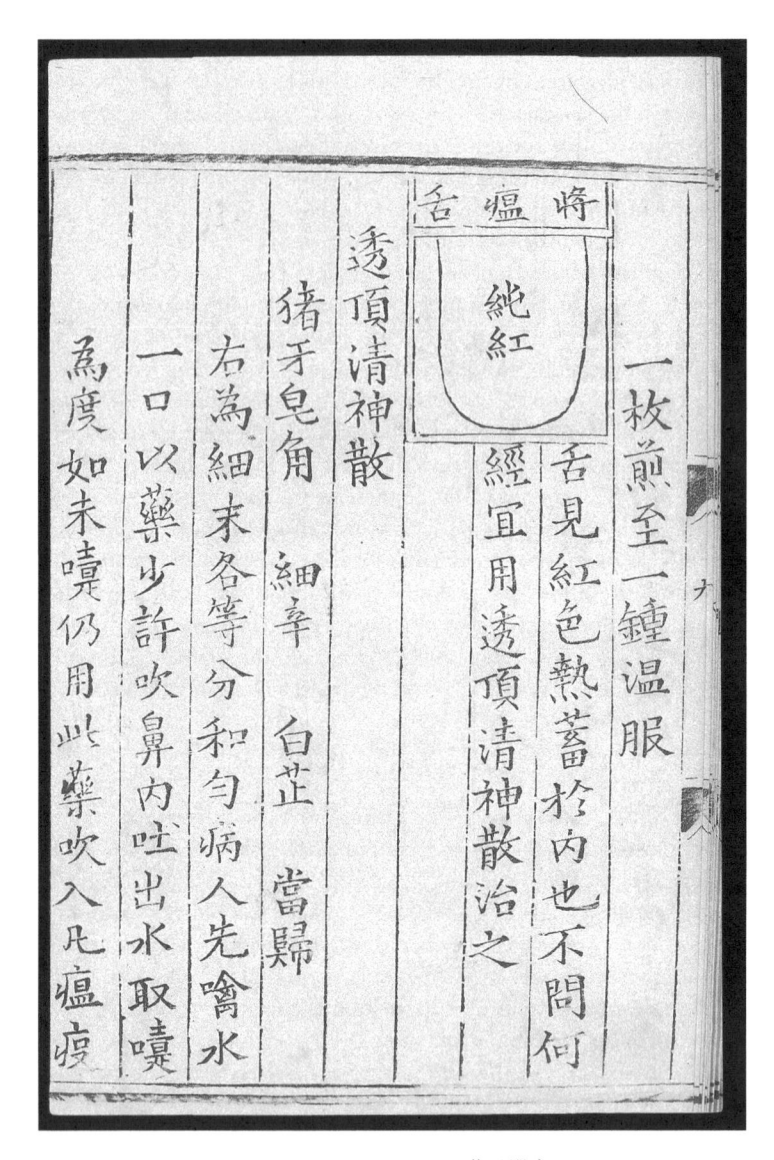

Figure 10.3 From a 1529 edition of Xue Ji's *Xueshi yi'an* 薛氏醫案, showing a pure red tongue and prescribing *Touding qingshen san* 透頂清神散 (Extremely Penetrating Clear-the-Spirit Powder)

Chinese tongue diagnosis diagram: 'Incipient plague' tongue. Wellcome Collection. Public Domain Mark.

invaluable to amateur physicians. However, despite this display, later evidence shows that the uptake of the tongue as a diagnostic indicator remained slow.

In 1531, the *Shishan yi'an* 石山醫案 (Stone Mountain Medical Case Records), a discrete collection of medical case records collected by disciples of Wang Ji 汪機 (1463–1539) (and compiled by Chen Jue 陳桷), showed the growing use of the tongue inspection within medical practice (Grant 2003: 20). Judging by the practice of physicians such as Sun Yikui 孫一奎 (1522–1619) and Cheng Congzhou 程從周 (style name Maoxian 茂先), at that time, the tongue was still only a site where the signs of illness were recorded as part of a more generalised diagnosis; it was not a system of diagnosis in itself. By the end of the sixteenth century, however, the role of the tongue in medical diagnostics was beginning to develop. In the (1591) *Mingyi lei'an* 名醫類案 (Classified Case Records by Famous Physicians), an encyclopaedic anthology which served both as a historical archive and a clinical reference of cases from famous physicians, past and present, we see a growing concentration on the tongue. The number of records of the tongue increases exponentially from the Ming to the Qing dynasty texts, mostly among those cases that included new discourses around *wenbing* 溫病 (warmth factor) theories. Warmth and heat diagnoses were indicative of the climate of epidemics and febrile illnesses (Holroyde-Downing 2017: 56–60, 136; Chapters 9, 16 and 17 in this volume). Fire was the principal root cause of illnesses attributed to cold, damp, wind and dryness (Holroyde-Downing 2017: 147–66; chapter 9 in this volume). Dai Tianzhang 戴天章 (fl. 1675–95), the author of the 1695 *Guang wenyi lun* 廣瘟疫論 (Expanded Treatise on Febrile Epidemics), provided a useful list of diagnostic criteria used to identify *wenyi* 溫疫 (febrile epidemics). The list included the quality of the patient's *qi*, their complexion, tongue, disposition and pulse (Liu 2005: 16; Hanson 2011: 112). Where heat played a significant role, physicians made use of the tongue. In the context of febrile illness, the heat dried the moisture throughout the body: the tongue was arguably the most easily observable indicator of dryness.

Within a century, the 1764 compilation of case records known as *Linzheng zhinan yi'an* 臨証指南醫案 (Medical Case Records as a Guide to Clinical Practice), which detailed a large, posthumously published compilation of a physician known as Ye Tianshi 葉天士 (1666–1745), provided systematic evidence of how tongue diagnosis was emerging as a system. A 'pioneer in the employment of aromatic stimulants for epidemic fevers', Ye had the ability to combine treatment strategies in the evolving treatment of febrile illnesses (Hummel 1943/44: 902; Holroyde-Downing 2017: 153). In the case detailed below, we see inspection of the tongue following the course of an illness with potential for altering the ingredients as the symptoms change; importantly, the case was diagnosed and treatment prescribed without reference to the pulse (Holroyde-Downing 2017: 154). Ye Tianshi is represented as prescribing *zhen wu tang* 真武湯 (True Warrior Decoction) for a syndrome known as Cold in the Four Extremities. The symptoms include cold hands and feet, a tight chest and white tongue; the white tongue, a symptom of the cold of the spleen. Treatment is succinct: Warm Spleen Yang. The prescription is:

> *cao guo ren* 草果仁 (Tsaoko Fruit Seed), *zhi fu zi* 製附子 (prepared Aconite), *sheng jiang* 生姜 (Fresh Ginger), *baifu ling* 白茯苓 (White Poria), *wu of case records known mei rou* 烏梅肉 (Mume, or Chinese Plum), *guang pi* 廣皮 (Tangerine Peel).

If there is waist pain when bending and the Kidney qi is exhausted, then also:

> *gou qi zi* 枸杞子 (Fructus Lyceum), *rou cong rong* 肉蓯蓉 (Cistanche), *fu zi* 附子 (Aconite), *sheng du zhong* 生杜仲 (Fresh Eucommia), *chuan shang jia* 穿山甲 (Pangolin Scales), *lu rong* 鹿茸.
>
> *(Cornu Cervi, or Deer Antler) (Weikeben Yeshi yi'an, 54)*

Here, the white tongue is used to diagnose cold in the spleen, without the corresponding use of the pulse. One can therefore deduce that by the eighteenth century, the appearance of the tongue had become a crucial medical diagnostic for some physicians in China. The use of the changing appearance of the tongue as one critical element in diagnosis was likely a reflection of the prevailing epidemic environment. There were also new social and cultural factors inspired by the arrival of foreign missions and foreign priorities in medicine.

The arrival of Jesuit missionaries

The expansion of sea routes, enabling the increase of commerce, led to the arrival of European missions in China – notably including the Jesuits – beginning around the late seventeenth century. They were tasked with the objective of spreading the word of the Catholic Church. Arguably, this might be most effectively achieved through the medium of a demonstrably superior intellectual tradition, including knowledge of anatomy.

French Jesuits Joachim Bouvet (1656–1730) and Dominique Parrenin (1665–1741) were privileged to instruct the Kangxi Emperor (reigned 1662–1722), who was interested in the human form as it was described in the most current anatomical knowledge in Europe (Walravens 1996: 365; Asen 2009: 40). Contemporary French anatomy was translated into Manchu in a book known in English as the 'Manchu Anatomy'. Anatomy was evidence of the work of 'divine creator' and lauded as a Christian endeavour (Osler 2001: 12). It became a potent vehicle for Jesuits to spread the gospel, as Florence Hsia aptly describes as 'science in service to religion' (Hsia 1999: 247).

Grounded in the new epistemological practices of Europe as 'the most certain of all parts of medicine' (Dionis 1703: 35), Europeans regarded anatomy as the only 'correct' conception of the human body. To Parrenin, Chinese medical knowledge would be incorrect without it (Le Comte 1698: 1). Here emerged a palpable tension. Anatomy conflicted with those Chinese conceptions of the body that were the basis of medical practice. Jesuit missionaries were quick to slander the Chinese conception of medicine as 'indifferent to anatomy' and 'confused, imaginative or completely false' (Asen 2009: 4; Chiang 2015: 25). Contemporary European medicine presumed superiority over the medicine in China.

Through illustrated anatomical works, the Jesuits endeavoured to 'correct' the Chinese idea of anatomy. The notion that various conceptions of the body could have existed or that the Chinese conception was 'normal' and aided practical and effective therapy within its own schema was of no interest to Europeans. The power of visual representation to demonstrate European cultural superiority became a way of influencing, convincing and enticing the Emperor. Jesuit missionaries used innovative techniques to ensure anatomy was palatable to the Chinese and employed illustrations, which were adapted for the Chinese through Chinese annotations and Chinese cultural representation. By doing so, Chinese identity was assimilated into anatomical illustrations through facial features and distinctive Chinese clothing. The purpose was to ensure anatomy could account for Chinese bodies as well as Europeans.

Figure 10.4 details the anatomy of a placenta and foetus featured in Manchu Anatomy.

The attention to detail through the use of intricate cross-hatching, shadowing and defined borders provides visual appeal through visuospatial dimension. Figurative representations took on anatomic verisimilitude and materiality, a concept that the Europeans assumed the Chinese lacked in their representation of the body. This heightened the power of visual representation. Europeans used the power of the eye to convince the Chinese of anatomy. Arguably, this was a two-prong assault whereby proving a 'correct' anatomy would endorse, through scientific and therefore cultural superiority, the 'correct' religion as purveyed by the Jesuits. This new culture of accurate visual representation was to prove fortuitous for tongue diagnosis.

Figure 10.4 The Foetus and placenta. Parrenin [1715?], v. XV, p. 254. Image permission of BnF

The rise of objectivity

The mid-nineteenth century witnessed a revolution in the visual interpretation of science. In Daston and Galison's historical analysis of *Objectivity* (2007), we discover a shift in the scientific focus of the portrayal of objects. Since the Renaissance 'images [had been] intentionally directive as well as descriptive' focusing the audience's gaze on certain aspects of the body's inner spaces and not others (Bivins 2017: 340). Eighteenth-century science had then depicted 'the characteristic, the essential, the typical', rather than what was to be considered the true specimen (Daston and Galison 2007: 20). The style had been true-to-nature; its role in science was performative; to train, teach and guide the eye of scientists on how to see. In contrast to the new nineteenth-century style of image making, the long-standing traditional culture had emphasised the artist's wilful intelligence to intervene and incorporate the essential and dismiss the accidental. But scientific visual interpretation that once celebrated the active role of the artist now embraced a passive role; atlases, botanical illustrations and snowflakes bore new faces which more faithfully reproduced objects as they were observed without any recourse to the imagination of a Platonian 'ideal type'. This, according to the authors, marked the birth of scientific objectivity.

This culture of objectivity encouraged artists to submit to the peculiarities of nature, rather than interpret the object through the filter of an artist's conception or a scientist's searching to demonstrate the ideal characteristics of a thing. Aptly described as a 'will to willessness' or a will not to will, drawings with as little human intervention as possible were encouraged (Ibid.: 38). To 'bear no trace of the knower' and forgo the ideal form became a new concept that preserved objects for their uniqueness (Ibid.: 17). In other words, objectivity was 'blind sight', a practice where the artist was encouraged to blind themselves to conscious interpretation that would skew faithful representation.

The rise of objectivity, so-described, and the importance of visual representation in Europe would certainly have privileged the tongue over the pulse. The cumbersome art of pulse diagnosis required training and experience, and was problematic for students who were acquiring large amounts of their knowledge through books and not personalised training. In this context, new notions of objectivity, encouraged by the democratisation of knowledge embedded in printing and in new modern networks of knowledge and its transmission, bound the untrained novice physician to a 'blind sight' approach. One could forego the difficulties of pulse diagnosis for the ease of learning the pathologies signified by the tongue from a book. Now, the trained haptic skill required in pulse diagnosis could be swapped for the virtually untrained eye. A white-coated tongue with a pale body could be much more easily communicated than a pulse that beat like 'the gathering of a flock of birds'. Even its individual peculiarities, whether it was tooth-marked, with a red tip or a black root, indicated a limited set of diagnostic possibilities. With an increasing focus on visual perception, visual representations began to take precedence. In the absence of the personal attention involved in apprenticeship, untrained physicians could simply replicate tongue diagnosis from a book or a manuscript, since many manuscript copies were circulating at that time (Chapter 18 in this volume).

Arthur Worthington (1852–1916), a British scientist, and pioneer of the science of objectivity wanted 'real (*objects*) as opposed to imaginary' (Daston and Galison, 2007: 36).[3] In China, the Chinese conception of the pulse was thought to be real. Across the seas, European scientists debated 'the realness' and therefore the value of the Chinese conception of the pulse, some drawing on the authority of its detailed descriptions to revive the old art of sphygmology, and others dismissing it as imaginary (Kuriyama 1999: 37; Jenner 2010: 650–1; Chapter 38 in this volume). Unsurprisingly, the pulse proved a topical debate. This raises important comparative issues which contextualises the rise in tongue diagnosis in China and subsequently worldwide. The pivotal point which led to the rise of tongue diagnosis, both in its availability and in its acceptability, was that there was no doubt that a tongue was indeed a tongue. It was a physical object that could be visualised, interpreted and communicated with significant ease compared to the pulse.

Conclusion

The pervasive influence of the Chinese medical classics leads us to believe that the pulse has always been the pre-eminent tool for diagnostics. Chinese medical diagnosis, however, remains a highly mobile feast that has utilised many different sensory-based techniques such as observing the five colours of the complexion, and the sounds of the voice. This chapter has analysed a range of contexts within which the tongue became an influential diagnostic tool, as an example of change in the hierarchy of diagnostic techniques. Tongue diagnosis was indeed a latecomer, with an initial slow uptake; several factors contributed to its eventual success, including the epidemic environment, and a change in scholarly career patterns. Finally, that it is used equally to the pulse today, is testament to the impact of European visual cultures of objectivity on medicine in China. With the modernisation, standardisation and scientisation of traditional Chinese medicine in the 1950s' academies of Chinese medicine, the apparently accurate reproducibility of tongue diagnoses, achieved with the help of imagery, has given it yet further credibility. Today, we find tongue diagnosis rivalling the pulse as the most common form of diagnosis in Chinese medical hospitals and clinics all around the world.

Notes

1 The *locus classicus* for the combined diagnostic techniques in the standard medical phrase *wang wen wen qie* 望聞問切 (visual inspection, auditory/olfactory examination, interrogation and palpation) is said to be *Gujin yiting daquan* 古今醫統大全 (Complete Collection of Ancient and Modern Medical Works, finalised in 1556). See also *Shengji jing* 聖濟經 (Classic of Divine Assistance, finalised in 1118) for an earlier reference.
2 Tanksuqnama-yi Ilkhan, Istanbul, Süleymaniye Ms. Aya Sofya 3596, fol.64b; and Rashid al-Din Tabib,Tansuqnama, ed. Minuvi, 126. 100 Istanbul, Süleymaniye Ms. Aya Sofya ff2180; see Terzioğlu, 'Ilkhanischen Krankenhäuser'.
3 Bivins (2000) recognises that the study of images is essential to understanding the processes that are passed from one culture to another. Important attention to Ten Rhijne's (1683): *Dissertation de Arthritide: Mantissa Schematica: De Acupunctura: Et Orationes Tres* shows the emergence of both Asian and European conceptions of the body through the illustrations of acupuncture in body maps,. These are routes and maps for needle placement that were indicative of a European visual culture being incorporated into Chinese theory and practice.

Bibliography

Pre-modern sources

Aoshi shanghan jinjing lu 敖氏傷寒金鏡錄 (Scholar Ao's Golden Mirror Record of Cold Damage), 1341, Du Qingbi 杜清壁 [1341] 1529 (ed.), Beijing: Library Archives of the Beijing Academy of Traditional Chinese Medicine.
Gujin yitong daquan 古今醫統大全 (Complete Collection of Ancient and Modern Medical Works) 1556, Xu Chunfu 徐春甫 (ed.) (1520–96) (1991), Beijing: Renmin weisheng chubanshe.
Hanfei zi 韓非子, mid 3rd c. BCE, atr. to *Hanfei zi* 韓非子(c. 280–233 BCE), in Wang Xianshen 王先慎 (ed.) (1998) *Hanfei zi jijie* 韓非子集解, Beijing: Zhonghua shuju.
Memoirs and Observations Topographical, Physical, Mathematical, Mechanical, Natural, Civil, and Ecclesiastical. Made in a late Journey Through the Empire of China, and Published in Several Letters, Le Comte, L. (1698, [1696]), London: printed for Benjamin Tooke at the Middle Temple Gate in Fleetstreet. Lettres.
Mingyi lei'an 名醫類案 (Classified Case Records by Famous Physicians) 1552, Jiang Guan 江瓘, Qiu Peiran 裘沛然 (ed.) (1994), *Zhongguo Yixue Dacheng Sanbian* 中國醫學大成三編 (Book 11), Changsha: Yuelu Shushe.
Nanjing 難經 (Classic of Difficult Questions) c. 2nd century CE.
Qianjin yifang 千金翼方(Annexes to Prescriptions Worth a Thousand Ducats) 681,
Sun Simiao 孫思邈 (1982), Beijing: Renmin weisheng chubanshe.
Shiji 史記 (Records of the Grand Historian) c. 86 BCE, Sima Qian 司馬遷 (1975), Beijing: Zhonghua shuju.
Tanksuqnama-yi Ilkhan, Rashid al-Din Tabib, Istanbul, Süleymaniye Ms. Aya Sofya 3596, fol.64b.
Wang Shu-ho (4th century A.D.) Pulse taking. Courtesy of the Wellcome Collection, London, https://wellcomecollection.org/works/wkqzmwfr, accessed 29/3/2020.
Weikeben Yeshi yi'an 未刻本葉氏醫案 (Case Histories from the Ye Family, Not Yet Printed) c.1746, Ye Tianshi 葉天士, Cheng Menxue 程門雪 (ed.) (2010), Shanghai: Shanghai kexue jishu chubanshe.

Modern sources

Asen, D. (2009) '"Manchu anatomy": anatomical knowledge and the Jesuits in seventeenth- and eighteenth-century China', *Social History of Medicine*, 22.1: 23–44.
Barrett, T. (2008) *The Woman Who Discovered Printing*, London: Yale University Press.
Bivins, R. (2000) *Acupuncture, Expertise and Cross-cultural Medicine*, Basingstoke: Palgrave.
Brokaw, C.J. and Chow, K.-W. (eds) *Printing and Book Culture in Late Imperial China*, Berkeley: University of California Press.
Brown, M. (2015) *The Art of Medicine in Early China: the ancient and medieval origins of a modern archive*, New York: Cambridge University Press.
Brownell, S. (ed.) (2011) *From Athens to Beijing: west meets east in the Olympic Games, volume 1: sport, the body, and humanism in ancient Greece and China*, New York: greek.works.com.
Cahill, J. (ed.) (1981) *Shadows of Mt. Huang: Chinese painting and printing of the Anhui school*, Berkeley: University of California Press.
Carter, T. (1955) *The Invention of Printing in China and Its Spread Westward*, New York: Ronald Press.
Chaney, T. (2016) *Living Health Integrated Medicine*, http://www.mylivinghealth.com, accessed 6/12/2017.
Chao, Y. (2009) *Medicine and Society in Late Imperial China: a study of physicians in Suzhou, 1600–1850*, New York: Peter Lang.
Chia, L. and De Weerdt, H.G.D. (eds) (2011) *Knowledge and Text Production in an Age of Print: China, 900–1400*, Leiden: Brill.
Chiang, H. (2015) *Historical Epistemology and the Making of Modern Chinese Medicine*, Manchester: Manchester University Press.
Clark, H. (1991) *Community, Trade, and Networks: southern Fujian province from the third to the thirteenth century*, Cambridge: Cambridge University Press.
Cook, H.J. (ed.) (2020) *Translation at Work: Chinese medicine in the first global age*, Leiden: Brill Rodopi.
Daston, L. and Galison, P. (2007) *Objectivity*, Cambridge, MA: MIT press.
Despeux, C. (2005) 'From prognosis to diagnosis of illness in Tang China, comparison of the Dunhuang manuscript P. 3390 and medical sources', in V. Lo and C. Cullen (eds) *Medieval Chinese Medicine: the Dunhuang medical manuscripts*, London; New York: Routledge, pp. 176–205.
——— (2018) 'Picturing the body in Chinese medical and Daoist texts from the Song to the Qing period (10th to 19th centuries)', in V. Lo and P. Barrett (eds) *Imagining Chinese Medicine*, Leiden: Brill, pp. 52–68.
Dionis, P. (1703) [1690] *The Anatomy of Humane Bodies Improv'd, According to the Circulation of the Blood, and all the Modern Discoveries*, London: printed for H. Bonwicke, W. Freeman, T. Goodwin, M. Wotton, B. Tooke, and S. Manship.
Farquhar, J. (1994) 'Multiplicity, point of view, and responsibility in traditional Chinese healing', in A. Zito and T.E. Barlow (eds) *Body, Subject and Power in China*, Chicago, IL: University of Chicago Press, pp. 78–99.
Floyer, J. (1649–1734) 'The physician's pulse-watch', *Center for the History of Medicine: OnView*, https://collections.countway.harvard.edu/onview/items/show/12605, accessed 28/12/2017.
Goldschmidt, A. (2009) *The Evolution of Chinese Medicine: Song dynasty, 960–1200*, London: Routledge.
Grant, J. (2003) *A Chinese Physician: Wang Ji and the Stone Mountain medical case histories*, London: Routledge Curzon.
Hanson, M. (2011) *Speaking of Epidemics in Chinese Medicine: disease and the geographic imagination in Late Imperial China*, New York: Routledge.
Hanson, M. and Pomata, G. (2020) 'Travels of a Chinese pulse treatise: the Latin and French translations of the *Tuzhu maijue bianzhen* 圖註脈訣辨真 (1650s–1730s)', in H.J. Cook (ed.) *Translation at Work: Chinese medicine in the first global age*, Leiden: Brill Rodopi, pp. 23–57.
Hao, Xue (1987) *On the Detection of Disturbances at a Distance in Arterial Pulse Waves*, PhD thesis, University of California, San Diego.
Harper, D. (1998) *Early Chinese Medical Literature: the Mawangdui medical manuscripts*, London: Kegan Paul International.
——— (2001) 'Iatromancy, diagnosis, and prognosis in early Chinese medicine', in E. Hsu (ed.) *Chinese Medicine and the Question of Innovation*, Cambridge: Cambridge University Press, pp. 99–120.
Hett, W.S. (trans.) (1957) *Aristotle on the Soul. Parva Naturalia. On breath*, Cambridge, MA: Harvard University Press.

Hinrichs, TJ (2011) 'Governance through medical texts and the role of print', in L. Chia and H.G.D. De Weerdt (eds) *Knowledge and Text Production in an Age of Print: China, 900–1400*, Leiden: Brill, pp. 217–38.

Hinrichs, TJ and Barnes, L. (2013) *Chinese Medicine and Healing*, Cambridge, MA; London: Belknap Press of Harvard University Press.

Holroyde-Downing, N. (2005) 'Mysteries of the Tongue', *Asian Medicine*, 1.2: 432–61.

——— (2017) *Tongues on Fire: on the origins and transmission of a system of tongue diagnosis*, PhD thesis, University College London.

Hsia, F. (1999) 'Jesuits, Jupiter's satellites, and the Académie Royale des sciences', in J.W. O'Malley, G.A. Bailey, S.J. Harris and T.F. Kennedy (eds) *The Jesuits: cultures, sciences, and the arts, 1540–1773*, Toronto, ON: University of Toronto Press, pp. 241–57.

Hsu, E. (ed.) (2001) *Innovation in Chinese Medicine*, Cambridge: Cambridge University Press.

Hummel, A.W. (ed.) (1943/1944) *Eminent Chinese of the Ch'ing Period, 1644–1912*, Washington, DC: U.S. Government Printing Office.

Hymes, R.P. (1987) 'Not quite gentlemen? Doctors in the Sung and Yuan', *Chinese Science*, 8: 9–76.

Jenner, M. (2010) 'Tasting Lichfield, touching China: Sir John Floyer's senses', *The Historical Journal*, 53.3: 647–70.

Kobayashi, H. and Sabin, S. (1981) 'The great age of Anhui printing', in J. Cahill (ed.) *Shadows of Mt. Huang: Chinese painting and printing of the Anhui school*, Berkeley: University of California Press, pp. 25–32.

Kuriyama, S. (1999) *The Expressiveness of the Body and the Divergence of Greek and Chinese Medicine*, New York: Zone Books.

Landau, A.S. and Adamjee, Q. (eds) (2015) *Pearls on String: artists, patrons, and poets at the great Islamic courts*, Baltimore, MD: Walters Art Museum; University of Washington Press.

Lei, S. (2002) 'How did Chinese medicine become experiential? The political epistemology of jing-yan', *Positions: Asia Critique*, 10.2: 333–64.

Li Jianmin 李建民 (2000) *Sisheng zhi yu: Zhou Qin Han maixue zhi yuanliu* 死生之域:周秦漢脈學之源流 (The Territory between Life and Death), Taipei: Zhongyang yanjiuyuan lishiyuyan yanjiusuo.

Liu, G. (2005) *Warm Pathogen Diseases: a clinical guide*, Seattle, WA: Eastland Press.

Lo, V. (2011) 'Training the senses through animating the body in ancient China', in S. Brownell (ed.) *From Athens to Beijing: West meets East in the Olympic Games, volume 1: sport, the body, and humanism in Ancient Greece and China*, New York: greek.works.com, pp. 67–85.

——— (2018) 'Imagining practice: sense and sensuality in early Chinese medical illustration', in V. Lo and P. Barrett (eds) *Imagining Chinese Medicine*, Leiden: Brill, pp. 69–88.

Lo, V. and Barrett, P. (eds) (2018) *Imagining Chinese Medicine*, Leiden: Brill.

Lo, V., Berlekamp, P. and Wang, Y. (2015) 'Administering art, history, and science in the Mongol Empire', in A.S. Landau and Q. Adamjee (eds) *Pearls on String: artists, patrons, and poets at the great Islamic courts*, Baltimore, MD: Walters Art Museum; University of Washington Press, pp. 53–85.

Lo, V. and Cullen, C. (eds) (2005) *Medieval Chinese Medicine: the Dunhuang medical manuscripts*, London; New York: Routledge.

Lo, V. and Wang, Y. (2018) 'Chasing the vermillion bird: late medieval alchemical transformations in "The Treasure Book of Ilqan on Chinese Science and Techniques"', in V. Lo and P. Barrett (eds) *Imagining Chinese Medicine*, Leiden: Brill, pp. 291–304.

McDermott, J. (2005) 'The ascendance of the imprint in China', in C.J. Brokaw and K.-W. Chow (eds) *Printing and Book Culture in Late Imperial China*, Berkeley: University of California Press, pp. 55–106.

Needham, J. (2000) *Science and Civilization in China: volume 6 (biology and biological technology)*, part VI (medicine), Cambridge: Cambridge University Press.

Osler, M. (2001) 'Whose ends? Teleology in early modern natural philosophy', *Osiris*, 16: 151–68.

Ozanam, C (1886) *La Circulation et le pouls: Histoire, physiologie, semiotique, indications therapeutiques*, Paris: Librarie J.B. Bailliere et Fils.

Parrenin (1715?) 'The Foetus and placenta', *Bibliothèque Nationale de France*, v.XV: 254.

——— (1715?) *Wargi namu oktosilame niyalma beye giranggi sudala nirugan-i gisun Xiyi renshen gumo tushuo* 西醫人身骨脈圖說 (Explanations and Illustrations of the Bones and Vessels of the Human Body in Western Medicine), Bibliothèque Nationale, Paris, FM 191.

Ten Rhyne, W. (1683) *Dissertatio de Arthritite; Mantissa Schematica; de Acupunctura; et Orationes tres: I. De Chymiae et Botaniae antiquitate & Dignitate; II. De Physionomia; III. De Monstris*, Londres: Chiswell.

Terzioğlu, 'Ilkhanischen Krankenhäuser' (The Treasure Book of the Ilkhan on Chinese Science and Techniques), dated AH 713 (1313) (Lo and Wang 2017: 291; Lo, Berlekamp and Wang 2015: 60).

Unschuld, P.U. (1986a) *Medicine in China: a history of pharmaceutics*, Berkeley: University of California Press.

——— (1986b) *Nan-ching: the classic of difficult issues*, 1st ed., Berkeley: University of California Press.

——— (1998) *Chinese Medicine*, Indiana: Paradigm Publications.

Unschuld, P.U., Tessenow, H. and Zheng, Jinsheng (2011) *Huang Di Nei Jing Su Wen: an annotated translation of Huang Di's Inner Classic – Basic Questions*, Berkeley; Los Angeles: University of California Press.

Walravens, H. (1996) 'Medical knowledge of the Manchus and the Manchu anatomy', *Etudes Mongoles et Siberiennes*, 27: 359–74.

Yang, S. (1997) *The Pulse Classic: a translation of the Mai Jing by Wang Shu-he*, Denver: Blue Poppy Press.

11
CASE RECORDS *YI'AN* 醫案

Nancy Holroyde-Downing

Case records, or the setting out of details of a medical encounter, have a long history in China, and offer a cornucopia of information about the practices of individual physicians.

The identification of the person seeking treatment and the date; the person consulted; the complaint and the assessment of its cause are the minimal facts of a case. As such, the roots of case records in China are arguably found in the inscriptions on the oracle bones of the Shang dynasty (c.1600–1046 BCE), as seen here.

> Day *renxu*. Crackmaking. Diviner: *Huan*. [The king] has a sick tooth.
> It is [a case of] *chi* [=attack by demonic agency].
>
> *(Cullen 2001: 302)*

To be sure, this is a minimalist account; yet, these components remain quite constant in subsequent case record structures. From these first vignettes inscribed on oracle bones to much later collections of cases documenting a physician's lifetime of practice, case records allow us to see not only what physicians did, but also what they valued in both the assessment of the patient before them and the role they saw themselves playing in the encounter.

Case records were sometimes written by physicians themselves, and sometimes compiled by students as a tribute to their teacher. As they developed into a distinct literary genre, they began to tell us a great deal more than the bare facts of the clinical encounter. They inform us about the social and cultural milieu in which they were written, reference the theoretical underpinnings of a physician's reasoning and argue for particular methods of practice.

However, the gathering of specific information to form a record was not a uniquely medical concern. While we translate *yi'an* 醫案 as medical case records, the character *an* 案 (case) is also a particularly significant term in law, and is found in the legal writings of the Han dynasty (206 BCE–220 CE). A tomb in Hebei, *Zhangjiashan* 張家山 tomb 247, excavated in 1983, held texts that suggest its occupant was an official skilled in law, medicine, military arts and calculation who retired in 194 BCE (Li and Xing 2001: 127). During the Han, '*an*' referred to 'on the spot investigation into a crime' (Furth *et al.* 2007: 5). Texts from the *Zhangjiashan* site include cases of livestock theft; robbery; eluding conscription;

murder, and even suffering from contagious illness could be considered a social offence. The fact that these cases are nearly identical to a collection entitled *Fengzhenshi* 封診式 found in an earlier burial site, *Shuihudi* 睡虎地 tomb 11, suggests that they were being copied from one set to another as models of legal practice used to establish norms.[1] Standardisation was an all-encompassing pursuit of the newly established Han empire. Attempts to conform the practices of law and medicine were an extension of this, and it is therefore no surprise that it was exactly in the first half of the first century BCE that we also see the first extant collection of a physician's cases, which were created to justify medical practice in the face of a criminal charge (Holroyde-Downing 2017).

In both legal cases and medical cases, a narrative of events displays the practices and assumptions of the record keeper. In groups of medical case records, a mosaic forms which reveals the workings of a tradition at a particular point in time and culture. We see established treatments and practices being upheld or contested, and innovations taking shape. On an individual level, we see physicians chronicling their clinical encounters to highlight their expertise.

The story of case records in China begins with the compilation of patient notes set before us by the Grand Historian and astrologer, Sima Qian 司馬遷 (c. 145–86 BCE) of the Western Han dynasty (202–9 BCE). He composed the first Chinese dynastic history, the *Shiji* 史記 (Records of the Grand Historian, c. 90 BCE). In the 105th chapter, we find twenty-five medical case records compiled by the physician Chunyu Yi 淳于意 (b. 215 BCE). Yi had been the Master of the Granary (*Canggong* 倉公), who gave up his official post to devote himself to the study of medicine. Having been accused of a crime, punishable by mutilation, by a patient he refused to treat, Yi was imperially summoned and imprisoned. Though eventually spared punishment through the intervention of his youngest daughter, who pleaded to the Emperor on his behalf, he had compiled a collection of case records during his imprisonment in defence of himself and his medical practice.

Chunyu Yi's medical knowledge and diagnostic practices resonate with what we find in the received texts of the Han dynasty such as the *Huangdi neijing* 黃帝內經 (Yellow Emperor's Inner Classic).[2] In his records, he makes use of the *mai* 脈 (vessels and/or movements in the vessels, or pulses); the concept of *qi* 氣; of *yin* and *yang* and the five viscera (*wuzang* 五藏) as he recounts his consultations (Chapters 1 and 2 in this volume). He also notes that he taught massage techniques to others. In his cases, we see his use of drug therapy, acupuncture and moxibustion. Indeed, in line 28 of case 6, he mentions all three modalities.

> In cases where the body form has been fading away, it is not fitting to apply cauterization (moxibustion) and needle therapy, and to make [the patient] drink potent drugs.
>
> (Hsu 2010: 233)[3]

Physician Yi did not call his accounts case records, but rather consultation records (*zhenji* 診籍). In his response to questioning about his practice, he stated that, 'In every case where your vassal has conducted a medical consultation, he has always made a consultation record' (Cullen 2001: 305). In his Memoir (*liezhuan* 列傳) and its twenty-five cases, Chunyu Yi gives us the first social history of medical practice in China. We learn about his education as he tells us about his teachers, his methods of learning, the skills he applies in practice and the books he owns. He offers his own assessment of his professional competence and the declaration that his prognostications have been correct. Even in cases in which the patient is beyond help, he knows when this is so, and his therapeutic successes are noted. He offers his opinion of the role of case records in his own medical learning and practice:

The reason I am able to differentiate disorders is because of what I accomplished with my teacher. Since my teacher died, I have set out 'consultation records' of disorders that I have diagnosed to predict the time [allotted] for life or death. I observe where my predictions are accurate or amiss and find the results in agreement with the pulse method. This is how I gain knowledge.

(Furth 2007: 127)

Record keeping in the Song (960–1278)

The Han educational model of medical learning that Chunyu Yi's writings describe was that of master and disciple. Following the Tang dynasty (618–907), a transformation occurred as Xie Guan 謝觀 points out in his 1935 *Zhongguo yixue yuanliu lun* 中國醫學源流論 (Origins and Development of Medicine in China):

中國醫術, 當以唐宋為一大界。自唐以前, 醫者多守專門授受之學, 其人皆今草澤 鈴醫之流, 其有士大夫而好研方書… 代不數人耳。自宋以後, 醫乃一變為士大夫 之業, 非儒醫不足見重於世。

In Chinese medical arts, the Tang and Song should be taken as an important boundary. Up through the Tang, most doctors clung to specialised learning transmitted [from master to disciple]. The literati among them who loved the study of formularies… numbered only a few. From the Song onward, medicine suddenly became a job for the literati. If a doctor was not a scholar physician, he was not worthy of recognition in the world.

(Xie Guan in Boyanton (trans.) 2015: 62–3)

There were various reasons why medical practice became a profession that the scholar elite took up in increasing numbers during the Song.[4] Among them was the accrual of wealth by the rising merchant class, which allowed them to fund the education of their sons. Greater numbers of scholars then took, and passed, the imperial examinations, though the posts requiring examination success did not become any more numerous. Not only did Song scholars increasingly outnumber any potential civil service positions. As aspiring medical practitioners, they also outnumbered experienced physicians with whom to apprentice. Consequently, the traditional structure of medical learning – that of master and apprentice – became an impracticable model.

The Song dynasty's encouragement of advances in printing, which created an unprecedented proliferation of texts, was a crucial factor in the upsurge of scholars turning to medicine. Though not yet on the scale that would later be seen in the Ming dynasty (1368–1644), the resulting wider distribution and availability of medical texts brought physicians out of the relative isolation in which they had previously practised. Canonical medical texts, treatises and government formularies were no longer confined primarily to the domain of apprenticeship models and solitary or family based physicians but sought after by a wider range of the population. The ensemble of literate scholars interested in medicine grew beyond practising physicians, to include government officials, literate gentry and students, ensuring a broad market for medical publications.

While scholars could master medical principles from their textual studies, the scrutiny of case records afforded the student – particularly those without an experienced practising mentor – a window onto the actual clinical practice of learned physicians. Consequently, while Tang literati had engaged largely in the study of formularies, from the Song dynasty onwards, case records became an increasingly useful genre of medical writing.

While not extant, written medical record keeping is implied in the bureaucratic demands of both the Song and Ming governments. The first demanded *yinzhi* 印紙 (stamped papers, or official records) of cases treated to be presented by students for grading (Cullen 2001: 304), and the second set out a hierarchy of rewards to physicians according to numbers of successful treatment outcomes (Goldschmidt 2009: 50).

The physician Xu Shuwei

One of the Song publications that presaged the establishment of the case history genre was Xu Shuwei's 許叔微 (1079–1184) *Shanghan jiushi lun* 傷寒九十論 (Ninety Discourses on Cold Damage). Scholars disagree as to whether this is a lineage text – one passed down from master to apprentice or kept within a medical family – or case records intended for public consumption,[5] though whatever its intended audience, the earliest record of its existence is from a Qing dynasty (1644–1911) publication (*Congshu jizheng xinbian* 45: 617; *Zhongguo yiji tongkao* 1: 279–80). What is certain is that it is a compilation of clinical cases with added discussion that appears to be aimed at knowledgeable individuals, most likely physicians, centring on the appropriate use of the drug recipes found in the *Shanghan lun*. An example of Xu's record keeping follows:

> Cao Sheng initially suffered from Cold Damage. After six or seven days, his abdomen was full and he was vomiting. He couldn't get food down. He had a fever, and his hands and feet were hot. His abdomen ached, and he was nauseated. The physicians called it excessive *yang*. [His family] still had misgivings about his hands and feet being hot, fearing that heat had amassed in the stomach causing vomiting and nausea, or, seeing the vomiting and diarrhoea, took it to be sudden turmoil [disease]. [They] asked me to diagnose [him]. His pulse was fine and sunken. I evaluated him saying, 'this is a Greater *Yin* pattern. In Greater *Yin* disease, there is abdominal fullness and vomiting, inability to get food down, severe spontaneous diarrhoea, and occasional spontaneous pain of the abdomen'. I used Regulate the Centre Pill (*Tiaozhong wan* 調中丸) to stop [the illness]. I used five or six pills the size of an egg yolk per day. Then I used Five Accumulation Powder (*Wuji san* 五積散). After several days [Cao] recovered.
>
> Discussion: I see common physicians diagnosing Cold Damage and only labelling them *yin* patterns and *yang* patterns. Zhongjing has three *yin* and three *yang* [diseases]. Even in one pattern, there are also leanings toward exuberance or insufficiency. What is necessary is to clearly differentiate in which channel [the illness is present]. The formula must correspond to the signs, and there are standards for the use of medicinals. Moreover, in the case of Greater *Yin*, Lesser *Yin*, and Reverting *Yin*, they have [situations which demand] either supplementing or draining. How can [they] stop at naming [the disease] a *yin* pattern?
>
> *(Boyanton (trans.) 2015: 158–9)*[6]

Xu provides us with the patient's name (though not his age), his symptoms and some sense of the duration of the illness. There are several other nuggets of information here. Reading that Xu is requested to give his diagnosis after other physicians had already done so reflects patient (and family) anxiety in a world of diverse medical practitioners, frequent epidemics (Chapters 16 and 17 in this volume; Goldschmidt 2009: 79) and often ineffective remedies. What is fascinating is that not only did his family have 'misgivings' about the diagnoses of the other physicians, but they had them around issues of diagnostic significance. In other words, they were medically learned enough to know that symptoms of heat (hot hands and feet) had significance in the functioning of the digestive processes of the body (vomiting and diarrhoea), that this should probably indicate a short illness, and in the absence of improvement after a week, called in Xu to give an additional opinion. Another arresting facet of this case record is the following discussion, in which Xu delivers a tirade concerning the diagnostic incompetence of the other medical personages attending (Holroyde-Downing 2017: 125). The case record as an argument for one's own medical expertise is apparent.

Ming dynasty case records

It was during the Ming dynasty that case records became an established form of medical literature in their own right. In addition to published cases of individual physicians, encyclopaedic collections, containing large numbers of case histories of different eminent physicians, appeared. Ming advances, such as the restructuring of the imperial examination system, improvements in agriculture supporting population growth, industrial expansion bringing about a newly wealthy merchant class, increased mobility due to improved transport links and the continued flourishing of publishing, contributed further to the upsurge in the literate gentry population. The number of available government posts continued to stagnate and, in the search for additional career opportunities, the practice of medicine as an option resonating with Confucian benevolence was very attractive. In the words of a Ming official,

> Today, in selecting a technique from among the professions, only medicine is close to benevolence. In practising it, one can save life and raise a family; one can spread kindness and save many lives.
>
> *(Brokaw 1991: 66)*

Compiling records of one's clinical encounters became a significant part of the practice of literati physicians, as membership of the elite could be based on establishing one's authority in medicine. Scholarly and literary flourishes, coupled with medical prescribing, were distinguishing features of the aspirational behaviour of Ming physicians. As Cullen puts it, 'what could be more useful as a quick route to apparent expertise than a collection of authoritative medical cases statements?' (Cullen 2001: 319).

An example of the role record keeping played in the social status of Ming physicians can be seen in the cases of Sun Yikui 孫一奎 (1522–1619). He was a scholar from Anhui, whose repeated attempts at advanced examination success were met with failure. He turned to medicine, and through his study of medical texts became a self-taught physician. His case records, published by his sons and disciples, demonstrate the particularly vivid style and use of descriptive detail employed by scholarly physicians in the Ming (Zeitlin 2007: 169–202). An example is the case of Wang Dong's husband who had a *yang* deficiency (*Sun Wenyuan yi'an* 孫文垣醫案 2012: 106). In it, we learn that the patient has just returned from a trip, that he was treated for a cold that was getting better, but the accompanying mouth ulcers were getting worse and he wanted another diagnosis. One Mr. Chen, a self-proclaimed throat specialist of lower status than physician Sun, is introduced, as well as the family chickens who become poisoned after eating the patient's vomitus. We are told that the patient himself is unaware of the coldness in his legs beneath his knees until this crucial piece of diagnostic information is pointed out to him by Sun. There is a description of the patient's family and friends – all of whom are impressed with the physician's diagnostic skills. We even learn that

Sun went for a two-hour walk with friends after breakfast in the midst of dealing with the case. The record is not confined to medical information but has all the page-turner quality of a good story. The use of such literary flourish in a case record was a method of underscoring one's scholarly pedigree at a time when competition in the field of medicine was significant and becoming greater. A beautifully crafted case record submitted to the patient's family was proof of eminence. Indeed, one renowned Ming physician, Cheng Congzhou 成從周 (1581–?), was reputed to have such polished literary skills that he read segments of his case records to his poetry group (Zeitlin 2007: 169–202).

Encyclopaedias of case records

By the end of the sixteenth century, *yi'an* writing had become an established medical form, and another innovative genre appeared: the encyclopaedic anthology of cases taken from the records of eminent physicians. The *Mingyi lei'an* 名醫類案 (Classified Case Records by Famous Physicians) was the first of these, and was compiled by Jiang Guan 江瓘 (father) and Jiang Yinsu 江應宿 (son), who gathered cases of 141 famous physicians. Published in 1591, it was both a historical archive and a clinical reference. This collection was expanded during the Qing dynasty by Wei Zhixiu 魏之琇 (1722–72), who augmented the first compilation by including the case records of more contemporary, well-known, physicians and by adding disease categories. This extended text was entitled *Xu mingyi lei'an* 續名醫類案 (Supplement to Classified Case Records by Famous Physicians) and published in 1770. The publication of these texts gave physicians an unprecedented window on the practice of celebrated physicians.

Qing dynasty case records

By the Qing dynasty, changes were taking place in both the style and content of medical record keeping. In terms of style, while the personal expertise of the practising physician was still on show, a more laconic approach to the telling of the encounter became the norm. In terms of content, Furth (2007: 146) maintains that Qing case records 'moved ever closer to the centre of the production of medical knowledge'. In other words, the case record can demonstrate how individual physicians might negotiate the differences between the authority of the medical canons on the one hand, and their actual experience of medical practice on the other.

The case records of one of the most famous Qing physicians, Ye Gui 葉桂, also known by his stylename, Ye Tianshi 葉天士 (1667–1747), continue to be admired and studied. They demonstrate a particularly sparse reporting of the encounter and illustrate Farquhar's (1994: 190) observation that 'much of the intellectual life of Chinese medicine revolves around the reading and writing of prescriptions'. Andrews offers an example of physician Ye's terse reporting, which comprised nothing more than a pulse quality and a medicinal recipe:

> *Zuocun shu* 左寸數
> *Shudi* 熟地, *tiandong* 天冬, *tianbei shashen* 甜北沙參, *fushen* 茯神, *huohu* 霍斛, *chaosong maidong* 炒松麥冬
> Left (pulse) *cun* (position) accelerated
> *Radix Rhemanniae Praeparata*; *Radix Asparagi*; sweetened Northern *Radix Gleheniae*; *Sclerotium Poriae Cocos Pararadicis*; *Herba Dendrobium off.*; roasted, loose *Radix Ophiogonis*.
> (*Andrews 2001: 325*)

In further contrast to many of the Ming case records which set forth magnificent recoveries, Ye's case records omit any report of outcomes.

The physicians' accounts of the clinical encounters with their patients have evolved over time and varied in both form and content among practitioners. In contrast to the embellished cases of Sun Yikui, or the minimalist record set down by Ye Gui, a case set out by the Republican-Era (1911–49) physician, He Bingyuan 和炳元 (1861–1929),[7] uses a format familiar to a more modern medical style of record keeping. He sets out his case with a series of headings: Patient name; Disease; Causes of disease; Symptoms; Diagnosis; Therapy; Prescription; Results. Written at a time when Western medicine was gaining prominence in areas of Chinese society, it is credible to assume that its templates of case recordings were influential. His model of case recording appeared at that time and it is therefore possible to assume that the use of this more modern template to document a case owed something to Western practice.

As we have seen, case records in China ranged from Chunyu Yi's self-justificatory collection, to the self-promotional writings of Sun Yikui and Cheng Congzhen. They were also teaching tools, in that the theory and practice embedded in the stories of the cases enabled readers to discern the theoretical underpinnings of a physician's actual treatments given in a clinical encounter at a particular time.

And yet, in all of these cases, from the Shang oracle bones onwards, we have a record of something very constant – a healer's intervention on behalf of a sufferer. As Scheid (2002: 102) describes it, whatever the differences, 'Each note is still, however, a recording of the clinical encounter as understood by the physician and is thus a reflection of his or her subjectivity'. As we have seen, this subjectivity reflects not only a personal understanding of medicine, but also the culture in which each physician practises.

Notes

1 For further discussion of the tomb manuscripts of the Han dynasty, see Lo and Cullen (2005).
2 The *Huangdi neijing corpus* consists of two treatises: *Suwen* 素問 (Fundamental Questions) and *Lingshu* 靈樞 (The Numinous Pivot), written down *c.* second century BCE, plus the *c.* seventh-century *Taisu* 太素 (Great Basis) recension, which overlaps with both. There is general scholarly consensus that the corpus came into being through a process of medical texts being combined and transmitted among physicians during *Qianhan* 前漢, Former Han (202 BCE–9 CE).
3 This book provides a translation and interpretation of ten of Chunyu Yi's cases.
4 For a discussion of the rise of the scholar physician during the Song dynasty, see Robert Hymes (1987: 9–76).
5 Furth (2007) argues the former, while Boyanton (2015) and Goldschmidt (2015) argue the latter.
6 The three *yang* and three *yin* are here referring to Zhang's six-channel framework, made up of the three *yang* categories (*taiyang* 太陽, *yangming* 陽明, *shaoyang* 少陽), and three *yin* categories (*taiyin* 太陰, *shaoyin* 少陰, *jueyin* 厥陰).
7 He Bingyuan published under his style name of He Lianchen 和廉臣. For a discussion of this case, see Andrews (2001: 324-36).

Bibliography

Pre-modern sources

Huangdi neijing suwen 黃帝內經素問 (The Yellow Emperor's Inner Classic: Basic Questions) (1963), Beijing: Renmin weisheng chubanshe.
Mingyi leian 名醫類案 (Case Histories Recorded by Famous Doctors) 1552, Jiang Guan 江瓘, in Qiu Peiran 裘沛然 (ed.) (1994), *Zhongguo yixue dacheng sanbian* 中國醫學大成三編 (book 11), Changsha: Yuelu shushe.

Shanghan jiushi lun 傷寒九十論 (Ninety Discourses on Cold Damage) 1132, Xu Shuwei 許叔微, in *Congshu jicheng xin bian* 叢書集成新編 (Vol. 45) (1986), Taipei: Xinwenfeng chuban gongsi.
Shiji 史記 (Records of the Grand Historian) c. 86 BCE, Sima Qian 司馬遷 (1975), Beijing: Zhonghua shuju.
Sun Wenyuan yi'an 孫文垣醫案 (Medical Case Records of Sun Wenyuan) c.1619, Sun Wenyuan 孫文垣 (2012), Beijing: Zhongguo yiyao keji chubanshe.
Xu mingyi leian 續名醫類案 (Supplement to Classified Case Records by Famous Physicians) 1770, Wei Zhixiu 魏之琇 (1957), Beijing: Renmin weisheng chubanshe.
Zhongguo yixue yuanliu lun 中國醫學源流論 (Origins and Development of Medicine in China) 1935, Xie Guan 謝觀 (2003), Fuzhou: Fujian kexue jishu chubanshe.
Zhongguo minbgyi yan'an leibian 中國名醫驗案類編 (Classified Case Histories by Famous Chinese Physicians) 1927, He Lianchen 和廉臣 (1969), Hong Kong: Minglang chubanshe.

Modern sources

Andrews, B. (2001) 'From case records to case histories: the modernisation of a Chinese medical genre, 1912–49', in E. Hsu (ed.) *Innovations in Chinese Medicine*, Cambridge: Cambridge University Press, pp. 324–36.
Boyanton, S. (2015) *The Treatise on Cold Damage and the Formation of Literati Medicine: Social, Epidemiological, and Medical Change in China, 1000–1400*, PhD thesis, Columbia University.
Brokaw, C. (1991) *The Ledgers of Merit and Demerit: social change and moral order in late imperial China*, Princeton, NJ: Princeton University Press.
Cullen, C. (2001) 'Yi'an 醫案 (Case Statements): the origins of a genre of Chinese medical literature', in E. Hsu (ed.) *Innovations in Chinese Medicine*, Cambridge: Cambridge University Press, pp. 297–323.
Farquhar, J. (1994) *Knowing Practice: the clinical encounter of Chinese medicine*, Boulder, CO: Westview Press.
Furth, C. (2007) 'Producing medical knowledge through cases: history, evidence and action', in C. Furth, J.T. Zeitlin and P.C. Hsiung (eds) *Thinking with Cases: specialist knowledge in Chinese cultural history*, Honolulu: University of Hawai'i Press, pp. 125–51.
Furth, C., Zeitlin, J.T. and Hsiung, P.C. (eds) (2007) *Thinking with Cases: specialist knowledge in Chinese cultural history*, Honolulu: University of Hawai'i Press.
Goldschmidt, A. (2009) *The Evolution of Chinese Medicine: Song dynasty, 960–1200*, London: Routledge.
——— (2015). 'Reasoning with cases: the transmission of clinical medical knowledge in twelfth-century Song China', in B. Elman (ed.) *Antiquarianism, Language, and Medical Philology, from Early Modern to Modern Sino–Japanese Medical Discourse*, Leiden/Boston: Brill, pp. 19–51.
Holroyde-Downing, N. (2017) *Tongues on Fire: on the origins and transmission of a system of tongue diagnosis*, PhD Thesis, University College London.
Hsu, E. (ed.) (2010) *Pulse Diagnosis in Early Chinese Medicine*, Cambridge: Cambridge University Press.
Hymes, R. (1987) 'Not quite gentlemen? Doctors in Sung and Yuan', *Chinese Science*, 8 (January): 9–76.
Li Xueqin and Xing Wen (2001) 'New light on the Early-Han code: a reappraisal of the Zhangjiashan bamboo-slip legal texts'. *Asia Major* xiv(1): 125–46.
Lo, V. and Cullen, C. (eds) (2005) *Medieval Chinese Medicine: the Dunhuang medical manuscripts*, Needham Research Institute Series, London: RoutledgeCurzon.
Scheid, V. (2002) *Chinese Medicine in Contemporary China: plurality and synthesis*, Durham, NC: Duke University Press.
Zeitlin, J.T. (2007) 'The literary fashioning of medical authority: a study of Sun Yikui's case histories', in C. Furth, J.T. Zeitlin and P.C. Hsiung (eds) *Thinking with Cases*, Honolulu: University of Hawai'i Press, pp. 169–202.

12
ACUPUNCTURE ILLUSTRATIONS

Huang Longxiang and Wang Fang

When we look into the history of Chinese acupuncture, a magical medical landscape seems to open up before us. But, as in the story of the blind men and the elephant, it is all too easy to mistake the part we first encounter for the whole. Where can we find the lamp that will light our way through the shadowy depths of China's medical history and traditions? For historians and practitioners alike, the extant diagrams and bronze statues depicting the channels and points of acupuncture and moxibustion (hereafter acu-moxa) are like flickering candles in our hands. Each picture may be worth a thousand words, but it only speaks to those who can decipher its visual codes. This chapter aims to provide the reader with some keys to reading acupuncture illustrations. It will establish three genres of illustrations, identify their contexts, recover historical details, which will help us understand important facts about traditional Chinese anatomy and physiology, and finally pinpoint some critical contemporary research issues.

Every picture in a pre-modern series of works on acupuncture encodes an iconographic continuity that functions as a kind of 'serial number'. Over the centuries, many of these illustrations have been scattered and some are lost forever. But once a series is reconstituted and, importantly, when its 'serial numbers' are decoded, it becomes possible to articulate a new value for each separate unit. In other words, it is only when all of the scattered images are strung back together in an historical sequence that the full significance of each one truly emerges. Collectors who own a single illustration can then easily establish its identity and its value in relation to the larger sequence.

Whenever Chinese acupuncture is mentioned, people tend to think of the theory of the channels and their collateral networks – matters of great mystery for those who have not grown up with the tradition, but also for Chinese medical historians and practitioners themselves. Without the diagrams and bronze acupuncture statues surveyed in this chapter, we would have little prospect of understanding the ancient textual accounts of the acu-moxa points.

DOI: 10.4324/9780203740262-15

Categories of extant ancient and mediaeval diagrams of channels and their collateral networks

Acu-moxa point diagrams

The traditional *Mingtang* 明堂 (Bright Hall) diagrams produced up until the Tang dynasty (618–907) were mostly general diagrams illustrating acu-moxa points of the whole body. They usually consisted of a set of three images of a human figure showing frontal, back and lateral views. In some circumstances, the term *Mingtang* became largely synonymous with 'acu-moxa points'. *Mingtang* diagrams, together with 'bronze statue diagrams' (*tongren tu* 銅人圖), constitute the majority of the extant acu-moxa diagram collections. They serve to show the locations of acu-moxa points on channels rather than illustrating the system of channels. The *Mingtang* genre can be divided into three major categories: general diagrams of acu-moxa points; diagrams of acu-moxa points of the fourteen channels; and sectional diagrams and diagrams showing acupoints categorised by type or by illness. A standard diagram belonging to the *Mingtang* genre was in wide circulation before the third century CE, but unfortunately the early versions have long since disappeared. The best evidence we have for them comes from the early Tang dynasty when the well-known physician and writer Sun Simiao 孫思邈 (581–682) first used polychrome ('five colour') technique to create colour-coded charts. As documented in *Prescriptions Worth a Thousand Gold* (*Qianjin yaofang* 千金要方, 652), he redrew the *Mingtang* diagram as three hanging charts (hereafter the triptych), showing front, back and lateral views: the acu-moxa points on the four limbs were arranged in the order of the channels (other acu-moxa points were not necessarily represented in this joined-up fashion).

Mingtang *triptychs (Tang dynasty)*

Unlike the earlier monochrome *Mingtang* diagrams, Sun Simiao's diagrams (Figure 12.1) were multicoloured, and the colours were keyed to the channels as they corresponded to *Wuxing* 五行 (the Five Agents). Sun Simiao tells us that the scale of the diagrams corresponded to half the size of an average person of his times, calculated according to a system of measurements known as *xiajia xiaochi* 夏家小尺, where one *chi* appears to be about 24.69 cm. He notes that the height of an average person as depicted in his diagrams is '7 *chi*, 6 *cun* and 4 *fen*' as in the *Mingtang jing* 明堂經 (Acupoints Canon of the *Mingtang*) rather than the '7 *chi*, 5 *cun*' recorded in *Lingshu jing* 靈樞經 (Miraculous Pivot) (Chapters 4, 5 and 7 in this volume). Sun Simiao's style of illustration exerted far-reaching influence over the evolution of *Mingtang* and bronze statue diagrams, including those drawn by Wang Tao 王燾 (conventional dates: 670–755) of the Tang dynasty and other physicians active during and after the Song dynasty. Sun's original diagrams have not been handed down to us, and it is, therefore, hard to study his acu-moxa point locations with precision today. However, we can gather important information from the acu-moxa point order in other extant *Mingtang* triptychs, such as those preserved in the *Supplement to the Essential Prescriptions worth a Thousand Gold* (*Qianjin yifang* 千金翼方, 682) and the *Illustrated Manual of Acu-moxa Points of the Bronze Figure* (*Tongren shuxue zhenjiu tujing* 銅人腧穴針灸圖經, 1206; hereafter *Illustrated Manual of Acupoints*). Sun's *Mingtang* triptych was like other such diagrams created up to the Tang dynasty in that only the acu-moxa points on the four limbs were arranged in the order of the channels (but not the other acu-moxa points). This meant that the *Mingtang* diagrams of that time did not include the full complement of acu-moxa point connection lines for the standard twelve or fourteen channels that we know today.

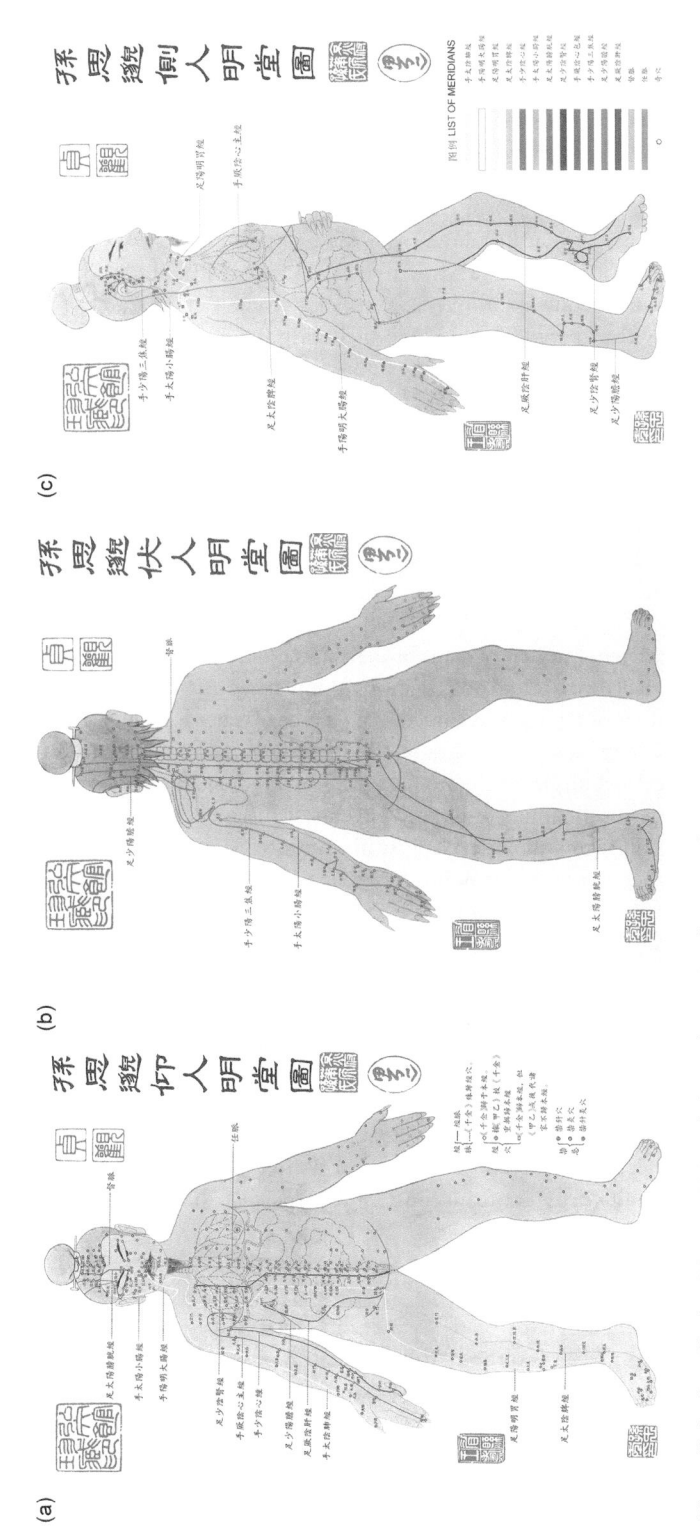

Figure 12.1 Sun Simiao *Mingtang* Triptych (reproduction). All images from Huang Longxiang (2003)

Acupuncture bronze statues and their diagrams

The whole-body acu-moxa diagrams as they were shown on bronze statues extant since the Song dynasty (960–1279) (Figure 12.2) exerted increasing influence in both medical education and medical practice, and have achieved iconic status in acupuncture history. The 'Bronze Man' is now a familiar symbol of the art of acupuncture. According to historical documents relating to the bronzes, the eminent Northern Song (960–1127) physician Shi Cangyong 石藏用 (birth year unknown) was the first person to draw whole-body diagrams of acu-moxa points. In 1474, in the Ming dynasty (1368–1644), Shi Su 史素 (birth year unknown) revised and reproduced in five-colour polychrome the diagrams of the points of the twelve channels with front and back views, including the points of the Governor and Conception Vessels. Later, during that dynasty under the Hongzhi 弘治 emperor (r. 1488–1505), Qiu Jun 丘濬 (1418–95) revised and recoloured their diagrams, adding the *zangfu* 臟腑 (viscera and bowels) and the skeleton. Until recently, there was an academic consensus that the above-mentioned bronze statue triptychs had been lost, although the design (encoding information from those triptychs) survived, thanks to a set of diagrams engraved and published by Zhao Wenbing 趙文炳 (1541–1602) during the reign period of Emperor Wanli 萬曆 of the Ming dynasty (r. 1572–1620), which were well known and widely disseminated. However, after much investigation and research, we find that Shi Su's and Qiu Jun's diagrams do still exist.

The problem of recasting an ancient bronze statue deserves our special attention. In the absence of a textual description of acu-moxa point locations directly linked to a specific bronze statue and set of charts, it is impossible to recreate any lost source, or indeed even to copy an existing bronze statue or diagram. This is evident since different physicians, whether or not they were working in the same period, portrayed the acu-moxa point locations on their bronze statues or diagrams differently, even when they consulted the same texts.

Bronze statue diagrams show striking similarities to *Mingtang* diagrams in a number of respects. Both types of acu-moxa point diagrams show two or three figures. However, the bronze statue diagrams are obviously produced from bronze statues, while *Mingtang* diagrams are derived from the acu-moxa point literature. It is hard for those who do not know the history of these two types of diagrams to distinguish between them. However, there are some fundamental differences that reflect the technique and context of production. When copying from a bronze statue, the artist is constrained by the fact that only the particular acu-moxa points that are visible from a single perspective can be shown. In contrast, a *Mingtang* diagram is a two-dimensional schematic drawing, so the artist need not consider perspective. Also, if the acu-moxa points on a bronze statue are not connected on channels in a linear fashion, the connection sequences of acu-moxa points in the corresponding bronze statue diagrams may be quite different from those in the *Mingtang* diagrams.

Acupoint diagrams of fourteen channels

The acu-moxa point diagrams of the fourteen channels (Figure 12.3) serve primarily to display the acu-moxa points on the channels, but also provide extra information about those channels. The way in which the acu-moxa points became associated in channel theory greatly influenced the course of the lines connecting them. Representative cases are the lines of acu-moxa points on the head; the acu-moxa points of the foot-*taiyang* channel on the lower limbs and the acu-moxa points of the foot-*shaoyin* channel on the foot. To best illustrate this argument, consider the lines connecting the five acupoints on the foot *shaoyin*

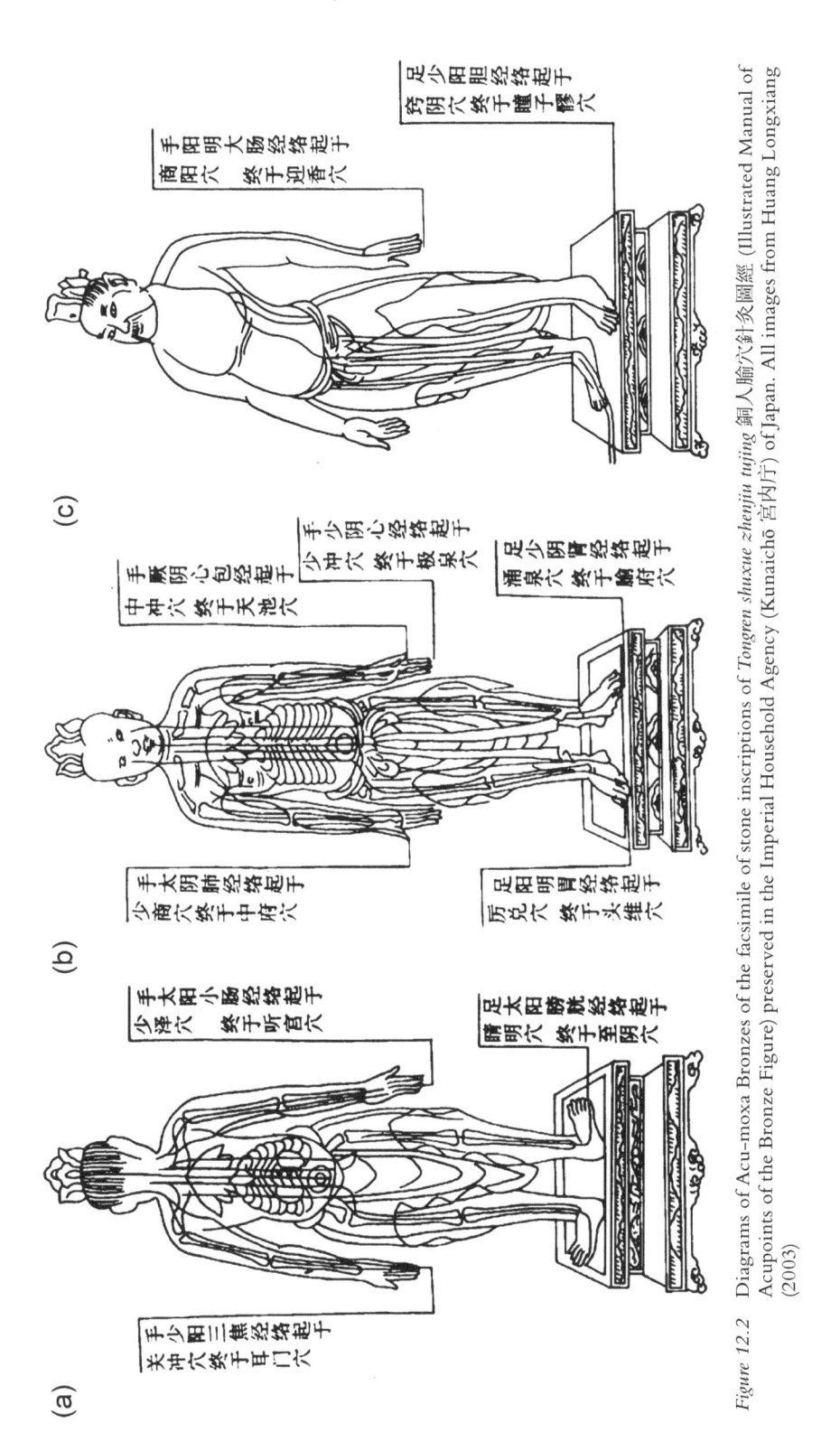

Acupuncture illustrations

Figure 12.2 Diagrams of Acu-moxa Bronzes of the facsimile of stone inscriptions of Tongren shuxue zhenjiu tujing 銅人腧穴針灸圖經 (Illustrated Manual of Acupoints of the Bronze Figure) preserved in the Imperial Household Agency (Kunaichō 宮内庁) of Japan. All images from Huang Longxiang (2003)

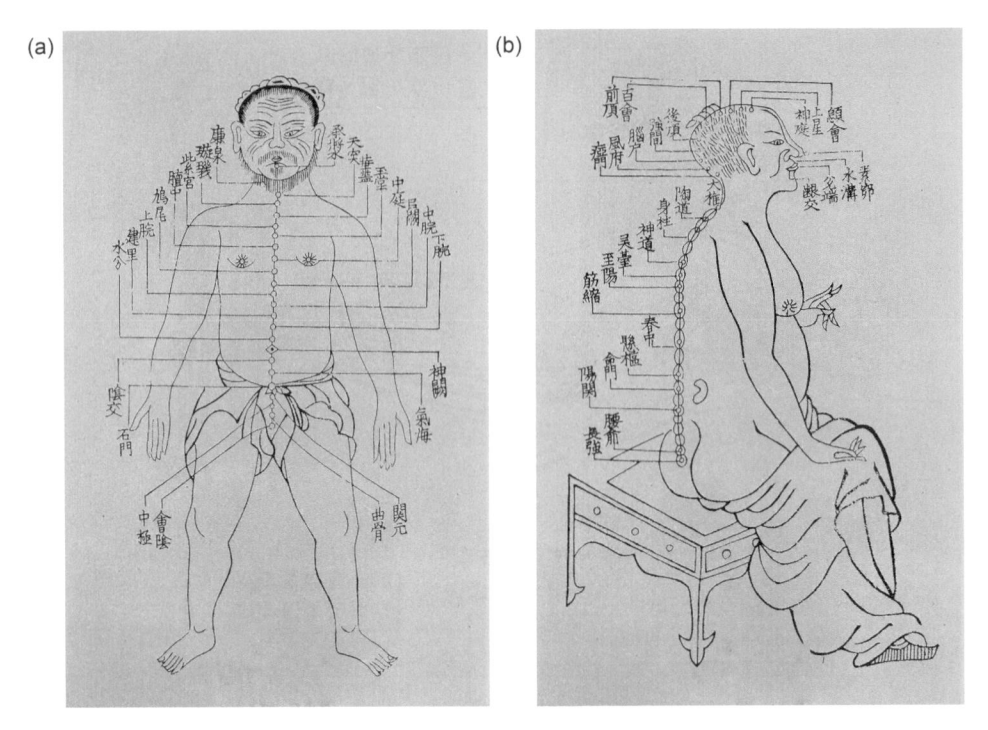

Figure 12.3 Hua Shou 滑壽, 1341, diagrams of the acu-moxa points of the fourteen channels in *An Elucidation*. Edition of the tenth year of Kansei era, Japan (1798) printed in *Zhenjiu yixue dianji daxi* 針灸醫籍典籍大系 (Great Compilation of Acupuncture and Moxibustion Medical Dictionary). All images from Huang Longxiang (2003)

channel: KI2 (*Rangu* 然骨), KI3 (*Taixi* 太溪), KI4 (*Dazhong* 大鐘), KI5 (*Shuiquan* 水泉) and KI6 (*Zhaohai* 照海), which circle around the ankle and do not describe a straight linear route.

All illustrations of acu-moxa points of the fourteen channels can be divided into three systems: the system originating from 'An Elucidation of the Fourteen Channels' (*Shisijing fahui* 十四經發揮; hereafter 'An Elucidation') by Hua Shou 滑壽 (1304–86) of 1341; the system from 'An Exemplary Collection of Acupuncture and Moxibustion and their Essentials' (*Zhenjiu juying* 針灸聚英; hereafter 'An Exemplary Collection') by Gao Wu 高武 (birth year unknown, fl. sixteenth century) of 1529 and the system from 'Illustrated Supplement to the Classified Canon' (*Leijing tuyi* 類經圖翼; hereafter 'Illustrated Supplement') by Zhang Jiebin 張介賓 (1563–1640) dating to 1624. In 'An Exemplary Collection', the figures in the acu-moxa point illustrations of the fourteen channels differ from those in 'An Elucidation'. In that text, Gao Wu notes, 'The diagrams of five *zang* and six *fu*', a feature which continued to appear in the *Mingtang* books of acupuncture from then on.

Channel diagrams

The channel diagrams illustrate the routes of the channels. From our own research, we have determined that there are only six types of extant diagrams of the channels that can be dated to the Song dynasty or earlier:

- 'Diagram of Ten Channels' of the *Chanjing* 產經 (Classic of Childbirth): Anonymous, Six Dynasties (220–589)
- 'Diagram of Twelve Channels and Acu-moxa Points': Wang Weiyi 王惟一 (987–1067) of the Northern Song dynasty
- 'Diagrams of Twelve Channels' in *Cunzhen huanzhong tu* 存真環中圖 (Diagram of Preserving the Truth and Diagrams of the Circulatory Course of Channels): Yang Jie 楊介 (1060–1113)
- 'Diagrams of the Twelve Channels' of *Neiwai erjing tu* 內外二景圖 (Diagrams of the Inner and Outer Views of the Body): Zhu Gong 朱肱 (1050–1125)
- 'Illustrations of Channels' of the *Huo ren shu* 活人書 (Lifesaving Book), date?
- 'Illustrations of Twelve Channels' of the *Ziwu liuzhu zhenjing* 子午流注針經 (Acupuncture Classics of Midnight-Midday and Ebb-Flow Doctrine; hereafter *Acupuncture Classic*): Yan Mingguan 閻明廣 (birth year unknown), Jin dynasty (1115–1234).

After the Song dynasty, there was a new trend towards combining diagrams of channels and acu-moxa points as two-in-one diagrams. Wu Qian 吳謙 (1689–1748), a medical official of the Imperial Medical Department of the Qing dynasty, was influenced by the illustrations of channels in the *Lifesaving Book* when he compiled *Yizong jinjian* 醫宗金鑒 (Golden Mirror of the Medical Tradition) in 1742 (hereafter *Golden Mirror*). For the first time, he distinguished the channel diagrams from composite channel and acu-moxa point diagrams.

Wang Weiyi's 'Diagram of Twelve Channels and Acu-moxa Points' is not a standard diagram of the channels in that Wang specifically combined illustrations of the channel acu-moxa points with illustrations of the channels themselves. The channel lines, as he depicts them on the surface of the body, are a synthesis of the connection lines of acu-moxa points on the body surface with the routes of the channels as described in the chapter *Jingmai* 經脈, hereafter 'Channels' – part of the *Lingshu* (Miraculous Pivot) book of the seminal Chinese medical classic *Huangdi neijing* (Yellow Emperor's Inner Classic, hereafter *Inner Classic*).[1] As a result of this approach, he shows us only the routes of channels on the body surface, but not the routes connecting to the inside of the body. Similarly, the courses of the three *yin* channels of the foot are only depicted as they appear on the thorax and abdomen. All of this differs from the description in *Channels*.

It is worth noting that tubular double lines were used to illustrate channels in the (monochrome) diagrams in the Song period *Lifesaving Book*, and that most of the *Mingtang* diagrams that were subsequently handed down used the same method to show the channels. Since the polychrome channel lines on Shi Cangyong's 石藏用 bronze statue diagrams were rather thick, it made sense to use tubular double lines to represent the channels when the thick, coloured lines were interpreted in black and white (e.g. in mass-market woodblock prints). In other words, the technology transfer involved a slight reinterpretation or misrepresentation of the earlier bronze statue diagrams. This kind of 'misrepresentation' was quite common.

Diagrams of ten channels

There are numerous citations from Six Dynasty writings, such as the *Classic of Childbirth* in the Japanese medical compilation *Ishinpō* 醫心方 (Prescriptions from the Heart of Medicine) by Tanba Yasuyori 丹波康頼 (912–995) in 984. The 'Diagrams of Ten Channels' recorded in the chapter titled *Ninpu myakuzu tsuki kinpo* 妊婦脈圖月禁法 (Charts of the Channels and Method of Monthly Prohibitions During Pregnancy) in volume 22 of *Ishinpō* has the following characteristics:

1 The diagrams of ten channels illustrate the relationship between the stages of pregnancy and the channels.
2 The channel lines run both on the body surface and inside the body and the corresponding organs are shown.
3 The relationship between the 'heart regular channel of the hand' and the heart itself is illustrated. The name of the channel differs from the more familiar one that is recorded in *Channels*, i.e. the significance of '*xinzhu* 心主' in the name of the channel, *shou xinzhu mai* 手心主脈, is that here the heart is regarded as 'the lord'. This concept is commonly seen in the other chapters of the *Inner Classic* and other classical works that date to the Han dynasty.
4 The hand-*shaoyang* channel pertains solely to the upper *jiao* 焦 (lit. 'burner', a tripartite division of the abdomen) rather than the three *jiao* in combination (the 'Triple Burner'). Thus, the hand-*shaoyang* channel as illustrated here is relevant only to the heart and the lung of the upper *jiao*, rather than the five *zang* and six *fu* as they reside in the upper, middle and lower *jiao*.
5 Only one line of the bilateral foot-*taiyang* channel is depicted lateral to the Governor Vessel on the back. This is different from the description in *Channels*, where a branch is derived from, and runs parallel with it.
6 The foot-*shaoyin* channel runs on the back, rather than on the abdomen. This characteristic is the same as in *Channels* (Figure 12.4).

Diagrams of the twelve regular channels

Huanzhong tu 環中圖 (Diagrams of the Circulatory Course of the Channels; hereafter *Diagrams of the Circulatory Course*) was written by Yang Jie 楊介 in 1113 (Song dynasty) (Figure 12.5). It is included here as the most typical of the complete channel diagrams.

The diagrams and texts of *Diagrams of the Circulatory Course* were first quoted in volume 44 of the Japanese compilation *Tonishō* 頓醫抄 (Book of the Simple Physician; by Shozen Kajihara 梶原性全 [1265–1337] in 1302–04 in the Kamakura period of Japan). Later, the contents of volume 44 were compiled into Shozen Kajihara's *Mananpō* 萬安方 (Myriad Relief Prescriptions), which was published in the second year of the Enkyo era of the Kamakura period (1309) and is now preserved in the Cabinet Library of the National Archives of Japan.

The diagrams of twelve channels in *Mananpō* show that *Diagrams of the Circulatory Course* was a standard set of channel diagrams. The circulatory courses of the channels included not only the main courses and branches but also the surface courses and internal courses. In addition, there were vivid images of the internal viscera and bowels (*zangfu* 臟腑) in some diagrams (such as the diagram of the hand-*yangming* channel). The only disadvantage of *Diagrams of the Circulatory Course* was that it was not easy to distinguish the surface courses from the internal courses because they were all produced as full lines, in the style of the *Classic of Childbirth*.

Illustration of channels in Lifesaving Book

There were many versions of Zhu Gong's Northern Song *Lifesaving Book* (1108). In fact, he revised his own book, correcting over one hundred mistakes and republishing the original

Acupuncture illustrations

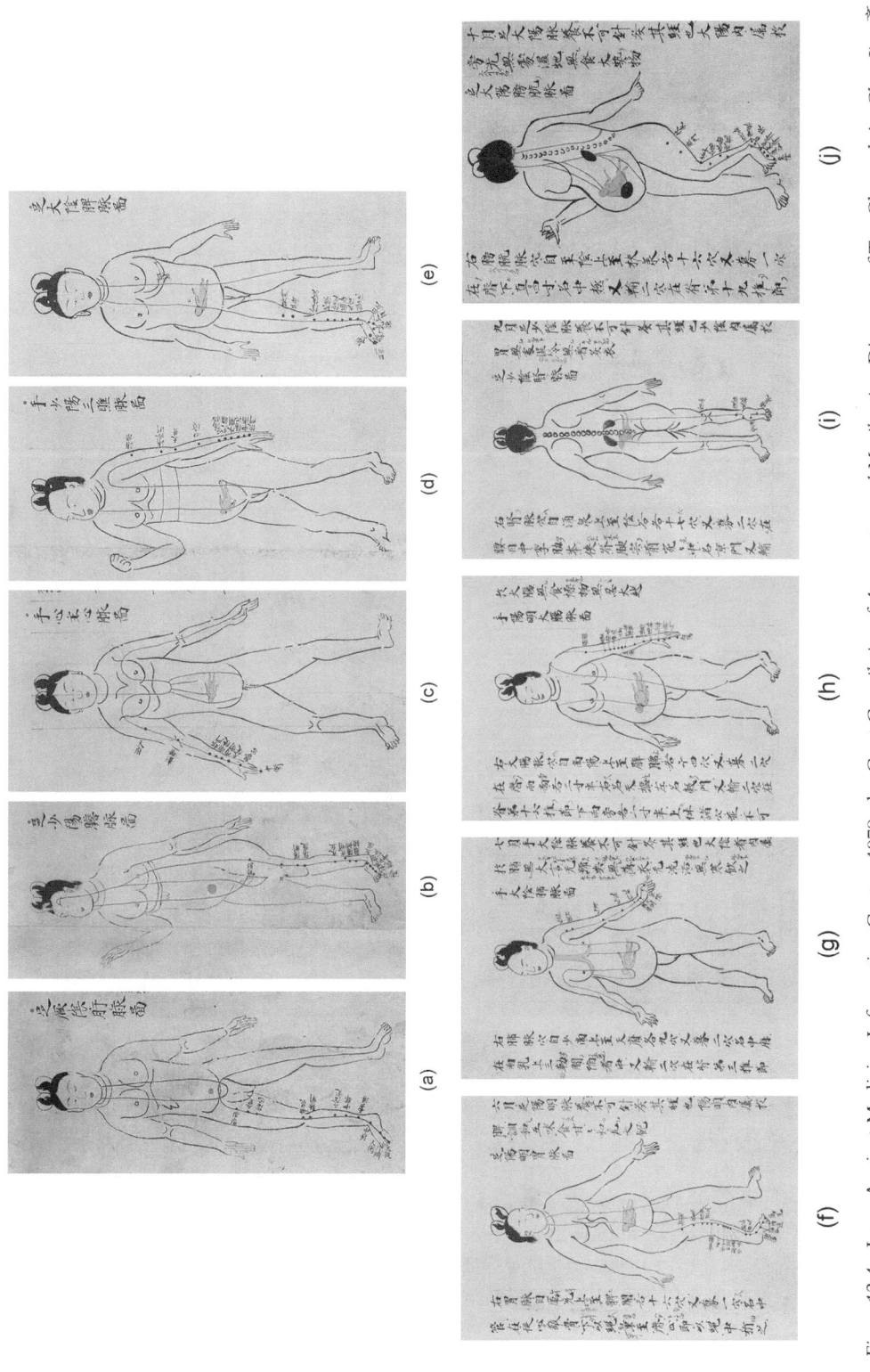

Figure 12.4 Japan Ancient Medicine Information Center, 1978, the *Great Compilation of Acupuncture and Moxibustion*, Diagrams of Ten Channels in *Chan Jing* 產經 (Classic of Childbirth); printed in *Zhenjiu yixue dianji daxi* 針灸醫學典籍大系 (Great Compilation of Acupuncture and Moxibustion Medical Dictionary). All images from Huang Longxiang (2003)

(a) (b) (c) (d)

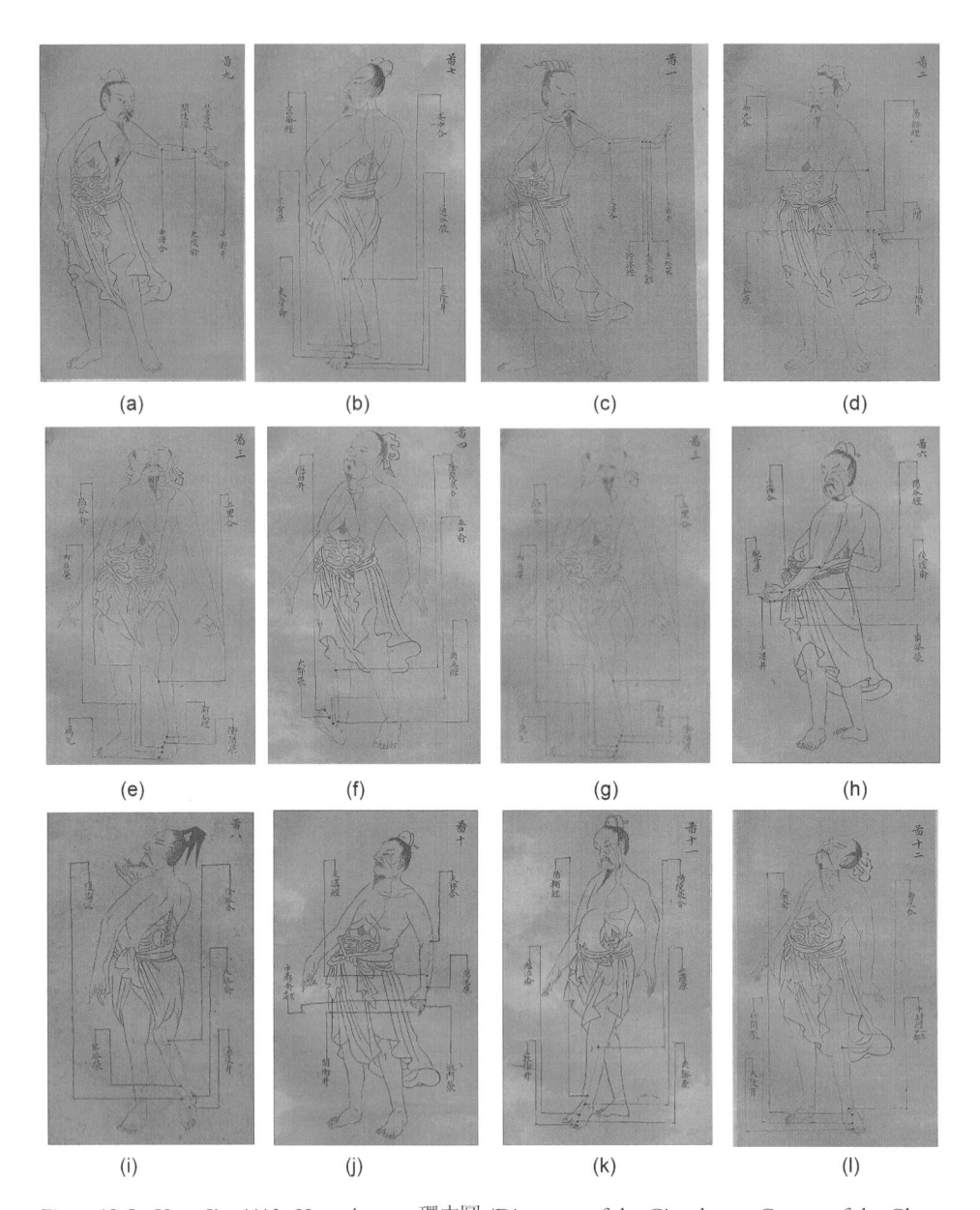

(e) (f) (g) (h)

(i) (j) (k) (l)

Figure 12.5 Yang Jie, 1113, *Huanzhong tu* 環中圖 (Diagrams of the Circulatory Course of the Channels), as cited in *Mananpō* 萬安方 (Myriad Relief Prescriptions), edition of the second year of the Enkyo era of the Kamakura period (1309), Japan, collected in the Cabinet Library of the National Archives of Japan. All images from Huang Longxiang (2003)

text with amendments in 1118. Later, in the Southern Song dynasty (1127–1279), Wang Zuosu 王作蕭 (birth year unknown) provided detailed explanatory notes for the texts, and changed the book title to *Zengshi nanyang huorenshu* 增釋南陽活人書 (Lifesaving Book with Explanatory Notes). Copies of the Song or Yuan dynasty (1271–1368) editions were so rare and hard to get hold of that the medical officials of the Qing dynasty (1636–1912) designed the illustrations of the twelve channels in the *Golden Mirror* on the basis of a version from

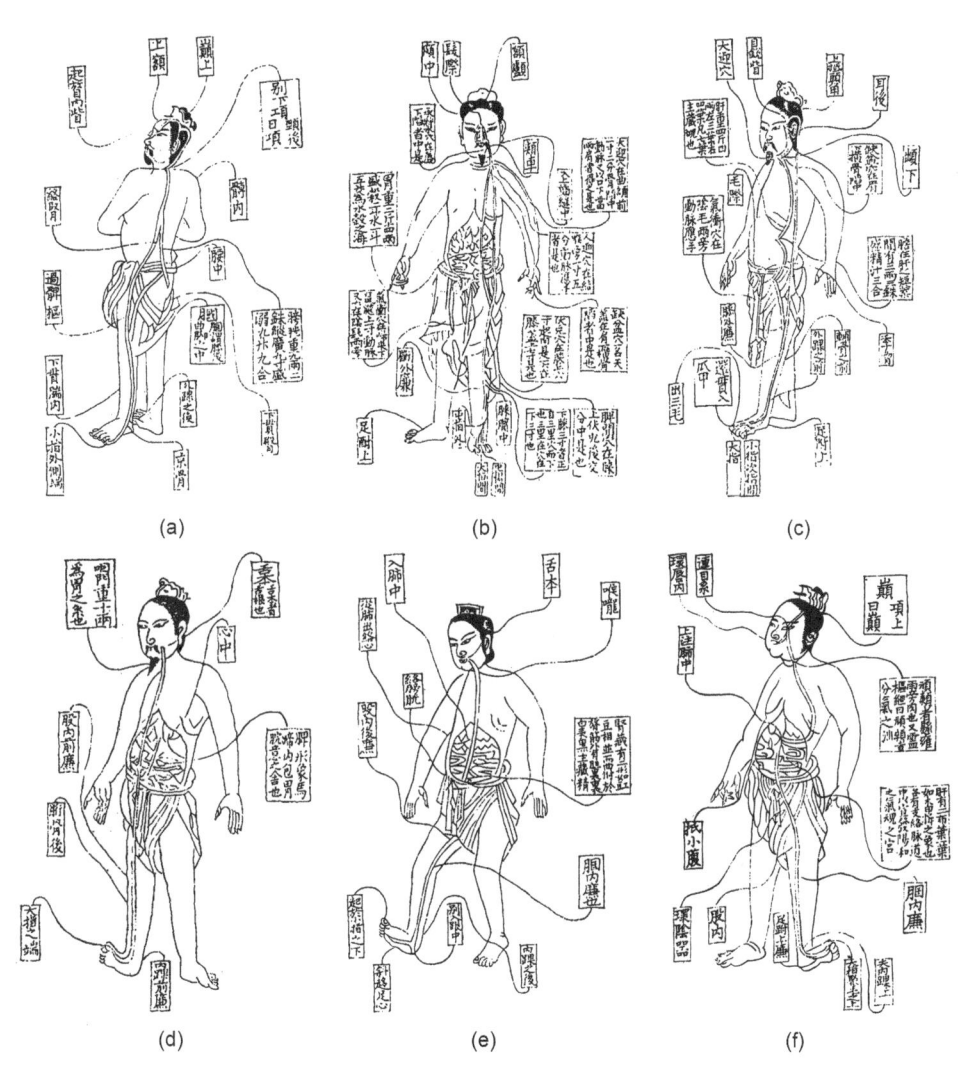

Figure 12.6 Zhu Gong 朱肱, 1118, Illustrations of channels in *Chongjiazheng huorenshu* 重校正活人書 (Revised Lifesaving Book), Song dynasty edition, preserved in Seikado Library, Japan. All images from Huang Longxiang (2003)

the Wanli reign of the Ming dynasty (1572–1620). Today, we have a Song version with a different title: *Chongjiazheng huorenshu* 重校正活人書 (Revised Lifesaving Book, 1118).

It is worth reiterating that tubular double lines were used to illustrate the channels in the *Lifesaving Book* in the Song dynasty, as in most of the *Mingtang* diagrams that have been handed down (Figure 12.6).

Eight Extra Channels (*Qijing bamai* 奇經八脈)

Apart from the twelve channels, there are Eight Extra Channels, namely the Governor Vessel (*dumai* 督脈), Conception Vessel (*renmai* 任脈), Thoroughfare Vessel (*chongmai* 衝脈), Belt Vessel (*daimai* 帶脈), Yin Heel Vessel (*yinqiaomai* 陰蹻脈), Yang Heel Vessel (*yinqiaomai* 陽蹻脈), Yin Link Vessel (*yinweimai* 陰維脈) and Yang Link Vessel (*yangweimai* 陽維脈).

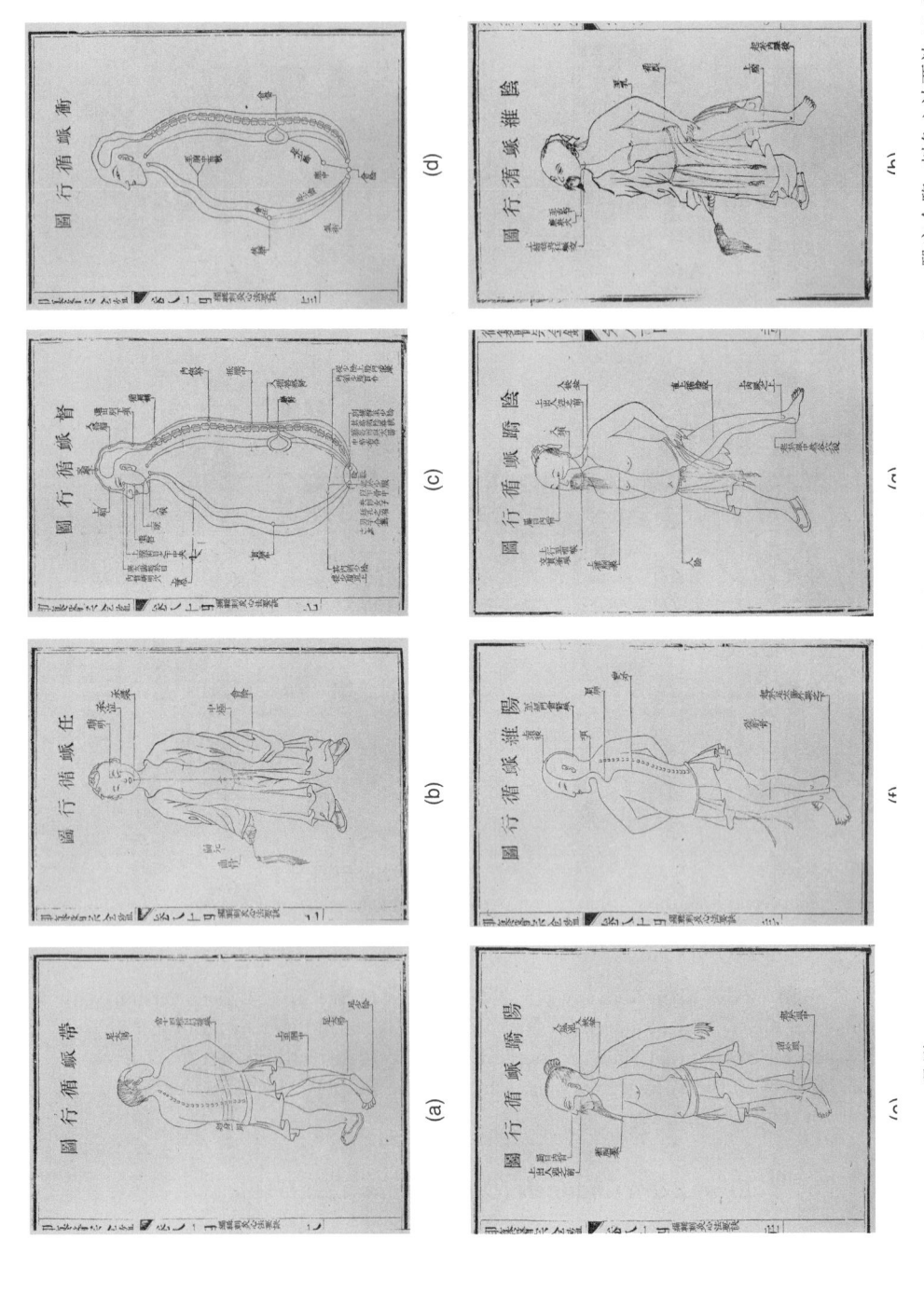

Figure 12.7 Wu Qian 吳謙, Qing, 1742, 'Diagrams of the Eight Extra Channels' in *Yizhong jinjian: Cijiu xinfa yaojue* 醫宗金鑒・刺灸心法要訣 (Golden Mirror of the Medical Tradition: Essential Teachings on Acupuncture and Moxibustion). All images from Huang Longxiang (2003)

They are different from the twelve channels because none of them relates to the *zangfu* (viscera and bowels). Apart from the Governor Vessel and Conception Vessel, these extra channels share their points with other channels. The earliest extant example of a typical diagram of the Eight Extra Channels is found in the *Golden Mirror*, compiled for 'universal use' in the Qing dynasty.

In the majority of the diagrams, the channel lines are simply drawn according to the intersecting points (*jiaohui xue* 交會穴) of the extra channels. Strictly speaking, these diagrams are not historically standardised channel diagrams. The theory of the Eight Extra Channels was apparently established at a relatively late date, since there are no literary descriptions of the course of the Yin Link Channel and Yang Link Channel (two of the Eight Extra Channels) in the earliest acupuncture sources. The first systematic textual descriptions of the terminology, routes and indications of the Eight Extra Channels are recorded in *Nanjing* 難經 (Classics of [81] Difficult Questions), probably compiled in the first or second century CE.

Influenced by the *Lifesaving Book*, Wu Qian compiled the *Golden Mirror* and drew not only the diagrams of the twelve channels, but also the diagrams of the Eight Extra Channels for the first time (Figure 12.7).

Diagrams of collaterals (luomai 絡脈), divergent channels (jingbie 經別) and the muscle region (pibu 皮部)[2] of the regular channels

In ancient literature, there were no diagrams of the routes of the 'collateral' branches of the regular channels, the 'divergent' channels or the 'muscle region of the regular channels'. In the 1961 *Zhenjiuxue jiangyi* 针灸学讲义 (Teaching Materials on the Science of Acupuncture and Moxibustion), an editorial committee of the Shanghai College of Traditional Chinese Medicine compiled textual descriptions of this complementary material and created a set of precise diagrams to elucidate the texts.

Huang (2001) pointed out that the so-called channel, collateral and muscle regions of the regular channels were in fact different terms employed to explain the same phenomenon according to the experience of a variety of different schools, and the concept of a divergent channel is simply an auxiliary hypothesis arising from the classic theory of the channels. Thus, it was found that a very similar conclusion was derived from these apparently conflicting doctrines, when the phenomenon was presented diagrammatically.

The collateral network, in fact, constitutes 50% of the channel and collateral structure. One cannot produce an internally coherent diagram of it. Even now, there are no charts of the collateral network in acupuncture teaching materials. It would not be possible to draw the routes of the collateral channels on to the standard channel diagram since the channels would overlap. By following classical descriptions, one could not define the routes of the collateral channels separately. In addition, after the Tang dynasty, the only acu-moxa points on the collateral network were the fourteen Luo-connecting points (*luoxue* 絡穴), which were classified as just one of the acu-moxa points of each of the fourteen corresponding channels. Therefore, since there were no points to illustrate, there was no practical purpose to be served by illustrating the channels and this is certainly why there was no demand for a diagram of the collateral network. For the same reason, one cannot find a diagram of the divergent channels: there is not a single point located on the divergent channels themselves, and the divergent channels were co-opted into channel theory a long time before *Zhenjiuxue jiangyi* was compiled in the 1960s. So, by this time, it was very clear that there was no need to illustrate the divergent channels independently.

Concerning the channel diagrams of the muscle region, in early and medieval China, the scope of acupuncture was very wide and many kinds of tools, devices and techniques were used (Chapter 13 in this volume). For example, the fire needling technique was very popular in the Tang dynasty and the relevant techniques and manipulations were explained in great detail. Moreover, the massage, pressure point and stretching and exercise techniques now known as *tuina/anmo* 推拿/按摩 were specialities in the medical repertoire of the Tang dynasty (Chapter 6 in this volume).[3] Since these involved whole channel therapy, diagrams of the muscle region were required. Later, perhaps the relevant techniques were lost, or the channel diagrams of the muscle region became redundant due to insufficient demand.

Research questions and reflection

From diagrams to theories

It is worth noting that, in the absence of visual aids such as the bronze statues and their associated two-dimensional diagrams, or the *Mingtang* diagrams, to ground research into the ancient locations of acu-moxa points, great attention should be paid to the classification of acu-moxa points in the ancient acu-moxa point literature. If one compares the acu-moxa point sequences on the 1443 Ming acupuncture bronze statue that was cast in the reign period of Emperor Zhengtong 正統 (r. 1435–49) with those in the *Illustrated Manual*, it is apparent that there are remarkable coincidences. For example, the *Illustrated Manual* was the work first to arrange GB4 (*Hanyan* 頷厭), GB5 (*Xuanlu* 懸顱) and GB6 (*Xuanli* 懸厘) on the lateral side of the head (unlike the pre-Song literature). The same arrangement of these three acu-moxa points can also be found on the bronze acupuncture statue cast in the reign of the Zhengtong Emperor. In documents published before the Song dynasty, the acu-moxa points located anterior to the ear, TE21 (*Ermen* 耳門) are situated above GB2 (*Tinghui* 聽會), which is the precise opposite of the locations in *Illustrated Manual*. They are identically situated on the acupuncture bronze statue dating to the reign of Emperor Zhengtong. In view of the aforementioned facts, when investigating the locations of acu-moxa points in the pre-Song writings for which no *Mingtang* diagrams of acu-moxa points and no bronze acupuncture statues are available, we should pay added attention to the description of the sequences of acu-moxa points.

Studies of the bronze acupuncture statues, bronze statue diagrams or *Mingtang* diagrams have academic value and significance in establishing the history of the locations of acu-moxa points. In fact, acu-moxa point locations have been determined retrospectively in light of the bronze statues and the acu-moxa point diagrams, reflecting a large degree of uncertainty when it comes to textual descriptions. In summary:

- With respect to literary descriptions of acu-moxa point locations, much attention should be paid to the relationship between acu-moxa points.
- As new national standards for the location of acupoints are drawn up, it would be useful to cast two figurines of the acupuncture channels and locations to reflect their three-dimensional relations. Such models both help to locate acu-moxa points and also represent all the divergent models and styles that are embraced within the national standard texts, embodying higher academic and historical values for our own times.

Nowadays, a set of three wall-charts represents the standard channel acu-moxa points, namely, the front, lateral and back views. These are derived from traditions that were embodied separately in sets of four or five wall-charts. The current sets are both more concise and more practical, and cover the contents of the ancient *Mingtang* diagrams of *zangfu* organs in front and back views more comprehensively.

From the ancient to the modern

When we compare charts of the channels to channel acu-moxa point diagrams, why are there fewer charts of the channels historically, and far fewer extant versions of these rare historical sources? The key extant source for the channel diagrams is the specialised book, *Diagrams of the Circulatory Course* which, as we saw above, dates to the Song dynasty. The situation is roughly the same today. In general, people are very unfamiliar with the diagrams of the extra channels and the channel diagrams of the muscle region. This may be because first, the significance and value of channel diagrams has arguably never been completely understood and, second, there was always scope to improve the performative value of the charts' design. Authors and artists failed to make full use of pictures and diagrams in the interpretation and application of theory. Thus, they lost an important opportunity to exploit what is surely the most powerful motivation that might drive forward the development of acupuncture imagery. Initially, modern medicine neglected the muscular fasciae and the connective tissue. So why has this kind of visual aid become so attractive in recent years? The recent rapid increase in demand for the diagnosis and treatment of myofascial pain, for instance, has seen a market for diagrams and books on the ancient Chinese channel theory of the muscle region.

A good visual design is better than thousands of words. Diagrams or pictures are not replaceable by words and text. *Myofascial Pain and Dysfunction: the Trigger Point Manual* (Travell *et al.* in Donnelly (ed) 2019) is regarded as a bible for the diagnosis and treatment of myofascial pain. Such a book could never have won such a wide audience without its excellent graphics and exquisite pictures. Much the same can be said of *Anatomy Trains: Myofascial Meridians for Manual and Movement Therapists* (Myers 2014), another extremely influential book in recent years, which presents a unique 'whole system' view of myofascial/locomotor anatomy in which the body-wide connections among the muscles within the fascial net are described in detail for the first time. Without the stunning pictures it contains, it would have been impossible for this book to stand out against the many other text-based books on acupuncture or have such a big impact. When this book was reviewed so favourably in China, it seems that everyone, including the author's colleagues, initially assumed that it was some sort of modern re-discovery and re-development of the theory of channels and collaterals from China. Comparisons were made between these depictions of the newly discovered myofascial chains and the ancient channel diagrams from China. But actually, this book demands comparison with the much overlooked Chinese textual accounts of the channel diagrams of the muscle region, even if the associated diagrams are no longer extant. Naturally, those who are unaware of the existence of such diagrams hundreds of years ago in China could not know of the historical authenticity of the modern text and illustrations.

With a little knowledge of the history of science, it is clear that illustrations facilitate the longevity of innovative knowledge. Mendeleev's extraordinary insights into chemistry might not have made such an impact without the Periodic Table (1869). Therefore, if those who constructed the theory of channel and collaterals could have designed a graphic or a table, their theories and practice might never have been forgotten.

Advanced visualisation tools and techniques are not the decisive factor in the design of better channel and collateral diagrams. Rather, the demands of clinical practice and teaching on the knowledge and transmission of acupuncture are essential motivations, as is a historically refined understanding of the theory of channels and collaterals. The collaborative efforts of scholars at home and abroad are necessary to make this theory more effective in its wider applications in a concerted attempt to respond to the demands of practice.

Notes

1 Much of *Huangdi neijing* was arguably compiled in the late Warring States period (see 'Introduction' to this volume for controversy over dating of HDNJ).
2 *Pibu* 皮部 (lit. cutaneous region) is defined by the WHO as 'the region of the skin reflecting the functional condition of a certain meridian' (WHO 2007: 34).
3 The term *tuina* 推拿 (Pushing and Grasping) was not used in the Tang period. It was not until the Ming dynasty (1368–1644 CE) that the term *tuina* was introduced to describe the specifically external manipulation of massage, often with reference to the paediatric massage of children (*xiao'er tuina* 小兒推拿).

Bibliography
Pre-modern sources

Chanjing 產經 (Classic of Childbirth) Six Dynasties (220–589), Anon., lost.
Chong jiazheng huorenshu 重校正活人書 (Revised Lifesaving Book) 1118, Zhu Gong 朱肱, Tokyo: Seikado Library.
Cijiu xinfa yaojue 刺灸心法要訣 (Essential Teachings on Acupuncture and Moxibustion) 1742, Wu Qian 吳謙, (2006), Beijing: Renmin weisheng chubanshe.
Cunzhen huanzhong tu 存真環中圖 (Diagrams of the Circulatory Course of the True Channels) 1113, Yang Jie 楊介, Beijing: Library of China Academy of Chinese Medical Sciences.
Huangdi neijing 黃帝內經 (The Inner Canon of the Yellow Emperor), Anon. c. Han dynasty (202 BCE–220 CE). *Suwen* 素問 (Plain Questions), Wang Bing 王冰 (c. 710–805) ed.; *Lingshu* 靈樞 (The Numinous Pivot) and *Mingtang jing* 明堂經 (Illuminated Hall Canon).
Huoren shu 活人書 (Lifesaving Book) 1108, Zhu Gong 朱肱, Tokyo: Seikado Library.
Ishimpō 醫心方 (Prescriptions from the Heart of Medicine) 984, Tanba Yasuyori 丹波康賴 (1993), Beijing: Huaxia.
Leijing tuyi 類經圖翼 (Illustrated Supplement to the Classified Canon) 1624, Zhang Jiebin 張介賓 (1965), Beijing: Renmin weisheng chubanshe.
Mananpō 萬安方 (Myriad Relief Prescriptions) 1309, Shozen Kajihara 梶原性全, Tokyo: Cabinet Library of the National Archives of Japan.
Qianjin yifang 千金翼方 (Supplementary Prescriptions Worth a Thousand Gold Pieces) 682, Sun Simiao 孫思邈 (d. 682), (1955), Beijing: Renmin weisheng chubanshe.
Shisijing fahui 十四經發揮 (An Elucidation of the Fourteen Channels) 1341, Hua Shou 滑壽 (1956), Shanghai: Shanghai weisheng chubanshe.
Tongren shuxue zhenjiu tujing 銅人腧穴針灸圖經 (Illustrated Manual of Acu-moxa Points of the Bronze Figure) 1207, Wang Weiyi 王惟一 (1987), Beijing: Zhongguo shudian.
Tonishō 頓醫抄 (Book of the Simple Physician) 1303, Shozen Kajihara 梶原性全(1986), Tokyo: Cabinet Library of the National Archives of Japan.
Yizong jinjian 醫宗金鑒 (Golden Mirror of the Medical Tradition) 1742, Wu Qian 吳謙 (1997) Shenyang: Liaoning kexue jishu chubanshe.
Zengshi nanyang huorenshu 增釋南陽活人書 (Lifesaving Book with Explanatory Notes) Southern Song Dynasty (1127–1279), Wang Zuosu 王作肅, Tokyo: Seikado Library.
Zhenjiu juying 針灸聚英 (An Exemplary Collection of Acupuncture and Moxibustion and Their Essentials) 1529, Gao Wu 高武 (2007), Beijing: Zhongguo zhongyiyao chubanshe.
Ziwu liuzhu zhenjing 子午流注針經 (Acupuncture Classics of Midnight-Midday and Ebb-Flow Doctrine) Jin Dynasty, 1115–1234, Jin Dynasty, Yan Mingguang 閻明廣 (1986), Shanghai: Shanghai zhongyi xueyuan chubanshe.

Modern sources

Donnelly, J.M. (ed.) (2019) *Travell, Simons & Simons' Myofascial Pain and Dysfunction: the trigger point manual*, Philadelphia: Wolters Kluwer.

Huang Longxiang 黄龙详 (2001) *Zhongguo zhenjiu xueshushi dagang* 中国针灸学术史大纲 (The Historical Development of Acupuncture), Beijing: Huaxia chubanshe.

——— (2003) *Zhongguo zhenjiushi tuijian* 中国针灸史图鉴 (Illustrated History of Chinese Acupuncture and Moxibustion), Qingdao: Qingdao chubanshe.

Japan Ancient Medicine Information Center (1978) *Great Compilation of Acupuncture and Moxibustion Medical Dictionary* (*Zhenjiu yixue dianji daxi* 針灸醫學典籍大系), Japan *Nihon ko-igaku shiryō sentā* 日本古医学資料センター, *Shinkyū igaku tenseki taikei* 鍼灸医学典籍大系, Tokyo: Shuppan Kagaku Sōgō Kenkyūjo.

Myers, T.W. (2014) *Anatomy Trains: Myofascial meridians for manual and movement therapists*, Edinburgh: Churchill Livingstone/Elsevier.

Shanghai zhongyi xueyuan zhenjiuxue jiaoyanzu 上海中医学院针灸学教研组 Shanghai College of Traditional Chinese Medicine (1961) *Zhenjiuxue jiangyi* 针灸学讲义 (Teaching Materials on the Science of Acupuncture and Moxibustion), Shanghai: Shanghai kexue jishu chubanshe.

World Health Organization Regional Office for the Western Pacific (2007) *WHO International Standard Terminologies on Traditional Medicine in the Western Pacific Region*, Manila: WHO Regional Office for the Western Pacific.

Zhonghua renmin gongheguo guojia biaozhun 中华人民共和国国家标准 (GB 12346-1990) (1990) *Jingxue buwei* 经穴部位 (Location of Acupoints), Beijing: Beijing biaozhun chubanshe.

13
ANATOMY AND SURGERY

Li Jianmin 李建民

Translated by Michael Stanley-Baker and Vivienne Lo

Waike 外科, literally 'the curriculum of external medicine', is a medical metonym established over a thousand years ago that mainly describes diseases of the material components of the body, most commonly the skin and flesh. *Waike* is also the modern Chinese term for 'surgery'. This dual meaning arises from the particular history of external diseases which was shaped by the interrelationship between fleshy, anatomical knowledge and surgical procedures. It therefore encompasses those Chinese perspectives on the medical body which are oriented towards its fleshy parts or, in other words, what one might call 'the muscular gaze' in Chinese medicine. The management and elimination of *nong* (膿; hereafter pus), that is, the suppurating pathological transformations of body fluids which cause flesh to fester, was the focus of recorded surgical practice in China. During the Song (960–1279) and Yuan (1271–1368) periods, anatomical knowledge became increasingly sophisticated and ultimately integrated with local cultures of self-cultivation, or *yangsheng* 養生, literally 'nourishing life' (Chapter 49 in this volume; Despeux 2018b). Yet, it was always difficult for external medicine to treat diseases which manifested pus. Operations on what we would now consider infected flesh persisted right into the twentieth century before there were either antibiotics or anaesthesia and were inevitably dangerous surgical procedures anywhere in the world before then. This chapter focuses on the changing boundaries of external medicine and *neike* (內科; hereafter internal medicine) as they have occurred since the Song period.

> That which we call *chuang* 瘡 are illnesses of broken and bruised flesh.
> Chen Shigong 陳實功
> *(1555–1636;* Waike zhengzong*: 14)*

Contrary to popular academic belief, Chinese surgical procedures *did* involve an 'anatomical gaze' (Chapter 38 in this volume) on the muscular and fleshy body and this is finally being recognised in the English language historical literature on China (Despeux 2005, 2018a, 2018b; Wu Yi-Li 2011, 2015a, 2015b, 2017, forthcoming; Hu 2018).[1] The *Shiji* 史記 (Record of the Grand Historian, completed 91 BCE) records early Chinese surgical activities such as *gepi jieji* 割皮解肌 (cutting the *pi* 皮 'skin' and releasing the muscles) and *juemai jiejin* 訣脈結筋 (severing the channels and knotting the tendons) (Li Jianmin 2007: 3). '*Pi*' 皮 refers to the superficial layer of the human body, and the *jirou* 肌肉 (hereafter muscular flesh) with the *jinrou* 筋肉 (hereafter tendinous flesh) were the subsequent anatomical layers as one penetrated more deeply into the body. The *ji* 肌 (muscly) part of the flesh can be seen at the surface of the body, particularly where it rises up and gathers when it is called the *jiongrou* 䐃肉 (bulky flesh).

Depictions of muscular flesh in the classical corpus of Chinese medicine, the *Huangdi neijing* 黃帝內經 (Yellow Emperor's Inner Canon, c. first century CE), hereafter *Inner Canon*, also employ the terms *fenrou* 分肉 (hereafter differentiated flesh), muscular flesh and other technical terms, all of which refer to corporeal forms (Zhao 2014: 255–6, 399; Chapter 7 in this volume). 'Differentiated flesh' refers to the different types of flesh and sinews that display a patterned structure and lie in proximity to the skeleton, such that it only becomes visible after dissection. One can map the tendinous flesh to the divisions of the twelve channels (Chapter 1 in this volume), and therefore an early Chinese anatomical system which embraces twelve tendons and related flesh. The tendinous flesh is the type of flesh that can exert physical force and is intimately associated with the physical structures and relationships that create movement (Li Ding 1998: 14).

The understanding of muscular flesh as it is explained in *the Inner Canon* is that it is responsible for bodily movement. The *Taiyin yangming lun pian* 太陰陽明論篇 (On the theory of Taiyin and Yangming) treatise of the *Suwen* 素問 describes the relationship between movement and human bodily fluids as follows:

> All the four limbs are supplied with *qi* by the stomach, yet it doesn't [thereby] get to the channels; the necessary factor in the supply is the spleen. Now, if the spleen is sick and is unable to move the fluids for the stomach and the four limbs are not supplied with the *qi* of water and grain, the *qi* weakens by the day; the channel paths do not connect up; the sinews and the bones, the muscles and the flesh, none of them has *qi* to survive. So, from this they no longer function.[2]

The activity of the limbs is due to the function of moistening and nourishment produced by the *qi* of the *piwei* 脾胃 (stomach and spleen). Thus, if the *piwei* is unable to perform the function of transporting bodily fluid, then the muscular flesh and related tendons and skeletal structures of the limbs lose their normal physical structures and relationships which create movement.

Fluids and muscular flesh have interrelated effects. Should there be some external injury or other cause which gives rise to weeping wounds, there were prescriptions aimed at regenerating muscles and flesh (*Liu Juanzi guiyifan*: 14, 30, 40). The story of the famous circa third-century doctor, Hua Tuo 華佗, cutting out a festering spleen, also conceives pathology as being of the 'fleshy' body (Fan 2004; Shang 2005: 129). Huang Longxiang 黃龍祥 has styled the pre-modern Chinese study of anatomy as 'skin-deep' or 'superficial', and as characteristically prioritising various bodily structures such as 'the beginning and ends of the musculature' and understandings of how humans were 'endowed with muscular function' (Huang and Huang 2007: 34; Chapter 12 in this volume).

For example, through observations of the superficial parts of the muscular flesh, diagnoses could anticipate pathogenic transformations in the internal organs (Shang 2005: 323). This is Huang's insight into the Eastern Han dynasty *Huangdi mingtang jing* 黃帝明堂經 (Yellow Emperor's Classic of the Bright Hall). Anatomical research and knowledge gained in this way was usefully employed in pre-modern surgical techniques (Huang and Huang 2007: 323). A decision as to whether or not to proceed with surgery involved an assessment of the extent to which the body's superficial flesh was festering. *The Yellow Emperor's Inner Canon* states that 'when flesh decays it makes pus', and the festering of the muscular flesh was articulated within the terms of an 'aetiology concerned with cold and heat causes of disease', with cold dominating early causal explanations and heat predominant after the Song and Yuan (Wang 2014).

Chinese surgical practices were thus grounded in an anatomical gaze. The abovementioned theory that 'the spleen governs the flesh' was given its fullest expression in Li Gao's 李杲 (1180–1251) 'Discourse on the Spleen and Stomach'. Li's doctrine gradually became mainstream during the Yuan, Ming and Qing dynasties (1271–1911). He pointed out that fatigue and irregularity in eating and drinking led to wasting away of the flesh. In particular, his emphasis on internal injury (*neishang* 內傷) caused later doctors to pay attention to internally caused diseases of the seven emotions (*qiqing* 七情) in their relation to external medicine as well (Liu 1993: 27–84). Therefore, recuperation of the spleen lay at the heart of surgery and external medicine in China.

In cases where festering superficial flesh could not be healed, Li Chan 李梴 opined in his *Yixue rumen* 醫學入門 (Introduction to Medicine, 1575) that, 'when wounds do not close, this is because flesh does not grow' (*Yixue rumen*: 467), and went on to state that surgical procedures and herbal medicines could stimulate regrowth of the festering flesh. Pathological changes of the spleen and stomach were therefore considered the underlying cause of gradual decay of the flesh. As stated by Yu Chang 喻昌 (style name Jiayan 嘉言, 1585–1664), 'when nutrient qi (*rongqi* 榮氣) decays and becomes turbid, then the flesh slowly festers. The flesh is governed by the stomach' (*Yuyi cao*: 67).

When the barber surgeon Ambroise Paré (1510–90) was designing surgical instruments for the Kings of France, the Chinese surgeon, Chen Shigong 陳實功 (1555–1636), was articulating his perspective on the fleshy body. He went so far as to argue that, 'external medicine is especially relevant and of critical importance' to the spleen and stomach, which he deemed inseparably related to the flesh (*Waike zhengzong*: 13–14). Pus fluids (*nongye* 膿液) caused by suppurating flesh were always difficult to heal. Chen Shigong said:

> And when at this time there is pus, yet it cannot be externally expressed, use a needle and hook to pull out the stiffened flesh to the surface of the body. Using a knife or scissors, cut about an inch or more at the main peak, enabling the pus to flow out. Try not to let the head of the wound become blocked up.
>
> (*Waike zhengzong*: 11)

This is a crucial piece of historical information. Chinese doctors used various instruments, such as needles, hooks and knives and scissors, to get rid of extraneous material, that is, the accumulated pus which is trapped inside. Moreover, as the position of the suppurating swelling gets deeper, there develops a 'pus conduit' (*nong guan* 膿管), and the process of drawing it becomes more difficult, increasing the risk of infection.

Pus is a sticky type of pathological fluid, which tends to be classified in the category of 'fluids' (*jinye* 津液) in Chinese medical pathology (Yu 2012: 113, 250). It was expressed in the classics of early China as 'pus and blood' (*nongxue* 膿血) or 'swelling blood' (*zhongxue* 腫血). When flesh rots, it was said to generate matter, including corrupted fluids of the internal organs. Interconnected with the concept of *qi* in Chinese medicine, fluids are closely tied up with surgical procedures. The most important treatments in Chinese medicine, such as sweating, upward and downward purgation (*han* 汗, *tu* 吐, *xia* 下), all involved the regulation and purging of pathological fluids.

Emphasising the centrality of the body fluids to Chinese medical theory, Chen Xiuyuan 陳修園 (1753–1823) identified that the highest priority was their preservation in cases of Cold Damage or febrile diseases: 'conserving the fluids, this is the true exposition' (Fang 2007: 379; Sun 2011: 17–34; Chapters 8, 16 and 17 in this volume). When external injuries bleed, this also causes overall disequilibrium of the bodily fluids, and leads to pathological changes.

Diseases that required surgery were always different from internal medicine diseases. Initially, external medicine did not rely on pulse diagnosis. The condition and colour of the muscular flesh, such as changes in swelling and redness, were subject to visual examination. Hardness and softness of lesions and wounds could be palpated. Among the many forms of external diagnosis popular in early Chinese medicine, there was once a form of diagnosis that involved touching the skin all over the entire body.

In one overlooked essay on medical history, Liao Ping 廖平 (1852–1932) collected early methods for diagnosing the flesh (*ji* 肌) and skin (*jifu* 肌膚), and found that the technical terms for this kind of external diagnostics, such as 'slippery' (*hua* 滑), 'rough, choppy' (*se* 澀), 'tight' (*jin* 緊), 'hard' (*jian* 堅) and others, had later become adopted as vocabulary for pulse diagnosis. He pointed out that:

> Since the *Canon of Difficulties* (*Nanjing* 難經; 1st or 2nd centuries CE) rather arbitrarily established a new technique that uniquely diagnosed through palpation of the 'two inches' (*liangcun* 兩寸), [that is, the radial pulse at the *cunkou* 寸口 position of the wrist], later writings on the pulse blatantly adopted vocabulary from skin diagnosis and applied it to the pulse.
>
> (Liao 2010: 125)

Additionally, the second-century *Jin'gui yaolue* 金匱要略 (Essential Prescriptions from the Metal Coffer) retains a method for skin diagnosis which assesses whether or not the pus has developed (Gao 1964: 253). Whether or not there was pus located at the surface of the skin, or deeper below, was also an important factor in deciding whether or not to use surgical techniques. This move from external to internal diagnosis (by a focus on pulsing) occurred very late in the history of external medicine in China. For example, in his *Waike jingi* 外科精義 (Essentials of external medicine), the fourteenth-century imperial physician of external medicine, Qi Dezhi 齊德之, questioned whether surgeons understood the pulse. Qi was of the opinion that:

> all those who practise medicine, should first refine their understanding of complexion and pulse [diagnostics]. This is even more so for traumatology; one that is not expert in it, even if intelligent, wise and broadly learned, he will not be fit to be entrusted with a commission.
>
> (Waike jingyi: 1)

In particular, he argued, it was necessary with illnesses such as 'internal sores' (*neichuang* 內瘡) and 'internal ulceration [possibly of a gangrenous type]' (*neiju* 內疽) where the internal organs could fester and rot. Types of swelling diseases which 'are not seen by the eye, the hand cannot come near, these are the most difficult, yet can be discerned by examining the pulse' (Ibid.: 8). Furthermore, in the Chinese surgical gaze, attention is also given to 'external ulcers' (*waiyang* 外瘍), which turn into 'internal ulcers' (*neiyang* 內瘍).

Li Jianmin 李建民

In addition to incorporating pulse diagnostics, the treatment methods of Chinese external medicine, in the hands of the scholarly practitioners who left records and therefore evidence of their work, started to use pharmacological decoctions to substitute for surgical and other external treatments. Angela Leung made the useful observation:

> when scholarly physicians consolidated traditional medicine, some ancient aspects were increasingly marginalised, particularly those that were considered technically more like 'handicraft' or were deemed superstitious; in particular those which were considered too technical and included acupuncture, eye surgery, and other external techniques that involved esoteric rituals.
>
> *(Leung 2011: 12)*

From the Song and Yuan periods onwards, there was an increasing trend towards integrating the treatment of illnesses that had formerly been in the domain of external medicine with pulse diagnostics and treatment according to the principles of drug prescription.

A scholar working at the end of the Yuan period, Wu Hai 吳海 (dates unknown), wrote that an external medicine physician by the name of Guo 郭 (dates unknown) had pointed out in his preface that 'what the world of physicians specialising in ulcers called external medicine was different to what was known as internal medicine' and that the two defining features of the latter were 'taking the pulse' and 'drinking decoctions' (*Wu Chaozong xiansheng wenguozhai ji*: 8). He said, 'even though Mr Guo called ulcers external, in fact they broke out internally, and should be first pursued at their root, and only then treatment should be applied to the ulcer' (Ibid.). Even though some external medicine illnesses could be observed at the surface of the body, the original site where the pathogenic processes, including [excesses of] joy and anger, first broke out was in the internal organs and therefore could be included in internal medicine (Chapter 2 in this volume). In the relationship between external medicine and internal medicine, the scope for treating illness of the latter sort was comparatively large, whereas that of surgery was rather limited. Xu Dachun 徐大椿 (1693–1771) believed that, 'the methods for ulcer treatment are all external treatment, and these manual techniques had to be transmitted within a lineage tradition' (*Yixue yuanliu lun*: 63). Yet, one had to be on the alert to all the dangers of surgical methods that employed the knife. Xu also stated, 'if you cut flesh that has not yet completely decayed, when blood comes out copiously then there will be instant death' (*Xuping waike zhengzong*: 10). Since cases of death by surgery were obvious and easily seen and the rotting of muscular flesh was frequently impossible to control effectively, Xu suggested using drug therapy in external medicine. For the treatment of muscular flesh, he stated that 'external medicine involves no more [internal medicine treatments] than purging toxins, cooling fire, and the various methods for generating muscles and flesh, that's it' (Ibid.: 28). Toxins and fire here can be interpreted as the insurmountable problems related to the phenomenon of 'infection'. The most dangerous syndrome in Chinese external medicine was the 'toxic mire' (*duxian* 毒陷) where the 'mire' refers to a syndrome where the entire body is subject to corruption by the gradual, unceasing spread of various disease-causing entities (Jin 1958). Xu's choice of wording identified the limitations of what external medicine therapies could do in clinical practice (Ibid.). Chen Xiuyuan 陳修園 (1753–1823) criticised the various techniques of external medicine, saying:

> Scholarly [doctors] are ill-informed and have little to say about it; and, so these techniques become less and less effective. In my youth when I encountered dangerous and contrary syndromes, external medicine practitioners' hands were tied, and they had no

therapeutic strategies. They were forced to choose life-threatening approaches to correct them, and [only] seven or eight in ten might be cured. None had any other techniques; they probably deduced them from the *Shanghan lun* (Discourse on Cold Disorders).

(Huang 1995: 149)

This represents a great change in external medicine theory. Chen Xiuyuan hoped to go back to the classic ancient decoction recipes. In fact, the therapies in external medicine did begin to include decoctions, as shown by Qing dynasty documents. In drug therapy, texts such as the *Yizong jinjian* 醫宗金鑑 (Golden Mirror of Medical Learning; published 1742), which was the instructional texts of the Imperial Academy of Medicine, contained a great many powerful and toxic drugs, quite different from internal medicine decoctions (Xie 2004). A fellow countryman and friend of the famous Qing dynasty doctor Wang Shixiong 王士雄 (style name Mengying 王孟英; 1808–68) from Qiantang 錢塘 (now Hangzhou) in Zhejiang, Guan Rongtang 管榮棠, made a comparison of changes between ancient and contemporary external medicine methods. Guan pointed out that the use of decoctions was merely a delaying tactic, performed as a rote response. He said:

Examining the ancient therapies, they did not distinguish internal (*nei*) and external (*wai*); they used all the methods of knives, needles, *bian* 砭 stones, piercing, cautery, hot pressing, and washes, and did not specialize in one branch of decoctions. Now transmission of all these methods has been lost, and sole reliance is placed on decoctions.

(*Wang Mengying yixue quanshu*: 431–2)

Yet, the taking of decoctions in external medicine therapy usually could not control festering of the flesh, nor could it heal pus and bloody wounds, such that illness conditions successively worsened. Guan Rongtang questioned whether decoctions could be relied on:

Giving up knives and needles, and not using methods for removing pus and fester, engaging only in the domain of prescriptions and decoctions, waiting for abscesses (*ju* 疽) to resolve on its own – this is indolence and irresponsibility.

(Ibid.: 432)

In relation to surgical procedures, decoction therapy was a passive approach. Pan Mingde 潘明德 (1867–1928) of Menghe 孟河, Jiangsu, thus created the term 'Zhongjing-[style] external medicine', referring to this gradual transformation of external medicine towards internal medicine within a scholarly context (Pan 2014: 36). Moreover, as time went on, surgical procedures and activities involving cutting the flesh became ever more distanced from scholarly and elite medicine in China and more and more exclusive to the domain of those with less prestigious manual skills.

In roughly the eleventh century, Chinese doctors performed several large-scale dissections, from which remain records such as *Ou Xifan's Diagrams of the Five Organs* (*Ou Xifan wuzang tu* 歐希範五藏圖) based on sketches of the organs of the executed rebel fighter Ou Xifan 歐希範, who was executed in 1041 CE, and the *Diagrams for Preserving Perfection* (存真圖 *Cunzhen tu*) (Zhang 2014: 121–3). These books of anatomical diagrams showed that the left kidney was slightly lower than the right kidney or described arteries and the oesophagus penetrating the diaphragm (Despeux 2018a, 2018b). However, knowledge of the inner organs became linked with the imagination of Daoist *yangsheng* 養生 (nurturing life regimes). Cultivating the *zigong* 子宮 (translated from modern Chinese as 'uterus') described in the roughly sixteenth- or seventeenth-century *Xunjing kaoxue bian* 循經考穴編 (Investigations into the Points along the Channels) was the core and target of such self-cultivation (Yan 1961: 7): Similarly, the seventeenth-century *Zangfu zhizhang tu shu* 藏府指掌圖書 (*Illustrated Guide to the Zangfu Organs*) by Shi Pei 施沛 (1585–1661) collected multiple diagrams of the body which bore no relation to anatomical dissection, but were closely related to an expanding health regimen culture (Li Jianmin 2010). Soon thereafter, a number of Western anatomical works arrived in China, which Wang Xuequan 王學權 (1728–1810) of Hangzhou in Zhejiang received and to which he added his own critique, evincing a staunch faith in the knowledge of the formless physiology in Chinese medicine: 'Dead material that has form can be seen, but the functions of the formless are invisible' (*Chongqingtang suibi*: 116). Research in Western medical anatomy never came any closer to Chinese medical surgery.

Moreover, the most important puzzle for Chinese surgery remained the interpretation of the internal pathological changes indicated by decaying flesh at the surface of the body. As a result, the government commissioned Wu Qian 吳謙 and others to compile the 1742 *Yizong jinjian* 醫宗金鑒 (The Golden Mirror of Medical Learning), which maintained the view that no matter how serious external illnesses might be, as soon as the flesh decayed one must consider surgery, 'decaying flesh was bad flesh'. It also stated, 'if one encounters a person with full *qi*, use a knife to cut them to be effective' (*Yizong jinjian*: 62. 630). Between the eighteenth and the nineteenth centuries, the external medicine physician from Qingpu 青浦 Jiangsu, Zhu Feiyuan 朱費元, believed that the onset of pathological changes in the flesh was dependent on whether *yuan qi* 元氣 (primordial *qi*) was functioning normally or not:

> When the circulation of blood and *qi* around an individual's entire body suffers dissipation day and night. The body, already depleted, suffers loss during circulation, and because of stagnation and hardening of damp-phlegm, stasis and build-up of blood accumulation, and festering corruption of the muscly flesh, pus and fatty-oil is produced.
>
> *(Zhu 2004: 133)*

With an increase in the flow of *nongzhi* (膿脂 pus and oil-fat) generated by the body, the festering of the muscular flesh would spread throughout the entire body. Even so, Zhu believed that surgery was not necessarily appropriate, and that it could result in too many complications, leaving:

> a chance in ten thousand; a [surgical intervention] may not be careful, may destroy the inner membranes, may harm the tendino-muscular channels. If it is serious then it will bring fatal harm to the body, if minor it will harm the limbs.
>
> *(Ibid.: 135)*

If surgical procedures directly harmed the inner organs and the tendinous flesh, then they would inflict permanent damage on the body.

Without exception, the 116 external medicine case histories that Zhu Feiyuan left for posterity all employed methods from internal medicine. He expressed the frank view that there were no benefits to surgical and other treatments (Ibid.: 80), 'do not be in the habit of using knife or needle so as not to fall short of the royal way' (Ibid.: 136). Crucial here is the phrase, 'the kingly way (*wangdao* 王道)'. The terminology comes from the discourse common among scholarly physicians about an idealised and gentle model of therapeutics and is opposite to the 'the way of the tyrant (*badao* 霸道)' (He 1998: 85–6). Surgery is violent, likened to punishment and the rule of terror.

Historically, surgery in China was not 'superficial surgery'. Take, for example, sores on the occiput (*duikou* 對口), or the back (*fabei* 發背), two serious types of external medicine symptoms that develop pus, and which are frequently treated with surgery (Ling 1957; Zhang 1960).

Attached to the above-mentioned set of case histories, the *Linzheng yide fang* 臨証一得方 (*Comprehensive Recipes for Clinical Syndromes*) by Zhu Feiyuan, there is a commentary written by Zhu's sons and grandsons. It states 'in recent times crude handworkers treating the two syndromes, occipital and back sores, frequently use the knife to cut away *e'rou* 惡肉 (flesh that has gone bad), and they boast of their handicraft' (Zhu 2004: 138). The so-called *e'rou* refers to muscular flesh that has changed to become suppurating and rotten. As scholars have explained: 'at first, superficially there emerges a hot flushing erythema, raised swelling, burning heat, broiling pain, and gradually it begins to suppurate' and so on as the external lesions go deeper and enter the tendons and bone (Liao 1962: 6–8).

Right through to late Qing times, surgical treatment was used to treat a broad range of serious complaints. According to his personal observations of the harm caused by surgery, Wang Yanchang 王燕昌 (1831-1895), who came from a family lineage of physicians from Gushi 固始 in Henan, spoke of:

> those suffering with their eyes, have to endure cutting by the knife. Those with pain in their arms and legs have to endure a hundred needles. Illnesses such as scrofula, pain in the throat, swelling of sores, choking on food, flatulence and abdominal distension, empty swellings, heart pain, infant wind-induced fright, jaundice and wasting away, they were [being treated with] chaotic needling and disorderly cutting, which resulted in death.
>
> *(Wangshi yicun jiaozhu: 145)*

From this list, we know that surgery was employed for a rather broad ranging set of conditions.

Damage from surgical procedures, in Wang Yanchang's view, resulted in the long-term trauma of sick people. The wife of an epigrapher, Wang Yirong 王懿榮 (1845–1900), suffered from a breast tumour, an affliction that went on for seven years; when she died, she was only thirty-seven years old. When Wang's wife had the breast tumour surgery, they 'were misled by a quack, and blood oozed out from the wound' (Lu 1999: 93). Since blood from surgery is hard to stem, there was no way to close the wounds. external medicine illnesses could be treated with internal medicine in China, but such cases would likely fail.

Unsuccessful surgery, and therapies that used decoctions that were sold with exaggerated claims of efficacy, both harmed the reputation of the medical tradition within China and affected its popularity in late imperial China. At the end of the Qing, the scholar Fang Renyuan 方仁淵 (1844–1926) stated, 'those who wish to become famous specialists in treating ulcers must read large quantities of internal medicine prescriptions' (*Wangxugao yian*: 295). The gradually increasing tendency for China's field of external medicine to turn to internal medicine methods was set against background sociocultural causes. Ma Peizhi 馬培之 (1820–1905), the renowned physician of external medicine from Menghe in Jiangsu, observed:

> Among the officially appointed [doctors] whom I have encountered, all give weight to internal medicine and slight external medicine, saying that specialists in treating sores do not take [account of] the pulse patterns. In cases of external sores they even extend to

using the ingested drugs of 'prescription and pulse specialists'; this trend is most extreme in Jiang[su] and Zhe[jiang] provinces.

(Wu Zhongtai 2010: 151)

In external medicine, they took the pulse and [prescribed] ingested drugs. The 'prescription and pulse specialists' referred to here are doctors of internal medicine. This culture of opposition to surgery in Southern China was very common during the Ming and Qing periods. Even until now, this current in Chinese medicine of 'giving weight to internal medicine and slighting external medicine' is still prevalent. Is it possible that in the future history of Chinese medicine, research into China's medical traditions could take a long awaited and highly necessary external medicine turn?

Notes

1 Translator's note: Translating the anatomical terminology is complex and can only ever approximate to the range of meanings of the original text which, in itself, would have been interpreted differently in different periods. Our decisions, hereafter, are provisional.
2 *Huangdi neijing Suwen jiaoshi* (2009: 320). See also Unschuld *et al.* (2011: 483).

Bibliography

Pre-modern sources

Chongqingtang suibi 重慶堂隨筆 (Random Jottings of the Chongqing Hall) 1808, Wang Xuequan 王學權 (a.k.a. Wang Bingheng 王秉衡) (2012), Beijing: Renmin junyi chubanshe.

Huangdi neijing Suwen jiaoshi 黃帝內經素問校釋 (Inner Canon of the Yellow Emperor – Basic Questions, Collated and Annotated), Shandong zhongyi xueyuan and Hebei yixueyuan (eds) (2009), Beijing: Renmin weisheng chubanshe.

Liu Juanzi guiyifang 劉涓子鬼遺方 (Liu Juanzi's Remedies Bequeathed by Ghosts) Jin Dynasty, Liu Juanzi 劉涓子, Gong Qingxuan 龔慶宣 (ed.) in 499 (2004), Tianjin: Tianjin kexue jishu chubanshe.

Waike jingyi 外科精義 (Essentials of External Medicine) Yuan dynasty, Qi Dezhi 齊德之, in He Qinghu *et al.* (eds) (1999), *Zhonghua yishu jicheng* 中華醫書集成 (Compilation of Chinese Medicine) Vol. 13, Beijing: Zhongyi guji chubanshe.

Waike zhengzong 外科正宗 (Orthodox Manual of External Medicine) 1617, Chen Shigong 陳實功 (2007), Beijing: Renmin weisheng chubanshe.

Wang Mengying yixue quanshu 王孟英醫學全書 (Complete Medical Works of Wang Shixiong) Qing dynasty, 王孟英 Wang Mengying (1808–1868), Cheng Dengxiu 盛增秀 et al. (eds) (1999), Beijing: Zhongguo yiyao chubanshe.

Wangshi yicun jiaozhu 王氏醫存校注 (Wang's Medical Work: Collated and *Annotated*) Qing Dynasty, Wang Yanchang 王燕昌, Cheng Chuanhao 程傳浩 and Wu Xinke 吳新科 (2014), Zhengzhou: Henan kexue jishu chubanshe.

Wangxugao yian 王旭高醫案 (Medical Cases by Wang Xugao), Qing dynasty, Wang Xugao 王旭高 (1798–1862) (2010), Shanghai: Shanghai kexue jishu chubanshe.

Wu Chaozong xiansheng wenguo zhaiji 吳朝宗先生聞過齋集 (The Wenguozhai Anthology of Mr. Wu Chaozong) Yuan dynasty, Wu Hai 吳海 (1963), Shanghai: Shangwu yinshuguan.

Xuping waike zhengzong 徐評外科正宗 (Xu's Commentary on 'Orthodox Manual of External Medicine') Qing dynasty, Xu Dachun 徐大椿 (a.k.a. Xu Lingtai 徐靈胎) (2014), Beijing: Zhongguo yiyao chubanshe.

Yixue rumen 醫學入門 (Introduction to Medicine) 1575, Li Chan 李梴 (1999), Beijing: Zhongguo zhongyiyao chubanshe.

Yixue yuanliu lun 醫學源流論 (Discussions on the Origin and Development of Medicine) Qing dynasty, Xu Lingtai 徐靈胎 (1693–1771) (2011), Beijing: Zhongguo yiyao keji chubanshe.

Yizong jinjian 醫宗金鑒 (Golden Mirror of Medical Learning) 1742, Wu Qian 吳謙 *et al.* (2011), Taiyuan: Shanxi kexue jishu chubanshe.

Yuyi cao 寓意草 (Notes that Indirectly Express My Intentions) 1643, Yu Jiayan 喻嘉言 (1585–1670) (2013), Shangshui: Shanghai pujiang jiaoyu chubanshe.

Zangfu zhizhang tushu 藏府指掌圖書 (*Illustrated Guide to the Zangfu Organs*) Ming dynasty, Shi Pei 施沛 (1585–1661), in Li Ding 李鼎 (ed.) (2007) *Zangfu jingxue zhizhang tu shisi jing hecan pingzhu* 藏府經穴指掌圖十四經合參評注 ('Illustrated Guide to the *Zangfu* Organs' and 'Synopses on the Fourteen Channels' with commentary and annotations), Shanghai: Shanghai kexue jishu chubanshe.

Modern sources

Despeux, C. (2005) 'Visual representations of the body in Chinese medical and Daoist texts from the Song to the Qing period (tenth to the nineteenth century)', trans. P. Barrett, *Asian Medicine: Tradition and Modernity*, 1.1: 10–52.

——— (2018a) *Taoism and Self Knowledge: The chart for the cultivation of perfection (Xiuzhen tu)*, Leiden: Brill.

——— (2018b) 'Picturing the body in Chinese medical and Daoist texts from the Song to the Qing period (10th to 19th Centuries)', in V. Lo and P. Barrett (eds) *Imagining Chinese Medicine*, Leiden: Brill, pp. 51–68.

Fan Ka-wai 范家偉 (2004) 'On Hua Tuo's position in the history of Chinese medicine', *The American Journal of Chinese Medicine*, 32.2: 313–20.

Fang Yaozhong 方藥中 (2007) *Yuxue sansijing qianshuo* 醫學三字經淺說 (A General Discussion on 'Medical Three Character Classic'), Beijing: Renmin weisheng chubanshe.

Gao Xueshan 高學山 (1964) *Gaozhu jingui yaolue* 高註金匱要略 (Gao's Annotations on 'Essentials of the Golden Casket'), Shanghai: Shanghai kexue jishu chubanshe.

He Shaoqi 何紹奇 (1998) '*Wangdao yu badao* 王道與霸道 (The Royal Way and the Hegemonic Way)', in He Shaoqi (ed.) *Dushu xiyi yu linzheng deshi* 讀書析疑與臨證得失 (Analysis of Problems from the Texts and Clinical Gains and Losses), Beijing: Renmin weisheng chubanshe, pp. 85–6.

Hu Xiaofeng 胡曉峰 (2018) 'A brief introduction to illustration in the literature of surgery and traumatology in Chinese medicine', in Lo and Barrett (eds) *Imagining Chinese Medicine*, Leiden: Brill, pp. 183–196.

Huang Longxiang 黃龍祥 and Huang Youmin 黃幼民 (2007) *Shiyan zhenjiu biaomian jiepoxue: zhenjiuxue yu biaomian jiepoxue yingxiangxue de jiehe* 實驗針灸表面解剖學: 針灸與表面解剖學影像學的結合 (Evidence-Based Surface Anatomy for Acupuncture: Acupuncture Integrated with Surface Anatomy and Imaging), Beijing: Renmin weisheng chubanshe.

Huang Jiexi 黃杰熙 (1995) *Nüke yaozhi qianzheng* 女科要旨箋正 (Commentary and Corrections to 'The Essentials of Gynaecology'), Taiyuan: Shanxi kexue jishu chubanshe.

Jin Bogong 金伯恭 (1958) *Waike duxian zhengzhi zhi taolun* 外科毒陷証治之討論 (Discussion on Syndrome and Treatment of Poison in External Medicine), *Zhongguo zazhi*, 8: 521–3.

Leung, Angela Ki-che 梁其姿 (2011) *Miandui jibing: Chuantong zhongguo shehui de yiliao guannian yu zuzhi* 面對疾病：傳統中國社會的醫療觀念與組織 (In the Face of Disease: Concepts and Institutions of Medicine in Traditional Chinese Society), Beijing: Zhongguo renmin daxue chubanshe.

Li Ding 李鼎 (1998) *Zhenjiuxue shinan* 針灸學釋難 (Explanation on the Perplexities in Acupuncture and Moxibustion), Shanghai: Shanghai zhongyiyao daxue.

Li Jianmin 李建民 (2007) *Faxian gumai: Zhongguo gudian yixue yu shushu shenti guan* 發現古脈：中國古典醫學與數術身體觀 (Discovering the Ancient *Mai*: The Numerological Conceptualisation of the Body in Classical Chinese Medicine), Beijing: Shehui kexue wenxian chubanshe.

——— (2010) *Zangfu zhizhang tushu de zangxiang guan ji guankan de shijian* 藏府指掌圖書的藏象觀及觀看的實踐 (The concept of '*zang* organs' in *Illustrated Guide to the Zangfu Organs* and the practice of observation), *Jiuzhou xuelin*, Winter: 45–81.

Liao Ping 廖平 (2010) *Liao Ping yishu heji* 廖平醫書合集 (A Compilation of Medical Books by Liao Ping), Tianjin: Tianjin kexue jishu chubanshe.

Liao Yinyuan 廖蔭元 (1962) '*Beibu chanshang de bianzheng lunzhi* 背部搶瘍的辨証論治' (Syndrome differentiation and treatment for back sore), *Jiangsu zhongyi*, 6: 8.

Ling Yunpeng 凌雲鵬 (1957) '*Fabei dashou de bianzheng yu zhiliao* 發背搭手的辨症與治療' (Syndrome differentiation and treatment for back sore), *Shanghai zhongyiyao zazhi*, 9: 40–2.

Liu Bingfan 劉炳凡 (1993) *Piweixue zhenquan* 脾胃學真詮 (Veritable Interpretation on Spleen-Stomach Theory), Beijing: Zhongyi guji chubanshe.

Lo, V. and Barrett, P. (eds) (2018) *Imagining Chinese Medicine*, Leiden: Brill.
Lu Weida 呂偉達 (ed.) (1999) *Wang Yirong ji* 王懿榮集 (Anthology of Wang Yirong's [works]), Jinan: Qilu shushe.
Pan Mingde 潘明德 (2014) *Yifa tiyao* 醫法提要 (Synopsis of Medical Treatment), Beijing: Xueyuan chubanshe.
Shang Qidong 尚啟東 (2005) *Hua Tuo kao* 華佗考 (Investigating Hua Tuo), Hefei: Anhui kexue jishu chubanshe.
Sun Xin 孫欣 (2011) *Huangdi neijing shuiyi mingci yanjiu* 黃帝內經水液名詞研究 (Research on the Naming of Body Fluids in the 'Inner Canon of the Yellow Emperor'), MA thesis, Liaoning University of Traditional Chinese Medicine.
Unschuld, P.U., Tessenow, H. and Zheng Jinsheng (2011) *Huang Di Nei Jing Su Wen: an annotated translation of Huang Di's Inner Classic – Basic Questions*, Berkeley: University of California Press.
Wang Fusheng 王伏聲 (2014) 'Zhongyi waikexue fanchozhong de "hanre" guan 中醫外科學範疇中的「寒熱」觀 (The concepts of "cold and heat" in the field of Chinese external medicine), *Zhongguo zhongyi jichu yixue zazhi*, 20.10: 1324–5.
Wu Yi-Li (2011) 'Body, gender, and disease: the female breast in late imperial Chinese medicine', *Late Imperial China*, 32.1: 83–128.
——— (2015a) 'Bodily knowledge and western learning in late imperial China: The case of Wang Shixiong (1808–68)', in H. Chiang (ed.) *Historical Epistemology and the Making of Modern Chinese Medicine*, Manchester: University of Manchester Press, pp. 80–112.
——— (2015b) 'Between the living and the dead: trauma medicine and forensic medicine in the Mid-Qing', *Frontiers of History in China*, 10.1: 38–73.
——— (2017) 'A trauma doctor's practice in nineteenth century China: The medical cases of Hu Tingguang', *Social History of Medicine*, 30: 2: 299–322.
——— (forthcoming) *The Injured Body: a social history of medicine for wounds in Late Imperial China* (under review, Berghahn Books).
Wu Zhongtai 吳中泰 (ed.) (2010) *Menghe Ma Peizhi yian lun jingyao* 孟河馬培之醫案論精要 (Essentials of Medical Cases by Ma Peizhi from Menghe), Beijing: Renmin weisheng chubanshe.
Xie Haizhou 謝海洲 (2004) 'Duyao yigong yishi: du *Yizong jinjian: Waike xinfa yaojue* 毒藥以供醫事：讀《醫宗金鑑・外科心法要訣》的啟示' (Poisons for medical purposes: reading 'the golden mirror of medical learning: knowledge and skills of external medicine'), *Tianjin zhongyiyao*, 21.4: 265–7.
Yan Zhen 嚴振 (1961) *Xunjing kaoxue bian* 循經考穴編 (Investigations into the Points along the Channels), Shanghai: Shanghai kexue jishu chubanshe.
Yu Yunxiu 余雲岫 (2012) *Gudai jibing minghou shuyi* 古代疾病名候疏義 (Explaining the Ancient Disease Names and Syndromes), Beijing: Xueyuan chubanshe.
Zhang Yupeng 張宇鵬 (2014) *Zhangxiang xinlun: Zhongyi zhangxiang de hexin guannian yu lilun fanshi yanjiu* 藏象新論：中醫藏象的核心觀念與理論範式研究 (New Theories on Visceral Manifestations: Research on the Core Concepts and Theoretical Paradigms of 'Visceral Manifestations' in Chinese Medicine), Beijing: Zhongguo zhongyiyao chubanshe.
Zhang Zanchen 張贊臣 (1960) 'Naoju zhengzhi 腦疽証治' (Syndrome and treatment for Cerebral Ulcers), *Shanghai zhongyiyao zazhi*, 5: 203–7.
Zhao Jingsheng 趙京生 (ed.) (2014) *Zhenjiu xue jiben gainian shuyu tongdian* 針灸學基本概念術語通典 (General Dictionary of Basic Concepts and Terms in Acupuncture and Moxibustion), Beijing: Remin weisheng chubanshe.
Zhu Feiyuan 朱費元 (2004) *Linzheng yidefang* 臨証一得方 (Comprehensive Recipes for Clinical Syndromes), Shanghai: Shanghai kexue jishu chubanshe.

14
HISTORY OF DISEASE
Pre-Han to Qing

Di Lu

Illness has always posed a major threat to human survival and well-being, but in a history of disease the ways in which bodily deterioration, accident and the experience of suffering have been understood, and the strategies deployed to address the consequences of ill health have to be placed within their specific historical and cultural contexts. The subject of disease and illness in pre-modern China is far too complex to relate in its entirety and this chapter, therefore, focuses on a few representative topics and an important tension between them. First, although the rise of classical medical concepts has been described in detail elsewhere in this handbook, for the purposes of this chapter, the salient point of contrast is the application of the theories of *yin*, *yang* and the five agents to all the individual permutations of illness and the experience of it. Second, the main thrust of this chapter traces the history of contagious disease in China – as representative of those kinds of disorders that affect populations in apparently identical ways. Styles of medicine that deal with general types of disease are quite separate from those that deal with individual expressions of suffering and illness and a historical analysis of this distinction in China has critical contemporary relevance; it brings into focus contemporary discourses about 'holism' and 'personalised medicine', which have been seen as a corrective to the impersonal biomedical models of public health practice and its search for the chemical 'magic bullet' for diseases (Mann 1999: 54–6; Scheid 2016: 66–86). Appreciating this distinction is essential to understanding the significance that Chinese medicine has taken on for patients in the late twentieth century, and the enduring global relevance of Traditional Chinese Medicine (TCM).

In recent decades, the relative definitions of disease and illness have attracted much academic attention (Kleinman 1988: 187–93; Cassell 1991: 81–93). Establishing a lexicon of disease types requires the classification of distinct sets of symptoms linked to the articulation of causes and discrete aetiologies, without particular reference to the individual sufferer. Illness, in contrast, refers to the unique experience of sub-optimal health and bodily disorder. Thus conceived, it is possible to identify coeval, yet polarising tendencies in the history of treatments for both diseases and illnesses in China, in an analysis that emphasises the plural medical environment that has existed from pre-imperial times to the beginnings of the twentieth century in China and beyond. The main thrust of this chapter is, however, concerned with the nature of disease in the history of Chinese medicine, since the background to classical Chinese medicine and its application to individual suffering is provided by Chen Yunju in this volume. There are very many other topics related to this history that are not approached or elaborated in this chapter, some of which, like the history of anatomy and surgery and diseases specific to women, are also dealt with elsewhere in this handbook (Chapters 13 and 23 in this volume).

A pervasive drive in histories of disease has been to match ancient to modern diseases to aid comprehension – a tendency which has gained pace and authority with the bio-anthropology of human remains. It has become possible to know a great deal more about pre-modern diseases in modern terms. This tendency is, however, dangerous historically, because of the many ways in which it distorts the ancient record. A modern disease term brings with it all the assumptions of causes and aetiologies framed in modern terms: we cannot now conceive of, for example, diabetes without thinking of high blood sugar, or cancer without thinking of malignant cells. One of the complex problems that has occupied physicians and the intellectual elite since the late Qing has been how to clarify the correspondence between Chinese and European language disease terms (Leung 2010: 25–50; Smith 2017: 139–60). For example, the term *huoluan* 霍亂, which can be found in classical medical treatises, generally refers to acute vomiting and diarrhoea in TCM, but it has often been used as a Chinese translation for cholera (an infectious disease caused by the bacterium *Vibrio cholerae*) since the 1830s, and vice versa (Dudgeon 1877: 44–5; Chen 1981: 29–31; Hanson 2011: 136; Unschuld and Tessenow 2011: 475). This problem not only concerns appropriate translation, but also reflects difficulties in twentieth- and twenty-first-century attempts to integrate Chinese medical and biomedical epistemologies of disease. However, in the science of history, it is not sustainable to simply match an ancient Chinese disease term to a modern one, if the intention is to understand the past on its own terms. It has become necessary to disambiguate disease from the symptoms of illness. Understanding disease in history involves refining our appreciation of the cultural systems that have generated the unique disease terminology, while it may be relatively more feasible to match symptoms of illness across time and in translation. A broken leg is a broken leg, frequent urination easily expressed and understood, but how well can we know how people 2,000 years ago experienced degrees and qualities of pain, or a depletion of *qi*? This is not an exact science and the identification of disease and its distinction from illness is always a process contingent on many factors, some of which I hope will become clear in this chapter.

Multidisciplinary studies on human remains from prehistory to the Shang dynasty (*c*. 1600–1046 BCE), recovered in a variety of archaeological sites in modern China, reveal that the communities of the ancient people who resided in the Yellow River plains had easily identifiable symptoms, such as dental caries; malocclusion; bone eburnation in relation to osteoarthritis and other visible bone disorders (Zhang 1982; Sakashita *et al*. 1997; Wang *et al*. 2012; Zhang *et al*. 2017). But with the discovery of oracle-bone inscriptions dating to Shang rule, we can recover the names of more than fifty types of diseases contracted by local people and can begin to interpret their nature, despite the abstruse script. It is possible to identify afflictions of different parts of the human body and the ways in which Shang accounts of disease conceptualised the body. Some inscriptions, for example, record nasal, dental and other unspecified diseases suffered by Lady Fu Hao (died *c*. 1200 BCE), wife of King Wu Ding. Shang diviners frequently refer to the power of ancestors and gods to intervene in human health and indicate the aetiology of the diseases informed by these spirits and sometimes followed by ritual remedies (Keightley 2001: 150–68; Song 2004). In this atmosphere, the role of drugs assumed minor significance, because whether a person would fall sick or recover from disease was primarily dependent on the intention of the spirits concerned, which therefore indicates a neglect of connections between disease and the physical body.

When the Zhou (c. 1046–256 BCE) succeeded the Shang rulers, the conquerors also maintained the custom of medical divination. The Duke of Zhou, namely Ji Dan 姬旦, once prayed to (the spirits of) three ancestors to take on the ailments of his sick brother, the King Wu of Zhou (?–1043 BCE) (Gren 2016: 300). This practice embodied a general belief about the transferability of disease or ominous matter from one person to another. For example, in 489 BCE, the King Zhao of Chu (r. 515–489 BCE) sent an emissary to enquire of the Zhou grand scribe about the clouds resembling a flock of red birds. The grand scribe replied that it meant the King would suffer from illness which, however, could be transferred to the chief minister or the supervisor of the military through sacrifice (Zuoqiu 2016: 1867). In 480 BCE, the ominous Fire Star moved to the celestial region that corresponded to the State of Song. As the Duke Jing of Song worried about a presaged disaster for him, his astrologer suggested that the disaster could be transferred to ministers, ordinary people or even annual harvests (Chang and Saussy 1999: 187). So sometimes, prognoses could also be made by glancing at the sky.

By the Warring States Period (475–221 BCE), we can identify the emergence of learned practitioners concerned specifically with the treatment of both diseases and illness. Their work seems to have been endorsed officially, although no doubt interlaced with ritual methods of treatment, and with ritualist specialists themselves. Members of this group also had specialist skills and were known for curing specific diseases such as *yin* 瘖, *bi* 疕 and *jie* 疥, which to a modern eye may look like dysphonia, porrigo and scabies, respectively. But the epistemic accuracy and value of the translations has always to be born in mind for every disease and illness term. We know about them from the discovery of their personal seals; each seal contains two or three characters, often constituting a combination of a physician's surname or the term 'treats' and a disease or symptom name (Chen 1988: 288–92). Before and around the unification of China in 221 BCE, the Qin state authorities established a strict set of rules relating to the reporting and diagnosis of contagious diseases such as *li* 癘 (possibly leprosy) and the segregation of infected patients; these rules and regulations were to be executed by officials and these specialist physicians (Leung 2009: 3–22).

The oldest extant Chinese treatise that is organised specifically around medical conditions, a silk manuscript that records over 280 prescriptions for fifty-two groups of diseases, was discovered in the Mawangdui tomb 3, a Western Han (202 BCE–9 CE) tomb constructed in 168 BCE. Considered to pre-date the tomb by about a century, the silk manuscript contains valuable relics of the medical knowledge of earlier periods. The prescriptions address the treatment of skin, inner-body, parasitic and other diseases (such as female haemorrhoid [*pinzhi* 牝痔], abscess [*yong* 癰] and dry itch [*gansao* 乾瘙]) by means of medicinal substances as well as ritual spells and incantations (Harper 1997: 14–5, 221–304; Brown 2015: 48–9). The use of spells, incantations and rituals in medical treatment in the Han dynasty was not without controversy: some Han intellectuals criticised people who relied solely on the ritual therapy of *wu* 巫 (controversially translated 'spirit mediums' or 'shamans', and referring to those using incantations, spells and various techniques for communication with ancestors, spirits and demons), rather than on the medicine of physicians (*yi* 醫) (Mair et al. 2005: 178). Ritual techniques only appear in more than thirty of the total prescriptions, but there is no evidence to show that their efficacy in curing serious diseases was doubted, for example, in the case of warts (*you* 疣); or that these methods were regarded as inferior to those using medicinal substances; diseases could still be attributed to ancestral spirits, or ghosts, which could be expelled by certain incantations or ritual pacing, such as the pace of Yu (*yubu* 禹步) (Harper 1997: 244, 290). The coexistence of therapeutic medicinal remedies with a lesser percentage of ritual techniques deployed by the *wu* indicates that spirit medicine remained an option, if apparently of secondary significance (given the ratio of the respective remedies) for patients within the newly emerging literate medical contexts of imperial China (Cho 2013). Even in contemporary China, patients and their families often pin their hopes for the sick upon divination and ritual, demonstrating a remarkable continuity in the plural medical environment.

Physicians of the Han dynasty (202 BCE–220 CE) increasingly differentiated their skills from ritualists, and diagnosed and treated diseases by applying medical measures. This trend is exemplified in Sima Qian's biography of Chunyu Yi 淳于意 (c. 205 BCE–?) as recorded in the first standard history of China, the Records of the Historian (Shiji 史記). Chunyu Yi mainly mentions medicinal substances and acumoxa among his therapeutic strategies, and does not refer to spirit medicine. In the eyes of Chunyu Yi and the myriad anonymous writers who contributed to the classical compilations of Chinese medicine over the course of the Han dynasty, ill health was closely related to excessive behaviour and inappropriate emotions as well as a result of losing synchronicity with the environment and astro-calendrical rhythms. Dietary habits, sexual activities and overwork were often implicated and reveal underlying moral censure of undesirable lifestyles at a time in the early empires when social disparity was becoming more pronounced. Exogenous illness-causing factors (e.g. wind [*feng* 風]) could invade the body and harm the organs where there was endogenous weakness or depletion. Such a conception also reflects more attention than ever being given to the body itself in Han physicians' understanding and explanation of disease and aetiology (Epler 1988; Kuriyama 1994: 26–7; Hsu 2010: 101–341).

Medical manuscripts organising prescriptions according to different types of diseases multiplied in the Han dynasty and thereafter, but a uniform classification system for illness was still lacking. Over the 400 years of the Han dynasty, however, medical authors who had inherited and developed technologies and technological terminology that had been emerging in previous centuries attempted to systematise them in a medical context: they appropriated and extended the theories of *qi*, *yin-yang* 陰陽, the five agents (*wuxing* 五行), viscera and organs (*zangfu* 臟腑) and channels (*jingmai* 經脈) to explain the mechanisms of ill health, and ultimately associated the nature of both illnesses and disease with the intrinsic qualities of medicinal substances creating a fully fledged classical medicine. The timing was not incidental to this process. Many scholars have noted that the concept of depletion, of the disruption of an essential integrity of the body by internal weakness, was conceptually and practically consistent with the new political realities of the early empire (Unschuld 2010 [1985]: 51–99). Concepts of bodily healthcare were contiguous with the proper organisation of the body politic: a strong empire would be based on a powerful administration, with seasonal rituals, and the free flow of resources around the newly unified geography of China. The political leaders were the link between heaven and earth. Likewise, the organs had to function together in a seasonal order so that a free flow of *qi* could power a strong and unified body and integrate it with the movements of heaven and earth (Lo 2001: 25–39; Sterckx 2006: 24–25; Brown 2015: 52–62, 125–127). The organs were even named for the officials of state, the heart having the function of the lord himself, the liver, the general, etc. Constitutional and inherited weaknesses could be aggregated with damage caused by excessive and immoral behaviour or by attack by external factors in the environment in the creation of a specific medical profile. It is this unique conception of medicine that allowed a physician to interpret the constellation of symptoms that made up an individual patient's illness and to tailor-make a solution, whether that be a carefully chosen set of acupuncture points or a custom-built remedy. Yet, even as this unique and enduring personalised medicine developed in China, the urgent need to find measures for treating contagious and epidemic diseases was gaining momentum.

A group of infectious febrile diseases called cold damage (*shanghan* 傷寒), for example, evident in *Huangdi's Inner Classic-Basic Questions* (*Huangdi neijing suwen* 黃帝內經素問, *c.* 100 BCE), hereafter *Suwen*, began to receive special attention from the scholar official, Zhang Ji 張機 (*c.* 150–219). Since the beginning of the Jian'an reign (196–220), almost half of Zhang's deceased relatives had died from epidemic diseases within a decade. This tragedy motivated Zhang to study contagious disease in depth and to record as many relevant prescriptions as possible (Goldschmidt 2009: 95–6; Hanson 2011: 4–5). By referring to the works of earlier scholars, he devised a classification system that simplified the common symptoms and pulse conditions of the sufferers of the epidemics into six progressive types according to the six *jing* 經 (three *yin* and three *yang* 'channels', each related to individual viscera and organs) in the body. The system thus described the changing condition of the pulse in the individual in order to place the individual at a stage in the course of these diseases unifying the diverse methods of diagnosis and linking the interpretations to a standard set of remedies (Mitchell et al. 1999). In other words, the individual was subordinated to the disease. Thus, the core method of diagnosing cold damage demanded a process of streamlining multiple symptoms. Zhang's theory of cold damage was strongly advocated by the reputed physician Sun Simiao 孫思邈 (*c.* 581–682), and has received extensive attention since the Tang dynasty (618–907), especially in Edo Japan (Daidoji 2013). It is clear that manuscript copies of cold damage treatises circulated widely after the Han period, although some argue that cold damage did not begin assuming anything like its later importance until the epidemics of the Song period (Lo and Cullen 2005: 385–6; Goldschmidt 2009: 2–3, 69–102; 141–72). Court physicians of that time were searching the archives for coherent medical theories that could be applied *en masse* to help the state address the cycles of contagious disease. Remarkably, the renowned healer Hua Tuo 華佗, a contemporary of Zhang Ji, would offer his patients a decoction of numbing powder and then surgical treatment if their internal bodily illness could not be cured by acupuncture or medication. But his medical expertise also includes psycho-emotional styles of therapy and disease prevention through the five-animal exercise invented by himself (Fu 2002), both of which reflect the concept that vulnerability to disease was closely tied to the strengths and weaknesses of an individual rather than to external factors such as cold damage.

At the end of the Eastern Han dynasty when Daoist institutions began to appear in west and central China, Daoists used the cure of disease as a tool for promoting their religion (Chapter 27 in this volume). They associated disease with a patient's personal guilt, and treated disease by using both community ritual and medical measures; a patient's recovery would be a sign of her or his individual merit and success in negotiations with the afterlife. In the Six Dynasties (229–589) when Daoism flourished, Daoist physicians contributed much to the knowledge of disease and treatment. For example, Ge Hong (284–*c.* 343), a distinguished writer and practitioner of the arts of transcendence loosely affiliated to Daoist culture, compiled a book that records many acute and chronic diseases as well as a greater number of related emergency prescriptions. One of the prescriptions for *nüe* 瘧 (intermittent fevers) involves the use of the plant *qinghao* 青蒿 (*Artemisia annua* or *Artemisia lancea*), which inspired the discovery of the antimalarial biomedical drug artemisinin from *Artemisia annua* in the twentieth century (Hsu 2006). In the eyes of Daoist physicians, however, disease was just a stumbling block on the road to longevity and immortality; and health and recovery were just first steps to these ultimate goals. Although almost all the theories of causes and cures of diseases in TCM can be found in forms of therapy described in Daoist texts, these therapies also reference a range of distinctive theories. One of these is characterised by the agents of destruction, the *sanshi* 三尸 (three corpse-worms) and *jiuchong* 九蟲 (nine worms). These worms were thought to live in different parts of the body, and to give rise to various desires which would prove damaging to people's health. They would also report their host's faults and guilt to the gods. By linking the cause of disease with desires, this Daoist theory combined practices involving powerful and effective medicinal materials to expel internal parasites with moral cures (Huang 2011). Buddhists also related disease causation to inappropriate desire and, despite living in different religious landscapes peopled by different pantheons of deities, Daoists and Buddhists both believed that diseases could be brought to the temporal world by spirits (Salguero 2014: 23–6). Both Daoist and Buddhist religious concepts of disease bred an element of fatalism which struck root in the hearts of the Chinese laity. A ninth-century text, for example, records that the famous Tang intellectual and official Pei Du 裴度 (765–839) once claimed that birth, ageing, illness and death were nothing but scheduled programmes (*Yinhualu*, vol. 2, [*c.* 853] 1979: 80–1), which possibly engages with the public belief that heavenly bureaucracies might keep tallies and registers of every individual's behaviour pre-determining their health outcomes (Kuriyama 2003: 52–9).

A profound medical revolution happened in the ephemeral Sui dynasty (581–618). In this short lived dynasty characterised by large-scale imperial projects involving both military expansion and building of canals and other infrastructure, the pedagogical function of the Imperial Medical Academy expanded. Together with an increased medical bureaucracy dedicated to the creation and transmission of medical knowledge, the Sui period was an important time for both the preservation of traditional texts and medical innovation (Needham and Lu 2000: 98–102; Hinrichs and Barnes 2013: 87–90). Chao Yuanfang 巢元方, a member of the Academy, summed up almost all previous knowledge of aetiology and the signs and symptoms of disease. His *Treatise on the Origin and Symptoms of Diseases* (*Zhubing yuanhou lun* 諸病源候論), hereafter *Yuanhou lun*, is the first comprehensive Chinese monograph on disease. The diseases are organised under sixty-seven rubrics (including 1,739 sub-types) and the book contains many new insights into the mechanics of epidemics, and diseases, which possibly can be associated with diabetes (*xiaoke* 消渴); pestilential diseases (*yili* 疫癘); urinary calculi (*shilin* 石淋); scabies (*jie* 疥); parasitoses (*jiuchong hou* 九蟲候) and motion sickness (*zhuchechuan* 注車船) (*Zhubing yuanhou lun*, vols. 5, 10, 14, 18, 35, 38, 40, [610] 1991: 155–6, 334–6, 441, 557–8, 1000–1, 1149).

Matching pre-modern disease categories with modern diseases, especially cross-culturally, is a serious teleological crime for historians who recognise the temporal nature of concepts of disease, especially those of our own time. However, this is not to say that it is impossible to *interpret* what ancient people suffered in modern terms, so long as one remains sensitive to the inevitable distortions of such an enterprise. Chao, for example, refuted the conventional view that infertility (*wuzi* 無子) was only caused by the female partner, and stressed that infertility would also occur if a man was unable to ejaculate or his semen (*jing* 精) looked as limpid as water or felt as cold as ice and iron. This gendered perspective on fertility matches modern knowledge in obstetrics and gynaecology, although Chao also associated infertility with not having offered sacrifices to dead ancestors (*Zhubing yuanhou lun*, vols. 3, 38, [610] 1991: 101, 1124).

Yuanhou lun follows the earlier *Suwen* when it states that wind (*feng*) is the 'master of one hundred diseases'. Wind diseases (*fengbing* 風病) and their related symptoms are his first and most important rubric in the book (*Zhubing yuanhou lun*, vol. 1, [610] 1991: 1–84). Wind and cold (*han* 寒); heat (*shu* 暑); wet (*shi* 濕); dryness (*zao* 燥) and fire (*huo* 火) are called the six *qi* (*liuqi* 六氣, i.e. six exogenous disease-causing factors) in *Suwen*. Wind was the primary cause of diseases; and wind disease could transform into other diseases (Unschuld and Tessenow 2011: 72, 625, 631; Chapters 4 and 5 in this volume). Compared with earlier medical writers, Chao was more inclined towards integrating theories of *yin-yang*, *qi*, *wuxing* into the mechanisms of wind diseases and other disease rubrics and, overall, his accounts were more methodical and detailed than any we know of before him, and his observations of symptoms much more careful. Due to its comprehensive coverage, *Yuanhou lun* has frequently been cited by later physicians, and is a first point-of-call for researchers into disease, and for clinicians (Pregadio 2008: 87–8). Meanwhile, the sheer number of symptoms and mechanisms of diseases related in this book has made it difficult for readers and researchers to get to grips with differential diagnosis.

Tang physicians also paid more attention to the recording of symptoms which was helpful to the diagnosis of diseases. Chao had only appended some therapeutic exercises and drug remedies to his disease categories, claiming that prescriptions concerning decoctions, hot packs, needles and stones are given elsewhere (Chapter 6 in this volume). Other medical writers, however, attached prescriptions to the disease rubrics, enabling their books to be more practical. The main reason given by Wang Tao 王燾 (*c.* 670–755) for compiling his *Secret Essentials from the Outer Terrace* (*Waitai miyao fang* 外臺秘要方), with a collection of over 6,800 prescriptions, was that he regretted the lack of prescriptions in Chao's book (*Waitai miyao fang*, preface, [752] 1993: 3–6). Sui and Tang physicians explored disease-causing factors and medical treatment in a variety of ways, many of which anticipate modern sensibilities. For example, Chao Yuanfang noticed the close relationship between acute vomiting and diarrhoea (*huoluan* 霍亂) and dietary factors (*Zhubing yuanhou lun*, vol. 22, [610] 1991: 648–9). The Tang physician Sun Simiao further pointed out that acute vomiting and diarrhoea were solely caused by dietary factors, stressing the irrelevance of the spirits (*guishen* 鬼神) (*Beiji qianjin yaofang*, vol. 20, [*c.* 652] 1998: 442). Sun is also considered by modern positivist commentators to be a Chinese expert in 'nutrient deficiency diseases', exemplified by one of his findings that some people suffering from what was called 'night blindness' (*yan mu wusuojian* 眼暮無所見) were helped by his remedy for the disease which was pig's liver, a food which is now proved rich in vitamin A (Kuhnlein and Pelto 1997: *Qianjin yifang*, vol. 11, [*c.* 682] 1998: 181–2; 113; Han et al. 2003: 301).

In the Sui and Tang dynasties, Sun was not the only physician who studied cold damage, but his angle was special: he categorised symptoms according to the treatment prescribed, and particularly the key ingredients in cold damage remedies (*Qianjin yifang*, vol. 9, [*c.* 682] 1998: 127–40). Cold damage epistemology prevailed in later periods, and was closely related to a trend for collecting and arranging prescriptions. This trend is exemplified in two of the earliest extant Chinese monographs on bone injuries and obstetrics, respectively, *Divinely Transmitted Secret Prescriptions for the Management of Wounds and Fractures* (*Xianshou lishang xuduan mifang* 仙授理傷續斷秘方, *c.* 846) and *Tested Treasure in Obstetrics* (*Jingxiao chanbao* 經效產寶, *c.* 852). The former book formulates a regular procedure for treating bone injuries (e.g. bone fractures and joint dislocation), and particularly warns that incisions should not be contaminated by wind and water when doing surgical operations because that might cause *poshangfeng* 破傷風 (lit. wind-syndrome due to open wounds, a term now associated in modern times with tetanus). The latter book mentions the mechanisms of pregnancy, birth and post-partum diseases. Traditional theories concerned with harmony were esteemed in both books, and how to harmonise blood and *qi* in the body became critical to the healing of bone injuries and obstetric diseases (Chapter 23 in this volume). In Tang times, spirits and ritual treatments occupied an inferior position in the cognition and treatment of diseases.

The trend for simplifying aetiology, diagnostics and treatment gained pace in the Song dynasty (960–1279), especially the Southern Song dynasty (1127–1279), perhaps as a corrective to the burgeoning number of encyclopaedic remedy books published at the instigation of the state. Such books include *Imperial Grace Formulary of the Great Peace and Prosperity Reign Period* (Taiping shenghuifang 太平聖惠方, 992), *Comprehensive Record of Sagely Beneficence* (Shengji zonglu 聖濟總錄, c. 1118) and *Prescriptions of the Public Pharmacy of the Era of Great Peace and of the Bureau of Medicines* (Taiping huimin hejiju fang 太平惠民和劑局方, 1151). Inspired by Zhang Ji's theories, the physician Chen Yan 陳言 (1121–90) in his book *Treatise on Three Categories of Pathogenic Factors* (Sanyin jiyi bingzheng fanglun 三因極一病證方論) proposed that disease-causing factors could be catalogued into three groups: inner factors (*neiyin* 內因, i.e. seven emotions [*qiqing* 七情]), external factors (*waiyin* 外因, i.e. six exogenous diseases-causing factors [*liuyin* 六淫]) and factors that were neither inner nor external (*buneiwaiyin* 不內外因, related to accidents, improper diet, fatigue, etc.) (*Sanyin jiyi bingzheng fanglun*, preface and vol. 2, [1174] 1957: 1, 19). Thereafter, medical writings and institutions regularly organised an otherwise eclectic array of symptoms into the three categories. This classification served to draw individual experience and systematic theory closer together, narrowing the gap between what historians now distinguish as 'illness' and 'disease', and it remains a popular set of distinctions in both modern institutional medicine and TCM, due to its simplicity and practical nature. Chen's student Wang Shuo 王碩 (late twelfth–early thirteenth century) went further. Despite the existence of several encyclopaedic official prescription books, he compiled the succinct *Simple Prescriptions* (*Yijian fang* 易简方, 1196) with some thirty prescriptions which were expected to cope with a wide variety of diseases (Smith 2017: 75–9). This book, which enjoyed an excellent reputation at the time, prompted the publication of similar books which served as a corrective to the complexities involved in practising within a plural and increasingly competitive medical environment.

In the Song dynasty, there had been new developments in ophthalmology and early forms of obstetrics (including gynaecology), both of which were first officially added into the medical education system at the Imperial Medical Academy (Furth 1999: 66–7; Chapter 23 in this volume). Eye diseases and a related seventy-two symptoms were synoptically divided into internal and external categories according to the sites of the symptoms; the internal eye diseases were usually attributed to the malfunction of organs, while the external ones were mostly associated with the invasion of the body by wind or fire (Kovacs and Unschuld 1998: 46–7; Goldschmidt 2009: 53). Song physicians also re-conceptualised women's diseases following the hypothesis that female physiology was based on blood and assisted by *qi*; henceforth, the central principle in understanding and treating the phases of female disease was the regulation of blood and *qi* (Furth 1999: 70–4). For example, Chen Ziming 陳自明 (*c.* 1190–1270) described a disease known as *ruyong* 乳癰 (possibly breast carbuncle), and attributed its emergence to obstruction of blood flow and *qi*. He assented to Chen Yan and other predecessors' opinions that women and men shared most of the aetiological factors, but women's diseases were ten times more difficult to cure than men's diseases. This was because women had more desires and thus became more susceptible to diseases; furthermore, the deeply ingrained emotional character of women also contributed to the difficulty in curing women's diseases (*Furen daquan liangfang*, vols. 2, 23, [1237] 1992: 63–4, 644; Introduction in this volume). Thus, the causes of female diseases were associated with emotional disorder, belonging to the inner factors in Chen Yan's classification of aetiological factors. Zhu Zhenheng 朱震亨 (1281–1358) and many later physicians held a similar gendered view that emphasised women's susceptibility to emotions (Hanson 2005; Wu 2011). Zhu also observed that breast lumps in their later stages could become incurable, as in terminal breast cancer in the eyes of some modern authors who look for medical intelligence of the ancients according to modern priorities (*Danxi xinfa*, vol. 5, [1347] 1994: 174; Yan 2013).

Song, Jin (1115–1234) and Yuan (1271–1368) physicians offered dynamic observations on a few contagious diseases such as pre-modern equivalents to the poxes: diseases likely to approximate measles, smallpox and chicken pox. Rough differentiations between the poxes were made according to the characteristics of eruptions; and the causes and aetiology of pox disease were often described in terms of foetal toxin (*taidu* 胎毒), a heat toxin derived from the foetus' maternal contaminated blood or liquid in the viscera (*zangfu* 臟腑) and stored in the foetus' gate of life (*mingmen* 命門) (Chang 2010: 23–38). In the late eighteenth century, Pierre-Martial Cibot (1727–80), a French Jesuit missionary in Beijing, at one time introduced some Chinese understanding of smallpox and this theory to France (Heinrich 2008: 24–9). Despite the poxes being described in culturally specific ways, many histories of disease are often framed according to the ways in which they anticipate modern epistemologies. The histories of smallpox are a case in point, since it seems that Chinese methods included, in practice, what we understand today as some basic forms of immunotherapy that took increasingly sophisticated shape in the Ming and Qing (Wong and Wu 1932: 215–6, 273–6; Chen 1981: 50–66; Needham and Lu 2000: 134–45).

Chinese medical records of smallpox can probably be traced back to the fourth century or about the turn of the sixth century when Ge Hong or Tao Hongjing 陶弘景 (456–536) mentioned marauders' pox (*luchuang* 虜瘡) and indicated that it was transmitted to China by means of wars between China and foreign nations (Needham and Lu 2000: 126–7). It remains controversial as to whether Song physicians invented human variolation. Certainly by the Ming (1368–1644), it is clear, however, that physicians were successfully introducing infected smallpox materials into the bodies of those who had not yet been infected for the purposes of creating immunity. The disease had always baffled Chinese physicians but in the early seventeenth century, human variolation (*zhongdou* 種痘) began to be spread from central China (Chiu 2019). Qing (1644–1912) physician, Zhang Lu 張璐 (1617–99), recorded a few prevention methods that introduced a smallpox lymph or scab into a healthy person, or by wearing the clothes that were worn by smallpox patients (*Zhang's Treatise on General Medicine* (*Zhangshi yitong* 張氏醫通, 1695; Zhang 1994: 629). The seventeenth century then saw variations in treatment techniques and conceptions. It must have been observed quite early on that people who had been infected by smallpox would not always die: and some of them could naturally acquire immunity after infection, as exemplified in the Qing period by the medical history of the emperor Kangxi 康熙 (1654–1722) himself (Chang 2002: 182). These methods were gradually refined and generalised, and encountered cowpox vaccination transmitted to China from the early nineteenth century which, however, did not drive human variolation away (Leung 2011: 5–12; Andrews 2014: 43). Smallpox was just one of the various epidemics that ravaged Ming and Qing China.

With respect to acute febrile epidemics in some areas of Qing China, for example, Jiangnan, the frequency of their outbreaks did not reduce but increased due to factors such as population growth and environmental deterioration, while their rampancy in local society declined due to medical advances and timely social responses. Nevertheless, praying to spirits remained one of the responses for local people in Jiangnan treating epidemics (Yu 2003: 187–93, 249–88, 344). Such behaviour had its origin in early China and remains significant today. In the ecologically diverse areas of Yunnan, some pervasive pathogenic or infectious factors were grouped under the name of *zhang* 瘴 (miasma) or *zhangqi* 瘴氣 (miasma *qi*) during the Qing dynasty, and are linked to rapid population growth and unprecedented regional agricultural and economic explorations in the areas (Zhou 2007: 238–406). However, new strategies to counter the effects of *zhangqi* and, for example, the *zaqi* 雜氣 (heterogeneous [pathogenic] *qi*) in Wu Youxing's 吳有性 (1582–1652) discourse of disease would not be unveiled until Western microbiology and immunology began to be transmitted to China from the end of the nineteenth century (Andrews 1997; Kuriyama 2010: 18–20).

Given the rich and plural traditions of medicine in China, the transmission of those traditions that passed for 'Western' medicine, especially after the late Ming and early Qing periods, was bound to be contested both by a powerful local medical community and by those looking for effective cures (Cunningham and Andrews 1997: 1–32). With the exception of stunning new surgery for cataracts, pain relief for operations and a few drugs like quinine for malaria, the knowledge and practice that pioneering European physicians deployed in Asia for treating disease and illness were not, initially, necessarily more effective than what the Chinese already had – and, indeed, the shady practices of the anatomists provoked backlashes (Hinrichs and Barnes 2013: 220–6; Luesink 2017: 7–16). But by the early twentieth century, the arrival of modern quarantine measures for controlling contagious diseases marked a profound improvement in healthcare (Wong and Wu 1932: 728–33; Andrews 2014: 156–8). Modern medicine has had a profound influence on Chinese society and increasing investment into cutting-edge medical research is seeing modern medical provisions in China rank very highly on the world stage. But due to a less rigid distinction between tradition and modernity in China, and a commitment to the integration of 'Chinese' and 'Western' medicine, the search for new treatments for disease in China includes mining the traditional *materia medica* for effective cures. The global success of artemisinin as a treatment for malaria is a well-known example of the Chinese tradition being re-invented to meet one of the biggest disease challenges of the twenty-first century. At the same time, in a medical world where patients are increasingly reduced to their syndromes, the 2,000-year-old tradition of treating illness and suffering with a form of personalised medicine, that is able to describe and respond to the patient as an individual with their unique expressions of illness, is gaining new momentum and significance worldwide.

Bibliography

Pre-modern sources

Beiji qianjin yaofang jiaoshi 備急千金要方校釋 (Collation and Annotations of Prescriptions for Emergencies Worth a Thousand Gold) *c.* 652, Sun Simiao 孫思邈, Li Jingrong 李景榮 et al. (eds) (1998), Beijing: Renmin weisheng chubanshe.

Danxi xinfa 丹溪心法 (Good Recipes from Danxi) 1347, Zhu Zhenheng 朱震亨, in Qiu Peiran 裘沛然 (ed.) (1994), *Zhongguo yixue dacheng sanbian* 中國醫學大成三編 (The Third Important Collection of Chinese Medical Texts), Changsha: Yüelu shushe.

Furen daquan liangfang 婦人大全良方 (Complete Good Prescriptions for Women) 1237, Chen Ziming 陳自明 (1992), Beijing: Renmin weisheng chubanshe.

Qianjin yifang jiaoshi 千金翼方校釋 (Collation and Annotations of Supplement to Prescriptions Worth a Thousand Gold) *c.* 682, Sun Simiao 孫思邈, Li Jingrong 李景榮 et al. (eds) (1998), Beijing: Renmin weisheng chubanshe.

Sanyin jiyi bingzheng fanglun 三因極一病證方論 (Treatise on Three Categories of Pathogenic Factors) 1174, Chen Yan 陳言 (1957), Beijing: Renmin weisheng chubanshe.

Waitai miyaofang 外臺秘要方 (Secret Essentials from the Outer Terrace) 752, Wang Tao 王燾(1993), Beijing: Huaxia chubanshe.

Yinhualu 因話錄 (Notes on Hearsays) *c.* 853, Zhao Lin 趙璘 (1979), Shanghai: Shanghai guji chubanshe.

Zhangshi yitong 張氏醫通 (*Zhang's Treatise on General Medicine*) 1695, Zhang Lu 張璐, in Qiu Peiran 裘沛然 (ed.) (1994) *Zhongguo yixue dacheng sanbian* 中國醫學大成三編 (The Third Important Collection of Chinese Medical Texts), Changsha: Yüelu shushe.

Zhubing yuanhoulun jiaozhu 諸病源候論校注 (Collation and Annotations of the Treatise on the Origin and Symptoms of Diseases) 610, Chao Yuanfang 巢元方, Ding Guangdi 丁光迪 (ed.) (1991), Beijing: Renmin weisheng chubanshe.

Modern sources

Andrews, B. (1997) 'Tuberculosis and the assimilation of germ theory in China, 1895–1937', *Journal of the History of Medicine and Allied Sciences*, 52.1: 114–57.

Andrews, B. (2014) *The Making of Modern Chinese Medicine, 1850–1960*, Vancouver: University of British Columbia Press.

Brown, M. (2015) *The Art of Medicine in Early China: the ancient and medieval origins of a modern archive*, Cambridge: Cambridge University Press.

Cassell, E.J. (1991) *The Nature of Suffering and the Goals of Medicine*, Oxford: Oxford University Press.

Chang, C.F. (2002) 'Disease and its impact on politics, diplomacy, and the military: the case of smallpox and the Manchus (1613–1795)', *Journal of the History of Medicine and Allied Sciences*, 57.2: 177–97.

—— (2010) 'Dispersing the foetal toxin of the body: conceptions of smallpox aetiology in pre-modern China', in L.I. Conrad and D. Wujastyk (eds) *Contagion: perspectives from pre-modern societies*, Aldershot: Ashgate, pp. 23–38.

Chang, K.S. and Saussy, H. (eds) (1999) *Women Writers of Traditional China: an anthology of poetry and criticism*, Stanford, CA: Stanford University Press.

Chen Shengkun 陳勝昆 (1981) *Zhongguo jibingshi* 中國疾病史 (History of Chinese Disease), Taipei: Ziran kexue wenhuashiye gongsi.

Chen Zhi 陳直 (1988) *Wenshi kaogu luncong* 文史考古論叢, Tianjin: Tianjin guji chubanshe.

Chiu Chung-lin 邱仲麟 (2019) 'Wanming rendoufa qiyuan jiqi chuanbo de zaisikao 晚明人痘法起源及其傳播的再思考' (Rethinking the origin of variolation and its spread in Ming-Qing China), *Taida lishi xuebao*, 64: 125-204.

Cho, P.S. (2013) 'Healing and ritual imagination in Chinese medicine: the multiple interpretations of *Zhuyou*', *East Asian Science, Technology, and Medicine*, 38: 71–112.

Conrad, L.I. and Wujastyk, D. (eds) (2010) *Contagion: perspectives from pre-modern societies*, Aldershot: Ashgate

Cunningham, A. and Andrews, B. (eds) (1997) *Western Medicine as Contested Knowledge*, Manchester: Manchester University Press.

Daidoji, K. (2013) 'The adaptation of the *treatise on cold damage* in eighteenth-century Japan: text, society, and readers'. *Asian Medicine*, 8.2: 361–93.

Dudgeon, J. (1877) *The Diseases of China: their causes, conditions, and prevalence, contrasted with those of Europe*, Glasgow: Dunn and Wright.

Durrant, S. et al. (2016) *Zuo Tradition Zuozhuan: commentary on the "spring and autumn annals"*, Seattle: University of Washington Press.

Epler, D.C. (1988) 'The concept of disease in an ancient Chinese medical text, the discourse on cold-damage disorders (Shang-han Lun)', *Journal of the History of Medicine*, 43.1: 8–35.

Fu, L.K.T. (2002) 'Hua Tuo, the Chinese god of surgery', *Journal of Medical Biography*, 10.3: 160–6.

Furth, C. (1999) *A Flourishing Yin: gender in China's medical history (960–1665)*, Berkeley: University of California Press.

Goldschmidt, A. (2009) *The Evolution of Chinese Medicine: Song dynasty, 960–1200*, London: Routledge.

Gren, M.R. (2016) 'The Qinghua 'Jinteng' manuscript: what it does not tell us about the Duke of Zhou', *T'oung Pao*, 102.4–5: 291–320.

Han, H. et al. (2003) *Ancient Herbs, Modern Medicine*, New York: Bantam.

Hanson, M. (2005) 'Depleted men, emotional women: gender and medicine in the Ming dynasty', *Nan Nü: men, women and gender in China*, 7.2: 287-304.

—— (2011) *Speaking of Epidemics in Chinese Medicine: disease and the geographic imagination in late imperial China*, New York: Routledge.

Harper, D. (1997) *Early Chinese Medical Literature: the Mawangdui medical manuscripts*, London: Kegan Paul International.

Heinrich, A.L. (2008) *The Afterlife of Images: translating the pathological body between China and the West*, Durham, NC: Duke University Press.

Hinrichs, TJ and Barnes, L.L. (eds) (2013) *Chinese Medicine and Healing: an illustrated history*, Cambridge, MA: The Belknap Press of Harvard University Press.

Hsu, E. (ed.) (2001) *Innovation in Chinese Medicine*, Cambridge: Cambridge University Press.

——— (2006) 'The history of *qing hao* in the Chinese *materia medica*', *Transactions of the Royal Society of Tropical Medicine and Hygiene*, 100.6: 505–8.

——— (2010) *Pulse Diagnosis in Early Chinese Medicine: the telling touch*, Cambridge: Cambridge University Press.

Huang, S.S. (2011) 'Daoist imagery of body and cosmos, part 2: body worms and internal alchemy', Journal of Daoist Studies, 4: 33–64.

Keightley, D.N. (2001) 'The "science" of the ancestors: divination, curing, and bronze-casting in late Shang China', *Asia Major*, 14.2: 143–87.

Kleinman, A. (1988) *The Illness Narratives: suffering, healing, and the human condition*, New York: Basic Books.

Kovacs, J. and Unschuld, P.U. (1998) *Essential Subtleties on the Silver Sea: the Yin-Hai Jing-Wei: a Chinese classic on ophthalmology*, Berkeley: University of California Press.

Kuhnlein, H.V. and Pelto, G.H. (eds) (1997) *Culture, Environment, and Food to Prevent Vitamin A Deficiency*, Boston, MA: International Nutrition Foundation for Developing Countries.

Kuriyama, S. (1994) 'The imagination of winds and the development of the Chinese conception of the body', in A. Zito and T.E. Barlow (eds) *Body, Subject & Power in China*, Chicago, IL: The University of Chicago Press, pp. 23–41.

——— (2003) 'Concepts of disease in East Asia', in K.F. Kiple (ed.) *The Cambridge World History of Human Disease*, Cambridge: Cambridge University Press, pp. 52–59.

——— (2010) 'Epidemics, weather, and contagion in traditional Chinese medicine', in L.I. Conrad and D. Wujastyk (eds) *Contagion: perspectives from pre-modern societies*, Aldershot: Ashgate, pp. 3–22.

Leung, A.K.C. (2009) *Leprosy in China: a history*, New York: Columbia University Press.

——— (2010) 'The evolution of the idea of *Chuanran* contagion', in A.K.C. Leung and C. Furth (eds) *Health and Hygiene in Chinese East Asia: policies and publics in the long twentieth century*, London: Duke University Press, pp. 25–50.

——— (2011) '"Variation" and vaccination in late imperial China, ca. 1570–1911', in S.A. Plotkin (ed.) *History of Vaccine Development*, New York: Springer, pp. 5–12.

Lo, V. (2001) 'The influence of nurturing life culture on the development of Western Han acumoxa therapy', in E. Hsu (ed.) *Innovation in Chinese Medicine*, Cambridge: Cambridge University Press, pp. 19–50.

Lo, V. and Cullen, C. (eds) (2005) *Medieval Chinese Medicine: the Dunhuang medical manuscripts*, London: Routledge.

Luesink, D. (2017) 'Anatomy and the reconfiguration of life and death in Republican China', *The Journal of Asian Studies*, 76.4: 1–26.

Mair, V.H. et al. (2005) *Hawai'i Reader in Traditional Chinese Culture*, Honolulu: University of Hawai'i Press.

Mann, J. (1999) *The Elusive Magic Bullet: the search for the perfect drug*, Oxford: Oxford University Press.

Mitchell, C., Ye, F. and Wisemman, N. (1999) *Shang Han Lun: on cold damage, translation and commentaries*, Brookline: Paradigm Publications.

Needham, J. and Lu, G.D. (2000) *Science and Civilisation in China: volume 6 part 6: medicine*, Sivin, N. (ed.), Cambridge: Cambridge University Press.

Pregadio, F. (2008) 'Chao Yuanfang', in H. Selin (ed.) *Encyclopaedia of the History of Science, Technology, and Medicine in Non-Western Cultures*, Berlin: Springer-Verlag, pp. 87–8.

Sakashita, R., Inoue, M., Inoue, N. Pan, Q. and Zhu, H. (1997) 'Dental disease in the Chinese Yin–Shang period with respect to relationships between citizens and slaves', *American Journal of Physical Anthropology*, 103.3: 401–8.

Salguero, C.P. (2014) *Translating Buddhist Medicine in Medieval China*, Philadelphia: University of Pennsylvania Press.

Selin, H. (ed.) (2008) *Encyclopaedia of the History of Science, Technology, and Medicine in Non-Western Cultures*, Berlin: Springer-Verlag.

Scheid, V. (2016) 'Holism, Chinese medicine and systems ideologies: rewriting the past to imagine the future', in A. Whitehead and A. Woods (eds) *The Edinburgh Companion to the Critical Medical Humanities*, Edinburgh: Edinburgh University Press, pp. 66–86.

Smith, H.A. (2017) *Forgotten Disease: illnesses transformed in Chinese medicine*, Stanford, CA: Stanford University Press.

Song Zhenhao 宋鎮豪 (2004) 'Shangdai de jihuan yiliao yu weisheng baojian 商代的疾患醫療與衛生保健' (Medical Care and Healthcare in the Shang Dynasty), *Lishi yanjiu* 歷史研究, 2: 3–26.

Sterckx, R. (2006) 'Sages, cooks, and flavours in Warring States and Han China', *Monumenta Serica*, 54.1: 1–46.

Unschuld, P.U. (2010 [1985]) *Medicine in China: a history of ideas*, Berkeley: University of California Press.

Unschuld, P.U. and Tessenow, H. (2011) *Huang Di nei jing su wen: an annotated translation of Huang Di's inner classic-basic questions*, Berkeley: University of California Press.

Wang, W. et al. (2012) 'Malocclusions in Xia dynasty in China', *Chinese Medical Journal*, 125.1: 119–22.

Whitehead, A. and Woods, A. (eds) (2016) *The Edinburgh Companion to the Critical Medical Humanities*, Edinburgh: Edinburgh University Press.

Wong, K.C. and Wu, L.T. (1932) *History of Chinese Medicine*, Tientsin: The Tientsin Press.

Wu, Y. L. (2011) 'Body, gender, and disease: the female breast in late imperial Chinese medicine', *Late Imperial China*, 32.1: 83–128.

Yan, S.H. (2013) 'An early history of human breast cancer: West meets East'. *Chinese Journal of Cancer*, 32.9: 475–77.

Yu Xinzhong 余新忠 (2003) *Qingdai jiangnan de wenyi yu shehui: yixiang yiliao shehuishi de yanjiu* 清代江南的瘟疫與社會：一項醫療社會史的研究 (Wenyi and Society in the Jiangnan Area of Qing China: A Social Study of Medicine), Beijing: Zhongguo renmindaxue chubanshe.

Zhang, H. *et al.* (2017) 'Osteoarthritis, labour division, and occupational specialization of the Late Shang China – insights from Yinxu (ca. 1250–1046 B.C.)', *PLoS ONE*, 12.5: E0176329, https://doi.org/10.1371/journal.pone.0176329, accessed 27/8/2018.

Zhang, Y.Z. (1982) 'Dental disease of Neolithic Age skulls excavated in Shaanxi province', *Chinese Medical Journal*, 95.6: 391–6.

Zhou Qiong 周瓊 (2007) *Qingdai yunnan zhangqi yu shengtai bianqian yanjiu* 清代雲南瘴氣與生態變遷研究 (A Study of Miasma and Ecological Changes in Yunnan in the Qing Dynasty), Beijing: Zhongguo shehui kexue chubanshe.

15
PRE-MODERN MADNESS

Hsiu-fen Chen

Introduction

In English, the term 'madness' often refers to various kinds of insanity. In addition to foolishness, enthusiasm and uncontrollable emotions in colloquial usages, it also refers to melancholy, mania, hysteria, hypochondria, neurasthenia, psychosis and a wide range of 'mental illness' that have been identified through time in the medical realm (Foucault 1964; Scull 2015).

Similar to the English counterpart of 'madness', in modern China, *diankuang* 癲狂, *fengkuang* 瘋狂 and *fengdian* 瘋癲 are probably the most common colloquial words in Chinese to describe the chaotic mind, disturbed and disturbing emotions and/or uncontrolled behaviour. For the purposes of this chapter, these terms serve as a convenient category equivalent to madness or insanity.[1] These compounds among other terms, such as *fengren* 瘋人 (madmen), *fengbing* 瘋病 (madness illness), *jingshenbing* 精神病 (mental illness) and *shenjingbing* 神經病 (nerve illness), have become widely used, within and outside psychiatric hospitals, since the early twentieth century (Shapiro 1995; Baum 2018).

Nevertheless, *dian* 癲, *kuang* 狂 and *feng* 瘋 had different, though not mutually incompatible, connotations in pre-modern times. Over time, there have been many interpretations of them. Generally, *kuang* 狂 could signify those who had peculiar characteristics, such as special virtues, specific talents and/or magical power. The social images of those deemed to be *kuang* were therefore not necessarily negative – they could refer to genius, the eminent, as well as the deviant, or strange (Chen 2003: 200–35, 2016: 80–1). In the eyes of physicians, however, both *kuang* and *dian* were often seen as correlative illnesses in relation to insanity in classic medicine, while *feng*, of which the etymology could be traced back to the conception of wind, had become a legal term for labelling insanity by the Qing dynasty (1644–1911) (Messner 2000: 91; Simonis, 2010: 408–9).

In his research on madness in late imperial China, Fabien Simonis states that, '[t]here were effectively no "types of madness" in Chinese medicine, just one madness – *kuang* or *diankuang* – with many symptoms'. Besides, '*dian* had mostly meant "falling sickness", an epileptiform disorder [which] characterized seizures and convulsions' (Ibid.: 22, 47). It is true that some Western missionaries and physicians, who had introduced biomedicine into China, identified *dian*, *xian* 癇 and *dianxian* 癲癇 as an illness of epileptic seizure or 'Epilepsia' in the early twentieth century (Yu 1953: 109–13; Mathews 1996: 916). But the historical connotations of *dian* were more diverse in history. In the late imperial period, in particular, certain renowned physicians invented a new category of 'State of Mind' (*shenzhi* 神志), in which *dian*, *kuang* and *xian* were distinguished as different disorders. Here, *dian* was redefined as a mental illness owing to frustration of unfulfilled desires. Women's *huadian* 花癲/顛 (flower *dian*) was another example to show the association of *dian* with love madness. In addition to tracing the medical history of *kuang* madness, therefore, this chapter is also an attempt to clarify *dian* as an illness rather than just 'falling sickness'.

Classification of disease

Kuang 狂

Etymology suggests that early medical concepts of *kuang* madness were indistinguishable from the image of a mad dog (or of rabies in an ancient form) (Yu 1953: 154–5). In the Mawangdui manuscript, given the modern title *Recipes for Fifty-two Disorders* (*Wushier bingfang* 五十二病方 *c*. 200 BCE), *kuang* is not linked to any human ailment but a 'mad dog's biting men' (*Wushier bing fang*: 43; Ma 1992: 379–80; Harper 1998: 234–5). It later came to refer to a person's wild and extravagant behaviour. According to the *Yellow Emperor's Inner Classic* (*Huangdi neijing* 黃帝內經, *ca.* the first century CE, hereafter *Inner Classic*), the earliest received medical canon in China, a sufferer from *kuang* may appear to be initially grieving, likely to be forgetful, angry and fearful due to distress and hunger. When *kuang* starts to develop, the sufferer rests less and loses appetite and may regard himself as a paragon of virtue, acute intellect or nobility. He is easily provoked to swear and curse ceaselessly.[2] If an incidence of *kuang* is originally induced by great fear, the sufferer may appear frightened, be inclined to laugh and sing joyfully, and run around rashly and restlessly. When the *kuang* is originally induced by insufficient *qi*, its symptoms involve visual and auditory hallucinations. Furthermore, a sufferer from *kuang* that is originally induced by great joy may eat much, tends to see ghosts and spirits, and be inclined to laugh yet hide from people.[3]

No less important is the *yin-yang* dualism in the explanations for *kuang* madness in the *Inner Classic*. In this renowned homological thought, both *yin* and *yang* always generate, cooperate and fight each other (Chapter 1 in this volume). For example, when *yin* stores the spirit, *yang* defends its exterior. As a *yang* disorder, *kuang* is believed to be caused by 'pathogens invading the *yang*'. It is also assumed that when *yin* is defeated by *yang*, the vessel flow fights diseases, and this process also leads to *kuang*.[4]

Damage to the *yangming* 陽明 (Yang Brightness) vessel is another disorder that is frequently paralleled or linked to *kuang* madness. In addition to physical ailments, this disorder may be accompanied by other symptoms of emotional disturbance, such as 'aversion for people and fire, panicked alarm when hearing wood agent sounds',[5] 'desire to close doors and windows and to dwell in isolation', 'a tendency to mount high places and sing, cast off clothing and run about', 'not eating for several days, climbing over walls and onto roofs' and 'speaking wildly and swearing at people no matter who they are'. They attack in particular when *yang* (*qi*) and *yin* (*qi*) fight each other.[6] This canonical view of madness was anticipated by earlier clinical practice. In one of his twenty-five medical case histories, the renowned doctor Chunyu Yi 淳于意 (205–150 BCE) explained clearly that 'when the *yangming* vessel is injured, one runs about wildly' (*yangming mai shang ji dang kuang zou* 陽明脈傷 即當狂走).[7] His view of *kuang* madness was later echoed in the *Inner Classic*. In short, *kuang* disorders often manifested themselves as imbalanced emotions, unrestrained behaviour and unusual capability in early medical thought.

Dian 顛, 巔, 瘨

Ancient usage of the term *dian* often made the following characters interchangeable – 顛, 巔 or 瘨 – revealing its etymological and etiological aspects. 顛 means either 'upside down' or 'falling down', while 巔, which contains the 'mountain' component, refers to the 'peak of a hill', implicitly referencing the 'crown of the head'. Their respective connotations suggest that *dian* is an illness in proximity to the head, which usually comes along with the symptom of feeling upside down and/or falling down. Furthermore, the fact that 瘨 was sometimes interpreted as *kuang* or *xian* seems to have muddled these implications.

In terms of *yin-yang* dualism, for example, the *Canon of Problems* (*Nanjing* 難經 *c*. the second century CE) indicates that 'a doubling of *yang* [influences results in] *kuang*; a doubling of *yin* [influences results in] *dian*'.[8] It further notes the distinction between *dian* and *kuang* based on their respective emotional reactions: when the disorders develop, a *kuang* sufferer is joyful, whereas a *dian* sufferer is 'unhappy' (*bule* 不樂). In this homological thought of *yin-yang* dichotomy, *dian* is thought passive and negative when *kuang* is active and positive.[9]

Both *dian* and *kuang* are like the two extremes of the same spectrum, which at a first glance appear to parallel the bipolar disorder in modern psychiatry. But *dian* is also defined as a foetal disorder, which originates in the womb of a mother when she is greatly frightened. According to the *Inner Classic*, it is a disorder of *qi* that suddenly goes up and cannot go down due to great fright.[10] When the disorder begins to develop, it is often accompanied with unhappiness. Then come symptoms such as a heavy and painful head, staring upwards with reddened eyes. When it becomes severe, one may feel vexed in the Heart/mind.[11] Other symptoms of *dian* may include shouting, panting and palpitating, rigidity and back pain.[12]

In the view of the *Inner Classic*, the struggle between *yin* and *yang* also gives rise to the *dian* disorder. When it occurs in the *yangming* channel, the *dian* disorder will drive the sufferer to run and shout; his abdomen becomes bloated and he has difficulty lying down; his face will be red and hot and he may also have wild vision and mad speech. In this case, the *dian* disorder may appear similar to the image of *kuang* depicted above. But the *dian* disorder is more often accompanied by symptoms of head ailments, in particular when one's '*qi* only goes upwards and yet not downwards'.[13]

The *yin-yang* analogy remained influential in later medical discourses. Zhang Ji 張機 (150–219), for instance, stated that 'the decline of *yin qi* 陰氣 leads to *dian*; the decline of *yang qi* leads to *kuang*'. In other words, *dian* is not caused by excess but decline of *yin qi*, while *kuang* is not caused by excess but decline of *yang qi* (*Jinkui yaolue lunzhu*, 11.150). This view is contrary to the teachings of the *Inner Classic*, which shows that physicians might have employed diverse interpretations when they applied early medical doctrines. It also reveals that the various symptoms related to *dian* and *kuang* had no single, standard, unified understanding in antiquity.

Pre-modern physicians not only paralleled *dian* and *kuang*, but confused *dian* and *xian* as well. The *Inner Classic* categorises the symptoms of the *xian* disorder based on the five Viscera. When the pulse of the Heart is full and large, or the pulse of the Liver is small and hurried, the symptom of 'convulsions and spasms' occurs. When the Heart is hot and the Liver is cold, the disorder of *xian* also attacks.[14] To be sure, the symptoms of *xian* are always associated with convulsive movements and therefore refer to what in modern terms is described as 'falling sickness' or 'epilepsy'. But the *Inner Classic* also describes one *dian* disorder with symptoms of full and bloated cheeks and muscles, stiffness in the bones, exhausted body, spasm, suddenly falling down and so forth, which could also indicate epilepsy.[15]

This similarity between the symptoms designated by the early terms has generated ongoing debates about the distinction between *dian* and *xian* in later periods. They became complicated in the newly popular category of Wind malady (*feng bing* 風病).

Wind malady

The concept of Wind as a pathogenic factor had prevailed in Chinese medical theories throughout the ages. Illnesses such as 'wind stroke' (*zhongfeng* 中風) as a paralysis disorder, and numbing wind (*mafeng* 痲瘋) or great wind (*dafeng* 大風) referring to leprosy, were all attributed to the attack of wind (*Jinkui yaolue lunzhu*, 5.71–8; Leung 1999: 399–438). In mediaeval China, 'Wind vertigo' (*fengxuan* 風眩) was paid much attention, since certain politicians and emperors suffered from it.[16] The *Recipes for Wind Vertigo* (*Fengxuan fang* 風眩方 501) was probably the first work exclusively for treating this illness. Xu Sibo 徐嗣伯 (c. the fifth to sixth century), the author, stated that *dian* and *xian* are the same disease at different stages: the disorder of adults is called *dian*; that of infants is called *xian* (*Beiji qianjin yaofang*, 14.201). Not long after, the *Treatise on the Origins and Symptoms of Medical Disorders* (*Zhubing yuanhou lun* 諸病源候論, 610) listed 'Wind malady' first before all to other diseases, and indicated Wind as a source of various types of stroke and insanity. Similar to Xu's view, this officially compiled medical work named *dian* as an ailment for people aged over ten and *xian* for people under ten. It further introduces an innovative, unprecedented typology: the 'five types of *dian*' under the category of 'Wind malady' (*feng bing* 風病), and the 'three types of *xian*' regarded as 'children's miscellaneous disorders' (*xiaoer zabing* 小兒雜病). They are differentiated based on symptoms (e.g. those of *dian* were similar to those sometimes shown by a horse, such as rolled-back eyes, clenched jaw, convulsion of limbs and the whole body heat), or on their causes (e.g. wind, fright and food).[17]

These works led later medical writers to associate *dian*, *kuang* and *xian* with the invasion of Wind: they often arise unexpectedly, spontaneously and irregularly, making harsh and abrupt shifts. Some later Tang (618–907) and Song dynasty (960–1271) medical texts followed the model of the *Treatise on the Origins and Symptoms of Medical Disorders*, but created different categories of *xian*, depending on either their origins or their symptoms.[18] Up until the thirteenth century, medical writers still had no agreement on the taxonomy and typology of *dian* and *xian*. It was only made clear that they were thought to be different from the *kuang* disorder.[19]

The category of Wind malady seemed to lose its privileged position in medical treatises from the fourteenth century onwards. Its relationship to the disorders/symptoms of madness was also reformulated. On the one hand, the illnesses such as *kuang*, *dian*, *xian* and fright (*jing* 驚) due to wind attacks continued to be recorded in medical texts during the fifteenth and sixteenth centuries. On the other hand, more physicians tended to portray a distinction between Wind malady and *dian*, *kuang* and *xian*.[20] An exception is the disorder of 'Heart wind' (*xinfeng* 心風). Dai Sigong 戴思恭 (1324–1405) believed that it is caused by 'phlegmatic *qi*', of which the symptoms often manifest themselves as trance, joy-anger alternation, silence and distraction. To some extent, it looks similar to the *dian* disorder, yet not as serious as the latter.[21] Xu Chunfu 徐春甫 regarded Heart wind as a colloquial name for *diangkuang* illness 癲狂病, claiming that when the Heart/Spirit is disturbed and chaotic, it gives rise to wind pathogen (*xinshen huailuan er you fengxie* 心神壞亂 而有風邪).[22] Gong Xin 龔信 (fl. sixteenth century) also indicated the close relationship between Heart wind and *diankuang*, attributing one of their causes to anger.[23] On the other hand, both Sun Yikui 孫一奎 (1522?–1619?) and

Wang Kentang 王肯堂 (1549–1613) asserted that Heart wind (or 'absent-mindedness due to wind' 失心風) is a colloquial term for *dian* disorder owing to long-term frustration and unfulfilled desires.[24] Despite their different interpretations, these physicians showed a common interest in emotional roots of *dian* and *kuang* disorders in the late imperial period.

As the influence of the character *feng* 風 with reference to wind faded, it was later replaced with the character *feng* 瘋 (wind with a 疒 character component referring to sickness). The *feng* 瘋 character became more frequently used in the Qing Empire particularly in legal and colloquial languages. Numerous examples of *fengren* 瘋人, *fengdian* 瘋癲 and *fengbing* 瘋病 used to refer to madmen and a variety of types of insanity are recorded in the officially compiled *History of the Ming dynasty* (*Mingshi* 明史 1739), the *Qing Code* (*Daqing lüli* 大清律例) and Imperial Archives of Grand Secretariat and Board of Punishment (*Mingshi*, 218.5760, 242.6280, 244.6343, 245.6351; Simonis 2010: 408–20; Gabbiani 2013: 115–41).

This usage suggests that 'legal and medical spokesmen across the centuries were discussing more or less the same phenomenon of insanity' (Chiu 1981: 76).

Fire, phlegm and the heart disorders

During the twelfth and thirteenth centuries when the northern tribes of Jurchen and Mongols established their regimes and invaded China, respectively, changing views towards health and disease also emerged amidst the political turmoil. The medical systems of different cultures challenged accepted understanding of the body. Localised and regional medical views were gradually shaped, and came to prevail in the medical theories and practices of later ages. It is through this route that phlegm-fire entered medical discourse as a specific and particular cause of medical disorders.

Some physicians at that time even conceptualised phlegm-fire as a cause of madness. The physicians who are nowadays commonly called 'the Four Masters' made great contributions to these issues (Chapter 9 in this volume). Of them, Zhu Zhenheng 朱震亨 (aka Zhu Danxi 朱丹溪, 1282–1358) is the most renowned. Zhu elaborated Liu Wansu's 劉完素 (1110–1200) theory of fire and heat and proposed phlegm as a pathology in his medical treatises on various disorders. For example, he explained that the symptoms of 'depletion disorders' and 'phlegm disorders' look similar and are often mistaken as 'demonic afflictions'. In these cases, they should be treated by replenishing depletion, clearing heat and dispelling phlegm with herbal recipes (*Gezhi yulun* in *Danxi yiji*, 23; Simonis 2014a: 632–3).

In the Ming dynasty (1368–1644), despite the remaining influence of the classic *yin-yang* doctrines in the medical discourses of madness, an increasing number of physicians began to attribute the causes of *dian*, *kuang* and *xian* to phlegm and fire, mostly owing to the wide circulation of Zhu Zhenheng's medical teachings. In Zhu's *Plumbing the Mysteries of the Golden Casket* (*Jingui gouxuan* 金匱鉤玄) collated and supplemented by Dai Sigong, a royal physician in the early Ming dynasty, it suggests that phlegm in the diaphragm will drive people to *diankuang* and forgetful (*Jinkui gouxuan*, 1.121). Meanwhile, three distinct editions of the *Essential Methods of [Zhu] Danxi* (*Danxi xin fa* 丹溪心法) compiled by different persons in the second half of the fourteenth century enabled Zhu Zhenheng's medical teaching to reach a larger spread of readers (Simonis 2010: 167–78). One of these editions elaborates Zhu's aetiology of *xian* and explains that 'when the Spirit cannot guard the body because of fright, then phlegm will accumulate within'. Similar principles can be found in the treatments for *diankuang*. 'If it is heat that has accumulated in the Heart Vessel, then [one should] clear the Heart and eliminate heat. If it is phlegm that has clouded the Heart orifices, then [one should] dispel phlegm and calm the emotions'.[25]

As Zhu's most renowned disciple, Dai Sigong also ascribed an aetiology of phlegm and emotions to *diankuang*. His own work points out that *diankuang* is caused by 'gloom of the seven emotions, which therefore induce phlegm and saliva which block and cloud the orifices of the Heart'. The resulting symptoms often include loss of consciousness; staring, wide-open eyes without blinking; wild speech; shouting and scolding and even going outside the walls of one's home and ascending rooftops while naked and beating people, etc.[26] Lou Ying 樓英 (1320–1400), another of Zhu Zhenheng's disciples, also deserves attention. In his medical treatises, Lou regarded *dian* and *xian* as the same and stated that *dianxian* can attack when phlegm is in the diaphragm. As such, one may feel dizzy yet not fall down; when phlegm overflows above the diaphragm, one may feel dizzy and then fall down and lose consciousness.[27] Therefore, in addition to the works attributed to Zhu Zhenheng, it is also through Zhu's disciples that his views of madness became renowned. Many physicians in the later eras often explained the aetiology of *dian*, *kuang* and *xian* in terms of either 'phlegm clouding the orifices of the Heart', or 'pathogenic phlegm reversing upwards', and so on.[28]

The medical perceptions of Heart wind, Heart fire and Heart blood echoed the contemporary concern with self-cultivation of the mind/Heart in Neo-Confucianism. Apart from phlegm, excessive fire of the Heart was also highlighted as the cause of madness. Frequently quoted terms include 'exuberant Heart fire', and 'restless Heart fire'.[29] Some physicians further integrated phlegm and fire, elaborating the aetiology in terms of phlegm-fire being 'replete and flourishing' (*tanhuo shisheng* 痰火實盛), 'blocked and flourishing' (*tanhuo yongsheng* 痰火壅盛) or 'internally flourishing' (*tanhuo nei sheng* 痰火內盛). In certain cases, factors such as 'depletion of Heart blood' (*xinxie buzu* 心血不足), or 'gradually-dried-out Heart blood' (*xinxie ri he* 心血日涸) were also indicated.[30]

In the late Ming, *xinji* 心疾 (Heart disorders) was a colloquial term for labelling various physical/mental disorders in relation to insanity. Xu Wei 徐渭 (1521–93), a scholar renowned for his multi-faceted talents in poetry, prose, drama, painting and calligraphy, had committed several self-harms through his life. What made him more notorious would be his killing of his third wife for suspected infidelity. According to the Ming Codes, if a man catches his wife and her lover in the act of adultery, then he would be exonerated for murdering them. (*Ming lu jijie fu li* 明律集解附例 19.803, 1461, 1476). From the death penalty imposed upon Xu, he would have failed to provide any evidence for defending his crime. Nor did the Ming Codes, just as the Qing Codes did, stated clearly whether one's sentence can be reduced in a fit of madness (Chiu 1981: 75–94; Ng 1990: 136–7). In any case, Xu's death penalty was never executed. He was released instead after imprisonment for six years, because of the amnesty due to the succession of Emperor Wanli 萬曆 (r. 1572–1620). Despite the imperial leniency, Xu's contemporaries remained suspicious of his mental states when killing his wife. Some believed that Xu's misconduct was due to *kuangji* 狂疾 (*kuang* illness) and *bingkuang* 病狂 (affliction of *kuang*). Others, on the contrary, claimed that Xu initially feigned his 'madness' in order to escape from political implication, but that it soon developed beyond his control and became real. Interestingly, Xu never regarded himself mad. He only admitted that he had suffered from *xinji* 心疾 (Heart disorders) and affliction of *yi* 易 (change; turn) in terms of *sui* 祟 (demonic calamity) (Chen 2016: 112). This case shows that pre-modern Chinese might employ a somatic or cosmic reason for explaining some unreasonable acts in relation to insanity.

Emotional disorders

The roles of emotion in the medical analyses of madness should not be overlooked. The Chinese have recognised the importance of emotions in health and sickness since antiquity. The

Inner Classic sets out the theory of the five major human emotions, i.e. anger (*nu* 怒), joy (*xi* 喜), pensiveness (*si* 思), sorrow (*you* 憂) and fear (*kong* 恐). If any of the five emotions reaches an extreme state, the functions of its corresponding visceral organ, i.e. the Liver, the Heart, the Spleen, the Lungs and the Kidneys, will be injured, too.[31] This view explains emotions in a typical body-mind holism in early Chinese medicine.

The classic medical theory of five emotions seemed largely unchanged before mediaeval China. However, both medical theories and practices with regard to emotional disorders underwent drastic changes during the twelfth and the fourteenth centuries. Certain physicians expanded the category of the five emotions into seven by including grief and fright. An important example is Chen Yan 陳言 (fl. twelfth century), who stated clearly that 'Seven Emotions' (*qiqing* 七情) were the 'inner causes' of various disorders.[32] The term 'Seven Emotions' has become widely used since then. It remained an important category for many physicians in the nineteenth century: Fei Boxiong 費伯雄 (1800–79) was just one example.[33]

Parallel to the increasing significance of the 'Seven Emotions' was the emergence of a new category for classification of illnesses originating in emotional disturbance. In the late Ming, 'emotions' or 'emotional disorders' became popular terms in medical texts. Evidence can be found in medical works by Wu Kun 吳昆 (1551–?) and Zhang Jiebin 張介賓 (1563–1640).[34] One century later, the great imperial encyclopaedia, the *Synthesis of Past and Present Books and Illustrations* (*Gujin tushu jicheng* 古今圖書集成, 1723), again demonstrates a wide range of medical treatises on 'emotions' throughout the ages (*Gujin tushu jicheng yibu quanlu*, 321.2368–73). During this development of emotion theories, certain physicians believed that the dysfunction of the body, no matter whether due to obstruction of phlegm, excessive fire or blood depletion, was often reflected in emotional disorders. Furthermore, they redefined *dian* and *kuang* in terms of 'gloom of the seven emotions', 'contorted [emotions] with no means of smoothing out, and anger with no means of being released' (*qu wu suo shen nu wu suo xie* 屈無所伸怒無所泄), 'when one has far-reaching ambition but [events] do not accord to one's wish' (*zhiyuan gaoda er bu sui suo yu* 志願高大而不遂所欲), or 'tangled-up thought' (*silu yujie* 思慮鬱結), and so on.[35] Fright was also often attributed as the emotional cause of *xian*.[36]

Of the late Ming physicians, Wang Kentang was remarkable for contributing a systematic and comprehensive survey of madness. In the *Standard for Diagnosis and Treatment* (*Zhengzhi zhunsheng* 證治準繩, 1602), his newly invented category of 'States of Mind' (*qingzhi* 情志) integrated important passages from medical canons into one all-inclusive argument accompanied by his critical comments. Under this category are listed three sub-categories entitled *diankuangxian* 癲狂癎, *fanzao* 煩躁 and *jingjikung* 驚悸恐, in which a wide range of mental disorders or emotional symptoms are detailed. In addition to *dian*, *kuang* and *xian*, they also include *xufan* 虛煩 (weakness vexation), *zao* 躁 (agitation), *zhanwang* 譫妄 (delirious mania), *xixiao bu xiu* 喜笑不休 (gleeful unceasing laughter), *nu* 怒 (anger), *shan taixi* 善太息 (frequent sigh), *bei* 悲 (grief), *jing* 驚 (fright), *ji* 悸 (palpitation), *kung* 恐 (fear) and *jianwang* 健忘 (forgetfulness), and so on.[37]

Mostly following Sun Yikui's arguments, Wang Kentang also stated that the distinction between *dian*, *kuang* and *xian* was largely based on their peculiar symptoms. As he put it:

> One [who suffers from] *dian* is either frantic or stupid, singing or laughing, grieving or weeping, as if drunken and fixated; his speech is incoherent; he is unable to tell difference between dirty and tidy; it cannot be cured for many years. This is commonly called 'Heart Wind'. It is most likely to happen to people who have far-reaching aims yet cannot attain what they want. As for *kuang*, when the illness occurs, [one may become] wild and violent as if going mad due to Cold damage, of which [one's] *yangming* Vessels

appear enormous and replete. [One will] swear at people no matter intimate or remote, even ascending to heights and singing, casting off clothes and running about, climbing over walls and scrambling up on roofs, doing something beyond his normal capability; or, to tell people something [one] has never seen, as if [one] has been possessed by Evil. When the disease of *xian* comes, one becomes dizzy and loses consciousness; falling down on the ground due to vertigo; unable to tell which way is which. In severe cases [one] may even convulse and tremble; eyes looking upwards; or mouth and eyes being wry, or making noises like livestock do.[38]

To Wang, the principle for grouping these three disorders is somehow different. *Dian* obviously refers to a likely chronic illness along with the symptoms of uncontrolled emotional imbalance, deranged state of mind, incoherent speech and thought. It is sometimes caused by frustration of unfulfilled desires. The attack of *kuang*, by contrast, usually appears suddenly in the shape of excessive violence, extravagant behaviour and delusion. By contrast, *xian* is not identified with any emotion-oriented symptoms, but solely indicates convulsions, including involuntary contraction of muscles, violent motion of the limbs and loss of consciousness.[39] In this regard, he viewed *dian* disorder as originating in emotional/mental disturbance, unlike *kuang* and *xian*. This notion echoes the increasing importance at that time of emotional aetiologies in medical discourses in general.

With the increasing attention paid in medical circles to the role and function of emotions, not only men's but women's emotional states were highlighted in diagnostics. While some male doctors complained that women's emotional disorders were particularly hard to cure, others went even further to attribute certain kinds of madness exclusively to women. The gendered madness of women, to some extent, was often associated with their sexual frustration and/or reproductive ailments (Chen 2003: 149–99; 2011: 51–82; Simonis 2010: 367–406). As Chen Shiduo 陳士鐸 (*c*. 1621–1711) suggested, 'flower *dian*' (*huadian* 花顛), a kind of love madness, can only attack a woman who 'desires a man and thus generates evil in the heart/mind' (*sixiang qi ren er xin xie* 思想其人而心邪). When her madness results from yearning for a man whom she cannot attain, she may suddenly go mad and hug any man she meets 'regardless of any sense of shame' (*wang shi xiuchi* 罔識羞恥). Her Liver fire will thus burn profusely and her Liver pulse may bulge from (the locus of) her left wrist pulse.[40] She is 'merry when meeting men and angry when seeing women'. Sometimes, she will even transgress the bounds of social orders, 'taking off clothes and staying naked'.[41]

During the sixteenth and the eighteenth centuries, the new classifications of illnesses involved emotions and the mind, suggesting that physicians and medical writers appeared to give new significance to emotions. Assessments of emotional states became frequently highlighted in both medical theories and clinical encounters. Under this scope, the *dian* disorder was certainly more than 'falling sickness'. Women's flower *dian* (love madness) also showed a gendering aspect of *dian* madness. Therefore, little credence can be given to a recent viewpoint that 'the symptomatic distinction between *dian* and *kuang* as two distinct kinds of madness only stabilised in the 19th century' (Simonis 2014a: 604).

Medical treatments

In early and mediaeval times, the seizure of madness was often viewed as a result either of natural dysfunction or of supernatural influence. In response to the explanations of 'demonic attack', a religious ritual such as exorcism was often recommended. This kind of treatment was especially popular among magicians and religious healers. The excavated manuscript

Recipes for Fifty-two Disorders records two seemingly magico-medical therapies for treating *dian* disorder. These recipes, which use materials such as a white chicken; dog faeces; dog tails (possibly a sort of grass appearing as a dog tail) and grain growing on the wall of an animal pen, appear to be a sort of magical healing (*Wushi'er bing fang*, 58–9; Ma 1992: 425–8; Harper 1998: on the recipe 246–7, on magic 148–72). By contrast, scholarly physicians, who in the main were inspired by medical classics, tended to treat *dian* and *kuang* in a rather naturalistic way. When they diagnosed such diseases as being caused by any reason – whether *yin-yang* imbalance, wind attack, phlegm, fire or emotional disturbance – most of the treatments they recommended were herbal remedies, acupuncture and/or moxibustion.[42] In the late imperial period, most of the physicians would still prescribe formulæ to dispel phlegm, to repress fire and to nourish blood (of *yin*) for symptoms of *dian* and *kuang*. Only in very extreme cases of 'demonic affliction', some of them might suggest ritual therapies for a treatment (Chen 2019: 132–3).

In addition, physicians also developed unique emotional counter-therapies for treating illnesses caused by emotional imbalance. One sixteenth-century physician stated that 'when emotions become extreme, no drug can cure [the resulting disorders]; [they] must be overcome by emotions [themselves]' (*qingzhi guoji fei yao ke yu xu yi qing sheng* 情志過極 非藥可愈 須以情勝); this certainly signified a new perspective on therapy for emotional disorders (*Yifang kao*, 3.201; Sivin 1995: 1–19). This treatment by appeal to emotional manipulation and talking cures referred to the principle of 'overcoming one emotional state by another' (*yi qing sheng qing* 以情勝情), which originally derived from the old doctrine of the *Inner Classic*. The most skilful physician in such emotional counter-therapy, or 'therapy by counter-affect', seems to have been Zhang Congzheng 張從正 (1156–1228). His remarkable healing arts of emotional catharsis were followed by a number of successful cases in the late imperial period, and numerous records appear in medical texts and popular literature (*Rumen shiqin in Zihe yiji*, 3.109; Simonis 2010: 88; Chen 2014: 42–5). This method of ritual 'talking cures' which 'announce the cause' (*zhuyou* 祝由), a form of incantation that was often accompanied by religious talismans, could even be combined with herbal recipes in the treatment of *diankuang* patients. In the treatments of 'inner injury' (*neishang* 內傷), for example, Wu Tang 吳塘 (c. 1758–1836) believed that a doctor should firstly detail the origin of the illness and make this known to the patient, so that the latter can avoid having another fit. The doctor should also examine how the illness develops in meticulous detail, as well as scrutinise the hidden feelings of 'mentally exhausted men' (*laoren* 勞人) and 'pensive women' (*sifu* 思婦) (Chen 2014: 41–2). Wu also emphasised the importance to healing process of making patients change their point of view of themselves. In addition to drug prescriptions, he also used *zhuyou* to treat one of his patients suffering from *diankuang*, reinterpreting the practice of *zhuyou* from an 'incantation' against spirits, to an 'exhortation' remonstrating the patient (Simonis 2014b: 68–9).

In general, medical treatments with an appeal to emotional manipulation and communication techniques seemed to have been positively received. Nevertheless, these examples created no need for a special classification of mental or emotional illness in pre-modern China. In most cases, drug/herbal remedies and methods other than psychological counselling were more commonly used (Chen 2014: 51). Similar to psychosomatic and somatopsychic disorders in certain modern contexts, there is no split between mind and body or between what moderns might call somatic and psychological symptoms in pre-modern medicine (Kleinman 1980: 77–8; Sivin 1995: 17). In this regard, the therapies for *dian*, *kuang* and *xian* look no different from those for other disorders.

Conclusion

In early China, *dian* and *kuang* are usually explained with an appeal to theoretical notions in medical classics. The pathologies of *dian* and *kuang* were often discussed in homological terms – *yin-yang* imbalance, or *qi* deficiency. Among them, *kuang* is considered *yang*, while *dian* is *yin*. In contrast to *kuang*, sometimes *dian* is comparable or even confused with *xian*.

When Wind became an overwhelming factor to account for various disorders in mediaeval China, *dian*, *kuang* and *xian* were no exception. During the twelfth and fourteenth centuries, however, Chinese medical learning underwent an unprecedented development mostly because of the Jin-Yuan masters. Their works led to new explanatory modes for diseases – such as the combined aetiology of phlegm-fire – which gradually became applied not only to diseases in general but also to madness in particular.

Furthermore, an increasing attention was paid to the relationship between emotional influences and the emergence of madness. Not only women's unstable emotional states, but their potential sexual frustration, in particular, were highlighted by some male physicians, either in the symptoms of love madness (e.g. excessive hugging) or of erotic dreams and 'demonic foetuses'. Medical judgement of this kind was strongly supported by dominant social attitudes towards sex and sexuality. These all helped to shape the diverse aetiology and pathology of *dian* and *kuang* in the late imperial period. When physicians increasingly paid more attention to the seven emotions in aetiology, the category of wind malady gradually faded away and partly reformulated itself into the new discourses of *feng* madness particularly in legal discourses.

Of those physicians who had paid attention to the emotional origins of *dian* and *kuang* disorders, Wang Kentang deserves special mention since he grouped *dian*, *kuang*, *xian* and several emotional disorders into an unprecedented category, 'States of Mind', carefully distinguishing their differences based on a detailed examination of their pathology. At a glance, what Wang revealed might be promptly recognised as the evidence for pre-modern Chinese 'mental diseases', as some modern scholars often claim (Liao 1997: 247–8). But I should emphasise that Wang's (and perhaps most of his contemporaries as well) conceptions of these illnesses rarely signified a 'mental illness' alone in the mind-body holism, since their psychic syndrome usually came along with a somatic one. Nor did they help to produce conditions that warranted the existence of the speciality of 'psychiatry', 'mad-doctors' and 'asylums for the insane' in pre-modern China, as their European counterparts introduced before the nineteenth century. After all, Chinese physicians tended to treat *dian*, *kuang* and all kind of emotional disorders with herbal remedies, acupuncture, moxibustion, etc., throughout the ages. Only very few of them had attempted to apply the healing arts of emotional therapy and talking cures deriving from *the Inner Classics* and the Jin-Yuan masters. Apart from the absence of Western mind-body dualism, the lack of medical institutions exclusively dedicated to insane patients reflects the fact that pre-modern Chinese healthcare systems remained loosely organised up until the late imperial period.

In any case, the medical history of *dian* and *kuang* will help to explain the reasons why modern Chinese, to some extent, like to express their mental illness in physiological terms. In addition to psychiatric and medical anthropological approaches to mental illness, e.g. the so-called 'somatisation' syndrome and 'culture-bound' syndrome, there is certainly more to be learned from the past.

Notes

1. For example, Michel Foucault's *Histoire de la folie à l'âge classique* (1961) was translated into Chinese as *Gudian shidai fengkuang shi* 古典時代瘋狂史 (1998). But its more renowned English abridged version *Madness and Civilization* (1964) was published as *fengdian yu wenming* 瘋癲與文明 (1992).
2. *Huangdi neijing lingshu jizhu* 黃帝內經靈樞集注 (The Yellow Emperor's Inner Classic: Divine Pivot, with Annotation), 3.178-180.
3. *Huangdi neijing lingshu jizhu*, 3.178-180. *Huangdi neijing taisu* 黃帝內經太素 (Yellow Emperor's Inner Classic: Grand Basis), 30.589-91.
4. *Huangdi neijing suwen jizhu* 黃帝內經素問集注 (The Inner Canon of the Yellow Emperor: Basic Questions, with Annotation), 11.100. For a full English translation of *Suwen*, see Unschuld and Tessenow (2011).
5. Editor's note: 'Wood agent sounds' refers to sounds that correspond to wood in the system of the five agents water, wood, fire, earth and metal (see Chapter 1 in this volume).
6. *Huangdi neijing suwen* 黃帝內經素問 (The Yellow Emperor's Inner Classic: Basic Questions), 49.270-1; *Huangdi neijing suwen jizhu*, 30.121-2, 49.188.
7. *Shiji* 史記 (Records of the Historian), 105.2802.
8. For a full English translation of this work, see Unschuld (2016).
9. *Nanjing benyi* 難經本義 (Canon of Problems: the Original Meanings), 20.78, 59.138-9; *Nanjing yizhu* 難經譯注 (Canon of Problems: Translation and Annotation), 20.110-2, 59.271-3.
10. *Huangdi neijing suwen jizhu*, 47.181; *Huangdi neijing taisu*, 30.588.
11. In this paper, I capitalise terms such as Heart, Liver, Wind and others to indicate that these terms translate pre-modern Chinese concepts, and are not to be confused with common English usage.
12. *Huangdi neijing lingshu jizhu*, 3.175; *Huangdi neijing taisu*, 30.222.
13. *Huangdi neijing suwen jizhu*, 4.100 and 118, 5.173-174, 9.375.
14. *Huangdi neijing suwen*, 5.264, n.9-10; *Huangdi neijing suwen jizhu*, 5.182.
15. *Huangdi neijing lingshu jizhu*, 3.177-178.
16. *Xinjiaoben houhanshu* 新校本後漢書 (New Collated Edition of the East Han History), 82.2738. *Taiping guangji* 太平廣記 (Broad Records of the Great Peace), 218.429.
17. *Zhubing yuanhou lun* 諸病源候論 (Treatise on the Origins and Symptoms of Medical Disorders), 2.55-60, 63-4, 1289-90.
18. *Beiji qianjin yaofang* 備急千金要方 (Essential Prescriptions Worth a Thousand for Urgent Need), 5.63, 66-7, 203; *Waitai miyao* 外臺秘要 (Arcane Essentials from the Royal Library), 15.411-14; *Taiping shenghui fang* 太平聖惠方 (Imperial Grace Formulary of the Great Peace Reign), 85.613.
19. *Xiaoer yaozheng zhijue* 小兒藥證直訣 (Chants for Drug Therapy and Symptoms of Children), 上.11; *San yin ji yi bing zheng fang lun* 三因極一病證方論 (The Three Causes Epitomized and Unified: the Treatise on Diseases, Symptoms and Prescriptions), 9.15b-16b, 18a; *Xin dacheng yifang* 新大成醫方 (The New All-inclusive Collection of Medical Recipes), 3.7a-7b.
20. *Puji fang zhulu* 普濟方注錄 (Annotation on Prescriptions for General Aid); *Yuji weiyi* 玉機微義 (Esoteric Meanings of the Jade Secrets) in *Liu Chun yixue quanshu* 劉純醫學全書 (Liu Chun's Complete Book of Medicine), 41.392; *Gujin yitong daquan* 古今醫統大全 (Great Compendium of Medical Tradition, Past and Present), 10.566.
21. *Michuan Zhengzhi yaojue* 秘傳證治要訣 (Secretly Transmitted Essential Formulae for Diagnosis and Treatment), 9.118.
22. *Gujin yitong daquan*, 49.1404.
23. *Gujin yijian* 古今醫鑑 (The Medical Mirror, Past and Present), 7.216.
24. *Yizhi xuyu* 醫旨緒餘 (Supplements to the Medical Theme), 43.73; *Zhengzhi zhunsheng* 證治準繩 (Standard for Diagnosis and Treatment), vol.5.275.
25. *Danxi xinfa* 丹溪心法 (Zhu Danxi's Heart Methods), 4.359-61.
26. *Michuan zhengzhi yaojue*, 9.117.
27. *Yixue gangmu* 醫學綱目 (Outlines for Medicine), 11.184.
28. *Mingyi zhizhang* 明醫指掌 (A Handbook for Eminent Physicians): 7.175-6; *Yixue rumen* 醫學入門 (A Guide into Medicine), 4.919; *Jingyue quanshu* (The Complete Medical Works of Zhang Jiebin), 34.735.
29. *Gujin yitong daquan*, 49.1405; *Yixue rumen* 醫學入門 (A Guide into Medicine), 4.919.
30. *Yixue zhengzhuan* 醫學正傳 (The Medical Orthodoxy), 5.267; *Gujin yijian*, 7.216; *Yixue rumen*, 4.919-20; *Gujin yitong daquan*, 49.1405; *Yizhi xuyu*, 上.73; *Yixue liuyao* 醫學六要 (Six Essentials for Medicine), 7.884-5.

31 *Huangdi neijing suwen*, 2.37–42; 6.124–5; 19.374–85.
32 *San yin ji yi bing zheng fang lun* 三因極一病證方論 (The Three Causes Epitomised and Unified: the Treatise on Diseases, Symptoms and Prescriptions), 2.6b–7a.
33 *Yichun shengyi* 醫醇賸義 (The Supplementary Meanings of Medical Essence), 2.195–6.
34 *Yifang kao* 醫方考 (*Research on Medical Formulas*), 3.201–4; *Lei jing* 類經 (Classified Canon), 12.31a–37a.
35 *Michuan Zhengzhi yaojue*, 9.117–8; *Gujin yitong daquan*, 49.1405; *Yizhi xuyu*, 上.72; *Zhengzhi zhunsheng*, vol.5.303; *Jingyue quanshu* 景岳全書 (The Complete Medical Works of Zhang Jiebin), 34.735.
36 *Mingyi zhizhang* 明醫指掌 (A Handbook for Eminent Physicians), 7.175.
37 *Zhengzhi zhunsheng* 證治準繩 (Standard for Diagnosis and Treatment), vol.5.275, 303. *Yizhi xuyu*, 275.
38 *Zhengzhi zhunsheng*, vol.5.303.
39 Ibid.: 309, 311, 313.
40 The Liver pulse normally manifests in the middle of the three pulse positions at the left wrist.
41 *Shishi milu* 石室秘錄 (Secret Records of the Stone Chamber), 6.295; *Bianzheng qiwen* 辯證奇聞 (Strange Records of Diagnoses), 4.138.
42 *Huangdi neijing taisu* 黃帝內經太素 (Yellow Emperor's Inner Canon: Grand Basis), 13.222–4; *Huangdi neijing suwen jizhu* 黃帝內經素問集注 (The Inner Canon of the Yellow Emperor: Basic Questions, with Annotation), 5.177.

Bibliography

Pre-modern sources

Beiji qianjin yaofang 備急千金要方 (Essential Prescriptions Worth a Thousand for Urgent Need) 650, Sun Simiao 孫思邈 (1996), Beijing: Huaxia chubanshe.

Bianzheng qiwen 辯證奇聞 (Strange Records of Diagnoses) 1687, Chen Shiduo 陳士鐸 (1995), Beijing: Zhongguo zhongyiyao chubanshe.

Danxi xinfa 丹溪心法 (Zhu Danxi's Heart Methods) attr. to Zhu Zhenheng 朱震亨, 1481 edn, in *Zhejiangsheng zhongyiyao yanjiuyuan wenxian yanjiushi* 浙江省中医药研究院文献研究室 (eds) (2013), *Danxi yiji* 丹溪医集 (Medical Collection of Zhu Danxi), Beijing: Renmin weisheng chubanshe, pp.185–462.

Gezhi yulun 格致餘論 (The Supplementary Treatise on Investigating Things and Attaining Knowledge) 1347, Zhu Zhenheng 朱震亨, in *Zhejiangsheng zhongyiyao yanjiuyuan wenxian yanjiushi* 浙江省中医药研究院文献研究室 (eds) (2013), *Danxi yiji* 丹溪医集 (Medical Collection of Zhu Danxi), Beijing: Renmin weisheng chubanshe, pp. 3–48.

Gujin tushu jicheng yibu quanlu 古今圖書集成醫部全錄 (Past and Present Synthesis of Books and Illustrations: Complete Records of Medical Division) 1723, Chen Menglei 陳夢雷 (ed.) (1979), Taipei: Xinwenfeng chuban gongsi.

Gujin yijian 古今醫鑑 (The Medical Mirror, Past and Present) 1577, Gong Xin 龔信 and Gong Tingxian 龔廷賢 (1997), Beijing: Zhongguo zhongyiyao chubanshe.

Gujin yitong daquan 古今醫統大全 (Great Compendium of Medical Tradition, Past and Present) 1556, Xu Chunfu 徐春甫 (1996), Beijing: Renmin weisheng chubanshe.

Huangdi neijing lingshu jizhu 黃帝內經靈樞集注 (The Yellow Emperor's Inner Classic: Divine Pivot, with Annotation) 1672, Zhang Zhicong 張志聰 (ann.) (2002), in *Huangdi neijing jizhu* 黃帝內經集注, Hangzhou: Zhejiang guji chubanshe.

Huangdi neijing suwen jizhu 黃帝內經素問集注 (The Inner Canon of the Yellow Emperor: Basic Questions, with Annotation) 1669, Zhang Zhicong 張志聰 (ann.) (1983), Tainan: Wangjia chubanshe.

Huangdi neijing suwen 黃帝內經素問 (The Yellow Emperor's Inner Classic: Basic Questions), Wang Bing 王冰 (ed.) (1996), Beijing: Renmin weisheng chubanshe.

Huangdi neijing taisu 黃帝內經太素 (Yellow Emperor's Inner Classic: Grand Basis) 666–83, Yang Shangshan 楊上善 (ed.), Lanling Tang 蘭陵堂 edn, Xiao Yanping 蕭延平 (ann.) (2000), Tainan: Wangjia chubanshe.

Jingyue quanshu 景岳全書 (The Complete Medical Works of Zhang Jiebin) 1624, Zhang Jiebin 張介賓 (1997), Beijing: Renmin weisheng chubanshe.

Jinkui gouxuan 金匱鉤玄 (Enquiring the Mysteries of the Golden Casket), Zhu Zhenheng 朱震亨, Dai Sigong suppl. and collated, Shenxiu Hall 慎修堂 edn, in *Zhejiangsheng zhongyiyao yanjiuyuan wenxian yanjiushi* 浙江省中医药研究院文献研究室 (eds) (2013), *Danxi yiji* 丹溪医集 (Medical Collection of Zhu Danxi), Beijing: Renmin weisheng chubanshe, pp. 113–84.

Jinkui yaolue lunzhu 金匱要略論註 (The Medical Essentials in Golden Casket: Annotations) 1671, Zhang Ji 張機, Xu Bin 徐彬 (ann.) (1993), Beijing: Renmin weisheng chubanshe.

Lei jing 類經 (Classified Canon), Zhang Jiebin 張介賓 (1624), Taipei: National Library, Microfilm.

Michuan zhengzhi yaojue ji leifang 秘傳證治要訣及類方 (Secretly Transmitted Essential Formulae for Diagnosis and Treatment and Analogised Recipes) 1443, 1601 edn, Dai Yuanli 戴原禮 (1998), Beijing: Zhongguo zhongyiyao chubanshe.

Ming lu jijie fu li 明律集解附例 (Ming Legal Codes with Commentaries), Wu Yuannian 吳元年 (1969), Taipei: Chengwen chubanshe.

Mingshi 明史 (History of the Ming) 1678–1739, Zhang Tingyu 張廷玉 *et al.* (1974), Beijing: Zhonghua shuju.

Mingyi zhizhang 明醫指掌 (A Handbook for Eminent Physicians) 1556, Huangfu Zhong 皇甫中 (1999), Beijing: Zhongguo zhongyiyao chubanshe.

Nanjing benyi 難經本義 (Canon of Problems: The Original Meanings) 1361, Hua Shou 滑壽 (ann.) (1982), Taipei: Jiwen shuju.

Nanjing yizhu 難經譯注 (Canon of Problems: Translation and Annotation), Niu Bingzhan 牛兵占 (ann.) (2004) Beijing: Zhongyi guji chubanshe.

Puji fang zhulu 普濟方注錄 (Annotation on Prescriptions for General Aid) 1390, Zhu Shu 朱橚, Teng Shuo 滕碩 *et al.* (ann.) (1996), Harbin: Heilongjiang jishu chubanshe.

Rumen shiqin 儒門事親 (Serving the Parents in a Confucian's Family) 1228, Zhang Congzheng 張從正, in Deng Tietao 鄧鐵濤 *et al.* (1994), *Zihe yiji* 子和醫集 (Medical Collection of Zhang Zihe), Beijing: Renmin weisheng chubanshe, pp. 1–384.

San yin ji yi bing zheng fang lun 三因極一病證方論 (The Three Causes Epitomized and Unified: the Treatise on Diseases, Symptoms and Prescriptions) 1174, Chen Yan 陳言 (1991), Taipei: Tailian guofeng chubanshe.

Shiji 史記 (Records of the Historian), Sima Qian 司馬遷 (1981), Taipei: Dingwen shuju.

Shishi milu 石室秘錄 (Secret Records of the Stone Chamber) 1689, Chen Shiduo 陳士鐸 (1998), Beijing: Zhongguo zhongyiyao chubanshe.

Taiping guangji 太平廣記 (Broad Records of Great Peace) 978, Li Fang 李昉 (ed.) (1987), Shanghai: Shanghai guji chubanshe.

Taiping shenghui fang 太平聖惠方 (Imperial Grace Formulary of the Great Peace Reign) 992, Wang Huaiyin 王懷隱 (1982), Beijing: Renmin weisheng chubanshe.

Waitai miyao 外臺秘要 (Arcane Essentials from the Royal Library) 752, Wang Tao 王燾 (1955), Beijing: Renmin weisheng chubanshe.

Wushier bing fang 五十二病方 (Recipes for Fifty-Two Disorders), *Mawangdui hanmu boshu zhengli xiaozu* 馬王堆漢墓帛書整理小組 (The Group to Collate the Manuscripts from the Han Tombs at Manwangdui, or MWD Group) (ed.) (1979), Beijing: Wenwu chubanshe.

Xiaoer yaozheng zhijue 小兒藥證直訣 (Chants for Drug Therapy and Symptoms of Children) 1119, Qian Yi 錢乙 (1955), Beijing: Renmin weisheng chubanshe.

Xin dacheng yifang 新大成醫方 (The New All-inclusive Collection of Medical Recipes) 1267, facsimile of the Song edn, Wang Yuanfu 王元福 (1991), Taipei: National Central Library.

Xinjiaoben houhanshu 新校本後漢書 (New Collated Edition of the East Han History) the fifth century, Fan Ye 范曄 (1965), Beijing: Zhonghua shuju.

Yichun shengyi 醫醇賸義 (The Supplementary Meanings of Medical Essence) 19th century, Fei Boxiong 費伯雄, in Cao Bingzhang 曹炳章 (ed.) (2000), *Zhongguo yixue dacheng xuji* 中國醫學大成續集 (The Great Collection of Chinese Medical Works: Sequel), vol. 29, Shanghai: Shanghai kexue jishu chubanshe.

Yifang kao 醫方考 (*Research on Medical Formulas*) 1584, 1586 ed., Wu Kun 吳昆 (1985), Jiangsu: Jiangsu kexue jishu chubanshe.

Yixue gangmu 醫學綱目 (Outlines for Medicine) 1396, Lou Ying 樓英 (1996), Beijing: Zhongguo zhongyiyao chubanshe.

Yixue liuyao 醫學六要 (Six Essentials for Medicine) 1609, Zhang Sanxi 張三錫 (2005), Shanghai: Shanghai kexue jishu chubanshe.

Yixue rumen 醫學入門 (A Guide into Medicine) 1575, Li Yan 李梴 (1999), Tianjin: Tianjin kexue jishu chubanshe.

Yixue zhengzhuan 醫學正傳 (The Medical Orthodoxy) 1515, Yu Chuan 虞摶 (1981), Beijing: Renmin weisheng chubanshe.
Yizhi xuyu 醫旨緒餘 (Supplements to the Medical Theme) 1573, Sun Yikui 孫一奎, (1985), Jiangsu: Jiangsu kexue jishu chubanshe.
Yuji weiyi 玉機微義 (Esoteric Meanings of the Jade Secrets) 1396, Xu Yongcheng 徐用誠 and Liu Chun 劉純, 1439 edn, in Jiang Dianhua 姜典華 (ed.) (1999), *Liu Chun yixue quanshu* 劉純醫學全書 (Liu Chun's Complete Book of Medicine), Beijing: Zhongguo zhongyiyao chubanshe.
Zhengzhi zhunsheng 證治準繩 (Standard for Diagnosis and Treatment) 1602, Wang Kentang 王肯堂 (1979), Taipei: Xinwenfeng chuban youxian gongsi.
Zhubing yuanhou lun 諸病源候論 (Treatise on the Origins and Symptoms of Medical Disorders) 610, Chao Yuanfang 巢元方 (1996), Beijing: Renmin weisheng chubanshe.

Modern sources

Baum, E. (2018) *The Invention of Madness: state, society, and the insane in modern China*, Chicago, IL: University of Chicago Press.
Chen, Hsiu-fen (2003) *Medicine, Society, and the Making of Madness in Imperial China*, PhD thesis, School of Oriental and African Studies, University of London.
——— (2005) 'Wind malady as madness in medieval China: some threads from the Dunhuang medical manuscript', in V. Lo and C. Cullen (eds) *Medieval Chinese Medicine: the Dunhuang medical manuscripts*, London: Routledge Curzon, pp. 345–62.
——— (2011) 'Between passion and repression: medical views of demon dreams, demonic fetuses, and female sexual madness in late imperial China', *Late Imperial China*, 32.1: 51–82.
——— (2014) 'Emotional therapy and talking cures in late imperial China', in H. Chiang (ed.) *Psychiatry and Chinese History*, London: Pickering and Chatto, pp. 37–54.
——— (2016) '"Zhenduan" Xu Wei: Wan ming dui yu kuang yu bing de duoyuan lijie 「診斷」徐渭：晚明社會對於狂與病的多元理解 ('Diagnosing' Xu Wei: pluralistic views of madness and illness in late Ming society), *Journal of Ming Studies*, 27 (Dec.): 71–121.
——— (2019) 'Ghostly encounters', in H. Chiang (ed.) *The Making of the Human Sciences in China: historical and conceptual foundations*, Leiden: Brill, pp. 124–41.
Chiu, M.L. (1981) 'Insanity in imperial China: a legal case study', in A.M. Kleinman and T.Y. Lin (eds) *Normal and Abnormal Behavior in Chinese Culture*, Dordrecht: D. Reidel, pp. 75–94.
Foucault, M. (1964) *Madness and Civilization: a history of insanity in the Age of Reason*, trans. R. Howard, New York: Vintage Books.
——— (1992) *Fengdian yu wenming* 瘋癲與文明 (Madness and Civilization: a history of insanity in the age of reason), trans. Liu Beicheng 劉北成 and Yang Yuanying 楊遠嬰, Taipei: Laurel Publishing.
——— (1998) *Gudian shidai fengkuang shi* 古典時代瘋狂史 (Histoire de la folie à l'âge Classique; The History of Madness in the Classic Age), trans. Lin Zhiming 林志明, Taipei: China Times Publishing.
Gabbiani, L. (2013) 'Insanity and parricides in late imperial China (eighteenth – twentieth centuries)', *International Journal of Asian Studies*, 10.2: 115–41.
Harper, D.J. (1998) *Early Chinese Medical Literature*, London; New York: Kegan Paul International.
Kleinman, A. (1980) *Patients and Healers in the Context of Culture: an exploration of the borderland between anthropology, medicine, and psychiatry*, Berkeley: University of California Press.
Leung, A.K.C. (1999) 'Mafeng bing gainian yanbian de lishi' 麻風病概念演變的歷史 (The evolution history of the concept of *Mafeng Bing*), *Zhongyang yanjiuyuan lishi yuyan yanjiusuo jikan* 中央研究院歷史語言研究所集刊 (Journal of the Institute of History and Philology, Academia Sinica), 70.2: 399–438.
Liao Yuqun 廖育群 (1997) 'The history of psychiatric diagnosis in traditional Chinese medicine', in Y. Kawakita, S. Sakai and Y. Otsuka (eds) *History of Psychiatric Diagnoses*, Tokyo: Ishiyaku Euro-America, pp. 239–51.
Ma Jixing 馬繼興 (1992) *Mawangdui guyishu kaoshi* 馬王堆古醫書考釋 (Ancient Medical Works Excavated from the Mawangdui Han Tombs: Exegesis and Annotation), Changsha: Hunan keji chubanshe.
Mathews, R.H. (1996) *The Mathews' Chinese-English Dictionary*, compiled for the China Inland Mission and first published in Shanghai in 1931, Cambridge, MA: Harvard University Press.

Messner, A.C. (2000) *Medizinische Diskurse zu Irreisein in China (1600–1930)*, Stuttgart: Franz Steiner Verlag.
Ng, V.W. (1990) *Madness in Late Imperial China: from illness to deviance*, Norman: University of Oklahoma Press.
Scull, A. (2015) *Madness in Civilization: a cultural history of insanity from the Bible to Freud, from the madhouse to modern medicine*, London: Thames and Hudson.
Shapiro, H. (1995) *The View from a Chinese asylum: Defining madness in 1930s Peking*, PhD thesis, Harvard University.
Simonis, F. (2010) *Mad Act, Mad Speech, and Mad People in Late Imperial Chinese Law and Medicine*, PhD thesis, Princeton University.
——— (2014a) 'Ghosts or mucus? Medicine for madness: new doctrines, therapies, and rivalries', in J. Lagerwey (ed.) *Modern Chinese Religion Part I: Song, Liao, Jin and Yuan (960–1368)*, Leiden: Brill, pp. 603–39.
——— (2014b) 'Medicaments and persuasion: medical therapies for madness in nineteenth-century China', in H. Chiang (ed.) *Psychiatry and Chinese History*, London: Pickering and Chatto, pp. 55–70.
Sivin, N. (1995), 'Emotional counter-therapy', in N. Sivin (ed.) *Medicine, Philosophy, and Religion in Ancient China*, Hampshire: Variorium, Part II, pp. 1–19.
Unschuld, P.U. (2016) *Huang Di Nei Jing Ling Shu: the ancient classic on needle therapy*, Berkeley: University of California Press.
Unschuld, P.U. and Tessenow, H. (2011) *Huang Di Nei Jing Su Wen: annotated translation of Huang Di's inner classic – basic questions*, Berkeley: University of California Press.
Yu Yan 余巖 (1953) *Gudai jibing minghou shuyi* 古代疾病名候疏義 (The Commentaries on Names and Symptoms of Ancient Diseases), Beijing: Renmin weisheng chubanshe.

16
LATE IMPERIAL EPIDEMIOLOGY, PART 1
From retrospective diagnosis to epidemics as diagnostic lens for other ends, 1870s to 1970s

Marta Hanson

The beginning of the 1640s marked a precipitous fall into chaos. In the wake of rebellions (Parsons 1970) and war, locusts and famine, floods and epidemics, husbands left their wives, parents abandoned their children, and worse. Under the 'Omens and Anomalies' (*xiang yi* 祥異) section of the *Gazetteer of Tongxiang County*, the scholar Chen Qide 陳其德 provided an account of the local situation in 1641 titled 'Record of Disasters and Famines' (*Zaihuang jishi* 災荒紀事). He wrote, 'If not from war, one died from famine. If not from famine, one died from epidemics'.[1] Responses to epidemics around this time varied by sectors of society. The government sometimes set up centres to distribute medicines, using the model of grain stations for famine relief (Will 1990) that harkens back to Mencius who argued that blaming poor harvests rather than governance for famine was no different from blaming the weapon rather than its bearer for killing someone (Lau 1970: 52). Some temples organised ceremonies every three years to appease the epidemic gods (Schipper 1985) or carried out community rituals to expel epidemic-carrying demons (Katz 1995). During the waning years of the Ming dynasty, local elites responded to long-term imperial neglect of actual medical relief during epidemics by increasing privately supported efforts at the local level (Leung 1987: 139, 1997). Officials continued to tabulate epidemics across the country (Imura 1936–37), while some local scholars recorded what they observed in more detail.

Chen further wrote, for example, that 'previous famines could usually be counted by the prefecture or by the province, [but] this [one] severely affected not only Zhejiang and the southern capital [Nanjing], but also as far north as the capital [Beijing] and as far south as Guangdong'. Chen proposed two causes for this catastrophe. The first appealed to a concept of divine population control: 'Is it that Heaven created too many people, so it spread this fierce scourge'? The second blamed human greed: 'Or is it because the people have already consumed so much that Heaven hates and discards them to such a degree'. Human indulgence, Chen concluded, had caused the people to stray from the correct way – the ultimate cause of the famine and epidemics of 1641. If only they could 'always regard rice as they do pearls, and wood as they do cassia' he pondered, 'then the bad qualities in their hearts would never develop'.[2]

Famine and war were the only environmental elements Chen recognised as contributing to the rise of epidemics; he did not mention poor living conditions, polluted water, the weakened constitutions of the impoverished, or the greater susceptibility of one group of people compared to another. Chen's explanation and solution remain within a conventional cosmo-moralistic framework in which Heaven sends floods, droughts, famines, and epidemics as justified divine retribution for excessive human indulgence and corrupt court politics (Hanson 1997: 117–22).

A man known only as Mr. Shen also wrote about the early 1640s epidemics in Zhejiang province.[3] Instead of the broader political-moral context, he focused on the steady decline in the local economy and social order in Gui'an County from 1640 up through the autumn of 1643.[4] First, there was flooding from a serious downpour in July, whereupon 'hoodlums' (i.e. refugees) banded together in groups of three to five demanding food. The local economy shut down as neighbourhoods discontinued markets and villagers closed their doors. Finally, a new County Magistrate captured the leader of the bandits (i.e. refugees) and calmed the people. However, the fields were still too wet to plant new rice sprouts until the end of July. During this time, the cost of a 'picul' (dan 石) of rice, or about 60 kilograms, reached 1 tael 5–6 copper cash, the equivalent of just over 2 ounces of silver.[5] The following year [1641], the village of Huangmei suffered a drought. It did not rain that spring or summer until July 16th. Only one to two tenths of the total crop of rice was planted, but the shoots planted after the July rain did not survive the cold of the fall and an early frost. A picul of rice jumped to more than twice the price of the previous year, to 3 tael 5–6 copper cash, almost 5 ounces (4.7–4.9 oz.) of silver.[6] Shen recorded that at this point people began to die of starvation or resort to cannibalism to survive.

A big snowstorm on the first day of the next year [1642] was taken initially as a good omen for the next year's harvest. Instead, a major epidemic swept through the region. Shen estimated that by the summer solstice, the epidemic had affected eight of ten households and sometimes wiped out entire families. Meanwhile, the government made the lives of the people worse by refusing to give them tax relief. Although in 1643 crops were successfully planted, when the farmers were about to plant their second crop, a massive flood turned the leveed areas into ponds. The gazetteer entry directly following Shen's account records that the famine in that year was so severe that people again resorted to cannibalism. In the spring of 1644, another major epidemic struck similar to the one two years before, but this time with victims vomiting blood before they died.

In contrast to Chen's moral-political interpretation, Shen located the cause of the epidemics in severe famines, worsened by excessive government taxes. As was the convention, he used the price of rice to gauge the severity of the famine and attributed these disasters to taxation, on the one hand, and crop failure due to vicissitudes of nature, on the other. His concluding questions – 'Who will pay the state taxes this year? And how will the debts be paid?' – reflect the primary concerns of the rural elite whose views he represented.

In the autumn of 1642 in neighbouring Jiangsu province, the physician Wu Youxing 吳有性 (c. 1582–1652) substantiated Chen Qide's account of the same year: that the epidemic had spread throughout the country both north and south. In the preface to his *Treatise on Febrile Epidemics* (*Wenyi lun* 瘟疫論, 1642), Wu wrote that the epidemics worsened in the fifth and sixth lunar months of 1641, until '[everyone in] entire households infected each other'.[7] The epidemic the previous year had swept through Shandong, Zhejiang, Hebei, Northern Zhili (Beijing region), and Southern Zhili (Nanjing region). Wu, however, did not speculate about moral or cosmological causes like Chen or political or economic causes like Shen. Rather, he sought a natural cause in the environment.

Wu also did not look back to the 1580s and 1630s epidemics across China, which had probably contributed to the crisis about which he was writing. Instead, he criticised conventional medical treatments:

During the initial onset, fashionable practitioners erroneously used Cold Damage methods to treat the disorder. I never saw a case of theirs that did not get worse. Some patients and their families mistakenly heeded the claims that by the seventh or fourteenth day it would heal itself. Because of this they were not treated. Some died from not being treated in time, or not taking the medicine in time. Others wrongly took drastic formulas, and by not following the normal sequence for attacking and then replenishing, died.

(Hanson 2011: 91)[8]

If some doctors were too aggressive, Wu thought others were too timid, using mild slow-working drugs for acute symptoms, prolonging the suffering. These inadequacies prompted him to attack the focus on climate that dominated traditional Chinese epidemiology up to that time with an argument that deviant qi (戾氣), pestilential qi (liqi 癘氣), and other non-climatic factors caused these epidemics (Qi 1981; Xiao 1987). Starting from the first sentence of the preface to his *Treatise*, Wu moved from the traditional medical view of correspondence between seasonal cycles and human illness to his new view of a specific kind of *anomalous qi* (*zaqi* 雜氣) separate from normal changes in seasonal *qi*.

The pathology of Warm epidemics is not that of Wind, Cold, Summer-Heat, or Damp [*qi*]. Rather it is stimulated by a type of anomalous *qi* in Nature [lit. 'Heaven and Earth,' *tiandi* 天地]. There are nine stages of transmission; these are the critical junctures for treating epidemic disorders. Why is it that from antiquity to the present no one has ever discovered (*faming*) this?

(Hanson 2011: 94–5)

Wu then argued that while his medical colleagues thought they were seeing cases of 'Cold Damage' (*shanghan* 傷寒), in fact, they were mostly dealing with what he called epidemics of 'Warm diseases' (*wenbing* 溫病). Warm disease is an umbrella term comparable to the modern-day concept of acute febrile diseases in that it encompassed a range of symptoms and epidemiological phenomenon, such as an illness spreading broadly and indiscriminately throughout a large population, that today would fall under acute infectious diseases. This Chinese disease concept, however, was first defined during the Han dynasty (202 BCE–220 CE) within a *configurationist* perspective that emphasised that something in the environment had become pathogenic.

Generally, this perspective means that something in the environment – a configuration of cold or hot air, weather, climate, mists, etc. – caused the outbreak, thus accounting for the fact that many people became sick at the same time with comparable symptoms. In classical Chinese medicine, the external causes were seasonal pathogenic factors such as Cold or Hot seasonal *qi* (Hanson 2011: 16–17). This is in contrast to a *contaminationist* perspective that emphasises human-to-human transmission via some kind of pathogen or contaminant. Both perspectives depended on a third *predispositionist* perspective that explained why some people do not become sick, others do but recover, and still others perish (Rosenberg 1992a: 195–6).

Because literate Chinese physicians were trained to think in terms of macro-microcosm relationships, one scholar has argued that they were more weather-conscious than contagion-conscious (Kuriyama 2000), though conceptions of contagion certainly circulated among the populace at the same time and had a complex history within classical Chinese medicine as well (Leung 2010; Lu 2021, Chapter 14 in this volume). Nonetheless, while the characteristic febrile symptoms of what were called Warm-factor epidemics (*wenyi* 瘟疫) did not change much over time, interpretations of their etiology, best therapeutic interventions, and appropriate institutional responses changed dramatically from the pre-modern period up through the 2002–2003 SARS epidemic (Hanson 2010).

What was the disease?

Although there is no question that widespread epidemics occurred in the early 1640s, the question about what caused them remains. From Chen's moral argument and Shen's political-economic analysis to medical debates about whether the epidemics were actually Cold Damage or Warm diseases or due to pathogenic climatic *qi* or as yet unidentified anomalous *qi* not related to the weather, it is clear that contemporaries had wide-ranging opinions on the subject. Over 400 years later, although medical historians concur that the end-of-Ming epidemics contributed to the Ming's fall (Zhao 2004), they still remain divided on what caused them, albeit on very different conceptual foundations and in the context of the broader epidemiology of infectious diseases in the Ming-Qing period (Fan Ka-wai *et al.* 2005). While one historian concluded that their cause was a mystery (Elvin 1973), others argued that mostly bubonic plague, and in some causes even pneumonic plague, was definitely the main cause (Cao 1997; Cao and Li 2006), though others noted that dysentery, typhoid, and malaria were also present (Mei and Yan 1996). Still others insist that bubonic plague cannot be conclusively determined in a context within which not only were other diseases occurring simultaneously but also at a time when historical actors' concepts of disease did not neatly fit one-to-one correspondences with modern-day disease concepts (Dunstan 1975; Benedict 1996a, 1996b, 1996c; Hanson 1997, 2011).

So what caused the late Ming epidemics? The present chapter on late imperial epidemiology in China, Part 1, uses this question as a heuristic device to write a broader historiography of epidemics in China and traditional Chinese epidemiology up through the 1970s. First, I present two diametrically opposite approaches to the history of disease that have emerged since the 1970s in the medical history practised in European-American institutions. One scholar has coined the terms 'naturalist-realist' and 'historicalist-conceptualist' to represent the two sides of the broader historical debate (Wilson 2000). Since these two approaches have also largely shaped the medical historiography of China from the late nineteenth century up to the present, next I sketch a historiography of the naturalist-realist method of retrospective epidemiology that dominated treatments of the historical evidence of epidemics in China from the 1870s to the 1930s. Third, I classify scholarship on epidemics in China between the 1940s and 1970s into two disciplinary trajectories: on the one hand, economic and social historians largely concerned about epidemics with respect to demographic transformations in the past and, on the other hand, anthropologists focused on current social and religious responses to epidemics. Despite the different questions scholars sought to answer based on extant records of past epidemics, or their observations of contemporary Chinese responses to them, both types of scholars nonetheless crafted answers to their questions within a shared natural-realist framework.

Two approaches to the history of disease

Answering the question of what caused the late Ming epidemics thus depends on how one approaches them. Should we cast out Chen's human greed and Shen's excessive taxation along with Wu's anomalous *qi* in favour of a list of modern biomedical disease categories that could have caused these epidemics? To do so requires drawing retrospective diagnoses, based on modern disease concepts, from historical sources on epidemics recorded within very different linguistic realms and cultural frames. Here, the distinction in cultural anthropology between emic and etic viewpoints is instructive: the emic view is the insiders' perspectives, how subjects perceive things, and so whatever distinctions are meaningful within their society; the etic view is the outsiders' perspectives, how observers analyse things in another culture, and so whatever analytical concepts they find meaningful to interpret any given society. The emic view, for example, would value the responses of Chen, Shen, and Wu within their historical milieus; the etic view, by contrast, would use the sources they wrote as evidence to make retrospective diagnoses of the cause of the epidemics according to modern disease concepts. Both approaches are valuable.

Historians have dealt with this fundamental contradiction between past and present interpretations of disease experience in different ways. The medical historian Adrian Wilson summarised the two abovementioned opposing approaches to the history of disease: the *naturalist-realist* and the *historicalist-conceptualist* (Ibid.). Efforts to make retrospective diagnoses of past disease experience based on present understanding exemplify what Wilson termed the naturalist-realist approach. What he designated as the historicalist-conceptualist approach, by contrast, represents scholarship that takes disease concepts themselves – past and present – as objects of historical analysis. From here on, we will refer to this approach more simply as 'historical-conceptual' and borrow from another scholar's shorthand reference to practitioners of the former as 'realists' and of the latter as 'historicists' (Packard 2016).

The realists take the modern disease concept as equivalent to 'natural reality' and thereby seek to connect historical sources about past disease experiences with the present-day modern consensus on disease etiologies. Historicists, by contrast, start from the premise that all knowledge (medical as well as scientific and humanistic) is socially constituted, socially maintained, and dynamic. Knowledge about diseases therefore changes over time depending on the social consensus that surrounds the phenomena the disease concepts refer to at any given time and place, and constitutes the most reliable kind of information about them. The naturalist-realist position also considers scientific and medical knowledge as transcending social context, whereas the historical-conceptual position assumes that even the most stable consensus on somatic disease concepts today have complex socially embedded histories. The naturalist-realist perspective considers that the only relevant past meanings of historical disease concepts are those that fit into a clear trajectory progressing towards a present definition. The historical-conceptual perspective, by contrast, examines the consensus-building process that stabilised past (as well as current) disease concepts. The historicist thus considers historical disease concepts relevant on their own terms with their own histories and logic separate from possible modern correlates (Duffin 2005; Packard 2016).

Randall Packard (2016) argued in his article '"Break-Bone" Fever in Philadelphia, 1780: Reflections on the History of Disease' that, although both approaches have been the basis of a rift within the medical history profession in the West since the 1970s, each approach can make distinct contributions to the history of disease. The influential social historian of American medicine, Charles Rosenberg, famously used the nineteenth-century cholera epidemics in the US as a heuristic device to illuminate major social, political, religious, and economic transformations (Rosenberg 1962). His essay 'Toward an Ecology of Knowledge' moved the field even more towards the sociology of medicine and science (Ibid. 1979). Just over twenty years later, his co-edited volume *Framing Disease* established the position that historicising disease concepts can be just as illuminating of present-day disease concepts that have been conventionally understood within what Wilson later termed the naturalist-realist position (one which Wilson equally considered a cultural construction) by situating them as emergent and changing as well within specific places, institutions, and time frames (Rosenberg and Golden 1992; Packard 2016).

As one of Rosenberg's students, I was also inspired by another one of my advisors who wrote the essay 'Topics for Research in Ch'ing History' (Naquin and Rawski 1987) and raised epidemics as an important new topic to pursue. It was from this socio-intellectual milieu that my dissertation 'Inventing a Tradition in Chinese Medicine' emerged (Hanson 1997). I also integrated into my analysis of Chinese disease concepts for febrile diseases, however, Andrew Cunningham's critique of retrospective epidemiology: the modern laboratory had so fundamentally changed the identity of infectious diseases, such as plague, that any one-to-one correspondence between modern and past disease concepts was not only futile but even worse, distorted the past (Cunningham 1992). David Harley followed suit along these lines in his article on 'Rhetoric and the Social Construction of Sickness and Healing' (Harley 1999). Taking the opposite position, Roger Cooter argued in '"Framing" the End of the Social History of Medicine' for what was lost in the shift from the social to cultural history of disease through the 'framing disease' metaphor especially related to pragmatic research on medical disparities and justice issues (Cooter 2004).

Since 2007, however, Rosenberg's 'Biographies of Disease' series (Johns Hopkins University Press monographs) fully represents both sides of this academic spectrum. Speaking of his own book in the series, Packard noted that *The Making of a Tropical Disease: a short history of malaria* (Packard 2007) took the *naturalist-realist* position arguing that malaria had existed for centuries as a distinct biological entity, whereas Steve Peitzman's *Dropsy, Dialysis, Transplant: a short history of failing kidneys* (Peitzman 2007) traces a historical-conceptual history of disease concepts on kidney failure. In 2011, I positioned my first book, *Speaking of Epidemics in Chinese Medicine*, firmly on the *historical-conceptual* side of the spectrum (Hanson 2011). In the same year, the paleopathologist and physical anthropologist Piers Mitchell published an influential essay that acknowledged the methodological limitations of a retrospective epidemiology that relied on completely different disease concepts from the past. Yet, he also argued that this approach is viable when researchers seek not to equate current disease concepts uncritically with past ones but rather seek to understand the micro-organism itself, how it spread, and who it infected through careful use of historical sources and archeological evidence (Mitchell 2011: 81–8; Packard 2016: 200).

Later, I published an article on the *naturalist-realist* approach to late imperial Chinese epidemiology through a history of the first fifty years of Western medical maps of diseases in China (Hanson 2017). I neither embraced the *naturalist-realist* position in analysing these disease maps nor applied Mitchell's recommendations for studying the transmission paths of micro-organisms that these same disease maps were often intended to visualise. Rather, I attempted both to situate the maps within the socio-historical milieu of the authors and publications that produced them and to interpret their range of rhetorical functions and how these changed over time revealing new political regimes and intended audiences.

Inspired by the scholars synthesising both sides of the realist-historicist divide in this chapter, I argue, however, that when done judiciously both approaches to past disease concepts can be productive for understanding late imperial Chinese epidemiology. Each perspective illuminates very different dimensions of Chinese medical history. *Historical-conceptual* methods are most useful for understanding the social, cultural, and religious contexts within which historical actors' categories and their own disease concepts made sense. However, *natural-realist* methods, when used judiciously, can be productive for answering questions about political-economic, ecological, and demographic transformations.

Packard wrote about his own work on diseases in colonial and postcolonial Africa, for example, that he 'employed the history of disease as a way of illuminating the complex ecological relationships that link social, economic, and biological processes together to produce disease states' (Packard 2016: 200). Taking a natural-realist approach also allows historians to follow the historical movement of a disease entity across boundaries by controlling for characteristic symptoms, such as the buboes of plague, despite multiple and multivalent disease concepts within one culture and even more variation cross-culturally (Hymes 2014). The modern genetic determination of the *Yersinia pestis* in combination with a scientific analysis of the remains of plague victims, for example, has opened new pathways of historical and scientific collaboration on the global transmission of plague (Green 2014).

In terms of the historiography of late imperial epidemiology in China, authors applied the *naturalist-realist* method of retrospective epidemiology well before they applied the *historical-conceptual* approach to analysing historical actors' categories on their own terms. Part I of this essay thus reviews the history of the former, whereas Part II addresses the history of the latter.

Historiography of retrospective epidemiology in China

Western physicians began to define the 'Diseases of China' from the mid-1800s according to their own medical training in the newly emergent laboratory-based approach to disease – the method that has most informed the *natural-realist* approach. Some notes on epidemics of this period can be found in the letters of the British official Robert Hart (1835–1911) while he was Inspector General of the Imperial Maritime Customs Service (IMCS) from 1868 to 1907 (Fairbank *et al.* 1975). He used the new infrastructure of the Customs Service as a clearinghouse for medical reports of disease among foreigners and Chinese alike across China. Some of the earliest examples of the *naturalist-realist* approach to interpreting epidemics can therefore be found in the forty years of *Medical Reports* in the *Customs Gazette* published from August 1871 to 1911 (Gordon 1884). For the first time, European and American physicians based in China and Japan had a central place to publish their medical observations without limitations in length. The Scottish physician John Dudgeon (1837–1901) based in Beijing and Tianjin during the last forty years of his life wrote several of the earliest medical reports between 1871 and 1875 (Dudgeon 1871a, 1871b, 1872a, 1872b, 1875). He also wrote the first book focused on *The Diseases of China: their causes, conditions, and prevalence contrasted with those of Europe* (Dudgeon 1877). This book illustrates well a transition period when British physicians and their Chinese counterparts had more in common conceptually than divided them, especially regarding disease etiology (Rogaski 2004). Dudgeon even argued, for instance, that the Chinese had a more moderate and healthier diet that Europeans should consider adopting to remedy their dietary and drinking excesses (Li 2010a).

In the late 1870s and early 1880s, the deepening Western medical consensus on disease classification (nosology), new developments in laboratory science in the direction of isolating causative agents, and related developments in medical statistics helped strengthen a *naturalrealist* interpretation of China's disease concepts (Hanson 2011: 151–2; 2017). Attempts to understand the epidemics spreading across Yunnan in the 1870s particularly exemplify the earliest natural-realist lens through which Western observers filtered their observations of China's epidemics. Even before Alexandre Émile Jean Yersin (1863–1943) for the first time identified under a microscope the specific *bacillus* that caused plague in his field laboratory during the 1894 bubonic plague epidemic in Hong Kong, a French official of the Imperial Chinese Maritime Customs, Émile Rocher, wrote, 'Notes sur la Peste au Yün-nan' about his observations of epidemics of the 1870s during the Muslim-led multiethnic Panthay rebellion (1856–1873) in Yunnan (Rocher 1879).

Recognising the significance of Rocher's account for bubonic plague studies, Sir Patrick Manson (1844–1922), considered the founder of tropical medicine based on research he largely did while living in China (Li 2002, 2004, 2012, 2018), translated into English most of Rocher's 'Notes sur la Peste' in his contribution to the *Medical Reports*. Manson also included the first map of the 1871–73 plague epidemics in Yunnan (Manson 1879); a coloured version of this map also appears in Rocher's book published the following year. What local Chinese called *Yangzi bing* 痒子病 – a disease characterised by severe itching of skin or *pruritus* – both Rocher and Manson, by this time, confidently identified as bubonic plague. Their consensus that the 1870s Yunnan epidemics were one thing – namely, bubonic plague even without laboratory confirmation – also contributed to their ability to map them. The following fifty years of Western maps of the diseases of China ranging from plague and cholera to beriberi and apoplexy (stroke) further represent this *natural-realist* approach (Hanson 2017).

The first edition of *The Diseases of China, Including Formosa and Korea*, for example, exemplified the scientific transformations in medicine over the turn of the nineteenth century and included national maps of the major diseases of the period according to Western disease classification (Jefferys and Maxwell 1910). Although the second edition no longer included maps, it brought the first edition up to date with new developments in the *natural-realist* understanding of disease concepts (Jefferys 1928). Another early representative of the *natural-realist* approach from the 1930s is in Wong and Wu's summary of China's historical epidemiology of plague during the 'Period 1894–99' in chapter ten of their *History of Chinese Medicine: being a chronicle of medical happenings in China from ancient times to the present period* (Wong and Wu 1936: 506–37).

Titled 'Describing (a) the spread of plague in China leading to the Great Outbreak at Canton and Hong Kong in 1894 and (b) further consolidation of medical efforts', the authors synthesised the primary sources they relied on to sketch an early history of plague in China. Noting that 'the chroniclers do not differentiate between the diseases met with', they first make the following clarification: 'Treating the subject with our present knowledge of epidemics in China, we can rule out a number of these pestilences as having nothing to do with true plague – especially those following in the wake of war, famine, floods and other catastrophes, which were in all probability typhus and relapsing fever – disease apparently rampant in China from the earliest time'. To determine what were true plague cases, they then followed two criteria: 'A smaller group of outbreaks on the other hand, seem to have been plague visitations, either because they occurred simultaneously with an established plague pandemic (e.g. the Black Death) or because they took place in regions where afterwards the existence of frequent plague epidemics or even endemicity was established (e.g. Mongolia, Shansi)' (Wong and Wu 1936: 506–7).

Because Yunnan no longer had plague by the time they wrote their history of Chinese medicine, Wong and Wu acknowledged the problem of using these earlier sources on the Yunnan epidemics for their history of plague. Because Yunnan was one of the 'regions now entirely free from plague but in the past suspected to have been endemic centres', they justified focusing on the region 'to reconstruct the course of events leading to the 1894 outbreak at Canton and Hongkong' (Wong and Wu 1936: 507). They thereby relied on the earliest accounts by Rocher and Manson discussed above in their retrospective epidemiology of plague in China. Wu Lien-teh's other publications on cholera in 1934 (Wu *et al*. 1934), plague in 1936 (Wu 1936a, 1936b), and his autobiography of 1959 similarly represent this *natural-realist* approach to the history of disease in China (Ibid., 1959). Wu was also the first to propose that the even earlier late-Yuan epidemics (1344–45, 1356–60, 1362) were plague, despite scarce evidence (Brook 2013: 65).

Also in the 1930s, the Japanese scholar Imura Kôzen published tabulations of epidemic diseases listed in local gazetteers for the Ming dynasty (Imura 1936–37). He thereby established a solid primary source foundation for carrying out retrospective epidemiology as Wong and Wu and their predecessors had exemplified. Because his tabulations preserved all the original disease concepts as they were recorded in the often very terse entries in the local gazetteers, he also made valuable source material more easily accessible for scholars who sought to understand traditional Chinese epidemiology on its own terms using historical-conceptual methods.

Functionalist anthropology and natural-realist demography

In the 1940s to 1960s, one sees a related divergence in how scholars interpreted the history of epidemics in China. On the one side, the anthropologist Francis Hsu wrote a functionalist interpretation of the medical and religious responses that townspeople in Yunnan province took to cholera epidemics in 1942 (Hsu 1952). At that time, Hsu could assume the identity of the epidemics as cholera (natural-realist) in his otherwise more contextual (albeit not historical) analysis of religious and scientific responses to the epidemic. Following what had become by then a well-established *natural-realist* line of reasoning, historians of China also started to refer to epidemics in their narratives of major demographic transformations during the first to eighth centuries (Bielenstein 1947) and the mid-fourteenth century through mid-twentieth century (Ho 1959).

Within China, researchers published *A Glossary on the Names and Symptoms of Ancient Disease* (Yu 1953) and *The Intellectual History of Preventive Medicine in China* (Fan 1955). From a *naturalist-realist* perspective, both scholars attempted to make one-to-one translations of historical disease concepts in line with modern ones, stripping away older meanings in the process. Yet because of the Chinese primary source material provided, their books can be used to understand late imperial Chinese epidemiology from either side of the spectrum. Another Chinese historian using the *naturalist-realist* perspective focused on *The Study of Contagious Diseases in China* (Shi 1956) and thus considered the late Ming physician, Wu Youxing quoted above, as a Chinese expert of contagious disease whose 'deviant qi doctrine' was comparable to modern germ theory (Ibid. 1957). Wu's deviant qi concept, though, was based within a *configurationist* not *contaminationist* perspective and so incommensurable with germ theory. Furthermore, all three examples of 1950s Chinese scholarship on late imperial epidemiology are themselves best understood as scholarly responses to the early PRC politicising of public health, disease prevention, and control of communicable diseases (Scheid 2002: 67–72; Rogaski 2004: 285–7).

Developing these 1950s Chinese precedents, Joseph Needham and Lu Gwei-djen published an article on 'Hygiene and Preventive Medicine in Ancient China' (Needham and Lu 1962) and a chapter on 'Records of Diseases in Ancient China' for a book on *Diseases of Antiquity* (Ibid. 1967). Both publications also fall on the *naturalist-realist* side of the spectrum, yet also provide useful primary source material for the *historical-conceptual* side. Meanwhile, the Song historian Robert Hartwell integrated records of epidemics and related demographic trends as essential dimensions of his analyses of social and economic transformations during the Song (Hartwell 1967, 1982).

An outlier in this 1960s scholarship, but no less influential, historian Carl Nathan published a book on *Plague Prevention and Politics in Manchuria, 1910–1931* (Nathan 1967). This book set the foundation for further analyses of the relationship between public health, epidemics, and politics through the example of Penang-born, Cambridge-University-educated, bacteriologist, and self-described 'Plague Fighter', Wu Lien-teh, whom we have already met as an author. Wu had established China's first public health system in response to the Manchurian pneumonic plague epidemic of 1910–11, a historical thread scholars have picked up again since the early 1990s (Benedict 1993; Flohr 1996; Fisher 1995/1996; Gamsa 2006; Lei 2010; Summers 2012; Lynteris 2016). He is also credited with having co-authored the first English *History of Chinese Medicine* published in 1936 (Wong and Wu 1936; Luesink 2009), though by that time they could rely on a foundation of earlier histories of Chinese medicine in Chinese (Xie 1935).

In the 1970s, new developments in ecological and medical history inspired historians of China to turn their attention to what caused the late Ming epidemics. In an ecological reinterpretation of late imperial Chinese history, Mark Elvin succinctly summarised the end-of-Ming epidemiological crisis as follows: 'In 1586–89 and 1639–44 China suffered from the two most widespread and lethal epidemics in her recorded history, although their medical nature remains a complete mystery' (Elvin 1973: 310). Based on Imura's data (1936–37), Elvin sketched the disaster's scope: ninety-two prefectures and counties across thirteen provinces were affected in the 1588–1592 epidemic; seventy-nine prefectures and counties across ten provinces were affected in 1641, the worst year of the 1640s epidemics; of the densely populated provinces, only Sichuan and Guangdong escaped.

Inspired by Elvin, Helen Dunstan attempted to identify the types of epidemic diseases that spread episodically across China during the last six decades of the Ming following the *natural-realist* approach of retrospective diagnosis (Dunstan 1975). Although she questioned whether it was possible to make a positive diagnosis, she nonetheless endeavoured to make one-to-one correspondences between Chinese disease concepts such as *wenyi* 瘟疫 ('febrile epidemics'), *wenbing* 溫病 ('warm disease'), and *shanghan* 傷寒 ('cold damage') and modern Western disease concepts, such as bubonic plague, meningitis, and typhoid. She also expressed scepticism that this was a viable approach with the following question: 'Making the perhaps rather large assumption that all these references from the early seventeenth century to the twentieth are to the actual disease entity, is it possible to arrive at any diagnosis?' (Ibid.: 26). Dunstan was the first historian to begin to analyse Wu Youxing's *Treatise on Febrile Epidemics* (1642) through a more historical-conceptual lens in order to explore the main issues in traditional Chinese epidemiology.

Meanwhile, Francis Hsu, the anthropologist who had previously examined what people actually did in response to the 1942 cholera epidemics in Yunnan province (Hsu 1952), was by 1975 taking notes during a plague prevention ritual that had been held every ten years since the 1870s in Shatin, Guangdong province. Hsu found that the analytic separation between religion and science then current in academic anthropology did not exist in practice. As late as the mid-twentieth century, Chinese responded equally in religious and scientific ways to protect themselves from cholera in 1942 and prevent plague in 1975 (Ibid. 1983). Hsu's functionalist analyses of how ordinary Chinese responded to cholera provided a model for studying epidemics anthropologically comparable to William McNeill's demonstration of the demographic impact of epidemics as he integrated the role of infectious disease as a key player in shaping world history in *Plagues and Peoples* (McNeill 1976). McNeill unified the ecological and cultural dimensions of human experience from the perspective of ecological determinism that supported, for example, his argument that infectious diseases (what he called 'microparasites') greatly facilitated one civilisation (i.e. 'macroparasite') conquering another, as he argued was the case with the European conquest of the Americas. Alfred Crosby developed this ecological determinism further in his *The Columbia Exchange* (Crosby 1972) and *Ecological Imperialism* (Crosby 1986).

With respect to China, McNeill also adopted Wu Lien-teh's earlier claim that plague was the cause of the late Yuan epidemics to support his argument that the Mongolian steppe was the origin of fourteenth-century's Black Death in Europe (McNeill 1976: 132–75; Brook 2013: 65). In one version of McNeill's hypothesis, the Mongol passage through Yunnan in the 1250s was crucial for the plague's transmission to Europe (McNeill 1976: 143–45; Hymes 2014: 287). The 'China origin of plague' hypothesis immediately engaged historians in debate, starting with Michael Dols' demonstration that even Middle Eastern accounts of plague assumed that it came from the East (Dols 1977). Norris took issue with this view (Norris 1977, 1978; Dols 1978). Both sides of this debate nonetheless remained within the *naturalist-realist* epistemology in that neither raised issues with the method of making retrospective diagnoses of plague for past epidemics with a more complex multi-causal epidemiology.

As for historians of China, Denis Twitchett evaluated the effect of epidemics on the population in his analysis of the uneven records for historical demography during the Tang dynasty (Twitchett 1979). While Dunstan could rely on local gazetteers of the Ming for her analysis of sixteenth- through seventeenth-century epidemics across large geographic distances, there were no comparable Tang local gazetteers. Instead, Twitchett had to rely on contemporary registered population records along with the Tang dynastic history's *Treatise on the Five Agents* (*Wuxing zhi* 五行志), where epidemics were listed along with other natural disasters. He used these sources to interpret fluctuations in the population as indicative of one of three things: varying administrative efficacy, changing methods of registration by the state (which were highly variable over time), and disasters of war, famine, and pestilence. In some cases, he argued that it was difficult to tell whether the decreased records of disasters, including epidemics in extant sources, reflected accurate records of fewer natural or manmade disasters or rather problems in the Tang administration's capacity to collect data about their occurrence. He concluded that this discrepancy in the historical records was most likely a combination of both factors. Within a decade, Chinese historical demographers integrated the then current knowledge of China's historical epidemiology with a broad historical overview of China's demography into *The History of Population in China* (Zhao and Xie 1988). This historical synthesis of Chinese demography may be productively understood within the broader 1980s socio-political context of the one-child policy, which was first introduced in 1979 as a means to control population growth in China (Greenhalgh 2008).

Conclusion

In the century following the 1870s, when lab-based bacteriology began to transform concepts of infectious disease in a way that improved tracking their transmission, to the 1970s, when records of epidemics themselves became useful for tracking other types of change, the *natural-realist* approach to late imperial Chinese epidemiology developed and came to dominate the interpretation of historical evidence as well as contemporary experience of epidemics. Even before germ theory transformed laboratory medicine, during the Panthay Rebellion of the 1870s in Yunnan, the diplomat Émile Rocher and English physician Patrick Manson had already agreed on a disease concept of plague that was based mostly on clinical symptoms, and that could facilitate determining its etiology and transmission pathways. Over the next forty years, academics and scientists such as Manson, Western physicians writing 'Medical Reports' for China's *Customs Gazette*, and the co-editors Jefferys and Maxwell of *The Diseases of China* (1910), also made some of the earliest attempts to translate the more multivalent Chinese disease concepts into the more narrowly defined Western medical equivalents.

For the next thirty years, the self-styled 'Plague Fighter', Wu Lien-teh, dominated scholarship on late imperial epidemiology. He had had a key role in first controlling the Manchurian pneumonic plague epidemic in 1910–11 and in managing cholera during the 1930s when he was the Director of the National Quarantine Service, helping track the transmission paths of the plague and cholera epidemics he was charged with controlling, his *natural-realist* perspective facilitated the first retrospective historical epidemiology of plague stretching back to the Yuan dynasty. This interpretation later became influential through McNeill's contested argument about the Chinese origin of the Black Death.

In the next thirty years from the early 1940s to the late 1960s, scholars began to study the history of epidemics in China as a tool for yet other ends. Hsu leveraged his analyses of a contemporary cholera epidemic to challenge the false separation of science and religion in anthropology. The historians Bielenstein, Ho, and Hartwell integrated historical records of epidemics to reveal other social, economic, and demographic transformations in Chinese society. Carl Nathan used the Manchurian plague epidemic as a lens to see how national political crises and global geopolitics intersected in the opening decade of the twentieth century.

In the 1970s, scholarship on late imperial Chinese epidemiology started to become more sceptical about making one-to-one translations between traditional Chinese and modern disease concepts. Following Elvin's doubts about being able to identify the late Ming epidemics based on primary sources, Dunston questioned whether this was even a viable approach. Nevertheless, she suggested that the epidemics were most likely a combination of plague, typhoid, dysentery, and meningitis. Tang records of epidemics were no more illuminating for equating ancient with modern disease concepts, but Twitchett used them effectively to speculate about varying administrative capacities, changing methods of state registration, and fluctuations in population.

Just as many modern-day scholars have used historical records of epidemics as a tool for other ends so too did some pre-modern Chinese commentators. The two accounts of the late Ming epidemics that began this chapter – Chen Qide's political-moral assessment and Mr. Shen's social-economic analysis – were rhetorically comparable to what anthropologists and historians began to do from the 1940s through the late 1960s. Both Chen and Shen, for instance, found the late Ming epidemics of their lifetimes to be useful means to illuminate what they perceived to be the underlying social, moral, political, and economic factors, fissures, and failures that gave rise to them.

Even physician Wu Youxing's criticism of how his predecessors and contemporaries defined febrile epidemics (*wenyi*) as Cold Damage (*shanghan*) rather than Warm diseases (*wenbing*) is rhetorically analogous to the *natural-realist* efforts to match past Chinese disease concepts with new biomedical interpretations. The latter translation strategy characterised one of the most important dimensions of the history of late imperial Chinese epidemiology just over two centuries later from the 1870s through the 1970s. Part II picks up the baton in the 1980s when the *historicist-conceptual* approach to the history of disease and epidemics started to develop within the scholarship of medical historians. This new approach shifted historians' attention to changing meanings of Chinese disease concepts and the therapeutic responses they legitimated both on their own terms and within unique histories deeply embedded within Chinese history. Many of these same disease concepts persist in present-day Chinese medical practices, namely because healers as well as patients continue to find them useful to frame illness experience and determine appropriate therapeutic responses.

Notes

1 *Zaihuang jishi* 災荒紀事 (Record of Disasters and Famines) by Chen Qide 陳其德 in *Tongxiangxian zhi* 桐鄉縣志 (Gazetteer of Tongxiang County), 20.9a.
2 For evidence of blaming human moral failures for widespread epidemics in Daoist thought of the late Han to early medieval period, see ch. 2 on 'Demonology and Epidemiology' in Strickmann (2002). For evidence of this in European medical history, see Rosenberg (1992a, 1992b).
3 For Mr. Shen's account of the epidemics in context of broader decline of the economy and social order from 1640 to 1643, see *Guianxian zhi*, 27.16b–17a.
4 For a punctuated version of this text, see appendix in Chen (1958: 289–91). Primary text in *Guianxian zhi*, 27.16b–17a.
5 A picul of rice equalled about 133.3 pounds or 60.4 kilograms. A tael of silver equalled about 1.3 ounces or 37.6 grams. One copper cash equalled a string of 100 copper coins. Although from about 1500 to 1645 the exchange ratio of copper coin to tael fluctuated between 500 and 750, the ratio skyrocketed after 1640 when in Beijing, and especially the Yangzi Delta, it rose above 1,000 to 2,000 copper coins per tael by 1645. For 1640, the exchange ration was 800 copper cash/tael (von Glahn 1996: 160).
6 This figure is also calculated on the 800 copper coins/silver tael exchange ratio with 1 tael to 1.3 ounces.
7 This quotation and those following by Wu Youxing come from his preface to *Wenyi lun*. See *Wenyi lun pingzhu*, 2; or *Wenbingxue quanshu*, 981.
8 This and the following translation come from Wu Youxing's preface to the *Wenyi lun*. See *Wenyi lun ping zhu*, 1–2; and *Wenbingxue quanshu*, 981.

Bibliography

Pre-modern sources

Guianxian zhi 歸安縣志 (Gazetteer of Guian County) 1882, Li Yixiu 李昱修 and Lu Xinyuan 陸心源 (eds) (1970), Taipei: Chengwen chubanshe. For Mr. Shen's account of the epidemics in context of broader decline of the economy and social order from 1640–1643, *juan* 27, 16b–17a.
Tongxiangxian zhi 桐鄉縣志 (Gazetteer of Tongxiang County) 1887, Yan Chen 嚴辰 (ed.) (1970), Taipei: Chengwen chubanshe.
Wenbingxue quanshu 溫病學全書 (Complete Works on the Studies of Warm Disorders), Li Shunbao 李順保 (ed.) (2002), 2 vols., Beijing: Xueyuan chubanshe.
Wenyi lun 瘟疫論 *Treatise on Febrile Epidemics*, 1642, Wu Youxing 吳有性 (c. 1582–1652), Critical edn. by Meng Shujiang 孟淑江 and Yang Jin 楊進. Beijing: Remin weisheng chubanshe, 1990.
Wenyi lun ping zhu 瘟疫論評注 (Critical Notes and Annotations on the 'Treatise on Febrile Epidemics'), Liu Fangzhou 劉方舟 (ed.) (1709), first reprint (1977), second reprint (1985), Beijing: Renmin weisheng chubanshe.
Zaihuang jishi 災荒紀事 (Record of Disasters and Famines) 1641, Chen Qide 陳其德 in *Tongxiangxian zhi* 桐鄉縣志 (Gazetteer of Tongxiang County) 1887, Yan Chen 嚴辰 (ed.) (1970), Taipei: Chengwen chubanshe, *juan* 20, 8a–9a.
Zaihuang youji 災荒又記 (Another Record of Disasters and Famines) 1642, Chen Qide 陳其德 in *Tongxiangxian zhi* 桐鄉縣志 (Gazetteer of Tongxiang County) 1887, Yan Chen 嚴辰 (ed.) (1970), Taipei: Chengwen chubanshe, *juan* 20, 9b–10a.

Modern sources

Bauer, W. (ed.) (1979) *Studia Sino-Mongolica: Festschrift für Herbert Franke*, Wiesbaden: Franz Steiner.
Benedict, C. (1988) 'Bubonic plague in nineteenth-century China', *Modern China*, 14.2: 107–55.
—— (1993) 'Policing the sick: plague and the origins of state medicine in late imperial China', *Late Imperial China*, 14.2: 60–77.
—— (1996a) *Bubonic Plague in Nineteenth-Century China*, Stanford, CA: Stanford University Press.
—— (1996b) 'Framing plague in China's past', in G. Hershatter *et al.* (eds) *Remapping China: fissures in historical terrain*, Stanford, CA: Stanford University Press, pp. 27–41.

—— (1996c) 'Epidemiology and history: an ecological approach to the history of plague in Qing China', *Chinese Environmental History Newsletter*, 3.1: 6–11.
Bielenstein, H. (1947) 'The census of China during the period 2–742 A.D.', *Bulletin of the Museum of Far Eastern Asia*, 19: 125–63.
Boroway, I. (ed.) (2009) *Uneasy Encounters: the politics of medicine and health in China 1900–1937*, Frankfurt: Peter Lang.
Brook, T. (2013) *The Troubled Empire: China in the Yuan and Ming dynasties*, Cambridge, MA: Belknap Press of Harvard University Press.
Brothwell, D. and Sandison, A.T. (eds) (1967) *Diseases in Antiquity: a survey of the diseases, injuries, and surgery of early populations*, Springfield: Thomas.
Cao Shuji 曹树基 (1997) 'Shuyi liuxing yu Huabei shehui de bianqian' 鼠疫流行与华北社会的变迁 (1580–1644) (Plague epidemics and social transformations in North China) *Lishi yanjiu* 历史研究 (Historical Research), 1: 17–32.
Cao Shuji 曹树基 and Li Yushang 李玉尚 (2006) *Shuyi: zhanzheng yu heping: Zhongguo de huanjing yu shehui bianqian, 1230–1960* 鼠疫: 战争与和平: 中国的环境 与社会变迁 (1230–1960) (Plague: war and peace: environmental and social transformations in China, 1230–1960), Ji'nan: Zhongguo huabao chubanshe.
Chen Hengli 陳恆力 (1958) *Bu nongshu yanjiu* 補農書研究 (Research on the 'Supplement to the Treatise on Agriculture'), Beijing: Nongye chubanshe.
Conrad, L.I. and Wujastyk, D. (eds) (2000) *Contagion: perspectives from pre-modern societies*, Aldershot: Ashgate.
Cooter, R. (2004) '"Framing" the end of the social history of medicine', in F. Huisman and J.H. Warner (eds) *Locating Medical History: the stories and their meanings*, Baltimore, MD: Johns Hopkins University Press, pp. 309–37.
Crombie, A.C. (ed.) (1963: 2nd edn 1977) *Scientific Change*, London: Heineman.
Crosby, A. (1972) *The Columbia Exchange: biological and cultural consequences of 1492*, Contribution to American Studies, no. 2, Westport, CT: Greenwood Press.
—— (1986) *Ecological Imperialism: the biological expansion of Europe, 900–1900*, Cambridge: Cambridge University Press.
Cunningham, A. (1992) 'Transforming plague: the laboratory and the identity of infectious disease', in A. Cunningham and P. Williams (eds) *The Laboratory Revolution in Medicine*, Cambridge: Cambridge University Press, pp. 209–44.
Cunningham, A. and Williams, P. (eds) *The Laboratory Revolution in Medicine*, Cambridge: Cambridge University Press.
Dols, M. (1977) *The Black Death in the Middle East*, Princeton, NJ: Princeton University Press.
—— (1978) 'Geographical origin of the Black Death: comment', *Bulletin of the History of Medicine*, 52.1: 112–13.
Dudgeon, J. (1871a) 'Report on the health of Peking for the half year ended 31st March, 1871', *Customs Gazette, Medical Reports*, No. 1, April–June, 6–15, Shanghai: The Customs Press. Chinese Imperial Maritime Customs Service (1871–1911) *Customs Gazette, Medical Reports*, https://www.hathitrust.org/, accessed 27/3/2020.
—— (1871b) 'Report on the physical conditions of Peking and the habits of the Pekinese as bearing upon health, first part', *Customs Gazette, Medical Reports*, No. 2, July–September, 73–82, Shanghai: The Customs Press. Chinese Imperial Maritime Customs Service (1871–1911) *Customs Gazette, Medical Reports*, https://www.hathitrust.org/, accessed 27/3/2020.
—— (1872a) 'Report on the health of Peking for the half year ended 30th September, 1871', *Customs Gazette, Medical Reports*, No. 3, January–March, 7–9, Shanghai: The Customs Press. Chinese Imperial Maritime Customs Service (1871–1911) *Customs Gazette, Medical Reports*, https://www.hathitrust.org/, accessed 27/3/2020.
—— (1872b) 'Report on the physical conditions of Peking and the habits of the Pekinese as bearing upon health, second part', *Customs Gazette, M* Chinese Imperial Maritime Customs Service (1871–1911) *Customs Gazette, Medical Reports*, https://www.hathitrust.org/, accessed 27/3/2020. *Medical Reports*, No. 4, April–September, Shanghai: The Customs Press, pp. 29–42.
—— (1875) 'Report on the health of Peking for the half year ended 31st March, 1875', *Customs Gazette, Medical Reports*, No. 9, October–March, 34–44, Shanghai: The Customs Press. Chinese Imperial Maritime Customs Service (1871–1911) *Customs Gazette, Medical Reports*, https://www.hathitrust.org/, accessed 27/3/2020.

—— (1877) *The Diseases of China: their causes, conditions, and prevalence contrasted with those of Europe*, Glasgow: Dunn and Wright. Chinese Imperial Maritime Customs Service (1871–1911) *Customs Gazette, Medical Reports*, https://www.hathitrust.org/, accessed 27/3/2020.

Duffin, J. (2005) *Lovers and Livers: disease concepts in history*, the 2002 Joanne Goodman lectures, Toronto, ON: University of Toronto Press.

Dunstan, H. (1975) 'The late Ming epidemics: a preliminary survey', *Ch'ing-shih wen-t'i*, 3.3: 1–59.

Elvin, M. (1973) *The Pattern of the Chinese Past: a social and economic interpretation*, Stanford, CA: Stanford University Press.

Fairbank, J.K., Bruner, K.F. and Matheson, E.M. (eds) (1975) *The I.G. in Peking: letters of Robert Hart Chinese maritime customs 1868–1907*, Cambridge, MA: The Belknap Press of Harvard University Press.

Fan Ka-wai, Yu Xinzhong, Cheung Hok-ming and Lao Sze-nga (2005) 'Studies on Ming dynasty infectious diseases', *Ming Qing Yanjiu*, Napoli, Italy: 133–50.

Fan Xingzhun 范行準 (1955) *Zhongguo yufang yixue sixiang shi* 中國預防醫學思想史 (The Intellectual History of Preventive Medicine in China), Beijing: Renmin weisheng chubanshe.

Fisher, C.T. (1995/1996) 'Bubonic plague in modern China: an overview', *Journal of the Oriental Society of Australia*, 27 and 28: 57–104.

Flohr, C. (1996) 'The plague fighter: Wu Lien-teh and the beginning of the Chinese public health system', *Annals of Science*, 53: 361–80.

Gamsa, M. (2006) 'The epidemic of pneumonic plague in Manchuria 1910–1911', *Past and Present*, 190 (February): 147–83.

Gordon, C.A. (1884) *An Epitome of the Reports of the Medical Officers to the Chinese Imperial Maritime Customs Service, from 1871 to 1882: with chapters on the history of medicine in China; materia medica; epidemics; famine; ethnology; and chronology in relation to medicine and public health*, London: Bailliere.

Green, M.H. (ed.) (2014) *Pandemic Disease in the Medieval World: Rethinking the Black Death*, TMG Occasional Volume 1, Kalamazoo; Bradford: Arc Medieval Press.

Greenhalgh, S. (2008) *Just One Child: science and policy in Deng's China*, Berkeley: University of California Press.

Hanson, M. (1997) *Inventing a Tradition in Chinese Medicine: from universal canon to local medical knowledge in south China, the seventeenth to the nineteenth century*, PhD thesis, Department of the History and Sociology of Science, University of Pennsylvania.

—— (2001) 'Robust northerners and delicate southerners: the nineteenth-century invention of a southern *wenbing* tradition', in E. Hsu (ed.) *Innovation in Chinese Medicine*, Cambridge: Cambridge University Press, pp. 262–92.

—— (2010) 'Conceptual blind spots, media blindfolds: the case of SARS and traditional Chinese medicine', in A. Leung and C. Furth (eds) *Health and Hygiene in Chinese East Asia: publics and policies in the long twentieth century*, Chapel Hill, NC: Duke University Press, pp. 369–410.

—— (2011) *Speaking of Epidemics in Chinese Medicine: disease and the geographic imagination*, London: Routledge Press.

—— (2017) 'Visualizing the geography of the diseases of China: Western disease maps from analytical tools to tools of empire, sovereignty, and public health propaganda, 1878–1929', *Science in Context*, 30.3: 219–80.

Harley, D. (1999) 'Rhetoric and the social construction of sickness and healing', *Social History of Medicine*, 12: 407–35.

Hartwell, R. (1967) 'A cycle of economic change in imperial China: coal and iron in northeast China, 750–1350', *Journal of Economic and Social History of the Orient*, 10: 102–59.

—— (1982) 'Demographic, political and social transformations of China, 750–1550', *Harvard Journal of Asiatic Studies*, 42.2: 365–442.

Ho Ping-ti (1959) *Studies in the Population of China 1368–1953*, Cambridge, MA: Harvard University Press.

Hsu, E. (ed.) (2001) *Innovation in Chinese Medicine*, Cambridge: Cambridge University Press.

Hsu, F.L. (1952; 2nd edn 1973) *Religion, Science, and Human Crises: a study of China in transition and its implications for the West*, London: Routledge and Kegan Paul; repr. Westport, CT: Greenwood Press.

—— (1983) *Exorcising the Trouble Makers: magic, science, and culture*, contributions to the Study of Religion, No. 11, Westport, CT: Greenwood Press.

Huisman, F. and Warner, J.H. (eds) (2004) *Locating Medical History: the stories and their meanings*, Baltimore, MD: Johns Hopkins University Press.

Hymes, R. (2014) 'Epilogue: a hypothesis on the East Asian beginnings of the Yersinia pestis polytomy', in M.H. Green (ed.) *Pandemic Disease in the Medieval World: rethinking the Black Death*, Kalamazoo; Bradford: Arc Medieval Press, pp. 285–308.

Imura Kôzen 井村哮全 (1936–37) 'Chihôshi ni kisaiseraretaru Chûgoku ekirei ryakkô', 地方志に記載せられたる中國疫癘略考 (Brief Study of Records of Chinese Epidemics in Local Histories), 8 parts, *Chûgai iji shimpô* 中外醫事新報 (*New Journal of Chinese and Foreign Medicine*), no. 1232, pp. 5–17; no. 1233, pp. 12–21; no. 1234, pp. 14–24; no. 1235, pp. 20–28; no. 1236, pp. 25–33; no. 1237, pp. 25–28; no. 1238, pp. 28–33; no. 1239, pp. 30–39.

Jefferys, W.H. (1928) *The Diseases of China, including Formosa and Korea*, 2nd edn, Shanghai: A.B.C. Press.

Jefferys, W.H. and Maxwell, J.L. (1910) *The Diseases of China, including Formosa and Korea*, Philadelphia, PA: P. Blakiston's Son.

Katz, P. (1995) *Demon Hordes and Burning Boats: the cult of Marshal Wen in late imperial Chekiang*, New York: State University of New York Press.

Kim, Yung Sik and Bray, F. (eds) (1999) *Current Perspectives in the History of Science in East Asia*, Seoul: Seoul National University Press.

Kiple, K.F. (ed.) (1993) *The Cambridge World History of Human Disease*, Cambridge: Cambridge University Press.

Kuriyama, S. (1993) 'Concepts of disease in East Asia', in K.F. Kiple (ed.) *The Cambridge World History of Human Disease*, Cambridge: Cambridge University Press, pp. 52–9.

——— (2000) 'Epidemics, weather, and contagion in traditional Chinese medicine', in L.I. Conrad and D. Wujastyk (eds) *Contagion: perspectives from pre-modern societies*, Aldershot: Ashgate, pp. 3–22.

Lau, D.C. (trans) (1970) *Mencius*, Harmondsworth: Penguin.

Lei Hsiang-lin S. (2010) 'Sovereignty and the microscope: constituting notifiable infectious disease and containing the pneumonic plague in Manchuria', in A. Leung and C. Furth (eds) *Health and Hygiene in Modern Chinese East Asia: policies and publics in the long twentieth century*, Durham, NC: Duke University Press, pp. 73–106.

Leung, Angela Ki Che [Liang Qizi 梁其姿] (1987) 'Organized medicine in Ming-Qing China: state and private medical institutions in the lower Yangzi region' *Late Imperial China*, 8.1: 134–66.

——— (1997) *Shishan yu jiaohua: Ming Qing de cishan zuzhi* 施善與教化:明清的慈善組織 (Bestow Charity and Transform via Education: charitable institutions in the Ming and Qing), Taipei: Lianjing chuban shiye gongsi.

——— (2010) 'Evolution of the idea of *chuanran* contagion in imperial China', in A. Leung and C. Furth (eds) *Health and Hygiene in Chinese East Asia: policies and publics in the long twentieth century*, Durham, NC: Duke University Press, pp. 25–50.

Leung, A. and Furth, C. (eds) (2010) *Health and Hygiene in Chinese East Asia: policies and publics in the long twentieth century*, Durham, NC: Duke University Press.

Li Shang-Jen 李尚仁 (2002) 'Natural history of parasitic disease: Patrick Manson's philosophical method', *Isis*, 93.2: 206–28.

——— (2004) 'The nurse of parasites: gender concepts in Patrick Manson's parasitological research', *Journal of the History of Biology*, 37.1: 103–30.

——— (2010a) 'Discovering "the secrets of long and healthy life": John Dudgeon on hygiene in China', *Social History of Medicine*, 23.1: 21–37.

——— (2010b) 'Eating well in China: diet and hygiene in nineteenth-century treaty ports', in A. Leung and C. Furth (eds) *Health and Hygiene in Chinese East Asia: policies and publics in the long twentieth century*, Durham, NC: Duke University Press, pp. 109–31.

——— (2012) *A Physician to Empire: Patrick Manson and his founding of British tropical medicine*, Taipei: Asian Culture Publishing.

——— (2018) 'Visualisation in parasitological research: Patrick Manson and his Chinese assistants', in V. Lo and P. Barrett (eds) *Imagining Chinese Medicine*, Leiden: Brill, pp. 457–66.

Lo, V. and Barrett, P. (eds) (2018) *Imagining Chinese Medicine*, Sir Henry Wellcome Asian Series, Volume 18, Leiden: Brill.

Lu, D. (2021) 'History of Epidemics in China: Some reflections on the role of animals', *Asian Medicine: Journal of the International Association for the Study of Traditional Asian Medicine*, 16.1.

Luesink, D. (2009) 'Wu Lien-teh and the history of Chinese medicine: empires, transnationalism and medicine in China, 1908–1937', in I. Boroway (ed.) *Uneasy Encounters: the politics of medicine and health in China 1900–1937*, Frankfurt: Peter Lang, pp. 149–76.

Lynteris, C. (2016) *Ethnographic Plague: configuring disease on the Chinese-Russian frontier*, London: Palgrave Macmillan.

Manson, P. (1879) 'Dr. Manson's report on the health of Amoy for the half-year ended 31st March 1878', *Customs Gazette, Medical Reports*, 2 (January–March 1878): 25–27. Chinese Imperial Maritime Customs Service (1871–1911) *Customs Gazette, Medical Reports*, https://www.hathitrust.org/, accessed 27/3/2020.

McNeill, W.H. (1976; 2nd edn 1998) *Plagues and Peoples*, New York: Anchor Books.

Mei Li 梅莉 and Yan Changgui 晏昌貴 (1996) '*Guanyu Ming dai chuanranbing de chubu kaocha*' 關於明代傳染病的初步考察 (A preliminary study of infectious disease in the Ming), *Journal of Hubei University*, 5: 80–88.

Mitchell, P. (2011) 'Retrospective diagnosis and the use of historical texts for investigating disease in the past', *International Journal of Paleopathology*, 1: 81–88.

Naquin, S. and Rawski, E.S. (1987) 'Topics for research in Ch'ing history', *Late Imperial China*, 8.1: 187–203.

Nathan, C.F. (1967) *Plague Prevention and Politics in Manchuria, 1910–1931*, Harvard East Asian Monographs, Cambridge, MA: Harvard University Press.

Needham, J. (1970) *Clerks and Craftsman in China and the West: lectures and addresses on the history of science and technology*, Cambridge: Cambridge University Press.

Needham, J. and Lu Gwei-djen (1962) 'Hygiene and preventive medicine in ancient China', *Journal of the History of Medicine and Allied Sciences*, 17, reprinted in J. Needham (1970), pp. 340–78; and J. Needham and N. Sivin (2000), pp. 67–94.

——— (1967) 'Records of diseases in ancient China', in D. Brothwell and A.T. Sandison (eds) *Diseases in Antiquity: a survey of the diseases, injuries, and surgery of early populations*, Springfield: Thomas, pp. 222–37; republished as 'Diseases of Antiquity in China', in K.F. Kiple (ed.) (1993), pp. 345–53.

Needham, J. and Sivin, N. (ed.) (2000) *Science and Civilization in China, Vol. 6, biology and biological technology, part VI: medicine*, Cambridge: Cambridge University Press.

Norris, J. (1977) 'East or West? The geographic origin of the Black Death', *Bulletin of the History of Medicine*, 51.1: 1–24.

——— (1978) 'Response' [to M. Dols (1978)], *Bulletin of the History of Medicine*, 52.1: 114–20.

Olson, A. and Voss, J. (eds) (1979) *The Organization of Knowledge in America 1860–1920*, Baltimore, MD: Johns Hopkins University Press.

Packard, R. (2007) *The Making of a Tropical Disease: a short history of malaria*, Baltimore, MD; London: Johns Hopkins University Press.

——— (2016) 'The fielding H. Garrison lecture: "Break-Bone" fever in Philadelphia, 1780: reflections on the history of disease', *Bulletin of the History of Medicine*, 90.2: 193–21.

Parsons, J. B. (1970) *Peasant Rebellions of the Late Ming Dynasty*, Tucson: University of Arizona Press.

Peitzman, S. (2007) *Dropsy, Dialysis, Transplant: a short history of failing kidneys*, Baltimore, MD; London: Johns Hopkins University Press.

Qi Tao 祁濤 (1981) 'Wu Youxing "liqi shuo" qiantan' 吳有性' 戾氣說' 淺談 ('A superficial discussion of Wu Youxing's doctrine of deviant *qi*'), *Yunnan zhongyi xueyuan xuebao* (The Journal of the Yunnan College of TCM), 4: 7–8.

Rocher, Émile (1879) *La Province Chinoise du Yün-nan* (The Chinese Province of Yunnan), Paris: E. Leroux.

Rogaski, R. (2004) *Hygienic Modernity: meanings of health and disease in treaty-port China*, Berkeley: University of California Press.

Rosenberg, C.E. (1962) *The Cholera Years: the United States in 1832, 1849, and 1866*, Chicago, IL: University of Chicago Press.

——— (1979) 'Toward an ecology of knowledge: on discipline, context, and history', in A. Olson and J. Voss (eds) *The Organization of Knowledge in America 1860–1920*, Baltimore, MD: Johns Hopkins University Press, pp. 440–55.

——— (1992a) *Explaining Epidemics and Other Studies in the History of Medicine*, Cambridge: Cambridge University Press.

——— (1992b) 'Framing disease: illness, society, and history', in C.E. Rosenberg and J. Golden (eds) *Framing Disease: studies in cultural history*, New Brunswick: Rutgers University Press, pp. xiii–xxvi.

Rosenberg, C.E. and Golden, J. (eds) (1992) *Framing Disease: studies in cultural history*, New Brunswick: Rutgers University Press.

Sandison, A.T. (ed.) (1967) *Diseases in Antiquity: a survey of the diseases, injuries, and surgery of early populations*, Springfield: Thomas.

Scheid, V. (2002) *Chinese Medicine in Contemporary China: plurality and synthesis*, Durham, NC; London: Duke University Press,

Schipper, K. (1985) 'Seigneurs royaux, dieux des épidémies', *Archives de Sciences Sociales des Religions*, 59.1: 31–40.

Shi Changyong 史常永 (1956) *Zhongguo chuanran bing xue* 中國傳染病學 (The Study of Contagious Diseases in China), Shanghai: Shanghai weisheng chubanshe.

—— (1957) 'Shilun chuanranbing xuejia Wu Youke ji qi liqi xueshuo' 試論傳染病學家吳又可及其戾氣學說 ('An essay on the scholar of infectious diseases, Wu Youxing, and his doctrine of deviant qi'), *Yixueshi yu baojian zazhi* 醫學史與保健雜誌 (Journal of Medical History and Health), 9.3: 180–6.

Strickmann, M., edited by B. Faure (2002) *Chinese Magical Medicine*, Asian Religions and Cultures, Stanford, CA: Stanford University Press.

Summers, W.C. (2012) *The Great Manchurian Plague of 1910–1911: the geopolitics of an epidemic disease*, New Haven, CT: Yale University Press.

Temkin, O. (1963; 2nd edn. 1977) 'The scientific approach to disease: specific entity and individual sickness', in A. C. Crombie (ed.) *Scientific Change*, London: Heineman, pp. 629–47; republished in O. Temkin (1977), pp. 441–5.

—— (1977) *The Double Face of Janus and Other Essays in the History of Medicine*, Baltimore, MD: Johns Hopkins University Press.

Twitchett, D.C. (1979) 'Population and pestilence in T'ang China', in W. Bauer (ed.) *Studia Sino-Mongolica: Festschrift für Herbert Franke*, Wiesbaden: Franz Steiner, pp. 35–68.

von Glahn, R. (1996) *Fountain of Fortune: money and monetary policy in China 1000–1700*, Berkeley: University of California Press.

Will, P. (1990) *Bureaucracy and Famine in Eighteenth-Century China*, Stanford, CA: Stanford University Press; translated by E. Forster, originally published as *Bureaucratie et famine en Chine au 18e siècle* (1980), Paris: École des Hautes Études en Sciences Sociales; Hague: Mouton.

Wilson, A. (2000) 'On the history of disease concepts: the case of pleurisy', *History of Sciences*, 38: 271–319.

Wong, K.C. and Wu Lien-teh (1936) *History of Chinese Medicine: being a chronicle of medical happenings in China from ancient times to the present period*, Shanghai: National Quarantine Service.

Wu Lien-teh (ed.) (1936a) *Manchurian Plague Prevention Service Memorial Volume: 1912–1932*, Shanghai: National Quarantine Service.

—— (1936b) *Plague: a manual for medical and public health workers*, Shanghai: Weishengshu National Quarantine Service.

—— (1959) *Plague Fighter: the autobiography of a modern Chinese physician*, Cambridge: W. Heffer.

Wu Lien-teh, Chun, J.W.H., Pollitzer, R. and Yu, C.Y. (1934) *Cholera: a handbook for the medical profession in China*, Shanghai: National Quarantine Service.

Xiao Defa 肖德發 (1987) 'Shilun Wu Youke *Wenyi lun* de zhuyao xueshu guandian', 試論吳又可瘟疫論的主要學術觀點 ('An essay on the principal scholarly views of Wu Youxing's *Treatise on Febrile Epidemics*'), *Jiangxi zhongyi yao*, 2: 1–3.

Xie Guan 謝觀 (1935; 2nd edn 1970) *Zhongguo yixue yuanliu lun* 中國醫學源流論 (On the Origins of Medicine in China), Taipei: Jinxue shuju.

Yu Xinzhong (ed.) (2004) *Wenyi xia de shehui zhengjiu: Zhongguo jin shi zhong da yiqing yu shehui fanying yanjiu* 瘟疫下的社會拯救:中國近世重大疫情與社會反應研究 (Social Salvation during Epidemics: studies of serious modern epidemics and social responses in China), Beijing: Zhongguo shudian.

Yu Yan 余巖 (Name Given as Yunxiu 雲岫) (1953) *Gudai jibing ming hou shu yi* 古代疾病名候疏義 (Glossary on the Names and Symptoms of Ancient Diseases), Beijing: Renmin weisheng chubanshe.

Zhao Wenlin 赵文林 and Xie Shujun 谢淑君 (1988) *Zhongguo renkou shi* 中國人口 史 (The History of Population in China), Beijing: Renmin chubanshe.

Zhao Xianhai 趙獻海 (2004) 'Mingmo da yi yu Ming wangchao di miewang' 明末大疫 與明王朝滅亡 (Epidemics at the end of the Ming and the fall of the Ming empire), in Yu Xinzhong (ed.) *Wenyi xia de shehui zhengjiu*瘟疫下的社會拯救 (Social Salvation during Epidemics), Beijing: Zhongguo shudian, pp. 49–101.

17
LATE IMPERIAL EPIDEMIOLOGY, PART 2

New material and conceptual methods, 1980s to 2010s

Marta Hanson

The previous chapter concluded with two clear themes: (1) historical actors as well as historians, demographers, and anthropologists have wrestled with retrospective diagnosis of epidemic diseases, revealing in the process that disease concepts had complex histories as much in the past as they do today; and (2) historical actors and modern scholars alike have found epidemics to be useful as a diagnostic lens on contemporary problems, whether as fault lines in the moral economy, as fissures in social order, or as failures in governance. While both of these key themes continue into the present, the scholarship on late imperial epidemiology from the 1980s to the present markedly differs from that of the previous century in terms of both conceptual and material methods. ('Material methods' refers to extant primary sources; 'conceptual methods' refers to how people interpret them.) From late nineteenth-century Western physicians to 1970s historians of Chinese demography, for example, the history of late imperial epidemiology was dominated conceptually by a naturalist-realist perspective. From the 1980s onwards, however, medical historians have increasingly explored late imperial Chinese epidemiology from the historical-conceptual side of the spectrum.

As for material methods, historical records of epidemics created by Chinese administrators (from dynastic histories to local gazetteers and jottings), religious leaders (from tracts on doctrines to rituals and liturgy), and medical authorities (from treatises to case records) constituted the primary-source foundation for the former period. From the 1980s onwards, however, geneticists began to develop means to extract human and bacterial DNA (aDNA) from ancient remains, and so bring new evidence into the conversations about the global history of epidemics. The resulting new field of paleomicrobiology since the 1990s has analysed ancient DNA (aDNA) in ways that confirmed, for example, the retrospective diagnoses of a range of infectious diseases from tuberculosis in ancient Egypt to plague in fourteenth-century Europe and influenza in the US during the 1918 Spanish Flu pandemic. Furthermore, scientific research on the history of non-human diseases and even viruses has developed additional evidence that medical historians may both historicise and integrate into new global histories of disease in non-human as well as human populations.

By 1997, one also finds the full range of conceptual methods applied to the history of late imperial Chinese epidemiology, when one medical historian applied retrospective

DOI: 10.4324/9780203740262-20

epidemiology to argue that the late Ming epidemics were mostly due to plague (Cao 1995, 1997), while I took a more historical-conceptual approach to the same epidemics in my PhD thesis and book (Hanson 1997, 2011).

Finally, this essay reviews the main transformations in scholarship in the past twenty years on both sides of the spectrum. The historical-conceptual side dug deeper into historical actors' categories, while the naturalist-realist side took a definitive genetic turn. Initially, the two approaches were considered incompatible (Wilson 2000); yet more recently, some historians are willing not only 'to wear both hats' (Packard 2016), but also to integrate aDNA isolation of the *Yersinia pestis* in plague victims in Europe into their histories of epidemics in medieval China, where comparable remains have yet to be found (Hymes 2014; Brook 2020). These historians demonstrate how syntheses of both sides of this methodological divide allow medical as well as socio-economic historians to switch judiciously between two sides of the spectrum of conceptual methods: namely contextualising past disease concepts on their own terms and the related responses to epidemics, and also using current scientific criteria not only to identify the infectious cause of past epidemics retrospectively but also for historical ends related to a deeper understanding of social, economic, and environmental transformations.

Historiography of Chinese late imperial epidemiology in the 1980s–1997

Within China during the 1980s, medical historians also began to integrate the history of disease more into general Chinese history, first in *A History of Disease in China* (Chen 1981) and then in the *New Significance of the History of Disease in China* (Fan 1989; on epidemics, see 161–94, 241–4). Both authors began to study the history of epidemics as means to explore questions in social, economic, and demographic history. Following more along the anthropological lines of enquiry about Chinese experiences of epidemics that Francis Hsu initiated from the 1940s to 1980s (Hsu 1952, 1983), Kristofer Schipper described a Taoist (Daoist) ceremony in Taiwan that occurred every three years to protect the community from epidemic diseases (Schipper 1985). While Hsu's work described one community's scientific as well as religious responses to a cholera epidemic in order to reveal problems with a religion-science binary then prevalent in anthropology, Schipper explained how three distinct levels of one Chinese community – the Taoist priesthood, local village chiefs, and the popular or vernacular – attempted to appease the epidemic gods within what he considered to be a coherent religious framework with long historical antecedents in mainland China that persisted in modern-day Taiwan. Both Hsu and Schipper thus explored Chinese epidemic responses in pursuit of other aims, such as demonstrating the inadequacies of Western models to do justice to the social meanings and functions of ritual within Chinese contexts. Their ethnographies furthermore record details of religious and social responses to epidemics that historical documents rarely record and that more naturalist-realist oriented authors largely disregard.

With regard to the history of modern public health in China, Kerrie MacPherson produced a detailed portrayal of the foreign settlement in Shanghai from 1843 to 1893 and the public health infrastructure foreign settlers initiated (MacPherson 1987; Chapter 47 in this volume). She showed that Western medical and sanitary principles were successfully implanted and later fused with Chinese reformist efforts in areas open to Western ideas, especially in Hong Kong and Peking during the 1890s. Taking the epidemiological model developed by British epidemiologist William Farr (1807–1883) as it was carried out in China as one example, and demonstrating that Westerners negotiated with Chinese social reformers, MacPherson refuted previous interpretations that emphasised imperialist exploitation.

Although she applied a symmetric approach to the social negotiation over Farr's model among Chinese and Westerners, she did not entertain the possibility that Chinese disease concepts or public health methods themselves might have been relevant to this interaction.

In the same year, however, Angela Leung published two articles related to China's public health history that did better justice to the *longue-durée* history of Chinese medical governance. The first article traced the transformation from late medieval ideals of state medical governance during the Song-Jin-Yuan period to more privatised medical and charitable institutions during the Ming dynasty in the Lower Yangzi Delta region (Leung 1987). The second article examined the related evolution of smallpox prevention measures in the Ming and Qing dynasties as a way to examine the broader phenomenon of privatisation of medical and charitable institutions (Leung 1987–88). Meanwhile, to explore the conceptual foundations of Chinese medical thought that arguably underlay this history of state and private medical governance, Nathan Sivin translated a 1970s textbook on Chinese medicine in order to analyse how the meanings of core medical terms had changed from classical times to the 1970s China (Sivin 1987).

When Susan Naquin and Evelyn Rawski wrote in their review article that the history of disease and epidemics was an important new research trajectory in Qing history (Naquin and Rawski 1987), they captured a shift in the China field that was well under way. In response to their question 'When epidemic diseases struck, what were the responses of the state and of communities to the crisis?', they proposed that the study of the transmission paths of epidemics in the past could help clarify communication and transport networks, migration routes, and levels of social interaction. This review exemplified how mainstream historians saw value in the natural-realist angle on China's disease history. Just a year later, Carol Benedict (1988) analysed the spread of epidemics of plague from Western Yunnan to the Southeast Coast microregion during the nineteenth century, applying William Skinner's macroregion model, predominantly used within Chinese social and economic history, for the first time to Chinese historical epidemiology (see the Appendix for map). But she also discussed Chinese disease concepts and various treatments for what clinically appeared to be plague, as well as social reactions to Western treatments and draconian quarantines. Published in the same year, the analysis of *The Epidemiological Transition in Hong Kong: Changes in Health and Disease since the Nineteenth Century*, however, took a public health approach to many of the same historical sources to narrate Hong Kong's epidemiological transition from acute infectious diseases, arising out of poverty, food scarcity, and poor sanitation, to predominantly chronic diseases under improved living conditions (Phillips 1988).

Applying instead a political-history perspective, Carney Fisher (1988) evaluated the influence of a 1550 smallpox epidemic on Ming-Mongol relations during the Jiajing 嘉靖 reign (1522–67). He demonstrated how smallpox epidemics influenced the Altan-qaghan's decision to accept tributary status with the Ming government, for instance, and discussed some of the healing methods for, and social taboos created around, smallpox. Chinese, for whom smallpox was an endemic childhood disease that granted immunity to those who survived it, were often asked during this period to help take care of Mongols for whom smallpox was epidemic and thus more deadly for adult Mongols, not previously exposed as were Chinese during childhood. For Fisher, the 1550 smallpox epidemic was an illuminating lens on unique dimensions of sixteenth-century Chinese-Mongol political and social relationships, rather than an occasion to problematise how to identify smallpox in China's past or contextualise how Chinese conceptualised it then.

Other historians nonetheless had grasped the baton of the historical-conceptual method and started to apply it to explain the cultural distinctiveness of Chinese concepts of disease.

Dean Epler analysed 'Cold Damage' (*shanghan* 傷寒) in the first known Chinese disease monograph, the *Treatise on Cold Damage and Miscellaneous Disorders* (*Shanghan zabing lun* 傷寒雜病論), by Han physician Zhang Ji 張機 (150–219 CE). Cold Damage referred to both its general aetiology due to cold (within a configurationist understanding of pathogenic climatic factors) and its major symptoms of excessive heat and cold aversion (Epler 1988). From then on, Cold Damage meant both an ontological cause (cold) and multiple physiological signs (heat and cold aversion). This multivalent meaning of Cold Damage fits well medical historian Owsei Temkin's (1977) famous distinction between disease as a specific entity (ontological) and as an individual's sickness (physiological) (Chapter 14 in this volume).

Several scholars followed suit along these lines of interpretation in the early 1990s with chapters for *The Cambridge World History of Human Disease* that explained East Asian disease concepts on their own terms (Jannetta 1993; Kuriyama 1993; Leung 1993). In a historical survey on 'Sexually transmitted diseases in modern China', another scholar discussed some Chinese terms for, and so understanding of, venereal diseases though within a natural-realist frame (Dikötter 1993). In a comparable vein, Chinese historian Hsiao Fan examined an even wider range of Chinese disease terminology from the early to late medieval period that illuminated how Chinese related the environment to local diseases, sometimes making one-to-one correspondences to modern disease concepts (Hsiao 1993). Medical historian Shigehisa Kuriyama, by contrast, took a fully historical-conceptual approach to the history of disease in his analysis of the earliest Chinese concepts of pathogenic wind, the human body, and an individuated self (Kuriyama 1994).

Continuing the legacies of anthropologists Hsu and Schipper, historian Kenneth Dean included a chapter about a cult to a medical god in *Taoist Ritual and Popular Cults of South-East China* (Dean 1993). Similarly, Paul Katz united disease and social history by situating the cult of Marshal Wen (Marshal of Epidemics) within religious communities of late imperial Zhejiang province (Katz 1995). Combining both the historical-conceptual and the natural-realist perspectives, Chang Chia-feng analysed Chinese 'strategies of dealing with smallpox' (*dou* 痘, lit. 'bean' with an illness radical) as part of a range of religious responses to 'fetal poison' (*taidu* 胎毒), the main Chinese disease concept that overlaps with the symptoms of modern-day smallpox (Chang 1995, 1996a, 1996b, 1996c). But Chinese conceptualised *dou* not as caused by an external infectious agent but rather as a type of congenital 'fetal poison' internal to all humans that needed to be expelled (*shang deng dou* 上等痘) in order for them to survive into adulthood, which made a great deal of sense considering that, for at least the Chinese population it had become a nearly universal childhood disease.

These developments in historical-conceptual approaches continued during the early 1990s alongside natural-realist approaches. Grasping the baton from Naquin and Rawski's proposal to integrate the history of epidemics into Qing history, Carol Benedict argued that the government's responses to plague epidemics during the New Policies period (1900–11) helped establish the first state medical institutions for modern China (Benedict 1993). Meanwhile, French scholars published a naturalist-realist overview of the effect of epidemics of plague, cholera, smallpox, and leprosy on China, which also explained Chinese medical approaches to understanding and treating them (Lu *et al*. 1995). In her book on plague in nineteenth-century China, Benedict approached from a more historical-conceptual perspective the earlier epidemics of late Ming (Benedict 1996a). She cautioned historians of disease not to assign the biomedical category 'bubonic plague' hastily to epidemics for which there was little historical evidence of either bubo-like swellings or death of plague-vector rodents. She combined modern-day scientific evidence of plague reservoirs – animal-based reservoirs of the plague bacillus – with Chinese descriptions of symptoms to determine which epidemics were likely bubonic plague. Two sources of evidence, one biological and the other theoretical, helped her reconstruct the diffusion of plague from Yunnan in the southwest to Fujian along the southeastern coastline. Developing upon her earlier publications, her book synthesised modern scientific research on natural plague reservoirs in China with G. W. Skinner's regional systems analysis of economic and social change to determine the diffusion patterns of plague in China from the late eighteenth through the early twentieth century.

Benedict (1996b, 1996c) also separately wrote articles on methodological approaches to epidemiology and history in modern China that clarified for medical historians of China the distinctions that Wilson (2000) would later call naturalist-realist and historical-conceptual approaches. During the same period, Carney Fisher reviewed primary sources and secondary scholarship on bubonic plague in China from the 1800s through the 1980s without, however, including Benedict's more nuanced mid-1990s scholarship that cautioned, as already noted, against retrospective diagnoses of bubonic plague without specific primary evidence (Fisher 1995–96).

Naturalist-realist interpretation of the end-of-Ming epidemics as plague

In 1995, the Chinese historian of demography, Cao Shuji, used McNeill's 'China origin of plague' argument to stitch together a narrative from actual Chinese primary sources to place the origin of plague in Mongolia itself and as the cause of the late Yuan epidemics of 1344–45, 1356–60, and 1362 (Cao 1995). Cao followed this article two years later in 1997 with another one that argued that during the late Ming catastrophes of 1587–88, 1639–41, and 1643–44, the widespread epidemics in China were also due to plague (Ibid. 1997). Cao thus reexamined the evidence for identifying the two waves of epidemics of the late Ming as bubonic plague and, in a few cases, pneumonic plague. His scholarship at that time exemplified what the naturalist-realist perspective could contribute to historical epidemiology. First, he argued that these epidemics were either bubonic or pneumonic plague based on accounts in local gazetteers and modern evidence of natural plague reservoirs in Inner Mongolia (Ibid.). Scholars had previously pointed out that it is both biomedically difficult (Dunstan 1975) and theoretically problematic to identify plague without laboratory evidence of the *Yersinia pestis* – the bacterial cause of both the bubonic and pneumonic forms of plague (Cunningham 1992: 211–19; Benedict 1996a: 8–9). Nonetheless from textual, ecological, and historical evidence, Cao argued for the retrospective diagnosis of bubonic plague in the northern provinces. The local gazetteer descriptions of symptoms, such as a 'node' or 'hard lump' (*he* 核), 'swollen neck' (*zhongxiang* 腫項), and boils/swelling disorder (*geda bing* 疙疸病), align with plague (though he included the symptoms of 'enlarged heads' (*datou* 大頭) and 'obstructed throat' (*houbi* 喉痺) that do not), and accounts of mortality rates of 40% to 90% and a two-to-three-day mortality span all suggested to him plague.

Cao used ecological and historical evidence to further support his hypothesis of plague in the north. A natural plague reservoir exists today on the Inner Mongolian Plateau (Benedict 1996a: 6). Cao argued that it spanned a much wider region during the Ming and Qing dynasties: this reservoir included Mongolian pasture lands just north, east, and west of the Great Bend section of the Yellow River and south along the Great Wall at the northern borders of Hebei and Shanxi (Cao 1997: 28). Although the tarbagan (Mongolian-Siberian marmot) also lives across Inner Mongolia, Cao focused on the Mongolian gerbil (*Meriones unguiculatus*) as the more important natural host for the Asiatic flea that carries the plague bacillus in this region (Ibid.: 27–9). Normally, according to Cao, the bacillus, flea, and Mongolian gerbil lived in a symbiotic host–parasite relationship on the Inner Mongolian and Shiliyn Boyd Uul

Plateaus, rarely changing the course of human history. Seen within this broader ecological, economic, and social setting, Cao used a broad brush to attribute the drastic demographic losses from 1580 to 1640 mostly to plague.

Cao pinpointed the beginning of this process of depopulation by disease in the early Jiaqing reign (1522–1566), when there was a large migration of Han settlers to the northern Shanxi region bordering Inner Mongolia from 1533 to 1534. The Han migrants began the extensive transformation of the region's environment from pastureland to agricultural fields that upset the ecological balance of the natural plague reservoirs (Ibid.: 28–9). He did not consider, however, the possibility of concurrent smallpox epidemics (Fisher 1988), to which we will return. A similar process occurred in late eighteenth-century Yunnan when a large influx of migrant workers arrived to mine mineral deposits and so expanded urban settlements, which led to an increase in intraregional trade and encroachment on the natural plague reservoirs. With newly opened agricultural lands came an increase in human settlement on the dry plateau and an expansion of markets.

This commercial expansion ultimately led to plague outbreaks from 1772 to 1830 in Yunnan (Benedict 1996a: 24–9). One hypothesis is that these socio-economic changes in the region combined to disrupt the rodents' normal habitats, forcing them into new interactions or, at least, proximity with humans or their domestic animals, increasing chances that people would come into more contact with plague-carrying rodents. In addition, a significant decline in yearly precipitation and the resulting droughts during the reigns of Wanli (1573–1620) and Chongzhen (1628–1644) led to regular famines that weakened human resistance to disease and further expanded the opportunities for rodent-human contact. In such years of dearth, both animals and humans forage for food in new regions, losing their resistance to disease and spreading it as they migrate. According to modern research, people in Shandong during droughts also dug into the nests of the plague-carrying gerbils seeking the grain they had stored there and, in some cases, even eating the gerbils, thereby increasing their chances of dying (Cao 1997: 29).

Although Cao did not adequately consider the range of disease possibilities, his focus on the effect of the two waves of epidemics in Shaanxi and the North China macroregion (Shanxi, Henan, Hebei, and Shandong) synthesised evidence on the northern origin of the 1580–88 and 1633–44 epidemics that clarified transmission patterns more broadly for epidemics throughout the North China macroregion. By applying a naturalist-realist method of retrospective epidemiology to a wide range of sources, Cao sketched an epidemiological picture in the North China macroregion in the 1580s, and again in the 1630s–1640s, that contributed to the history of epidemics in China. Cao significantly contributed to clarifying transmission pathways, temporal duration, local severity, and range of underlying factors that contributed to the late Ming outbreak epidemics, even if the historical evidence does not yet conclusively support his bubonic-plague hypothesis.

A naturalist-realist critique of the end-of-Ming epidemics as only plague

Clearly, severe droughts and famines, combined with continued Mongol raids, peasant uprisings from 1628 to 1644 (Parsons 1970), and Manchu invasions starting again in 1629 (Atwell 1988), led to a second wave of epidemics in the North China macroregion that were even more complex, widespread, and destructive to the population than the earlier 1588–89 epidemics (Cao 1997: 25–6). Manchu military forces finally broke through the Shanhai Pass on the far eastern shore of Hebei province in December 1629, continued their raids through the early 1630s, entered the northern Zhili region in the summer of 1636, and so approached the capital in Beijing (Atwell 1988: 629). Despite clear military successes over the previous seven years, the Manchus then suddenly withdrew from the region. There was in 1635–36 an epidemic outbreak at Shanhai Pass. Despite Cao's argument that this too was plague, smallpox was a significant concern within the Manchu military (Chang 2002), as it was also for the Mongols in the previous century (Fisher 1988).

Furthermore, beginning with the initial rise of Manchu power in the 1610s well into the consolidation period after 1644, smallpox played a significant role in Manchu military and political decisions (Chang 1996a: 169–92). Since at least 1613, the Manchus organised their campaigns to avoid contact with smallpox patients; they also distinguished the Mongol and Manchu princes who had survived smallpox from those who had not experienced it. In 1613, the Jurchen chieftain Nurhaci (1559–1626) decided not to withdraw his troops from Usu, a Yehe Manchu city, despite a smallpox epidemic raging there, because he wanted to avoid spreading it to his other troops. In 1627, and again in 1633, the new Jurchen leader Hong Taiji 皇太極 (1592–1643) of the Later Jin dynasty (r. 1626–1636) decided to withdraw Mongol and Manchu princes who had not experienced smallpox from campaigns in both Korea and China (Ibid.: 172, 179).

As for the epidemic in 1636 on the northern Shaanxi border in Yulin prefecture, the following year it spread further south to Yan'an prefecture in north central Shaanxi. Datong in northern Shanxi also had an epidemic in 1637. These prefectures were all sufficiently close to the horse markets along the northern frontier and the natural plague reservoirs on the Inner Mongolian Plateau to have possibly experienced plague outbreaks. For the next three years leading up to 1640, there was a hiatus in epidemics in three of the provinces of the North China macroregion, excluding Shandong. The shift of the peasant rebellions to the southwest in Shaanxi and Sichuan, and south to Henan and Hubei, left Shanxi, Hebei, and Shandong provinces virtually free of rebel activity from 1637 to 1640. The decline in political upheaval that reduced the numbers of dislocated, moving, and homeless people largely explains the break in epidemics (see Maps 10–13 in Parsons 1970: 59, 61, 67, 73).

In 1639, however, another epidemic ravaged the population in Shandong in the region surrounding the provincial capital Jinan. Cao curiously did not discuss this epidemic in his tally of late-Ming plague epidemics. The connection between this epidemic and a Manchu retreat similarly weighs against a plague diagnosis. The epidemic broke out shortly after Hong Taiji crossed the Great Wall in the winter of 1638 and raided Shandong by the first month of 1639. Shortly afterwards, the Manchu armies penetrated deep into the Central Plain and reached Jinan, Shandong, where epidemics broke out in the surrounding region at Qihe, to the northwest of Qihe in Yucheng, and also in Linqing to the east. The historian Frederic Wakeman wrote that it had not been determined 'whether the Manchus brought the illness with them, or turned back because of it' (Wakeman 1985: 143, fn. 171; Dunstan 1975: 27–8; Spence 1990: 24). Yet, he still cited the General Zuo Maodi (1601–1645), who sent a memorial in 1643 from Linqing in Shandong reporting that he estimated 30% died from starvation, 30% from smallpox, and the remaining 40% were forced into banditry in order to survive (Wakeman 1985: 155, fn. 216).

The Manchus thought that smallpox was one of their greatest enemies; as a natural barrier, it proved more formidable than the Great Wall for Ming military forces. The regent Dorgon of the child emperor Shunzhi (r. 1638–1661) went to great lengths to protect the emperor from contact with smallpox. Even before the Manchu victory, the child was made to spend long periods in a pox isolation centre (*bidousuo* 避痘所) to quarantine him from contact with people who had smallpox. Once the Qing forces had settled in Beijing, in 1645, they forced the Chinese who suffered from smallpox to move a dozen or so miles out of the capital (Chang 1996a: 174; Wakeman 1985: 465–6). Nonetheless, the year after the Manchu entered the capital on June 6th, 1644, a major smallpox epidemic broke out there (Dunstan 1975: 27; Chang 2002). Long familiar with the symptoms of smallpox, the Manchus had policies to protect those in the ruling family and upper military ranks who had not experienced it as children and therefore had not acquired immunity. It was only after their conquest of China, however, that they began to adopt Chinese variolation methods to protect themselves. Despite such precautions against smallpox infection, the Shunzhi emperor died from it in 1661.

As for the Lower Yangzi microregion, the severe epidemic that erupted in 1639 for the first time may have also been due in part to the same smallpox epidemic that hit the Jinan region in Shandong the same year. The 1639 epidemics in Jiangnan do not appear, however, to have been extensions of the northern epidemics of 1633–1637, but rather were the consequence of the combined effects of floods in northern Zhejiang, bad weather in Jiangsu, and locusts in Anhui, all of which contributed to severe famines (Hanson 1997: 350–4). The tax increase during that same summer and dearth of foreign silver bullion since 1636 also made it difficult for peasants to purchase rice, even when it was available (Atwell 1988: 632).

In 1640, another major epidemic hit Shandong, but available primary sources do not specify its nature; nor do they indicate that a smallpox epidemic continued through the following year. An epidemic erupted again the same year in the northwest and central regions of the North China macroregion, with greater virulence than before. This epidemic hit Fengxiang prefecture to the far northwest of Shaanxi, Hejian prefecture to the east of Beijing, Shunde prefecture in southwest Hebei, and Zhangde and Huaiqing prefectures in northern Henan. Instead of coming out of north and central Shanxi, as the Wanli epidemics did, the 1641 epidemic spread out simultaneously from Hebei and Henan provinces. It subsided in 1642, but struck again in Hebei, Henan, and Shanxi provinces from 1643 to 1644 when, coupled with internal warfare and the Manchu invasion, it took its most devastating toll (Cao 1997: 23–7).

The campaigns of rebel-leader Li Zicheng (1605?–1645) in Henan during the spring of 1641 could not have facilitated the spread of the northern epidemics south of Kaifeng prefecture (where there was a severe epidemic) through Anhui to Jiangsu and Zhejiang in the Lower Yangzi macroregion. Although the rebel army, and fleeing Chinese in their wake, may well have spread the epidemic through major thoroughfares in northern provinces, the rebels neither went far enough out of southern Henan, nor deeply enough into northern Anhui, to connect them with the epidemics in the Lower Yangzi macroregion (Atwell 1988: 633).

The records describe buboes-like symptoms in northern provinces during the mid-1640s epidemics: 'boils disorder' in the 1643 epidemic in Beijing and surrounding region to the northwest; and 'hard lumps under the armpits' in the 1644 epidemic in Lu'an prefecture of central Shanxi. The 'enlarged head epidemics' also reached northern Shandong's Binzhou County the same year (Hanson 1997: 358). Although Cao examined the transmission paths of the 1633–1637 epidemics throughout North China, he falsely assumed that once the epidemic arrived by 1641 in northern Henan and southern Hebei, it rapidly spread south to Jiangnan (Cao 1997: 27). Evidence of buboes-like symptoms, for example, does not appear in the primary sources for the Jiangnan region, though other types of symptoms do. Such clinical symptoms are only found in gazetteers from northern not southern provinces. The concentration in the northern provinces of buboes-like symptoms indicates that, if these epidemics were indeed plague, they probably did not spread south to Jiangnan. The range of symptoms described in the local gazetteers indicates that physicians had not reached a consensus on the character of these epidemics, unlike smallpox and other known diseases. When editors compiled the local gazetteers for each region, they may have borrowed the designation for the epidemic given in an earlier local gazetteer.

Furthermore, several accounts of the 1640–44 epidemics in the southern Zhejiang and Jiangsu gazetteers refer variously to 'Sheep's Wool heat epidemic' (*yangmaowen* 羊毛瘟), 'Sheep's Wool clove-like sores' (*yangmao ding* 羊毛丁), and 'Sheep's Wool papules' (*yangmao zhen* 羊毛疹) – a cluster of disease concepts unique to this time and place that does not appear in any northern records. This phrase appears first in Zhejiang during a 1640 epidemic in Huzhou prefecture south of Lake Tai, and then again in Jiangsu during the 1644 epidemic in Zhenjiang prefecture just northwest of Lake Tai. The gazetteer sources are from Wucheng County in Zhejiang and from Dantu, Danyang, and Jintan counties in Jiangsu (Imura, 1936–37: no. 7, 20–1). The *Gazetteer of Wucheng County*, for example, records: 'In 1640 an epidemic raged; the symptoms were unusual and incomprehensible. It was called Sheep's Wool epidemic. Strands of sheep's wool would suddenly come up out of foods and fruits. All those who accidentally ate [this contaminated food] would immediately get sick and die' (Imura 1936–7, no. 7: 21; Dunstan 1975: 24–7; Hanson 1997: 109–10).

With only these scant textual sources, it is not possible after over 300 years to biomedically identify the Sheep's Wool epidemics in 1640 Huzhou and 1644 Zhenjiang prefectures as plague or any other modern disease category. Nevertheless, clearly contemporaries understood their experience to be unprecedented; nor did they borrow from preexisting medical terms for buboes. We also find that smallpox epidemics clearly raged at the end of the Ming in different regions of China, especially judging from how Manchus made considerable efforts to avoid contact with smallpox victims. Furthermore, in the context of severe famines and far from the natural plague reservoirs in Inner Mongolia, some other possible candidates for the late Ming epidemics in the Lower Yangzi macroregion are also typhus, one of the most lethal famine fevers in human history (Zinsser 1934: 159–60; Cartwright 1972: 18–65), and other water-borne infectious diseases such as dysentery and typhoid.

To sum up, Cao did not apply a macroregional systems analysis to examine the possibility of multiple epidemic diseases as did Carol Benedict (Benedict 1993, 1996a). Nor was he engaged with her scholarship when he published his. Furthermore, Chang's scholarship on the Manchu fears of smallpox provided further evidence that smallpox not plague erupted in Shandong in 1639 and 1643, and then in Beijing in 1645. Cao did not determine whether the same epidemics spread southwards through the cities along the canal and the Yangzi River throughout the Lower and Middle Yangzi macroregions. As we have just seen, smallpox certainly and other possibilities seem more likely (Hanson 1997: 109).

This revised naturalist-realist account of the late Ming epidemics argues that a combination of infectious diseases – smallpox certainly, something that may never be identified associated with 'sheep's wool' in Jiangsu and Zhejiang, and possibly because of the famine and war of the period typhus, dysentery, and typhoid – better explains the range of possibilities for epidemics along the lower reaches of the Yangzi River in the early 1640s. This multifactorial assessment is more complex and convincing than the easier to grasp, but methodologically flawed, retrospective diagnosis of plague for most of the late Ming epidemics. Furthermore, by taking a more historical-conceptualist approach to the primary sources of the period, we can learn a great deal about how southern Chinese wrote about their experience of the 1640s epidemics from cosmo-moralistic judgements to political-economic interpretations. One physician's response, in fact, reveals a crisis of confidence in the conventional cosmology that held together Chinese society as profound as their experience of the political convulsions of the last years of the Ming.

Historical-conceptual perspectives on end-of-Ming epidemics

A switch to a historical-conceptual lens on medical responses to the epidemics from the same period also opens up new avenues of interpretation aiming at very different objectives. As we saw in Part 1, figures like Chen Qide and Mr. Shen had interpreted epidemics in cosmological, social, and moral terms. Their contemporary, the Suzhou physician Wu Youxing, by contrast, took a strictly medical approach to the epidemics of 1641. Instead of interpreting them as manifestations of agrarian crises, excessive taxation, political corruption, moral failures, or the breakdown of the social order, Wu explained them in his preface to the *Treatise on Febrile Epidemics* completely within a naturalist framework. While Chen sought moral solutions and Shen worried about paying off debts, Wu criticised conventional treatment methods, promoted experimentation with new drug therapies, and launched an attack on traditional epidemiology through a new understanding of 'warm diseases' (*wenbing* 溫病), distinct from the Cold Damage framework within which Warm diseases were originally understood to be spring illnesses due to latent cold acquired during the winter.

If we situate his *Treatise* within the social-intellectual issues of his time, Wu Youxing's criticism of traditional epidemiology illuminates aspects of late imperial Chinese social and intellectual history. The broad trend of cosmological criticism since the sixteenth century clearly informed Wu Youxing's critique of traditional medical cosmology in the seventeenth (Henderson 1984: 163–4; Hanson 2011: 92–3). Wu redefined 'febrile epidemics' (*wenyi* 瘟疫), to which 'warm diseases' belonged, as being caused by a kind of 'deviant *qi*' (*liqi* 戾氣) that did not follow normal seasonal cycles of *qi*. While Chinese classical medicine understood epidemics to be the result of anomalies in the seasonal *qi* within a universal, agrarian cosmology based on the cyclical changes of seasons, Wu used 'deviant *qi*' to explain epidemics as the contingent consequence of the local and unclassified *qi* of a particular time and place, independent of predictable seasonal change.

Wu thereby challenged the most basic assumption in Chinese medical theory: the assumption of a system of correspondence between cosmological phases, seasonal cycles, and individual health. His new epidemiology represented a major shift in medical thinking among certain Ming and Qing physicians from a universal-cosmological to a local-environmental framework. While reading his *Treatise on Febrile Epidemics* as a naturalist-realist would find little evidence of bubonic plague beyond mention of the subcategory 'boils/swelling disorder epidemics' (*geda wen* 疙瘩瘟) mentioned earlier in the review of Cao's research, approaching it as a historical-conceptualist opens up many questions about the medical world within which Wu Youxing wrote his *Treatise* in response to the late-Ming epidemics.

For instance, Wu was part of a long historical tradition of medical scepticism towards classical medical texts among his predecessors, who found them spatially and temporally limited. The concept of anomalous disorders proved useful for thinking through problems within traditional cosmological categories. But instead of identifying the epidemics he witnessed with climatic pathogens out of season, due to northern rather than southern climates, or possibly the pathogenic environment of the Far South, Wu newly attributed the epidemics to pathogenic local *qi*.

Furthermore, the chaos of the mid-seventeenth century challenged the intellectual elite, compelling them to re-evaluate traditional cosmology and forcing them to become sceptical towards a regular and comprehensible universe (Henderson 1984: 171). Comparable to the astronomers of the same period, many of whom had read Western astronomical works that had challenged their Chinese training, Wu also thought that there were fundamental imponderables and irregularities in the world that were not simply due to the limitation of knowledge, though in his case without any known Western exposure (Ibid.: 248). If boundaries such as those which distinguished each season could not contain the activity of pestilential *qi*, then such phenomena simply changed, irrespective of a regular sequence, and were thus temporally and spatially contingent. The order of the universe, according to Wu, was fundamentally varied and unpredictable. Epidemics exposed the weakness of the system of correspondence based on seasonal configuration and opened up the realm of discourse to other ways of explaining epidemics.

Henceforth, for those who followed these new arguments within the febrile discourse, the identity of Warm diseases became tied to that of epidemics and separated from their previous association as a type of Cold Damage. Those who continued to align themselves with the Cold Damage tradition and the 'canonical formula current' (*jingfang pai* 經方派) associated with it, however, never accepted this distinction. They continued to understand Warm diseases as merely one of several possible seasonal transformations of an original underlying Cold Damage disorder. In the eighteenth century, physicians who aligned themselves with 'Han learning' (*Hanxue* 漢學) or the 'Return to antiquity' (*fugu* 復古) movement would, in fact, harden their position as a defence against the spread of the 'contemporary formula' (*jinfang* 今方) or 'modern formula' (*xinfang* 新方) currents of learning with which Wu Youxing would later become associated.

Nonetheless, Wu's conceptualisations of heterogeneous, pestilential, anomalous, and deviant *qi*, as well as his thoughts on the limitations of knowledge, occurred during a period of profound scepticism among educated elites. The traditional medical framework was found to be just as insufficient as the corrupted Confucian classics that formerly legitimated state ideology. Wu's critique of medical cosmology had profound effects on later physicians, because it exposed the limitations of the cosmology generally and introduced a new way of defining diseases, particularly epidemic ones. It thus became the basis for a new discourse on epidemics in the succeeding eighteenth through nineteenth centuries (Hanson 1997, 1998). Asking only natural-realist questions about what caused the late-Ming epidemics not only limits answers to medical causes, despite even contemporary Chinese accounts to the contrary of multiple socio-political-moral causes, but worse, ignores the richer history of Chinese medical scepticism and the broader socio-intellectual trends within which Chinese physicians articulated solutions to perceived limitations within their received medical traditions.

Historical-conceptual scholarship on the history of Chinese medicine 1997 onwards

Several other scholars have contributed to an expanding trend to contextualise historically Chinese disease concepts. French historians Catherine Despeux and Frédéric Obringer published a book analysing concepts of the 'cough' (*ke* 咳, *sou* 嗽) that focused on actors' categories in ancient and medieval Chinese medicine (Despeux and Obringer 1997). In the history of modern medicine in China, Bridie Andrews analysed how Chinese physicians first assimilated as well as questioned germ theory from the 1850s to 1930s through their various translations of the Western disease concept 'tuberculosis' (*fei jiehe* 肺結核 'lung tubercule'), including the Chinese disease concept literally meaning 'exhaustion disorder' (*laobing* 癆病) that corresponded well with the English 'consumption' (Andrews 1997). Angela Leung carried out a historical-conceptual analysis of the changing meanings of the Chinese disease concept 'numbing wind' (*mafeng* 痲瘋), which had many overlapping symptoms with those subsumed under the modern disease concept Hansen's disease or the outdated leprosy (Leung 1999).

The following year, medical historians published a cross-cultural synthesis of concepts of contagion (Conrad and Wujastyk 2000) that included three relevant chapters: one on 'Epidemics, weather, and contagion' argues that although there was 'contagion-consciousness' among ordinary people the scholarly Chinese medical tradition was more weather-conscious because of its broader cosmological frame (Kuriyama 2000); a chapter on the 'Threatening Stranger' (*kewu* 客忤) unpacks a unique concept of contagion in Chinese paediatrics (Cullen 2000); and a third chapter on the concept of 'Dispersing the Foetal Toxin of the Body' demonstrates how this therapeutic strategy was integral to treating smallpox as a congenital not contagious disease (Chang 2000). Furthermore, many contributions to Elisabeth Hsu's (2001) edited volume *Innovation in Chinese Medicine* can be considered representative of historical-conceptual methods. One chapter relates the history of doctrines underlying Chinese epidemiology to Northern-Song intellectual and political history (Despeux 2001). Another presents the history of medical case records as a distinct genre (Cullen 2001), within which physicians also recorded their responses to epidemics. My contribution charted the nineteenth-century development of a new discourse on epidemics in the Qing dynasty that took a regional as well as empirical turn (Hanson 2001).

Meanwhile, Leung cast her net even more broadly to examine local medical treatments (Leung 2001) and the relationship between concepts of disease and concepts of locality (Ibid. 2002). Her later articles on the popularisation of medicine in the Ming-Qing period (Ibid. 2003a), medical learning from the Song to the Ming (Ibid. 2003b), and vaccination in nineteenth-century Canton (Ibid. 2008) were not directly on the history of Chinese disease concepts but essential reading nonetheless for understanding their broader context. A decade after her first article on 'numbing wind', Leung completed *Leprosy in China: A History* (Ibid. 2009). This is a historical-conceptual *longue-durée* 'biography' of the disease concepts 'numbing wind' and 'skin afflictions with ugly sores' *li/lai* 癩 that Leung argued often resembled the symptoms of Hansen's disease but that had their own unique historical trajectories over Chinese medical history. Just as in the nineteenth century, the modern disease concept bubonic plague became equated with the late nineteenth-century neologism 'rat epidemics' (*shuyi* 鼠疫) (Benedict 1996a) and beriberi was equated with the fourth-century 'foot *qi*' (*jiaoqi* 腳氣), despite covering a much wider range of symptoms, since then (Smith 2008a, 2008b); modern leprosy has also found equivalents in the Chinese disease concepts *li/lai* 癩 and *mafeng* 痲瘋, despite their historically more multivalent and complex meanings.

Leung chose to write the opening chapter to the book she co-edited with Charlotte Furth on *Health and Hygiene in Chinese East Asia* (Leung and Furth 2010) on the 'Evolution of the Idea of *Chuanran* Contagion in Imperial China'. With this chapter, she effectively brought insights from her previous biographies of Chinese disease concepts to bear on the *longue-durée* history of *chuanran* 傳染 – the Chinese term chosen in the early twentieth century to translate contagion and infection. Since the early seventh century, the second character used the metaphor 'to dye' to signify 'to contaminate', similarly to infection's roots in *infecire* 'dye'. In the twelfth century, the first character meaning 'transmission' is first found combined with 'dye', first to mean contamination via sexual exchange and then to human-to-human disease transmission. But Leung argues that *chuanran* cannot be reduced to modern 'contagion' because over more than a millennium, it accrued multiple layers of meaning. These include non-human-to-human contamination, contact with pathogenic environmental *qi*, non-epidemic as well as epidemic diseases, and even sexually transmitted diseases. Her historical archeology of *chuanran*'s multivalency uncovers a finely granulated spatio-temporal terrain of how Chinese articulated disease transmission that cannot be reduced to modern-day contagion.

This methodological criticism of one-to-one translations holds as well for the retrospective diagnoses that translate Chinese disease concepts with equally complex historical legacies into modern disease concepts. Yet, major breakthroughs in the genetics of aDNA have forged a new path towards doing just this, opening new avenues of research in the process that have brought scientists and historians together in productive new ways.

Scientific transformations in the history of plague in China since 2011

During the first decade of the twenty-first century, scientists applied new methods to determine the genetics of the plague bacillus, *Yersinia pestis*, work that proved also to transform historical interpretations of the plague's global trajectory. A major genetic turn occurred in the global history of plague with the full gene sequencing of *Y. pestis* in 2011 based on aDNA that was gathered from human remains at East Smithfield (London's Black Death cemetery). This breakthrough solidified arguments that *Y. pestis* was one of the causative factors in the first plague outbreak in London from 1348 to 1350 and transformed historical scholarship on medieval plague thereafter (Green 2014, 2020). It also stimulated new lines of genetic research on the history of plague in China.

Shortly afterwards, Chinese geneticists made more precise claims about the evolution and origin of plague that are particularly important for the 'Chinese origin hypothesis' (Cui et al. 2013). Summing up this research, historian Robert Hymes wrote that their research 'signals a new departure in the cumulative study of the genetics of the bacillus over the preceding fifteen years' (Hymes 2014: 285). Based on genetic evidence, they made bold historical claims that sometime between 1142 and 1339, there was a 'polytomy' (namely, the simultaneous or nearly simultaneous genetic divergence of multiple lineage branches) from which most of today's strains of *Y. pestis* developed. Furthermore, they argued that the Qinghai-Tibet Plateau was where the bacillus originated.

Extending further their research, Hymes argued that the polytomy – some call it the 'Big Bang' – that yielded most of *Y. pestis*'s current strains 'can by placed in space and time in historical sources, too: that the polytomy first manifests itself historically in the long destruction, by the Mongols under Cinggis-Qan (Genghis Khan), of the Xia state of the Mi or "Tangut" people in the early 1200s, and continues with the movement of the Mongols into north China, south China, and much of Eurasia' (Ibid.). Recent hypotheses place the polytomy further west and a bit earlier during the Mongol conquest of Kara Khitai (Kyrgyszstan) (Green forthcoming). The new aDNA methods have strengthened the naturalist-realist approach to the history of plague in China and placed the *Y. pestis* polytomy in the Qinghai-Tibet Plateau between the mid-twelfth and mid-thirteenth centuries.

Hymes synthesised this modern genetic research with thirteenth- to fourteenth-century Chinese historical and medical records of epidemics in Central Asia and China to which he began to apply a historical-conceptual approach to how physicians recorded their experience with what they perceived to be a newly virulent epidemic disease, an angle he has pursued more deeply since (Hymes 2021). By combining arguments from both sides of the natural-realist and historical-conceptualist spectrum, Hymes was able to situate temporally the 'beginnings of the Black Death' – arguably one of the most transformative events for global human history – to the 1210s–1220s period when Mongols repeatedly attacked the Tangut people of the state of Xia and successfully conquered 'a state that sat cheek-by-jowl with what the new genetic evidence is telling us was probably the first home of plague' (Ibid.: 287). Hymes contributed not only an indispensable Central Asian-Chinese perspective to the groundbreaking book on *Pandemic Disease in the Medieval World: Rethinking the Black*

Death (Green 2014) but also an exemplary model of cross-fertilisation between natural-realist and historical-conceptual methods of analysis of the global history of pandemics.

Conclusion

Despite the current resurgence of naturalist-realist approach to history of epidemics in China with recent breakthroughs on genetics of *Y. pestis*, excellent scholarship from the historical-conceptual camp continues to be produced. Hilary Smith's book *Forgotten Disease: Illnesses Transformed in Chinese Medicine*, for example, narrates a long-durée history of the multivalent meanings of the Chinese disease concept 'foot *qi*' (*jiaoqi*) since the fourth century to demonstrate that the nineteenth-century conflation of *jiaoqi* with the modern disease concept beriberi distorts both China's past medical history and its encounters with Western imperialism in the past two centuries (Smith 2017). Smith asks the reader to consider what does unpacking the history of this 'Forgotten disease' Foot *qi* reveal about the social, political, and economic changes that Western imperialism brought to East Asia? The following answer that she provided illustrates the pitfalls of retrospective epidemiology: 'The history of foot *qi* suggests that beriberi outbreaks of the nineteenth century were, in fact, a new phenomenon that reflected the rise of imperialism and industrialisation just as surely as did the large-scale epidemics of smallpox, typhus, tuberculosis, and malaria in the same period'. Reframing beriberi as but one change in foot *qi*'s long story highlights the novelty of those nineteenth-century conditions. In this way, the study contributes to the more complex and more accurate understanding that is currently forming of how the modern distribution of power helped shape the global disease burden and how Western medicine reinforced imperial hierarchies at the same time that it relieved some illnesses (Ibid.: 22).

In other words, the modern experience of beriberi was not a feature of East Asian morbidity waiting to be discovered, as many Western physicians had previously argued, but rather a direct result of the spread of Japanese imperialism in East Asia from 1882 to the mid-1920s. As another historian wrote, 'the history of beriberi in Japan provides a good example of the socio-political issues at stake in the definition of disease' and 'how political contexts shape the production of scientific knowledge about disease, as well as the institutional dynamics that determine policy' (Peckham 2016: 18).

Furthermore, the *Visual Representations of the Third Plague Pandemic* project that Christos Lynteris directed for five years (2013–2018) has resulted in a series of stimulating monographs and edited volumes that combine the history of human with that of non-human epidemics into a new global history of pandemics. The titles alone illustrate their historical and contemporary relevance from the monographs *Ethnographic Plague: Configuring Disease on the Chinese-Russian Frontier* (Lynteris 2016) and *Human Extinction and the Pandemic Imaginary* (Lynteris 2020) to the edited volumes *Histories of Post-Morten Contagion* (Lynteris and Evans 2018), *Framing Animals as Epidemic Villains* (Lynteris 2019), *The Anthropology of Epidemics* (Kelly et al. 2019), and *Plague and the City* (Engelmann, Henderson, and Lynteris 2019). Collectively, these books offer some insights into our pandemic present by contributing fresh perspectives on the history of epidemics in late imperial and modern China while opening up new vistas onto the history of pandemics in the world. The full range of material as well as conceptual methods being applied now to the global history of pandemics, in fact, is as inseparable from the history of epidemics in China as it is likely to be inspirational for even more innovative research into the history of late imperial Chinese epidemiology.

Bibliography

Andrews, B. (1997) 'Tuberculosis and the assimilation of germ theory in China, 1895–1937', *Journal of the History of Medicine and Allied Sciences*, 52.1: 114–57.
Atwell, W. (1977) 'Notes on silver, foreign trade, and the late Ming economy', *Ch'ing-shih-wen-t'i*, 3.8: 1–33.
—— (1982) 'International bullion flows and the Chinese economy circa 1530–1650', *Past & Present*, 95: 68–90.
—— (1986) 'Some observations on the "seventeenth-century crisis" in China and Japan', *Journal of Asian Studies*, 65.2: 223–44.
—— (1988) 'The T'ai-ch'ang, T'ien-ch'i, and Ch'ung-chen reigns, 1620–1644', in F. Mote and D. Twitchett (eds) *The Cambridge History of China*, Vol. 7, *The Ming Dynasty, 1368–1644*, Part I, Cambridge; New York: Cambridge University Press, pp. 585–640.
Bates, D. (ed.) (1995) *Epistemology and the Scholarly Medical Traditions*, Cambridge: Cambridge University Press.
Bello, D. (2005) 'To go where no Han could go for long: malaria and the Qing construction of ethnic administrative space in frontier Yunnan', *Modern China*, 31.3: 283–317.
Benedict, C. (1988) 'Bubonic plague in nineteenth-century China', *Modern China*, 14.2: 107–55.
—— (1993) 'Policing the sick: plague and the origins of state medicine in late imperial China', *Late Imperial China*, 14.2: 60–77.
—— (1996a) *Bubonic Plague in Nineteenth-Century China*, Stanford, CA: Stanford University Press.
—— (1996b) 'Framing plague in China's past', in G. Hershatter, E. Honig, J.N. Lipman and R. Stross (eds) *Remapping China: fissures in historical terrain*, Stanford, CA: Stanford University Press, pp. 27–41.
—— (1996c) 'Epidemiology and history: an ecological approach to the history of plague in Qing China', *Chinese Environmental History Newsletter*, 3.1: 6–11.
Bielenstein, H. (1947) 'The census of China during the period 2–742 A.D.', *Bulletin of the Museum of Far Eastern Asia*, 19: 125–63.
Boyanton, S. (2015) *The Treatise on Cold Damage and the Formation of Literati Medicine: social, epidemiological, and medical Change in China, 1000–1400*, PhD thesis, Columbia University.
Bretelle-Establet, F. (2002) *La santé en Chine du Sud (1898–1928)*, Asie Orientale, Paris: CNRS Editions.
Brook, T. (2010) *The Troubled Empire: China in the Yuan and Ming dynasties*, Cambridge, MA: Belknap at Harvard University Press.
—— (2019) *Great State: China and the world*, London: Profile Books.
—— (2020) 'Comparative pandemics: the Tudor-Stuart and Wanli-Chongzhen years of pestilence, 1567–1666', *Journal of Global History*, 15.3: (note: need #s).
Buell, P.D. (2012) 'Qubilai and the rats', *Sudhoffs Archiv*, 96.2: 127–44.
Cao Shuji 曹树基 (1995) 'Dili huanjing yu Song-Yuan shidai de chuanranbing' 地理环境与宋元时代的传染病 (Geographic environment and contagious diseases in the Song-Yuan period), *Lishi yu dili* 历史与地理 (*History & Geography*), 12: 183–92.
—— (1997) 'Shuyi liuxing yu Huabei shehui de bianqian' 鼠疫流行与华北社会的变迁(1580–1644) (Plague epidemics and social transformations in North China) *Lishi yanjiu*历史研究, 1: 17–32.
Cao Shuji 曹树基 and Li Yushang 李玉尚 (2006) *Shuyi: Zhanzheng yu heping – Zhongguo de huanjing yu shehui bianqian, 1230–1960* 鼠疫: 战争与和平:中国的环境与社会变迁 (1230–1960) (Plague: War and Peace – Environmental and Social Transformations in China), Ji'nan: Zhongguo huabao chubanshe.
Cartwright, F.F., in collaboration with M.D. Biddiss (1972) *Disease and History: the influence of disease in shaping the great events of history*, New York: Thomas Crowell.
Chang Chia-feng 張嘉鳳 (1995) 'Strategies of dealing with smallpox in the Qing imperial family', in K. Hashimoto, C. Jami and L. Skar (eds) *East Asian Science: tradition and beyond: papers from the 7th international conference on the History of Science in East Asia, Kyoto, 2–7 August, 1993*, Osaka: Kansai University Press, pp. 199–205.
—— (1996a) *Aspects of Smallpox and its Significance in Chinese History*, PhD thesis, University of London, School of Oriental and African Studies.
—— (1996b) '*Qing chu de bi dou you cha dou zhidu*' 清初的避痘與查痘制度 (Eradicating and Diagnosing Smallpox during the Early Qing Dynasty), *Hanxue yanjiu*, 14.1: 135–56.

―――― (1996c) 'Qing Kangxi huangdi cai yong rendou fa de shijian yu yuanyin shi tan' 清康熙皇帝採用人痘法的時間與原因史探 (Historical exploration of the reasons and period when Emperor Kangxi chose to use human variolation methods), *Zhonghua yishi zazhi*, 26: 30–2.

―――― (2000) 'Dispersing the foetal toxin of the body: conceptions of smallpox aetiology in pre-modern China', in L.I. Conrad and D. Wujastyk (eds) *Contagion: perspectives from pre-modern societies*, Aldershot: Ashgate, pp. 23–38.

―――― (2001) '"Jiyi" yu "xiangran" – yi *Zhubing yuan hou lun* wei zhongxin shilun Wei Jin zhi Sui Tang zhijian yiji de jibing guan' '疾疫'與'相染'以 <諸病源候論> 為中心試論魏晉至隋唐之間醫籍的疾病觀 (Epidemics and Contagion: using the 'Treatise on the Origins and Symptoms of Various Diseases' to discuss the medical perspective on illness from the Wei and Jin to the Sui-Tang period), *Taida lishi xuebao*, 27: 37–82.

―――― (2002) 'Disease and its impact on politics, diplomacy, and the military: the case of smallpox and the Manchus (1613–1795)', *Journal of the History of Medicine and Allied Sciences*, 57.2: 177–97.

Chen Shengkun 陳勝崑 (1981) *Zhongguo jibing shi* 中國疾病史 (A History of Disease in China), Taipei: Ziran kexue wenhua shiye gongsi chubanbu.

Conrad, L.I. and Wujastyk, D. (eds) (2000) *Contagion: perspectives from pre-modern societies*, Aldershot: Ashgate.

Crombie, A.C. (ed.) (1963) *Scientific Change*, London: Heineman.

Cui, Y., Yu, C., Yan, Y., Li, D., Li, Y., Jombart, T., Weinert, L.A., Wang, Z., Guo, Z., Xu, L., Zhang, Y., Zheng, H., Qin, N., Xiao, X., Wu, M., Wang, X., Zhou, D., Qi, Z., Du, Z., Wu, H., Yang, X., Cao, H., Wang, H., Wang, J., Yao, S., Rakin, A., Li, Y., Falush, D., Balloux, F., Achtman, M., Song, Y., Wang, J. and Yang, R. (2013) 'Historical variations in mutation rate in an epidemic pathogen, *Yersinia pestis*', *Proceedings of the National Academy of Science*, 110.2: 577–82.

Cullen, C. (2000) 'The threatening stranger: kewu in pre-modern Chinese paediatrics', in L.I. Conrad and D. Wujastyk (eds) *Contagion: perspectives from pre-modern societies*, Aldershot: Ashgate, pp. 39–52.

―――― (2001) '*Yi'an* (Case statement): the origins of a genre of Chinese medical literature', in E. Hsu (ed.) *Innovation in Chinese Medicine*, Cambridge; New York: Cambridge University Press, pp. 297–323.

Cunningham, A. (1992) 'Transforming plague: the laboratory and the identity of infectious disease', in A. Cunningham and P. Williams (eds) *The Laboratory Revolution*, Cambridge: Cambridge University Press, pp. 209–44.

Cunningham, A. and Williams, P. (eds) (1992) *The Laboratory Revolution in Medicine*, Cambridge: Cambridge University Press.

Dean, K. (1993) *Taoist Ritual and Popular Cults of South-East China*, Princeton, NJ: Princeton University Press.

Despeux, C. (2001) 'The system of five circulatory phases and the six seasonal influences (*wuyun liuqi*), a source of innovation in medicine under the Song', in E. Hsu (ed.) *Innovation in Chinese Medicine*, Cambridge; New York: Cambridge University Press, pp. 121–65.

Despeux, C. and Obringer, F. (1997) *La Maladie dans la Chine Médiévale: la toux*, Recherches Asiatiques, Paris: Editions L'Harmattan.

Dikötter, F. (1993) 'Sexually transmitted disease in modern China: a historical survey', *Genitourin Medicine*, 69: 341–5.

Dunstan, H. (1975) 'The late Ming epidemics: a preliminary survey', *Ch'ing-shih wen-t'I*, 3.3: 1–59.

Engelmann, L., Henderson, J. and Lynteris, C. (eds) (2019) *Plague and the City*, The Body in the City Series, London: Routledge.

Epler, D.C. (1988) 'The concept of disease in an ancient Chinese medical text: the discourse on cold-damage disorders (Shang-han Lun)', *The Journal of the History of Medicine and Allied Sciences*, 43.2: 8–35.

Fan Xingzhun 范行準 (1986) *Zhongguo yishi shilue* 中國醫學史略 (Historical Summary of Chinese Medical Learning), Beijing: Zhongguo guji chubanshe.

―――― (1989) *Zhongguo bingshi xinyi* 中國病史新義 (New Significances for the History of Disease in China), Beijing: Zhongyi guji chubanshe.

Fisher, C.T. (1988) 'Smallpox, salesmen, and sectarians: Ming-Mongol relations in the Jiajing reign (1522–67)', *Ming Studies*, 25: 1–23.

―――― (1995/1996) 'Bubonic plague in modern China: an overview', *Journal of the Oriental Society of Australia*, 27 and 28: 57–104.

Green, M.H. (ed.) (2014) Inaugural Double Issue of *Medieval Globe*, 'Pandemic Disease in the Medieval World: Rethinking the Black Death,' Volume 1, Kalamazoo, MI; Bradford: Arc Medieval Press.

——— (ed.) (2014) *Pandemic Disease in the Medieval World: rethinking the Black Death*, TMG Occasional Volume 1, Kalamazoo, MI; Bradford: Arc Medieval Press.

——— (2020) 'The four black deaths', *The America Historical Review* 125.5: 1601–31.

Hanson, M. (1997) *Inventing a Tradition in Chinese Medicine: From universal canon to local medical knowledge in South China, the seventeenth to the nineteenth Century*, PhD Thesis, University of Pennsylvania.

——— (2001) 'Robust northerners and delicate southerners: the nineteenth-century invention of a southern *wenbing* tradition', in E. Hsu (ed.) *Innovation in Chinese Medicine*, Cambridge: Cambridge University Press, pp. 262–92.

——— (2008) 'Hand mnemonics in classical Chinese medicine: texts, earliest images and arts of memory', *Festschrift Issue in Honor of Nathan Sivin, Asia Major*, series 3, 21.1: 325–57.

——— (2010) 'Conceptual blind spots, media blindfolds: the case of SARS and traditional Chinese medicine', in Angela Ki-Che Leung and C. Furth (eds) *Health and Hygiene in Chinese East Asia: publics and policies in the long twentieth century*, Chapel Hill, NC: Duke University Press, pp. 369–410.

——— (2011) *Speaking of Epidemics in Chinese Medicine: disease and the geographic imagination*, London: Routledge Press.

——— (2017) 'Visualizing the geography of the diseases of China: western disease maps from analytical tools to tools of empire, sovereignty, and public health propaganda, 1878–1929', *Science in Context*, 30.3: 219–80.

Hashimoto, K., Jami, C. and Skar, L. (eds) (1995) *East Asian Science: tradition and beyond: papers from the 7th international conference on the History of Science in East Asia, Kyoto, 2–7 August, 1993*, Osaka: Kansai University Press.

Henderson, J.B. (1984) *The Development and Decline of Chinese Cosmology*, New York: Columbia Press.

Hershatter, G., Honig, E., Lipman, J.N. and Stross, R. (eds) (1996) *Remapping China: fissures in historical terrain*, Stanford, CA: Stanford University Press.

Hsiao Fan 蕭璠 (1993) '*Han Song jian wenxian suojian gudai zhongguo nanfang de dili huanjing yu difang bing ji qi yingxiang*' 漢宋間文獻所見古代中國南方的地理環境與地方病及影響 (Extant sources from the Han to Song on the environment and local diseases of ancient South China and their Influence), *Zhongyang yanjiuyuan Lishi yuyan yanjiusuo jikan*, 63.1: 67–171.

Hsu, E. (2008) 'The experience of wind in early and medieval Chinese medicine', in E. Hsu and C. Low (eds) *Wind, Life, Health: anthropological and historical perspectives*, Oxford: Blackwell Publishing, pp. 111–27.

——— (ed.) (2001) *Innovation in Chinese Medicine*, Needham Research Institute Series no. 3, Cambridge: Cambridge University Press.

Hsu, E. and Low, C. (eds) (2008) *Wind, Life, Health: anthropological and historical perspectives*, Oxford: Blackwell Publishing.

Hsu, F.L. (1952; 2nd edn 1973) *Religion, Science, and Human Crises: a study of China in transition and its implications for the west*, London: Routledge and Kegan Paul, repr. Westport: Greenwood Press.

——— (1983) *Exorcising the Trouble Makers: magic, science, and culture*, contributions to the study of religion, No. 11, Westport, CT: Greenwood Press.

Huang Ko-Wu 黃克武 (ed.) (2002) *Xingbie yu yiliao: disan jie guoji hanxue huiyi lunwenji lishi zu* 性別與醫遼: 第三屆國際漢學會議論文集歷史組 (Gender and Medicine: Collected Papers from the Third International Conference on Sinology, History Section), Taipei: Institute of Modern History, Academia Sinica.

Hymes, R. (2015) 'Epilogue: a hypothesis on the East Asian beginnings of the Yersinia pestis polytomy', in M.H. Green (ed.) *Pandemic Disease in the Medieval World: rethinking the Black Death*, Kalamazoo, MI: Arc Medieval Press, pp. 285–308.

——— (2021) 'A tale of two sieges: Liu Qi, Li Gao, and epidemics in the Jin-Yuan transition', *Journal of Song-Yuan Studies*, 50: 295–363.

Imura Kôzen 井村哮全 (1936–37) `Chihôshi ni kisaiseraretaru Chûgoku ekirei ryakkô' 地方志に記載せられたる中國疫癘略考 (Brief study of records of Chinese epidemics in local histories), 8 parts, *Chûgai iji shimpô* 中外醫事新報 (New Journal of Chinese and Foreign Medicine), 1232–1239: 30–39, 263–275, 316–325, 366–376, 414–422, 459–467, 505–507, 550–555.

Jannetta, A.B. (1987) *Epidemics and Mortality in Early Modern Japan*, Princeton, NJ: Princeton University Press.

——— (1993) 'Diseases of the early modern period in Japan', in K.F. Kiple (ed.) *The Cambridge World History of Human Disease*, Cambridge: Cambridge University Press, pp. 385–9.

——— (2006) *The Vaccinators: medical knowledge and the 'opening' of Japan*, Stanford, CA: Stanford University Press.

Katz, P.R. (1995) *Demon Hordes and Burning Boats: the cult of Marshal Wen in late imperial Chekiang*, Albany, NY: SUNY Press.
Kelly, A., Keck, F. and Lynteris, C. (2019) *The Anthropology of Epidemics*, Routledge Studies in Health and Medical Anthropology, London: Routledge.
Kim, Yung Sik and Bray, F. (eds) (1999) *Current Perspectives in the History of Science in East Asia*, Seoul: Seoul National University Press.
Kiple, K.F. (ed.) (1993) *The Cambridge World History of Human Disease*, Cambridge: Cambridge University Press.
Kuriyama, S. (1993) 'Concepts of disease in East Asia', in K.F. Kiple (ed.) *The Cambridge World History of Human Disease*, Cambridge: Cambridge University Press, pp. 52–9.
——— (1994) 'The imagination of winds and the development of the Chinese conception of the body', in A. Zito and T.E. Barlow (eds) *Body, Subject and Power in China*, Chicago, IL: University of Chicago Press, pp. 23–41.
——— (1999) *The Expressiveness of the Body and the Divergence of Greek and Chinese Medicine*, New York: Zone Books.
——— (2000) 'Epidemics, weather, and contagion in traditional Chinese medicine', in L.I. Conrad and D. Wujastyk (eds) *Contagion: perspectives from pre-modern societies*, Aldershot: Ashgate, pp. 3–22.
Leung, A.K.C. [Liang Qizi 梁其姿] (1987) 'Organized medicine in Ming-Qing China: state and private medical institutions in the lower Yangzi region', *Late Imperial China*, 8.1: 134–66.
——— (1987–88) 'Ming Qing yufang tianhua cuoshi zhi yanbian' 明清預防天化措施之演變 (History of the Evolution of Smallpox Preventive Measures in the Ming and Qing), in Yang Liansheng 楊聯陞 (ed.) *Guoshi shi lun: Tao Xisheng xiansheng jiu zhi rongqing zhushou lunwen ji* 國史釋論: 陶希聖先生九秩榮慶祝壽論文集 (Essays on Chinese History: Festschrift for Professor Tao Xisheng to Celebrate His Ninetieth Birthday), Taipei: Shihuo chubanshe, pp. 239–53.
——— (1993) 'Diseases of the premodern period in China', in K.F. Kiple (ed.) *The Cambridge World History of Human Disease*, Cambridge: Cambridge University Press, pp. 354–62.
——— (1997) *Shishan yu jiaohua: Ming Qing de cishan zuzhi* 施善與教化:明清的慈善組織 (Bestow Charity and Transform via Education: Charitable Institutions in the Ming and Qing), Taipei: Lianjing chuban shiye gongsi.
——— (1999) 'Mafeng bing gainian yanbian de lishi' 麻風病概念 演變的歷史 (The history of the evolving concepts of Leprosy), *Zhongyang yanjiu yuan lishi yuyan yanjiu suo jikan*, 70.2: 399–438.
——— (2001) 'Song Yuan Ming de difang yiliao ziyuan chu tan' 宋元明的地方醫療資源初探 (A preliminary survey of local medical treatments during the Song, Yuan, and Ming), *Zhongguo shehui lishi pinglun*, 3: 219–37.
——— (2002) 'Jibing yu fangtu zhi guanxi: Yuan zhi Qing jian yijie de kanfa' 疾病與風土之關係:元至清間醫界的看法 (The relations between diseases and locality: views of physicians from the Yuan to the Ming), in Huang Ko-Wu 黃克武 (ed.) *Xingbie yu yiliao: disan jie guoji hanxue huiyi lunwenji lishi zu* 性別與醫遼: 第三屆國際漢學會議論文集歷史組 (Gender and Medicine: Collected Papers from the Third International Conference on Sinology, History Section), Taipei: Institute of Modern History, Academia Sinica, pp. 165–212.
——— (2003a) 'Medical instruction and popularization in Ming-Qing China', *Late Imperial China*, 24.1: 130–52.
——— (2003b) 'Medical learning from the Song to the Ming', in P.J. Smith and R. von Glahn (eds) *The Song-Yuan-Ming transition in Chinese history*, Cambridge, MA: Harvard University Asia Center, pp. 374–98.
——— (2008) 'The business of vaccination in 19th-century Canton', *Late Imperial China*, 29.1: 7–39.
——— (2009) *Leprosy in China: a history*, New York: Columbia University Press.
——— (2010) 'Evolution of the idea of *chuanran* contagion in imperial China', in A. Leung and C. Furth (eds) *Health and Hygiene in Chinese East Asia*, Durham, NC: Duke University Press, pp. 25–50.
Leung, A. and Furth, C. (eds) (2010) *Health and Hygiene in Chinese East Asia: policies and publics in the long twentieth century*, Durham, NC: Duke University Press.
Lu Dong, Ma Xi, and Thann, F. (1995) *Les maux épidémiques dans l'empire chinois*, Paris: Editions L'Harmattan.
Lynteris, C. (2016) *Ethnographic Plague: configuring disease on the Chinese-Russian frontier*, London: Palgrave Macmillan.
——— (2020) *Human Extinction and the Pandemic Imaginary*, London: Routledge.

Lynteris, C. (ed.) (2019) *Framing Animals as Epidemic Villains: histories of non-human disease vectors*, Medicine and Biomedical Sciences in Modern History Series, London: Palgrave Macmillan.

Lynteris, C. and Evans, N.H.A. (eds) (2018) *Histories of Post-Mortem Contagion: infectious corpses and contested burials*, Medicine and Biomedical Sciences in Modern History Series London: Palgrave Macmillan.

MacPherson, K.L. (1987; 2nd edn 2002) *A Wilderness of Marshes: the origins of public health in Shanghai, 1843–1893*, Lanham, MD: Lexington Books.

Mote, F. and Twitchett, D. (eds) (1988) *The Cambridge History of China*, Vol. 7, *The Ming Dynasty, 1368–1644*, Part I, Cambridge; New York: Cambridge University Press.

Naquin, S. and Rawski, E.S. (1987) 'Topics for research in Ch'ing history', *Late Imperial China*, 8.1: 187–203.

Packard, R. (2016) 'The Fielding H. Garrison lecture: "Break-bone" fever in Philadelphia, 1780: reflections on the history of disease', *Bulletin of the History of Medicine*, 90.2: 193–21.

Parsons, J.B. (1970) *Peasant Rebellions of the Late Ming Dynasty*, Ann Arbor, MI: Association for Asian Studies.

Peckham, R. (2016) *Modern Epidemics in Asia*, New Approaches to Asian History no. 15, Cambridge: Cambridge University Press.

——— (2019) 'Plague Views: epidemics, photography, and the ruined city', in Engelmann, L., J. Henderson and C. Lynteris (eds) *Plague and the City*, London: Routledge, pp. 91–115.

Phillips, D.R. (1988) *The Epidemiological Transition in Hong Kong: changes in health and disease since the nineteenth century*, Hong Kong: University of Hong Kong.

Schipper, Kristofer (1985) 'Seigneurs royaux, dieux des épidémies', *Archives de Sciences Sociales des Religious*, 59.1: 31–40.

Sivin, N. (1987) *Traditional Medicine in Contemporary China*, Science, Medicine, and Technology in East Asia, 2, Ann Arbor: Center for Chinese Studies, University of Michigan.

Smith, H.A. (2008a) *Foot Qi: history of a Chinese disorder*, PhD Thesis, University of Pennsylvania.

——— (2008b) 'Understanding the *jiaoqi* experience: the medical approach to illness in seventh-century China', *Asia Major*, Third Series, 21.1: 273–92.

——— (2017) *Forgotten Disease: illnesses transformed in Chinese medicine*, Studies of the Weatherhead East Asian Institute, New York: Columbia University Press.

Smith, P.J. and von Glahn, R. (eds) (2003) *The Song-Yuan-Ming transition in Chinese history*, Cambridge, MA: Harvard University Asia Center.

Spence, J. (1990) *The search for modern China*, New York: W. W. Norton.

Temkin, O. (1963; 2nd edn 1977) 'The scientific approach to disease: specific entity and individual sickness', in A.C. Crombie (ed.) *Scientific change*, London: Heineman, pp. 629–47, republished in Temkin 1977, pp. 441–55.

——— (1977) *The double face of Janus and other essays in the history of medicine*, Baltimore, MD: Johns Hopkins University Press.

Wakeman, F. Jr. (1985) *The great enterprise: the Manchu reconstruction of imperial order in seventeeth-century China*, Berkeley: University of California Press.

Wilson, A. (2000) 'On the history of disease concepts: the case of pleurisy', *History of Sciences*, 38: 271–319.

Yang Liansheng 楊聯陞 (ed.) (1987–1988) *Guoshi shi lun: Tao Xisheng xiansheng jiu zhi rongqing zhushou lunwen ji* 國史釋論：陶希聖先生九秩榮慶祝壽論文集 (Essays on Chinese History: Festschrift for Professor Tao Xisheng to Celebrate his Ninetieth Birthday), Taipei: Shihuo chubanshe.

Zinsser, H. (1934; 2nd edn 1963) *Rats, lice, and history: the biography of a bacillus, a bacteriologist's classic study of a world scourge*, Boston: Little, Brown, and Company.

Zito, A. and Barlow, T.E. (eds) (1994) *Body, subject and power in China*, Chicago, IL: University of Chicago Press.

18
FOLK MEDICINE OF THE QING AND REPUBLICAN PERIODS

A review of therapies in Unschuld's Berlin manuscripts

Nalini Kirk

Introduction[1]

Healing in China has always encompassed many more thoughts and practices than those transmitted in the printed literature of literati medical traditions. Medical manuscripts offer a unique view into the fluid boundaries between various styles of practice and types of medical practitioners otherwise obscured by the received tradition. More than 40 years ago, Paul and Ulrike Unschuld began to collect medical manuscripts from local markets in China. Preserved now in the Berlin State Library and the Ethnological Museum, this collection encompasses 1,000 handwritten texts, the majority of which date to the nineteenth and early twentieth centuries.[2] These common writings by no-name authors are of immeasurable value for the historian interested in medicine of late imperial and early Republican China. Apart from containing abundant material for literary, linguistic, and socio-historical studies, they reveal dimensions of medical practice that have so far only rarely received scholarly attention.

The main part of this chapter is a review of the various therapies described in the manuscripts. I start by providing a summary of groups of authors and their purposes, and a definition of what folk medicine consisted of at the time. The review itself is largely based on Paul Unschuld and Zheng Jinsheng's catalogue (Unschuld and Zheng 2012), which serves as a starting point for any research on the collection's material. This I have supplemented with my own findings from MS 8051, the *Collected Notes on Medical Recipes* (*Yifang jichao* 醫方集抄) by the nineteenth-century Guangdong physician Tang Tingguang 唐廷光 and his successors.[3] Readers interested in particular therapies will find references to exemplary manuscripts and selected secondary literature consisting of historical studies, anthropological works, or books on practical application – after all, similar forms of therapy are applied in China and Taiwan even today.[4]

Apart from Unschuld and Zheng, few people have yet published any research on the collection. He Bian has used a text by a wholesale medicinal merchant (MS 8808) in an analysis of late imperial interregional drug trade and questions of authenticity (Bian 2014: 212–9) and Andrew Schonebaum has examined various manuscripts as places of intersection of medicine and vernacular literature (Schonebaum 2016).

Other scholars have meticulously studied older manuscripts, and a comparison with the texts discussed in their works might well shed some light on the history of the therapies presented in the Berlin collection. Taking us right back to the third and second centuries BCE is Donald Harper's translation of the medical texts of Mawangdui (Harper 1998). Vivienne Lo and Christopher Cullen have edited sixteen essays on the Dunhuang medical manuscripts (Lo and Cullen 2005), and more information on the same text corpus is provided in Catherine Despeux' volumes (Despeux 2010a).

The scarcity of secondary literature on folk medicine in late imperial China is perhaps due to the lack of primary medical sources. There are some, however: household almanacs constitute a window to domestic healthcare; there are apotropaic manuals; texts written or inspired by itinerant healers; recipe books by hereditary doctors and amateurs, or *materia medica* like the *Systematic Materia Medica* (*Bencao gangmu* 本草綱目) of Li Shizhen 李時珍 (1518–1593) that integrate many elements of folk medicine.[5]

A few scholars have directed their focus away from the field of late imperial literati medicine: Yili Wu's studies on popular gynaecology and traumatology touch upon certain aspects of what one might consider folk medicine (Wu 2000, 2015, 2017). Linda Barnes and T.J. Hinrichs have edited a comprehensive overview of the history of medicine in China, which does not hesitate to affirm folk practices as an integral part of public healthcare (Hinrichs and Barnes 2013). Their volume provides brief essays (including bibliographical information) on various topics discussed in this chapter. Other authors have investigated the world of medicine and healing encountered in non-medical sources such as late imperial fiction (Cullen 1993; Schonebaum 2016; Berg 2000, 2001) or legal case studies (Unschuld 1977; Sommer 2010).

Healing practices not unlike those encountered in the manuscripts were also observed by anthropologists in modern Taiwan and China (Kleinman *et al.* 1975; Kleinman 1980; Zhang 1989) only a few decades later. As for practical application, several Chinese monographs and journal articles discuss contemporary folk therapies (mostly those of ethnic minorities) but unfortunately simply list treatment methods and indications without providing any background information whatsoever (Liu and Liu 1992a; Wang 1998, 2000).

Folk medicine and its practitioners

The authors and their purposes[6]

Printed medical texts, even case studies, were to at least some extent intended to establish medical and scholarly authority and therefore cannot be assumed to represent unadorned medical reality. In comparison, most authors of the Berlin manuscripts only put down in writing what they considered useful for *practical application*. High expenses for medicines or limited access to physicians and pharmacies made laypersons record affordable treatment methods they considered effective. Medical experts transmitted and safeguarded knowledge within their group by compiling secret manuscripts. This established a sense of group identity, ensured exclusivity and economic benefit, while less honourably, secrecy also prevented deceptive methods from being leaked to outsiders. Finally, some authors copied excerpts from printed texts for personal use, and a few apparently planned to publish their own writings.

Always keeping the great diversity of actors in the medical marketplace in mind, the authors of the Berlin collection can roughly be divided into the following groups:

1. *Non-medical authors* collected pharmaceutical and dietary recipes, acumoxa therapies, or apotropaic techniques (*zhuyou* 祝由) for their own personal use at home. Their manuscripts follow no systematic order, and recipes are simple and include many substances of rural or local origin rarely found in conventional recipe literature.
2. *Folk healers'* manuscripts list the same therapies as those of lay authors but also include heterodox practices and specialist knowledge on external medicine, paediatrics, gynaecology, and abortion. In comparison to regular physicians' writings, these manuscripts are characterised by the relative absence of theoretical reasoning or by ideas differing from regular medical theory. Many texts Unschuld and Zheng attribute to folk healers, however, reveal that the authors had undergone professional training to varying extents, were at least somewhat familiar with regular medical theory and printed texts, and practised medicine for a living.[7]
3. This makes them comparable to *regular physicians*, who had received literary and professional medical training of some sort.[8] Physicians' manuscripts were for the most part based on excerpts from printed medical books or consisted of their own writings intended for publishing. We find comments on recipes, case studies reflecting personal experience, theoretical elaboration, and clinical specialisation. However, even the medicine of well-educated scholar-physicians was not free of folk influences. Others have already noted the inclusion of popular knowledge into literati medicine (Scheid 2005: 97), and the manuscripts provide ample proof that the boundaries between folk and literati medicine were indistinct at best. MS 8051 is just one example of a text that lists complex recipes from printed literature side by side with recipes containing local herbs or single-ingredient recipes, with bloodletting and wick cauterisation, or ritual instructions for the preparation of medicinals.
4. *Itinerant healers'* manuscripts contained rhetorical and practical tricks to attract customers and make maximum profit in addition to various therapies. While their theoretical ideas, though scanty and simplified, were sometimes based on regular medical theory, itinerant healers' therapeutic interventions differed considerably from literati medicine. According to the *Refined Therapies of Itinerant Physicians* (*Chuanya* 串雅) of Zhao Xuemin 趙學敏 (c. 1720–1805), itinerant healers possessed skills in acupuncture and massage; they were specialists at 'pulling teeth; dabbing [medicinals] onto macules; removing eye shades and grasping worms' and performed apotropaic rituals (*Chuanya*, prefaces: 9–10, 13). Because therapies were required to be cheap (*jian* 賤), effective (*yan* 驗), and convenient (*bian* 便), they consisted mostly of cure-all plaster remedies and pharmaceutical recipes made from few or even single ingredients.[9]
5. Although we find apotropaic therapies scattered throughout the collection, a number of texts solely devoted to divination and magical healing suggest that these were written by *magico-religious healers* who specialised in some form of prognostics and ritual therapy. Their methods were manifold and reflect the potpourri of hemerological, Daoist, Buddhist, and local cultic beliefs that characterises religion in China.
6. *Practitioners of martial arts*, naturally, took a special interest in healing injuries. Manuscripts of such authors are mainly concerned with traumatology and include diagnostic and prognostic procedures, pharmaceutical recipes, and descriptions of bone setting techniques. These were often based on physiological ideas differing from those of orthodox medicine.
7. *Pharmacists* constitute another important group of authors. Their manuscripts were medication lists for commercial purposes and not intended to record actual therapies; therefore, they will not be discussed here.

Folk medicine in the manuscripts

Folk medicine was therapeutic practice for everybody, no matter where a person lived or to which social stratum he or she belonged. Anderson and Anderson use the term 'folk' to refer to 'the behaviour of ordinary community member, the man-in-the-street, as opposed to the medical experts [···] who are carriers of a classical or great tradition, or some similar body of highly specialized lore' (Anderson and Anderson 1975: 143). However, healthcare in China encompassed multiple layers of therapeutic action and expertise in the field expanding from the 'ordinary community member' to the 'carriers of a classical or great tradition': common people applied simple home remedies; literati amateurs prescribed elaborate medical recipes. A sick person's family could employ the services of elite scholar-physicians, hereditary physicians, monks, shamans, bone setters, wound doctors, female healers, itinerant drug-peddlers, acupuncturists, or massage therapists. Healers from the lower echelons of society crossed the boundaries in between these groups by acquiring knowledge and gaining fame as specialists in their field of expertise. Knowledge and practices travelled by personal and literary exchange from one sphere of healing to another. This fluidity is reflected in the Berlin manuscripts and reveals that even the division of what was called Chinese folk medicine by anthropologists of the 1970s to 1980s into a 'sacred' and 'secular' sector is tenuous (Ahern 1975; Kleinman 1980: 63–5).[10] Reviewing the manuscripts, one finds not only that most authors, regardless of their literary or social origin, did not limit themselves to one type of therapy, but also that the secular and sacred spheres were not as far apart as Kleinman observed decades later. Pharmaceutical texts included acumoxa or scraping techniques, texts on traumatology or paediatrics listed pharmaceutical recipes side by side with manual methods or lamp wick cauterisation. Manuscripts otherwise devoted entirely to pharmaceutical recipes contained apotropaic therapies, and apotropaic manuscripts listed herbal formulæ. Nevertheless, in all this hotchpotch we need to make an assessment as to the gradual differences that qualify certain manuscripts as more 'on the folk side' in comparison to those that are representative of classical literati medicine. I propose the following criteria:

1. the level of literary education of the author and the amount of material copied from printed works;
2. the proportion of simple, cheap recipes to complex medical recipes from printed texts (and the theoretical reasoning associated with these);
3. the proportion of substances of domestic, local, or rural origin to substances available on the urban, transregional drug market;
4. the proportion of sacred medicine, heterodox practices, and surgical and manual techniques to secular medicine and orthodox pharmacology.

For this review, I will focus on manuscripts characterised by a relatively low level of literary education of the authors, by an emphasis on simple and cheap recipes, pharmaceutical substances of local or rural origin, magico-religious or heterodox practices, and on surgical or manual techniques. This does not exclude specialist knowledge on certain fields of expertise such as traumatology or paediatrics. However, while classical medicine was mostly employed by the urban, literate elite, the therapies preserved in these texts were available to the 'man in the street' as well.

Therapies

Pharmacotherapy

Pharmaceutical substances

Regional plant and animal products fitted the 'easy-to-obtain' requirement of common healers. This included victuals such as vegetables, grains, fruit, or meats but also non-edible indigenous plants. Provided that a manuscript can be pinpointed to one locality, an identification of the medicinal plants used there appears a fascinating field of research.[11]

Non-herbal substances included objects from everyday life such as chopsticks, coins (MS 8121), or needle shards (MS 8051), and the various types of dust or carriage-axle grease (MS 8198) and drugs of human or animal origin might be regarded with disgust by the modern observer. Traditional *materia medica* such as the *Bencao gangmu* include entries on human or animal excretions, head or body grease, hair, fingernails, and the like, but the brevity with which these substances are treated indicates that they were not used widely in literati medicine. In folk medicine, however, they frequently served as medicinals in their own right or were used for the preparation of other drugs. MS 8051 recommends the faeces of boys, cattle, rats, and cats in internal or external recipes, and elsewhere we find human sperm, pigeon's droppings, insect and donkey's dung (MS 8121), or even a human skull for the treatment of a wide variety of diseases.[12]

External and simple recipes

Throughout the manuscripts, we find advice on the external applications of medicinals. This is to be expected in the treatment of sores and traumatic injuries, but plasters and pastes applied on the navel, acupuncture points, or other parts of the body were used for the treatment of fright-wind (*jingfeng* 驚風), cough, and other internal diseases as well.

Simple recipes (applied internally or externally) consisted of a combination of a few medicinals with 'dirty' substances or foods such as eggs or innards. Even easier to apply were single-ingredient recipes consisting of only one pharmaceutical drug or foodstuffs like salt or egg yolk.[13]

Fakes and tricks

Pharmaceutical recipes were not only required to be simple and effective, practitioners also needed to earn their livelihood and therefore sought financial profit. For this purpose, itinerant healers relied on rhetoric, sleights of hand, and the faking of medicinals (Zheng 1996). Traditional *materia medica* frequently warn of fake substances and describe methods to expose them. In itinerant healers' manuscripts, we find first-hand information recorded by the adulterators themselves (MS 8722, MS 8253).

However, a regular physician's practice was not free of trickery either: the authors of MS 8051 increased the value of cheap recipes by adding other substances to change their colour or odour, by crushing them into a pulp or powder, or by providing well-known remedies with new, fine-sounding names (Unschuld and Zheng 2012: 62–4). This clearly shows that in medical practice, the ingredients and form of administration of medical recipes did not only depend on theoretical considerations and pharmaceutical effects but served economic purposes as well.

Pharmaceutical substances in magico-religious therapies

Many effects attributed to pharmaceutical drugs were derived from belief in magical correspondences, but certain substances were considered magically effective in themselves.[14] People ingested the ash of burnt amulets with medicinal decoctions, used medicinal plants for fumigation, kept or planted them in or around the house, performed ritual baths, or carried them on the body. Certain woods considered magically effective (e.g. peachwood) were carved into ritual devices, and rituals were performed before or during the preparation of medicinals (MS 8051). The fact that we find these methods not only in apotropaic manuscripts but in pharmaceutical ones as well shows us that a strong connection between pharmacotherapy and ritual healing still prevailed in Qing and Republican China.

Piercing, cauterisation, and surgery

Marginalised practices

Cauterisation, piercing (I will use this term here to designate bloodletting and acupuncture), and the lancing of abscesses were closely related in ancient times (Harper 1998: 92–4; Lo 2002). As theory-based pharmacology evolved, these hands-on techniques were disdained by many literati physicians. There is only a small amount of specialised acumoxa manuscripts in the Berlin collection, most of which date from the 1920s to 1950s, this scarcity probably reflecting the marginalisation of acupuncture during the Qing dynasty (Unschuld and Zheng 2012: 126). They were mostly copied from printed books and reveal only limited information on the clinical practice of acumoxa in folk medicine. More promising in this regard seems an assessment of the few manuscripts that cannot be traced back to printed works (8065) and of the numerous piercing and cauterisation techniques interspersed into recipe texts by practising physicians. These, along with the striking absence of theory they reveal, might be an indicator that acumoxa was still practised as the 'quick and easy' healthcare Vivienne Lo detected in the Dunhuang moxibustion charts (Lo 2005).

Piercing and cauterisation in magico-religious context

Some scholars have emphasised the connection between ritual and heat therapy in ancient medicine (Harper 1998: 95–6; Yamada 1998: 67–72); others have surveyed acumoxa in medieval religious healing, Strickmann tentatively suggesting a connection between exorcistic 'sword methods' (see below) and acupuncture (Strickmann 2002: 241–3, 150; Stanley-Baker 2013: 76–109). The Berlin collection provides valuable material for an investigation of the relationship between acumoxa and the magico-religious practices of the time: piercing techniques are not only listed side by side with apotropaic methods, but some manuscripts combine bloodletting with pharmaceutical recipes strongly reminiscent of apotropaic formulæ (MS 8178) or even integrate acupuncture into exorcistic rituals (MS 8798).

Similarly, prohibitions and indications for the stimulation of anatomical locations at certain times must be seen in the context of hemerological calculation, magico-religious ideas, and life-cultivation practices. As Vivienne Lo has shown, this holds true for early and medieval China (Lo 2001), and the manuscripts bear witness of the survival of such recommendations until the twentieth century.

Lamp wick cauterisation

Descriptions of cauterisation with a hot cord appear in medical texts as early as the Warring States period (Harper 1998: 95). MS 8036, a manuscript devoted mostly to lamp wick cauterisation (*denghuo fa* 燈火法, a therapy based on the scorching of certain points on the body with a burning lamp wick or oil-soaked rush pith),[15] traces this practice back to the *Yellow Thearch's Inner Classic* (*Huangdi neijing* 黃帝內經), and laments its loss in later acupuncture literature. However, lamp wick cauterisation is very much present in the Berlin collection. Several specialised manuscripts include exact descriptions and illustrations of when and where to apply such treatment in cases of smallpox or fright-wind (MS 8004, MS 8102, MS 8119, MS 8494, MS 8764). Wick cauterisation apparently formed an integral part of late imperial folk paediatrics (Unschuld and Zheng 2012: 111, 113) but was in no ways limited to the treatment of children, as various other manuscripts show. MS 8051 proposes cauterisation of red macules erupting in the course of acute enteritis, a treatment that Li Shizhen, 300 years earlier, considered significant enough to include in the 'lamp wick' entry of the *Bencao gangmu* (*Bencao gangmu* 6: 421) (Figure 18.1).

Comment: This illustration from MS 8036 depicts 'black sand fright wind' (*wusha jingfeng* 烏沙驚風), one of twenty-six conditions of fright-wind characterised by black sand-like eruptions on the entire body, blue veins on the belly, bloated abdomen, black lips and mouth, vomiting, and diarrhoea. The text recommends cauterising the area below the throat, the navel, the 'tiger mouth' (*hukou* 虎口) points, and the blue veins above the 'back seam' (*beifeng* 背縫) points with a lamp wick.

Bloodletting and minor surgery[16]

Kuriyama has emphasised the decline of bloodletting from late Han times onwards (Kuriyama 1995: 15), but this view only holds true for literati medicine. Again, the *Bencao gangmu* repeatedly refers to bloodletting techniques, and the Berlin manuscripts reveal that the average physician had no problem with applying such methods in medical practice. Even today, certain forms of bloodletting are still conducted in folk practice and acupuncture clinics.[17]

Similarly, the lancing of abscesses (now carried out with needles and knives instead of lancing stones), though frequently disapproved of by literati authors, remained a necessary intervention in Qing medicine (Chapter 13 in this volume). Other minor surgeries included the opening of pox papules (MS 8221), the removal of arrowheads (MS 8269), the stitching up of wounds (MS 8111), and tooth extraction (MS 8166). Pharmaceutical substances were sometimes applied externally for anaesthesia during such procedures.[18]

The scraping method (*guafa* 刮法, i.e. scraping the skin with a blunt instrument in order to break the capillaries and cause petechiae) is placed in close proximity to piercing and cauterisation for the treatment of sand-like skin eruptions (*sha* 痧)[19] (MS 8358, MS 8397). It seems that scraping was intended to cause bruising or possibly bleeding of papulous exanthema, which implies that it was considered a bloodletting technique rather than a type of massage as it is nowadays (Liu and Liu 1992b: 127–31).

Bone setting and massage

Bone setting (*zhenggu* 正骨) and massage (*anmo* 按摩, from Ming times on *tuina* 推拿) constitute important elements of popular healthcare until today.[20] Though *tuina* has now been re-integrated into Chinese medical institutions, this was not always the case. The Song

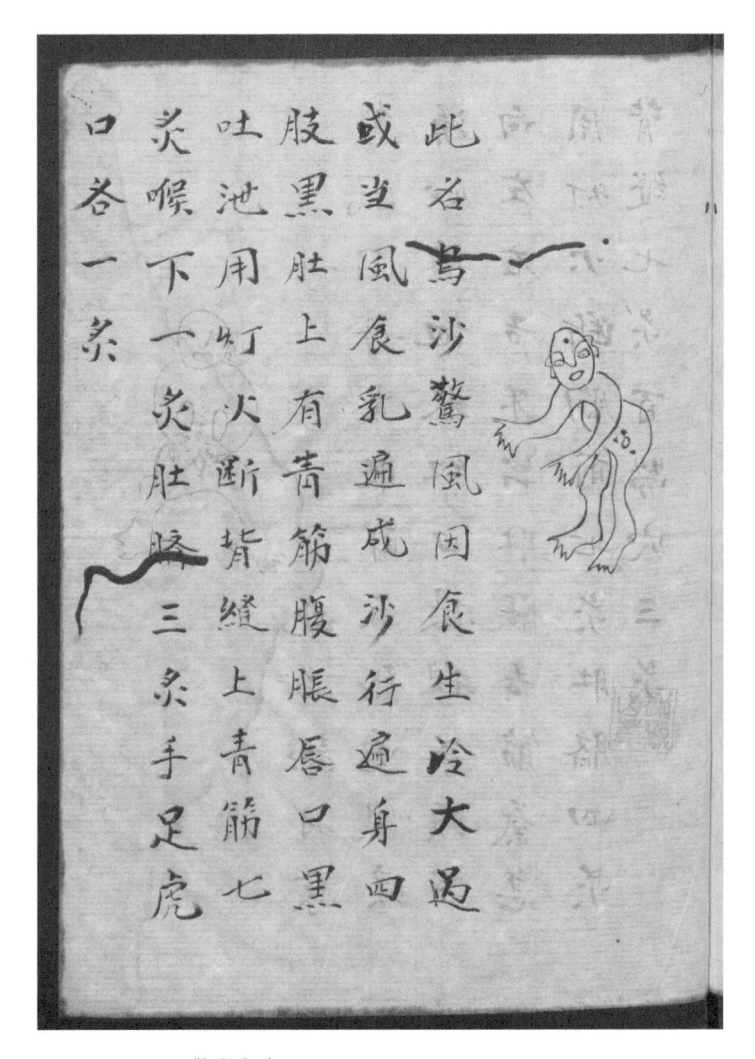

Figure 18.1 Jingzheng huoshi 驚症火式 (Fire Approach to Pathoconditions of Fright, MS 8036), p. 15. Staatsbibliothek zu Berlin – PK

government disbanded the Tang massage department (Chapter 6 in this volume), and with the rise of literati medicine and its focus on pharmaceutics massage was increasingly marginalised. Nevertheless, it remained part of folk medicine and rose in popularity again during the Ming and Qing dynasties as a simple, cheap, and safe way of treating children, possibly in place of more invasive procedures like acupuncture. All *tuina* manuscripts in the collection were copied from printed texts (Unschuld and Zheng 2012: 111–3) and concern themselves almost exclusively with paediatrics. This, of course, does not mean that only children enjoyed such treatments. Influential books like the *Golden Mirror of Medical Learning* (*Yizong jinjian* 醫宗金鑒) generally recommend massage for diseases of muscles, tendons, and joints (*Yizong jinjian*: 87.803), and at least MS 8131 indicates that the author practised his art as an alternative to potentially dangerous pharmaceutics.

Various manuscripts in the collection deal with bone setting and trauma medicine. MS 8167 includes a manual inspired by Japanese orthopaedics and Western anatomy on how to reposition dislocated joints, while other manuscripts discuss traumatology in general. Alternative physiologies of channels and openings on the body through which blood was thought to flow according to a specific timetable served as a basis for diagnosis, prognosis, and manual treatment (Unschuld and Zheng 2012: 120–4). Sometimes illustrated in diagrams (MS 8111, MS 8027, MS 8041), these notions were derived from martial arts rather than regular medical theory (Figure 18.2).

Comment: This illustration from MS 8111 shows 'holes' (*xue* 穴) on the front of the head and trunk, the injury of which can be fatal. Names and locations of these holes, such as the 'blood sack hole' (*xuenang xue* 血囊穴) below the right twelfth rib, are specific to trauma medicine.

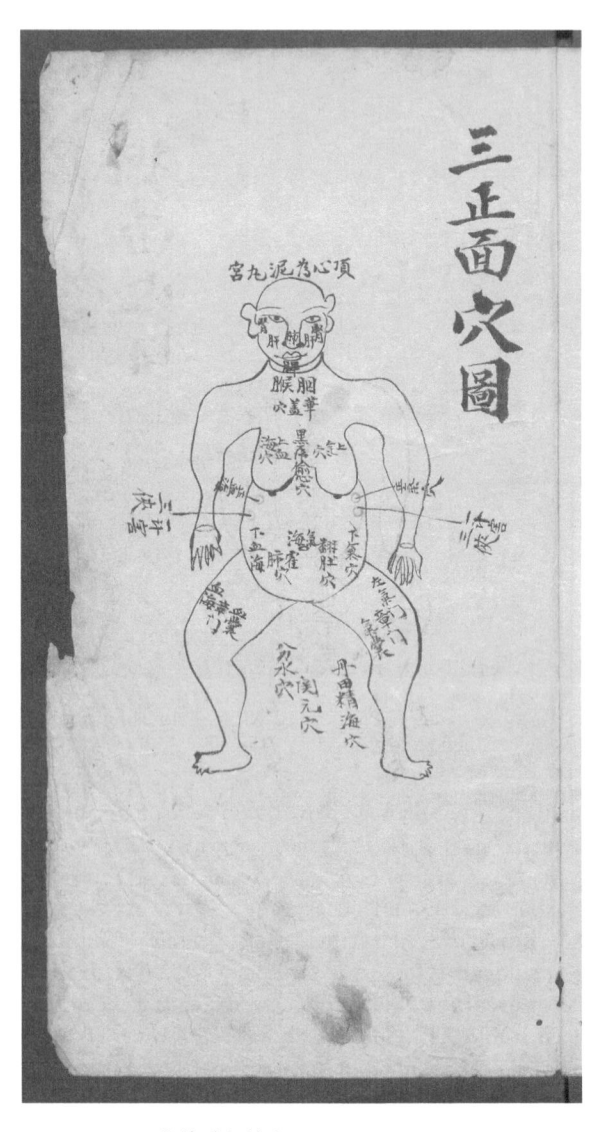

Figure 18.2 Michuan dieda bofang 秘傳跌打鈸方 (Secretly Transmitted Recipes [Treating Injuries from] Knocks and Falls, [Strung Together like] Cymbals, MS 8111), p. 7. Staatsbibliothek zu Berlin – PK

Childbirth practices and abortion

Childbirth practices

The scholarly impact on women's medicine in imperial China was less apparent in the field of hands-on obstetrics, which remained largely the responsibility of female experts. While printed gynaecology books contain extensive erudite treatises on pregnancy and postpartum care, chapters on parturition are, in comparison, rather brief and permeated with folk practices. They include birthing charts, instructions on amulet application, single-ingredient and apotropaic recipes, or advice on manual techniques for birth complications.[21] Obstetric treatments in the Berlin manuscripts mainly consist of pharmacotherapy and magic but also include some descriptions of hands-on manipulations (MS 8014). The vast amount of apotropaic recipes for childbirth complications indicate that the abandonment of ritual instructions Furth has identified in late imperial printed literature (Furth 1999: 178) did not occur in folk medicine. Just as a practice like pricking the foetus with a needle to make it turn (MS 8538) had apparently survived since pre-Tang times (Lee 2006: 141), so were amulets, spells, or objects that magically corresponded to childbirth still applied to hasten delivery by the authors of the manuscripts (MS 8121).

Abortion

Zheng Jinsheng has analysed the methods of abortion in the Berlin collection and compared them to ones from printed literature (Zheng 2013). The percentage of such recipes is significantly higher in the manuscripts, and recipes consist of highly toxic substances that were often applied externally through the vagina instead of being ingested orally. Examining legal case studies, Matthew Sommer concluded that, far from being a widespread, safe instrument of birth control, abortion in Qing China was potentially incriminating for the practitioner, unreliable, and bore considerable risks for a woman's health (Sommer 2010). His findings may explain the caution and criticism concerning the termination of pregnancies that become apparent in printed literature, but the evidence in the manuscripts indicates that abortion did indeed constitute a considerable part of popular gynaecology.

Dietetics

Ute Engelhardt (2001: 173) has noted the lack of differentiation between foodstuffs and medicinal drugs in the Mawangdui manuscripts and has identified the emergence of a distinction between drug and dietary therapy in Tang dynasty medical literature. The fact that there is not one specialised manuscript on dietetics in the Berlin collection indicates that such 'literati dietetics' was not part of folk medicine. In the 1960s, the Andersons still observed that 'food graded into medicine and medicine graded into food' in Chinese folk practice, and diet and medicine were closely interconnected (Anderson and Anderson 1975: 154). Thus, in addition to proper therapeutic food recipes (MS 8148), we find recipes that could qualify as dietetic ones in various medical recipe manuscripts. MS 8051 lists preparations with candle wax to treat hunger and therapeutic food recipes such as soup of pork belly and toad, carp stuffed with salt and medicinals to be eaten with gruel, or chives boiled in pig's blood side by side with pharmaceutical recipes. At the same time, medical recipes are often flanked with dietary recommendations and prohibitions. The origins of these instructions may be traced back to notions about magical correspondences (Engelhardt 2001: 183) or the thermic qualities of foodstuffs (Anderson and Anderson 1975: 144–8; Zhang 1989: 149–57), but some, like the prohibition of salt in cases of abdominal drum-distension (*guzhang* 鼓脹), were apparently based on practical observation rather than theory.

Magico-religious practices

Apotropaic manuscripts have three main purposes: banning evil (that is, pathogenic influences), divination in order to predict the outcome of a disease and to identify its cause, and healing illness. It might be a point of discussion whether hemerological and calendrical calculations should be counted as magic, but they are so closely interwoven with demonological ideas in the manuscripts that it seems suitable to do so here.

The eclectic apotropaic practices revealed by the Berlin collection correspond with the mutual influences between beliefs and healing rituals of Buddhism, Daoism, and folk religion Strickmann and Sivin identified in Song and pre-Song texts (Strickmann 2002, Sivin 2015).[22] In several manuscripts, we encounter figures from local popular cults, Daoist rituals, Buddhist phrases and deities (MS 8012, MS 8390, MS 8408, MS 8564, MS 8721), or ideas from Indian astrology (MS 8651).[23] We find practices resembling those described in ancient manuscripts (Harper 1998: 148–83), Song medical works, and late imperial *zhuyou* manuals (Cho 2005: 47–53, 71–9, 2013: 78–84, 94–9; Lin 2013), as well as those performed by ritual specialists in modern Taiwan (Gould–Martin 1975; Kleinman 1980: 210–42; Zhang 1989: 74–82, 89–90). Obviously, the magico-religious therapies recorded by the authors of the collection were neither new, nor did they fade into oblivion in modern times.

Rituals

Healing rituals usually included the application of amulets, spells and incantations, and ritual movements, but some also showed distinctive features: as in earlier texts, Daoist healing rituals in the manuscripts projected secular bureaucratic processes onto the spirit world (MS 8650), and the appropriate treatments consisted of petitions, lawsuits, and the like.[24] Rituals involved the use of fire, nails, or a sword made of wood with magical properties, or, less elaborately, the burning of paper money or the removal of objects from the house (MS 8184, MS 8794).

Harper advocates the connection between ancient southern traditions of 'spitting and spouting' and Qing local exorcistic traditions (Harper 1998: 179–83). Healing rituals involving amulets and spitting were reported from a 1970s Taiwanese village as well (Gould-Martin 1975: 121). Thus, manuscripts that list similar methods (MS 8407) provide another clue for the transmission of ancient apotropaic techniques through the centuries.

Amulets

Maybe the most important magical therapy described in the manuscripts is the application of amulets. Amulets were usually written on paper, burnt and ingested, sometimes together with decoctions of pharmaceutical substances (MS 8070). People also placed such charms in the house (e.g. nailing them to the door or keeping them underneath the pillow), carried them on the body, or wrote them directly onto the skin. The manuscripts reveal a wide variety of amulets, ones with written magical characters and spells, ones without any fixed form of writing at all, some that consist of diagrams, and one with images of star constellations (MS 8407). The application of an amulet was usually accompanied by a spell or incantation, in some cases also with the performance of ritual movements.

Spells and incantations

Spells and incantations usually consisted of requests for assistance from numinous powers such as spirits, animals, celestial bureaucrats, or natural forces, or of interdictions directed towards harmful influences. While spells were written on amulets or other places, incantations were recited during amulet application, acupuncture, the performance of physical postures and gestures, or any other sort of ritual.

Ritual movements

The hand gestures illustrated in some manuscripts (Unschuld and Zheng 2012: 158–60) are strongly reminiscent of Indian *mudrās*. Such finger seals (*yin* 印) had apparently been transmitted through the ages since medieval times (Strickmann 1995; Strickmann 2002: chs. 4 and 6; Cho 2005: 73–5). Different types of gestures were believed to ban evil or invoke favourable influences (MS 8070, MS 8806), and some were used for divination (MS 8565).[25]

The feet of the religious specialist served as ritual instruments as well. Just one example is the 'step of Yu' (*yubu* 禹步, MS 8588) which we find referred to as early as the Warring States (Harper 1998: 167–9), thus serving as another proof for the ancient origin of twentieth-century ritual acts.

Calendrics, divination, and demonology

The Berlin collection reveals abundant information on practices that combined calendrical calculation, divination, and demonological ideas.[26]

Specific demons were associated with certain days of the year that were calculated according to the sixty-day cycle of the stems and branches (MS 8184, MS 8554, MS 8309), the thirty days of the month (MS 8184, MS 8563), or the eight trigrams (MS 8610, MS 8611). The birth date, date of consultation, or the time of onset of an illness were used to determine the demonological cause of a disease and to choose the appropriate ritual treatment. The diviner might gain additional information by the assessment of cardinal directions, the five transitional phases, dreams and bodily manifestations (MS 8268), or the random tossing of glossy objects (MS 8656). The application of the 'emolument and horse diagram' (MS 8439) as well as the calculation of auspicious and inauspicious years according to Indian astrology (MS 8651) are other methods of divination found in the manuscripts (Unschuld and Zheng 2012: 162–74).

Summary

We have now identified some aspects of folk therapies in the Berlin collection: simple pharmacotherapy; piercing and cauterisation; surgical and manual techniques; obstetrics and abortion; folk dietetic and magical healing. In comparison with literati medicine and its asserted emphasis on elaborate cosmological and pathophysiological theory, the constituting factors for folk medicine and its therapies were practical and magico-religious: treatments had to be readily available and profitable, and effectiveness was assessed by experience, magical belief, or (sometimes) alternative theory. Each practice deserves closer attention in itself, but the interconnections between different therapies and traditions appear to be a fascinating field of research as well. Not only were acumoxa and pharmaceutics closely connected to magical healing, folk and literati medicine as a whole reveal large areas of intersection. Even the daily practice of regular physicians, though grounded in varying degrees of literary medical education, was permeated with practical and economical considerations or magico-religious notions, thus including the use of local drugs, dirty substances, single-ingredient recipes, piercing techniques, trickery, or ritual. These practices appear in various genres of literature, even, if read closely, in printed medical books, and the manuscripts provide a valuable additional source for a 'bottom up' approach to the history of medicine in China.

Notes

1 The open access publication of this chapter was funded by the Federal Ministry of Education and Research BMBF project SimLeap (FKZ 13GW0226A) at Charité - Universitätsmedizin Berlin.
2 Note that the vast majority of manuscripts are held at the Berlin State Library (Staatsbibliothek zu Berlin) and only few at the Ethnological Museum (Ethnologisches Museum).
3 For an extensive description of MS 8051, see Unschuld and Zheng (2012: 58–65, 835–57).
4 The limited space of this chapter neither allows for an exhaustive listing of manuscripts on certain therapies, nor for a complete bibliography of relevant primary and secondary literature. For further information on individual manuscripts, the reader is referred to the corresponding entries in the catalogue, for additional primary and secondary sources to the bibliographies in the literature listed here.
5 On almanacs as medical sources, see Schonebaum (2016: 25–7); on apotropaic manuals, see Cho (2005: esp. 4–5, 60–9). On sources associated with itinerant physicians, see Unschuld and Zheng (2012: 103–6); on popular medical publishing, see Wu (2010: 57–9).
6 This classification is based on Unschuld and Zheng's overview of Chinese medical manuscript literature (Unschuld and Zheng 2005, 2000). I have included additional information wherever I considered it necessary.
7 Unschuld and Zheng's definition of 'folk doctors' as people who 'were not professional physicians who practised for a living, but people who possessed medical and pharmaceutical knowledge and who treated patients without being bound to regular hours or a specific place' (Unschuld and Zheng 2005: 32) is slightly inconsistent with their later attribution of specialised texts by evidently professional practitioners to 'folk healers' in the catalogue. I have therefore amended their definition here.
8 On transmission of medical knowledge in early imperial China, see Sivin (1995); in modern China, Farquhar (1996) and Hsu (1999).
9 For the translation of *yan* 驗 as 'effective' in this context, see *Chuanya*, preface: 11: 'Itinerant doctors rely on a three-word-principle: The first [word] is "cheap," which means that they only use pharmaceutical substances that are not expensive. The second [word] is "effective," which means that by swallowing [the medicine] the illness can be removed [immediately]. The third [word] is "convenient," which means that even in secluded places in the mountains and forests one must be able to obtain it easily'.
   For extensive information on the language, theories, and practices of itinerant physicians, see Unschuld and Zheng (2012: 73–102). See also Wang (2013); Wu (2012); Unschuld (1978); Unschuld (1995); Zheng (1996).
10 Although Kleinman distinguishes between literati physicians (that is, modern TCM doctors) and folk healers, he nevertheless lumps together the two in his definition of secular folk medicine.
11 Bian (2017) provides valuable insights on the inclusion of eighteenth-century local knowledge and medicinal substances into the *Supplement to Materia Medica* (*Bencao shiyi* 本草拾遺, pr. 1871) by the *Chuanya's* author Zhao Xuemin. On local medicinals used by ethnic minorities today, see, for example, Zhang (2007); Zhao (2013).
12 See Unschuld and Zheng (2012: 22–6) for the history of 'dirty' substances.
13 See Liu and Liu (1992d) for present-day folk recipes.
14 For various forms of application of pharmaceutical substances in magico-religious contexts, see Strickmann (2002); for detailed and contextualised examples of drug intake in medieval Daoist hagiographies, see Stanley-Baker (2013: ch. 6). Cho (2005: 64–5) identifies a significant increase of drug application in exorcistic rituals in the Ming and Qing dynasties.
15 The term 'lamp wick cauterisation' is introduced by Unschuld in the catalogue; see Unschuld and Zheng (2012: 113).
16 See Li (2011) for an illustrated history of external medicine (*waike* 外科) and surgery.

Folk medicine

17 See Liu and Liu (1992c) for present-day folk practices of bloodletting.
18 Some cases in the *Compilation of Teachings on Traumatology* (*Shangke huizuan* 傷科匯纂, 1815) by the trauma medicine specialist Hu Tingguang 胡廷光 offer fascinating insights into late imperial surgery. Apart from setting broken bones and repositioning dislocated joints, Hu amputates a fingertip, opens up a male urethra, and stitches scrotums back on (*Shangke huizuan* 6: 230–239). Wu (2017) provides an overview of Hu's work and medical cases.
19 *Yizong jinjian*: 52.489 describes *sha* as white, millet-like skin rashes erupting during epidemic diseases.
20 See Wu (2015, 2017) for nineteenth-century bone setting. Liu and Liu (1992b) provide a monograph on contemporary folk *tuina*, which includes various methods of massage, bone setting, scraping, and other techniques.
21 On early imperial childbirth practices, see Lee (2006); for later literati views on obstetrics, see Furth (1999: 106–19, 174–7); Wu (2010: 147–87).
22 See also Lin (2008, 2011).
23 On early medieval Daoism and medicine, see Stanley-Baker (2013). On medieval Buddhist medicine, see Salguero (2010); Despeux (2010b); Fang (2010); Chen (2010). On *zhuyou* exorcism of the twelfth through eighteenth centuries, see Cho (2005).
24 On the development of exorcistic and apotropaic rituals involving talismans and petitions, see Bumbacher (2012).
25 For a detailed analysis of hand gestures in a Song treatise, see Hanson (2008).
26 On divinatory calculation in the Dunhuang manuscripts, see Kalinowski (2005); Arrault (2010). On iatromancy in the Dunhuang manuscripts, see Harper (2005: 148–83); in Warring States texts, see Harper (2001).

Bibliography

Pre-modern sources

Bencao gangmu 本草綱目 (Systematic Materia Medica) 1596, Li Shizhen 李时珍 (2007), 2nd edn, 2 vols, Beijing: Renmin weisheng chubanshe.

Chuanya neiwai bian 串雅内外编 (Refined Therapies of Itinerant Healers – Inner and Outer Chapters) 1759, Zhao Xuemin 赵学敏 (2004), Beijing: Xianzhuang shuju.

Jingzheng huoshi 驚症火式 (Fire Approach to Pathoconditions of Fright, MS 8036) late Qing/1940s to 50s, Xie Jikang 謝吉康 and Xie Guanshan 謝觀山, manuscript held by the Berlin State Library (Staatsbibliothek zu Berlin – Preußischer Kulturbesitz), https://digital.staatsbibliothek-berlin.de/werkansicht/?PPN=PPN3346229858, accessed 14/4/2020.

Michuan dieda bofang 秘傳跌打鈸方 (Secretly Transmitted Recipes [Treating Injuries from] Knocks and Falls, [Strung Together like] Cymbals, MS 8111) late Qing, Yao Yangsan 姚仰三 and Huang Tingxuan 黃廷選, manuscript held by the Berlin State Library (Staatsbibliothek zu Berlin – Preußischer Kulturbesitz), https://digital.staatsbibliothek-berlin.de/werkansicht/?PPN=PPN3346230155, accessed 14/4/2020.

Shangke huizuan 傷科彙纂 (Compilation of Teachings on Traumatology) 1815, Hu Tingguang 胡廷光, manuscript held by the Beijing University Library, facsimile reprint in Xu xiu si ku quan shu 續修四庫全書, Shanghai: Shanghai guji chubanshe.

Yifang jichao 醫方集抄 (Collected Notes on Medical Recipes, MS 8051) 19th to first half of the 20th century, Tang Tingguang 唐廷光 et al., manuscript held by the Berlin State Library (Staatsbibliothek zu Berlin–Preußischer Kulturbesitz), https://digital.staatsbibliothek-berlin.de/werkansicht/?PPN=PPN3346229882, 14/4/2020.

Yizong jinjian 醫宗金鑒 (Golden Mirror of Medical Learning) 1742, Wu Qian 吳謙 *et al.* (1997), vol. 5, Shenyang: Liaoning kexue jishu chubanshe.

Modern sources

Ahern, E.M. (1975) 'Sacred and secular medicine in a Taiwan village: a study of cosmological disorders', in A. Kleinman *et al.* (eds) *Medicine in Chinese Cultures: comparative studies of health care in Chinese and other societies*, Washington, DC: U.S. Government Printing Office, pp. 91–113.

Anderson, E. and Anderson, M.L. (1975) 'Folk dietetics in two Chinese communities and its implications for the study of Chinese medicine', in A. Kleinman, *et al.* (eds) *Medicine in Chinese Cultures: comparative studies of health care in Chinese and other societies*, Washington, DC: U.S. Government Printing Office, pp. 143–75.

Arrault, A. (2010) Activités médicales et méthodes hémérologiques dans les calendriers de Dunhuang du IXe au Xe siècle: esprit humain (*renshen*) et esprit du jour (*riyou*)', in C. Despeux (ed.) *Médecine, religion et société dans la Chine médiévale: étude de manuscrits chinois de Dunhuang et de Turfan*, vol. 1, Paris: Collège de France Institut des hautes études chinoises, pp. 285–332.

Berg, D. (2000) *Perceptions of Lay Healers in Late Imperial China, Durham East Asian papers*, vol. 15, Durham, NC: Dept. of East Asian Studies, University of Durham.

—— (2001) 'Bell doctors in the late imperial Chinese novel *Xingshi yinyuan zhuan*', *Monumenta Serica*, 49: 57–70.

Bian, He (2014) *Assembling the Cure: materia medica and the culture of healing in late imperial China*, PhD thesis, Harvard University.

—— (2017) 'An ever-expanding pharmacy: Zhao Xuemin and the conditions for new knowledge in eighteenth-century China', *Harvard Journal of Asiatic Studies* 77.2: 287–319.

Bumbacher, S.P. (2012) *Empowered Writing: exorcistic and apotropaic rituals in medieval China*, St. Petersburg, FL: Three Pines Press.

Chen Ming (2010) 'Maladies infantiles et démonologie bouddhique. Savoirs locaux et exotiques dans les manuscrits de Dunhuang', in C. Despeux (ed.) *Médecine, religion et société dans la Chine médiévale: étude de manuscrits chinois de Dunhuang et de Turfan*, vol. 2, Paris: Collège de France Institut des hautes études chinoises, pp. 1095–127.

Cho, P.S. (2005) *Ritual and the Occult in Chinese Medicine and Religious Healing: the development of zhuyou exorcism*, PhD thesis, University of Pennsylvania.

—— (2013) 'Healing and ritual imagination in Chinese medicine: the multiple interpretations of zhuyou', *East Asian Science, Technology, and Medicine*, 38: 71–113.

Committee on Scholarly Communication with the People's Republic of China (U.S.), National Institute of Health (eds) (1981) *Rural Health in the People's Republic of China: report of a visit by the rural health systems delegation, June 1978*, Washington, DC: U.S. Government Printing Office.

Cullen, C. (1993) 'Patients and healers in late imperial China: evidence from the *Jinpingmei*', *History of Science*, 31: 99–150.

Despeux, C. (ed.) (2010a) *Médecine, religion et société dans la Chine médiévale: Étude de manuscrits chinois de Dunhuang et de Turfan*, 3 vols, Paris: Collège de France Institut des hautes études chinoises.

—— (2010b) 'Pratiques magico-religieuses bouddhiques', in C. Despeux (ed.) *Médecine, religion et société dans la Chine médiévale: étude de manuscrits chinois de Dunhuang et de Turfan*, vol. 2, Paris: Collège de France Institut des hautes études chinoises, pp. 899–999.

Engelhardt, U. (2001) 'Dietetics in Tang China and the first extant works of *materia dietetica*', in E. Hsu, (ed.) *Innovation in Chinese Medicine*, Cambridge: Cambridge University Press, pp. 173–91.

Fang, Ling (2010) 'Sutras apocryphes et maladie', in C. Despeux (ed.), *Médecine, religion et société dans la Chine médiévale: étude de manuscrits chinois de Dunhuang et de Turfan*, vol. 2, Paris: Collège de France Institut des hautes études chinoises, pp. 1001–94.

Farquhar, J. (1996) *Knowing Practice: the clinical encounter of Chinese medicine*, Boulder, CO: Westview.

Furth, C. (1999) *A Flourishing Yin: gender in China's medical history, 960–1665*, Berkeley; Los Angeles; London: University of California Press.

Gould-Martin, K. (1975) 'Medical systems in a Taiwan village: *Ong-ia-gong* (Wang-yeh-kung 王爺公), the Plague God as modern physician', in A. Kleinman et al. (eds) *Medicine in Chinese Cultures: comparative studies of health care in Chinese and other societies*, Washington, DC: U.S. Government Printing Office, pp. 115–41.

Hanson, M. (2008) 'Hand mnemonics in classical Chinese medicine: texts, earliest images, and arts of memory', *Asia Major, Third Series*, 21.1: 325–57.

Harper, D.J. (1998) *Early Chinese Medical Literature: the Mawangdui medical manuscripts*, London; New York: Kegan Paul International.

—— (2001) 'Iatromancy, diagnosis, and prognosis in early Chinese medicine', in E. Hsu (ed.) *Innovation in Chinese Medicine*, Cambridge: Cambridge University Press, pp. 99–120.

—— (2005) 'Dunhuang iatromantic manuscripts', in V. Lo and C. Cullen (eds) *Medieval Chinese Medicine: the Dunhuang medical manuscripts, Needham Research Institute series*, London; New York: RoutledgeCurzon, pp. 134–64.

Hinrichs, T J and Barnes, L.L. (eds) (2013) *Chinese Medicine and Healing: an illustrated history*, Cambridge, MA: Belknap Press of Harvard University Press.

Hsu, E. (1999) *The Transmission of Chinese Medicine*, Cambridge: Cambridge University Press.

——— (ed.) (2001) *Innovation in Chinese Medicine*, Cambridge: Cambridge University Press.

Kalinowski, M. (2005) 'Mantric texts in their cultural contexts', in V. Lo and C. Cullen (eds) *Medieval Chinese Medicine: the Dunhuang medical manuscripts, Needham Research Institute series*, London; New York: RoutledgeCurzon, pp. 109–33.

Kang Bao 康豹 and Liu Shufen 劉淑芬 (eds) (2013) *Di si jie guoji Hanxuehui yilun wenji - Xinyang, shjian yu wenhua tiaoshi* 第四屆國際漢學會議論文集 - 信仰、實踐與文化調適 (Papers from the Fourth International Meeting of Sinological Studies – Faith, Practice, and Cultural Adaptation), Vol. 2, Taipei: Zhongyang yanjiuyuan.

Kleinman, A. (1980) *Patients and Healers in the Context of Culture: an exploration of the borderland between anthropology, medicine, and psychiatry*, Berkeley; Los Angeles: University of California Press.

——— (1981) 'Traditional doctors', in Committee on Scholarly Communication with the People's Republic of China (U.S.) (ed.) *Rural Health in the People's Republic of China: report of a visit by the rural health systems delegation, June 1978*, Washington, DC: U.S. Government Printing Office, pp. 63–74.

Kleinman, A. et al. (eds) (1975) *Medicine in Chinese Cultures: comparative studies of health care in Chinese and other societies*, Washington, DC: U.S. Government Printing Office.

Kuriyama, S. (1995) 'Interpreting the history of bloodletting', *Journal of the History of Medicine and Applied Sciences*, 50.1: 11–46.

Lee, Jen-der (2006) 'Childbirth in early imperial China, in A.K.C. Leung (ed.) *Medicine for Women in Imperial China*, Leiden; Boston, MA: Brill, pp. 108–78.

Leung, A.K.C. (ed.) (2006) *Medicine for Women in Imperial China*, Leiden; Boston, MA: Brill.

Li Jianmin 李建民 (2011) *Hua Tuo yincang de shoushu: Waike de zhongguo yixue shi* 華佗隱藏的手術: 外科的中國醫學史 (The Secret Surgeries of Hua Tuo: a Chinese medical history of external medicine), Taipei: Dongda tushu gongsi.

Lin Fushi 林富士 (2008) *Zhongguo zhong gu shiqi de zongjiao yu yiliao* 中國中古時期的宗教與醫療 (Religion and Healing in Ancient China), Taipei: Lianjing chubanshe gongsi.

——— (2011) *Zongjiao yu yiliao* 宗教與醫療 (Religion and Healing), Taipei: Lianjing chubanshe gongsi.

——— (2013) 'Zhuyou yixue yu daojiao de guanxi – Yi *Shengji zonglu fujin men* wei zhu de taolun「祝由」醫學與道教的關係 —以《聖濟總錄. 符禁門》為主的討論' (On the relationship between apotropaic medicine and Daoist religion: a discussion based on the talisman chapter in the *Shengji zonglu*), in Kang Bao 康豹 and Liu Shufen 劉淑芬 (eds) *Di si jie guoji Hanxuehui yilun wenji : xinyang, shjian yu wenhua tiaoshi* 第四屆國際漢學會議論文集: 信仰、實踐與文化調適, Taipei: Zhongyang yanjiuyuan, pp. 403–48.

Liu Shaolin 刘少林 and Liu Guangrui 刘光瑞 (1992a), *Zhongguo minjian erliao tujie* 中国民间儿疗图解 (Chinese Folk Paediatrics Illustrated and Explained), Chengdu: Sichuan kexue jishu chubanshe.

——— (1992b) *Zhongguo minjian tuina shu* 中国民间推拿术 (The Art of Chinese Folk Tuina), Chengdu: Sichuan kexue jishu chubanshe.

——— (1992c) *Zhongguo minjian cixue shu* 中国民间刺血术 (The Art of Bloodletting in Chinese Folk Medicine), Chengdu: Sichuan kexue jishu chubanshe.

——— (1992d) *Zhongguo minjian xiaodan fang* 中国民间小单方 (Simple and Single Ingredient Recipes of Chinese Folk Medicine), Chengdu: Sichuan kexue jishu chubanshe.

Lo, V. (2001) 'Huangdi Hama jing (Yellow Emperor's Toad Canon)', *Asia Major, Third Series*, 14.2: 61–99.

——— (2002) 'Spirit of stone: technical considerations in the treatment of the Jade Body', *Bulletin of the School of Oriental and African Studies*, 65.1: 99–128.

——— (2005) 'Quick and easy Chinese medicine: the Dunhuang moxibustion charts', in V. Lo and C. Cullen (eds) *Medieval Chinese Medicine: the Dunhuang medical manuscripts, Needham Research Institute series*, London; New York: RoutledgeCurzon, pp. 227–51.

Lo, V. and Cullen, C. (eds) (2005) *Medieval Chinese Medicine: the Dunhuang medical manuscripts, Needham Research Institute series*, London; New York: RoutledgeCurzon.

Salguero, C.P. (2010) *Buddhist Medicine in Medieval China: disease, healing and the body in crosscultural translation (2nd to 8th centuries CE)*, PhD thesis, Johns Hopkins University.

Scheid, V. (2005) 'Restructuring the field of Chinese medicine: a study of the Menghe and Ding Scholarly currents, 1600–2000', Part 2, *East Asian Science, Technology, and Medicine*, 23: 79–130.

Schmidt-Glintzer, H. (ed.) (1995) *Das andere China: Festschrift für Wolfgang Bauer zum 65, Geburtstag*, Wiesbaden: Harrassowitz.

Schonebaum, A. (2016) *Novel Medicine: healing, literature, and popular knowledge in early modern China*, Washington, DC: University of Washington Press.

Sivin, N. (1995) 'Text and experience in classical chinese medicine', in D. Bates (ed.) *Knowledge and the Scholarly Medical Traditions*, Cambridge: Cambridge University Press, pp. 177–204.

——— (2015) *Health Care in Eleventh-Century China*, New York: Springer.

Sommer, M.H. (2010) 'Abortion in late imperial China: routine birth control or crisis intervention?', *Late Imperial China*, 31.2: 97–165.

Stanley-Baker, M. (2013) *Daoists and Doctors: the role of medicine in Six Dynasties Shangqing Daoism*, PhD thesis, University College London.

Strickmann, M. (1995) 'Brief note: "the seal of the jungle woman"', *Asia Major, Third Series*, 8.2: 147–53.

——— (2002) *Chinese Magical Medicine*, Stanford, CA: Stanford University Press.

Tan Xunyun 覃迅云 et al. (2002) *Zhongguo Yao yaoxue* 中国瑶药学 (Pharmacology of the Yao People in China), Beijing: Minzu chubanshe.

Unschuld, P.U. (1977) 'Arzneimittelmissbrauch und heterodoxe Heiltätigkeit im kaiserlichen China: Ausgewählte Materialien zu Gesetzgebung und Rechtsprechung', *Sudhoffs Archiv*, 61.4: 353–85.

——— (1978) 'Das Ch'uan-ya und die Praxis chinesischer Landärzte im 18. Jahrhundert', *Sudhoffs Archiv*, 62: 378–407.

——— (1995) 'Der chinesische Wanderarzt und seine Klientel im 19. Jahrhundert. Rekonstruktion eines Dialogs', in H. Schmidt-Glintzer (ed.) *Das andere China: Festschrift für Wolfgang Bauer zum 65. Geburtstag*, Wiesbaden: Harrassowitz, pp. 129–57.

Unschuld, P.U. and Zheng, Jinsheng (2000) 'Handschriften als Quelle Chinesischer Medizingeschichte', *Monumenta Serica*, 48: 471–99.

——— (2005) 'Manuscripts as sources in the history of Chinese medicine', in V. Lo and C. Cullen (eds) *Medieval Chinese Medicine: the Dunhuang medical manuscripts*, London; New York: Routledge-Curzon, pp. 19–44.

——— (2012) *Chinese Traditional Healing: the Berlin collections of manuscript volumes from the 16th through the early 20th century*, Leiden; Boston, MA: Brill.

Wang Jing 王静 (2013) 'Qingdai zoufangyi de yishu chuancheng ji yiliao tedian 清代走方医的医术传承及医疗特点' (The transmission of medical skills and the characteristics of medical treatments of itinerant physicians), *Yunnan shehui kexue* 云南社会科学, 3: 161–5.

Wang Wen'an 王文安 (1998) *Minjian yishu jinghua – Nei wai ke* 民间医术精华–内外科 (The Quintessence of the Art of Folk Medicine – internal and external medicine), *Zhongguo yishu mingjia jinghua* 中国医术名家精华, vol. 4, Xi'an: Shijie tushu chuban gongsi.

——— (2000) *Zhongguo shaoshu minzu yishu juezhao* 中国少数民族医术绝招 (Unique Skills of the Medical Art of Chinese Ethnic Minorities), Hohhot: Yuanfang chubanshe.

Wu Xiaoming 吴小明 (2012) 'Cong "Chuanya" kan Zhejiang minjian zoufangyi de zhibing tedian 从《串雅》看浙江民间走方医的治病特点' (The characteristics of treatments of Zhejiang itinerant folk-healers, as found in the "*Chuanya*"), *Zhejiang zhongyiyao daxue xuebao* 浙江中医药大学学报, 36.1: 9–10.

Wu, Yi-Li (2000) 'The bamboo grove monastery and popular gynecology in late imperial China', *Late Imperial China*, 21.1: 41–76.

——— (2010) *Reproducing Women: medicine, metaphor, and childbirth in late imperial China*, Berkeley; Los Angeles: University of California Press.

——— (2015) 'Between the living and the dead: trauma medicine and forensic medicine in the mid-Qing', *Frontiers of History in China*, 10.1: 38–73.

——— (2017) 'A trauma doctor's practice in nineteenth-century China: the medical cases of Hu Tingguang', *Social History of Medicine*, 30.2: 299–322.

Yamada, Keiji (1998) *The Origins of Acupuncture, Moxibustion, and Decoction*, Kyoto: Nichibunken International Research Center for Japanese Studies.

Zhang Shaoyun 张绍云 et al. (2007) 'Lahuzu minjianyi zhiliao waishang yaoyong zhiwu de tiaocha 拉祜族民间医治疗外伤药用植物的调查' (Survey of medicinal plants used by Lahu folk healers to treat external injuries), *Yunnan zhongyi xueyuan xuebao* 云南中医学院学报, 30.6: 18–20.

Zhang Xun 張珣 (1989) *Jibing yu wenhua – Taiwan minjian yiliao renlei xue yanjiu lunji* 疾病與文化：台灣民間醫療人類學研究論集 (Disease and Culture – Collection of Anthropological Studies on Medical Folk Treatments in Taiwan), Taipei, Banqiao: Daoxiang chubanshe.

Zhao Jingyun 赵景云 *et al.* (2013) '*Achangzu minjianyi changyong de ji zhong duxing yaocai* 阿昌族民间医常用的几种毒性药材' (A few toxic medicinal herbs commonly used by Achang folk healers), *Zhongguo minzu yiyao zazhi* 中国民族医药杂志, 2.2: 37–8.

Zheng Jinsheng 郑金生 (1996) '*Zoufangyi weiyao chutan* 走方医伪药初探' (An Initial Study of Fake Pharmaceutical Substances Used by Itinerant Physicians), *Zhongyaocai* 中药材, 11: 587–90.

——— (2013) 'Über *da tai* 打胎 – Abtreibung in alten volksmedizinischen Handschriften Chinas', *Sudhoffs Archiv*, 97.1: 102–20.

PART 3

Food and sex

19
WHAT NOT TO EAT – HOW NOT TO TREAT
Medical prohibitions[1]

Vivienne Lo and Luis F-B Junqueira

As this Handbook amply demonstrates, we now know a great deal about what people did to maintain their health and cure illnesses in China; yet, very little research has been focused on the subject of medical prohibitions; what not to eat, how not to treat and why. Since prohibitions have shaped the Chinese medical traditions for 2,000 years and more, why are they among the most neglected of topics? As part of their healing and curative functions, foods, drugs and medical techniques always have socio-cultural and symbolic potency, as a result of a combination of religious, secular and scientific beliefs and practices. Indeed, it is precisely this potency of meaning which intensifies their nutritive and healing action but, equally, the severity of the consequences if one transgresses the established medical and nutritional codes that surround them. 'Do not bury talismans in places where chickens and dogs may trample them or dig them up', an early twentieth-century popular manual warns us, as doing so 'will offend the gods and make talismans ineffective' (Yu Zhefu 1924: 3–4).

The codes have generated both therapeutic strategies and prohibitions aimed at staying well as an individual or community and keeping other people well who are in one's care. Yet, the range and reach of the prohibitions, interdictions and taboos in China became so vast that they easily seem incomprehensible, inconsistent and irrational; and they are therefore an easy target for those looking to debunk Asian medical traditions. With the exception of a handful of scholars, it has therefore not been in the interest of practitioners or historians alike to focus on this most controversial and complex of subjects, when the study of prohibitions is widely construed as the least promising of scientific innovation, and most likely to draw critical attention.

Nevertheless, given the diversity of the prohibitions, it is premature to make any kind of judgement about their nature and value without preliminary definitions and analysis of how they came into being. In order to make sense of the vast range of medical prohibitions in China, it is helpful to ask at the outset whether the categories of prohibition we identify are emic or etic: i.e. were they categories applied in the past by those using the prohibitions in their daily lives, the insider's perspective; or have they been constructed retrospectively in the social sciences? In the latter case, we have to be aware of how our academic disciplines of historical, scientific or anthropologic analysis, or some combination of these, distort understanding of the past.

DOI: 10.4324/9780203740262-23

Given the lack of secondary literature about prohibitions, the concepts of emic and etic are easier to differentiate when analysing positive medical strategies. Many scholars seek precedents for modern science in early Chinese empirical knowledge. Nutritionists, for example, are not often concerned that their research into the history of the Chinese diet might be deemed anachronistic by cultural historians: for they have reason to believe that something in the Chinese diet prevented beri-beri, goitre, night blindness and rickets – even if the illnesses were not identified by these modern disease terms in the past and there has never been one single Chinese diet to speak of. They would contest that there is enough similarity across Chinese culture and the framing of diseases throughout time to argue, for example, in the case of beri-beri, that the Chinese diet as represented in a fourteenth-century Sino-Mongol manual was sufficiently high in Vitamin D, particularly given the use of unpolished rice, to prevent deficiency diseases. Thus, pre-modern Chinese knowledge 'that proper feeding may eliminate certain diseases was vindicated [in modern scientific terms] after six hundred years' (Lu and Needham 1951: 19; Huang 2000: 578–86, 588–91 *et passim*).

When it comes to the patterns of interdiction, many nutritional prohibitions also seem to anticipate modern universal recommendations, where common prejudices arise within a transcendent and shared realm of sensory observation. So, for example, the bad or unappealing odours of spoiled meat or fruit and vegetables out of season protect us from poison and indigestion. A piece of 'meat or fish that dogs do not eat, or birds do not peck' will cause sickness and even death. Yet, a 'white horse whose head is black' or an animal which died unexpectedly, 'its body bent over to the floor with the head facing north' must be avoided altogether (*Jingui yaolüe fanglun*, 25). Bernard Read (1931) lists forty-one kinds of poisonous and prohibited meats from Li Shizhen's 李時珍 (1518–93) *Bencao gangmu* 本草綱目 (Systematic *Materia Medica*). Most of the prohibitions caution against the anomalous where what is considered strange and inedible is not just diseased or old, rank or ruined, but the prohibitions extend to meat from animals with deformities or with inauspicious colourings and other culturally specific indicators of bad fortune (Read: rubric 348).

Mary Douglas' ground-breaking work on the nature of prohibition, *Purity and Danger* (1966) was the first to offer an analysis of the symbolic nature of why different peoples have chosen to prohibit the use of certain substances and not others (Douglas 1984: 3–58). Transcending simplistic moralistic or disciplinary readings of what has been labelled clean or unclean, polluting or cleansing, she emphasised that it was necessary to conduct an in-depth investigation into the social and cultural history of the target populations to gain any insight into their shared behaviour.

Her major contributions to the field include concentrating our attention on strictures applied to the consumption of boundary-crossing animals with the example of the abominations of Leviticus, the cloven-hoofed pig that does not chew the cud, and water creatures that do not have fins and scales; later, she was to prioritise the power of the ritual body of any chosen community to shape what was considered sacred or abominable: when people eat what they sacrifice, their violence against an animal, rather than the animal itself, might be deemed abominable (Ibid.: 42–58). Douglas' assumptions about the danger of eating liminal and sacred animals are germane to the Chinese context, but it is totally inadequate to capture the great diversity of Chinese prohibitions.

Other commentators suggest that people only prohibit the behaviours of 'other' people who perhaps threaten their own communities (Blackman *et al.* 1933), citing interdictions against the eating of pork, a domestic farm animal, among herding peoples of various religious persuasions. Whether one is convinced by such theories where prohibitions are deemed to originate in the alterity of ethnic exceptionalism, or the ritual homologies of body and altar, these authors opened the way for a wave of approaches to prohibition studies, most vividly in the analysis of dietary cultures, but also in medicine and every other aspect of everyday life (Chapter 20 in this volume). We will see how those instincts of alterity, of sanctity, purity and ritual aimed at avoidance of harm, all produced pervasive and enduring prohibitions in China.

There were also moral medical prohibitions against eating too much, indolence, or profiteering from patients' suffering all of which have different kinds of modern resonance. It is in Sun Simiao's 孫思邈 seventh-century *Beiji qianjin yaofang* 備急千金要方 (Prescriptions Worth a Thousand in Gold for Every Emergency) that we find the moral strictures that make up the well-known medieval Chinese ethical code for doctors: 'a doctor should not question rank or wealth, age or beauty, nor should he have personal feelings towards that person, his race, or his mental capacities. He should treat all his patients as equal, as though they were his own closest relatives' (*Beiji qianjin yaofang*, 1.7–8; trans. Pregadio 2008: 2048–9). Similar warnings resonate in occult texts concerned with talismans, prayers, astrology, spirit-writing and divination that were published as late as the twentieth century (see below).

These prohibitions against the health dangers of eating rotten meat or restraints on professional conduct have often been selected because they showcase modern sensibilities. But they do not get us any closer to understanding the mentalities and historical processes of past cultures of prohibition, which is the aim of this chapter. We can begin to determine those emic categories through examining how medical interdictions were packaged in their own time on their own terms: as changing moral instructions to doctors and diviners, in antagonisms between the ingredients of recipes and remedies, in dietary regimens, in day books, occult texts and almanacs.

Hemerology

Many medical prohibitions concern the avoidance of inappropriate times for medical treatment, especially acupuncture and cauterisation. In ancient China, core patterns of auspicious and inauspicious days in the calendar were keyed to ritual cycles of the spirit world when an ancestral, calendrical or planetary spirit might have a benign or malevolent influence on human activities depending on the day (Chapters 4 and 5 in this volume).

This discourse of lucky and unlucky days (hemerology) operated from as early as the second millennium BCE, when the spirit ancestors of the kings required sacrifices appropriate to the sixty-day Shang calendar, the Heavenly Stems *tiangan* 天干 and Earthly Branches *dizhi* 地支 (hereafter ganzhi; Chapters 4 and 5 in this volume). The relationships between kings and ancestors were therefore reciprocal, and the appropriate sacrifices would ensure plentiful harvests, the success of military campaigns and smooth births, as well as good fortune in more mundane projects. If the rituals were not adjusted to the calendar and the ancestors were not happy, disasters would ensue.

Fast forward a thousand or more years to early imperial times and the calendar linking the *ganzhi* cycle and the deceased Shang kings remained in use, even if the kings themselves had long lost their power to intervene. By then, there were also other spatio-temporal structures guiding prohibitions that related to the newly ascendant powers of *yin*, *yang* and the Five Agents, as well as a number of competing and complementary divinatory systems (Chapter 1 in this volume). There were also physiological theories of the body that mirrored the celestial circulation of Taiyi 太乙 (Grand Unity), a deity thought to live on Polaris and therefore to circle around the North Pole. Associated imperial cults that had originated in Chu, such as the one to Taiyi (Sukhu 1999: 154–7), became increasingly important during early

Han (202 BCE–9 CE). Taiyi's celestial circulation was embodied in the practice of observing contraindicated days for applying acupuncture therapy to certain parts of the body where the deity was supposed to reside (to prevent disturbing his celestial spirit) (Woolf 'Iatromancy' forthcoming; Lo 2001).

Medical interdictions of this kind are found scattered through textual miscellanies of the *rishu* 日書 (daybooks) of the late Warring States and the early imperial period, where they are listed according to the calendar, together with proscriptions and prescriptions for all manner of ritual and religious behaviour, life stages, dealings with the government and legal matters, and socio-cultural activities, including everyday ritual and household matters (Kalinowski 2017: 141). There were also alternative calendrical systems such as the *jianchu* 建除 (establish/remove) system, which assigned twelve-day qualifiers to months and days, *jian* being the first, in order to mark the fortune of a particular day for a particular activity. Each of the days was related through this system to the mythical sites and divine and astronomical entities that were later known as *shensha* 神煞 (calendar spirits), which had the power to influence human events (Ibid.: 171).

Here, two entries from the Jiudian 九店 (JD) and Fangmatan 放馬灘 (FMTA) day books give a flavour of how the discourse of auspicious days also naturally generated prohibitions.

JD.1
Jian (Establish) days: Greatly auspicious. Favourable for taking a wife, for sacrifices and cults, for building a house, for erecting altars to the gods of the soil and grain, and for wearing one's sword and ceremonial cap.

FMTA.1
Jian (Establish) days: A good day. Permitted to be an overseer; permitted for prayers and cults; permitted for breeding domestic animals. *Not permitted for bringing common people into the home.*

(Ibid.: 139)

Medieval texts, including those copied by and for religious institutions, inherited many of the prohibition traditions from earlier times, extending their range and detail. The concept of a *renshen* 人神, 'human spirit' or 'spirits' circulating around the body, further medicalised the physiology of roaming inner body spirits and appears in the *Lingshu* 靈樞 (Numinous Pivot) recension of the *Huangdi* 黃帝 (Yellow Emperor) corpus (Chapter 7 in this volume), the *locus classicus* for Chinese medicine first compiled some time in late Han. These theories of a circulation of spirits therefore pre-dated a mature theory of the circulation of *qi* and linked with the ancient Taiyi divination traditions (Chapter 2 in this volume). Among the medieval manuscripts discovered at the Buddhist Mogao shrines in Dunhuang, manuscripts assigned the numbers P. 2675 V° and S. 5737 set out different takes on the 'human spirit' traditions. Just as in the Taiyi system, these contain a list of body locations of the spirit in question, *renshen* (or sometimes *renqi* 人氣 'human *qi*'), where it takes up overnight lodging according to calendrical cycles, in this case the days of the lunar calendar. See, for example, the *Hama jing* 蛤蟆經 (Toad Canon), a series of charts and texts describing moxibustion prohibition of mostly medieval provenance with elements that date to the Han period (Lo 2001; Lo and Yoeli-Tlalim 2018). The charts follow the course of human *qi* as it moves around the body according to the waxing and waning of the moon and thereby illustrate the forbidden days and locations. The continuities of these cultures of lucky and unlucky days through to the present day are one demonstration of how a range of different systems and traditions of prohibition adapted to changing historical contexts.

Yin, Yang and seasonal restraint

In the late Warring States, and in the years preceding the imperial unification of 221 BCE, behaviour throughout the year was increasingly structured to the temporal cycles of *yin* and *yang*. Consider the ways in which a civilised gentleman was advised to deport himself in *Lüshi chunqiu* 呂氏春秋 (Spring and Autumn annals of Master Lü), a ritual encyclopaedia compiled in 239 BCE by one of the key advisors to, and sponsors of, the First Emperor of China (r. 221–210 BCE).

> A gentleman fasts and observes vigils, makes sure to stay deep inside his house, and keeps his body utterly still. He refrains from music and sex, eschews association with his wife, *maintains a sparse diet, and avoids use of piquant condiments.* He settles the vital energies of his mind, maintains quietude within his various bodily organs and engages in no rash undertakings. *He does all these things in order to assure completion of the first traces of yin.*
> (Lüshi chunqiu, 5.42; trans. Knoblock and Riegel 2000: 135, our emphasis)

This passage is concerned with practices of restraint to calm and quieten the body, in order to align the body to the cosmic cycles operating at the onset of *yin*, the coolest, most inactive part of the year in winter. This was the time when one's behaviour should follow the spirit of the season and be most conservative. Yet, with the seasonal prohibition of pleasure and indulgence, there was also a strong moral censure at work, and it is customary to find interdictions against excess regardless of the season. No doubt new disparities of wealth in early imperial times were in part responsible for the censure of excessive behaviour. Among the scanty evidence of real patients from the Western Han period (202 BCE–9 CE), the twenty-five case histories of Chunyu Yi 淳于意 are remarkable in that they indict either sex or alcohol or both in the aetiology of some three-quarters of the male deaths (*Shiji*, 105.2794–820; Grant 2003: 128–9; Hsu 2010: 52–3). In fact, a parallel medical biography of the semi-legendary patron of medicine, Bian Que 扁鵲, implies that many of the prohibitions up to the Han period were related to moral considerations, such as arrogance, greed, and not prioritising the care of one's body. Care of one's body involved both social rituals like appropriate dress and comportment, but also ensured that one's *yin*, *yang* and the *qi* of one's inner organs would not lose harmony with the season, or the constitution would not lack the strength to take medicines. Finally, there was a growing prohibition against engaging spirit mediums and unorthodox healers for medical treatment, marking the ascendency of a particular class of orthodox medical practitioners (*Shiji*, 70.2279–303).

By Han times, there was clearly an awareness of how over-eating, neglect of the signs of physical breakdown and lack of appropriate medical care could damage one's health. In the earliest extant manuscript devoted to the *mai* 脈 (trans. variously as channels, vessels, conduits), a very clear understanding of the link is shown between certain types of pain, over-eating, overweight and diseases of the blockage of the blood flow, which could result in a sudden attack on the heart. If left untreated, 'the sound of weeping will be heard', that is, the condition would lead to immediate death.

> Now the six pains all exist in the body yet there is not the wisdom to treat them, so if the gentleman becomes fat and neglects boundaries, this is an unbearable burden for the stomach, muscles and bone. His *qi* is too plentiful, his blood is excessive. Qi and blood decay and fester, the hundred joints all sink. The twenty extremities clog and rebelling

it goes to the heart. If these cannot be treated in advance then the sound of weeping will be heard.

(*Maishu*; Lo 1998)

Excess itself became a fundamental category of the prohibitions. There were also twelve injunctions against giving acupuncture treatment at certain times related to excessive or extreme conditions (after sex, over-eating/drinking, over exertion, emotional intensity, when hungry, thirsty) (*Lingshu*, 9; Unschuld 2016: 175–6).

In the first century CE, the Grand Historian of the Western Han period, Sima Qian 司馬遷 (c.145–86 BCE), complained that the 'techniques of *yin* and *yang* are too detailed, and multiply prohibitions, restraining [people] and exacerbating the things they are afraid of' (*Shiji* 130.3288–99). Even at that time, close to their origin, the different ritual systems had their supporters, habitual users and detractors. Wang Chong's 王充 (27–100 CE) *Lunheng* 論衡 provides the most well-known critique of the widespread culture of *zeri* 擇日 (lit. 'selecting days', i.e. hemerology) for misleading people across all the divisions of society, so that they lived in fear of the inauspicious intervention of less than well-meaning spirits in their lives (Li 2017: 274).

At first sight, it is easy to sympathise with Wang Chong and Sima Qian, since the complexity of prohibitions is indeed bewildering. Angus Graham damningly concluded that the early imperial period saw the end of a golden age of intellectual innovation. In his view, the rise of the kind of hemerology and cosmological correspondence thinking towards the end of the first millennium BCE, which underpinned much of the prohibition culture, heralded a low point in the debasement of Chinese thought (Graham 1989: 382). Nevertheless, sympathetic magic, avoidance of substances and activities which might act as a vehicle for demonic infestation, enrage the spirits, or poison from decay or damage due to eating or taking food and medicines on the wrong day or in the wrong season are all so well represented in both everyday medical and administrative literatures that they should not be ignored.

The Five Agents (Wood: Fire: Earth: Metal: Water)

Understanding the Five Agents (Introduction and Chapter 1 in this volume) is the key to a significant element of the imperial prohibition cultures that have survived into the modern world. The Five Agents structured everyday and state activities according to two standard cycles: the *xiangsheng* 相生 cycle of generation (which followed the sequence of the seasons: Spring; Summer; Late Summer; Autumn; Winter) and the earlier *xiangke* 相克 control/conquest cycle, which posited that every Agent overcame another as Water naturally overcomes Fire and in the following sequence: Wood > Earth > Water > Fire > Metal. By the late Warring States, the *xiangke* (conquest) cycle had grown into a ritual theory which, during Han times, was used to endorse the succession of the new ruling house.

There is some controversy over the extent to which the legitimacy of the ritual theories was actually evoked in practice. Nevertheless, the Han ruling house had certainly decided by 104 BCE to rule by virtue of Earth, and its power was portended by reported sightings of golden dragons in 165 BCE (Loewe 1994: 55–61); and it used the ritual colour of the yellow earth, symbolic also of China, in its insignia. Yellow/gold corresponded with Earth and Earth was the victor over Water. It had been argued that Water and its ritual colour of black/blue were the power behind the previous regime, Qin (221–206 BCE), and so the political succession was naturalised as aligning with the principles of Heaven and its cosmological cycles.

All of the recensions of the *Huangdi* 黃帝 (Yellow Emperor) corpus are permeated with prohibitions, which reflect the medical manifestations of these cycles. Each of the Five Agents was associated with different qualities (flavours, organs, directions, colours, seasons, etc.) that constantly generated and/or dominated one another, in turn. Their cycles governed all worldly phenomena, including the medical functions of the body and even the organs and bowels themselves. Each was associated with an inner organ, the heart, liver, kidneys, lungs and spleen, which also formed basic textual units. One treatise of the *Huangdi neijing suwen* 黃帝內經素問 (Yellow Emperor's Inner Canon: Basic Questions) recension, for example, gives us an oblique and rather alarming insight into how the empirical aspects of acupuncture knowledge came about by trial and error according to this standard set of five organs. It lists the damage one can do by inserting sharp instruments inappropriately and to dangerous depths:

> When a piercing hits the heart, death occurs within one day. If the [heart] was [merely] excited, this causes belching. When a piercing hits the liver, death occurs within five days. If the [liver] was [merely] excited, this causes talkativeness. When a piercing hits the kidneys, death occurs within six days. If the [kidney] was [merely] excited, this causes sneezing. When a piercing hits the lung, death occurs within three days. If the [lung] was [merely] excited this causes coughing. When a piercing hits the spleen, death occurs within ten days. If the [spleen] was [merely] excited, this causes swallowing.
> (*Suwen*, 52; trans. Unschuld and Tessenow 2011: 744–5)

There was a kind of seasonal homoeopathy where things that were deemed alike resonated sympathetically. Recommendations therefore existed to eat foods that supported the season: thus, in spring, liver *qi* and the 'acid' or 'sour' flavour predominated, so one should eat foods from these categories and less of others (*Zhouli*, 5:72; Unschuld 1986: 206). There were also seasonal interdictions according to an individual's constitution and illnesses. The Yellow Emperor Corpus notes that the five flavours (pungent, salty, bitter, sweet and sour) were prohibited for patients with diseases of the physical substrates: *qi*, *xie* 血 (blood), *gu* 骨 (bones), *rou* 肉 (flesh) and *jin* 筋 (sinews), respectively. The corresponding pairs (pungent-*qi*; sweet-flesh, etc.) roughly corresponded with the Five Agents system and, in contradistinction to the *Zhouli* recommendations indicated above, proscribed the eating of the flavour of the corresponding physical substrate for fear of creating excess. Eating the sour flavour would therefore harm the body in springtime.

Prohibitions were increasingly associated with the classic incompatibilities according to an emerging lore of the Five Agents. Eating the flavour that was antagonistic to the physical substrate through the *xiangke* system could also be dangerous: excessive consumption of salty flavours would cause harmful changes to the *mai* 脈; bitter to the *pi* 皮 (skin); pungent to the sinews; sour to the flesh and sweet to the bones and hair (*Suwen*, 10 and 23). To take just one of these cases Wang Bing 王冰 (eighth century), one of the most famous commentators on the *Suwen*, evokes the *xiangke* and *xiangsheng* cycles in this way when explaining the bitter phenomenon. Bitter was the flavour associated with the Agent Fire. Since the bitter flavour has the nature of hardness and dryness, it nourishes the organs associated with Earth, the stomach and spleen but, when consumed in excess, it dominates the lung (Metal). The lung is also associated with the skin and the hair on the skin and, when it is weakened, the former dries out, while the latter falls out – due to the nature of the bitter flavour (*Suwen*, 3 and 10; trans. Unschuld 2003: 80–1, 186–7). In other words, one must not ingest any of the Five Flavours in excess, and the proper amount must be assessed on an individual basis, taking into account not only the individual, but also the environment, time of the year and so forth. It is also interesting to note that the most obvious concerns about flavours expressed in the *Suwen* related to prohibitions against excess, and less is made of deficiency.

Combining both *xiangsheng* and *xiangke* systems, one must not, in contradistinction to the homoeopathic style of *Zhouli* recommendations, 'eat liver in spring'. Spring and the liver were associated with the Agent Wood, and it was in spring that the *qi* of the liver reached its peak and the *qi* of the spleen (Earth) was the weakest. Therefore, by eating liver in spring, the person's liver might become over-nourished and further deplete an already weak spleen, rendering the disease incurable (*Jingui yaoliie fanglun*, 24). This is because Wood dominates Earth. The same principle applies to each of the Five Agents. Following this rationale, and avoiding excess, one should also be wary of eating animals that resonate with the season that one is passing through.

At the same time, the Five Agent cycles were used pervasively throughout society to narrow and focus individual choices about what to do, and where and when to do it, if at all. The harmonious or antagonistic relations between the Agents and their relationship to the passage of time meant that they could be used to pinpoint auspicious and inauspicious moments for health, as well as everyday activities. Raising crops, animal husbandry, and managing the household, washing one's hair or simply going on an excursion were shaped by these cycles in everyday life, as much as decisions concerning the governance of the empire. All this knowledge would keep running through Chinese society for millennia, as reflected in nineteenth- and twentieth-century almanacs.

While the Five Agents were a core part of early acupuncture theory and structured knowledge of dietary importance, it was some time before this theory was systematically applied to pharmacological theory and recipe literature. Zhang Zhongjing's 張仲景 famous *Jingui yaoliie fanglun* 金匱要略方論 (Essential Prescriptions from the Golden Chamber, hereafter *Essential Prescriptions*) devotes two chapters to the matter of food and drug prohibitions divided into (1) 'birds, beasts, fishes and insects' and (2) 'fruits, nuts, vegetables and grains'. Following the *Suwen*, many of the prohibitions in the *Essential Prescriptions* are associated with the flavour and nature of substances as described above. Invoking the Five Agents framework, the text begins by warning readers not to consume certain flavours when treating diseases related to the five organs.

The *Essential Prescriptions* lists an extensive number of foodstuffs that must not be combined – e.g. 'do not consume fish together with chicken meat' (*Jingui yaoliie fanglun*, 24) – or should not be eaten in excess – e.g. 'eating too many plums causes tooth decay' (Ibid.: 25). People should also refrain from eating certain kinds of food at particular times: 'in the 4th and 8th months', for example, 'do not eat coriander as it harms the spirit' and 'if a pregnant woman eats [too much] ginger, an extra finger will grow in her child' (Ibid.). The reasons behind most of the prohibitions contained in the *Essential Prescriptions* are left without explanation, but clearly include empirical evidence about spoiled or strong foods, and avoidance of excessive consumption.

The presence of prohibitions is far less obvious in the 365 entries of herbs, animals and minerals recorded in the *Shennong bencao jing* 神農本草經 (The Divine Farmer's *Materia Medica*; hereafter: Divine Farmer). It is only in a short preface at the beginning of the text that we learn that substances in a recipe must be coordinated in terms of *yin* and *yang*. While some drugs can be used alone, others must be combined, and the healer must attend closely to those that reinforce, 'fear' or 'kill' the effects of each other, avoiding the use of substances that 'are mutually averse or mutually clash' (*Shennong bencao jing*, preface, trans. Yang, ix–xvi). Their application must follow the basic allopathic principle that 'to treat cold diseases, one should use hot drugs' and 'to treat heat diseases, one must use cold drugs'. Curiously, none of this information is provided in the 365 individual entries, so that the reader must rely either on tacit knowledge or on expanded editions. Likewise, it is only much later, in editions of the *Shennong bencao jing* with commentaries, that the Five Agents theory was incorporated and began to serve as a guide to the selection of drugs. Drug preparation was also subject to regulations where some herbs were best made into pills, powders, decoctions or wines, while others were not, and the healer 'must not violate these rules' (Ibid.: xii).

By Tang times (618–907 CE), what one should consume, or prescribe to a sick person, was governed by a plurality of theories about the interactions between the Five Agents. Sun Simiao puts it succinctly when he opines that bad combinations will make the body impure:

> If the *qi* of different foods is incompatible, the vital essence will be damaged. The body achieves accomplishment thanks to the flavours that nourish it. If the different flavours of foods are not harmonised, the body becomes impure. This is why the sage starts off by obeying the alimentary prohibitions in order to preserve his nature; if that proves ineffective, he has recourse to remedies for sustaining life.
>
> (*Qianjin fang*, 26: 465–75)

The householder and agriculturalist, Jia Sixie 賈思勰, writing in *Qimin yaoshu* 齊民要術 (Everyman's Essential Arts, c. 540 BCE), claimed that 'it suffices to know the great principles [without need] to follow the meanderings in their smallest details', a comment which reveals strategies for dealing with the many systems of prohibition in practice (Sabban 1996: 333). In order to do justice to the enormous quantity of proscriptions that make up historical medical literature and everyday practice, and to understand why these traditions were able to survive into the twentieth and twenty-first centuries, these great principles of prescription and prohibition, together with some of the major 'meanderings', are discussed.

Despeux's (2007) survey of medical prohibitions suggested that one-third of all prohibitions related to incompatibilities between food items, one-sixth to times and seasons. She noted that the larger part of the dietary prohibitions was related to foods of animal origin, and that there were biological, dietetic, symbolic and religious reasons underlying the interdictions. The seasonal prohibitions, she believes, were primarily based on practical agricultural experience and observational knowledge and, from the earliest records, there is plenty to substantiate everyday knowhow about the optimum time to sow, harvest and preserve.

Liu Zenggui 劉增貴 (2007) drew out more specifically the complexities of Qin and Han period prohibitions and taboos as Five Agent theories were grafted on to early ritual belief; and developed in particular ways in specialist medical contexts, such as in pregnancy and reproduction. Liu divided taboos into three cultural categories that pervaded society: those governed by local concepts of the seasons and the passage of time, including the twenty-four *jieqi* 節氣 (solar periods) and the times of the day; those governing social relations and customs, particularly at the points of critical life-transitions, such as birth, marriage or death, but also interdictions related to food, clothes, the household and travel; and third prohibitions related to ghosts and gods, and rituals for lifting related calamities. Taboo objects and prohibited behaviour, and their powers, both brought about inauspicious consequences that could only be counteracted with the kind of knowledge mediated by specialists in ritual or, ultimately, from more modern times, through printed books.

The Chinese medical lexicon offers an excellent vantage point for us to understand the fundamental role prohibitions and taboos have played in China. Based on around a thousand medical texts compiled between the Han (202 BCE–220 CE) and late Qing dynasties (1644–1912), our preliminary survey (Table 19.1) indicates the core phrases and

Table 19.1 Survey Based on the Database *Beijing Airusheng shuzihua jishu yanjiu zhongxin* (2016)

Individual Characters or Words	General English Translation	Number of Times It Appears
bu 不[a]	No/Not/Do not	1,295,711
buke 不可	Cannot	108,517
mo 莫	No/Not/Do not	14,185
moneng 莫能	Cannot/Must not	1,254
wu 勿	Do not	33,148
wuling 勿令	Do not let	5,970
ji 忌	Taboo[b]	39,198
jin 禁	Prohibition	23,402
fan 犯	Transgression	14,230
bi 避	Avoidance	6,217
Total	–	**1,541,832**

[a] The *bu* category includes *buke*. Similarly, *mo* includes *moneng* and *wu* includes *wuling*.
[b] The characters *ji* 忌, *jin* 禁, *fan* 犯 and *bi* 避 can be read as nouns or verbs depending on the context.

words whose meanings are associated with prohibitions with taboos and the frequency of occurrence.

The survey draws mostly upon printed medical literature published in the Ming (1368–1644) and Qing (1644–1912) dynasties; manuscripts, almanacs and specialised occult texts are not included in the count. It is hard, however, to identify changes in the proportion of prohibitions across texts dating from different periods of time as most extant early Chinese medical texts were either compiled or survived only in late imperial editions so that the original dates are obscure.

Some of the prohibitions are gathered together into specialist sections of the medical literature. This is most easily traced through the literature on women's health in pregnancy and childbirth: what a woman should eat at different stages of pregnancy, what she should allow herself to see or listen to and how she should deport herself. One of the earliest manuscript collections devoted to medicine and matters pertaining to the gestating body was buried in Mawangdui tomb 3 in 168 BCE and is full of prohibitions to protect the development of the foetus:

> In the first month it is called 'flowing in the form'. Food and drink must be finest: the sour boiled dish must be thoroughly cooked. Do not eat acrid or rank foods. This is called 'initial fixture'. In the second month it first becomes lard. Do not eat acrid or stinking foods. The dwelling place must be still. For a boy there must be no exertion, lest the hundred joints all ail. This is called 'first deposition'. In the third month it first becomes suet and has the appearance of a gourd. During this time it does not have a fixed configuration, and if exposed to things it transforms. For this reason lords, sires and great men must not employ dwarves. Do not observe monkeys. Do not eat onion and ginger; and do not eat a dish of boiled rabbit.
>
> (*Taichan shu*; trans. Harper 1998: 378–9)

Similar interdictions restricting and shaping the behaviour of women while pregnant are copied into, and extended, in later literature. The treatises on women's illnesses in Sun Simiao's *Beiji qianjin yaofang* (*juan* 2–4) are very clear that a woman's behaviour has to be modest and exemplary. The rationale for many of the prohibitions are not made explicit, but it is clear that a majority concern the purity, quietude and moderation of the expectant mother, with a particular taboo against sexual relations and having fun. Knowledge of complex sets of prohibitions is often assumed. She should not use the latrine, only a chamber pot; she should not indulge herself in whims which will offend the spirits; she should not look at or eat things which may have qualities that will affect the pregnancy or the babies. In the different months of pregnancy, it was also prohibited to use the acupuncture points of the channels which were nourishing the foetus at that moment.

A much extended list of interdictions to do with behaviour during pregnancy is to be found in the Mongol period *Yinshan zhengyao* 飲膳正要 (Principles of Correct Diet), which is the earliest text to include both theoretical monographs about dietary substances and integrated recipes and cooking instructions. Donkey meat, for example, will make for a stubborn birth; eating turtles will give the baby a short neck (Buell and Andersen 2010: 264–5).

As Table 19.2 demonstrates, our survey mostly corroborates Despeux's (2007) findings; although a substantial proportion of medical sources also warn against certain actions and the use of certain objects, drugs and healing methods, taboos around sex, women and animals

Table 19.2 Survey Based on the Database *Beijing Airusheng Zhongyi dianhai*, 1st edition, 2016

Terms in Chinese	General English Translation	Number of Entries
buke yong 不可用/*bu ke yi* 不可以/*moyong* 莫用/*jiyong* 忌用/*jinyong* 禁用/*wuyong* 勿用	Must not use	16,127
buke shi 不可食/*mo shi* 莫食/*jishi* 忌食/*jinshi* 禁食/*wushi* 勿食	Must not eat	7,355
bu ke ling 不可令/*moling* 莫令/*jinling* 忌令/*jinling* 禁令/*wuling* 勿令	Must not let	6,915
buke fu 不可服/*mofu* 莫服/*jifu* 忌服/*jinfu* 禁服/*wufu* 勿服	Must not ingest	4,495
buke yu 不可與/*moyu* 莫與/*jiyu* 忌與/*jinyu* 禁與/*wuyu* 勿與	Must not [...] with/ Must not give	3,247
buke gong 不可攻/*mogong* 莫攻/*jigong* 忌攻/*jingong* 禁攻/*wugong* 勿攻	Must not interfere/ use/cure	2,143
buke ci (zhen) 不可刺(針)/*jici (zhen)* 忌刺(針)/*jinci (zhen)* 禁刺(針)/*wuci (zhen)* 勿刺(針)	Must not pierce/use needles	1,924
buke jiu 不可灸/*mojiu* 莫灸/*jijiu* 忌灸/*jinjiu* 禁灸/*wujiu* 勿灸	Must not use moxa	1,705
jitie 忌鐵/*jintie* 禁鐵/*wufantie* 勿犯鐵	Must not [use] metal [utensils]	1,673
buke jin 不可近/*mojin* 莫近/*jijin* 忌近/*jinjin* 禁近/*wujin* 勿近	Must not get close to	1,549
buke jian 不可見/*mojian* 莫見/*jiian* 忌見/*jinjian* 禁見/*wujian* 勿見	Must not look at	1,335
fanghi buke 房事...不可/*jifangshi* 忌房事/*jinfangshi* 禁房事/*jiefangshi* 戒房事/*fanfangshi* 犯房事	Must not have intercourse	1,303
buke yan 不可言/*moyan* 莫言/*jiyan* 忌言/*jinyan* 禁言/*wuyan* 勿言	Must not speak	967
jiquan buke 雞犬...不可/*jijiquan* 忌雞犬/*jinjiquan* 禁雞犬/*wuling jiquan* 勿令雞犬	Must not let chickens or dogs...	921
buke yin 不可飲/*moyin* 莫飲/*jiyin* 忌飲/*jinyin* 禁飲/*wuyin* 勿飲	Must not drink	812
fu buke 婦不可/*fu bude* 婦不得/*jifu* 忌婦/*jinfu* 禁婦/*wuling fu* 勿令婦/*fumo* 婦莫	Must not let women...	712
buke dong 不可動/*modong* 莫動/*jidong* 忌動/*jindong* 禁動/*wudong* 勿動	Must not move	603
buke ru 不可入/*moru* 莫入/*jiru* 忌入/*jinru* 禁入/*wuru* 勿入	Must not enter	511
Total		**54,297**

are equally common. In the category 'must not eat', for example, most of the entries are related to fish and meat, followed by herbs and vegetables with strong flavours, like onion, garlic and mustard. Remedies, medicinal herbs, healing techniques and utensils are the most seen in 'must not use' and, overall, 'must not let women' and 'must not let chickens or dogs' can be counted together, as these categories often only change the order of who comes first.

Much has been written about the reasons that led to the creation of taboos around women – a phenomenon common to all cultures worldwide across time (Knight 1991). Analysing the Chinese case, Emily Ahern sees women pollution beliefs as justification for their marginal position within the patriarchal kinship system (Ahern 1978). At first, the prohibitions that align women, chickens and dogs might seem random and alarming, and reinforce the lower status of women. But they are also categories found combined in the art of spells and charming and suggest a different kind of agency (Liu 2005: 167–9). After all, as a group, women, birds and dogs are both ubiquitous and inclined to inspire intense emotions on account of a certain dependence on them for reproduction, domestic harmony, nutrition and protection. This undoubtedly imbues them with a kind of magic, and the potential to be dangerous domestically. They are all also associated with a very powerful blood taboo, as discussed below.

The wide range of prohibitions contained in Sun Simiao's *Beiji qianjin yaofang* are nicely illustrated by its volumes on gynaecology (trans. Wilms 2007). Prohibitions are scattered throughout the text and its prescriptions and they can be placed in already familiar categories, such as those organised by spatio-temporal markers, by the kind of activities engaged in, by excess and extreme emotional states, and by the actions and qualities of the substances themselves. To secure a successful pregnancy, the couple should observe the Five Agents cycle and not engage in sexual intercourse during periods of time deemed inauspicious. The woman must avoid eating certain flavours at particular times: nothing acrid or rancid in the first month. Certain kinds of meat, such as mountain goat, mule or turtle, were to be avoided, as well as sex or excessive work. A woman in labour must remain relaxed and calm, and the delivery tent should not be polluted by blood or death. After delivery, the place to bury the placenta must follow the design of the birth charts so as not to offend the spirits. Such charts remain pervasive in modern-day almanacs. Postpartum concerns proscribe harsh purgatives, and suggest preventative and curative measures, warning time and again against sexual activities, consumption of cold foods and exposure to wind, all of which deplete the woman's five organs. Given the unique vulnerabilities of pregnancy, birth and blood loss, Sun Simiao considered women's diseases far more difficult to treat than men's. Hence the very large number of prohibitions for women in the *Beiji qianjin yaofang*; perhaps this is, as Ahern would suggest, a way of limiting and restraining women's behaviour and therefore their social position, but it was also clearly better to prevent (by observing the lists of prohibitions) than to lament.

Medical prohibitions in the twentieth century

Prohibitions have been at the foundation of talismanic culture, which has been integral to the occult healing arts for millennia (Junqueira 2018). The efficacy of healing talismans relies not only on the talismans themselves – their provenance and proper construction – but especially on moral codes both healers and patients should observe. A wave of practically oriented occult manuals swamped China between the late nineteenth and early twentieth centuries, and these often dedicated entire sections to the subject of prohibitions, explaining their rationale and how their transgression affects efficacy. In a talismanic manual, *Xuanyuan beiji yixue zhuyou shisanke* 軒轅碑記醫學祝由十三科 (The Yellow Thearch's Stele of the Thirteen Medical Disciplines of Zhuyou), published in 1885, four out of the five 'don'ts' are related to general moral standards – 'those who do not cultivate a sincere and respectful heart cannot heal/be healed' and 'those who care for profit more than saving lives cannot heal/be healed'. Only one is specifically associated with the writing of talismans, 'if talismans and spells do not comply with [the disease] or are incomplete, you cannot heal/be healed' (*Xuanyuan beiji yixue zhuyou shisanke*). It is unclear if the rules applied only to the healer, the patient or both.

The lists of prohibitions contained in later manuals grew in length and rationale. The principle of moral cleanliness remained critical. Making an illicit profit from talismanic arts, not healing those who cannot pay for your service or instead healing robbers, murderers and rapists were all condemned, but by far the major concern was around sex, an action morally incompatible with the practice of occult arts. 'Don't go after (female) prostitutes', 'avoid carnal pleasures' and 'don't conduct rituals during your honeymoon' are warnings pervasive among occult texts, ramifications of the age-old, almost universal, prohibition of blood contamination (*xuewu* 血污), 'the only thing that terrifies deities and immortals' (Xu 1924; Foyin jushi 1926: 36–8).

Blood contamination referred to the slaughtering of animals, sex and menstruation; to avoid breaking this powerful taboo and ruining 'decades of physical training and spiritual cultivation', the healer, always a male, must never 'write talismans in the presence of women', 'let them touch the talismans' or 'conduct rituals where women have recently given birth' (Yu 1924: 3–4; Xu 1924: 5–6). Not surprisingly, these women-related prohibitions are often accompanied by another warning: 'do not bury talismans in places where chickens and dogs may trample them or dig them up' (Yu 1924: 3–4). As we might have noticed, none of these taboos are entirely new, which shows the relevance of this *longue durée* perspective to our subject of medical prohibitions.

Almanacs were a crucial driving force for the popularisation of prohibition knowledge from the Tang period onwards and performed some of the same functions as the ancient day book manuscripts, but were produced in far larger numbers. They were unofficial, privately produced, calendrical works that had for centuries coexisted alongside the official calendars promulgated by the imperial court (Smith 2017: 336–7). But, contrary to occult texts, which emphasised the healer's moral conduct, the prohibitions contained in almanacs could be observed by anyone. They mostly concern what not to do, or when and where to do it. Almanacs satisfied people's need to live their lives in accordance with the cosmic patterns of Heaven and Earth and, intertwined with occult knowledge, their calendrical information was supposed to structure, and thereby protect, all aspects of an individual or household's life (Ibid.: 340). While banned by the Imperial, Republican and Communist states alike, millions of almanacs circulated freely (Nedostup 2009: 230–8; Smith 2017: 353); in fact, almanacs were the most commercially successful publications in pre-Communist China (Smith 2017: 353; Brokaw 2007: 452–4), which makes them valuable sources for the study of popular medical prohibitions in the Late Imperial and Republican periods (c. 1368–1949).

Time-sensitive, the content of almanacs is mostly related to a specific year, but their format and guidance about auspicious and inauspicious days remained remarkably consistent over the centuries (Brokaw 2007). Almost all the dozens of early twentieth-century almanacs we have consulted combined the Gregorian and Chinese lunisolar calendars, and for each day there were several appropriate (*yi* 宜), prohibited (*ji* 忌) and inappropriate (*bujiangyi* 不將宜) activities. There is no set of activities that is consistently designated as appropriate, inappropriate or prohibited: these always vary from day to day and from year to year. The list of activities is broad, including offering sacrifices, eating dog meat, cutting hair, practising divination, planting, praying, taking medicine and receiving acupuncture. While one is not obliged to do an activity on its respective auspicious day, doing an inauspicious or prohibited activity, when warned not to do so directly, offends calendrical spirits, causing environmental disasters, disease and death. Each day is also accompanied by a list of hours – according to the *ganzhi* 干支 system – when a woman must not give birth.

There were many ways to prepare for a good birth. A method based on the sexagenary system of pregnancy names the days a pregnant woman should not visit parts of the household or touch certain objects associated with the Placenta Spirit (*Taishen* 胎神); doing otherwise would provoke a miscarriage, pain or an inauspicious birth. A set of twelve talismans, one for each month of the year that a child was born in, was followed by an exposition about how their birth offended certain spirits associated with the directions of the *ganzhi* system. Almanacs also emphasised other negative effects of the wrathful calendar spirits (*sha* 煞) and how they would influence a person during their lives. Graphic depictions of twenty-six 'barriers' (*guan* 關, lit. passes) illustrated the perils a child born in a certain month or hour would encounter as it moved through key stages into adulthood. Each one is associated with offences against one of the twenty-six *sha*-spirits; to get through them, the child must abstain from doing certain activities. The parents could find an occultist to help them select a proper name or perform household rituals for the infant. An almanac from 1950 also offered two talismans the pregnant woman should ingest to protect her child's fate, undo any pregnancy-related taboos or hasten parturition – possibly to comply with an auspicious time of birth (Self-Reliance Press 1950). Most striking, these almanacs, with all their prohibition knowledge, remain best-sellers in Taiwan and Hong Kong bookstores.

The ritual systems are so tenacious that they continue to structure people's everyday health choices, particularly in relation to dietary care in elderly communities. Older people with smartphones share messages, memes and short videos every day with detailed content relating to what they should and shouldn't eat or do. And intergenerational conflicts rage on social media as to the value and science of food combinations and prohibitions (Wang and Lo 2019). The Shang calendar still operates, albeit without its ghostly kings. Structuring time, and the discourse of auspiciousness, is evidently one of the relatively secure elements of the *longue durée* of a distinctive Chinese culture where geographical and environmental conditions have sustained the continuity of certain technologies and practices (Braudel 1949, 1972–3; Keightley 1990). The ritual use of the calendar for selecting days for undertaking particular activities has been largely ignored in academic writing until relatively recently – perhaps because the discourse of lucky and unlucky days is redolent of a medieval magic that Chinese scholars are trying to shed, along with the perception of Chinese culture as having been millennia old and in terminal decline until the revolutions of the twentieth century. But we should be wary of assuming that cultures of prohibition have ever been either uniform or unchanging; they have adapted to new technologies and clearly serve new purposes.

Fundamentally, cultures of prohibition in China have served to negotiate the complex nature of plural Chinese identities. They have undoubtedly sustained a sense of tradition in communities of the Yellow River and Yangtze valleys and among those who self-identify as Chinese in diaspora. What one chooses not to eat and not to do for one's health nurtures a sense of a distinct and exceptional community, and simultaneously makes it quite clear who does and does not belong to that community. Should we have researched the prohibition literature and traditions of some of the 55 'minorities' currently formally identified by the Chinese government as living within China (Chapter 37 in this volume), no doubt we would have discovered that, in reality there were both shared and alternative sets of prohibitions operating within and across the population of China. While proper cultures of prohibition varied across different regions, social classes and periods of time, it is also remarkable to see the universality of such taboos as those relating to women, blood and sex, which may indicate certain enduring human prejudices.

Over three millennia, the accumulated traditions of divination, diagnosis and fortune telling in China have meant that there have been many competing systems and their advocates, but that there was always also a range of choices with which individuals could negotiate some freedom of manoeuvre. In other words, there was enough restraint to know who you were and how to go about your life, but there was also sufficient freedom to negotiate individual and community identities within the overall tradition. It was clearly not necessary to follow a single system of prohibition since Chinese prohibitions, like its culinary traditions, were and remain famously eclectic and often contradictory. As Jia Sixie wrote in 540 CE, you just need to know the basic principles of the prohibitions. In the same way, knowledge of the basics of a great culinary tradition allows a flexibility in practice. With expertise in the basic principles of prohibitions, one might manage effectively the confusing array of possible pathways that life presents at any one time. And perhaps that very flexibility within the tradition is the key to the extraordinary continuity of a Chinese identity over two millennia. If you avoid burying your talismans in places where chickens and dogs may trample them or dig them up, it provides a certain certainty about your identity and your future. It works, it lasts and, crucially, it is a sufficiently flexible reality to embrace multiple lifestyles and identities.

Note

1 This work was supported by the Wellcome Trust [References: 201616/Z/16/Z and 217661/Z/19/Z].

Bibliography

Pre-modern sources

Beiji qianjin yaofang 備急千金要方 (Prescriptions Worth a Thousand in Gold for Every Emergency), 7th century, Sun Simiao 孫思邈 (1955), Beijing: Renmin weisheng chubanshe.

Huangdi neijing lingshu 黃帝內經靈樞 (Yellow Thearch's Inner Canon: Spiritual Pivot) c. 3rd century CE, (1995) *Xuxiu siku quanshu*, Vol. 981, Shanghai: Shanghai guji chubanshe.

Huangdi neijing suwen 黃帝內經素問 (Yellow Thearch's Inner Canon: Basic Questions) c. 3rd century CE, (1995) *Xuxiu siku quanshu*, Vol. 980, Shanghai: Shanghai guji chubanshe.

Jingui yaolüe fanglun 金匱要略方論 (Prescriptions of the Golden Chamber) 2nd century CE, Zhang Zhongjing 張仲景 (1965) *Sibu congkan chubian* 四部叢刊初編, Vol. 21, Taipei: Shangwu yinshuguan.

Lüshi chunqiu 呂氏春秋 (Master Lü's Spring and Autumn Annals) c. 3rd century BCE, Lü Buwei, (1986) *Jingyin wenyuange siku quanshu*, Vol. 848, Taipei: Shangwu yinshuguan.

Nanjing 難經 (The Classic of Difficult Issues) c. 3rd century CE, (1986) *Jingyin wenyuange siku quanshu*, Vol. 733, Taipei: Shangwu yinshuguan.

Qimin Yaoshu 齊民要術 (Everyman's Essential Arts) c. 540 CE, Jia Sixie 賈思勰, (1936), Shanghai: Zhonghua shuju.

Shennong bencaojing 神農本草經 (The Divine Farmer's *Materia Medica*) c. 3rd century CE, (1986) *Jingyin wenyuange siku quanshu*, Vol. 785, Taipei: Shangwu yinshuguan.

Shiji 史記 (Records of the Historian) 90 BCE, Sima Qian 司馬遷 (1972), Beijing: Zhonghua shuju.

Xuanyuan beiji yixue zhuyou shisanke 軒轅碑記醫學祝由十三科 (The Yellow Thearch's Stele of the Thirteen Medical Disciplines of Zhuyou) 1885, unpublished woodblock edition, Luis F.B. Junqueira's personal collection.

Zhouli zhushu 周禮註疏 (Rites of Zhou: With Commentary), 1st century CE, Ruan Yuan 阮元 (ed.) (2001), Taipei: Yiwen yishuguan.

Modern sources

Ahern, E.M. (1978) 'The power and pollution of Chinese women', in A.P. Wolf (ed.) *Studies in Chinese Society*, Stanford, CA: Stanford University Press, pp. 269–90.
Beijing Airusheng shuzihua jishu yanjiu zhongxin 北京愛如生數字化技術研究中心 (2016) *Zhongyi dianhai* 愛如生中醫典海 (Ocean of Chinese Medical Texts), database.
Blackman, A.M., Hooke, S.H., Gadd, C.J., Hollis, F.J. and James, O.E. (1933) *Myth and Ritual: essays on the myth and ritual of the Hebrews in relation to the culture pattern of the ancient East*, London: Oxford University Press.
Braudel, F. (1949) *The Mediterranean and the Mediterranean World in the Age of Philip II*, trans. S. Reynolds (1972–1973), London: Collins.
Brokaw, C.J. (2007) *Commerce in Culture: the Sibao book trade in the Qing and Republican periods*, Cambridge, MA: Harvard University Press.
——— (2017) 'The dance of 'old' and 'new' in Chinese print culture, 1860s–1955', *Science in Context*, 30.3: 281–324.
Buell, P.D. and Anderson, E.N. (2010) *A Soup for the Qan: Chinese dietary medicine of the Mongol era as seen in Hu Sihui's Yinshan zhengyao*, Leiden: Brill.
Chang, K.C. (ed.) (1977) *Food in Chinese Culture: anthropological and historical perspectives*, New Haven, CT: Yale University Press.
Despeux, C. (2007) 'Food prohibitions in China', trans. P. Barrett, *The Lantern: A Journal of Traditional Chinese Medicine*, 4.1: 22–32.
Douglas, M. (1966/1984) *Purity and Danger: an analysis of the concepts of pollution and taboo*, London: Routledge.
Foyin jushi 佛穩居士 (1926), *Yuanguang mijue qishu* 圓光真傳秘訣 (Transmitted Secrets of Yuanguang), Shanghai: Zhongxi shuju.
Goossaert, V. (2005) *Le Tabou du Bœuf en Chine. Agriculture, éthique et sacrifice*, Paris: Collège de France, Institut des Hautes Études Chinoises.
Graham, A. (1989) *Disputers of the Tao: philosophical argument in ancient China*, Chicago, IL: Open Court.
Grant, J. (2003) *A Chinese Physician: Wang Ji and the Stone Mountain medical case histories*, New York: Routledge.
Harper, D.J. (1998) *Early Chinese Medical Literature: the Mawangdui medical manuscripts*, London: Kegan Paul.
——— (2005) 'Ancient and medieval Chinese recipes for aphrodisiacs and philters', *Asian Medicine*, 1.1: 91–100.
Hsu, E. (2010) *Pulse Diagnosis in Early Chinese Medicine: the telling touch*, Cambridge: Cambridge University Press.
Huang, H.T. (2000) 'Fermentations and food science', in J. Needham (ed.) *Science and Civilisation in China, Volume 6, Biology and Biological Technology, Part 5*, Cambridge: Cambridge University Press.
Junqueira, L.F.B. (2018) *Mingqing zhuyou zhi wenti kaoshi* 明清「祝由」之問題考釋 (The Origins of Talismanic Healing According to Ming and Qing Sources), MA thesis, Fudan University.
Kalinowski, M. (2017) 'Hemerology and prediction in the daybooks: ideas and practices', in D. Harper and M. Kalinowski (eds) *Books of Fate and Popular Culture in Early China: the daybook manuscripts of the Warring States, Qin, and Han*, Leiden: Brill, pp. 128–206.
Keightley, D.N. (1990) 'Early civilization in China: reflections on how it became Chinese', in P.S. Ropp (ed.) *Heritage of China: contemporary perspectives on Chinese civilization*, Berkeley: University of California Press, pp. 15–34.
Kieschnick, J. (2005) 'A history of Buddhist vegetarianism in China', in R. Sterckx (ed.) *Of Tripod and Palate: food, politics and religion in traditional China*, New York: Palgrave-Macmillan, pp. 186–212.
Knight, C. (1991) *Blood Relations: menstruation and the origins of culture*, New Haven, CT: Yale University Press.
Knoblock, J. and Riegel, J. (trans.) (2000) *The Annals of Lü Buwei: a complete translation and study*, Stanford, CA: Stanford University Press.
Lévi, J. (1983) 'L'abstinence des céréales chez les taoïstes', *Études Chinoises*, 1: 3–48.
Li Ping (2017) 'The Zidanku silk manuscripts', in D. Harper and M. Kalinowski (eds) *Books of Fate and Popular Culture in Early China: the daybook manuscripts of the Warring States, Qin, and Han*, Leiden: Brill, pp. 249–77.

Liu Lexian 劉樂賢 (2005) 'Love charms among the Dunhuang manuscripts' in V. Lo and C. Cullen (eds) *Medieval Chinese Medicine: the Dunhuang medical manuscripts*, London: Routledge, pp. 165–75.

Liu Tseng-kuei 劉增貴 (2007) 'Jinji Qinhan Xinyang de yige cemian 禁忌 – 秦漢信仰的一個側面' (Taboo: an aspect of belief in the Qin and Han), *Xinshixue*, 18.4: 1–70.

Lo, V. (1998) *The Influence of Yangsheng Culture on Early Chinese Medical Theory*, PhD thesis, SOAS, University of London.

—— (2001) 'Huangdi Hama jing (Yellow Emperor's Toad Canon)', *Asia Major*, 14.2: 61–69.

—— (2005) 'Pleasure, prohibition and pain: food and medicine in China', in R. Sterckx (ed.) *Of Tripod and Palate: food, politics and religion in traditional China*, New York: Palgrave-Macmillan, pp. 163–86.

Lo, V. and Barrett, P. (2005) 'Cooking up fine remedies: on the culinary aesthetic in a sixteenth-century Chinese *materia medica*', *Medical History*, 49.4: 395–422.

Lo, V. and Cullen, C. (eds) (2005) *Medieval Chinese Medicine: the Dunhuang medical manuscripts*, London: Routledge.

Lo, V. and Yoeli-Tlalim, R. (2018) 'Travelling light: Sino-Tibetan moxa-cautery from Dunhuang', in V. Lo and P. Barrett (eds), *Imagining Chinese Medicine*, Leiden: Brill, pp. 271–90.

Loewe, M. (1994) *Divination, Mythology, and Monarchy in Han China*, Cambridge: Cambridge University Press.

Lu, G.D. and Needham, J. (1951) 'A contribution to the history of Chinese dietetics', *Isis*, 42.1: 13–20.

Nedostup, R. (2009) *Superstitious Regimes: religion and the politics of Chinese modernity*, Cambridge, MA: Harvard University Press.

Pregadio, F. (2008) 'Sun Simo', in H. Selin (ed.) *Encyclopaedia of the History of Science, Technology, and Medicine in Non-Western Cultures*, Dordrecht: Springer, pp. 2048–9.

Read, B.E. (1931) *Chinese Materia Medica: animal drugs*, Beijing: Peking Natural History Bulletin.

Sabban, F. (1996) 'Follow the season of the heavens: household economy and the management of time in sixth century China', *Food and Foodways*, 6.3/4: 329–49.

Self-Reliance Press (1950) *Zili shuju* 自力書局 (Self-Reliance Press [Almanac]), Guangzhou: Zili shuju.

Smith, R. (2017) 'The legacy of daybooks in late imperial and modern China', in D. Harper and M. Kalinowski (eds) *Books of Fate and Popular Culture in Early China: the daybook manuscripts of the Warring States, Qin, and Han*, Leiden: Brill, pp. 336–72.

Sukhu, G. (1999) 'Monkeys, shamans, emperors and poets: images of Chu during the Han dynasty', in C. Crook and J. Major (eds) *Defining Chu: image and reality in ancient China*, Honolulu: University of Hawai'i Press, pp. 145–65.

Unschuld, P.U. (trans.) (1986) *Nan-ching: the classic of difficult issues*, Berkeley: University of California Press.

—— (2003) *Huang Di Nei Jing Su Wen: nature, knowledge, imagery in an ancient Chinese medical text*, Berkeley: University of California Press.

—— (2016) *Huang Di Nei Jing Ling Shu: the ancient classic on needle therapy*, Berkeley: University of California Press.

Unschuld, P.U., Tessenow, H. and Zheng, Jinsheng (trans.) (2011) *Huang Di Nei Jing Su Wen: an annotated translation of Huang Di's inner classic – basic questions*, Berkeley: University of California Press.

Wang Xinyuan and Lo, V. (2019) 'Food-related yangsheng: short videos among the retired population in Shanghai', in V. Lo, C. Berry and L. Guo (eds) *Film and the Chinese Medical Humanities*, London: Routledge, pp. 226–41.

Wilms, S. (trans.) (2007) *Bei Ji Qian Jin Yao Fang: prescriptions worth a thousand in gold for every emergency, 3 volumes on gynecology*, Portland, OR: Chinese Medicine Database.

Woolf, D. (forthcoming) 'Iatromancy', in S.N. Kory (ed.) *Handbook of Divination Techniques*.

Xu Jinghui 徐景輝 (1924) *Zhuyouke zhibing qishu* 祝由科治病奇書 (Occult Book of Zhuyou), Shanghai: Zhongxi shuju.

Yang Shou-zhong (trans.) (2008) *The Divine Farmer's Materia Medica: a translation of the Shen Nong Ben Cao Jing*, Boulder, CO: Blue Poppy Press.

Yu Zhefu 余哲夫 (1924) *Fuzhou yanjiufa* 符咒研究法 (Research Methods of Talismans and Spells), Shanghai: Jingling xueshe.

20
CHINESE TRADITIONAL MEDICINE AND DIET[1]

Vivienne Lo

With its rapidly changing economy and increasing urban wealth, China is at a critical moment when global cultures are changing traditional patterns of healthcare, often for the worse. Despite extraordinary public health achievements in the twentieth century, evident in most of the standard health indicators, the country is now facing an epidemic of obesity among its urban children and a marked increase in first world diseases such as diabetes. Where once China's railway stations were surrounded by a multitude of small vendors hawking every kind of snack, albeit of dubious sanitation, suddenly the ultra-modern concourses are home exclusively to MacDonalds, Kentucky Fried Chicken and equivalent Chinese fast food chains. Nevertheless, at home and abroad, the older generation still keep traditional food cultures alive and, as in other Asian cultures, that includes paying attention to seasonal eating, the nutritional qualities of ingredients and their mutual interactions according to ancient precepts.

Culinary technology in China, through the manipulation of the potencies attributed to flavour, links to a history of nutritional ideas that begins in pre-imperial times and has echoes today in everyday life for over a billion people, and many more if we are to count those Chinese living overseas (Chapter 21 in this volume). People in China are still inclined to have a view about the effect on their bodies of what they eat, and to have an opinion about the techniques of adjusting the individual ingredients in a dish to the individual constitution and appetites of the consumer (Farquhar 2002; Wang and Lo 2019; Chapter 49 in this volume). They think about their food in unique ways. This chapter is therefore dedicated to understanding how this knowledge came about, to explaining the background assumptions that have informed the collective practices that surround us. It is not only a short historical and ethnographic survey aimed at understanding 'other' people. It also considers how the legacy of Chinese nutritional ideas might remain relevant today for those who are interested in Chinese cuisine and Chinese medicine, and among those who care about the relationship between local and global attitudes to health and patterns of consumption.

It is in the concepts and associated technologies of *qi* (the stuff of life that is) thought to power the body and *wei* 味 (the flavours, which came to have medical potency) that I will focus this account of potency and flavour in China (Buell and Anderson 2000: 575–91; Lo 2005). The story behind this ethnically distinctive culinary world begins in the centuries before the unification of China in 221 BCE. It is a two and a half thousand year story of ever-changing dietary rules and regulations that have shaped complex and pervasive nutritional practices, which survive to this day in many forms. Embedded in this story, there are certainly practices that are offensive and reprehensible today, from the consumption of rare wild animals to resorting to eating human flesh during famines (actually a worldwide practice in pre-modern times). But there are also attractive ideas that we can use to think with, that can model how both individuals and communities can remain well connected with the environment, and nourished socially and psychologically through what they eat and the way they eat it, suggesting how diets can be tailored to individual and community constitutions.

In the ancient world, it was not only the inhabitants of the Yellow River valley that developed a science to guide dietary choices to fit their unique constitutions. The ancient Greeks also designed foods that were deemed suitable for various states of humoral imbalance according to food potencies and flavours. As Ken Albala observes, 'most complex societies codify their foods investing in them a significance beyond satisfying hunger. In the West, at least since the ancient Greeks, this significance has been medical' (Albala 1994: 1–2). For China, this codification, which involved assigning qualities to food such as heating, cooling, the power to support the functions of the different organs and their associated physiological systems, was a process that can be seen coming to fruition as an integrated practice over a period of a millennium or more.

The notion of potent flavours was enshrined in the creation of imperial culture from the very beginning of empire (221 BCE). According to the foundational myths of Chinese civilisation, of the five culture bringers of a Golden Age in pre-history, the Yellow Emperor and the Divine Husbandman imparted knowledge of the agricultural seasons, and the cultivation and relative potencies of food and medicines. Shennong 神農 (the Divine Farmer), in particular, is famous for his self-experimentation. He tasted one hundred herbs to find out which were edible, which poisonous and according to *Huainanzi* (Master Huainan c. 139 BCE), he 'thrashed every single plant with a rust-coloured whip. In the end he learned their characteristics, the bland, the toxic, the cool, and the hot, taking their smell and taste as a guide' (Birrell 1993: 49). Aroma and flavour were by Han times (202 BCE–220 CE) the key to the potencies of foodstuffs and medicine. Thus, the altruistic physical suffering of this culture hero simultaneously embodied the work of empire and the empirical spirit at the foundation of medical and nutritional knowledge.

The story of the Shennong is often cited by those who hold the first of the following two extreme views of Chinese pharmacology. First, many Chinese medical practitioners suggest that Chinese drug therapy as we know it now is the product of long empirical experience. They believe that, just like Shennong, early physicians tested all the substances in the *materia dietetica* for their efficacy and toxicity and recorded their findings for posterity. That by today's count would be a dizzying array of products, a list of foodstuffs that runs today into 1,500 species, varieties and cultivars (Hu 2005).

However, cynics will argue that the drug and food classification system is an overworked theoretical elaboration lacking any coherent threads. It is probable that both views have some elements of truth. First, it is clear that theory does play an explicit and important role in guiding therapy. It is however also clear that even in the most hostile view, Chinese drugs and foods contain many pharmacologically active substances, which in many cases do have effects of the kind appropriate to the condition for which they are prescribed (the most immediately obvious are the purgatives!).

Earliest evidence

Historically, food and drug culture in China was not always limited to curing illness; the potential for herbal, mineral and animal products to fortify and stimulate the body was well known in ancient China. Food and drugs were an integral part of regimen aimed at longevity, immortality and even the preservation of the body beyond death. Early evidence for drug therapy appears in *Wushier bing fang* 五十二病方 (Fifty-two Remedies, tomb closed 168 BCE), a manuscript written on silk that was excavated from the tomb of the son of the Lady Dai in modern Changsha, Hunan at the Mawangdui burial mound (Harper 1998). This text records a wide range of illnesses treated with an equally wide range of preparations and magical rituals. These early remedies often combined foodstuffs and common herbs, human and animal excreta, ground insects and household objects in simple preparations and techniques to heal specific illnesses, such as abscess; haemorrhoids; scabies; gynaecological problems; fright; convulsions or to increase sexual vigour. They included medicinal powders prepared from dried and ground substances, and from charred raw foods and drugs that were consumed as pills, pastes or ointments, often using wine, vinegar and fats as binders but also honey, date pulp, birds eggs, blood and resins.

It took some while, however, before each individual foodstuff was systematically assigned medical efficacy according to classical Chinese scientific categories. These categories had been much more easily integrated with acupuncture theories, since the work of attributing nutritional qualities to the myriad foodstuffs was a much more complex and lengthy task than establishing the *yin* and *yang* qualities of the twelve acupuncture channels.

Yin and *yang* represent an ancient tendency to categorise the world in terms of polarities. Something is not to be spoken of as *yin* or *yang* either absolutely or in isolation. They derive their status from their relation to other things – thus, a man may be *yang* in relation to his wife, but *yin* in relation to his father. The 'complementary opposition' of *yin/yang* is often expressed in the expectation that *yin/yang* shall be in temporal alternation, *cf.* the examples of day/night and the warm/cold seasons of the year. When *yang* reaches an extreme, it transforms into *yin*. For the properties of foodstuffs, the oppositions of heating and cooling, or upwards and downwards (purgatives and emetics) are most often encountered (Lo 2012: 31–64).

The basic category assigned to a substance was its *wei*. Historians of Chinese medicine often translate *wei* 'sapors', rather than 'flavour', the common translation, to distinguish the medical rather than culinary virtues of the term in this context. *Wuwei* 五味, the five flavours, often cited in early Chinese literature, refer generally to the range of sensual pleasures that one might dream of eating, but in a medical context they were also specific flavours that could stimulate the movement of *qi*, the essential 'stuff of life' that animated and invigorated the body, making a person astute, healthy and effective. Carrying all the potency of their association with the five agents (wood, fire, earth, metal, water), the five flavours or 'sapors' extended a framework of correlative knowledge that had been used to organise political, ritual and medical life from the beginning of imperial times.

In some of the earliest ritual writings, for example, a pentic system of correspondences structured the emperor's diet according to a calendrical schedule (Knoblock and Reigel 2000: Books 1–12). According to the season, the emperor was advised to change the colour of his ceremonial robes, his rituals and the food that he would eat. The following is a sample of the kind of basic correspondences established by the second century BCE (Table 20.1).

Joseph Needham called the traditional Chinese world view exemplified by this correspondence thinking 'Organismic'; that is one in which all the phenomena of the universe are interconnected and each and every level of existence is an exact analogue of every other: the whole cosmos, including man, is one integrated organism (Needham 1956). The human body was a map of the cosmos, the empire and the emperor's palace; its rhythm followed patterns immanent in the universe. Cosmos and body correspond part to part and function to function.

Table 20.1 Five Agent Correspondences with Flavours, Seasons, Grains and Meat

Earth	Sweet	Season	Pannicled Millet	Beef
Metal	Pungent/acrid	Autumn	Sorghum	Dog
Water	Salty	Winter	Millet	Pork
Wood	Sour	Winter	Millet	Mutton
Fire	Bitter	Summer	Beans	Fowl

The medieval synthesis

The flavours attributed to food were the same set as were used for marking the nature of drug efficacy in medieval pharmacological treatises (Engelhardt 2001). Of all the associations with the five agents, flavour was the most important to a dietary regime. Controlling the flavours in diet meant more than the judicious combination of tastes for gustatory pleasure. The exact combinations were linked to specific therapeutic effects through the system of correspondences with the five agents. To summarise the main points of the mature theory: foods, like drugs, were classified according to their thermostatic qualities (hot, warm, neutral, cool and cold), by flavour (pungent, sweet, salt, sour and bitter), and by the directions in which they induced movements within the body and by the organ system that they supported as well as by the types of illness that they could influence.

Sweet, for example, was thought slightly yang in nature and promoted an upward and outward movement. It entered the stomach and spleen channels. Mildly sweet foods, such as grains, nuts, fruits and many vegetables, should form the main bulk of any diet. Stronger sweet flavours have a very warming and nourishing effect, but should be avoided by people with signs of damp. Those who crave sweet food would be well advised to choose from the list of mildly sweet foods such as the grains. Sour and salty were slightly *yin* in flavour and were therefore cooling. They promoted a downward movement in the body. Salt moistened the body, while sour gathered and contracted, cleansing the body and moving the blood. Salt entered the kidneys and sour the liver. Bitter was the most *yin* of flavours. It caused contraction and made *qi* descend and move inwards reducing fever and calming agitation. It was also drying and therefore good for dampness. Bitter entered the heart clearing heat and calming the spirit.

In general, medical priorities demanded restraint with a distinct moral overtone aimed at the leisured classes: too much strong meat, spice, oil and fat would create excess heat, while raw vegetables and cold food and water were indigestible and harmed the stomach (Chapter 19 in this volume). Balance, harmony and the careful intervention of the chef in cooking were the keys to a healthy diet.

If we focus on the physiological effects of the foodstuffs and the mix of food and drug stuffs in the pharmacological treatises, there is little to distinguish food from medicine. In practice, medical advice from the medieval period advised beginning with the gentler dietary therapy: Sun Simiao 孫思邈 (581?–682? CE) wrote:

> A good doctor first makes a diagnosis, and having found out the cause of the disease, he tries to cure it first by food. When food fails, then he prescribes medicine. The nature of drugs is hard and violent, just like that of imperial soldiers. Since soldiers are so savage and impetuous, how could anybody dare to deploy them recklessly? If they are deployed inappropriately, harm and destruction will result everywhere. Similarly, excessive damage is the consequence if drugs are thrown at illnesses.
>
> *(Qianjin yaofang, 26.554–5)*

Thus, the best doctors diagnosed and treated the body with dietary advice before illness manifested itself. By attributing thermostatic, and other physiological effects to every foodstuff, medical authors embraced nutrition within a larger framework of knowledge, thereby providing the rationale for what became an enormous and influential tradition of food combinations and prohibitions (Chapter 19 in this volume). Sun Simiao also emphasised the importance of dietary prohibitions in his chapter on dietetics (Despeux 2007):

> If the *qi* of different foods are incompatible, the essence will be damaged. The body achieves completion from the flavors that nourish it. If the different flavours of foods are not harmonized, the body becomes impure. This is why the sage starts off by obeying the alimentary prohibitions in order to preserve his nature; if that proves ineffective, he has recourse to remedies for sustaining life.
>
> *(Qianjin yaofang, 26.557)*

He emphasised that correct patterns of eating protected the body's essences, and that medical or nutritional remedies were only appropriate when food combinations and prohibitions failed. In practice, the origins of any specific prohibition might be religious, particularly Buddhist, seasonal or simply related to hygiene.

It was not until the fourteenth century that a dietary treatise integrated information on each individual foodstuff with recipes. *Yinshan zhengyao* 飲膳正要 (Proper and Essential Things for the Emperor's Food and Drink) integrated theories about specific foodstuffs with recipes in a woodblock print book presented to the Mongol Emperor at the Yuan court in 1330 by the court dietary physician Hu Sihui 忽思慧. Demonstrating the diversity of Chinese dietary culture, it presents a blend of Mongol food, Muslim spice and often ethnically central Chinese cooking methods, overlaid with Chinese medical philosophy. The recipes themselves are sandwiched between substantial tracts on longevity, foods to avoid or not to combine, how to behave in pregnancy and miscellaneous precepts for hygiene and deportment taken from some well-known early Chinese sources as well as lists that are un-attributed and therefore presented as general knowledge.

The majority of the recipes 'supplement the centre and increase *qi*' and can be classified as fortifying. We also see how new foods were embraced within the local nutritional system of codification. In the description of a Mongolian favourite, Roast Wolf Soup, the dilemma is explicit:

> ancient *materia medica* do not include entries of wolf meat. At present we state that its nature is heating.........it warms the five internal organs and warms the centre.
>
> *(Buell and Anderson 2000: 295)*

New food classifications were clearly established by their similarities to known foods or were loosely classified according to the obvious resonance between remedy and ailment, i.e. boiled sheep's heart, treats heart energy agitation, while the loins treat lumbago.

Old foodways for new appetites?

With the globalisation of Cantonese culinary culture in our time, ready-made selections of restorative foods are commonly available on the shelves of overseas Chinatown supermarkets. Middle-aged Chinese at home and abroad are generally able to tell you that crab must be eaten with ginger to counteract its 'cold' qualities since ginger is warming. Walnuts are a tonic for the brain and the kidneys and increase your sexual potency, jujube enriches the blood, chrysanthemum tea and steamed fish are cooling, mandarin oranges are heating, but oranges, cut rather than peeled, are cold. Some of the information is subject-specific: 'round white (coleslaw) cabbage' will make the head heavy and the feet light, so a child might fall. Many recommendations concern pregnancy: if you eat fish soup, the child will blow bubbles, if you look at a hare, the child will be born with a split lip, and steamed carp increases the flow of breast milk. Chinese dietary practice demands attention to timing and knowledge of your own and other's constitutions: if your liver is hyperactive, don't eat sour flavours in spring time; if there's heat in your body, avoid chilli; if you're suffering from cold in the stomach, eat warming foods. Others appear simply to be related to culinary preference: for instance, bean curd should not be eaten with vinegar. Superstitious precepts survive, such that strong spices like garlic or ginger on the breath will attract the ghosts who will want to lick your lips.

Where China differs, for example, from Europe is that so many of those ancient ideas and practices survive in popular culture and are increasingly the subject of modern scientific research and commercial exploitation. If you sit in any public place in China and listen to the conversations that surround you, before long you are most likely to overhear a casual exchange about food that references the past. Often, there will be warnings about what will be bad for your health, will boost your vitality, calm your stress or clear your headache. Everyone has an opinion.

We can pick up shrink-wrapped packs of *goujizi* 枸杞子 (Chinese wolfberry) to strengthen blood and *yin*, *hongzao* 紅棗 (jujube) and chrysanthemum tea in every cosmopolitan city of the world and, whether or not it is legal to print the traditional belief about their potencies on the packaging, we buy these products amid a growing, and often baffling, network of nutritional advice from family, friends, advertising and multi-media, in our information-soaked world.

Exacerbating the problem is the current fashion for super or power foods, those products that come with the promise of increased vitality, sexual vigour, youth and longevity. Many of these claim an evidence base in modern nutritional science. Chinese green tea is advertised as rich in anti-oxidants good for combatting cancer, garlic prevents bacterial infections, epimedium (*yinyang huo* 淫羊藿: also known as horny goat's weed) has the same active ingredient as Viagra (Lo and Re'em 2018). Perhaps the latter have grains of truth and will continue to be popular, but many of the power foods, like *gouji* berry, enjoy a very short celebrity status outside China, as the global market for superfoods is nothing if not faddish.

Modern nutritional justifications do much to undergird past assumptions, pointing, for example, to the iron content of lamb and *danggui* 當歸 (Chinese Angelica) as the effective ingredient in the dish's traditional use as a blood tonic for post-partum mothers. Equally, old beliefs lose much of their efficacy when they are no longer culturally relevant. While many millions of elderly men are prepared to drink their rice spirits laced with *gouji* berries because it tonifies their kidneys (and therefore their virility), it is unlikely that those with swollen scrotums will be prepared to take the following cure from the Mawangdui manuscripts: 'soak a woman's (menstrual cloth). Boil meat in the liquid. Eat it and drink the liquid' (*Wushier bing fang*, tr. Harper1998: 261).

Other products, mostly prepared for the aphrodisiac market of ageing Asian men, have come up against the World Wildlife Fund and other bodies concerned with the protection of endangered species. Rhinoceros and deer horns, tiger bone, bear gall, are all animal products that carry with them a certain symbolism associated with the power of the animals themselves. These kinds of links to sympathetic magic persist in the popular imagination, but are clearly out-dated and inappropriate today.

Are the myriad dietary prohibitions and recommendations the stuff of old wives tales, the vestiges of a vanishing and irrational past? To the modern eye, Chinese dietary traditions might seem burdened with a history of ritual, religion, sexual lore and magic. Yet, what one believes in is hardly the point. Chinese dietary lore is not so much a set of beliefs, but of shared social practices within which ordinary people can claim a certain expertise. The older generations of Chinese still care about what they eat, as do those younger people who have not been totally seduced by fast food culture and who buy in to a new-look tradition of gastronomic healthcare. Lavish imperial-style banqueting aside, ordinary people still eat plenty of grain, green vegetables, steamed food and little meat. Therapeutic diets involve rice and millet congees, many different types of brassica and wood ear soups with berries, slow cooked foods that are easy on elderly digestions. A traditional Chinese diet not only pays attention to individual constitutions, and allows people a framework within which to test what diet suits them; it is also a diet that is sustainable in global terms.

Our modern medical obsession with clinical or laboratory 'evidence' as the only criterion for public health recommendations should not lead to dismissing traditional dietary cultures. When the whole evidence-based medical tradition has only managed to come up with the universal diktat of 'five fruit and vegetables a day', or worse a plate full of vitamins and minerals for breakfast, and cannot demonstrate that public health campaigns actually change behaviour, we have to look elsewhere for efficacy – for nutritional ideas that can be successfully translated into practice, and aim beyond just keeping our bodies functioning.

Note

1   This work was supported by the Wellcome Trust [References: 201616/Z/16/Z].

Bibliography

Pre-modern sources

*Beiji qianjin yaofang jiaoshi*备急千金要方校释 (Collation and Annotations of Prescriptions for Emergencies Worth a Thousand Gold) c. 652, Sun Simiao 孫思邈, Li Jingrong李景荣*et al.* (eds) (1998), Beijing: Renmin weisheng chubanshe.

Modern sources

Albala, K (1994) 'Dietary regime in the Renaissance', *Malloch Room Newsletter*, Jan. 1994, **7**: 1–2.
Anderson, E.N. (1988) *The Food of China*, New Haven, CT: Yale University Press, http://www.krazy-kioti.com/articles/296/, accessed 27/9/2019.
Birrell, A. (1993) *Chinese Mythology*, Baltimore, MD: John Hopkins University Press.
Buell, P.D. and Anderson, E.N. (2000) *A Soup for the* Qan, London; New York: Kegan Paul International.
Chang, K.C. (ed.) (1977) *Food in Chinese Culture: anthropological and historical perspectives*, New Haven, CT: Yale University Press.
Despeux, C. (2007) 'Food prohibitions in China', trans. P. Barrett, *The Lantern: A Journal of Traditional Chinese Medicine*, 4.1: 22–32.
Engelhardt, U. (2001) 'Dietetics in Tang China and the first extant works of materia dietetica', in E. Hsu (ed.) *Innovation in Chinese Medicine*, Cambridge: Cambridge University Press, pp. 173–91.
Farquhar, J. (2002) *Appetites: food and sex in post-socialist China*, Durham, NC; London: Duke University Press.
Harper, D. (1998) *Early Chinese Medical Literature*, London; New York: Kegan Paul International.
Hinrichs, TJ and Barnes, L. (eds) (2012) *Chinese Medicine and Healing*, Cambridge, MA: Harvard University Press.

Hsu, E. (ed.) (2001) *Innovation in Chinese Medicine*, Cambridge: Cambridge University Press.
Hu Shiu-ying (2005) *Food Plants of China*, Hong Kong: The Chinese University of Hong Kong.
Huang, H.T. (2000) *Science and Civilisation in China, Vol. VI, Part 5: fermentations and food science*, Cambridge: Cambridge University Press.
Knoblock, J. and Riegel, J. (2000) *The Annals of Lu Buwei*, Stanford, CA: Stanford University Press.
Lloyd, G.E.R. and Zhao, J.J. (eds) (2018) *Ancient Greece and China Compared*, Cambridge: Cambridge University Press.
Lo, V. (2005) 'Pleasure, prohibition and pain', in R. Sterckx (ed.) *Of Tripod and Palate*, London: Palgrave MacMillan, pp. 163–86.
––––––– (2012) 'Han' in TJ Hinrichs and L. Barnes (eds) *Chinese Medicine and Healing*, Cambridge, MA: Harvard University Press, pp. 31–64.
Lo, V. and Barrett, P. (2005) 'Cooking up fine remedies: on the culinary aesthetic in a sixteenth-century Chinese materia medica', *Medical History*, 49.4: 395–422.
Lo, V., Berry, C. and Guo Liping (eds) (2019) *Film and the Chinese Medical Humanities*, London: Routledge.
Lo, V., Kadetz, P., Datiles, M.J. and Heinrich, M. (2015) '*Potent substances* – an introduction', *Journal of Ethnopharmacology*, 167: 2–6.
Lo, V. and Re'em E. (2018) 'Recipes for love in the ancient world', in G.E.R. Lloyd and J.J. Zhao (eds) *Ancient Greece and China Compared*, Cambridge: Cambridge University Press, pp. 326–52.
Needham, J. (series ed.) (1956) *Science and Civilisation in China, Vol 2, History of Scientific Thought*, Cambridge: Cambridge University Press.
––––––– (2000) *Science and Civilisation in China*, Cambridge: Cambridge University Press.
Sterckx, R. (ed.) (2005) *Of Tripod and Palate: food, politics and religion in traditional China*, New York; Basingstoke: Palgrave Macmillan.
Wang, Xinyuan and Lo, V. (2019) 'Food-related yangsheng short videos among the retired population in Shanghai', in V. Lo, C. Berry and Guo Liping (eds) *Film and the Chinese Medical Humanities*, London: Routledge, pp. 226–241.

21
FOOD AND DIETARY MEDICINE IN CHINESE HERBAL LITERATURE AND BEYOND

Paul D. Buell

Herbal medicine is ages old, starting literally before history. Unfortunately, almost everything about healing before the written record is lost since it is hard to trace healing practices from archaeological evidence alone. In China, the problem is made worse by the fact that writing was only invented in the mid-second millennium BCE, rather late when compared to Egypt or Babylonia. When we do have reliable evidence, primarily but not exclusively herbal, drug therapy included *materia medica*, also derived from animal and even mineral sources. As part of this tradition, recipes and information about individual *materia medica* (simples) were written down and gradually formed into specialised texts, perhaps from Han 漢 (202 BCE–220 CE) times onwards. From the early imperial period onwards, *materia medica* in China became increasingly framed by concepts drawn from Chinese ritual and state philosophy, and particularly merged with streams of thought that became associated with Daoism. Here, early alchemical traditions, both physical alchemy using mineral and other substances, and physiological alchemy, which aimed to refine the physiological substances of the body, were key and dominant (Chapters 2, 29 and 30 in this volume; Needham *et al.* 1976, 1980). Also important for the broad medical context which we have to understand, food and dietary medicine reflected medical traditions, but with their own validity, and were not so caught up in herbal medicine *per se*. The texts making up the three parallel versions of the *Yellow Emperor's Inner Classic* are but one example of classical medical traditions which also shaped medical foodways (Chapter 7 in this volume). At least part of this corpus of work is pre-Han and may be based upon much older material, now lost. Existing recensions date to the Tang 唐 (618–907) and Song 宋 (960–1279) periods, a time when the entire tradition was codified more systematically than ever before (Unschuld 2003).

From the beginning of a codification of foods and medicines as part of larger traditions, the nutritional and the therapeutic become virtually indistinguishable in our texts. Thus, the histories of Chinese dietary and herbal medicine not only interpenetrate, but both the specialised and the general traditions reflect Chinese food and food history as well as medical history. Interestingly, the revolutions in content that we find in early texts, and those from later times, closely reflect technical revolutions in food production as well as changes in the way people thought about medicine and the use of food to treat the body.

In this connection, the earliest Chinese herbal literature that survives, the text known as the *Shennong bencao jing* 神農本草經 (*Materia Medica* Canon of Shennong), written slightly after Han in its present form, already contains entries for individual *materia medica* which we might think of purely as foods, since they are primarily domestic today (Unschuld 1986: 11).[1] Later, the food as medicine tradition already present in the Chinese herbal literature was considerably strengthened from an outside source – Buddhist medicine that was adapted from Indian traditions, which strongly associated foods and health, including specific foods thought to act on specific conditions (Salguero 2014).

Both in Buddhist and in Chinese medicines, the actions on the body of specific *materia medica* and, for that matter, purely *materia dietetica* (if there is such a thing) were identified through a classification system structured in terms of humoral properties (Chapters 20, 28 and 32 in this volume). These properties were, in turn, matched against categories of prevention, and with illnesses, whose nature and prognosis were also expressed in humoral terms. Such classification was apparently already well-established by the time that it first appeared in Chinese documents, but large elements of that system may not be of Chinese origin at all. Where these ideas ultimately came from is unclear, but remarkable parallels exist between similar classification systems in, for example, the so-called Galenic system, and later the Galen-influenced systems of the Islamic world, and of China. Quite probably, the European ideas arrived in China with exotic medicinals (and exotic foods), increasingly sought for beyond the borders of imperial China as China's connections with the outside world expanded. How profound the cross-cultural influences in question became can easily be seen from an expansion of the first herbals from coverage of a few hundred *materia medica*, to thousands, and even more if we consider all the variants listed in a text such as the sixteenth-century *Bencao gangmu* 本草綱目 (*Materia Medica* Organised by Headings), for example, on which more is described below.

During China's medieval period (post-Han through Tang), a time of disunity and foreign invasion, the *materia medica/dietetica* texts as we have received them in printed form became more complex. In many ways, this complexity reflected changes within Chinese society itself, with a real shift of the Chinese centre of gravity southwards. The corpus of writing associated with Daoist master Tao Hongjing 陶弘景 (456–536) is one example of a new medieval complexity focused on the south (Unschuld 1986: 28–44 and *passim*). By his time, Daoism and its alchemical traditions had become a major fertiliser for Chinese medicine in general, but particularly for its *materia medica* traditions. As already noted, a complicating influence was the arrival of Buddhist medicine in China which brought not only its own humoral traditions, but substantial Indian and Western *materia medica* as well, along with specifically Buddhist methods of treatment, including a well-developed dietary medicine that was part of the appeal of Buddhism in China.

Following Indian and Buddhist influence on Chinese foodways, the Tang period also saw the first introduction of Middle Eastern medical lore and *materia medica*, along with many Middle Eastern foods, including the whole *samosa* tradition. In some cases, new medicinals and theories were introduced by persons of Middle Eastern extraction actually present in China. What resulted was a considerable enrichment of the Chinese tradition of *materia medica*, including one herbal largely devoted to Middle Eastern medicinals (and many foods) by an individual of Persian extraction (Chen Ming 2007). At this time, Islamic lore was already spreading by sea, as well as through Central Asia as the Islamic faith and lifestyles spread beyond Arabia and Persia, including to what is now Indonesia.

By Tang times, the manuscript tradition of the herbal was well developed, including major new works such as the *Xinxiu bencao* 新修本草 (Newly Revised *Materia Medica*), compiled by several authors, and published in 659 in fifty-four *juan* 卷 (scrolls/chapters) with 850 drug monographs (Unschuld 1986: 44–5). This was a much larger work than any previously published. This text was still firmly in the *Shennong bencao jing* tradition as revised by Tao Hongjing, even if much expanded over previous *materia medica*. Also appearing during Tang were the first illustrated herbal texts. They became increasingly well developed as time passed. Efforts were also made to add illustrations to older texts originally lacking them, such as Tao Hongjing's definitive version of the *Shennong bencao*, the *Shennong bencao jing jizhu* 神農本草經集注 (The Shennong *Materia Medica* Canon with Collected Commentaries and Notes) (Ibid.: 46). Illustrations can be particularly important since they can, if they are realistic and accurate enough, facilitate identification of foods and medicines, at least more so than in purely textual description. However, illustrations tend to deteriorate content-wise when they are copied and recopied. Note that the illustrated herbal was very important in the European and Islamic worlds, thanks to the Dioscorides tradition, and there may possibly have been distant influence.

Illustration became a very strong tradition after the first experiments with printing took place around the eighth and ninth centuries (Barrett 2008). For the first time in the history of any culture, *materia medica* texts could be mass produced for wide dissemination. The real flowering of this initiative came in Northern and Southern Song 宋 times (960–1279), particularly the former, when commercial printing of medical texts soon superseded official government publications as the major sources for wide distribution (Chia 2002).

More than any other dynasty, before or after, Song dynasty administrations sought to standardise texts of every kind, including those with medical content (Goldschmidt 2009). And, most important, the imperial government endeavoured to produce great editions of standardised texts. These remain among the most beautiful of China's *shanben* 善本 (fine printings, fine books) since they were carefully printed and illustrated. Many examples survive today because, among other reasons, the Song also printed on the highest quality paper. Books from this time have not only been appreciated, but have also survived.

Song dynasty administrations, as is known, thus produced standard editions of the Confucian classics and related works, but also of the medical classics then regarded as part of a Confucian as well as medical tradition. The version of the *Huangdi neijing suwen* 黃帝內經素問 (The Yellow Emperor's Inner Classic: Simple Questions) that circulates today, for example, is the version of the Song editors (Unschuld, 2003: 1–7). Although their text has been reprinted many times since, it has not been extensively re-edited. They also produced editions of earlier herbals such as the *Shennong bencao jing* and the works of Tao Hongjing. The latter work comprised principally Tao's expanded version of the *Shennong bencao jing*, in three juan with 730 drug monographs, twice the number of the original text, and his *Bencao jing jizhu* 本草經集注 (*The Materia Medica Canon with Collected Commentaries and Notes*), in seven juan. (Unschuld 1986: 30–43).[2] As was the case with most later herbals, this text had far more detail than was typical of the early traditions.

Among many original Song medical works, the most important are a series of official herbals which I list in Table 21.1. These, by and large, became the foundations for later dietary as well as mainstream herbal medicines.

Major Southern Song herbals included still another version of Tang Shenwei's 唐慎微 ((fl. eleventh to twelfth centuries) illustrative work, the *Chongxiu zhenghe jingshi zhenglei beiyong bencao* 重修正和經史證類備用本草 (Revised *Materia Medica* Proven and Classified Throughout History for Use from the Zhenghe 正和 Period), published in 1249 in thirty *juan* with 1,746 monographs (Unschuld 1986: 81–2). This latter text was profusely illustrated and subsequently, illustration became the standard rather than the exception. And in the cases of such texts, the same illustrations were used again and again, being retraced from earlier versions for re-cutting as part of later editions, usually with a steady deterioration in

Table 21.1 Song dynasty (960–1279) Official Herbals

Dates	Title	Author	juan	Monographs
973	Kaibao xin xiangding bencao 開寶新詳定本草 *Materia Medica* Newly Revised in Detail during the Kaibao 開寶 (968–976) Period	Numerous authors/editors listed	21	983
974	Kaibao chongding bencao 開寶重定本草 The Re-revised *Materia Medica* from the Kaibao Period (Unschuld 1986: 55–8)			
1061	Jiayou buzhu Shennong bencao 嘉祐補注神農本草 Annotated Shennong *Materia Medica* from the Jiayou 嘉祐 (1056–1063) period (Ibid.: 60–4)	Various authors	21	1084
Comp. 1080–1107; officially published in 1108 in a much-expanded version	Jingshi zhenglei beiji bencao 經史證類備急本草 *Materia Medica* for Emergencies Proven and Classified through History (Ibid.: 70–1)	Tang Shenwei 唐慎微	31	1744
Revision of the entry above	Jingshi zhenglei Daguan bencao 經史證類大觀本草 *Materia Medica* Proven and Classified through History from the Daguan 大觀 period (1107–1110) (Ibid.: 72–7)			
	Zhenghe xinxiu jingshi zhenglei beiyong bencao 正和新修經史證類備用本草 (Ibid.: 77)		30	1748
1062	Tujing bencao 圖經本草 (Ibid.: 64–8)	Su Song 蘇頌 et al.	21	634 drug descriptions from *Xinxiu bencao*

quality. Compare, for example, the surviving pages of the Yuan 元 (1271–1368), first edition of the imperial dietary manual *Yinshan zhengyao* 飲膳正要 (Proper and Essential Things for the Emperor's Food and Drink), against the Ming edition of 1456.[3] The latter is still of high quality in its illustrations but the original edition is clearly better.

The large *materia medica* as first published during Song times continued to constitute the main stream for Yuan and Ming 明, although a growing number of specialised works appeared as well. Jin 金 (1115–1234) authors of *materia medica* books based their work on Song books, smuggled or imported from the south, but also developed their own traditions. In large part, this was a response to changing medical theories, namely the rise of a fully developed correspondence theory based upon juxtapositions of *yin* 陰, *yang* 陽 and Five Agents,

331

in medicine, *e.g.* as seen in the thirteenth-century *Tangye bencao* 湯液本草 (*Materia Medica* of Soups and Decoctions) by theorist Wang Haogu 王好古 (thirteenth century), in three *juan* with 224 monographs (Unschuld 1986: 108–17). Little is known about Wang but his work is nonetheless a theoretical watershed.

A great specialised work from this time, and the most important work of dietary medicine *per se* in its time and long thereafter, was the aforementioned *Yinshan zhengyao* 飲膳正要. Since appearing, this text has become the most influential of all Chinese texts on dietary medicine and *materia dietetica*. This three *juan* text was submitted to the Mongol court in China in 1330 by Hu Sihui 忽思慧, possibly an Uighur or some other kind of Turk, but one thoroughly familiar with Chinese culture. The *Yinshan zhengyao* subsequently has constituted a major source for our understanding of foreign influences in Chinese dietary and general medicine. It remains popular today. One of the text's three *juan* is devoted to an illustrated collection of *materia dietetica*. Many examples occur under Mongolian or other non-Chinese names. It was another officially sponsored work, although much of the Yuan original edition, other than a few pages, is now lost, its publication history, including an official new 1456 Ming edition.

The *Yinshan zhengyao* was certainly not the first Chinese work devoted to *materia dietetica* although much of the earlier tradition is now lost. Sun Simiao 孫思邈 (581?–682?), for example, included a chapter on dietary treatment in his important *Beiji qianjin yaofang* 備急千金要方 (Essential Recipes for Need worth a Thousand in Gold) (Engelhardt 2001: 176–84). All of the characteristics of later dietary manuals are present in this chapter and re-occur, in expanded forms in later dietaries, including the *Yinshan zhengyao*, which quotes material from the Sun Simiao text, although often in paraphrase, mixing it with observations and substances derived from the Middle East.

After Sun Simiao, others wrote on dietary medicine as can be seen from fragments of lost works quoted in the Japanese anthology *Ishimpō* 醫心方 (Recipes at the Heart of Medicine), and also noticed in other sources such as the bibliographies of the standard dynastic histories the *Hanshu* 漢書 (Book of Han) and the *Suishu* 隋書 (Book of Sui),[4] but the first known dedicated work devoted specifically to dietetics was the Tang-era *Shiliao bencao* 食療本草 (*Food* Therapy *Materia Medica*). It was composed between 721 and 739 by one Zhang Ding 張鼎 and based upon an earlier work by Meng Shen 孟詵, the *Buyang fang* 補養方 (Recipes to Tonify and Nourish). The *Buyang fang* does not survive but a large fragment of the *Shiliao bencao* was found at Dunhuang 敦煌 in 1907 (Unschuld 1986: 208–11; Engelhardt 2001).

The tradition of food as medicine continued as did the tradition closely associating dietary medicine and Daoist alchemy. The late Northern Song *Jingshi zhenglei beiji bencao* includes a number of quotations from the *Shiliao bencao*, but also from two completely lost works, the *Shixing bencao* 食性本草 (*Materia Medica* of Food Qualities) in ten *juan*, written by one Chen Shiliang 陳士良 at the end of the ninth century, and the *Shizhi tongshuo* 食治通說 (Explanation of Dietary Therapy) in one *juan*, written by Lou Juzhong 婁居中 in the twelfth century (Unschuld 1986: 211–13).

As far as can be determined, the *Yinshan zhengyao* is keyed into the main traditions of herbal medicine described above, except that the Mongol-era text goes far beyond anything previously attempted. Not only does it include miscellaneous foreign material, not just *materia medica*, but even what seems to be an extensive adaptation of text passages from Ibn Sīnā on wet nursing. At least, there is really nothing else like the material in other Chinese sources. There are also many parallels in the *Yinshan zhengyao* between Ibn Butlan's twelfth-century *Kitab Taqwim as-sihha* (Almanac of Health), known in Europe as the *Tacuinum Sanitatis* (The *Taqwim* of Health). But other Arabic works may have been involved as well since Ibn Butlan wrote only at the end of a long tradition (Arano 1976; Ullmann 1978: 99; Buell and Anderson 2021: 115–16).

The *Yinshan zhengyao* also makes a strong effort to quote earlier literature and directly identifies works that are very much in the Daoist alchemical traditions. These include the fourth-century *Baopu zi* 抱朴子 by Ge Hong 葛洪, the name of the work coming from the author's nickname, Master of Embracing Simplicity. Also quoted were the *Shenxian zhuan* 神仙傳 (Traditions of the Beneficent Immortals), attributed to Ge Hong but now lost, the *Liexian zhuan* 列仙傳 (Tradition of the Ranks of immortals) also lost; and another lost work called the *Yao jing* 藥經 (Canon of Medicines) apparently recommending dosages with particular herbs, the very similar *Zhen zhong ji* 枕中記 (Record Kept in the Pillow), attributed to Sun Simiao, and the mysterious *Taiqing zhu bencao* 太清諸本草 (Various *Materia Medica* of the Taiqing 太清 era (547–9)). This is presumably a version of the *Shennong bencao*. Also present in the *Yinshan zhengyao* are many unattributed passages from the *Yellow Emperor's Classic*.

There is thus considerable Daoist lore in the text of the *Yinshan zhengyao*, but also present alongside traditional dietary theory and *materia dietetica* are many entirely new elements. China has, over time, undergone many shocks of the new in its food systems, the coming of wheat in late Zhou 周 times, for one; later, in medieval and early modern times, the rise of rice, which became the major grain in most areas; and then the coming of tea, which transformed society. The *Yinshan zhengyao*, in its turn, represents another change as massive amounts of new foods and food cultures are first introduced in this text (Laudan 2013). The text, above all, shows the need to incorporate the new foods into China's existing food and dietary cultures. Although many of the foods in question remained little more than exotics outside court circles, the *materia dietetica* of the text has, by and large, become an important part of Chinese dietary medicine. The importance of the text lies in its legacy in the dietary traditions and explains why it was reprinted during the Ming dynasty in 1456, nearly 130 years after first being presented at court. Many editions exist today, some quite popular in orientation.

Ming contributions to dietary medicine were not limited to reprinting the *Yinshan zhengyao*, but the thrust of publication was aimed at a broader audience, reflecting the rapid development of commercial publishing from Song times onwards. There existed a popular appetite for medical texts of every sort by Ming times. This included what was the greatest herbal of its time, the late sixteenth-century *Bencao gangmu* of Li Shizhen 李時珍 (1518–93). The book was not fully published in the author's lifetime,[5] but subsequently enjoyed a long-life having been published commercially in many illustrated editions since his time, and there have recently been a number of full or partial translations.

Li Shizhen, the author of the *Bencao gangmu*, was from Hubei 湖北 (Needham *et al*. 1986: 308–11; Unschuld 1986: 145–63; Métailié 2001: 221–61; Nappi 2006). His grandfather had been an itinerant doctor. His father, by contrast, although clearly a practising doctor, had been awarded the *xiucai* 秀才 (flowering talent degree) at the lowest level of the Ming imperial exam system, and had enough land so that he did not have to live from hand to mouth. Or at least so goes the traditional biography. By virtue of his qualifications, and the influence of physicians in general, Li Shizhen's father later served as a medical officer at court. From Li Shizhen's citations in the *Bencao gangmu*, we know that, in addition to being a practising doctor, he was also the author of a number of medical works, all of which are now lost. Shizhen, himself, passed the primary level examinations at age fourteen, but ultimately failed the imperial exam. This was in spite of his rigorous classical education and broad learning.

Li Shizhen was trained in medicine, but was very widely read outside medicine. This fact is shown strongly in the range of material that he drew into the composition of the *Bencao gangmu*. His truly encyclopaedic *Bencao* is not only based on a wide range of literature still available today, but also cites many rare and now lost works. His broad approach not only makes the *Bencao gangmu* a medical compendium, but also a much broader array of cultural facts, including extensive regional information and minutiae about food and diet: and this includes rich information about the history of distillation as summarised by Needham (Needham *et al.* 1980: 132 ff and *passim*). Distillation was already old in China by the time of the Mongols and was particularly associated with the Daoists and their alchemy, but the Mongols were associated with the introduction of lighter and more portable stills and, most importantly, widely disseminated distilled beverages of every sort. These beverages are usually recorded under the Arabic name *arkhi* (*arak*). This term first occurs in the *Yinshan zhengyao*, which shows clearly the revolution affected by the new distillates. Distillation represents another major dietary change of the era, and one that has persisted.

Otherwise, Li Shizhen frequently quotes the *Yinshan zhengyao* and mentioned most of the new, exotic *materia dietetica* of that text. And, like the *Yinshan zhengyao*, the medicine of the *Bencao gangmu* has strong alchemical roots.

Li Shizhen spent decades writing his classic work. He set out to systematise knowledge associated with *materia medica* of every sort, and this effort he based upon extensive reading (he cites 952 authors, many lost and cited at second hand), although most of his material comes from major *materia medica* texts such as the various versions of the *Jingshi zhenglei bencao* discussed above. He also travelled to places where *materia medica* were produced and collected specimens.

Following Li Shizhen's example, subsequent *materia medica* also attempted to be systematic and closely followed his approach in form and structure. His influence extends to the treatises on dietary medicine written after his time. The most important of these was Yao Kecheng's 姚可成 seventeenth-century *Beikao shiwu bencao gangmu* 備考食物本草綱目 (*Materia Medica* Organised by Headings with Individual Listings for Foodstuffs for Consideration) (Unschuld 1986: 225 ff). This great work has 1,699 monographs. As its title implies, this is a *Bencao gangmu* for foods.

Not so extensive, but among other dietary works of the Ming was the *Shiwu bencao* 食物本草 (*Materia Medica* of Foodstuffs), published by Lu He 廬和 in 1571 with 391 monographs. Although the emphasis in Lu's work is pragmatic and practical, the author is influenced by the correspondence theory of Wang Haogu and others (Ibid.: 221–2). Another dietary text of the same name is a revision of Lu He's work in seven *juan* originally by Wang Ying 汪穎, but then revised by Qian Yunzhi 錢允治, and published in 1620 (Ibid.: 224–5). Not a *materia medica*, but offering dietary medicine implicitly from a Daoist perspective is the *Yunlin tang yinshi zhidu* 雲林堂飲食制度 (The Drink and Food System of the Cloudy Forest Temple) of literatus and painter Ni Zan 倪瓚 of early Ming. Although the stress is on dietary alchemy, the recipes are quite varied and delicious and show the sophisticated environment from which Ni Zan came. Interestingly, there is little foreign material in them (Buell *et al.* 2010: 58–60).

Another famous dietary (and medical) work of Ming times, this one written before Li Shizhen's time is the *Jiu huang bencao* 救荒本草 (*Materia Medica* for Aid in Famines), published in 1406. This text is unique in combining information about famine foods *per se* as foods, with parallel information stressing the medical importance of the same foods. Altogether some 400 food/drugs from trees and herbs are listed with excellent illustrations so that the trees and plants in question could be easily identified and used (Unschuld 1986: 221). Numerous editions of this work, some of them modified, survive. So many editions survive, in theory, because of the great usefulness of the information in the text to people experiencing hard times.

Although not strictly speaking dietary texts, China also has a rich cookbook literature, which is often quite specialised. This literature is of interest here because such works often include recipes that are more or less medicinal from our perspective; little distinction is made between the medical and the dietary in the way that we define the boundaries nowadays in law and practice. Perhaps typical of such material, although not, strictly speaking, a cookbook, is the early Ming household encyclopaedia *Jujia biyong shilei* 居家必用事類 (Various Categories that Must be Used Living at Home). There are several versions (collections) but all contain many recipes and 'how to' guides for producing things such as fermentation starters, as well as descriptions of still technology to support the new beverages. One of these stills is specifically labelled as an import from the Islamic world of the Indian Ocean trading system. Unlike Ni Zan's text, the shock of the new is readily apparent in the collection's many foreign recipes, including an important collection of Turkic recipes, some recorded under their original Turkic names (Buell 1999: 200–23).

An important part of dietary medicine is, last but not least, Chinese folk traditions, which dictate not only the traditional, medical values of foods but even the desired colour balance to the dishes of a meal. And if the colours don't work, foods can be treated with food colouring. As long as the yin and yang and food properties are balanced, even if only in theory, their nutritional value might be deemed efficacious. Thus, the pervasive outcome of thousands of years of dialogue about food and medicine in China grows out of complex terroirs and environments, where foods act as medicines, and medicines as foods, connecting multiple social contexts and dietary variations. Influences on food and dietary medicine as discussed in Chinese herbal literature are not just limited to the Yellow River valley, but embrace broader cultures that, by the medieval times, stretched westwards as far as Persia and beyond – and the practices engendered transculturally survive today to appropriate the authority and practices of modern science to an ever-emerging and living ancient tradition.

Notes

1 Note the early association of Shennong 神農 with a lost book on dietetics. For an introduction to China's early dietaries and dietary information in more general sources, see Engelhardt (2001: 173–91).
2 During Tang times, there was an official edition of this book, the first of its kind, edited by Su Jing 蘇敬 (599–674) and published in the middle of the seventh century. An attempt was made to add colour illustrations to this text but we do not know how successful the request of the publishers was. Colour printing had not been invented yet but soon was.
3 Portions of both the Ming and the Yuan editions are reproduced in Buell, Anderson and Perry (2010).
4 Early lost texts are discussed in Unschuld (1986: 206–8).
5 In an email to the author of 19 September 2019, Paul Unschuld notes that 'the Chinese now believe that the *Bencao gangmu* was published in 1593 – maybe Li Shizhen saw it on his deathbed'. He also notes that the compendium is unlikely to be directed at practising doctors, and that he is not sure for whom it was intended.

Bibliography

Modern sources

Amitai, R. and Morgan, D.O. (eds) (1999) *The Mongol Empire and Its Legacy*, Amsterdam: E. J. Brill.
Arano, L.C. (1976) *The Medieval Health Handbook: tacuinum sanitatis*, New York: George Braziller.
Barrett, T.H. (2008) *The Woman Who Discovered Printing*, London: Yale University Press.

Buell, P.D. (1999) 'Mongol empire and turkicization: the evidence of food and foodways', in R. Amitai and D.O. Morgan (eds) *The Mongol Empire and Its Legacy*, Amsterdam: E. J. Brill, pp. 200–23.

Buell, P.D. and Anderson, E.N. (2021) *Arabic Medicine in China: tradition, innovation and change*, Leiden; Boston, MA: E. J. Brill.

Buell, P.D., Anderson, E.N. and Perry, C. (2010) *A Soup for the Qan: Chinese dietary medicine of the Mongol era as seen in Hu Sihui's Yinshan zhengyao: introduction, translation, commentary, and Chinese text*, Leiden: Brill.

Chen Ming 陳明 (2007) 'The transmission of foreign medicine via the silk roads in medieval China: a case study of the *Haiyao Bencao* 海藥本草', *Asian Medicine, Tradition and Modernity*, 3: 241–64.

Chia, L. (2002) *Printing for Profit, the Commercial Publishers of Jianyang, Fujian (11th–17th Centuries)*, Cambridge MA: Harvard University Press.

Engelhardt, U. (2001) 'Dietetics in Tang China and the first extant works of *material dietetica*', in E. Hsu (ed.) *Innovation in Chinese Medicine*, Cambridge: Cambridge University Press, pp. 173–91.

Goldschmidt, A. (2009) *The Evolution of Chinese Medicine, Song Dynasty, 960–1200*, London; New York: Routledge.

Hsu, E. (ed.) (2001) *Innovation in Chinese Medicine*, Cambridge: Cambridge University Press.

Laudan, R. (2013) *Cuisine and Empire, Cooking in World History*, Berkeley; Los Angeles; London: University of California Press.

Métailié, G. (2001) 'The *Bencao gangmu* of Li Shizhen: an innovation in natural history?', in E. Hsu (ed.) *Innovation in Chinese Medicine*, Cambridge: Cambridge University Press, pp. 221–61.

Nappi, N. (2006) *The Monkey and the Inkpot: natural history and its transformations in early modern China*, PhD thesis, Harvard University.

Needham, J., Ho, P.Y. and Lu, G.-D. (1976) *Science and Civilization in China, Volume 5, Chemistry and Chemical Technology, Part III: spagyrical discovery and invention: historical survey, from cinnabar elixirs to synthetic insulin*, Cambridge: Cambridge University Press.

Needham, J., Ho, P.Y., Lu, G.-D. and Sivin, N. (1980) *Science and Civilization in China, Volume 5, Chemistry and Chemical Technology, Part IV: spagyrical discovery and invention: apparatus, theories and gifts*, Cambridge: Cambridge University Press.

Needham, J., Lu, G.-D. and Huang, H.T. (1986) *Science and Civilization in China, Volume 6, Biology and Biological Technology, Part I: botany*, Cambridge: Cambridge University Press.

Salguero, C.P. (2014) *Translating Buddhist Medicine in Medieval China*, Philadelphia: University of Pennsylvania Press.

Ullmann, M. (1978) *Islamic Medicine*, Edinburgh: Edinburgh University Press.

Unschuld, P.U. (1986) *Medicine in China: a history of pharmaceutics*, Berkeley; Los Angeles; London: University of California Press.

——— (2003) *Huang Di Nei Jing Su Wen: nature, knowledge, imagery in an ancient Chinese medical text*, Berkeley; Los Angeles; London: University of California Press.

22
THE SEXUAL BODY TECHNIQUES OF EARLY AND MEDIEVAL CHINA – UNDERLYING EMIC THEORIES AND BASIC METHODS OF A NON-REPRODUCTIVE SEXUAL SCENARIO FOR NON-SAME-SEX PARTNERS

Rodo Pfister

The sexual body techniques of early and medieval China are treated heuristically to form a sexual scenario for non-same-sex partners that is discussed in (1) textual sources dating from approximately 200 BCE to 1000 CE. These texts were transmitted and reformulated throughout this period as part of the wider sexual knowledge culture of imperial China (Wells and Yao Ping 2015; Yao Ping 2018). Minimal referential series of short extracts taken from such primary sources will be presented in a historical order to illustrate some fairly consistent basic ideas, concepts, theories and practical advice documented therein.

This concise review discusses (2) general aspects of the sexual scenario of early and medieval China in which gender-specific roles during the sexual encounter must be emphasised. As 'essence' is considered to be the most precious generative fluid in the human body, men are advised to (3) deal with male essence as a scarce good, and thus learn to avoid emission and ejaculation during a sexual encounter. In stark contrast to this male preoccupation with containment, women are thought to be a superior source of nourishment. (4) Repeated female ejaculation provides the 'female essence' that can be absorbed by the man. (5) Performing a sexual encounter means mutual stimulation to this end during foreplay and onset phase, followed by a series of penetrative 'advances' with 'intermissions', and culminating in a 'grand finale'.

Textual sources to a sexual knowledge culture, 200 BCE to 1000 CE

Tangible *textual evidence* for sexual body techniques comes early in China – about five hundred years before the Indian *Kāmasūtra*, which is considered to date from circa 300 CE (transl. Doniger and Kakar 2002) – in the bamboo and silk manuscripts excavated from Tomb 3 at Mawangdui 馬王堆 (near Changsha, in today's Hunan province) that date to the late third, or the early second century BCE. (See the archaeological reports of Fu Juyou and Chen Songchang 1992, and He Jiejun ed. 2004. For source texts, see Mawangdui Hanmu Boshu Zhengli Xiaozu 1985; Ma Jixing 1992; Qiu Xigui ed. 2014; Ōgata 2015; Chapter 3 in this volume).

Two of the recovered bamboo manuscripts – *Uniting Yin and Yang* (**He yinyang* 合陰陽)[1] and *Discussion of the Utmost Method Under the Sky* (*Tianxia zhidao tan* 天下至道談) – treat exclusively a sexual scenario for non-same-sex partners. A third – *Ten Interviews* (**Shiwen* 十問) – includes additional advice for general health, breathing techniques and wellness. The first two manuscripts mentioned are made up of small modular text sequences that are organised as a hypertext whose links the reader mentally constructs by jumping from module to module (Pfister 2013). The third text consists of ten dialogues in a question-and-answer format. These early sources are written in a concise, often metrically regulated language. They use lists and rhymes to facilitate memorisation, and develop a special technical vocabulary that now needs to be reconstructed and interpreted as its terminology became obsolete early on, or underwent considerable changes in transmission. (See the translations in Wile 1992; Harper 1998; Pfister 2003; Ōgata 2015; compare Harper 1987; Li Ling and McMahon 1992; Li Ling 2006.)

The sexual scenario falls within the overall topic of nurturing life techniques (*yangsheng*), which receive further treatment in silk manuscripts. *Recipes for Nurturing Life* (**Yangsheng fang* 養生方) includes prescriptions to cure various sex-related health issues, to increase arousal and to stimulate sexual performance, as well as a line drawing of the vulva labelled with the names of outer aspects of the genital area and locations inside the vagina (Pfister 2016; Chapters 25, 26, 30 and 49 in this volume).

During the Later Han dynasty (25–220 CE), 'arts of the bedchamber' (*fangzhong shu* 房中術, in Japanese *bōnaijutsu*) became a bibliographic label for various sexual body techniques, which was set apart from medical and nurturing life writings, but this distinction was not followed in later Sui dynasty (581–618 CE) listings (Chen Guofu 1963: 365–9; Okanishi 2010, vol. 2: 1183–8; Lin Fu-shih 2008: 335–7; Li Ling 2011a: 207–10).

In 984 CE Tamba no Yasuyori 丹波康賴 (912–995 CE) presented a monumental work to the Japanese Emperor Enyū 円融: The *Core Prescriptions of Medicine* (*Ishinpō* 醫心方). This work comprised a large compilation of thirty scrolls, and gives a comprehensive view of Chinese medicine as understood in the 10th century CE (Tamba [1955]2000; Liu Xiuqiao ed. 1976; Gao Wenzhu *et al.* 1996; compare Ōta Tenrei 1976; Society for the Commemoration of the One Thousandth Anniversary of the 'Ishimpo' 1984; Triplett 2014).

Tamba excerpted mainly medical texts brought to Japan from China, and meticulously noted the title of each fragment throughout. Many of the titles mentioned were subsequently lost in part or entirely in China. Scroll 28 is entitled 'On the Chamber-Intern' (*bōnei* 房内 in Japanese, or *fangnei* in Chinese). *Nei* 内 means both the 'inner quarter' reserved for women in a major household, and the women themselves thereby addressed as 'inmates'. Thus, the chapter title politely points to the sexual encounters of a patriarch with his several women in their closed sphere; however, the process of them 'being brought in (the household)' (*na* 納) is not discussed. Issues of reproduction are treated separately in Scroll 24 (Chapters 23, 24 and 30 in this volume).

Scroll 28 contains thirty subchapters or rubrics that deal with various aspects of sexual body techniques (Table 22.1; Pfister 2013). This topical categorisation allows us to assess the relative importance of the sources quoted therein. Most prominent is *Secret Decisions in the*

Jade Chamber (*Yufang mijue* 玉房祕決), as its passages are quoted in most rubrics before any other source text, and several rubric titles are probably derived from it (Table 22.1). It is followed by *Master Dong Xuan* (*Dongxuan zi* 洞玄子), *Essentials of the Jade Chamber* (*Yufang zhiyao* 玉房指要), *The Book of the Dark Woman* (*Xuannü jing* 玄女經) and several others. (Translations include Van Gulik [1951]2003, [1961]2004; Umayahara Shigeo *et al.* 1967; Ishihara and Levy 1969; Hsia *et al.* 1989; Wile 1992. For critical reviews of Van Gulik, see Furth 1994; 2005, and Hinsch 2005.)

Even though *The Book of the Plain Woman* (*Sunü jing* 素女經) is quoted only twice in *Ishinpō* 28.1 and 28.5, and despite Tamba no Yasuyori's explicit titling of all citations, Ye Dehui 葉德輝 (1903: 1a–11b) assembled all the *Ishinpō* passages mentioning the interlocutor 'Su nü 素女' ('Plain Woman') into a composite work of his own making. By assigning those passages a single title, this made-up '*Sunü jing*' conflates several sources, including cases when passages contradict each other, or otherwise do not fit. Yet, Ye's *rifacimento Sunü jing* of 1903 was received as the major work on sexual techniques in the twentieth century (trans. Mussat 1978; Wile 1992: 84; 85–94, *The Classic of Su nü* (sic); compare Rocha 2015).

Tapping into the same early medieval texts as the *Ishinpō*, the famous medical writer Sun Simiao 孫思邈 (581–682) includes in his voluminous work of 652 CE – titled the *Essential Prescriptions Worth a Thousand in Gold for Urgent Cases* (*Beiji qianjin yaofang* 備急千金要方 27.8: 488–91) – a chapter on 'Replenishing and Benefitting in the Bedchamber' (*Fangzhong buyi* 房中補益; trans. Wile 1992: 114–1, and Wouters 2010: 73–8). This chapter shares its composite character, its intertextual relationship with predecessors and the combination of health-related and Daoist religious concerns with 'Losses and Benefits in Steering Women' (*Yunü sunyi* 御女損益), Chapter 6 in the *Records on Nourishing the Disposition and Prolonging the Mandate of Life* (*Yangxing yanming lu* 養性延命錄) of unknown date and origin (trans. Wile 1992: 119–22; textual history, Stanley-Baker 2006).

The sexual rites of Celestial Master (*Tianshi* 天師) Daoism – 'merging the pneumas' (*he qi* 合炁) – were distinguished from the arts of the bedchamber by this religious community itself, which criticised the latter, and they appear to have been socially oriented *rites de*

Table 22.1 The Thirty Rubrics of Scroll 28 *On the Chamber-Intern* in *Ishinpō*

1 Culminant Principles	16 Eight Increases [of *Qi*][b]
2 Nourishing *Yang*	17 Seven Decreases [of *Qi*][b]
3 Nourishing *Yin*[b]	18 Reverting the Essence
4 Harmonise Strivings	19 [Male] Emission and Ejaculation
5 Approaching the Ride	20 Treating Damages[b]
6 Five Constants[b]	21 Seeking Offspring
7 Five Proofs[b]	22 Attractive Women
8 Five Desires[b]	23 Repulsive Women
9 Ten Stimulations[b]	24 Interdictions and Avoidances
10 Four Optima [of male erection][a]	25 Interception of Ghost Intercourse[b]
11 Nine *Qi* [of the woman][a]	26 Use of Herbs and Minerals
12 Nine Procedures[a]	27 Small Jade Stalk [Penis]
13 Thirty Procedures[a]	28 Wide Jade Gate [Vagina][a]
14 Nine States[a]	29 Pains of Young Women
15 Six Positions[a]	30 Injuries of Grown-up Women

a Rubrics that do not quote the Secret Decisions in the Jade Chamber.
b Rubrics quoting exclusively the Secret Decisions in the Jade Chamber.

passage rather than focussing on the sexual encounters of couples (see Maspero 1937: 401–13; Kalinowski 1985; Raz 2008, 2012; Kleeman 2014; Mollier 2016). The relationship of the sexual body techniques with various Daoist religious movements awaits further clarification (on the latter, see Maspero 1937; Needham 1983; Hidemi 1991; Zhu Yueli 2002; Lin Fu-shih 2001, 2008; De Meyer 2006: 345–74; Hudson 2007; Chapter 27 in this volume).

Literary treatments that use the vocabulary of sexual body techniques, like the *Rhapsody on the Great Pleasure in the Mutual Joys of Heaven and Earth, Yin and Yang* (*Tiandi yinyang jiaohuan dale fu* 天地陰陽交歡大樂賦, trans. Idema 1983; see Umekawa 2005a; Harper 2010; Yao Ping 2013; McMahon 2019), the art of charming (Li Jianmin 1996; Zhang Hanmo 2013), aphrodisiacs (Harper 2005; Umekawa 2005b; Lo and Re'em 2018) and treatment strategies for sexual disorders or sexual medicine (Liu Jie 1995; Fan Youping *et al.* eds 2007), as well as later developments of the art of the bedchamber (see Van Gulik [1951]2003, [1961] 2004; Kobzev 1993; Li Ling 2006; Sakade and Umekawa 2003; Marié 2007; Tsuchiya Eimei 2008; Chen Hsiu-fen 2009: 83–128; Yao Ping 2015; Wells and Yao Ping 2015; Umekawa and Dear 2018), lie outside the topical or temporal focus of the present entry (Chapters 23, 24, 30 and 49 in this volume).

A note of caution and critique is due regarding the twentieth-century reception of the textual sources of the arts of the bedchamber: not only did translation of this ancient knowledge culture pose exceptional difficulties (as amply documented by Wile 1992 in his notes, albeit not without adding *Verschlimmbesserungen*), but also the conceptional frameworks used did not attempt to recover or reconstruct emic perspectives on body, consciousness and disease concepts of their times. (On the emic-etic distinction, see Headland, Pike, Harris eds 1990; de Sardan 1998.) Instead, various kinds of then new, normative approaches to sexological issues were anachronistically inserted without discussion of their being etical terminology. 'Orgasm' may serve as an example: it was considered as being synonymous with male ejaculation or used to translate a medieval Chinese term for male and female 'satisfaction' or 'jouissance' (*kuai* 快, see Pfister 2012), but came without reflection on the shifting technical meaning and conceptual development 'orgasm' underwent in modern sexology. (For critical remarks on orgasm paradigm, and its development, see Walter 1999; Lewandowski 2001; Janssen 2007.) Some terms like 'injaculation' and 'sexual energy' are applied to translate terms or to paraphrase concepts. But while this might appeal to a fashionable *Zeitgeist*, it does not faithfully render the Chinese wording and emic perspective. (General overviews of the field include Wile 2018, Wells and Yao Ping 2015 and Yao Ping 2015, 2018. For insights into the reception process and hybrid popularisations, see Rocha 2011, 2012; Chapters 25 and 26 in this volume.)

General aspects of the sexual scenario

The *sexual scenario* of the arts of the bedchamber does not discuss same-sex relations, and it is necessarily based on a two-sex model, wherein men are classified as *yang* and male, and women as *yin* and female. This applies also to haptic pleasures during sexual acts: 'What is categorised as male, rubs the outside; what is categorised as female, rubs the inside. This is called the skill of *yin* and *yang*' (*Tianxia zhidao tan*, slip 65). During non-same-sex encounters, the superior female potency (Sherfey 1966) wins out against male impotence and weakness: 'If the woman defeats the man, it is like water extinguishing fire' (*Yufang mijue*, *Ishinpō* 28.1: 2a). In most texts, male fragility is emphasised and foregrounded; it receives a phantasmatic treatment in the sexual techniques and is medicalised prominently in the recipe literature. In contrast, female strength and capacity for nourishment forms an implicit background against which the sexual techniques seek to address men's health issues, especially those of men past forty (*Beiji qianjin yaofang* 27.8: 488b). Thus, *yin* and *yang* are used to describe a fundamental and dynamic *asymmetry* in female and male sexual responses, and each gender is advised to adopt a specific kind of learned behaviour – or sexual body technique – to play out the basic programme that unites *yang* with *yin* (on the concept of body techniques, see Mauss 1935, and Crossley 2005). The *yin-yang* language obscures to a certain extent hierarchies and asymmetric conceptual features as applied to 'sexual intercourse' – or *jiaojie* 交接, an 'exchange' (*jiao* 交) (of bodily fluids) and 'contact' (*jie* 接) (by touch).

To create norms for sexual behaviour is conceptually naturalised: 'When humans are born only two things are not to be learned: the first is breathing, the second is eating. Besides these two, there is nothing that has not to be learned and practised. Because what reproduces life is eating and what diminishes life is sensuality (*se* 色), sages must have standards (*ze* 則) for uniting men and women' (*Tianxia zhidao tan*, slips 40–41). Such teaching offers guidance in potentially life-threatening situations. As humans leak, the 'nine openings' of the body – the two of the lower body, anus and urethral meatus, and the seven of the sense organs – are problematic spheres of in- and outflows. They are inroads for diseases, and the generative fluid or 'essence' (*jing* 精) escapes during ejaculatory coitus. The learnt practice is meant to overcome men's most serious problem of losing a scarce good necessary for life-maintenance. From an outsider perspective and psychologically speaking, sexual body techniques open detailed standardised ways for phantasmatic self-affection (Lohmar 2008), but emically and androcentrically these are seen as a self-reproductive process: 'The Daoist considers essence (*jing*) as a treasure. If spent, it engenders others; if retained, it engenders oneself' (*Yangxing yanming lu* 6: 8b; a similar formulation is found in *Wangwu zhenren koushou yindan mijue lingpian*, *Yunji qiqian* 64: 19a).

What may be considered as non-reproductive sexual behaviour (Wundram 1979) between non-same-sex partners in early and medieval Chinese texts is indeed more than recreational sex – one of its twentieth-century interpretations (Rocha 2012). Earlier thought communities held that sexual intercourse should nourish men in need for essence, and thereby 'heal humans with humans' (*Beiji qianjin yaofang* 27.8: 488b). The sexual arts are presented as a healing method, and were assumed to have an enormous therapeutic value for needy men. Androcentric self-reproduction – viewed as a therapeutic process – serves to resolve the considerable male feeding envy. Grafting – a widespread metaphor for sexual intercourse (Taiz and Taiz 2017: 299–301) – is used to express linguistically this idea of nourishment. In *Shiwen, several interviews describe methods titled with the phrase 'to graft the privates' (*jie yin* 椄陰), implying a form of intimate contact in which the man – conceived as scion – draws up the sap from the woman's slit – forming the rootstock, and thereby 'feeds his spiritual flow' (*shi shenqi* 食神氣). The eighth-century text *Numinous Tablets of the Secret Instructions on the Yin Elixir Orally Transmitted by the Perfected Person of Wangwu* (*Wangwu zhenren koushou yindan mijue lingpian* 王屋真人口授陰丹祕訣靈篇) takes up the idea as follows: 'The arts of prolonging life are similar to the grafting of trees (*jie shu* 接樹), as one develops one life mandate with another life. (...) As one begins to open the jade gate, it should have the signs of blood [*xue hou* 血候]; the initally stopped up embankment veins (*chengli* 塍理) start to become permeable [i.e. the vasocongestion of the labia minora and suburethral region sets in like the mud ridges or "embankments" are opened when fields are watered]. When *yin* and *yang* stimulate another, it is the time when conception is going to occur. Whether one acts in such a way as to let the life mandate of descendants ripen [i.e. one begets children], or develops the life mandate of us forebears [i.e. one engenders oneself], are matters of one and the same category. The principle is clear as daylight, too. In this moment it is only important to not let [the essence] leak out (*wu xie* 勿洩)' (*Yunji qiqian* 64: 19a; reject emendation *cheng* 塍 to *cou* 腠!). The late imperial *Exposition of Cultivating the True by the Great Immortal of the Purple Gold Splendour* (*Zijin guangyao daxian xiuzhen yanyi* 紫金光耀大仙修真演義, hereafter *Exposition of Cultivating the True*, fol. 111, transl. Wile 1992: 144) gives a slightly more eloquent description of the matter in section 18 'Renewed Blossoming by Grafting on the Decayed (Rootstock)', including the motto: 'Worldly people who do not know the principle of prolonging life merely have to look at the grafted pear in the mulberry'.

Conceptually speaking, this phantasmatic absorption theory is not presented as a negatively connoted 'sexual vampirism' (revamped by Goldin 2006), but rather as a men-centred optimisation process of self-growth whose possible adverse effects on women do not enter the picture. The subjective phantasmatic experiencer perspective visualises textually the permeable, excited bodies, and assumes that as soon as their body images merge, the female essences can be transported from the female to the male system.

However, a concrete guide to literally drawing up female fluids by the penis could not be found in our Chinese sources. This kind of exploration of the humanly possible has apparently only been executed by Bengal yogins in Tārāpīṭh applying the *vajrolī mudrā* (lightning seal) for seminal retention and absorption of female fluids during sexual intercourse. They use rubber or silver catheters to train their uro-genital musculature in order to develop urethral suction (Roşu 2002; Darmon 2002; Mallinson 2018). It remains unclear what emic perspective then explains what should happen with the fluids once they arrive in the urinary bladder.

Taken at face value, one might conceive of the rubric 3 'Nourishing *Yin*' as the female counterpart to rubric 2 'Nourishing *Yang*' (Table 22.1), where the *Secret Decisions in the Jade Chamber* advises menfolk to change partners frequently in order to benefit more (from the female partner's fluids), to choose, preferably, young women – not under fourteen and up to thirty years old – who have not given birth, and to keep the sexual arts secret from them. Despeux and Kohn (2003: 36–40) claimed that three passages of the *Secret Decisions in the Jade Chamber* promote the 'power of yin' by embracing the woman's female sexuality, 'giving encouragement to all women who follow her'. However, their partial translations omitted all those passages wherein female adepts are judged negatively. We read that the Queen Mother of the West 'copulates once with a man, and this establishes damaging diseases'; and that she likes to have intercourse with young men, which is said 'not to be admissible to teach to the world; why was the Queen Mother like that?' The explicit call is for moderation and restriction; we read that an early exhaustion of a woman's '*yin* essence' (*yin jing* 陰精) should be 'adequately restrained and observed' (specifically when, on hearing that the man has had sex with others, she becomes jealous and agitated, so that her 'essence juices' flow out by themselves), and we are told that a woman who 'knows the art of nourishing *yin*' might 'transform into a man'. Clearly, theories of equal benefit were secondarily emerging, derived theories that did not even attempt to smooth out the theoretical inconsistencies with their forerunners. An early example for such an equal benefit scenario can be found in the *Biography of Lord Pei, the Realised Person of Pure Refinement* (*Qing Ling zhen ren Pei jun zhuan* 清靈真人裴君傳; see Raz 2012: 185–6).

Dealing with male essence: on male ejaculation and how to avoid it

'Essences' (*jing* 精) are fatty, whitish and slimy fluids in the body that get dispersed inside the body and into the 'hair-fine vessels' (*mao mai* 毛脈). The *Secret Decisions in the Jade Chamber* contrasts the healthy 'manifest essence' (*yang jing* 陽精) – which is thick and coagulates –

with five kinds of spoilt ejaculated fluids, and their respective damaged body constituents, all of which are results of hectic intercourse, leading to abrupt and violent emissions (*Ishinpō* 28.20: 25a). Compare also the seven afflictions and seven injuries to the male uro-genital system as given by Marquis of White Waters in the Wuwei medical tablets 85A–85B (trans. in Yang and Brown 2017: 293–5), and in *Bei ji qian jin yao fang* (19.8: 354a).

'Essence' as such constitutes a scarce good. In 1335 CE, Master Shang Yang calculates in his *Great Essentials on the Golden Elixir by Master Shang Yang* (*Shangyangzi jindan dayao* 上陽子金丹大要 3: 1a–2b) that at age sixteen, the 'essence' in the body only amounts to about 1,520 ml (1 *sheng* 升 6 *ge* 合) of liquid volume, which according to the text weighs 633 g (1 *jin* 斤). By a continent lifestyle, it can be continuously augmented up to about 2,850 ml (3 *sheng*) at best. But each ejaculation results in a loss of about 47.5 ml (half a *ge*), which is of great concern!

Li Shizhen 李時珍 (1518–1593) cites the calculation without mentioning the source in his *Materia Medica Arranged According to Monographs and Technical Criteria* (*Bencao gangmu* 本草綱目, *juan* 52; vol. 2: 1932). By his time, the total volume of essence in the body would vary from about one at sixteen years of age to about three litres in a continent life, and the ejaculation would amount to 50 ml. However, if Master Shang Yang had used earlier standards of measurement, all these volumes would have been slightly lower than indicated above.

Using the voice of the Selected Woman (Cai *nü* 采女), the everyman's question 'wherein lies the fun of preventing ejaculation?' is answered by Ancestor Peng: 'If the essence is emitted, trunk and body parts become sluggish and limp, the ears are bitterly (painfully) buzzing, the eyes can hardly be kept open, the throat dries out, and the bones and articulations loosen and decay. Even though there is occasionally a short-term satisfaction, in the end it is not pleasurable. If one stimulates and does not ejaculate, the force of the *qi* has a surplus, trunk and body parts can be at ease, ears and eyes are sharp and bright. Even though one restrains and calms oneself, imagination and (loving) care are emphasised even more. It is constantly as if it were not enough, how can this not be pleasurable?' (*Yufang mijue, Ishinpō* 28.18: 22a; see Pfister 2012: 52–3).

Thus, ejaculation – the 'short-term satisfaction' (*zan kuai* 暫快) – is contrasted with an intermediate state of bliss, echoing prepubescent boys' orgasmic experiences. (On the principal and experiential separation of ejaculation and orgasm, see Marcuse 1922; Haeberle 1985: 266–8; Kothari 1990.) Several texts elaborate on the profitable health outcome of a series of prolonged sexual intercourse performed without ejaculation: from improved sight and hearing to a glowing skin, to strengthened spine and bones, to free-flowing waterways (of the uro-genital system), to a hard and strong erection; and as one's strivings become untameable (a feeling of boldness) (see *Tianxia zhidao tan*, slips 22–4; *He yinyang, slips 112–15; *Shiwen, slips 19–22). This culminates in an altered state of consciousness, during which subjective light experiences emerge as he 'follows the heavenly blossom', or his 'spirit brightens up'. Clearly, the sexually stimulated inner transports of the 'essences' benefit overall fitness and well-being – at least from the experiencer's perspective – and contribute to agreeable psychological states, and a feeling of bodily lightness. (See Pfister 2006a on spirit brightening; Pfister 2012 on phosphenes and Hsu 2012 on the feeling of lightness. On altered states of consciousness during sexual encounters, see Swartz 1994; Cohen and Lévy 1996; Meston *et al.* 2004; Safron 2016.)

The basic methods to avoid emission and ejaculation are (a) the free-handed method, (b) the urethral pressure method and (c) combinations thereof with visualisations. Ejaculation is not completely avoided, but simply decreases with age (Maspero 1937; Needham 1983; Wile 1992; Pfister 1995; Karamanou *et al.* 2010).

Rubric 18 of the *Ishinpō* 28 is named 'reverting the essence' (*huan jing* 還精), a term that denotes internal dispersion of the essence after high excitation during sexual intercourse and avoiding emission and ejaculation. However, the meditative texts of the Shangqing tradition use the term in solo practices that move around visualised contents and circle coloured *qi* inside the adept's body (Maspero 1937: 379–80; Robinet 1984, I: 175 n. 1; Hudson 2008: 413–30).

a   The Mawangdui texts report a *free-handed method* closely linked with an altered state of consciousness: 'The matter of the spirit brightening consists in what is locked away; cautiously handle the jade lock (*yu bi* 玉閉), and spirit brightening will arrive' (*Tianxia zhidao tan*, slips 18–19). While maintaining a light and regular breathing pattern throughout and avoiding vocalisations, the man reduces thrusting movement and 'contracts the ring' (*xi zhou* 翕州, i.e. the anal sphincter) to maintain the 'jade lock' (and thus avoid emission and ejaculation), that is, by locking away his own 'essence', which is made one whole (*yi* 壹) that will be shifted (*qian* 遷) upwards in his body (Ibid., slip 22). The use of anal sphincters and genital musculature is in no way tabooed or restricted (on anal pleasures, see Lo and Barrett 2012). Insight into pelvic muscle function might have been furthered by squatting behaviour, and sitting on mats, which trains relevant musculature on the go (Paciornik 1985).

Before each 'intermission' (*yi* 已), he stops moving, sucks air in and presses the penis downwards, waiting for some time in order to 'retain the surplus', or washes the penis with an aphrodisiac lotion, so that it begins to erect itself anew; this action is called 'stabilising the tilting' (Ibid., slips 31, 33–6). In all, there are eight ways to 'increase (bodily) flow constellations' (*yi qi* 益氣) and overall stamina of the man (compare Pfister 2006b: 90–7). *Yufang zhiyao* (*Ishinpō* 28.18: 22b) adds some elements to guide the man's attention away, stating: 'When the essence is strongly aroused, you quickly raise the head, open your eyes wide and look to the left and to the right, upwards and downwards; you contract the lower body, hold the breath, and the essence is stopped by itself'.

b   The *urethral pressure method* is considered by some a beginner's practice, which should, after some training, be replaced by the free-handed procedure. 'The method to revert the essence and to replenish the brain (marrow) (*huanjing bunao zhi dao* 還精補腦之道): as the essence is greatly aroused during sexual intercourse, and about to be emitted, press quickly to drive it back with the two middle fingers of the left hand behind the privates' bag [scrotum] and in front of the great opening [anus]. Press it during its bristling activity, prolongedly eject the breath while clapping the teeth ten times. Do not block off the breath, as in that case your essence will be released. Whereas if the essence is not allowed to be emitted, it returns and reverts in the jade stalk [penis], and moves upwards into the brain (marrow)' (*Yufang zhiyao Ishinpō* 28.18: 22b).

More than a thousand years later, the method was integrated without further ado into planned parenthood tracts of the twentieth century, but now called the 'urethral pressure method to prevent conception' (*niaodao yapo fa* 尿道壓迫法, or *yapo niaodao biyunfa* 壓迫尿道避孕法; see Yang Geng 1964; Han Xiangyang 1972; Edwards 1976).

Retrograde ejaculation of semen into the bladder is an undesired outcome of cases when emission was already on its way, and urethral pressure applied a trifle too late. It would become apparent as cloudy urine in the lacquerware urinals in the form of a tiger (*huzi* 虎子) that were used by men in well-to-do households in early China (Huang Gangzheng 1986). Ancestor Peng explains it to the Selected Woman: 'One forcefully presses [the essence in the urethra], and closes it off tightly; it being difficult to hold and easily lost, this causes one's essence to leak and the urine to be turbid' (*Yangxing yanming lu* B.6: 9a; *Beiji qianjin yaofang* 27.8: 490a; Maspero 1937: 382). The late imperial *Exposition of Cultivating the True* (fol. 104, section 12) describes the difficulties encountered by beginners who want to 'lock and bar the mysterious trigger', and quantifies the occurrence of reflux: 'Wanting to forcibly close it off (*qiǎng bì* 強閉), the wasted essence (*bai jing* 敗精) necessarily silts up (*ni* 泥) and enters the urinary bladder and the kidney bags. (...) Out of a number of five thousand (sexual) excitations (*shān gu* 肩〈扇〉皷), one only about once emits wasted essence' (Revising Wile 1992: 140, 265 n. 3). By this estimation, retrograde ejaculation is considered an occasional and relatively rare event, which can be avoided by using a more gentle approach to prevent emission, the texts' own elaborate free-handed method.

Female ejaculation and female essence

At the end of the nineteenth century, *female ejaculation* became a controversial topic in the then new research field of sexual science (*Sexualwissenschaft*, sexology), and it remains controversial to this day (Stifter 1988; Korda *et al.* 2010). However, the topic was prominently present in several ancient cultural spheres of Afroeurasia, e.g. in circum-Mediterranean cultures of antiquity (Andò 2009), ancient India (Syed 1999) and China (Pfister 2006b, 2007). Ancient and medieval Chinese texts on sexual body techniques integrate female potency and ejaculation into the body image of all women. A woman is capable of ejaculation, actively provoked by herself, solo or in combination with adequate stimulation by a partner or a sex toy.

The sixth of the Nine Procedures in *Xuannü jing* is called 'phoenixes soaring': the man kneels between the raised legs of the woman, supporting himself with his hands on the mat, thereby reducing physical contact with the genital area; he penetrates hard and hot, piercing the deepest vaginal portion of her cervix or 'descendant's pebble' (*kun shi* 昆石), and then drags out the penis along the upper vaginal wall. 'Let the woman move herself and perform the technique of three times eight: [three times] pressing her buttocks tightly against each other, and [eight times] opening and relaxing the female privates so as to drive out herself the essence juices (*jing ye* 精液)'.

This position maximises friction for both. The woman is able to provoke ejaculation by repeatedly firmly pushing outwards (as if to force urination) for some time, and then relaxing her pubococcygeal musculature (Sundahl 2003: 86–7, 102–3, 121–4).

Sex training tools both for women and for men have been archaeologically recovered in tombs of elite households from early China: bronze dildoes to cover the man's penis, stone eggs to train the vaginal musculature, as well as bronze dildoes with bony rubbing devices aimed at maximal stimulation of the clitoral complex and upper vaginal wall (Eggebrecht 1994: Cat. 95; Chen Hai 2004; Khayutina ed. 2013: 291, fig. 198; Li Ling 2006: 343–70; Li Ling 2011b).

The **He yinyang* module eloquently describes, in 'proofs of the ten intermissions', the consistency and smell of the female emission as being variable, ranging from being emitted clear and fresh, like groats, smelling like cooked bones, smelling fishy or like grain, being creamy, slippery, viscous, tallowy, sticky, clotty; 'having clotted, it becomes slippery again, is again emitted clear and fresh; and that is called the 'grand finale' (*da zu* 大卒) [that includes carpopedal spasms, elevating the buttocks from the mat, pale lips, and a sweaty nose].'

(Slips 129–32; the parallel passage of *Tianxia zhidao tan*, slips 56–8, ends with 'floodlike flow constellations are only now emitted'; see Pfister 2006b: 98).

*He yinyang is unique in claiming a debit by postulating that a women owes the man her ejaculate: 'In the evening the essences of the man are provided for (*jiang* 將) [by prolonged sexual intercourse without emission and ejaculation], in the morning the essences of the woman are demanded (*ze* 責) [i.e. emitted and ejaculated], and thereby my essence will be nourished by your essence' (slip 127). At the time of the 'grand finale', body boundaries are lost during the experience (Meston et al. 2004: 177), and this allows for the phantasmatic transport of essence from woman to man (slips 128, 133; Pfister 2006b: 97–100).

Performing a sexual encounter

The scenario of sexual performance is developed in text modules. Such a presentation strategy parallels the breaking down of the apparent tumult of a sexual encounter into modular units, which, in turn, are cognitively manageable by the aspiring male participant and observer. The modules facilitate the learning of a specific descriptive and technical description language, and guide attention to observing and interpreting specific female bodily signs and kinetic reactions. He receives advice for adequate interaction, which in each case must be adapted to a variety of situational constellations.

a   Foreplay, or the 'playful ways' (*xi dao* 戲道), sees the partners breathing upon each other's bodies, embracing, snuggling, touching and arousing each other in a leisurely way. The man learns the five proofs of her fivefold desire, which are rising bodily flows and a heated face, hardened nipples and a sweating nose, descent of bodily saps and wet thighs, a dry throat and swallowing saliva. Only after all the proofs are seen, he is allowed to mount. A rich vocabulary is used to describe the caressing of her erogenous zones from the wrist, over neck, to 'mount constancy' (*chang shan* 常山), massaging her by breath, touch and body weight (*He yinyang, slips 102–11; see Harper 1987; Li and McMahon 1992; Middendorf 2007).

b   The onset phase, called 'approaching the matter' (*lin shi* 臨事), or 'approaching the ride' (*lin yu* 臨御) – despite being a regular rubric in the literature (*Ishinpō* 28.5) – has so far been overlooked by virtually all modern commentators. Onset is differentiated from foreplay – when fingers and tongue arouse the couple – by its use of the penis in order to promote frictional pleasure and to further tumescence of the erectile tissues of both parties. Here, the distinction between 'outer' (*wai* 外) and 'inner' (*nei* 內) aspects of the female genitals is technically important. It is paralleled by the male actions of 'mounting' (*shang* 上) the vulva and *mons veneris*, and 'entering' (*ru* 入, *na* 內 or 納) the vaginal canal itself, where thrusting can be directed to both areas, varied and alternated in either a vulval or 'shallow' (*qian* 淺), or a vaginal or 'deep' (*shen* 深) fashion. During onset – that is, before any 'deep' penetration occurs – special attention is given for some time to the clitoral complex. (On the conception of the clito-urethro-vaginal complex, see O'Connell et al. 2008; Foldès and Buisson 2009; Jannini et al. 2014; Mazloomdoost and Pauls 2015; Levin 2018.) In *He yinyang, he takes the penis or 'jade whip' to 'hit upwards' (*shang zhen* 上揕) without penetration in order to cause her bodily flows to arrive (*zhi qi* 致氣), producing 'warmth' (slips 108–9). In *Sunü jing*, the woman reclines, and the man positions himself between her legs; he uses his lips on the 'mouth' and sucks the 'tongue' (clitoris) of her genitals, and by holding his 'jade stalk' then 'beats' (*ji* 擊) both sides of her 'jade gate' (aiming at the clitoral bulbs) for about the 'time of a meal' (*shi qing* 食頃;

Ishinpō 28.5: 11b). Even more detailed is the description in *Dongxuan zi*: 'The jade stalk drags at the mouth of the jade gate. (...) He then attacks and hits with the *yang* blade to and fro, or storms downwards to the jade streaks (*yu li* 玉理) [fourchette], or rams upwards to the golden ditch (*jin gou* 金溝) [pudendal cleft]; he hits and pierces the sides of the ring wall (*bi yong* 辟雍, read 璧廱) [the side areas around the urethral meatus, or periurethral glans (Levin 1991)], or rests at the right of the reddish jade platform (*xuan tai* 璿臺) [elevation of glans clitoridis]'. An added interlinear commentary specifies: 'The above is wandering outside, and not yet copulation inside'. To stimulate the clitoris, the clitoral bulbs and the area around the urethral meatus for quite some time by hitting with the penis guided by the hand are known in Rwanda as *gukubita rugongo*, 'hit the clit', and on the Chuuk Islands and the Ulithi Atoll, it's called *wechewechen Chuuk*, 'Trukese striking or prodding' (see Vincke 1991: 175; Bizimana 2005: 64; 2008: 60–64; Swartz 1958: 477–8, 481–3; Lessa 1966: 87). All three cultures prefer wet sex, and thus our last text concludes the passage with: 'The woman's lustful juices must spill from the cinnabar grotto, thereafter you throw your *yang* blade into the children's palace [i.e. the vagina]' (*Ishinpō* 28.5: 11a). The onset activity mimicks ways of same-sex and solo stimulation that further erection and tumescence. The deliberate insertion of such an onset phase, which specifically stimulates the clitoral complex, modifies the influential, but often criticised four-phase model of the sexual response cycle – consisting of excitation, plateau, orgasm and resolution phases – established by the modern sexologists Masters and Johnson (1966) (see also Haeberle 1985: 65–7; for criticisms, see Tiefer 1991; Levin 2001, 2008). It is a research desideratum to analyse ancient Chinese recipes to tighten the vagina and other vaginal practices regarding the question if they further lubrication (wet sex preference), or rather dry up the vagina (dry sex preference) (Levin 2005).

c The linking passage in *Tianxia zhidao tan* elucidates what should be at one's disposition during the encounter: 'Hold ready the ten embellishments, arrange conveniently the ten positions, and vary the eight ways [of thrusting]' (slip 47). Purposefully using some adapted military terms, the sexual interplay is cadenced by ten 'advances' (*dong* 動), each followed by an 'intermission' (*yi* 已). (On metaphors of fighting and warfare, see Van Gulik [1951]2004: 68, 158–9; [1961]2003: 76, 157, 278–80, 320; Wile 1992: 35.) The scenario thus relinquishes the one-climax-structure for a ten-fold, longer and thoroughly modulated one. The ten positions have animal names; one needs to know in what position toads or dragonflies copulate to get the point of the list. Later texts list in dense description up to thirty variations (*Dongxuan zi*, *Ishinpō* 28.13: 16a–18a). Some positions are considered therapeutic, especially those to increase the man's weakened *qi*, where the woman straddles over the reclined man and provokes the emission of her 'essence juices' that 'overflow to the outside', coming 'like rain' or a 'spring' (*Xuannü jing*, *Ishinpō* 28.12: 14b–15b). The sources transmit several partonymic sets of vulvo-vaginal locations, which allow insights into the prototypical cognitive representation of the female genitals (Pfister 2007, 2016; Middendorf 2007). The observance of the distinction between shallow and deep thrusting addresses stimulation to both vulval (clitoral) and vaginal locations in changing rhythms, which are organised into a series of thrusting, punctuated by intermissions (on the role of rhythmic stimulation, see Safron 2016; Levin 2018).

The male participant observer interprets her 'five sounds' or vocalisations to guide his activity: when her breathing is throaty or she catches her breath – she's inwardly tight; when gasping – she reaches delight; when continuously wailing – insert the jade whip and the nourishing starts; when breathing with a *hmaj sound – pleasant sweetness is extreme; if she's grinding her teeth – he should wait for her. By carefully distinguishing the sounds, he recognises where her 'heart' (attention) is located; likewise, he knows by her 'eight ways of [involuntary] movement' where pleasure passes through. Four examples out of 'eight observations' may serve: if she reaches out for him with her hands, the bellies should draw close; if she extends her elbows, the hammering of the upper vaginal wall is wanted, while he supports himself with his hands on the bed; if she crosses her thighs, the piercing is greatly overdone; if she's shivering, it's excellent (*Tianxia zhidao tan*, slips 50–53, 63–4; compare *He yinyang*, slips 120–26; see Levin 2006).

Slowing down and lingering without losing persistence is the essential advice given to men at the end of the *Discussion of the Utmost Method Under the Sky*, adding a description of emotional afterglow: 'For the teasing entertainment it is important to endeavour to linger on and hold out; if one is capable to linger on and hold out, a woman greatly rejoices. She will be attached to you as to her younger and older brothers, and take care of you as of her father and mother. Everyone who can go this way is called a heavenly master' (slips 66–7).

Note

1 Manuscript titles are starred * when their title was added by modern editors. Translations of titles and excerpts of source texts are those of the present author throughout.

Bibliography

Primary sources

Beiji qianjin yaofang 備急千金要方 (Essential Prescriptions Worth a Thousand in Gold for Urgent Cases), 652 CE, Sun Simiao 孫思邈 1955(1994), Beijing: Renmin weisheng.

Bencao gangmu 本草綱目 (Materia Medica Arranged According to Monographs and Technical Criteria), 1596 CE, Li Shizhen 李時珍 (comp.), Liu Hengru 劉恒如 and Liu Shanyong 劉山永 (coll., comm.) (1998) «*Bencao gangmu*» *xinjiao zhuben* 《本草綱目》新校注本, *shang xia ce* 上下冊, Lidai zhongyi mingzhu wenku 歷代中醫名著文庫, Beijing: Huaxia.

Dongxuan zi 洞玄子 (Master Dong Xuan), unknown collective; 12 fragments in *Ishinpō* 28.

**He yinyang* 合陰陽 (Uniting Yin and Yang), scroll, slips 102–33. Unknown collective; Mawangdui bamboo manuscript, title by modern eds, in Mawangdui Hanmu Boshu Zhengli Xiaozu (1985: 99–103 repr., 153–6 transcr.); Qiu Xigui (ed.) (2014, 2: 212–4 col. repr.; 6: 153–8 transcr.).

Ishinpō • *Yixin fang* 醫心方 (Core Prescriptions of Medicine), 984 CE, Tamba no Yasuyori 丹波康賴 1955(2000), Beijing: Renmin weisheng chubanshe; Liu Xiuqiao 1976; Gao Wenzhu *et al.* 1996.

Qingling zhenren Peijun zhuan 清靈真人裴君傳 (Biography of Lord Pei, the Realised Person of Pure Refinement), before 5th c. CE (?), in *Yunji qiqian* 雲笈七籤 105 (*Daozang* 道藏, fasc. 698, DZ 1032).

Shangyangzi jindan dayao 上陽子金丹大要 (Great Essentials on the Golden Elixir by Master Shang Yang), preface 1335 CE, by Chen Zhixu 陳致虛 (fl. 1289–1335), in *Daozang* 道藏, fasc. 736–8, DZ 1067.

**Shiwen* 十問 (Ten Interviews), scroll, slips 1–101. Unknown collective; Mawangdui bamboo manuscript, title by modern eds, in in Mawangdui Hanmu Boshu Zhengli Xiaozu (1985: 87–97 repr., 143–152(transcr.); Qiu Xigui (ed.) (2014, 2: 203–11 col. repr.; 6: 139–52 transcr.).

Sunü jing 素女經 (The Book of the Plain Woman), unknown collective; 2 fragments in *Ishinpō* 28.

Tiandi yinyang dale fu 天地陰陽交歡大樂賦 (Rhapsody on the Great Pleasure in the Mutual Joys of Heaven and Earth, Yin and Yang), in Wei, Tongxian 魏同賢 (ed.) (2001) *Facang Dunhuang xiyu wenxian* 法藏敦煌西域文獻 (Dunhuang and other Central Asian Manuscripts in the Bibliothèque Nationale de France), Shanghai: Shanghai guji and Faguo guojia tushuguan, vol. 15: 229–233.

Tianxia zhidao tan 天下至道談 (Discussion of the Utmost Method Under the Sky), scroll, slips 12–67. Unknown collective; Mawangdui bamboo manuscript, in Mawangdui Hanmu Boshu Zhengli

Xiaozu (1985: 109–15 repr., 161–67 transcr.); Qiu Xigui (ed.) (2014, 2: 216–20 col. repr.; 6: 163–72 transcr.).

Wangwu zhenren koushou yindan mijue lingpian 王屋真人口授陰丹祕訣靈篇 (Numinous Tablets of the Secret Instructions on the Yin Elixir Orally Transmitted by the Perfected Person of Wangwu), after 780 CE, in *Yunji qiqian* 雲笈七籤 64: 14a–19a (*Daozang* 道藏, fasc. 690, DZ 1032).

Xuannü jing 玄女經 (The Book of the Dark Woman), unknown collective; 7 fragments in *Ishinpō* 7, and 28.

Yangsheng fang 養生方 (Recipes for Nurturing Life), MS III, cols. 1–219, drawing, content list; fragments. Mawangdui bamboo manuscript, title by modern eds, in Mawangdui Hanmu Boshu Zhengli Xiaozu (1985: 53–70 col. repr., 97–119 transcr.); Qiu Xigui (ed.) (2014, 2: 108–27 col. repr.; 6: 35–72 transcr.; 7: 237–55 orig. col. repr.).

Yangxing yanming lu 養性延命錄 (Records on Nourishing the Disposition and Prolonging the Mandate of Life), in *Daozang* 道藏, fasc. 572, DZ 838.

Yufang mijue 玉房祕決 (Secret Decisions in the Jade Chamber), unknown collective, Six Dynasties (?); 28 fragments in *Ishinpō* 7, 13, 21, 24, and 28.

Yufang zhiyao 玉房指要 (Essentials of the Jade Chamber), unknown collective; 7 fragments in *Ishinpō* 28.

Zijin guangyao daxian xiuzhen yanyi 紫金光耀大仙修真演義 (Exposition of Cultivating the True by the Great Immortal of the Purple Gold Splendor), 1594 CE, preface by Deng Xixian 鄧希賢, in Van Gulik 1951(2004), vol. II: 105–122.

Secondary references

Andò, V. (2009) 'Sogni erotici e seme femminile nella antica medicina greca', *Medicina nei secoli*, 21.2: 663–91.

Bizimana, N. (2005) *Weiblicher Orgasmus und Weibliche Ejakulation Dank Afrikanischer Liebeskunst*, Norderstedt: Books on Demand.

―――― (2008) *Le Secret De L'amour à L'africaine. La Caresse Magique Que Chaque Homme Devrait Connaître*, Paris: Leduc.S Editions.

Chen Guofu 陳國符 (1963) *Daozang yuanliu kao* 道藏源流考 (Original Sources of the Daoist Canon), Beijing: Zhonghua shuju.

Chen Hai 陳海 (2004) 'G-dian yu Xihan nüyong xingwanju kao G點與西漢女用性玩具考 – Research on G-spot and female sexual toys of Western Han dynasty', *Kaogu yu wenwu*, 3: 62–7, figs. 1–4 (back cover).

Chen Hsiu-fen 陳秀芬 (2009) 'Yangsheng yu xiushen 養生與修身: wan Ming wenren de shenti shuxie yu shesheng jishu 晚明文人的身體書寫與攝生技術 (Nourishing Life and Cultivating the Body: writing the literati's body and techniques for preserving health in the late Ming), Taipei: Dawhsiang.

Cohen, H. and Lévy, J. (1996) 'L'orgasme comme état de conscience modifié', *Sexologies*, 5.21: 5–8.

Crossley N. (2005) 'Mapping reflexive body techniques: on body modification and maintenance', *Body and Society*, 11.1: 1–35.

Darmon, R.A. (2002) '*Vajrolī Mudrā*: La rétention séminale chez les yogis *vāmācāri*', in V. Boullier and G. Tarabout (eds) *Images du corps dans le monde hindou (Monde Indiens – Sciences sociales 15e–20e siècle)*, Paris: CNRS éditions, pp. 213–240.

De Meyer, J. (2006) *Wu Yun's Way: life and works of an eighth-century Daoist master*, Leiden: Brill.

De Sardan, J.-P.O. (1998) 'Émique', *L'Homme*, 38.147: 151–66.

Despeux C. and Kohn L. (2003) *Women in Daoism*, Cambridge, MA: Three Pines Press.

Doniger, W. and Kakar, S. (trans.), Vatsyayana Mallanaga (comp.) (2002) *Kamasutra*, Oxford: Oxford University Press.

Edwards, J.W. (1976) 'The concern for health in sexual matters in the "old society" and "new society" in China', *Journal of Sex Research*, 12.2: 88–103.

Eggebrecht, A. (ed.) (1994) *China, eine Wiege der Weltkultur: 5000 Jahre Erfindungen und Entdeckungen*, Mainz: Verlag Philipp von Zabern.

Fan Youping 樊友平, Zhu Jiaqing 朱佳卿, Li Jiarong 李佳蓉 and Gong Xiaoying 公曉穎 (eds) (2007) *Zhonghua miniao nankexue gudian jicheng* 中華泌尿男科學古典集成 (Collection of Ancient Texts on Chinese Urologic Andrology), Beijing: Zhongyi guji.

Foldès, P. and Buisson, O. (2009) 'The clitoral complex: a dynamic sonographic study', *Journal of Sexual Medicine*, 6: 1223–31.

Fu Juyou 傅舉有 and Chen Songchang 陳松長 (1992) *Mawangdui Hanmu wenwu* 馬王堆漢墓文物 (The Cultural Relics Unearthed from the Han Tombs at Mawangdui), Changsha: Hunan Publishing House.

Furth, C. (1994) 'Rethinking Van Gulik: sexuality and reproduction in traditional Chinese medicine', in C.K. Gilmartin, G. Hershatter, L. Rofel and T. White (eds) *Engendering China: women, culture and the state*, Cambridge, MA: Harvard University Press, pp. 125–46.

———— (2005) 'Rethinking van Gulik again', *Nan Nü*, 7.1: 71–8.

Gao Wenzhu 高文鑄 (1996) '«Yixin fang» yinyong wenxian kaolüe 《醫心方》引用文獻考略' (Brief study of the works cited in *Ishinpō*), in Gao Wenzhu 高文鑄 *et al.* (1996) «*Yixin fang*» *jiaozhu yanjiu*《醫心方》校注研究, Beijing: Huaxia chubanshe, pp. 725–806.

Gao Wenzhu 高文鑄 *et al.* (1996) «*Yixin fang*» *jiaozhu yanjiu* 《醫心方》校注研究 (Annotation and Study of the *Ishinpō*), Beijing: Huaxia chubanshe.

Goldin, P.R. (2006) 'The cultural and religious background of sexual vampirism in ancient China', *Theology and Sexuality*, 12.3: 285–307.

Haeberle, E.J. (1985) *Die Sexualität des Menschen*, Berlin: De Gruyter.

Han Xiangyang 韓向陽 (ed.) (1972, 1973) *Jihua shengyu zhishi wenda* 計劃生育知識問答 (Questions and Answers on Family Planning Knowledge), Beijing: renmin weisheng chubanshe.

Harper, D. (1987) 'The sexual arts of ancient China as described in a manuscript of the second century B.C.', *Harvard Journal of Asiatic Studies*, 47.2: 539–93.

———— (1998) *Early Chinese Medical Literature: the Mawangdui medical manuscripts*, London; New York: Kegan Paul International.

———— (2005) 'Ancient and medieval Chinese recipes for aphrodisiacs and philters', *Asian Medicine*, 1.1: 91–100.

———— (2010) 'La littérature sur la sexualité à Dunhuang', in C. Despeux (ed.) *Médecine, Religion et Société Dans la Chine Médiévale: Étude de manuscrits chinois de Dunhuang et de Turfan*, Paris: Collège de France, Institut des Hautes Études Chinoises, pp. 871–98.

He Jiejun 何介鈞 (ed.) (2004) *Changsha Mawangdui er, san hao Hanmu* 長沙馬王堆二三號漢墓, *di yi juan* 第一卷: *tianye kaogu fajue baogao* 田野考古發掘報告 (Tombs 2 and 3 of the Han Dynasty at Mawangdui, Changsha Report on Excavation, volume 1: with abstracts in English and Japanese), Beijing: Wenwu.

Headland, Th., Pike K. and Harris M. (eds) (1990) *Emics and Etics: the insider/outsider debate*, Newsbury Park; London; New Delhi: Sage.

Hinsch, B. (2005) 'Van Gulik's sexual life in ancient China and the matter of homosexuality', *Nan Nü*, 7.1: 79–91.

Hsia, E.C.H., Veith, I. and Geertsma, R.H. (trans.) (1986) *The Essentials of Medicine in Ancient China and Japan. Yasuyori Tamba's Ishimpō* 醫心方, Leiden: Brill.

Hsu, E. (2012) '"Feeling lighter": why the patient's treatment evaluation matters to the health scientist', *Integrative Medicine Research*, 1.1: 5–12.

Huang Gangzheng 黃綱正 (1986) 'Changsha chutu de Zhanguo huzi ji youguan wenti 長沙出土的戰國虎子及有關問題 (The Warring States urinals excavated in Changsha, and related questions)', *Wenwu*, 9: 86–91.

Hudson, W.C. (2007) 'Daoism (Taoism)', in F. Malti-Douglas (ed.) (2007) *Encyclopedia of Sex and Gender*, New York: Macmillan, pp. 373–6.

———— (2008) *Spreading the Dao, Managing Mastership, and Performing Salvation: the life and alchemical teachings of Chen Zhixu*, PhD thesis, Indiana University.

Idema, W.L. (1983) *Bai Xingjian (775–826): Het hoogste genot*, Leiden: Uitgeverij De Lantaarn.

Ishida, Hidemi 石田秀實 (1991) 'Shoki no bōchū yōsei shisō to sensetsu 初期の房中養生思想と僊說 (On the Notion of Nurturing Life through Sexual Practice – with special reference to the early hsien cult)', *Tōhō shūkyō*, 77: 1–21.

Ishihara A. and Levy H.S. (transl.) (1968, 1969, 1973) (Tamba Yasuyori) *The Tao of Sex. An Annotated Translation of the 28th Section of the Essence of Medical Prescriptions (Ishimpō)*. Yokahama (Japan): Shibundō; 1969 Yokohama: General Printing Co.; 1973 New York: Harper and Row.

Jannini E.A., Buisson O. and Rubio-Casillas A. (2014) 'Beyond the G-spot: clitourethrovaginal complex anatomy in female orgasm', *Nature Reviews Urology*, 11: 531–8.

Janssen, D.F. (2007) 'First stirrings: cultural notes on orgasm, ejaculation, and wet dreams', *Journal of Sex Research*, 44.2: 122–34.

Kalinowski, M. (1985) 'La transmission du dispositif des Neuf Palais sous des six dynasties', in M. Strickmann (ed.) *Tantric and Taoist Studies*, Vol. 3, Bruxelles: Institut Belge des Hautes Études Chinoises, pp. 773–811.

Karamanou, M., Demetriou, T., Liappas, I. and Androutsos, G. (2010) 'Techniques de contrôle de l'éjaculation dans l'ancienne Chine – ejaculation control techniques in ancient China', *Andrologie*, 20: 266–72.

Khayutina, M. (ed.) (2013) *Qin. The Eternal Emperor and His Terracotta Warriors*, Zurich: NZZ Libro.

Kleeman, T.F. (2014) 'The performance and significance of the merging the pneumas (*heqi*) rite in early Daoism', *Daoism: Religion, History and Society*, 6: 85–112.

Kobzev, A.I. (ed.) (1993) *Kitajskij Ėros: naučno-chudožestvennyj sbornik*, Moskva: SP Kvadrat.

Korda, J.B., Goldstein, S.W. and Sommer, F. (2010) 'The history of female ejaculation', *Journal of Sexual Medicine*, 7: 1965–75.

Kothari, P. (1990) 'Orgasmic disturbances: a new classification', in F.J. Bianco and R.H. Serrano (eds) *Sexology: an independent field*, Amsterdam: Excerpta Medica, Elsevier, pp. 193–206.

Lessa, W.A. (1966) *Ulithi: a micronesian design for living*, New York: Holt, Rinehart and Winston.

Levin, R.J. (1991) 'VIP, vagina, clitoral and periurethral glans: an update on human female genital arousal', *Experimental and Clinical Endocrinology and Diabetes*, 98.2: 61–9.

——— (2001) 'Sexual desire and the deconstruction and reconstruction of the female sexual response model of Masters and Johnson', in W. Everaerd, E. Laan and S. Both (eds) *Sexual Appetite, Desire and Motivation: energetics of the sexual system*, Amsterdam: Royal Netherlands Academy of Arts and Sciences, pp. 63–93.

——— (2005) 'Wet and dry sex – the impact of cultural influence in modifying vaginal function', *Sexual and Relationship Therapy*, 20.4: 465–74.

——— (2006) 'Vocalised sounds and human sex', *Sexual and Relationship Therapy*, 21.1: 99–107.

——— (2008) 'Critically revisiting aspects of the human sexual response cycle of Masters and Johnson: correcting errors and suggesting modifications', *Sexual and Relationship Therapy*, 23: 393–9.

——— (2018) 'The clitoral activation paradox – claimed outcomes from different methods of its stimulation', *Clinical Anatomy*, 31.5: 650–60.

Lewandowski, S. (2001) 'Über Persistenz und soziale Funktionen des Orgasmus (paradigmas)', *Zeitschrift für Sexualforschung*, 14: 193–213.

Li Jianmin 李建民 (1996) 'Furen meidao' kao – zhuantong jiating de chongtu yu huajie fangshu 「婦人媚道」考 – 傳統家庭的衝突與化解方術 (The art of charming in Han China), *Xin shixue*, 7.4: 1–32.

Li Ling 李零 (2006) *Zhongguo fangshu zhengkao* 中國方術正考 (Corrected Study of the Chinese Divinatory and Medical Arts), Beijing: Zhonghua shuju.

——— (2011a) *Lan tai wan juan* 蘭臺萬卷. *Du «Han shu • yi wen zhi»* 讀《漢書•藝文志》 (The Ten Thousand Scrolls of the Orchid Platform: Reading the 'Treatise on Skills and Literature' in the 'History of Han Dynasty'), Beijing: Sheng huo, dushu, xinzhi sanlian shudian.

——— (2011b) 'Jiao mao kao – kaogu faxian yu Ming-Qing faxian xiaoshuo de bijiao yanjiu 角帽考 – 考古發現與明清小說的比較研究 (Study of the horn hat: comparing archaeological finds and Ming–Qing fiction)', *Qinshi huangdi ling bowuguan*, 2011: 213–28.

Li Ling and McMahon, K. (1992) 'The contents and terminology of the Mawangdui texts on the arts of the bedchamber', *Early China*, 17: 145–85.

Lin Fu-shih 林富士 (2001) 'Lüelun zaoqi daojiao yu fangzhongshu de guanxi 略論早期道教與房中術的關係 (Taoism and Sexual Arts in Medieval China), *Bulletin of the Institute of History and Philology, Academia Sinica*, 72.2: 233–300.

——— (2008) 'Lüelun zaoqi daojiao yu fangzhongshu de guanxi 略論早期道教與房中術的關係 (Taoism and Sexual Arts in Medieval China), in Lin Fu-shih (ed.) *Zhongguo zhonggu shiqi de zongjiao yu yiliao* 中國中古時期的宗教與醫療 (Religion and Medicine in Middle Period China), Taipei: Lianjing 聯經, pp. 333–402.

Liu Jie 劉杰 (1995) *Zhongguo xingyixue* 中國性醫學 (Chinese Sexual Medicine), Beijing: Zhongguo renkou.

Liu Xiuqiao 劉修橋 (ed.) (1976) *Yixin fang (fu Zhong Ri wen jieshuo yi ce)* 醫心方 (附中日文解說一冊) (*Ishinpō* [With one Volume of Chinese and Japanese Explanations]), Taipei: Xinwenfeng chuban gongsi.

Lo, V. and Barrett P. (2012) 'Other pleasures? Anal sex and medical discourse in pre-modern China', in R.A.G. Reyes and W.G. Clarence-Smith (eds) *Sexual Diversity in Asia, c. 600–1950*, London; New York: Routledge, pp. 25–46.

Lo, V. and Re'em E. (2018) 'Recipes for love in the ancient world', in G.E.R. Lloyd and J.J. Zhao (eds) *Ancient Greece and China Compared*, Cambridge: Cambridge University Press, pp. 326–52.

Lohmar, D. (2008) *Phänomenologie der schwachen Phantasie. Untersuchungen der Psychologie, Cognitive Science, Neurologie und Phänomenologie zur Funktion der Phantasie in der Wahrnehmung*, Dordrecht: Springer.

Ma Jixing 馬繼興 (1992) *Mawangdui gu yishu kaoshi* 馬王堆古醫書考釋 (Examination and Explanation of the Ancient Mawangdui Medical Texts), Changsha: Hunan kexue jishu chubanshe.

Mallinson, J. (2018) 'Yoga and sex: what is the purpose of vajrolīmudrā?' in K. Baier, Ph.A. Maas and K. Preisendanz (eds) *Yoga in Transformation: historical and contemporary perspectives*, Vienna: Vienna University Press, pp. 181–222.

Marcuse, M. (1922) 'Orgasmus ohne Ejakulation', *Deutsche Medizinische Wochenschrift*, 48.35: 1171–3.

Marié, É. (2007) 'L'art de la chambre à coucher dans la Chine ancienne', *Revue d'études japonaises du CEEJA, Benkyô-kai*, II: 33–60.

Maspero, H. (1937) 'Les procédés de 'nourrir le principe vital' dans la religion taoïste ancienne', *Journal asiatique*, 229 (avril–juin): 177–252; (juillet–septembre): 353–430.

Masters, W.H. and Johnson, V.E. (1966) *Human Sexual Response*, Boston, MA: Little, Brown.

Mauss, M. (1935) 'Les techniques du corps', *Journal de psychologie normale et pathologique*, 32.3–4: 271–93.

Mawangdui Hanmu Boshu Zhengli Xiaozu 馬王堆漢墓帛書整理小組 (ed.) (1985) *Mawangdui Hanmu boshu (si)* 馬王堆漢墓帛書 (肆) (The Silk Manuscripts from the Mwangdui Han Tomb [vol. 4]), Beijing: Wenwu chubanshe.

Mazloomdoost, D. and Pauls, R.N. (2015) 'A comprehensive review of the clitoris and its role in female sexual function', *Sexual Medicine Reviews*, 3.4: 245–63.

McMahon, K. (2019) 'The art of the bedchamber and *Jin Ping Mei*', *Nan Nü*, 21: 1–37.

Meston, C.M., Levin, R.J., Sipski, M.L., Hull, E.M. and Heiman, J.R. (2004) 'Women's orgasm', *Annual Review of Sex Research*, 15.1: 173–257.

Middendorf, U. (2007) *Resexualizing the Desexualized: the language of desire and erotic love in the classic of Odes*, Pisa, Roma: Accademia Editoriale.

Mollier, C. (2016) 'Conceiving the embryo of immortality: "seed-people" and sexual rites in early Taoism', in A. Andreeva and D. Steavu (2016) *Transforming the Void: embryological discourse and reproductive imagery in East Asian religions*, Leiden; Boston, MA: Brill, pp. 87–110.

Mussat, M. (prés.) and Leung, Kwok Po (trad.) (1978) *Sou Nü King: La sexualité taoïste de la Chine ancienne*, Paris: Édition Seghers.

Needham, J. (1983) *Science and Civilisation in China, vol. 5, part V: spagyrical discovery and ivention: physiological alchemy, sexuality and the role of theories of generation*, Cambridge: Cambridge University Press.

O'Connell H.E., Eizenberg N., Rahman M. and Cleeve J. (2008) 'The anatomy of the distal vagina: towards unity', *Journal of Sexual Medicine*, 5: 1883–1891.

Ōgata, Tōru 大形徹 (2015) *Taisansho • Zakkinhō • Tenka shidōdan • Gōin'yōhō • Jūmon* 胎産書 • 雜禁方 • 天下至道談 • 合陰陽方 • 十問 (Book of the Generation of the Fetus • Recipes for Various Charms • Discussion of the Utmost Method Under the Sky • Uniting Yin and Yang • Ten Interviews), Tōkyō-to Chiyoda-ku: Tōhō Shoten.

Ōta, Tenrei 太田典礼 (1976) 'Guanyu yixin fang 關於醫心方 (On the *Ishinpō*)', in Liu Xiuqiao 劉修橋 (ed.) (1976) *Yixin fang* 醫心方 (*Fu zhongriwen jieshuo yice* 附中日文解說一冊), Taipei: Xinwenfeng chuban gongsi, pp. 1–270.

Okanishi, Tameto 岡西為人 (author), Guo, Xiumei 郭秀梅 (collation) (2010) *Song yiqian yiji kao* 宋以前醫籍考 (Research on Medical Works Before the Song Dynasty), Beijing: Xueyuan 學苑.

Paciornik, M. (1985) *Come partorire accoccolate: la posizione migliore per madre e bambino*, Palermo: I.P.S.A. editore.

Pfister, R. (1995) 'Bodies and nature in ancient Chinese sexual body technique: three illustrations', in H. Johannessen, S.G. Olesen and J.Ø. Andersen (eds) *Studies in Alternative Therapy 2: body and nature*, Gylling (Denmark): Odense University Press and International Network for Research on Alternative Therapies (INRAT), pp. 82–105.

——— (trans.) (2003) *der beste weg unter dem himmel: sexuelle körpertechniken aus dem alten china: zwei bambustexte aus mawangdui*, Zürich: Museum Rietberg Zürich.

——— (2006a) 'The production of special mental states within the framework of sexual body techniques – as seen in the Mawangdui medical corpus', in P. Santangelo, D. Guida (eds) *Love, Hatred and Other Passions: questions and themes on emotion in Chinese civilisation*, Leiden: Brill, pp. 180–94.

——— (2006b) 'The jade spring as a source of pleasure and pain: the prostatic experience in ancient and medieval medical and Daoist texts', in H.U. Vogel, C. Moll-Murata and X. Gao (eds) *Studies on Ancient Chinese Scientific and Technical Texts. Proceedings of the 3rd ISACBRST. March 31–April 3, 2003 Tübingen, Germany*, Zhengzhou: Elephant Press, pp. 88–106.

——— (2007) 'Der Milchbaum und die Physiologie der weiblichen Ejakulation: Bemerkungen über Papiermaulbeer- und Feigenbäume im Süden Altchinas', *Asiatische Studien Études Asiatiques*, LXI.3: 813–44.

——— (2012) 'Phosphenes and inner light experiences in medieval Chinese psychophysical techniques: an exploration', in G. Tamburello (ed.) *Concepts and Categories in East Asia*, Roma: Carrocci editore, pp. 38–70.

——— (2013) 'Gendering sexual pleasures in early and medieval China', *Asian Medicine*, 7.1: 34–64.

——— (2016) 'On the partonymy of female genitals in Chinese manuscripts on sexual body techniques. 28p. Addendum to 2016' (Shanghai 上海) *Chutu Yixue Wenxian Yanjiu Guoji Yantaohui Lunwenji* 出土醫學文獻研究國際研討會論文集, Shanghai: Shanghai University of Traditional Chinese Medicine.

Qiu Xigui 裘錫圭 (main ed.), Hunansheng bowuguan 湖南省博物館 and Fudan daxue chutu wenxian yu guwenzi yanjiu zhongxin 復旦大學出土文獻与古文字研究中心 (2014) *Changsha Mawangdui Hanmu jianbo jicheng* 長沙馬王堆漢墓簡帛集成 (Collection of Bamboo and Silk Manuscripts from the Mawangdui Han Tombs in Changsha), Beijing: zhonghua shuju.

Raz, G. (2008) 'The way of the yellow and the red: re-examining the sexual initiation rite of Celestial Master Daoism', *Nan Nü*, 10.1: 86–120.

——— (2012) *The Emergence of Daoism: creation of tradition*, London; New York: Routledge.

Robinet, I. (1984) *La Révélation du Shangqing Dans L'histoire du Taoïsme*, 2 vols., Paris: École Française d'Extrême-Orient.

Rocha, L.A. (2011) 'Scientia sexualis versus ars erotica: Foucault, van Gulik, Needham', *Studies in History and Philosophy of Science Part C: Studies in History and Philosophy of Biological and Biomedical Sciences*, 42.3: 328–43.

——— (2012) 'The way of sex: Joseph Needham and Jolan Chang', *Studies in History and Philosophy of Science Part C: Studies in History and Philosophy of Biological and Biomedical Sciences*, 43.3: 611–26.

——— (2015) 'Translation and two Chinese "sexologies": double plum and sex histories', in H. Bauer (ed.) *Sexology and Translation: cultural and scientific encounters across the modern world*, London: Temple University Press, pp. 154–73.

Roşu, A. (2002) 'Pratiques tantriques au regard de l'andrologie médicale', *Journal Asiatique*, 290.1: 293–313.

Safron, A. (2016) 'What is orgasm? A model of sexual trance and climax via rhythmic entrainment', *Socioaffective Neuroscience and Psychology*, 6: 31763.

Sakade, Yoshinobu 坂出祥伸 and Umekawa, Sumiyo 梅川純代 (2003, 2006²) *"Ki" no Shisō kara Miru Dōkyō no Bōchūjutsu – Ima ni Ikiru Kodai Chūgoku no Seiai Chōju Hō* 「気」の思想から見る道教の房中術 – いまに生きる古代中国の性愛長寿法 (Daoist Sexual Techniques from a point of view of Qi – Chinese ancient sexual techniques which are still available), (Kokoro to Kyōyō Shirīzu 心と教養シリーズ, 3), Tokyo: Goyō Shobō 五曜書房.

Sherfey, M.J. (1966, 1972) *The Nature and Evolution of Female Sexuality*, New York: Vintage; 1972: Random House.

Society for the Commemoration of the One Thousandth Anniversary of the 'Ishimpo' (ed.) (1984) *Dawning of Japanese Medicine – Ishimpo*, Kyoto: Tsumura Juntendo.

Stanley-Baker, M. (2006) *Cultivating Body, Cultivating Self: a critical translation and history of the Tang dynasty Yangxing yanming lu (records of cultivating nature and extending life)*, MA thesis, Indiana University, Bloomington.

Stifter, K.F. (1988) *Die Dritte Dimension der Lust. Das Geheimnis der weiblichen Ejakulation*, Frankfurt am Main: Ullstein.

Sundahl, D. (2003) *Female Ejaculation and the G-spot, Not Your Mother's Orgasm Book!*, Alameda, CA: Hunter House.

Swartz, L.H. (1994) 'Absorbed states play different roles in female and male sexual response: hypotheses for testing', *Journal of Sex and Marital Therapy*, 20.3: 244–53.
Swartz, M.J. (1958) 'Sexuality and aggression on Romonum, Truk', *American Anthropologist*, 60.3: 467–86.
Syed, R. (1999) 'Zur Kenntnis der "Gräfenberg-Zone" und der weiblichen Ejakulation in der altindischen Sexualwissenschaft', *Sudhoffs Archiv*, 83: 171–90.
Taiz, Lincoln and Taiz, Lee (2017) *Flora Unveiled: the discovery and denial of sex in plants*, New York: Oxford University Press.
Tiefer, L. (1991) 'Historical, scientific, clinical and feminist criticisms of "the human sexual response cycle" model', *Annual Review of Sex Research*, 2.1: 1–23.
Triplett, K. (2014) 'For mothers and sisters: care of the reproductive female body in the medico-ritual world of early and medieval Japan', *Dynamis*, 34.2: 337–56.
Tsuchiya, Eimei 土屋 英明 (2008) *Chūgoku no seiaijutsu* 中国の性愛術 (Erotic Art of China), Tokyo: Shinchō Shu.
Umekawa, Sumiyo 梅川純代 (2005a) 'Tiandi yinyang jiaohuang dalefu and the art of the bedchamber', in V. Lo and C. Cullen (eds) *Medieval Chinese Medicine: the Dunhuang medical manuscripts*, London; New York: Routledge Curzon, pp. 252–77.
——— (2005b) 'III.8 – Biyaku: Chūgoku sei gihō ni okeru shoku 媚薬: 中国性技法における〈食〉 (Aphrodisiacs: Diet in Chinese Sexual Techniques), in Suzuki, Akihito 鈴木晃仁 and Ishizuka, Hisao 石塚久郎 (eds) *Shokuji no gihō* 食餌の技法. (Shintai i bunka ron 身体医文化論, IV) Tōkyō: Keiō gijuku daigaku shuppankai 慶応義塾大学出版会, pp. 193–216.
Umekawa, Sumiyo 梅川純代 and Dear, D. (2018) 'The relationship between Chinese erotic art and the art of the bedchamber: a preliminary survey', in V. Lo and P. Barrett (eds) *Imagining Chinese Medicine*, Leiden; Boston, MA: Brill, pp. 215–26.
Umayahara, Shigeo 馬屋原成男 (*kanshū* 監修); Ishihara, Akira 石原明, Takada, Shōjirō 高田正二郎 (*kaisetsu* 解說) and Iida, Yoshirō 飯田吉郎 (*kundoku* 訓読) (1967) *Ishinpō: maki dainijūhachi, bōnai: kunaichō shoryōbu zōhon* 醫心方: 卷第廿八, 房內: 宮內庁書陵部蔵本 (*Ishinpō*, scroll 28, The Chamber-Intern: Edition of the Imperial Household Agency, Archives and Mausolea Division), Tōkyō: Shibundō.
Van Gulik, R.H. 1951(2003) *Erotic Colour Prints of the Ming Period, with an Essay on Chinese Sex Life from the Han to the Ch'ing Dynasty, B.C. 206 – A.D. 1644*, Leiden: E.J. Brill.
——— 1961(2004) *Sexual Life in Ancient China: a preliminary survey of Chinese sex and society from ca. 1500 B.C. till A.D. 1644, with a new introduction and bibliography by Paul R. Goldin*, Leiden: Brill.
Vincke, É. (1991) 'Liquides sexuels féminins et rapports sociaux en Afrique centrale', *Anthropologie et Sociétés*, 15.2–3: 167–88.
Walter, T. (1999) 'Plädoyer für die Abschaffung des Orgasmus: Lust und Sprache am Beginn der Neuzeit', *Zeitschrift für Sexualforschung*, 12.1: 25–49.
Wells, S. and Yao, Ping (2015) 'Discourses on gender and sexuality', in C. Benjamin (ed.) *The Cambridge World History, Volume 4: A World with States, Empires and Networks 1200 BCE–900 CE*, Cambridge: Cambridge University Press, pp. 154–78.
Wile, D. (1992) *Art of the Bedchamber: the Chinese sexology classics including women's solo meditation texts*, Albany: State University of New York Press.
——— (2018) 'Debaters of the bedchamber: China reexamines ancient sexual practices', *JOMEC Journal*, 12: 5–69.
Wouters, L.C.S. (2010) *Sun Simiao's Qianjinfang* 千金方 (Voorschriften Ter Waarde Van Duizend Goudstukken) *Fangzhong buyi di ba* 房中補益第八 (Het Herstellende van de Slaapkamer[kunsten]: deel acht), MA thesis, Universiteit Gent.
Wundram, I.J. (1979) 'Nonreproductive sexual behavior: ethological and cultural considerations', *American Anthropologist*, n.s. 81.1: 99–103.
Yang Geng 楊耕 (ed.) (1964) *Tantan jihua shengyu* 談談計劃生育 (Talking about Family Planning), Tianjin: Tianjin renmin chubanshe.
Yang, Yong and Brown, M. (2017) 'The Wuwei medical manuscripts: a brief introduction and translation', *Early China*, 40: 241–301.
Yao, P. (2013) 'Historicizing great bliss: erotic in Tang China (618–907)', *Journal of the History of Sexuality*, 22.2: 207–29.
——— (2015) 'Changing views on sexuality in early and medieval China', *Journal of Daoist Studies*, 8: 52–68.

——— (2018) 'Between topics and sources: researching the history of sexuality in imperial China', in: H. Chiang (ed.) *Sexuality in China: histories of power and pleasure*, Seattle: University of Washington Press, pp. 34–49.

Ye Dehui 葉德輝 (1903) *Shuangmei jingan congshu* 雙梅景闇叢書 (The Shadow of the Double Plum Tree Anthology), Changsha: Yeshi kan 葉氏刊.

Zhang, H. (2013) 'Enchantment, charming, and the notion of the femme fatale in early Chinese historiography', *Asian Medicine*, 8.2: 249–94.

Zhu Yueli 朱越利 (2002) 'Fangxiandao he Huang-Lao dao de fangzhongshu 方仙道和黃老道的房中術 (The Arts of the Bedchamber of Magic and Immortality Tradition and Huang–Lao Tradition)', *Zongjiaoxue yanjiu*, 1: 1–11.

23
SEXING THE CHINESE MEDICAL BODY

Pre-modern Chinese medicine through the lens of gender

Wang Yishan

Generally speaking, the development of a feminist scholarship of Chinese medical history can be traced back to the late 1990s with the publication of Charlotte Furth's pioneering and prodigious monograph *A Flourishing Yin: Gender in China's Medical History, 960–1665* (1999). Furth's work represents a conceptual shift in research on Chinese medicine that echoed a larger tendency of the late twentieth century to add 'gender' to the vocabulary of one or another discipline in the Humanities. Gender, as Scott (1986) affirmed, is a 'useful category of analysis'. But gender can only be 'freed to do its critical work' (Scott 2010: 12) if we first cast scrutiny on the word itself. The distinction of gender and sex as first articulated by Gayle Rubin (1975) has been arguably the most crucial framework for feminist theory. Nevertheless, the seemingly clear-cut separation of gender from sex – the former being the socially and culturally constructed difference between men and women, while the latter denotes a biological sexual classification – has come to be contested and reconceived by feminist theorists (Butler 1990, 1993; Errington 1990; Fausto-Sterling 1992, 2000).

Indebted to the feminist deconstruction of the sex-gender binary and sexual difference in particular, an array of historians, including Schiebinger (1989), Laqueur (1990), Cadden (1993) and Duden (1998), have confronted the naturalist epistemologies of sexuality in medicine. Their work demonstrates that neither the body nor sexuality is an objective entity with transhistorical existence, but that these are concepts that need to be studied in the specific historical contexts in which they gained saliency. In a similar vein, historians such as Bray (1997), Furth (1999), Leung (2006a,b), Lee (2012), to name just a few, have attempted to contextualise feminist critiques[1] of gender, sex and the body in relation to Chinese medicine; and their efforts have been repaid with a burgeoning of historical scholarship on Chinese medicine through the lens of gender.

But what does it mean to adopt gender as a 'category of analysis' in the historical investigation of Chinese medicine? In light of the feminist disruption of the taken-for-grantedness of sexual dimorphism, Scott revisited her argument of gender as a 'useful category of analysis' in a more recent article, 'Gender: Still a Useful Category of Analysis?' (Scott 2010). Here, she points out that the potential efficacy of taking a gender perspective still pertains only if the focus shifts to 'the construction of sexual difference itself' rather than 'the roles assigned to women and men' (Ibid.: 10). My intention then is to identify what a gender approach, as advocated by Scott, has contributed, and might further contribute, to the field of Chinese medicine. In this introduction to the subject, I will restrict my focus to studies exploring the ways in which sexed bodies have been brought into coherence and crisis within the context of Chinese medicine from the early imperial period to the Qing dynasty (1644–1912). I suggest that opening up the sexed binary for interrogation might represent a desirable direction in which historical research on Chinese medicine and the body could be usefully pursued.

The *Yellow Emperor's* body: rethinking sexual difference

Furth began her gender analysis of Chinese medical history with the idea of the 'Yellow Emperor's body', a metaphor developed as an interpretative tool for studying the medical body as represented in the classic *Inner Canon of the Yellow Emperor* (*Huangdi neijing* 黃帝內經), hereafter the *Inner Canon*. This is a corpus compiled c. first century CE (Chapter 7 in this volume) from anonymous manuscript texts that were written down during the previous several centuries (Chapter 3 in this volume). Furth's finely nuanced analysis of the 'Yellow Emperor's body' paved the way for subsequent feminist writers who have grappled with the subject of sexual difference in Chinese medicine.

It seems that Furth took her inspiration essentially from Laqueur's deconstructive interpretation (Laqueur 1990) of early European conceptualisations of a 'one-sex' body, which portrayed the female body as an inferior male body. The Yellow Emperor's body was Furth's point of departure in a search for 'making sex' in classical Chinese medical knowledge. However, what she discovered was not a 'one-sex' body but a 'truly androgynous' body with *yin* and *yang* as 'the most important signifier[s] of bodily gender' (Furth 1999: 46). This characterisation of the body later conjured up ideas of 'an androgynous womb' (Wu 2010: 98) and the lactating male scholar of Ming-Qing medicine (Furth 1999: 221).

Furth's conclusion that the body of classical Chinese medical texts is androgynous, surprising though it may seem, is not hastily drawn; it is solidly based on her keen observation that the Yellow Emperor's body is a cosmologised entity that expresses the unity of the human being with the heavenly macrocosm. This 'body of generation', as Furth aptly names it, encompasses and is encompassed by the dyad of *yin* and *yang* which also orders the cosmos at large (Ibid.: 48).

According to Furth, the dyad of *yin* and *yang* was organised according to two main patterns, both of which were incorporated into Chinese medical discourse (Ibid.: 28). In the first model, the relationship of *yin* and *yang* is complementary and equivalent. This pattern evokes the complementarity of male and female partners as a *yin-yang* pair with parallel and homologous generative capacities and bodies (Ibid.: 41). Furth explains this by citing the *Inner Canon*'s classic account of the sexual maturation of boys and girls, which starts with the development of the Kidney[2] and the abundance of 'Kidney *qi*'. When the girl is seven, the *Inner Canon* says:

> The *qi* of the Kidneys abounds. The first teeth are substituted and the hair grows long. With two times seven [at the age of 14], the heaven *gui* 癸 arrives, the controlling vessel is passable and her thoroughfare vessel is abundant; her menses flow regularly and she is able to bear children'.

(adapted from Unschuld et al. 2011: 36–8)

The corresponding description for boys goes like this:

> With two times eight [at the age of 16], his *qi* of Kidneys is abundant and the heaven *gui* arrives, his seminal essence overflows; he is able to unite *yin* and *yang* so as to beget sons.
>
> *(idem, 39)*

It is clear that for both boys and girls, reproductive ability depends on the Kidney and Heaven *gui* (*tiangui* 天癸).[3] Simply put, the master system here is the Kidney, 'the seat of primary life vitalities' (Furth 1999: 45), controlling the generative function for both sexes. Furth suggests that sexual dimorphism, which presumes natural and innate differences between the two sexes, is not evidenced in this passage, and notes the absence of anatomical secondary sexual characteristics such as breasts or facial hair. Furth's reading echoes Bray's earlier interpretation of the same paragraphs, where Bray argues that the womb, a distinctive anatomical organ on which explanations of sex differences were centred in the Greek medical tradition, 'is not a synecdoche for woman, nor a ruling feature of her constitution' (Bray 1995: 237). What matters here is how puberty marks girls' and boys' complementary roles as procreators, which resonate with the mirroring positions of *yin* and *yang* in cosmogenesis, rather than marking their bodies with gendered anatomical signs (Furth 1999: 45).

The second model establishes a hierarchy of the *yin-yang* pair, which served as a point of reference for the hierarchical and encompassing relationship of Blood[4] and Essence (*jing* 精), 'the sexual power of the body' (Ibid.: 47). At the highest level, heavenly *qi* or primordial *qi*, permeating both the universe and the body, was first divided into Essence, which was *yin*, and *qi*, which was *yang*. Moving down through the hierarchy from the level of the universe to that of the generative body,[5] we see that *yin* Essence bifurcated into a further dyad of *yin* Blood and *yang* Essence. This *yang* Essence was the '*qi* aspect' of Essence (Ibid.: 48), while *yin* Blood at this level denoted all the sexual fluids, including male seminal Essence and female menstrual Blood as well as the generative vitalities underlying them. At the 'most material' level, which was concerned with the reproductive power of the individual body, male Essence (semen) was *yang* and female Blood (female reproductive fluids, including breast milk and menses) was *yin* (Ibid.: 47–8). Not only were Blood and Essence attributed flexibly to *yin* and *yang*, but they also formed a hierarchy analogous to that of Earth and Heaven in the cosmos, and woman and man in society (Figure 23.1).

Furth's abstraction of the Yellow Emperor's body has been well received; it links with studies of the unique conception of the body in Chinese medicine as 'functional and processual' (Bray 1995: 236) rather than anatomical, while at the same time, it has seeded a rethinking of the notion of the idealised body that simply transcended essentialised differences between the two sexes in the historiography of Chinese medicine (Chapters 22, 24–26 and 30 in this volume).

Bray was perhaps the first to stress that the difference between male and female bodies was viewed in pre-modern China not so much in terms of 'two essentialised separate categories', as in terms of a 'continuum of probability' along which men and women took different positions according to, say, their ages (Ibid.). Simply put, her argument is that sexual difference in Chinese medicine was understood as a matter of 'degree rather than one of essential nature' (Bray 1997: 317). In a similar vein, Raphals, discussing the concept of *yin-yang* in medical texts, argues that in the *Inner Canon,* the body was analogised with a cosmos of *yin* and *yang* and the male and the female body were conceived of as symmetrical systems of *yin* and *yang*. She points out that, with the exception of a few diseases related to childbirth, neither in diagnosis nor in treatment was gender difference emphasised in early medical theorisation

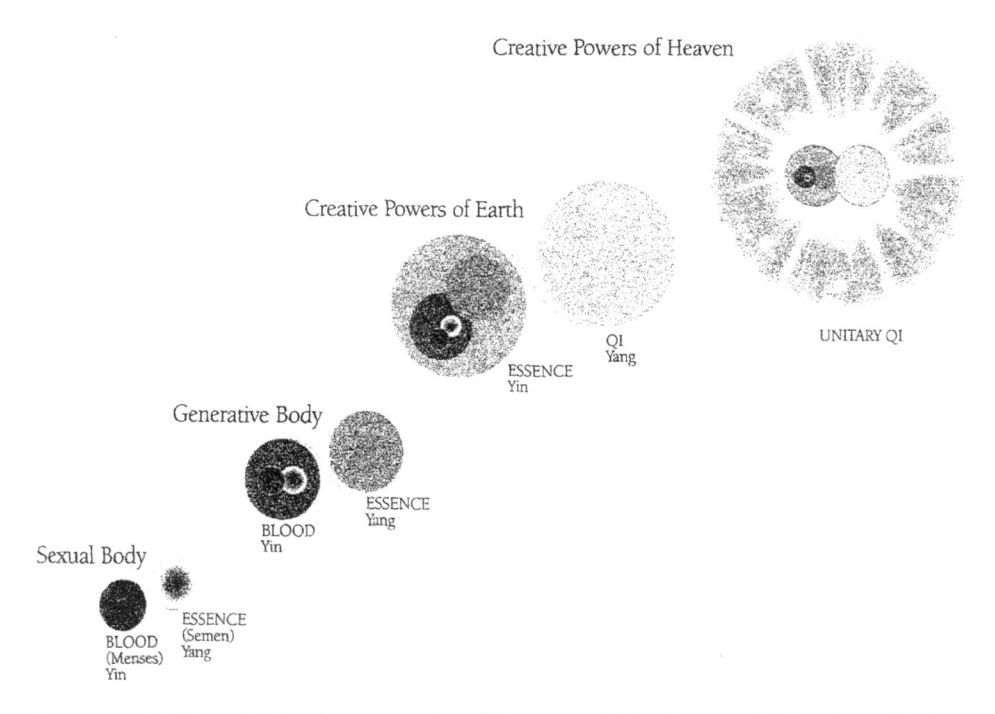

Figure 23.1 Different levels of *yin-yang* pairs of Essence and Blood. Reproduction from Charlotte Furth (1999) *A Flourishing yin*, p. 49. Credit: University of California Press

(Raphals 1998, ch. 7). Wilms takes her inspiration from Furth and also finds that none of the early medical texts, e.g. the *Classic of Difficult Issues* (*Nanjing* 難經), which was composed in the middle of Eastern Han (25–220 CE), recognised essential differences between the two sexes (Wilms 2006). Wilms also refers to the text of the *Inner Canon*, in particular the account of sexual development to illustrate her point (Ibid.: 80–1).

It appears that the only diverging opinion comes from the Taiwanese Chinese scholar Li Jianmin 李建民, who suggests that Chinese physicians recognised a 'two-sex model' of human bodies, which grounded theories of sexual difference in different *qi* and vessel functions (Li Jianmin 2005). He examines the Superintendent channel (*dumai* 督脈) and its significant role in constructing a unique vision of the male body in early China. The Superintendent channel starts in the lower abdomen, crossing male or female genitals and travelling up the spine. Connecting the Kidney and the brain, it was imagined as the main vessel responsible for the physiology of practices involved in nourishing the male body. During these practices, the male practitioner refined his own body through *coitus reservatus*, avoiding ejaculation during intercourse, while at the same time attempting to obtain the energies of the female partner(s), which would result for him in an accumulation of the *yin* Essence, thought to be secreted by women when they reached orgasm. The ultimate goal of such techniques was to 'return the essence to nourish the brain' (*huanjing bunao* 還精補腦) (Chapters 22, 25 and 26 in this volume). The Superintendent channel was certainly crucial here, as the marrow-like fluid in the brain and the Essence stored in the Kidney could only supplement each other via the Superintendent channel. Li thus concludes that the Superintendent channel was associated with the conception of a distinctive male body despite the fact that it was common to both men and women (Ibid.: 21–2).

While Furth bases her reading of the Yellow Emperor's body almost solely on the *Inner Canon*, the *locus classicus* of Chinese *medicine,* Li collates a heterogeneity of materials, a significant portion of which belong to the tradition of the art of the bedchamber (*fangzhong shu* 房中術). It should be noted that the early bedchamber art manuals, such as those found in the *Mawangdui* 馬王堆 manuscripts[6] of the early Han – which as Harper (1998: 7) argues, 'reveal more of what medicine was like' in Early China than the *Inner Canon* – were indeed preoccupied with a gender-specific gaze on human bodies. For example, Pfister shows that the Mawangdui manuscript *Discussion Of the Culminant Way In Under-Heaven* (*Tianxia zhidao tan* 天下至道談) is rich in terminology that describes the female genitalia with great clarity and specificity (Pfister 2016; see also Lo and Re'em 2018). Li (2005: 253) points out that in the *Inner Canon* too, descriptions of physical difference between the sexes (e.g. having a beard or not) are abundant. However, Li makes it clear that it was the functions of certain vessels and related organs within the body, more so than anatomical differences as such, that marked the Chinese medical body as male or female. We have yet to see how Li's view might be brought into conversation with Furth's reading of the androgynous Yellow Emperor's body. Nevertheless, both Furth's and Li's works have made admirable contributions to challenging 'what is in fact "natural" in our own as well as other cultures' understandings of gender and the body' (Furth 1994: 250).

To sum up, Furth's understanding of the androgynous body of the Yellow Emperor's is a fine example of re-reading of the medical body from a gender perspective. Her imagination of the Yellow Emperor's body teases out the gendered code embedded in the Chinese medical body, which was associated with the high traditions of Chinese medicine. For instance, while ideas like Blood, *qi* and Essence are nothing new to historians analysing medicine in China, Furth is probably the first who proposed a most innovative interpretation of them as a three-tiered system of bodily power that 'participate[s] in the gendered hierarchical' of the human microcosm and heavenly macrocosm (Furth 1999: 48).

At the same time, her feminist approach to the Yellow Emperor's body is path-breaking in the way in which it undermines the 'reality' of sexual difference in the context of Chinese medicine, and introduces an alternative way to understand 'the Chinese model of bodily gender difference' (Ibid.: 27). The Yellow Emperor's body conceptualised in earlier Chinese medical texts was androgynous in the sense that it did not serve as an *a priori* locus of reference for sexual difference; its ontology was metaphorical and non-foundational. In the context of correlative thinking (Chapter 1 in this volume), men and women as *yang* and *yin* are homologous and complementary pairs in their participation in the cosmic process of creation. Thus, the process of gendering is accomplished through differentiated *yin/yang* positions and roles within different social contexts, but is not necessarily enacted on clearly sexed bodies. In this sense, not only is the distinction between gender (*qua* culture) and sex (*qua* nature) uncritical, but 'gender is the key to sex' (Scott 2010: 13).

Gendering the history of Chinese medicine: a brief review of feminist historiographies of medicine for women

The richness of Chinese medical theories makes us sceptical of any easy claim for the continuity of an enduring master narrative. Indeed, what we usually encounter throughout the history of Chinese medicine is a cacophony of voices about multiple bodies across historical periods. It is thus not surprising that at a certain, much debated, point in the history of Chinese medicine, as we will see, the notion of the androgynous Yellow Emperor's body was conjoined with an emerging conception of a distinctive female body. This new female body came into view as, in the medical encounter, physicians started to systematically theorise the thorny problems of difference that their female patients presented them with in the matters of gestation, reproduction, etc. To understand when and how the Chinese medical body was ultimately 'sexed' and especially the medical envisioning of female uniqueness in Chinese medicine, we have to turn to the story of *fuke* 婦科,[7] a subfield within Chinese medicine which is exclusively for women, yet, at least in its medical conceptualisation, mostly dominated by male physicians.

According to Leung, historical research on medicine for women as an integral part of the general history of Chinese medicine is not a new area, but the feminist approach to the history of *fuke* only began to receive sustained attention from the late 1990s onwards (Leung 2006a). Acknowledging the potential of the female body as a primary site through which medical discourses are expressed, an increasing number of scholars have attempted to mobilise gender as an interpretative tool to study the rich and copious literature of Chinese *fuke*. Feminist historiographies of *fuke* set out not simply to map the historical development of *fuke*, but also to unpick the process through which the idea of the female body as distinct from the male one was substantiated in the area of medicine for women.

Post-Han to the Song dynasty (960–1279): the construction of the sexed body

Despite their various chronological focuses, all the works I shall review shortly share an interest in the 'gendering of Chinese medicine', a term that undoubtedly bears the hallmark of a feminist framework for understanding the history of *fuke*. In all these works, gender is deployed as an 'useful category of analysis' by Scott's standard, as they all set the aim to 'think critically about how the meanings of sexed bodies are produced in relation to one another' (Scott 2010:10).

Scholars agree that Chinese medicine was 'gendered' at a certain point, coinciding with the emergence of a medical conception of the female body that was clearly separated from the androgynous ideal of the Yellow Emperor's body. Despite the ongoing debate about the exact periodisation of Chinese *fuke*, the general consensus is that the seventh century was a watershed in which male physicians started to systematically theorise and categorise pathologies of female disorders so that a separate division of medicine was deemed necessary (Furth 1999; Lee 2003; Wilms 2006; Yates 2006).

Given that the professionalisation of medicine for women stemmed from an '*a priori* assumption that women had unique illnesses requiring unique therapeutic strategies' (Wu 2010: 42), tracing the trajectory of *fuke* in this context sheds light on the medical narrative of the female body and by extension the notion of womanhood. As such, at the centre of feminist scholars' enquiries sit the following questions: How was the female body envisioned? Was it homologous to the male as the ancient medical classic taught or different from the male body? Or in the context of *fuke*, 'what aspects of the female body were regarded as specifically female' (Wilms 2006: 76) such that they would precipitate a new field of medical knowledge dedicated to female diseases alone?

To start with, historians have unveiled a stereotypical depiction of the female body, which was seen as innately sicklier and constituently different from the male one due to female reproductive functions. Women's susceptibility, physicians theorised, was directly associated with their essentialised roles as child-bearers. Though men and women alike were afflicted with many of the same diseases, it gradually came to be believed that women were specifically prone to more disorders and that those disorders were more difficult to cure; thus, separate recipes for women were rationalised. Historical studies of *fuke* without exception stress the importance of the work of Sun Simiao 孫思邈 (581–682), who is believed to have set forth the earliest explicit justification of the need for 'separate prescriptions' (*biefang* 別方) and gender-specific treatment for women in his *Essential Prescriptions For Every Emergency Worth a Thousand Pieces of Gold* (*Beiji qianjin yaofang* 備急千金要方, 652).[8] Indeed, as Furth notes, the single fact that Sun's account of the female body was cited in 'almost every important writer on the subject of *fuke* in the Song dynasty' (Furth 1999: 72) attests to the impact of Sun's legacy on the consolidation of Chinese *fuke*.

With particular interest in medical techniques of childbirth from early imperial China to the Tang period (618–907), Lee (2003), among others, opens up the discussion of how medical discourse endorsed the entrenched view that reproduction weakened women's bodies. In another discussion of reproductive medicine in China in late antiquity and early medieval times, she explains that an increasing number of fertility recipes for women became mainstream, marginalising the previous advice and methods to beget sons found in the literature of the art of the bedchamber, which solely addressed a male readership (Lee 1997). A corollary of this was that the female body was increasingly envisioned as weaker and more fragile than its male counterpart, thus necessitating gender-specific treatment.

Another exemplary feminist approach to the history of Chinese *fuke*, especially in the Tang-Song era, is to address the medical narrative of femaleness in tandem with the medicalisation of menstruation, through which sexual difference was also conceptualised. Several authors (Furth 1999; Wilms 2006; Yates 2006) maintain that the 'hallmark' of Song *fuke* (Wilms 2006: 101) lies in its prioritisation of female Blood. When Song physicians repeatedly cited the aphorism 'In women, Blood is the leader', a teaching that was indebted to Sun Simiao, they in fact highlighted the primacy of Blood as the 'primary bodily signifier of female difference' (Furth 1999: 77). Interestingly, Chinese physicians' preoccupation with a vulnerable and fluid female body is analogous to Johann Storch's (1681–1751) concern for the periodic blood flow of his female patients, which was explicitly identified with femaleness in eighteenth-century Germany (Duden 1998).

In particular, Furth's re-interpretation of Blood and *qi* in the context of Song *fuke* exemplifies the depth of entanglement of gender ideology with medical discourses. She illustrates how, in Song doctors' theorisations, Blood was 'feminized as secondary and dependent in the hierarchy of bodily energies' (Furth 1999: 74) – Blood after all cannot stands alone, but follows *qi*. The hierarchy and encompassing relationship between Blood and *qi*, according to Furth, mirrored the gender hierarchy at the societal level. I would argue that what makes gender a constructive analytical tool here lies in the ways in which Furth problematises the asymmetry itself, which is deeply embedded both at the level of medical thinking (hierarchy of *qi* and Blood) and at the level of the structure of medical practice (institutionalised medicine for women versus the absence of medicine for men), more so than the actual hierarchical relationship between men and women in the larger society.

Post-Song to Qing: deconstructing female difference

As we move away from the flourishing of *fuke* in the Song dynasty, gender also proves an insightful perspective to scrutinise medical revisionism from the post-Song period well into the end of the Qing era, during which time some particularly striking paradigmatic shifts revised discourses of sexual difference.

To be sure, feminist historians note that the female body within the context of *fuke* was not linear nor 'well-gendered' (Leung 2006a: 5). They reveal that what confronts us in the post-Song *fuke* is no longer an emphasis on female difference constructed around Blood, but rather the revivification of the classic androgynous body, albeit not without revision.

Furth first highlights that the axiom of the 'leadership of Blood' in women in Song *fuke* was gradually eclipsed by the mainstream medical thinking of the Ming, which seemed to 'return to the classically androgynous body' (1999: 142). The centrality of Blood as the marker of femaleness was thus rejected along with the Song construct of sexual difference.

Specialising in Ming-Qing medicine, Wu (2010) makes it her aim to understand the way in which late imperial doctors defined women's reproductive bodies. The overarching thesis throughout her book is that the distinctive feature of the late imperial *fuke* is its 'de-exoticization of female difference and… increasingly benign view of female reproductive bodies' (Ibid.: 14). Similarly to Furth, she also points out the significant changes revealed in Xue Ji's 薛己 (1487–1559) revision of *Comprehensive Compendium of Good Formulas for Women* (*Furen daquan liangfang* 婦人大全良方) (1237), originally composed by Chen Ziming 陳自明 (1190–1270). Particularly noteworthy among these changes is Xue's deliberate omission of the rationalisation of separate prescriptions for women, which was indebted to Sun Simiao. What is more, Wu offers a fascinating comparative reading of Sun's *Essential Prescriptions For Every Emergency Worth a Thousand Pieces of Gold* with the *Imperially Commissioned Golden Mirror of Medicine* (*Yuzuan yizong jinjian* 御纂醫宗金鑒, 1742), hereafter the *Golden Mirror*. Sun's rationale of separate prescriptions for women was based on his conception of women as sicklier and more emotional than their male counterparts. The *Golden Mirror*, in contrast, opens with this remark: 'The diseases of women are fundamentally no different from those of men' (婦人之病本與男子無異) (translation in Wu 2010: 48). Wu thus concludes that the *Gold Mirror* adopted the 'same as men' perspective in the sense that disorders relevant to childbirth were portrayed as 'an exception to the rule that men and women were the same, rather than as a master narrative of female difference' (Wu 2010: 44).

There are two compelling examples that Wu adopts to support her argument which I find most fascinating. The first is her detailed discussion of how doctors in late imperial China came to accept the idea that men like women had wombs.[9] Basing her analysis on Zhang Jiebin's 張介賓 (1563–1640) medical treatise *The Complete Works of Jingyue* (*Jingyue quanshu* 景岳全書, 1624), Wu concludes that by analogising a unique female organ in men, Zhang in fact subsumed the womb into the 'broader discourse of generative vitalities' (Ibid.), evoking the ideal body of the Yellow Emperor as aspiring to androgyny.

In an important article on the medical rhetoric of the female breast in late imperial China, Wu (2011) points out that Xue notably de-emphasised the connection between female reproductive function and breast diseases, thus opening up possibilities for arguing for a category of breast disease that afflicted both men and women. In sum, Wu argues in the same vein as Furth that late imperial Chinese *fuke* was characterised by an emphasis on 'the common dynamics of male and female bodies' as well as a narrowing of 'the scope of gender difference' (Ibid.: 107–8).

Finally, Joanna Grant's gender analysis (Grant 2003) of the medical practice and theories of Ming physician Wang Ji 汪機 (1463–1539) is a fine addition to the feminist historiography of medicine for women. Her book stands out among studies on *fuke* by virtue of its clear aim to examine female bodies as well as male bodies within the context of one set of case histories, *The Stone Mountain Medical Case Histories* (*Shishan yi'an* 石山醫案) by Wang Ji. Grant shows that gender ideology is embedded in every aspect of Wang Ji's medical encounters, including diagnosis, aetiology, illness syndromes, treatment and outcome, corresponding to five subsections of the fourth chapter of her book, 'A gender analysis of *The Stone Mountain Medical Case Histories*'. The compelling conclusion she draws is that Wang stressed the 'essential sameness of understandings of male and female bodies' (Ibid.: 152). It seems that for Wang, the differences between women whose disorders were associated with reproduction and those who were suffering from non-reproductive disorders were greater than the dissimilarities between the two sexes (Ibid.: 147).

In summary, the aforementioned authors reach a consensus that in the post-Song period, Chinese *fuke* experienced an epistemological shift, reversing the pendulum swing from the emphasis on female difference to one of sameness between male and female bodies.

Two questions remain to be settled. First, how should we account for this quite dramatic shift in the history of Chinese *fuke*? When, after the Song dynasty, physicians gradually returned to favouring the classic androgynous body, does this imply that the learned tradition of Song *fuke* was forgotten and discredited all together?

Furth indicates that one salient feature of *fuke* after the Song period and up until the late Ming dynasty was its revisionism, a term that she uses to denote the continuity of preceding medical traditions. Some transformations in *fuke*, she maintains, 'added complexity more than transformed its foundation' (Furth 1999: 183). In other words, the *fuke* knowledge that survived from previous dynasties was not merely vestigial but still maintained a strong presence, entwined with the evolved medical theories of later periods, and this coexistence of old and new knowledge was presumably one of the reasons why Chinese medical theories are strewn with contradictions.

This leads to a further question: If the imagination of a unique female body persisted into the Song period and onwards, how then should we understand the bewildering view whereby the similarities between the two sexes are highlighted which seems to epitomise late imperial *fuke* knowledge? I prefer Wu's interpretation of the 'same as men' perspective as 'women's illnesses were the *same* as men's, except those *related* to childbirth' (Wu 2010: 43, italics in original text). It is not a question of whether women and men are different, because they clearly are and no one can deny the fact that women give birth but men do not. Perhaps we should rephrase the question like this: Do reproductive functions define women as qualitatively different from men or are such differences only 'a question of degree' (Bray 1997: 317)? When late imperial physicians theorised that reproduction aside, men's and women's diseases were no different, they in fact implied that the male and the female bodies were the same in nature. This fundamental similarity stands in stark contrast to the medical thinking of the Song period when femaleness was constructed around Blood, indicating that the female body, gestational or not, was envisioned as constituently and categorically different from the male one, hence entailing the absolute need for separate prescriptions.

At this point, it might be useful to ponder the constant tension between the androgynous body of generation revealed in the high tradition of Chinese medicine versus the female body of gestation in *fuke*. Furth encapsulates the development of *fuke* as a conceptual oscillation between two separate poles representing androgyny and sexual difference. In contrast, Wu, using a linguistic analogy, explains the male and female bodies in classical medicine as two ways of conjugating the same 'infinitive body' which 'serves as the basis of all human bodies' (Wu 2010: 232). As long as the generative body continues to serve as the metaphoric embodiment of the creative powers of Heaven and Earth, the seemingly unique female body remains simply one morphological variant of the infinitive body. In this context, the maturation of *fuke* then can be read as a process through which the unique female body of gestation was subsumed into the 'infinitive' body, whose multiple conjugated forms gradually become intelligible as medical knowledge advances. Wilms' view seems to be a hybrid of the arguments of Furth and Wu. She concludes that while the master narrative reflects the androgynous ideal of the Yellow Emperor's body, the fluidity of the notion of the Chinese medical body allows for the integration of a distinctive female body into the 'overarching paradigm of androgyny' (Wilms 2006: 107).

Conclusion

> Gender was a call to disrupt the powerful pull of biology by opening every aspect of sexed identity to interrogation, including the question of whether or not male/female, masculine/feminine was the contrast being invoked.
>
> *(Scott 2010: 12)*

Feminist scholarship broadens the scope of the historiography of Chinese medicine, with an agenda that includes but is not limited to the portrayal of female practitioners of medicine ranging from midwives to female healers, the development of medicine for women, medical discourses of the sexed body and gendered bodily techniques such as nourishing life (*yangsheng* 養生). In the light of Scott's statement, at the core of the analytical work involved in gendering medical history lies the question: In what way should we probe how differences between men and women were constructed and given credibility? Ultimately, it is the scholarly efforts to bring into relief questions about what Scott calls 'the essential "truth" about sexual difference' (Scott 2010: 12) that marks the studies I have reviewed in this chapter as feminist historiographies of Chinese medicine par excellence. All of them labour at a dual task: To offer a feminist re-interpretation of the rich repertoire of knowledge and practices of Chinese medicine, as well as to problematise the Chinese medical construct of the sexed body where the biomedical category of sex does not find a ready home.

In a word, at the heart of the feminist historiography of Chinese medicine lies an impulse to expose the taken-for-grantedness of the body and so to historicise it; and to question how, if at all, the female body is envisioned to live, function and suffer in different ways from the male body, and under what various social and discursive conditions. After all, the body itself, no less than ideas of sex and gender difference, is 'the product of a certain history and not one we should consider inviolate' (Ibid.: 12).

Notes

1 Furth in particular acknowledges her debt to the insights of scholars such as Roy Porter, Barbara Duden and Thomas Laqueur. See Furth (1999: 4).
2 I have adopted the convention of capitalising names of organs where the reference is to functional systems in Chinese medicine rather than anatomical structures. See Furth (1999: 23, n. 5).
3 The meaning of *tiangui* has been contested by many scholars. One interpretation I tend to agree with holds that it is 'the *yin* essence' generated from an accumulation of primordial *qi* stored in the Kidney. See Unschuld *et al.* (2011). It was believed to be a substance responsible for the promotion of the growth and reproduction of the human body. For a more detailed review of *tiangui*, also see Leo (2011: 140–2).
4 I hereafter use Blood with capital 'B' throughout to distinguish it from biomedical blood. As well as the red liquid that circulates through the veins and nourishes the foetus of a pregnant woman, Blood can also take the form of breast milk. See Furth (1986).
5 The generative body here bridges the higher level of cosmogenesis and the lower level of human reproduction as it relates the universal origins of human life and growth.
6 For more information about the *Mawangdui* texts, see Li and McMahon (1992), Harper (1998) and Wile (1992), as well as Chapters 3 and 22 in this volume.

7 The Chinese term *fuke* (in modern Chinese, the term for gynaecology) literally means 'a discipline of medicine for women'. I hereafter will adopt '*fuke*' to refer to the learned tradition of medicine for women which today is identified with gynaecology and obstetrics.
8 For a translation, see Wilms (2007).
9 For a brief discussion of the womb in Chinese medicine, see Wu (2010: 92–3). The Chinese word for 'womb' is '*zigong*'(子宫) which literally means 'child palace'.

Bibliography

Atkinson, J.M. and Errington, S. (eds) (1990) *Power and Difference: gender in island Southeast Asia*, Stanford, CA: Stanford University Press.
Bates, D.G. (ed.) (1995) *Knowledge and the Scholarly Medical Traditions*, Cambridge: Cambridge University Press.
Bray, F. (1995) 'A deathly disorder: understanding women's health in late imperial China', in D.G. Bates (ed.) (1995) *Knowledge and the Scholarly Medical Traditions*, Cambridge: Cambridge University Press, pp. 235–50.
—— (1997) *Technology and Gender: fabrics of power in late imperial China*, Berkeley: University of California Press.
Brownell, S. and Wasserstrom, J.N. (eds) (2002) *Chinese Femininities/Chinese Masculinities: a reader*, Berkeley: University of California Press.
Butler, J. (1990) *Gender Trouble: feminism and the subversion of identity*, New York: Routledge.
—— (1993) *Bodies that Matter: on the discursive limits of 'sex'*, New York: Routledge.
Cadden, J. (1993) *Meanings of Sex Difference in The Middle Ages: medicine, science, and culture*, Cambridge: Cambridge University Press.
Duden, B. (1998) *The Woman Beneath the Skin: a doctor's patients in eighteenth-century Germany*, Cambridge, MA: Harvard University Press.
Errington, S. (1990) 'Recasting sex, gender, and power: a theoretical and regional overview', in J. M. Atkinson and S. Errington (eds) *Power and Difference: gender in island Southeast Asia*, Stanford, CA: Stanford University Press, pp. 1–58.
Farquhar, J. (1994) 'Multiplicity, point of view, and responsibility in traditional Chinese healing', in Zito, A. and Barlow, T. E. (eds) *Body, subject and power in China*, Chicago, IL: University of Chicago Press, pp. 78–99.
Fausto-Sterling, A. (1992) *Myths of Gender: biological theories about women and men*, New York: Basic Books.
—— (2000) *Sexing the Body: gender politics and the construction of sexuality*, New York: Basic Books.
Furth, C. (1986) 'Blood, body, and gender: medical images of the female condition in China, 1600–1850', *Chinese Science*, 7: 43–66; reprinted in S. Brownell, S. and J.N. Wasserstrom (eds) (2002) *Chinese Femininities/Chinese Masculinities: a reader*, Berkeley: University of California Press, pp. 291–314.
—— (1994) 'Ming-Qing medicine and the construction of gender difference', *Jindai Zhongguo funü shi yanjiu*, 2: 229–50.
—— (1999) *A Flourishing Yin: gender in China's medical history, 960–1665*, California: University of California Press.
Grant, J. (2003) *A Chinese Physician: Wang Ji and the stone mountain medical case histories*, London: RoutledgeCurzon.
Harper, D.J. (1998) *Early Chinese Medical Literature: the Mawangdui medical manuscripts*, London: Kegan Paul International.
Laqueur, T. (1990) *Making Sex – body and gender from the Greeks to Freud*, Cambridge, MA: Harvard University Press.
Lee, J.D. (Li Zhende) 李貞德 (1997) 'Han Tang zhijian qiuzi yifang shitan – jian lun fuke lanshang yu xingbie lunshu 漢唐之間求子醫方試探—兼論婦科濫觴與性別論述' (Reproductive medicine in late antiquity and early medieval China: gender discourse and the birth of gynaecology), *Bulletin of the Institute of History and Sinology, Academia Sinica*, 68.2: 283–367.
—— (2003) 'Gender and medicine in Tang China', *Asia Major* (3rd series), 16.2: 1–32.
—— (2012) *Xingbie, shenti yu yiliao* 性別，身体与医疗 (Gender, Body and Medicine), Beijing: Zhonghua shuju.

Leo, J. (2011) *Sex in the Yellow Emperor's Basic Questions: sex, longevity, and medicine in early China*, Dunedin: Three Pines Press.

Leung, A.K.C. 梁其姿 (2006a) 'Recent trends in the study of medicine for women in imperial China', in Leung (ed.) *Medicine for Women in Imperial China*, Leiden: Brill, pp. 2–18.

——— (ed.) (2006b) *Medicine for Women in Imperial China*, Leiden: Brill.

Li Jianmin 李建民 (2005) 'Dumai yu Zhongguo zaoqi yangsheng shijian: Qijing ba mai de xin yanjiu zhi er 督脉与中国早期养生实践: 奇经八脉的新研究之二' (The Superintendent Channel and the practice of nourishing life in early China: new research on the eight Extraordinary Channel pulses, part 2), *Bulletin of the Institute of History and Philology, Academia Sinica*, 76.2: 249–314.

Li Ling 李零 and McMahon, K. (1992) 'The contents and terminology of the Mawangdui texts on the arts of the bedchamber', *Early China*, 17: 145–85.

Lloyd, G.E.R. and Zhao J.J. (eds) *Ancient Greece and China Compared*, Cambridge: Cambridge University Press.

Lo, V. and Re'em E. (2018) 'Recipes for love in the ancient world', in G.E.R. Lloyd and J.J. Zhao (eds) *Ancient Greece and China Compared*, Cambridge: Cambridge University Press, pp. 326–52.

Pfister, R. (2016) 'On the partonymy of female genitals in Chinese manuscripts on sexual body techniques', addendum to *Chutu yixue wenxian yanjiu guoji yantaohui lunwenji* 出土醫學文獻研究國際研討會論文集, Shanghai: Shanghai University of Traditional Chinese Medicine.

Raphals, L.A. (1998) *Sharing the Light: representations of women and virtue in early China*, Albany, NY: State University of New York Press.

Reiter, R.R. (ed.) (1975) *Toward an Anthropology of Women*, New York: Monthly Review Press.

Rubin, G. (1975) 'The traffic in women: notes on the "political economy" of sex', in R.R. Reiter (ed.) *Toward an Anthropology of Women*, New York: Monthly Review Press, pp. 157–210.

Scott, J.W. (1986) 'Gender: a useful category of historical analysis', *American Historical Review*, 91.5: 1053–75.

——— (2010) 'Gender: still a useful category of analysis?', *Diogenes*, 57.1: 7–14.

Schiebinger, L.L. (1989) *The Mind Has No Sex?: women in the origins of modern science*, Cambridge, MA: Harvard University Press.

Unschuld, P.U., Tessenow, H. and Zheng Jinsheng (2011) *Huang Di Nei Jing Su Wen an Annotated Translation of Huang Di's Inner Classic – Basic Questions*, 2 vols, Berkeley: University of California Press.

Wile, D. (1992) *Art of The Bedchamber: the Chinese sexual yoga classics including women's solo meditation texts*, Albany: State University of New York Press.

Wilms, S. (2006) '"Ten times more difficult to treat": female bodies in medical texts from early imperial China', in A.K.C. Leung (ed.) *Medicine for Women in Imperial China*, Leiden: Brill, pp. 74–107.

——— (2007) Sun Simiao 孙思邈, *Bèi Jí Qiān Jīn Yào Fāng – Essential Prescriptions Worth A Thousand Gold for Every Emergency*, volumes 2–4 on Gynecology, Portland, OR: Chinese Medicine Database.

Wu, Y.-L. 吳一立 (2010) *Reproducing Women: medicine, metaphor, and childbirth in late imperial China*, Berkeley: University of California Press.

——— (2011) 'Body, gender, and disease: the female breast in late imperial Chinese medicine', *Late Imperial China*, 32.1: 83–128.

Yates, R. (2006) 'Medicine for women in early China: a preliminary survey', in A.K.C. Leung (ed.) *Medicine for Women in Imperial China*, Leiden: Brill, pp. 19–73.

Zito, A. and Barlow, T.E. (eds) (1994) *Body, Subject and Power in China*, Chicago, IL: University of Chicago Press.

24
GYNECOLOGY AND OBSTETRICS FROM ANTIQUITY TO THE EARLY TWENTY-FIRST CENTURY

Yi-Li Wu

Introduction

Scholarship on the history of gynecology and obstetrics in China has been nurtured by the intersection of three fields: 'traditional Chinese medicine' (TCM), women's and gender studies, and science and technology studies. After the People's Republic of China was established in 1949, the government sponsored the creation of TCM, a modernized form of historical practices that was formally integrated into the healthcare system. The growth of TCM has been accompanied and facilitated by a proliferation of studies explaining its foundational ideas and their historical development. This includes histories of the medical subfield known as 'medicine for women' (*fu ke* 婦科) (Ma 1992; Zhang 2000). Meanwhile, scholars have also analyzed Chinese gynecology and obstetrics as a 'technology of gender' (Bray 1997) whose methods and ideas were intertwined with prevailing gender norms and gendered practices (Furth 1987, 1998; Bray 1995, 1997; Leung, ed. 2006; Lee 2008; Wu 2010). While diverse in chronological scope and methodology, these analyses of gender and medicine share an interest in three interrelated sets of questions. The first addresses epistemological issues, namely how people defined what a 'woman' was, how the female body differed from the male, and what methods were most appropriate for treating a woman's special medical needs. In particular, how were beliefs about pregnancy and childbirth related to social and medical norms of womanhood? A second set of questions addresses the socio-institutional patterns that shaped healing activity, especially the relative roles that female and male practitioners played in providing care to women. Also important in this regard is the status of lay healing within the Chinese household, which historically was a woman-managed 'inner sphere' and the place where healing activities typically took place. Finally, a third set of questions pertains to cultural-experiential issues, namely how women have experienced their own bodies, and how those experiences were shaped by gender norms and medical practices. In earlier historical periods, much of this information must be inferred from male-authored sources, but sources expressing women's own voices become much more available beginning in the early twentieth century, whether published in the popular press (Judge 2015; Lin 2013) or in ethnographic accounts and oral histories (Idema and Grant 2004: 542–66; Hershatter 2007, 2011; Ahn 2011).

This chapter surveys scholarship on the epistemological, socio-institutional, and cultural-experiential aspects of women, gender, and medicine during the long and varied history of the polity now known as 'China'. After examining the broad outlines of gynecological and obstetrical care in imperial China from antiquity to the nineteenth century, I discuss the advent of biomedicine and how women's reproductive bodies have been implicated in modern state-building under Republican and Communist regimes in the twentieth and early twenty-first centuries.

The imperial period

A main focus of research has been to understand the massive corpus of extant historical writings on women's diseases. The most comprehensive bibliography of extant Chinese medical books held in Chinese libraries lists over 300 titles of specialized works on women's diseases and childbearing produced up to 1900 (Xue, ed. 2007). Thanks to the growth of TCM, numerous historical texts have been reprinted in modern facsimile or typeset editions, many have been digitized and are full-text searchable on internet sites or in proprietary databases, and some have also been translated into Western languages (Wilms 2007). These texts include both printed works and manuscripts, written and circulated by a range of men (and some women), for reasons ranging from the desire to promote a particular medical theory, to the philanthropic impulse to make useful remedies widely available (Furth 1998; Wu 2010). The tone and content of these texts range from esoteric discussions of *yin-yang* theory to lists of drug recipes meant to be used by laypeople. Thus while the overall corpus tends to foreground classical gynecology and obstetrics as practiced by elite doctors, these texts are sufficiently eclectic and diverse that scholars have also used them to investigate popular and folk practices, as well as forms of religious and ritual healing (Furth 1998; Leung ed. 2006; Lee 2008; Wu 2010). Scholars have also expanded understandings of gender and medicine by reading medical texts in conjunction with literature, legal cases, and historical chronicles (Cullen 1993; Furth 1998; Schonebaum 2004; Lee 2008; Sommer 2010).

The emergence of 'separate prescriptions for women' from the Han (202 BCE–220 CE) to the Tang (618–907) dynasties

China's earliest written records—the 'oracle bones' used by the kings of the Shang dynasty (1766–1054 BCE)—include numerous inscriptions meant to divine the outcome of royal pregnancies (Zhang 2000, 12–16). At some point, people began recording methods for managing gestation and childbirth. The manuscripts excavated from the Han dynasty tombs at Mawangdui include a silk document titled *Book of The Generation of the Fetus* (*Taichan shu* 胎產書), believed to date from the late third or early second century BCE (Harper 1998). It featured diagrams showing where to bury the afterbirth and for calculating the neonates's fate, and discussed conception, monthly fetal development, recommended or prohibited activities during pregnancy, and how to influence the sex of the fetus. Doctors of antiquity also theorized about menstruation, fertility, and female health. The canonical *Yellow Emperor's Inner Classic, Plain Questions* (*Huangdi neijing suwen* 黃帝內經素問), which recorded teachings dating from the first to second centuries BCE, described menarche and menopause as the beginning and end points of a woman's reproductive life. The case records of physician Chunyu Yi 淳于意 (c. 170–150 BCE) show him treating female patients, including one woman with menstrual stoppage (Raphals 1998; Hsu 2010: 84). The physician Zhang Ji 張機 (style name Zhongjing 仲景, late second to early third centuries) wrote about the ailments of pregnancy and postpartum as well as women's 'miscellaneous illnesses' (zabing 雜病) in his *Essentials of the Golden Cabinet* (*Jingui yaolue* 金匱要略). These miscellaneous illnesses, he explained, were caused by pathogenic cold that caused the menses to stop and blood to accumulate in the opening of the womb.

Although many early writings on women's diseases and childbirth are no longer extant (Yates 2006: 27, 35), the great interest in women's diseases is evident in works such as the *Treatise on the Origins and Manifestations of the Myriad Diseases* of 610 (*Zhubing yuanhou lun* 諸病源候論), conventionally credited to Chao Yuanfang 巢元方, an 'Erudite' (*taiyi boshi* 太醫博士) in the state medical service during the Daye reign (605–618) of the Sui dynasty (581–618). It included information on over 1,700 illness manifestations, including problems related to pregnancy, birth, postpartum, and 142 'miscellaneous diseases of women' ranging from noxious vaginal discharges to febrile disorders (Lee 2008). However, the earliest author to articulate an explanation of female bodily difference was the seventh-century polymath and medical expert Sun Simiao 孫思邈 (d. 682). In his influential *Prescriptions Worth a Thousand Gold Pieces for Urgent Situations* (*Beiji qianjin yaofang* 備急千金要方, 652), Sun began his discussion of illnesses with three sections on 'formulæ for women' (*furen fang* 婦人方). 'The reason that there are separate prescriptions for women', Sun began, 'is [because] they differ in experiencing pregnancy, childbirth, and injury from [blood] collapse. This is why the illnesses of women are ten times harder to treat than those of men' (*Beiji qianjin yaofang*, 2.1a-b). Sun's explanation of women's special constitutional tendencies and bodily experiences also depicted them as vulnerable to pathological accumulations of *yin qi*—including reproductive blood—as well as prone to intense and uncontrolled emotions. Thus, he said, 'the roots of [women's] disorders are deep and they are difficult to cure'.

Texts on 'bedchamber techniques' (*fangzhong shu* 房中術) had long taught men how to harness their sexual energy and that of their female partner(s) to enhance fecundity. Between the fifth and seventh centuries, however, the center of attention shifted to the female body, exemplified by an increase of medical writings on promoting female fertility (Lee 2008). Sun's special formulæ for women embody this trend. In addition to beginning with the topic of 'seeking children' (*qiuzi* 求子), Sun explained that promoting female reproductive health was an important aspect of 'nourishing life' (*yangsheng* 養生), a realm of practice conventionally focused on male vitality and fertility (Wilms 2005). In essence, Sun melded attention to the female role in human generation with concern for women's special medical needs.

In such writings, male doctors claimed expert knowledge of women's diseases. The Tang imperial state included an expansive medical bureaucracy that employed both male and female medical practitioners, although the latter were considered subordinate and inferior (Lee 2008). In addition to female palace doctors, Daoist priests and nuns were also active as healers of women. In this competitive environment, the learned male physician constructed his authority through mastery of medical literature. Sun's *Prescriptions* incorporated numerous earlier and contemporary works, and its eclectic range of therapies reflected the rich diversity of Tang dynasty healing practice. For example, the influence of Buddhism and Indian medical ideas appears in Sun's discussion of how to change a female fetus to a male one (Chen 2005). Worries about birthing pollution, which owed much to religious beliefs, co-existed with medical concerns about depletion of female vital energies. While such diversity also colors later writings, Sun's successors focused on expanding the medical approaches to women's reproductive health.

Neo-Confucian gynecology in the Song dynasty (960–1279)

The Song dynasty was characterized by the expansion of scholarly medicine, which also fostered new ways of conceptualizing women's bodies. Emperors and officials with a personal interest in medicine sought to reform medical learning by infusing it with the norms of classical scholarship. The government compiled authoritative versions of medical classics and promoted literate medicine as part of a broader agenda of cultural reform (Goldschmidt 2009; Hinrichs forthcoming). In 1060, 'the division of childbearing' (*chanke* 產科) became one of nine departments in the Song Imperial Medical Bureau. Although elites generally viewed medicine as a lowly technical occupation, a culture of amateur medical connoisseurship flourished (Chen Yuan peng 1996). Upper-class women could also wield medical knowledge, suggesting that such information circulated readily in elite circles (Cheng 2012).

The growth of Neo-Confucianism encouraged literate men—including physicians—to investigate how the primordial patterns of the cosmos manifested themselves in observable phenomena. Neo-Confucian thinkers also taught that human society should conform to these universal principles, and they promoted strict norms of female virtue and chastity. It was in this context that Song doctors sought to understand what made women unique and how best to treat women's diseases. The centerpiece of Song gynecological thinking was a model of gender difference summed up in the saying, 'Women take Blood as the main principle' (*furen yi xue wei zhu* 婦人以血為主). This formulation was rooted in *yin-yang* cosmology, where the *yin* bodies of women had a special affinity for the *yin* form of *qi*—namely blood. At the same time, it focused attention on the menses, now conceptualized as the crux of female health and fertility (Furth 1998). During the Tang, Sun Simiao had placed menstrual ailments under the heading of 'girdle discharges' that evoked the polluting nature of female flows. Now, by contrast, the Song rhetoric of 'regulating menses' integrated female flows into a larger discourse of bodily harmony (Wilms 2006). Notably, the category of blood-related 'women's diseases' also encompassed ailments that were not directly related to pregnancy and birth.

These orientations are most fully elaborated in Chen Ziming's 陳自明 (style name Liangfu 良父; fl. early to mid-thirteenth century) *Comprehensive Compendium of Worthy Prescriptions for Women* (*Furen daquan liangfang* 婦人大全良方) of 1237, which synthesized earlier and contemporary writings with Chen's own experiences as a doctor (Furth 1998). Chen organized his discussion into eight topical categories, each linked to the next in a chain of therapeutic logic. Chen began with 'regulating menses' (*tiaojing* 調經) explaining that 'when providing medical treatment to women, the first necessity is to regulate the menses'. The second category was ailments caused by menstrual irregularity. Only then did Chen address 'seeking descendants' (*qiusi* 求嗣, the topic that Sun Simiao had addressed first), for conception was impossible without menstrual health. The remaining categories addressed 'fetal education' (*taijiao* 胎教, how to produce a high-quality child); the diseases of pregnancy; preparing for delivery; difficult childbirth; and postpartum ailments.

Notably, however, male doctors did not attempt to develop a field of expertise in practical obstetrics. This is apparent in the historical transmission of Yang Zijian's 楊子建 (style name Kanghou 康侯, fl. ca. 1100) 'Discussion of the Ten Forms of Childbirth' (*Shichan lun* 十產論), an essay that survives only because it was included in Chen Ziming's compilation (Furth 1998; Wu 2010; Ng 2013). Yang first described normal, uncomplicated birth, in which the child was in the proper head down position and was delivered smoothly. His subsequent enumeration of birthing complications included four entries on different forms of malpresentation where he explained how to correct the baby's position. Yang's intended readers, however, would not be handling the baby themselves: Yang explained what instructions should be given to the 'birthing attendant' (*kan sheng zhi ren* 看生之人, presumably the midwife), and he warned that disaster would result if the attendant did not have clever hands. Later medical discussions of how to deliver a child generally just quoted Yang, and to the extent that they added new therapeutic information, it was about medicinal formulæ that a doctor could recommend to ease delivery. In addition to this division of labor with the midwife, the Song male doctors shared a stage with the spirits. Chen Ziming's section on preparing for childbirth included charms and hemerological calculations meant to protect the mother and child from demons and inauspicious cosmological alignments. Over the subsequent centuries, as doctors applied new medical doctrines to women's ailments, this attention to ritual medicine diminished in learned texts (Furth 1998).

Rethinking blood and gender in the Jin (1115–1234), Yuan (1271–1368) and Ming (1368–1644) dynasties

In 1127, the Jurchen Jin dynasty conquered the northern half of the Song empire, compelling the Song state to reconsolidate itself in the south. After the Mongols conquered both the Jin (in 1234) and the southern Song (in 1279) in the name of the Yuan dynasty, the reunification of north and south stimulated the interregional exchange of medical ideas. Prominent doctors who lived under Jin or Yuan rule created new, competing doctrines that sought to identify which dynamics were the most important in maintaining health and treating disease. Although they did not write specialized texts on women's diseases, their ideas provided new ways of understanding female blood that profoundly influenced Ming dynasty writings on gynecology (Furth 1998). Particularly influential in this respect was Li Gao 李杲 (style name Dongyuan 東垣, 1180–1251), who argued that the stomach and spleen were the root of all bodily vitalities, since they were responsible for transforming food and drink into qi and blood. Also important was Zhu Zhenheng 朱震亨 (style name Danxi 丹溪, 1282–1358), whose doctrinal innovation was to argue that the body had an inherent tendency toward *yin* deficiency.

When the famous Ming doctor Xue Ji 薛己 (sobriquet Lizhai 立齋 1487–1559) composed a redacted version of Chen Ziming's *Comprehensive Compendium of Worthy Remedies for Women*, he used Li and Zhu's teachings to rework Chen's therapeutic recommendations (Furth 1998; Chapter 9 in this volume). While Chen Ziming favored remedies that acted directly on female blood, Xue's strategy was to treat blood by regulating the spleen and stomach. Xue also deleted the non-reproductive ailments that Chen had gendered as rooted in female blood, thus circumscribing the scope of 'women's diseases' to those issues pertaining directly to women's perceived reproductive functions (Furth 1998; Wu 2010). In effect, Xue and other Ming writers narrowed the perceived scope of female difference in favor of a more androgynous model of the body. These views arose at a time when medical thinkers were taking a renewed interest in 'nourishing life' and enhancing men's generative and procreative energy (Furth 1998, ch. 6). The influence of nourishing life culture in late sixteenth- and early seventeenth-century medicine is especially evident in the belief that men and women both had 'wombs' serving as centers of reproductive energy, an idea that also subsequently appears in some gynecological discussions (Wu 2010; deVries 2012).

But even as male doctors expanded their investigations, female healing experts continued to play important roles. For example, the Ming imperial government included the 'Lodge of Ritual and Ceremony', an office that recruited female doctors, midwives, and wet-nurses to care for court women (Cass 1986). The fact that the Lodge did not train healers but only recruited women who already had expertise shows the wide dissemination of female medical skills. One female doctor whose writings survive was Tan Yunxiang 談允賢 (1461–1554), daughter of a scholarly medical family who wrote a book of medical cases subsequently published by her son (Furth 1998). While elite men often complained about the worrisome influence of the 'three aunties and six grannies' (*sangu liupo* 三姑六婆, female religious specialists and healers, including midwives) on their womenfolk, such complaints also show that female healers were a routine source of care for women (Furth 1998).

Popularizing classical gynecology in the Qing dynasty (1644–1912)

Beginning in the late Ming, a vast expansion of printing helped to popularize scholarly culture, including classical medicine, among a growing urban readership. Many medical works were aimed at literate men who turned to medicine as an alternative profession when they failed to advance in the civil service examinations. The Qing period also saw the proliferation of popular works that readers could use to treat themselves or their relatives. A standard complaint was that norms of gender segregation and female modesty made it difficult for male doctors to interact with female patients. While men did in fact succeed at treating women for various gynecological conditions (Furth 1998, Grant 2003), people also promoted popular gynecology guides as valuable resources for women who were unable or unwilling to see a male doctor (Wu 2000, 2010). One of the most widely circulating Qing texts was the *Treatise on Easy Childbirth* (*Dasheng bian* 達生編) of 1715, which taught readers how to dispense with both doctors and midwives. Authored by a lower-level literatus who identified himself only with the pseudonym 'Lay Buddhist Jizhai', *Easy Childbirth* argued that childbirth was an inherently trouble-free process that recapitulated the spontaneous ease with which the cosmos generated all things (Wu 2010). The woman simply needed to remain calm and bear her labor pains steadfastly, and childbirth would take care of itself. No drugs were needed, Jizhai argued, and the midwife's only role was to catch the baby as it came out. Jizhai was more extreme than other writers in arguing against intervention in childbirth. However, his teachings echoed the medical view that disharmony of maternal qi and blood was a chief cause of difficult labor, and thus that difficult childbirth could be prevented beforehand by proper medical and behavioral regulation (Wu 2010).

While the gynecological literature was meant to promote childbearing, people might also draw on it in an attempt to prevent births. Doctors warned about drugs that would provoke miscarriage, and formulæ designed to 'clear menstrual blockage' (*tongjing* 通經) might be repurposed to end a suspected pregnancy. Although earlier studies of medical literature hypothesized that such techniques were successfully used for fertility control (Bray 1995, 1997; Furth 1998), an investigation of Qing legal cases involving attempted abortion suggests that these methods were ineffective (Sommer 2010). Medical beliefs about the generative power of women's blood also resonated with bodily practices that aimed to transcend reproduction. A distinctive Qing development was the growth of 'female inner alchemy' (*nüdan* 女丹) over the seventeenth to the nineteenth century (Chapter 29 in this volume). Inner alchemy had long been practiced by male adepts, who used visualization and meditation techniques to concentrate and transform essence/semen within their bodies to attain immortality. The female version worked by retaining blood, rather than semen, and one of the signs that the adept was on the path to enlightenment was that her menses ceased altogether. In contrast to pollution narratives that portrayed female blood as defiling, inner alchemy celebrated it as a vital source of transcendence (Valussi 2008).

European medicine in China

During the seventeenth century, Jesuit missionaries at the Qing court compiled texts on European science and medicine aimed at the emperor and high officials. Their aim was to demonstrate the superiority of Western knowledge, and thereby of Western civilization and its Christian religion. The Protestant missionaries that came to China in the decades following the Sino-British Opium War (1839–42) were even more ambitious, employing medical care as a vehicle for evangelizing the masses. Cognizant of Chinese gender norms, missionary societies recruited female physicians and nurses to tend to Chinese women. Furthermore, both female and male medical missionaries established hospitals and dispensaries, including dedicated facilities for women and children which offered perinatal and obstetrical services. The Chinese Christians who studied abroad under missionary auspices also included female medical students. The most famous were Mary Stone (Shi Meiyu) and Ida Kahn (Kong Aide) who obtained their medical degrees from the University of Michigan in 1896 and returned to China to run missionary hospitals for women and children (Shemo 2011).

Protestant missionaries also compiled books on Western medicine, used for training Chinese personnel to work in mission institutions and to disseminate Western knowledge more broadly. The earliest such work was published in 1851 by Benjamin Hobson of the London Missionary Society. Working with Chinese scholars who helped him render his ideas into Chinese (Chan 2012), Hobson eventually produced four medical treatises, including one on 'midwifery and the diseases of children'. In it, Hobson pointed out that in Europe, medical men trained in anatomy were in charge of childbirth, but that the Chinese relied on midwives. While Chinese male doctors were fascinated by Hobson's texts (Wu 2015), there is no evidence that he inspired them to pursue obstetrical practice themselves. During the latter half of the nineteenth century, however, imperialist pressures on China were intensifying just as Western medicine was being transformed by advances in aseptic surgery and microbiology. As Chinese reformers sought to save their country and people from 'racial extinction', they focused on women's roles as producers of Chinese citizens. Promoting successful reproduction became a matter of national importance, and Western biomedical science increasingly appeared to be an essential means to this end.

State-building in the biomedical age

The 1910s witnessed the end of China's imperial age, a failed attempt to establish a democratic republic, and the beginning of intense cultural experimentation and iconoclasm. Drawing on scientific ideas from Europe and North America, Chinese reformers sought to revise older ideas and practices pertaining to the female body, reproduction, sexuality, and childbirth. Only by changing old habits and instilling new scientific practices could a healthy, modern citizenry be created (Dikötter 1995, 1998; Rocha 2010; Zhou 2010). One venue for promoting these ideas was the new literary genre of women's magazines, founded by reformers who sought to create a new Chinese womanhood. In articles, advice columns, advertisements, and images, these journals presented frank biomedical information about menstruation, childbirth, and other reproductive issues (Judge 2015). They exhorted their readers to discard older attitudes that saw such topics as shameful and called on them to talk openly about their bodies in a more 'scientific' way. New ideas included the concept of 'female hygiene' which both borrowed from and effaced older ideas about menstrual regulation. A central pillar of female hygiene was the use of sanitary napkins during menses, promoted by commercial firms such as Kotex as well as by social reformers (Zhou 2010; Lin 2013).

Racial strengthening discourse was also accompanied by renewed attention to the ancient practice of 'fetal education'. While doctors disagreed about the modern utility of fetal education per se, they concurred that maternal thoughts and actions would influence the quality of the unborn child (Richardson 2012, 2015).

Midwifery reform in the Nanjing decade (1927–37)

Beyond worrying about individual women's behavior during pregnancy, reformers sought to create a scienized institutional framework for childbirth. They lamented China's high maternal and infant mortality rate, which they blamed on the ignorance and malpractice of the midwives who delivered China's babies (Johnson 2008, 2011). After the Nationalist Party established nominal control over China in 1927, ushering in a decade of concerted state-building, midwifery reform was included in an expansive program of public health initiatives. Its chief architect was the Chinese-born, Western-trained obstetrician-gynecologist Marion Yang Chongrui (Johnson 2008, 2011). Holder of a Johns Hopkins University medical degree, Yang taught at the Peking Union Medical College. Because there were far too few biomedical physicians in China, Yang focused on creating a corps of modern midwives. In practical terms, modern midwifery meant sterile techniques of delivery and neonatal care. Yang was instrumental in creating the government's National Midwifery Board in 1929, serving as its head. She established the First National Midwifery School in Beijing and hoped to create a network of sixty schools nationwide (Johnson 2008). Unfortunately, the Japanese invasion of China in 1937 and subsequent years of international and civil war curtailed these initiatives.

The People's Republic of China (1949–present)

From its beginning, the Communist government was ideologically committed to improving the lives of the people, which included reducing maternal and infant mortality. Public health was now also linked to the political goal of creating new socialist citizens. The government recruited female healthcare workers to promote prenatal care, retrained midwives in scientific (sterile) childbirth, and sought to convince women to employ new practices and abandon old ones. Simultaneously, officials sought to replace local, woman-centered, family-based practices with a new 'ideal of socialist childbirth' carried out by medical cadres in the name of state-building (Goldstein 1998). Midwives were now subordinated to government medical workers and institutions. Practices such as 'fetal education' were deemed to be feudal superstition. Scholars have sought to understand how such campaigns affected birthing practices or shaped the experiences of practitioners and parturient women (Goldstein 1998; Hershatter 2007, 2011; Ahn 2011, 2013). For example, Communist Party propaganda decried midwives as a feudal legacy of the old society. In practice, however, midwives were treated with respect, and government health workers tried to learn from them even while attempting to reform them. The gruesome complications of birth as well as childbirth-related deaths were a fact of life for the older generation, but midwifery reform succeeded in making these less and less common (Hershatter 2011).

The politicization of childbirth reforms is epitomized by Chinese attempts to adopt 'Soviet psychoprophylactic painless childbirth' (popularized in Europe and North America as the Lamaze method) (Goldstein 1998; Ahn 2013). This method taught that normal childbirth was not inherently painful, but that women had been socially conditioned to associate labor contractions with pain. In properly trained women, therefore, pain would disappear from normal birthing experiences, appearing only if childbirth became complicated. Beginning in 1952, the Chinese Ministry of Health ordered obstetrical facilities to promote this Soviet method. Patronage of 'bourgeois' Western medicine was already politically sensitive, and it became especially fraught after 1952, when China accused the U.S. of using biological warfare in Korea. In this context, Soviet techniques provided an 'ideological weapon' for discrediting Western science and affirming the power of socialist knowledge (Ahn 2013). But Chinese advocates exaggerated the claims that the Soviets had originally made for psychoprophylactic painless childbirth and claimed that it would be universally effective, even with only perfunctory training, as long as practitioners and patients had sufficient faith in socialist medicine. The cadres charged with promoting it had no medical expertise, and obstetricians and midwives seem to only have paid lip service to the practice, which faded after Sino-Soviet relations soured in 1958 (Ahn 2013).

Obstetrics and gynecology in the period of economic reforms (1978– present)

In 1978, the post-Mao Chinese leadership initiated economic reforms to decollectivize, decentralize, and privatize large sectors of the socialist economy. These reforms undergirded three decades of rapid growth that transformed China into the world's second largest economy. Reform-era policies also had significant consequences for women's health. Alarmed by what they perceived to be a looming overpopulation crisis that would cripple economic growth, Chinese leaders rejected more moderate policies and adopted the seemingly 'scientific' models of cyberneticists who argued that all couples should be limited to one child (Greenhalgh 2008). A one-child policy was formally codified in 1980 and remained in effect, with very few allowable exceptions, until concerns over demographic distortions compelled the government to replace it with universal two-child policy, effective as of January 1, 2016 (Wang et al. 2016). During its lifetime, the human costs of the one-child policy fell heavily on the bodies of women and girls. Women felt enormous social pressure to terminate non-authorized pregnancies or to be sterilized after their one child was born, and officials were known to coerce reluctant women into such procedures (Nie 2005; Wang et al. 2016: 967). Son-preference also led to female infanticide and sex selective abortion, starkly illustrated in years of abnormal sex ratios at birth (SRB). While the expected SRB is 106 (the number of boys born for every 100 girls), China's SRB reached a high of 121 in 2004. Although it declined during 2008–2014 to 115.8, the SRB remains a concern (National Health and Family Planning Commission 2015).

Meanwhile, economic reforms included dismantling the communes which had been the major source of rural medical care. The central government also cut its funding to hospitals and medical centers which now became dependent on local tax revenues as well as their own ability to generate income (Blumenthal and Hsiao 2005). This led to rapidly widening health disparities between richer, urban localities, and poorer, rural ones, including in the areas of prenatal and emergency obstetrical care. A report co-authored by China's Ministry of Health reported that in 2004, the rural rates of maternal mortality were over three times higher than urban ones. Furthermore, it estimated that about three-fourths of these maternal deaths could have been prevented with access to the necessary medical care (Ministry of Health et al. 2006). Since the mid-1990s, the government's key strategy for reducing infant and maternal mortality has been to increase the percentages of births taking place in hospitals. This policy was accompanied by the discontinuation of midwifery training programs and campaigns against home births. By 2008, close to 95% of births were hospital births, except in the remotest rural regions (Feng *et al.* 2011: 433).

While the hospital birth policy has reduced overall maternal and infant mortality rates (Feng *et al.* 2011; Liang *et al.* 2012), there are still persistent regional differences in the quality and accessibility of maternal care. Furthermore, hospital births often involve interventions that carry their own risks. Rates of caesarian delivery surged during the 1990s, and a WHO study reported that China's rate of caesarian delivery in 2007 was 46.2% of all births (Lumbiganon *et al.* 2010). A main factor is the high percentage of elective caesarians (Feng *et al.* 2011; Hellerstein *et al.* 2015). Women and doctors alike share the belief that caesarian deliveries are more efficient and predictable—and thus safe—than vaginal births, a compelling argument when families could only have one child. The state-mandated fee structure and insurance system also makes caesarians more profitable than vaginal births. To lower the high caesarian rates, numerous researchers and institutions are seeking ways to promote vaginal delivery and to mitigate the over-medicalization of childbirth in hospitals (Liu *et al.* 2016).

In the wake of the WHO report, and as the government was phasing in a two-child policy, it also revived its support of midwifery, establishing university degree programs in the subject as part of a broader campaign to increase the number of trained obstetrical staff (National Health and Family Planning Commission 2014). State media also praised midwifery as a way to reduce the caesarian delivery rate (China Daily 2014). Meanwhile, the universal two-child policy means that some 90 million additional couples are now entitled to have a second child, straining existing medical resources (Cheng and Duan 2016). How women and policy makers negotiate and influence healthcare choices in this new era remains to be seen. However, the twenty-first century is already proving to be another thought-provoking chapter in the ongoing history of gender, medicine, and society in China.

Bibliography

Pre-modern sources

Beiji qianjin yaofang 備急千金要方 (Prescriptions Worth a Thousand Gold Pieces for Urgent Situations) 652, Sun Simiao 孫思邈, reprinted in *Siku quanshu* (Complete Libraries of the Four Treasuries), Wenyuange edition, 1782.

Modern sources

Ahn, B. (2011) *Modernization, Revolution, and Midwifery Reforms in Twentieth-Century China*, PhD thesis, University of California Los Angeles.
——— (2013) 'Reinventing scientific medicine for the socialist republic: the Soviet psycho-prophylactic method of delivery in 1950s China', *Twentieth-Century China*, 38.2: 139–55.
Bates, D. (ed.) (1995) *Knowledge and the Scholarly Medical Traditions*, Cambridge: Cambridge University Press.
Blumenthal, D. and Hsiao, W. (2005) 'Privatization and its discontents–the evolving Chinese health care system', *New England Journal of Medicine*, 353.11: 1165–70.
Bray, F. (1995) 'A deathly disorder: understanding women's health in late imperial China', in D. Bates (ed.) *Knowledge and the Scholarly Medical Traditions*, Cambridge: Cambridge University Press, pp. 35–50.
——— (1997) *Technology and Gender: fabrics of power in late imperial China*, Berkeley: University of California Press.
Brown, J. and Pickowicz, P.G. (eds) (2007) *Dilemmas of Victory: the early years of the People's Republic of China*, Cambridge, MA; London: Harvard University Press.

Cass, V.B. (1986) 'Female healers in the Ming and the lodge of ritual and ceremony', *Journal of the American Oriental Society*, 106.1: 233–40.

Chan, M.S. (2012) 'Sinicizing western science: the case of *Quanti xinlun*', *T'oung Pao*, 98: 528–66.

Chen Ming 陳明 (2005) 'Zhuan nü wei nan turning female to male: an Indian influence on Chinese gynecology', *Asian Medicine: Tradition and Modernity*, 1.2: 315–34.

Chen Yuanpeng 陳元朋 (1996) *Liangsong de 'shangyi shiren' yu 'ruyi': Jianlun qi zai Jin Yuan de liubian* 兩宋的「尚醫士人」與「儒醫」:兼論其在金元的流變 ('Elites Who Esteemed Medicine' and 'Confucian Doctors' in the Northern and Southern Song Dynasties, with a Discussion on Their Spread and Transformation during the Jin and Yuan Dynasties), MA thesis, Taiwan National University.

Cheng, H.-W. (2012) *Traveling Stories and Untold Desires: female sexuality in Song China, 10th–13th Centuries*, PhD thesis, University of Washington.

Cheng, P.-J. and Duan, T. (2016) 'China's new two-child policy: maternity care in the new multiparous era', *BJOG*, 123.53: 7–9.

Chiang, H. (ed.) (2015) *Historical Epistemology and the Making of Modern Chinese Medicine*, Manchester: University of Manchester Press.

China Daily (2014) 'Program would solve shortage of midwives', June 12, www.china.daily.com.cn/china/2014-06/12/content_17580601.htm, accessed 25/9/2016.

Cullen, C. (1993) 'Patients and healers in late imperial China: evidence from the *Jinpingmei*', *History of Science*, 31: 99–150.

de Vries, L. (2012) *The Gate of Life: before heaven and curative medicine in Zhao Xianke's Yiguan*, PhD thesis, Universiteit Gent.

Dikötter, F. (1992) *The Discourse of Race in Modern China*, Stanford, CA: Stanford University Press.

——— (1995) *Sex, Culture and Modernity in China*, Honolulu: University of Hawai'i Press.

——— (1998) *Imperfect Conceptions: medical knowledge, birth defects, and eugenics in China*, New York: Columbia University Press.

Feng, X.-L., Xu, L., Guo, Y. and Ronsmans, C. (2011) 'Socioeconomic inequalities in hospital births in China between 1988 and 2008', *Bulletin of the World Health Organization*, 89: 432–41.

Furth, C. (1986) 'Blood, Body, and Gender: medical images of the female condition in China', *Chinese Science*, 7: 48–51.

———. (1987) 'Concepts of pregnancy, childbirth, and infancy in Ch'ing dynasty China', *Journal of Asian Studies*, 46.1: 7–35.

——— (1998) *A Flourishing Yin: gender in China's medical history, 960–1665*, Berkeley: University of California Press.

Goldschmidt, A. (2009) *The Evolution of Chinese Medicine: Song dynasty, 960–1200*, Abingdon: Routledge.

Goldstein, J. (1998) 'Scissors, surveys, and psycho-prophylactics: prenatal health care campaigns and state building in China, 1949–1954', *Journal of Historical Sociology*, 11.2: 153–84.

Grant, J. (2003) *A Chinese Physician: Wang Ji and the 'Stone Mountain medical case histories'*, London; New York: Routledge Curzon.

Greenhalgh, S. (2008) *Just One Child: science and policy in Deng's China*, Berkeley: University of California Press.

Harper, D. (1998) *Early Chinese Medical Literature: the Mawangdui medical manuscripts*, London; New York: Keegan Paul International.

Hellerstein, S., Feldman, S. and Duan, T. (2015) 'China's 50% caesarean delivery rate: is it too high?', *BJOG*, 122: 160–5.

Hershatter, G. (2007) 'Birthing stories: rural midwives in 1950s China', in J. Brown and P.G. Pickowicz (eds) *Dilemmas of Victory: the early years of the People's Republic of China*, Cambridge, MA; London: Harvard University Press, pp. 337–58.

——— (2011) *The Gender of Memory: rural women and China's collective past*, Berkeley; London: University of California Press.

Hinrichs, TJ (forthcoming) *Shamans, Witchcraft, and Quarantine: the medical transformation of governance and southern customs in mid-imperial China* (Harvard East Asia Series).

Hsu, E. (2010) *Pulse Diagnosis in Early Chinese Medicine: the telling touch*, Cambridge: Cambridge University Press.

Idema, W. and Grant, B. (2004) *The Red Brush: writing women of imperial China*, Cambridge, MA: Harvard University Asia Center and Harvard University Press.

Johnson, T.P. (2008) 'Yang Chongrui and the first national midwifery school: childbirth reform in early twentieth-century China', *Asian Medicine*, 4: 280–302.

——— (2011) *Childbirth in Republican China: delivering modernity*, Lanham, MD: Lexington Books.

Judge, J. (2015) *Republican Lens: gender, visuality, and experience in the early Chinese periodical press*, Berkeley: University of California Press.

Li Zhende 李貞德 (J.D. Lee) (2008) *Nüren de zhongguo yiliao shi: Hantang zhi jian de jiankang zhaogu yu xingbie* 女人的中國醫療史：漢唐之間的健康照顧與性別 (A Women's History of Chinese Medicine: health care and gender from the Han to the Tang dynasties), Taipei: Sanmin shuju.

Leung, A.K.C. (ed.) (2006) *Medicine for Women in Imperial China*, Leiden: Brill.

——— (1999) 'Women practicing medicine in premodern China', in H.T. Zurndorfer (ed.) *Chinese Women in the Imperial Past: new perspectives*, Leiden: Brill, pp. 101–34.

Liang, J., X. Li, L. Dai et al. (2012) 'The changes in maternal mortality in 1000 counties in mid-western China by a government-initiated intervention', *PLOS ONE*, 7.5: e37458.

Lin, S.-T. (2013) '"Scientific" menstruation: the popularization and commodification of female hygiene in Republican China, 1910s–1930', *Gender and History*, 25.2: 294–316.

Liu, X, Lynch, C.D., Cheng, W.W. and Landon, M.B. (2016) 'Lowering the high rate of caesarean delivery in China: an experience from Shanghai', *BJOG*, 123: 1620–28.

Lumbiganon, P., Laopaiboon, M., Gülmezoglu, A.M. *et al.* (2010) 'Method of delivery and pregnancy outcomes in Asia: the WHO global survey on maternal and perinatal health, 2007–08', *The Lancet*, 375: 490–99.

Ma Dazheng 马大正 (1992) *Zhongguo fuchanke fazhan shi* 中国妇产科发展史 (The History of the Development of Gynecology and Obstetrics in China), Taiyuan: Shanxi kexue jiaoyu chubanshe.

Ministry of Health of the People's Republic of China, UNICEF, World Health Organization, and United Nations Fund for Population Activities (2006) *Joint Review of the Maternal and Child Survival Strategy in China*, Beijing: Ministry of Health.

National Health and Family Planning Commission of the People's Republic of China (2014) 'Guojia weisheng jisheng wei guanyu zuo hao xin xinshi xia fu you jian kang fuwu gongzuo de zhidao yijian 国家卫生计生委关于做好新形势下妇幼健康服务工作的指导意见' (The National Health and Family Planning Commission's Guiding Opinions on How to Properly Carry Out Maternal and Child Health Services in the New Situation), June 19, http://www.nhfpc.gov.cn/fys/s3581/201406/fb7b841983a2460e92b350bef532a6a1.shtml, accessed 25/9/2016.

——— (2015) 'Woguo chusheng renkou xingbie bi shixian "Liu lian jiang" 我国出生人口性别比实现"六连降"' (China's Sex Ratio at Birth Has Experienced 'Six Continual Years of Decline'), 3 February, http://www.nhfpc.gov.cn/jtfzs/s3578/201502/ab0ea18da9c34d7789b5957464da51c3.shtml, accessed 25/9/2016.

Ng, W.-S. M. (2013) *Male Brushstrokes and Female Touch: medical writings on childbirth in imperial China*, PhD thesis, McGill University.

Nie, J.-B. (2005) *Behind the Silence: Chinese voices on abortion*, Lanham, MD: Rowman and Littlefield.

Raphals, L. (1998) 'The treatment of women in a second-century medical casebook', *Chinese Science*, 15: 7–28.

Richardson, N. (2012) 'The nation in utero: translating the science of fetal education in Republican China', *Frontiers of History in China*, 7.1: 4–31.

——— (2015) *A Nation in Utero: pregnancy and fetal education in early Republican China, 1912–1937*, PhD thesis, University of California, Davis.

Rocha, L.A. (2010) *Sex, Eugenics, Aesthetics, Utopia in the Life and Work of Zhang Jingsheng (1888–1970)*, PhD thesis, University of Cambridge.

Schonebaum, A.D. (2004) *Fictional Medicine: diseases, doctors, and the curative properties of Chinese fiction*, PhD thesis, Columbia University.

Shemo, C.A. (2011) *The Chinese Medical Ministries of Kang Cheng and Shi Meiyu, 1872–1937: on a cross-cultural frontier of gender, race, and nation*, Bethlehem, PA: Lehigh University Press.

Sommer, M. (2010) 'Abortion in late imperial China: routine birth control or crisis intervention?' *Late Imperial China*, 31.2: 97–165.

Valussi, E. (2008) 'Blood, tigers, dragons: the physiology of transcendence for women', *Asian Medicine: Tradition and Modernity*, 4.1: 46–85.

Wang, Z., Yang, M., Zhang, J-M. and Chang, J. (2016) 'Ending an era of population control in China: was the one-child policy ever needed?' *American Journal of Economics and Sociology*, 75.4: 929–79.

Wilms, S. (2005) 'The transmission of medical knowledge on 'nurturing the fetus' in early China', *Asian Medicine: Tradition and Modernity*, 1.2: 276–314.

—––– (2006) '"Ten times more difficult to treat": female bodies in medical texts from early imperial China', in A.K.C. Leung (ed.) *Medicine for Women in Imperial China*, Leiden: Brill, pp. 74–107.

—––– (2007) *Sun Simiao, Bei Ji Qian Jin Yao Fang: essential prescriptions worth a thousand gold for every emergency, volumes 2–4 on gynecology*, Portland, OR: The Chinese Medicine Database.

Wu, Y.-L. (2000) 'The bamboo grove monastery and popular gynecology in Qing China', *Late Imperial China*, 21.1: 41–76.

—––– (2010). *Reproducing Women: medicine, metaphor, and childbirth in late imperial China*, Berkeley: University of California Press.

—––– (2015) 'Bodily knowledge and western learning in late imperial China: the case of Wang Shixiong (1808–68)', in H. Chiang (ed.) *Historical Epistemology and the Making of Modern Chinese Medicine*, Manchester: University of Manchester Press, pp. 80–112.

Xue Qinglu 薛清录 (ed.) (2007) *Zhongguo zhongyi guji zongmu* 中国中医古籍总目 (General Catalogue of Ancient Medical Books on Chinese Medicine), Shanghai: Shanghai cishu chubanshe.

Yates, R. (2006) 'Medicine for women in early China: a preliminary survey', in A.K.C. Leung (ed.) *Medicine for Women in Imperial China*, Leiden: Brill, pp. 19–73.

Zhang Zhibin 张志斌 (2000) *Gudai zhongyi fuchanke jibing shi* 古代中医妇产科疾病史 (History of Diseases in Ancient Chinese Gynaecology and Obstetrics), Beijing: Zhongyi guji chubanshe.

Zhou Chunyan 周春燕 (2010) *Nüti yu guozu: qiang guo qiang zhong yu jindai Zhongguo de funü weisheng (1895–1949)* 女體與國族: 強國強種與近代中國的婦女衛生 (Women's Bodies and the Nation: strengthening the country, strengthening the race, and female hygiene in modern China), Taipei: Guoli zhengzhi daxue lishixue xi.

Zurndorfer, H.T. (ed.) (1999) *Chinese Women in the Imperial Past: new perspectives*, Leiden: Brill.

25
THE QUESTION OF SEX AND MODERNITY IN CHINA, PART 1
From *xing* to sexual cultivation

L. A. Rocha

The discourse of *Xing* 性: modern terms for modern ideas

Perhaps the best starting point for thinking about the question of sex and modernity in China is the very word 'sex' (Dikötter 1995: 68; Zhong 2000: 54; Farquhar 2002: 250–5; Sang 2003: 102–6; Li 2004; Rocha 2010a). In modern Chinese, the character *xing* 性 is most frequently used to denote matters related to sexual practices, gender difference, and reproductive anatomy and physiology. Open any present-day Chinese dictionary and one would be able to find numerous compounds containing *xing* 性, such as *xingjiao* 性交 (sexual intercourse or coitus); *xingbie* 性別 (sexual difference or gender); *nüxing* 女性 (woman or female); *tongxinglian* 同性戀 (same-sex love or homosexuality) and *xing qiguan* 性器官 (sex organs or genitalia). Often overlooked, however, is the historical fact that in classical Chinese the character *xing* 性 generally referred to an innate 'human nature' or essence, which had no necessary connection to sexuality. It was only around the early twentieth century that the character *xing* 性 came to signify simultaneously sex *and* human nature. This was a watershed development in the discourse on sexuality in China because, effectively, sex became part of human nature, something built-in to all human beings, as basic as the appetite for food and drink. It stood in contrast to Confucian philosophy whereby, for Confucius and Mencius, for example, human nature is defined as the disposition to perform good deeds.

As with most neologisms – in this case a new meaning becoming attached to an existing word – it is difficult to pinpoint exactly who first used the Chinese character *xing* 性 to mean sex. It is most likely to have originated in Japanese intellectuals' encounter with Western science and medicine, specifically German *Sexualwissenschaft*, which was imported into Japan from the late Meiji period (1868–1912) onwards (Oda 1996: 54–67; Frühstück 2003: 83–115; Sang 2003: 102–3). This usage then travelled back to China via Chinese translations of Japanese translations of European sexological writings, a process that historians and linguists have labelled 'return graphic loan' (Liu 1995: 32–4). By the late 1910s and early 1920s, quantities of books and articles written by Chinese intellectuals began to use *xing* 性 to refer to human sexuality and reproduction in an objective, respectable, value-neutral, matter-of-fact, consciously modern fashion. Older words that carried connotations of excess or obscenity (such as *se* 色 and *yin* 淫), or euphemistic terms that contained rich literary and historical allusions, were gradually displaced in the intellectual, medico-scientific realm.

This linguistic shift could be situated squarely in the context of the 'May Fourth New Culture Movement' (*Wusi xin wenhua yundong* 五四新文化運動), that took place around the mid-1910s to mid-1920s (Chow 1960; Schwarcz 1986; Doleželová-Velingerová and Král 2000; Mitter 2004; Chow *et al.* 2008). The Movement was named after the mass demonstration in Tian'anmen Square in Beijing on 4 May 1919. In the aftermath of the First World War, regions in China occupied by the Germans were, under the Treaty of Versailles, handed over to the Allies and the Japanese. The protesters, led by Chinese intellectuals and students, voiced their anger and frustration at China's repeated 'humiliation' by the foreign powers. To rescue China from the forces of imperialism and colonialism, many Chinese thinkers of the May Fourth Movement embarked on an audacious, nationalist mission to build a 'New Culture'. Concurrently, they wanted to demolish 'Confucianism'; 'feudalist morality' and the traditions and 'superstitions' that were, according to these iconoclasts, responsible for China's backwardness *vis-à-vis* the West. In terms of body and sexuality, probably the most visible sign of the brutality of China's 'Old Culture' was the subjugation of women through foot-binding (Wang 2002; Ko 2005). For this 'May Fourth' project, Chinese intellectuals translated, creatively appropriated and distributed a tremendous range of new knowledge from Europe, United States and Japan – Lydia Liu called this 'translated modernity' (Liu 1995).

The 'May Fourth' generation of thinkers frequently endorsed Social Darwinism; racial science and eugenics. They believed that the quality of China's population had to be urgently improved; otherwise, the Chinese would risk further degeneration or even extinction (Pusey 1983; Dikötter 1992, 1998; Chung 2002; Sakamoto 2004). Numerous proposals and solutions were debated in intellectual magazines and newspapers: Chinese women had to be emancipated from traditional family structures and values; sexual customs and ethics had to be thoroughly scrutinised and fundamentally reconfigured; and reproduction had to be regulated in some manner, possibly through birth control or the implementation of eugenic policies, to stop the 'weak' and 'pathological' people from procreating. Some Chinese intellectuals even suggested intermarriage with the white race, so that the Chinese could 'ascend' the racial pyramid (Teng 2013: 112–34). The adoption of the character *xing* 性 reflected the conceptualisation of sex as something central to human nature, a basic instinct that had been denied in the 'Old Culture' of China. The emergence of the discourse of *xing* 性 at this historical juncture coincided with the intensification of the proliferation of 'truthful' knowledge about human sexuality by Chinese intellectuals in their endeavours to renew and modernise the nation. To borrow Michel Foucault's phrasing, sex in early twentieth-century China became regarded as 'a kind of natural given which power tried to hold in check' and 'an obscure domain which knowledge tried gradually to uncover' (Foucault 1978: 105).

The case of 'doctor sex': the modernisation of intimacy

To illustrate further these cultural and political dynamics, we can take the emblematic case of Zhang Jingsheng 張競生 (1888–1970), one of the most notorious Chinese intellectuals from the Republican period (Leary 1994; Peng 1999; Zhang Peizhong 2008; Rocha 2010b, 2015). Zhang was a philosopher who received his doctoral degree from the University of Lyon for his dissertation on Rousseau's pedagogical thought, and was a professor at Peking University in the 1920s. In 1926, Zhang published *Sex Histories* (*Xingshi* 性史), one of the most sensational and controversial books in the Republican period, which earned him the nickname 'Doctor Sex' (*Xing boshi* 性博士). *Sex Histories* contained seven sexually explicit autobiographical confessions that Zhang Jingsheng solicited from the public. As befitting of a disciple of Rousseau, Zhang believed that the modernisation of the Chinese psyche had to begin with the liberation of the sexual self via confession. At the end of each confession in *Sex Histories*, Zhang appended his sexological commentaries, written in the vein of British sexologist Havelock Ellis' (1859–1939) magnum opus *Studies in the Psychology of Sex* (1897–1928, 6 volumes). Zhang presented his book as a self-help manual that would teach China about sexual hygiene. He prescribed a 'correct path of sex' (*xing de zhenggui* 性的正軌) that, he claimed, would lead to a fulfilling marital relationship; produce quality offspring and prevent the spread of 'perverted' and 'deviant' acts, such as masturbation and homosexuality, that were corrupt wastages of a person's vitality (Zhang 2006, 2008: 154; Chapter 22 in this volume). In short, *Sex Histories* was supposed to push its readers towards modernity and civilisation by revolutionising their intimate practices.

Zhang Jingsheng's 'correct path of sex' revolved around his concept of the 'Third Kind of Water' (*disanzhong shui* 第三種水). Zhang suggested that a woman's genitals could secrete three kinds of fluids: from the labia, the clitoris and the Bartholin's glands. He argued that a woman had to release all three kinds of fluids in copious amounts, via extensive foreplay and penetration. All sexual secretions had to be absorbed for their health-promoting effects; a woman had to absorb a man's semen and the man had to absorb the precious 'Third Kind of Water' from his female partner. Moreover, Zhang recommended that a man had to discipline himself to delay ejaculation so that his orgasm could coincide exactly with the woman's climax and her release of the 'Third Kind of Water'. A child conceived at the moment when a man and a woman reached simultaneous orgasm, with the ovum and the sperm surrounded by the 'Third Kind of Water', would be physically stronger and more intelligent. By regularly engaging in such a 'perfect', pleasurable form of intercourse, Zhang argued, couples would feel physically and psychologically satisfied, and husbands would, therefore, be less likely to seek out prostitutes and to contract venereal diseases, like syphilis, while wives would become less prone to developing 'hysteria'. In one stroke, Zhang connected 'good sex' (always already heterosexual and heteronormative) with a couple's eugenic duty to improve China's racial stock, and with public health and the elimination of social ills (Zhang 1927, 2006: 80–4).

As bizarre and pseudoscientific as Zhang Jingsheng's theory might sound, the 'Third Kind of Water' could be placed in the global genealogy and traffic of sexological ideas. Zhang claimed that he found inspiration for his ideas from Marie Stopes' (1880–1958) famous work from 1918, *Married Love* (Stopes 2004; Zhang 2006: 80–1). In Stopes' version of the 'ideal' marital sex, there would be penile-vaginal penetration over a lengthy period of time, culminating in mutual and simultaneous orgasm. After the climax, the couple had to continue to lie in a 'coital embrace' or 'locking position' so that the man and the woman could both absorb all the highly beneficial sexual secretions (Jackson 1994: 138–9; Stopes 2004: 54–5; Jagose 2013: 58–9). References to the eugenic function of synchronised orgasm could also be found in a wide range of marriage manuals from late nineteenth- and early twentieth-century Europe and America (Gordon 1971; McLaren 1999). Nevertheless, Zhang Jingsheng's contemporaries, for instance, the science populariser and journalist Zhou Jianren 周建人 (1888–1984) and the sociologist and eugenicist Pan Guangdan 潘光旦 (1899–1967), were extremely hostile towards the theory of the 'Third Kind of Water'. Essentially, they argued that Zhang Jingsheng's ideas were an odd mash-up of 'Daoist' practices which were totally contrary to modern scientific understanding of the human body. They questioned Zhang Jingsheng's credentials and his motives for publishing *Sex Histories*; Zhang was attacked for peddling pornographic nonsense for financial gain (Leary 1993; Lee 2007: 186–219; Larson 2009: 55–9).

One might point out the similarities between Zhang Jingsheng's theory of the 'Third Kind of Water' and what came to be construed by twentieth-century authors as 'Daoist' inner alchemy. For instance, a cornerstone of sexual cultivation practice was the so-called 'coitus reservatus' (Van Gulik 1961; Needham 1983; Wile 1992; Rocha 2012; Chapter 22 in this volume). Some forms of this involved a male practitioner having intercourse with as many young maidens as possible in order to absorb their *yin* essence (*yinqi* 陰氣), and at the same time preventing ejaculation and 'returning the semen to the brain' (*huanjing bunao* 還精補腦). Another phrase to describe this practice was 'harnessing the *yin* essence [from women] to nourish the *yang* essence [of the man]' (*caiyin buyang* 採陰補陽). The difference between Zhang's ideal sex and *coitus reservatus* was that Zhang emphasised the mutual, profitable transaction of genital fluids between men and women. Moreover, 'bedchamber techniques' (*fangzhong shu* 房中術) were centred on macrobiotics, specifically the promotion of health and longevity of the *male* practitioner, while sexual pleasure and reproduction were secondary if not peripheral issues (on Female Alchemy, see Chapter 30 in this volume). Zhang Jingsheng was adamant that his ideas were based on cutting-edge sexological knowledge in the West, and that the 'Third Kind of Water' was not a synthesis or attempt at reconciling Chinese traditions and modern science. The circulation of knowledge around the world is always embroiled with the politics of deciding what is 'old' ('traditional', 'indigenous', 'Chinese', etc.) and what is 'new' ('modern', 'scientific', 'Western', etc.).

The fate of tradition: sexual alchemy in modern China

This leads to the question of the complicated fate of these practices of inner alchemy and sexual cultivation in modern China, which were misunderstood to be 'Daoist' although the historical sources we have are in the main non-sectarian (Chapter 22 in this volume). The most crucial textual source of these practices, until the discovery of the Mawangdui 馬王堆 silk texts in 1973, was compiled in the early twentieth century. This was the *Shadow of the Double Plum Tree Anthology* (*Shuangmei jing'an congshu* 雙梅景闇叢書, 1903–1917) by the late Qing literatus Ye Dehui 葉德輝 (1864–1927) (Du and Zhang 1986; Zhang Jingping 2008; Rocha 2015). Ye was a member of the Hunanese elite, who had a short-lived career as a Beijing official. Controlling major agricultural and commercial enterprises in his native Changsha, Ye amassed considerable wealth that financed one of the largest collections of rare books and manuscripts at the time in China. He fashioned himself as a stalwart of Chinese traditions, and staunchly defended the social and political organisation of the Qing dynasty.

Arguing that everything that was wise and useful had already been said millennia ago by Chinese sages, the decay and decline of China for Ye was a result of the Chinese people's deviation from tradition. Therefore, Ye advocated a return to the classic Chinese texts and the rejection of Western knowledge *tout court*, in sharp contrast to his intellectual nemeses, the reformists Kang Youwei 康有為 (1858–1927) and Liang Qichao 梁啟超 (1873–1929) who called for the strategic appropriation of Western knowledge to strengthen China. Ye Dehui's conservative, anti-Western, mentality animated all his scholarly projects, which included the so-called 'elementary studies' (*xiaoxue* 小學) and 'bibliographic studies' (*banben mulu xue* 版本目錄學). The former combined phonetics, semantics and etymological analysis to determine the correct pronunciation; appearance and meanings of words. This in Ye's intellectual outlook would ensure the accurate transmission of ideas. The latter was concerned with the authentification, restoration and preservation of classical Chinese texts. Ironically, these endeavours were enabled in Ye's case by modernity itself, which made possible his long-range network of book-collectors, traders, Sinologists and Sinophiles that stretched as far as Japan.

In the early 1900s, Ye serendipitously discovered the oldest surviving medical work from Japan, called *The Core Prescriptions of Medicine* (in Japanese *Ishinpō* 医心方 or in Chinese *Yixinfang* 醫心方, 984). Edited by Tamba Yasuyori 丹波康賴 (912–995), *Ishinpō* was a compilation containing fragments and quotations from various Chinese texts (Gao *et al.* 1996; Pfister 2012). And from *Ishinpō*, Ye Dehui reconstructed four texts – *Classic of the Plain Girl* (*Sunü jing* 素女經); *Secrets of the Jade Chamber* (*Yufang mijue* 玉房秘訣); *Essentials of the Jade Chamber* (*Yufang zhiyao* 玉房指要) and *Master Dongxuan* (*Dongxuan zi* 洞玄子) – which could be dated to the Sui and Tang Dynasties (581–907) or even as early as the Western Han dynasty (202 BCE–9 CE). For Ye, the sexual alchemy described in these treatises was superior to European and American sexologies and anticipated eugenics and birth control; his chief motive for publishing these texts as part of the *Shadow of the Double Plum Tree Anthology* was precisely to resist modern science and Western medicine (Ye 1903–1917; Furth 1994: 130).

It is difficult to assess what kind of impact Ye Dehui's *Shadow of the Double Plum Tree Anthology* had in the late Qing and Republican periods and, without available sources, impossible to say how many seriously followed these bedchamber techniques. It appears that Daoist adepts at that time generally favoured the sublimation or suppression of carnal desire, and marginalised or even attacked the tradition of sexual cultivation. Take, for example, Chen Yingning 陳攖寧 (1880–1969), arguably the most influential Daoist adept in early twentieth-century China. Chen claimed that the production of semen – voluntarily during intercourse or involuntarily through nocturnal emissions – diminished the body's limited stock of primordial *qi* 氣 available for alchemical transformation. He advocated 'cutting off desire' (*duanyu* 斷慾) altogether by abstaining from the sources of mental and physical stimulation. Chen regarded 'pure and quiet solo practice' (*qingjing gongfu* 清靜功夫), which involved a combination of meditation; gymnastics and breathing exercises, as the privileged pathway towards transcendence and immortality. There is little room for sexuality in Chen Yingning's Daoist practices (Liu Xun 2009: 151–2; Liu Xun 2012).

Nevertheless, sexual cultivation would continually surface in and outside China throughout the twentieth century. Ye Dehui's *Shadow of the Double Plum Tree Anthology* became an important source for a number of mid-twentieth-century European Sinologists who studied the history of religion and medicine in China. One of them was Joseph Needham (1900–1995), inaugurator of the monumental *Science and Civilisation in China* project, who stumbled upon Ye's compilation in the early 1950s and labelled it 'the greatest Chinese sexological collection' (Needham 1977; Rocha 2012). For Needham, Ye's bedchamber texts offered undeniable proof that there existed a sophisticated eroticism and refined sensibility in ancient China that was supposedly healthier and more 'normal' than Western Judeo-Christian sexuality. He believed that the 'Daoist' practitioners concentrated on the enhancement of intensity and ecstasy, as opposed to the control of reproduction and gender. Needham wanted to construct a universal sex ethics, a syncretism between Daoist 'naturalism' and the mysticism that he found in the work of William Blake (1757–1827); D.H. Lawrence (1885–1930); Edward Carpenter (1844–1929) and others (Needham 1974). If Ye Dehui misread Chinese sexual cultivation, perhaps wilfully, as an indigenous predecessor for Western eugenics and reproductive hygiene to further his anti-modern agenda, then Needham romantically imagined an ancient China in which people sought pleasure purely for the sake of pleasure. For Needham, China could somehow be an antidote, a panacea even, for the hypocrisies and ills of modernity. In the next chapter, we explore the 1960s and 1970s more closely, with ancient Chinese sexuality becoming something valorised by postmodern thinkers and New Age gurus, while sexuality during the Cultural Revolution period was seen as both repressed and perverse.

Bibliography

Bauer, H. (ed.) (2015) *Sexology and Translation: cultural and scientific encounters across the modern world*, Philadelphia: Temple University Press.

Chang, J. (ed.) (1977) *The Tao of Love and Sex: the ancient Chinese way to ecstasy*, London: Wildwood House.

Chow, Kai-wing, Tze-ki Hon, Hung-yok Ip and Price, D.C. (eds) (2008) *Beyond the May Fourth Paradigm: in search of Chinese modernity*, Lanham, MD: Rowman and Littlefield.

Chow, Tse-tsung (1960) *The May Fourth Movement: intellectual revolution in modern China*, Cambridge, MA: Harvard University Press.

Chung, J.Y. (2002) *Struggle for National Survival: eugenics in Sino-Japanese contexts, 1896–1945*, London: Routledge.

Dikötter, F. (1992) *The Discourse of Race in Modern China*, London: C. Hurst.

—— (1995) *Sex, Culture, and Modernity in China: medical science and the construction of sexual identities in the early Republican period*, London: C. Hurst.

—— (1998) *Imperfect Conceptions: medical knowledge, birth defects, and eugenics in China*, London: C. Hurst.

Doleželová-Velingerová, M. and Král, O. (eds) (2000) *The Appropriation of Cultural Capital: China's May Fourth project*, Cambridge, MA: Harvard University Press.

Du Maizhi 杜迈之 and Zhang Chengzong 张承宗 (1986) *Ye Dehui pingzhuan* 叶德辉评传 (A Critical Biography of Ye Dehui), Changsha: Yuelu shushe.

Farquhar, J. (2002) *Appetites: food and sex in postsocialist China*, Durham, NC: Duke University Press.

Foucault, M. (1978) *History of Sexuality Volume 1: the will to knowledge*, trans. R. Hurley, New York: Vintage.

Frühstück, S. (2003) *Colonising Sex: sexology and social control in modern Japan*, Berkeley: University of California Press.

Furth, C. (1994) 'Rethinking van Gulik: sexuality and reproduction in traditional Chinese medicine', in C.K. Gilmartin, G. Hershatter, L. Rofel and T. White (eds) *Engendering China: women, culture, and the state*, Cambridge, MA: Harvard University Press, pp. 125–46.

Gao Wenzhu et al. 高文铸 (1996) *Yixinfang jiaozhu yanjiu* 医心方校注研究 (Studies on the Collation and Annotation of *The Core Prescriptions of Medicine*), Beijing: Huaxia Chubanshe.

Gilmartin, C.K., Hershatter, G., Rofel, L. and White, T. (eds) (1994) *Engendering China: women, culture, and the state*, Cambridge, MA: Harvard University Press.

Gordon, M. (1971) 'From an unfortunate necessity to a cult of mutual orgasm: sex in American marital education literature, 1930–1940', in J.M. Heslin (ed.) *Studies in the Sociology of Sex*, New York: Appleton-Century-Crofts, pp. 53–77.

Heslin, J.M. (ed.) (1971) *Studies in the Sociology of Sex*, New York: Appleton-Century-Crofts.

Jackson, M. (1994) *The Real Facts of Life: feminism and the politics of sexuality 1850–1940*, London: Routledge.

Jagose, A. (2013) *Orgasmology*, Durham, NC: Duke University Press.

Ko, D. (2005) *Cinderella's Sisters: a revisionist history of footbinding*, Berkeley: University of California Press.

Larson, W. (2009) *From Ah Q to Lei Feng: Freud and revolutionary spirit in twentieth-century China*, Stanford, CA: Stanford University Press.

Leary, C. (1993) 'Intellectual orthodoxy, the economy of knowledge and the debate over Zhang Jingsheng's *Sex Histories*', *Republican China*, 18.2: 99–137.

—— (1994) *Sexual Modernism in China: Zhang Jingsheng and 1920s urban culture*, PhD thesis, Cornell University.

Lee, H. (2007) *Revolution of the Heart: a genealogy of love in China, 1900–1950*, Stanford, CA: Stanford University Press.

Li, Xiaojiang [as Li Xiao-Jian] (2004) '*Xingbie* or gender', in N. Tazi (ed.) *Keywords: gender*, New York: Other Press, pp. 87–103.

Liu, L.H. (1995) *Translingual Practice: literature, national culture, and translated modernity — China, 1900–1937*, Stanford, CA: Stanford University Press.

Liu Xun 劉迅 (2009) *Daoist Modern: innovation, lay practice, and the community of inner alchemy in Republican Shanghai*, Cambridge, MA: Harvard University Press.

—— (2012) 'Scientising the body for the nation: Chen Yingning and the reinvention of Daoist inner alchemy in 1930s Shanghai', in D.A. Palmer and Liu Xun (eds) *Daoism in the Twentieth Century: between eternity and modernity*, Berkeley: University of California Press, pp. 154–172.
McLaren, A. (1999) *Twentieth-Century Sexuality: a history*, Oxford: Blackwell.
Mitter, R. (2004) *A Bitter Revolution: China's struggle with the modern world*, Oxford: Oxford University Press.
Needham, J. (as Henry Holorenshaw) (1974) 'The making of an honorary Taoist', in M. Teich and R.M. Young (eds) *Changing Perspectives in the History of Science: essays in honour of Joseph Needham*, London: Heinemann, pp. 1–20.
—— (1977) 'Foreword', in J. Chang (ed.) *The Tao of Love and Ssex: the ancient Chinese way to ecstasy*, London: Wildwood House, pp. 9–10.
—— (1983) *Science and Civilisation in China, Volume V: chemistry and chemical technology, part 5: spagyrical discovery and invention: physiological alchemy*, Lu Gwei-Djen (collab.), Cambridge: Cambridge University Press.
Oda, Makoto 小田亮 (1996) *Ichigo no jiten: sei* 一語の辞典: 性 (*Dictionary of One Word: sex*), Tokyo: Sanseido.
Palmer, D.A. and Liu Xun 劉迅 (eds) (2012) *Daoism in the Twentieth Century: between eternity and modernity*, Berkeley: University of California Press.
Peng, Hsiao-Yen (1999) 'Sex histories: Zhang Jingsheng's sexual revolution', in *Critical Studies*, 18: 159–77.
Pfister, R. (2012) 'Gendering sexual pleasures in early and mediaeval China', in *Asian Medicine: Tradition and Modernity*, 7: 34–64.
Pusey, J.R. (1983) *China and Charles Darwin*, Cambridge, MA: Harvard University Press.
Rocha, L.A. (2010a) 'Xing: the discourse of sex and human nature in modern China', *Gender and History*, 22: 603–28.
—— (2010b) *Sex, Eugenics, Aesthetics, Utopia in the Life and Work of Zhang Jingsheng (1888–1970)*, PhD thesis, University of Cambridge.
—— (2012) 'The way of sex: Joseph Needham and Jolan Chang', *Studies in History and Philosophy of Biological and Biomedical Sciences*, 43: 611–26.
—— (2015) 'Translation and two 'Chinese sexologies': *double plum* and *sex histories*', in H. Bauer (ed.) *Sexology and Translation: cultural and scientific encounters across the modern world*, Philadelphia: Temple University Press, pp. 154–73.
Sakamoto, Hiroko (2004) 'The cult of 'love and eugenics' in May Fourth Movement discourse', *Positions: East Asia Cultures Critique*, 12: 329–76.
Sang, D.T. (2003) *The Emerging Lesbian: female same-sex desire in modern China*, Chicago, IL: University of Chicago Press.
Schwarcz, V. (1986) *The Chinese Enlightenment: intellectuals and the legacy of the May Fourth Movement of 1919*, Berkeley: University of California Press.
Stopes, M. (2004) *Married Love: a new contribution to the solution of sex difficulties*, Ross McKibbin (ed.), Oxford: Oxford University Press.
Tazi, N. (ed.) (2004) *Keywords: gender*, New York: Other Press.
Teich, M. and Young, R.M. (eds) (1974) *Changing Perspectives in the History of Science: essays in honour of Joseph Needham*, London: Heinemann.
Teng, E. (2013) *Eurasian: mixed identities in the United States, China, and Hong Kong, 1842–1943*, Berkeley: University of California Press.
Van Gulik, R.H. (1961) *Sexual Life in Ancient China: a preliminary survey of Chinese sex and society from ca. 1500 BC till 1644 AD*, Leiden: Brill; reprinted and edited in 2003, with an introduction by Paul Rakita Goldin, Leiden: Brill.
Wang, Ping (2002) *Aching for Beauty: footbinding in China*, Minneapolis: University of Minnesota Press.
Wile, D. (1992) *Art of the Bedchamber: the Chinese sexual yoga classics including women's solo meditation texts*, Albany, NY: State University of New York Press.
Ye Dehui 葉德輝 (1903–1917) *Shuangmei jing'an congshu* 雙梅景闇叢書 (Shadow of the Double Plum Tree anthology), Changsha: Yeshi xiyuan.
Zhang Jingping 张晶萍 (2008) *Ye Dehui shengping ji xueshu sixiang yanjiu* 叶德辉生平及学术思想研究 (Research on Ye Dehui's Life, Scholarship, and Thought), Changsha: Hunan shifan daxue chubanshe.

Zhang Jingsheng 張競生 (1927) 'Disanzhongshui yu luanzhu ji shengji de dian he yousheng de guanxi, huo mei de xingyu' 第三種水與卵珠及生機的電和優生的關係, 或美的性慾 (The Relationship between the Third Kind of Water and the Ovum, and Vital Electricity and Eugenics, or Beautiful Sexual Desire), *Xin Wenhua* 新文化, 1.2: 23–48.

—— (2006) *Xingshi* 性史 (Sex Histories), Taipei: Dala.

—— (2008) *Fusheng Mantan: Zhang Jingsheng suibi xuan* 浮生漫谈：张竞生随笔选 (Reveries on a Floating Life: Zhang Jingsheng's selected miscellaneous essays), Zhang Peizhong 张培忠 (ed.), Beijing: Sanlian shudian.

Zhang Peizhong 张培忠 (2008) *Wenyao yu xianzhi: Zhang Jingsheng zhuan* 文妖与先知张竞生传 (The Monster and the Prophet: a biography of Zhang Jingsheng), Beijing: Sanlian shudian.

Zhong, Xueping (2000) *Masculinity Besieged?: issues of modernity and male subjectivity in Chinese literature of the late twentieth century*, Durham, NC: Duke University Press.

26
THE QUESTION OF SEX AND MODERNITY IN CHINA, PART 2
From new ageism to sexual happiness

L. A. Rocha

Spirituality of the postmodern age: Chinese sex as modernity's antidote

The previous chapter ends with Joseph Needham's romantic vision of ancient Chinese sexuality. As a perfect example of the global, circuitous and sometimes surprising flow of ideas, Needham's vision impacted Michel Foucault's theorisation in *History of Sexuality Volume 1* (Foucault 1978, first published in French in 1976), one of the foundation texts, of course, in the field of gender and sexuality studies. In that book, Foucault postulated the famous distinction between Western *scientia sexualis* and Eastern *ars erotica*. Western *scientia sexualis* allegedly emerged in nineteenth-century Europe via the secularisation and scientisation of the Catholic ritual of confession, yielding a modern sexuality that served as a point of anchorage for the governance of the nation-state – the surveillance of the population; the maintenance of productivity; the disciplining of bodies and behaviour; the regulation of reproduction and so forth. By contrast, Eastern *ars erotica* was simply about the intensification of manifold pleasures, the enhancement of sensual delight. Even though in the 1980s Foucault distanced himself from this problematic distinction, he nevertheless remained convinced that Chinese civilisation, in particular, had an *ars erotica*, a configuration of sex and desire that was diametrically opposite to that of the West (Rocha 2011).

Foucault did not know Chinese and he relied exclusively on the work of Dutch Sinologist Robert van Gulik (1910–1967), particularly his *Sexual Life in Ancient China* (originally published in English as Van Gulik 1961; French translation in 1971). Van Gulik's work was enthusiastically received by the 1968 generation of Parisian intellectuals; they generally regarded his presentation of Chinese eroticism as offering a promising avenue to think one's way out of Western sexuality and modernity. In the early 1950s, before writing *Sexual Life in Ancient China*, van Gulik actually held a fairly negative view of certain elements of Chinese sexual practice. Daoist techniques of sexual cultivation, especially the aforementioned *coitus reservatus*, were troublesome for van Gulik because it appeared to treat young women as reservoirs of vital *qi* 氣 that could be 'harvested' by the male adept in his quest for enlightenment and rejuvenation. In fact, van Gulik thought that this was a kind of exploitation or 'sexual vampirism' (van Gulik 1951: 11; Furth 1994; Furth 2005; Goldin 2006; Chapters 22 and 49 in this volume).

Joseph Needham, however, believed that everything about Chinese sexuality was wholesome and good, and that Daoist sex could be one of China's most important 'exports' to the modern world. He pressured van Gulik to alter his views when he penned *Sexual Life in Ancient China* in the 1960s (Rocha 2011). Therefore, both Needham and van Gulik ended up painting an idealised, if not Orientalist, picture of Chinese eroticism, which Foucault uncritically assimilated in *History of Sexuality Volume 1* and endowed with the name *ars erotica* (Foucault 1978). Of course, not all Western scholars contemporaneous with Needham and van Gulik agreed with this characterisation of Chinese sexuality. For instance, Sinologists Jacques Pimpaneau (1935–) and Kristofer Schipper (1934–) argued that there was nothing particularly 'superior' about Chinese erotic culture (Pimpaneau 1969; Schipper 1993: 145–50). For Schipper and others, esoteric Daoist practices were worth analysing, not because they carried something useful that might be revived for present consumption but for what they revealed of the conceptualisations of body, health and longevity in early China.

By the late 1960s and 1970s, a number of practitioners of Chinese heritage began to promote Daoist sexual techniques in Europe and America. Two well-known personalities were Jolan Chang (or Chang Chung-lan 章鍾蘭, 1917–2002) and Mantak Chia (or Xie Mingde 謝明德, 1944–). Chang was the Chinese-Canadian-Swedish author of *The Tao of Love and Sex: The Ancient Chinese Way to Ecstasy* (Chang 1977), which was highly praised by Joseph Needham who penned the preface and the afterword. Chang fashioned himself as a follower of the Dao who mastered the 'cultivation of life' (*yangsheng* 養生), and claimed that he would have sex several times a day as part of his health regime. He also suggested that Daoist techniques helped him to prolong ejaculation and led to more satisfying and pleasurable sex with his partners. *The Tao of Love and Sex* went on to be translated into sixteen languages and continues to be in print today; Chang became something of a curious celebrity in Sweden and later produced a sequel entitled *The Tao of the Loving Couple: True Liberation Through the Tao* (Chang 1983; Rocha 2012).

Mantak Chia, however, was the founder of the 'Healing Tao', one of the most widespread forms of popular Daoism in the West. Born in Thailand to Chinese parents, Chia claimed that he was mentored in Hong Kong by a Daoist master called Yi Eng (*Yiyun* 一雲) who taught him the 'Seven Formulae for Immortality'. In the late 1970s, Chia moved to New York City and opened a healing centre where he practised acupuncture and disseminated his teachings, now known as the 'Universal Healing Tao System'. He became a household name through a series of bestselling books. Those that discussed sexuality included: *Awaken Healing Energy through the Tao: The Taoist Secret of Circulating Internal Power*; *Taoist Secrets of Love: Cultivating Male Sexual Energy*; *Healing Love Through the Tao: Cultivating Female Sexual Energy* (Chia 1983, 1984, 1986). More recently, Chia and his collaborators published *The Multi-Orgasmic Man: Sexual Secrets Every Man Should Know*; *The Multi-Orgasmic Woman: How Any Woman Can Experience Ultimate Pleasure and Dramatically Enhance Her Wealth and Happiness* and *The Multi-Orgasmic Couple: Sexual Secrets Every Couple Should Know* (Chia and Abrams 1996, 2005; Chia et al. 2001). Chia prescribed various methods for controlling ejaculation and menstruation in order to transform the 'sexual energy' residing in everyone into *qi* 氣 and spiritual forces (Siegler 2012). Chia's numerous works, like Jolan Chang's, are essentially a *bricolage* of Daoist techniques, popular sexology and New Age spirituality, underpinned by a philosophy of self-discovery; individualism; entrepreneurship and experimentation that resonated with a Western, 'postmodern' audience eager for alternative styles of thinking and living (Clarke 2000: 128–35).

Repression of sexuality? Communist China and the Cultural Revolution

While Western intellectuals and practitioners investigated and publicised Daoist practices, what was happening to sexuality in China during the time of communism? The Hungarian-British journalist Paul Tabori (1908–1974), writing in 1967 under the pseudonym Peter Stafford, claimed that the Communist regime under Mao enforced a 'self-denying, masochistic, almost inhuman philosophy of the supreme importance of the state and the total insignificance of the individual' (Tabori [as Stafford] 1967: 13). Echoing some of the sensationalist rhetoric of the Cold War, Tabori stated that all Communist societies (China, Albania, Bulgaria, Czechoslovakia, East Germany, Hungary, Poland, Romania, the Soviet Union and Yugoslavia) were necessarily totalitarian and always led to the brutal and dehumanising suppression of all desires – in China's case reducing ordinary people into an unfree, indistinguishable mass of uniformed workers or 'blue ants' (Tabori [as Stafford] 1967: 13). Moreover, the uncompromising puritanism and sexual repression, Tabori argued, resulted in the manifestation of aberrant, criminal and 'perverted' behaviour such as sadomasochistic torture, rape and child abuse – as if these things never happened in capitalist nations.

The French philosophers Roland Barthes (1915–1980) and Julia Kristeva (1941–), as well as other literary figures associated with the *Tel Quel* group, visited China for three weeks between April and May 1974. Witnessing the failure of the May 1968 revolution in France, these Parisian intellectuals turned to Maoist philosophy as the source of hope and renewal. As they travelled across China, the country was in the final phase of the so-called 'Cultural Revolution' (*wenhua dageming* 文化大革命) decade (1966–1976), which involved the 'Criticise Lin Biao and Criticise Confucius' campaign (*Pi Lin pi Kong yundong* 批林批孔運動, 1973–1976). In his travel diaries, Barthes remarked that China was a 'desert of flirtation' where eroticism was 'abolished' through the 'total uniformity of clothes' (Barthes 2012: 9). Engaging in conversation with physicians at the Shanghai Second People's Hospital, Barthes reported that mental illnesses were seen by Chinese doctors as having physical and hereditary causes. Psychoanalysis was rejected because of Freud's 'sexualism' and because the Chinese insisted that 'reality [was] not sexual' (Ibid.: 34). Sexual tensions among the young were sublimated through 'effort, study, work'; sexual freedom before marriage was 'considered as debasing' and 'not accepted by young people' (Ibid.).

If Barthes' observations on sexuality in China under communism were scattered and ambivalent – at one point, Barthes seemed simply disappointed to have missed the chance to see 'the willy of a single Chinese man' (Barthes 2012: 100) – his fellow traveller, Julia Kristeva, saw Chinese women as offering nothing less than the possibility of the radical reconstruction of Western female subjectivity. In her deeply controversial *About Chinese Women* (Kristeva 1977), Kristeva 'speculated that the matrilineal, orgasmic, truthful language that she felt Chinese women had spoken in the archaic past had somehow survived into the present. She "theorised" that the Chinese Communist Party had emancipated this archaic *jouissance* when it liberated women from Confucianism and returned them to native prehistory' (Barlow 2004: 311). Postcolonial thinkers, including Gayatri Spivak and Rey Chow, criticised Kristeva for reproducing tired stereotypes regarding 'Oriental' women and femininity as primitive and unchanging, turning China into grist for her own theoretical mill, while ignoring historical realities and exhibiting a classic 'colonial benevolence' symptomatic of Western, first-world, feminism (Spivak 1981; Chow 1993).

In stark contrast to these earlier Western (and often Eurocentric) gazes at China, more recent scholarship from Chinese History and Gender Studies paint a complex and nuanced picture of sexuality in the Mao era – the work of Harriet Evans, Gail Hershatter and

Emily Honig, for example (Hershatter 1996, 2007, 2011; Evans 1997; Honig 2000, 2002, 2003). Investigating sex during the Cultural Revolution, Honig challenged in her seminal article the conventional story of 'state silencing and popular submission or of state prohibition and popular resistance' (Honig 2003: 145). This was not to deny that romance and sexuality became taboo subjects, references to which were removed from official publications. Individual authors who wrote on such subjects were indeed imprisoned and made to confess their 'errors' and 'reform' their thinking. Ordinary folk engaging in non-marital relations were subjected to 'struggle sessions' and public humiliation, and, in turn, 'bourgeois class enemies' were often accused of sexual immorality. However, the role of the party-state was highly ambiguous: issues surrounding sexuality were not a priority during the Cultural Revolution, and the government did not lay down any policies and declarations specifically prohibiting romantic liaisons and sexual relations. Honig argued that it would be more appropriate to describe the Cultural Revolution 'as a period characterised by a profound conflation of political and sexual impurity' (Honig 2003: 154).

From interviews and personal memoirs, Emily Honig found that youths sent down to the countryside or those remaining in the cities in the 1960s and 1970s – often freed from parental supervision and familial control – had a range of new opportunities and sexual experiences from brief encounters to prolonged cohabitation with youths of the same or the opposite sex. There were indeed records of abuse and harassment from local cadres but also accounts of consensual sex. Even though some youths were reprimanded or punished for pursuing romantic relationships, and young women were particularly vulnerable to sexual attack and scapegoating, repression and subjugation were only part of a wider landscape of Chinese sexuality during the Cultural Revolution decade. People might have been censured for their 'transgressions' of sexual norms but at the same time such 'transgressions' were commonplace and ultimately difficult to police. The presence of a discourse of purity was as much a symptom of the fact that people's behaviour was often 'impure'. None of this was to make light of the sufferings of victims of sexual assault and persecution but to highlight the flaws of narratives that simplistically reduced sexuality in Communist China to nothing but state oppression. Such simplistic narratives served political purposes: to legitimise the post-socialist economic reforms in the Deng Xiaoping era by condemning the Cultural Revolution as a period of total darkness and 'sexual blindness' (*xingmang* 性盲), or more generally to reinforce the ideology that capitalism, unlike socialism, was the only social configuration that did not run contrary to 'human nature'.

Just how did ordinary folk in China learn about human sexuality and reproduction? One of the most widely read texts on the subject during the Communist period was a seventy-eight-page booklet entitled *Sexual Knowledge* (*Xing de zhishi* 性的知識). Written by three physicians – Wang Wenbin 王文彬, Zhao Zhiyi 趙志一 and Tan Mingxun 譚銘勛 – the manual was first published in 1956, a few years after the promulgation of the New Marriage Law (*Xin hunyin fa* 新婚姻法) that afforded women legal equality with men (Wang, Zhao and Tan: 1956). By the late 1950s and early 1960s, there were millions of copies in circulation across China. *Sexual Knowledge* imparted basic facts on male and female reproductive anatomy and physiology, as well as information on contraceptive methods, venereal diseases and sexual hygiene. It also prescribed an ideal of sexuality, specifically between the heterosexual couple in a marital setting. This did not involve the sublimation of sexual pleasure, nor the flattening of sexuality into reproduction, nor the complete dissolution of the boundaries between the private and the public. Instead, *Sexual Knowledge* promoted the management of pleasure for the sake of familial and social harmony – an ideology of marital sex that would make a comeback in the post-socialist period. As Harriet Evans pointed out, however, books on sexual health that were endorsed by the Chinese government in the 1950s and 1960s ceased to be published during the Cultural Revolution (Evans 1997: 41–4; Honig 2003: 146), but *Sexual Knowledge* was passed secretly among people and sometimes copied by hand. Evidently, more research, especially the collection of oral histories, is needed on how those growing up in the Mao era came to know – or failed to know – the 'facts of life'.

Modernising Chinese sexuality again: of heritage and happiness

The initiation of Deng Xiaoping's (1904–1997) 'Open Door Policy' in the late 1970s ushered in Western ideals of liberty, freedom and individual rights. The short-lived 'Anti-Spiritual Pollution Campaign' (*Qingchu jingshen wuran yundong* 清除精神污染運動), launched by the conservative factions within the Communist Party in 1983, was largely unsuccessful in eliminating the supposed onslaught of 'obscene'; 'barbarous'; 'bourgeois' and 'vulgar' materials from the West. With the introduction of the Family Planning (*Jihua shengyu* 計劃生育) Policy (so-called 'One Child Policy') in 1979, with the growing wealth of city dwellers from industrial investment and the migration of millions of people from villages to urban areas, with the 'information explosion' and the concomitant proliferation of images and discussions on sexuality in public culture and with the global HIV/AIDS crisis unfolding in the early 1980s, it became imperative for the Chinese state to choreograph the emerging 'sexual revolution' (*xing geming* 性革命) in the early to mid-1980s (Ruan 1991; Farquhar 2002; Farrer 2002; Jeffreys 2004; McMillan 2006; Evans 2008; Rofel 1999, 2007; Zhang 2011; Burger 2012; Davis and Friedman 2014; Jeffreys and Yu 2015). The Chinese government was interested in maintaining social stability via the strengthening of familial 'harmony' (*hexie* 和諧); disseminating modern knowledge on sexual health and reproduction and cultivating 'appropriate' kinds of desires. A key part of this endeavour was the dramatic expansion of the social sciences in higher education institutions and the inauguration of academic sexological research – in other words, ways to investigate the lives of ordinary individuals; develop authoritative works on sex education and reproductive medicine for the general populace and generate policy recommendations for the Chinese state's experiments in economic reform and social liberalisation (Ruan and Lau 2004). This involved collaboration with European and American universities and experts (such as the Kinsey Institute at Indiana University Bloomington), as well as funding from international non-government organisations and international agencies (for instance, the Ford Foundation).

The professional trajectory of sociologist and sexologist Liu Dalin 劉達臨 (1932–) is illuminating in this regard (Liu 2001, 2008, 2011). Liu was a Professor at Shanghai University and spearheaded the Shanghai Sex Sociological Research Centre's *National Sex Civilisation Survey* (*Zhongguo xing wenming diaocha baogao* 中國性文明調查報告, 1989–1990). Liu and his team collected responses to some 200 questions from almost 20,000 individuals across China. The report of the survey, commonly referred to as the 'China Kinsey Report', was published in 1992 and translated into English in 1997 (Liu *et al.* 1992). In 1994, Liu was the recipient of the Magnus Hirschfeld Medal, awarded by the German Society for Social-Scientific Sexuality Research (DGSS) for his contributions to the development of sexual science in China. In his later career, Liu established the China Sex Museum (*Zhonghua xingwenhua bowuguan* 中華性文化博物館) in Shanghai in 1995, which relocated to Tongli in Jiangsu Province in 2003 (permanently closed by spring 2017). The China Sex Museum showcased the art, manuscripts and other artefacts belonging to a supposedly nature-oriented, frank, open Chinese people whose sexual instincts were repeatedly repressed over the past centuries by Confucian propriety and, more recently, by Communist puritanism, and now a healthy, normal and thoroughly modern sexual subject could finally emerge in the late twentieth century.

Liu Dalin also produced a number of popular histories on ancient Chinese sexual life (Liu 2001, 2004; Liu and Hu 2008), which sat alongside a whole series of publications by other authors that emerged in the 1990s and 2000s highlighting China's 'erotic heritage' and 'hedonism'. Take, for instance, the 'Mawangdui Silk Texts' (*Mawangdui boshu* 馬王堆帛書). In 1973, Chinese archaeologists excavated the Mawangdui (or King Ma's Mound) in Changsha, Hunan Province, which consisted of two saddle-shaped hills housing the tombs of an aristocratic family from the Western Han dynasty (202 BCE–9 CE). The tombs were sealed around 168 BCE, and buried in Tomb Number 3 were the so-called Mawangdui Silk Texts – philosophical and medical works written on silk, some of which were even older than the ones reconstructed by Ye Dehui 葉德輝 (1864–1927) in the early twentieth century from Japanese sources. The texts from Mawangdui contained recipes for aphrodisiacs, as well as discussions on the cultivation of health and vitality via sexual techniques (for scholarly treatments, see Li and McMahon 1992; Harper 1998). Arguments concerning the erotic cultures of the distant past continue to affect debates on modern Chinese people's attitude to sexual pleasure under the socialist market economy.

Another key personality in the field of Chinese sex research is Pan Suiming 潘綏銘 (1950–), formerly professor at Renmin University of China in Beijing. Pan became renowned in the 1990s and 2000s for his sociological scholarship on Chinese people's sexual customs and mentalities via nationwide surveys, as well as his path-breaking studies on red-light districts, sex workers and the sex industry in China (Pan 1995, 1999, 2000; Pan *et al.* 2004). For many of these projects, including the foundation of the Institute for Research on Sexuality and Gender at Renmin University for which Pan served as the director, Pan received generous sponsorship from global philanthropic organisations chiefly the Ford Foundation (Kaufman *et al.* 2014). Following the Programme of Action devised at the 1994 International Conference on Population and Development (ICPD) in Cairo, the Ford Foundation rebranded its population stream to incorporate investigations into sexual behaviour and sexual cultures. Ford directors and officers promoted a holistic conception of sexual well-being that not only involved the absence of reproductive diseases and problems but also an individual's fulfilment in intimate life. The Ford Foundation's programmes on sexuality and sexual health disseminated a sexual script that invoked the vocabulary of rights; accountability; fairness; respect; dignity and mutual exchange of pleasure. The most recent iteration is the idea of 'sexual happiness' (*xingfu* 性福), an affirmative outlook on sexuality, which maintains that each individual has a fundamental entitlement to sexual happiness, and to which Pan Suiming and his colleagues subscribe (Pan and Huang 2011). The best possible social and political configuration, according to this discourse, allows for the fullest expression of sexuality (including lesbian, gay, bisexual, transgender and queer subjectivities) which then leads to the maximisation of human potential.

★ ★ ★

In these two short chapters, I have shown how a diverse, intriguing cast of historical actors have attempted to do things *to*, or do things *with*, 'Chinese sexuality'. We have had:

i Reformist literati in the twilight of the Qing Empire who accepted Western science and medicine (by arguing that it was urgently needed to rescue China from the brink of collapse and/or arguing that Western and Chinese knowledges were compatible); conservative thinkers who rejected Western science and medicine (by claiming that Chinese science and medicine were superior, e.g. Ye Dehui's philological reconstruction of Chinese 'bedchamber classics' from Japanese sources).

ii Western-educated intellectuals in Republican China, who argued that China could only step into modernity by fundamentally revolutionising the Chinese people's sexual mores and practices, and by adopting racial science, eugenics and sexology from Europe and America (e.g. Zhang Jingsheng and many of his 'May Fourth' contemporaries).

iii Groups of Western Sinologists in the mid-twentieth century, who investigated the sexual cultures of early China and argued that Chinese sexuality might offer a kind of counter-discourse to the 'repressive', alienating, Judeo-Christian sexuality, a kind of therapy for modernity (e.g. Joseph Needham, Robert van Gulik); Chinese practitioners and populariser of Daoist sexual techniques who were welcomed by a Western, 'postmodern' audience searching for alternative lifestyles and spiritualities (e.g. Jolan Chang, Mantak Chia).

iv Cold War commentators who regarded sexuality as an index to social and political freedom and who argued that China and other Communists regimes brought about the repression of all of human beings' natural instincts (e.g. Paul Tabori); the 1968 generation of Parisian, Leftist philosophers, who saw something in Chinese sexuality that allowed them to rethink gender relations and sexual desire in the West (e.g. Julia Kristeva, Michel Foucault).

v   A transnational network of contemporary social researchers and sociologists; sex educators and sexologists; physicians and social workers; archaeologists and museum curators, as well as officers and directors of non-government organisations and global philanthropic foundations supporting projects in post-socialist China (e.g. Liu Dalin, Pan Suiming and the Ford Foundation). The Chinese state is invested in supervising its ongoing 'sexual revolution', Chinese intellectuals are interested in liberalising society through their sociological and sexological scholarship and the transnational agents propagate a (neo-) liberal sexual script and ethics invoking the language of rights, accountability and happiness.

Even though all these different actors held and hold widely divergent agenda, their projects inevitably revolved around a constellation of questions: Is the West more liberated or repressed than China? Does traditional China have a refined sexual culture to be emulated, or do the Chinese need to be brought in line with 'civilisation'? Is the reform of sexuality a privileged way to transform society, or are sexual revolutions the inevitable side-effect of economic reforms? How does one define China's position *vis-à-vis* 'modernity' and the 'modern world'? Does there exist a code of sexual ethics and norms and regimes of education that could be adopted universally? Is there something characteristically 'Chinese' about the ways that desires and pleasures have been pursued in China? What are the connections between sexuality and socio-political configurations? Fundamentally, the question of Chinese sexuality and modernity reflects and shapes the entanglements between China and the Western world.

Bibliography

Barlow, T. (2004) *The Question of Women in Chinese Feminism*, Durham, NC: Duke University Press.
Barthes, R. (2012) *Travels in China*, trans. A. Herschberg-Pierrot and A. Brown, Cambridge: Polity.

Beurdeley, M. (ed.) (1969) *The Clouds and the Rain: the art of love in China*, trans. D. Imber, Fribourg: Hammond, Hammond and Co.
Brownell, S. and Wasserstrom, J.N. (eds) (2002) *Chinese Femininities/Chinese Masculinities*, Berkeley: University of California Press.
Burger, R. (2012) *Behind the Red Door: sex in China*, Hong Kong: Earnshaw Books.
Chang, J. (1977) *The Tao of Love and Sex: the ancient Chinese way to ecstasy*, London: Wildwood House.
—— (1983) *The Tao of the Loving Couple: true liberation through the Tao*, New York: E.P. Dutton.
Chia, M. (1983) *Awaken Healing Energy through the Tao: the Taoist secret of circulating internal power*, New York: Aurora Press.
—— (1984) *Taoist Secrets of Love: cultivating male sexual energy*, New York: Aurora Press.
—— (1986) *Healing Love through the Tao: cultivating female sexual energy*, Huntington, NY: Healing Tao Books.
Chia, M. and Abrams, D. (1996) *The Multi-Orgasmic Man: sexual secrets every man should know*, San Francisco, CA: Harper.
Chia, M. and Abrams, R.C. (2005) *The Multi-Orgasmic Woman: how any woman can experience ultimate pleasure and dramatically enhance her wealth and happiness*, Emmaus, PA: Rodale.
Chow, R. (1993) *Writing Diaspora: tactics of intervention in contemporary cultural studies*, Bloomington: Indiana University Press.
Clarke, J.J. (2000) *The Tao of the West: western transformations of Taoist thought*, London: Routledge.
Davis, D.S. and Friedman, S.L. (eds) (2014) *Wives, Husbands, and Lovers: marriage and sexuality in Hong Kong, Taiwan, and urban China*, Stanford, CA: Stanford University Press.
Evans, H. (1997) *Women and Sexuality in China: dominant discourses of female sexuality and gender since 1949*, Cambridge: Polity.
—— (2008) *The Subject of Gender: daughters and mothers in urban China*, Lanham, MD: Rowman and Littlefield.
Farquhar, J. (2002) *Appetites: food and sex in post-socialist China*, Durham, NC: Duke University Press.
Farrer, J. (2002) *Opening Up: youth sex culture and market reform in Shanghai*, Chicago, IL: University of Chicago Press.
Foucault, M. (1978) *History of Sexuality Volume 1: the will to knowledge*, trans. R. Hurley, New York: Vintage.
Francoeur, R.T. and Noonan, R.J. (eds) (2004) *The Continuum Complete International Encyclopaedia of Sexuality*, London: Continuum.
Furth, C. (1994) 'Rethinking van Gulik: sexuality and reproduction in traditional Chinese medicine', in C.K. Gilmartin, G. Hershatter, L. Rofel and T. White (eds.) *Engendering China: women, culture, and the state*, Cambridge, MA: Harvard University Press, pp. 125–46.
—— (2005) 'Rethinking van Gulik again', *Nan Nü: men, women and gender in early and imperial China*, 7: 71–8.
Gilmartin, C.K., Hershatter, G., Rofel, L. and White, T. (eds.) (1994) *Engendering China: women, culture, and the state*, Cambridge, MA: Harvard University Press.
Goldin, P.R. (2006) 'The cultural and religious background of sexual vampirism in ancient China', *Theology and Sexuality*, 12: 285–307.
Gutwisle, B. and Henderson, G. (eds) (2000) *Redrawing the Boundaries of Work, Households, and Gender*, Berkeley: University of California Press.
Harper, D. (1998) *Early Chinese Medical Literature: the Mawangdui medical manuscripts*, London: Routledge.
Hershatter, G. (1996) 'Sexing modern China', in G. Hershatter, E. Honig, J.N. Lipman and R.Stross (eds), *Remapping China: fissures in historical terrain*, Stanford, CA: Stanford University Press, pp. 77–93.
—— (2007) *Women in China's Long Twentieth Century*, Berkeley: University of California Press.
—— (2011) *The Gender of Memory: rural women and China's collective past*, Berkeley: University of California Press.
Hershatter, G., Honig, E., Lipman, J.N. and Stross, R. (eds) (1996) *Remapping China: fissures in historical terrain*, Stanford, CA: Stanford University Press.
Honig, E. (2000) 'Iron girls revisited: gender and the politics of work in the Cultural Revolution', in B. Gutwisle and G. Henderson (eds) *Redrawing the Boundaries of Work, Households, and Gender*, Berkeley: University of California Press, pp. 97–110.

——— (2002) 'Maoist mappings of gender: reassessing the Red Guards', in S. Brownell and J.N. Wasserstrom (eds) *Chinese Femininities/Chinese Masculinities*, Berkeley: University of California Press, pp. 255–68.

——— (2003) 'Socialist sex: the cultural revolution revisited', *Modern China*, 29: 153–75.

Jeffreys, E. (2004) *China, Sex and Prostitution*, London: Taylor and Francis.

Jeffreys, E. and Yu, H. (2015) *Sex in China*, Cambridge: Polity.

Kaufman, J., Burris, M.A., Lee, E.W. and Jolly, S. (2014) 'Gender and reproductive health in China: partnership with foundations and the United Nations', in J. Ryan, L.C. Chen and T. Saich (eds) *Philanthropy for Health in China*. Bloomington: Indiana University Press, pp. 155–74.

Kleinman, A., Yan, Y., Jun, J., Lee, S. and Zhang, E. (eds) (2011) *Deep China: the moral life of the person*.

Kristeva, J. (1977) *About Chinese Women*, trans. A. Barrows, London: Marion Boyers.

Li, L. and McMahon, K. (1992) 'The contents and terminology of the Mawangdui texts on the arts of the bedchamber', *Early China*, 17: 145–85.

Liu Dalin 刘达临 (2001) *Wo de xingxue zhi lu* 我的性学之路 (My Path to Sexology), Beijing: Zhongguo qingnian chubanshe.

——— (2004) *Zhongguo qingse wenhua shi* 中国情色文化史 (The History of Erotic Culture in China), Beijing: Renmin ribao chubanshe.

——— (2008) *Wo yu xing wenhua* 我与性文化 (Sex Culture and I), Shanghai: Dongfang chuban zhongxin.

——— (2011) *Zou xiang xing wenming* 走向性文明 (Stepping towards Sexual Civilisation), Shanghai: Shanghai renmin chubanshe.

Liu Dalin 刘达临, Wu Minlun 吴敏伦 and Qiu Liping 仇立平 (1992) *Zhongguo dangdai xing wenhua: Zhongguo liangwan li "xing wenming" diaocha baogao* 中国当代性文化: 中国两万例「性文明」调查报告 (Sexual Behaviour in Modern China: a report of the nationwide "sex civilisation" survey on 20,000 subjects in China), Shanghai: Sanlian shudian.

Liu Dalin 刘达临 and Hu Hongxia 胡宏霞 (2008) *Lishi de da yinsi: Zhonghua xing wenhua shi ershi jiang* 历史的大隐私: 中华性文化史二十讲 (The Greatest Secrets in History: twenty lectures on the history of Chinese sexual culture), Zhuhai: Zhuhai chubanshe.

McMillan, J. (2006) *Sex, Science and Morality in China*, London: Routledge.

Palmer, D.A. and Liu Xun 劉迅 (eds) (2012) *Daoism in the Twentieth Century: between eternity and modernity*, Berkeley: University of California Press.

Pan Suiming 潘绥铭 (1995) *Zhongguo xing xianzhuang* 中国性现状 (The Current State of Sex in China), Beijing: Guangming ribao chubanshe.

——— (1999) *Cunzai yu huangmiu: Zhongguo dixia "xing chanye" kaocha* 存在与荒谬:中国地下"性产业"考察 (Existence and Absurdity: a survey of China's underground "sex industry"), Beijing: Qunyan chubanshe.

——— (2000) *Zhongguo dangdai daxuesheng de xing guannian yu xing xingwei* 中国当代大学生的性观念与性行为 (Contemporary Chinese University Students' Sexual Concepts and Sexual Behaviour), Beijing: Commercial Press.

Pan Suiming 潘绥铭 and Huang Yingying 黄盈盈 (2011) 'The rise of rights and pleasure: towards a diversity of sexuality and gender', in K. Zhang (ed.) *Sexual and Reproductive Health in China: reorienting concepts and methodology*, Leiden: Brill, pp. 215–62.

Pan Suming 潘绥铭, Parish, W., Wang Aili 王爱丽 and Lauman, E. (2004) *Dangdai Zhongguoren de xing xingwei yu xing guanxi* 当代中国人的性行为与性关系 (Sexual Behaviour and Relationships in Contemporary China), Beijing: Shehui kexue chubanshe.

Pimpaneau, J. (1969) 'The erotic novel in China', in M. Beurdeley (ed.) *The Clouds and the Rain: the art of love in China*, trans. D. Imber, Fribourg: Hammond, Hammond and Co., pp. 79–134.

Rocha, L.A. (2011) '*Scientia sexualis* versus *Ars erotica*: Foucault, van Gulik, Needham', *Studies in History and Philosophy of Biological and Biomedical Sciences*, 42: 328–43.

——— (2012) 'The way of sex: Joseph Needham and Jolan Chang', *Studies in History and Philosophy of Biological and Biomedical Sciences*, 43: 611–26.

Rofel, L. (1999) *Other Modernities: gendered yearnings in China after socialism*, Berkeley: University of California Press.

——— (2007) *Desiring China: experiments in neoliberalism, sexuality, and public culture*, Durham, NC: Duke University Press.

Ruan, F. (1991) *Sex in China: studies in sexology in Chinese culture*, New York: Plenum Press.

Ruan, F. and Lau, M.P. (2004) 'China', in R.T. Francoeur and R.J. Noonan (eds) *The Continuum Complete International Encyclopaedia of Sexuality*, London: Continuum, pp. 182–209.

Ryan, J., Chen, L.C. and Saich, T. (eds) (2014) *Philanthropy for Health in China*. Bloomington: Indiana University Press.

Schipper, K. (1993) *The Taoist Body*, trans. K.C. Duval, Berkeley: University of California Press.

Siegler, E. (2012) 'Daoism beyond modernity: the "Healing Tao" as postmodern movement', in D.A. Palmer and Liu Xun (eds) *Daoism in the Twentieth Century: between eternity and modernity*, Berkeley: University of California Press, pp. 274–92.

Spivak, G.C. (1981) 'French feminism in an international frame', *Yale French Studies*, 62: 154–84.

Tabori, Paul (as Peter Stafford) (1967) *Sexual Behaviour in the Communist World: an eyewitness report of life, love and the human condition behind the Iron Curtain*, New York: Julian Press.

Van Gulik, R.H. (1951) *Erotic Colour Prints of the Ming Period: with an essay on Chinese sex life from the Han to the Ching dynasty, BC 206–AD 1644*, 3 volumes, Tokyo: Private printing; reprinted and edited in 2004, with introductions by James Cahill, W.L. Idema and S. Edgren, Leiden: Brill.

—— (1961) *Sexual Life in Ancient China: a preliminary survey of Chinese sex and society from ca. 1500 BC till 1644 AD*, Leiden: Brill; reprinted and edited in 2003, with an introduction by P.R. Goldin, Leiden: Brill.

Wang Wenbin 王文彬, Zhao Zhiyi 赵志一 and Tan Mingxun 谭铭勋 (1956) *Xing de zhishi* 性的知识 (Sexual Knowledge), Beijing: Renmin weisheng chubanshe.

Zhang, E.Y. (2011) 'China's sexual revolution', in A. Kleinman, Y. Yan, J. Jun, S. Lee, E. Zhang (eds) *Deep China: the moral life of the person*, Berkeley: University of California Press, pp. 106–51.

Zhang, K. (ed.) (2011) *Sexual and Reproductive Health in China: reorienting concepts and methodology*, Leiden: Brill.

PART 4

Spiritual and orthodox religious practices

27
DAOISM AND MEDICINE[1]

Michael Stanley-Baker

A popular aphorism states, 'Daoism and medicine emerged from the same source' (*yidao tongyuan* 醫道同源). They both share a common ancestry in the Shang dynasty (ca. 1600–1046 BCE), when diviners (*wu* 巫) asked the gods and ancestors about the disease of the kings, recording them in the earliest Chinese writings (Keightley 2001). After that time emerged a broad variety of disparate popular medical practices (Cook 2013), including drug recipes, divination and exorcistic magic (Harper 1998). The aphorism also alludes to how the common scientific and cosmological principles, such as *yin–yang*, the five agents (*wuxing* 五行) and *qi*, which lie at the base of classical medicine (Chapters 1 and 2 in this volume) formalised in the Han dynasty (202 BCE–220 CE) were also the basis for Daoist ritual and physiology which emerged towards the end of this period.

Taken together, these common origins and expressions might lead one to imagine the Han medical community as intrinsically open to 'spiritual' health practices. However, medical practitioners and classical texts also demonstrate an antipathy towards spirit-based notions of disease, and towards spirit mediums. This can be seen in another aphorism attributed to the mythical physician Bianque 扁鵲 that one of the six obstacles to treatment is 'believing in spirit mediums, and not in physicians' (*xin wu bu xin yi* 信巫不信醫), a statement recorded and circulated in the early Han dynasty text, *Records of the Grand Historian* (*Shiji* 史記, 105.2794). The late Han dynasty 'Yellow Emperor's Inner Classic' (*Huangdi neijing* 黃帝內經, hereafter *Inner Classic*) also criticised medical incantation (see *zhuyou* below) as an outdated practice from bygone times, when bodies were more subtle and diseases less penetrating, arguing that acupuncture was more suited to recent eras (Unschuld and Tessenow 2011: *Suwen* 13.219–22). Some scholars understand these and other passages to represent a rejection among physicians of their competitors who practised spirit medium medicine, the *wu* (巫) and the emergence of a medicine oriented towards natural, not spiritual laws (Unschuld 2010, 2016: 15–19). Others acknowledge that the situation was nuanced, and that anachronistic frameworks of rationality, empiricism or individualism do not reflect the epistemological admixture of early sources (Lloyd and Sivin 2002, 239–51; Cook 2013; Stanley-Baker 2014). Early competition amidst groups was not grounded on such terms. It should also be borne in mind that given the time period of these Han dynasty and earlier sources, opponents identified in these physicians' texts were not 'Daoists' per se, as the religion was not created until 142 CE. Rather, they were spirit mediums, a term which by the Han dynasty functioned as an epithet of exclusion much as the term 'witch' did in medieval Europe, marking off social or epistemological out-groups without consistency or any attempt to define those practitioners on their own terms. The Daoist Church itself was also antagonistic towards *wu*, but competed with them on different terms (Stein 1979), even as it also sought to distinguish itself from physicians, transcendents and other contemporary practitioners, even supressing medical techniques such as acupuncture, herbs and moxibustion.

This chapter examines the current state of scholarship in order to discuss what classical Chinese medicine and practices used by Daoists for health, longevity and transcendence have in common. It argues, in addition, that conditions at the end of the Han dynasty influenced the Daoist religion, as it first emerged in an institutionalised form, to produce a distinctively medicalised imagination of the workings of the cosmos and individual destiny. With no single solutions to the complex problems of political and military chaos, famine and widespread epidemics of that time, Daoists produced composite ritual programmes that responded to the total situation. It was also from this period that religion and healing began to emerge as co-related, but increasingly distinct, domains of knowledge.

The state, epidemics and individual destiny: the complex problem of Ming 命

The Eastern Han dynasty (25–220 CE) was a watershed period which profoundly influenced the religious imagination of China. The same period saw the crystallisation of the medical classics (Chapter 7 in this volume); the *materia medica* tradition – the systematic identification and definition of individual medicinal products and their properties (Chapter 8 in this volume); the rise in popularity of transcendents (*xian* 仙) – practitioners whose self-cultivation exercises and ingestion of rare minerals and herbs were thought to bestow magical powers and supernatural longevity; as well as the birth of the Daoist religion – an organised body with ranks of ordained priests who performed communal and individual rites for healing and salvation. Taken together, these developments point to an inextricably entangled religious and medical imagination, with continuities between self-cultivation, morality and notions of bodily well-being, cure and disease. On the one hand, these continuities were the result of a sophisticated Han state coalescing its philosophical, technical and spiritual rationales about the body and disease. On the other, they were also produced by the collapse of the Han: internal political corruption, extensive military conflict and widespread epidemics led to a search for stable communities and systems of meaning and pragmatic responses that integrated various fields of activity.

This broader context is important for understanding the emergence of Daoism and its particularly medical imagination, and distinguishes the Daoist medical landscape from the two other major religions which spread medical knowledge at different times in China: Buddhism, which arrived in China during the first century CE, roughly at the same time that the Daoist religion was forming, and Christianity which came much later with the Jesuits during the seventeenth century (Chapters 16, 17 and 28 in this volume). All of them used medicine as a means to promote religious enterprise, as expressions of compassion, and conceived of their teachings as a kind of healing that solved the problem of worldly existence (Choa 1990; Salguero 2018). However, Christian and Buddhist teachings were based on disembodied concepts of salvation and disparaged the body as an obstacle to the realisation of religious goals. Daoist salvation, on the other hand, integrated a medical framework where the body is seen as the vehicle and locus of salvation, rather than an obstacle. To understand better how this came about, it is useful to understand the context within which it emerged.

The collapse at the end of the Han dynasty is commonly attributed to a series of weak child emperors, the rise of the empresses' families and corrupt eunuchs restricting access to power at court, resistance to which produced various outbreaks of military conflict and dissent in multiple regions across the country (Beck 1986). These skirmishes caused human migration as peasants fled war-torn areas, leaving crops untended and unprotected from raiding militias and armies, which, in turn, led to crop failure and famine. As these undernourished, stressed and exposed populations moved across the countryside in search of food and shelter, they brought disease with them, plagues which spread like wildfire across the land (Lin Fushi 2008a, 2008c). Plagues (*yibing* 疫病) covered the land with such speed they were considered to be ghost-borne, or ghost infixation (*guizhu* 鬼注), the result of the unquiet dead, improperly buried far from home, or bearing grudges against the living (Li Jianmin 2009; Nickerson 1997; Strickmann 2002). The constant impact and stress of epidemics should not be understated. These epidemics were so frequent that imperial histories record outbreaks on average every 7.6 years, with drastic mortality rates ranging between 45% and 85% of the local population (Li Wenbo 2004: 1; Chapters 16 and 17 in this volume).

To describe these interlinking catastrophes as simply problems of 'the state' would be to neglect the scale and complexity of the problem – the Han dynasties covered the largest land mass in Chinese recorded history by that time, a territory which was commonly referred to as 'all under Heaven' (*tianxia* 天下). This collapse constituted a 'complex problem', also known as a 'wicked' problem (Rittel and Webber 1973), that is, a situation which brooks no single causal explanation and no unitary solution. This confluence of problems unravelled the very fabric of the world as it was known, and could only be satisfactorily explained on a scale that was cosmic.

This complexity is perhaps best encapsulated as a problem of *ming* 命 (Stanley-Baker 2014), a multi-valent word that is usually translated as 'life' or 'destiny'. Its connotations, however, extended to the political, military, social, bodily and the cosmic (Lupke 2005: see especially chapters by Campany and Bokenkamp; Verellen 2019). The concept of *ming* reflects the potential for creating a fundamental order within these different domains. *Mingling* 命令 can mean 'to command', as in to give orders to troops, while the divine authority of the monarch to rule, the heavenly mandate, is *tianming* (天命). *Ming* refers to one's individual lifespan, as well as one's individual destiny – that is, one's overall trajectory or career, as well as the vicissitudes and minor misfortunes of daily life. These are influenced by one's astrological designation or natal destiny (*benming* 本命), which is based on the day and time of birth, and for which there are rituals to adjust one's chances in life.

The collapse of the Han dynasty was a threat to all these forms of *ming*, and saw a proliferation of methods, institutions and ideologies which tackled this complex problem in various ways. Medical texts, such as the *Inner Classic*, specifically addressed physical disease in the language of political rebellion which had to be controlled (Chapter 7 in this volume). Transcendents were people who exercised a power not only over the length of their biophysical lives; they transcended the social boundaries of the human order, rising beyond to different orders of being. They are portrayed as being literally 'above' the rule of the emperor. The hagiography of the Sire who Dwells by the Riverside (*Heshang gong* 河上公), for example, who floats above the earth when the emperor comes to visit, demonstrating that he has risen above earthly rule (Campany 2002: 91). Late Han self-cultivation texts, such as the commentary on the *Daode jing* 道德經 named after the aforesaid Sire, equated bodily self-mastery and spiritual self-governance with the mystical power to govern the country. Exercises lengthened life and staved off disease, while grain-fasting diets enabled one to live independently from an agricultural, settled economy. The Daoist ritual programmes initiated by the Celestial Master (*Tianshi* 天師) and Great Peace (*Taiping* 太平) movements conceived of disease as a manifestation of incorrect alignment with the moral order of the universe, while attempting to found an alternate state. In different ways, each responded to the complex problem of *ming*.

Overlapping goals: preventing disease, long life, delaying death, transcending mortality

Between the third and first centuries BCE, the cosmological principles of *yin* and *yang*, *qi* and the five agents coalesced into a coherent foundation for scientific practice, most fully expressed in the *Inner Classic*, while multiple styles of therapy remained in use (Harper 1999; Lo 2013; Chapter 1 in this volume). It has been argued that stages of this development may have taken place within a mixed current of ideas and technical practices that articulated a coherent, ethical universe governed by observable predictable laws, which the virtuous sage should learn through observation, and to which he should align himself to maintain personal and cosmic harmony. Some (Peerenboom 1993, among others) identify this as a movement named *Huanglao* 黃老, a term used by the father of the great Han historian Sima Qian 司馬遷 (ca. 145–ca. 86 BCE). The name is taken from the sagely figures of Laozi 老子 and the Yellow Emperor (Huangdi 黃帝), who are invoked in the titles of many related works, notably the *Inner Classic* itself. However, more recent scholarship argues that no social communities or 'schools' identified with the term, and that conceptual boundaries between philosophies such as Legalism and Daoism, reconstructed by later scholars, are inconsistent with the primary sources (Wang 2000, 182–83; Csikszentmihalyi and Nylan 2003). Such historiographic concerns aside, Han dynasty works such as the *Huainanzi* 淮南子 (139 BCE), which contains a plethora of chapters on astronomy, geography, philosophy and science in addition to embodied self-cultivation (Major *et al.* 2010), as well as manuscripts on silk and on bamboo strips excavated from an early Han dynasty tomb in Mawangdui, in Changsha, which was sealed in 168 BCE, exhibit common interests. These include legal culture, cosmogenesis and philosophies that came to be retroactively associated with 'Legalism' and 'Daoism', as well as in technical notions such as the five agents, and the emotional or spiritual aspects of the five organs (*wuzang* 五臟). In these ways, such texts were contiguous with the *Inner Classic*, which articulates disease and cure in political terms of rebellion (*ni* 逆) and chaos (*luan* 亂) vs control (*zhi* 治), emphasising a homology between individual bodily cultivation and managing the state.

> 是故聖人不治已病, 治未病;不治已亂, 治未亂。
>
> This is why the sage does not cure disease when already manifest, but cures it before it manifests; he does not control that which is already chaotic, but governs before chaos manifests.
>
> *(Huangdi neijing: Suwen 2.12)*

The aspiration to anticipate disease, and regulate the body so as to avoid falling ill and to lengthen one's years, took form in a vigorous culture of bodily cultivation aimed at 'nourishing life' (*yangsheng* 養生).[2] This broad set of exercises, diet, sexual cultivation, attention to daily rhythms and seasonal changes was an integral part of early medicine, and the textual works on the subject influenced the formation of the *Inner Classic* (Lo 2001). Attested to in texts dating as early as the fourth century BCE, such as the *Guanzi* 管子 and *Zhuangzi* 莊子, this culture flourished in the early Han dynasty, and is exemplified in multiple texts and images excavated from Mawangdui (Chapter 6 in this volume). The relationship to 'Daoism', that is to the philosophical and spiritual practices of quietude, inwardness and meditative breath practice, was differently marked in these texts and others, showing an ambivalence about their role in spiritual cultivation. Where the *Guanzi* affirmed the value of longevity practices and the closeness of spiritual self-cultivation to bodily cultivation, the *Zhuangzi* decried such unworthy huffings and puffings as beneath the attention of the true aspirant (Stanley-Baker 2019a).

It was those who aspired to become transcendents who made it their primary study to cultivate the body to become not only impervious to disease, but even death itself. Myths and hagiographies describe these figures as human beings possessed of knowledge of rare and exotic plant and mineral drugs, special fasting diets, and skills with talismans and spells which enabled them to ward off disease and death. Transcendent bodies acquired magical powers and spectacular longevity, being able to hear and see for thousands of miles, living from two to three hundred to thousands of years or even avoiding the tomb entirely. Myths and stories date back to 400 BCE, and texts on methods to achieve such powers can be found in the Western Han Mawangdui collection, but it was in the Eastern Han dynasty that they became dramatically more popular, during roughly the same period when the *Inner Classic* was compiled. Tales in works such as the 'Arrayed Biographies of Transcendents' (*Liexian zhuan* 列仙傳) att. to Liu Xiang 劉向 (77 BCE–6 CE), and the 'Biographies of Divine Transcendents' (*Shenxian zhuan* 神仙傳) by Ge Hong 葛洪 (283–343 CE), describe these individuals prominently displaying knowledge of drugs and curative arts in the marketplace (Ogata Toru 大形徹 2015), and transitioning from their careers as transcendents to those of doctors, and vice versa (Lin Fushi 2008b).[3] While these narratives are not factual historical accounts, they are not entirely fictitious either, and accurately communicate a cultural understanding of medicine and transcendence as intimately related domains of knowledge.[4]

The continuity between transcendence and medicine is further attested to in two important historical records of medicine in the Han dynasty. The first is the catalogue of the imperial library, the *Hanshu yiwen zhi* 漢書藝文志, which includes one section on 'Recipes and Techniques' (*fangji* 方技) that lists four parallel sections: 'Medical Classics' (*yijing* 醫經); 'Classic [drug] Recipes' (*jingfang* 經方); '[Arts of the] Bedchamber' (*fangzhong* 房中) and 'Divine Transcendence' (*shenxian* 神僊) (*Hanshu*, 30.1776–81). Although transcendence is listed last, indicating a lower hierarchy of importance (which is corroborated by the compiler's annotations), there is clearly a perceived continuity between these types of technical expertise (Stanley-Baker 2019a).

Second, the organisation of the earliest layers of the *materia medica* literature is influenced by transcendent drug culture. The 'Divine Farmer's *Materia Medica*' (*Shennong bencao jing* 神農本草經), produced in the first or second centuries CE, divides up known drugs into three hierarchical categories, where superior drugs benefit one's lifespan (*yangming* 養命), and lighten the body and prevent ageing; middling ones nourish inner nature (*yangxing* 養性) and strengthen the body against disease; and the lowest merely cure disease (*zhibing* 治病).[5] This triple-layered hierarchy was echoed in many transcendent texts from the period which distinguished transcendence, robust health and curing disease as three stages of spiritual cultivation, or classes of practice. The religious overtones of this ranking are unmistakable in this excerpt from the 'Four Classics of the Divine Husbandman' (*Shennong sijing* 神農四經).[6] Here, superior drugs do not merely lighten the body and lengthen the lifespan; they have far more dramatic effects:

上藥令人身安命延，昇為天神，遨遊上下，使役萬靈，體生毛羽，行廚立至。中藥養性，下藥除病，能令毒蟲不加，猛獸不犯，惡氣不行，眾妖併辟。

Superior drugs put people's bodies at ease and lengthen their lives; they ascend and become heavenly spirits, freely roaming above and below, and command myriad spirit minions. Their bodies grow feathers and wings, and they can immediately summon the travelling canteen. Middling drugs cultivate inner nature. Inferior drugs purge disease, and can cause poisonous insects not to multiply, wild animals not to offend, noxious *qi* not to circulate, and demon hordes to flee en masse.[7]

(*Baopuzi neipian* 11.196)

This family of related practices can be, and sometimes was, articulated in a graded scale of aspirations of health practices, which created a continuity between the goals in the *Inner Classic*, excavated recipe texts and transcendent literature. This scale ranged from curing manifest disease; to curing disease prior to manifestation; the use of drugs to strengthen the body and prevent disease; to gain miraculous command over bodily health and longevity; and to postpone death indefinitely, roam the heavens and earth, and gain dominion over spirits and the natural world.

This continuity was not merely one of common cosmology or shared health goals, but drew from a common repertoire of related techniques. Many of the same or similar cultivation practices used by transcendents to attain the heights of their spiritual goals were also used by mere mortals for the simple goals of preventing or curing disease, or extending a normal lifespan. The difference lay in the aspirations of the practitioner and claims made about the effects of the practice (Stanley-Baker 2006: 34–47).

Illness theodicy, disease transmission and the bureaucratic imagination: composite ritual solutions in the early Daoist Church

The rich medical imagination of Daoism, and its profusion of health-related practices is perhaps best understood by examining the emergence of the early church in the midst of the vicious cycle at the end of the Eastern Han dynasty, with power waning on the state's periphery and increasing epidemics, famine and military unrest all around. The Celestial Master movement began in roughly 142 CE, as some hagiographies relate the tale, when a spirit-medium healer Zhang Daoling 張道陵 (also Zhang Ling 張陵) had a vision of Laozi 老子 in the form of Taishang Laojun 太上老君 atop Crane-call Mountain (*Heming shan* 鶴鳴山) in the hills just west of the semi-independent border capital of Shu 蜀 (modern-day Chengdu). While accounts vary – some say he was given talismanic healing practices there, some say he went there to make alchemical drugs – the tradition agrees that during his revelation from the god, he established with him the Covenant of the Powers of the Orthodox Unity (*Zhengyi mengwei* 正一盟威). This contract, a bond based on righteous sacrifice, granted him healing powers and a new ritual of confession which, according to contemporary historical accounts, he then used to attract huge crowds of followers in the Sichuan basin (Kleeman 2016: 21–62). These settled into twenty-four small communities, or dioceses (*zhi* 治), which maintained their health through communal rites. Members re-established their covenant with the Dao through thrice-yearly communal gatherings, at which they tithed infrastructural support – the fixing of roads (a play on the words for 'the Way of Governance' *zhidao* 治道) and donating rice to travelling way stations or 'charity lodges' (*yishe* 義舍) in the region. In this way, the ritual programme functioned to some degree as a public health programme, responding to the issues of migration and famine. The healing rite of confession invited further health measures: tithing grain, a three-day period of fasting, purification and retreat in a chamber of silence, followed by a formal written confession of sin, submitted by a priest with prayers to the gods of the four directions (Strickmann 2002: 1–17). It is worth considering that these interventions would have functioned as rudimentary forms of quarantine, convalescence, dietary therapy and psychological counselling. The later Great Peace (*Taiping* 太平) movement sought to overturn the corrupt Han government, and conceptualised the plagues of the time as the result of a corrupt state, thus aligning political, spiritual, health and military goals. Like the Celestial Masters, they employed talismans and prayers, but also used needles and drugs to preserve health and stave off disease (Toshiaki 2008b).

Daoist illness theory resolved two important conceptual problems during this time of widespread epidemics, namely the problem of communicable illness, and the problem of evil. At the time, there was no theory of human-to-human disease transmission in classical Chinese medicine. Etiological theory in the *Inner Classics* focused on the body's own strength to resist external meteorological forces, such as wind, damp, dryness and so on (see the six *qi* in Chapters 1 and 2 in this volume), and internal emotions, including anger, joy, sadness worry and so on, but did not offer a theory for how disease was transmitted between humans. Terms for epidemics in imperial histories referred to their fast flowing nature (*yi* 疫), and described their suddenness and speed (*ji* 疾), but nothing about the means by which this occurred.

Daoist aetiological theory, however, considered that the fault for the sin which caused disease could be shared between people, across families, down generations, within communities: it was even meted out to members of a corrupt state. It was communicated by the ghosts of the wronged dead, who wreaked their vengeance on the living. Their fast movement across the land explained the rapid spread of disease, following vectors of military and famine migration. Furthermore, this theory also resolved the tacit connection between military conflict, mass migration and displacement of communities, and the rapid spread of disease. The disease that entered the body, even if it originated in one's own, or one's family's sins, was first processed and calculated by bureaucratic denizens of the afterlife, who then meted it out on living individuals via ghost infixation (*guizhu* 鬼注), that is the pouring of vengeful ghosts or their substance into the body, where it then manifested as disease. Disease thus formed an interlocking part of the breakdown of state, society and civil and military order.

Daoist disease theory resolved the problem of evil – why bad things happen to good people and vice versa (Lagerwey 2007). It provided what I term an 'illness theodicy', an explanation for why virtuous people without sin could be struck by disease that was the result of moral failure. Because punishment was meted out down the family line, on those who did not commit evil deeds, and who had no obvious, visible connection to them (Strickmann 2002: 1–59; Tsuchiya 2002; Toshiaki 2008a), disease could strike those who had not done anything observably wrong to merit it, offering a resolution to the unpredictability of epidemic disease transmission. In this way, Daoist disease notions had the potential to settle a sense of random threat and offered an outlet whereby the bereaved and the diseased could respond with proactive measures, such as ritual confession. This aetiology provided a way of comprehending the passing of loved ones, as can be seen in family records, which provide detailed descriptions of sepulchral lawsuits from deceased family members, as well as enemies, and the ritual prescriptions to ameliorate them (Nickerson 1997; Bokenkamp 2007: 130–57). It may by extension have also provided a means to cope with survivor's guilt amidst the random, wide spread of multiple pandemics. Read with close attention to the context and motivations for Daoist ritual, early sources show that many rituals were employed in the case of illnesses so conceived. In making these assertions, the Daoist religion maintained, like medical theory, that the universe operated on a series of predictable, regular, impersonal laws. Daoists distanced themselves, just as physicians did, from the mediumistic *wu*, whose deities were capricious and based their actions on personal favour and corrupt bargains rather than on a just and regular order (Kleeman 1994). Daoists theorised a moral universe in which the apparently random strike of disease was the result of explicable, but hidden, causes originating in human behaviour and foibles. The primary difference was that, for physicians, the relevant laws were purely natural laws of material forces, whereas Daoists modelled their spiritual world on an ideal spiritual bureaucracy which operated according to moral laws.

Despite these differences, both classical medicine and early Daoism shared a similar vision of a coherent, predictable world, which could be managed through regularised practices.

In this way, the unpredictable strike of epidemic disease could be explained, and ritual solutions offered to cure it in the present and prevent it in the future. Daoist aetiology did more than explain and produce observable order, i.e. biological or social continuity across generations – patterns discerned and produced by medical theory and Confucian norms. It resolved a fundamental problem of *dis*order that had become especially pertinent in the context of epidemics.

Blended practices

From the Han dynasty onwards, we find a profusion of health-related practices circulating across different communities and genres of writing, many of which have been claimed to be Daoist or related to Daoist aspirations. It must be recalled, however, that these practices were not exclusively or even intrinsically 'Daoist', even though they have come to be identified as such by some.

'Guiding and pulling' (*daoyin* 導引) is a genre of stretching and movement exercises mentioned as early as the fourth century BCE, and appearing in excavated Han dynasty practical manuals. During the Six Dynasties (220–589 CE), it became prominently adopted by Daoist and transcendent practitioners, but by the Sui and Tang dynasties (581–907 CE) was assimilated into court medicine (Despeux 1989; Engelhardt 1989). The *Treatise on the Causes and Symptoms of All Diseases* (*Zhubing yuanhou lun* 諸病源侯論) lists over a hundred 'methods to nurture life with guiding and pulling' (*yangsheng fang daoyin fa* 養生方導引法) as cures for individually specified syndromes. These are usually some form of exercises to stretch the body or circulate *qi* (*xingqi* 行氣) to nourish health or draw the illness out. Such practices also formed a central element in the Imperial Medical Academy, where the majority of staff appointments were masters of such methods. Many of the practices and texts used were attributed to Daoist masters during the Six Dynasties, but were assimilated into court medicine in the Sui/Tang period (Stanley-Baker 2006; Kohn 2012; Yang 2018, Chapter 6 in this volume).

'Arts of the Bedchamber' (*fangzhongshu* 房中術), or sexual cultivation, was a variety of methods which stimulated the female partner and enabled the male partner to retain semen during sex, circulating its nourishing powers throughout the body. These practices were highly regarded for preventing old age. The literature circulated widely beyond Daoist contexts: they are found in the Mawangdui cache, and a huge quantity is listed in the Han dynasty imperial catalogue (Harper 1987; Wile 1992). It also became an important theme in Six Dynasties Daoist cultivation (Eskildsen 1998: 76–8; Kirkland 2008). Related to these were the sexual initiation rites of early Celestial Masters, for which they were harshly criticised, perhaps unfairly, by their contemporaries. While the Celestial Masters were keen to distinguish their practice from mundane bodily cultivation (Mollier 2008; Raz 2008), they deserve comparison as rites for self-preservation and transformation through sexual congress. By the late imperial period, the relation of sexual cultivation to inner alchemy was a subject of heated debate in male-oriented inner alchemy practice, while the newly emergent female alchemy advocated cognate, but novel, physiological claims in relation to sexual abstinence (Valussi 2009, Chapter 30 in this volume; Hudson 2008).

Diet (*fushi* 服食) more generally included a wider set of practices of which grain-fasting was only a part. In this period, it encompassed a broader range of substances than is normally considered 'food' today, including the ingestion of herbs, rare minerals, alchemical products, talismans, different grades of *qi* 氣 and even not eating at all. These 'cuisines' were attributed physiological, social or spiritual hierarchies, the more refined products producing more refined bodies, sensibilities and miraculous ability (Campany 2005). Passages from numerous texts, such as the *Huainanzi* 淮南子, *Shennong jing* 神農經, *Dadai liji* 大戴禮記 and *Soushen ji* 搜神記, classify inhabitants of the natural world by their diet.[8] Contents vary but the scale is generally of the same ilk: eaters of meat were vicious but powerful; eaters of vegetables, calm and docile; grains, intelligent; lithic potions (minerals) could prevent ageing; those who lived on *qi* could not die. The Supreme Clarity (*Shangqing* 上清) sect of Daoism in particular, but others as well, practised visualisations of stellar, solar and lunar *qi*, which entered the body and refined it to the extent that the practitioner could rise up to the heavens, whether in meditation or bodily form was a matter of skill and extent of practice.

'Fasting from grains' (***bigu*** 辟穀) was nominally the avoidance of cultivated crop grains in preference for a reliance on wild-crafted herbs and plants, with various objectives: to achieve transcendence; as a staged practice among others on the path to become a Perfected (*zhenren* 真人), an adept even higher than a transcendent; and to purify the body from the Three Corpses and Nine Worms (*sanshi jiuchong* 三尸九蟲), demonic parasites which infested the body (Kohn 1995; Eskildsen 1998: 43–68; Campany 2002: 22–4; Despeux 2008a;

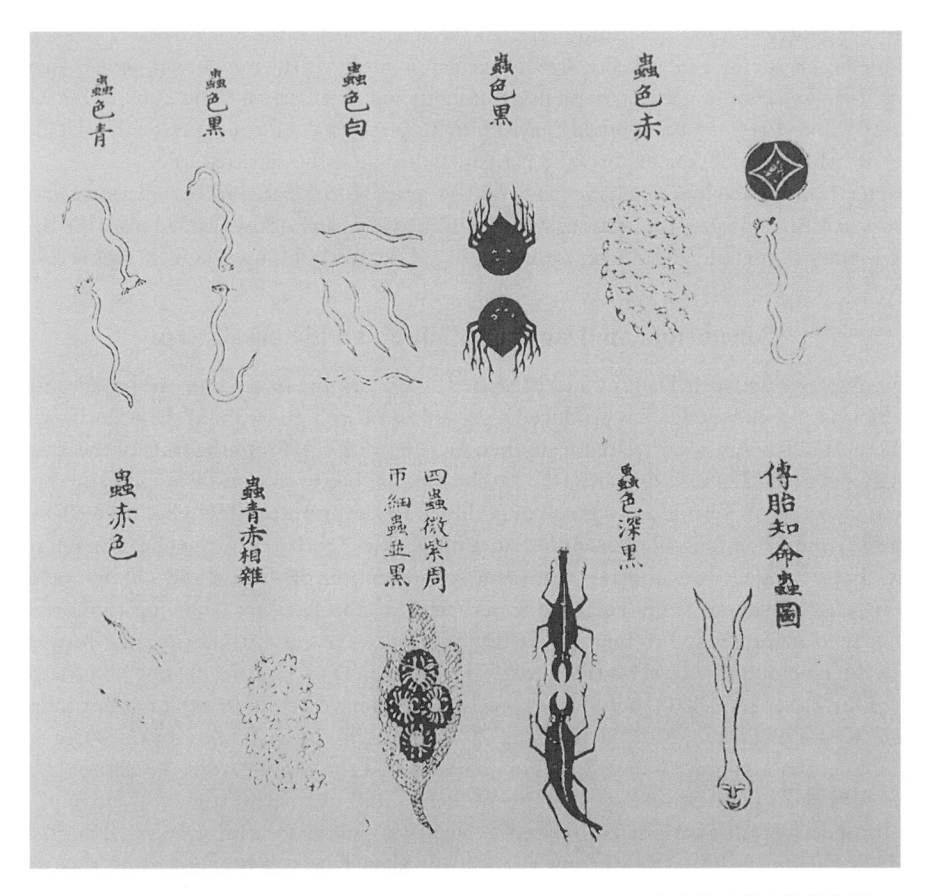

Figure 27.1 Parasites from *Taishang chu sanshi jiuchong baosheng jing* 太上除三尸九蟲保生經 (Most High Scripture on Conserving Life and Expelling the Three Corpses and Nine Worms) DZ 871 9b15a, Five Dynasties (907–960 CE)

Tadd 2012; Arthur 2013). Despite the phantasmagoric properties attributed to these parasitic beings, they were not purely figments of the imagination – many of their physiological effects were described in ways that evoke images of parasites and bugs recognised by modern public health authorities (Kohn 1993), and images in the Daoist canon, while mixed, are clearly partly derived from empirical experience (Figure 27.1).

Although earlier manuals are extant, the rise in popularity of grain-fasting after the Eastern Han gives the impression that the widespread famines at the end of the Han played a role in bolstering its appeal (Despeux 2008a). Living without cultivated food was not simply a low-carb fasting diet, but tantamount to an attempt to 'get off the grid' of agricultural life and out of civilised society, modes of life that could be starkly precarious during times of political and military strife (Campany 2005).

Talismans and Incantations (*fuzhou* 符咒) had the power to command beings from the spirit world (Bumbacher 2012). These magically scripted writs were a hallmark of emergent Han dynasty Daoist sects, such as the Celestial Masters (*Tianshi* 天師) and Great Peace (*Taiping* 太平) movements, and were often prepared, presented or ingested along with a spell or prayer. This means of addressing the spiritual causes of disease was similar to the practice of '**incantations to the origin [of the disease]'** (*zhuyou* 祝由), and together these formed a broad category of practices (Lin Fushi 2012).[9] While imperial medical texts distanced themselves from such practices, they never fully rejected the spirit realm – the Song dynasty (960–1279 CE) medical academy established *zhuyou* as the rubric for its thirteenth bureau, which focused on spells and charms in general, including talismans and incantations (Cho 2005). One Ming dynasty (1368–1644) doctor adopted *zhuyou* into his practice as a form of personalised 'talking cure', in which the doctor exhorts the patient, rather than the spiritual cause of the disease (Chapter 15 in this volume). Manuscripts on this practice circulated into the later twentieth century at least, and a good collection are fully scanned and can be downloaded from the Berlin State Library (Unschuld and Zheng Jinsheng 2012; Chapter 18 in this volume) (Figure 27.2).

Conclusion: on 'Daoist Medicine' as a historical term

The homology between Daoism and Medicine, their origins in a common, broad pool of technicians and knowledge has produced a great deal of ambiguity about the ways in which medicine and Daoism were related from then on (Figure 27.2). From the time of the Eastern Han dynasty (25–220 CE) onwards, the Daoist religion began to transform radically the religious landscape of China, and a range of bodily practices and attitudes to bodily well-being began to emerge and circulate in different communities, with various medical or religious aims. These ranged from disease treatment, to prevention of disease and old age, and to bodily self-cultivation; from 'curing disease before it manifests' to achieving immortality. Earlier scholarship tended to treat the Daoist religion as clearly distinct from the history of classical medicine and vice versa (Porkert 1974; Lu Gwei-Djen and Needham 1980; Maspero 1981; Unschuld 1985/2010), but more recently scholars have begun to explore their interrelation (Katz 1995; Davis 2001; Strickmann 2002; Hinrichs 2003; Stanley-Baker 2013; Sivin 2015). Scholars and practitioners in China have coined the term 'Daoist medicine' (*daojiao yixue* 道教醫學) to encapsulate a total phenomenon and distinctive historical current, and also to underscore the early scientific endeavours that emerged within Daoism (Hu Fuchen 1995; Gai Jianmin 2001; Xi Zezong et al. 2010; Shen Chen 2020).[10] In using this term, these latter studies have produced very useful synoptic views of the broad variety of healing and self-cultivation practices used by Daoists, covering large swaths of material. However, they have not accounted for the fact that 'Daoist medicine' is itself a modern term and

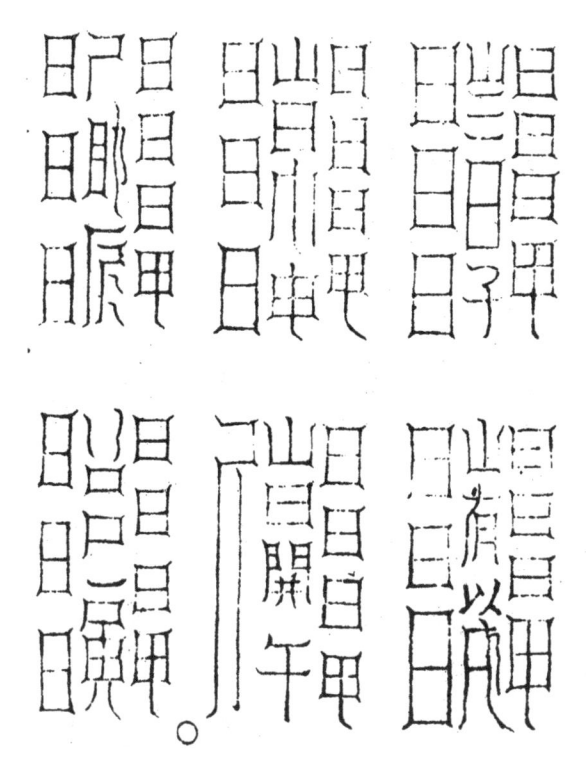

Figure 27.2 Talisman from *Taishang chu sanshi jiuchong baosheng jing* 太上除三尸九蟲保生經 (Most High Scripture on Conserving Life and Expelling the Three Corpses and Nine Corpse-worms) DZ 871 17a, Five Dynasties (907–960 CE)

that it introduces certain biases. This word (in Chinese), and its medieval Chinese analogue *daoyi* 道醫, is nowhere to be found in the Daoist canon itself, nor in other major historical collections (Stanley-Baker 2019b). This neologism privileges early 'science', a move expressly intended by Gai Jianmin 盖建民 (Gai 2001: 6–11), who argued that it could be used to promote more research and prove the relevance of Daoism in the modern scientific world. This is fine and laudable but has the unintended effect of tacitly separating out ritual as a discrete domain, in ways that do not account for how early practitioners organised knowledge. No historical figure or writer used this term before the twentieth century, and this fact alone indicates that it is a modern, academic, retrospective, analytical category, but not a term of art from the past. While it is useful for scholars and practitioners today to understand the interrelation of these different therapies and therapists, it has the effect of erasing complex relations between early sects of Daoism and how they thought about healing. Many of these studies leave unmentioned the Celestial Master programmes of ritual confession, considered the foundational core of liturgical Daoism, to say nothing of the fact that the early Church forbade the use of classical medical practices: acupuncture, herbs and moxibustion.

The object lesson here is that critical attention is necessary when applying organisational categories – we must pay attention to modern assumptions that they may contain. When approaching Daoist methods of healing, it is important to hold in abeyance familiar modern categories, like 'medicine', 'religion', 'science', 'ritual' and 'politics', in order to situate Daoists more accurately in their time and place, and understand how they organised and employed therapeutic knowledge.

Notes

1 Funding for the research, writing and publication of this paper was generously provided by the Max Planck Institute for the History of Science Dept. III, under the project 'Charting Interior and Exterior Worlds'; the Kolleg Forschungsgruppe 'Multiple Secularities' at the University of Leipzig, and Nanyang Technological University, under the project 'Situating Medicine, Religion and *Materia Medica* in China and Beyond'.
2 Although used earlier on to refer to other meanings, the term did not cover this general category of practice until the third century CE (Despeux 2008; Chapters 6 and 49 in this volume).
3 On these primary sources and the provenance of the versions which survive, see Kaltenmark (1953), Kubo Teruyuki 久保輝幸 (2011), Campany (2002). Campany contains a full translation of the *Shenxian zhuan*, including its variant editions.
4 See Campany (2009) on fictionality and the social roles of transcendence.
5 The dating of this text to the first or second century CE is agreed on by Schmidt (2006) and Harper (1998: 34) although Ma Jixing 馬繼興 (1990) argues that it dates as early as the third century BCE. Earlier *bencao* have been cited, but none survive. No singular, original, edition of this text survives either, but it has been reconstructed from multiple citations in later Tang and Song works. These, however, require some interpretation as the citations vary considerably. For an overview on the bibliographic history of this text, see Despeux (2015).
6 The *Sijing* no longer survives, but is cited in the famed alchemical work *Baopuzi neipian* 抱朴子內篇 (11.196). For a comparison of how the *Shennong sijing* and the *Shenong bencao* describe these categories, see Stanley-Baker (2013: 149–50, 273).
7 The travelling canteen, or mobile kitchen (*xingchu* 行廚), referred to here was a magical banquet summoned at will by ascetic masters who could produce exotic food and drink in resplendent table settings (Campany 2002: 29, 221–2; 2005: 46–7). This term also refers to large communal feasts which accompanied Celestial Masters feasting and fasting rituals (Mollier 2008; Stein 1979).
8 *Dadai liji* 大戴禮記, juan 81; *Huainan Honglie jie* 淮南鴻烈解 DZ 1176, 7.8b. An independent *Shennong sijing* does not survive with this passage intact, but it is cited in the fourth-century *Yangsheng yaoji*, which survives today as the *Yangxing yanming lu* 養性延命錄 DZ 838, 1.4b (Stanley-Baker 2006: 69–70, 130–1). For a translation of the *Soushen ji* passage, see DeWoskin and Crump (1996: 142).
9 On *zhuyou* across historical periods, see Cho (2005). For an example of a combined use of incantation with acupuncture, see Kleeman (2009).
10 These scholars, in particular Gai Jianmin and Xi Zezong *et al.*, refer to arguments by the historian of science in China, Joseph Needham, who regarded Daoism as a foundational philosophy for Chinese science. However, Needham's position has been roundly critiqued on the grounds that the term 'Daoism' is often unclearly defined and inconsistently used, and he privileges the earlier 'philosophy', but denigrates the religious tradition as a debasement of the 'purer' philosophical strands (Sivin 1968, 1978, 1995). Sivin called for more refined arguments, which attended specifically to the Daoist elements in any scientific endeavour. Xi Zezong 席澤宗, Jiang Sheng 姜生 and Tang Weixia 汤伟俠 (2010: 7–9, 469–594) acknowledge Sivin's call, and laudably attempt to isolate rationales for a particular brand of 'Daoist medicine', covering a wide range of useful material. However, their tacit, modern emphasis on 'science', as intrinsically distinct from religion, leads into simple but yawning categorical traps. For example, they arbitrarily separate out categories of literature as 'medical' and 'Daoist' when analysing imperial catalogues of technical literature (Ibid.: 478–9), despite the fact that the early catalogue itself makes no such distinction. It includes all four categories equally as subsets of technical literature, literally 'techniques and recipes' (*fangji* 方技). Furthermore, their study ignores the Celestial Masters (*tianshi* 天師), the founding sect of the Daoist religion and the basis of its enduring liturgical core.

Bibliography

Pre-modern sources

Baopuzi neipian 抱朴子內篇 (The Master Who Embraces Simplicity: inner chapters) 317 CE, Ge Hong 葛洪, in Wang Ming 王明 (ed.) (1981) *Baopuzi jiaoyi* 抱朴子內篇校釋 (Annotated Critical Edition of The Master Who Embraces Simplicity: inner chapters), Taipei: Liren shuju.
Da Dai Liji 大戴禮記 (Records of Ritual Matters by Dai the Elder) Western Han (202 BCE–9 CE), Dai De 戴德, in Gao Ming 高明 (ed.) (1984) *Da Dai liji jinzhu jinyi* 大戴禮記今註今譯 *(Dadai Liji* with Modern Commentary and Translation), Taipei: Taiwan Shangwu yinshuguan.

Hanshu 漢書 (Records of the Han) 111, Ban Gu 班固, Yan Shigu 顏師古 and Ban Zhao 班昭 (1962), Beijing: Zhonghua shuju.

Huainan Honglie jie 淮南鴻烈解 (Explanations on the *Great Achievements* by Huainan), DZ 1184.

Huangdi neijing suwen jiaozhu yushi 黄帝内经素问校注语译 (Yellow Emperor's Inner Classic: Plain Questions – Critically Compared, Annotated and Translated), Guo Aichun 郭靄春 (ed.) (1981), Tianjin: Tianjin kexue jishu chubanshe.

Lie xian zhuan 列仙傳 (Arrayed Traditions of Transcendents), DZ 294.

Shiji 史記 (Records of the Grand Historian) 78 CE, Sima Qian 司馬遷, in Pei Yin 裴駰, Yang Shuda 楊樹達 and Yang Jialuo 楊家駱 (eds) (1981) 史記集解 (Records of the Grand Historian, Collected Annotations), Taipei: Dingwen shuju.

Taishang chu sanshi jiuchong baosheng jing 太上除三尸九蟲保生經 (Life-Preserving Scripture from the Most High on Expelling the Three Corpses and Nine Corpse-worms), DZ 871.

Yangxing yanming lu 養性延命錄 (Records of Nourishing Inner Nature and Lengthening Life), DZ 838.

Modern sources

Arthur, S. (2013) *Early Daoist Dietary Practices: examining ways to health and longevity*, Lanham, MD: Lexington Books.

Beck, B.J.M. (1986) 'The fall of Han', in D. Twitchett and M. Loewe (eds) *The Cambridge History of China: Volume 1: the Ch'in and Han empires, 221 BC–AD 220*, Cambridge: Cambridge University Press.

Bokenkamp, S.R. (2007) *Ancestors and Anxiety: Daoism and the birth of rebirth in China*, Berkeley: University of California Press.

Bumbacher, S.P. (2012) *Empowered Writing: exorcistic and apotropaic rituals in medieval China*, St. Petersburg, FL: Three Pines.

Campany, R.F. (2002) *To Live as Long as Heaven and Earth: a translation and study of Ge Hong's traditions of divine transcendents*, Berkeley: University of California Press.

——— (2005) 'The meanings of cuisines of transcendence in late classical and early medieval China', *T'oung Pao*, 91: 1–57.

——— (2009) *Making Transcendents: ascetics and social memory in early medieval China*, Honolulu: University of Hawai'i Press.

Cho, P.S. (2005) *Ritual and the Occult in Chinese Medicine and Religious Healing: the development of zhuyou exorcism*, PhD thesis, University of Pennsylvania.

Choa, G.H. (1990) *'Heal the Sick' Was Their Motto: the protestant medical missionaries in China*, Hong Kong: Chinese University Press.

Cook, C.A. (2013) 'The Pre-Han Period', in T.J. Hinrich and L.L. Barnes (eds) *Chinese Medicine and Healing: an illustrated history*, Cambridge, MA: Belknap Press of Harvard University Press, pp. 5–30.

Csikszentmihalyi, M. and Nylan, M. (2003) 'Constructing lineages and inventing traditions through exemplary figures in early China', *T'oung Pao*, 89.1/3: 59–99.

Davis, E. (2001) *Society and the Supernatural in Song China*, Honolulu: University of Hawai'i Press.

Despeux, C. (1989) 'Gymnastics: the ancient tradition', in L. Kohn and Y. Sakade (eds) *Taoist Meditation and Longevity Techniques*, Ann Arbor: Center for Chinese Studies, University of Michigan, pp. 223–61.

——— (2008a) '*Bigu* 辟穀', in F. Pregadio (ed.) *The Encyclopedia of Taoism*, London: Routledge, pp. 233–4.

——— (2008b) '*Yangsheng* 養生', in F. Pregadio (ed.) *The Encyclopedia of Taoism*, London: Routledge, pp. 1148–50.

——— (2015) '*Shennong Bencao Jing* 神農本草經', in C.L. Chennault, K.N. Knapp, A.E. Dien and A.J. Berkowitz (eds) *Early Medieval Chinese Texts: a bibliographical guide*, Berkeley: Institute of East Asian Studies, pp. 264–68.

DeWoskin, K.J. and Crump, J.I. (trans.) (1996) *In Search of the Supernatural: the written record*, Stanford, CA: Stanford University Press.

Engelhardt, U. (1989) 'Qi for Life: longevity in the Tang', in L. Kohn and Y. Sakade (eds) *Taoist Meditation and Longevity Techniques*, Ann Arbor: Center for Chinese Studies, University of Michigan, pp. 263–96.

Eskildsen, S. (1998) *Asceticism in Early Taoist Religion*, Albany: State University of New York Press.

Gai Jianmin 盖建民 (2001) *Daojiao yixue* 道教医学 (Daoist Medicine), Beijing: Zongjiao wenhua chubanshe.
Harper, D. (1987) 'The sexual arts of ancient China as described in a manuscript of the second century B.C.', *Harvard Journal of Asiatic Studies*, 47.2: 539–93.
——— (1998) *Early Chinese Medical Literature: the Mawangdui medical manuscripts*, London; New York: Kegan Paul.
——— (1999) 'Warring states: natural philosophy and occult thought', in M. Loewe and E.L. Shaughnessy (eds) *The Cambridge History of Ancient China: from the origins of civilization to 221 B.C.*, Cambridge: Cambridge University Press, pp. 813–84.
Hinrichs, TJ (2003) *The Medical Transforming of Governance and Southern Customs in Song Dynasty China (960–1279 C.E.)*, PhD thesis, Harvard University.
Hu Fuchen 胡孚琛. (1995) 'Daojiao yiyao xueshu yao' 道教医药学述要 (Summary of Studies of Daoist Medicine), *Zhongguo zhongyi jichu yixue zazhi*, 1.4: 17–18.
Hudson, C. (2008) *Spreading the Dao, Managing Mastership, and Performing Salvation: the life and alchemical teachings of Chen Zhixu*, PhD thesis, Indiana University.
Kaltenmark, M. (1953) *Le Lie-Sien Tchouan: biographies légendaires des immortels Taoïstes de l'antiquité*, Pékin: Centre d'études sinologiques de Pékin.
Katz, P.R. (1995) *Demon Hordes and Burning Boats: the cult of Marshal Wen in late imperial Chekiang*, Albany: State University of New York Press.
Keightley, D.N. (2001) 'The "science" of the ancestors: divination, curing, and bronze-casting in late Shang China', *Asia Major*, XIV.2: 143–88.
Kirkland, R. (2008) '*Fangzhong shu* 房中術', in F. Pregadio (ed.) *The Encyclopedia of Taoism*, London: Routledge, pp. 407–11.
Kleeman, T.F. (1994) 'Licentious cults and bloody victuals: sacrifice, reciprocity, and violence in traditional China', *Asia Major*, 7.1: 185–211.
——— (2009) 'The ritualized treatment of stroke in early medieval Daoism and the secret incantation of the Northern Thearch', in F.C. Reiter (ed.) *Foundations of Daoist Ritual: a Berlin symposium*, Wiesbaden: Harrassowitz, pp. 227–38.
——— (2016) *Celestial Masters: history and ritual in early Daoist communities*, Cambridge, MA: Harvard University Asia Center.
Kohn, L. (1993) 'Kōshin: a Taoist cult in Japan: part I: contemporary practices', *Japanese Religions*, 18.2: 113–39.
——— (1995) 'Kōshin: a Taoist cult in Japan: part II: historical development', *Japanese Religions*, 20.1: 34–55.
——— (2012) 'Daoyin among the Daoists: physical practice and immortal transformation in Highest Clarity', in V. Lo (ed.) *Perfect Bodies: sports, medicine and immortality: ancient and modern*, London: British Museum, pp. 111–20.
Kohn, L. and Sakade, Y. (eds) (1989) *Taoist Meditation and Longevity Techniques*, Ann Arbor: Center for Chinese Studies, University of Michigan.
Kubo Teruyuki 久保 輝幸 (2011) '"Retsu sen den" no bōshitsu shita senden ni soku nitsuite' 「列仙伝」の亡失した仙伝2則について, *Jinbun gaku ronshū* 人文学論集, 29: 109–28.
Lagerwey, J. (2007) 'Evil and its treatment in early Taoism', in J.D. Gort, H. Jansen and H.M. Vroom (eds) *Probing the Depths of Evil and Good*, Leiden: Brill, pp. 73–86.
Li Jianmin 李建民 (2009) 'They shall expel demons: etiology, the medical canon and the transformation of medical techniques before the Tang', in J. Lagerwey and M. Kalinowski (eds) *Early Chinese Religion, Part One: Shang through Han (1250 BC–220 AD)*, Leiden; Boston, MA: Brill, pp. 1103–50.
Li Wenbo 李文波 (2004) *Zhongguo chuanranbing shiliao* 中国传染病史料 (Historical Records of Infectious Diseases in China), Beijing: Huaxue gongye.
Lin Fushi 林富士 (2008a) 'Zhongguo zhonggu shiqi de wenyi yu shehui' 中國中古時期的瘟疫與社會 (Epidemics and Society in Medieval China), in Lin Fushi (ed.) *Zhongguo zhonggu shiqi de zongjiao yu yiliao* 中國中古時期的宗教與醫療 (Medicine and Religion in Medieval China), Taipei: Lianjing chubanshe, pp. 3–28.
——— (2008b) 'Zhongguo zaoqi daoshi de "yizhe" xingxiang: yi *Shenxian zhuan* wei zhu de chubu tantao' 中國早期道士的「醫者」形象：以《神仙傳》為主的初步探討 (The image of Doctors among Early Medieval Daoists in China: preliminary explorations of the *Traditions of Divine Transcendents*), in Lin Fushi (ed.) *Zhongguo zhonggu shiqi de zongjiao yu yiliao* 中國中古時期的宗教與醫療 (Medicine and Religion in Medieval China), Taipei: Lianjing chubanshe, pp. 277–302.

——— (2008c) 'Donghan shiqi de jiyi yu zongjiao' 東漢時期的疾疫與宗教 (Epidemics and Religion in the Eastern Han), in Lin Fushi (ed.) *Zhongguo zhonggu shiqi de zongjiao yu yiliao* 中國中古時期的宗教與醫療 (Medicine and Religion in Medieval China), Taipei: Lianjing chubanshe, pp. 29–84.

——— (2012) '*Zhuyou* shiyi: yi *Huangdi neijing - Suwen* wei hexin wenben de taolun' 「祝由」釋義：以《黃帝內經‧素問》為核心文本的討論 (Interpretation of the term 'chanting to the origins' (*zhuyou*): discussion based on 'The Inner Canon of the Yellow Emperor: Plain Questions'), *Zhongyang yanjiu yuan lishu yuyan yanjiu suo jikan* 中央研究院歷史語言研究所集刊, 83.4: 671–738.

Lloyd, G. and Sivin, N. (2002) *The Way and the Word: science and medicine in early China and Greece*, New Haven, CT: Yale University Press.

Lo, V. (2001) 'The influences of nurturing life culture on the development of Western Han Acumoxa Therapy', in E. Hsu (ed.) *Innovation in Chinese Medicine*, Needham Research Institute Studies, Cambridge: Cambridge University Press, pp. 19–50.

——— (2013) 'The Han period', in T.J. Hinrich and L.L. Barnes (eds) *Chinese Medicine and Healing: an illustrated history*, Cambridge, MA: The Belknap Press of Harvard University Press, pp. 31–64.

Lu Gwei-Djen and Needham, J. (1980) *Celestial Lancets: a history and rationale of acupuncture and moxa*, Cambridge: Cambridge University Press.

Lupke, C. (ed.) (2005) *The Magnitude of Ming: command, allotment, and fate in Chinese Culture*, Honolulu: University of Hawai'i Press.

Ma Jixing 马继兴 (1990) *Zhongyi wenxian xue* 中医文献学 (Bibliographic Studies in Chinese Medicine), Shanghai: Shanghai kexue jishu chubanshe.

Major, J.S., Queen, S.A., Roth, H.D., Meyer, A.S., Puett, M. and Murray, J. (trans.) (2010) *The Huainanzi: a guide to the theory and practice of government in early Han China*, New York: Columbia University Press.

Maspero, H. (1981) *Taoism and Chinese Religion*, trans. F.A. Kierman, Amherst: University of Massachusetts Press.

Mollier, C. (2008) 'Chu 廚 "Cuisines"', in F. Pregadio (ed.) *The Encyclopedia of Taoism*, London: Routledge, pp. 539–44.

Nickerson, P.S. (1997) 'The great petition for sepulchral plaints', in S.R. Bokenkamp (ed.) *Early Daoist scriptures*, Berkeley: University of California Press, pp. 230–60.

Ogata Toru 大形徹 (2015) 'Rensen den ni miru doutokuteki sennin no houga' 「列仙傳」にみる道德的仙人の萌芽 (Emergent ethics in the *Liexian zhuan*), *Jinbun kagaku ronshyu* 人文学論集, 33: 29–38.

Peerenboom, R.P. (1993) *Law and Morality in Ancient China: the silk manuscripts of Huang-Lao*, New York: State University of New York Press.

Porkert, M. (1974) *The Theoretical Foundations of Chinese Medicine: systems of correspondence*, Cambridge, MA: MIT Press.

Raz, G. (2008) 'The way of the yellow and the red: re-examining the sexual initiation rite of Celestial Master Daoism', *Nan Nü*, 10.1: 86–120.

Rittel, H.W. and Webber, M.M. (1973) 'Dilemmas in a general theory of planning', *Policy Sciences*, 4.2: 155–69.

Salguero, C.P. (2018) '"This fathom-long body": bodily materiality and ascetic ideology in medieval Chinese Buddhist scriptures', *Bulletin of the History of Medicine*, 92.2: 237–60.

Schmidt, F.R.A. (2006) 'The textual history of the materia medica in the Han period: a system-theoretical reconsideration', *T'oung Pao*, 92: 293–324.

Shen Chen 申琛 (2020) *Daojiao yixue* 道教医学, Zhengzhou: Zhongzhou guji chubanshe.

Sivin, N. (1968) *Chinese Alchemy: preliminary studies*, Harvard Monographs in the History of Science, Cambridge, MA: Harvard University Press.

——— (1978) 'On the word 'Taoist' as a source of perplexity: with special reference to the relations of science and religion in traditional China', *History of Religions*, 17.34: 303–30.

——— (1995) *Medicine, Philosophy and Religion in Ancient China*, Aldershot: Variorum.

——— (2015) *Health Care in Eleventh-Century China*, New York: Springer.

Stanley-Baker, M. (2006) *Cultivating Body, Cultivating Self: a critical translation and history of the Tang dynasty Yangxing yanming lu* (Records of Cultivating Nature and Extending Life), MA thesis, Indiana University, Bloomington.

——— (2013) *Daoists and Doctors: the role of medicine in six dynasties Shangqing Daoism*, PhD thesis, University College London.

——— (2014) 'Drugs, destiny, and disease in medieval China: situating knowledge in context', *Daoism: Religion, History and Society*, 6: 113–56.

——— (2019a) 'Health and philosophy in pre- and early imperial China', in P. Adamson (ed.) *Health: a history*, New York: Oxford University Press, pp. 7–42.

——— (2019b) '*Dao*ing medicine: practice theory for considering religion and medicine in early imperial China', *East Asian Science Technology and Medicine*, 50: 21–66.

Stein, R.A. (1979) 'Religious Taoism and popular religion from the second to the seventh centuries', in H. Welch and A. Seidel (eds) *Facets of Taoism*, New Haven, CT: Yale University Press, pp. 53–81.

Strickmann, M. (2002) *Chinese Magical Medicine*, Stanford, CA: Stanford University Press.

Tadd, M. (2012) 'The power of parasites and worms', in B.S. Walter and M. Tadd (eds) *Parasites, Worms, and the Human Body in Religion and Culture*, New York: Peter Lang, pp. ix–xxxv.

Toshiaki, Yamada (2008a) 'Chengfu 承負', in F. Pregadio (ed.) *The Encyclopedia of Taoism*, London: Routledge, pp. 265–6.

——— (2008b) '*Taiping* 太平: great peace; great equality', in F. Pregadio (ed.) *The Encyclopedia of Taoism*, London: Routledge, pp. 937–8.

Tsuchiya, Masāki 土屋晶明 (2002) 'Confessions of sins and awareness of self in the *Taiping Jing*', in L. Kohn and H.D. Roth (eds) *Daoist Identity: history, lineage, and ritual*, Honolulu: University of Hawai'i Press, pp. 39–57.

Unschuld, P.U. (1985/2010) *Medicine in China: a history of ideas*, Berkeley: University of California Press.

——— (2010) 'When health was freed from fate: some thoughts on the liberating potential of early Chinese medicine', *East Asian Science Technology and Medicine*, 31: 11–24.

——— (2016) *Huang Di Nei Jing Ling Shu: the ancient classic on needle therapy*, Berkeley: University of California Press.

Unschuld, P.U. and Tessenow, H. (2011) *Huangdi Neijing Suwen: an annotated translation of Huang Di's inner classic – basic questions*, Berkeley: University of California Press.

Unschuld, P.U. and Zheng Jinsheng (2012) *Chinese Traditional Healing: the Berlin collection*, Leiden; Boston, MA: Brill.

Valussi, E. (2009) 'Female alchemy: an introduction', in R.R. Wang and L. Kohn (eds) *Internal Alchemy: self, society, and the quest for immortality*, St. Petersburg, FL: Three Pines Press, pp. 141–62.

Verellen, F. (2019) *Imperiled Destinies: the Daoist quest for deliverance in medieval China*, Boston, MA: Harvard University Press.

Wang, A. (2000) *Cosmology and Political Culture in Early China*, Cambridge; New York: Cambridge University Press.

Wile, D. (1992) *Art of the Bedchamber: the Chinese sexual yoga classics: including women's solo meditation texts*, Albany: State University of New York Press.

Xi Zezong 席澤宗, Jiang Sheng 姜生 and Tang Weixia 汤伟侠 (2010) *Zhongguo daojiao kexue jishu shi: Nanbeichao Sui Tang Wudai* 中国道教科学技术史：南北朝隋唐五代卷 (History of Chinese Daoist Science and Technology: Volume 'Southern and Northern Dynasties; Sui; Tang and the Five Dynasties'), Beijing: Kexue chubanshe.

Yang, D. (2018) *Prescribing 'Guiding and Pulling': the institutionalisation of therapeutic exercise in Sui China (581–618 CE)*, PhD thesis, University College London.

28
BUDDHIST MEDICINE
Overview of concepts, practices, texts, and translations

Pierce Salguero

'Buddhist medicine' (Ch. *foyi* 佛醫 or *fojiao yixue* 佛教醫學, Jp. *bukkyō igaku* 仏教医学) is a modern term commonly used by East Asian scholars and devotees alike to refer to a body of medical knowledge that was introduced to East Asia via the transmission of Buddhist texts (Salguero 2015). These texts were translated and composed in China between the second and the eleventh centuries CE, based on source materials imported from many parts of South, Southeast, and Central Asia. Despite the ongoing efforts of historical and contemporary East Asian exegetes to present Buddhist medicine as a coherent system of medical knowledge, the perspectives preserved in these diverse texts do not represent a single point of view. They are best thought of as a series of snapshots indicative of many local variations on a central theme. For historians, the very heterogeneity of the source base makes it invaluable as evidence of the development of medical thought in India, the reception of foreign medicine in China, and the cross-cultural exchange of medicine across first-millennium Eurasia.

Many of the main concepts underpinning Buddhist medicine ultimately derive from the Indo-European intellectual context, and for this reason some of its central doctrines bear some similarities with Indian, Greco-Roman, Islamic, and other Eurasian medical traditions. For example, the human body is generally said to be composed of the Great Elements (Sk. *mahābhūta*; Ch. *sida* 四大), Earth, Water, Fire, and Wind (to which are frequently added Space and Consciousness). Among the most important causes or symptoms of disease are the so-called Three Humours or Three Defects (Sk. *tridoṣa*; Ch. *sandu* 三毒 or *sanbing* 三病), Wind, Bile, and Phlegm. While reminiscent of Greco-Roman humoral medicine, these concepts are even more closely related to the principal doctrines of Āyurveda, a form of Indian medicine whose foundational texts are datable to between the third century BCE and about 600 CE (see Wujastyk 2003). However, since Buddhist texts often present variant formulations of even the most basic Āyurvedic doctrines, it is probable that they reflect separate streams of medical thought current among distinct interpretive communities (Zysk 1998; Mazars 2008).

Similarities with other traditions notwithstanding, the medical ideas presented in Buddhist texts are usually framed in ways that are uniquely Buddhist and that feed into Buddhist religious and philosophical discourses. The Great Elements, for example, are introduced as objects of meditation and are connected with the Buddhist virtues of impermanence and non-attachment (Salguero 2014: 71–3, 2018b). The Three Defects are commonly used as metaphors for the mental poisons of Greed, Aversion, and Delusion (Demiéville 1985: 69–71).

Ideas about foetal development are introduced within narratives that focus on karmic retribution and the need to escape the cycle of rebirth (Kritzer 2014; Salguero 2014: 74–6). Bathing and personal hygiene are explicitly and implicitly related to moral virtue and spiritual purity (Heirman and Torck 2012; Salguero 2014: 76–8, 112–16). Always, healing activities of any kind were in Mahāyāna Buddhism considered integral parts of the practice of compassion and skilful means that should be exhibited by a devotee (Salguero 2018a).

In addition to new doctrines, Buddhism also introduced China to a pantheon of Indian deities and semi-divine heroes who were reputed to have potent healing powers (Birnbaum 1989a). One of the most widely venerated Buddhist figures in East Asia, the bodhisattva Guanyin 觀音 (Sk. Avalokiteśvara), began to appear in popular Chinese tales about miraculous cures as early as the fourth century (Campany 1993, 1996, 2012a: 49–51; Kieschnick 1997: 103–5; Yü 2001: 58–84). The Buddha of Infinite Light or Infinite Life (Sk. Amitābha or Amitāyus, Ch. Wuliangguang fo 無量光佛 or Wuliangshou fo 無量壽佛) was also credited with performing medical miracles in Buddhist writings from the early medieval period. By the middle of the Tang dynasty (618–907), the Master of Medicines Buddha (Ch. Yaoshifo 藥師佛; Sk. Bhaiṣajyaguru) had become a major focal point of Buddhist worship in many segments of society (Birnbaum 1989b; Ning 2004; Shi 2020). In the Song dynasty (960–1279), the deities Ucchuṣma (Ch. Wushusemo 烏芻瑟摩 or Wuchushamo 烏芻沙摩) and Nāgārjuna (Longshu 龍樹) became objects of cultic devotion among certain groups of Buddhist ritual healers (Davis 2001). Buddhists throughout history recognized numerous other divine figures who could be called upon to heal, to protect against disease, or to ensure the safety of the state from epidemics and other calamities.

Rituals calling upon the power of these deities for healing and protection ranged in size and complexity from merit-making by individual devotees to massive imperially sponsored ceremonies involving thousands of monks (see Zhiru 2020, trans. Salguero et al. 2017: 286–89). Such measures consisted of exoteric practices – for example, giving offerings (Ch. *gongyang* 供養; Sk. *pūjā*), praying to deities, reciting scriptures, performing repentance rituals, patronizing Buddhist monks known as capable healers, or donating bathhouses to local monasteries – which could be engaged in by devotees of all types in order to purify their karma and positively affect their health (Birnbaum 1989a, 1989b; Salguero 2013, 2014: 76–86; Lowe 2014; trans. Salguero 2017: ch. 25, 26, 31). Buddhist therapeutics also included the occult rites of Tantric or Esoteric Buddhism (*mijiao* 密教) – such as reciting healing incantations (Sk. *dhāraṇī*; Ch. *zhou* 咒), invoking or channelling deities, creating protective seals or talismans; consecrating water, medicines, or healing implements, and constructing mandalas to purify the body and mind – which almost always required specialized knowledge and initiation, and which were thought to have profoundly transformative effects (Satirajan Sen 1945: 85–95; Davis 2001; Strickmann 2002; Mollier 2008; Despeux 2010; McBride 2011; Salguero 2014: 86–92; trans. Salguero 2017: ch. 28, 29, 30, 45).

In addition to rituals, Buddhist texts describe a wide variety of other types of healing practices. Some provide advice on maintaining a healthy diet or making seasonal adjustments to one's regimen (e.g. trans. Satirajan Sen 1945: 76–84; trans. Salguero 2017: ch. 4). Others introduce meditations to maintain health or alleviate disease (trans. Greene 2021: 249–300; trans. Salguero 2017: ch. 36, 37). Examples of the latter include concentration exercises, breathing techniques, absorption meditations intended to manipulate the balance of the Great Elements, and the visualization of deities performing procedures such as massage or surgery on one's body. Simply reflecting on the ultimate wisdom that the physical body is an illusory mental construct is also said to eradicate spontaneously all diseases (Salguero 2017: 387).

While there are a few extant Chinese Buddhist texts that explain how to perform ophthalmological surgical procedures (Deshpande 1999, 2000; trans. Salguero 2017: ch. 54), abdominal and cranial surgeries are more frequently described in narratives about Buddhist deities and semi-divine healers. These became objects of considerable fascination for devotees. Tales about Buddhist healers with all sorts of wondrous healing powers began to appear in copious numbers in the fifth to sixth centuries (Wright 1948; Fu and Ni 1996; Kieschnick 1997; Salguero 2009, 2014: 121–40, 2020a; trans. Campany 2012a; Salguero 2017: ch. 21). Such narratives often centre around foreign masters who are said to have studied medicine as part of their monastic training before arriving in China. Whatever their origins, these heroes are depicted performing miraculous feats of healing for important individuals and the general population by means of a wide range of Indian and Chinese therapeutic techniques. By the Tang period, the mysterious and potent 'eminent monk' (*gaoseng* 高僧) had become something of a stock character in the Chinese literary landscape. In many of these tales, Buddhist monks are compared favourably to rival Daoist adepts, doctors, spirit mediums, and other healers. Such stories reflect the real-world competition for patronage among religious and medical sectarians in medieval China, and the importance of healing in their contests for cultural capital (Campany 2012b; Salguero 2014: 59–65, 2020a).

While healing narratives are reflections of the medieval Chinese literary imagination, we also have more reliable historical evidence about certain Chinese monastics who studied, practised, or wrote about Buddhist medicine. The most important among these figures, in terms of the quantity of information he left behind, is the pilgrim Yijing 義淨 (635–713). A native of China, Yijing travelled to northeastern India to learn at the monastic university Nālandā, and while doing so acquired some knowledge about Indian Buddhist medical traditions. Among Yijing's copious writings is a travelogue that reports upon many facets of life in the Indian monastery, giving historians a unique window onto the medical and hygienic practices of the residents (T. 2125; Heirman & Torck 2012; Salguero 2014: 112–16; trans. Li 2000; Salguero 2017: ch. 16). Apart from Yijing's writings, there are few other accounts of Buddhist monastic healing practices in the medieval period which are both reliable and detailed. However, what materials are available in the historical records attest that Tang rulers sought out Indian medical knowledge, that Buddhist monks were commonly employed as ritual healing specialists, and that Buddhist medicine became fashionable in Tang China (Tansen Sen 2001; Chen Ming 2013; Despeux 2017). We also know that monastic complexes in the Sui-Tang period often included not only bathhouses but also infirmaries, medical dispensaries, and hospice facilities (Despeux 2010, 2020). These institutions were important resources for the laity as well as for the resident monastics, a social function that intensified as Buddhist charities were increasingly absorbed into the nascent public health apparatus of the state (Liu 2008).

Though the bulk of the translation into Chinese of texts related to Buddhist medicine took place between the fifth and eighth centuries, discourses about healing can be found in virtually all Buddhist genres from every period of translation activity (see, e.g., trans. Salguero 2017). Relevant materials are found in even the most revered Buddhist scriptures. The *Vimalakīrti Sutra*, for example, presents an extended argument about the illusory nature of the body and disease in a chapter entitled 'Inquiring about illness' (T. 475; trans. Watson 1997: 64–74; Richter 2020). The *Lotus Sutra*, one of the most celebrated Buddhist scriptures in East Asia, dedicates a chapter to extolling the Medicine King Bodhisattva (Ch. Yiwang pusa 醫王菩薩) (T. 262; trans. Watson 1993: 280–9). The *Sutra of Golden Light*, a text primarily concerned with Buddhist models of kingship, contains a chapter that outlines Indian medical theory (T. 665.24; trans. Salguero 2017: ch. 4) as well as a chapter that describes ritual bathing in medicated water for strength and protection (T. 665.15; trans. Skjaervø 2004: 172–181). The various monastic disciplinary codes translated into Chinese over the early medieval period contain sections on medicine in which they outline the allowable and non-allowable therapeutic procedures, rules on the storage of medicines and protocols for interacting with the sick (T. 1421, 1425, 1428, 1435, 1448; trans. Salguero 2017: ch. 13, Salguero et al. 2017: 281–83). There are even two Indian medical treatises embedded within the Chinese Buddhist Tripitaka, texts which have clear connections with Āyurvedic medicine and demonology (T. 1330, 1691; trans. Bagchi 1941, 2011). The most well-known translated scripture pertaining to healing in the corpus, however, is the *Sutra of the Master of Medicines Buddha*, the core text devoted to the principal deity of healing (Birnbaum 1989b). This text exists in multiple Chinese translations (T. 449–51, T. 1331.12) and is accompanied by a series of ritual manuals (T. 922–8; trans. Salguero 2017: 299–301), leaving us in no doubt about its centrality in Buddhist healing practice in medieval China.

Medical topics are well-represented in domestic Chinese Buddhist writings such as commentaries, compilations, reference works, and 'apocryphal sutras' (i.e. pseudotranslations composed anew in China that purport to be authentic translations of Indic texts). Among the more important examples available in English translation are admonishments to preserve morality in promise of divine protection (Lowe 2014; Goble 2017), as well as ritual instructions that synthesize Buddhist and indigenous Chinese methods (Salguero 2017: ch. 45). Additionally, a set of sixth-century meditation manuals by Zhiyi 智顗 (538–597), the founder of the Tiantai School 天台宗, describes both Chinese and Indian models of diagnosis and meditative therapies (T. 1911, 1915, 1916; trans. Salguero 2017: ch. 37). The travelogue by Yijing (T. 2125) has already been mentioned. Also of interest are multiple Chinese versions of the biography of Jīvaka, the Buddhist physician par excellence, in which the hero is depicted performing a number of Indian therapies, including abdominal and cranial surgeries (T. 553, T. 1428: 851–4, T. 2121: 166–170; Salguero 2009). Finally, an influential pair of encyclopaedias compiled by Daoshi 道世 (?–683) collected together scriptural passages from across a wide range of Buddhist sutras, disciplinary texts, commentaries, and miracle tales, offering a more comprehensive picture of Chinese Buddhist medical knowledge (T. 2122.95, 2123.29; Salguero 2014: 109–12; Hsu 2018).

While hard and fast distinctions between these many categories of sources cannot be made, broadly speaking, Buddhist discourses on healing across the vast Chinese corpus tend to be presented differently in genres intended mainly for monastics versus those intended for wider lay audiences (Salguero 2014: 67–95). The former – including philosophical treatises, meditation manuals, and monastic disciplinary treatises, among other genres – tend to connect Buddhist medicine with ascetic Buddhist doctrines such as emptiness and non-self (Salguero 2018b). The latter types of texts, however, tend to celebrate the healing powers of Buddhas and bodhisattvas, the ability of lay devotees to rectify their karma, and the efficacy of Buddhist rituals in vanquishing disease (Salguero 2020a). The more complex composite texts with varied histories of circulation among different communities, of course, intermix both registers in complicated ways.

Another way to analyse the sprawling corpus of Chinese texts on Buddhist medicine is to distinguish on the basis of translation strategies. Buddhist writings on medicine are often infused with foreign metaphors and translation tactics such as transliteration and neologisms, and thus stand in stark contrast to mainstream Chinese medical discourses about *qi*, *yin-yang*, and the Five Agents. At the same time, other Buddhist translators and commentators often attempted to resituate foreign medical knowledge in domestic cultural and social contexts, explaining Indian medical doctrines using Chinese medical vocabularies. The use of these

Buddhist medicine

approaches was influenced by the authors' sociopolitical contexts, intended audiences, and individual biographical circumstances (Salguero 2014).

Efforts to build conceptual and linguistic bridges between Indian and Chinese medical thought peaked in the period from the Sui to the mid-Tang, when creative cross-cultural mediation allowed for the absorption of many aspects of Indian medical doctrine in China and Chinese medical doctrine into Buddhist texts. The received Buddhist literature from this period, the extant writings of Sui-Tang physicians; the competing writings of Daoist sectarians, and the recovered manuscripts from medieval sites such as Dunhuang and Turfan collectively leave us with little doubt that Buddhist practitioners played a major role in the contemporary medical marketplace or that Indian ideas and practices were important features of the Chinese medical landscape (Zhu 1999; Lo and Cullen 2005; Chen Ming 2005a, 2005b, 2013; Li and Shi 2006; Mollier 2008; Despeux 2010, 2020; trans. Salguero *et al.* 2017: 290–93).

Why, then, did Buddhist medicine not play a larger role in later Chinese medical history? One significant reason is the classicist movement that began to gather steam in the late eighth to ninth centuries. Rising nativist and xenophobic sentiments culminated in the violent repression of Buddhist institutions and clerics in 842–845. While Buddhism's cultural and social significance eventually recovered from this blow, interest in the Indo-Sinitic medical syncretism of the early medieval period began to decline. In the Song, the government-led reformation of medicine elevated texts from the pre-Buddhist era (Goldschmidt 2009). By that time, certain elements of Indian religion and medicine had been inseparably integrated into Chinese thought and practice. Certain fields of medicine – such as pharmacology, ophthalmology, and embryology – exhibited many traces of Indian medical influence (Unschuld 1998; Deshpande 1999, 2000, 2003, 2008; Xue 2002; Chen Ming 2005c, 2007; Li and Shi 2006; Deshpande and Fan 2012; Salguero 2017: ch. 52, 53, 54). However, in the Song, knowledge expressed in Indian medical vocabularies was sidelined from official Chinese medical discourses, increasingly replaced by a neo-classical orthodoxy.

While Indian medical doctrines receded in importance in official circles, however, other aspects of Buddhist medicine continued to be widely practised across all layers of East Asian society in the post-medieval period and eventually became durable features of the East Asian medical landscape. Throughout Chinese history, monks continued to be celebrated as healers, monasteries continued to offer health services to their communities, the devout continued to organize medical charities, and Buddhist deities continued to receive the prayers of the sick (Wu 2000; Chen 2008; Liu 2008; Huang 2009). Today, Buddhist ritual, literature, and lore persist as important fonts of popular healing knowledge in Chinese communities worldwide (Salguero 2020b: ch. 9, 14, 15, 26, 27). Although in the long run Buddhist medicine was not to become as formative in China as it was in Southeast Asia or Tibet, it remains a significant minor theme without which our picture of the history of Chinese medicine cannot be complete (e.g. trans. Salguero 2020b: ch. 3).

List of Chinese texts on Buddhist medicine

As mentioned above, there are many references to medical topics scattered throughout the Chinese Buddhist corpus. As it is impossible to list every relevant text in the allotted space, this section includes only the most important and it is by no means comprehensive (see also Salguero 2018c). For convenience, the received texts below are subdivided by the traditional designations of exoteric/esoteric scripture, monastic discipline, and commentary, followed by a list of archaeologically recovered manuscripts. They are listed in order of their reference numbers in the *Taishō-Era Newly Revised Tripitaka* (Jp. *Taishō shinshū daizōkyō*; Ch. *Dazheng xinxiu dazang jing* 大正新脩大藏經). The latter is the most widely cited collection of historical Buddhist sources, available in print in Takakusu and Watanabe (1924–35) or in corrected digitized form at http://cbeta.org and http://21dzk.l.u-tokyo.ac.jp/SAT.

Exoteric scriptures

T. 150A, *Foshuo qichu sanguan jing* 佛說七處三觀經 (*Sutra on the Seven Points and Three Contemplations*). Attributed to An Shigao, fl. 148–170.

T. 219, *Foshuo yiyu jing* 佛說醫喻經 (*Sutra on the Medical Simile*). Translated by Dānapāla, late tenth century.

T. 262, *Miaofa lianhua jing* 妙法蓮華經 (*Lotus Sutra*), Chapter 23. Translated by Kumārajīva, 406. [Alternate translations T. 263.10, T. 264.22.]

T. 293, *Da fangguang fo huayan jing* 大方廣佛華嚴經 (*Flower Ornament Sutra*), pp. 710–12. Translated by Prajñā, ca. 800.

T. 317, *Foshuo baotai jing* 佛說胞胎經 (*Sutra on the Embryo*). Translated by Dharmarakṣa, 281 or 303. [Alternate translations T. 310.13, T. 310.14, T. 1451: 251a-262a.]

T. 374, *Da banniepan jing* 大般涅槃經 (*Mahāparinirvāṇa-sūtra*), pp. 378–79. Translated by Dharmakṣema, ca. 421. [Alternate translation T. 375.3.]

T. 451, *Yaoshi liuli guang qifo benyuan gongde jing* 藥師瑠璃光七佛本願功德經 (*Medicine Buddha Sutra*). Translated by Yijing, 707. [Alternate translations T. 449, T. 450, T. 1331.12.]

T. 475, *Weimojie suoshuo jing* 維摩詰所說經 (*Vimalakīrti Sutra*), Chapter 5. Translated by Kumārajīva (344–413). [Alternate translations T. 474.5, T. 476.5.]

T. 554, *Foshuo Nainü Qipo jing* 佛說奈女耆婆經 (*Āmrapālī and Jīvaka Sutra*). Unknown author, fifth century. Translation misattributed to An Shigao (fl. 148–70). [Alternate translations T. 553, T. 1428: 851–4, T. 2121: 166–170.]

T. 620, *Zhi chanbing miyao fa* 治禪病祕要法 (*Secret Essential Methods for Treating the Maladies of Meditation*). Translated or compiled by Juqu Jingsheng, 455.

T. 665, *Jin guangming zuisheng wang jing* 金光明最勝王經 (*Sutra of Golden Light*), Chapters 15, 24. Translated by Yijing, 703. [Alternate translations in T. 663, T. 664.]

T. 701, *Foshuo wenshi xiyu zhongseng jing* 佛說溫室洗浴眾僧經 (*Sutra on Bathing the Sangha in the Bathhouse*). Translation misattributed to An Shigao (fl. 148–70).

T. 793, *Foshuo foyi jing* 佛說佛醫經 (*Sutra on the Buddha as Physician*). Translated by Zhu Lüyan and Zhiyue, after 230.

Esoteric scriptures and ritual manuals

T. 922, *Yaoshi liuliguang rulai xiaozai chu'nan niansong yigui* 藥師琉璃光如來消災除難念誦儀軌 (*Ritual Procedure for Recitation to the Master of Healing, the Lapis Lazuli Radiance Tathāgata, for Eliminating Disaster and Escaping Hardships*). By Yixing (684–727). [Related texts T. 923–928.]

T. 1028A, *Foshuo hu zhu tongzi tuoluoni jing* 佛說護諸童子陀羅尼經 (*Dhāraṇī for the Protection of All Children*). Translated by Bodhiruci (d. 727).

T. 1028B, *Tongzi jing niansong fa* 童子經念誦法 (*Method of Recitation of the Sutra on the [Dhāraṇī for the Protection of All] Children*). Translated by Śubhakarasiṃha (637–735).

T. 1043, *Qing Guanshiyin pusa xiaofu duhai tuoluoni zhoujing* 請觀世音菩薩消伏毒害陀羅尼咒經 (*Sutra of the Dhāraṇī Spell to Ask Guanyin Bodhisattva to Absorb Poisons*). Translated by Zhu Nanti (fl. ca. 419).

T. 1059, *Qianshou qianyan Guanshiyin pusa zhibing heyao jing* 千手千眼觀世音菩薩治病合藥經 (*Sutra on the Use of Medicinal Herbs for Healing Illness by the Thousand-eyes, Thousand-hands Avalokiteśvara*). Translated by Bhagavatdharma, Tang dynasty. [Related texts T. 1060, T. 1070.]

T. 1161, *Foshuo guan Yaowang Yaoshang er pusa jing* 佛說觀藥王藥上二菩薩經 (*Sutra on the Contemplation of the Two Bodhisattvas King of Medicine and Supreme Medicine*). Translated by Kālayaśas (fl. 424–42).

T. 1323, *Chu yiqie jibing tuoluoni jing* 除一切疾病陀羅尼經 (*Sutra on the Dhāraṇī to Eliminate All Illnesses*). Translated by Amoghavajra (705–74).

T. 1324, *Nengjing yiqie yan jibing tuoluoni jing* 能淨一切眼疾病陀羅尼經 (*Sutra on the Dhāraṇī That Can Clear Up All Eye Ailments*). Translated by Amoghavajra (705–74).

T. 1325, *Foshuo liao zhibing jing* 佛說療痔病經 (*Sutra on the Treatment of Sores*). Translated by Yijing (635–713).

T. 1326, *Foshuo zhou shiqi bing jing* 佛說呪時氣病經 (*Sutra on the Spell for the Illness of Seasonal Qi*). Uncertain authorship, possibly Zhu Tanwulan (fl. 381–395).

T. 1327, *Foshuo zhou chi jing* 佛說呪齒經 (*Sutra on the Spell for the Teeth*). Translated by Zhu Tanwulan (fl. 381–95).

T. 1328, *Foshuo zhou mu jing* 佛說呪目經 (*Sutra on the Spell for the Eyes*). Uncertain authorship, possibly Zhu Tanwulan (fl. 381–395).

T. 1329, *Foshuo zhou xiaoer jing* 佛說呪小兒經 (*Sutra on the Spell for Children*). Uncertain authorship, possibly Zhu Tanwulan (fl. 381–395).

T. 1330, *Luomona shuo jiuliao xiaoer jibing jing* 囉嚩拏說救療小兒疾病經 (*Sutra on the Cure of Childhood Diseases Spoken by Rāvaṇa*). Translated by Faxian (d. 1001).

T. 1691, *Jiaye xianren shuo yi nüren jing* 迦葉仙人說醫女人經 (*Sutra on Women's Medicine Spoken by the Sagely Kāśyapa*). Translated by Faxian (d. 1001).

Monastic disciplinary texts

T. 1421, *Mishasaibu hexi wufen lü* 彌沙塞部和醯五分律 (*Five-part Vinaya of the Mahīśāsaka School*), Section 3.7. Translated by Buddhajīva and Daosheng, 423–24.

T. 1425, *Mohesengqi lü* 摩訶僧祇律 (*Vinaya of the Mahāsāṃghika School*), pp. 455a–457b. Translated by Buddhabhadra and Faxian, 416.

T. 1428, *Sifen lü* 四分律 (*Four-part Vinaya [of the Dharmaguptaka School]*), Section 3.4. Translated by Buddhayaśas and Zhu Fonian, 408–13.

T. 1435, *Shisong lü* 十誦律 (*Ten Recitations Vinaya [of the Sarvāstivāda School]*), Section 4.9.6. Translated by Puṇyatāra and Kumārajīva, 404–6.

T. 1448, *Genben shuoyiqieyoubu pinaiye yaoshi* 根本說一切有部毘奈耶藥事 (*Mūlasarvāstivāda Vinaya: Medical Matters*). Translated by Yijing, 703.

T. 1804, *Sifenlü shanfan buque xingshi chao* 四分律刪繁補闕行事鈔 (*Emended Commentary on Monastic Practices from the Dharmaguptaka Vinaya*), Chapters 18, 26. By Daoxuan, 626–30.

Commentaries and reference works

T. 1793, *Wenshi jing yiji* 溫室經義記 (*Commentary on the Bathhouse Sutra*). By Shi Huiyuan (523–92).

T. 1911, *Mohe zhiguan* 摩訶止觀 (*Great [Treatise on] Śamatha and Vipaśyanā Meditation*), Section 7.3. By Zhiyi (538–97), transcribed by Guanding (561–632). [Related text T. 1916.4.4.]

T. 1915, *Xiuxi zhiguan zuochan fayao* 修習止觀坐禪法要 (*Essentials of Practicing Śamatha and Vipaśyanā Meditation*), Chapter 9. By Zhiyi, 575–85.

T. 2122, *Fayuan zhulin* 法苑珠林 (*Forest of Pearls in the Garden of the Dharma*), Chapters 60, 95. Compiled by Daoshi, 668. [Related text T. 2123.]

T. 2125, Nanhai *jigui neifa zhuan* 南海寄歸內法傳 (*Record of Buddhist Practices Sent Home from the Southern Seas*), Chapters 4–8, 18, 20, 23, and 27–29. By Yijing, 691.

Recovered manuscripts incorporated into *Taishō Tripitaka*

T. 2766 *Yaoshi* jing *shu* 藥師經疏 (*Commentary on the Master of Medicines Sutra*). Unknown authorship. [Related text T. 2767]

T. 2780, *Wenshi jing shu* 溫室經疏 (*Commentary on the Bathhouse Sutra*). By Huijing, seventh century.

T. 2865, *Hu shenming jing* 護身命經 (*Scripture on Saving and Protecting Body and Life.* Unknown authorship, likely sixth century. [Related text T. 2866.]

T. 2878, *Foshuo* jiuji *jing* 佛說救疾經 (*Sutra on Deliverance from Disease*). Unknown authorship.

T. 2916, *Quanshan jing* 勸善經 (*Sutra Urging Goodness*). Unknown authorship.

Bibliography

Bagchi, P.C. (1941) 'New materials for the study of the Kumāratantra', *Indian Culture*, 7.3: 269–86.

——— (2011) 'A Fragment of the Kāśyapa-saṃhitā in Chinese', in B. Wang and T. Sen (eds) *India and China: a collection of essays by professor Prabodh Chandra Bagchi*, London: Anthem Press, pp. 75–85.

Birnbaum, R. (1989a) 'Chinese Buddhist traditions of healing and the life cycle', in Lawrence E. Sullivan (ed.) *Healing and Restoring: Health and medicine in the world's religious traditions*, New York; London: Macmillan, pp. 33–57.

——— (1989b) *The Healing Buddha*, Boulder: Shambhala.

Campany, R.F. (1993) 'The real presence', *History of Religions*, 32.3: 233–72.

——— (1996) 'The earliest tales of the Bodhisattva Guanshiyin', in Donald Lopez (ed), *Religions of China in Practice*, Princeton, NJ: Princeton University Press, pp. 82–96.

——— (2012a) *Signs from the Unseen Realm: Buddhist miracle tales from early medieval China*, Honolulu: University of Hawai'i Press.

——— (2012b) 'Religious repertoires and contestation: a case study based on Buddhist miracle tales', *History of Religions*, 52.2: 99–141.

Chen Ming 陳明 (2005a) *Dunhuang chutu huhua yidian Qipo shu yanjiu* 敦煌出土胡話醫典《耆婆書》研究, Hong Kong: Xinwenfeng chuban.

——— (2005b) *Shufang yiyao: chutu wenshu yu xiyu yixue* 殊方異藥: 出土文書與西域醫學, Beijing: Peking University Press.

——— (2005c) 'Zhuan nü wei nan: turning female to male, an Indian influence on Chinese gynaecology?', *Asian Medicine: Tradition and Modernity*, 1.2: 315–34.

——— (2007) 'The transmission of foreign medicine via the silk roads in medieval China: a case study of Haiyao Bencao', *Asian Medicine: Tradition and Modernity*, 3.2: 241–264.

——— (2013) *Zhonggu yiliao yu wailai wenhua* 中古医疗与外来文化, Beijing: Peking University Press.

Chen Yunü (2008) 'Buddhism and the medical treatment of women in the Ming Dynasty: a research note', *Nan Nü* 10: 279–303.

Davis, E.L. (2001) *Society and the Supernatural in Song China*, Honolulu: University of Hawai'i Press.

Demiéville, P. (1985) *Buddhism and Healing: Demiéville's Article 'Byō' from Hōbōgirin,* trans. M. Tatz, Lanham, MD; London: University Press of America.

Deshpande, V. (1999) 'Indian Influences on early Chinese ophthalmology: glaucoma as a case study', *Bulletin of the School of Oriental and African Studies* 62.2: 306–22.

——— (2000) 'Ophthalmic surgery: a chapter in the history of Sino-Indian medical contacts', *Bulletin of the School of Oriental and African Studies* 63.3: 370–88.

——— (2003) 'Nāgārjuna and Chinese medicine', *Studia Asiatica* 4–5: 241–57.

——— (2008) 'Glimpses of Āyurveda in medieval Chinese medicine', *Indian Journal of History of Science* 43.2: 137–61.
Deshpande, V. and Fan Kawai (2012) *Restoring the Dragon's Vision: Nagarjuna and medieval Chinese ophthalmology*, Hong Kong: City University of Hong Kong.
Despeux, C. (ed.) (2010) *Médecine, religion, et société dans la Chine médiévale: Étude de manuscrits chinois de Dunhuang et de Turfan*, Paris: Collège de France, Institut des Hautes Études Chinoises.
——— (2017) 'Chinese medical excrement: is there a Buddhist influence on the use of animal excrement-based recipes in medieval China?' *Asian Medicine: Journal of the International Association for the Study of Traditional Asian Medicine* 12.1/2: 139–69.
——— (2020) 'Buddhist healing practices at Dunhuang in the medieval period', in C.P. Salguero and A. Macomber (eds) *Buddhist Healing in Medieval China and Japan*, Honolulu: University of Hawai'i Press, pp. 118–59.
Fu Fang 傅芳 and Ni Qing 倪青 (1996) *Zhongguo foyi renwu xiaozhuan* 中国佛医人物小传, Xiamen: Lujiang chubanshe.
Goldschmidt, A.M. (2009) *The Evolution of Chinese Medicine: Song dynasty, 960–1200*, London: Routledge.
Goble, G.C. (2017) 'Three Buddhist texts from Dunhuang: the scripture on healing diseases, the scripture urging goodness, and the new Bodhisattva scripture', *Asian Medicine: Journal of the International Association for the Study of Traditional Asian Medicine* 12.1/2: 265–78.
Greene, E.M. (2021) *The Secrets of Buddhist Meditation: visionary meditation texts from early medieval China*, Honolulu: University of Hawai'i Press.
Heirman, A. and Torck, M. (2012) *A Pure Mind in a Clean Body: bodily care in the Buddhist monasteries of ancient India and China*, Gent: Academia Press.
Hsu, A.O. (2018) *Practices of Scriptural Economy: compiling and copying a seventh-century Chinese Buddhist anthology*, Ph.D. diss. University of Chicago.
Huang, C.J. (2009) *Charisma and Compassion: Cheng Yen and the Buddhist Tzu Chi Movement*, Cambridge, MA: Harvard University Press.
Kieschnick, J. (1997) *The Eminent Monk: Buddhist ideals in medieval Chinese hagiography*, Honolulu: University of Hawai'i Press.
Kritzer, R. (2014) *Garbhāvakrāntisūtra: the Sūtra on entry into the womb*, Tokyo: The International Institute for Buddhist Studies.
Li Rongxi (2000) *Buddhist Monastic Traditions of Southern Asia: a record of the inner law sent home from the south seas*, Berkeley: Numata Center for Buddhist Translation and Research.
Li Yingcun 李應存 and Shi Zhenggang 史正剛 (2006) *Dunhuang fo-ru-dao xiangguan yishu shiyao* 敦煌佛儒道相关医书释要, Beijing: Minzhu chubanshe.
Liu Shufen 劉淑芬 (2008) 'Tangsong shiqi sengren, guojia he yiliao de guanxi: cong yaofang dong dao huimin ju 唐、宋時期僧人、國家和醫療的關係: 從藥方洞到惠民局', in Li Jianmin 李建民 (ed.) *Cong yiliao kan Zhongguo shi* 從醫療看中國史, Taipei: Lianjing chuban gongsi, pp. 145–201.
Lo, V. and Cullen, C. (eds) (2005) *Medieval Chinese Medicine: the Dunhuang medical manuscripts*, London; New York: RoutledgeCurzon.
Lowe, B.D. (2014) 'The scripture on saving and protecting body and life: an introduction and translation', *Journal of Chinese Buddhist Studies* 27: 1–34.
Mazars, S. (2008) *Le bouddhisme et la médecine traditionnelle de l'Inde*, Paris: Springer.
Mollier, C. (2008) *Buddhism and Taoism Face to Face: scripture, ritual, and iconographic exchange in medieval China*, Honolulu: University of Hawai'i Press.
Ning Qiang (2004) *Art, Religion and Politics in Medieval China: the Dunhuang cave of the Zhai family*, Honolulu: University of Hawai'i Press.
McBride, R.D. (2011) 'Esoteric Buddhism and its relation to healing and demonology', in C.D. Orzech, H.H. Sorensen and R.K. Payne (eds) *Esoteric Buddhism and the Tantras in East Asia*, Leiden; Boston, MA: Brill, pp. 208–214.
Richter, A. (2020) 'Teaching from the sickbed: ideas of illness and healing in the *Vimalakīrti Sūtra* and their reception in medieval Chinese literature', in C.P. Salguero and A. Macomber (eds) *Buddhist Healing in Medieval China and Japan*, Honolulu: University of Hawai'i Press, pp. 57–90.
Salguero, C.P. (2009) 'The Buddhist medicine king in literary context: reconsidering an early example of Indian influence on Chinese medicine and surgery', *History of Religions*, 48.3: 183–210.
——— (2013) 'Fields of merit, harvests of health: some notes on the role of medical karma in the popularization of Buddhism in early medieval China', *Asian Philosophy*, 23.4: 341–9.

––––– (2014) *Translating Buddhist Medicine in Medieval China*, Philadelphia: University of Pennsylvania Press.
––––– (2015) 'Reexamining the categories and canons of Chinese Buddhist healing', *Journal of Chinese Buddhist Studies*, 28: 35–66.
––––– (ed.) (2017) *Buddhism & Medicine: an anthology of premodern sources*, New York: Columbia University Press.
––––– (2018a) 'Healing and/or salvation? The relationship between religion and medicine in medieval Chinese Buddhism', *Working Paper Series of the HCAS: Multiple Secularities — Beyond the West, Beyond Modernities* 3.
––––– (2018b) '"This Fathom-long body": bodily materiality & ascetic ideology in medieval Chinese Buddhist scriptures', *Bulletin of the History of Medicine*, 92: 237–60.
––––– (2018c) 'A "missing link" in the history of Chinese medicine: a research note on the medical contents in the Chinese Buddhist Taishō Tripiṭaka', *East Asian Science, Technology, and Medicine*, 47: 93–117.
––––– (2020a) '"A multitude of ghosts bursting forth and scattering": healing narratives in a sixth-century Chinese Buddhist hagiography', in C.P. Salguero and A. Macomber (eds) *Buddhist Healing in Medieval China and Japan*, Honolulu: University of Hawai'i Press, pp. 23–56.
––––– (ed) (2020b) *Buddhism & Medicine: an anthology of modern and contemporary sources*, New York: Columbia University Press.
Salguero, C.P., Toleno, R., Giddings, W.J., Capitanio, J., Bingenheimer, M. (2017) 'Medicine in the Chinese Buddhist canon: selected translations', *Asian Medicine: Journal of the International Association for the Study of Traditional Asian Medicine* 12.1/2: 279–94.
Sen, S. (1945) 'Two medical texts in Chinese Translation', *Visva-Bharati Annals*, 1: 70–95.
Sen, T. (2001) 'In search of longevity and good karma: Chinese diplomatic missions to middle India in the seventh century', *Journal of World History*, 12.1: 1–28.
Shi, Z. (2020) 'Lighting lamps to prolong life: ritual healing and the Bhaiṣajyaguru cult in fifth- and sixth-century China', in C.P. Salguero and A. Macomber (eds) *Buddhist Healing in Medieval China and Japan*, Honolulu: University of Hawai'i Press, pp. 91–117.
Skjaervø, P.O. (2004) *This Most Excellent Shine of Gold, King of Kings of Sutras: the Khotanese Suvarṇabhāsottamasūtra*, Cambridge, MA: Harvard University Department of Near Eastern Languages and Civilizations.
Strickmann, M. (2002) *Chinese Magical Medicine*, Stanford, CA: Stanford University Press.
Takakusu, Junjirō, and Watanabe Kaikyoku (eds) (1924–1935) *Taishō shinshū daizōkyō* 大正新修大藏經, 85 vols., Tokyo: Issaikyō kankōkai.
Unschuld, P.U. (1998) *Essential Subtleties on the Silver Sea*, Berkeley: University of California Press.
Watson, B. (trans.) (1993) *The Lotus Sutra*, New York: Columbia University Press.
––––– (1997) *The Vimalakirti Sutra*, New York: Columbia University Press.
Wright, A.F. (1948) 'Fo-Tu-Têng: a biography', *Harvard Journal of Asiatic Studies*, 11.3/4: 312–71.
Wu, Y.-L. (2000) 'The bamboo grove monastery and popular gynecology in Qing China', *Late Imperial China*, 21.1: 41–76.
Wujastyk, D. (2003) *The Roots of Ayurveda*, London: Penguin Books.
Xue Gongchen 薛公忱 (2002) *Ru-dao-fo yu zhongyiyao xue* 儒道佛與中醫藥學, Beijing: Zhongguo shudian.
Yü, Chün-fang (2001) *Kuan-yin: the Chinese transformation of Avalokiteśvara*, New York: Columbia University Press.
Zhu Jianping 朱建平 (1999) 'Sun Simiao "Qianjin fang" zhong de fojiao yingxiang 孙思邈《千金方》中的佛教影响', *Zhonghua yishi zazhi* 中华医史杂志, 4: 220–2.
Zysk, K.G. (1998) *Asceticism and Healing in Ancient India: medicine in the Buddhist monastery*, Delhi: Motilal Banarsidass.

29
TIME IN CHINESE ALCHEMY

Fabrizio Pregadio

Introduction

Time plays a major role in the doctrines and practices of both main branches of Chinese alchemy. The first of these branches, *waidan* 外丹, or external alchemy (documented from the mid-second century BCE), is the art of elixir-making aimed at immortality. The second branch, *neidan* 內丹, or internal alchemy (documented from the early eighth century CE), understands the elixir as the inner body essences, *qi* 氣, and *shen* 神 (spirits) and cultivates them through a variety of meditative and exercise practices. Within both branches, time is understood in two main aspects. First, the cosmos is generated by the Dao (the primordial principle usually translated as 'the Way') through a number of stages. Time here is meant in a metaphoric sense: as these stages precede the emergence of time, they occur in a state of timelessness. Time as we ordinarily perceive it begins at the conclusion of that sequence. Under this second aspect, time is manifested in the cyclical alternation of *yin* and *yang*, visible, for instance, in the succession of the diurnal cycles of day and night, of the moon phases of the month, and of the four seasons of the year.

Neidan is grounded in embodied practice. Health and illness are critical concerns for the adept who, in *Waidan*, processes minerals and, in *Neidan*, processes bodily essences, *qi* and *shen* (Chapter 27 in this volume). Interest in the ingestion of mineral substances waned after the Tang dynasty (618–907) when the new internal meditative practices of *neidan* became more widespread. The literature detailing both practices shares language and terminology, as well as cosmological principles expressed through numerology (Chapters 4 and 5 in this volume). This numerology draws together the spatiotemporal practices of medicine and alchemy.

In a significant portion of *waidan*, and in virtually the whole of *neidan*, both aspects of time are described and represented by classical Daoist concepts and by emblems drawn from the system of Chinese cosmology. As in other Daoist traditions (Schipper and Wang 1986), these concepts and emblems represent both 'regressive' or 'upward' sequences, in which time is traced backwards (*ni* 逆, or 'inverting the course'); and in 'progressive' or 'downward' sequences, which reproduce the course of the major time cycles (*shun* 順, or 'following the course'). In the first case, a sequence of pre-cosmic stages, prior to the emergence of material form, provides a framework for gradual reabsorption from the state of material existence into previous primordial stages by means of *waidan* or *neidan* practices. In the second case, the stages identified within the daily, monthly, and yearly time cycles provide templates for the 'refining' (*lian* 鍊) of natural substances (in *waidan*) or of the person's bodily components (in *neidan*).

'Inverting the Course': from the post-celestial to the pre-celestial

Chinese alchemy uses two main numerical sequences to illustrate the process through which the Dao generates the cosmos. Both sequences describe an ontology (by displaying the hierarchy among those stages) and a cosmogony (by representing those stages as succeeding one another in a metaphoric time).

The *Daode jing* Sequence. The first sequence is Dao → 1 → 2 → 3. This sequence has its *locus classicus* in the *Daode jing* 道德經 (Book of the Way and Its Virtue):

> The Dao generates the One, the One generates the Two, the Two generate the Three, and the Three generate the ten thousand things.
> *(Book of the Way and Its Virtue, section 42)*

According to one of the several ways in which this passage has been understood within and outside Daoism (Robinet 1995b: 198–203 and *passim*), One, Two, and Three, respectively, stand for the state of Unity, the emergence of *yin* and *yang*, and the product of their reconjunction. The 'ten thousand things' are the sum of entities and phenomena generated by the continuous reiteration of this three-stage process. In addition, the sequence of the *Book of the Way and Its Virtue* is also associated with three parallel states or stages that the Dao takes on or generates in its self-manifestation: Dao → Spirit (*shen* 神) → Breath (*qi* 氣) → Essence (*jing* 精). After the last of these stages, the Dao gives birth to the cosmos through its own Essence (*jing*) (*Book of the Way and Its Virtue*, section 21). While all these stages are contained within the Dao, their completion marks the shift from the pre-celestial (or pre-incarnate, to the post-celestial or incarnate domains (*xiantian* 先天 to *houtian* 後天).

The *Yijing* 易經 Sequence. The *Yijing* or *Book of Changes* is a divinatory text dating to the ninth century BCE. It is based on a system of 64 hexagrams (*gua* 卦), which are permutations of six broken (*yin*) or solid (*yang*) lines (*yao* 爻) probably derived from numerical symbols. The second sequence used is [Dao →] 1 → 2 → 4.[1] Its *locus classicus* is the *Xici* 繫辭 (Appended Sayings) appendix of the *Yijing* (Chapter 1 in this volume):

> Therefore in change (*or*: in the *Changes*) there is the Great Ultimate (*taiji* 太極, Unity). This generates the two principles (*liangyi* 兩儀). The two principles generate the four images (*sixiang* 四象).
> *(Book of Changes, Appended Sayings, section A.11)*

This sequence—which continues with the generation of the eight trigrams and the sixty-four hexagrams—intends to show that the modes of change represented by lines, trigrams, and hexagrams are issued from the state of Unity and are ultimately contained within it: these emblems portray the progressive unfolding of Unity into multiplicity.

In the traditional interpretation of the *Book of Changes*, the two principles of Pure Yang (━) and Pure Yin (╌) are referred to by unbroken and broken single lines. The four images contain two lines: Minor Yang (⚍), Greater Yang (⚌), Minor Yin (⚎), and

Greater *Yin* (☷). In alchemy, the 'two' are understood in the same way, even though they are usually called True *Yin* (*zhenyin* 真陰) and True *Yang* (*zhenyang* 真陽). The 'four', instead, are understood differently as different states or qualities of *yin* and *yang* in the pre-celestial and post-celestial domains. Pre-celestial True *Yang*, represented as three unbroken lines, is Qian ☰; pre-celestial True Yin is Kun ☷. The other two principles, Kan ☵ and Li ☲, respectively, stand for post-celestial *Yin* (two broken lines) containing True *Yang* (a single solid line) and for post-celestial *Yang* (two solid lines) containing True *Yin* (a single broken line).

Application to Alchemy. Both sequences outlined above are used as templates for the 'reversion to the origin' (*huanyuan* 還元) that is performed in both *waidan* and *neidan*. In the *waidan* works that do not use the emblems of cosmology to describe the alchemical process of smelting the minerals—these texts date up until c. the seventh century—the elixirs are usually compounded by heating the ingredients in a hermetically closed vessel that reproduces the inchoate state (*hundun* 混沌) prior to the emergence of the cosmos. The final product is often called the 'Essence' (*jing* 精) of the ingredients. A source dating from ca. 650 equates this Essence to the one mentioned in the *Book of the Way and Its Virtue* (section 21) as the seed of existence (see Pregadio 2006: 78). However, since many *waidan* elixirs hold the *yin* and *yang* principles reverted to the state of Unity, they incorporate all three stages, the generation of Spirit, Breath, and Essence.

The three stages of the *Book of the Way and Its Virtue* are even more important in the meditative practices of *neidan*. Its doctrinal discourse represents the inversion from the post-celestial to the pre-celestial by different sets of cosmological emblems, but its practice consists in inverting the three-stage sequence of generation. Accordingly—as shown in more detail below—the exemplary *neidan* practice consists of three stages: (a) refining Essence into Breath; (b) refining Breath into Spirit; and (c) refining Spirit to return to Emptiness, or the Dao. Here, the emblematic numbers 1, 2, and 3 also represent the progressive reduction of the components: 3 (Essence, Breath, Spirit) → 2 (Breath and Spirit) → 1 (Spirit) → 0 (Emptiness).

With regard to the sequence of the *Book of Changes*, the main principles are equally shared by both *waidan* and *neidan*. Pre-celestial True *Yang* and True *Yin* are first extracted from the respective post-celestial counterparts, namely Kan ☵ and Li ☲, and their placements are exchanged. This restores Qian ☰ and Kun ☷, which are then joined to one another in order to re-establish their Unity (also represented by Qian, which now stands for the stage prior to Yin and Yang). This process can be represented as shown in Figure 29.1. A chart drawn by Li Daochun 李道純 (late thirteenth century), a Daoist master from Jiangsu, makes clear that this representation of the alchemical process also applies to the refining of Essence, Breath, and Spirit in *neidan* (see Figure 29.2).

In all these cases, the elixir is a token of the successive states taken on by the Dao as it gives birth to the cosmos, reverted to the state prior to their inception—a visible token in *waidan*, an invisible one in *neidan*. Certain *neidan* masters emphasise an additional point. The backward movement of 'inverting the course' is, in fact, an upward movement that leads the alchemist from the cosmos to the Dao by means of 'doing' (*youwei* 有為, here meaning 'doing' the practice). The alchemical work, however, is entirely accomplished only if the course is completed by an opposite movement of descent, performed by 'non-doing' (*wuwei* 無為). Therefore, after the three stages of the alchemical practice have been completed, the alchemist should return to the domain from which he (or she) had departed, and realise the Unity and identity of Dao and cosmos, or timelessness and time.

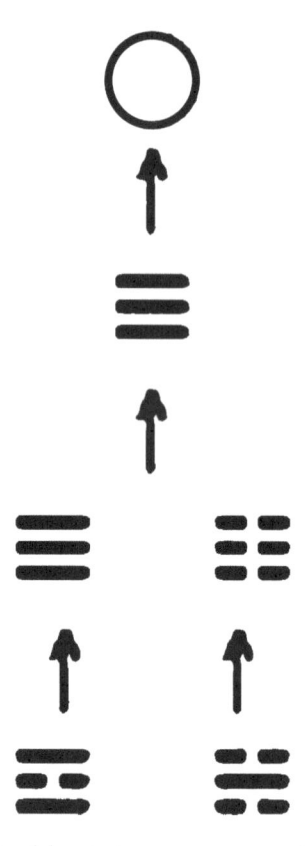

Figure 29.1 The three stages of the alchemical process represented by means of trigrams of the *Book of Changes*

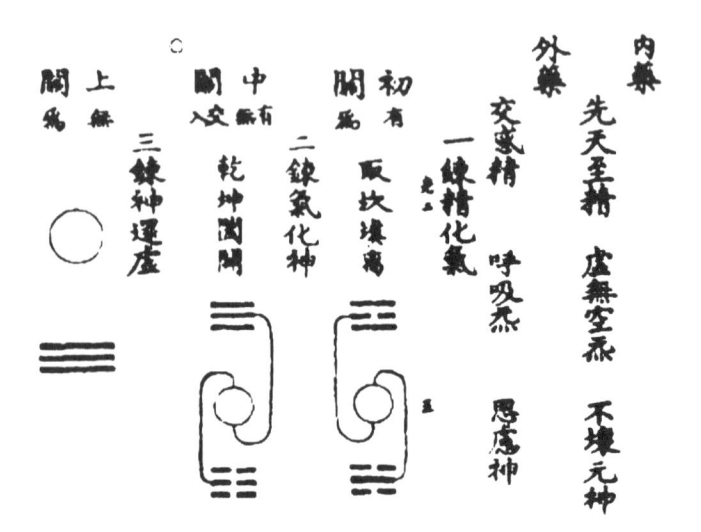

Figure 29.2 The refining of Essence, Breath, and Spirit and the corresponding trigrams of the *Book of Changes*. Li Daochun (fl. 1288–92), *The Harmony of the Centre*, 2.6a–b

'Following the Course': emblematic time cycles

The two 'regressive' sequences seen above, respectively based on the *Book of the Way and Its Virtue* and on the *Book of Changes*, show how, to use Nathan Sivin's words, 'Chinese alchemical theories were essentially numerological' (Sivin 1976: 521; see also Robinet 2011: 66–72). This feature is even more visible in the three main 'progressive' sequences used in alchemy to illustrate the cyclical flow of time. Although, for the sake of clarity, these sequences are described here in a separate section, it should not be forgotten that in both *waidan* and *neidan*, the 'progressive' time cycles are embedded in the 'regressive' time cycles.

These sequences became relevant to alchemy through the *Zhouyi cantong qi* 周易參同契 (*Seal of the Unity of the Three, in Accordance with the Book of Changes*, hereafter *Unity of the Three*), a work dating, in its present form, from not earlier than the mid-fifth century (and possibly from one or even two more centuries later; Pregadio 2011: 11–26). Under the influence of this work, which changed the history of *waidan* and gave origin to *neidan*, the trigrams and hexagrams of the *Book of Changes*—and with them the whole basic repertoire of cosmological emblems and terminology—entered the field of alchemy. As far as we know, this occurred for *waidan* approximately in the eighth century, the same period in which *neidan* also begins to be documented in extant sources.

The cosmological portions of the *Unity of the Three* describe three emblematic time cycles: the day, the month, and the year. In the view of the *Unity of the Three*, these cycles manifest the operation of the One Breath (*yiqi* 一氣) of the Dao in the cosmos. All of them—but especially the third one, as shown in the next sections—became models of *waidan* and *neidan* practices.

Sixty Hexagrams: The Daily Cycle. The first cycle concerns the thirty days of the lunar month (*Unity of the Three*, sections 3 and 45; references are to the annotated translation in Pregadio 2011). During each day, the *yang* principle prevails at daytime, from dawn to dusk, and the *yin* principle prevails at night time, from dusk to dawn. The two parts of the day are ruled by a pair of hexagrams: a *yang* hexagram presides over the first half, and a *yin* hexagram presides over the second half. Accordingly, sixty of the sixty-four hexagrams are distributed among the thirty days of the month, following one another in the order in which they are arranged in the *Book of Changes* and are described in its 'Hexagrams in Sequence' (*Xugua* 序卦) appendix. Zhun 屯 ䷂ and Meng 蒙 ䷃, the first and second hexagrams after Qian and Kun, respectively, correspond to daytime and night time on the first day of the month. Jiji 既濟 ䷾ and Weiji 未濟 ䷿, the next-to-last and last hexagrams, respectively, correspond to daytime and night time of the month's last day. The remaining four hexagrams, namely Qian 乾 ☰, Kun 坤 ☷, Kan 坎 ☵, and Li 離 ☲, reside at the centre. While they are not part of the time cycles, they enable them to occur.

Further, the rise and decline of *yin* and *yang* during the day are marked and measured by the twelve lines of the ruling pair of hexagrams for that day. Time in China was calculated according to ten heavenly stems (*tiangan* 天干) and twelve earthly branches (*dizhi* 地支) which when combined produced a cycle of sixty units (Chapter 4 in this volume). Each of the twelve double hours (*shi* 時) in a day was associated with one of the twelve branches. The six lines of the first hexagram are represented by the first six branches (*zi* 子, *chou* 丑, *yin* 寅, *mao* 卯, *chen* 辰, and *si* 巳), and the six lines of the second hexagram are represented by the last six branches (*wu* 午, *wei* 未, *shen* 申, *you* 酉, *xu* 戌, and *hai* 亥) (Chapters 4 and 5 in this volume).

'Matching Stems': The Monthly Cycle. The second cycle concerns the six stages of the lunar month (*Unity of the Three*, sections 13 and 49). This cycle is represented by the device known as *yueti najia* 月體納甲 (Matching Stems of the Moon), which is ascribed to

Table 29.1 The *yueti najia* (Matching Stems of the Moon) device

node	day	phase	trigram	stem and direction
(1–5)	3	beginning of waxing (*shuo* 朔)	*zhen* 震	*geng* 庚 W
(6–10)	8	first quarter (*shangxian* 上弦)	*dui* 兌	*ding* 丁 S
(11–15)	15	full moon (*wang* 望)	*qian* 乾	*jia* 甲 E
(16–20)	16	beginning of waning (*jiwang* 既望)	*xun* 巽	*xin* 辛 W
(21–25)	23	last quarter (*xiaxian* 下弦)	*gen* 艮	*bing* 丙 S
(26–30)	30	end of cycle (*hui* 晦)	*kun* 坤	*yi* 乙 E

Yu Fan 虞翻 (164–233). The month is divided into six periods ('nodes', *jie* 節) of five days each: 1–5, 6–10, 11–15, 16–20, 21–25, and 26–30. Each period is portrayed by one trigram and one of the celestial stems (*tiangan*; see Table 29.1). The sequence of trigrams and stems is Zhen ☳ (*geng* 庚) → Dui ☱ (*ding* 丁) → Qian ☰ (*jia* 甲) → Xun ☴ (*xin* 辛) → Gen ☶ (*bing* 丙) → Kun ☷ (*yi* 乙). These trigrams and stems are matched to nodal days in the waxing and waning of the moon: the 3rd (middle day of the first node), the 8th (middle day of the second node), the 15th (last day of the third node), the 16th (first day of the fourth node), the 23rd (middle day of the fifth node), and the 30th (last day of the sixth node). As shown by the sequence of the trigrams, the first half of the lunar cycle is governed by the *yang* principle (represented by the solid line), which grows until it culminates in the middle of the month (☰). The second half is governed by the *yin* principle (the broken line), which similarly grows until it overcomes the *yang* principle at the end of the month (☷).

The most significant aspect of this representation is the symbolic event that occurs in the night between the end of a month and the beginning of the next (*Unity of the Three*, sections 10 and 48). On the 30th day, the *yang* principle is entirely obscured and Kun ☷ (pure Yin) dominates over the entire cosmos. However, during that night, the Sun, represented by Li ☲, and the Moon, represented by Kan ☵, meet at the centre of the cosmos and exchange their essences. Their conjunction replicates in the post-celestial domain the conjunction of Qian ☰ and Kun ☷ in the pre-celestial domain. The monthly conjunction of the Sun (*ri* 日) and the Moon (*yue* 月) regenerates the light (*ming* 明): Kun performs her motherly function and gives birth to her first son, Zhen ☳, the initial trigram in the new lunar cycle, whose lower Yang line represents the rebirth of light. After an instant of suspension, time again begins to flow, and the next month begins.

'Twelve-Stage Ebb and Flow': The Yearly Cycle. The third cycle concerns the twelve months of the year (*Unity of the Three*, section 51). Usually called Twelve-stage Ebb and Flow (*shi'er xiaoxi* 十二消息), this cycle represents change by the twelve 'sovereign hexagrams' (*bigua* 辟卦). This representation ultimately derives from Meng Xi's 孟喜 (fl. 69 BCE) cosmological device known as Breaths of the Hexagrams (*guaqi* 卦氣). Meng Xi assigns four hexagrams, namely Zhen 震, Li 離, Dui 兌, and Kan 坎, to the four seasons, and each of their lines to one of the twenty-four 'nodal breaths' (*jieqi* 節氣) of the year. In a development of this system attributed to Jing Fang 京房 (77–37 BCE), the remaining sixty hexagrams are related to the twelve months in five sets of twelve. The 'sovereign hexagrams' are one of the five sets.

Analogously to the Matching Stems, here too the solid and broken lines flow first upwards and then downwards (see Table 29.2). Beginning with Fu 復, which stands for the first stage

Time in Chinese alchemy

Table 29.2 The twelve 'sovereign hexagrams' (*bigua*) and their relation to other duodenary series: earthly branches (*dizhi*), bells and pitch-pipes (*zhonglü*), months of the year, and 'double hours' (*shi*)

䷗	䷒	䷊	䷡	䷪	䷀	䷫	䷠	䷋	䷓	䷖	䷁
復	臨	泰	大壯	夬	乾	姤	遯	否	觀	剝	坤
fu	lin	tai	dazhuang	guai	qian	gou	dun	pi	guan	bo	kun
子	丑	寅	卯	辰	巳	午	未	申	酉	戌	亥
zi	chou	yin	mao	chen	si	wu	wei	shen	you	xu	hai
黃鐘	大呂	太蔟	夾鐘	姑洗	仲呂	蕤賓	林鐘	夷則	南呂	無射	應鐘
huangzhong	dalü	taicou	jiazhong	guxi	zhonglü	ruibin	linzhong	yize	nanlü	wuyi	yingzhong
11	12	1	2	3	4	5	6	7	8	9	10
23–1	1–3	3–5	5–7	7–9	9–11	11–13	13–15	15–17	17–19	19–21	21–23

of the growth of *yang* at the winter solstice, each hexagram represents one lunar month. The twelve-stage sequence also establishes correspondences with other duodenary series: the earthly branches (*dizhi* 地支), the bells and pitch-pipes (*zhonglü* 鐘呂), and the double hours (*shi* 時) of the day.

Time in *Waidan*

Waidan alchemists shared with their companions in other cultures the idea that their work reproduces the processes by which minerals and metals are transmuted into gold within the earth's womb. In their way of seeing, the elixir compounded in the alchemical laboratory has the same properties as the Naturally Reverted Elixir (*ziran huandan* 自然還丹) which nature refines in a cosmic cycle of 4,320 years. This number corresponds to the total sum of the twelve double hours contained in the 360 days that form one year according to the lunar calendar (Sivin 1976: 515–6, 1980: 264–6). The alchemical work, therefore, reproduces in a relatively short time the same process that requires an entire cosmic cycle to occur naturally. The *Insights on the Purport of the Alchemical Treatises* (*Danlun juezhi xinjian* 丹論訣旨心鑑) states:

> The Naturally Reverted Elixir is formed when Flowing Mercury, embracing Sir Metal (i.e. lead), becomes pregnant. Wherever there is cinnabar there are also lead and silver. In 4320 years the elixir is finished. Realgar to its left, orpiment to its right, cinnabar above it, malachite below. It embraces the *qi* of Sun and Moon, *yin* and *yang*, for 4320 years; thus, upon repletion of its own *qi*, it becomes a Reverted Elixir for immortals of the highest grade and celestial beings. When in the world below lead and mercury are subjected to the alchemical process for purposes of immortality, [the elixir] is finished in one year.
>
> (*Insights on the Purport of the Alchemical Treatises*;
> trans. based on Sivin 1976: 516, and 1980: 232)

The reduction of an extended cosmic cycle to one year—in other words, from 4,320 years to 4,320 double hours—is achieved by phasing the heating of the elixir ingredients according to suitable time patterns. These patterns are provided by the system of the Fire Phases (*huohou* 火候, also rendered as Fire Times or Fire Regime), which is modelled on the 'Twelve-Stage Ebb and Flow' of the *Unity of the Three*. In agreement with this model, firing is progressively increased during the first six stages, and then decreased during the last six stages. The *Unity of the Three* hints at the application of this system to alchemy by saying about the elixir:

> Watch over it with heed and caution: inspect it attentively and regulate the amount of its warmth. It will rotate through twelve nodes, and when the nodes are complete, it will again need your care.
> (*Unity of the Three*, section 40; Pregadio 2011: 91, 176–77)

Several varieties of the Fire Phases system exist in *waidan* (Sivin 1980: 266–79), but a particularly important example is provided by Chen Shaowei 陳少微, who lived in the early eighth century (Sivin 1976: 519–20, 1980: 272–3). Starting at the midnight of the first day of a sixty-day cycle in the month of the winter solstice (the eleventh month, which marks the beginning of the yearly cycle), fire is progressively increased by feeding growing amounts of charcoal to the furnace through its six *yang* doors, at intervals of five days for each door—that is, for six times and altogether one month. Heating then is progressively decreased for another month by placing lower amounts of charcoal in the furnace through its six *yin* doors. This operation also is repeated six times, so that the whole process takes one year. When this procedure is read in the light of the time cycles described in the *Unity of the Three*, it appears clear that Chen Shaowei intended to model the heating process not only on the cycle of the year, but also on the cycle of the month, which is made up of six periods of five days.

The compression of time achieved by means of the Fire Phases is a crucial aspect of *waidan*, but the creation of the alchemical microcosm also requires a smaller-scale representation of space. Chen Shaowei's furnace provides an example: its square shape represents the four directions, and its twelve doors are arranged in three tiers corresponding to Heaven (top), Earth (bottom), and Humanity (middle). Several other spatial correspondences may be embodied in the furnace, the reaction vessel, and the arrangement of the laboratory itself (Sivin 1980: 279–97). By placing himself in the spaceless centre of this microcosmos, the alchemist was able to observe the unfolding of a complete cosmic cycle. As bringing time to its end is the same as bringing it to its beginning, the alchemist was able to see the state of timelessness in the elixir.

Time in *Neidan*

As mentioned above, through meditation, breath cultivation, and inner body visualisations, *neidan* purports to restore the state prior to the shift from the pre-celestial to the post-celestial domains, which occurs as the Dao gives birth to the cosmos through its own Spirit (*shen*), Breath (*qi*), and Essence (*jing*). Accordingly, the model *neidan* practice consists of a preliminary stage, followed by a gradual sequence in which each of the three elements is re-integrated into the previous one. The preliminary stage serves to replenish Essence, Breath, and Spirit in the human body and to clear the vessels along which they are circulated. The respective functions of the three main stages are refining the Essence in order to conjoin it with Breath; refining that Breath in order to conjoin it with Spirit; and refining that Spirit in order to 'return to Emptiness', or the Dao. As part of the first stage, *neidan* includes a practice that bears several analogies to the Fire Phases of *waidan* (Extended descriptions of the *neidan* practice in different times and subtraditions are found in Despeux 1979: 48–82; Baldrian-Hussein 1984: 59–193; Robinet 1995a: 103–45; Wang Mu 2011).[2]

The Lesser Celestial Circuit. At the end of the preliminary stage, Essence, Breath, and Spirit are collected in the lower Cinnabar Field. Cinnabar Fields (*dantian* 丹田) were defined during the third to fourth century as three regions of the inner body located below (or behind) the navel, in the heart, and behind the forehead. In the 'living *zi* hour' (*huo zishi* 活子時, which marks the beginning of the daily time cycle, but is so called in order to distinguish it from the ordinary *zi* 'hour', corresponding to 11pm–1am), the External Medicine (*waiyao* 外藥) emerges as 'Original Essence' (*yuanjing* 元精). The first stage of the practice consists in circulating this Essence along a route called Lesser Celestial Circuit (or Lesser Celestial Orbit, *xiao zhoutian* 小周天) by means of repeated breathing cycles. This route is named 'lesser' in contrast to the Greater Celestial Circuit (or Greater Celestial Orbit, *da zhoutian* 大周天), which is used in the second stage of the practice (Despeux 1979: 57–63; Baldrian-Hussein 1984: 88–105; Robinet 1995a: 120–31; Neswald 2009: 35–7; Wang Mu 2011: 71–86).[3]

The Lesser Circuit is based on two of the eight 'Extraordinary Vessels' (*qijing* 奇經), namely the *dumai* 督脈 (variously translated as Function, Governor, or Superintendent Vessel) and the *renmai* 任脈 (Control or Conception Vessel). The *dumai* vessel runs along the

Figure 29.3 The twelve stages of the Lesser Celestial Circuit (*xiao zhoutian* 小周天)

back of the body, from the Meeting of Yin cavity (*huiyin xue* 會陰穴) near the coccyx to the Mouth Extremity cavity (*duiduan xue* 兌端穴) above the upper lip. The *renmai* vessel runs along the front of the body, from the Meeting of *Yin* cavity to the Receiver of Fluids cavity (*chengjiang xue* 承漿穴) below the lower lip. In the *neidan* view, the 'circuit' itself is actually completed by the two Magpie Bridges (*queqiao* 鵲橋), which conjoin the two vessels: the upper Bridge, located at the Meeting of *Yin* cavity, and the lower Bridge, located between the palate and the tongue (or identified with the tongue itself). Through the conjunction of the two vessels, the Essence can be circulated in a way contrary to its ordinary downward flow: first, it rises to the upper Cinnabar Field (crossed by the *dumai*), then it descends again to the lower Cinnabar Field (crossed by the *renmai*).

The Lesser Circuit is further subdivided into twelve segments, which are designated by any of the duodecimal series of cosmological emblems—such as the twelve 'sovereign hexagrams' and the twelve earthly branches—but are especially tied to the twelve double hours of the days (see Figure 29.3). This correspondence is often said to be symbolic and not to be understood in a literal sense (Chapters 4 and 5 in this volume): the twelve segments only serve to determine the stages of the Fire Phases, or the varying intensity of heat to be applied in order to refine the Essence. As in *waidan*, a progressively stronger 'Martial Fire' (*wuhuo* 武火) is used in the first six stages, which correspond to the ascent of the Essence along the *dumai* vessel (in terms of the earthly branches, from *zi* 子 to *si* 巳); and a progressively weaker 'Civil Fire' (*wenhuo* 文火) is used in the last six stages, corresponding to the descent of the Essence along the *renmai* vessel (from *wu* 午 to *hai* 亥). These two main stages of the Fire Phases are called 'advancing the *Yang* Fire' (*jin yanghuo* 進陽火) and 'withdrawing the *Yin* Tallies' (*tui yinfu* 退陰符, i.e. responding to the progressively stronger heating by a progressively weaker heating, in order to moderate and temper the *yang* of the first half of the cycle). Fire corresponds in *neidan* to Spirit, and the active faculty of Spirit (or True Intention, *zhenyi* 真意) leads the entire process: 'Spirit leads Breath and refines the Essence' (Wang Mu 2011: 52). At the two intermediate points of the Lesser Circuit (represented by the branches *mao* in the

Figure 29.4 The Waterwheel (*heche*). *Chart of the Inner Warp* (*Neijing tu*), detail

back and *you* in the front of the body), one should 'bathe' (*muyu* 沐浴), that is, temporarily suspend the heating process.

With regard to time, the Fire Phases of *neidan* enable the simultaneous operation of two antithetical sequences: the 'progressive' sequence of ordinary time, represented by the twelve segments, is encased in a 'regressive' sequence, represented by the circulation of the Essence in a way contrary to its ordinary downward flow. This process, in turn, is the first part of the inversion of time to timelessness performed in *neidan*.

As it does in *waidan*, the *neidan* practice of the Fire Phases also requires the use of a spatial framework. This framework is provided in the first place by the 'circuit' itself. In the lower part of its course is found the Waterwheel, or *heche* 河車 ('water-raising machine', Needham 1983: 250); this instrument is pictured in the well-known *Chart of the Inner Warp* (*Neijing tu* 內經圖), where it inverts the downward flow of the Essence and enables it to begin its upward course (Figure 29.4). In another sense, the *heche* is understood as the River Chariot, the vehicle that drives the Essence along the Lesser Circuit (Figure 29.5). Due to its symbolic importance, moreover, *heche* is also used as a synonym of the Lesser Circuit and the related practices.[4]

The 'chariot' metaphor and the spatial features of the Lesser Circuit reappear in relation to the three barriers (or passes, *sanguan* 三關), which are three key points in the back of the body: (1) the Caudal Funnel (or Tail Gate, *weilü* 尾閭), located at the base of the spine; (2) the Spinal Handle (or Spinal Straits, *jiaji* 夾脊), located in the middle of the spine, across from the heart; (3) the Jade Pillow (*yuzhen* 玉枕), located at the level of occipital bone, across from the mouth. As shown by the expression 'three fields in the front, three barriers in the back' (*qian santian, hou sanguan* 前三田後三關), the three barriers are seen as corresponding to the three Cinnabar Fields in the front of the body. The barriers are said to be arduous to overcome. Drawing on the metaphor of the three Vehicles of gradual liberation that prefigure

Figure 29.5 The River Chariot (*heche*). Xiao Tingzhi, *The Great Achievement of the Golden Elixir: An Anthology* (*Jindan dacheng ji*), 9.3b

the Highest Vehicle in the Buddhist *Lotus Sutra* (Watson 1994: 56–62), the practitioner is enjoined each time to proceed as if riding a cart, loaded with the Essence and driven first by a goat (that is, lightly and slowly), then by a deer (quickly and in a lively way), and lastly by an ox (strongly and powerfully). (See Despeux 1994: 80–7, 149–51; Neswald 2009: 42–5; Wang Mu 2011: 34–6, 83–4.)

Finally, it should be noted that the Lesser Circuit has, for some of its aspects, two main antecedents, both of which pertain to *yangsheng* 養生 (Nourishing Life) practices. The first one is the variety of breathing methods first documented in the 'Circulating Breath' (*xingqi* 行氣) inscription on jade, dating from the Warring States (Needham 1983: 142; Harper 1998: 125–6; Chapter 2 in this volume). The second is the method of 'reverting the essence to replenish the brain' (*huanjing bunao* 還精補腦, where 'essence' means the male semen) of the sexual practices (*fangzhong shu* 房中術), also known to have existed in the same period (Despeux 2008) (Chapters 22, 30, and 49 in this volume). In both cases, however, there are essential differences compared to the *neidan* practice. The method outlined in the jade inscription, and all those modelled on the same pattern require practitioners to circulate the Breath first downwards and then upwards—exactly the opposite of the *neidan* pattern. The 'reversion of the essence' in the sexual practices, instead, does follow an ascending path, but does not comprise the crucial descending half of the cycle, which coagulates the Essence into the seed of the Internal Elixir. In both cases, we have examples of a recurrent pattern in *neidan*, which draws from earlier methods or ideas, but adjusts them to suit its own views and purposes.

The Greater Celestial Circuit. The operation of the two contrasting sequences mentioned above results, after repeated cycles (sometimes said to be 300 or 360), in the formation of the Internal Medicine (*neiyao* 內藥). As soon as it is formed, the Internal Medicine should be conjoined with the External Medicine in order to generate the Great Medicine (*dayao* 大藥) in the lower Cinnabar Field. The Great Medicine is also called 'Mother of the Elixir' (*danmu* 丹母). After a further seven days of refining (called 'entering the enclosure', or *ruhuan* 入環), it conceives the Embryo (*tai* 胎).[5]

The Embryo is equivalent to the Breath (theoretically denoted by the homophonous graph「炁」instead of 「氣」, although this rule is not always followed) (Chapter 2 in this volume) and is formed by the conjunction of Essence and Breath. The second stage of the practice consists in nourishing it between the middle and the lower Cinnabar Fields for ten metaphoric months (the time required for gestation in the Chinese reckoning) by means of the Greater Celestial Circuit. However, this second stage is unrelated to time in the ordinary sense of the word. While the 'living *zi* hour' of the Lesser Circuit symbolises the beginning of a time cycle reproduced by the alchemist in practice, the Great Medicine is said to appear in the 'true *zi* hour' (*zheng zishi* 正子時, lit., 'correct *zi* hour'). Notwithstanding the allusion to the beginning of a time cycle (the *zi* hour), this term only denotes 'a state or a condition; one could call it a sign that the Great Medicine has been completed' (Wang Mu 2011: 107). While the absence of an exemplary time pattern, such as the one provided by the Fire Phases, results in different sources giving different descriptions of the Greater Circuit, its practice involves all of the eight 'Extraordinary Vessels', with no subdivisions into sequences or stages. The stages of 'bathing' also have no temporal correspondences, even symbolic: 'bathing' now consists in 'washing the mind and cleansing the thoughts, steaming them with the True Breath (*zhenqi* 真氣), observing subtle silence and brightness with the eyes, and preventing the mind from wandering around unrestrained and becoming unstable' (Wang Mu 2011: 103).

All this shows that the task of reverting ordinary time to its origins is essentially performed in the first stage of the practice. From the second stage onwards, a different time scale applies, unrelated to time as we ordinarily understand and measure it. In the third and final stage, the Embryo is moved to the upper Cinnabar Field and is finally delivered through the sinciput. The Infant (often called Red Child, chizi 赤子) is then first 'breast-fed' (buru 哺乳) and later nourished for nine symbolic years (the time that, according to tradition, Bodhidharma spent in meditation facing a wall after he transmitted Chan Buddhism from India to China, but also the number that represents Great Yang). As the practitioner 'returns to Emptiness', the Infant—an immortal replica of him or herself—roams throughout space-lessness and timelessness.

Time as a metaphor

The Fire Phases are said to be one of the most important aspects of *neidan*, but also one of its most carefully guarded secrets. Statements found in different sources indicate that one of the main issues is knowing when to terminate the Fire Phases and proceed to the higher stages of the practice. However, the whole discourse is framed in a way that both emphasises their importance and warns that the main points are left unsaid. Zhang Boduan 張伯端 (987?–1082), who is placed at the origins of the Southern Lineage (Nantong 南宗, one of the two main lineages that emerged in the history of *neidan* between the eleventh and the thirteenth centuries), says at first in his *Awakening to Reality* (*Wuzhen pian* 悟真篇, c. 1075):

> Even if you discern the Vermilion Sand and the Black Lead,
> it will be useless if you do not know the Fire Phases.
>
> (*Awakening to Reality, Jueju* 絕句 27)

Then it adds, with a reference to the *Seal of the Unity of the Three*:

> The *Seal* and the treatises, the scriptures and the songs expound ultimate reality,
> but do not commit the Fire Phases to writing.
> If you want to know the oral instructions and comprehend the mysterious points,
> you must discuss them in detail with a divine immortal.
>
> (Ibid., *Jueju* 28)

A later Nanzong master, Xue Daoguang 薛道光 (1078?–1191), is ascribed with a similar statement often quoted in later sources: 'The sages transmit the Medicine*, but do not transmit the Fire' (*Huandan fuming pian* 還丹復命篇, *Qiyan jueju* 七言絕句 11).

Within this context, one of the main underlying points is how to deal with the ordinary onward progression of time while concurrently undertaking the return to timelessness. The Fire Phases provide a valuable model, as they make it possible to begin the practice by following a progressive time sequence while submitting the Essence (the most basic component of the human being) to a course contrary to its ordinary flow. In the next stages, the practice continues by entering a different time frame (represented by the 'true *zi* hour') and is concluded with the return to timelessness. This procedure could not succeed if the time cycles of the Fire Times were followed in a literal way, as the practitioner would not be released from time as it occurs in the cosmic domain. The time of the Fire Phases is by its very nature a metaphoric time.

This issue has been repeatedly approached during the history of *neidan*. The Zhong-Lü 鐘呂 texts—written from the second half of the Tang period onwards, and belonging to the earliest identifiable tradition of *neidan*—seem to interpret various emblematic macro-microcosmic correspondences, including the Fire Phases, in quite literal ways. Examples of this understanding are found in the main Zhong-Lü doctrinal treatise, the *Zhong-Lü chuandao ji* 鍾呂傳道集 (Anthology of the Transmission of the Dao from Zhongli Quan to Lü Dongbin; Section 'Lun heche 論河車' [Discussion of the Waterwheel]); and in the main text devoted to the practice, the *Lingbao bifa* 靈寶畢法 (Secret Methods of the Numinous Treasure; see Baldrian-Hussein 1984: 237–59, and her explanations, 116–59).

This attitude changes in the Southern Lineage texts and in the later *neidan* tradition, to such an extent that warnings about a literal understanding of the time sequences are probably more numerous—and certainly more authoritative—than descriptions of the sequences themselves. With a clear reference to the 'Twelve-Stage Ebb and Flow', Zhang Boduan advises against strictly patterning one's practice on any time course established by sequences of cosmological emblems:

> The whole world delusively clings to the [hexagram] images:
> they practise the 'breaths of the hexagrams' (*guaqi*) and hope thereby to rise in flight.
>
> *(Awakening to Reality, Jueju 37)*

The preface to another work attributed to Zhang Boduan, the *Four Hundred Words on the Golden Elixir* (*Jindan sibai zi* 金丹四百字), describes several macro-microcosmic correspondences, saying, for instance, that the 30,000 'quarters of an hour' (or 'intervals', *ke* 刻) contained in the ten months of the gestation of the Embryo correspond to a cosmic cycle of 30,000 years. It adds, though, that the whole alchemical work actually occurs in the One Opening (*yiqiao* 一竅), the non-material centre of the human being where ordinary time and space do not apply:

> If one is able to understand this Opening, then the winter solstice, the Medicine, the Fire Phases, the bathing, the coalescing of the Embryo, and the delivery of the Embryo are all found there.
>
> *(Four Hundred Words on the Golden Elixir, Preface)*

One of the poems found in this work is famous for saying:

> The Fire Phases do not depend on the hours,
> and how could the winter solstice be at *zi*?
> As for the method of bathing,
> the times of *mao* and *you* are empty similitudes.
>
> *(Ibid., poem 13)*

Many later works quote an analogous statement attributed to Bai Yuchan 白玉蟾 (1194–1229?): 'The True Fire fundamentally has no phases'.

Li Daochun (late thirteenth century) gives an extended description of the Fire Phases, but laments that while they are only meant to provide a template for the practice, many understand them in a literal way. In his *Harmony of the Centre: An Anthology* (*Zhonghe ji* 中和集), he reiterates that the alchemical work takes place in the One Opening, and reminds us that the practice has nothing to do with the year, the month, or the day: 'The birth of the Medicine★ has its times, but these are not the time of the winter solstice, the time of the birth of the moon, or the time of the *zi* hour' (section '*Zhao Ding'an wenda* 趙定菴問答' [Questions and Answers with Zhao Ding'an]).

The later Longmen 龍門 (Dragon Gate) tradition shows another way of dealing with the same issue. The famous *Secret of the Golden Flower* (*Taiyi jinhua zongzhi* 太一金華宗旨, c. 1700), placed by Min Yide 閔一得 (1748–1836) at the source of the Jin'gai 金蓋 lineage of Longmen, is mainly devoted to the practice of 'reversing the light' (*huiguang* 回光) within the practitioner's person (Wilhelm 1962). About this practice, the *Secret* says: 'The reversion of the light is the same as the Fire Phases' (section 3). Min Yide, however, also proposed a different view. In one of his works, he refers to the Lesser Circuit by calling the *dumai* vessel the Black Path (*heidao* 黑道), and the *renmai* vessel the Red Path (*chidao* 赤道). In addition, he describes a Yellow Path (*huangdao* 黃道), which is situated between them. While this term ordinarily is a synonym of the Lesser Circuit, in Ming Yide's view, the Yellow Path overrides the other two vessels and directly connects the three Cinnabar Fields to one another. This 'central path', also called the 'Path of the Immortals' (*xiandao* 仙道), allows a practitioner to achieve the whole alchemical work in one instant, without depending on gradual stages or on time sequences (Esposito 2001: 209–13).

Many other views of time in *neidan* deserve attention; in particular, those of Liu Huayang 柳華陽 (1735–99) (Wilhelm 1962: 71–4; Wong 1998: 29–35), and those of Zhao Bichen 趙避塵 (1860–after 1933) (Lu 1964: 35–7 and *passim*; Despeux 1979: 55–63, 106–10). However, perhaps nothing better than a statement by Liu Yiming 劉一明 (1734–1821), found in his *Xiuzhen houbian* 修真後辨 (Further Discriminations in Cultivating Reality, section '*Zi, wu, mao, you*'), summarises the way ordinary time is seen in the context of the *neidan* tradition as a whole. Mentioning the four cardinal earthly branches, which represent the quarters of the day and the seasons of the year, Liu Yiming says: 'Alas! Those are the *zi, wu, mao*, and *you* of Heaven: what do they have to do with me? Heaven has Heaven's time, I have my own time'.

★Editor's note: 'Medicine' here refers to an inner alchemical product which is not, in the strict sense, a 'curative' as normally conceived; although it may have curative effects, its ultimate goals are extraordinary. Note also the use of acupoints along the channels for cultivation purposes.

Notes

1. I place 'Dao' within brackets only because it is not explicitly mentioned in the passage quoted immediately below. In the eyes of an alchemist, however, the whole sequence can only begin with the Dao.
2. *Neidan* practices include several varieties, but many of them are based on the pattern outlined above. It should also be mentioned that these or similar sequences—which are said to give initial priority to the cultivation of *ming* 命 (one's existence or embodiment)—do not exhaust the field of *neidan*. In other cases, the practice initially gives emphasis to *xing* 性 (one's inner nature) and consists in removing the causes of its obfuscation. These two emblematic modes of *neidan* practice are typically merged in the 'conjoined cultivation of nature and existence' (*xingming shuangxiu* 性命雙修).
3. In Chinese astronomy, 'Lesser Circuit' defines the year cycle made up of twelve months, and 'Greater Circuit' defines the Jupiter cycle made up of twelve years. As used in *neidan*, however, the two terms only refer to the lower and the higher stages of the practices in which the 'Lesser' and the 'Greater' Circuits are used, respectively.
4. In fact, the term *heche* is even more complex. In its earliest sense, found in the *Unity of the Three* (section 22), this term means the metal lead, which in *neidan* corresponds to the Essence. The metaphors represented by the *heche*, therefore, include at the same time the Essence, the instrument that accomplishes its inversion, the vehicle that transports it along the Lesser Circuit, and the Lesser Circuit itself.
5. Seven is an important number in *neidan*, as it represents the rebirth of the *yang* principle (the first solid line of Fu 復) from the *yin* principle (the six broken lines of Kun 坤). As for the term *ruhuan*,

it appears to derive from, or to be related to, the Quanzhen meditation practice of retirement in the *huandu* 環堵, 'enclosure', which originally lasted one hundred days or three years, and later was performed for shorter periods. The Quanzhen *huandu* retirement, in turn, bears analogies with the Buddhist practice of *biguan* 閉關, or 'confinement' in a solitary cell.

Bibliography

Pre-modern sources

Danlun juezhi xinjian 丹論訣旨心鑑 (Insights on the Purport of the Alchemical Treatises), Tang, DZ 935.

Daode jing 道德經 (Book of the Way and Its Virtue), 4th century BCE, *Sibu beiyao* 四部備要 ed., 1936.

Huandan fuming pian 還丹復命篇 (Reverting to the Mandate by the Reverted Elixir), Xue Daoguang 薛道光 (1078?–1191), DZ 1088.

Jindan dacheng ji 金丹大成集 (The Great Achievement of the Golden Elixir: An Anthology), Xiao Tingzhi 蕭廷芝 (fl. 1260–64), in *Xiuzhen shishu* 修真十書 (Ten Books on the Cultivation of Reality, DZ 263, *juan*. 9–13.

Jindan sibai zi 金丹四百字 (Four Hundred Words on the Golden Elixir), attr. Zhang Boduan 張伯端 (ca. 987–1082), DZ 1081.

Lingbao bifa 靈寶畢法 (Secret Methods of the Numinous Treasure), 11th century (?), DZ 1191.

Taiyi jinhua zongzhi 太一金華宗旨 (Ultimate Teachings on the Golden Flower of Great Unity), c. 1700, *Daozang xubian* 道藏續編 ed., 1834.

Wuzhen pian 悟真篇 (Awakening to Reality), Zhang Boduan 張伯端 (c. 987–1082). Text in Wang Mu 王沐 (1990) *Wuzhen pian qianjie (wai san zhong)*『悟真篇』淺解（外三種）(A Simple Explanation of the *Wuzhen pian* and Three Other Works), Beijing: Zhonghua shuju.

Xiuzhen houbian 修真後辨 (Further Discriminations in Cultivating Reality). In *Daoshu shi'er zhong* 道書十二種, Huguo an 護國庵 ed., 1819, reprinted in *Zangwai daoshu* 藏外道書, vol. 8, Chengdu: Bashu shushe, 1992.

Yijing 易經 (Book of Changes). Text in the *Zhouyi yinde* 周易引得 (A Concordance to Yi Ching), Peking: Harvard-Yenching Institute, 1935.

Zhong-Lü chuandao ji 鍾呂傳道集 (Anthology of the Transmission of the Dao from Zhongli Quan to Lü Dongbin), 11th century (?), in *Xiuzhen shishu* 修真十書 (Ten Books on the Cultivation of Reality, DZ 263, *juan*. 14–16.

Zhonghe ji 中和集 (The Harmony of the Centre: An Anthology), Li Daochun 李道純 (fl. 1288–92), DZ 249.

Zhouyi cantong qi 周易參同契 (Seal of the Unity of the Three, in Accordance with the Book of Changes), 5th/7th century, Text in Pregadio (2006).

Modern sources

Baldrian-Hussein, F. (1984) *Procédés secrets du joyau magique: traité d'alchimie taoïste du XIe siècle*, Paris: Les Deux Océans.

Despeux, C. (1979) *Zhao Bichen: Traité d'Alchimie et de Physiologie taoïste (Weisheng shenglixue mingzhi)*, Paris: Les Deux Océans.

—— (1994) *Taoïsme et corps humain: Le Xiuzhen tu*, Paris: Guy Trédaniel.

—— (2008) 'Huanjing bunao', in F. Pregadio (ed.) *The Encyclopedia of Taoism*, London: Routledge, pp. 514–5.

Esposito, M. (2001) 'Longmen Taoism in Qing China: doctrinal ideal and local reality', *Journal of Chinese Religions*, 29: 191–231.

Fraser, J.T, Lawrence, N. and Haber, F.C. (eds) (1986) *Time, Science, and Society in China and the West*, Amherst: University of Massachusetts Press.

Harper, D.J. (1998) *Early Chinese Medical Literature: the Mawangdui medical manuscripts*, London; New York: Kegan Paul International.

Kohn, L. and Wang R.R. (eds) (2009) *Internal Alchemy: self, society, and the quest for immortality*, Magdalena, NM: Three Pines Press.

Lu K'uan Yü (Charles Luk) (1964) *The Secrets of Chinese Meditation: self-cultivation by mind control as taught in the Ch'an Mahayana and Taoist schools in China*, London: Rider.

Needham, J. and Lu Gwei-Djen (1983) *Science and Civilisation in China*, vol. V: *chemistry and chemical technology*, part 5: *spagyrical discovery and invention: physiological alchemy*, Cambridge: Cambridge University Press.

Neswald, S.E. (2009) 'Internal landscapes', in L. Kohn and R.R. Wang (eds) *Internal Alchemy: self, society, and the quest for immortality*, Magdalena, NM: Three Pines Press, pp. 27–52.

Pregadio, F. (2006) *Great Clarity: Daoism and alchemy in early medieval China*, Stanford, CA: Stanford University Press.

—— (ed.) (2008) *The Encyclopedia of Taoism*, London: Routledge.

—— (2011) *The Seal of the Unity of the Three: a study and translation of the Cantong qi, the source of the Taoist way of the golden elixir*, Mountain View, CA: Golden Elixir Press.

Robinet, I. (1995a) *Introduction à l'alchimie intérieure taoïste: de l'unité et de la multiplicité*, Paris: Éditions du Cerf.

—— (1995b) 'Un, deux, trois: les différentes modalités de l'un et sa dynamique', *Cahiers d'Extrême-Asie*, 8: 175–220.

—— (2011) 'Role and meaning of numbers in Taoist cosmology and alchemy', in *The World Upside Down: essays on Taoist internal alchemy*, Mountain View: Golden Elixir Press, pp. 45–73.

—— (trans. and ed.) (2011) *The World Upside Down: essays on Taoist internal alchemy*, Mountain View: Golden Elixir Press.

Schipper, K.M., and Wang Hsiu-huei (1986) 'Progressive and regressive time cycles in Taoist ritual', in J.T. Fraser, N. Lawrence, and F.C. Haber (eds) *Time, Science, and Society in China and the West*, Amherst: University of Massachusetts Press, pp. 185–205.

Sivin, N. (1976) 'Chinese alchemy and the manipulation of yime', *Isis* 67: 513–27; reprinted in N. Sivin (ed.) (1977) *Science and Technology in East Asia: articles from Isis, 1913–1975*, New York: Science History Publications, pp. 109–22.

—— (ed.) (1977) *Science and Technology in East Asia: articles from Isis, 1913–1975*, New York: Science History Publications.

—— (1980) 'The theoretical background of elixir alchemy', in J. Needham (ed.) *Science and Civilisation in China*, vol. V: *chemistry and chemical technology*, part 4: *spagyrical discovery and invention: apparatus, theories and gifts*, Cambridge: Cambridge University Press, pp. 210–305.

Wang Mu (2011) *Foundations of Internal Alchemy: the Taoist practice of neidan*, Mountain View: Golden Elixir Press.

Watson, B. (1994) *The Lotus Sutra*, New York: Columbia University Press.

Wilhelm, R. (1962) *The Secret of the Golden Flower: a Chinese book of life*, trans. C.F. Baynes, London: Routledge and Kegan Paul.

Wong, E. (1998) *Cultivating the Energy of Life*, Boston, MA; London: Shambhala.

30
DAOIST SEXUAL PRACTICES FOR HEALTH AND IMMORTALITY FOR WOMEN

Elena Valussi

This chapter will discuss Daoist views on sexuality, health and immortality in relation to gender. The sources will be Daoist practice manuals, which in large part are non-gender specific, and in small part directed specifically at women. While I will give an overview of early practices, this chapter will mainly cover the Late Imperial period, since the early period is covered in Pfister Chapter 22 in this volume.

In Daoism, references to the female principle (*yin* 陰) abound. Scholars have interpreted these references as indication that women had a larger and more positive role in the Daoist tradition than in other religious traditions like Buddhism, which criticises female bodies and sexuality for being a distraction from detachment from worldly affairs (Ames 1981). However, while this may be true in theory, it is not always the case in practice; looking at a variety of texts describing physical practices for health and immortality will help us to understand the multiplicity of views, positive and negative, on the female body.

There is indeed evidence that in the early Daoist communities, women had high ritual and teaching roles; however, their presence and engagement decreased with the shift towards a more segregated society, where women's area of activity was increasingly defined as inside the home.[1] In the Late Imperial period, women who made space for their religious practice were described as unusual and out of the norm, mainly because this would detract from fulfilling their roles as wives and mothers. Often, though, women of Daoist and Buddhist religious persuasion did take up religious practices once their reproductive duties were completed.

As in most religious and non-religious literature, Daoist practice manuals were almost always written by males. Furthermore, though these texts do not often distinguish by gender, it is clear that they are generally directed at men, since women had little opportunity to follow the suggestions provided in them. These included wandering far and wide in search of the teachings of different masters, recruiting wealthy sponsors to fund the building of a secluded refuge where the practice could take place, finding attendants to support during the more intense parts of the practice, and often finding female sexual companions.

Thus, while it is true that references to feminine elements are quite common in Daoist scriptures, the role of women in the Daoist religious and intellectual community was not as central as those references might lead us to believe; the roles of women in practices of health and immortality, sexual and non-sexual, were also quite complex and varied.

While in the early period there is evidence that women did participate in sexual practices that benefitted both partners, in the Late Imperial period some sexual practices became exploitative of and harmful to the female body. Non-sexual techniques, on the other hand, historically did not distinguish by gender, and were generally directed at males; by the Qing dynasty (1644–1911) when a tradition of solo practices specifically for women developed, the female body became the central vehicle of the practice.

Sexual practices in the early period

Sexual practices have a long history in Chinese healing, as detailed by Donald Harper in his work on the Mawangdui sexual arts manuscripts (Harper 1987), which describe in detail sexual techniques for the improvement of health and for longevity, and the health benefits of correct sexual intercourse. While early sexual techniques were aimed towards spiritual cultivation and immortality, they did not indicate a specific religious affiliation. There are examples of Daoist sexual techniques from the second to third century CE and, by the Song dynasty, sexual techniques have a subsection under the category of 'Section on Gods and Immortals 神仙類', in the *History of the Song* (*Songshi* 宋史) (*Songshi*, 205.4).

The *Huangting jing* 黃庭經 (Scripture of the Yellow Court), dated probably to the second or third century CE, is the oldest Daoist text to discuss sexual cultivation (Robinet 2006). Sexual techniques were also used in the earliest Celestial Masters Daoist movement (Tianshi dao 天師道), active during the late Han dynasty, as a means to bring about great peace (*taiping* 太平) through the uniting of *qi* (*heqi* 合氣). We know of these practices by the Celestial Masters mainly through their detractors: in the *Annals of the Three Kingdoms* (*Sanguo zhi* 三國志) they are described as 'demonic'; in Buddhist treatises they are also criticised as demonic and false. Here is a well-known passage by Zhen Luan 甄鸞 from the treatise *Laughing at the Dao* (*Xioadao lun* 笑道論):

> When I was twenty years old I was fond of Daoist methods, so I enrolled in a Daoist monastery to study. First, I was taught the practice of merging pneumas[2] (*heqi* 合氣) according to the *Huangshu* 黃書 (Yellow Writ) and the three, five, seven, nine method of sexual intercourse. In pairs of 'four eyes and two tongues' we practiced the Dao in the Cinnabar Field. Those who practice this thereby overcome obstacles and prolong their lives. Husbands were instructed to exchange wives merely for carnal pleasure. Practitioners had no shame, even before the eyes of their fathers and elder brothers. This they called the 'perfect method of concentrating pneuma (*jingqi* 精氣)'. At present, Daoists regularly engage in these practices, in order thereby to attain the Way. The basis for this is unclear.[3]

The *Huangshu* mentioned above might be related to a text found in the *Daozang* 道藏, the *Shangqing Huangshu guodu yi* 上清黃書過度儀 (Initiation Rite of the Yellow Writ of Highest Clarity, DZ 1294), dated by Gil Raz to the late fourth century CE. According to Raz, this text does describe a ritual including sexual union, but this is not the main aim of the ritual:

> I argue that the focus on the ritualised intercourse in the initiation rite of the *Huangshu guodu yi* has blinded us to its real significance, which is, in fact, to transcend the mundane realm, symbolised by the sexual act, and to attain the primordial undifferentiated oneness, beyond sexual division.

(Raz 2008: 90)

A careful reading of the ritual instructions reveals that the ritualised intercourse is not in fact the climax of the ritual procedure. Rather, the climax is the production of a perfected being within the body of the initiate through the visualisation, realisation, and coagulation of the three primordial pneumas (*yuanqi* 元氣). The ritualised intercourse that follows the production of this homunculus is the first stage in the ritual reconstruction of the cosmos.

(Raz 2008: 91)[4]

The description of the ritual reveals that the sexual union was highly codified, involved meditation and visualisations, followed rules set by the contemporary understanding of cosmology, and resulted in the production of an immortal embryo. Even though the text does not specify the outcome by gender, we can surmise that both men and women benefitted from it.

The later Shangqing 上清 tradition criticised the use of sexual intercourse in Daoist ritual on the basis that it was dangerous, and recast the sexual union into individual practices in which the adept would work within his/her own body to produce the 'immortal embryo'.[5] Also, as Raz explains, 'the practice of Merging Pneumas was remade into a "pure" marriage rite, called Pairing Radiances (*oujing* 偶景) between male adepts and female deities' (Raz 2008: 94). Here we see a shift from an actual pairing between a man and a woman in the tradition of the Celestial Masters, to the pairing between the male adept and a female deity. In this case, it is the male adept who benefits from the union with the female Perfected; no mention is made of female adepts uniting with male deities.

Ge Hong 葛洪 (283–363 CE), in his *Baopuzi neipian* 抱朴子內篇 (Book of the Master Who Embraces Simplicity), mentions the existence of sexual practices for the achievement of immortality, and even though he does not dismiss them altogether, he categorises them among other *yangsheng* exercises (Chapter 6 in this volume), and forcefully asserts that it is folly to think of them as the only path to immortality (Chapter 8, section 2; trans. Wile 1992: 24). His criticism reveals to us their presence within the panorama of Daoist practices.

A much later criticism towards using sexual techniques to achieve immortality is that of Zeng Zao 曾慥 (?–1155), who included the *Rongcheng pian* 容成篇 (Chapter on Rongcheng) in his collection of Daoist works, the *Daoshu* 道樞 (Pivot of the Dao, DZ 1017). In this chapter, written by Zeng Zao himself, he criticises the *Ruyao jing* 入藥鏡 (Mirror for Compounding the Medicine), for its supposed descriptions of sexual techniques for immortality. Zeng Zao rejects the notion that sexual intercourse is needed for the achievement of immortality and that in fact it may even be a hindrance on that path, and harmful to the health of the practitioners, who would lose their *jing*.[6]

Even though there is scant evidence, from the few examples above we can see that sexual techniques for health and immortality were indeed present from a very early time in China, that they developed within the Daoist ritual and alchemical tradition, and that they also had detractors from within and without Daoism. The tension between sexual and non-sexual Daoist practices highlighted above is a constant theme within Daoism; with the emergence of an alchemical tradition of immortality, we see the clear development of both dual and solo alchemical practices.

Sexual alchemy in the Ming (1368–1644) period

By the time of the Ming dynasty, there was a flourishing Daoist literature openly advocating the use of sexual techniques for the achievement of health and immortality within the already well-developed *neidan* 內丹 (inner alchemy) school; many of these texts are related to the cult of the Daoist Immortal Zhang Sanfeng 張三丰.

The most representative of these texts are the *Jindan zhenchuan* 金丹真傳 (True Transmission of the Golden Elixir) by Sun Ruzhong 孫汝忠 (fl. 1615) and the *Jindan jiuzheng pian* 金丹就正篇 (Folios on Seeking the Proper Understanding of the Golden Elixir)[7], preface dated 1564, by Lu Xixing 陸西星 (1520–1601?). Other well-known works on sexual alchemy are attributed to the Immortal Zhang Sanfeng and found in the *Sanfeng danjue* 三丰丹訣 (Zhang Sanfeng's Alchemical Instructions), published as part of the Qing dynasty compilation *Daoshu Shiqizhong* 道書十七種 (Seventeen Books on the Dao), edited by Fu Jinquan 傅金銓: they are the *Jindan jieyao* 金丹節要 (Synopsis of the Golden Elixir; trans. Wile 1992, 169–78), the *Caizhen jiyao* 採真機要 (Essentials of the Process for Gathering the True; trans. Wile 1992, 178–88), and the *Wugen shu* 無根樹 (*The Rootless Tree*; trans. Wile 1992, 188–92), a text allegedly transmitted by Zhang Sanfeng to Lu Xixing, accompanied by an extensive commentary by Li Xiyue 李西月 (1806–1856) and Liu Wuyuan 劉悟元 (1734–1821).[8]

The sexual practices described in these texts, in a very similar way to all other alchemical texts, are all directed at restoring the prenatal level of energies that every human is endowed with at birth, but that is squandered throughout life, through excessive sexual intercourse, excessive mental and physical exertion, and, specifically to women, the menstrual period. However, it is clear that, while both men and women are mentioned in these texts, the techniques only benefit men. The alchemical texts, some more explicitly than others, describe the use of women as tools for the restoring of health and seeking of immortality for men. In them, women are often referred to as 'cauldrons' (*ding* 鼎), and described as tools for men to deposit their *jing* and to refine it, before extracting it and using it to replenish their own vital energies. The term 'internal alchemy' was developed from external (operational) alchemy to indicate the transformation of substances akin to the smelting of ore in a receptacle, but this time inside the body of the adept, resulting in the production of the Golden Elixir (*jindan* 金丹) of immortality. In Daoist sexual practices, it is the female body that is used as the cauldron for the transformation of substances and the production of pure *Yang*, which the male can then pluck and utilise for his own realisation. Below are excerpts from the *Jindan zhenchuan* by Sun Ruzhong; this section describes the original well-being of the just born baby, full of energies that are then squandered through life; in order to restore these vitalities, it is necessary to utilise the body of another human being:

> A human being is endowed with the father's *jing* [精 essence] and the mother's blood, and from this the body is formed. After a process of combination and gestation, it gradually reaches the stage of manifesting as physical form. The father's *jing* is stored in the kidneys, the mother's blood is stored in the heart. The heart and kidneys are connected by a channel, and following the mother's breathing, the *jing* and blood are produced together. (...) When accumulation proceeds for one year, it reaches two *liang* 兩 (1/16th of a *jin* 斤) of *jing*, three *liang* at two years, and one *jin* 斤 (600 grams)[9] at fifteen. Now the Tao of the male is complete. At this moment, the *jingqi* is complete and the state of pure *Qian* is realised. This is called 'the highest power'. If he receives enlightening instructions from an adept, then his foundation may be secured by itself, and there will be no need to engage in exercises as 'strengthening the *qi*', 'strengthening the blood', 'obtaining the medicine', or 'returning the elixir'. (...) However, at this stage in his life, a man's life knowledge arises and emotions are born. When the *jing* is full, one is unable to control oneself: when the spirit is

complete one cannot remain stable. (…) They are busy by day and active at night, all of which harms the *jing* and damages the blood. As a result our pure bodies become corrupt. (…) Therefore, one must employ methods for repairing and returning. (…) If bamboo breaks, bamboo is used to mend it; **if a human being suffers injury, another human being may be used to repair the damage.**

(Wile 1992: 157)

In the following section, even though the language is cryptic, it seems clear that the body of a woman is used to restore the vitalities of the male practitioner:

The medicine is the external medicine produced in the postnatal crucible. 'Obtaining the medicine' refers to gathering **the postnatal crucible's** external medicine, taking it into my body and combining it with the *qi* and blood that I have finished refining… After this, the body and spirit are both whole.

(Wile 1992: 158)

The postnatal (*houtian* 後天) crucible is the female body.[10] The *Jindan Jieyao* is much more specific about what is a crucible and how to select it:

Chapter 8 – choosing the crucible

The crucibles are the 'True Dragon' and the 'True Tiger'. First choose a beautiful tiger with clear eyebrows and lovely eyes. You must find one with red lips and white teeth. There are three grades of crucibles. The lowest are twenty-five, twenty-four or twenty-one. Although they belong to the 'postnatal', they may be employed in 'practicing fire work', 'nourishing the weak dragon', 'adding oil to supplement the lamp', and 'enriching the nation and ordering the people'. Those of middle grade are twenty, eighteen or sixteen, who have never engaged in intercourse, but already have had their first menses. Because they have never given birth, their placenta has never been broken; and they may be used to extend life and achieve 'human immortality'. The highest grade are 'medicine material' of fourteen. Their condition precedes the division of 'primal unity'(···). These are called the 'true white tigers'. A Qian dragon of sixteen who has never lost his 'true *ching*' is a 'true Green Dragon'

(Wile 1992: 174).[11]

The text describes not only the sexual act itself, but also the most auspicious circumstances that are needed to bring it to successful completion, which include a secluded chamber, several 'crucibles', attendants and, most importantly a 'Yellow Dame', an older woman who will attend to and direct the young women participating in the practice.

From the above passages, and from analysing other sexual alchemy treatises, it is clear that women are utilised as 'crucibles' for the smelting of energies that are then plucked by male practitioners and united in their bodies with their own energies, in order to repair health and possibly achieve immortality. Xun Liu has proven that these texts did advocate for the well-being of the young women involved: 'the emotional harmony and reciprocity were clearly defined as the precondition for engendering rising of the vital and generative fluids within the female body and for ensuring the successful give-and-take during the practice' (Liu 2009: 139); nonetheless, their well-being was understood as the best possible setting for male plucking of female vitalities.

The Qing dynasty: non-sexual practices and *nüdan*

The more reactionary climate of the Qing, together with the widening of the audience for inner alchemy, and the development of a state-mandated chastity cult, made the above-described sexual practices much less acceptable. This changing social climate resulted in two developments: 1. the reinterpretation of sexual alchemy in non-sexual terms, and 2. the solidifying of specific solo techniques for women that did not involve men, called *nüdan* 女丹, or female alchemy.

Non-sexual alchemy

Clarke Hudson discusses the non-sexual reinterpretation of Ming sexual techniques by Qing Daoist masters (Hudson 2010). He mentions two examples in particular. The first is a text by the Chan monk and Daoist master Liu Huayang 柳華陽 (1735–99). In the *Huiming jing* 慧命經 (Scripture of Wisdom and Life) Liu claims that, while humans unite sexually to conceive children, and thereby lose part of their life energy, 'transcendents or buddhas intermingle or copulate using their spirits or *qi*' (Hudson 2010:6). In this way, he asserts that sexual union only results in the loss of vital energies, and that immortality is only achieved by uniting spiritually. The second example is Quanzhen Daoist master Min Yide 閔一得 (1758–1836), who, referring to the above mentioned *Jindan zhenchuan*, recasts corporeal sexual alchemy into spiritual sexual alchemy, where a man and a woman meditate together, joining their *qi* but not their bodies:

> The inner partners who copulate as dragon and tiger are people united in mind and intention. This is not muck [alchemy], but is in fact a method whereby *qi* and spirit virtuously unite. What you gain from it is grain after grain of celestial treasure. … If your partner is the type of partner who can set down a seed, then there [need be] no parting of clothing, nor untying of belts. There are just one dragon and one tiger, who join their pure *qi* and spirit together to penetrate void emptiness.[12]

In both cases, physical sexual union is recast into the union of *qi* and spirits. In this way, the female body is no longer used as a tool for the realisation of male immortality, but as a locus of immortality itself.

Wayward women

Despite the more reactionary attitude during the Qing dynasty and the recasting of sexual into non-sexual alchemy discussed above, we find that men and women were sometimes still engaged in sexual alchemy. This is confirmed by negative reactions to these practices throughout the Qing, especially as it pertains to women. A compelling example of the 'wrong' behaviours which women were engaging in appears in a list written by He Longxiang 賀龍驤 in Chengdu in 1906; this list is included in the preface to a large collection of female solo alchemy, or *nüdan*, the *Nüdan hebian* 女丹合編 (Collection of Female Alchemy); this preface lists the behaviours which women were said to be prone unless they engaged in female alchemy. These include everything from a faulty mastery of the Daoist scriptures to a mistaken reliance on Buddhist practices. In addition,

> … there are those who do not discriminate between Qian and Kun and do not know that there are differences [between female alchemy] and male alchemy (*nandan* 男丹);

there are those who know alchemical books for men but do not know that there are books for women …

(Nüdan hebian, Preface: 1b–2a)

These misguided beliefs and practices can, according to He, have dire results not only for individual women, but for society as a whole:

There are those [women] who mistakenly become involved in heterodox sects and do not know the correct way. (…) Others are lured into chambers where lewd [activities take place]. There are [those] who secretly seduce good girls into serving as human cauldrons (rending 人鼎), and [some] themselves serve as Yellow Dames (huangpo 黃婆), [13] as a result of which [the girls] lose their reputations and integrity. [Then] there are those good women who do what palace ladies like to do; they are fond of serving as cauldrons in their search for immortality, [but in the end] do nothing but lose their reputations and integrity. There are those [women] who go on pilgrimage, enter the temple and throw themselves in a disorderly manner at Buddhist and Daoist monks; others plant the seed of passion in male teachers of good schools.

(Nüdan hebian, Preface: 3a)

It is clear that the target of the above invective is the sexual alchemy advocated by the late Ming texts described above. The preface continues by suggesting that women do not practise in the correct way because they do not know that there is a correct way designed just for them. This assertion is important because it gives a solid basis to the claim that women do indeed require a separate practice from men, and that inner alchemy, historically non-gendered, is in fact (and always was) directed to males. In this way, the writers make space for nüdan.

The issue of misunderstanding the difference between male and female practices is also connected to that of morally wrong and physically dangerous practices. Among women who do not know of nüdan, it is said, some simply never progress in their path, because they are following practices that are not specific to the female body. Others, however, incur greater dangers because they follow heterodox practices such as becoming a human cauldron. In order to better understand the trend that was directly attacked by He Longxiang in his preface, we will turn to the Pangmenlu 旁門錄 (Record of the Heterodox Schools), a text He Longxiang includes in his collection. This scripture includes poems attributed to Buddhas and heavenly worthies, annotated by various gods and immortals. These poems were regarded as a guide for correct practice, and they refer to such things as the difficulty in distinguishing between proper and improper instructions and the diffusion of heterodox practices. In his post-face to the Nüdan hebian, He Longxiang states that he includes the Pangmenlu in his collection as a warning to women practitioners. An excerpt from the text exemplifies the kinds of problems He was afraid women would encounter:

It is wrong to think that 'you and I'[14] can make you realised.
Don't you know that yin and yang reside in your own body? Buy concubines, sleep with prostitutes, practise the 'plucking battle',[15]
in this way you steal the original qi and supplement your spirit.
How is it then that there are immortal guests on Penglai Isle
that decry debauched practitioners who cherish lewd practices?

The master says:

> ... Now, the heterodox schools say that 'I' (*wo* 我)' is the man and 'you (*bi* 彼)' is the woman. Because of this, they buy beautiful women and rear them, they use an 'external' Yellow Dame to gather their menstrual flow; then, they engage in 'the plucking battle' in order to steal the original *qi* .... All of these are wrong understandings of the two characters 'you' and 'I'. They violate the principal heavenly rules and the kingly ways; they cannot escape the sternest of punishments. How can they possibly hope to attain immortality?
>
> (*Pangmenlu*: 1a)

The two paragraphs translated above reflect a difference of opinion as regards the practice of gathering the original *qi* (*yuanqi* 元氣) to create the immortal embryo. Those who practise sexual union believe they can gather the original *qi* from the body of a woman. The authors of the *Pangmenlu*, and He Longxiang with them, insist that only by gathering one's own *yuanqi* can one attain immortality. This is further explained in a later passage in the same text:

> The *yuanqi* within the body is the elixir of immortality. Heterodox schools and wayward ways go against reason in seeking for it outside the body; by refining the elixir and ingesting it they hasten their own death (...). Men and women come together and then extract the spoiled *jing*. How is this different from [what is done by] cattle and horses, dogs and pigs? They say that by eating the spoiled *jing*, one can become immortal; why then, do cattle and horses, dogs and pigs not ascend to the heavenly hall? This is a case of misreading the two words 'ingest and eat' (*fushi* 服食) found in alchemical books. When studying alchemical books, one must find a teacher to explain them so that one clearly understands the principles behind them.
>
> (*Pangmenlu*: 4b)

Nüdan and female physiology

The new school of *nüdan* thus focusses on the female body as the sole locus for the immortality of women. The texts associated with it describe in much detail female physiology and the processes for its transformation. References to the specificity of the female body can be found in inner alchemical texts as early as the Song period, but not until the first *nüdan* texts started to appear in the seventeenth century, and especially until the first collections at the end of the eighteenth century, do we find descriptions of the need for a completely separate path for women practitioners. This is therefore a retrofitting of the history of inner alchemy, and its new (but purportedly old) category is now called *nandan*. The physiological language, which points at locations and processes of the female body with great clarity, lent even more weight to this development.

As in any practice towards health and immortality, the goal of female practice is to refine one's constitution, reclaiming the energies one received at birth and slowing their loss, thus delaying or eliminating death. But while the standard course of refinement in alchemy proceeds from *jing* 精 (essence) to *qi*, from *qi* to *shen* 神, and from *shen* to emptiness, women need to refine their blood (*xue* 血), not their essence, into *qi*. The physical starting point for female practice is the *qi* cavity, a point between the breasts. Through breast massage and visualisations, the blood that has previously descended from the *qi* cavity to the infant's palace (uterus) is sent upward in a backwards motion. The infant's palace, also called the sea of blood (*xuehai* 血海), is located three and a half inches below the navel. It is not to be confused with the lower elixir field, where male practice begins.

Unlike a man, a woman needs first and foremost to refine her exterior form, her bloody and impure constitution and her sexual characteristics (i.e. breasts). This attention to her exterior form directly relates to the structure of the female cosmological and physiological body that is *yin* and impure in nature. Through breast massages the blood that would flow down from the heart to the womb, and outside the body as menstruation, is prevented from doing so and accumulates in the *qi* cavity between the breasts. Practitioners repeat this process many times, achieving the thinning and eventual disappearance of the menstrual flow, a process called Beheading the Red Dragon (*zhan chilong* 斬赤龍). When this happens, other sexual characteristics change: the breasts shrink and the body becomes more androgynous. At this point the woman has completed the first stage of the practice (Valussi 2008a, 2008b: 155).

Within Late Imperial Daoist practices for health and immortality that involve women, we therefore observe a shift from sexual, to non-sexual, to solo practices specifically intended for women. While the shift away from being used as a tool in the quest for health and immortality of males may be perceived as positive for women, in *nüdan* texts and practice, female physiology and the pre-eminence of blood are clearly described as a hindrance for women's refinement. Here is an example:

> It is said that blood is the energetic basis of the woman. Her nature is inclined toward the *yin*, and the nature of *yin* is to enjoy freshness. If a woman does not avail herself of massage, by which she can help the *qi* mechanism to circulate subtly, she will easily sink into pure *yin*. *Yin* is cold, and cold is ice-like. The failure to activate it by means of circulating movements, may result in illnesses such as congestion and blood obstruction, which would make the practice [of alchemy] difficult to implement.
>
> *(*Niwan Li Zushi nüzong shuangxiu baofa 泥丸李祖師女宗雙修寶筏*)*
> *(Precious Raft on Paired Cultivation of Women by Master Li Niwan)*

Thus blood is central for the health and immortality of women. Dealing with blood and refining it into a more ethereal substance is described in these texts as much more difficult than refining *jing* is for men.

In modern times, both sexual and solo health and immortality techniques have been 're-discovered', in China and in the West. Mantak Chia is the practitioner who first popularised these practices in the West. In his books on Daoist health and sexuality, Chia reinterpreted sexual alchemy as offering an equal path to men and women (Chia, 2005 [1986]). Solo women's techniques have also been reinterpreted by contemporary *qigong* practitioners, concerned more about health than about immortality, and defining *nüzi qigong* 女子氣功 (Women's *Qigong*) as an eminently healing practice for women (Liu 2015; Valussi 2008a). Interestingly, menstrual regularity and a healthy flow are still central to both of these practices, indicating that blood is still the most concerning and difficult to deal with element of female physiology.

Notes

1 For a description of women's roles in early Daoism, see Despeux and Kohn (2005), especially Part 2. For a description of the shift towards a more segregated society, see Ebrey (1993).
2 Editor: *Qi* is often rendered in earlier scholarship as the Greek *pneuma*, and this translation is kept in quoted translations.

3 Zhen Luan, *Xiaodaolun* (compiled 570 CE) in *Guang hongming ji* 廣弘明記, T.52.2103.152; translated in Kohn (1995: 147–50).
4 Christine Mollier (2015: 87–110) describes these early sexual rituals in a different manner, for procreation of the community, in 'Conceiving the Embryo of Immortality: "Seed-People" and Sexual Rites in Early Taoism', in A. Andreeva and D. Steavu (eds) *Transforming the void: embryological discourse and reproductive imagery in East Asian Religions*, Leiden: Brill.
5 Early traces of this critique date back to the founding of the sect, in *j*. 6 of the *Zhen'gao* 真誥 DZ1010 (Stanley-Baker 2013: 101–5, 288 ff.; Strickmann 1981: 179–95).
6 See *Daoshu* 3.4b–7b. A discussion of this is found in Wile (1992: 26–27), and Baldrian-Hussein (2006: 330).
7 In *Fanghu waishi* 方壺外史 (The Untold History of Master Square Pot), reprinted in *Zangwai daoshu* 藏外道書 (Daoist Texts Outside the Canon), vol. 5: 208–375.
8 *Zangwai daoshu*, vol. 5: 578–603.
9 Weight measures changed over time. This conversion refers to Song dynasty measures. Reference in Wilkinson (2013: 556).
10 See Chapter 29 in this volume on *houtian*.
11 On the Qian hexagram, Ibid.
12 *Xiuzheng biannan canzheng* 修真辨難參證, *juan* 1. Translation in Hudson (2010).
13 The term Yellow Dame has multiple meanings. Here it refers to the person who oversees the sexual union, the aim of which is to produce pure *yang* by transferring the woman's internal *yang* to the man.
14 The terms 'you and I' (*biwo* 彼我) are very commonly used in alchemical texts of the Ming Qing era. In texts with a clear sexual overtone, they indicate the male (*wo*) and the female (*bi*), and they are used in the context of a transfer of energies from the female to the male by means of a sexual encounter. In texts that are not overtly sexual, they still refer to the male and female principle, but usually as found within the body of an individual practitioner. In the present case, the terms are used as a metaphor for sexual techniques, which, in turn, is criticised for being both misguided and fruitless.
15 Sexual intercourse.

Bibliography

Pre-modern sources

Baopuzi neipian 抱朴子內篇 (Book of the Master Who Embraces Simplicity), Ge Hong 葛洪 (283–343), DZ 1185.
Bianhuo lun 辨惑論 (To Discriminate Errors), 5th century, Shi Xuanguang 釋玄光, T. 2102, 48a–49c.
Caizhen jiyao 採真機要 (Essentials on the Mechanism for Gathering the True), attributed to Zhang Sanfeng 張三丰, in Hu Daojing 胡道靜 *et al*. (eds) (1992–1994) *Zangwai daoshu* 藏外道書 (Daoist Texts Outside the Canon), vol. 11, pp. 348–51, trans. D. Wile (1992), pp. 178–88.
Daoshu 道樞 (Pivot of the Dao), compiled by Zeng Zao 曾慥 (?–1155), DZ 1017.
Daoshu shiqizhong 道書十七種 (Seventeen Books on the Dao), ca. 1842, Fu Jinquan 傅金銓, in Hu Daojing 胡道靜 *et al*. (eds) (1992–1994) *Zangwai daoshu* 藏外道書 (Daoist Texts Outside the Canon), vol. 11, pp. 1–720.
Fanghu waishi 方壺外史 (The External Secretary of Mount Fanghu), 1571, collected by Lu Xixing 陸西星 (1520–1601 or 1606), in Hu Daojing 胡道靜 *et al*. (eds) (1992–1994) *Zangwai daoshu* 藏外道書 (Daoist Texts Outside the Canon), vol. 5, pp. 208–375.
Huiming jing 慧命經 (The Book of Wisdom and Life), 1794, Liu Huayang 柳華陽 (ca. 1736) in Hu Daojing 胡道靜 *et al*. (eds) (1992–1994) *Zangwai daoshu* 藏外道書 (Daoist Texts Outside the Canon), vol. 5, pp. 875–918.
Jindan jieyao 金丹節要 (Synopsis of the Golden Elixir) n.d., collected in *Sanfeng danjue* 三丰丹訣 (Elixir Intructions of Zhang Sanfeng) in Hu Daojing 胡道靜 *et al*. (eds) (1992–1994) *Zangwai daoshu* 藏外道書 (Daoist Texts Outside the Canon), vol. 11, pp. 336–43, trans. D. Wile (1992), pp. 169–78.
Jindan jiuzheng pian 金丹就正篇 (Treatise on receiving correct instructions on the golden elixir), Lü Xixing 陸西星 (1520–1601 or 1606), in Hu Daojing 胡道靜 *et al*. (eds) (1992–1994) *Zangwai daoshu* 藏外道書 (Daoist Texts Outside the Canon), vol. 5, pp. 368–70.

Jindan zhenchuan 金丹真傳 (True Transmission of the Golden Elixir), Sun Ruzhong 孫汝忠 (1574–?), 1615, in Hu Daojing 胡道靜 *et al.* (eds) (1992–1994) *Zangwai daoshu* 藏外道書 (Daoist Texts Outside the Canon), vol. 11, pp. 860–76.

Niwan Li Zushi nüzong shuangxiu baofa 泥丸李祖師女宗雙修寶筏 (Precious Raft on Paired Cultivation of Women by Master Li Niwan), n.d., in Min Yide 閔一得 (ed.) (1834), *Gu Shuyinlou cangshu* 古書隱樓藏書 (Collection of the Ancient Hidden Pavilion of Books), Wuxing: Jingai Chunyang gong cangban.

Nüdan hebian 女丹合編 (Combined Collection of Female Alchemy), 1906, He Longxiang 賀龍驤 (ed.), Chengdu: Er xian'an.

Pangmenlu 旁門錄 (Record of the Heterodox Schools), in He Longxiang 賀龍驤 (ed.) (1906) *Nüdan hebian* 女丹合編 (Combined Collection of Female Alchemy), Chengdu: Er xian'an, pp 23.1a–7a.

Rongcheng pian 容成篇 (Chapter on Rongcheng), Zeng Zao 曾慥 (?–1155), in *Daoshu* 道樞 (Pivot of the Dao), DZ 1017, 3.4b–7b.

Ruyao jing 入藥鏡 (Mirror for Compounding the Medicine), attr. to Cui Xifan 崔希范 (ca. 880–940), in Yan Yonghe 閻永和 *et al.* (eds) (1906, 1985) *Chongkan daozang jiyao* 重刊道藏輯要, Chengdu: Bashu shushe, vol. 4, pp. 485–94.

Songshi 宋史 (History of the Song), 1346, in *Chizaotang Siku quanshu huiyao* 摛藻堂四庫全書薈要 (1773), *juan* 129–39 (1985), Taipei: Shijie shuju.

Wugen shu 無根樹 (The Rootless Tree), attr. to Zhang Sanfeng 張三丰, in Hu Daojing 胡道靜 *et al.* (eds) (1992–1994) *Zangwai daoshu* 藏外道書 (Daoist Texts Outside the Canon), vol. 5, pp. 578–603.

Xiaodaolun 笑道論 (Laughing at the Dao) 570, Zhen Luan 甄鸞, in *Guang hongming ji* 廣弘明記 (The Extended Collection of Great Clarifications), T.52.2103.152.

Xiuzhen biannan canzheng 修真辨難參證 (Revision of the Discussions on the Cultivation of Perfection), Liu Yiming 劉一明 (1734–1821) and Min Yide 閔一得 (1989), Shanghai: Shanghai guji chubanshe.

Zangwai daoshu 藏外道書 (Daoist Texts Outside the Canon), Hu Daojing 胡道靜 *et al.* (eds) (1992–1994), Chengdu: Bashu Shushe.

Modern sources

Ames, R. (1981) 'Taoism and the androgynous ideal', in R.W. Guisso and S. Johannesen (eds) *Women in China*, Youngstown: Philo Press, pp. 21–45.

Andreeva, A. and Steavu, D. (eds) (2015) *Transforming the Void: embryological discourse and reproductive imagery in East Asian religions*, Leiden: Brill.

Baldrian-Hussein, F. (2006) 'Daoshu', in F. Pregadio (ed.) *The Encyclopedia of Taoism*, London: Routledge, vol. 1, pp. 329–31.

Chia, M. (2005) [1986] *Healing Love through the Dao: cultivating female sexual energy*, Rochester: Destiny Books.

Despeux, C. and Kohn, L. (2005) *Women in Daoism*, Magdalena, NM: Three Pines Press.

Ebrey, P. (1993) *The Inner Quarters: marriage and the lives of Chinese women in the Sung period*, Berkeley: University of California Press.

Guisso, R.W. and Johannesen, S. (eds) (1981) *Women in China*, Youngstown: Philo Press.

Harper, D. (1987) 'The sexual arts of ancient China as described in a manuscript of the second century B.C.', *Harvard Journal of Asiatic Studies*, 47.2: 539–93.

Hsi, L. (2001) *The Sexual Teachings of the White Tigress: secrets of the female Taoist master*, Rochester: Destiny Books.

Hudson, C. (2010) 'Man does not part his robes, Woman does not loosen her belt, controversy over non-contact sexual alchemy', paper presented at the meeting of the AAR, 2010.

Kohn, L. (1995) *Laughing at the Dao*, Princeton, NJ: Princeton University Press.

Kohn, L. and Wang, R. (eds) (2009) *Internal Alchemy: self, society, and the quest for immortality*, Magdalena, NM: Three Pines Press.

Liu, X. (2009) 'Numinous father and holy mother: late Ming duo-cultivation practice', in L. Kohn and R. Wang (eds) *Internal Alchemy: self, society, and the quest for immortality*, Magdalena, NM: Three Pines Press, pp. 121–40.

Liu Yafei 劉亞非 (2015) *Nüzi qigong* 女子氣功 (Women's Qigong), DVD, Sierksdorf: AVR Verlag.

Mollier, C. (2015) 'Conceiving the embryo of immortality: "seed-people" and sexual rites in early Taoism', in A. Andreeva and D. Steavu (eds) *Transforming the Void: embryological discourse and reproductive imagery in East Asian religions*, Leiden: Brill, pp. 87–110.

Pregadio, F. (ed.) (2006) *The Encyclopedia of Taoism*, London: Routledge.

Raz, G. (2008) 'The way of the yellow and the red: re-examining the sexual initiation rite of Celestial Master Daoism', *Nan Nü*, 10.1: 86–120.

Robinet, I. (2006) '*Huangting jing*', in F. Pregadio (ed.) *The Encyclopedia of Taoism*, London: Routledge, vol. 1, pp. 511–4.

Stanley-Baker, M. (2013) *Daoists and Doctors: the role of medicine in Six Dynasties Shangqing Daoism*, PhD thesis, University College London.

Strickmann, M. (1981) *Le Taoïsme du Mao Chan: chronique d'une révélation*, Paris: Collège de France.

Valussi, E. (2008a) 'Women's alchemy: an introduction', in L. Kohn and R. Wang (eds) *Internal Alchemy: self, society, and the quest for immortality*, Magdalena, NM: Three Pines Press, pp. 141–62.

Valussi, E. (2008b) 'Men and women in He Longxiang's *Nüdan hebian* (Collection of Female Alchemy)', *Nannü*, 10.2: 242–78.

Wile, D. (1992) *Art of the Bedchamber: the Chinese sexual yoga classics, including women's solo meditation techniques*, Albany: State University of New York Press.

Wilkinson, E.P. (2013) *Chinese History: a new manual*, Cambridge, MA: Harvard University Asia Centre.

31
NUMINOUS HERBS
Stars, spirits and medicinal plants in Late Imperial China[1]

Luis Fernando Bernardi Junqueira

Introduction

Before the creation of an evidence-based Traditional Chinese Medicine (TCM) (Chapter 45 in this volume) in the last century, astrology, divination, talismans, spells and incantations, often combined with herbs, had constituted the core of the healthcare culture in Chinese society. Extracted from a seventeenth-century talismanic manuscript, the following recipe illustrates this intertwined relationship between herbs and occult arts in Late Imperial China.

> At the midday of the 5th day of the 5th month, sit beneath a white poplar tree (*bai yangshu* 白楊樹) and wait until a leaf falls on your side. Remove its stalk and recite the following spell: "Your home was hanged upside down, and I am heartbroken for you; [but] I am willing to fly under the sky with you". Whittle the stalk and out of it make a person figurine around three *cun* tall,[2] offer it wine and preserved fruits for forty-nine days and nights, and then wrap it with a white cloth; use [this amulet] during an emergency so that people will not be able to see you.
>
> (*Zhuyouke mijue qishu*, 2.35–36)

Invoked through herbal incense, gods and spirits were active agents in the processes of healing, while demons, beasts and insects could be chased away with such household herbs as artemisia, peach flowers and ginger. Past studies have persistently dismissed this ritual knowledge of herbs as 'magical' or 'superstitious', and the worldview wherein this knowledge is rooted as 'ludicrous'. Focussing on manuscripts and printed books produced between the sixteenth and early twentieth centuries and drawing on fieldwork with folk healers in Shanghai and Hong Kong, I will instead approach the longstanding relationship between herbs and occult arts in traditional China from historical and anthropological perspectives. What is the place of occult medical knowledge in Chinese medical literature? To what extent does this knowledge relate to scholarly medicine, its theories and practices? What was its significance for everyday life? And who were the actors mostly involved with the ritual use of medicines?

Just as in other traditional medical systems worldwide, herbs constitute the essence of Chinese healing culture. The colossal 'Compendium of *Materia Medica*' (*Bencao gangmu*

本草綱目), compiled in the sixteenth century, contains 1,892 entries, over half of which is dedicated to medicinal herbs alone (Unschuld 1984: 148–9). Even though such works as the Compendium also include hundreds of minerals, animals and other substances, the Chinese term for *materia medica*, or *bencao* 本草 (lit. 'roots and grasses'), has for millennia been used to summarise the whole spectrum of therapeutic substances used in Chinese pharmacotherapy. Early dictionaries and traditional histories explain the character for medicinal drug, or *yao* 藥, as 'the herbs which heal disease' (*Shuowen jiezi*, 1.15), praising them as the first healing methods primordial deities bestowed upon the Chinese (*Lidai mingyi mengqiu*, upper.1). It therefore comes as little surprise that the ritual use of herbs in such fields as astrology, talismans and incantations has thrived over the past few centuries – a fact medical histories of Late Imperial China have only recently begun to acknowledge (Unschuld and Zheng 2012).

Alternatively known as the 'Continent of Spirits' (*Shenzhou* 神州), China is conceived of as a divine land inhabited by numinous forces (Lagerwey 2010: 17). In both traditional and modern China, the ritual use of herbs has therefore been based on the holistic cosmology wherein everything in the universe is animated, sacred and interconnected. From spirits, stars and rivers to plants, humans and even written characters, everything is on a continuum, with the body seen as a microcosmic image of the universe itself (Lo 2001).

As a result, herbs were never expected to be used in isolation. They were picked, cut, decocted, ground, smashed, burned, shaped, hung on walls, carried as charms, mixed with talismans and even made sacred. The time, place and direction in which they grew could affect their healing efficacy as much as their ritual use. In fact, ritual action was sometimes more crucial than the selection of particular herbs (Sax 2010), so much that modern botanical understandings, which break down herbs into chemical components and active agents, would sound absurd to the pre-twentieth-century Chinese. It was the intense interaction between spirits, stars, herbs and humans, rather than their mere existence in isolation, which sustained the Chinese medical practices that are analysed here. Whereas the sources examined in this chapter are textual, my primary concern is with practice, or the culture-specific ways the Chinese have transformed 'natural herbs' into 'cultural artefacts' infused with human and divine meanings (Hsu 2010). Looking at their ritual use in three distinct yet correlated dimensions, namely, astrology, talismans and incantations, the following pages will try to illuminate the role of herbs as powerful mediums between the human, natural and spiritual worlds in Late Imperial China.

Herbs and astrology

The astrological use of herbs constituted the foundation of scholarly and folk healing traditions in imperial China (Unschuld 1985: 55–99; Lo 2008). From their planting and harvesting to processing and consumption, the life cycle of herbs was intrinsically bound to the motions of the stars in the sky and the ritual actions of humans on earth. A herb picked at a certain time of the year, month, week, day or hour had its healing efficacy enhanced or featured miraculous therapeutic properties that would otherwise be concealed. (On these timings, see Chapter 4 in this volume.) Geomantic principles, such as the position a herb grows in or the place where it is kept after collection, also affected its healing and apotropaic qualities. In other words, an ordinary herb could metamorphose into a powerful weapon against demons, insects and epidemic diseases depending on its time of collection and ritual manipulation.

Whereas pervasive in all aspects of Chinese medicine, astrological elements are particularly evident at certain dates and behind the use of certain herbs. In medical texts, the

Duanwu 端午, known more commonly in English as Dragon Boat Festival, is by far the most popular day for the manipulation of herbs. Falling on the fifth day of the fifth month of the Chinese lunar calendar – which means any time between late May and late June in the Gregorian calendar – the *Duanwu* is considered the hottest day of the year. In traditional terms, it marks the apex of *yangqi* 陽氣 and the rock bottom of *yinqi* 陰氣 (*Lüshi Chunqiu*, 4.2). On the one hand, the *Duanwu* is the day when poisonous insects, evil spirits and epidemic diseases are most active; so much that the fifth month as a whole is also known as the 'evil month' (*eyue* 惡月) or 'month of a hundred poisons' (*baiduyue* 百毒月) (*Jingchu suishi ji*, n.p.; *Qingjia lu*, 5.1–2). On the other hand, the healing and apotropaic qualities of herbs were also at their greatest on that day, especially at noon. On the *Duanwu*, people employed the 'five auspices', namely, the mugwort, pomegranate, calamus, garlic and *longchuanhua* 龍船花 (*Ixora chinensis*) to fight off the 'five poisons', represented by the centipede, snake, scorpion, toad and gecko (*Suishi guangji*, 21.8, 16, 21).

Encyclopaedias and local gazetteers record a wide variety of herb-based rituals performed on the *Duanwu*. An early seventeenth-century gazetteer of the Taizhou 泰州 county, Jiangsu province, states that on that date:

> [locals] eat rice dumplings, brew hot medicinal wine made of realgar and calamus, and eat cured meat; both men and women wear talismans written in cinnabar, tie threads of the five colours around their arms, and pin their hair with flowers and mugwort twigs.
> (*Chongzhen Taizhou zhi*, 1.21)

These rituals, the text concludes, are not restricted to Taizhou but conspicuous everywhere in China. Around 200 years later, a gazetteer dedicated to the same locality added boat racing and bath with the 'decoction of a hundred herbs' (*baicaotang* 百草湯) as two other rituals locals performed on the *Duanwu*. While the herbs for this decoction varied from region to region, it often included mugwort and mulberry leaves, wild chrysanthemum flowers, peach twigs, calamus and fish mint (*yuxingcao* 魚腥草). The herbs should be picked in the early morning of the *Duanwu*, decocted and then used as a herbal bath later in the afternoon (*Daoguang Taizhou zhi*, 5.3–4). The ritual use of some of these herbs, intended to attract good luck and chase away the evil, can be traced back to periods as early as the second century BCE (Yan 2017: 215, 230–1).

Mugwort and calamus have for millennia been praised as powerful apotropaic herbs whose potency was enhanced on the *Duanwu*. A calendrical treatise compiled between the sixth and seventh centuries urges members of the same clan to pick those mugwort leaves, which resemble the image of a tiger or cut tiger shapes out of the same leaves; these would then be known as 'mugwort tigers' (*aihu* 艾虎) (*Suishi guangji*, 21.14–15). Almost a thousand years later, the practice of collecting mugwort leaves on the *Duanwu* and cutting shapes out of them persisted; now, the leaves should be collected before cockcrow, have the shape of humans and be hung on top of the main entrances to chase away venomous creatures (*Bencao gangmu*, 15.10). A recipe to wipe out demons living inside one's body calls on the sick to collect mugwort leaves on the *Duanwu*, mix them with musk and honey, and then prepare small pills to be taken with ginger soup (*Huisheng ji*, 2.37).

Praised as a 'celestial herb' (*lingcao* 靈草) in Daoist texts (Bokenkamp 2015), the calamus was believed to increase longevity, improve memory and 'heal a thousand diseases'. A traditional method of collecting and preparing calamus leaves into pills combines astrological and geomantic principles:

The time to pick calamus falls on the 3rd day of the 3rd month, 4th day of the 4th month, 5th day of the 5th month, 6th day of the 6th month, 7th day of the 7th month, 8th day of the 8th month, 9th day of the 9th month or 10th day of the 10th month. You should pick the plants growing in pure water which flows from the south; those [growing in water which flows] from the north are not auspicious.

(*Shenxian fushi lingcao changpu wanfang zhuan*, n.p.)

Calamus picked on the *Duanwu* and mixed with honey to form pills was a particularly powerful way to boost one's vitality. After taking twenty-five pills in the early morning, twenty-five at noon and thirty late at night for a whole year, demonic creatures would leave the body and the person would enjoy long life. Daoist immortals and hermits carried calamus leaves with them on their journeys to sacred mountains, while the plant had for centuries been cultivated inside imperial palaces as a reminder of immortality (*Shenxian fushi*, n.p.). The sharply pointed shape of its leaves assigned it the name of 'calamus swords' (*pujian* 蒲劍), another effective apotropaic weapon (*Qingjia lu*, 5.7).

Recognised as the hallmarks of the *Duanwu*, mugwort tigers combined with calamus swords could drive off ghostly and poisonous beings. A ubiquitous ritual in traditional Chinese households consisted of hanging up mugwort twigs and calamus leaves, tied with a red thread, on the top of the main entrance of the home. Local variations included combining those herbs with bitter fleabane (*peng* 蓬), peach twigs and strings of garlic, and then hanging the bundle on the top of the bed (*Qingjia lu*, 5.7). Descriptions of such rituals are pervasive in novels like the seminal eighteenth-century 'Dream of the Red Chamber' (*Hongloumeng* 紅樓夢; trans. Joly 2010: ch. 31), while legends about the sacred relationship between mugwort and calamus continue to circulate in China today. Oral accounts collected among the people of Yuanling 沅陵 county, Hunan province, in the early 2000s, tell how in high antiquity a water monster was trying to submerge farmlands so as to put them under his control. Afraid that this would harm the locals, celestial immortals employed mugwort and calamus as weapons against the dreadful monster. The immortals won and spared the creature under the condition that he would never again cause trouble to their descendants,[3] marking the locals' houses with the same mugwort and calamus leaves they employed in the battle. Then, the *Duanwu* day came. Riding a wave all the way in, to his chagrin the ravenous monster only found a few unmarked houses. Later, people learnt that mugwort and calamus terrified diabolic monsters, wicked bandits and poisonous creatures, and transmitted these rituals to posterity (Liu 2014: 165–6).

The *Qingming* 清明 has for millennia been another important date for the astrological use of herbs. Known in English as the Tomb-Sweeping Day, on the *Qingming* Chinese families clean the tombs of their ancestors, worshipping and making ritual offerings to them. Contrary to the *Duanwu*, whose date is calculated according to the cycles of the moon, the *Qingming* falls on the first day of the fifth solar term. In the Gregorian calendar, it translates as the fifteenth day after the spring equinox, on 4, 5, or 6 April each year. Just as the mugwort and calamus are associated with the *Duanwu*, on the *Qingming* it is the willow tree which takes centre stage. An eighteenth-century gazetteer describes some rituals families from the Yangwu 陽武 county, Henan province, performed on the *Qingming*. As an act of communal worship, in the early morning locals would move spirit tablets and statues from their private shrines to open altars especially set up in front of willow trees, hanging up bundles of willow twigs on the top of the main entrance of their residences. The purpose of these rituals was to protect one's home against evil in all its forms (*Yangwu xian zhi*, 3.3). Also known as *guibumu* 鬼怖木, or 'the tree that terrifies ghosts', willow wood could be carved into apotropaic amulets, a practice that can be traced back to the sixth century (*Qimin yaoshu*, 5.12–13), or into an arsenal of ritual weapons employed against demonic creatures on hungry ghost festivals. Its sacred qualities have been consolidated in the gracious image of Guanyin 觀音, the most beloved bodhisattva in Chinese Buddhism, using a willow branch to sprinkle blessing water upon humankind.

Other important astrological dates included the Spring Festival, or Chinese New Year, as well as the third day of the third month, sixth day of the sixth month, seventh day of the seventh month and ninth day of the ninth month of the Chinese lunisolar calendar. The Spring Festival lasts from the evening preceding the first day of the year until the Lantern Festival, which falls on the eighteenth day, and it is marked by the beginning of the new moon between late January and late February each year. By taking a bath with a decoction of wolfberries on the New Year's Eve, and then repeating the procedure on the first, second, eighth, thirteenth, fifteenth and twentieth days of the New Year's first month, one could prevent disease and attract good luck in the upcoming year (*Yashang zhaizun shengba jian*, 5.23). And the ritual burning of weed boats, believed to have the power of 'burning up' bad karma, had to be performed in a river on the eighteenth day of the New Year's first month (*Yueling cuipian*, 4.35). Given their antiquity, it is hard today to make firm claims about why and how certain herbs became associated with particular dates of the Chinese traditional calendar, although some believe those dates mark the death of legendary or historical figures (Yan 2017: 231–44).

Herbs and talismans

> [...] and, respectfully abiding by the Heavenly principles and for the terror of ghosts and demons, the Yellow Thearch suddenly realised that characters are numinous.
>
> (*Zhuyouke zhibing qishu* 1927: 2)

This passage embodies the essence of talismanic healing: characters are sacred entities imbued with healing power. Known as *fu* 符 in Chinese, talismans are artefacts composed of characters and symbols written in a certain fashion and used for healing or apotropaic purposes. Talismans can assume multiple forms and are not supposed to be read by humans; rather they serve as a writ through which humans communicate with, or give commands to, the spirit world (Bumbacher 2014; Junqueira 2021). Recent archaeological discoveries have shown that talismans and spells – the verbal aspect of talismans – have been used in China for, at least, the past two millennia (Harper 2015). When the first Chinese imperial medical institutions emerged around the sixth century CE, talismans and spells were soon incorporated as officially sponsored medical disciplines alongside drugs and massage, a status they enjoyed until the late sixteenth century (Junqueira 2018a; Cho 2005). Despite the fierce debates around the legitimacy of talismanic healing and its place vis-à-vis orthodox medical traditions, talismans and spells remained pervasive among both scholarly and folk healers in traditional China (Junqueira 2021, 2018a).

Given the prominence of herbs in post-sixteenth-century Chinese medical traditions (Leung 2003), it is not surprising that talismans developed an intimacy with the herbal world. A mid-nineteenth-century talismanic manuscript preserved at Shanghai Library offers some of the most comprehensive accounts on the connection between talismans and herbs in Chinese medicine. Before creating a talisman, the healer should hold his brush or seal and, with a pure and sincere heart, invoke the Heavenly Physician

(*Tianyi* 天醫) through herbal incense. Breathing in pure *qi*, he then draws talismans on rice or bamboo paper while uttering secret spells. The talismans are then burned, and their ashes mixed with a decoction prepared with pure water and three *qian* of each herb stated in the recipe.[4] Herbs, it concludes, are used to guide (*yin* 引) the talismanic energy inside the patient's body, helping deliver its occult message to internal and external deities (*Zhuyouke*, 3).

Prior to creating a talisman, the healer should identify if the disease belongs to *yin* or *yang*. When the patients have trouble in speaking and their feet and hands are cold, this is the sign of a *yin*-syndrome, and one should create the talisman mixing ginger juice with black ink. When the person's body itches and he or she feels alternations of pain, this is the sign of a *yang*-syndrome, and the ink must instead be mixed with rice vinegar. Should the person be afflicted by skin diseases, the healer could create the talisman directly onto the injured area, drawing a circle around it and then applying lard and herbs on its top. If after drinking the talismanic potion the person vomits, feels hot or acts as if possessed, these are all signs the treatment was successful (*Zhuyouke*, 3). The inclusion of herbal decoctions in talismanic treatments and the admonition to identify syndromes based on the principles of *yin-yang* might have been attempts by post-sixteenth-century healers to reintegrate talismans into orthodox medical traditions.

The writing style of healing talismans can be divided into two main categories: secret characters (*mizi* 秘字) and elaborate talismans (*huafu* 花符) (Figures 31.1–2). Secret characters are talismans often composed of the radicals *shang* 尚 (to be in charge of) or *yu* 雨 (thunder) at the top, and *gui* 鬼 (ghost) or *shi* 食 (to eat) at the bottom. Other characters can be added depending on the disease the talisman is expected to cure. For a difficult labour, a talisman composed of the radicals *shang*, *ge* 革 (to change) and *shi* 食 should be burned into ashes and taken by the woman together with a decoction of rice or motherwort (*yimucao* 益母草) (*Zhuyouke*, 5). And for nocturnal emissions, a talisman formed of *shang*, *yu* 羽 (arrow), *yuan* 元 (primordial) and a variant of the character *gui* could be employed; here, the talismanic ashes should be ingested with a decoction of soft rush (*dengxinocao* 燈芯草), a herbaceous flowering plant traditionally used as a repellent for mosquitoes and flies or as wicks in household lamps (*Zhuyouke*, 27).

Certain secret characters work as invocatory spells. An instruction to heal *nüe* 瘧 (this term for feverish epidemics is now used to translate 'malaria'), attributed to evil spirits, requires that one takes seven tangerine leaves and on each leaf writes, in cinnabar, one of the seven characters *kui* 魁, *zhuo* 魆, *huan* 䰠, *xing* 魓, *bi* 魓, *fu* 魖, and *piao* 魒; each of these characters refers to the secret name of an astral deity from the Northern Dipper. The leaves should then be dried by fire and ground into powder; when the disease breaks out, the victim should take it with plain soup (*Chuanya waibian*, 1.n.p). This method thus serves as a contrast to the use of artemisia against *nüe* by the Daoist alchemist Ge Hong, whose cold-washing technique was an inspiration for the modern nobel prize-winner Tu Youyou 屠呦呦 (Chapter 51 in this volume).

Contrary to secret characters, elaborate talismans have no fixed pattern. They are usually created by the combination of stylish characters, ancient scripts and occult symbols, and then activated by mudras and verbal spells. Whereas the creation of elaborate talismans often requires lengthy rituals of purification, consecration and invocation, their application does not differ much from secret characters. A rare five-volume manuscript once kept in the Emperor Qianlong's (1711–1799) imperial library, and now held in the National Central Library, Taiwan, records hundreds of elaborate healing talismans. A ritual against plague begins with the purification of ritual implements – yellow paper, brush and cinnabar – followed by the

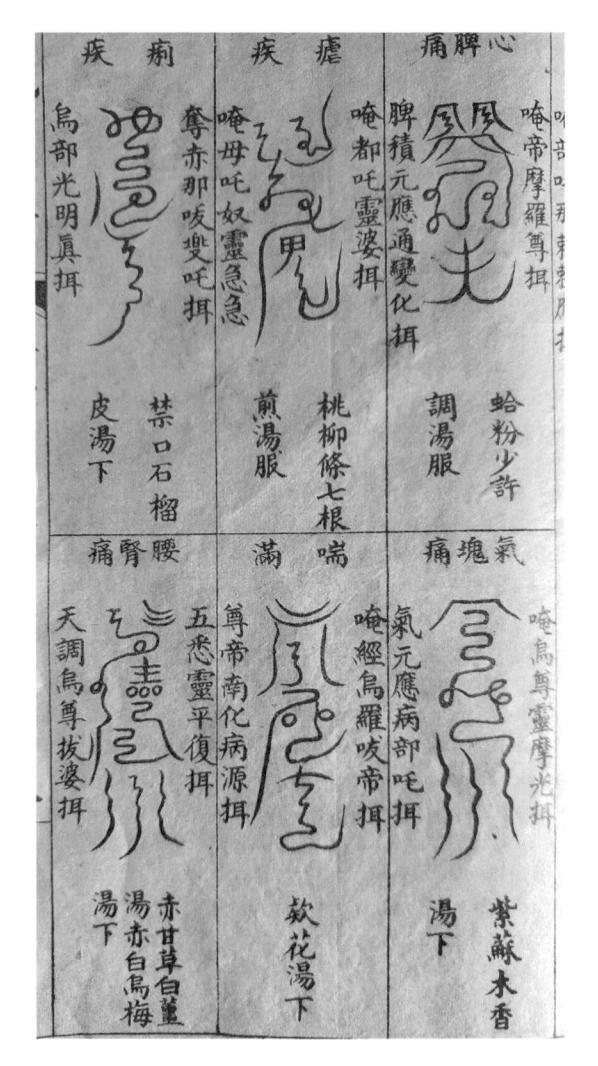

Figure 31.1 Example of twelve 'secret characters' (on the left). *Miben fuzhou quanshu* 秘本符咒全書 (Compendium of Secret Talismans), Shanghai: Jingzhi tushuguan, 1912–1949, n.p

invocation of the Heavenly Generals Ding Chengzong 丁成宗 and Xu Shiheng 徐士衡 using two talismans. The healer should then draw, consecutively, three other talismans: one to suppress, one to confine and one to slash plague demons. While the last two talismans must be pasted on the home's walls and main entrance, the former three had to be burned and swallowed with a decoction of Chinese dates or tea leaves (*Zhuyouke mijue qishu*, 1.94–5).

The manuscript's third volume divides elaborate talismans into thirteen categories of illness – like children, adults, sense organs and so forth.[5] Each category is ruled by a deity invoked through herbal incense, talismans and spells prior to the creation of each healing talisman. After the invocatory ritual, the pattern to draw talismans remains virtually unchanged: each talisman must be written in cinnabar on strips of yellow paper, activated by a spell, burned to ashes and mixed with herbs. A talisman for congested throat, for instance, should be taken with a decoction of mugwort (*ai* 艾), while for *huoluan* 霍亂 another talisman should be ingested with a decoction of cloves and Korean mint (*huoxiang* 藿香) (*Zhuyouke*

Numinous herbs

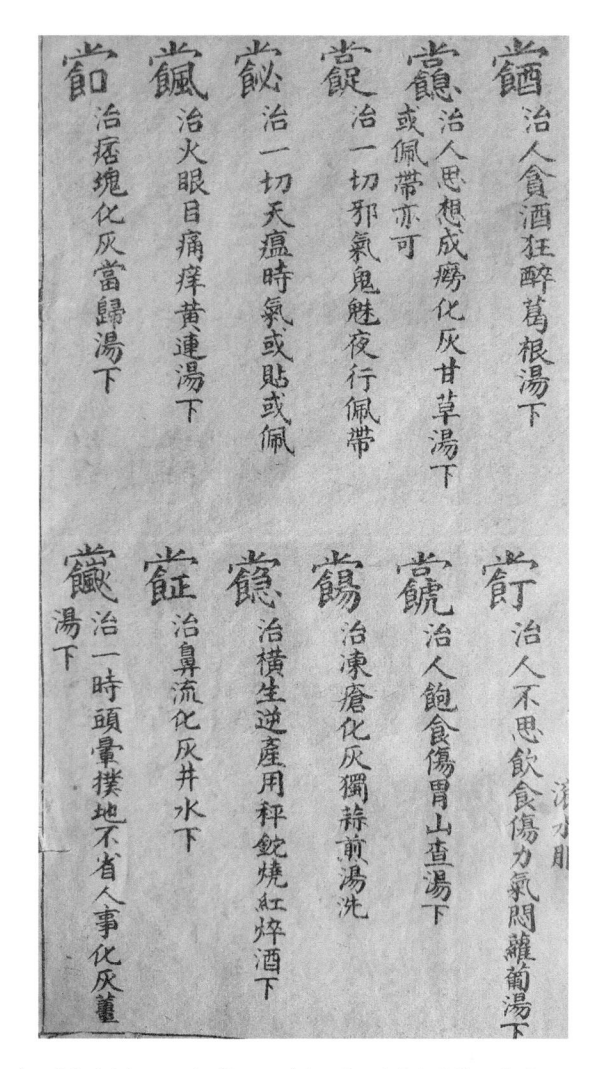

Figure 31.2 Example of six 'elaborated talismans' (on the right). *Miben fuzhou quanshu*, n.p

mijue qishu, 3.57, 30). For 'dripping' (*linyi* 淋溢), a condition in which the individual feels 'a frequent urge to urinate, with the urine dribbling and failing to completely leave the body' (Zhang and Unschuld 2015: 316), talismanic ashes should be ingested with the juice of three-leaf clover flowers, whereas for mouth ulcers, it should be combined with a decoction of apricot kernel and used for rinsing (*Zhuyouke mijue qishu*, 3.64, 59).

In the same way that herbal decoctions assisted the circulation of talismanic essence inside the body, talismans could also maximise the healing power of herbs. A ritual pervasive in talismanic manuscripts involves invoking the Heavenly Physician (*Tianyi* 天醫) through the burning of incense and talismans; at midnight, the God, accompanied by his divine troops, would then descend on the altar dedicated to him and bless the herbs displayed thereon (Figure 31.3) (*Zhuyouke*: preface, n.p.). Another ritual comprises drawing, printing or carving talismans in clay bottles, peach wooden boxes or calabashes; the physician then fills the container with medicine, whose therapeutic potency is increased by the talismans' numinous power (Figure 31.4).

463

Figure 31.3 Altar for the Heavenly Physician. The sixteenth-century statue at the centre is Shennong, the God of Medicine and Agriculture. He is surrounded by talismans, medicinal jars and ritual objects, with herbs displayed in front of him for blessing. Photo by author 14 November 2019, private altar, Shanghai

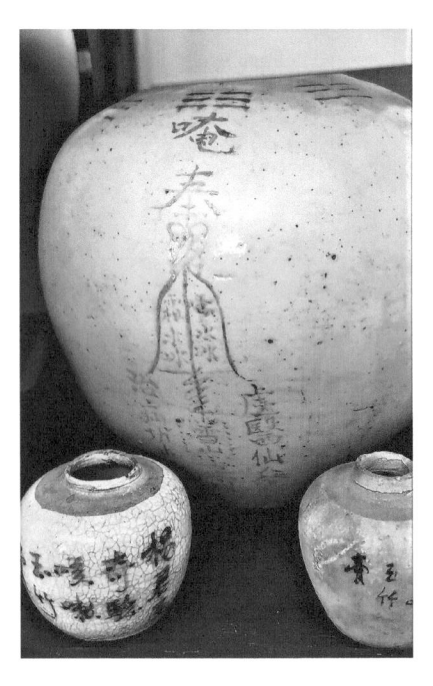

Figure 31.4 Nineteenth-century medicinal jar inscribed with talismans. Such jars are filled with medicinal herbs and then placed in the altar for the Heavenly Physician to bestow upon them His blessings. This ritual, still pervasive among Chinese folk healers today, is believed to maximise the healing power of herbs. Photo by author 14 November 2019, private altar, Shanghai

Numinous herbs

Herbs did not act simply as guides but also as a physical support for the creation of talismans. First, most of the talismans should be written on rice or bamboo paper. Second, as a divine writing, talismans had their origin ascribed to deities, who bestowed them upon humans by means of spirit-possession, dreams or revelations in the natural realm – which, in traditional China, was itself holy. Local gazetteers show that herbs have long served as a material support for the divine revelation of talismans. The seventeenth-century *Tanghu qiaoshu* 倘湖樵書 (Book of Tanghu), a collection of personal notes compiled by Lai Jizhi 來集之 (1604–1682), describes how in Leshan 樂山, Sichuan province, there was once a tree whose leaves grew veins in the form of talismanic writing, while in a neighbouring area, Mount Rongzi 容子, there was another sacred plant whose leaves resembled those from lychee trees yet their veins featured the pattern of the worm-seal, a script commonly used for talismans. Locals from Guangdong and Hubei provinces also discovered that the 'talismanic leaves' (*yefu* 葉符) of bamboos, which had been found emerging

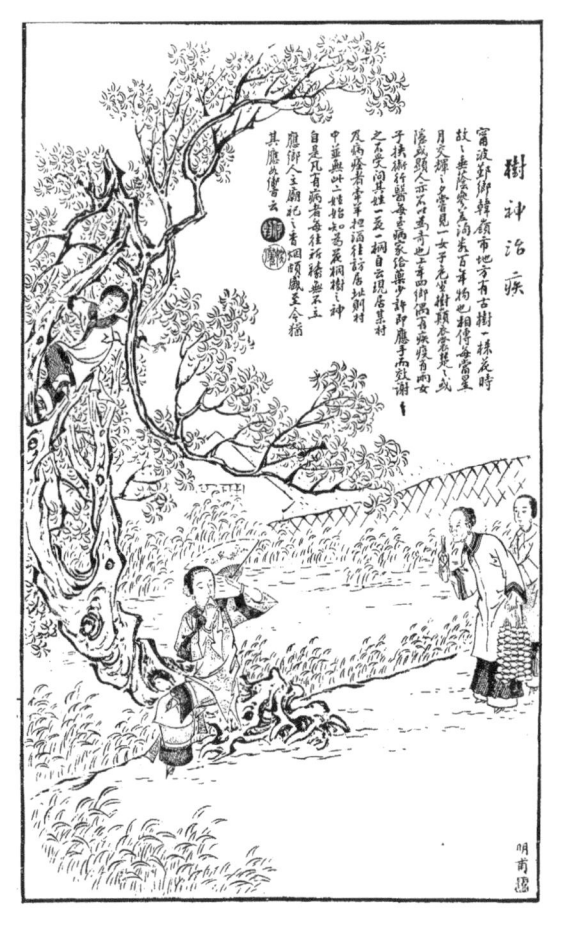

Figure 31.5 'Divine Tree that Heals Disease'. This drawing depicts an ancient divine tree and its two girl-spirits (on the left) in Hanling 韓嶺 village, Ningbo, in the late nineteenth century. When falling sick, villagers would pray and burn incense to the tree. Its wondrous healing powers led locals to erect a special temple for its worshipping. *Dianshizhai huabao* 點石齋畫報 (Illustrated Lithographer), vol. 4 upper, (1884–1889): 1

from sacred altars, stones and caves could serve as weapons to suppress the evil, ward off snakes or heal disease (*Tanghu qiaoshu*, 12.29–33) (Figure 31.5).

The carving of talismans in peach wood has a long history (Lu 2012; Yan 2017: 230–1). For centuries, the Chinese have praised the peach tree as the *primus inter pares* among the 'five kinds of trees' – namely, the peach, elm, mulberry, locust and willow (*Taiping yulan*, 967.2). While its wood was used to carve talismans, sacred statues and exorcistic weapons, its kernels were effective for a difficult labour – the *Chuanya waibian*, a late eighteenth- or early nineteenth-century recipe manuscript, describes how one simply had to split a peach pit into two, write the characters *ke* 可 (can) on the one side and *chu* 出 (out) on the other, and give it to the pregnant woman; the baby would then be delivered immediately (*Chuanya waibian*, 1.n.p.). Peach roots, leaves, resin and fruits all also possessed apotropaic qualities. Peach twigs were hung on walls or the tops of windows, doors and gates so as to chase away malevolent forces and, when boiled with water, could serve as a preventive bath against epidemics (*Chuanya waibian*, 1.n.p.). Peach trees stuck by lightning were cherished for their exceptional quality of warding off the evil; the talismans carved from their wood, known as 'peach talismans' (*taofu* 桃符), embodied the numinous essence of the thunder gods. These talismans could then be worn as protective amulets, transformed into ritual seals or prepared as decoctions to treat heart and abdominal pain, spirit-possession and blood-related ailments (*Suishi guangji*, 21.21).

Herbs and incantations

In Chinese medicine, *jinshu* 禁術 or *jinfa* 禁法 (incantations) comprise a broad category of healing techniques underpinned by the view that there are occult powers in the world that people can harness and manipulate in a myriad of ways. Distinguishing from the same Chinese term *jin* 禁 which elsewhere in this volume specifically refers to, and we translate as, 'prohibitions' (Chapter 19 in this volume), in this context *jin* refers to the category of 'incantations'. 'Incantations' include a wide range of written talismans, as well as verbal spells and household rituals in which herbs – in their various forms – often constitute the central component for healing. Stories about the manipulation of the occult power of herbs, for instance, pervade traditional medical histories. The eighteenth-century *Mingyi lei'an* 名醫類案 (Medical Cases from Famous Physicians) records the story of Gongsun Tai 公孫泰, a wealthy old man from 'ancient times' whose back suffered from *yongju* 癰疽 (abscess and ulceration), a skin condition that conventional drugs and acupuncture were unable to treat. Xue Bozong 薛伯宗, who had mastered the esoteric arts of 'disease transfer' (*yibing* 移病), then used his own *qi* to close the wound and transferred it onto a willow tree nearby. The next day, the carbuncle began to vanish, while a tumour of the size of a fist appeared in the willow. A couple of weeks later, thick yellowish-red pus emerged from the tumour as if it were festering; the tree then withered and Gongsun was finally healed (*Mingyi lei'an*, 10.5). While Shennong 神農, the God of Medicine and Agriculture, ingested a herb that made his belly transparent so as to observe the effect of other herbs on internal organs, the divine physician Bian Que 扁鵲 received a secret recipe that allowed him to see through walls and the human body. Today, folk healers continue to celebrate Xue's art of 'disease transfer', as well as Shennong and Bian Que's 'see-through vision' with highest esteem and as examples of the antiquity, ubiquity and efficacy of incantations in Chinese society (Junqueira 2018a, 28–29, 33–34, 50–52, 57–59, 70–77).

While the rationale behind incantations is not always clear, the constant criticism they received from scholar-physicians leads us to presume incantations did not comply with the orthodox interpretation of medical classics. Recent studies have demonstrated that the likelihood of finding incantations in folk medical manuscripts is far higher than in printed medical books (Chapter 18 in this volume; Unschuld and Zheng 2012: 154–75). This fact should not come as a surprise considering that printed books are the final product of careful compilation and editing efforts by Chinese intellectuals who often condemned, if not simply neglected, the existence of incantations (Junqueira 2018b). Manuscripts, therefore, offer the most fascinating records for the study of healing incantations.

The *Chuanya* 串雅 (Corrected Recipes of Itinerant Healers), a nineteenth-century recipe text, represents a scholarly attempt to incorporate certain folk incantations into orthodox medicine. A substantial portion of the *Chuanya* comprises recipes and techniques collected from folk healers (*caoze yi* 草澤醫), literally 'physicians of grasses and swamps'. Divided into two parts, the second volume dedicates roughly one-third of its content to incantations, categorised as 'incantations with drugs' (*yaojin* 藥禁), 'arts and incantations' (*shujin* 術禁) and 'incantations with characters' (*zijin* 字禁). Another chapter, 'tricks with drugs' (*yaoxi* 藥戲), comprises techniques itinerant healers employed to attract their clientele's attention, including potions to suppress appetite when undertaking hard work, wines to foresee the future and powders to create dreams.

Whereas 'incantations with characters' resembles the talismans analysed in the previous section of this chapter, 'arts and incantations' includes verbal spells and simple household rituals against demons, beasts and insects. To prevent or heal *tui* 㿉, a disease that affects genital organs, hang a couple of eggplants on top of the main door and always glance at them when entering or leaving the house. And a recipe to chase away mosquitoes requires that in the midday of the *Duanwu*, one stares at the sun and, while holding a bunch of soft rush, recites 'in Heaven there is a golden chicken who eats the brain and marrow of mosquitoes'; after this, the person breathes in the *qi* from the sun, blows it on the soft rushes and lights them at night (*Chuanya waibian*, 1.n.p.).

The section 'incantations with drugs' accounts for recipes wherein minerals, animals and herbs, ritually empowered, are the main components. To keep mosquitoes away from a new home, for instance, during the construction period one must bury a cattail leaf fan (*pushan* 蒲扇) underneath the four main pillars that sustain the house, whereas to ward off evil spirits, stone tiles should be buried beneath the four corners of the house and then beaten on their top with seven peach pits (*Chuanya waibian*, 1.n.p.). Most of the recipes in this section are relatively easy to put into practice and only require everyday herbs. To make a child stop crying at night, a father should pick herbs located close to a well and hide them, without the mother knowing, under his child's mat; alternatively, the well's herbs could be replaced with twigs and leaves from a chicken coop or pigsty. And to heal a child's abdominal pain, the father must pick any herbs growing from a tree's hollow and stealthily place them on the roof of the house (*Chuanya waibian*, 1.n.p.). These last two recipes demonstrate that the exact kind of herb was not as important as the ritual manipulation of the same.

The oral nature of folk healing knowledge makes it difficult to clarify the original rationale behind the ritual use of the herbs contained in the *Chuanya*. The reason is that, just as in medieval and early-modern Europe (Davies 2007), in traditional China the image of folk and itinerant healers remained rather controversial. Praised by some as living repositories of medical knowledge, others despised them as chief vectors of medical heterodoxy who only cared about how to make a profit out of the naïve (Wang 2013). Traditional biographies depict the divine physician Bian Que, considered the father of Chinese medicine, and Zhang Congzheng 張從正 (1156–1228), known as one of the greatest physicians of the Jin-Yuan period (1115–1368), as itinerant healers. Their stories raise questions about the social competition between learnt and itinerant physicians, and whether they were composed thus as rhetorical attempts to criticise the arrogance of scholar-physicians, rather than a celebration of folk healers.

The controversial picture of folk healers is best manifested in the notes of the nineteenth-century physician Mao Duishan 毛對山 regarding the fish seller Li Bo 李跛. When Li was a child, a couple of tiny lumps appeared on the back of his feet but, since the pain was not severe, he simply ignored them. On a hot summer day, however, Li's feet began to turn so reddish and swollen that it caught the attention of an itinerant healer who was passing by. The healer then warned Li:

> This is not *xuanjie* 癬疥-illness (a general term for parasitic skin rash). The poison has infected the *yangming* 陽明 channels for very long. It is now time to treat it with external medicine, otherwise the disease will turn into intestine ulcers (*changju* 腸疽) and you will lose your feet.
>
> (Mao Duishan yihua, n.p.)

Tragically, Li Bo's father did not believe the itinerant healer. A few days passed, the disease broke out again and, because conventional drugs were unable to heal him, Li Bo lost his feet. Mao concluded this anecdote arguing that, although most itinerant healers were charlatans, their drug and talismanic knowledge was worth preserving due to its proven therapeutic effectiveness.

The widespread condemnation of itinerant healers overshadowed their sporadic appraisal in printed medical literature (Junqueira 2018b; Unschuld and Zheng 2012: 73–106). Scholar-physicians criticised them for not understanding the true principles of medical classics – or even worse, for distorting their 'real meaning'– and for adopting all kinds of illicit means to make a profit: from producing and selling fake medicine to bragging about mysterious recipes that promised wondrous results. Indeed, the *Chuanya* records several of these recipes. To reduce thirst when walking long miles, one should grind sugar, *bai fuling* 白茯苓 (white *Poria cocos*), mint (*bohe* 薄荷) and liquorice (*gancao* 甘草) into powder, mix it with honey, and mould the mixture into pills to be taken once a day. Pills made of hemp seeds (*damazi* 大麻子), calamus and parturition blood, if taken every morning for a hundred days, allowed one to see ghosts. Rubbed on one's hands, a fragrant potion made of dried flowers and liquid honey was believed to attract a kaleidoscope of butterflies around the healer, a shimmering spectacle used to captivate potential customers by showing the healer's wondrous powers (*Chuanya waibian*, 1.n.p.).

Yet, a few records also illustrate how itinerant healers employed poisonous herbs to induce disease in someone who was initially healthy: ashes of 'snake mushroom' (*shexun* 蛇蕈) to create skin ulcers, wine of datura metel (*mantuojiu* 曼陀酒) to induce insanity (*liudian* 留癲), and in remedies supposed to improve the spleen, they instead added dew of pawpaw (*mugualu* 木瓜露) to weaken the patient's body (*Chuanya neibian*, personal preface). Folk medical manuscripts record dozens of secret drug recipes itinerant healers used to cheat, attract or gain the trust of their clientele, behaviour scholar-physicians abominated (Unschuld and Zheng 2012: 9–13, 73–106; Andrews 2014: 25–51).

Spirit-possession and its myriad of herbal treatments are central aspects of folk healing. But here, again, scholarly and folk healers did not agree on its pathological nature. The folk belief – which, we might assume, most of the Chinese embraced – associated possession with entering old, abandoned temples or being suddenly attacked by the evil while wandering around forests, mountains and valleys (*Yixue zhengzhuan*, 1.30). Scholar-physicians, on the other hand, explained the aetiology of spirit-possession from the perspective of the medical classics. The scholar-physician Xu Dachun 徐大椿 (1693–1771), for example, did not reject the idea that ghosts and demons could harm humans but rather argued that only those with a weak essence, spirit and *qi* would be affected. Whereas talismans and spells could occasionally prove effective, physicians should rather employ drugs to heal the root cause of spirit-possession – which, he believed, resulted from an internal body disharmony (*Yixue yuanliulun*, 1.39–40; trans. Unschuld 1998: 127–8). Xu's view summarises much of the scholarly medical debates on spirit-possession between the sixteenth and twentieth centuries (Junqueira 2018a; Chen 2008).

The recipe collection 'Sea of Medicine' (*Yixue huihai* 醫學匯海) illustrates such debates. Published in 1826, it dedicates an entire chapter to *xiesui* 邪祟, a category of diseases attributed to demons and poisonous insects. After echoing Xu's words, the chapter lists over twenty recipes against *xiesui*, all of which are based on minerals, animals and especially herbs. Pills to treat illnesses caused by fox-spirits who dwell in valleys are made of a complex mixture of thirteen ingredients, nine of which are herbs, including ginseng, young peach fruits and burning bush (*guijianyu* 鬼箭羽). After being wrapped in gold foil, the patient takes a few pills, together with a decoction of *muxiang* 木香 (*Radix Aucklandiae*), just before going to bed; but to chase away evil spirits, five or seven pills must be kept in a small crimson bag and hung on top of the bed. For madness caused by malicious ghosts or fox-spirits, a recipe combines principles of acupuncture and herbal medicine: the patient's hands must be tied up by the thumb, while dried mugwort is burned onto the area between the nails and cuticle up to seven times; this area, the text reminds us, is known as *guikuxue* 鬼哭穴, or 'the point where ghosts sob' (*Yixue huihai*, 34.38–48).

Conclusion

During most of the twentieth century, attempts to modernise Chinese medicine have encouraged the marginalisation of most of the therapies and worldviews explored in this chapter. Astrology, talismans and incantations are now alien to most Chinese medical students. Old medical texts, selected for present-day publication, are carefully edited so as to eradicate their most obvious ritual aspects and shape Chinese medicine as a modern scientific discipline. The only edition of the *Chuanya* published with a commentary so far, for instance, warns readers that the original text 'contained content at odds with scientific values': as a result, from hundreds of incantations, the editors included only eighteen in their book, despising all rituals and astrological knowledge as 'primitive religious non-sense' (Fujiansheng 1977: 1–4).

Although the 'obsession with scientific accuracy' is a recent phenomenon in modern Chinese medicine (Li 2012: 5; Chapter 45 and Chapter 51 in this volume), the tendency not to transmit certain kinds of knowledge through print is not entirely new in China. The occult use of herbs is more conspicuous in folk manuscripts than in orthodox medical books. Likewise, while deemed as an awkward subject in TCM, the ritual manipulation of herbs has remained essential in folk healing traditions today, with herbs celebrated as mediators between the human, natural and divine worlds. In the contemporary Chinese religious revival, such practices as hanging up mugwort twigs on the top of doors on the *Duanwu* or making exorcistic weapons out of peach wood are gradually being brought back to life. In late imperial times, herbal ritual knowledge constituted the core of Chinese life, and it is unlikely this will change in the near future.

Notes

1 This work was supported by the Wellcome Trust [Reference: 217661/Z/19/Z]. I am indebted to Vivienne Lo, Gao Xi 高晞, Andrew Wear, Michael Stanley-Baker and Ephraim Ferreira Medeiros. I could not have written this piece without the invaluable help of Lo Ping Kwan 盧炳坤, and especially masters Chen 陳 and Pau Hin Tsun 鮑顯震. For supplementary materials for this chapter, including further illustrations, please visit https://luisfbj.com/publications/.
2 One *cun* corresponds to 9.9cm.
3 Locals consider themselves descendants of those immortals.
4 One *qian* accounts for 3.125g.
5 The thirteen categories are *dafangmai* 大方脈 (internal medicine), *zhufeng* 諸風 (wind-illnesses), *taichang* 胎產 (foetus and childbirth), *yanmu* 眼目 (eyes), *xiao'er* 小兒 (children illnesses), *kouchi* 口齒 (mouth and teeth), *zhuwai* 諸外 (external medicine), *shangzhe* 傷折 (physical injuries), *er'bi* 耳鼻 (ear and nose), *chuangzhong* 瘡腫 (wounds and swelling), *jincu* 金鏃 (injuries by weapons), *shujin* 書禁 (written incantations), *bianzhen* 砭鍼 (acupuncture and moxibustion) (*Zhuyouke mijue qishu*, 3.3).

Bibliography

Pre-modern sources

NCLT: *Guojia tushuguan* 國家圖書館 (National Central Library, Taiwan).
NJUTCM: *Nanjing zhongyiyao daxue* 南京中醫藥大學 (Nanjing University of Traditional Chinese Medicine).
NLC: *Zhongguo guojia tushuguan* 國家圖書館 (National Library of China).
SL: *Shanghai tushuguan* 上海圖書館 (Shanghai Library).
Bencao gangmu 本草綱目 (Compendium of Materia Medica) 1596, Li Shizhen 李時珍, printed by Zhang Dingsi 張鼎思 in 1603, woodblock edition kept in NLC.
Chongzhen Taizhou zhi 崇禎泰州志 (Records of Taizhou from the Chongzhen Reign-Period) 1633, Li Zici 李自滋 and Liu Wanchun 劉萬春, woodblock edition collected in *Zhongguo shuzi fangzhiku* 中國數字方志庫 (database), Beijing guji xuan tushu shuzi jishu youxian gongsi.
Chuanya neibian 串雅內編 (Corrected Recipes of Itinerant Healers: Inner Volume) 1759, Zhao Xuemin 趙學敏, copied by Shao Yi 少伊 in 1859, manuscript kept in NJUTCM.
Chuanya waibian 串雅外編 (Corrected Recipes of Itinerant Healers: Outer Volume) 1759, Zhao Xuemin, copied by Shao Yi in 1859, manuscript kept in NJUTCM.
Daoguang Taizhou zhi 道光泰州志 (Records of Taizhou from the Daoguang Reign-Period) 1827, Wang Youqing 王有慶 and Chen Shirong 陳世鎔, woodblock edition collected in *Zhongguo shuzi fangzhiku*.
Huisheng ji 回生集 (Formulas for Bringing Back to Life) 1789, Chen Jie 陳杰, printed in the late 18th century, woodblock edition kept in SHL.
Jingchu suishi ji 荊楚歲時記 (A Record of the Seasonal Customs in the Jing-Chu Region) 6th century, Zonglin 宗懍, printed in the Wanli reign-period (1573–1620), woodblock edition kept in NLC.
Lidai mingyi mengqiu 歷代名醫蒙求 (Clarifying the History of Famous Physicians) 11th century, Zhou Shouzhong 周守忠, printed in the 12th century, woodblock edition collected in (1932) *Tianlu linlang congshu* 天祿琳琅叢書, 1st edition, vol. 21–22, Beijing: Gugong bowuguan.
Lüshi chunqiu 呂氏春秋 (Master Lü's Spring and Autumn Annals) 2nd century BCE, Lü Buwei 呂不韋, commented by Gao You 高誘, printed in the Wanli reign-period, woodblock edition kept in NLC.
Mao Duishan yihua 毛對山醫話 (Medical Records of Mao Duishan) 1903, Mao Duishan, printed in 1924, Hangzhou: Sansan yishe.
Mingyi lei'an 名醫類案 (Medical Cases from Famous Physicians) 1770, Jiang Guan 江瓘, woodblock edition kept in SHL.
Qimin yaoshu 齊民要術 (Essential Techniques for the Common People) 5th or 6th century, Gu Sixie 賈思勰, printed in 1524, woodblock edition kept in SHL.
Qingjia lu 清嘉錄 (Records of Beauty and Praise) early 19th century, Gu Lu 顧祿, printed in the Daoguang reign-period (1821–1850), woodblock edition collected in (1995) *Xuxiu siku quanshu* 續修四庫全書, vol. 1262, Shanghai: Shanghai guji chubanshe.

Shenxian fushi lingcao changpu wanfang zhuan 神仙服食靈草菖蒲丸方傳 (Divine Herbal and Calamus Pills Transmitted by the Immortals), anonymous author, printed in the (1923–1926) *Zhengtong Daozang* 正統道藏, vol. 573, Shanghai: Shangwu yinshuguan.

Shuowen jiezi 說文解字 (Explaining Graphs and Analysing Characters) 2nd century, Xu Zhen 許慎, printed in 1598, kept in NLC.

Suishi guangji 歲時廣記 (Record of Seasonal Customs) 13th century, Chen Yuanjing 陳元靚, printed in the Guangxu reign-period (1875–1908), woodblock edition collected in (1995) *Xuxiu siku quanshu*, vol. 885.

Taiping yulan 太平御覽 (Imperial Readings of the Taiping Era) late 10th century, Li Fang 李昉, printed in the 11th or 12th century, woodblock edition collected in (1935–1936) *Sibu congkan* 四部叢刊, vol. 234–369, Taipei: Shangwu yinshuguan.

Tanghu qiaoshu 倘湖樵書 (Book of Tanghu) late 17th century, Lai Jizhi 來集之, printed in the Kangxi reign-period (1661–1722), woodblock edition collected in (1995) *Xuxiu siku quanshu*, vol. 1195–1196.

Yangwu xian zhi 陽武縣志 (Records from the Yangwu County) 1936, Dou Jingkui 竇經魁, woodblock edition collected in *Zhongguo shuzi fangzhiku*.

Yashang zhaizun shengba jian 雅尚齋遵生八牋 (Eight Treatises on Following the Principles of Life) late 16th century, Gao Lian 高濂, printed in 1591, woodblock edition kept in SHL.

Yixue huihai 醫學匯海 (Sea of Medicine) 1820, Sun Derun 孫德潤, printed in 1826, woodblock edition kept in SHL.

Yixue zhengzhuan 醫學正傳 (Orthodox Transmission of Medicine) early 16th century, Yu Tuan 虞摶, printed in the Jiajing reign-period (1522–1566), woodblock edition collected in (1995) *Xuxiu siku quanshu*, vol. 1019.

Yueling cuibian 月令粹編 (Compilation of the Months and Seasons) 17th century, Qin Jiamo 秦嘉謨, printed in 1812, woodblock edition collected in (1995) *Xuxiu siku quanshu*, vol. 855.

Zhuyouke mijue qishu 祝由科秘訣奇術 (Occult Techniques of Talismanic Healing) 1674, anonymous author, manuscript kept in NCLT.

Zhuyouke zhibing qishu 祝由科治病奇書 (Occult Book of Talismanic Healing) 1927, Xu Jinghui 徐景輝, Shanghai: Zhongxi shuju.

Zhuyouke 祝由科 (Book of Talismanic Healing) late 19th century, anonymous author, manuscript kept in SHL.

Modern sources

Andrews, B. (2014) *The Making of Modern Chinese Medicine, 1850–1960*, Vancouver: UBC Press.

Bokenkamp, S.R (2015) 'The herb calamus and the transcendent Han Zhong in taoist literature', *Studies in Chinese Religions* 1.4: 293–305.

Bumbacher, S.P. (2014) *Empowered Writing: exorcistic and apotropaic rituals in medieval China*, St. Petersburg, FL: Three Pines Press.

Chen Hsiu-fen 陳秀芬 (2008) 'Dang bingren jiandao gui shilun Mingqing yizhe duiyu xiesui de taidu 當病人見到鬼：試論明清醫者對於「邪祟」的態度' (When Patients Met Ghosts: a preliminary survey of scholarly doctors's attitudes towards 'Demonic Affliction' in Ming-Qing China), *Guoli zhengzhi daxue lishi xuebao*, 30: 43–86.

Cho, P.S. (2005) *Ritual and the Occult in Chinese Medicine and Religious Healing: the development of zhuyou exorcism*, PhD thesis, University of Pennsylvania.

Davies O. (2007) *Popular Magic: cunning-folk in English history*, London: Hambledon Continuum.

Fujiansheng yiyao yanjiusuo 福建省醫藥研究所 (ed.) (1977) *Chuanya waibian xuanzhu* 串雅外編選注 (Selections and Commentaries to 'Corrected Recipes of Itinerant Healers: Outer Volume'), Beijing: Renmin weisheng chubanshe.

Harper, D. (2015) *Early Chinese Medical Literature: the Mawangdui medical manuscripts*, New York: Routledge.

Hsu, E. (2010) 'Qing hao 青蒿 (*Herba Artemisiae annue*) in Chinese *materia medica*', in E. Hsu and S. Harris (eds) *Plants, Health and Healing: on the interface of ethnobotany and medical anthropology*, New York: Berghahn Books, pp. 83–130.

Joly, H.B. (trans.) (2010) *The Dream of the Red Chamber*, Tokyo: Tuttle Publishing.

Junqueira, L.F.B. (2021) 'Revealing secrets: talismans, healthcare and the market of the occult in early twentieth-century China', *Social History of Medicine*, 34.4: 1068 -93. https://doi.org/10.1093/shm/hkab035, accessed 07/1/2022.

——— (2018a) *Mingqing zhuyou zhi kaoshi wenti* 明清「祝由」之問題考釋 (The Origins of Talismanic Healing According to Ming and Qing Sources), MA thesis, Fudan University.

——— (2018b) 'Popular healing in printed medical books: the compilation and publication of the *Chuanya* from the late Qing through the Republican period', *Monumenta Serica*, 66.2: 391–436. https://doi.org/10.1080/02549948.2018.1534357.

Lagerwey, J. (2010) *China: a religious state*, Hong Kong: Hong Kong University Press.

Leung, A.K.C. (2003) 'Medical learning from the Song to the Ming', in P. Smith and R. Von Glahn (eds) *The Song-Yuan-Ming Transition in Chinese History*, Cambridge, MA: Harvard University Asia Center, pp. 374–98.

Li Jianmin 李建民 (2012) 'Foreword,' in Li Jianmin (ed.) *Cong yiliao kan zhongguoshi* 從醫療看中國史 (Chinese History through Medicine), Beijing: Zhonghua shuju.

Liu Changlin 劉昌林 (2014) *Minsu fengqing* 民俗風情 (Chinese Popular Customs), Beijing: Zhongguo wenshi chubanshe.

Lo, V. (2001) 'Huangdi hama jing', *Asia Major*, 14: 61–99.

——— (2008) 'Heavenly bodies in early China: astro-physiology in context', in A. Akasoy, C. Burnett and R. Yoeli-Tlalim (eds) *Astro-Medicine: astrology and medicine, east and west*, Florence: Sismel, 143–188.

Lu Xixing 陸錫興 (2012) 'Kaogu faxian de taogeng yu taoren 考古發現的桃梗與桃人' (Recent Discoveries of Peach Stalks and Human Figurines), *Kaogu* 12: 78–85.

Sax, W.S. (2010) 'Ritual and the problem of efficacy', in W.S. Sax, J. Quack and J. Weinhold (eds) *The Problem of Ritual Efficacy*, Oxford: Oxford University Press.

Unschuld, P.U. (1984) *Medicine in China: a history of pharmaceutics*, Berkeley: University of California Press.

——— (1985) *Medicine in China: a history of ideas*, Berkeley: University of California Press.

——— (1998) *Forgotten Traditions of Ancient Chinese Medicine*, Brookline, MA: Paradigm Publications.

Unschuld, P.U. and Zheng Jingsheng 鄭金生 (eds) (2012) *Chinese Traditional Healing: the Berlin collections of manuscript volumes from the 16th through the early 20th century*, Leiden: Brill.

Wang Jing 王靜 (2013) 'Qingdai zoufangyi de yishu chuancheng ji yiliao tedian 清代走方醫的醫術傳承及醫療特點' (Itinerant healers in the Qing dynasty), *Yunnan shehui kexue*, 3: 161–5.

Yan Changui (2017) 'Daybooks and the spirit world', in D. Harper and M. Kalinowski (eds) *Books of Fate and Popular Culture in Early China: the daybook manuscripts of the Warring States, Qin, and Han*, Leiden: Brill, pp. 207–47.

Zhang Zhibin and P.U. Unschuld (2015) *Ben Cao Gang Mu Dictionary: Chinese historical illness terminology*, Berkeley: University of California Press.

PART 5

The world of Sinographic medicine
A diversity of interlinked traditions

32
TRANSMISSION OF PERSIAN MEDICINE INTO CHINA ACROSS THE AGES

Chen Ming 陳明

Translated by Michael Stanley-Baker[1]

I Travellers, Traders and Texts

Cultural exchanges between China and Persia have a deep-rooted history, the earliest traces of which date back to the Warring States, Qin and Han Dynasties (550 BCE–220 CE). Berthold Laufer's *Sino-Iranica* (1919), Aly Mazahéri's *La Route de la Sole* (1983) and other works have shown the close links between the two cultures. Medicine provided an important element in Sino-Persian exchange. Song Xian's 宋峴 *Gudai posi yixue yu Zhongguo* 古代波斯醫學與中國 (Ancient Persian medicine in China) seems to be the sole book-length work devoted to the subject. However, the transmission of Persian medicine into China is extremely complex, and warrants more intensive research.

In the thirteenth year of He Di's 和帝 Yongyuan 永元 reign (101 CE), King Pacorus II (*Manju* 滿屈, Pers. *Manūchihr*) of the ancient Persian Kingdom of Parthia (*Anxi* 安息) sent envoys to China bearing tribute of one lion and 'a great bird', an ostrich, constituting the first reliable record of contacts between China and Persia (*Houhan shu*, 88.2909–34). After the end of the Eastern Han (9–220 CE) a number of disciples associated with Persia arrived in China, following the north-western route of Buddhist transmission from India into China. The most well-known of these was the first translator in the history of the translation of Buddhist scripture into Chinese, An Shigao 安世高, style name Qing 清. He was a Buddhist from Parthia who, because of family and political troubles, came to China at the end of the second century to spread the teachings. According to biographical information in the *Chu sanzang jiji* 出三藏記集 (Compilation of Notes on the Translation of the *Tripiṭaka*) T.2145 by the monk Sengyou 僧祐 (445–518), and the *Gaoseng zhuan* 高僧傳 (Biographies of Eminent Monks) T. 2059 by Hui Jiao 慧皎 (497–554), An Shigao was widely learned, and possessed a refined understanding of foreign classics, and many other forms of knowledge, extending to the seven heavenly bodies,[2] medical recipes and other arcane arts. Moreover, his medical skills were excellent; his prescriptions could be relied upon to effect a cure, and he was skilled at needling the vessels and diagnosing illnesses by observing the complexion, and by many other such methods. Looking at his intellectual background, it is undoubtable that the medical arts he studied included those from Persia and India and that, through the display of his breadth of knowledge and talent, he had gained great esteem in the countries west of China. The Buddhist scriptures that An Shigao translated included new medical knowledge from outside the region. Chapter 5 on 'Five Kinds of Success and Failure' in the *Daodi jing* 道地經 (Skt. *Yogâcāra-bhūmi*, Sutra of the Path of Stages of Cultivation) T.607.232a–235b, translated by An Shigao, and Chapter 1 of the *Xiuxing daodi jing* 修行道地經 (Sutra of the Path of Stages of Cultivation) T.606.183b–189b, entitled 'Number 5: Categories of Success and Failure produced by the Five Skandhas' (*Wuyin chengbai pin diwu* 五陰成敗品第五), translated by Dharmarakṣa (Zhu Fahu 竺法護, 230?–316), are variant translations of the same text. Both contain descriptions of Indian medicine, and thus it can be said that An Shigao made a contribution to the transmission of Indian medical knowledge to China.

The Jin, Southern and Northern Dynasties (265–589) saw a gradual increase in exchanges between China and Persia. Official histories of China began to contain extensive accounts of Persia, describing Persian local customs and social conditions, while frequently referring to Persian products, including a number of *materia medica*. The section on 'Western Regions' in the Records of the Wei (*Weishu: Xiyu zhuan*, 102.2270–1) mentions materials from the sixth century, including minerals like tutty (*toushi*, MC *thuw dzyek* 鍮石, Pers. *tutiya*),[3] great pearl (*dazhenzhu* 大真珠), cinnabar (*zhusha* 硃砂), mercury (*shuiyin* 水銀), verdigris (*yanlü* 鹽綠/碌 Pers. *zinjar*,) and orpiment (*cihuang* 雌黃), as well as plant drugs such as frankincense (*Xunlu/ruxiang* 熏陸/乳香), saffron (*yujinxiang* 鬱金香), storax or oriental sweetgum (*suhe xiang* 蘇合香), slender Dutchman's pipe root or *Aristolochia debilis* Siebold & Zucc. (*qing muxiang* 青木香), black pepper or *Piper nigrum* L. (*hujiao* 胡椒, Skt. *marica*), *pippali* or long pepper (*bibo* 蓽撥), dates (*qiannian zao* 千年棗, Pers. *khṛmā*), nutgrass galingale root (*xiangfuzi* 香附子), yellow myrobalan or *Terminalia chebula* Retz. or (*helile* 訶黎勒, Skt.*harītāki*) and Aleppo gall, produced by wasps on *Quercus infectoria* G.Olivier (*wushizi/meishizi* 無食子/沒食子). A number of healing practices and beliefs from Persian religions also entered the Chinese cultural field through a variety of means. A tomb of Anjia 安伽 in Xi'an city, Shaanxi province, which was constructed during the Northern Zhou dynasty, was found to contain a number of very fine screens and a stone funerary bed, decorated with Zoroastrian murals from Persia. The lintel to the stone door at the entrance to the tomb was decorated with a mural, with scenes of two sacrificants wearing gauze masks on their faces in the midst of a Zoroastrian sacrifice, including images of the sacred plant *haoma*. An image of the plant can be found on the headboard of the stone bed, now stored in the Miho museum in Japan. These testify to the adherence of foreigners to their Zoroastrian faith, while in China, and their transmission of immortality beliefs around *haoma* in the central Chinese plains.

Although the Sassanid Empire of Persia was superseded in the seventh century by the Tajiks (*dashi* 大食, Pers. *tāzīk*), throughout the Tang and Five Dynasties, China never discontinued her exchange with Persia. On the contrary, they became much closer than in the Six Dynasties through increased trade along both the Silk Road and sea routes. Impressions of the Islamic world can be found in what remains of Du Huan's 杜環 (fl. eighth century) *Jing xing ji* 經行記 (Records of Travels and Experiences) in the Tang dynasty *Tongdian* 通典 (Comprehensive Institutions) by Du You 杜佑 (735–812). The lost portions of this work possibly contained further descriptions by Du Huan of customs and habits originating from Persia. *Juan* 971.11408a of the *Cefu yuangui* 冊府元龜 (Models from the Archival Bureau) records that in the first lunar month of the year 730, the Persian prince Ji Hu Suo 繼忽娑 came to Chang'an, to pay tribute of incense, rhinoceros horn and other ritual items. Among the foreigners coming from outside China, who included traders, doctors, soldiers, envoys, and disciples of the 'three foreign teachings' (that is, Nestorianism, Zoroastrianism and Manicheism), there were many who hailed from Persia. Along the rivers of Guangzhou 廣州 during the Tang dynasty (628–907) were anchored countless boats trading with India, Persia and southeast Asia, teeming with cargos of incense and treasures. There are many records of Persian exports in historical literature. The 'Biography of the Western Regions' (*Xiyu zhuan* 西域傳) of the *Xin tangshu* 新唐書 (New Tang History 221–2.6213–65) records that Persian products such as Aleppo gall, nutgrass galingale root, yellow myrobalan, black pepper, *pippali*, dates, graphite (*shimo* 石墨) and sweet-dew peaches (*ganlu tao* 甘露桃) could all be used in medicine. *Juan* 193 of the *Tongdian* 通典 (Comprehensive Institutions) records these Persian exports: incenses such as mastic resin (*xunlu* 薰陸), turmeric (*yujin* 鬱金), storax, slender Dutchman's pipe root (*qingmu* 青木), black pepper, evaporated cane juice (*shimi* 石蜜, Skt. *śarkarā*), dates, nutgrass root, *Terminalia bellirica* fruit, Aleppo galls, verdigris, orpiment and even the beautiful Queen of the Night flower, *Epiphyllum oxypetalum* (*youbotan* 優鉢曇).

II Importing Knowledge Through Translations and Tales

As part of the continuous expansion of the professionalisation of Chinese pharmacology, not only were foreign *materia medica* absorbed into the corpus, but foreign pharmacological works were also translated into Chinese. The *Xinxiu bencao* 新修本草 (Newly Revised Pharmacopoeia), edited in the early Tang by Su Jing 蘇敬 (599–674) *et al.*, included 850 entries, of which 114 were newly added. During the Kaiyuan 開元 reign (713–741), Chen Zangqi 陳藏器 (687–757) compiled the *Bencao shiyi* 本草拾遺 (Supplement to the Pharmacopoeia), incorporating foreign-derived medicinals, recording hundreds of *materia medica* that were missing from the earlier work, including many foreign imports. After the Kaiyuan reign period, more texts describing foreign pharmacopoeia emerged, such as Zheng Qian's 鄭虔 (691–759) *Hu bencao* 胡本草 (Western Pharmacopoeia). In these Chinese pharmacological works, many Persian *materia medica* are to be found. According to statistical measures, the *materia medica* appearing in the *Newly Revised Pharmacopoeia*, such as verdigris, lead oxide (*mituoseng* 密陀僧), shellac (*zikuang* 紫礦), dragon's blood or resin of *Calamus draco* Willd. [Arecaceae] (*qilin jie* 麒麟竭), sal ammoniac (*naosha* 硇砂), Euphrates poplar or *Populus euphratica* Oliv. (*hutonglei* 胡桐淚), Aucklandia root or *Aucklandia costus* Falc. [Asteraceae] (*muxiang* 木香), asafoetida or latex *Ferula narthex* Boiss. (*awei* 阿魏, Skt. *hingu*), amber (*hupo* 虎魄), Chinese eaglewood or *Aquilaria sinensis* (*chenxiang* 沉香, Skt. *aguru*), storax, benzoin (*Anxixiang* 安息香 lit. 'Parthian aromatic', *Styrax tonkinensis* (Pierre) Craib ex Hart.), Chebulae Fructus, Aleppo gall, theriac (*diyejia* 底野迦, Pers. *diryaq*), evaporated cane juice, Parthian pomegranate (*anshiliu* 安石榴), were all more or less closely related to pre-Islamic Persian culture, and greatly augmented Chinese pharmacopoeic knowledge. The recipes collected in Wang Tao's 王燾 (670–755) *Waitai miyao fang* 外臺秘要方 (Recipes based on the Secret Essentials of the Imperial Archives), and those from works such as Sun Simiao's 孫思邈 (581?–682 CE) *Qianjin yaofang* 千金要方 (Essential recipes worth a Thousand Gold) and the *Jinxiao fang* 近效方 (Quick-Working Recipes, auth. unknown), used Persian ingredients such as: white evaporated cane juice (*baishimi* 白石蜜), Persian verdigris (*Bosi yanlü* 波斯鹽綠), Persian salt (*bosi yan* 波斯鹽), true Persian indigo (*zhen bosi qing dai* 真波斯青黛), Carpesium fruit Persian *Carpesium abrotanoides* L. (*Bosi heshi* 波斯鶴虱), and imported *pippali* among others. Apart from being used as a cosmetic, Persian indigo, also known as *Murex trunculus* indigo (*luozi dai* 螺子黛), also had medicinal applications for smallpox-like diseases, such as *tianxing fachuang* 天行發瘡 and *wandou paochuang* 豌豆皰瘡. Persian verdigris was used for many kinds of eye problems. The section entitled 'Recipes for drinking milk, which energises listlessness and breaks *qi* [stagnation] (*Fu niuru buxu poqi fang* 服牛乳補虛破氣方)' in the Sun Simiao's *Qianjin yifang* 千金翼方 (Supplementary Prescriptions Worth a Thousand Gold Pieces) uses cow's milk and *pippali* to remove all kinds of *qi* [stagnation]. This recipe, sometimes known as '*pippali* fried in milk' (*rujian bibo fang* 乳煎畢撥方), was highly regarded in Persia and

Byzantium: Chinese sources refer to it as a 'decoction from *Bei* (which could mean Persian) powder' (*beisan tang* 悖散湯). According to tradition, this recipe was used by the Tang emperor Taizong 太宗 (598–649), and was so renowned that Hu Sihui 忽思慧 (fl. fourteenth century) included it in *j*. 2 of his *Yinshan zhengyao* 飲膳正要 (Proper and Essential Things for the Emperor's Food and Drink), entitled 'Dietary cures for all illnesses' (*Shiliao zhubing* 食療諸病), where it is called 'Method for sauteeing *pippali* in milk' (*niunai zi jian bibo fa* 牛奶子煎蓽撥法), testifying to the broad influence of the recipe. Medical and religious texts from the Chinese central plains frequently contain laudatory statements about medicinals like 'Persian ones are of good quality', 'Persian ones are superior' or 'Its nature is not as good as Persian ones', emphasising their reliability and efficacy.

Chinese translations of Buddhist scriptures record many drugs or products of Persian origin, including white Persian evaporated cane juice, glass and glass objects, alfalfa or *Medicago sativa* L. [Fabaceae] (*muxuxiang* 苜蓿香), root of *Triarrhena sacchariflora* (Maxim.) Nakai (*digeng* 荻根), Persian dates and other items which could all be used as *materia medica*. Persian white evaporated cane juice also appeared in ritual instructions for esoteric Buddhist rites. The translations also include Persian *materia medica*. Commenting on a Chinese translation for juice from the wild Persian date (*keshuluo* 渴樹羅, Skt. *kharjūra*, Pers. khrmā), the Buddhist monk and master translator Yi Jing 義淨 (635–713) made the following observation: 'It is shaped like a small jujube, it is astringent and sweet, and comes from Persia. It is also found in the central regions of India, but its flavour is slightly different. The tree grows singly, resembling a palm tree, bearing lots of fruit. When one is about to arrive in Fanyu 番禺 (modern-day Guangzhou), people call it Persian date. It tastes very similar to dried persimmons'.[4] Buddhist dictionaries also include Persian medicines, such as the *Yiqie jing yi yinyi* 一切經音義 (Sounds and Meanings [of all words in] the Scriptures) by Hui Lin 慧琳 (737–820), which states: '*Biba* 蓽茇: Pronunciation of the first Chinese character is *bi*, Sanskrit *pippali*, name for drug from Western country, originally from Persia and India, it looks like mulberry fruit, thin and long, it is extremely acrid and spicy'.[5] Chan Buddhist masters from the central Chinese plains are also recorded as using phrases such as 'Persians eat black pepper' in answer to questions raised by their disciples.

Narratives in Tang dynasty biographies record a number of stories of Persian merchants, describing them as skilled at differentiating precious stones, and as superior in the medical arts. For example, 'Essence of red cornetfish (*shaoyu* 鮹魚)' was renowned for its miraculous effects; simply pasting it on a patient's abdomen could reduce abdominal masses. Although these miracle tales and stories are not reliable as historical data, at the least they reflect the high regard people of the time held for foreign Persian products. Tang dynasty literati enjoyed reading pharmacopoeia and other encyclopaedic works, in order to increase their knowledge, expand their written works and increase their quality of life. Authors of Tang encyclopaedias also enjoyed writing about exotic foreign cultural products, thereby increasing the transmission of foreign medical products in the central plains. The most important Tang encyclopaedia is the short work, the *Youyang zazu* 酉陽雜俎 (Miscellaneous Morsels from Youyang). The first eighteen *juan* record twenty-two different plants from the same territories and regions, among which are included the following Persian medical products: Borneo camphor wood *Dipterocarpaceae* (*longnao xiangshu,* 龍腦香樹 – from which borneol is produced), benzoin, Aleppo gall, shellac (*zimao* 紫鉚), asafoetida, *poso* tree 婆娑樹, Persian dates, sweet almonds or *Prunus persica* (L.) Batsch (*piantao* 偏桃), the *pannüse* tree 槃砮穄樹,[6] *qitun* wood 齊暾樹, *bie qi* 酺齊, Persian honey locust or *Gleditsia caspia* Desf. (*zaojia* 皂莢), myrrh trees, jasmine (*yeximi* 野悉蜜),[7] and *di'er* fruit 底欄實. The incorporation of foreign *materia medica* in the *Youyang zazu* 酉陽雜俎, is predominantly due to the Grand Councillor's son, Duan Chengshi 段成式, style name Kegu 柯古 (803–863), who, relying on his family's social status, mingled with the educated classes and foreign dignitaries in Chang'an. The *Youyang zazu* records five foreigners' names, the Persian emissaries Wuhai 烏海 and Shalishen 沙利深, the Chenla[8] emissary and Commandant of the Assault-resisting Garrison, the monk Shanibato 沙尼拔陁, the Byzantine monk Wan 彎, and the monk Deva (*Tipo* 提婆) from Magadha, ancient India. There is no record of the religious background of Wuhai and Shalishen, they may well have been Zoroastrian. The culture and lifestyle of Persia, including its religious customs, had a profound impact and influence on Tang society. The Persian *materia medica* entering China was highly valued, and was frequently used by officials and the affluent.

Among the medical manuscripts and fragments unearthed from Dunhuang are records of the Persian *materia medica*, sulphur and shellac. The remnants of the *Shiliao bencao* 食療本草 (Materia Dietetica) from Dunhuang S076R describes evaporated cane juice (Pers. *sarkara*; Skt. *śarkarā*) as being cold, and primarily governing hot swelling in the upper and lower abdomen and thirst. It emphasises that among all the extant varieties, the Persian was of the highest quality, and that those from Shu (modern Sichuan) and Eastern Wu (modern Jiangsu and Zhejiang) along the coast were not equal to it. Not only documents in Chinese, but also Tibetan-language manuscripts from Dunhuang (S.756) attest to the use of Persian materials in medical practice, such as Persian brocade, Persian paper and so on. In chapter 91 of the earliest extant Tibetan medical work, the 'Medical Method of the Lunar King' (Tib. *sman dpyad zla ba'i rgyal po*, skt. *somarāja*),[9] one also finds descriptions of theriac (Tib. *dar ya kan*), the famed product of ancient Greece, brought via Persian traders and passing through the Tibetan kingdom of Zhang Zhung. One 'Recipe for Massaging the Crown of the Head' from the Sui dynasty opthalmological text, *Longshu pusa yanlun* 龍樹菩薩眼論 (Nagarjuna's Discourse on the Eye),[10] uses theriac from Western regions in the shape of a camel's gall bladder, which could also be a Persian product.

III Local Texts and Uses of Western Drugs

During the Five Dynasties (907–960), the single most influential work of Persian medicine was the *Haiyao bencao* 海藥本草 (Overseas *Materia Medica*) by Li Xun 李珣, style name De Run 德潤 (?855–930?).[11] Li Xun was a 'local-born Persian' in China. His ancestors were Persians, who accompanied the Tang emperor Xi Zong 僖宗 (r. 873–888) in his flight from Chang'an into the province of Shu in 874, at one time occupying the post of Commandant of the Eastern Palace Guard Command. Living in the ninth and tenth centuries, Li Xun's talent as a scholar made him stand out among his peers and led to his achieving office in the Former Shu dynasty (907–925), after which he travelled through or resided from some time in Guangzhou. Li Xun was among the representative poets of the 'Amidst the Flowers' school, penning *ci*-style poems such as the *Nanxiang zi* 南鄉子 (Southern Lad) and other renowned works. His poems describe strange landscapes, a disdain for fame and glory, and his artistic style was clear and limpid, sparkling with wit. One depiction by Li Xun of a southern landscape painting transformed it from an object of beauty to an image of a wild desert riddled with miasmic plagues. Li Xun is one of the foreign-born figures of great importance in medieval Chinese literature, his greatest contribution to which was his descriptions of southern scenery, which expanded the scope of the *ci* poetic genre. Li Xun's younger brother, Li Xuan 李玹 inherited the characteristics of Persian merchants, taking up selling incense for a career. He also had a liking for the Chinese strategic board-game *weiqi* 圍棋, as well as for *yangsheng*, in particular the arts of Daoist inner alchemy. In his later years he spent his entire family fortune on alchemical products. Li Xun's little sister, Li Shunxian 李舜絃, is the only recorded foreign-born female poet of the period, leaving works such as *Shugong yingzhi shi* 蜀宮應制詩 (Commissioned Poems for the Shu Palace), *Diaoyü bude shi* 釣魚不得詩 (A Poem on Fishing Unrequited) and *Suijia you Qingcheng shi* 隨駕遊青城詩 (A Poem While Riding to Qingcheng Mountain). Li Xun's family personifies Sino-Persian cultural exchange.

Li Xun's greatest contribution to medical history is his *Overseas Materia Medica*. This work is a pharmacopoeia which focusses exclusively on drugs imported into China along the maritime routes, but the framework and style of writing are completely modelled on native Chinese pharmacopoeic literature. The *Overseas Materia Medica* relies on the style and structure of the *Xinxiu bencao* (Newly Revised Pharmacopoeia) and explains foreign-derived drugs according to the progressive descriptions traditional in Chinese medicine. This is an example of cultural fusion, and reveals the reception and transformation of foreign materials within Chinese medicine.

According to statistics in Shang Zhijun's 尚志鈞 edited *Overseas Materia Medica*, the products referred to as produced in Great Qin 大秦國 (Eastern Roman Empire) include these five items: Persian alum (*Bosi fan* 波斯礬), *wufeng duyao cao* 無風獨搖草, nutmeg (*roudoukou* 肉豆蔻), fragrant rosewood (*jiangzhen xiang* 降真香) and seeds of the large-fruited elm, or *Ulmus macrocarpa* Hance (*wuyi* 蕪荑). Those related to Persia include seventeen items, such as: gold-streaked alum (*jinxian fan* 金線礬), silver shards (*yingxie* 銀屑), green salt (*lüyan* 綠鹽), Euphrates poplar, betel pepper fruit (*jujiang* 蒟醬), dill (*shiluo* 蒔蘿), benzoin, myrrh (*moyao* 沒藥), Aleppo gall, marking nut or *Semecarpus anacardium* L. (*poluode* 婆羅得, *poluole* 婆羅勒, Skt. *Bhallātaka*, Pers. *Balādur*, Tocharian B *Bhallātak*), litchi (*lizhi* 荔枝), white Persian alum (*Bosi baifan* 波斯白礬), frankincense (*rutou xiang* 乳頭香), resin of Persian pine (*bosi shuzhi* 波斯樹脂), pistachio (Pers. *ayuehun*, *wumingzi* 無名子), *Semen Ulmus macrocarpa* Hance or Persian elm seeds (*bosi wuyi* 波斯蕪荑), bark of *Lithocarpus glaber* (*keshu pi* 柯樹皮) and haritaka (*helile* 訶梨勒). One text cited within the *Overseas Materia Medica*, the *Bie bao jing* 別寶經 (Classic on Distinguishing Treasures) was probably a surviving trace of foreign Persians, and their traditional specialisation in identifying gemstones. The betel nut recipe in the *Overseas Materia Medica* described as coming from Great Qin is one rarely seen in native Chinese sources. Together with this work, the *Nanhai yaopu* 南海藥譜 (Treatise on Drugs from the Southern Seas) and the *Haiyao lun* 海藥論 (Discourse on Overseas Drugs) all deal with the importation of foreign drugs. The *Overseas Materia Medica* was lost earlier, and is mostly recorded in Song period Chinese works on *materia medica* and, together with common folk lore, complemented the accumulated knowledge within traditional Chinese medicine in this way gradually influencing the East Asian region.

Within China there were many Persians who worked in medicine. The Persian Li Miyi 李密醫 who made the journey across from China to Japan in 736, the twenty-fourth year of Tang Emperor Xuan Zong 唐玄宗, may well have been a medical doctor.[12] In 812, the seventh year of the Yuanhe reign, the ship's captain Li Mohe 李摩訶 who made an offering of a recipe for *Psoralea corylifolia* or babchi (*buguzhi* 補骨脂) to the Prefect of Guangzhou, was also likely to have been a Persian merchant. The *Beimeng suoyan* 北夢瑣言 (Fragmentary Sayings of a Northern Dream), an important tenth-century historical record, describes the Persian Mu Zhaosi 穆昭嗣, who as a child was fond of medicinal arts, and who, because his drugs were effective, took up a position in his local government. It is possible that this 'Persian' Mu Zhaosi was in fact from the land of Mu 穆 in the Sogdian states. One document from Dunhuang, S.1366, contains a late tenth-century mention of which local government donated comestibles, such as flour and oil, to a Persian monk (perhaps a Nestorian) who had contributed medicine to local authorities. The Collected Essays of Li Deyu 李德裕, towards the end of the Tang dynasty, records one 'monk from Great Qin, who was a specialist in treating the eyes' and also a Nestorian, as active in Chengdu (in modern-day Sichuan) (*Li Wenrao wenji*, ch.12). In his tenth-century *Kitāb al-Fihrist* (Categorical Index of Collected Writings), Ibn al-Nadim records a Chinese youth from Baghdad who learnt Galenic medicine from the Persian doctor Mohammad-e Zakariā-ye Rāzi. Using rapid mnemonic methods, he memorised the Galenic classics in six months before returning to his homeland. However, the work does not leave us this Chinese student's name. Nevertheless, it indicates that medical scholars travelled far into Baghdad and wider Persia to learn Islamic medical knowledge.

Along with Persian food culture travelling East came Persian wines, jams, honey, candies such as rock sugar, dates, almonds, *qitun* bark, figs (*wuhuaguo* 無花果), Parthian pomegranates, dill (*shiluo* 蒔蘿), beetroot (*tiancai* 甜菜, also *junda* 軍達or 莙蓬), spinach (*lit*. Persian greens, *bosi cai* 波斯菜), which became part of the categorical records (*pulu* 譜錄) of common Chinese people, and were ingredients frequently employed in Chinese food and drink. The term *zhujunda* 諸軍達, encountered in the Dunhuang text *Zaji shiyao yongzi* 雜集時要用字 (Miscellany of Important Terms Used in Daily Life) P.3391, originally derives from the New Persian term *čugundur* or *čugonder*, and refers to the plant used to make sugar, beetroot, which formed an important commodity of cultural exchange between China and Samarkand (Yu Xin 2013). Because medieval Persian and Indian cultures were closely linked at this time, Indian medicaments and their use were also transported to Persia, and thence into China, and in this way Persian medicine exerted a particular type of influence in medieval China. The juice of *triphalā* (Skt. for 'three fruits', *sanle jiang* 三勒漿) was a fruit-based liquid imported from Persia, but originally sourced from India, made from yellow myrobalan or *Terminalia chebula* Retz. (*helile* 訶梨勒, Skt. *harītāki*), beleric myrobalan or *Terminalia bellirica* (*pilile* 毗梨勒, Skt. Vibhitaka), and emblic myrobalan or *Phyllanthus emblica* L. (*anmole* 庵摩勒 Skt. *amalika*). Following the fashion in the higher echelons of Tang society, during the Song dynasty related medicinals from southern regions, such as yellow myrobalan decoction (*hezi tang* 訶子湯) and beleric myrobalan decoction (*yuganzi tang* 余甘子湯), enjoyed widespread fame in the north. The Yuan dynasty doctor Xu Guozhen 許國禎, style name Jinzhi 進之, temporarily renewed the fashion for *triphalā*, which had declined in the Ming dynasty to the extent that it was hardly made any more. However, texts containing related knowledge were passed down the generations, so that people today still attempt to replicate it in memory of the historical Tang type. The three fruits held an important place in Indian medicine and cuisine, and there exists matching mythography concerning them. Persian and Arabic medical texts and the Mongolian *Huihui yaofang* 回回藥方 (A Collection of Muslim Prescriptions) describe the use of the 'three fruits' in both medicine and drink, and mention their higher popularity in Persian and Arabic regions as quite different from their dissemination in the central plains of China, because of the difference between Persian, Chinese and Indian cultures. Acting as a midway point, Persian trade routes not only made for a secondary mechanism of external migration of Indian dietary custom, but cloaked them in Persian-style cultural trappings when entering Chinese soil. In this way, the customs of food and drink of China, India and Persia embodied variety, difference and mutual exchange between the cultures and religions of the three regions.

In the pursuit of longevity, medieval alchemists and experts in the Daoist arts looked to Western merchants dealing in drugs from Western regions. Persian drugs were frequently used by Chinese alchemists, as described in the work by the Late Tang and Five Dynasties Li Guangxuan 李光玄 from Bohai 渤海 (modern-day Binzhou in Shandong), the *Jinye huandan baiwen jue* 金液還丹百問訣 (Explanations of the 'Hundred Questions' on the Cyclically

Transformed Elixir of Liquefied Gold, DZ 266). He writes: 'That referred to as Numinous Elixir is not from here, it is said the ultimate drug is produced overseas, one should look to the interior of Persia, and from there seek white alum (*baifan* 白礬) and purple alum (*zifan* 紫礬). Or turn to the Uighur lands, and ask for diamonds and shards of jade'. Of the foreign drugs that could be used for refining elixirs referred to in alchemical texts in the *Daoist Canon*, many came from Persia, India, Khotan, Nanhai 南海 (modern-day Guangzhou), Silla, Beiting 北庭 (modern-day Ürümuqi). These were primarily mineral drugs, and plant drugs were in the minority. Frequently mentioned minerals from Persia include: Persian chalcopyrite (*Bosi toushi* 波斯鍮石), 'true superior Persian brass with the hue of a horse tongue' (*zhen Bosi mashe se shang tou* 真波斯馬舌色上鍮), Persian *aurichalcite* (*Bosi zhetou* 波斯折鍮), Persian lead (*Bosi qian* 波斯鉛), Persian Verdigris, white lead powder (*hufen* 胡粉), borax (*da peng sha* 大鵬砂), lithargyrum or lead oxide (*mituoseng* 密陀僧), sulphur ore (*shiliu huang* 石硫黃), naptha (*shinao* 石腦), iron sulphate (*jiang fan* 絳礬), 'chicken dung' alum (*jishifan* 雞屎礬),[13] heaven's brilliance sand (*tianming sha* 天明砂), yellow floriate ore (*huang hua shi* 黃花石), asbestos (*buhui mu* 不灰木), rock salt, sulphur imported by sea (*boshang liuhuang* 舶上硫磺), northern calamine (*bei lüganshi* 北盧甘石, Pers. *tūtiya*), pyrolusite (*wumingyi*, 無名異), Persian alum, Persian white alum, potassium chrome alum (*zi fan* 紫礬), Persian refined lead (*Bosi qian jing* 波斯鉛精), Persian red salts (*Bosi chiyan*, 波斯赤鹽), Persian silver ore (*xigezhi*, 悉悋脂), indigo (*qingdai* 青黛) and *Carpesium abrotanoides* (*heshi* 鶴虱).

IV Chinese Institutions and Markets

From the Song dynasty onwards, the unimpeded commerce along the maritime silk route accelerated trade in foreign drugs, so that Persian traders in Quanzhou, Yangzhou, Siming (modern-day Ningbo) and other such places continued their trade in incense and drugs. Early in the Song dynasty, at least forty-four varieties of ferula incense are recorded as being imported, but by 1133 they rose to more than two hundred types. The record of goods on trading vessels in the *Siming Zhi* 四明志 (Siming Gazetteer) during the *Baoqing* period (1125–1228) describes numerous 'fine goods' such as ferula resin, myrrh, Aleppo gall, aloe (*lühui* 蘆薈), rosewater (*qiangwei shui* 薔薇水), shellac and others, many of which originated in Persia. Song dynasty pharmacological texts included many more foreign products than earlier works. The *Kaibao bencao* 開寶本草 (*Materia Medica* of the Kaibao Reign), the *Bencao tujing* 本草圖經 (/*Tujing bencao* 圖經本草, Illustrated *Materia Medica*), the *Bencao yanyi* 本草衍義 (Further Discussion on *Materia Medica*) and Tang Shenwei's 唐慎微 (fl. eleventh to twelfth centuries) great compendium the *Zhenglei bencao* 証類本草 (*Classified Materia Medica*) all included many foreign ingredients. The main ones in the *Classified Materia Medica* include indigo, aloe, babchi, pistachios, myrrh, *cile* resin (*yuancile* 元慈勒), dates (*wulouzi* 無漏子), gold-streaked alum, sulphur ore, lead oxide, coral (*shanhu* 珊瑚), ferula resin and *pippali*. In the main, it can be said that they definitely became integrated as important components in Chinese pharmacology.

Following the westward march of Mongolian armies and the entry of Muslim scholars from Central Asia and the Middle East, the trade in Chinese and Persian medicine in the Yuan dynasty was in an unprecedented situation. For the most part, the upper echelons of Yuan society mainly used officially recommended medicines, approved by institutions such as the Medical Bureau, the Imperial Pharmacy, the Office of Broad Grace (*Guanghui si* 廣惠司), which provided West Asian medical service, the Islamic Pharmaceutical Bureau (*Huihui yaowu yuan* 回回藥物院), the Islamic Pharmaceutical Dispensary (*Huihui yaowu ju* 回回藥物局) and the Pharmaceutical Dispensary of Gracing the People (*Huimin yaoju* 惠民藥局),

otherwise Persian doctors offered their services. They not only brought medicinal products with them, they also translated a number of Persian and Islamic medical texts into Chinese. The Yuan dynasty Director of the Palace Library preserved the *Tebi yijing shisan bu* 忒畢醫經十三部 (Classic of Tibb Medicine in Thirteen Volumes), because it was the most practical and important translation of this Islamic medical work. During Kublai Khan's *zhiyuan* 至元 period (1264–1295), he collected medical classics from around the world, and had the *Dayuan bencao* 大元本草 (Pharmacopoeia of the Great Yuan) compiled. Although this work has since been lost and it is difficult to know what it contained, it is very likely that it would have included Persian and other foreign medicinals. In *juan* seven of *Nancun chuogeng lu* 南村輟耕錄 (Records from the Southern Hamlet of Setting the Hoe Aside) in the Yuan dynasty, Tao Zongyi 陶宗儀 records a variety of Persian gems and minerals, such as corundum (Pers. *yakut*, *yagu* 鴉鶻) under the term 'Muslim minerals' (*huihui shitou* 回回石頭), some of which had medicinal applications. The most prominent foreign drink was sherbet (*shelibie* 舍利別, *shelibai* 攝里白, Pers. *šarba*, *šarbat*), a concentrate made from high-grade fruit mixed with sugar or honey and then diluted with water,[14] and there were also similarly named fruit wines such as *šarāba* (*shelibi* 舍剌必) and grape wine. As sherbet entered the local market, it became known as 'slake-thirst drink' (*jiekeshui* 解渴水), and developed into ten or more different varieties. The *Yinshan zhengyao* 飲膳正要 (Important Principles of Food and Drink) by Hu Sihui 忽思慧 (fl. 1314–1330) is among the most important works on dietetics and medicinal curing (Buell and Anderson 2000). This work not only refers to the successes of Chinese and Mongolian folk dietary cures, but also contains elements of Persian and other Western dietary cultures.

During the early Ming dynasty, *Huihui yaofang* 回回藥方 (A Collection of Muslim Prescriptions) exemplifies Yuan dynasty medical exchange, but this is not an independent work written by some ethnically Hui medical scholars, but is more likely to be an edited translation of an Islamic medical work, including selections from one or more Persian or Arabic medical encyclopaedias, closely drawn from Avicenna's *Canon of Medicine* (Arab. *al-Qānūn fī aṭ-Ṭibb*). The original *Hui Medicinal Recipes* (*Huihui yaofang*) contained thirty-six *juan*, most of which have been lost. Only four now survive, and these are stored in the rare book section of the Chinese National Library. The *Huihui yaofang* contains a number of ancient recipes and the names of medical doctors from Persia. These include one of the three great Persian doctors, Muḥammad ibn Zakariya al-Rāzī (*Mahe made [ben] zakeliya* 馬哈麻的[本]咱可里牙, 864–925/932); the Abbasid doctor of Persian descent from the Arabian empire, Isa ibn Saharbakht (*Sahe'er baheite* 撒哈爾八黑忒); the ninth to tenth-century Nestorian doctor from Baghdad, originally born in the Persian city of Marv, Abu Yaḥyā al-Marwarruzi (*Ma'er waji* 麻而瓦即); together with the famed Persian doctor from Jundi Shahpur, Sābūr ibn Sahli (*Shabu'er sanheli* 沙卜而撒哈里) and the greatest of the three major Persian doctors, Ibn-Sīnā (*Bu'ali* 卜阿里, Abu alisanna 阿卜阿里撒納). The title *Kelimei wenshu* 可里眉文書 refers to *The Complete Book of the Medical Art also known as Kitāb al-Malikī* (Pers. *Kitāb Kāmil al-ṣinā'ah al-ṭibbīyah*) by al-Majūsī (*Maijuxi* 麥朱西). Furthermore, the *Huihui yaofang* acted as a Persian mediator to transmit ancient Greek medical knowledge to China. The *Huihui yishu* 回回醫書 (A Collection of Muslim Medical Works) was translated by Mashayihei Mohamed 馬沙亦黑馬哈麻. This book is similar to the *Huihui yaofang* mentioned above, and was collected in *juan* 1426–1464 of the massive Ming dynasty encyclopaedia, the *Yongle dadian* 永樂大典, including twenty-six *juan* on internal medicine, six on external medicine and seven on drug recipes.

According to preliminary statistics by Song Xian 宋峴, about five hundred foreign drugs are included in Song, Yuan and Ming Chinese medical works, and the majority of these are recorded in the *Huihui yaofang*. The Western drugs which appear in the *Pujifang* 普濟方 (Recipes for Universal Relief) by Zhou Dingwang 周定王 from 1406 CE include: zinc oxide or tutty (*duotiya* 朵梯牙, Pers. *tutiyā*,), *anzarout* tree resin (*anzalu* 安咱蘆, Pers. *anzarūt*), gum tragacanth (*ketiela* 可鐵剌, Pers. *kateerā*,), opium (*afeiyong* 阿飛勇, Pers. *afyoon*), plum tree gum (*lizi shujiao* 李子樹膠), which refers to gum Arabic (*sanyi* 三亦, Pers. *samgh*), among others. Practical prescriptions from Persia, for example, using wine to disperse toxins and cure bones, blowing through the nose to remove cataracts, using vinegar medicinally to ease childbirth, and distillation methods for producing rosewater and other medicated waters and wines, all appear in Chinese recipe texts or notebooks. There are many large pharmacopoeic works in the Ming dynasty, among which the most important is Li Shizhen's 李時珍 (1518–93) *Bencao gangmu* 本草綱目 (Categorical and Itemised Pharmacopoeia). Apart from the *Huihui yaofang*, which was inaccessible as it was stored within the imperial library, Li Shizhen incorporated all the Persian drug tracts available to him in his *Pharmacopoeia*, including opium which entered China from Western Asia. The two officially compiled major drug compendia, the *Bencao pin hui jingyao* 本草品匯精要 (Collected Essentials of *Materia Medica* Species) and the *Buyi Leigong paozhi bianlan* 補遺雷公炮製便覽 (Supplement to Master Lei's Guide to Drug Preparation), contain beautiful diagrams produced by palace painters, depicting the external appearance of the drugs and of the methods for preparing them. The eight pictures of foreign drugs show examples belonging to foreigners from Persia and elsewhere, and the ways in which they prepared and presented drugs, thus portraying the image Han Chinese intellectuals had of the world abroad (Chen Ming 2018: 305–14).

V Lasting Influence

Even more deserving of notice are the studies of by Western scholars of the plants used by Persians in their medicine, their treatment methods and customs, the translations of which into Chinese further influenced medicine in China. Although this knowledge only represents a second-hand influence of Persian medicine in China, it should still not be overlooked. The *Manual of Materia Medica and Therapeutics*, co-authored by the English medical doctors John Forbes Royle and Frederick William Headland (1879), recorded quite a few ancient Persian and Arabic doctors, their *materia medica* and related medical knowledge. A number of Persian works are cited in the Chinese edition of the *Manual of Materia Medica and Therapeutics*, the *Xiyao dacheng* 西藥大成, translated by John Fryer and Zhao Yuanyi 趙元益, including references to 'Persian writers', 'Persian *Materia Medica*', 'Persian works on *Materia Medica*', 'F. Gladwin's *Ulfaz Udwieh*' and 'The Persian works on *Materia Medica* in use in India'. The work also refers to numerous ancient Persian, Syrian and Arabic famous scholars, such as 'The ancient botanist Avicenna' (*Afeisena* 阿非色那), 'Rhases, Rhazis/Zakariya al-Rāzī' (*Laxisi* 拉西司), 'the botanist Serapion/Yahya ibn Sarafyun' (*Saila pi'en* 塞拉披恩) and the Arab 'Geber/Jābir ibn Hayyān' (*Qiba* 奇巴) or 'the chemist' (*Qiba* from the Middle East).

Concentrating only on medical systems, Persian *materia medica* were the dominant among all those which found their way into China, although medical theory and practical therapeutic methods had limited influence. What should not be overlooked is that, since the Ming and Qing Dynasties onwards, the strongest source of Persian medicine in China has been Uyghur medicine, which basically relies on Islamic medicine as a model, including ancient Persian medical theories, texts, use of *materia medica*, therapeutic methods and traditional treatments. Quite a few Persian medical works, which have been translated into Chagatai, Uyghur and other languages, continue to be circulated in Xinjiang and neighbouring Central Asian countries. Naturally, Sino-foreign exchange is two-directional. From one aspect, we can see the historical export of Persian medicine to China; from another, Chinese *materia medica* and related knowledge were also transmitted into Persia. During the Ilkhanate period (1256–1353), Chinese tea (Pers. *tchay*) was used a medicine in the Ilkhan. Until the mid-nineteenth century, tea was commonly drunk by people in Persia for medicinal reasons. A fair number of Chinese *materia medica* (such as cinnamon, ginger, rhubarb root, bodhi seeds, musk and others) were also commonly used by Persian doctors. Even more representative is the work edited by the Ilkhanate minister, Rashīd al-Dīn Fażl Allāh Hamadānī (1247–1318), the *Tānksūqnāmeh* (Treasure Book of the Ilkhan on Chinese Science and Techniques). Rashīd al-Dīn established a centre for technological culture near Tabriz, in the town of Rob'-e Rashīdī, to which came a number of Chinese doctors, who introduced local Iranians to acupuncture and moxibustion, pulse diagnosis, the making of herbal preparations and other traditional Chinese therapeutic methods, as well as a number of medical works by famous Chinese doctors. The *Tānksūqnāmeh* was originally arranged in four sections, including such works as the *Maijue* 脈訣 (Explanations of the Pulse) by Wang Shuhe 王叔和 (180–270), a famous literati from Gaoyang 高陽, the *Tongren shuxue zhenjiu tujing* 銅人腧穴針灸圖經 (Illustrated Classic on Points for Acupuncture and Moxibustion on the Copper Man), the *Shennong bencao jing* 神農本草經 (Divine Farmer's Pharmacopoeia) and the *Taihe yijing* 太和易經 (Great Harmony Classic of Changes). A copy of the *Tānksūqnāmeh* currently survives in the Aya Sophia in Istanbul, of which only *Wang Shuhe's Explanations of the Pulse* remains, a comprehensive testament to Chinese perspectives on medicine, pulse diagnosis and their explanations of the internal organs.[15] The diagrams in the *Tānksūqnāmeh* also show signs of the influence of Chinese painting on Persian painting. Therefore, the mutual connections between Persia and China should be the subject of further research, in order to bring to light the complex and rich history of cultural exchanges between these two lands.

Appendix: translated terms for Persian *materia medica*

Common name	Pinyin	Chinese	Sanskrit	Scientific	Persian
Aleppo gall	wushizi meishizi	無食子 沒食子		Quercus infectoria G.Olivier	
alfalfa	muxuxiang	苜蓿香		Medicago sativa L.	
aloe	lühui	蘆薈		*Aloe vera* (L.) Burm.f.	
amber	hupo	虎魄			
anzaroot tree resin	anzalu	安咱蘆		Astragalus sarcocolla Dymock	anzarūt
asafoetida	awei	阿魏	hingu	Ferula narthex Boiss.	anguza
asbestos	buhui mu	不灰木			
Astragalus gum gum tragacanth	ketiela	可鐵剌		Astragalus gummifer Labill.	kateerā
Aucklandia root	muxiang	木香		Aucklandia costus Falc.	

(Continued)

Common name	Pinyin	Chinese	Sanskrit	Scientific	Persian
beetroot	tiancai	甜菜		Beta vulgaris L.	čugonder
	junda	軍達 / 若蓬			
	zhujunda	諸軍達			
beleric myrobalan	pilile	毗梨勒	vibhitaki	Terminalia bellirica (Gaertn.) Roxb.	
beleric myrobalan decoction	yuganzi tang	余甘子湯			
benzoin	Anxixiang	安息香		Styrax tonkinensis (Pierre) Craib ex Hart	
black pepper	hujiao	胡椒	marica	Piper nigrum L.	
	bie qi	䤵齊			
borax	da peng sha	大鵬砂			
Borneo camphor wood	longnao xiangshu	龍腦香樹		Dryobalanops aromatica C.F.Gaertn.	
carpesium fruit	Bosi heshi	波斯鶴蝨		Carpesium abrotanoides L. Persiana	
Carpesium fruit	heshi	鶴蝨		Carpesium abrotanoides L.	
chicken dung alum	jishifan	雞屎礬			
Chinese eaglewood	chenxiang	沉香	aguru	Aquilaria sinensis (Lour.) Spreng.	
Chinese tea	cha	茶		Camellia sinensis (L.) Kuntze	tchay
cile resin	yuancile	元慈勒			
cinnabar	zhusha	硃砂			
coral	shanhu	珊瑚			
corundum	yagu	鴉鶻			yakut
dates	qiannian zao,	千年棗	kharjūra	Phoenix dactylifera L.	khrma
Persian date	keshuluo,	渴樹羅			
	wulouzi	無漏子			
di'er fruit	di'er shi	底欄實			
dill	shiluo	蒔蘿		Anethum graveolens L.	
dragon's blood	qilin jie	麒麟竭		Calamus draco Willd.	
dried sugar cane, white dried sugar cane	shimi, baishimi	石蜜 白石蜜	śarkarā	Saccharum officinarum L.	
Dutchman's pipe cactus Queen of the Night	youbotan	優鉢曇		Epiphyllum oxypetalum (DC.) Haw.	
emblic myrobalan	anmole	庵摩勒	amalika	Phyllanthus emblica L.	

Common name	Pinyin	Chinese	Sanskrit	Scientific	Persian
Euphrates poplar	hutonglei	胡桐淚		Populus euphratica Oliv.	
figs	wuhuaguo	無花果		Fructus Fici	
fragrant rosewood	jiangzhen xiang	降真香		Dalbergia odorifera T.C.Chen	
frankincense	xunlu	熏陸		Boswellia carteri Birdw.	
	ruxiang	乳香			
	rutou xiang	乳頭香		Boswellia sacra Flück.	
fruit wine	shelibi	舍剌必			Šarāba
fruit-spike of betel pepper	jujiang	蒟醬		Piper betle L.	
gold-streaked alum, fiboferrite	jinxian fan	金線礬			
graphite	shimo	石墨			
great pearl	dazhenzhu	大真珠			
green salt	lüyan	綠鹽			
gum Arabic	lizi shujiao	李子樹膠		Senegalia senegal (L.) Britton	
'Plum tree gum'	sanyi	三亦			samgh
heaven's brilliance sand	tianming sha	天明砂			
indigo	luozi dai	螺子黛		Murex trunculus	
	qingdai	青黛			
Japanese oak	keshu pi	柯樹皮		Cortex Lithocarpus glaber (Thunb.) Nakai	
jasmine	yeximi	野悉蜜		Jasminum grandiflorum L.	
	suxinhua	素馨花			
juice of three fruits	sanle jiang	三勒漿	triphalā		
large-fruited (Persian) elm seeds	wuyi, bosi wuyi	蕪荑 波斯蕪荑		Ulmus macrocarpa Hance	
lead oxide lithargyrum	mituoseng	密陀僧			
litchi	lizhi	荔枝		Litchi chinensis Sonn.	
long pepper	bibo	蓽撥	pippali	Piper longum L.	
	biba	荜茇			
marking-nut	polude	婆羅得	Skt. bhallātaka Toch. B bhallātak	Semecarpus anacardium L.	balādur
mastic resin tears of Chios	xunlu	薰陸		Pistacia lentiscus L.	
mercury	shuiyin	水銀			
myrrh	moyao	沒藥			
naptha	shinao	石腦			

(Continued)

Common name	Pinyin	Chinese	Sanskrit	Scientific	Persian
northern calamine tutty	bei lüganshi	北盧甘石			tūtiya
nutgrass galingale root	xiangfuzi	香附子		Cyperus rotundus L.	
nutmeg	roudoukou	肉豆蔻		Myristica fragrans Houtt.	
opium	afeiyong	阿飛勇		Papaver somniferum L.	afyoon
orpiment	cihuang	雌黃			
Parthian pomegranates	anshiliu	安石榴		Punica granatum L.	
Persian alum	Bosi fan	波斯礬			
Persian aurichalcite	Bosi zhetou	波斯折鍮			
Persian honey locust	zaojia	皂莢		Gleditsia caspia Desf.	
Persian lead	Bosi qian	波斯鉛			
Persian pine resin	bosi shuzhi	波斯樹脂			
Persian red salts	chiyan	赤鹽			
Persian refined lead	Bosi qian jing	波斯鉛精			
Persian salt	bosi yan	波斯鹽			
Persian silver ore	xigezhi	悉悋脂			
Persian verdigris	bosi yanlü	波斯鹽綠			
Persian verdigris, verdigris	Bosi yanlü, yanlü	波斯鹽綠鹽綠/碌			zinjar
Persian white alum	Bosi baifan	波斯白礬			
pistachio	wumingzi	無名子		Pistacia chinensis subsp. integerrima (J.L.Stewart) Rech.f.	ayuehun
pomegranate	anshiliu	安石榴		Punica granatum L.	
poso tree	poso shu	婆娑樹			
psoralea fruit	buguzhi	補骨脂	babchi	Cullen corylifolium (L.) Medik.	
purple alum, chrome alum chromium potassium sulphate	zifan	紫礬			
pyrolusite	wumingyi	無名異			
qitun tree	qitun shu	齊暾樹			
red alum iron sulphate	jiang fan	絳礬			
red cornetfish	shaoyu	鮹魚			
rosewater	qiangwei shui	薔薇水			
saffron	yujinxiang	鬱金香		Crocus sativus L.	
sal ammoniac	naosha	硇砂			

Common name	Pinyin	Chinese	Sanskrit	Scientific	Persian
shellac	zikuang	紫礦			
	zimao	紫鉚			
sherbet	shelibie	舍利別			šarba,
'slake-thirst drink'	shelibai	攝里白			šarbat
	jiekeshui	解渴水			
silver shards	yingxie	銀屑			
slender Dutchman's pipe root	qingmu	青木		Aristolochiae debilis Siebold & Zucc.	
spinach, Persian greens	bosi cai	波斯菜		Spinacia oleracea L.	
storax styrax oriental sweetgum	su he(xiang)	蘇合(香)		Liquidambar orientalis Mill.	
sulphur imported by sea	boshang liuhuang	舶上硫磺			
sulphur ore	shiliu huang	石硫黃			
sweet almonds	piantao	偏桃		Semen Prunus persica L. Batsch	
sweet-dew peaches	ganlu tao	甘露桃			
sweetgum oil benzoin	anxixiang	安息香		Styrax tonkinensis Pierre Craib ex Hart.	
the *pannüse* tree	pannüse	槃笯穡樹			
theriac	diyejia	底野迦			diryaq
true Persian indigo	zhen bosi qing dai	真波斯青黛			
true superior Persian brass with the hue of a horse tongue	zhen Bosi mashe se shang tou	真波斯馬舌色上鍮			
turmeric	yujin	鬱金		Curcuma longa L.	
white alum	baifan	白礬			
white lead powder	hufen	胡粉			
	wufeng duyao cao	無風獨搖草			
yellow floriate ore	huang hua shi	黃花石			
yellow myrobalan decoction	hezi tang	訶子湯			
yellow/black/chebulic myrobalan	helile	訶黎勒	harītakī	Terminalia chebula Retz.	
zinc oxide tutty Persian chalcopyrite	toushi (MC thuw dzyek) duotiya Bosi toushi	鍮石 朵梯牙 波斯鍮石			tutiyā
	digeng	荻根		radix Triarrhena sacchariflora Maxim. Nakai	

Notes

1 Translator's note: For the sake of simplicity, botanical names here refer to the most commonly used, or normative equivalences to the Chinese term. These are listed in the appendix. However, it should be noted that, even in contemporary use, there is a high degree of ambiguity – many Chinese medical plant terms cover multiple species, sometimes dozens, and there is currently no concise way to reference this. The recent volume by botanists Christine Leon and Lin Yulin, based on years of fieldwork, not only attends to Chinese medical functions of the plants, but the botanical variation and habits of substitution in Chinese markets (Leon and Lin 2017). Native Chinese systems of plant identification themselves varied across history (Métailie and Needham 2015). Historical substance and region names were subject to even more ambiguity and change. The term *Bosi* 波斯, for example, here translated ubiquitously as 'Persia', at times referred to a southeast Asian region of unsure identity (Laufer 1919: 486 ff.). While some scholars agree this refers to the thirteenth to sixteenth-century north Sumatran kingdom of Pasai, this cannot account for fifth-century Chinese references to a SE Asian region by the same name (Kotyk forthcoming). It is possible the term refers to colonies of Persian merchants in SE Asia. Thus when considering items like *Anxixiang* 安息香 (lit. Parthian aromatic), derived from the SE Asian species *Styrax tonkinensis* Craib ex Hartwich, we need to consider that these may have been products which came *through* Persia, but were not *from* Persia. For more detailed work on individual products, see the author's other works (Chen 2007, 2013, 2018). Useful sites for this work include the *Global Biodiversity Information Facility* https://gbif.org, Kew Gardens' *Medicinal Plant Name Services* https://mpns.science.kew.org/mpns-portal/, and *Zhongyi shijia* 中醫世家 http://www.zysj.com.cn/zhongyaocai/index.html.
2 The *qiyao* 七曜 literally refer to the sun, moon, Mercury, Venus, Jupiter, Mars and Saturn, but broadly refer to astronomy, and often astrology.
3 MC refers to Middle Chinese pronunciation. The reference used is from Baxter and Sagart (2014).
4 *Genben shuo yiqie youbu baiyi jiemo* 根本說一切有部百一羯摩 (Skt. Mūlasarvāstivāda ekaśataka karman), CBETA, T. 1453, p. 478.a20.
5 CBETA, T54, no. 2128, p. 710, c8.
6 This item remains unidentified in recent scholarship. For variant names, see Santos (2010: 225–6).
7 *Zhonghua bencao* lists this as an alternate name for *suxinhua* 素馨花.
8 The Chenla kingdom controlled Indochina prior to the Khmer empire between 550–706 CE.
9 On this text, see the unpublished paper by F. Meyer, 'Syncrétisme médical en Haute-Asie d'après un texte Tibetain censé avoir été introduit de Chine au VIIIème siècle', and R. Yoeli-Tlalim (2012) 'Re-visiting 'Galen in Tibet'(1)', *Medical History*, 56.3: 355–65.
10 On this work, see Needham (1974: 163 f.) and Deshpande and Fan (2012).
11 Now lost, this work has been reconstructed in Shang Zhijun (1997).
12 The term *yi* in Li Miyi's name may mean medicine or doctor.
13 Translator's note: of unstable identity, this ore was named for its mixed yellow and black colouring, and probably contained various minerals composited over time. See Han Jishao (2011: 69).
14 Translator's note: not to be confused with other sweets and ices for which the names 'sherbert', 'sherbet' and 'sorbet' later came to be used in English.
15 Cf. Shi Guang (2016).

Bibliography

Pre-modern sources

Bencao gangmu 本草綱目 (Categorical and Itemised Pharmacopoeia) 1596, Li Shizhen 李時珍 (1979), Beijing: Renmin weisheng chubanshe.

Bencao shiyi 本草拾遺 (Supplement to the Pharmacopoeia) 739, Chen Zangqi 陳藏器, in Shang Zhijun 尚志鈞 (ed.) (2002) *Bencao shiyi jishi* 本草拾遺輯釋 (Supplement to the Pharmacopoeia with Annotations), Hefei: Anhui kexue jishu chubanshe.

Cefu yuangui 冊府元龜 (Models from the Archival Bureau) 1013, Song dynasty, Wang Qinruo 王欽若 (ed.) (2003), Beijing: Zhonghua shuju.

Chu sanzang jiji 出三藏記集 (Compilation of Notes on the Translation of the Tripiṭaka), Sengyou 僧祐, CBETA T55 n2145.

Gaoseng zhuan 高僧傳 (Biographies of Eminent Monks), Hui Jiao 慧皎 (497–554), CBETA T50 n2059.

Haiyao bencao 海藥本草 (Overseas *Materia Medica*) Five Dynasties, Li Xun 李珣 (855?–930?), in Shang Zhijun 尚志鈞 (ed.) (1997) *Haiyao bencao jijiaoben* 海藥本草 輯校本 (Overseas *Materia Medica*, Compiled and Edited), Beijing: Renmin weisheng chubanshe.

Hou han shu 後漢書 (Book of the Later Han) 445, Fan Ye 范曄, Li Xian 李賢, Sima Biao 司馬彪 and Liu Zhao 劉昭 (1965), Beijing: Zhonghua shuju.

Hu bencao 胡本草 (Western Pharmacopoeia) 740, Zheng Qian 鄭虔, lost.

Jinxiao fang 近效方 (Quick-Working Recipes), lost.

Jinye huandan baiwen jue 金液還丹百問訣 (Explanations of the 'Hundred Questions' on the Cyclically Transformed Elixir of Liquefied Gold), Li Guangxuan 李光玄, DZ 266.

Li Wenrao wenji 李文饒文集 (Collected Essays of Li Wenrao, namely Li Deyu 李德裕) Tang dynasty, Li Deyu 李德裕 (787– 849) in Fu Xuanchong 傅璇琮 and Zhou Jianguo 周建國 (ed.) (2018) *Li deyu wenji jiaojian* 李德裕文集校箋 (Collected Essays of Li Deyu with Annotations), Shijiazhuang: Hebei jiaoyu chubanshe.

Nancun chuogeng lu 南村輟耕錄 (Records of Setting the Hoe Aside from the Southern Hamlet) Yuan dynasty, Tao Zongyi 陶宗儀 (1329–1412) (2004), Beijing: Zhonghua shuju.

Qianjin yaofang 千金要方 (Essential Recipes Worth a Thousand Gold Pieces) 652, Sun Simiao 孫思邈, in Li Jingrong 李景榮 et al. (eds) (1997), *Beiji qianjin yaofang jiaoshi* 備急千金要方校釋 (Essential Recipes Worth a Thousand Gold Pieces, Punctuated and Annotated), Beijing: Renmin weisheng chubanshe.

Qianjin yifang 千金翼方 (Supplementary Prescriptions Worth a Thousand Gold Pieces) 682, Sun Simiao 孫思邈 in Zhu Bangxian 朱邦賢 et al. (eds) (1999), *Qianjin yifang jiaozhu* 千金翼方校注 (Supplementary Prescriptions Worth a Thousand Gold Pieces, Punctuated and Annotated), Shanghai: Shanghai guji chubanshe.

Tong dian 通典 (Comprehensive Institutions) 801, Du You 杜佑 (1984), Beijing: Zhonghua shuju.

Waitai miyao fang 外臺秘要方 (Recipes based on the Secret Essentials of the Imperial Archives) 752, Wang Dao 王燾 (670–755), Gao Wenzhu 高文鑄 (ed.) (1993), Beijing: Huaxia chubanshe.

Weishu 魏書 (Wei Dynasty History) 554, Beiqi Dynasty, Wei Shou 魏收 (1997), Beijing: Zhonghua shuju.

Xin tangshu 新唐書 (New Tang History)1060, Song Dynasty, Ou Yangxiu 歐陽修 and Song Qi 宋祁 (eds) (1975), Beijing: Zhonghua shuju.

Xinxiu bencao 新修本草 (Newly Revised Pharmacopoeia) 659, Su Jing 蘇敬, Shang Zhijun 尚志鈞 (ed.) (2005), Hefei: Anhui kexue jishu chubanshe.

Yinshan zhengyao 飲膳正要 (Important Principles of Food and Drink) 1330, Hu Sihui 忽思慧 (1986), Beijing: Renmin weisheng chubanshe.

Yiqie jing yi yinyi 一切經音義 (Sounds and Meanings [of All Words in] the Scriptures), Tang Dynasty, Hui Lin 慧琳 (1986), Shanghai: Shanghai guji chubanshe.

Youyang zazu 酉陽雜俎 (Miscellaneous Morsels from Youyang), Tang Dynasty, Duan Chengshi 段成式 (1981), Beijing: Zhonghua shuju.

Modern sources

Akasoy, A., Burnett, C. and Yoeli-Tlalim, R. (eds) (2011) *Islam and Tibet: interactions along the musk routes*, Surrey: Ashgate.

––––––– (eds) (2013) *Rashid al-Din: agent and mediator of cultural exchanges in Ilkhanid Iran*, London: Warburg Institute.

Allsen, T.T. (2001) *Culture and Conquest in Mongol Eurasia*, Cambridge: Cambridge University Press.

Baxter, W.H. and Sagart, L. (2014) *Old Chinese: a new reconstruction*, New York: Oxford University Press.

Buell, P.D. (2007) 'How did Persian and other western medical knowledge move East, and Chinese West? A look at the role of Rashīd al-Dīn and others', *Asian Medicine: tradition and modernity*, 3.2: 279–95.

Buell, P.D. and Anderson, E.N. (2010) *A Soup for the Qan: Chinese dietary medicine of the Mongol era as seen in Hu Szu-Hui's Yin-shan Cheng-yao*, Leiden: E.J. Brill.

Chen Ming 陳明 (2007) 'The transmission of foreign medicine via the Silk Road in medieval China: a case study of *Haiyao Bencao*', *Asian Medicine: Tradition and Modernity*, 3.2: 241–64.

––––––– (2013) *Zhonggu yiliao yu wailai wenhua* 中古醫療與外來文化 (Foreign Medicine and Culture in Medieval China), Beijing: Beijing daxue chubanshe.

——— (2018) 'Fanciful images from abroad: picturing the other in *Bencao pinhui jingyao*', in V. Lo and P. Barrett (eds) *Imagining Chinese Medicine*, Leiden: E.J. Brill, pp. 305–14.

Deshpande, V.J. and Fan, J. (2012) *Restoring the Dragon's Vision: Nagarjuna and medieval Chinese ophthalmology*, Hong Kong: Chinese Civilisation Centre, City University of Hong Kong.

Han Jishao 韓吉紹 (2011) *Liandanshu zhong de wailai fanshi* 煉丹術中的外來礬石 (Exotic Alumite in Chinese Alchemy), *Hongdao* 弘道, 47.2: 67–72.

Kotyk, J. (2020) On the identification and use of anxi-xiang 安息香/安悉香/安西香 in Medieval China. in E. Nissan (ed.) *Quaderni di Studi Indo-Mediterranei XII: For the centennial of Berthold Laufer's Classic Sino-iranica (1919)*, Milan: Mimesis Edizioni, pp. 519–28.

Laufer, B. (1919) *Sino-Iranica: Chinese contributions to the history of civilization in ancient Iran, with special reference to the history of cultivated plants and products*, Chicago, IL: Field Museum of Natural History.

Leon, C., and Lin, Y.-L. (2017) *Chinese Medicinal Plants, Herbal Drugs and Substitutes: an identification guide*, Chicago, IL: Kew Publishing, Royal Botanic Gardens Kew.

Lo, V. and Barrett, P. (eds) (2018) *Imagining Chinese Medicine*, Leiden: E.J. Brill.

Mazahéri, A. (1983) *La Route de la Sole*, Paris: Papyrus.

Métailie, G. and Needham, J. (2015) *Science and Civilisation in China: biology and biological technology-traditional botany: an ethnobotanical approach, Vol. 6. Part IV*, Cambridge: Cambridge University Press.

Meyer, F. (unpublished) 'Syncrétisme médical en Haute-Asie d'après un texte Tibetain censé avoir été introduit de Chine au VIIIème siècle'.

Needham, J. (ed.) (1974) *Science and Civilisation in China: historical survey, from cinnabar elixirs to synthetic insulin*, vol. 5, Chemistry and Chemical Technology, pt. 3, spagyrical discovery and invention, Cambridge: Cambridge University Press.

Royle, J.F. and Headland, F.W. (1879–1894) *Xiyao dacheng* 西藥大成 (A Manual of *Materia Medica* and Therapeutics: including the preparations of the pharmacopoeias of London, Edinburgh, and Dublin, with many new medicines), trans. J. Fryer (Fu Lanya 傅蘭雅) and Zhao Yuanyi 趙元益 (1887), Shanghai: Jiangnan zhizaoju.

Santos, D.M. (2010) 'A note on the Syriac and Persian source of the pharmacological section of the Youyang zazu', *Collectanea Christiana Orientalia*, 7: 217–29.

Schafer, E.H. (1985) *The Golden Peaches of Samarkand: a study of T'ang exotics*, Berkeley; Los Angeles: University of California Press.

Schottenhammer, A. (2010) 'Transfer of *xiangyao* 香藥 from Iran and Arabia to China: a reinvestigation of entries in the *Youyang zazu* 西陽雜俎 (863)', in R. Kauz (ed.) *Aspects of the Maritime Silk Road: from the Persian Gulf to the East China Sea*, Wiesbaden: Harrassowitz Verlag, pp. 117–49.

——— (2013) 'Huihui medicine and medicinal drugs in Yuan China', in M. Rossabi (ed.) *Eurasian Influence on Yuan China*, Singapore: Institute for Southeast Asian Studies, pp. 75–102.

Shang Zhijun 尚志钧 (ed.) (1997) *Haiyao bencao jijiaoben* 海藥本草輯校本 (Overseas *Materia Medica*, Compiled and Edited), Beijing: Renmin weisheng chubanshe.

——— (ed.) (2002) *Bencao shiyi jishi* 本草拾遺輯釋 (Supplement to the Pharmacopoeia with Annotations), Hefei: Anhui kexue jishu chubanshe.

Shi Guang 時光 (2016) *Yili hang zhongguo keji zhenbao shu jiaozhu* 伊利汗中國科技珍寶書校注 (Research and Translation of Tānksūqnāmeh – The Ilkhanate Treasure Book of Khatay Sciences and Technologies), Beijing: Beijing daxue chubanshe.

Song Xian 宋峴 (2000) *Huihui yaofang kaoshi* 回回藥方考釋 (The Textual Study on a Collection of Muslim Prescriptions), Beijing: Zhonghua shuju.

——— (2001) *Gudai Bosi yixue yu Zhongguo* 古代波斯醫學與中國 (Persian Traditional Medicines in Imperial China), Beijing: Jinjing ribao chubanshe.

Yoeli-Tlalim, R. (2012) 'Re-visiting 'Galen in Tibet' (1)', *Medical History*, 56.3: 355–65.

Yu Xin 余欣 (2013) 'Zhonggu shidai de caishu yu wailai wenming: zhujunda de yilang yuanyuan 中古時代的菜蔬與外來文明：諸軍達的伊朗淵源' (Vegetables in the medieval ages and foreign culture: the Iranian origin of Zhujunda), *Fudan xuebao* 復旦學報, 6: 71–7.

33
VIETNAM IN THE PRE-MODERN PERIOD

Leslie de Vries

This chapter focusses on Chinese medicine in Vietnam between the period of Chinese domination (almost continuously from 111 BCE to 938 CE) and the end of the nineteenth century, when the French took over full control of the country in 1885. Throughout history, Vietnam was known by many names, including *Đại Việt*, *An Nam*, *Đại Nam*, and *Nam Việt*. *Việt* (Ch. *Yue* 越) refers to the main ethnic group, the Việt people (also known as *Kinh*); *Nam* (Ch. *Nan* 南) refers to the south. Vietnam has defined itself and was defined as southern in relation to its big northern neighbour, China. The name 'Vietnam' appears in few sources of the pre-modern era. The Chinese Qing dynasty emperor Jiaqing 嘉慶 (r. 1796–1820) imposed Vietnam as an official name on Gia Long (r. 1802–20), the founder of the Nguyễn dynasty (1802–1945), Vietnam's last dynasty. Nevertheless, the Vietnamese court continued to refer to itself as *Đại Nam* (Woodside 1971: 120–1). 'Vietnam' only became commonly used for the country during the early twentieth century in nationalist circles (Taylor 2013: 398). For matters of convenience, I anachronistically refer to Vietnam by its modern name for convenience. Việt culture originated in the northern part of what is now Vietnam. During the second millennium the Việt expanded their territory southwards where they replaced former Cham and Khmer governance. It was only in the nineteenth century that Vietnam took the geographical form it has today.

The Red River delta, the northern part of what is now Vietnam and the legendary homeland of the Việt people, was inhabited from early prehistoric times onwards (Higham 1996, Kim 2015). Yet Vietnamese identity and language were shaped through interactions between earlier inhabitants and non-Han ethnic people (Yue) of southern China, who were pushed further southwards by Chinese expansion during the Qin dynasty (221–206 BCE) (Brindley 2015). The millennium-long Chinese occupation of Vietnam during the subsequent Han dynasty left strong imprints on Vietnamese culture and language. In this period Vietnam became largely Sinicised (Holcombe 2001). After its independence, Vietnam maintained a tributary relationship with China until the nineteenth century. Throughout most periods during the second millennium CE, Vietnam's ruling elite looked up to China (Kelley 2005). They had a strong demand for things Chinese and also preferred Chinese medicine, also known as 'northern medicine' (*thuốc bắc*) in Vietnam.

Medicine in Vietnam

Northern medicine was practised alongside indigenous Vietnamese medicine, or 'Southern medicine' (*thuốc Nam*), and various forms of religious healing (e.g. Marr 1987). Southern medicine, also known as 'our medicine' (*thuốc ta*) in the twentieth century (Wahlberg 2014: 53), includes various forms of folk medicine, which rely almost entirely on local medicinal products (Marr 1987: 196). Whereas northern medical literature is mainly written in Chinese (*Hán*), most texts of Southern medicine are in Vietnamese vernacular script (*Nôm*). Southern medicine is said to be less theoretical and more pragmatic than northern medicine, and also served the lower social classes (Thompson 2015). A third main category of medicine Marr distinguished, religious healing, primarily dealt with diseases caused by harmful spirits and required experts like 'Buddhist monks, Taoist priests, sorcerers and mediums' (1987: 172–3). The boundaries between these three main kinds of Vietnamese medicine are not clear-cut. During the anti-colonial struggle in the twentieth century, different forms of Vietnamese medicine (mainly Northern and Southern medicine) became known as 'Vietnamese traditional medicine' (*y học cổ truyền Việt Nam*) and 'Eastern medicine' (*Đông y*). These terms distinguish Vietnamese traditional medicine from Western (bio)medicine (*thuốc Tây*) (Monnais et al. 2012: 2–3; Wahlberg 2014). Finally, the many non-Việt ethnic groups in Vietnam practised and still practise their own forms of medicine.

Northern (and southern) visions of the south

During the period of Chinese colonisation, the region of the Red River delta in northern Vietnam became part of Lingnan, made up parts of the modern Chinese provinces Guangdong, Guangxi, Guizhou, Yunnan, Hunan, and Hainan. In the Chinese 'geographical imagination', the local *qi* of this hot and humid region deviated from the annual agricultural cycles of the northern plains, the cradle of Chinese civilisation. Lingnan, notorious for its toxic miasmic mists, served as a location where convicts were sent into exile. Many of them fell ill in this malarial ridden area and died early (Schafer 1967: 37–44, 130–34; Hanson 2011).

The term 'miasma' (*zhang* 瘴) features in the biography of the Han dynasty general Ma Yuan 馬援 (d. 49 CE) in the *History of the Han Dynasty* (*Hanshu* 漢書). In 42 CE, Emperor Guangwu 光武 (r. 25–55) commissioned general Ma to pacify the Yue people. After returning from a successful campaign, many of his soldiers fell ill and died of a 'miasmic epidemic' (*zhangyi* 瘴疫) they contracted in the south. The miasmatic *qi* of the far south became a trope in later Chinese medical texts (Hanson 2011; Chen 2015). Although we are not sure of which disease Ma Yuan's soldiers died (*zhang* has been associated with a wide range of diseases, including malaria, syphilis and Hansen's Disease), Ma's campaign became the origin story of smallpox during the Ming dynasty (1368–1644), and is recounted as such in many Chinese medical texts. After a short and failed occupation of Vietnam (1407–27), Chinese rulers refrained from further attempts to include the region in the Chinese empire. Miasmic mist became an explanation for the political frontier or natural barrier of China. The Chinese word for 'miasma' is etymologically related to the word 'barrier' (*zhang* 障). The 'miasmic climate' of Vietnam was understood as a 'deadly "barrier" that set limits for military garrisons and Han settlements' (Hanson 2011: 67).

Vietnamese medical authors acknowledged the specific climatic characteristics of their region. Lê Hữu Trác 黎有晫 (1724?–1791), for instance, famously argued that fevers relating to the disease category Cold Damage (*thương hàn*, Ch. *shanghan* 傷寒), belonging to the harsh winters in the Northern planes of China, do not occur in Lingnan. He further warned his readers to be careful with signature Cold Damage formulæ, which induce sweat and harm the bodies of people living in hot and damp environments (Hoang *et al.* 1993: 23-4). Although this erudite doctor referred to complex cosmological principles, modern authors have underlined the rather pragmatic approach in the adaptation of Northern medicine in Vietnam. In the conclusion to her *Vietnamese Traditional Medicine: a social history*, C. Michele Thompson writes, for instance: 'Almost everyone who studies Vietnamese medicine concludes that the contributions to the practice of Sino-Vietnamese medicine have been *practical* rather than *theoretical*' (2015: 140). Furthermore, local doctors had a keen eye for the richness of the indigenous ecology, which provided necessary medicines to deal with local diseases. These adaptations have been eulogised in modern, nationalist Vietnamese historiography as evidence of the pragmatism and intelligence of Vietnamese doctors (e.g. Dương 1947–50: 1; Hoang *et al.* 1993: 23–4).

Traces of medical history

Our views on Vietnam's medical past prior to the nineteenth century are heavily blurred. Due to unfavourable conditions, like the hot and damp climate, prolonged periods of warfare, and the absence of large-scale commercial printing until the twentieth century, only a small number of Vietnamese texts made it to the twentieth century (Cadière and Pelliot 1904; Henchy 1998; McHale 2007; Mayanagi 2010b). Most Vietnamese medical texts date from after the short Ming occupation of the early fifteenth century. Most scholars point the finger at Ming occupiers for confiscating and destroying earlier Vietnamese texts (Dương 1947–50: 38–39; Hoang *et al.* 1993: 12; Thompson 2015: 18). Also later Vietnamese rulers controlled the production and diffusion of printed knowledge. Most notorious was Minh Mạng's (r. 1820–41) campaign against texts written in Nôm (Thompson 2010).

Although inventories of books in official libraries were made throughout history, we have no information on the numbers of books and their contents in catalogues before the Nguyễn dynasty. Inventories like that of the Tụ Khuê library, founded under Minh Mạng, date only from the first decade of the twentieth century (Trần Nghĩa 1993: 51–2). The modern collecting and cataloguing of old Vietnamese texts was started by French colonial researchers. In 1958, the École française d'Extrême-Orient (EFEO) handed their collections over to the Vietnamese authorities, who afterwards further expanded them (Henchy 1998).

Some 400 'ancient' medical texts, written in Chinese (Hán) and the Vietnamese vernacular (Nôm), of which about 150 authors are known, survived. These texts include works by Vietnamese authors, Chinese texts copied or printed in Vietnam, and imports from China. The largest collection of ancient texts can be found in the Institute of Hán-Nôm Studies in Hanoi. A good starting point to look for medical and pharmacological texts is the index to the bilingual *Di sản Hán Nôm Việt Nam: Thư mục đề yếu / Catalogue des livres en Han Nôm* edited by Trần Nghĩa and François Gros (1993: 283–4), containing works in the Institute of Hán-Nôm Studies and several French libraries. In this catalogue, C. Michele Thompson counts 366 entries of texts exclusively dealing with medicine and pharmacy, and another 166 non-medical texts which discuss medical and pharmaceutical topics. Looking at texts written by Vietnamese authors only, Thompson concludes that 40.8% of them are written in Hán, 16.4% in Nôm and 42.8% in a mixture of Hán-Nôm (Thompson 2006: 258–9). A Taiwanese catalogue, although in a different arrangement, makes the information in Trần Nghĩa and François Gros's work available to a Chinese language readership (Liu *et al.* 2002). The most extensive catalogue of medical texts preserved in Vietnam is *Research on Vietnamese Traditional Medical Texts (Tìm hiểu thư thịch y dược cổ truyền Việt Nam)* edited by Lâm Giang (2009).

This catalogue counts 394 titles of medical works in the Institute of Hán-Nôm Studies and lists another 192 works preserved in various libraries in Hanoi, including the National Library of Vietnam. Some of the texts in the aforementioned catalogues are listed under various titles, and are counted twice; other works listed as separate texts are also preserved in larger works. We should thus be careful with counts based on indices of these catalogues as a way of obtaining a definitive number of preserved medical texts in Hanoi. A more elaborate statistical analysis of Hán-Nôm medical texts preserved in the National Library of Vietnam, and the libraries of the Institute of Hán Nôm Studies and the Hà Tây Museum can be found in an article written in Vietnamese by Nguyễn Thị Dương (2009a). The best description of the medical collections in both the Institute of Hán-Nôm Studies and the National Library of Vietnam is in Japanese by Mayanagi Makoto, who provides an extensive list of the texts divided into categories. He describes the external physical characteristics of the works and comprehensive information on their contents. Mayanagi further concludes that the majority of preserved medical texts in Vietnam consist of handwritten manuscripts, not older than the nineteenth century (Mayanagi 2006, 2011, 2012–15).

A number of medical texts have been made available online through the joint efforts of the National Library of Vietnam and the Vietnamese Nôm Preservation Foundation (Shih and Chu 2006), and can be consulted through the online catalogues on their websites (National Library of Vietnam, 30/5/2018; Vietnamese Nôm Preservation Foundation, 30/5/2018). Smaller collections of Hán-Nôm texts are held in France, the United States, Japan, the Netherlands, and the United Kingdom. Some of these collections, such as the *Southeast Asian Digital Library of the University of Northern Illinois* are accessible online and contain medical texts. Unfortunately, the Hán-Nôm medical texts online are not searchable or machine-readable.

Chinese medical texts were a sought-after commodity in Vietnam. Many texts reached the country through the Nagasaki trade (Li 2011). Other texts were brought back by emissaries as official presents. Ming loyalist refugees may have imported medical texts as well. Makoto Mayanagi's study of the circulation of books throughout East Asia in the Early Modern period provides an overview of Chinese medical texts transmitted to Vietnam. He concludes that far fewer editions of popular Chinese medical texts can be found in Vietnam than in Japan and Korea. Mayanagi lists only fifteen Vietnamese reprints of Chinese medical texts, all written in the Ming and Qing dynasties. Fifteen is a significantly lower figure than the approximate 315 Japanese and 93 Korean reprints of Chinese texts Mayanagi counts. Moreover, all these 15 Vietnamese editions of Chinese medical text date from the nineteenth century. As Mayanagi points out, we should consider the conclusions of his survey with care since many books were lost in Vietnam. References to Chinese sources in important Vietnamese medical texts, such as Lê Hữu Trác's *Intuitive Understandings of Hai Thuong's Medical Lineage* (*Hải Thượng y tông tâm lĩnh* 海上醫宗心領, 1770–86) also give clues as to which Chinese medical texts were available in Vietnam (Mayanagi 2010b).

In an essay on medicine in the collection *Essays Written during the Rain* (*Vũ trung tùy bút* 雨中隨筆), the scholar Phạm Đình Hồ (1768–1839) makes a distinction between external and internal medicine. Phạm mentions three famous family traditions of external medicine (*ngoại khoa* 外科), all sharing the same family name: Nguyễn. Most of his attention goes to internal medicine (*nội khoa* 内科), however. He writes that texts by Li Chan 李梴 (sixteenth to seventeenth centuries), Gong Tingxian 龔廷賢 (1522–1619), Zhang Jiebin 張介賓 (1563–1642), and Feng Zhaozhang 馮兆張 (seventeenth to eighteenth centuries) were the most popular medical texts at his time (Chapter 9 in this volume). Phạm further complains about a schism between medical practitioners in his day. On the one hand, there were doctors who always 'nourished and supplemented' (*tẩm bổ* 滋補); on the other hand, there were those who only 'attacked and dispersed' (*công tán* 攻散). According to Phạm, the first group followed texts by Cảnh Nhạc 介賓 ([Zhang] Jiebin) and Phùng Thị 馮氏 (Feng-shi, i.e. Feng Zhaozhang); the other group took *Y học* (*Yixue* [*rumen*] 醫學 [入門], by Li Chan 李梴 [?]) and *Hồi xuân* ([*Wanbing*] *huichun* [萬病]回春, by Gong Tingxian 龔廷賢 [1522–1619]) as examples. Although not himself a physician, Phạm criticised the irreconcilability between these two therapeutic stances, and pointed out the dangers of stubbornly sticking to one approach only. He argued for a middle position, and believed that a doctor, like a good statesman, needs to know when to reward and when to punish. In the last part of his essay, Phạm devotes attention to the most influential doctor in his day: Lê Hữu Trác (Nguyễn 1973). The names of Li Chan, Gong Tingxian, Zhang Jiebin, and Feng Zhaozhang also feature prominently in Mayanagi's survey of important Chinese medical texts in Vietnam (compare Mayanagi 2010b).

Scattered information on medicine during the second millennium can also be retrieved from steles, genealogical books, historiography and literature, including Chinese sources. For instance, one of the earliest surviving Vietnamese histories *Abridged Records of An Nam* (*An Nam chí Lược*, Ch. *Annan zhilüe* 安南志略) written in 1250 and included in the *Complete Library of the Four Treasuries* (*Siku quanshu* 四庫全書, SKQS), provides information on medicine during the Trần dynasty (1225–1400) (Dương 1957–50: 37). The fourteenth-century *Wonders Picked up in Vietnam* (*Lĩnh Nam chích quái*, Ch. *Linnan zhiguai* 嶺南摭怪) recounts legendary accounts of the origin of moxibustion (Durand 1953). Descriptions by Christoforo Borri of Nguyễn ruled southern 'Cochinchina' and Samuel Baron on Trịnh-ruled northern 'Tonkin' offer Western views on medicine practised in Vietnam during the seventeenth century (Dror and Taylor 2006). A rich autobiographical account by the doctor Lê Hữu Trác gives valuable insights on how medicine was practised at the end of the eighteenth century (Nguyễn 1972). Also the above-mentioned essay by Phạm Đình Hổ provides general information on medicine from a Vietnamese scholarly point of view in the early nineteenth century (Nguyễn 1973).

Historiography

French scholars, like the biomedical doctor Anatole Mangin (1887) and the archaeologist and ethnographer George Dumoutier (1887), ignited modern research on Sino-Vietnamese or Sino-Annamese medicine, as Vietnamese indigenous medicine was called during the colonial period (1887–1954). Doctors such as Henry and Vialet observed local practice and were positive about the empirical knowledge of local doctors concerning plants, but criticised their ignorance of scientific methodology (Monnais 2012: 63). Some later colonial doctors also showed an interest in the historical background of traditional medicine. The military doctor Albert Sallet (1930), for instance, wrote an essay on Lê Hữu Trác, one of the founding fathers of Sino-Vietnamese medicine. Pierre Huard, professor of surgery and anatomy and medical historian, stimulated his PhD students to write dissertations on Vietnamese traditional medicine in the period he acted as dean of Faculté française de médicine de Hanoi from 1947 to 1954. Dương Bá Bành's history of Vietnamese medicine (1947–50), based on his PhD dissertation with Huard, is referred to in most of the postcolonial English and Vietnamese language scholarship discussed hereafter. One of Huard's other students, Nguyễn Trần Huân (1921–2001), translated parts of the texts of the famous doctors Tuệ Tĩnh 慧靖 (1330–ca. 1389) and Lê Hữu Trác in his dissertation (1951). Nguyễn combined his career as biomedical doctor with philological studies of Vietnamese and Chinese texts, and published extensively on literature and Chinese and Vietnamese medical history.

An important introduction to traditional medicine in Vietnam in English is Marr's seminal essay *Vietnamese Attitudes Regarding Illness and Healing* (1987). Various Vietnamese scholars provide overviews in English as well: one of the most cited introductions is Hoang Bao Chau, Pho Duc Thuc and Huu Ngoc's 'Overview of Vietnamese traditional Medicine' in their *Vietnamese Traditional Medicine* (1993). A general introduction to the field can be found in Chu Lan Tuyet's 'An Introduction to the History of Traditional Medicine and Pharmaceutics in Vietnam' (2002). Chu's essay also contains an overview of scholarship in Vietnamese and of texts translated in modern Vietnamese. Chu also mentions the works of three doctors who authored medical texts in the Nguyễn ruled south during the eighteenth century (2002: 269). Lâm Giang's (2009) above-mentioned catalogue of Vietnamese medical texts includes introductory essays and extensive references to scholarship in Vietnamese.

Only a few contemporary scholars have gone beyond descriptive overviews of medicine in Vietnam. *Southern Medicine for Southern People*, a recent edited volume by Monnais *et al.* (2012), is an important contribution to the field. Although this volume offers a broad view on medicine in Vietnam, only the editors' 'Introduction' (2012) and C. Michele Thompson's 'Setting the Stage' (2012) discuss medicine before the nineteenth century. Thompson is one of the few scholars who has published on pre-modern Vietnamese medicine in English. Her recent *Vietnamese Traditional Medicine: A Social History* is the only academic monograph devoted on traditional medicine in pre-modern Vietnam (Thompson 2015). However, her study is mostly based on post-eighteenth-century sources and her focus is foremost on smallpox in Vietnam. Thompson emphasises the importance of Nôm texts and Southern medicine, but her book does not give a comprehensive overview of the history of Chinese medicine (or northern medicine) in pre-modern Vietnam.

Echoing Liam C. Kelley's (2006) conclusion about Confucian studies in Vietnam, the field of medical history in Vietnam in the pre-modern period is virtually non-existent. Most Western postcolonial research tends to take earlier French colonial and Vietnamese research on Vietnamese traditional medicine for granted. Shawn F. McHale warns us, however, that we have to approach the primary sources critically in order to avoid falling into the trap of a historiography of grand narratives, informed by ideological agendas such as Confucianism, Marxism, Colonialism, and Nationalism. McHale points out, for instance, how Dương Bá Bành's (1947–50: 37–8) and Hoang *et al* (1993: 11–2). Confucian portrayals of the fourteenth-century doctor Trâu Canh do not correspond to the original account in *A Short History of Annam*. McHale's case-study of Trâu Canh reminds us how history gets distorted (sometimes by mixing up two historical persons), and how constructed stories become accepted as facts (McHale 1999).

Famous doctors

No history of Vietnamese medicine is complete without mentioning Tuệ Tĩnh and Lê Hữu Trác (also known as Mister Lazy of Hai Thuong, Hải Thượng Lãn Ông 海上懶翁). The names of these two founding fathers of traditional medicine ring a bell in the ears of the Vietnamese public, as streets, hospitals, and schools are named after them. Raised as an orphan in a Buddhist pagoda, the fourteenth-century Tuệ Tĩnh prepared for official examinations, but he chose to remain a monk and to practise medicine. In his fifties, he was sent to China as a living tributary present to the Ming emperor. At the imperial court in Nanjing, Tuệ Tĩnh wrote his famous book *Miraculous Drugs from the South* (*Nam dược thần hiệu* 南藥神效) in which he explained Southern medicine in Chinese for a Chinese audience (Thompson 2017). Another work attributed to Tuệ Tĩnh, partly written in Nôm verses, is entitled *Master Hong Nghia's Medical Writings* (Hồng Nghĩa giác tư y thư 洪義覺斯醫書) (Mayanagi 2010a). Lê Hữu Trác, the second father of Vietnamese medicine, hailed from a family of high officials, but chose a life in seclusion in Hương Sơn. His numerous rejections of an official career earned him the nickname Mister Lazy or Mister Lazy from Hải Thương. Lê Hữu Trác's 'medical encyclopedia', *Intuitive Understandings of Hai Thuong's Medical Lineage*, in which he synthesised and elaborated the ideas of Chinese doctor Feng Zhaozhang, is revered as the opus magnum of Vietnamese medicine (Sallet 1930; Huard and Durand 1953, 1956; Bates and Bates 2007; Mayanagi 2010a; De Vries 2017). His autobiographic *Account of the Journey to the Capital* (Thượng kinh ký sự 上京記事, 1784) recounts his travel and stay at the capital in 1782, where he was summoned to treat members of the ruling Trịnh family. This text not only gives insights into his medicine, but also documents daily life at the court and that of the common people, and contains a rich collection of Lê Hữu Trác's poetry (Nguyễn 1972).

Although McHale (1999) questions the Confucian motives of Vietnamese doctors during the Trần dynasty, and Thompson (2015) is sceptical about the connections between Confucianism and medicine before the Minh Mạng reforms in the nineteenth century, other scholars have highlighted the interconnection between Confucianism and medicine in Vietnam (e.g. Chu 2002; Nguyễn Thị Dương 2009). Mayanagi has pointed out that many famous doctors in Vietnamese history held the degree of 'advanced scholar' (tiến sĩ, Ch. *jinshi* 進士) or were related through family to the highest degree holders. Therefore, these doctors were not only trained in medicine but also in the Confucian classics (Mayanagi 2010a). Famous medical texts attributed to important scholars-cum-physicians are Chu Văn An's 朱文安 (?–1370) *Essential Explanations about the Study of Medicine* (Y học yếu giải 醫學要解), Nguyễn Đại Năng's 阮大能 (active during the Hồ dynasty, 1400–7), *Songs on the Swift Efficacy of Acu-Moxa* (Châm cứu tiệp hiệu diễn ca 針灸捷效演歌, in Nôm verses); Phan Phu Tiên's 潘孚先 (1370–1482) *Comprehensive Collection of Materia Medica and Food* (Bản thảo thực vật toản yếu 本草食物纂要, 1428; the oldest book on pharmacology and dietetics preserved in Vietnam); Nguyễn Trực's 阮直 (1417–1474) *Efficacious Formulas for Protecting Infants* (Bảo anh lương phương 保嬰良方, 1455; the oldest Vietnamese text on paediatrics); Hoàng Đôn Hoà's 黃敦和 *Collecting Essentials for Saving Lives* (Hoạt nhân toát yếu 活人撮要; later expanded and supplemented by the court physician Trịnh Đôn Phác, 1692–1762); and various works on gynaecology, paediatrics, and epidemics by Nguyễn Gia Phan 阮嘉璠 (1749–1829), a tiến sĩ degree holder who lived through three dynasties. More comprehensive lists of doctors with information on their works can be found in Dương 1947–50; Hoang *et al.* 1993; Chu 2002; Lâm Giang 2009; Nguyễn Thị Dương 2009; Mayanagi 2010a.

Further research

Much remains to be studied about the history of medicine in pre-modern Vietnam. A small but substantial amount of sources, written in Hán and representative of Chinese (or northern) medicine, dating back to the period between the fourteenth and nineteenth centuries has survived, but attracted hardly any attention from English-language scholars. The study of these texts in combination with other surviving sources, and in comparison with dynamics in medicine in other parts of East Asia, may yield invaluable insights into how Chinese medicine was adopted and adapted in Vietnam. Such an approach will offer more complex narratives and promises to go beyond the history of grand narratives that has dominated the field.

Bibliography

Aung-Thwin, M.A. and Hall, K.R. (eds) (2012) *New Perspectives on the History and Historiography of Southeast Asia: continuing explorations*, London; New York: Routledge.

Bates, A. and Bates, A.W. (2007) 'Lãn Ông (Lê Hữu Trác, 1720–91) and the Vietnamese medical tradition', *Journal of Medical Biography*, 15.3: 158–64.

Bretelle-Establet, F. (ed.) (2010) *Looking at It from Asia: the processes that shaped the sources of history of science*, Boston, MA: Springer.

Brindley, E.F. (2015) *Ancient China and the Yue: perceptions and identity on the southern frontier, c. 400 BCE–50 CE*, Cambridge: Cambridge University Press.

Cadière, L. and Pelliot, P. (1904) 'Première étude sur les sources annamites de l'histoire d'Annam', *Bulletin de l' École Française d'Extrême Orient*, 4.3: 617–71.

Chan, A.K.L., Clancey, G.K. and Loy, H.-C. (eds) (2002) *Historical Perspectives on East Asian Science, Technology and Medicine*, Singapore: Singapore University Press and World Scientific.

Chen, Y.-J. (2015) *Zhang ("Miasma"), Heat, and Dampness: the perception of the environment and the formation of written medical knowledge in Song China (960–1279)*, PhD thesis, University of Oxford.

Chu, L.T. (2002) 'An introduction to the history of traditional medicine and pharmaceutics in Vietnam', in A.K.L. Chan, G.K. Clancey and H.-C. Loy (eds) *Historical Perspectives on East Asian Science, Technology and Medicine*, Singapore: Singapore University Press and World Scientific, pp. 264–75.

De Vries, L.E. (2017) 'The Đồng Nhân pagoda and the publication of Mister Lazy's medical encyclopedia' in C.P. Salguero (ed.) *Buddhism and Medicine: an anthology of premodern sources*, New York: Columbia University Press, pp. 569–74.

Dror, O. and Taylor, K.W. (Introduction and Annotation) (2006) *Views of Seventeenth-Century Vietnam: Christoforo Borri on Cochinchina and Samuel Baron on Tonkin*, Ithaca, NY: Cornell Southeast Asia Program Publications.

Dumoutier, G. (1887) *Essai sur la pharmacie annamite: Détermination de 300 plantes et produits indigènes, avec leur nom en annamite, en français, en latin et en chinois et l'indication de leurs qualités thérapeutiques d'après les pharmacopées annamites et chinoises*, Hanoi: Imprimerie typographique F.–H. Schneider.

Dương Bá Bành (1947–50) *Histoire de la Médecine du Việt-Nam*, Hanoi: école française d'extrême-orient.

Durand, M. (1953) 'L' introduction des moxas en Vietnam d'après une légende de Linh-Nam Trich Quai', *Bulletin de la Société des Études Indochinoises*, 28.3: 295–302.

Hanson, M.E. (2011) *Speaking of Epidemics in Chinese Medicine: disease and the geographic imagination in late imperial China*, London: Routledge.

Henchy, J. (1998) *Preservation and Archives in Vietnam*, Washington, DC: Council on Library and Information Resources.

Higham, C. (1996) *The Bronze Age in Southeast Asia*, Cambridge: Cambridge University Press.

Hoang Bao Chau, Pho Duc Thuc and Huu Ngoc (eds) (1993) *Vietnamese Traditional Medicine*, Hanoi: Gioi Publishers.

——— (1993) 'Overview of Vietnamese traditional medicine', in Hoang Bao Chau, Pho Duc Thuc and Huu Ngoc (eds) *Vietnamese Traditional Medicine*, Hanoi: Gioi Publishers, pp. 5–29.

Holcombe, C. (2001) *The Genesis of East Asia 221 B.C.–907 A.D.*, Honolulu: Association for Asian Studies and University of Hawai'i Press.

Huard, P. and Durand, M. (1953) 'Lãn-Ông et la médecine sino-viêtnamienne', *Bulletin de la Société des Études Indochinoise*, 28.3: 222–93.

——— (1956) 'Un traité de médecine sino-vietnamienne du XVIIIe siècle: La compréhension intuitive des recettes médicales de Hai-Thuong', *Revue d'histoire des Science*, 2: 126–49.

Idema, W. (ed.) (2007) *Books in Numbers*, Hong Kong: Hong Kong University Press.

Kelley, L.C. (2005) *Beyond the Bronze Pillars: envoy poetry and the Sino-Vietnamese relationship*, Honolulu: Association for Asian Studies and University of Hawai'i Press.

——— (2006) '"Confucianism" in Vietnam: a state of the field essay', *Journal of Vietnamese Studies*, 1.1–2: 314–70.

Kim, N.C. (2015) *The Origins of Ancient Vietnam*, Oxford; New York: Oxford University Press.

Lâm Giang (ed.) (2009) *Tìm hiểu thư tịch y dược cổ truyền Việt Nam* (Research on Vietnamese Traditional Medical Texts), Hà Nội: Nhà xuất bản khoa học xã hội.

Li, T. (2011) 'The imported book trade and Confucian learning in seventeenth- and eighteenth-century Vietnam', in M.A. Aung-Thwin and K.R. Hall (eds) *New Perspectives on the History and Historiography of Southeast Asia: continuing explorations*, London and New York: Routledge, pp. 167–82.

Liu Chun-Yin 劉春銀, Wang Xiaodun 王小盾 and Trần Nghĩa 陳義 (eds) (2002) *Yuenan hannan wenxian mulu tiyao* 越南漢喃文獻目錄提要 (Annotated Catalogue of the Vietnamese Materials in Hán and Nôm), Taipei: Institute of Chinese Literature and Philosophy, Academica Sinica.

Marr, D.G. (1987) 'Vietnamese attitudes regarding illness and healing', in N.G. Owen (ed.) *Death and Disease in Southeast Asia: explorations in social, medical and demographic history*, Singapore: Oxford University Press, pp. 162–86.

Mangin, A. (1887) *La medicine en Annam*, PhD thesis, Faculté de Médicine de Paris.

Mayanagi Makoto 眞柳誠 (2006) 'Bedonamu Kokka toshokan no ko-iseki shoshi ベドナム國家圖書館の古醫籍書誌' (The Bibliography of Old Medical Books in the National Library of Vietnam), *Ibaraki daigaku jinbun gakubu kiyō: Jinbun gakka ronshū*, 45: 1–16.

——— (2010a) 'Bedonamu igaku keisei no kiseki ベトナム医学形成の軌跡' (Traces of the Development of Medicine in Vietnam), in Mayanagi Makoto (ed.) *Dainikai Ni Chū Kan igakushi gakkai gōdō shinpojiumu ronbun-shū* (Proceedings of the Second Joint Symposium of the Japanese, Chinese and Korean Societies for the History of Medicine), Mito: Dai 111 kai Nihon igakushi gakkai jimukyoku, pp. 65–75.

——— (2010b) 'Ni Kan Etsu no igaku to chūgoku isho 日韓越の医学と中国医書' (Japanese, Korean and Vietnamese medicine and Chinese medical books), *Nihon ishigaku zasshi*, 56.2: 151–9.

——— (ed.) (2010) *Dainikai Ni Chū Kan igakushi gakkai gōdō shinpojiumu ronbun-shū*, Mito: Dai 111 kai Nihon igakushi gakkai jimukyoku.

——— (2011) 'Bedonamu Kokka toshokan no ko-iseki shoshi hoi ベドナム國家圖書館の古醫籍書誌補遺' (The Bibliography of Old Medical Books in the National Library of Vietnam: Addendum), *Ibaraki daigaku jinbun gakubu kiyō: Jinbun komyunikēshon gakka ronshū*, 10: 21–39; 11: 51–73.

——— (2012–15) 'Bedonamu Kannomu kenkyūjo no ko iseki-shoshi ベドナム漢喃研究所の古醫籍書誌' (The Bibliography of the Old Medical Books in the Institute of Hán-Nôm Studies of Vietnam), *Ibaraki daigaku jinbun gakubu kiyō: Jinbun komyunikēshon gakka ronshū*, 12: 19–42; 13: 53–76; 14: 39–65; 15: 71–97; 16: 17–46; 17: 51–76; 18: 17–41.

McHale, S. (1999) 'Texts and bodies: refashioning the disturbing past of Tran Vietnam (1225–1400)', *Journal of the Economic and Social History of the Orient*, 42.4: 494–518.

——— (2007) 'Vietnamese print culture under French colonial rule: the emergence of a public sphere', in W. Idema (ed.) *Books in Numbers*, Hong Kong: Hong Kong University Press, pp. 377–412.

Monnais, L. (2012) 'Traditional, complementary and perhaps scientific? Professional views of Vietnamese medicine in the age of French colonialism', in L. Monnais, C.M. Thompson and A. Wahlberg (eds) *Southern Medicine for Southern People: Vietnamese medicine in the making*, Newcastle upon Tyne: Cambridge Scholars, pp. 61–84.

Monnais, L., Thompson, C.M. and Wahlberg, A. (2012a) 'Introduction: southern medicine for southern people', in L. Monnais, C.M. Thompson and A. Wahlberg (eds) *Southern Medicine for Southern People: Vietnamese medicine in the making*, Newcastle upon Tyne: Cambridge Scholars, pp. 1–19.

——— (eds) (2012b) *Southern Medicine for Southern People: Vietnamese medicine in the making*, Newcastle upon Tyne: Cambridge Scholars.

National Library of Vietnam, *Thư Viện Quốc Gia Việt Nam*, http://hannom.nlv.gov.vn, accessed 31/5/2018.

Nguyễn Thị Dương (2009a) 'Short introduction to Vietnamese medical texts in Hán Nôm (Giới thiệu khái quát về mảng thư tịch y dược Hán Nôm Việt Nam)', *Tạp chí Hán Nôm*, 92.1: 29–40.

——— (2009b) 'Vài nét về đóng góp của nho sĩ Việt Nam đối với việc tạo lập và truyền bá văn bản y học Hán Nôm' in Trịnh Khắc Mạnh, Phan Văn Các and Chu Tuyết Lan (eds) *Confucian Thoughts in Viet Nam: studies from an interdisciplinary perspective (Nghiên cứu tư tưởng nho gia Việt Nam: Từ hướng tiếp cận liên ngành)*, Hà Nội: Nhà xuất bản Thế giới, pp. 269–94.

Nguyễn Trần Huân (1951) *Contribution à l'étude de l'ancienne thérapeutique vietnamienne*, PhD thesis, Hanoi: École française d'Extrême-Orient.

——— (trans. and anno.) (1972) *Lãn-Ông: Thượng Kinh Ký-Sự* (Relation d'un voyage à la capitale), Paris: École française d'Extrême-Orient.

——— (trans.) (1973) 'La médicine Vietnamienne au XVIIIe siècle: texte original de Phạm Đình Hổ extrait du «Vũ Trung Tùy Bút»', *Bulletin de l' École française d'Extrême-Orient*, 60.1: 375–84.

Owen, N.G. (ed.) (1987) *Death and Disease in Southeast Asia: explorations in social, medical and demographic history*, Singapore: Oxford University Press.

Salguero, C.P. (ed.) (2017) *Buddhism and Medicine: an anthology of premodern sources*, New York: Columbia University Press.

Sallet, A. (1930) 'Un grand médecin d'Annam: Hai-Thuong Lan Ong (1725–1792)', *Bulletin de la Société Française d'Histoire de la Médecine*, 24: 170–8.

Schafer, E.H. (1967) *The Vermilion Bird: T'ang images of the south*, Berkeley; Los Angeles: University of California Press.

Shih, Virginia Jing-Yi and Tuyết Lan Chu (2006) 'The Han Nom digital library', in Trịnh Khắc Mạnh et al. (eds) *Ngiên cứu chữ Nôm: Kỷ yếu Hội nghị Quốc tế về chữ Nôm*, Hà Nội: Nhà xuất bản khoa học xã hội, pp. 366–78.

Stanley-Baker, M. and Salguero, C.P. (eds) (forthcoming) *Situating Religion and Medicine Across Asia*, Manchester: Manchester University Press.

Taylor, K.W. (2013) *A History of the Vietnamese*, Cambridge: Cambridge University Press.

Thompson, C.M. (2006) 'Scripts and medical scripture in Vietnam: Nôm and classical Chinese in the historic transmission of medical knowledge in pre-twentieth century Vietnam', in Trịnh Khắc Mạnh et al. (eds) *Ngiên cứu chữ Nôm: Kỷ yếu Hội nghị Quốc tế về chữ Nôm*), Hà Nội: Nhà xuất bản khoa học xã hội, pp. 255–70.

——— (2010) 'Sinification as limitation: Minh Mạng's prohibition on use of Nôm and the resulting marginalization of Nôm medical texts', in F. Bretelle-Establet (ed.) *Looking at It from Asia: the processes that shaped the sources of history of science*, Boston, MA: Springer, pp. 393–412.

——— (2012) 'Setting the stage: ancient medical history of the geographic space that is now Vietnam', in L. Monnais, C.M. Thompson and A. Wahlberg (eds) *Southern Medicine for Southern People: Vietnamese medicine in the making*, Newcastle upon Tyne: Cambridge Scholars, pp. 21–60.

——— (2015) *Vietnamese Traditional Medicine: a social history*, Singapore: National University of Singapore Press.

——— (2017) 'Selections from Tue Tinh (c 1330–c 1389) miraculous drugs of the south, the importance of local materia medica to Vietnamese traditional medicine', in C.P. Salguero (ed.) *Buddhism and Medicine: an anthology of premodern sources*, New York: Columbia University Press, pp. 561–8.

Trần Nghĩa (1993) 'Introduction générale', in Trần Nghĩa and F. Gros (eds) *Di sản Hán Nôm Việt Nam: Thư mục đề yếu/Catalogue des livres en Han Nôm* (vol. 1), Hà Nội: Éditions sciences sociales/Nhà xuất bản khoa học xã hội, pp. 48–77.

Trần Nghĩa and Gros, F. (eds) (1993) *Di sản Hán Nôm Việt Nam: Thư mục đề yếu/Catalogue des livres en Han Nôm* (3 vols), Hà Nội: Éditions sciences sociales/Nhà xuất bản khoa học xã hội.

Trịnh Khắc Mạnh et al. (eds) (2006) *Ngiên cứu chữ Nôm: Kỷ yếu Hội nghị Quốc tế về chữ Nôm*, Hà Nội: Nhà xuất bản khoa học xã hội.

Trịnh Khắc Mạnh, Phan Văn Các and Chu Tuyết Lan (eds) (2009) *Confucian Thoughts in Viet Nam: studies from an interdisciplinary perspective* (*Nghiên cứu tư tưởng nho gia Việt Nam: Từ hướng tiếp cận liên ngành*), Hà Nội: Nhà xuất bản Thế giới.

University of Northern Illinois, Southeast Asian Digital Library, http://sea.lib.niu.edu, accessed 30/5/2018.

Vietnamese Nôm Preservation Foundation, Digital Library of Hán-Nôm, http://lib.nomfoundation.org/collection/1/?uiLang=en, accessed 30/5/2018.

Wahlberg, A. (2014) 'Herbs, laboratories, and revolution: on the making of a national medicine in Vietnam', *East Asian Science, Technology and Society*, 8: 43–56.

Woodside, A.B. (1971) *Vietnam and the Chinese Model: a comparative study of Vietnamese and Chinese government in the first half of the nineteenth century*, Cambridge, MA: Harvard University Press.

34
HISTORY AND CHARACTERISTICS OF KOREAN MEDICINE

Yeonseok Kang

Translated by Jaehyun Kim

This is a summary and revision of 'The Characteristics of Korean Medicine Based on Time Classification' (Kang 2011). In the paper, I classified the history of Korean medicine into five periods: the period of loss of medical texts (the twelfth century and earlier); the period of local Korean herbal medicine (thirteenth to fifteenth centuries); the period of compilation of East Asian medicine (fifteenth to early seventeenth centuries); the period of independent development of Korean medicine (seventeenth to nineteenth centuries); and the period of exchanges between Asian traditional and Western medicine (twentieth to twenty-first centuries). Here I would like to introduce the latter four periods, which have had great influence in shaping contemporary Korean medicine.

Period of local Korean herbal medicine (thirteenth to fifteenth centuries)

Local Korean herbal medicine (*Hyangyak uihak* 鄉藥醫學) is a type of Korean medicine (KM) that prescribes mainly local Korean herbs.[1] The period which covers late Goryeo (thirteenth to fifteenth centuries) and early Joseon (fifteenth century) period was a time of active fusion of local medicine using domestic herbs with medical knowledge from overseas. The earliest medical text concerned with local Korean herbs is *Emergency Prescriptions Made with Local Korean Herbs* (*Hyangyak gugeupbang* 鄉藥救急方) published in 1236, which is also the oldest medical text in existence in Korea (Shin 1995; Suh 2017). Many books published since then have 'local Korean herbs' in their names, which shows that it was an important task for medical society at that time.

Local Korean herbal medicine has several characteristics:

a it mainly uses prescriptions made with local Korean herbs.
b it uses only one or two herbs.
c it uses raw and fresh materials.
d it focuses on using herbs that are quick and easy to get, and inexpensive.
e it aims for a high utilization rate in medically underserved areas.
f it applies the method of food therapy (Kang 2006).

In the thirteenth century, from 1231 to 1270, there was a war between Goryeo and Mongolia. Since importing medicinal herbs from the enemy country was not easy, *Emergency Prescriptions Made with Local Korean Herbs* was published to treat patients with herbal resources available in Korea.

With the arrival of the fourteenth century, circumstances changed: Goryeo began to adapt to a new world order of the Yuan dynasty (1271–1368). As Yuan culture became popular among the Goryeo people, the increase in imports of costly goods worsened the economy of the declining country. A medical official Bang Saryang, who was deeply involved in the medical policies of late Goryeo and early Joseon, advocated strengthening national power by stabilising its finances, in particular by banning luxury imports from China. Bang accordingly emphasised local Korean herbs and helped compile the *Compendium of Life-Saving Prescriptions Made with Local [Korean] Herbs* (Hyangyak jesaeng jipseongbang 鄉藥濟生集成方) in 1399. This book was intended to stabilise state finances, as well as to establish a national medical system during the foundation of the Joseon kingdom (Ahn *et al.* 2015b: 38–41; Chapter 8 in this volume).

The stabilisation of the Joseon dynasty in the fifteenth century brought another change in circumstance. King Sejong (r. 1418–50) ordered a survey of all goods, including domestically produced medicinal herbs, in order to levy taxes on them. For this purpose, he asked scholars to write geographical treatises which listed of these herbs, based on previous works of local Korean herbal medicine. He also endeavoured to build a national health system, by assigning doctors to underserved regions and encouraging medical education. The objective of local Korean herbal medicine in this period was to establish a nationwide medical system.

The *Compendium of Prescriptions Made with Local Korean Herbs* (Hyangyak jipseongbang 鄉藥集成方) was published in 1433, and covered 959 diseases, 10,706 prescriptions, 703 local Korean herbs, 211 ways of processing herbs, and 1,476 kinds of acupuncture and moxibustion methods. The vast amount of information in the book is the result of an active introduction of a number of Chinese prescriptions into local Korean herbal medicine. This use was at the same time very selective. All of the prescriptions in the work could be made up from a selection of 703 local Korean herbs (Kang 2006). For example, no matter how many Chinese prescriptions are cited in the book, no prescriptions include ephedra, cinnamon bark, aconite root, or liquorice root – all very popular ingredients in Chinese recipes but not local to Joseon.

The characteristics of *Compendium of Prescriptions Made with Local Korean Herbs* are summarised as follows:

a It shows how local medicine accepted and readjusted foreign medical knowledge.
b It is a compilation of East Asian prescriptions that are completely achievable with local Korean herbs.
c It contributed to the establishment of a nationwide medical system.
d It was in part adapted for military medicine, in that it was designed for treating patients without imported medications, and was republished frequently in times of war.

After the sixteenth century, medical texts specialising exclusively in local Korean herbs stopped appearing. This did not reflect a decline in local Korean herbal medicine, but rather, that there was no longer any perceived social need to exclude foreign *materia medica*. The prescriptions in the *Synopsis of the Medical World* (Uilim chalyo 醫林撮要), an early version of the comprehensive medical books that appeared later in the sixteenth and seventeenth centuries are more similar to local Korean herbal medicine than other comprehensive books

(Kim 2001). All herbology texts published after *Compendium of Prescriptions Made with Local Korean Herbs*, including the 'Herbology and Pharmacology' section of the *Treasured Mirror of Eastern Medicine* (*Dongui bogam* 東醫寶鑑), mark imported *materia medica* with the letter 'Dang' (唐), which means China. They clearly inherit this distinction from local Korean herbal medicine.

Period of compilation of East Asian medicine (fifteenth to early seventeenth centuries)

This third period was a time of compiling various treatment methods and knowledge from across East Asia. At the beginning of the period, from the fifteenth to sixteenth centuries, Joseon was an advanced, stable country with a highly developed economy, culture, and technology, until war with Japan (1592–98) brought many changes.

The representative text from this period is the *Classified Collection of Medical Prescriptions* (*Uibang yuchi* 醫方類聚), one of the most important projects carried out by the early Joseon dynasty. The book was completed in 1445 with 365 volumes, but not published right away. Rather, it underwent a long period of proofreading until 1477. During this proofreading period, about seventy officials were punished for not providing satisfactory results, while those who successfully finished their work were rewarded, indicating the high importance laid on the work by the government. It was finally published with 266 *juan* in eighty-seven sections, and is considered to have the greatest quantity of information on East Asian medicine among Korean medical works.

What's interesting about the *Classified Collection of Medical Prescriptions* is that each chapter is divided into sections on Theory, Prescription Methods, Food Therapy, Taboos, Guiding-pulling Exercises, and Acupuncture and Moxibustion, indicating that the authors tried to encompass as many treatment methods as possible. The sections on Food Therapy and Acupuncture and Moxibustion are especially worthy of notice. All three major medical books of Korea, the *Compendium of Prescriptions Made with Local Korean Herbs*, the *Classified Collection of Medical Prescriptions*, and the *Treasured Mirror of Eastern Medicine* include Acupuncture and Moxibustion sections in every chapter. The *Treasured Mirror of Eastern Medicine* has a Simple Prescription (*danbang* 單方) section in every chapter, which includes material on local Korean herbal medicine and food therapy.

Another characteristic of the *Classified Collection of Medical Prescriptions* is that the medical texts that it cites are arranged in chronological order, with very clear identification of the source names. It therefore serves as a very useful reference for the history of East Asian medicine, as many Chinese texts omit references to their sources. This work can be used both to identify parallel passages and to provide comparative dating of the texts. The book has been a great help in clarifying the publication dates of old medical texts, as well as in restoration of lost ones (Ahn 2000: 36–40). The *Compendium of Prescriptions Made with Local Korean Herbs* and *Treasured Mirror of Eastern Medicine* also notes the primary source in every sentence, demonstrating the academic precision of Korean medical writers.

The interest of neo-Confucianists in scientific knowledge and methods played a significant role in Joseon society at that time (Ahn *et al.* 2015a: 281–7, 353–64, 407–14, 433–55, 463–99, 2015b: 13–20). In the early fifteenth century, young neo-Confucian officials assigned to the Hall of Worthies (Jip-hyeon jeon 集賢殿) were engaged in a large compilation project covering all fields of study, including medicine. They organised these collections in a regularized bibliographic style that neo-Confucianists regarded as the proper academic method, called the 'division and subdivision system' (*Gangmok chae* 綱目體). The *Classified*

Collection of Medical Prescriptions, one of the products of the project, and other two major books which preceded and succeeded it, the *Compendium of Prescriptions Made with Local Korean Herbs* and the *Treasured Mirror of Eastern Medicine,* were all written following the format.

During the latter half of the fifteenth century, many medical texts dealing with communicable diseases or emergency diseases were published (Ahn 2000). These many publications on limited diseases in specialised fields were the result of systematic attempts by the Joseon government to control the outbreak of epidemics.

The *Annals of the Joseon Dynasty* records a severe epidemic that lasted from 1451 to 1452, which was so severe that the king had to consider moving the capital when it swept through Seoul. The chain of responsibility for monitoring and responding to this epidemic was allocated initially to local offices, then to the district minister, and from the minister to the prime minister as the epidemic became widespread. The control centre only despatched doctors in the first instance, but after the epidemic spread, the following measures were taken:

a isolation of patients in the Office for Saving People (*Hwalminwon* 活民院);
b provision of bath and poulticing facilities, heating, food, and medication for the patients;
c an information campaign informing the people that patients should go to the Office for Saving People;
d punishment of officials if people were found to be unaware of these instructions (Kang 2009).

The central government despatched doctors to observe a given epidemic, after which the doctors would return to report the symptoms. The royal doctors would then publish a customised medical text for the epidemic and the government would distribute it to the stricken area. Since these texts had to be published as soon as possible in order to treat a given disease, they inevitably dealt with specific diseases and were small in volume. This system of controlling epidemics became well-established and it may have been one of the reasons 'warm disease studies' (*wenbing* 溫病), which developed in China during the seventeenth and nineteenth centuries, were unpopular in Korea.

In the late sixteenth century, Yang Yesu published the *Synopsis of the Medical World*. This book and the *Treasured Mirror of Eastern Medicine* were closely related to the medical ideas of the Yuan and Ming, while the *Compendium of Prescriptions Made with Local Korean Herbs* and *Classified Collection of Medical Prescriptions* were more related to the medical ideas of the Song and Jin Dynasties (Chapters 8 and 9 in this volume). *The Synopsis of the Medical World* combined the contents of Korea's *Compendium of Prescriptions Made with Local Korean Herbs* and China's *Orthodox Transmission of Medicine* (*Yixue zhengchuan* 醫學正傳), *Introduction to Medical Studies* (*Yixue rumen* 醫學入門), and *[Treatments for] All Ailments and Restoring Youth* (*Wanbing huichun* 萬病回春).

The medical innovations from this period can be summarised as follows:

a Medicine became an important part of national policy;
b Neo-Confucianists led the establishment of a national health system and compilation of medical texts;
c This led to the use of the scholastic Division and Subdivision system in the organisation of medical texts;
d Medical texts clearly referenced earlier sources;
e Various treatment methods were collected in these works, ranging from prescriptions, acupuncture and moxibustion to food therapy and guiding-pulling exercise;

f The compilations of medicine of this period encompass all medical knowledge of the East Asian region of the time;
g Joseon developed a well-established epidemics control system.

Period of independent development of Korean medicine (seventeenth to nineteenth centuries)

The fourth period of the history of KM was a time for publication and rearrangement of the *Treasured Mirror of Eastern Medicine* as well as exploration of a new medicine. After wars with Japan (1592–98) and with the Qing empire (1627, 1636–37), Joseon became hostile to Qing culture, even though they were close in political terms and continued medical exchanges. There was little interchange with the outer world except the Qing, leaving the Joseon to form its own unique culture in the eighteenth century. It was not until the nineteenth century that direct exchanges with the West began.

Heo Jun published a book in his later life (1613) that had more effect on Korean medicine than any other medical texts, the *Treasured Mirror of Eastern Medicine*. In his 'Note to Readers', Heo explains the name of the book. He divides Chinese medicine into two groups: Northern Medicine of the Dongyuan 東垣 sect, and Southern Medicine of the Danxi 丹溪 sect (Chapter 9 in this volume). In addition, he named the medicine of Joseon 'Eastern Medicine' (*Dongui* 東醫) because he thought it to be a distinctive domain of medicine with a long indigenous history of its own making. He compared his book to a 'Treasured Mirror', since it brightly mirrors the true nature of diseases and the body. Since the compilation of this book, Korean medicine has become known as 'Eastern medicine', a title still used in North Korea today. The far reach of the *Treasured Mirror*'s reputation can be seen by the fact that it was published over thirty times overseas.

The *Treasured Mirror of Eastern Medicine* is classified into the following sections: Internal Bodily Elements; External Bodily Elements; Miscellaneous Disorders; Herbology and Pharmacology; and Acupuncture and Moxibustion. Diseases are classified into Internal or External Bodily Elements section by whether they occur inside or outside the body. The Miscellaneous Disorders section describes various diseases that cannot be explained using bodily elements. This kind of disease classification was first attempted by Heo, and many subsequent Korean medical texts followed this format.

The *Treasured Mirror of Eastern Medicine* is considered by Korean scholars and practitioners to have one of the finest classification systems in traditional East Asian medicine. The first three sections are divided into chapters according to: Elements of the Internal Body; Parts of the External Body; and Names of Miscellaneous Disorders respectively. Each chapter organises disorders with similar symptoms, and each disease includes subcategories such as Disease Terms, Theories and Disputations, Pulse Diagnosis, Treatment Methods, Prescriptions, Methods for Preserving Life, Acupuncture and Moxibustion, and Simple Prescriptions. This level of systematic classification is highly regarded and applicable for clinical and theoretical use.

Heo was well aware of the originality of his book's hierarchical structure. He devoted two of the twenty-five volumes to the table of contents, simply listing the names of chapters and subcategories (Kim *et al.* 2016: 18–21). This allows readers to sketch the overall character and structure of diseases and of the human body simply by familiarising themselves with the table of contents of the book.

The 'Miscellaneous Disorders' section begins with an introduction to the diagnosis and treatment of disorders. Interestingly, diseases and symptoms of diseases are addressed simultaneously, with instructions to consulting physicians that 'detection of a disease' should precede the 'differentiation symptoms' when forming a diagnosis. These instructions contradict a common misconception that traditional medicine concentrates on symptom differentiation and does not or need not confirm the disease.

The Herbology and Pharmacology section deals with 1,212 kinds of water, soil, animals, plants, and stones of medicinal use. Foreign materials are marked with the term 'Dang'. There are no specific indications for local Korean herbs, but Korean names of 637 medicinals are recorded.

The *Treasured Mirror of Eastern Medicine* made significant innovations:

a The detailed table of contents introduced a novel and sophisticated reorganisation of East Asian medicine;
b It advocates life-nurturing theory and practices from within a neo-Confucian outlook;
c It classifies diseases according to body parts (Ibid.: 19);
d It contains very precise observations of particular disorders (Cho 2009: 153–6);
e It performs a balanced overview of competing theories from various sects, laying a slightly greater emphasis on the theory of Yin Nourishment (Cha 2000);
f From this balanced overview, it presents reasoned arguments for the best theory on each particular disease, settling arguments and contradictions between the competing theories.

The *New Edition of Universal Relief* (*Jaejung sinpyeon* 濟衆新編 1799) and the *Compilation of Formulas and Medicinals* (*Bangyak hap-pyeon* 方藥合編 1884) both follow the disease classification of the *Treasured Mirror*. The former in particular was a summary of essential information from the immense *Treasured Mirror*. Both books replaced the vast amount of information in the Herbology and Pharmacology section with 'The Song of Medicinal Properties' (*Yakseong ga* 藥性歌), which comprised heptametrical poems, but still retained the *Treasured Mirror*'s practice of distinguishing between domestic and imported herbs.

While, on the one hand, medical authors in this period were occupied with reconstructing and editing the *Treasured Mirror*, others attempted to create new medicine to exceed its achievements. Lee Gyujun (1855–1923), a neo-Confucian scholar in the nineteenth century in Gyeongsang Province, wrote *Reconstructing Treasured Mirror of Eastern Medicine* (*Uigam jungma* 醫鑑重磨) and *Essentials of the Plain Questions* (*Somun daeyo* 素問大要). In the former, Lee strongly criticised the parts of the *Treasured Mirror* that emphasise the theory of Yin Nourishment over other theories. Also, in *Essentials of Plain Questions*, he suggests 'Yang Support Theory' (*buyang lon* 扶陽論), which focusses on the fire of the heart (also the mind, *shim* 心). This is considered to be a reinterpretation of the *Internal Classic* from a Korean perspective (Kwon 2010: 35–41). Followers of Lee's medical theories today still utilise his unique pulse diagnoses and prescriptions, and have formed a nationwide modern academic society of practitioners, based in Gyeongsang Province (Kim 2010). It may be worthwhile to carry out a comparative study on this school of thought with Xin'an 新安 Medicine, a local medical system in China during the Ming and Qing dynasties, because the two share many characteristics in common (Kwon 2010).

Lee Jema (1838–1900) founded a form of constitutional medicine, which emphasised a person's inner nature (心性) in his work, *Longevity and Life Preservation in Eastern Medicine* (*Dongui suse bowon* 東醫壽世保元). Like Lee Gyujun, Lee Jema emphasised the heart. He categorised constitutions according to the sizes of the lungs, spleen, liver, and kidneys. He excluded the heart from the five viscera, not because it is unimportant, but because it occupies a higher status than the others. His system is called 'Four Constitutional Medicine' or 'Sasang Constitutional Medicine' (*Sasang uihak* 四象醫學), and is the focus of many academic societies and books today (Kim 2010: 14–26). Lee regarded Heo Jun as one of the most important medical figures, along with Zhang Zhongjing 張仲景 (Chapter 8 in this volume). As well as being a constitutional medicine, his system can also be considered a Korean reinterpretation of the *Treatise on Cold Damage* (*Shanghan lun* 傷寒論), as it reanalyses and transforms many cold damage prescriptions.

Korean medicine began to develop in a very different direction from Chinese medicine (CM) after the publication of the *Treasured Mirror*. The tendency to reject Qing culture may have partly acted as an external factor, while confidence in the perfection of *Treasured Mirror* was an internal one. These resulted in some significant directions within Korean medicine.

First, the *Treasured Mirror* replaced Chinese classics of CM, such as the *Internal Classic* or *Treatise on Cold Damage*, and assumed their status as medical canon. This phenomenon inherits some trends in earlier KM texts. Neither the *Compendium of Prescriptions Made with Local Korean Herbs* nor the *Classified Collection of Medical Prescriptions* regarded these Chinese canons as important references, but primarily referred to works of the Tang and Song periods. In China, the Chinese classics went through a process of Evidentiary Verification (*kaozheng* 考證) in the fifteenth to seventeenth centuries, around the same time that the *Treasured Mirror* gained its elite status.

Second, the canonical status of the *Treasured Mirror* over the last 400 years has meant that it served as a ground for the standardisation of Korean medicine. It provided standards not only for disease classification, but also for the composition and quantity of drug prescriptions. Many prescriptions with the same name in Korea and China often have different compositions, quantities, or processing methods. These indicate that the canonical status of the recipes in China did not carry over during their transmission to Korea, where they were subject to further revision.

Third, there was a strong emphasis on pragmaticism in Korean medical writing and pedagogy. Confucian-influenced medical authors in China preferred to engage in cosmological reflection, and theoretical disputation in their explanations of the nature of the world. They emphasised mastery of the Confucian classics prior to studying medicine. By contrast, Joseon medical texts stress the description of diseases and symptoms over theoretical argument. Both the 'Note to Readers' (*Jip rye* 集例) in the *Treasured Mirror*, and the *Annals of the Joseon Dynasty* (*Joseon Wangjo Sillok* 朝鮮王朝實錄, 1415) – which records national policy on medical education – stress 'the study of *materia medica*' as a first step.

Fourth, Korean medicine placed an increasingly strong emphasis on categorising diseases according to their bodily locations. Instead of classifying them by divisions of external and internal pathogens, as in East Asian medicine, the opening chapters of the *Treasured Mirror* organise these locations into 'Internal Views' (*naegyeong* 內景) and 'External Forms' (*oehyeong* 外形). Diseases which East Asian texts normally catalogue as External Influences, such as the six pathogens (*liuqi* 六氣, e.g. wind, cold, summer-heat, dampness, dryness, and fire (Stanley-Baker and Lo Introduction in this volume) are instead sorted into the 'Miscellaneous Disorders' (*japbyeong* 雜病) section. This trend of emphasising the bodily regions and structures became more dominant in the nineteenth century, with the publication of *Reconstructing Treasured Mirror of Eastern Medicine*, which emphasises *yang qi* in the body (Kwon 2010) and *Longevity and Life Preservation in Eastern Medicine*, which is concerned with differentiating constitutions.

Fifth, research on cold damage or warm diseases was not popular as it was in the rest of East Asia. Instead, Korea held to the national system of controlling epidemics that had been stabilised in the early Joseon, and to the tradition of applying different treatment measures to different epidemics. During the same time, the study of warm disease was very popular in China, and research on *Cold Damage Treatise* made a great progress in Japan. The application of cold damage or warm disease theory to an epidemic means using standardised symptom differentiation and prescriptions. This stood in contrast to the Joseon government's policy of applying different methods to different epidemics. Additional research on this topic would be helpful, especially into the fact that Europeans landed later in Korea than in China and Japan; whether epidemics in Korea differed from those in other countries and, if so, how this influenced the emergence of a distinctive Korean tradition. A further point of difference is the strong emphasis on internal healthy *qi* rather than external pathogenic factors.

Period of exchanges between Asian traditional and Western medicine (twentieth to twenty-first centuries)

We now come to the fifth period of the history of KM. This period has seen times of struggle for KM to obtain a legal status in the national health system, and also to adapt itself to the field of modernised research. After 1905, Japan gained sovereignty over Joseon, and finally colonised it from 1910 to 1945. True modernisation began after the Second World War, and Korea has made startling economic progress over the past fifty years.

In 1900, before Japan took over the country, the *Regulations for Medical Doctors* (*Uisa gyuchik* 醫士規則) mandated that all those who learnt KM or Western medicine could be regarded as medical doctors. However, once Japan took over Joseon, doctors schooled in KM were no longer allowed to work in national hospitals, starting from 1907. This was because the Japanese Parliament had already decided in 1895 to eradicate traditional medicine from their national medical system (Chapter 35 in this volume). But the high cost of medical training made it prohibitive for Japan to produce Western medical doctors in colonised Joseon; this led to the promulgation of the *Regulations for Medical Students* (*Uisaeng gyuchik* 醫生規則) in 1913, which set up regulations for KM doctors. While Western medical doctors were referred to as 'doctors', Korean medical doctors were referred to as 'medical students'. The colonisation of Joseon widened the social gap between these two roles, leading to a mutual hostility that still exists in Korea today. The policy of bias against KM was reflected in educational institutions as well, and KM was not included in university education until independence in 1945 (Kim *et al.* 2006: 471–516).

Fortunately, the Korean National Assembly reintegrated KM into the national medical system in 1951, and 1952 saw the foundation of an educational institute of KM. Currently, there are twelve colleges of Korean medicine, eleven in private universities and one in a national university. There were 22,074 Korean medical doctors, over 600 professional researchers, 22,074 Korean medical clinics, and 234 Korean medical hospitals as of 2014. Approximately 850 Korean medical doctors graduate each year.

In 1987, KM was included in medical insurance coverage thanks to the constant demand for young KM doctors in the 1980s, when the medical insurance system was being stabilised. KM treatments continue to make up around 4–5 percent of total medical insurance coverage, despite its continued steady demand.

The unequal status of KM and Western medicine did not improve until the 1990s. In the 1980s, as more and more students with high grades began applying to Korean medical colleges, young students and practitioners began to criticise the lack of government policies on Korean medicine. The period from 1993 to 1997 saw numerous campaigns which targeted the government, such as long-term boycotts of lectures and street assemblies led by students and practitioners.

The government subsequently clarified its intention to raise the status of KM and enforced various policies. The Korea Institute of Oriental Medicine, the only government-funded research institute of Korean medicine, was founded in 1994. The government also established a bureau that deals with KM policies exclusively. Since 1998, KM doctors have been assigned as public health doctors in community health centres alongside medical doctors. The role of KM became even more prominent with the passing of the 'Bill for Fostering Korean Medicine' in 2003, and the appointment by President Roh Moo-Hyun of a doctor of Korean medicine as his family doctor. In 2007, a KM college was founded for the first time in a national university, at Pusan National University.

With the arrival of the twentieth century, the influence of Japanese medicine on KM was paradoxically stronger than that of Chinese medicine up until the 1970s. In the 1980s, the texts and education system of modern Traditional Chinese Medicine (TCM) were introduced to Korea. In addition, Western medicine was included in the six-year curriculum of Korean medical colleges, taking up more than half of the allotted class time. The status of *Treasured Mirror of Eastern Medicine*, the standard of KM for the last 400 years, is therefore at stake.

Since the 1990s, various debates and disputes have been carried out within and outside of KM circles on the topics of modernisation (standardisation, scientific experimentation, industrialisation) and globalisation of KM. First, this has led to increased conflict rather than collaboration among Korean medical doctors, biomedical doctors and pharmacists, because of limited resources. Second, there is a major argument ongoing between general practitioners of KM in clinics and specialists in hospitals, as well as between the younger and older generations, on how research and education should be oriented in this time of co-existence between traditional medicine and modern medicine and science. Third, competition exists between TCM and KM over who has the right to speak for East Asian medicine at various international forums. Fourth, disputes exist among the followers of *Treasured Mirror of Eastern Medicine*, Four Constitution (*sasang*) Medicine, and those who follow various assertions made by TCM or Japanese medicine within the realm of KM. If Korean medical doctors resolve conflicts between professions and generations reasonably, the various debates ongoing in Korean medical circles would be able to become a foundation for the birth of a new medicine.

Note

1 Local Korean herbs (*Hyangyak* 鄉藥) refers to medicinal herbs produced or cultivated in Korea. Some readers might think that this is a mistranslation since its literal meaning is 'countryside' or local herbs. In Korea, however, the term was used in contrast with 'Chinese herbs' (*Dangyak* 唐藥) to distinguish domestic herbs from imported ones.

Bibliography

Ahn Sangwu (2000) *A Bibliographical Researches on EuiBang-YooChui*, PhD thesis, Kyunghee University.
Ahn Sangwu, Lee Seona, Kang Yeonseok, Kim Namil, Kim HongGyun, Park SeongGyu, Shin Dongwon, Yu Hoseok, Lee Beongwuk, Lee Huihwan, Jeong Jaeseo, Cha WungSeok and Hong Seyeong (2015a) *The Founders of Korean Medicine 1*, Seoul: Munsacheol Publisher.
——— (2015b) *The Founders of Korean Medicine 2*, Seoul: Munsacheol Publisher.
Cha Wung Seok (2000) 'Comparative studies on EuiHakIpMun and DongEuiBoGam', *Journal of Korean Medical History*, 13.1: 111–28.

Cho Sun Young (2009) *The Medical History of SoGal* (消渴), PhD thesis, Kyunghee University.

Kang Yeonseok (2006) *The Study of HyangYak Medicine through HyangYakJipSeongBang*, PhD thesis, Kyunghee University.

——— (2009) Rewriting the history of Korean medicine: endeavors of Korean medicine to fight epidemics, *Ohmynews*, November 4th, http://www.ohmynews.com/NWS_Web/Mobile/at_pg.aspx?CNTN_CD=A0001252862, accessed 22/3/2017.

——— (2011) 'The characteristics of Korean medicine based on time classification', *China Perspectives*, 3: 33–43.

Kim Giuk, Kim Namil, Kim Dohun, Kim Yongjin, Kim HongGyun, Kim Hun, Maeng Woongjae, Park Gyeongnam, Park Hyeonguk, Bang JeongGyun, Shin Yeongil, Ahn Sangwu, Eon Seokgi, Eun Seokmin, Lee Beongwuk and Cha Wung-Seok (2006) *General History of Korean Medicine*, Seoul: Daeseong Publisher.

Kim HongGyun (2001) 'The influence of HyangYakJipSungBang on later generations', in Ahn Sangwu, Choi Hwansu, Kim Hongjun, Lee Jeonghwa and Kang Yeonseok (eds) *Establishment of Database on HyangYakJipSungBang*, Daejeon: Korea Institute of Oriental Medicine, pp. 105–86.

Kim Namil, Yoshida kazuhiro 吉田和裕, Sakai shizu 酒井シズ, Kang Yeonseok, Kim Ho, Kim Seonsu, Bang Seonhye, Jang Wuchang, Jo Namho, Jeong Wujin, Jeong Jihun and Flowers, J. (2016) *Knowledge System of 'Treasured Mirror of Eastern Medicine' and Medical Sciences in East Asia*, Seongnam: The Academy of Korean Studies.

Kim Taewoo (2010) *Medicine without the Medical Gaze: theory, practice and phenomenology in Korean medicine*, PhD thesis, State University of New York, Buffalo.

Kwon OhMin (2010) *A Study on Shukgok Lee Gyu-Jun's Medical Ideas*, PhD thesis, Kyunghee University.

Shin Youngil (1995) *Research on Compendium of Prescriptions from the Countryside*, PhD thesis, Kyunghee University.

Suh, S. (2017) *Naming the Local: medicine, language, and identity in Korea since the 15th century*, Cambridge, MA; London: Harvard University Asia Center.

The Annals of the Chosun Dynasty (1415), The annals of the Joseon Dynasty, Jan. 16th, http://sillok.history.go.kr/id/kca_11501016_001, accessed 22/3/2017.

35
CHINESE-STYLE MEDICINE IN JAPAN

Katja Triplett

Chinese-style medicine became the dominant style of medicine in Japan from the period of early state-building in the fifth century CE onwards. It is only since the middle of the nineteenth century that it has been called *Kanpō* 漢方, literally the 'Han method, or formulæ' of medicine. The history of Chinese-style medicine in Japan can be roughly divided into three periods. These periods are: early encounters with Chinese-style medicine and initial adaptations; the development of autonomous medical traditions; and the decline and revival of Chinese-style medicine in modern Japan (Michel-Zaitsu 2017). Medical ideas and practices reflect those in China but as they reached Japan in an irregular if not erratic fashion, Japan developed independent and unique approaches (Rosner 1989). Periods of intense exchange between Japan and her neighbours alternated with phases of near complete termination of these exchanges following the breakdown of diplomatic and trade relations in times of war and strife. Adaptation of ideas from European doctors and scientists in the early modern era, primarily from the seventeenth century onwards (Otori 1964; Bowers 1970), resulted in an even more hybrid system of medicine in Japan. These developments can, in part, be explained by the location and geography of the Japanese island empire.

The global trade conducted in Nagasaki, Ōsaka, Kōbe, Edo (today's Tōkyō) and other places has always included the import of raw materials for Chinese-style medicine as practised in Japan. Many of the plants, animal products and minerals necessary for practising this form of medicine had to be imported because they could not be found widely or at all on the Japanese archipelago. While attempts to cultivate the requisite plants in the seventeenth century were somewhat successful, Japan continued to depend on imported raw materials for medicines.

From the fifth century, immigrants, including doctors, monks and nuns, smiths and other craftsmen, settled in Yamato. The first doctors came from the Korean kingdoms of Silla, Baekje and Goguryeo. As part of early state-building, the Yamato monarchs, who called themselves 'heavenly rulers' (*tennō* 天皇), introduced Chinese-style forms of bureaucracy. New institutions were established including a Bureau of Medicine (Ten'yakuryō 典薬寮, renamed Yakuin 薬院 in the sixteenth century), a medical school and medicinal gardens (Chapters 6 and 7 in this volume). The eighth-century legal codes and their commentaries mention doctors who were to practise Chinese-style acupuncture, moxibustion, herbology, exorcism and massage (Hattori 1945).

Another important development in the history of medicine in Japan was the introduction of Buddhism from Baekje (Chapter 28 in this volume). Following a period of conflict with practitioners of Shintō 神道, the established religion devoted to the worship of local deities, Mahāyāna Buddhism became the dominant religion in Japan (Bowring 2005: 15–35). The Japanese started to practise Buddhism in its East Asian version in combination with Shintō. Indian medical ideas arrived in Japan with Buddhist writings that were often direct translations from Sanskrit into Chinese. There is no convincing record of direct encounters between Indian and local doctors in Japan, as there is for medieval China (Unschuld 1998: 64). Indian medical ideas were studied but were generally not put into practice. An exception here is that Indian-style eye surgery (cataract couching) was practised, and indeed became widely popular in medieval and early modern Japan (Mishima 2004: 65; Triplett 2017, Triplett 2019: 80–91). Buddhism remained a prevailing cultural force until the end of the sixteenth century and has continued to be influential in Japanese culture and society. As centres of learning and erudition, Buddhist temples in Japan offered training in Chinese-style medicine while state institutions also educated doctors. Secular court doctors were organised in family lineages. Both groups of medical practitioners, Buddhist monastics and court doctors, engaged in healing members of the ruling elite while Buddhist monastics also treated other members of medieval and early modern society (Hattori 1982; Shinmura 2013; Andreeva 2017; Triplett forthcoming).

Professionalisation through private and state academies in the Neo-Confucian style as well as an increasing differentiation between the religious and the secular sphere in Japanese society resulted in the emancipation of qualified doctors from the seventeenth century onwards. The new government that came to power in 1868 decreed an official separation of Buddhist and medical practice in 1874 in order to separate these two spheres more strictly. The introduction of Western medical systems and state licensing in this period of 'modernisation' led to the decline of traditional medicine in Japan. *Kanpō* medicine has enjoyed a revival since the 1950s after various efforts to obtain official recognition.

Early encounters with Chinese-style medicine in Japan and first adaptations

When the central state of Yamato became fully established in the eighth century CE, the *tennō* ordered the composition of official chronicles following the Chinese model. The *Record of Japan* (*Nihon shoki* 日本書紀, 720) mentions the activities of doctors from the Korean Peninsula. During the reign of Kinmei tennō 欽明天皇 (509–571, r. 539–571), contact with the Asian continent intensified. The *tennō* invited a doctor, herbalists and others specialising in various sciences from Baekje to Japan to exercise their skill and train local students. The Chinese-style medical arts practised in the Korean kingdoms were greatly admired but medical texts and medicines also reached Japan directly from China. Treatises such as the Sui-dynasty *Treatise on the Origins and Symptoms of All Disorders* (*Zhubing yuanhou lun* 諸病源候論, compl. 610), the *Arcane Essentials from the Imperial Library* (*Waitai miyao*, or *Waitai biyao* 外臺秘要, 752) by Wang Tao 王燾 as well as Sun Simiao's 孫思邈 (581–682) *Important Formulas Worth a Thousand* (*Qianjin yaofang* 千金要方, 650–659) had an important impact on Japanese medical culture (Chapter 8 in this volume). The early eighth-century legal codes – the no longer extant Taihō Code 大寶律令 and the later Yōrō Code 養老律令 that is preserved in the commentary *Ryō no gige* 令義解 of 833 (transl. Dettmer 2010) – regarding the medical training of official medical personnel closely followed the *Penal Code of the Tang* (*Tanglü shuyi* 唐律疏義). The Japanese codes prescribed several standard medical textbooks (*Ryō no gige* book 8, part 24, sections 3, 4, 11 and 13; Dettmer 2010: 438–42): the third-century *Numbered Classic of the [Yellow Emperor]* ([*Huangdi*] *jiayijing* 黃帝甲乙經), the as yet unidentified *Pulse Classic* (*Maijing* 脈經) from the same period and *Materia Medica* (*Bencao* 本草), the latter probably being the Tang-dynasty *Revised Materia Medica* (*Xinxiu bencao* 新修本草). Works that students had to study in Japan included many books that are now lost or unknown. Others are still standard works for medical students today. The legal code lists the following books: the *Lesser Grade Remedies* (*Xiaopinfang* 小品方), *Collection of Experiential Formulas* (*Jiyanfang* 集驗方) and *Luminous Hall* (*Mingtang* 明堂) – all now lost – the *Needling Classic of the Yellow Emperor* (*Huangdi zhenjing* 黃帝針經), which may refer to the *Lingshu*, as well as the extant *Basic Questions* (*Suwen* 素問) and *Pulse Diagnostics* (*Maijue* 脉訣). The latter was most probably an alternate title for *Stratagems for Taking the Pulse* (*Maijingjue* 脈經訣) (Dettmer 2010: 441n26). Students were also to memorise the images used in the *Flowing Commentary* (*Liuzhu* 流注), a work that cannot be identified today, and in the equally unknown *Images from the Side while Lying Down* (*Yancetu* 偃側圖). The list is concluded by the now lost *Divine Needling Classic in Red and Black [Ink]* (*Chiwu shenzhenjing* 赤烏神針經). Later Japanese sources indicate that the educational material varied in some respects but that, overall, the training closely resembled Tang-dynasty education in this early period. Bureaucratic organisation developed in a different way from that intended by the *tennō*. In Japan, the practice of medicine remained for the most part hereditary and tied to family clans not to offices obtained by intellectual or moral achievement.

The Bureau of Medicine was also in charge of training exorcists who applied their Daoistic incantations within the field of medicine. However, Daoist institutions did not take hold in Japan and Daoist-style rituals were largely absorbed into Buddhist rituals (Lomi 2014). Interestingly, the medical regulations of the early legal codes prohibited Buddhist monks and nuns from using Daoist incantations but allowed Buddhist healing spells. Monastics strove to study medicine as one of the traditional 'five sciences' according to the Indian tradition. Charismatic monks frequently appear as healers in Japanese sources (Kleine 2012). In the Buddhist world, the training in the 'five sciences' was thought to be necessary for compassionate acts in society. In Japan, monks and nuns particularly conducted healing rituals for the ruling elite. The fierce competition between monastic healers and the court doctors in regard to caring for the elite is attested to in diaries of courtiers in the twelfth and thirteenth centuries (Drott 2010).

Empress Kōmyō 光明皇后 (701–760), who is known to have emulated the Chinese (female) ruler Wu Zetian 武則天 (625–705), piously supported Buddhist institutions and founded a dispensary for medical drugs. The dispensary offered free medicines for all. She also donated a number of precious raw materials to the magnificent temple Tōdaji 東大寺 her husband Shōmu tennō 聖武天皇 (701–756, r. 724–749) had established. Nearly forty of the sixty materials that were originally donated survived over 1200 years in the Shōsōin 正倉院, the treasure house of the temple. Modern surveys have established that some of these drugs came from Persia, India, China, Thailand, Korea and other places in Asia (Kunaichō and Shibata 2000, Torigoe 2005). In times when epidemics ravaged the land, demand for free medicines increased to such an extent that a sufficient amount could not be sourced from the medicinal garden in Nara and imports from abroad. The government purchased plants gathered in the wild and those grown in the provinces in order to meet demand (Ueda 1930: 7–8). Textual records such as the collection of rules and procedures for governmental agencies, the *Procedures of the Engi Era* (*Engishiki* 延喜式, 927), list over fifty different medicinal herbs to be gathered from the wild or cultivated in the provinces and sent to the capital (which was Kyōto since 794). This list of over 200 drugs makes this work an important source on early medieval pharmaceutical knowledge in Japan (Hattori 1955).

The tenth century also saw the compilation of a *materia medica* compendium *Japanese Names for the Materia Medica* (*Honzō wamyō* 本草和名, 901–923) by the court physician Fukane Sukehito 深根輔仁 (dates unknown) (Karow [1948] 1978) and an early medical compilation in Japan, *Essentials of Medical Treatment* (*Ishinpō* 医心方) by the court physician Tanba Yasuyori 丹波康頼 (912–995) which he presented to the *tennō* in 984. Tanba Yasuyori belonged to one of the two most important family clans of doctors, the second one being the Wake 和気. Both the Tanba and the Wake attempted to keep the coveted knowledge gleaned from Chinese sources and their own clinical notes secret for centuries. *Essentials of Medical Treatment* is a compendium of a wide range of Chinese medical sources – many of them now lost – with comments for the local readership in Japan, making it an important source for the reconstruction of early Chinese medicine. Tanba Yasuyori also included Japanese names for plants to enable doctors and herbalists to identify medicinal plants in nature. The *Shingon* 真言 monk Fujiwara Ken'i (or Kenni, 藤原兼意, 1072–ca. 1169) compiled important compendia of precious substances, incense, drugs and cereals (Triplett 2012). These, often exotic, substances were used in Buddhist rituals and exact knowledge of their identity was indispensable for the proper execution of the rituals.

Song-dynasty works reached Japan in the twelfth century but these did not immediately exert much influence there. One notable exception was the monastic physician Kajiwara Shōzen 梶原性全 (or Jōkan 浄観, 1265–1337) (Goble 2009, 2011) who belonged to a reforming Buddhist movement founded by the charismatic monk Eison 叡尊 (1201–1290) and known for its humanitarian engagement. This movement, the Shingon Ritsu school (Shingon Risshū 真言律宗), established a large temple hospital and other care facilities for both humans and animals in Kamakura, at the seat of the military ruler at the time (Hattori 1964). Contact with China flourished and new influences reached Japan via trade, personal visits and study stays. Myōan Eisai 明菴栄西 (or Yōsai, 1141–1215), one of the Japanese who spent an extended period in China gained the support of Hōjō Masako 北条政子 (1156–1225), a powerful female leader of the military regime in Kamakura. With her support, he built the first temple propagating Meditation, or Zen (Ch. Chan 禪) Buddhism in Kamakura, the seat of the de facto government at the time, was established. Eisai not only propagated Zen Buddhism but also recommended the consumption of green tea. Among his writings is the *Record on Drinking Tea for Nourishing Life* (*Kissayōjōki* 喫茶養生記, 1211) (trans. Benn 2015: 157–71).

Takeda Shōkei 竹田昌慶 (1338–1380), who spent nearly ten years in China, studied under a Daoist named Jin Weng 金翁 and returned not only with his teacher's daughter as his wife but also with an ingenious Chinese invention: a doll for the practice of needling, usually called 'bronze doll' (*dōningyō* 銅人形) or 'channels doll' (*keiraku ningyō* 經絡人形) (Chapter 12 in this volume). He founded an influential school of physicians and rose to become the personal doctor of a member of yet another clan of military rulers: the Ashikaga. During the Ashikaga period, the seat of the de facto government returned to Kyōto and there was constant civil strife between local rulers in the provinces. Still, in this period, Japan developed several unique medical traditions that increasingly departed from the Chinese model although the basic outlook and the taxonomies remained intact during the late medieval and the early modern eras (sixteenth to nineteenth centuries) (Hattori 1971).

Tashiro Sanki 田代三喜 (1465–1537), who spent twelve years in China, was responsible for making Jin-Yuan dynasty medicine popular in Japan, particularly the theories of the scholar Li Dongyuan 李東垣 (or Li Gao 李杲, 1180–1251) and the works of the physician Zhu Danxi 朱丹溪 (1281–1358) (Chapter 9 in this volume). Both emphasised therapies of balancing and calming the *qi*. These therapies were adapted and established in Japan, especially by Tashiro Sanki's student Manase Dōsan 曲直瀬道三 (1507–1594) and his successors

(Machi 2014). This school of medicine is today called the 'School of the medical methods of the later epoch' (Goseihōha 後世方派) because its proponents departed from earlier Song-dynasty medical theory and concentrated on the later epoch of the Jin-Yuan dynasties.

Development of autonomous medical traditions

After many decades of war between local factions in Japan and failed attempts to conquer Korea, a period of political stability began in 1600 with a new military ruling clan, which had its seat in Edo. The shōgun (military ruler) and his military court attracted advisors and scholars in the Neo-Confucian tradition of Zhu Xi 朱熹 (1130–1200). Stagnation of *qi* became a major concern in this phase in Japanese medicine as state control began to dominate all areas of life and stagnation and blockages were part of one's daily experience (Michel-Zaitsu 2017: 185). Most doctors and thinkers advocated staying within the tightly controlled social spaces and recommended that people did their utmost to preserve or increase their vitality in order to fulfil their social role in an increasingly Confucian society (Ahn 2012).

One of the shōgun's most influential advisors was Hayashi Razan 林羅山 (or Dōshun 道春, 1583–1657) (Marcon 2015). He purchased Li Shizhen's 李時珍 (1518–1593) monumental *Compendium of Materia Medica* (*Bencao gangmu* 本草綱目) in Nagasaki and wrote a glossary for Japanese readers. The *Bencao gangmu* by Li Shizhen and the *Cure of Myriad Diseases* (*Wanbing huichun* 萬病回春, 1587) by Gong Tingxian 龔廷賢 (or Yunlin 雲林, active 1577–1593) were other more comprehensive medical works from China that were printed and adapted in Japan.

The shōgunate ordered plant explorations in Japan to gather or complement knowledge on the natural resources of the country. There was also an effort to cultivate valuable ginseng and other useful plants. Japan's official policy to become independent from imports resulted in the revival and the establishment of medicinal gardens in many parts of the country. The government also ordered herbals and other books from Europe. While Rembert Dodoens' (1517–1585) herbal *Cruijdeboeck* (1554) was not held in very high regard in Europe itself, a copy of this book incited much interest in Japan (Vande Walle and Kasaya 2001). 'Dutch studies' (*rangaku* 蘭學), as the study of European medicine and sciences was called, became firmly established in this period.

The influx of new ideas in the field of medicine was not limited to those developed in Europe. The eighth shōgun Yoshimune 吉宗 (1684–1751) who was a particularly active supporter of the advancement of medicine and the natural sciences, ordered that the Korean compendium *Treasured Mirror of Eastern Medicine* (*Dongui bogam* 東醫寶鑑, 1613) be adapted for a Japanese readership and printed in Japan (Chapter 34 in this volume). The resulting work was printed in 1724 and 1730, with a second edition printed in 1799 (Mayanagi 2004). Scholars in Japan also published new *materia medica* compendia and other medical works (Hübner 2014). One of the pioneers in this field, the Neo-Confucian Kaibara Ekiken 貝原益軒 (or Ekken, 1630–1714), published the herbal *Japanese Materia Medica* (*Yamato honzō* 大和本草, 1708–1709). Kaibara Ekiken also wrote on 'nourishing life' (*yōjō* 養生), or self-cultivation, which was an exceptionally popular topic in the early modern period. His Japanese herbal was based on Li Shizhen's work but introduced a different classification system. The most comprehensive Japanese herbal, *Illumination of the Principles and Varieties of Materia Medica* (*Honzōkōmoku keimō* 本草綱目啓蒙, 1803), was compiled by the doctor and scholar Ono Ranzan 小野蘭山 (1729–1810) who was knowledgeable in both East Asian and European medicine.

Another famous doctor of this dynamic period, Nagata Tokuhon 永田徳本 (1513–1630), emphasised the importance of the Han-dynasty classic *Treatise on Cold Damage* (*Shanghan lun* 傷寒論) by the doctor Zhang Zhongjing 張仲景 (150–219) (Chapter 8 in this volume). Japanese physicians such as Nagoya Gen'i 名古屋玄医 (1628–1696), Gotō Konzan 後藤艮山 (or Gonzan, 1659–1733) and several others even saw the approaches in the *Treatise on Cold Damage* as the only authoritative medicine. Interestingly, proponents of this school of medicine in Japan, later to be called the 'School of the old medical methods' (Kohōha 古方派), criticised Zhang Zhongjing, the author of this important Chinese medical work. This school distanced itself from some fundamental principles such as concepts of yin and yang and the channels. Gotō Konzan and others can be said to have 'Japanised' Chinese medicine (Otsuka 1976: 328).

Misono Isai's 御園意斎 (1557–1616) unique acupuncture style of applying the needle to the abdomen with a small hammer, the 'tapping acupuncture insertion technique' (*dashin-hō* 打鍼法), originally invented by the sixteenth-century Zen monk Mubun 無分 (dates unknown), is practised to this day (Michel-Zaitsu 2017: 79–80). Mubun envisioned the abdominal region as a map of the internal organs. Misono Isai developed a specific diagnostic technique of palpating the abdominal region, based on Mubun's ideas. In general, abdominal palpation (*fukushin* 腹診) departs from Chinese models and is thus a Japanese-specific technique.

Sugiyama Wa'ichi 杉山和一 (1610–1694) relied heavily on palpation, touching and feeling when examining his patients because he was blind. The masseur and acupuncturist is the inventor of a technique that involves using a small tube to direct and stabilise the acupuncture needle. Sugiyama's 'guide tube acupuncture insertion method' (*kanshinhō* 管鍼法) and other needling techniques such as using very fine needles are widely employed to this day. He founded a school to train the visually impaired in acupuncture and became a doctor of high renown. Before Sugiyama, most visually impaired individuals who wanted to work in the medical field could only train in Chinese-style massage (*anma* 按摩).

Understanding of the body was also altered by anatomical knowledge imported from Europe and local observations via autopsy and dissection. This knowledge especially served surgeons such as Takashi Hōyoku 高志鳳翼 (fl. eighteenth century) whose illustrated treatise *Precious Notes on the Medical Therapy of Bone-setting* (*Honetsugi ryōji chōhōki* 骨継療治重宝記, 1746) can be said to be the foundation of osteopathy in Japan. It combines traditional Chinese and European knowledge but also includes Takashi's personal observations (Michel-Zaitsu 2017: 170–1).

The above-mentioned Gotō Konzan who explored new avenues in the medical field suggested that one of his students, Yamawaki Tōyō 山脇東洋 (1706–1762) dissect an otter to clarify the ambiguities about the inner organs encountered in the various Chinese sources. Yamawaki, however, was not satisfied by the results and obtained permission to organise the dissection of the body of an executed man. His exploration of the organs of the man opened the way for other autopsies in Japan.

Yoshimasu Tōdō 吉益東洞 (1702–1773) who developed innovative diagnostics and medical therapies is another remarkable doctor of this epoch, and is even regarded as the 'father' of modern *Kanpō* medicine (Trambaiolo 2015). He followed a radical path that initially alienated him from other doctors until Yamawaki Tōyō started to support him. Yoshimasu Tōdō claimed that 'all diseases have one poison as their single cause' (*manbyō ichidoku* 萬病一毒). According to him, this 'poison' triggered the various diseases by attacking different parts of the body. A doctor must localise the 'poison' in the individual body of the patient and treat it with 'poison'. He therefore started classifying the traditional Chinese-style *materia medica* according to their degree of 'toxicity', clearly departing from Zhang Zhongjing who considered the effect of the entire formula and not of its individual ingredients.

The surgeon Hanaoka Seishū 華岡青洲 (1760–1835) used careful observation to look for useful narcotic substances. Hanaoka is known to have engaged in experiments on both animals and humans. He sought the Han-dynasty Chinese doctor Hua Tuo's 華佗 (d. 220) legendary formula for inducing general anaesthesia, the enigmatic 'hemp boiled powder' (Ch. *mafeisan* 麻沸散). The decoction Hanaoka eventually used, the 'Powder for Communicating with Transcendents' (*tsūsensan* 通仙散), consisted of a mixture of extracts from parts of highly toxic plants. He used it in hundreds of operations, mainly related to breast cancer. The formula remained within the closed circle of his students.

Other eighteenth-century developments include an increased interest in bloodletting and bathing therapies, in the production of simplified and easily accessible medical drugs and in the printing of material on medical therapies and self-cultivation ('nourishing life') directed at a more general audience. The flourishing book market saw new editions, sometimes pioneering print editions of authoritative medical texts from the medieval period. Most personal medical manuscript notes, however, remained within the schools or lineages of transmission, although some students did publish their notes on their master's teachings. The students' master was often their father.

Medical practice was not a purely male pursuit. As women of the elite normally possessed a higher education in the Chinese and Japanese classics, calligraphy, poetry and other subjects, they contributed to intellectual life although they were dissuaded from taking a public role. Sources on female medical activities are scarce but some women's lives are better documented than others. We know that the *haikai* 俳諧 (short poem, haiku) poet Shiba Sonome 斯波園女 (1664–1726), a student of the famous poet Matsuo Bashō 松尾芭蕉 (1644–1694), for instance, practised ophthalmology alongside her husband. The learned Nonaka En 野中婉 (1660–1725) worked as a doctor after she was released in 1703 following four decades under house arrest due to kin punishment (Sanpei and Nihon joishi hensai iinkai 2008).

While many doctors continued to practise within a Buddhist framework, the numbers of those who were tonsured but not ordained and those without any clerical rank increased throughout the Edo period (1600–1868) (Hattori 1978). Tensions between these groups rose and controversies intensified over the extent to which scholars, including medical doctors, were to follow or depart from China as the perennial model. Interestingly, critical philological research blossoming in eighteenth-century China also took root in Japan. The government Institute of Medicine (Igakkan 医学館) in Edo followed it and started systematically to collect and also edit medical texts. This philological trend also inspired nativist movements.

Decline and revival of Chinese-style medicine in modern Japan

The new state under the 'reinstated' Meiji tennō 明治天皇 (1852–1912) saw the introduction of an official qualification system for medical doctors trained at institutions modelled after German clinics and universities. Doctors who were influenced by the 'Dutch studies' medicine or the empiricist Kohōha promulgated and easily adapted to the new system. However, the new developments led to the marginalisation of traditional Japanese medicine that was based on Chinese-style medicine but, as we saw above, was of a highly pluralistic and heterogeneous nature. Nativist trends after the pervasive Meiji-period reforms resulted in nationalistic, xenophobic, even racist forms of medicine. These had roots in an earlier movement called the 'way of the ancient [Japanese] medicine' (*ko'idō* 古医道) (Karow and Weller 1954). As the Meiji state established a form of state ritual for the *tennō*, the Emperor and his divine ancestors, the Society for the Preservation of Knowledge (Onchisha 溫知社), founded in 1879, received nationwide attention and support despite the fact that European biomedicine was already firmly established in Japan. A more politically active association, founded in 1891, fought for the full acceptance of traditional Japanese Chinese-style medicine, which tended to be called 'imperial Chinese [-style] medicine' (*kōkan igaku* 皇漢医學), and a revision of the qualification law. However, the Sino-Japanese War of 1894–1895 that ended in China's defeat – allegedly caused by the backwardness of the once greatly admired neighbour – ended these activities.

Some doctors who were in favour of the Western-style approbation law also engaged in the preservation of tradition by, for instance, trying to develop a curriculum for the study of Japanese Chinese[-style] medicine (*wakan igaku* 和漢医學) adapted to the Western-style programme. This attempt failed because of the inherent differences between the European and Chinese medical systems and their foundational paradigms (Oberländer 1995: 71–2). There were also attempts to harmonise 'traditional' and 'modern' knowledge in order to improve the health of the people in Japan, and even reach an international audience. Traditional concepts of self-cultivation ('nourishing life') and nutrition as expounded by Kaibara Ekiken played an outstanding role in these novel formulations. Sakurazawa Yukikazu 桜沢如一 (or Nyoichi, 1893–1966) alias George Ohsawa who developed the 'macrobiotic diet' must be mentioned in this context. Sakurazawa translated into French carefully selected parts of an influential 1927 work by the ultra-nationalist Nakayama Tadanao 中山忠直 (1895–1957) for the Sinologist and writer George Soulié de Morant (1878–1955). The resulting book, *Acupuncture et médecine chinoise vérifiées au Japon* (Acupuncture and Chinese medicine as verified in Japan, 1934), and de Morant's own writings made acupuncture more widely known in France and the West.

Meanwhile, in the Japanese Empire and her colonies, Nakayama Tadanao and others fruitfully propagated a nationalistic form of *Kanpō* medicine and successfully sold 'traditional' medicinal drugs (Michel-Zaitsu 2017: 292–8, Oberländer 1995: 202–3). After the First World War, mass-produced Chinese medicines also paved the way for the revival of *Kanpō* medicine. Commercial success and an increase in credibility due to evidence-based analysis were not the only factors in the revival (Shin 2016). The idea that *Kanpō* is a holistic and gentle medicine and has a beneficial role in the treatment of chronic diseases eventually led to its official recognition. This idea goes back to the generation of early twentieth-century doctors who had not been trained in Chinese-style medicine but admired and appreciated the tradition. The post-Second World War association The Japan Society for Oriental Medicine (Nihon Tōyō Igakkai 日本東洋医学会), founded in 1950, remains a leading body in the promotion and advancement of *Kanpō* medicine as well as in the study of Chinese-style medicine in Japanese history, although there are numerous similar associations. *Kanpō* (or *Kampō*) is now largely considered to be pharmacotherapeutics. Moxibustion and acupuncture, bone setting, massage and related arts developed autonomously and, as extra-medical activities, were less impeded by legal pressures (Otsuka 1976: 337–8). Today, *Kanpō* medicine is generally accepted as complementary medicine and practised alongside modern biomedicine in Japan.

Japan inherited and developed Chinese medicine, not in isolation from other regions in Asia but in occasionally frequent and fruitful exchange among medical practitioners, bureaucrats as well as traders in medical books or pharmaceuticals. The Japanese history of Chinese-style medicine unfolded in the various political and social contexts, and while departing from Chinese models the new or divergent medical institutions, theories, therapies and formulæ that evolved in Japan never completely lost their connection to styles current on the Asian continent. The developments at the turn of the nineteenth to the twentieth century were the most dramatic and at the same time threatening to the Japanese medical traditions that were based on Chinese medicine. Eventually new forms of what is now referred to as Kanpō emerged that, along with treatments such as acupuncture, moxibustion and various styles of massage, contribute to a vibrant hybrid medical culture in Japan and beyond.

Bibliography

Pre-modern sources

Cruijdeboeck, 1554, Rembert Dodoens, Antwerp: J. van der Loe.

Dongui bogam 東醫寶鑑 (Treasured Mirror of Eastern Medicine) 1613, Heo Jun 許浚, Heo Jun 許浚 (compiler) and Minamoto Gensō 源元通 (1724) *Teisei tōi hōkan 25 kan* 訂正東醫寶鑑 25 巻 (Revised Treasured Mirror of Eastern Medicine in 25 Fascicles), n/a: Hayashi Tsugai Tōbei.

Engishiki 延喜式 (Procedures of the Engi Era) 927, Torao Toshiya 虎尾俊哉 and Shintō taikei hensankai 神道大系編纂会 (1991), in Shintō taikei hensan 神道大系編纂 (ed.) *Shintō taikei* 神道大系 (Important Shintō Sources) vol. 11–12, Tōkyō: Shintō taikei hensankai.

Honetsugi ryōji chōhōki 骨継療治重宝記 (Precious Notes on the Medical Therapy of Bone-setting) 1746, Takashi Hōyoku 高志鳳翼 (2006–2007), in Nagatomo Chiyoji 長友千代治 (ed.) *Ihō – yakuhō* 医方・薬方 (Medical and Medicinal Treatments) vol. 3, (Chōhōki shiryō shūsei 重宝記資料集成 vol. 25). Tōkyō: Rinsen shoten.

Honzō wamyō 本草和名 (Japanese Names for the Materia Medica) 901–923, Fukane Sukehito 深根輔仁 (1926) (Nihon koten zenshū 日本古典全集, vol. 1) Tōkyō: Nihon koten zenshū kankōkai.

Honzōkōmoku keimō 本草綱目啓蒙 (Illumination of the Principles and Varieties of Materia Medica) 1803, Ono Ranzan 小野蘭山 (1991–1992) (Tōyō bunko 東洋文庫 vols. 531, 536, 540, 552) Tōkyō: Heibonsha.

Ishinpō 医心方 (Essentials of Medical Treatment) 984, Tanba Yasuyori 丹波康頼 (1935), *Nihon koten zenshū* 日本古典全集, vol. 7, Tōkyō: Nihon koten zenshū kankōkai; (1993), in Maki Sachiko 槇佐知子 (ed.) *Ishinpō: Maki Sachiko zenshaku seikai* 医心方: 槇佐知子全訳精解 (Essentials of Medicine: complete translation with explanations by Maki Sachiko), Tōkyō: Chikuma shobō.

Kissayōjōki 喫茶養生記 (Record on Drinking Tea for Nourishing Life) 1211, Myōan Eisai 明菴栄西 (2000), in Furuta Shōkin 古田紹欽 (ed.) (Kōdansha gakujitsu bunko 講談社学術文庫, vol. 1445). Tōkyō: Kōdansha.

Nihon shoki 日本書紀 (Record of Japan) 720, Toneri shinnō 舎人親王, in Sakamoto Tarō 坂本太郎, Ienaga Saburō 家永三郎, Inoue Mitsusada 井上光貞 and Ōno Susumu 大野晋 (eds) (1965–1967), *Nihon koten bungaku taikei* 日本古典文學大系 (Complete Works of Japanese Classical Literature) vols. 67–8, Tōkyō: Iwanami shoten.

Wanbing huichun 萬病回春 (Cure of Myriad Diseases) 1587, Gong Tingxian 龔廷賢.

Xinxiu bencao 新修本草 (Revised Materia Medica) Tang-dynasty, Su Jing 蘇敬 (1983) Tosho ryōsōkan, Kunaichō shoryōbu 圖書寮叢刊 / 宮内庁書陵部 (eds), Tōkyō: Kunaichō shoryōbu.

Yamato honzō 大和本草 (Japanese Materia Medica) 1708–1709, Kaibara Ekiken 貝原益軒 (1911) Ekiken-kai 益軒会 (ed.) *Ekiken zenshū* 益軒全集, vol. 6, Tōkyō: Ekiken zenshū kankōbu.

Modern sources

Ahn, J.Y. (2012) 'Worms, germs, and technologies of the self – religion, sword fighting, and medicine in early modern Japan', *Japanese Religions* (Special Issue: Religion and Healing in Japan, ed. by Christoph Kleine and Katja Triplett), 37.1 and 2, 93–114.

Andreeva, A. (2017) 'Explaining conception to women? Buddhist embryological knowledge in the *Sanshō ruijūshō* 産生類聚抄 (Encyclopedia of Childbirth, ca. 1318), '*Asian Medicine*, 12, pp. 170–202.

Benn, J.A. (2015) *Tea in China: a religious and cultural history*, Honolulu: University of Hawai'i Press.

Bowers, J.Z. (1970) *Western Medical Pioneers in Feudal Japan*, Baltimore, MD: John Hopkins Press.

Bowring, R.J. (2005) *The Religious Traditions of Japan, 500–1600*, Cambridge: Cambridge University Press.

Dettmer, H.A. (2010) *Der Yōrō-Kodex: Die Gebote; Übersetzung des Ryō no gige, Teil 2 Bücher 2–10* (Veröffentlichungen des Ostasien-Instituts der Ruhr-Universität Bochum 55, 1–4), Wiesbaden: Harrassowitz.

Drott, E.R. (2010) 'Gods, Buddhas, and organs: Buddhist physicians and theories of longevity in early medieval Japan', *Japanese Journal of Religious Studies*, 37.2: 247–73.

Goble, A.E. (2009) 'Kajiwara Shōzen (1265–1337) and the medical silk road: Chinese and Arabic influences on medieval Japanese medicine,' in Goble, Andrew Edmund, Robinson, Kenneth R. and Wakabayashi, Haruko (eds) *Tools of Culture: Japan's cultural, intellectual, medical and technological contacts in East Asia, 1000–1500s*, Ann Arbor, MI: Association for Asian Studies, pp. 231–57.

────── (2011) *Confluences of Medicine in Medieval Japan: Buddhist healing, Chinese knowledge, Islamic formulas, and wounds of war*, Honolulu: University of Hawai'i Press.

Hattori Toshiyoshi (Toshirō) 服部敏良 (1945) *Nara jidai igaku no kenkyū* 奈良時代医学史の研究 (Studies in the Medical History of the Nara Period), Tōkyō: Tōkyō-dō (reprint Yoshikawa kobunkan 1988).

────── (1955) *Heian jidai igaku no kenkyū* 平安時代医学の研究 (Studies in the Medical History of the Heian Period), Tōkyō: Kuwana bunseidō.

────── (1964) *Kamakura jidai igakushi no kenkyū* 鎌倉時代医学史の研究 (Studies in the Medical History of the Kamakura Period), Tōkyō: Yoshikawa kōbunkan.

────── (1971) *Muromachi Azuchi Momoyama jidai igaku-shi no kenkyū* 室町安土桃山時代医学史の研究 (Studies in the History of Medicine in the Muromachi, Azuchi, Momoyama Periods), Tōkyō: Yoshikawa kōbunkan.

────── (1978) *Edo jidai igakushi no kenkyū* 江戸時代医学史の研究 (Studies in the Medical History of the Edo Period), Tōkyō: Yoshikawa kōbunkan.

────── ([1968] 1982) *Bukkyō kyōten o chūshin toshita Shaka no igaku* 仏教経典を中心とした釈迦の医学 (Shakyamuni's Medicine According to Buddhist Classical Texts), Nagoya: Reimei shobō.

Hübner, R.B. (2014) *State Medicine and the State of Medicine in Tokugawa Japan: Kōkei saikyūhō (1791), an emergency handbook initiated by the Bakufu*, PhD thesis, University of Cambridge.

Karow, O. ([1948] 1978) 'Das Honzō-Wamyō. Eine japanische Pharmakopoe der Heian-Zeit, und seine Bedeutung für die Geschichte der altjapanischen Medizin', *Archiv für Ostasien*, 1.1: 72–4 (repr. in Dettmer, Hans Adalbert and Endreß, Gerhild (eds) *Opera Minora*, 1978, pp. 195–201).

Karow, O. and Weller, F. (1954) 'Das Daidōruijuhō und die Kokugaku', *Monumenta Nipponica*, 10: 45–64.

Kleine, C. (2012) 'Buddhist monks as healers in early and medieval Japan', *Japanese Religions* (Special Issue: Religion and Healing in Japan, ed. by Christoph Kleine and Katja Triplett) 37.1 and 2: 13–38.

Kunaichō Shōsōin jimusho 宮内庁正倉院事務所 and Shibata Shōji 柴田承二 (2000) *Zusetsu Shōsōin yakubutsu* 図説正倉院薬物 (The Medicinals of the Shōsōin, Illustrated), Tōkyō: Chūōkōron.

Lomi, B. (2014) 'Dharanis, talismans, and straw-dolls ritual choreographies and healing: strategies of the Rokujikyōhō in medieval Japan', *Japanese Journal of Religious Studies*, 41.2: 255–304.

Machi, S. (2014) 'The evolution of "learning" in early modern Japanese medicine', in M. Hayek and A. Horiuchi (eds) *Listen, Copy, Read: popular learning in early modern Japan*, Leiden; Boston: Brill, pp. 164–204.

Marcon, F. (2015) *The Knowledge of Nature and the Nature of Knowledge in Early Modern Japan*, Chicago, IL; London: University of Chicago Press.

Mayanagi Makoto 真柳誠 (2004) 'Kankoku dentō igaku bunken to Nichi-Chū-Kan no sōgo denpa 韓国伝統医学文献と日中韓の相互伝播, (Books on Medicine in the Korean Tradition and Their Distribution between Japan, China and Korea)' *Kankoku gengo bunka kenkyū* 韓国言語文化研究 (九州大学韓国言語文化研究会) 6: 41–8.

Michel-Zaitsu, W. (2017) *Traditionelle Medizin in Japan von der Frühzeit bis zur Gegenwart*, München: Kiener Verlag.

Mishima, S. (2004) *The History of Ophthalmology in Japan*, Oostende: Jean-Paul Wayenborgh.

Nakayama, T. (1934) *Acupuncture et médecine chinoise vérifiées au Japon, traduites du japonais par T. Sakurazawa et G. Soulie de Morant et precedees d'une preface de G. Soulie de Morant*, Paris: Editions Hippocrate.

Oberländer, C. (1995) *Zwischen Tradition und Moderne: Die Bewegung für den Fortbestand der Kanpō-Medizin in Japan*, Stuttgart: Steiner.

Otori, R. (1964) 'The acceptance of western medicine in Japan', *Monumenta Nipponica*, 19.3 and 4: 254–74.

Otsuka, Y. (1976) 'Chinese traditional medicine in Japan', in C. Leslie (ed.) *Asian Medical Systems*, Berkeley: University of California Press, pp. 322–40.

Rosner, E. (1989) *Medizingeschichte Japans* (Handbuch der Orientalistik Abt. 5: Japan Bd 3 Abschnitt 5), Leiden: E.J. Brill.

Sanpei Kōko 三瓶孝子, and Nihon joishi hensai iinkai 日本女医史編纂委員会, eds (2008) *Hataraku josei no reikishi: Nihon joishi* 働く女性の歴史。日本女医史 (History of the Working Woman: history of the Japanese women doctors), Tōkyō: Kuresu shuppan.

Shin, Chang-Geon (2016) 'The formation and development of the self-image of Kampō medicine in Japan: the relationship between Showa-period Kampō and science', in O. Kanamori (ed.) *Essays on the History of Scientific Thought in Modern Japan*, Christopher Carr and M.G. Sheftall (trans), Tokyo: Japan Publishing Industry Foundation for Culture, pp. 235–67.

Shinmura Taku 新村拓 (2013) *Nihon bukkyō no iryōshi* 日本仏教の医療史 (History of Japanese Buddhist Medical Therapeutics), Tōkyō: Hōsei daigaku shuppankyoku.

Torigoe Yasuyoshi 鳥越泰義 (2005) *Shōsōin yakubutsu no sekai: Nihon no kusuri no genryū wo saguru* 正倉院薬物の世界: 日本の薬の源流を探る (The World of Medicinals of the Shōsōin: searching for the origins of Japanese medicines) (Heibonsha shinsho 平凡社新書 296), Tōkyō: Heibonsha.

Trambaiolo, D. (2015) 'Ancient texts and new medical ideas in eighteenth century Japan', in B. Elman (ed.) *Antiquarianism, Language, and Medical Philology: from early modern to modern Sino-Japanese medical discourses*, Leiden: Brill, pp. 81–104.

Triplett, K. (2012) 'Magical medicine? Japanese Buddhist medical knowledge and ritual instruction for healing the physical body', *Japanese Religions* (Special Issue: Religion and Healing in Japan, ed. by Christoph Kleine and Katja Triplett), 37.1 and 2: 63–92.

———— (2017) 'Using the golden needle: Nāgārjuna bodhisattva's *ophthalmological treatise* and other sources in the *essentials of medical treatment*', in P. Salguero (ed.) *Buddhism and Medicine: an anthology of premodern Sources*, New York: Columbia University Press, pp. 543–8.

Triplett, K. (2019) *Buddhism and Medicine in Japan: a topical survey (500–1600 CE) of a complex relationship*, Berlin: De Gruyter.

Ueda Sanpei 上田三平 ([1930] 1972) *Nihon yakuen-shi no kenkyū* 日本薬園史の研究 (Study of the History of Medicinal Gardens in Japan), Takada-machi: n/a. (repr. Tōky`: Watanabe shoten).

Unschuld, P.U. (1998) *Chinese Medicine*, Brookline, MA: Paradigm.

Vande Walle, W. and Kasaya, K. (2001) *Dodonaeus in Japan: translation and the scientific mind in the Tokugawa period*, Leuven; Kyoto: Leuven University Press; International Research Center for Japanese Studies.

36
A BRIEF HISTORY OF CHINESE MEDICINE IN SINGAPORE

Yan Yang 杨妍

Singapore is a multiracial and multicultural country with ethnic Chinese (74%), Malays (13%) and Indians (9%) accounting for most of its population. It was a British colony from 1819 to 1963. During this time, it attracted waves of Chinese immigrants from China and around the Malay Peninsula from the nineteenth century onwards. Chinese medicine was brought into Singapore with the influx of Chinese immigrants. In the absence of a municipal system of medical care and poor relief, the Chinese medical delivery system comprising medical institutions, medical halls, clan-based recuperation centres and individual Chinese physicians, who operated from temples, market place or their own homes, filled a much-felt void in the lives of the Chinese plebeian classes (Yeoh 1991: 38–9). Chinese medicine remains relevant today. Singapore was recently praised by a World Health Organization (WHO) officer for pioneering good policies that protect against the perceived dangers of Traditional Chinese Medicine (TCM) resulting from poor integration with biomedical safety checks (Chan 2016).

Although Western medicine is the main form of healthcare in contemporary Singapore, Chinese medicine continues to enjoy considerable popularity. Chinese medicine practice in Singapore is confined to outpatient care. Chinese medicine practitioners are also called Chinese physicians, herbalists or therapists, but not doctors, as 'doctor' is reserved for registered Western medicine practitioners. According to a survey conducted by the Singapore government in 1994, 45% of Singaporeans had consulted a TCM practitioner at least once, and 19% of the population had visited TCM clinics in the last year. The proportion who had sought care from a TCM practitioner was highest among Chinese (54%), with 16% of Indians and 8% of Malays surveyed having also attended at least once. TCM was favoured slightly more by the elderly. The main reasons for consulting TCM practitioners were sprains, aches and pain, and the common cold (Ministry of Health 1995: 12).

There are some anthropological and sociological studies of health systems in Singapore, focussing on the phenomena of Chinese medicine practices and practitioners (Wu 1987: 71–94; Quah 1990: 122–59; Sinha 1996; Smith 2018). Historical research on Chinese medicine in Singapore is relatively scarce. Several Chinese physicians wrote about the important persons, institutions and events in the history of its development from the internal historical approach (Lee 1983, 1986; Tan 2001, 2007; Wong 2012). Some historians have emphasised the importance of Singapore's context in discussing the development of Chinese medicine by examining the linkages between the socio-political-cultural environments and the complicated modernisation process affecting Chinese medicine in Singapore (Foo 2000; Yang 2018). Although only limited literature has been published, there are copious amounts of primary sources available from various stages of Singapore's history. The sources include government records, reports and related documents about TCM in Singapore, annual reports and anniversary commemorative magazines of the Chinese medical organisations; academic journals published by these organisations, newspaper reports and op-eds on Chinese medicine, a series of oral history interviews conducted by the National Archive of Singapore with those key persons leading the development of Chinese medicine in Singapore, memoires and biographies of some influential Chinese physicians, manuscripts and collections passed down in some famous Chinese physicians' families.[1]

This chapter gives a brief introduction to the history of Chinese medicine in Singapore from the early colonial period to the present day. The discussion is divided into three sections according to the three different agents acting towards the development of Chinese medicine in Singapore. The three agents are governments of different time periods, Chinese medical professional associations and religious organisations. By showcasing how these agents have led and influenced the development of Chinese medicine (institutionally and practically), this article highlights the unique characteristics and the circuitous modernisation process of Singapore's Chinese medicine.

State laws and policies

The Medical Registration Ordinance of 1905, which was the first medical law under the British administration, set a framework for the control of medicine in Singapore. Article 21 of this ordinance states:

> Nothing contained in this Ordinance shall be construed to prohibit or prevent practice of native systems of therapeutics according to Indian, Chinese or other Asian Method.
>
> *(Straits Settlements 1920: 276)*

This law has had a profound and long-lasting impact on the control of Chinese medicine in Singapore. It has three implications. First, it guaranteed that the government would not prevent or prohibit TCM practices, leaving a space for Chinese medicine to prosper. Second, it showed that the only official healthcare system is based on Western medicine, not Asian medicine which is portrayed as 'native' and traditional, in contrast to the image of Western medicine as a modern, translocal science. Third, it forbad the integration of Western medicine and Asian medicine, unlike current practice in China today. It set a clear boundary to separate the two different medical systems, one which has lasted until now. Due to the exclusion posed by the law, Chinese medical practices were considered as local businesses, so that Chinese physicians had to pay for business registration fees and signboard taxes.

During the Japanese occupation of Singapore (1942–1945), all Chinese physicians were requested to obtain a certificate of registration under the Japanese administration. The new government announced that all doctors, dentists and native doctors, including practitioners of Chinese, Malay and Indian medicine, must register before they start their profession. This was the first time that Chinese medicine was directly controlled under the government. Chinese physicians had to attend exams assessing their abilities. Only those who passed could register and were given quotas of Chinese herbs. The purpose of this was to control the herb supply strictly and hence to tighten control over the Chinese community.

After the war, the British returned and started the Medical Plan expanding healthcare services to the whole local population on the island. This marked the beginning of the dominant role of Western medicine, which had lasting effects on Singapore's current healthcare system until now. While modernising its medical system, the Singapore government also initiated the indirect control of Chinese medicine through the Medicine (Advertisement and Sale) Act 1955. Since Chinese medical practices were considered business undertakings, the government restricted the selling and advertising of all medicine, Chinese medicine in particular. The law prohibited exaggerated and unscientific advertisements of Chinese medicine by banning key words such as 'guaranteed to cure' in reference to certain diseases and conditions including blindness, cancer, cataract, dangerous drug addiction, deafness, diabetes, epilepsy or fits, hypertension, insanity, kidney diseases, leprosy, menstrual disorders, paralysis, tuberculosis, sexual function, infertility, impotency, frigidity, conception and pregnancy (Singapore Government Printing Office 1955: 1–3). There were lots of such advertisements for Chinese medicine in Chinese newspapers. The target of the law was obvious. Unfortunately, the law was not effectively enforced as Chinese advertisers found ways to change the key words without changing the meaning. For instance, they used 'waist (*yao* 腰)' to replace 'kidney (*shen* 肾)' since *shen* 肾 was explicitly prohibited in the law, but kidneys are well-known to govern the waist and lower back in Chinese medical theory. Nevertheless, this indirect control of Chinese medicine, together with the underlying logic to erase unscientific elaboration and to emphasise the importance of science is interesting to note in the light of later changes.

In 1965, the Republic of Singapore was established. One of the colonial legacies was the laws on medicine and related policies were retained. Chinese medicine was still not under the direct legal control of the government. In order to gain an understanding of Chinese medicine and to introduce direct control, the government invited doctors to conduct scientific comparative research on Chinese medicine with the cooperation of Chinese physicians (Gwee *et al.* 1969). The aim of the study was to see if there was any correlation between the two systems of medicine in the way of approach and diagnosis, and to discover the similarities and the differences between the two medicines. The roots of undertaking these investigations within a modern scientific framework can be traced to colonial period medical governance. The rationale behind this action also explains later measures taken by the government. The findings of the study emphasised the incommensurability between Chinese and scientific medicine, a difference which informed later conflicts, even though the study also found some common elements across the two diagnostic methods.

This epistemological difference came to a head in the prohibition on the use of berberine (*huangliansu* 黄连素), implemented under the Poisons Act in 1978. The drug advisory committee, which was made up of doctors and appointed by the government, announced that *huanglian* was to be banned because Western medical tests had proved it harmful to babies and pregnant women. The Singapore Chinese Physicians' Association (SCPA) appealed against the ban, writing to the minister for health and through newspaper editorials, claiming that *huanglian* was an important and useful drug in Chinese medicine (Singapore Chinese Physicians' Association 1995: 767). The committee insisted on doing scientifically controlled trials and finding statistically significant results to show the importance of *huanglian*. The SCPA listed out research papers by Chinese physicians in China as contradictory proof.

These papers were not accepted by the committee as they were not regarded as scientific enough because they were grounded in TCM theory. This showed that the government's stance on Chinese medicine was to allow the practice of Chinese medicine, but only on the basis of Western scientific criteria. This case was the first time when biomedical argumentation about scientific method was relied on in a case of law to regulate Chinese medicine. This was the beginning of using biomedical standards to control and regulate Chinese medicine in an open manner.

The Traditional Chinese Medicine Practitioners Act was enacted in 2000, marking a transition from regulating Chinese medicine through the Poisons Act, as in the case of berberine above, to the start of state regulation of TCM as a whole. The framework of this act is exactly the same as the acts on Western medicine, including the establishment of a Traditional Chinese Medicine Practitioners Board (TCMPB). Thus, the governance of TCM was copied directly from the governance structure of Western medicine. The role of the TCMPB is similar to the Singapore Medical Council, but targeting Chinese physicians instead of doctors. The ethical code and guidelines for Chinese physicians were also modified from the code and guidelines for biomedical doctors. The investigation procedures and penalties for misconduct by physicians are modelled on regulations for doctors. Continuing past practice, TCM practitioners are not allowed to use Western diagnostic equipment or dispense Western pharmaceuticals. Chinese Proprietary Medicines must not be adulterated with Western drugs.

In addition, the government implemented a new framework for administering clinical practice, teaching and research on Chinese medicine, based on the existing biomedical model, in order to complete a top-down modernisation of Chinese medicine. Here, further discrepancies arose due to the differentials of epistemology and institutional power between the two medicines. For example, the registration of Chinese physicians was different from the registration of doctors. Since the situation was complicated, the government learnt from the experience of Hong Kong and took a phased approach by invoking the grandfather clause during the transitional period. Those considered as older practitioners were grouped into four categories according to the years of experience and the certificates they had obtained. Some were fully or partially exempted from taking the standardised qualifying exam.

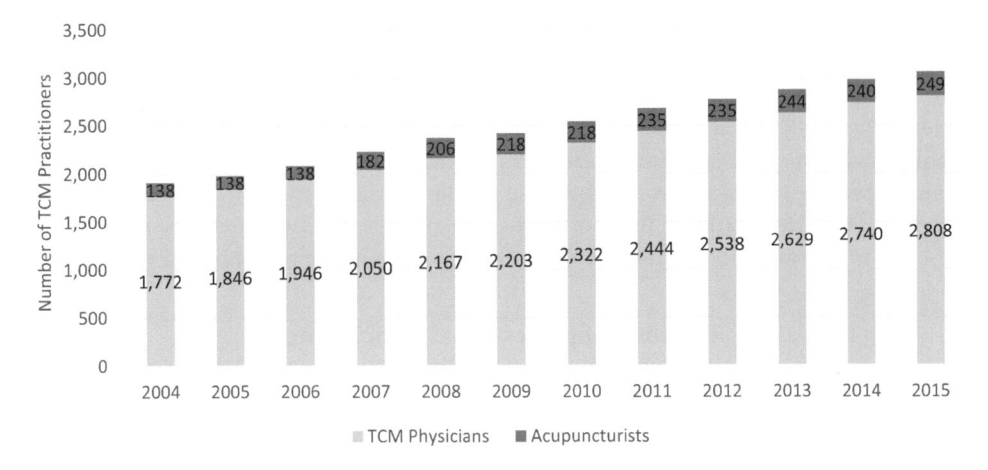

Figure 36.1 Number of TCM Practitioners on the Register, 2004–2015

From 2004 onwards, the transitional period was over. All physicians have to pass the national qualifying exam to get registered (Figure 36.1).

The training approved by the government duplicated the training system in Singapore medical schools. In the training aspect, the government copied the system for Western medicine doctors: introducing the new undergraduate double degree programme and continuing education programmes for Chinese physicians to upgrade their qualifications. The five-year double degree programme in Biomedical Sciences and Chinese Medicine is jointly organised by Nanyang Technological University in Singapore and Beijing University of Chinese Medicine in China. Within this framework, the curricula and the degrees offered by the university are subject to approval by the Ministry of Education. The continuing education programmes are organised mainly by professional associations with permission from the Ministry of Health.

In the research aspect, government encourages collaboration between Chinese physicians and Western medicine doctors and scientists in carrying out clinical research to provide scientific evidence. More and more TCM research grants were given to support research using scientific methodology.[2] All of these measures show the government's emphasis on biomedicine and the dominance of biomedical system.

Apart from the laws, the policies related to Chinese medical institutions show that the government is keen to promote Chinese medicine because of its potential to alleviate government healthcare cost. This support for Chinese medicine can be traced to the colonial period. Unlike in Hong Kong, Chinese medicine practices were endorsed and supported by the colonial government via land grant and tax exemption for TCM charity clinics. Although there were lots of complaints and criticisms from Western medical officers, the colonial government found that there was a need to allow Chinese medicine for healing the Chinese due to the scarcity of medical resources (Mugliston 1893/1894: 72). Thong Chai Medical Institution (*Tongji yiyuan* 同济医院), founded in 1867, was the earliest charitable organisation to provide free Chinese medical treatment and free Chinese herbs to help the needy sick. The Colonial Secretary of the Straits Settlements praised Thong Chai for its charitable work and granted a piece of land for Thong Chai to build its own premise. Other TCM charitable institutions, including the Chung Hwa Medical Institution (*Zhonghua yiyuan* 中华医院), enjoyed other benefits, such as exemption from paying property tax or licence fees.

These benefits continued after Singapore became an independent republic. Citing the ageing population and increasing burden of healthcare cost as their reason, the government continued providing incentives to support TCM charity clinics (Thong Chai Medical Institution 1979: 22–3). These expanded rapidly, which was in line with the government's community development masterplan. In each district, there is an average of six charitable TCM clinics providing free or cheap TCM service for the community, making it convenient for the elderly who live nearby. It is important to note that the tradition of charity practised by these Chinese medical institutions encourages Chinese medicine to flourish in Singapore.

The roles of professional associations

Professional associations have played a key role in developing Chinese medicine in Singapore since the government promoted self-regulation by the TCM community before implementing statutory regulation. There are eight self-funded non-governmental professional associations in Singapore (Table 36.1). Thong Chai is an exception as it takes the form of a medical institution rather than a professional association, yet it plays a role similar to the associations in building the training and research system. These grassroots efforts have been extremely

Table 36.1 TCM Associations in Singapore

Association	History	Affiliated Organisations	Members
Singapore Chinese Medical Union (Zhongyi zhongyao lianhehui 中医中药联合会)	Founded in 1929 under the leadership of Li Bo Gai 黎伯概 and Eu Yan Sang Medical Hall 余仁生药行, suspended in 1942, resumed in 1947. The current president is Guo Jun Xiang 郭俊翔.	TCM Cultural College (Zhongyiyao wenhua xueyuan 中医药文化学院)	Around 200 members in 1954, 141 members in 1989, around 200 members in the 2000s.
Singapore Chinese Physicians' Association (the SCPA, Zhongyishi gonghui 中医师公会)	Founded in 1946 under the leadership of Wu Rui Fu 吴瑞甫, You Xing Nan 游杏南 and Zeng Zhi Yuan 曾志远.	Chung Hwa Medical Institution (previous name as Chung Hwa Free Clinic), Singapore College of Traditional Chinese Medicine (Xinjiapo zhongyi xueyuan 新加坡中医学院, previous name as Singapore Chinese Physicians' Training College), Chung Hwa Medical Research Institute (Zhonghua yiyao yanjiuyuan 中华医药研究院), Chung Hwa Acupuncture Research Institute (Zhonghua zhenjiu yanjiuyuan 中华针灸研究院)	Largest organisation by number of members, around 600 members in 1980, 879 members in 1990, 1200 members in 1994, 1169 members in 2006, 1600 members in 2018.
Association for Promoting Chinese Medicine (Zhongyiyao cujinhui 中医药促进会)	Founded in 1957 under the leadership of Fang Zhan Lun 方展绘, previous name was Zhongyiyao tilianyaoye cujinhui 中医药提炼药业促进会. Re-established in 1971 under the leadership of Chen Zhan Wei 陈占伟.	Public Free Clinic Society (Dazhong yiyuan 大众医院), Institute of Chinese Medical Studies (Zhongyixue yanjiuyuan 中医学研究院), Research Institute of Chinese Medicine and Acupuncture (Zhongyiyao zhenjiu yanjiuyuan 中医药针灸研究院)	500 members in 2007.
Singapore Acupuncture Association (Zhenjiu yixue xiehui 针灸医学协会)	Founded in 1977 under the leadership of five doctors and five acupuncturists including Zhang Yu Hua 张玉华, renamed as Xinjiapo zhenjiu xuehui 新加坡针灸学会 in 1998.	School of Acupuncture (Zhenjiu xueyuan 针灸学院), Acupuncture Charitable Center (Xinjiapo zhenjiu shizhenzhongxin 新加坡针灸施诊中心)	85 members in 1995, 87 members in 2018.

(Continued)

Association	History	Affiliated Organisations	Members
Society of Chinese Medical Research (Xinyi xiehui 新医协会)	Founded in 1984 under the leadership of Cai Yu Quan 蔡玉泉.	Kang Ming Chinese Physicians College (Kangmin yiyao zhenjiu xueyuan 康民医药针灸学院), Kang Ming Clinic (Kangmin yiyuan 康民医院)	Small organisation as the members are from the graduates of Kang Ming college only. Number of members is not mentioned.
Xinhua Medical Society (Xinhua zhongyiyao xiehui 新华中医药协会)	Founded in 1994 under the leadership of Chen Pan Xu 陈磐绪, merged into Singapore Chinese Medical Union in 2000.	Jian Min Medical Hall (Jianmin yaofang 健民药房), Jian Min College of Chinese medicine (Jianmin zhongyi xueyuan 健民中医学院)	Small organisation as the members are from the graduates of Jian Min college only. Number of members is not mentioned.
Singapore Chinese Physicians' Training College Alumni Association (Zhongyi xueyuan biyeyishi xiehui 中医学院毕业医师协会)	Founded in 1958 under the leadership of student council in Singapore Chinese Physicians' Training College, renamed as Singapore College of TCM Alumni Association in 2012.	Singapore College of Traditional Chinese Medicine	Around 80 members in 1959, 282 members in 2017.
Society of Traditional Chinese Medicine (Xinjiapo zhonghua yixuehui 新加坡中华医学会)	Founded in 2000 under the leadership of Lee Kim Leong 李金龙	Specialist Traditional Chinese Medicine Center (Zhuanke zhongyiyuan 专科中医院), Academy of Traditional Chinese Medicine (Zhuanke zhongyi yanjiuyuan 专科中医研究院)	Around 200 members in 2018, the member must possess a master or doctorate degree in Chinese Medicine from TCM universities in China.

important in the institutionalisation and modernisation of Chinese medicine in Singapore, not only regulating the profession, but also leading the teaching and research of Chinese medicine without governmental guidance prior to 2000.

As shown in the table, these professional associations have a similar organisational structure, which is a three-in-one or four-in-one integrated structure. Each association is like a conglomerate assembling three or four subsidiaries under one umbrella. These subsidiaries, including clinics, schools and colleges and research institutes, focus on the areas of clinical practice, teaching and research. These umbrella-shaped institutions form a kind of guild which regulates the physicians. The highlight of this integrated structure is the interconnections among these entities that allow a better integration of different important aspects (clinical practice, education and research), essential to the modernisation of TCM. This organisational structure is also self-sufficient as it consists of all the essential aspects which provide support and sources to each other. Because of this unique structure, these associations played the key role of leading the development of Chinese medicine without help from the state before 2000.

Taking the SCPA as an example, the SCPA started its role as a gatekeeper and regulator in 1947, fulfilling a vacuum left by the colonial government. The SCPA announced that it followed the Physicians Act (*Yishifa* 医师法) of the Republic of China (ROC) promulgated in 1943. The SCPA considered that Chinese physicians overseas were still under the control of the ROC as they were Overseas Chinese. According to the act, SCPA was supposed to be the only legitimate local professional association in the region. The association started the work of registering physicians and regulating the profession according to the act. After Singapore gained independence, this situation changed completely as these physicians were no longer considered Chinese, but Singaporean/Malaysian physicians, and ROC laws no longer applied to them. Therefore, the SCPA started drafting its own principles and regulations. It referred to the previous Physicians Act and formulated new rules according to the local situation. These rules and regulations, including registration requirements, practice guidelines and disciplinary procedures, were revised several times to fit in with the local context and to accommodate the needs of different periods (Yang 2018: 128–49). For instance, the registration of physicians with the SCPA was initially open only to graduates from the SCPA-affiliated training college. Outsiders who wished to join the association had to be recommended by SCPA members and pass an exam in order to register. When the government called for the formation of one unified body to oversee all physicians, an initiative that lasted from 1980 to 1992, the association changed its entry requirement to recruit other physicians who had other certificates or equivalent experience without exam or member recommendation. The government ultimately gave up the endeavour, due to ongoing internal conflicts between various professional associations.

Membership in the SCPA was a symbol of professionalism. Being registered with the SCPA meant the physician was a qualified Chinese medicine practitioner certified by the SCPA and held the licence issued by the SCPA. The physician had to abide by the rules and regulations of the SCPA. Offenders would be investigated and punished according to society rules. The SCPA also introduced new measures such as issuing medical certificates to its members, an approach they learnt from the Western medical system. These efforts at self-regulation were effective, bringing a normalisation of practice and excluding poor practitioners to some extent, and had a significant impact on the Singapore community before 2000, when the Ministry of Health stepped in.

Besides these regulatory practices, the SCPA also established broad and extensive networks with overseas organisations, including TCM universities, international associations and government bodies in China. These networks have been helpful in raising TCM standards in Singapore, as they facilitate the exchange of resources and promote the interaction and collaboration between organisations. Members of the SCPA enjoy the privileges of attending workshops, and going for exchanges and internships to partner institutions. The SCPA also invites distinguished foreign experts to deliver lectures and conduct courses for its members.

Under the SCPA umbrella, the clinics, colleges and research institutes affiliated to the SCPA share resources and take advantage of collective international networks. The principal and teachers in the colleges are specialists or senior physicians who also work in associated clinics. They are also leaders of research groups in Chung Hwa Medical Research Institute and Chung Hwa Acupuncture Research Institute. Some of them also sit on the council of the SCPA. Students of the college are sent to SCPA clinics to observe and for internships. They have to participate in research groups in research institutes, where the projects are based on patient records and therapeutic practices used in the clinics. After graduation, they become members of the SCPA, and can continue to participate in activities hosted by affiliated organisations. The facilities and resources, such as the library and the labs, are also shared among these institutions.

This self-sufficient four-in-one structure is only possible because of the philanthropy of the physicians and the public. It is important to note that the school teachers, the researchers, the physicians working in the clinics and the council members in the SCPA are all volunteers. The clinics affiliated to the SCPA are all charitable organisations which rely on generous donations from the public. The clinics provide free resources for training and research.

The college affiliated to the SCPA is the largest TCM training school in Singapore. The history of this college shows the transformation of the training system of Chinese medicine from the traditional diverse acquisition methods to the modern standard college system. The college was founded in 1953. Because of restrictions imposed by the immigration ordinance issued by British colonial government in 1952, fewer physicians were allowed to enter Singapore. Facing a lack of successors, the SCPA initiated a plan to set up a TCM training school. Before this, there were different traditional acquisition methods such as apprenticeship training and self-learning. At the beginning stage when the school just opened, the key figures in the SCPA were the teachers. They wrote their own handouts and compiled their own textbooks. Their way of teaching was more like a master-apprentice model as there was no standard syllabus or curriculum. The situation changed in the 1960s when the school started to adopt the standard textbooks from China. Referring to the courses and curricula offered in TCM colleges in China, the teachers made some adaptations to meet local students' needs and designed a similar curriculum which was more suited to the Singaporean context.

Enrolment requirements also changed over the years. In the beginning, there were no specific degree requirements. From 1980 onwards, the entry requirement was changed to allow only those who had completed their junior college education. Those who did not meet this requirement had to take extra bridging courses before enrolment. Compared to China, students in Singapore's TCM colleges are from quite diverse social backgrounds. Most of them are middle aged and older, from different industries and income groups. About half of the students quit half-way through their training, and only 50% of graduates go on to pursue a professional career. The others study for personal benefit, such as learning self-care. This should be viewed in the light of the fact that TCM was not regulated as a discrete branch of medicine under the national healthcare system until 2000.

The main differences between the training in Singapore and China are,

1 Singaporean colleges only offer part-time courses;
2 clinical practice usually starts from year one onwards; [3]
3 across similar subjects (such as *Inner Canon of the Yellow Emperor*, Warm Diseases 溫病, Injury from Cold 傷寒, etc.), the instruction takes more class hours in Singapore as theoretical teaching is integrated with clinical practice. This is in line with the college mission to train TCM professionals with an emphasis on clinical capabilities and experience;
4 the focus on biomedicine is much less than in China as Singapore forbids the integration of Chinese medicine and Western medicine. Thus, colleges teach very few biomedical or science subjects.

After China opened up in the 1980s, China had a greater and more direct influence on TCM education in Singapore, while more China experts were invited to give lectures and to conduct training courses. More and more Singaporean physicians went to China to undertake postgraduate training. From 1994 onwards, local colleges started joint degree programmes with renowned TCM universities in China. These joint degree programmes were considered to have high standards of teaching and gained public recognition. This recognition was important as the government had yet to assess the standards of training since it had not legislated concerning TCM.

From the aspect of research, Singaporean professional associations have carried out research studies of TCM from the beginning to now. Before the 1950s, physicians tried multiple diverse research methods and methodologies. These included not only traditional approaches, such as textual research and annotation of medical classics, but also different scientific methods, including basic chemistry, keeping statistical results, systematic observation and logical deduction. Singapore has been creating a new modern industrial society since the 1950s, in which the government emphasises the study of mathematics and science. English is used as the medium of instruction, which also helps to spread Western influence. As a result, this environment has stimulated scientific research into TCM. A major hindrance faced by physicians in Singapore is the prohibition of integration of Chinese medicine and biomedicine. Because of this, researchers can only use the mathematical and scientific approaches from physics, chemistry and biology to do analysis and to conduct experiments. Researchers also focus more on the dialectical methods of pattern differentiation and treatment determination (*bianzheng lunzhi* 辨證論治) and traditional theories of TCM, which are different from the Westernised approach adopted by some TCM researchers in China.

The research institutes affiliated to the professional associations organise workshops and symposia, publish academic journals and set up specialty groups to promote academic research. Local physicians usually collaborate with experts in China who are their teachers or classmates. They consult these experts for ideas and theories while carrying out clinical research in Singapore on the local population. Sometimes they ask for technical support from overseas institutions due to the lack of facilities, as these research institutes are self-funded and small-scale.

Overall, these professional associations have made invaluable contributions to the development of TCM in Singapore, in such areas as the self-regulation and licencing of TCM practice, training of TCM practitioners and leading research on TCM (Yang 2018: 101–273).[4] They initiated a bottom-up modernisation by selectively learning from China's experience and adapting to fit into Singapore's context, implementing measures, such as forbidding the integration of Chinese and Western medicine, and assisting in the development of government-directed training programmes. The complicated process of development shows indigenous innovation and continuous adaptation. After the legislation on Chinese medicine, the government and the professional associations, via negotiation and collaboration, continued to modernise Chinese medicine.

As a response to the government's request, representatives from the first seven of the associations listed in Table 36.1 and Thong Chai formed the Singapore Traditional Chinese Medicine Organisations Coordinating Committee (*Zhongyi tuanti xietiao weiyuanhui* 中医团体协调委员会) in 1995. This committee, representing the TCM community, played a key role in negotiations with the government in the transition phase. It drafted the 'Guiding Principles Memorandum for Singapore Chinese Medical Education' and 'Practice Guidelines and Ethical Codes for Singapore TCM Practitioners'. It also certified the physicians and published the 'List of TCM Practitioners in Singapore'. Their work was based on their previous experience and provided important input for the government.

The actions taken by religious organisations

Apart from scholarly medicine, folk medicine (Topley 2011: 451–3) has coexisted in Singapore until now. Before the mid-twentieth century, some physicians even used Daoist rituals to treat patients or included Daoist philosophical mores in their prescriptions. This phenomenon started disappearing from the 1930s onwards after the physicians began to draw a strict line between scholarly and folk medicine and tried to eradicate what they saw as mere superstition.

Nevertheless, folk ideas and methods have been preserved and practised by religious professionals in temples even until now. We can still observe the combination of ritual and medical treatments in such places of worship as the Temple of Baosheng Emperor (*Zhenren gong* 真人宫), Singapore Teochew Charity Hall (*Xiude shantang yangxinshe* 修德善堂养心社) and Kiew Lee Tong Temple (*Jiuli dong* 九鲤洞). The common feature of these temples is that the Gods of Chinese medicine such as Hua Tuo 华佗, Wu Tao 吴夲 or Ji Gong 济公 are worshipped in these temples. Patients who seek treatment at these temples are considered to be seeking treatment from divine physicians. The seeking of treatment at these temples is understood to be by patients seeking help from the divine physicians. The types of healing include asking for prescription divinations (*Yaoqian* 药签) and getting prescriptions through the writing of spirit-diviners (*Fuji* 扶乩) after communication with the deity.

Over the years, there have been two prominent trends in the development of folk medicine. The religious organisations which provide these treatments started taking two steps. One is to re-explain and validate the rituals/activities by looking for scholarly proof. The other is to develop and institutionalise secular medicine by setting up and expanding affiliated free TCM clinics. Some physicians in these clinics may promote TCM knowledge as an expression of their religious ethics. These changes have both been influenced by the rationalisation of the country.

Conclusion

The Singapore case offers an interesting example of how Chinese medicine has been developed in an overseas context, affording a view of how Chinese medicine has adapted to conform to local conditions. Due to both internal and external influences, Chinese medicine has undergone a unique modernisation which is different from TCM in China. The non-governmental organisations, including professional associations and religious groups, were instigators of the professionalisation from below. The government, via collaboration with the professional associations, completed the top-down modernisation. The development of Chinese medicine in Singapore is a metaphor of Singapore, illustrating a blend of East and West embodying the interaction of Chinese influence and the British colonial legacy.

Notes

1 For a full list of these primary sources, refer to Yang Yan (2018: 350–61).
2 For the list of research projects funded by the government, refer to https://www.moh.gov.sg/docs/librariesprovider5/research-grants/application-guide-for-tcmrg-(jan-2017).pdf, accessed 8/10/2018. For information on how government funded research projects and how their terms and conditions follow a biomedical frame, see Yang (2018: 93–4). Two examples will suffice here to show the predominance of the biomedical method in government-funded research:

> K.H. Kong (2018) 'Acupuncture in the Treatment of Fatigue in Parkinson's Disease: A Pilot, Randomized, Controlled, Study' and Singapore General Hospital (2017) 'Traditional Chinese Medicine for Treatment of Irritable Bowel Syndrome'.

3 Colleges started to offer full-time courses only after the legislation of TCM under the supervision of TCMPB.
4 One example of new research is the innovative new proprietary Chinese medicine, *Bushen Yijing Wan* 补肾益精丸. Thong Chai Medical Research Institute invented this pill and successfully got the patent. For the research paper related to this development, see Lu (2015).

Bibliography

Chan, M. (2016) 'The contribution of traditional Chinese medicine to sustainable development', Keynote address at the International Conference on the Modernization of Traditional Chinese Medicine, http://www.who.int/dg/speeches/2016/chinese-medicine-sustainable/en/, accessed 17/8/2018.

Foo, D.S.L. (2000) *A Short History of a Long Tradition: the resilience of Chinese medicine in Singapore*, Singapore: Honour thesis, National University of Singapore.

Gwee, A.L., Lee, Y.K. and Tham, N.B. (1969) 'A study of Chinese medical practice in Singapore', *Singapore Medical Journal*, 10: 2–7.

Kong, K.H. (2018) 'Acupuncture in the treatment of fatigue in Parkinson's disease: a pilot, randomized, controlled, study', *Brain and Behavior*, 8.1: e00897.

Lee Kim Leong 李金龙 (1983) *Xinjiapo zhongyiyao de fazhan 1349–1983* 新加坡中医药的发展 1349–1983 (The Development of Chinese Medicine in Singapore), Singapore: Educational Publishing.

——— (1986) *Xinjiapo zhenjiu shiye de fazhan 1936–1986* 新加坡针灸事业的发展 1936–1986 (The Development of Acupuncture in Singapore), Singapore: Xinhua Cultural Enterprise Pte Ltd.

Lu, Jinyu (2015) 'Clinical observation of bushen yijing wan for treatment of male infertility with syndrome of kidney Yin deficiency', *Journal of New Chinese Medicine*, 47.4: 125–7.

Ministry of Health (1995) *A Report by the Committee on Traditional Chinese Medicine*, Singapore: Ministry of Health.

——— (2017) *Application Guide for Traditional Chinese Medicine Research Grant*, https://www.moh.gov.sg/docs/librariesprovider5/research-grants/application-guide-for-tcmrg-(jan-2017).pdf, accessed 8/10/2018.

Mugliston, T.C. (1893/1894) 'Unqualified practice in Singapore', *Journal of the Straits Medical Association*, 5: 68–85.

Quah, S.R. (1990a) 'The best bargain: medical options in Singapore', in S.R. Quah (ed.) *The Triumph of Practicality: tradition and modernity in health care utilization in selected Asian countries*, Singapore: Social Issues in Southeast Asia, Institute of Southeast Asian Studies, pp. 122–59.

——— (ed.) (1990b) *The Triumph of Practicality: tradition and modernity in health care utilization in selected Asian countries*, Singapore: Social Issues in Southeast Asia, Institute of Southeast Asian Studies.

Singapore Chinese Physicians' Association (1995) *Xinjiapo zhongyishi gonghui chengli wushi zhounian jinxi jinian tekan* 新加坡中医师公会成立50周年金禧纪念特刊 (A Commemorative Collection to Celebrate the Singapore Chinese Physicians' Association's Golden Jubilee), Singapore: Singapore Chinese Physicians' Association.

Singapore General Hospital (2017) *Traditional Chinese Medicine for Treatment of Irritable Bowel Syndrome*, https://clinicaltrials.gov/ct2/show/study/NCT03135821, accessed 21/12/2018.

Singapore Government Printing Office (1955) *Medicines (Advertisement and Sale) Act*, Singapore: Government Printing Office.

Sinha, V. (1996) *Theorizing the Complex Singapore Health Scene: reconceptualizing "medical pluralism"*, Ann Arbor, MI: University Microfilms International.

Smith, A.A. (2018) *Capturing Quicksilver: the position, power and plasticity of Chinese medicine in Singapore*, New York: Berghahn Books.

Straits Settlements (1920) *The Laws of the Straits Settlements Vol. 2, 1901–1907*, London: Waterlow and Sons.

Tan Hong Leng 陈鸿能 (2001) *Xinjiapo zhongyixue xianqu renwu yu yiyao shiye fazhan* 新加坡中医学先驱人物与医药事业发展 (Pioneers of Chinese Medicine in Singapore and Their Contributions), Singapore: Society of Traditional Chinese Medicine.

——— (2007) *Huaren yu xinjiapo zhongxiyixue 1819–1965* 华人与新加坡中西医学：从开埠1819年到建国1965年 (Chinese and the Developments of Chinese Medicine and Western Medicine in Singapore 1819–1965), Singapore: Lingzi Media.

Thong Chai Medical Institution (1979) *Tongji yiyuan dasha luocheng jinian tekan* 同济医院大厦落成纪念特刊 (A Collection Commemorating the Inauguration of the New Thong Chai Building), Singapore: Thong Chai Medical Institution.

Topley, M. (2011) *Cantonese Society in Hong Kong and Singapore: gender, religion, medicine and money*, Hong Kong: University of Hong Kong Press.

Traditional Chinese Medicine Practitioners Board (2015) *Traditional Chinese Medicine Practitioners Board Annual Report 2015*, Singapore: Traditional Chinese Medicine Practitioners Board.

Wong Peng 王平 (2012) *Xinglin xingzhi lu* 杏林行知录 (Essays on Development of Chinese Medicine in Singapore and Southeast Asia), Singapore: Singapore Chinese Physicians' Training College Alumni Association.

Wu, David Y.H. (1987) 'Traditional Chinese medicine in Singapore: cultural identity and social mobility', *Journal of the South Seas Society*, 42: 71–94.

Yang Yan 杨妍 (2018) *Zhongyi zai xinjiapo 1867–2014* 中医在新加坡 1867–2014 (Chinese Medicine in Singapore 1867–2014), PhD thesis, National University of Singapore.

Yeoh, B. (1991) *Municipal Sanitary Surveillance, Asian Resistance and the Control of the Urban Environment in Colonial Singapore*, Oxford: University of Oxford.

37

MINORITY MEDICINE

Lili Lai 賴立里 *and Yan Zhen* 甄艳

We two authors use the term 'minority medicine' (*shaoshu minzu yixue* 少数民族医学, lit. ethnic medicine)[1] to differentiate the concept from the majority Han 汉 medicine, commonly designated as 'Chinese medicine' (*zhongyi* 中医), which is by default the *national* medicine. This term refers to the medicines of all the other so-called nationalities apart from the so-called Han people, where the term Han is a self-designation for those who consider themselves members of the majority population. 'Minority medicine' is a somewhat different idea from the concept of 'ethnic medicine' that one might encounter, for example, in academic ethnographic studies, because it has been openly recognised that the Chinese state played a significant role in the construction of the country's officially recognised *minzu* (nationalities, ethnic groups), a practice of active identity-making which is different from the formation of ethnic groups as understood in an ethnological sense (Mullaney 2011). Between 1953 and 1954, following Mullaney's analysis, China's 'Ethnic Classification Project' (*minzu shibie* 民族识别) successfully compressed over 400 potential categories of *minzu* identity to under sixty, and over the course of the subsequent three decades, fifty-five of these were officially recognised. Now, including the majority Han Chinese category, there are fifty-six officially recognised categories that have become increasingly reified and ubiquitous. As Mullaney argues in his discussion of the 'terminological chaos' of *minzu*, the efforts made to standardise the blurry boundaries of minority medicines can be understood as 'a fundamental part of the history of the social sciences, the modern state, and the ongoing collaboration there-between' (Mullaney 2011: 15). This chapter surveys and evaluates the national policies concerned with the institutionalisation of China's minority medicine since 1951, and the development of minority medicine as an academic field since the 1980s.[2]

Before 1949, the only Chinese terms for 'minority medicines' were expressions referring to particular regions or cultures with their very own 'long histories', such as Tibetan medicine (*zang yi* 藏医), Mongolian medicine (*meng yi* 蒙医), Uyghur medicine (*weiwuer yi* 维吾尔医), and Korean medicine (*chao yi* 朝医), that is, those places and peoples with distinctive medicines that can be traced through the written record. The very first mention of *minzu yi* (民族医, doctors of minority medicine) as an overarching term appeared in the official document of the 'Programme for the Public Health Work of Minority Nationalities' (*Quanguo shaoshu minzu weishengg gongzuo fang'an* 全国少数民族卫生工作方案) issued in December 1951, which stated that 'the doctors of minority medicine employing herbs and folk medicine shall be brought together into groups and their work elevated to the greatest possible extent' (Zhu Guoben 2006: 2). It was only after 1949 that minority medicine began to develop alongside the state policies on minority nationalities (*shaoshu minzu* 少数民族).

On October 11, 1958 Mao Zedong famously declared: 'Chinese medicine is a great treasure house to be diligently explored and elevated' (Mao Zedong 1999, vol. 7: 423). In the same year, the National Workshop on Chinese Medicines and Drugs (*Quanguo zhongyi zhongyao gongzuo huiyi* 全国中医中药工作会议) exhorted the masses 'to respect and treasure many minority (Mongolian, Tibetan, etc.) medicines and work seriously on exploration and

Table 37.1 Policies and Projects Issued by the Government of China

Name	Pinyin	Hanzi	Issued Departments	Issued Date
Programme for the Public Health Work of Minority Nationalities	Quanguo shaoshu minzu weisheng gongzuo fang'an	全国少数民族卫生工作方案	Ministry of Health	December 1951
Suggestions for Promoting Minority Medicine	Guanyu jicheng fayang minzu yiyao xue de yijian	关于继承发扬民族医药学的意见	Ministry of Health and State Ethnic Affairs Commission	1983
Guidelines of Effectively Strengthening the Development of Minority Medicines	Guanyu qieshi jiaqiang minzu yiyao shiye fazhan de zhidao yijian	关于切实加强民族医药事业发展的指导意见	Jointly issued by eleven state departments, including State Administration of Traditional Chinese Medicine, State Ethnic Affairs Commission, Ministry of Health	October 2007
The Twelfth Five-year Plan on Informationization of Chinese Medicine	Zhongyiyao xinxihua jianshe shierwu guihua	中医药信息化建设'十二五'规划	State Administration of Traditional Chinese Medicine	July 2012
Program of Building Main Hospitals of Minority Medicines	Zhongdian minzu yi yiyuan jianshe xiangmu	重点民族医医院建设项目	State Administration of Traditional Chinese Medicine	Since 2007
Sorting All-China Minority Medicine Classics	Quanguo minzu yiyao wenxian zhengli gongzuo	全国民族医药文献整理工作	State Administration of Traditional Chinese Medicine	2001 and 2002
Demonstrative Research on Key Techniques of Minority Medicine Development	Minzu yiyao fazhan guanjian jishu shifan yanjiu	民族医药发展关键技术示范研究	Ministry of Science and Technology	2009
Minority Medicine Literature Collation and Appropriate Technology Selection and Propagation	Minzu yiyao wenxian zhengli yu shiyi jishu shaixuan tuiguang	民族医药文献整理与适宜技术筛选推广	State Administration of Traditional Chinese Medicine	2010

Minority medicine

Table 37.2 State Administration Departments Mentioned in the Article

Name	Pinyin	Hanzi
Ministry of Health	*Weisheng Bu*	卫生部
State Administration of Traditional Chinese Medicine	*Guojia Zhongyiyao Guanli Ju*	国家中医药管理局
State Ethnic Affairs Commission	*Guojia Minzu Shiwu Guanli Ju*	国家民族事务委员会

promotion', thus encouraging the development of minority medicines (Cheng Zhaosheng and Fang Mingjin 2012). However, this did not mean preserving medical traditions in their old forms, but exploiting them for how they could serve 'modern' medicine and the population at large. In this context minority medicine could be reformed to serve the interests of the socialist state.

State policy has not always been positive towards minority medicine and there have been constant tensions between local ethnic identities that are reflected in medicine and, of course, in religious life. Obviously, there are pros and cons involved in bringing local medicines under state administration. On the one hand, local medical practices have received official recognition and legal support; on the other, this has meant that campaigns against superstition and religion, especially during the Cultural Revolution (1966–76), greatly discouraged the development of traditional medicines across the country. This chapter also provides important background for further investigation into what changes centralisation and standardisation have made to (a) the social organisation of medicine and medical lineages; (b) the standardisation of prescriptions; (c) the ownership of the medicines and medical knowledge; (d) ritual and religious practices; and it offers food for (e) speculation about how standardisation and institutionalisation may or may not inhibit the dynamic nature of traditions, and their ability to innovate (Tables 37.1 and 37.2).[3]

Institutionalisation of minority medicines

Before 1950, practitioners of local medicines among ethnic groups living within the political boundaries of China were basically left to their own devices. Despite, or perhaps because of, the establishment of Republican China in 1912 and its commitment to modernisation and modern medicine, and due to the continuous warfare that characterised the subsequent thirty-five years, minority medicine survived only far away from institutional control without official support or management. Meanwhile, different groups, with diverse histories, cultural backgrounds and geographies, and uneven distributions of written/oral local traditions, all contributed to multiple approaches to healing in China. The most readily distinguishable of these medicines are those preserved in written traditions, notably Tibetan, Mongolian, and Uyghur medicine, all three of which have rather coherent theoretical systems grounded in a classical medical literature. In contrast, local medicines in the borderlands of the southwest were scattered, and as a rule were only practised and passed on by word of mouth. Written records are therefore scarce.

In the early years of the PRC, when the fifty-six-*minzu* model was still being constituted, there was not much effort put into constructing the terms within which *shaoshu minzu yixue* 少数民族医学 (minority medicine) was to be developed. The period between 1951 and 1966 was the time when minority medicine was first developed alongside *Zhongyi* 中医 (Chinese medicine). The year of 1951 witnessed the moment when minority medicine was first officially recognised in the 'Programme for the Public Health Work of Minority Nationalities', while 1966 was the year when the Culture Revolution began. More specifically, during the early 1960s, the famous Eight-Character Guiding Principles (*bazi fangzhen* 八字方针) were also gradually made applicable to minority medicine and there was a substantial government investment in minority nationality areas.[4] These Guiding Principles involved adjusting (*tiaozheng* 调整), strengthening (*gonggu* 巩固), enriching (*chongshi* 充实), and improving (*tigao*, 提高), principles that had already been integral to the project to develop Chinese medicine. The Cultural Revolution (1966–76) saw a suspension of these initiatives to foster China's ethnic medical diversity and an insistence on an assimilation to the majority medicine (Mullaney 2011: 125).

During the period when medical plurality was still encouraged, Mongolian and Tibetan medicines were also incorporated into higher education. The Department of Chinese and Mongolian Medicine (*Zhong meng yi xi* 中蒙医系) was set up in the Medical College of Inner Mongolia (*Neimenggu yixueyuan* 内蒙古医学院) in 1958, inaugurating one of the first majors in Minority Medicine that Chinese biomedical students could sign up for. Looking back at the unique historical trajectory of the Department of Chinese and Mongolian Medicine, it is evident that it was animated by a group of Mongolian intellectuals who had been inspired by the establishment of the very first four Chinese medicine colleges (Beijing 北京, Shanghai 上海, Guangzhou 广州, Chengdu 成都) in 1956. They soon gained support from the provincial government of the Inner Mongolian Autonomous Region, and in 1958, the provincial Party Committee approved the establishment of the department. Today the names of the department and institution have changed slightly; it is now the College of Mongolian Medicine (*Mengyiyao xueyuan* 蒙医药学院) at the Medical University of Inner Mongolia (*Neimenggu yike daxue*, 内蒙古医科大学). In 1978, another Mongolian medical college was established in Tongliao City 通辽市, in the Autonomous Region of Inner Mongolia: Jirim League Medical College (*Zhelimu meng yixueyuan* 哲里木盟医学院). This has now evolved into the College of Mongolian Medicine of the Minzu University of Inner Mongolia (*Neimenggu minzu daxue mengyiyao xueyuan* 内蒙古民族大学蒙医药学院).

During that time, the set-up of the Department mimicked the experience of the central Chinese medicine colleges, integrating Mongolian medicine and biomedicine in a ratio of 6 to 4, with Chinese Medicine not listed at all among the compulsory courses. The language of the entrance examination was also Mongolian, and the first forty-eight students recruited and enrolled in 1958 all held Mongolian national (ethnic) identity.

After what has been known in China as the Democratic Reform of Tibet (*Xizang minzhu gaige* 西藏民主改革) of 1959, a Lhasa Traditional Tibetan Hospital (today's Tibetan Medicine Hospital) was set up combining the old Sman rtsi khang (Institute of Medicine and Astrology) with the original Lcag po ri 'gro phan gling (Iron Hill Benefiting Institute). In 1983, the very first modern Tibetan Medical School was founded in Lhasa, followed by a Department of Tibetan Medicine at the University of Tibet in 1985. In 1989, the two institutions merged into the Tibetan Traditional Medical College of the Autonomous Region of Tibet (Bod ljong bod lugs gso rig slob grwa chen mo), which has become the professional training centre for Tibetan medicine.

Before the open-door policy of 1978, knowledge and practice of local practitioners of minority medicines (other than Tibetan and Mongolian) working with *materia medica* was mostly integrated into the 'vast treasure house' of Chinese medicine. At this point, despite the many and varied potentially healing interventions made by local doctors, 'minority medicine doctor' (*shaoshu minzu yi* 少数民族医) became a catch-all term for any doctors from recognised nationality groups other than the Han Chinese.

The year 1977 saw a Research Office for Minority Medicine (*Minzu yiyao diaoyan bangongshi* 民族医药调研办公室) set up in the Autonomous Prefecture of Xishuangbanna of Yunnan Province to 'salvage and sort' (*fajue zhengli* 发掘整理) Dai medicine (*Daiyi* 傣医), protecting the pre-modern medical literature of the Dai people and promoting the development of the tradition. 'Salvage and sort' has been the standard technique of governance for these initiatives since Mao Zedong's historic address on October 11, 1958 in Beijing on the subject of developing traditional medicine: it has involved surveys, collaboration meetings, and focus groups; scholarly and social research; professional education and training, regulation of practitioners and other agents (such as drugs and plantations); and subsidies for services needed. In the case of Dai medicine, the support of the local government of the Dai 'autonomous' prefecture played a significant role in the relative success of 'salvaging and sorting' in Dai medicine projects. Such was the case also for Zhuang medicine in Guangxi. By 1979 the Institute of Minority Medicine of the Dai Autonomous Prefecture of Xishuangbanna (*Xishuangbanna Daizu zizhizhou minzu yiyao yanjiusuo* 西双版纳傣族自治州民族医药研究所) was officially established. In 1984 the Ministry of Health officially recognised Dai medicine. Since then, Tibetan, Mongolian, Uyghur, and Dai medicines have become the four major minority medicines of China.

From the early 1980s 'minority medicines' have witnessed dramatic changes (Zhu 2006). The 1983 document 'Suggestions for Promoting Minority Medicine', for the first time openly stated that 'minority medicines are important components of our nation's traditional medicine' (Ministry of Health 1983). In the following year, 1984, the inaugural meeting of the All-China Minority Medical Work (*Diyijie quanguo minzu yiyao gongzuo huiyi* 第一届全国民族医药工作会议) in Hohhot was initiated by the Ministry of Health and the State Ethnic Affairs Commission. This meeting is widely considered to have been the launch of the national programme for the development of minority medicines, since subsequent years saw the establishment of a series of institutions dedicated to local cultures of medical care, and their education and research (Zhu Guoben 2006: 2). In particular, the inauguration of the China Medical Association of Minorities (*Zhongguo minzu yiyao xuehui* 中国民族医药学会, CMAM), founded in 1994, signalled an important new stage in the creation of a platform to represent China's minority medicine.

In 1995, the State Administration of Traditional Chinese Medicine and the State Ethnic Affairs Commission convened the Second Meeting of the All-China Minority Medical Work (*Di'er jie quanguo minzu yiyao gongzuo huiyi* 第二届全国民族医药工作会议), where the '316 Project' for developing minority medicine was inaugurated. The name of the project comprises three separate numerals, where '3' refers to the thirty major institutions to be selected nationwide; '1' refers to the one hundred academic leaders in clinical and technological practices; and '6' refers to the sixty minority medicine bases that were to coordinate the collection, manufacturing, and sale of minority medicinal drugs (Guojia zhongyiyao guanliju 1997: 791).

Entering the twenty-first century, the tenth five-year state plan (from 2001 to 2005) set out the task of 'salvaging and sorting' *all* minority medicines. In these five years, the State Administration of Traditional Chinese Medicine invested 3 million RMB to 'salvage and sort' nineteen minority medicines including Tibetan, Mongolian, Uyghur, Dai, Zhuang, Hui, Chao, Miao, Yi, Buyi, Yao, Kazakh, Tujia, Qiang, Dong, Mulao, Maonan, and Manchu. The work began by collating eighty-three medical documents and oral records and describing their contents, including historical development, basic theory, clinical medicine, drugs, and formularies (Zhu 2008: 363). From then on, the national project for developing minority medicines has remained on the action list of the State Five-Year Plan, right up until the current thirteenth plan (2016 to 2020).

Among such national administrative policies – both supportive and controlling – the official 'Guidelines for Effectively Strengthening the Development of Minority Medicines' has served as a schema for developing minority medicine and become a milestone in the history of China's minority medicine.

In July 2012, 'The Twelfth Five-year Plan on the Informatisation of Chinese Medicine' set out, in the item 'Platform Construction for Standard Information of Chinese Medicine' (*Zhongyiyao biaozhun xinxi pingtai jianshe* 中医药标准信息平台建设), to 'research into the standardisation of Mongolian, Tibetan, and Uyghur medical information and to promote the informatisation of minority medicines'.

At the 1984 All-China Meeting it was decided to establish a set of Minority Medicine hospitals in the minority nationality areas. By 2016, there were already 266 Minority Medicine hospitals, including ninety-nine Tibetan medicine hospitals (two grade A tertiary hospitals), seventy-two Mongolian medicine hospitals (four grade A tertiary hospitals), forty-five Uyghur medicine hospitals (two grade A tertiary hospitals), one Dai medicine hospital, and forty-nine other minority medicine hospitals. Most Chinese medicine hospitals now serving eighteen nationality groups in the minority nationality areas, such as township hospitals and some general hospitals, have opened a department of 'minority medicines'. Village clinics and some community health service centres in minority nationality areas also provide minority medical services.

The Construction Project of Major Minority Medicine Hospitals was launched in 2017, with a plan to build twenty-two major hospitals. These twenty-two hospitals now play a leading advisory role in the construction and development of China's minority medicine hospitals. So far China's minority medicine hospitals have a total of 16,255 beds, with 1,981 certified minority medical practitioners.

Academic development

First and foremost, the academic development of modern minority medicine began with literature studies. Before 1984, there were only isolated studies of the classical literature of Tibetan medicine, which were published in Tibetan (not Chinese) and therefore aimed exclusively at a Tibetan audience.

The 1984 meeting made it clear that 'to salvage and sort the heritage of minority medicines', it would be necessary to initiate a five-year project dedicated to the publication in Chinese of classical minority medicine literature. This publication project involved twenty-one classic works including eight on Tibetan medicine, seven on Mongolian medicine, two on Uyghur medicine, one on Dai medicine, two on Korean medicine, and one on miscellaneous minority medicines. It was the very first effort by the government to organise a large-scale collation of classic minority medicine works, to translate them into Chinese and to make them widely available for research and teaching.

In 2001 and 2002 respectively, two rounds of 'Sorting All-China Minority Medicine Classics' were launched and a series of fifty-nine books published. The series was called *Collections of Collation of Minority Medicine by the SATCM* (*Guojia zhongyiyao guanli ju minzu yiyao wenxian zhengli congshu* 国家中医药管理局民族医药文献整理丛书). This represents a major publication of important minority medicine classics, which also features summaries of the medical techniques and formularies of some specific minority medicines.

In general, the medicines of the so-called '55 minority nationalities' were divided into two large categories according to whether the medicines had a written tradition or not. In the first category were Tibetan, Mongolian, Uyghur, Dai, Korean, and Yi medicine. With their own written languages and a recognised corpus of medical literature, they could lay claim under the new administration to having a discrete medical system. In comparison, the local traditions with knowledge that had hitherto been passed on only through oral traditions were more difficult to 'bring under control'. With only random written records scattered throughout literature written by Chinese officials or literati, these local medicines resisted standardisation. By the time of writing, over twenty years of effort has gone into 'the salvaging and sorting' of these medicines through government initiatives. Many minority nationality groups such as the Zhuang 壮, Miao 苗, Yao 瑶, Tujia 土家, Dong 侗, Gelao 亿佬, Buyi 布依, Lahu 拉祜, She 畲, Qiang 羌, and Maonan 毛南 have also published their own books by summarising history, formula, drugs, and even the theoretical foundations for medical practices. This is considered a significant achievement in the state management of China's inheritance of oral and intangible cultural heritage (Zhu 2006). According to Tian Huayong 田华咏 of the Xiangxi Minority Medicine Research Institute (*Xiangxi minzu yiyao yanjiusuo* 湘西民族医药研究所), approximately thirty minority nationality groups in southern China alone, mainly living in the provinces of Lingnan,[5] Yunnan, Guizhou, Sichuan, Hunan and Hubei, can be sub-divided into three major cultural spheres based on their language families, ethnicity, and totem systems: the Panhu 盘瓠 medical cultural sphere represented by Miao and Yao medicine (including Miao, Yao, She, Mulao, Gelao); the Baiyue 百越 ethnic cultural sphere represented by Zhuang and Dong medicine (including Zhuang, Dong, Dai, Buyi, Shui, Maonan); the Diqiang 氐羌 medical cultural sphere represented by Yi 彝 and Tujia 土家 medicine (including Yi, Tujia, Bai, Naxi 纳西, Hani 哈尼, Lahu 拉祜, Pumi 普米, Achang 阿昌, and Lisu 傈僳) (Tian Huayong and Tian Di 2006). Research conducted into minority medicine regularly combines approaches from anthropology, ethnology, and social research, and uses cultural comparison in order to achieve a recognised public profile for the local medical assemblage which makes up the tradition. The tradition, meanwhile, is continuously emerging.

In 2009, the state initiative 'Demonstrative Research on Key Techniques of Minority Medicine Development' (*shaoshu Minzu yiyao fazhan guanjian jishu shifan yanjiu* 少数民族医药发展关键技术示范研究) approved a new focus on collecting and collating the medical literature and knowledge of the ten little-known minority medicines of the Lisu 傈僳, Blang 布朗 Deang 德昂, Nu 怒, Achang 阿昌, Hani 哈尼, Wa 佤, Mulao 仫佬, Man 满, and Ewenki 鄂温克 groups, most of whom live in the southwest regions (the Man and Ewenki live in the northeast of China). Information about these local medicines has been published by the TCM Ancient Books Publishing House.

In 2010, the state project for 'Minority Medicine Literature Collation and Appropriate Technology Selection and Propagation' (*Minzu yiyao wenxian zhengli yu shiyi jishu shaixuan tuiguang* 民族医药文献整理与适宜技术筛选推广) instigated research and collation of the literature of 150 forms of medical practice of twenty-eight minority nationalities (Zhen Yan and Hu Yingchong 2014). For the first time the medicines of several groups including the Bai 白, Kirgiz 柯尔克孜, Tatar 塔塔尔, and Tajik 塔吉克 were to be 'salvaged and sorted'. The largest of these ethnic groups is the Bai, with the majority living in Yunnan Province, while the others mainly live in Xinjiang Province.

In 2013, with the support of the National Funds for Publication (*guojia chuban jijin* 国家出版基金), the Tibetan People's Publishing House published the 30-volume *China's Traditional Tibetan Medical Texts* (*Zhongguo zang yiyao yingyin guji zhenben* 中国藏医药影印古籍珍本) including over 100 precious ancient books and manuscripts of Tibetan medicine, astronomy and calendrical calculation. Most ancient medical literature preserved in the Potala Palace in Lhasa has now been published with photo facsimiles of the originals. This has great significance for the preservation and increased awareness of Tibet's unique cultural heritage and for promoting research into the traditional knowledge of minority medicine.

Other publications include a Chinese *Materia Medica* that subsumes volumes on the national minorities medicines of the Mongolian, Miao, Tibetan, Uyghur, and Dai people into the more general rubric of 'Chinese' medicine. So far out of China's fifty-five minority medicines, a staggering 1,000 books represent research into thirty-nine of the groups. This prolific publishing record fully indicates the richness and plural characteristics of the new all-embracing rubric of Chinese medicine.

With regard to higher education, as mentioned above, so far Tibetan, Mongolian, and Uyghur medical colleges have been set up, and medical majors specialising in Dai, Korean, Zhuang, Kazakh, and Yi medicine have also been made available in Chinese medical colleges. All the above-mentioned groups have published textbooks and have instituted minority medicine licencing examinations. It was only in 1998 that the term *shaoshu minzu yixue* 少数民族医学 (Minority Medicine) first appeared in the 'List of Undergraduate Majors in Higher Education' as a second-level subject, ratifying its status as a working academic definition and a general state rubric for all minority medicines. In the second edition of the 'Subject Classification and Code of the PRC' (*Zhonghua renmin gongheguo xueke fenlei yu daima guojia biaozhun* 中华人民共和国学科分类与代码国家标准) of 2009, Minority Medicine was ranked as a second-level subject along with Chinese Medicine, Chinese *Materia Medica*, and Integrative Chinese and Western Medicine. Now, Minority Medicine has become an official discipline in China's higher education system.

There are thirty-five institutes for research into Minority Medicine above the county level employing over 1,000 researchers, including twelve minority groups. Of these institutes, six specialise in Tibetan medicine, nine in Mongolian, two in Uyghur, two in Dai, and two in Zhuang medicine. The others are engaged in research on Korean, Yao, Dong, Tujia, Hui, Miao, and Yi medicines.

The third Chinese Medicine Resource Survey, which ran from 1983 to 1987, identified more than 8,000 minority medicine drugs, 70% of the whole nation's *materia medica* resources. As regards to pharmaceutical development, the *Pharmacopeia of the People's Republic of China* (*Zhonghua renmin gongheguo yaodian* 中华人民共和国药典) has included minority medicines since 1977, and the 2015 edition added thirty-nine minority medicine drugs, of which twenty-one are identified as Tibetan medicine, eleven as Mongolian, two are used in both Tibetan and Mongolian medicine, two are Dai medicines, and two are Miao medicines and one is Yi medicine. So far 1209 minority medicine drugs have been put on the list issued by the Ministry of Health on Drug Standards, including 210 patent Tibetan medicines, 151 patent Mongolian medicines, 93 patent Uyghur medicines, 26 patent Dai medicines, 142 patent Miao medicines, and 71 patent Yi medicines (Cheng Yangyang et al. 2017: 804–8).

Conclusion

It has proved easiest for the above-mentioned government initiatives to 'salvage and sort' the medical heritage, clinical knowledge, and theoretical foundations for minority peoples that have their own written language and medical literatures. In particular, the teaching of the 'big four minority medicines', Tibetan, Mongolian, Uyghur, and Dai medicine, has entered the realms of higher education and provincial-level research institution and medical provision have been established. These minority medicine institutions, run by the state, cover clinical services and higher education, and sponsor research, and their staff have the status of public employees. In a word, these four medicines have been successfully institutionalised and have a secure academic status within the Chinese state system.

Those oral traditions with no written language remain a part of China's intangible cultural heritage, with some effort being made by researchers to materialise the immaterial through recording and translation. The key issue faced by the practitioners of these groups, nevertheless, is that they are not certified officially. In other words, there still exists the question of recognition by the health department of the government. Under the current system, if a medical tradition or practice is not recognised by the institutions and under the policies described above, there may well be restrictions placed on its practitioners, so that the tradition will run the risk of being marginalised or even eventual extinction. This is especially the case for the twenty-eight minority nationality groups with relatively small populations.

In summary, this policy review has made it clear that in China today, forms of traditional medicine and ethnic identifications are proliferating. The state-led research and regulatory projects to 'salvage and sort' the cultures and knowledges of the nation's fifty-five minority medicines, to a certain degree, accord with a prominent global turn in the social sciences, business, and government towards the management of indigenous knowledge, complementary and alternative medicine, intangible 'ethnic' cultural heritage, and intellectual property rights. The result is a new high value – commercial, legal, scientific, and cultural – attached to practices once neglected as mere folklore. The new climate of discovery and regulation, salvaging and sorting, also brings with it a certain risk: some traditions could be prohibited in the interest of consumer protection; others could be plundered for private gain by transnational big business.

These are issues worldwide, but China's situation displays special characteristics due to the country's unique history of medical pluralism and ethnic multiculturalism. The traditional Chinese medicine (TCM) system has long enjoyed official recognition and played a major role in public health and medical services; it has been both marketised and tightly state-regulated. The official importance of TCM offers opportunities for different, little-understood ethnomedicines to gain institutional and symbolic ground as healthcare, commodities, and cultural heritage. On the other hand, in China today, especially in the arena of traditional medicines, the line between buried knowledge (not qualifying for serious attention) and disqualified knowledge (denounced as naive, unscientific lore) is still under constant dispute, putting other elements of both unqualified and disqualified knowledges in the shade. The Chinese state-led project to salvage and sort minority medicine is thus a battleground for the truth that has much to teach us about how culture and authoritative knowledge grow, change, and disappear.

Notes

1 In this chapter, we follow Cai Jingfeng's translation, instead of the other common translation 'minority nationality medicines', as a general term for all the medicines of the 55 officially recognised ethnic groups in contrast to the default 'Chinese medicine' or 'Traditional Chinese Medicine' (TCM). See Cai Jingfeng (1998).
2 We do recognise the official English translation of 'medicine of minority ethnic groups' in the newly implemented *Law of the People's Republic of China on Traditional Chinese Medicine* (*Zhonghua renmin gonghe guo zhongyiyao fa* 中华人民共和国中医药法) on July 1, 2017, where 'Traditional Chinese Medicine' is adopted as the general term for 'medicine of *all* Chinese ethnic groups including the Han *and* the minority ethnic groups' [Article 2] (our italics). However we maintain that, as Mullaney has made clear, 'ethnic groups' does not convey the complicate history and great efforts expended in the 1950's 'Ethnic Classification Project', nor is the term accurate given what happened on the ground, so we stay with 'minority medicine' adopted by Cai and Zhen. The

same strategy goes for our usage of 'minority nationalities' instead of 'minority ethnic groups', but 'ethnic' does appear from time to time when it is needed to refer to a specific group with its very own local history.

3 Judith Farquhar and Lai Lili have been conducting a large-scale anthropological research project on the 'minority' medicine of seven groups in Southwest China since 2010. One of their research goals is to observe the impact and effect of the state-led projects for developing minority medicines (Farquhar 2017; Lai and Farquhar 2015; Farquhar and Lai 2014).

4 Zhongguo zhongyang pizhuan guojia jiwei dangzu 中共中央批转国家计委党组 (1960) 'Guanyu yijiu liuyi nian guomin jingji jihua kongzhu shuzi de baogao 关于一九六一年国民经济计划控制数字的报告' (Report of the Numbers of Planning and Controlling National Economy of the Year of 1961), http://www.china.com.cn/guoqing/2012-09/11/content_26747044.htm, accessed 7/10/2019.

Ministry of Health of the People's Republic of China 中华人民共和国卫生部 (1979) *Weishengbu guanyu yiyao weisheng keji gongzuo shiqie 'Bazi fangji' de yijian* 卫生部关于医药卫生科技工作贯彻'八字方针'的意见 (Suggestion on Thoroughly Implementing the Eight-character Guiding Principles in Medical, Health and Technology Work), https://wenku.baidu.com/view/15df-1d4a81eb6294dd88d0d233d4b14e84243e37.html, accessed 9/3/2020.

5 The area to the south of the Five Ridges, including Guangdong and Guangxi provinces.

Bibliography

Cai Jingfeng 蔡景峰 (1998) 'Zhengque lijie minzu yixue de hanyi 正确理解"民族医学"的涵义', *Chinese Journal of Medical History*, 28.3: 129–30.

Cheng Yangyang 程阳阳, Yu Jiangyong 于江泳, Lin Ling 林灵, Li Yanwen 李彦文 and Li Zhiyong 李志勇 (2017) 'Zhongguo yaodian shouzai minzuyao chengfang zhiji de tongji yu fenxi《中国药典》收载民族药成方制剂的统计与分析', *Zhongchengyao* 中成药, 39.4: 804–8.

Cheng Zhaosheng and Fang Mingjin (2012) 'Woguo zhongyiyao zhengce gaishu (xiapian) 我国中医药政策概述(下篇)', *Zhongguo zhongyiyao bao* 中国中医药报, 17/5/2012 (003).

Farquhar, J. and Lai, L. (2014) 'Information and its practical other: crafting Zhuang nationality medicine', *East Asian Science, Technology and Society*, 8: 417–37.

——— (2021) *Gathering Medicines: nation and knowledge in China's mountain south*, Chicago; London: University of Chicago Press.

Farquhar, J., Lai, L. and Kramer, M. (2017) 'A place at the end of a road: a yin-yang geography', *Anthropologica*, 59: 216–27.

Lai, L. and Farquhar, J. (2015) 'Nationality medicines in China: institutional rationality and healing charisma', *Comparative Studies of Society and History*, 57.2: 381–405.

Mao Zedong (1999) *Maozedong wenji diqijuan* 毛泽东文集第七卷, Beijing: Renmin chubanshe.

Ministry of Health and National Ethnic Affairs Commission (1983) *Weishengbu, Guojiaminwei guanyu jicheng fayang minzuyiyaoxue de yijian* 卫生部、国家民委关于继承、发扬民族医药学的意见 (Guidelines on Inheriting and Developing Nationality Medicines), http://china.findlaw.cn/fagui/p_1/361612.html, accessed 11/3/2020.

Mullaney, T.S. (2011) *Coming to Terms with the Nation: ethnic classification in modern China*, Berkeley; Los Angeles; London: University of California Press.

Tian Huayong 田华咏 and Tian Di 田莳 (2006) 'Miaoyao yu minzu yiyao wenhua tanyuan 苗瑶语民族医药文化探源' (Seeking out national medicine with culture of the Miao language), *Journal of Medicine and Pharmacy of Chinese Minorities*, 5: 1–3.

Ministry of Health of the People's Republic of China 中华人民共和国卫生部 (1979) *Weishengbu guanyu yiyao weisheng keji gongzuo shiqie "Bazi fangji" de yijian* 卫生部关于医药卫生科技工作贯彻 "八字方针" 的意见 (Suggestion on Thoroughly Implementing the Eight-character Guiding Principles in Medical, Health and Technology Work), https://wenku.baidu.com/view/15df1d4a81eb6294dd88d0d233d-4b14e84243e37.html, accessed 9/3/2020.

Zhen Yan and Hu Yingchong (2014) 'The collection of early texts and the establishment of databases for Chinese nationality medicines', *East Asian Science, Technology and Society: An International Journal*, 8.4: 461–78.

Zhongguo zhongyang pizhuan guojia jiwei dangzu 中共中央批转国家计委党组 (1960) *Guanyu yijiu liuyi nian guomin jingji jihua kongzhu shuzi de baogao* 关于一九六一年国民经济计划控制数字的报告

(Report of the Numbers of Planning and Controlling National Economy of the Year of 1961), http://www.china.com.cn/guoqing/2012-09/11/content_26747044.htm, accessed 07/10/2019.

Zhu Guoben 诸国本 (2006) *Zhongguo minzu yiyao sanlun* 中国民族医药散论 (Essays on China's Nationality Medicines), Beijing: Zhongguo yiyao keji chubanshe.

——— (2008) *Yilin zhaomu* 医林朝暮 (Days and Nights of the Medical World), Beijing: Zhongyi guji chubanshe.

PART 6

Wider diasporas

38
EARLY MODERN RECEPTION IN EUROPE
Translations and transmissions

Eric Marié

The first transmissions of medical knowledge from China to Western Europe were initiated as early as the end of the sixteenth century,[1] but the most important developments happened in the second half of the seventeenth century and during the eighteenth century. Strictly speaking, I refer to the importation of theories and practices from China to the Europhone world. Since the publication of the first treatises on Chinese medicine in Western languages, they continuously aroused interest, so much so that they gave rise to intellectual and doctrinal conflicts between 'sinophile' and 'sinophobe' physicians. At that time, pulse diagnosis was a recurring question. It would even result in a five-year correspondence (1784–89) between a French and a Chinese practitioner, transmitted thanks to the Jesuits of the French mission of Beijing (Mission française de Pékin). The texts which mediated these exchanges also constitute a catalogue of the various problems that followed the confrontation between two different systems of body representation, as physicians from across the globe tried to understand one another.

This study aims to show how medical knowledge circulated between China and Europe in the early modern period; to estimate how it was received; to reveal the difficulties of translating the terminology and their consequences, and finally, to analyse the confrontation between the two types of knowledge; its repercussions on medical practice and its general consequences for history and epistemology.

The two main categories of sources we have access to are testimonials from travellers and scholarly treatises. Most of these documents come from Catholic missionaries (Jesuits) and physicians from naval companies.

Tales and testimonials from travellers

During the seventeenth century, a great deal of information about China reached Europe, brought by Catholic missionaries and merchants. Although their testimonials are of great anthropological interest, their views on the medical system were more from the perspective of curious neophytes or occasional patients than that of specialists in the medical field. Three observers are illustrative of this contribution to the introduction of Chinese medicine in Europe.

The Spanish missionary Don Francisco de Herrera Maldonado, writer of a book printed in Madrid in 1621 and translated into French, then published in Paris the following year, expressed his admiration while giving some practical information: unlike European physicians, Chinese practitioners neglected the observation of urine but focussed more on the palpation of the pulse, which they classified, according to Herrera, into seventy different forms that were felt on different areas of the body (Herrera 1622: 169–70).

Johan Nieuhof (1618–76) dedicated to the practice of medicine in China two pages of a voluminous work which records the impressions of the representatives of the Dutch East India Company. Father Alexander of Rhodes (1591–1660), a French missionary who had travelled in China, including hard-to-reach areas, wrote with admiration the skill of the Chinese physicians which he witnessed as a patient, having received particularly efficient care preceded, naturally, by a long and thorough examination of the pulse. His detailed and personal description evokes particularly well the process of a consultation in seventeenth-century China.

> One will laugh at these people if I say that he becomes a Physician who wants to, and one will believe there's no good in trusting people that deceive their patients. But I, who have been in their hands, and who witnessed their abilities, I can say they easily bear comparison with our Physicians and even surpass them in some ways. (...) They particularly excel in knowledge of the pulse, from which they must learn all the subtleties. As soon as the Physician comes to see the sick, he feels his pulse and reflects on it for more than a quarter of an hour, then he has to tell to the patient where he is ailing and all the incidents which occurred since he fell ill.
>
> *(Rhodes 1653: 189–94)*

In addition to their interest as testimonials, these accounts from travellers hold some technical information. Thus the division of the radial pulse into three segments and the anatomical correspondence attributed to each part are partially described, even though there is some confusion: Alexander of Rhodes remembered little more than a vertical repartition in three parts from up to down, without perceiving more subtleties. The amount of seventy different pulses, according to both Herrera and Nieuhof, is surprising and inconsistent with older and contemporary Chinese sources. On that point, the similarity between the two writings is also surprising. Did Nieuhof draw on the account of Herrera to complete his description? At any rate, the impact of this information was to be felt for a long time in European medical circles, some influential members of which would sometimes consider Chinese sphygmology impracticable because of the extreme diversity of its pulse. Another source of interest is the allusion to peripheral pulses and the central role of palpation in clinical examination. The length of time devoted to palpation that they report, excessively long, is probably of a subjective nature. However, the European physicians who read them took the information they held as strictly correct. This helps to understand how some myths about Chinese medicine emerged in scholarly circles, up to the eighteenth century, with two opposing standpoints: for some, it was a scholarly medicine; for others, an empirical medicine, without philosophy or theoretical foundations, like some sort of worker's know-how learnt by practice.

First scholarly writings on Chinese medicine

Several monographs about theoretical and practical aspects of Chinese medicine, the first written by Westerners, were published during the second half of the seventeenth century.

Their influence was to be decisive for the interest and knowledge of Far Eastern health practices in Europe, as well as for the initial opinions, favourable or unfavourable, and consequently for the first medical debates on the value, purpose, validity and possible exploitation of Chinese theories and methods in Western medicine.

The first work was composed by Jacob de Bondt (1592–1631), a practitioner working for the Dutch East India Company. Lu Gwei-Djen and Joseph Needham quote this book, adding that it was the first real treatise written by a Western author on Chinese medicine and acupuncture:

> [...] The very first writer, so far as we can see, who spoke about acupuncture, was the Dane Jacob de Bondt who, in his capacity as surgeon-general for the Dutch East India Company at Batavia, had come into contact with Chinese and Japanese physicians [...].
> (Lu and Needham 1980: 260)

Jacob de Bondt had already written *De Medicina Indorum*, published in 1642, but the work in question here is another posthumous treatise, *Historiae Naturalis et Medicae Indiae Orientalis*, dated from 1658.

Another practitioner from the Dutch East India Company, Wilhem Ten Rhyne (1647–1700), having also lived in Batavia from 1673, made mention of that last work from Jacob de Bondt in his own book, published in 1683 in London, La Haye (the Hague) and in Leipzig.

These two authors, especially Ten Rhyne, are worth keeping in mind as they would be used as sources by physicians of the nineteenth century (Dujardin, for example) in their works on Chinese medicine, most notably when it came to acupuncture.

The first monograph specifically dedicated to pulse diagnosis was published in Grenoble (France) in 1671. It is an anonymous book entitled *Secrets of the Medicine of the Chinese that lie in the perfect knowledge of the Pulse, Sent from China by a French, A man of great merit*. The identity of the author is not the only enigma that surrounds this book. The origin of the knowledge constitutes another question of major interest. It does not contain any quotation or reference, and also no information concerning its sources. For more than three centuries, historians and sinologists have tried to solve this question with various conclusions. I endeavoured to study this little book, to try to figure out its author and the circumstances of its creation and, more importantly to me, to identify its main Chinese source, unknown at that time.[2] Actually, the main part of the book is a translation, sentence by sentence, of some parts of the *Zhenjia shuyao* 診家樞要 (Conducting principles of the Masters of diagnosis), written in 1359 by Hua Shou 滑壽 (1304–86), also known by his style name Hua Boren 滑伯仁.

The compilation of a travelling physician

At that time, the most able Jesuit regarding Chinese medicine was Michel Boym. The son of the first physician of Sigismond III (1566–1632), King of Poland, he chose, at the age of seventeen, to join the Society of Jesus where he undertook his studies instead of embracing the career of his father, although he kept an interest in the medical arts and developed an excellent knowledge of this discipline. His advanced mastery of Chinese language and culture allowed him to gather documents and to compile several works on botany and medicine.

In Siam in 1658, one year before his death, Michel Boym finished writing his work on Chinese medicine and entrusted his manuscripts, all written in Latin, to his companion Philippe Couplet who handed them to the Jesuits of Batavia so they could be published in Europe. But the documents were confiscated by the Dutch of the Dutch East India Company by way of reprisal for the position of the Jesuits in China, whom they accused of causing prejudice against their commercial interests. Boym died in 1659, at the frontier between Tonkin and Guangxi, without any news about the fate of his writings.

In 1681, Philippe Couplet was sent to Europe. He left Macao on the fifth of December but had to stop at Bantam (or Banten, on Java island) because of a storm. He probably met there a famous character who would play an important role in the spread of Chinese medicine in Europe: Andreas Cleyer (1615–90). This Prussian surgeon had lived for a long time in Batavia, studying the local flora and pharmacopoeia. He probably never set foot in the Chinese empire. Corresponding with several European scholars, including the sinologist and physician Christian Mentzel (1622–1701) and Georg Eberhard Rumphius (1628–1702), Cleyer worked for years on gathering documents about Eastern medicine. When they met, Philippe Couplet handed over to him some of Michel Boym's works he had kept with him. Cleyer integrated these new writings into a compilation he addressed to Mentzel which was published in 1682 in Frankfurt, entitled *Specimen medicinae Sinicae*. The fact that Boym wasn't quoted, although a part of his work was probably used, led some historians to accuse Cleyer of plagiarism. The polemic started at the beginning of the eighteenth century with the sinologist Bayer, when he discovered the borrowing by the Prussian physician. This was followed and amplified by Chabrié, then by Rémusat, who both called it a scandal, accusing Cleyer of being a plagiarist, or even a thief. This accusation of plagiarism deserves, with the benefit of hindsight more exhaustive analysis, some nuancing and even requalification, for Cleyer always introduced himself as editor and not as author of the *Specimen medicinae*.[3] It's precisely due to him that the treatise of Boym was published four years later, under another title. Furthermore, it's obvious that the two books are very different, as much in their general composition as in their style.

Strictly speaking, the *Specimen medicinae Sinicae* isn't a treatise but a collection of different texts on several subjects. The composite nature of the publication is clearly revealed by the variations in style of its different parts. It's probable that Cleyer simply put together several writings on oriental medicine. Six parts can be formally identified.

De Pulsibus Libros quatuor e Sinico translatos, De eplanatione pulsuum regulae

The first two booklets of this part are dedicated to the rules of the palpation of the pulse, attributed by Cleyer to 'Wam Xo Xo'. It's easy to guess that the latter is Wang Shuhe, author of the *Maijing* 脈經 (Pulse Classic). The pulses are associated with pharmacopoeia formulæ, most of them taken from the *Shanghan lun* 傷寒論 (Treatise on Cold Damage) (Chapter 9 in this volume).

Tractatus de Pulsibus ab erudito Europaeo Collectus

This part is a sort of commentary on the 'Nuy kim' ([*Huang di*] *Neijing* [黃帝]內經 [Inner Canon]), including a system of correspondence between pulse segments and the viscera. Mostly a compilation of various theories, the text is sometimes difficult to identify due to the lack of references and Chinese characters.

Fragmentum Operis Medici ab erudito Europaeo conscripti

This selection is a collection of texts and commentaries on diverse subjects, taken from the *Neijing*, the *Maijue* 脈訣 (Secret of the Pulses), based on some commentaries derived from the

Shanghan lun and various other writings. The content is very disparate, including explanations of the seasons and the Chinese calendar, indications of specific pulses or combinations of pathological pulses, references to pharmacopoeia treatments, various aphorisms on fever, diarrhoea, life and death and so on.

Excerpta Literis eruditi Europaei in China

Four letters written in Guangdong in 1669–70 are presented in this chapter. The first one is dated February 12, 1669. It notably contains information on the differences between the pulses of men and women, a comparison of the correspondence between viscera and pulse segments as they were previously mentioned, on the one hand, and as they appear in the *Inner Classic*, on the other hand,[4] an interesting analysis of the concepts of *mingmen* 命門 (gate of life) and *sanjiao* 三焦 (three burners) and a presentation of the number of pulsations in a day and its ensuing physiological consequences. The second letter (1669, October 20) mainly consists of a diagram of 'blood circulation' in the twelve vessels according to the twelve Chinese hours. The third one (1670, November 5) includes a presentation of the system of correspondences of the Five Agents and a study of the pulses of vital prognosis according to various illnesses. The fourth letter (1670, November 15) presents the theory of the fifty uninterrupted pulsations: the pulse of a healthy person must beat regularly and without break for at least fifty pulsations, thus meaning that the five organs are healthy. It then deals with two pulses located on the foot, 'chum yam' (*chongyang*, 42nd point of the meridian *zuyangming* of the stomach, on the dorsalis pedis artery) and 'tai hi' (*taixi*, third point of the meridian *zushaoyin* of the kidneys, on the posterior tibial artery), the absence of which means death. A sketch of the theory connecting radial and carotid pulse, 'ki keu' *(qikou)* / 'gin ym' *(renying)* is also mentioned. Finally, a last extract originates from an undated letter without place, which contains another, more explicit, schematic on the circulation in the vessels according to the hours of the day. Cleyer indicates that the author of the letter quotes the *Neijing* (written *Nuy Kim* in the document), without any more precision but, according to the content of the passage, we can assume that this theory originates from *Lingshu* 靈樞 (*Divine Pivot*), 18. This part of Cleyer's work ends with a succinct presentation, which he quotes as taken from the *Neijing*, on the 'eight extraordinary vessels'.

Schemata ad meliorem praecedentium Intelligentiam

This fifth part consists of eight pages of schematics recapitulating diverse information on the pulses, in a synoptic and fairly well-structured way, and of thirty medical illustrations mainly depicting viscera and vessels. These are of Chinese style and can be found, in almost identical form, in some treatises edited in China towards the beginning of the seventeenth century; it's probable that Cleyer simply copied these illustrations while adding annotations and captions in Latin. Two illustrations constitute an exception, being of European style. The first shows the positions of the fingers for taking the pulse (Figure 38.1); the other represents a body annotated with strategic medical points. In the same part of the book, there is a compendium of Chinese drugs, described by phonetic equivalents of their Chinese names (Figure 38.2).[5]

De Indiciis morborum ex Linguae coloribus & affectionibus

This last part is dedicated to the examination of the tongue and its coating, notably on the basis of their different colorations, with the main corresponding pathological indications (Chapter 10 in this volume).

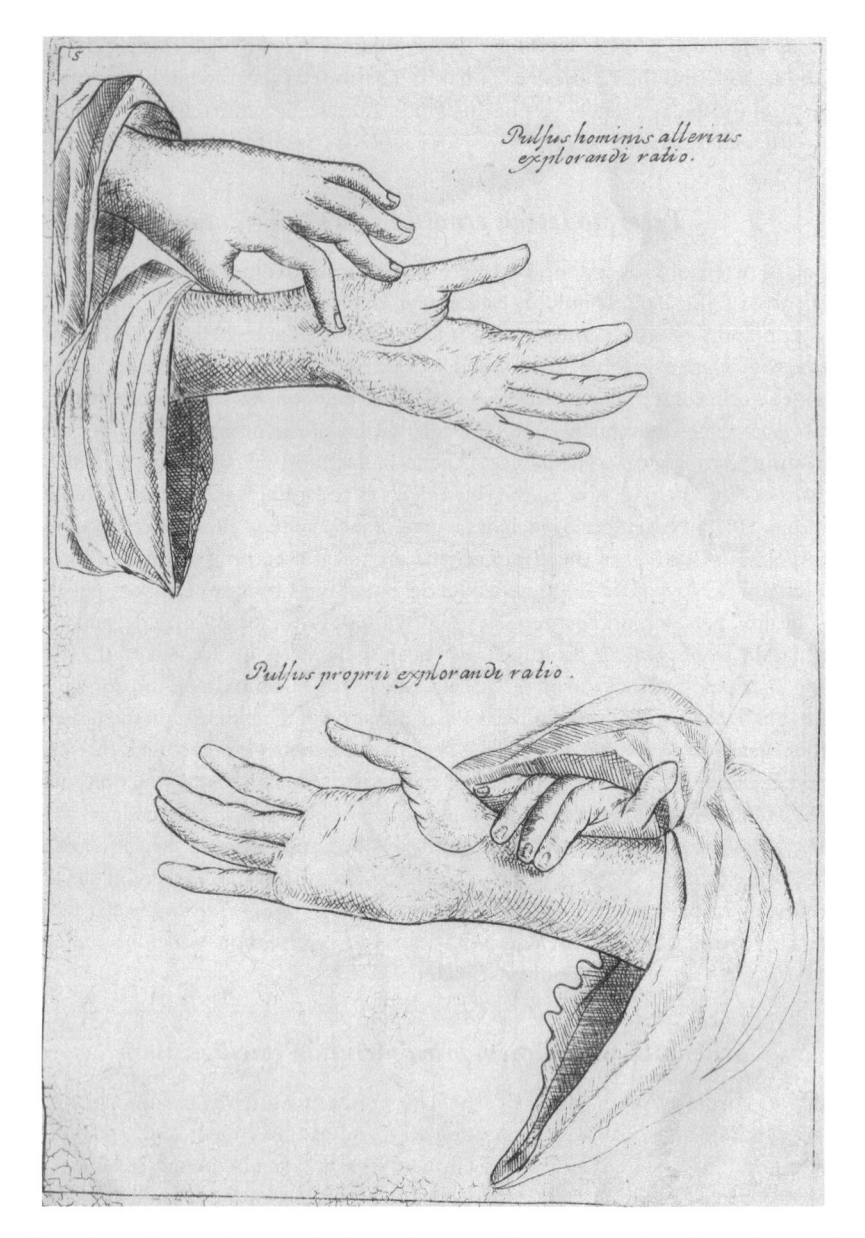

Figure 38.1 Pulse diagram, *Specimen Medicinae Sinicae*, p. 20. Courtesy of the New York Academy of Medicine Library

The treatise written or, more precisely, 'compiled' by Cleyer is of major interest, especially because of its composite nature, as it combines information from his own researches but also with fragments of studies by sinologists living in Europe and one anonymous writer living in Guangdong: it defines a state of knowledge on several aspects of Chinese medicine. Among the three mentioned treatises on Chinese sphygmology, it is the only one to have been produced by a renowned professional. It is probable that Cleyer gave more authority to Chinese medicine. His work rapidly became a reference and had a substantial influence among physicians and scholars interested in China, until the nineteenth century.

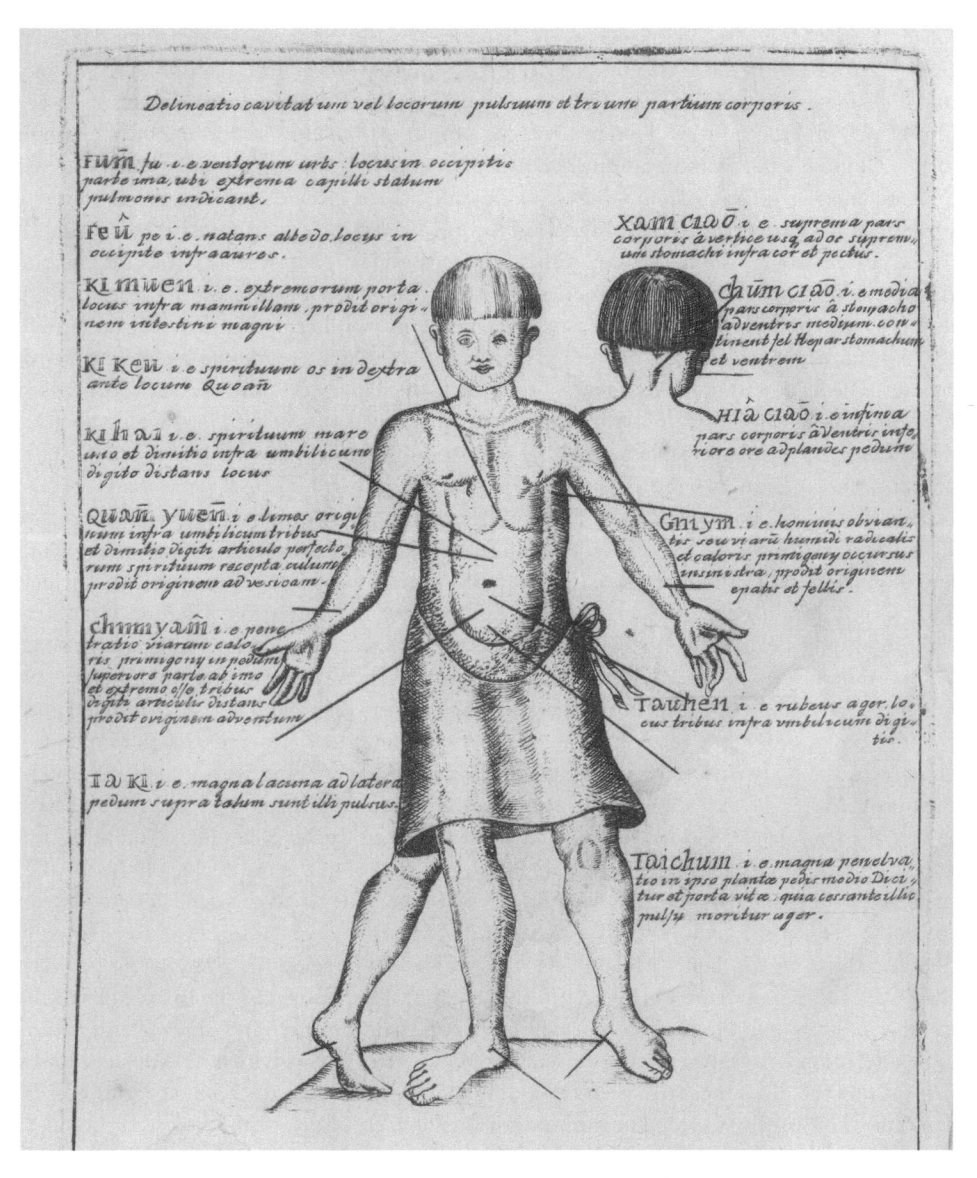

Figure 38.2 Conflation of acupoints and pulse points, Specimen Medicinae Sinicae, p. 68. Courtesy of the New York Academy of Medicine Library

A scholarly treatise on Chinese sphygmology

In 1658, shortly before his death, Michel Boym completed a book entitled *Clavis medica ad Chinarum Dotrinam de pulsibus*. After many tribulations, it would eventually be edited in Nuremberg, in 1686, paradoxically several years after the two other previously mentioned texts. Despite attempts to reconstitute the text through historical research, a certain mystery still hovers over the circumstances of its belated evolution. Boym's work is a true scholarly treatise, being constructed in a rigorous and coherent manner and redacted in a uniform style which distinguishes it clearly from the mosaic of disparate texts compiled by Cleyer.

The text mainly consists of seventeen chapters (actually eighteen, two distinct consecutive chapters bearing the same number). The first one presents an unusual (folk?) etymology of the character *ren* 人 (man, human being), stating that the left oblique stroke expresses the idea of *yang* and the right stroke represents the notion of *yin*. Thus, according to the author, for the Chinese, man would be defined as the reunion of the *yinyang*. This may be an abstruse interpretation originating from Chinese calligraphic theory: the first stroke, *pie* 撇 (left descending), thick at the top and thin at its end is considered as *yang* and the second stroke, *na* 捺 (right descending), thin at the top and thick at its end is considered as *yin*. The second chapter defines, in a succinct but meticulous way, the notion of 'cam & fu' (*zangfu* 臟腑), that is to say the Chinese theory of visceral physiology, while the third chapter deals with the classification of the vessels, presented as ways of circulation inside the body. The fourth chapter is dedicated to the Five Agents. From the fifth chapter, the author delves deeper into sphygmology, starting judiciously by explaining the movements of the *qi*, to which he gives the name of 'spiritus', probably by making an inappropriate semantic link with the Galenic concept of *vital spirit*, and those of the *xue* 血 (blood) which he translates as 'sanguis' (On Galenic medicine, Ballester 2002). The sixth chapter contains slightly surprising data: the author enunciates twelve major body locations for palpating the pulses, which he seems to have conflated with a set of major acupoints as, for some of them, there is no palpable artery. This set is reproduced by Cleyer in one of his anatomical illustrations (Figure 38.2). The seventh chapter deals with the theory of the Three Burners, presented as an anatomical division of the viscera in three levels, from top to bottom, which the author links to Heaven, Earth and Man; then he returns on other aspects of the *yin-yang* and the vessels. In the eighth chapter, Boym evokes the relation between the viscera and their respective orbs of influence, notably on the body tissues, by giving examples rather than dealing with it exhaustively. In the ninth chapter, he explains that the pulses cannot be taken only on the left wrist, but have to be palpated bilaterally; then he broaches the influence of time cycles (daily, monthly and annual) on the circulatory movements in the human being. In the ninth chapter, he justifies the reasons for radial palpation by clarifying how it is possible to perceive the general state of the body solely at the areas of the wrists, a reading which he deepens in the tenth chapter by defining the three segments. Chapters eleven to thirteen develop the relationship between the pulses and the measure of time, with the stages of circulation and the calculation of the number of daily pulsations. The fourteenth chapter mentions the physiological and pathological relations between respiratory rhythm and pulse frequency, which is completed in the fifteenth by the differences between adults and children, men and women, according to fat or slender constitutions, etc. The author explains how to distinguish the constitutional or physiological variations and the pathological alterations. The sixteenth chapter contains a list and descriptions of the twenty-four pulses, according to the well-known classification of the seven externals, the eight internals and the nine ways, first established by Wang Shuhe in the *Maijing*, and widely used since. The treatise then turns to the question of the functions of the *qijing bamai* 奇經八脈 (eight extraordinary vessels), that the author integrates into the pulses theory. The second 'chapter sixteen' gives various instructions on the interpretation of the pulses at the three segments of the radial artery. Finally, the seventeenth and last chapter presents the modifications of the pulses according to seasonal and climatic variations.

It appears clear, from reading his book, that Boym had carefully studied the Chinese medical treatises. However, on several occasions, we also notice knowledge based on oral transmission. The iconography illustrating his tractate is precise and explicit enough to be employed for practical purposes. Finally, Boym insists on the autonomy and completeness of the Chinese system which is totally independent from ancient Greek medicine, even though, in some places, he fails to escape from the risk of syncretism between the two doctrines, such as when he translated *yang* (written *yam*) into *calor primigenius* (primordial heat) and *yin* (written *in*) into *humidus radicalis* (radical humidity).

European scholars inspired by Chinese medicine

The above-mentioned writings inspired several European scholars to take an interest in Chinese medicine. We can mention a translation of Nieuhof's book from the Dutch original into English by John Ogilby, and the almost complete translation of the *Specimen Medicinae Sinicae* by William Wotton (1660–1726). Like Wotton, Sir William Temple (1628–99) and Isaac Vossius (1618–89) drew inspiration from this book. But their enthusiasm did not build on clinical experience. However, at least two British physicians, who endeavoured to understand and test Chinese medical practice, must be mentioned.

David Abercromby (1621–95) took an interest in the Chinese theories and drew inspiration from them and attempted to experiment with them. In his book, he states that the Chinese pulses bring an 'insight of the illness' and do so that the doctor, 'like a lynx, will scrutinize every change in the patient'.

Another British physician devoted himself to the study of Chinese pulse diagnosis, with an undeniable enthusiasm. Sir John Floyer (1649–1734) was a complex blend of conservative gentleman, erudite physician (he studied medicine and sciences for sixteen years in Oxford), prolific baroque writer and nonconformist researcher. This picturesque personality, passionate for the ancient civilisations of the Far East, contacted several sinologists, searched the libraries, and ended up discovering Chinese medicine and more especially, its pulse doctrine. He obviously knew the *Specimen medicinae Sinicae* (it was even probably his main source) of which he composed an English adaptation in the addenda of his most important publication on the matter: *The Physician Pulse Watch*. The explicit objective of Floyer was to propose a synthesis of Galenic and Chinese methods, completed by some personal contributions. In fact, he gives prominence to the Chinese method, considering that it was 'more evident, surer and more concise' than the Greek.

In France, it was mostly the physicians of the Vitalist school who collected and exploited the theories of Chinese medicine, most notably Jean-Jacques Menuret de Chambaud (1733–1815) and Henri Fouquet (1727–1806), who analysed Chinese pulse diagnosis in their writings.

The contribution of the Jesuits in the eighteenth century

Interest in Chinese medicine reached its peak after the 1730s. One of the main actors in this development was a French Jesuit, even though he never saw China: Jean-Baptiste Du Halde. He received from his superiors the task of collecting and redacting all the scientific reports of the missionaries sent to China, which he fulfilled with the publication of a remarkable encyclopaedic work in four folio volumes, ornate with a great number of engravings and containing, in the third volume, some information on Chinese medicine. It should be mentioned that some of them came from another Jesuit, Julien Placide Herviue (1671–1746), who arrived in China in 1701 and stayed there for forty-five years, until his death. His contribution to Du Halde's work is a partial translation of the *Tuzhu maijue bianzhen* 圖註脈訣辨真 (Discerned Truth of the Secrets of the Pulses, Illustrated and Commented), written during the Ming by Zhang Shixian 張世賢 (dates unknown), also known by his style name Zhang Tiancheng 張天成 and published in China in 1565. It contains information on methods of palpation, on the description of a number of pathological pulses and on the prognosis of various illnesses by the pulses. Throughout his explanations, Du Halde interprets Chinese theories and transcribes them into Galenic medical terminology. For him, *yang* corresponded to 'vital heat', *yin* to 'radical humidity', *k'i* (*qi*) to 'pneuma', etc. In this way, Far Eastern medicine became easily understandable for the European reader, but only under a dim light. The work of Du Halde, translated and published in English, German and Russian, would have a considerable influence on European physicians until the nineteenth century, who would borrow his descriptions and opinions, neither verifying them nor submitting them to arbitration of their own judgement. In this way the argument was often repeated, that the Chinese were ignorant of and had no interest in anatomy, to justify the opinion that their medicine could not ever reach the level of the European's one.

Towards the end of the eighteenth century, the contribution of the Jesuits would take another shape. At that time, Chinese medicine was not solely of interest to practitioners. It is sufficient to read some articles from the *Encyclopédie* of Diderot and d'Alembert, notably the one entitled 'Pouls' whose contents can also be read in a book, *Nouveau traité du pouls* (New treatise on the pulse, 1768) by Menuret de Chambaud on the subject: an important part of the book is dedicated to Chinese theories and practices. One can also observe in Parisian scholarly circles a deep curiosity for all that comes from Oriental civilisations. A high official, Henri Bertin (1720–92), at that time Controller-General of Finances and Secretary of State, initiated a correspondence between Paris and Beijing. Surrounded by a team of scientific collaborators, among whom Louis Oudard de Bréquigny (1761–95), a member of the French Academy, should be particularly mentioned, Bertin permitted an important quantity of data from China to be gathered, more precisely collected by the best special envoys at his disposal: the Jesuits of the French mission in Beijing, and especially Jean Joseph Amiot (1718–93). The regular reports they addressed to Paris are the basis of fifteen volumes of *memoirs*, without taking into account a number of unpublished letters. The mail that sent from France was added to by a Parisian physician who wished to delve deeper into the Chinese pulse method with which he had experimented for twenty years. He was Charles Jacques Saillant (1747–1814), who in his youth authored a thesis on pulse diagnosis, and went on to become Docteur-régent of the Faculty of Medicine of Paris and a member of the Royal Society of Medicine. On November 16, 1784, he wrote a letter to the missionaries of Beijing in which he interrogated them on precise points of Chinese pulse diagnosis. He notably mentions the confusion resulting from the various interpretations transmitted by works written or translated by Europeans.

The answer would reach him only in 1787, on the basis of the explanations of a Chinese doctor in Beijing interviewed by Amiot. At first, he specified the attributions of the various locations of the pulse. Encouraged by this first answer, Saillant continued his epistolary relationship with his informer by addressing him, this time in a memorandum composed of personal thoughts dated from September 22, 1787, under the title 'New questions about the pulse'. The French physician proposed an equivalent system between the Chinese pulses and those used by Europeans.

Amiot answered on June 26, 1789. His letter is interesting because it contains a personal account of an illness of which he was cured by Chinese medicine. He does his best to translate the explanations of his Chinese physician while expressing the difficulties he meets in the process. As soon as he comes to delve into technical subtleties, the epistemological gap between the two medical doctrines becomes obvious: how can a European physician understand the meaning of a '*shanghan* pulse' and how can a Chinese physician answer the questions about the pulses of the 'crisis' in reference to theories of medical vitalism of eighteen-century France?

We learn in the rest of Amiot's letter that the Chinese physician sent to Saillant a medicinal powder for treating headaches. The Jesuit missionary ends his message by repeating a general history of the development of medicine in China and by strongly criticising the translations of Chinese medical treatises made by Europeans. The epistolary relationship between those two practitioners could probably have continued and led to readjustments and exchanges of a major interest. Unfortunately Amiot, sick and ageing, complains that he is unable to continue to act as an intermediary.

The difficult confrontation between two medical cultures

At the beginning of the seventeenth century, Chinese medicine was seen as an exotic knowledge that aroused the curiosity of scholars. The first travellers to witness Chinese medicine were not interested in its theories. However, they bore witness to its practical efficiency. A physician reading their accounts could not fail to be intrigued, especially if sensitive to the exotic attraction of these foreign lands.

One should not conclude that Europe was ready to welcome Chinese medicine and to spontaneously adopt it. There was at the same time a strong resistance and a real attraction. On one side, there was no question of giving up the idea of the pre-eminence of the European medical model. Travellers returning from China thought that Chinese physicians were efficient but that they did not have any academic education; although they thought that their know-how was excellent, they erroneously assumed that their theories did not rest on any philosophy. However, some Europeans ended up wondering if it was reasonable to think that such a good practice was possible without theoretical foundations. This investigation, first stirred up by travellers' accounts, played out in more sophisticated ways through the theoretical considerations and translations in the three books that have been introduced.

The publication of the *Secrets de la médecine des Chinois* constituted the first practical contribution as a book of popular science. Furthermore, just because a subject was fascinating in its own right, it did not provide a reason to overthrow Western medical reasoning. It is clear from this that a European physician could learn these 'admirable secrets' without having to question the epistemological foundations of his own system. The theoretical approaches that fascinated Europeans were those that were contingent on existing medical debates in Europe, or easily understood, and thus this was the reason for their selection of the 'pulse', 'wind', 'vessels' (Bivins 2000). On the other hand, elements that were spectacularly foreign, also attracted attention, albeit for other regions. These two modes of analysis and selection meant that European reception would never directly mirror the medicine being transmitted.

When the *Specimen medicinae Sinicae* of Cleyer was printed, the discourse evolved. This compilation, written in Latin by an eminent surgeon, pointed to a much vaster knowledge, with its own representations of the body, its physiological conceptions, its diagnostic and treatment methods, its pharmacopoeia etc. Through this work, Europe discovered the scholarly nature of Chinese medicine.

Finally, after the publication of Boym's *Clavis medica*, written in a fluid and rigorous style which revealed the sophisticated understanding and philosophical approach which underpinned Chinese medicine, the European reception deepened to the point where it generated reactions from scholars far beyond the medical field. Boym's feat was showing that a good Christian can learn and use Chinese medicine, and that it possessed a real epistemological autonomy.

This analysis reveals the beginning of thinking in Europe about the relative and subjective nature of the theoretical value of medical systems. From the end of the seventeenth century, Chinese medicine found supporters as well as detractors. More importantly, it began to be considered as a comprehensive, systematic approach rather than a mere collection of techniques and exotic recipes. However, the confrontation between the two medicines wasn't easy. One can imagine the difficulty that Amiot and his Chinese physician, whose name is regrettably unknown, had encountered when translating medical terminology and, above all, concepts that were impossible to transpose from one medical system to the other. The confusions and mistranslations that occur are mainly the result of this impossibility of the transfer. Terms like 'crisis', which had an unambiguous and obvious meaning for a Western physician of the eighteenth century, were practically untranslatable in Chinese, except by using paraphrases assorted with explanations that only a specialist of both medicines could have produced. The opposite problem was met with medical expressions like *shanghan* which specifically belong to Chinese nosology. Amiot was conscious of this translating difficulty: he mentions and apologises for it on several occasions. However, this apparent linguistic obstacle hides another, even more pernicious one, less obvious at first glance: behind some terms whose translations appear easy, are hidden completely different notions of meaning. Translating *xin* 心 as 'heart' and *wei* 胃 as 'stomach' is literally accurate when speaking in terms of Western medicine. But these words and many other are as such *false friends* when talking about Chinese medicine. To facilitate the scientific exchange between Saillant and his Chinese correspondent, it would have been necessary to know not only both languages, but also both medical systems. Two centuries later, this problem is still topical in many medical exchanges between China and the West (Chapter 43 in this volume).

Notes

1 There were European travellers in China long before this period and their early impressions of Chinese medicine aroused Western curiosity and imagination (Barnes 2005: 8–35), but these observers did not contribute to what might be precisely defined as a transmission of medical knowledge.
2 The method that allowed me to identify this source led me to a better understanding of how the Chinese medical texts were translated, understood and exploited by Europeans during that time. For more details about the method and the demonstration, *cf.* Marié (2011: 296–304).
3 The title page of *Specimen medicinae Sinicae* states 'Edidit Andreas Cleyer'. The word 'editor', or 'redactor', in its contemporary meaning, is probably the closest to how Cleyer presents himself.
4 The restitution of the contents of the *Neijing* is perfectly accurate here, which shows that the writer knew very well this part of the text, which corresponds to a part of the chapter *Mai yao jingwei lun* 脈要精微論 (Treatise on the Main Subtleties of the Pulses), *Suwen*, 17.
5 The editors wish to thank the New York Academy of Medicine for rights to use these images.

Bibliography

Pre-modern sources

Zhenjia shuyao 診家樞要 (Key Points for Diagnosticians) 1359 CE, Hua Shou 滑壽, in Gao Wenzhu 高文鑄 (ed.) (1997), *Yijing bingyuan zhenfa mingzhu jicheng* 醫經病源診法名著集成, Beijing: Huaxia chubanshe.

Modern sources

Abercromby, D. (1685) *De Variatione, ac Varietate Pulsus Observationes*, Londini: S. Smith.
Allemand, L.-A. (1671) *Les Secrets de la Médecine des Chinois consistant en la parfaite connaissance du Pouls, Envoyez de la Chine par un François, Homme de grand mérite*, Grenoble: chez Philippes Charvys.

Barnes, L.L. (2005) *Needles, Herbs, Gods and Ghosts*, Cambridge, MA; London: Harvard University Press.
Bivins, R. (2000) *Acupuncture, Expertise, and Cross-Cultural Medicine*, London: Palgrave.
Bontii, J. (1642) *De Medicina Indorum Lib. IV*, Lugduni Batav.: Apud Franciscum Hackium.
—— (1658) 'Historiae Naturalis et Medicae Indiae Orientalis', in G. Pisonis (ed.) *De Indiae utriusque re naturali et medica libri quatuordecim*, Piso: Willem, Apud L. et D. Elzevirios.
Boym, M. (1686) *Clavis Medica ad Chinarum doctrinam de pulsibus*, Nuremberg.
Bréquigny, L.O. and Batteaux, C. (eds) (1776–1791) *Mémoires concernant l'histoire, les sciences, les arts, les mœurs, les usages etc. des Chinois, par les missionnaires de Pékin*, Paris: Nyon.
Brewer, L.A. (1983) 'Sphygmology through the centuries. Historical notes', *American Journal of Surgery*, 145.6: 696–702.
Cleyer, A. (1682) *Specimen Medicinae Sinicae, sive Opuscula Medica ad Mentem Sinensium*, Francofurti: J.P. Zubrodt.
Diderot, D. et Le Rond d'Alembert, J. (1751–1780) *Encyclopédie ou Dictionnaire raisonné des Sciences, des Arts et des Lettres*, Paris: Briasson.
Du Halde, J.B. (1735) *Description géographique, historique, chronologique, politique et physique de l'Empire de Chine et de la Tartarie chinoise*, Paris: P. G. Lemercier.
Floyer, J. (1707) *The Physician's Pulse-Watch, or, an Essay to Explain the Old Art of Feeling the Pulse, and to Improve It by the Help of a Pulse-Watch*, London: S. Smith and B. Walford.
García Ballester, L. and Arrizabalaga, J. (2002) *Galen and Galenism: theory and medical practice from antiquity to the European renaissance*, Aldershot: Ashgate Variorum.
Herrera, (de) F.M. (1622) *Nouvelle Histoire de la Chine*, Paris: chez la veuve de Ch. Chastellain.
Kuriyama, S. (1999) *The Expressiveness of the Body and the Divergence of Greek and Chinese Medicine*, New York: Zone Books.
Lu, G.-D. and Needham, J. (1980) *Celestial Lancets: a history and rationale of acupuncture and moxas*, Cambridge: Cambridge University Press.
Marié, E. (2008) *Précis de médecine chinoise*, St-Jean-de-Braye: Dangles.
—— (2011a) *Le diagnostic par les pouls en Chine et en Europe*, Paris; Berlin; Heidelberg; New York: Springer.
—— (2011b) 'The transmission and practice of Chinese medicine: an overview and outlook', *China Perspectives*, 2011.3: 5–13.
Menuret de Chambaud, J.-J. (1768) *Nouveau traité du pouls*, Amsterdam: Vincent.
Nieuhof, J. (1665) *L'Ambassade de la Compagnie orientale des Provinces Unies vers l'Empereur de la Chine ou Grand Cam de Tartarie, faite par les Srs. Pierre De Goyer et Jacob De Keyser*, Leyde: Jacob de Meurs.
Rhodes, A. (1653) *Divers voyages et missions du P. Alexandre De Rhodes en la Chine, et autres Royaumes de l'Orient, avec son retour en Europe par la Perse et l'Arménie*, Paris: S. et G. Cramoisy.
Saillant, C.J. (1770) *An ex vario variorum arteriarum motu variae diagnosci possint hominum diaqesij*, Paris: Quillan.
Shinnick, P. and Omura, Y. (1985) 'Difference in the location of finger placement on the artery for pulse diagnosis in the orient; and, 15th to 18th century occidental rare books on pulse diagnosis', *Acupuncture and Electrotherapeutics Research*, 10.4: 309–24.
Ten Rhyne, W. (1683) *Dissertatio de Arthritite; Mantissa Schematica; de Acupunctura; et Orationes tres : I. De Chymiae et Botaniae antiquitate & Dignitate; II. De Physionomia; III. De Monstris*, Londres: Chiswell.
Wotton, W. (1694) *Reflections upon Ancient and Modern Learning*, London: Printed by J. Leake for P. Buck.

39

THE EMERGENCE OF THE PRACTICE OF ACUPUNCTURE ON THE MEDICAL LANDSCAPE OF FRANCE AND ITALY IN THE TWENTIETH CENTURY

Lucia Candelise

The remarkable spread of Chinese medical practices in Europe from the first half of the twentieth century up to the present is embedded in a much broader social, cultural and political context, that of the circulation of knowledge and techniques on a global scale. Observed on this scale, the spread and reworking of Chinese medical knowledge, as it developed and took root on almost every continent from its territory of origin, can be described as a phenomenon or a process of globalisation. As a result, Chinese medicine – or rather Chinese medicines[1] and their techniques – can be defined as 'global' medical knowledge, since it has resulted in both medical knowledge and medical practices that are now pursued in the majority of countries in the world. If one accepts the definition of globalisation as the processes of intersection between the strategies of nation-states and the initiatives of transnational actors (Beck 2001: 11), it is interesting to analyse the processes by which this movement of knowledge becomes differentiated and localised. This can be done by looking at the conditions of the emergence of Chinese medical practices in the particular space of a nation-state. In the cases discussed in this chapter, that is, Italy and France, social networks can be seen to have supported the movements of 'deterritorialisation' and 'reterritorialisation' (Pordié and Simon 2013: 19–20) via the social integration of key figures in the circulation and adoption of Chinese medical knowledge beyond China's borders.

Knowledge in progress and the differentiated conditions of practice

Acupuncture is mainly known in Europe and the 'Western' world as a treatment technique from China that forms part of a wider set of medical practices, such as the administration of pharmacopoeia remedies, massage (*tuina* 推拿), martial arts (*qigong* 氣功 is frequently considered as a component of Chinese therapeutic practices) and dietetics. The practice of acupuncture known in the West today, particularly in France, is derived from a process of appropriation and reinterpretation of knowledge defined in China during the twentieth century. This is, in turn, the result of a complicated process of elaboration on the part of the

Chinese Communist government from 1949, negotiating demands for reform and standardisation and influences from the West, which shaped, as Kim Taylor has shown, what is now known as Traditional Chinese Medicine (TCM) (Taylor 2004; Chapter 40 in this volume).

In the course of the twentieth and twenty-first centuries, one finds, beyond China's borders, an extremely nuanced range of practices and therapeutic treatments that go under the name of Chinese medicine (or Traditional Chinese Medicine). These numerous healthcare resources are distinguished not only by their techniques (acupuncture in the majority of cases in Europe, and also pharmacopoeia, massage or martial arts), but also by the training or status of practitioners (physicians, midwives, physical therapists or people trained in one or more techniques of Chinese medicine), by the teachings and status of the institutions in which they are provided (frequently these are private schools, but they are also sometimes accommodated within universities), and by the stages of recognition and legalisation (for example, in France, Spain and Italy, the practice of acupuncture is restricted to those with a university medical degree, while in other European countries, such as Great Britain, Germany or Switzerland, acupuncture can be practised by non-medical personnel). This explains how one can speak of 'Chinese medicines' in the plural to designate the nebula of these numerous and multifarious medical practices and frameworks.

In this differentiated context, the most widespread healthcare practice relating to Chinese medical theories, within the European medical establishment, is without a doubt acupuncture. In this chapter, I propose an analysis of the process of globalisation of acupuncture through the analysis of its 'localisation' in France and Italy. The practice of acupuncture in France, which appeared in clinical practice relatively early in the twentieth century, is first and foremost carried out by general practitioners. In this chapter, I will show how this knowledge and expertise inspired, to a remarkable degree, the emergence and integration of acupuncture practices within the Italian medical community and how, in Italy in the 1990s, this knowledge was reworked and enriched by additional networks and other points and frames of references.

The evolution of the practice of acupuncture in France

In France, the first practitioners to declare an interest in Chinese medicine and acupuncture treatment were homeopathic physicians, belonging to a circle who defined themselves as neo-Hippocratic physicians. The history of the dissemination of this therapeutic technique is thus marked by the names of physicians of the 1930s such as Maxime Laignel-Lavastine (1875–1953), Marcel Martiny (1897–1982) and Paul Desfosses (1869–1944), and members of the Paris medical elite such as Professors Paul Carnot (1869–1957), Maurice Loeper (1875–1961) and René Leriche (1979–1955) [2].

As regards acupuncture in particular, its popularity in France in the 1930s is attributable to an infatuation with exoticism and orientalism (Segalen 1978; Said 1979) and probably also to the curiosity aroused by a treatment using needles. It should also be said that, at that time, acupuncture was not viewed as a true 'medicine' with its own corpus of medical knowledge and coherent system of interpreting the body and its functions. Rather it was perceived as a means of treatment, and perhaps among certain physicians, as a site for theories yet to be constructed. It was seen as an intriguing and astonishing method which had nothing to do with Western medicine, especially when it was taught and 'practised' by a non-physician, George Soulié de Morant, in 1930.

The figure of the acupuncture specialist

George Soulié de Morant was an eclectic, maverick character. Fascinated by the Chinese language from his early youth, he embarked on a diplomatic career in order to become more knowledgeable about China, and somewhat later, he became an expert in one aspect of Chinese medicine: acupuncture. Born into a bourgeois family in Paris in 1878, Soulié de Morant briefly pursued a course of study which equipped him for employment at the age of eighteen as a secretary for the Banque Lehideux and then the *Compagnie du Sud-Est Africain*. His linguistic skills enabled him to move into a diplomatic career with the Ministry of Foreign Affairs and to serve for some ten years in China as an interpreter.[3] In 1917, he was forced to abandon his career in diplomacy.[4] This proved to be a decisive step for his subsequent professional life. The decade spent in China and the knowledge that he acquired there would indeed be valuable to him. After leaving the Ministry of Foreign Affairs, he pursued a career as a 'learned specialist' on China between 1918 and 1929. Until 1929, however, he never addressed the theme of acupuncture in his writings. He wrote on music, literature, the history of art, on law, history and geography, on Confucius, and on Mongol grammar. He was also the author of seven novels inspired by China or by Chinese translations.[5]

Soulié de Morant's interest in acupuncture became evident at the end of the 1920s. In 1927, quite by chance, he met Paul Ferreyrolles (1880–1955), who showed an interest in his alleged knowledge of China and treatment with needles. A close collaboration between the two men ensued over several years, leading to the publication of Soulié de Morant's first article on acupuncture in 1929 (Soulié de Morant and Ferreyrolles 1929). From that point onwards, he dedicated himself entirely to this field and could thus lay claim to the status of 'sinologist and acupuncturist'. This major turning point in his career was largely due to the enthusiasm that his work aroused among certain French physicians, most of whom were neo-Hippocratic homeopathic practitioners, such as Marcel (1887–1982) and Thérèse Martiny (?–1979) as well as Paul Ferreyrolles. The success of Soulié de Morant as an acupuncturist was greatly facilitated by his membership in the circle of physicians that he succeeded in gathering around him, as he himself states in a text published in 1934 (Soulié de Morant 1934: 7–9). His links with clinicians furthermore offered him the opportunity to carry out work with hospital patients. For several years, he collaborated with Thérèse Martiny. Several physicians attended their consultations to learn the art of treatment with needles that Soulié de Morant carried out at the Bichat, Léopold Bellan, Saint-Louis, Saint-Jacques, Hahnemann and Foch hospitals in and around Paris.

Around 1935, after having worked in hospitals alongside physicians, Soulié de Morant sought to shift his status from scholar to practitioner of Chinese medicine. In this role, he practised at his home, 19 Boulevard d'Argenson in Neuilly, receiving patients from the Parisian bourgeoisie whom he treated with acupuncture. Among his patients were celebrities including the composer Maurice Ravel, the writers Antonin Artaud and Jean Cocteau, the psychoanalyst Marie Bonaparte, and the diplomat Maurice Peyrefitte. However, the fact that he was not a qualified physician put Soulié de Morant in a difficult position, and he was prosecuted for the illegal practice of medicine. This legal action began with a conflict with one of his students, Roger de la Füye (1890–1961), a homeopath and acupuncture physician.[6] In 1951, de la Füye filed a complaint against Soulié de Morant on behalf of the physicians' union and attempted to have him prosecuted. They eventually resorted to arbitration out of court, so the case did not go beyond the stage of pre-trial investigation, thus preventing a full-fledged trial; however, this incident deeply affected Soulié de Morant, who died in 1955.

Soulié de Morant's most important contribution to the establishment of acupuncture in France was *L'Acupuncture Chinoise*, published posthumously in 1957. In fact, it was his son and daughter who published the compendious manuscript on which Soulié de Morant had worked for more than twenty years, probably aided by Thérèse Martiny.[7] Soulié de Morant described his magnum opus as the embodiment of a scientific approach that could convey Chinese medical concepts to Western physicians. Thanks to the legitimacy he had acquired through his years spent in China, Soulié de Morant was able to position himself as the mediator or cultural broker of Chinese medical practices to Western physicians. The diversity of principles and theories at the heart of the two medicines would seem to have found a kind of harmony thanks to this learned scholar of China, who proposed a version of Chinese medicine using the language of Western medicine. The knowledge that he was able to bring together and organise had most likely been transmitted to him by a Japanese practitioner (T. Sakurasawa, otherwise known as Georges Ohsawa) rather than stemming from the clinical experience which he claimed to have acquired during his trips to China (Nguyen 2012). Nevertheless, he was able to establish himself as an authority on acupuncture in response to requests from physicians themselves as well as interest from segments of the French public at the beginning of the twentieth century. The charges brought against Soulié de Morant were ultimately dismissed and did not put an end to his career as a practitioner. Thus the demise of George Soulié de Morant marked the beginning of a period when acupuncture was championed as a medical practice by the physicians who carried it out – a period of moves towards institutionalisation and legalisation, when acupuncture was disseminated, and the teaching of acupuncture was organised, on a national and international scale.

The institutionalisation of knowledge and practices: the birth of acupuncture societies

The years 1944–1945 saw the birth of the two most important French acupuncture societies. Even though acupuncture had entered the world of French medicine a decade earlier, the Second World War disrupted its spread and institutional expansion in hospitals, as well as the realm of private practice. When the war was over, physicians already practising acupuncture took a more public stance, and established societies dedicated to the dissemination and the promotion of acupuncture. The physicians who had accompanied, supported or even attacked Soulié de Morant created and managed these associations and engaged, on the one hand, in propaganda activities, and on the other, in defending acupuncture against challenges from the public and institutions. Nonetheless, they were unable to prevent the rise of conflicts among rival factions. The first to emerge was the *Société Française d'Acupuncture* (SFA), founded by Roger de la Füye, followed by the *Société d'Acupuncture* (SA), which was set up by a group of physicians close to Soulié de Morant: Paul Ferreyrolles, Marcel and Thérèse Martiny, Hubert Khoubesserian, Paul Mériel and a number of others. These two societies would remain the twin poles of attraction for the development and defence of acupuncture (de la Füye also created the *Syndicat National des Médecins Acupuncteurs de France*), and for acupuncture teaching in France up until the 1960s.

The *Société Française d'Acupuncture*

The *Société Française d'Acupuncture* was established on October 15, 1943, at the behest of Roger de la Füye with the declared goal of 'the scientific development of acupuncture in France' (de la Füye 1949: 57). De la Füye had begun to take an interest in acupuncture at a time when homeopathy had already won the confidence of a certain sector of the public and had gained some credibility within the French medical community. In his book, *Traité d'acupuncture, la synthèse de l'acupuncture et de l'homéopathie, l'homéosiniatrie diadermique* (1956), de la Füye presented 'homeosiniatry', a theory aimed at the harmonisation of diagnosis, homeopathic remedies and acupuncture. His advocacy of acupuncture and his work to win it official recognition was entirely predicated on asserting the 'veracity' of that method. De la Füye advocated a synthetic approach to disease, which he attempted to legitimise by 'scientific' reasoning. For de la Füye, the possibility of integrating acupuncture into institutionalised medicine would guarantee its legitimacy within the community of practitioners. He argued that acupuncture's 'scientific' merits could be demonstrated simply by its ability to be incorporated within modern medicine, without regard to any epistemological differences. Its history of thousands of years was a testimony that it could be performed alongside and in addition to other therapies.

The work of Roger de la Füye was a significant contribution to the evolution of French acupuncture. His organisational capacities and managerial skills enabled not only the creation and advancement of the *Société Française d'Acupuncture*, but also the establishment of the *Société Internationale d'Acupuncture* (SIA) in 1946 and the *Syndicat National des Médecins Acupuncteurs de France* in 1947 as well as the founding of the journal *Revue Internationale d'Acupuncture* in 1949. These organisations and channels of communication, which still exist today, became the most widely accredited and widely recognised venues for the spread, teaching and defence of acupuncture in France. Their activities would remain centred on the person of de la Füye practically until his death in 1961.

The *Société d'Acupuncture*

The *Société d'Acupuncture* was created in 1945, just after the Second World War, and in reaction to the societies established by de la Füye. It included all the physicians formerly trained by George Soulié de Morant and belonging to his circle. In 1950, the *Société d'Acupuncture* began to publish the *Bulletin de la Société d'Acupuncture*. The majority of the articles in the *Bulletin de la Société d'Acupuncture* were original contributions by French physicians, and occasionally foreigner practitioners, presenting the results of their clinical research. It should be emphasised that relations between the physicians of the *Société d'Acupuncture* and the *Société Française d'Acupuncture* were subsequently marked by a climate of open hostility, for the two societies endorsed concepts that were hard to reconcile and became the source of factional differences. The physicians of the *Société d'Acupuncture* supported, practised and promoted a form of acupuncture that was as intended to be as close as possible to an 'Oriental reality', as rich as possible in ancient references, and as open as possible to all therapeutic 'mysteries'. At the same time, however, they maintained close attention to the scientific validation of their results. 'To practise good acupuncture, one must be imbued with oriental philosophy' said Ferreyrolles (Cassin 1955: 11), who thus rejected the dogmatic assertions of de la Füye against 'all the orientalisms' that misconstrued the truths of real acupuncture – acupuncture without mysteries. There was a growing desire to translate and analyse primary sources, with a view to a debate between the first Western works on acupuncture and Chinese medical theory, especially among those military physicians who had spent time in French Indochina and who had become interested, in various ways, in the medicine that they had observed there.[8] These physicians met at the *Société d'Acupuncture* to discuss their experiences and their differences with regard to the interpretation of texts.

Albert Chamfrault, translation of ancient medical texts

Among this group of military physicians was Albert Chamfrault, who was the first main figure after Soulié de Morant who showed an interest in the translation of medical texts and sought to draw directly from Chinese medical manuscripts. He played an important role in the transformation of the practice of acupuncture in France between the 1950s and the 1960s. Indeed, his achievements established him in both the theoretical and political arena – theoretical because of his contributions to the translation of texts and the publication of texts on acupuncture and Chinese medicine in Western languages, and political because he also succeeded in linking the two French acupuncture societies (the *Société Française d'Acupuncture* and the *Société d'Acupuncture*).

After the Second World War, Chamfrault's career as a military physician took him to Indochina for three years. Having already developed an interest in unofficial medical practices[9] during a period spent in Tonkin, he tried to better understand the 'curious science' that was called acupuncture in France (Chamfrault 1954: 7). With the assistance of a friend, the Chinese scholar Ung Kan Sam, he devoted an enormous amount of energy to the translation of the classic texts of Chinese medicine that would later prove useful to him in writing his own work after his return to France. Between 1954 and 1969, he published the six volumes of his *Traité de Médecine Chinoise*, the last volume in collaboration with the Vietnamese physician Nguyen Van Nghi (Chamfrault 1954; 1957; 1959; 1961; 1963; 1969).

In his translations, Chamfrault espoused a very particular methodological approach. Firstly, he underscored the importance of preparing translations under the direction of a physician (who 'recorded the translation') so that the translated text would remain as faithful as possible to the form of the Chinese source text, while also ensuring that it could be correctly understood by French readers. His principles and his work subsequently contributed to the creation of a style, a method, a conception and a 'medicine' – that of acupuncture physicians following a 'traditionalist' approach to acupuncture. In fact, from the second half of the 1950s until the beginning of the 1990s, French physicians such as Albert Chamfrault were very much interested in the translation, revision and even the interpretation of Chinese texts in order to construct a body of knowledge and a Chinese medical practice entirely based on this work of exegesis. In contrast, contacts with China and the practice of Chinese medicine in China came relatively late and were sometimes undertaken reluctantly. The intervention of a French physician in the translation of a Chinese medical text verified by a Chinese native speaker (the case of Chamfrault and Ung Kan Sam is not the only one that we have encountered)[10] entailed a certain degree of reinterpretation of the Chinese text under the influence of Western medical knowledge. This approach led to the construction of a particular style of acupuncture characteristic of France, that of 'traditionalist French acupuncture' (Candelise 2008) of which Albert Chamfrault was the creator.

Failure of the merger

A decade after the founding of the two acupuncture societies, the deaths between 1955 and 1961 of the first acupuncture physicians and of the main supporters and promoters of acupuncture in France had repercussions for relations between the *Société Française d'Acupuncture* and the *Société d'Acupuncture*. In fact, after almost ten years of competition, and even rivalry in the organisation of events (conferences, publications, teaching, membership activities, etc.), the two societies decided to merge in late 1964 to form a single body, the *Association Scientifique des Médecins Acupuncteurs de France* (ASMAF). The two societies shared the common goal of advancing the practice of acupuncture and significantly increasing the number of acupuncture physicians. This project could be best achieved with a merger but only if it was grounded in a reconciliation. While the death of Roger de la Füye certainly helped to mitigate the hostility between the two societies that had prevented any reconciliation up to that point, the peace nevertheless did not last long. As it turned out, the decision to create a single association was sudden and ill-prepared, and genuine differences of approach to acupuncture continued to divide French acupuncture physicians. This conflict resulted in a fresh split. Thus, at the end of 1966, the physicians who had been members of the *Société d'Acupuncture* retained the name of the *Association Scientifique des Médecins Acupuncteurs de France*, with Marcel Martiny becoming the president. Meanwhile, the former *Société Française d'Acupuncture* became the *Association Française d'Acupuncture* (AFA) with Albert Chamfrault assuming the presidency.

Duality between 'traditionalists' and 'modernists'

The reorganisation of the former *Société Française d'Acupuncture* in a new direction, confirmed by the arrival of a new president, Albert Chamfrault, and signalled by a new name, *Association Française d'Acupuncture*, but also by a revised set of statutes, was accompanied by internal political changes. While Roger de la Füye had campaigned for a 'modernised' acupuncture, the approach to acupuncture championed by the *Association Française d'Acupuncture* starting in 1966 and up until the 1980s was completely inspired by the work of Albert Chamfrault and Nguyen Van Nghi. Chamfrault was the champion of an acupuncture grounded in ancient texts and efforts at interpretation conducted by or directed by acupuncture physicians. Immediately after the publication of the first four volumes of Albert Chamfrault's works, Nguyen Van Nghi entered the field of French acupuncture by offering a translation of the basic texts of 'Traditional Chinese Medicine' that were produced in China at the end of the 1950s.[11]

The contribution of Nguyen Van Nghi (1909–1999) brought about an evolution in the concept of acupuncture in France. A Vietnamese physician trained in conventional medicine in France, Nguyen Van Nghi united in his person the conditions defined by Albert Chamfrault for the medical theory underpinning acupuncture; that is, the translation of original sources under the supervision of a conventional physician and in accordance with Chinese tradition. The collaboration between Nguyen Van Nghi and Chamfrault, which lasted until the end of the 1960s, contributed to the construction of another understanding of acupuncture. Their ideas would later be circulated in the 1980s by the Groupe Lacretelle and then by Jean-Marc Kespi and other physicians close to them.[12] The traditionalist approach to acupuncture, promoted by the *Association Française d'Acupuncture* from this time onwards, was rooted in the conviction that all knowledge of true acupuncture must be based on 'the Chinese tradition'. Kespi asserted that the reading and understanding of ancient texts, above and beyond their archaisms, 'showed another view of man, another physiology, another medicine' (Kespi 1981: 5).

It was also Kespi who wrote of the 'turnaround in the thinking and point of view' (retournement de la pensée et du regard) that characterised the theoretical orientation of these physicians (Kespi, Ibid.). Indeed the discourse of the acupuncture physicians of the *Association Française d'Acupuncture* emphasised the reversal of a 'scientific' understanding of the patient and his or her illness and disease in favour of an approach inspired by 'tradition'. This discourse and these values translated into a clear stance on Chinese therapeutics vis-à-vis conventional medicine: acupuncture is a form of medicine that is different from scientific medicine. Its goal is 'not to analyse the infinite phenomena of life, but to understand its principles and general laws', and furthermore 'the field of study of traditional medicine is total and complete, while modern science only considers a very limited field of reality' (Andrès 1999: 5–6). The global, ecological and qualitative vision specific to acupuncture is opposed to the quantitative and specialised approach to human beings and disease characteristic of conventional medicine. By their propositions, these acupuncture physicians demonstrated that they remained sceptical of the practice of Chinese medicine in current China and of any form of syncretism between Chinese medicine and Western medicine.

A group of physicians from the *Association Scientifique des Médecins Acupuncteurs de France* had begun to meet after the attempt to merge with the *Association Française d'Acupuncture*, the former *Société Française d'Acupuncture*. In 1969 they launched a movement under the aegis of the memory of the 'master' George Soulié de Morant. But beyond their homage to George Soulié de Morant, their work was essentially founded on two basic components: the demonstration of the scientific basis of acupuncture and the teaching of acupuncture. In their field of study, the acupuncture physicians attached to the *Association Scientifique des Médecins Acupuncteurs de France* distanced themselves from the physicians of the *Association Française d'Acupuncture*. Consequently, the 'modernist' or 'scientific' tendency was born out of the opposition to a vision of acupuncture focussed on and faithful to a presumed Chinese 'tradition'. For acupuncture physicians, being a modernist thus meant being able at the same time to demonstrate the scientific character of acupuncture, to establish correspondences between elements of conventional medicine – representing modern science – and the implementation of acupuncture, and to evaluate the positive and empirical effectiveness of acupuncture. Between 1970 and 1980, military physicians such as Jean Borsarello (1926–2007) sought in their work to provide a justification for the therapeutic prospects of acupuncture, continuing the work of Jacques-Emile Henri Niboyet (1913–1986).

Later, attention focussed less on the scientific demonstration of acupuncture than on establishing possible analogies, links and equivalences between certain aspects of Chinese medicine and pathologies or points of view specific to Western medicine. Jean Bossy in particular, but also other acupuncture physicians, on the one hand, pursued research on Chinese physiology according to scientific medicine, and, on the other, took an interest in the possible correlations between the two medicines and the impact of the use of acupuncture in clinical practice. This twofold approach can be seen in a number of articles in the journal *Méridiens*, the official journal of the *Association Scientifique des Médecins Acupuncteurs de France*.[13] Finally, from the 1990s up to the present, the question has been raised of the evaluation or the validation of experimental research in acupuncture following the example of other European and American countries as well as China itself. In effect, the *Association Scientifique des Médecins Acupuncteurs de France* is now the association closest to contemporary China and to the Chinese medicine practised there.

This duality of approaches to acupuncture in France is probably not as marked today as it was a few decades ago[14]. Nevertheless, the tendencies that we have highlighted have contributed to defining a very specific medical practice and conception of the human organism which I call 'French traditionalist acupuncture'. This approach to Chinese medicine and this medical trend certainly best characterise French acupuncture in the eyes of other nations. The contributions of the acupuncture physicians of the *Association Française d'Acupuncture* – and also of other institutions working on the interpretation of the classic texts of Chinese medicine[15]– have greatly contributed to the traditionalist character that distinguishes acupuncture in France, which we do not find in other European countries or North America.

The practice of acupuncture in Italy

Thirty years later than in France, clinical acupuncture spread widely in Italy in the 1970s. In 1973, influenced by the work of the *Association Française d'Acupuncture*, Alberto Quaglia Senta and Luciano Roccia founded the *Società Italiana di Ricerca in Agopuntura ed Auricoloterapia* (SIRAA) and the *Giornale Italiano di Riflessoterapia ed Agopuntura* in Turin. Around the same time, Ulderico Lanza began to develop an acupuncture treatment practices as conceived, formulated and supported in France from the 1940s onwards. With the help of Nguyen Van Nghi, Lanza had already founded the *Scuola Italiana di Agopuntura* and the *Società Italiana di Agopuntura* (SIA) in 1968. Acupuncture therapy in Italy had to be developed from the ground up, and those two organisations had given significant support, the former by providing locales where acupuncture physicians could be trained, and the latter – the SIA, which still exists today – by creating an institution for the research, development and maintenance of acupuncture knowledge in Italy. The *Società Italiana di Agopuntura* was actually a division of the *Société Internationale d'Acupuncture*, founded by Roger de la Füye, which was managed by Jean Schatz in the 1970s. Through the two societies that he founded, Ulderico Lanza began the trend towards exchanges between Italian and French acupuncture physicians. Nguyen Van Nghi and Albert Gourion would often visit the *Scuola Italiana d'Agopuntura* to give seminars, as would Yves Requena a few years later.

The 1970s saw the emergence of a considerable number of private schools in Italy, principally in northern Italy, and in a few isolated cases in the south. Among others, the *Centro Studi So Wen* was founded in Milan in 1974, the *Gruppo di Studio Società e Salute* in Bologna at the end of the 1970s, and the Scuola di Agopuntura Tradizionale della Città di Firenze (in Florence) in 1980, while the *Scuola Mediterranea di Agopuntura* opened in Catania in 1979. Each of these schools also represented a centre for research and investigations concerning Chinese medicine and acupuncture. Both the teaching and the thinking that took place in these centres were inspired by 'French traditionalist acupuncture'. French acupuncture physicians were regularly invited to Italian schools, including the most renowned representatives of the traditional approach to acupuncture in France. Chief among them was Nguyen Van Nghi, who was appointed curriculum director of the *Scuola di Agopuntura Tradizionale della Città di Firenze* and the *Scuola Mediterranea di Agopuntura* in Sicily, when both these institutions were founded. On the other hand, the *Gruppo di Studio Società e Salute* in Bologna and the *Centro Studi So Wen* in Milan were linked to the *Association Française d'Acupuncture* and the *Société Internationale d'Acupuncture* in their research and teaching methods. Physicians from the *Association Française d'Acupuncture*, such as Jean-Marc Kespi, Gilles Andrès and Jean Schatz, president of the *Société Internationale d'Acupuncture*, regularly visited these schools, organising many seminars each year as well as co-ordinating teaching activities.

Italian acupuncture departs from 'French traditionalist acupuncture'

However, during the 1990s, some twenty years after acupuncture began to develop in Italy, Italian acupuncture physicians diverged from the French acupuncture model that had hitherto been their sole reference. We can observe an opening-up by several Italian acupuncture schools and associations towards sources other than the French ones, concerned with Chinese medicine and acupuncture. More precisely, in order to respond to other specifically Italian demands connected to acupuncture, the work of Italian acupuncture physicians no longer fell under the exclusive intellectual influence of the French. In fact, the development of relationships with the acupuncture world in the UK and North America began at almost the same time as the establishment of links with China, the country of origin of the therapeutic method. The transition from reliance on a single European source, that is, French medical acupuncture knowledge ('French traditionalist acupuncture') to a situation of dialogue and exchange of knowledge among several countries took place in Italy in two almost simultaneous stages.

First, during the 1980s, a number of Italian acupuncture physicians felt the need to extend their theoretical knowledge and their clinical experience beyond acupuncture alone. A growing interest in the Chinese pharmacopoeia led these acupuncture physicians to make contact with British and American practitioners who were familiar with Chinese *materia medica* and experienced in using them. Therefore, these Italian acupuncture physicians went to England to attend seminars and to be trained in the use of the Chinese pharmacopoeia. At the same time from the 1980s onwards, it was easier to interact and establish links with China, so a number of Italian physicians went to China in order to be trained in Chinese medicine as it was taught there. This period also witnessed real changes in the teaching and diffusion of acupuncture and Chinese medicine in Italy. Several schools set up links and agreements with universities in China. We can cite among the Italian acupuncture schools that established ties with Chinese universities in the 1990s the *Medicina* school and *Centro Studi So Wen* in Milan and the *Fondazione Matteo* Ricci in Bologna.

This period was marked by a paradigm shift in the teaching and growth of acupuncture and Chinese medicine in Italy. The existence of Chinese medical practices in Italy in the 1980s was still largely due to the activity of a few physicians who were deeply involved in these practices as well as a number of schools that represented them. Thus certain physicians based in north Italy, such as Graziella Rotolo and Caterina Martucci, with the help of Alessandra Guli from Rome, opened up the field to other teaching and practice contexts for Chinese medicine. Thanks to these physicians, connections were established with milieus and practitioners in the UK such as Giovanni Maciocia, and American practitioners and researchers such as Ted Kaptchuk. But these same acupuncture physicians were also in the process of forging relationships with Chinese institutions which they had come to know by visiting China. During the 1990s, the Italian acupuncture schools mentioned above signed agreements with Chinese universities. This enabled them to regularly invite Chinese teachers to Italy for a few weeks each year and to send their students to go to China to undergo training in Chinese hospitals.

Conclusion

As we have shown, the emergence, implementation and transmission of medical knowledge between China and France, between France and Italy at a later date, and also directly between Italy and China, and between Italy and other European countries (for example, the UK) and even the Americas, provide evidence of non-linear movements of medical knowledge and practices among China and Europe and the rest of the world. In seeking to shed light on the processes of the differentiation and localisation of this movement of knowledge, we have seen that the case of acupuncture in France and Italy reveals a landscape of intersections between colonial politics and individual motivations, and is also characterised by distinctive personalities as well as an interest in relatively ancient Chinese medical knowledge and practices. This analysis highlights national strategies and the dynamics of exchange or the movements of transnational actors. The overlapping and the interaction of social and cultural factors (Werner and Zimmermann 2004) mean that these two fields of enquiry are unique sites in the context of the globalisation of this healthcare practice.

Notes

1 Distinctions in the reception and adoption of Chinese medical knowledge and practices lead us to consider that it would be more correct to speak of 'Chinese medicines' in the plural.
2 I use the terms 'Western medicine' (or 'Western physician') to indicate scientific medicine or biomedicine.
3 He spoke English (his mother was American), Spanish, Italian and Chinese. See the Annual notes of the Ministry of Foreign Affairs, Archives of the Ministry of Foreign Affairs, Dossier Georges C. Soulié, no. 1444.
4 His reasons for definitively abandoning his ambition to become Consul in China remain obscure.
5 For a complete bibliography of George Soulié de Morant, see Candelise (2010).
6 I distinguish between 'acupuncture physicians' who practised both Western medicine and acupuncture, and 'acupuncturists' who practised only acupuncture.
7 A preliminary, incomplete version of *L'Acupuncture chinoise* was published during his lifetime (vol. 1 in 1939 and vol. 2 in 1941).
8 I refer here to Albert Chamfrault and later, in the 1960s, Georges Cantoni and Jean-François Borsarello.
9 Albert Chamfrault was one of the first students of Roger de la Füye.
10 George Soulié de Morant is said to have worked with Chinese physicians to complete the writing of his medical dictionary, and Roger de la Füye is also said to have worked with a Chinese collaborator, while Jacques-Emile Henri Niboyet is supposed to have been trained by a mysterious Chinese physician living in France.
11 On this subject, a family friend, Nguyen Van Than, a pharmacist and cadre in the Ministry of Health, sent him several documents concerning Chinese medicine in the 1950s. The text translated by Nguyen Van Nghi was the translation into Vietnamese of the *Zhongyixue gailun* 中医学概论, a work first published in 1958 by Nanjing University of Chinese Medicine, which was translated by the Ministry of Health of Vietnam in that same year.
12 The 'Groupe Lacretelle' was a circle of Parisian acupuncture physicians, members of the *Association française d'Acupuncture*, who worked together on the traditional aspects of acupuncture. They took their name from the address where they met in rue Lacretelle in Paris, at the apartment of a member of the group.
13 Jean Bossy was professor of anatomy at the University of Nîmes. With other acupuncture physicians including Denis Colin, he worked to establish acupuncture teaching at the university from 1989. Nowadays the university's acupuncture diploma is the only one formally recognised in France.
14 The creation of the *Fédération des Acupuncteurs pour leur Formation Médicale Continue* (FAFORMEC) in 1997 brought together twenty-six acupuncture associations in France, which had very different approaches to acupuncture.
15 I am thinking especially of the *Institut Ricci* and the *Ecole Européenne d'Acupuncture*, which, although they are not associations of physicians, have worked on translations of and commentaries on Chinese medical texts since the beginning of the 1970s. If the works which inform the traditionalist approach of French acupuncture do not all answer to the academic criteria required by a purely sinological approach, this often reflects the facts that these were interpretations of medical texts by physicians, sometimes with the help of a Chinese collaborator, working outside a philological and historical framework.

Bibliography

Andrès, G. (1999) *Principes de la Médecine selon la Tradition*, Paris: Éditions Devry Annual notes of the Ministry of Foreign Affairs, Archives of the Ministry of Foreign Affairs, Dossier Georges C. Soulié, no. 1444.
Beck, U. (2001) *What Is Globalization?* Cambridge: Polity Press.
Candelise, L. (2008) 'Construction, acculturation et diffusion de l'"acupuncture traditionaliste française" au xxe siècle', *Documents pour l'histoire des techniques*, Paris, December, pp. 76–88.
—— (2010) 'George Soulié de Morant, le premier expert français en acupuncture', *Revue de Synthèse*, 131.3: 373–99.
Cassin, G. (1955) 'Sur la mort', *Bulletin de la Société d'Acupuncture*, 18: 11–13.

Chamfrault, A. (1954) *Traité de Médecine Chinoise: acupuncture, moxas, massage, saignées*, vol. 1, Angoulême: Coquemard.
―――― (1957) *Traité de Médecine Chinoise: livres sacrés de médecine Chinoise*, vol. 2, Angoulême: Coquemard.
―――― (1959) *Traité de Médecine Chinoise: pharmacopée*, vol. 3, Angoulême: Coquemard.
―――― (1961) *Traité de Médecine Chinoise: formules magistrales*, vol. 4, Angoulême: Coquemard.
―――― (1963) *Traité de Médecine Chinoise: de l'astronomie à la médecine Chinoise*, vol. 5, Angoulême: Coquemard.
Chamfrault, A. and Nguyen, V.N. (1969) *Traité de Médecine Chinoise: l'énergétique humaine en médecine Chinoise*, vol. 6, Angoulême: Imprimerie de la Charente.
De la Füye, R. (1949) 'L'avenir de l'Acupuncture', *Revue Internationale d'Acupuncture*, July–September: 57.
―――― (1956) *Traité d'acupuncture, la synthèse de l'acupuncture et de l'homéopathie, l'homéosiniatrie diadermique*, Paris: Le Franc,ois.
Kespi, J-M. (1981) 'Editorial', *Revue Française d'Acupuncture*, 27 (July–September).
Nguyen, J. (2012) *La Réception de l'Acupuncture en France: une biographie revisitée de George Soulié de Morant (1878–1955)*, Paris: L'Harmattan.
Pordié, L. and Simon, E. (eds) (2013) *Les Nouveaux Guérisseurs: biographies de thérapeutes au temps de la globalisation*, Paris, Éditions de l'École des Hautes Études en Sciences Sociales.
Said, E. (1979) *Orientalism*, New York: Vintage Books.
Segalen, V. (1978) *Essai sur L'exotisme: une esthétique du divers*, Paris: Fata Morgana.
Soulié de Morant, G. (1934) *Précis de la Vraie Acupuncture Chinoise*, Paris, Mercure de France.
Soulié de Morant, G. and Ferreyrolles, P. (1929) 'L'acupuncture en Chine vingt siècles avant J.-C. et la réflexothérapie moderne', *Homéopathie Française*, June: 403–17.
Taylor, K. (2004) 'Divergent interests and cultivated misunderstandings: the influence of the west on modern Chinese medicine', *Social History of Medicine*, 17.1: 93–111.
Werner, M. and Zimmermann, B. (eds) (2004) *De la Comparaison à l'Histoire Croisée*, Paris: Le Seuil.

40
ENTANGLED WORLDS
Traditional Chinese medicine in the United States

Mei Zhan

The history and context of 'traditional Chinese medicine' in the United States is a set of irreducibly complex and continuously unfolding stories with multiple points of entry and trajectories. Let me start this chapter with a story of my own. In 1995 I received my bachelor's degree in anthropology from Scripps College, part of the Claremont Colleges consortium in southern California. I had been an international student there for four years. Aside from majoring in anthropology, I also completed two minors one of which was in biology. During that time my biology advisor, Dr. David Sadava, was working on a laboratory research project on genetic therapies for lung cancer. Trying to draw me into the world of biology (for which I am eternally grateful), Dr. Sadava asked if I would be interested in conducting a laboratory research project testing the effects of Chinese herbal extracts on small-cell lung cancer cell cultures. It was an opportunity I could not possibly refuse. We began our research by driving to Monterey Park, a southern Californian city known for its concentration of Chinese American communities, and purchased three different kinds of herbs – *baihuashe shecao* 白花蛇舌草 (*odenlandia diffusa*), *banzhi lian* 半枝蓮 (*scutellaria barbata*), and *kunbu* 昆布 (*laminaria*) – which, Dr. Sadava told me (rather than the other way around), were commonly used in Chinese herbal prescriptions for cancer treatment.

The rest was history, so to speak. The research project grew over the next few years, bringing in other participants and resulting in a publication in a mainstream medical research journal (Sadava *et al.* 2002). But this experience also meant something else to me: a point of entry into both traditional Chinese medicine and the anthropological studies of science and medicine. Instead of choosing biology as my vocation, I became fascinated by the ways in which scientific experiments were actually conducted and scientific knowledge was produced in the wet lab, which is full of contingency, invention, and interpretation beyond textbook science. At the same time, it was clear that in order to understand how traditional Chinese medicine entered biology laboratories and thrived in the United States, we needed a cultural analysis of knowledge production through translocal encounters and entanglements that would go beyond the taken-for-granted ideas and practices of 'science' or 'medicine'.

The anecdote above is not an origin story: when I set my foot in the herbal store in Monterey Park, and when I applied herbal extracts to cell cultures, I was already caught up in the middle of the translocal movements and refigurations of traditional Chinese medicine:

a set of 'worlding' projects and processes that simultaneously produce multiple and dynamic forms of traditional Chinese medicine, as well as particular visions, understanding, and practices of what makes up the world and our place in it (Zhan 2009). In what follows I draw on my own fieldwork in California and other colleagues' research to recount some of the key events, routes, features, and debates in the world of traditional Chinese medicine in the United States. To do so I focus on three interwoven storylines: transnational movements and migrations; encounters with biomedicine; knowledge creation and world making.

Becoming 'traditional Chinese medicine'

Traditional Chinese medicine in the United States is by no means an overseas variation of the original. As historians and anthropologists have long argued, 'traditional Chinese medicine' as such did not acquire a distinctive professional identity until close encounters with modern biomedical institutions and practices in the late nineteenth and early twentieth centuries (Farquhar 1994; Scheid 2002; Zhan 2009; Andrews and Bullock 2014; Lei 2014; Rogaski 2014). The professionalization of traditional Chinese medicine in China at this time took place with a perhaps an unintended helping hand from American institutions. As they competed against other therapeutic practices (such as homeopathy) and rose to the status of a 'sovereign profession' in the United States and Europe (Starr 1982), biomedical colleges and hospitals also entered China around the turn of the twentieth century, often through missionary, philanthropic, and educational organizations. The Rockefeller Foundation and Johns Hopkins University, for example, played key roles in establishing the Peking Union Medical College – one of the oldest and most prestigious of its kind in China – in the 1910s. As biomedical institutions came to dominate healthcare in urban China, scholar-practitioners of herbal medicine and acupuncture found their collective livelihoods under intense scrutiny and threat. Alternately called 'old medicine' (*jiuyi* 舊醫) or 'national medicine' (*guoyi* 國醫), traditional Chinese medicine emerged as a key symbol in the intellectual and political debates of superstition and science, tradition and modernity, and backwardness and progress in a young Chinese nation-state struggling with colonialism, imperialism, and cultural identity. In response, these scholar-healers and their allies, especially those in treaty ports and cosmopolitan centers such as Shanghai, began setting up small professional schools and academies, compiling textbooks and curricula, and engaging in debates about the legitimacy of their knowledge and practice.

However, it was not until after the founding of the People's Republic of China that traditional Chinese medicine became standardized and institutionalized. In the mid-1950s, the science-based 'Traditional Chinese Medicine' (TCM) emerged as part of the socialist state's efforts to create a distinctively Chinese science and healthcare system that served the masses. (Chapter 45 in this volume). Biomedical professionals were enlisted to study with traditional healers and also played leading roles in founding and administering large TCM colleges and hospitals. In addition, biomedical professionals and traditional healers worked together to invent a body of basic TCM theories and novel TCM techniques that were then tested in laboratory and clinical trials and integrated into everyday medical practice.

Acupuncture anesthesia was one of these new inventions, which not only brought acupuncture onto the surgical table but also pushed it onto the geopolitical world stage. In July 1971, a few months before Richard Nixon's historic trip to China, the American journalist James Reston visited Beijing. During his trip, Reston had to undergo emergency appendectomy, and was offered acupuncture as a possible mechanism for post-surgery pain relief. He took the offer, and wrote a New York Times front-page article 'Now, About My Operation in Peking'. In it Reston gave a vivid firsthand account of his encounter with acupuncture:

> I was in considerable discomfort if not pain during the second night after the operation, and Li Chang-yuan, doctor of acupuncture at the hospital, with my approval, inserted three long thin needles into the outer part of my right elbow and below my knees and manipulated them in order to stimulate the intestine and relieve the pressure and distension of the stomach. That sent ripples of pain racing through my limbs and, at least, had the effect of diverting my attention from the distress in my stomach. Meanwhile, Doctor Li lit two pieces of an herb called *ai*, which looked like the burning stumps of a broken cheap cigar, and held them close to my abdomen while occasionally twirling the needles into action. All this took about 20 minutes, during which I remember thinking that it was a rather complicated way to get rid of gas in the stomach, but there was noticeable relaxation of the pressure and distension within an hour and no recurrence of the problem thereafter.
>
> *(1971:1; quoted in Zhan 2009:2)*

This encounter took place at the Anti-Imperialist Hospital, which was founded as the Peking Union Medical College Hospital more than half a century earlier and renamed at the beginning of the Cultural Revolution in 1966. Even as Reston's narrative emphasized the exotic and sensational aspects of his experience with acupuncture and moxibustion, he suggested to his American readership, most of whom was well educated and cosmopolitan, that this medicine might in fact be effective. He reported that herbal medicine and acupuncture were not only used for pain relief, arthritis, and paralysis, but also were cures for deafness and blindness (Reston 1971: 6). A frenzy over Chinese medicine quickly swept through the scientific and biomedical communities in the United States. Delegations of research scientists and biomedical professionals traveled to China to examine acupuncture and herbal medicine, and soon thereafter started publishing field reports and research articles on these subjects (see, for example, American Anesthesia Study Group 1976; American Herbal Pharmacology Delegation 1975).

Becoming 'legal'

In addition to attracting the attention from bioscientific research and biomedical communities, the 'acupuncture frenzy' galvanized advocates of traditional Chinese medicine in the United States, especially acupuncturists and herbalists who had been practicing behind closed doors in Chinese American communities. During my fieldwork in California, local acupuncturists and activists cited the publication of Reston's article as a pivotal moment when acupuncture broke out of Chinatowns and embarked on the process of legalization.

It is perhaps no surprise to anyone that the history of acupuncture and herbal medicine in the United States is inseparable from – though irreducible to – the history of Chinese immigration. According to the anthropologist Linda Barnes, the earliest arrivals took place along trade routes where Chinese laborers were employed (2013:286). By 1848 the Chinese had entered at least eleven East Coast cities in addition to Mexico, California, and the Pacific Northwest (2005: 234). Coming mainly from the southeast coast of China and speaking a variety of dialects including Cantonese, Hakka, and Fujianese, they came as 'sailors, servants, merchants, students, performers, and stage exhibits' (Ibid.).

Entangled worlds

The historians Jeffrey Barlow and Christine Richardson wrote about two of the earliest practitioners of Chinese medicine in Oregon:

> One thing we do know is that Chinese medicine arrived in the U.S. through the doctors who emigrated here, some as early as the 1800s⋯ The first documented acupuncturists and herbalists in the U.S. were Ing Hay and Lung On. In 1887, Ing Hay arrived in the mining town John Day, in Eastern Oregon. He soon began practising Chinese medicine with fellow practitioner Lung On, serving both the Chinese and Caucasian community there. Doctors Hay and On were both arrested several times for practising medicine without a license but due to their popularity in the community, each case brought against them was dismissed.

The theme of arrest and dismissal was to be repeated during the course of legalizing acupuncture in the United States in the 1970s. As acupuncture gained visibility through mainstream media, it also attracted controversies and increased resentment especially from biomedical communities. Miriam Lee (1926–2009), an acupuncturist in Palo Alto, California, was arrested for practicing medicine without a license in 1974. Lee was trained as a nurse and a midwife in China before becoming an acupuncturist. Like a number of other prominent Chinese American acupuncturists at the time; for example, Ding Zhongying, the second son of the Menghe School healer Ding Ganren (1866–1926), who is widely considered the 'father' of modern traditional Chinese medicine; Lee emigrated to Singapore after the communist takeover of China in 1949 and later to California (On the Menghe school, Scheid 2007). A devout Christian, Lee firmly believed that acupuncture was an effective and kind therapeutic practice which should be made available to all. Her clinic in Palo Alto, where she also mentored aspiring local acupuncturists, became wildly popular. Lee was arrested at her clinic on the early morning of April 16, 1974, one day after Governor Ronald Reagan vetoed a bill that would have legalized acupuncture in California. Her subsequent trial galvanized the efforts to legalize acupuncture. Her patients, some of whom arrived supported on the arms of their relatives, testified in court on the effectiveness of Lee's treatment. Local politicians, attorneys, and activists of acupuncture were mobilized in her support. Lee was eventually found not guilty and her court record was stricken. She went on to practice acupuncture and train students (see Zhan 2009 for discussion of Miriam Lee's case). Governor Reagan subsequently allowed acupuncture as an experimental procedure. Lee's court battle was one of the first victories on the road to legalize acupuncture in California and other US states.

What counts as 'legalization', however, is not as straightforward as it might appear. The key debates and battles centered on to what extent the biomedical profession should have the authority – and what kind of authority – over acupuncture and traditional Chinese medicine. Who is authorized to practice acupuncture and traditional Chinese medicine? Do medical doctors need additional training before inserting acupuncture needles into patients? Do acupuncturists need the supervision of biomedical physicians? Nevada was the first US state to acknowledge traditional Chinese medicine as a 'learned profession' (Barnes 2013) and to legalize acupuncture and oriental medicine through the 'Chinese Medicine Act' in 1973. In California, a series of state assembly and senate bills defined and redefined the legality and scope of acupuncture with a focus on its relations with the biomedical establishment and the insurance industry. The California Senate Bill 86 legalized acupuncture in 1975 on the condition that it was performed with prior diagnosis or referral by a physician. This precise language was the result of intense negotiations among legislators, attorneys, acupuncture activists, and biomedical representatives. While some insisted that acupuncture must require 'prior diagnosis *and* referral' by a physician, others argued that 'prior diagnosis *or* referral' was sufficient. The final language of the bill, opting for 'or' rather than 'and', essentially allowed patients to choose acupuncture without the authorization of a biomedical physician. However, it was not until 1997 – more than twenty years after the passing of SB 86 – that California Senate Bill 212 finally included 'acupuncturist' within the definition of 'physician' and placed acupuncture within the coverage of worker's compensation. Around the same time, in 1996, the US Food and Drug Administration reclassified acupuncture needles as accepted medical instruments instead of experimental devices. According to the data provided by the National Certification Commission for Acupuncture and Oriental Medicine (NCCAOM), by 2018 Alabama, Oklahoma, and South Dakota were the only states where acupuncture was not licensed.

Becoming 'alternative'

When I began conducting fieldwork on traditional Chinese medicine in California in the 1990s, many of my friends were eager to tell me that I should interview *their* acupuncturists. Since its legalization in Nevada, acupuncture has emerged as part of the diverse set of therapeutic practices institutionalized in the form of Complementary and Alternative Medicine (CAM). In 1993, David M. Eisenberg at Harvard Medical School and his colleagues published the first comprehensive article on the use of 'unconventional medicine' in the United States in the prestigious journal New England Journal of Medicine. In 1998, Eisenberg and his team published a follow-up article entitled 'Trends in Alternative Medicine Use in the United States'. This article, too, was supplemented by a later survey in 2003, 'The Evolution of Complementary and Alternative Medicine (CAM) in the USA over the Last 20 Years'. This set of surveys is an evolution in itself, from 'unconventional medicine' to 'alternative medicine' to the formalized 'complementary and alternative medicine (CAM)'.

The 1998 finding, for example, was based on a telephone survey on the use of alternative medicine. Among the sixteen therapies covered by the survey were such diverse methods as acupuncture, acupressure, chiropractic therapy, homeopathy, Ayurveda, biofeedback, guided imagery, relaxation techniques, high-dose vitamin therapy, and prayer. It involved 2,055 adults in various parts of the United States. The results of the survey revealed an increase in the number of medical conditions for which alternative medicines were used, with back trouble, allergies, arthritis, and digestive issues listed as some of the main problems (Eisenberg *et al.* 1998). The survey results also showed that between 1990 and 1997 visits to practitioners of alternative medicine increased by 47% from 427 to 629 million, and expenditure increased by 45% to $ 21.2 billion, with $ 12.2 billion out-of-pocket expenses (cited in Zhan 2009:80).

The nature and scope of CAM have shifted over the years, but acupuncture and Chinese herbal medicine remains the most systematically and widely practiced among a variety of therapies including homeopathy, biomedicine's old alternative and competitor. Today, having taught at the University of California, Irvine, for more than fifteen years, I myself have witnessed the founding of the Susan Samueli Center of Integrative Health on my home campus. Here, classes and student-organized study groups in acupuncture, *taiji* and traditional Chinese medicine have become an integral – even if elective – part of their premedical or medical school training. It is no exaggeration to say that traditional Chinese medicine has gone 'mainstream' in the United States. In addition to acupuncture and herbal medicine, other techniques and practices have also entered the public view. At the 2016 Rio Olympics, images of the champion swimmer Michael Phelps with cupping marks on his body caused a 'cupping frenzy'. While reminiscent of the acupuncture frenzy of the 1970s, the cupping frenzy captured a much wider and younger audience than the educated, primarily white, urbanites of the 1970s.

The last forty years of traditional Chinese medicine in the United States, then, have witnessed the processes by which it became simultaneously 'mainstream' and 'alternative'. How could this be? To begin, we need to revisit the heady days of the 1970s. A close look at the practitioners of acupuncture and traditional Chinese medicine reveals the fact that it was not just the Chinese American communities and immigrants from Asia who were practicing acupuncture at the time. One of my interlocutors during fieldwork, Barbara Bernie, was an advocate of legalizing acupuncture and one of the first licensed acupuncturists in California. Having grown up in an upper middle-class Russian Jewish family in New York, she came down with a mysterious debilitating illness in 1970. There was no biomedical diagnosis and therefore no treatment even after many laboratory tests: 'chronic fatigue syndrome', which likely ailed her at the time, was not recognized as a 'disease' until 1988 (Zhan 2009). She was treated by an acupuncturist – an immigrant from Singapore – on the advice of a friend and recovered. Bernie then pursued training in acupuncture first in Britain, where she was influenced by Jack Worsley's Five Element Acupuncture, and then in the Bay Area, where she learned from Miriam Lee. In the 1980s Bernie founded the American Foundation of Traditional Chinese Medicine, a nonprofit organization that promoted the legalization, education, and dissemination of traditional Chinese medicine in the United States.

Bernie's story is telling in several ways. Upon its introduction to those outside of Chinese immigrant communities, traditional Chinese medicine was primarily used for illnesses for which biomedicine was less effective or ineffective, or diseases and illnesses that were undiagnosed or undiagnosable by biomedicine (Zhan 2009). Indeed, one of the key moments for acupuncture and herbal medicine to enter the biomedical mainstream was during the HIV/AIDS crisis in the 1980s. Before bioscience could identify the cause of the disease, acupuncture entered community clinics as an experimental treatment (Zhan 2009). A national survey in the early 2000s indicated that 44% of those who used acupuncture did so because conventional biomedical treatments would not help their conditions, whereas 52% thought that it would be interesting to try (Barnes *et al.* 2004). These conditions include, for example, allergies and asthma, insomnia, certain pain syndromes, stress and depression, certain types of cancers that were resistant to biomedical therapies, and other chronic illnesses (Ibid.; Eisenberg *et al.* 1993, 1998; National Institute of Health 1997; Ni *et al.* 2002). It is noteworthy that many of these conditions – though by no means all – were closely associated with living an urban middle-class lifestyle. Through this association, acupuncture and Chinese herbal medicine came to be identified as a 'preventive medicine' for the cosmopolitan urban middle class (Zhan 2009).

Bernie's career choice was also representative of many US-born acupuncturists. Many began as patients who benefited from acupuncture or herbal medicine, and then decided to pursue studies in traditional Chinese medicine. These patient-turned-acupuncturists often noted that their decisions were not only based on their own medical experiences, but also on what they described as the sterile attitude, exorbitant cost, and sometimes perceived incompetence of biomedicine and biomedical physicians. According to the anthropologist Sonya Pritzker, who conducted fieldwork in acupuncture schools and colleges in California in the 2000s, about 70% of the students were Caucasian, 20% Asian (American), and 10% were of African-American, Latino or Persian descent (2014:4). About 60% of the faculty were native English speakers, mostly Caucasian. The other 40% were Asian, mostly trained in

Chinese and Taiwanese institutions (Ibid.). In my own fieldwork, over 50% of practitioners of acupuncture in California were of Caucasian descent, and the rest consisted mainly of those of East Asian descent such as Chinese, Korean, Vietnamese, and Japanese (Zhan 2009). This is not an invariably rosy picture of inclusiveness and diversity. As Linda Barnes (1995, 2005) and Tyler Phan (2017) have argued, the mainstreaming of acupuncture and traditional Chinese medicine in the United States has taken place in a racialized framework in which early immigrant practitioners were discriminated against on the basis of their racial profiles whereas today's white practitioners feel no qualms in claiming a place in the world of traditional Chinese medicine.

Interestingly, women also take up a large proportion of the students and practitioners of acupuncture and traditional Chinese medicine in the United States. By Pritzker's estimation, about 70% of current students in 2014 were women. This in part has to do with the fact that acupuncture and traditional Chinese medicine were perceived as the feminine, caring, naturalistic, and holistic counterpart of biomedicine. Even though this perception draws on conventional narratives of gender and nature, East-and-West, culture and science, it nevertheless challenges the dominant status and singular authority of biomedicine in the US healthcare system and changes it from within.

However, not all non-Chinese-American practitioners of acupuncture and traditional Chinese medicine entered their profession as patients. The counterculture movement in the 1960s and 1970s, for example, played an important part in changing the landscape of American mainstream culture and healthcare. During the course of my fieldwork on traditional Chinese medicine in the San Francisco Bay Area, I encountered a generation of non-Chinese and non-Asian American acupuncturists who came of age in the 1960s. In contrast to their younger colleagues who pursued a career in traditional Chinese medicine because of an interest in alternative medicine or personal struggles with illnesses, their entry into acupuncture, herbal medicine, and meditation came out of fascinations with 'Eastern' or 'Oriental' religions and philosophies as a way out of the ills of modern capitalism, the industrial military complex, war, the destruction of the environment, and the lack of civil liberty. Concepts such as *dao* and *yin-yang* (See Chapters 1 and 29 in this volume) – which opened the door to ideas and practices such as cosmic unity, harmony, irreverence, and escapism – became their spiritual guidance. Although many of my interlocutors disagreed with each other over which specific American or Eastern 'guru' inspired their own spiritual practice and life choices, virtually all agreed on the importance of Daoist texts such as *Laozi* and *Zhuangzi* in formulating their own naturalistic and spiritual thinking and experience. As the counterculture movement reshaped the American social and cultural landscape (Zhan 2009; Clarke 2000), Daoist ideas and idioms also entered mainstream discourses in the United States, especially in the age of consumerism and new planetary consciousness, contrary to some popular discourses about the 'hippies' having been 'washed up'. (See Chapter 27 in this volume.) Their presence, voice, and vision were in many ways derived from the *geist* and technologies making up cosmopolitan white-collar America at that time and, beyond that, an aspiring though profoundly precarious transnational middle class (Zhan 2009).

The mainstream and the alternative, then, are mutually inclusive rather than exclusive. Biomedical institutions and practitioners have, in the words of many practitioners of acupuncture and herbal medicine, 'turned around'. Major Bay Area hospitals and Health Maintenance Organizations (HMOs), such as the Chinese Hospital, St. Luke's Hospital, Kaiser Permanente, and the University of California, are eager to include acupuncturists in their hospitals and clinics. Major health insurance companies now offer coverage in acupuncture and other kinds of CAM such as chiropractice. Many practitioners of traditional Chinese medicine are bemused if not troubled by the turnaround of the biomedical mainstream. It is not exactly a dream come true. Although many celebrate the newly found status of legitimacy, they also worry about the appropriation by biomedical mainstream and commercial interests. As one acupuncturist and longtime activist put it, 'It is a love-hate relation. How can we push acupuncture into the mainstream without entering and competing in the market?'

Continuous unfoldings

By the late 2010s, there were approximately 30,000 licensed acupuncturists in the United States. Many trained in traditional Chinese medicine colleges in the United States rather than in China or Britain like the early generations. Some have traveled to China and other parts of East Asia to further their studies and, as I have learned from my current research on medical entrepreneurship in China, to teach and practice traditional Chinese medicine. Unlike the Rockefeller Foundation more than 100 years ago, this new generation of US-trained acupuncturists and herbalists were in urban China offering something different from both TCM and biomedicine and in so doing completing the trans-Pacific circle through which traditional Chinese medicine has traveled and constantly morphed. The trans-Pacific route is one of the many through which Chinese medicine has globalized. The vibrant forms of traditional Chinese medicine in the United States are part of the larger dynamic unfolding which through this set of lively medical practices continues to change our ideas and practices of who we are and the worlds we inhabit.

Bibliography

American Anesthesia Study (1976) *Acupuncture Anesthesia in the People's Republic of China: a trip report of the American Anesthesia Study Group, submitted to Committee on Scholarly Communication with the People's Republic of China*, Washington, DC: National Academy of Sciences.

American Herbal Pharmacology Delegation (1975) *Herbal Pharmacology in the People's Republic of China: a trip report of the American Herbal Pharmacology Delegation, submitted to Committee on Scholarly Communication with the People's Republic of China*, Washington, DC: National Academy of Sciences.

Andrews, B. and Bullock, M.B. (2014) *Medical Transitions in Twentieth-Century China*, Bloomington: Indiana University Press.

Barlow, J. and Richardson, C. (1979) *China Doctor of John Day*, Portland, OR: Binford and Mort.

Barnes, L. (1995) *Alternative Pursuits*, PhD thesis, Havard University.

Barnes, L. (2003) 'The acupuncture wars: the professionalizing of American acupuncture – a view from Massachusetts', *Medical Anthropology*, 22.3: 261–301.

—— (2005) *Needles, Herbs, Gods, and Ghosts: China, healing and the west to 1848*, Cambridge, MA: Harvard University Press.

Barnes, L. and Hinrichs, TJ (eds) (2013) *Chinese Medicine and Healing: an illustrated history*, Cambridge, MA: The Belknap Press of Harvard University Press.

Clarke, J.J. (2000) *The Tao of the West: Western transformations of Taoist thought*, London: Routledge.

Eisenberg, D., Davis, R., Ettner, S., Appel, S., Wilkey, S., Van Rompay, M. and Kessler, R. (1998) 'Trends in alternative medicine use in the United States, 1990–1997', *Journal of American Medical Association*, 280.18: 1569–75.

Eisenberg, D., Hufford, D., Jonas, W. B. and Crawford, C. (2013) 'The evolution of complementary and alternative medicine (CAM) in the USA over the last 20 years', *Forschende Komplementarmedizin*, 20.1: 65–72.

Eisenberg, D., Kessler, R., Foster, C., Norlock, D., Calkins, F. and Delbanco, T. (1993) 'Unconventional medicine in the United States', *New England Journal of Medicine*, 328: 246–52.

Farquhar, J. (1994) *Knowing Practice: the clinical encounter of Chinese medicine*, Boulder, CO: Westview Press.

Furth, C. (2011) 'The AMS/Paterson lecture: becoming alternative? modern transformations of Chinese medicine in China and in the United States', *Canadian Journal of Modern History*, 28.1: 5–41.

Lee, M. (1992) *Insights of a Senior Acupuncturist*, Boulder, CO: Blue Poppy Press.

Lei, Hsiang-lin (2014) *Neither Donkey nor Horse: medicine in the struggle over China's modernity*, Chicago, IL: University of Chicago Press.

National Institutes of Health (1997) 'Consensus development statement: acupuncture', *Consensus Statement Online 1997 Nov 3–5, 15.5:1–34*, https://consensus.nih.gov/1997/1997acupuncture107html.htm, accessed 1/4/2020.

Phan, Tyler (2017) *American Chinese Medicine*, PhD Thesis, University College London.

Pritzker, S. (2014) *Living Translation: language and the search for resonance in U.S. Chinese medicine*, Oxford; New York: Berghahn Books.

Reston, J. (1971) 'Now, about my operation in Peking', *The New York Times*, July 26: 1, https://www.nytimes.com/1971/07/26/archives/now-about-my-operation-in-peking-now-let-me-tell-you-about-my.html, accessed 1/4/2020.

Rogaski, R. (2014) *Hygienic Modernity: meanings of health and disease in treaty-port China*, Berkeley: University of California Press.

Sadava, D., Ahn, J., Zhan, M., Pang, Mei-lin, Ding, J. and Kane, S. E. (2002) 'Effects of four Chinese herbal extracts on drug-sensitive and multi-drug-resistant small-cell lung carcinoma cells', *Cancer Chemotherapy and Pharmacology*, 49:4: 261–6.

Scheid, V. (2002) *Chinese Medicine in Contemporary China: plurality and synthesis*, Durham, NC: Duke University Press.

——— (2007) *Currents of Tradition in Chinese Medicine, 1626–2006,* Seattle, WA: Eastland Press.

Starr, P. (1982) *The Social Transformation of American Medicine: the rise of a sovereign profession and the making of a vast industry*, New York: Basic Books.

Zhan, M. (2009) *Other-Worldly: making Chinese medicine through transnational frames*, Durham, NC: Duke University Press.

41

THE MIGRATION OF ACUPUNCTURE THROUGH THE *IMPERIUM HISPANICUM*

Case studies from Cuba, Guatemala, and the Philippines

Paul Kadetz

Introduction

It might be assumed that migration of 'culture', or, more accurately, of a given group's *order and meaning*, is primarily, or even solely, a consequence of the migration of people. However, the transmission of health practices and the understandings of health from one group to another may also be linked to factors other than migration. Political economy, similarities of praxis, and revolutionary representations of universal health coverage, are all factors that may help to explain the migration of acupuncture to Cuba, Guatemala, and the Philippines respectively; all former colonies of the Spanish Empire.

The factors of political economy, trade, foreign aid, and other forms of international soft power may lay a foundation for such migrations of order and meaning from one place to another. Sino-Latin American trade and economic relations have developed at a 'spectacular pace' in the late twentieth century (Dominguez 2006: 1). From 1990 to 2004, China's imports from Latin American countries increased from US$1 to $20 billion; with Chinese exports to Latin America increasing from US$432 million to $15 billion (Ibid.). Chinese companies have invested 'at least $25 billion in Latin America since 2005' (Gallagher 2010: 10). At the first ministerial meeting of the 'Forum of China and the Community of Latin American and Caribbean States', held in Beijing in January 2015, President Xi Jinping's goal for the next five years included US$500 billion in trade with the Latin American and Caribbean region and US$250 billion of direct investment (Dollar 2017).

Factors of political economy can also be identified in the antipathy that many Latin American countries share with China for American hegemony and imperialism (Dominguez 2006). Some Latin American governments have even viewed the People's Republic of China (PRC) as a means towards 'independence from the political influence of the United States and the economic dominance of Western institutions' (Ellis 2009: 1). At present, China continues to be an important economic and diplomatic force in Latin America.

In order to identify the factors that may have facilitated the migration of acupuncture to the countries of the former *Imperium Hispanicum*, this chapter will examine the three case studies of Cuba, Guatemala, and the Philippines.

The migration of acupuncture to Cuba: a lesson in political economy

In December 2010, the Cuban postal service issued four stamps to commemorate the fifty years of diplomatic relations between the PRC and Cuba. The PRC has provided a vital source of funding for Cuba, and has been Cuba's second largest trading partner and exporter of goods after Venezuela, with annual bilateral trade reaching approximately US$2.4 billion in 2015 (Kuo 2016).

Since 2002, the majority of physicians in Cuban medical schools received training in acupuncture, and all Cuban hospitals offered acupuncture anaesthesia for surgical procedures (Lo 2011). The use of acupuncture anaesthesia in surgery occurred in nearly 10% 'of the 336,622 major surgeries performed in 2008' (Stafford 2010: 45). However, the migration of acupuncture to Cuba pre-dated these recent economic ties and was locally supported, irrespective of Sino-Cuban relations; which in actuality were quite sporadic throughout the Cold War. Furthermore, the formal migration of acupuncture to Cuba had seemingly little to do with the presence of Chinese in Cuba since the nineteenth century.

Indentured servitude: the Chinese in Cuba

Mass emigrations from China that occurred between the nineteenth and mid-twentieth centuries are known collectively as the Chinese diaspora. Many of these migrations to Spanish colonies were due to economic deprivation in China and the promise of a better life in new lands. From 1847 to 1873 more than 200,000 Chinese (primarily farmers from Canton) arrived in Cuba with promises of employment that would lead to fortunes; though in reality they faced indentured servitude (Delgado Garcia 1995). This particular chapter of the Chinese diaspora was a result of numerous international changes, including the Treaty of Nanking that ended the Opium Wars and resulted in the 'creation of Hong Kong as an outpost of British imperialism'; facilitating access to labourers from South China (McKeown 1999: 313). The resulting increase in Chinese labour migration was often made via Hong Kong (Ibid.). Early nineteenth-century abolitionist campaigns in the United States and England that resulted in the restructuring of slave labour prepared the way for a new wave of Chinese migration and facilitated the Spanish colonists' procurement of Chinese workers (Dabney 2006).

The Spanish colonists in Cuba already had experience with the Chinese, having worked alongside Chinese traders over several centuries in the Philippines (Dolan 1991). The Chinese who ultimately made their way to Cuba were predominantly poor peasants from Southeast China, who believed that labour on sugar or tobacco plantations would eventually bring them wealth (Dabney 2006). Although the Chinese contracts stated that they were wage labourers, their contracts were usually sold to plantation owners for the equivalent of US$60.36 (Ibid.). Subsequently, the Chinese workers were paid a minimal amount and, for the most part, treated as indentured servants (Ibid.).

The Chinese maintained their community in Cuba in an area called *Sagua la Grande*. This area eventually housed a Chinese theatre, and numerous service businesses including tailors, barbers, candy stores, gambling houses, opium shops, and Chinese physicians (Dabney 2006). The Spanish colonists were reported to have held the Chinese physicians in higher esteem than their own physicians, 'whom nobody trusted' (Ibid.: 178).

The embrace of modernity ...at the expense of other 'local' practices

Along with Chinese medical practices, an abundance of plural ethnomedicines were practised throughout nineteenth century Cuba. However, just as Chinese medical practices were practised almost entirely within Chinese communities, the practices of the indigenous Guanahatabey and Ciboney, the Taino Indians, African slaves, and immigrants from other Caribbean islands were also not integrated into Cuban State healthcare of the nineteenth to mid-twentieth centuries (Volpato *et al.* 2009).

Similar to the trajectory of local and indigenous practices in China – which were initially dismissed by Mao Zedong and the Chinese Communist Party as 'old medicine' before being embraced by Mao as a 'national treasure', that could be used to redress rural health disparities – (Taylor 2005), Fidel Castro originally rejected all indigenous, non-biomedical practices as archaic (Rosenthal 1981; Cochetti 2008). Both Cuban and Chinese Communism embraced modernity and scientism while rejecting the past (Kadetz Chapter 42 in this volume). Modernisation was the discourse that fuelled the early Cuban communist State (Cochetti 2008).

Following the Cuban revolution (1953–1959), local healthcare practices, including religious practices, were actively discouraged. For example, beginning in 1961, Chinese pharmacies in Cuba were systematically closed until only one remained open (Stafford 2010). If biomedical practitioners used or referred patients for non-biomedical practices, they would receive sanctions from the Ministry of Public Health (Brotherton 2005). Hence, the State's initial rejection of local non-biomedical healthcare was further supported by Fidel Castro's renewed emphasis on the biomedical modernisation of Cuba, and the centrality of modern biomedicine to the communist State.

In Castro's 1953 speech, *La Historia me Absolverà* ('History will Absolve me'), the health of the population is proclaimed to be an important indicator of the success of governance. And in the 1975 Cuban constitution, health was incorporated as a right of Cuban citizens and the responsibility of the State (Official Gazette of the Republic of Cuba 2003: 3). As a result of this constitutional change, Castro's government instituted the Ministry of Public Health along with 'a national health system, the nationalization of private clinics and pharmaceutical companies and social and rural health care services at little or no cost' (Stafford 2010: 42). The goal of these changes was to provide universal access to biomedicine.

Interestingly, though, it was during this early post-revolutionary period that acupuncture started being utilised in place of specific biomedical procedures, such as acupuncture analgesia during the mid-1970s. However, the teaching lineages of acupuncture in Cuba cannot be traced to the Chinese migrants living in Cuba, nor indeed to China specifically (Padron Caceres and Perez Vinas 2005). In 1962, after a seminar in Havana was hosted by a physician who founded the Medical Institute of Acupuncture in Argentina, Floreal Carballo, acupuncture was formally, albeit minimally, introduced into the Cuban healthcare system (Acosta Martínez 2000). Further acupuncture courses during the 1980s and 1990s employed American acupuncturists (Padron Caceres and Perez Vinas 2005). Despite its acceptance in certain circles, acupuncture would not be fully accepted by the Cuban government and medical community until 1988, when three consultants from military academies in North Korea, North Vietnam, and the PRC were invited by the Minister of Military Affairs of Cuba to organise a speciality of traditional military medicine (Ibid.). Acupuncture would then go on to become part of a medical speciality that was incorporated into the national healthcare system under the rubric of *Medicina Tradicional y Natural* (MTN).

MTN was formally integrated into the Cuban healthcare system in 1992 through a national mandate (Applebaum *et al.* 2006). By 1995, a State commission was formed by the

Ministry of Health for the development of MTN throughout Cuba (Ibid.). In 1999, the National Program for MTN approved the use of acupuncture, homeopathy, chiropractic, and native medicinal plants into the national health system (Stafford 2010). But what can explain this marked departure from the State's initial embrace of biomedical modernity and rejection of non-biomedical practices?

The mother of invention

The Cuban government's reversal in its acceptance of acupuncture was largely an outcome of the constitutional commitment to universal healthcare that was severely challenged by the US embargo to Cuba. The United States has held an ongoing trade embargo against Cuba since 1961; the longest such embargo in modern history (Garfield and Santana 1997). However, the first three decades of the embargo had negligible effects on Cuba and Cuban healthcare, which were protected by maintaining 70% to 90% of international trade (which included pharmaceuticals and medical supplies) with Soviet bloc countries (Ibid.).

According to Garcìa (2002), the Cuban government was well aware that the blockade against Cuba could become global and, therefore, planned for self-sufficiency, particularly in terms of healthcare. In the late 1980s, Cuban military policy acted on the need for self-sufficiency and opened the Central Laboratory of Herbal Medicine at the Higher Institute of Military Medicine, thereby taking the first steps towards the formal integration of traditional medicine (MTN) into the Cuban healthcare system (Ibid.; Stafford 2010). The military's preparatory steps towards the widespread use of MTN were unerringly prescient, for within a matter of years, Cuba's protection from the ongoing US embargo vanished with the dissolution of the Soviet bloc.

Cuba had produced sugar for the Soviet Union, which paid higher than world market prices for Cuban exports, and exported Soviet goods in return (Garfield and Santana 1997). After the collapse of the Soviet bloc in 1989, Cuba lost 85% of its foreign trade; with an 80% reduction in Cuban exports and a 70% reduction in Soviet bloc exports to Cuba from 1989 to 1993 (Ibid.). Cuba's Gross National Product 'declined by 35%; and the value of imports from all sources declined from US$8 billion to $1.7 billion' (Ibid.: 15). Even though the PRC became Cuba's second largest trading partner (with an increase in Cuban imports from 4.3% in 1996 to 10.9% in 1991), with the dissolution of the Soviet Union, Cuba had not only lost its most important trading partner, but also its primary source for the raw materials for pharmaceutical production (Dominguez 2006: 10; Stafford 2010).

To add to Cuba's plight, the United States, at this juncture, strategically exacerbated trade restrictions in its embargo against Cuba (Garfield and Santana 1997: 15). In 1992, the U.S. embargo was made even more punitive with the passage of the (ironically titled) 'Cuban Democracy Act', whereby all US subsidiary trade to Cuba, including trade in food and medicines, was prohibited (Stafford 2010). Furthermore, any ships docking in Cuba were prohibited from docking at U.S. ports for six months thereafter, even if their Cuban cargoes were solely humanitarian goods (Garfield and Santana 1997: 15). It could be argued that the perverted logic of the Cuban Democracy Act (ie., that democracy in Cuba would be promoted by completely isolating the island economically) was, in fact, a direct reversal of the logic of neoliberalism; namely, that democracy is only fully promoted through free markets (Friedman 2002), and moreover a complete rejection of the democratic principles of sovereignty and autonomy upon which the United States was founded.

Thereafter, the United States' application of substantial pressure on other countries to cease both trading with, and providing humanitarian goods to Cuba, resulted in a near-global trade blockade against Cuba. In terms of biomedicine, the Cuba Diplomacy Act strictly prohibited trade of 'medicines or medical devices with 10% or more of their components made by a US company or foreign subsidiary of a US corporation [...] medical supplies for humanitarian aid can be sent to Cuba only after the Cuban government holds free and fair elections' (Stafford 2010: 43).

Such restrictions created severe issues in accessing pharmaceuticals and the materials for the domestic production of pharmaceuticals in Cuba, particularly considering that the global pharmaceutical industry was predominantly controlled by U.S. transnational corporations. The dollar value of imports for health to Cuba fell from US$227 million in 1989 to US$67 million in 1993 (Garfield and Santana 1997: 18). Many of the hard-won benefits of the Cuban healthcare system were quickly reversed, despite the State's organised targeting of limited resources to vulnerable populations, including the elderly, children, and women (Applebaum *et al.* 2006). Overall, the resulting shortage of pharmaceuticals was associated with a 67% increase in deaths due to infectious and parasitic diseases; a 77% increase in deaths due to influenza and pneumonia from 1989 to 1993; and a 48% increase in tuberculosis deaths from 1992 to 1993 (Garfield and Santana 1997: 17).

Hence, it was in this context that in 1992 the Cuban government completely reversed its position on the use of non-biomedical practices; reminiscent of Mao's need to reverse his position disparaging the use of Chinese medical practices in order to redress access to rural healthcare more than twenty years earlier (Applebaum *et al.* 2006). Following specific orders from Fidel Castro, certain standardised non-biomedical healthcare practices[1] were promoted as an important element of Cuban healthcare (Cooper, Kennelly, Ordunez-Garcia 2006). Suddenly, several non-biomedical practices, including acupuncture, that had hitherto been slowly and, at best, selectively and reservedly promoted or researched by the Cuban State could now scarcely be integrated into the national healthcare system quickly enough.[2]

The migration of acupuncture to Cuba was, therefore, an outcome of a very specific political economic context, while the migration of acupuncture to Guatemala involved a quite different set of conditions.

Remembrance of things past: the migration of acupuncture to Guatemala

Unlike Cuba, which historically housed a Chinese population that could have potentially paved the way for the migration of Chinese medicine, no Chinese population was ever present in Guatemala. In fact, just as the formal migration of acupuncture to Cuba was initiated by an Argentinian, the migration of acupuncture to Guatemala was initiated by a group of American acupuncturists. In the Northeast region of Guatemala known as Peten, American licenced acupuncture volunteers from the non-governmental organisation, The Guatemala Acupuncture and Medical Aid Project (GUAMAP), trained community health workers to treat their communities with basic acupuncture since 1994 (interview with GUAMAP CFO 2007). The ongoing success of this programme is a source of curiosity. Why would acupuncture be successfully introduced to the rural indigenous Maya of Guatemala? To attempt to answer this we need first to examine Mayan medicine.

Pre-colonial Mayan health practices

Contextualisation of pre-colonial Mayan health practices provides a basis for understanding current Mayan medical pluralism and the reception of acupuncture by indigenous Guatemalans. A marked disruption in Mesoamerican local order occurred with the Spanish conquests of the sixteenth century. After the conquests, indigenous practices, including local health practices, were markedly influenced by the Spanish colonists.

The lack of literature on indigenous health practices pre-dating the conquests presents challenges in disentangling Spanish influences from indigenous practices. Two works that may provide the most accurate depiction of pre-Columbian life are the *Popul Vuh* and the *Chilam Balam*. The *Popul Vuh*, written in the mid-sixteenth century, describes the early history of the Quiche' people and is one of the earliest surviving documents regarding Mayan illness classification and therapies. The *Chilam Balam*, a grouping of disparate writings named for the particular town in which they were found, and containing historical, medical, and divinatory knowledge (Roys 1933; Kunow 2003: 33), was believed to have been written by indigenous authors immediately after the conquest (Garcia, Sierra, Balam 1999). The Chilam Balam contains such information as the proper days for bleeding and purging, practices that were also affiliated with contemporary European and Chinese medical practices; which raises the question of whether Mayan practices were merely examples of 'travelling medicine' or forms of transcreation and hybridisation (Lo and Yoeli-Tlalim 2018) or, rather, continuities resulting from simultaneous, yet discrete, creation (Roys 1933; Gubler and Bolles 2000: 5).

Similar to the logic of myriad local, non-biomedical healthcare practices across the globe, ancient Mayans framed health as an outcome 'of living in harmony with social and natural laws; [whereas] disease is the result of transgressing these laws' (Garcia, Sierra, Balam 1999: xxxv). Medical practices are both a part of, and a product of, a given group and their specific system of order. According to anthropologist Mary Douglas (1966) all 'cultural' groups order the world and phenomena in a way that is meaningful to that particular group and that facilitates a common group understanding. Similar to Chinese medical understandings of order, categorical affiliations existed in Mayan medical order between chronology, the winds, animals, the cardinal directions, and illness (Chapters 5 and 6 in this volume; Roys 1933). Illness was diagnosed in pre-colonial Mayan society using palpation, divination, and consultation with the patient, and was primarily treated with medicinal plants, prayer, and bleeding in an attempt to restore harmony between 'the physical, social, and cosmological order and internal bodily processes' (Huber and Sandstrom 2001: 109). In describing the pre-Columbian classification system of illness, Orellana (1987: 28) differentiates between two categories of aetiology; illness caused by non-human beings – or by human beings who possessed supernatural powers – and natural causes, such as accidents, deficiencies, or excesses. In summary, religion, medicine, and morality were closely interwoven in pre-colonial Mayan medical practices (Ibid.: 257).

The influence of Spanish colonisation on Mayan medical order

Colonisation resulted in myriad changes in Mayan health practices by virtue of colonial religion, colonial health practices, and colonial policy aimed at replacing indigenous understandings and practices with those of the Spanish colonial order. In the mid-sixteenth century, Dominican missionaries in Guatemala opened the *Hospital de San Alexo*, treating local Maya with European medicine (Huber and Sandstrom 2001). Catholicism was inextricably linked to colonial medicine in Guatemala. Priests, friars, and missionaries operated the Spanish hospitals and the clergy promulgated the concepts of Spanish medicine to Mayans hoping to replace Mayan medical beliefs with European beliefs, which included Catholicism (Ibid.). Colonial policy, that was enforced by a *Protomedicato* responsible for licensure of medical personnel, resulted in the standardisation and complete legal control of Mayan medical practices by the State (Ibid.).

A fundamental outcome of colonisation and the ensuing epidemics the Spanish brought, was the hybridisation of European humoral medicine into constructions of Mayan medical order (Huber and Sandstrom 2001). There is strong support in the literature, concerning the history of humoral medicines in Latin America, that the humoral system is not an indigenous system (Currier 1966; Orellana 1987; Foster 1994; Huber and Sandstrom 2001). For example, Trevino identifies that the *Relacio'n de Tiripiti'o* (written in 1579) mentions Mayan physicians discussing illnesses in terms of 'the Galeno-Hippocratic humoral medicine that Spanish doctors practiced' (Huber and Sandstrom 2001: 51). Concurrently, indigenous medicine was attacked by the Spanish protomedicos for not following 'the doctrines of Hippocrates and Galen' (Ibid.: 55). According to Baer, Singer and Susser (1997: 307), 'humans universally have developed health care systems that reflect their living conditions and resources'. Hence, humoral medicine may not only have been a European system of order that was forced on the Mayans, but may also have been embraced by Mayans in order to survive the new infectious diseases the colonists introduced, against which the Maya were defenceless with their own interventions. Thereby, the Greco-Persian-Arabic humoral medicine of Hippocrates, Galen, and Avicenna may have been absorbed into Mayan medicine.

The centrality of the conception of equilibrium to Mayan cosmology may have provided a fertile ground for humoral theory (Orellana 1987: 260). In humoral theory, illness is believed to be caused by a disturbance resulting in an imbalance of hot and cold in the body, and remedied by hot and cold therapies that restore the imbalance (Wilson 1995; Adams and Hawkins 2007). However, in the Mayan adaptation of humoral medicine, hot and cold are not equivalent endpoints. Heat is considered more advantageous and positively associated with emotional support and affection (Wilson 1995). This meaning is illustrated in the hot classifications of fertility, pregnancy and menstruation, a powerful pulse, elders and ancestors, officials, and military officers (Ibid.). In contrast, cold is associated with factors that threaten one's survival, including sterility, weakness, rejection, and withdrawal (Ibid.).

Given that the humoral model is structured by life stages, gender, interpersonal relationships, and social hierarchies, this model can be understood to serve as a symbolic system onto which social anxieties can be projected and from which social desires can be fulfilled (Currier 1966). However, it should be understood that the meanings of hot and cold systems throughout Mesoamerica 'vary according to indigenous group and linguistic area' (Wilson 1995: 131). Hence, it is unlikely that a single interpretation of humoral medicine was adopted. For example, there is relatively little agreement as to which foods are hot and which are cold, not only between geographical areas of Mesoamerica, but even among the members of a single community (Currier 1966).

In order to understand why only certain elements of medical theories and practices are translatable and adopted between one group and another, it is essential to know who mediated and controlled these translations.

Reflections or illusions: Chinese medicine and Mayan medicine

The goal of the dynamic dualisms of humoral medicine, embedded in the order of Mayan medicine, is harmony; a central tenet of both Mayan medicine and traditional Chinese medicine (TCM); though this concept is relative to each system and we should not imagine that they are simply interchangeable. For example, in TCM, harmony of both the nutritive elements of the body (or simplistically stated, the more feminine or yin elements) and the functional (or more masculine or yang elements) is sought for ideal health (Ni 1995). In Mayan understandings, women and children are considered weak, whereas men are strong. Men are believed to be hotter, women colder. Men are believed to have stronger blood than women. Hot blood is considered healthier and stronger than cold blood. Hence, the concept of harmony that provides ideal health in Mayan medicine actually places stronger emphasis on masculine associations.

Furthermore, the general hot/cold duality of the adopted Mayan humoral concept can be perceived as similar to the concepts of hot/cold; yin/yang and myriad other dynamic binaries in Chinese medical systems. For example, in Chinese medical theory if there is a deficiency of yin, yang predominates and deficient heat ensues. The same is true of lack of cold resulting from yang deficiency. As death follows from the separation of hot and cold in Mayan medicine, so too is it a consequence of the separation of Yin and Yang (Orellana 1987: 29; Ni 1995). Despite the healthier and stronger nature of masculine attributes in Mayan medicine, according to Adams and Hawkins, one elderly male Mayan informant identified: 'If there was no cold we would die' (Adams and Hawkins 2007: 77). Similarly, *Tonalli,* or life force in Mayan medicine, *appears* to resemble the Chinese concept of life force or *qi*; as both can result in illness or death if deficient or absent from the body (Orellana 1987: 29; Ni 1995).

Both medical systems echo tensions with Western pharmaceuticals. Modern biomedicines are usually considered dangerously hot by Mayas, and as consuming yin in TCM (Wilson 1995). According to Mayan logic 'Vaccinations and strong medicines burn humans by creating excess amounts of heat damaging the body' (Adams and Hawkins 2007: 77).

Parallels can be further identified between Mayan and Chinese cosmology, mythology, understandings of chronicity and health, aetiological explanations for illness, diagnostic methods, therapeutics, and management. Garcia, Sierra, and Balam (1999: xxxii) report 'from the moment we began to introduce Asian techniques like acupuncture into our community health work, we were received quite differently than when we had presented ourselves as Western doctors'. They maintain that one reason for this reception may be that Mayan medicine had the most developed 'autochthonous system of acupuncture in Mesoamerica' (Ibid.: 109). Although, we can challenge whether what Mayans practised could actually be deemed an 'autochthonous system of acupuncture', Garcia *et al.* identify two indigenous variations of medical treatment utilising body piercings, that clearly resemble the manual techniques and body topology of Chinese acupuncture and the superficial surgical practices of *waike* (外科). Known as *jup* and *tok,* these practices have been located in Maya communities of the Yucatan peninsula.

Purportedly the first documentation of these therapies appear in a book of the late sixteenth century; *The Ritual of the Bacabes*, in which illness is treated by '*ix hun pudjub kik* (the needle which bleeds) and *ix hun pudjub olom* (the needle which frees the blood)'. The skin is punctured 'with the spines and thorns of several varieties of plants and several animals' (Garcia *et al.* 1999: 109). *Jup* is a technique in which discrete points on the body are pricked several times in rapid succession, but blood is not drawn (Ibid.). *Jup* is performed in two ways; the same point can be punctured three times at approximately 1 centimetre depth, or the same point on the body can be punctured repeatedly until a response of inflammation to the area is achieved (Ibid.). In *Tok,* a point on the body is punctured in order to draw blood (often facilitated with cupping over the point-another shared practice between the two medical systems) (Ibid.: 110).

Similar in practice to *Jup*, is the 1000-year-old Chinese practice of Plum Blossom Needle, named for the five needles bound together to resemble a plum blossom, in which the instrument is repeatedly struck over an acupuncture point or an area resulting in inflammation or micro-bleeding (Zhong 2007). Although in acupuncture, a needle is usually inserted in a specific point on the body and is typically retained for fifteen to twenty minutes, in the technique known as *ci xue liao fa* 刺穴疗法, bloodletting will occur (often with cupping) as practised in *tok* (Cheng 1996: 110).

In TCM, more than 350 acupuncture points are identified that can be utilised in treatment; a reflection of the original astromedical conception of the continuities between the number of days of the year of the macro world and the number of points on the body of the micro world (Cheng 1996). Garcia *et al.* (1999: 112) have identified fifty body points used in *tok* and *jup*, all of which they have correlated with the location of acupuncture points. For example, a common and important point on the body in both TCM (*yintang* 印堂) and *Tok/jup* (*tok lu ni*) is located midway between the eyebrows. The skin is lifted and pierced in both medical systems. In *tok/jup* it is both a point punctured in children to prevent future illness and in adults to stop 'evil wind attacks' that may cause headaches (Ibid.). Acupuncture at this same point location can be used for headaches, as well as for 'clearing external wind-heat' pathogens, pathology in the brain and nose, calming the spirit of the patient, and hypertension (Cheng 1996).

There are numerous instances of other correlations in both anatomical location and function between the two medical systems, even though these practices developed quite independently from one another. Yet, despite the differences in cosmology, understanding, meaning, and order, embedded in Chinese and Mayan medical systems, the similarities in the practice of these two distinct medical therapies may offer one explanation for the reception of acupuncture in some Mayan communities. But, there may also be other factors to help explain these 'cultural migrations' to both Mayan Guatemala and Cuba that can be best understood via a popular representation of acupuncture as a kind of revolutionary or 'rebel medicine'; for acupuncture represents the medicine of a people who have never had any colonial history, but who have been associated with socialist revolutions in Latin America.

Rebel medicine: acupuncture in the Philippines

Beyond political economic circumstances and beyond the similarities of foreign medical practices and local practices, there may be yet another reason that acupuncture migrated to both Guatemala and Cuba, as well as to many other similar contexts throughout the Hispanic world. A discussion between Mao Zedong and the Cuban party secretary, Blas Roca Calderio, in September 1960, marked the first instance of Sino-Latin American diplomatic relations. This bond was purportedly built upon the Cuban revolution itself. According to Lee (1964: 1132) 'When Che Guevara was interviewed in 1959 by a group of Chinese correspondents, he made clear his admiration for Mao'. Guevara purportedly said, 'When we were engaged in guerrilla war, we studied his [Mao's] theory on guerrilla warfare. Mimeographed copies of his work circulated widely among our commanders at the front. It was called 'the food from China' (Ibid.)'. Although this bond reportedly cooled during the Cold War period – when the Cuban government backed the Soviet Union in Sino-Soviet disputes – the significance of Maoist doctrine as a blueprint for revolution and rebellion, may have secured its symbolic value (Dominguez 2006). Similarly, though there have not been significant diplomatic relations between the PRC and Guatemala, Mayan Guatemalans, who survived a thirty-six-year civil war and continue to face structural violence on a daily basis, may find meaning in a doctrine of rebellion that can be symbolically represented through acupuncture; a means to achieve equitable healthcare access.

Thus, although Guatemala and Cuba represent two completely different contexts for the migration of acupuncture, it has been identified in both contexts as an equitable solution for healthcare access, and a resolution for the potential inequities embedded in market-driven biomedicine. After all, it was the Chinese barefoot doctor, who practised both TCM and biomedicine, and whom the World Health Organization globalised as the symbol for equitable 'health for all' in the Declaration of Alma Ata in 1978 (Chapter 44 in this volume). Thus, embedded in these global and local representations, is a symbol of traditional Chinese medicine as a socialised form of medicine that can redress social inequities. This depiction may not only be identified in the Cuban military's involvement in the development of MTN, and the self-organisation of vulnerable and underserved Mayan communities after the 36-year civil war, but also in another former Spanish colony halfway around the world: in the rural Philippines.

The groupings of grassroots healthcare programmes in the Philippines, known as Community-based Health Programs (CBHP), were born out of the poverty and political repression of the late 1960s. These programmes began in reaction to poor healthcare access: 'At its inception, the health program was a reaction to existing approaches to healthcare, which did not reach the people most needing them – the poor, especially in rural areas' (Council for Health and Development 1998: 11). In research I conducted in the Philippines, in both 2010 and 2020, informants discussed both 'community building' and 'nation building' as a primary desired outcome of CBHP groups. These groups trace their origin to the Rural Missionaries of the Philippines (ca. 1969) organised by the Association of the Major Religious Superiors of Women in the Philippines (Ibid.: 12). Informants from the various Community Health Worker (CHW) programmes of the CBHP groups identified the barefoot doctors programme of China as the template for CBHP. A CHD manual identifies CHWs as volunteer members of communities that 'are chosen by the community' and 'trained in basic health skills, such as prevention and treatment of common diseases, first aid, use of herbal medicine acupuncture and acupressure and dental hygiene and tooth extraction' (Ibid.: 8–9, 31). This CBHP curriculum was standardised in 1984 (Ibid.: 44).

Acupuncture was an early addition to the CBHP programmes. According to informants, China and the Philippines engaged in a cultural exchange during the Marcos administration (c. 1965–1986). As part of these exchanges, China offered both formal and informal short training courses in acupuncture in the Philippines, primarily for physicians. Acupuncture was quickly identified as an important cost-effective intervention for CBHPs. However, the development of CBHP organisation was soon viewed by the Marcos administration as 'conscientizing activities considered crimes by the government' and linked to revolutionary communist groups (Council for Health and Development 1998: 15). CHWs 'were targeted as subversives, arrested, and killed' (Ibid.: 30). However, the work of the CBHP's spread to other political groups.

The *Partido Komunista ng Pilipinas* (PKP), founded in 1930, is the 'oldest leftist party in the Philippines' (Quimpo 2008: 56). In 1968, a group of 'young communists' broke away from the PKP and started the Communist Party of the Philippines (CPP) (Ibid.: 58). The CPP actively denounced 'U.S. imperialism, feudalism, and bureaucratic capitalism' calling for 'the revolutionary overthrow of the reactionary Philippine State' (Ibid.). Several months later, in 1969, the CPP acquired a 'guerilla army' known as the New People's Army or NPA (Ibid.). This group became aligned with a larger 'revolutionary united front' in which several – but according to informants, not all – CBHP groups were aligned (Ibid.: 56).

It was during the period of martial law at the end of the Marcos administration, beginning in 1972, that any members of groups believed to have any affiliation with the CPP or NPA were routinely 'arrested, tortured, detained, and/or killed' (Quimpo 2008: 56). Several informants, including university professors and physicians, described years of torture and imprisonment for any suspected, though unproven, link with the NPA. 'Thousands disappeared without any trace, while many others were summarily executed' (Council for Health and Development 1998: 42). Several informants stated that neither one's political affiliation, nor the fact that equitable healthcare was the group's sought goal, was relevant to the government. The fact that community level organising was involved, in and of itself, constituted a threat to the government.

The NPA became a significant national movement, especially through forging bonds with the rural poor (Jones 1989). A sixty-two-year-old female physician informant in Manila reported: 'It seemed that communism was the way that Filipinos could be free of colonialism, and the NPA was fighting for a revolution to change the system'. Healthcare was a primary concern of the NPA and many young medical students became affiliated with both the NPA and/or CBHP groups in this capacity. According to informants, it was specifically the Philippine armed forces who began to affiliate acupuncture with both the NPA and CBHP groups. Jones (1989: 233) notes the NPA had their own medics who were 'trained to perform acupuncture and acupressure treatments and were skilled in the use of herbal medicines'. One male NPA informant reported: 'We learned [acupuncture] from books, from Chinese books, and [from] some trainings from community doctors. Basic training was three days'. However, other informants identified that the CPP was able to send some members to China to study acupuncture. Another NPA informant stated: 'We use acupuncture because it is cheap and does not require much [sic] materials'.

Interviews with current and former NPA-identified informants revealed that the NPA have utilised acupuncture as a means to recruit NPA members in rural communities. Informants explained that recruitment was initiated through free acupuncture treatments and later through acupuncture trainings. Thereby, the CBHPs, and especially the practice of acupuncture, became intensely politicised in the Philippines from the early 1970s. In fact, the mere presence of an acupuncture needle was considered sufficient evidence to identify one as having NPA affiliation. One female family medicine physician informant in Manila reported: 'When we were travelling in Mindanao, we would travel with [acupuncture] needles inside of pens, because we were afraid that the army was there, at a checkpoint, and you would be checked for needles. If they found them, you would be tortured – at best – or more likely killed'.

Healthcare providers were not the only group harassed. Patients were also harassed for seeking treatment from CBHP groups. One nurse informant who worked with several CBHP groups identified: 'The military were always watching us, and our patients were afraid to come in and be treated because of them, so they often would have no treatment'. Harassment of CBHPs did not cease after Marcos, but actually intensified with ['Auntie'] Corazon Aquino's administration (1986–1992), resulting in the cessation of CBHP groups in 267 communities and disruption in an additional 305 communities (Council for Health and Development 1978: 53).

In 2010, during the final days of the Macapagal-Arroyo administration, a group of forty-three healthcare workers undergoing training in disaster management and acupuncture, known as the 'Morong 43', were arrested by the military and detained for ten months on charges of illegal possession of firearms, explosives, *and* acupuncture needles (Department of Health 2015). According to an informant working with the Morong 43: 'They were illegally detained and tortured, while the military attempted to force them to admit they were part of the NPA'. The group filed a civil case for damages against former president Arroyo and high ranking military officials. This informant's allegations were corroborated by the 2015 report of the Philippines Commission on Human Rights that confirmed the illegal arrest and torture of the Morong 43 (Mateo 2015).

Hence, the case of acupuncture in the Philippines illustrates how the migration of acupuncture can be politicised for its provision of cost-effective accessible healthcare for all, and how it can be represented as a rejection of profit-driven healthcare. Thereby, acupuncture has served as a mechanism for political grassroots organisation.

Conclusion

In this chapter, we have discussed the migration of the Chinese 'cultural product' of acupuncture to countries of the *Imperium Hispanicum* that has not, in general, been accompanied by a marked influx of Chinese migrants, and in some cases, such as Guatemala, there has been no Chinese migration at all to support such 'cultural migrations'. In other words, we have examples of the migration of a people's systems of order and meaning without the migration of its people. One could argue that the globalisation of the outputs of the West, particularly biomedicine, has been achieved in numerous contexts without an accompanied influx of Western colonists or migrants, however Western 'cultural products' are clearly part of a dominant political economy and authoritative knowledge[3] of the West.

These three cases illustrate how the migration of acupuncture may have more likely been influenced by factors such as political economic circumstances and similarities to existing local systems order, than by any migration of Chinese. Forms of economic deprivation and structural violence[4] directed towards a group – or as in the case of Cuba, towards an entire country – that thwarts access to healthcare, are a common theme across these three case examples.[5]

Furthermore, the cases reviewed illustrate how the reception of Chinese medicine in a given local context can be influenced by the existing plurality of healthcare practices at any local level, which in itself is a function of social, cultural, and environmental impacts. However, the migration of Chinese medical practices cannot be reduced to a simple grand analysis. Migrations of a given group's order are specific to geographies, chronologies, political economies, and ways of knowing. Hence, by virtue of unique factors particular to a given time, place, and set of political, economic, and other social circumstances, each context may be uniquely prepared to adopt and incorporate some elements of Chinese medical practice into their local healthcare.

Notes

1 Particular globally standardised practices that fall under the rubric of Complementary and Alternative Medicine (CAM), such as acupuncture, chiropractic, and homeopathy were integrated under the National Program for MTN in 1999 (Stafford 2010). However, although the sale of some medicinal plants used in local practices, such as Santería (stemming from the Yoruba religion of West-African slaves to Cuba), was no longer prohibited, local non-biomedical practices were not commonly integrated into MTN. As of May 2015, ten non-biomedical practices (including: acupuncture, ozonotherapy, homeopathy, medical hydrotherapy, helio-thalassotherapy, Bach Flower Remedies, phytotherapy, apitherapy – utilising the products of bees, yoga, and 'naturist nutritional orientation') were formally recognised as a medical specialty requiring that 'treatments be provided only by duly trained and certified health professionals' (San Diego Union Tribune 2015). Yet, it is interesting to note that none of these practices were indigenous to Cuba.
2 Again, it is important to note that few of these 'traditional' practices, beyond the use of local herbs, were actually local or traditional to Cuba.
3 Anthropologist, Brigitte Jordan, notes how though 'equally legitimate parallel knowledge systems exist' that people may 'move easily between [often] one kind of knowledge gains ascendency and

legitimacy' (1997: 56). She identifies this phenomenon as the domination of 'authoritative knowledge', whose consequence 'is the devaluation, often dismissal, of all other kinds of knowing as backward, ignorant, naive' (Ibid.).
4 According to anthropologist, Paul Farmer (2004), with structural violence 'social inequality fuelled by bias can become embedded in social institutions (or structures) to effectively marginalise groups and exclude them from social benefits' (Kadetz 2018: 293).
5 It is interesting to note, and possibly not coincidental, that many of the political economic circumstances influencing the reception of acupuncture in these countries -from the Marcos regime in the Philippines, to the thirty-six-year civil war in Guatemala, to the nearly sixty-year embargo of Cuba– were all very much supported by the United States.

Bibliography

Acosta Martínez, B. (2000) Editorial: Palabras de recibimiento al III Congreso Internacional de Medicina Tradicional, Natural y Bioenergética, celebrado en la Facultad de Ciencias Médicas de Holguín del 7 al 11 de Júnio del 2000, *Correo Científico Médico de Holguín*, 4.1.

Adams, W.R. and Hawkins, J.P. (2007) *Health Care in Maya Guatemala: confronting medical pluralism in a developing country*, Norman: University of Oklahoma Press.

Applebaum, D., Kligler, B., Barrett, B. et al. (2006) 'Natural and traditional medicine in Cuba: lessons for U.S. medical education', *Academic Medicine*, 81.12: 1098–103.

Baer, H., Singer, M. and Susser, I. (1997) *Medical Anthropology and the World System: a critical perspective*, Westport, CT: Bergin and Garvey.

Brotherton, P. (2005) 'Macroeconomic change and the biopolitics of health in Cuba's special period', *Journal of Latin American Anthropology*, 10.2: 339–69.

Castro, F. (1953) *History Will Absolve Me*, https://www.marxists.org/history/cuba/archive/castro/1953/10/16.htm, accessed 25/6/2019.

Cheng Xinnong 程莘农 (1996) *Chinese Acupuncture & Moxibustion*, Beijing: Foreign Language Press.

Cochetti, C.S. (2008) *Integrating "Traditional" and "Scientific" Medicine in Contemporary Cuba*, Unpublished MSc Dissertation, University of London.

Cooper, R., Kennelly, J. and Ordunez-Garcia P. (2006) 'Health in Cuba', *International Journal of Epidemiology*, 35: 817–24.

Council for Health and Development (1998) 25 years of commitment and service to the people, Manila: *Council for Health and Development*.

Currier, R. (1966) 'The hot-cold syndrome and symbolic balance in Mexican and Spanish-American folk medicine', *Ethnology*, 5.3: 251–63.

Dabney, T. (2006) 'Cuba and China: labor links', in M. Font (ed.) *Cuba Today: continuity and change since the 'Periodo Especial'*, New York: Bildner Center for Western Hemisphere Studies, The Graduate Center, The City University of New York, pp. 171–83.

Delgado Garcia, G. (1995) *La medicina China y su presencia en Cuba, Cuadernos de Historia de la Salud Públic*, 95, http://bvs.sld.cu/revistas/his/his%2095/hist0595.htm, accessed 14/6/2019.

Department of Health, Republic of the Philippines (2015) *Free Morong 43*, https://web.archive.org/web/20150620184435/http://www.doh.gov.ph/node/1323.html, accessed 30/1/2020.

Dolan, R. (ed.) (1991) *Philippines: a country study*, Washington, DC: GPO for the Library of Congress.

Dollar, D. (2017) *China's Investment in Latin America. Order from Chaos: foreign policy in a troubled world*, Washington, D.C.: Brookings Institute.

Dominguez, J. (2006) *China's Relations with Latin America: shared gains, asymmetric hopes*, Inter-American Dialogue, China Working Paper, June 2006, http://www10.iadb.org/intal/intalcdi/PE/2007/00110.pdf, accessed 28/6/2019.

Douglas, M. (1966) *Purity and Danger: an analysis of concepts of pollution and taboo*, Abingdon: Routledge.

Ellis, R.E. (2009) *China in Latin America: the whats and wherefores*, Boulder, CO: Lynne Rienner Publishers.

Farmer, P., Bourgois, P., Fassin, D., Green, L., Heggenhougen, H.K., Kirmayer, L. and Wacquant, L. (2004) 'An anthropology of structural violence', *Current Anthropology*, 45.3: 305–25.

Foster, G. M. (1994) *Hippocrates' Latin American Legacy: humoral medicine in the new world* (Vol. 1), Langhorne: Gordon and Breach.

Friedman, M. (2002) *Capitalism and Freedom*, Chicago, IL: University of Chicago Press.

Gallagher, K.P. (2010) 'China discovers Latin America', *Berkeley Review of Latin American Studies*, 8–13.
Garcìa, M. (2002) 'Green medicine: an option of richness' in F. Funes and D. Maria (eds) *Sustainable Agriculture and Resistance: transforming food production in Cuba*, Oakland, CA: Food First Books, pp. 212–19.
Garcia, H., Sierra, A. and Balam, G. (1999) *Wind in the Blood: Mayan healing and Chinese medicine*, Berkeley, CA: North Atlantic Books.
Garfield, R. and Santana, S. (1997) 'The impact of the economic crisis and the US embargo on health in Cuba', *American Journal of Public Health*, 87.1: 15–20.
Gubler, R. and Bolles, D. (2000) *The Book of Chilam Balam of Na*, Lancaster, CA: Labyrinthos.
Huber, B. and Sandstrom, A. (2001) *Mesoamerican Healers*, Austin: University of Texas.
Jones, G. (1989) *Red Revolution: inside the Philippine Guerrilla movement*, Boulder, CO: Westview Press.
Jordan, B. (1997) 'Authoritative knowledge and its construction', in R. Davis-Floyd and C. Sargent (eds) *Childbirth and Authoritative Knowledge*, Berkeley: University of California Press, pp. 55–79.
Kadetz, P. (2018) 'Collective efficacy, social capital and resilience: an inquiry into the relationship between social infrastructure and resilience after Hurricane Katrina', in M.J. Zakour, N. Mock and P. Kadetz (eds) *Creating Katrina, Rebuilding Resilience*, Oxford: Butterworth-Heinemann, pp. 283–304.
Kunow, M. A. (2003) *Maya Medicine: traditional healing in Yucatan*, Albuquerque: University of New Mexico Press.
Kuo, M. (2016) 'China-Cuba relations: assessing US stakes', *The Diplomat*, https://thediplomat.com/2016/12/china-cuba-relations-assessing-us-stakes/, accessed 20/8/2021
Lee, J.J. (1964) 'Communist China's Latin American policy', *Asian Survey*, 4.11: 1123–34.
Lo, V. (2011) 'The Cuban Chinese medical revolution', in V. Scheid and H. Macpherson (eds) *Integrating East Asian Medicine into Contemporary Healthcare: authenticity, best practice and the evidence mosaic*, London: Elsevier, pp. 215–27.
Lo, V. and Yoeli-Tlalim, R. (2018) 'Travelling light: Sino-Tibetan Moxa-cautery from Dunhuang', in V. Lo and P. Barrrett (eds) *Imagining Chinese Medicine*, Leiden: Brill, pp. 253–71.
McKeown, A. (1999) 'Conceptualising Chinese diasporas, 1842 to 1949', *The Journal of Asian Studies*, 58.2: 306–37.
Mateo, J. (2015) 'CHR confirms torture of Morong 43', *ABS-CBN News*, https://news.abscbn.com/nation/06/20/15/chr-confirms-torture-morong-43, accessed 30/1/2020.
Ni, M. (1995) *The Yellow Emperor's Classic of Medicine*, London: Shambala.
Official Gazette of the Republic of Cuba (2003) *Constitution of the Republic of Cuba*.
Orellana, P. (1987) *Indian Medicine in Highland Guatemala: the pre-Hispanic and colonial periods*, Albuquerque: University of New Mexico Press.
Padron Caceres, L. and Perez Vinas, M. (2005) *Integración de las prácticas de la medicina tradicional y Natural al Sistema de Salud*, IDEASS, www.undp.org.cu/pdhl/ideass/BrochureMTNesp.pdf, accessed 20/6/2019.
Quimpo, N.G. (2008) *Contested Democracy and the Left in the Philippines after Marcos*, New Haven, CT: Yale University Press.
Rosenthal, M. (1981) 'Political process and the integration of traditional and western medicine in the People's Republic of China', *Social Science and Medicine*, 15A: 599–613.
Roys, R. (1933) *The Book of Chilam Balam of Chumayel*, Norman: University of Oklahoma.
San Diego Union Tribune (2015) 'Cuba recognizes traditional medicine as a medical specialty', May 26, 2015, https://www.sandiegouniontribune.com/en-espanol/sdhoy-cuba-recognizestradition-al-medicine-as-a-medical-2015may26-story.html, accessed 30/1/2020.
Stafford, L. (2010) 'The rich history, current state, and possible future of natural and traditional medicine in Cuba', *HerbalGram* (American Botanical Council), 85: 40–49.
Taylor, K. (2005) *Chinese Medicine in Early Communist China, 1945–1963: a medicine of revolution*, London: Routledge.
Volpato, G., Godínez, D. and Beyra, A. (2009) 'Migration and ethnobotanical practices: the case of Tifey among Haitian immigrants in Cuba', *Human Ecology*, 37: 43–53.
Wilson, R. (1995) *Maya Resurgence in Guatemala: Q'eqchi' experiences*, Norman: University of Oklahoma.
Zhong, M. (2007) *Chinese Plum Blossom Needle Therapy*, Beijing: People's Medical Publishing House.

42
LONG AND WINDING ROADS
The transfer of Chinese medical practices to African contexts

Paul Kadetz

Local practices that are transplanted from one context to another are dynamic processes influenced by local and international factors whose outcomes are specific to a given place and time. Local healthcare practices, for example, are not merely local in character, but may also be impacted by practices and understandings from further afield that have been altered *en route* through the processes involved in that transplant: namely, translation and adaptation to other social and 'cultural' contexts, particularly via the global commodification of 'wellness'. Chinese medical practices and products embrace multiple aspects of healthcare and combine prophylactic and curative approaches including: acupuncture, herbs and medical substances, tui-na (bodywork), longevity exercises, and dietary therapies. These interventions are a primary transfer of the global wellness industry and an example of globalisation. Zhan (2009: 33) describes this phenomenon as the making and remaking (or 'worlding') of Chinese medicine 'through multi-sited, multi-directional, and socio-historically contingent projects and processes' (Chapter 40 in this volume). In addition to the continents of Asia, Europe, and the Americas, Chinese medical practices[1] have found a home in many of the countries of African, though no single pathway can explain this phenomenon. This chapter will attempt to delineate the many winding roads along which Chinese medical practices have been transferred to African contexts, as exemplified by the case of Madagascar.

Though much has been written about China's political and economic motivations behind its diplomacy in Africa, little research has specifically concerned the transplantation of Chinese medical practices to African contexts. Chinese medical teams sent as part of China's health aid to African countries have likely been one of the largest sources of both state and popular acceptance of Chinese medical practices throughout Africa (Jennings 2005; Hsu 2007a, 2007b, 2012; Huang 2010; Zhang 2010; Freeman and Boynton 2011; Li 2011; Liu *et al*. 2014). Though Chinese medical practices may have been introduced in many of these countries through formal Chinese state mechanisms of health diplomacy, this introduction has often been followed by non-state (private) Chinese entrepreneurial migrants who practise acupuncture and/or sell Chinese medicinal formulæ (Jennings 2005; Hsu 2007a, 2012). Furthermore, the global endorsement by the World Health Organization (WHO) of the Chinese antimalarial herb, *qinghao* 青蒿 (*artemisia annua*), and its synthesised compound *artemisinin*, has proven to be particularly important in African contexts beleaguered by malaria (Willcox and Bodeker 2004; White 2008; Hsu 2009; zu Biesen 2010; Primer 2011).

This chapter is based on research that includes a review of the literature concerning Chinese medical practices in Africa, as well as qualitative research gathered at the 2013 Roundtable for China-Africa Health Cooperation in Beijing, and fieldwork in Madagascar. A targeted sample of 135 pertinent stakeholders, including: practitioners and instructors of Chinese medicine, members of the Malagasy Ministry of Health, Chinese medical teams, and Malagasy patients receiving Chinese medicines participated in semi-structured interviews from 2012 to 2016 in Beijing, and in and around Antananarivo, the capital of Madagascar.

The roads Chinese medical practices have taken to Africa

Trade and the historical diaspora of the Chinese to Africa

China has a long history of trade with the African continent that is often ignored in Western historical accounts. It has been suggested that the earliest record of trade between China and Africa dates to the first century CE (Smidt 2001). The development of trading routes throughout central Asia certainly provided a means for trade between China and Africa (Ibid.). The first cultural exchanges between China and Africa resulted from exchanges of trade, and there is more substantial archaeological evidence from the Islamic Indian Ocean trade of late medieval times (van der Veen 2014). A Chinese cartographer, Zhu Siben 朱思本 (1273–1337), for example, assembled one of the first maps of the Southern region of the African continent in 1320 (Ibid.)

Some of the first Chinese to arrive in colonial Africa in the seventeenth century, particularly to the Dutch-occupied Cape, were convicts or ex-convicts, banished to the Cape from Batavia in the former Dutch East Indies (Armstrong 1997). The convicts who chose to remain in Africa after the completion of their jail terms were classified as 'free blacks' (Yap and Man 1996: 6). Chinese labour migrants also settled in Africa. Similar to the diaspora of Chinese migrants to other contexts, nineteenth-century labour migrants were contracted through colonial empires. 'After the Opium Wars of the 1840s and 1850s, China was forced by the colonial powers to reduce restrictions on Chinese emigration, which saw the beginning of large-scale movements of Chinese overseas in the form of the "coolie trade"' (Yap and Man 1996; Mohan and Tans-Mullins 2009: 592). The migrants came to Africa believing in the prospect of a better financial future, even though hardship, and sometimes indentured servitude, awaited poor Chinese migrants.

South Africa, Madagascar, Mauritius, and Reunion have had long-standing Chinese communities dating back to the colonial period (Mohan and Tan-Mullins 2009: 598). These nations, 'which experienced earlier waves of Chinese migration, witnessed higher integration between the Chinese and the local communities, both socially or politically' (Ibid.: 598). Chinese migrants to Africa during the Cold War period, were believed to have 'contributed in diverse ways to African development, as they did not only enlarge the local economy, but were also more integrated and accepted into the local communities' (Ibid.: 601). Though Chinese migrants may have brought Chinese medical practices to their new countries, neither the literature reviewed nor the interviews conducted identify a cross-fertilisation of Chinese medical practices, beyond the enclaves of Chinese migrants. However, the presence of Chinese migrants and Chinese medical practices may have, at least, fostered a familiarity with these practices.

Madagascar's population of approximately 70,000 to 100,000 Chinese (Tremann 2013) is the second largest in the African Union, after South Africa.[2] Current estimated Chinese populations of over 30,000 on the African continent are identified in Ethiopia (Cook *et al.* 2016), Angola (Bloomberg 2017), Mauritius (US Dept. of State 2010), and Algeria (Reuters 2009) respectively.

The People's Republic of China's diplomacy on the African continent

The People's Republic of China (PRC) has engaged in diplomacy in post-colonial African states since the 1960s. Sino-African diplomacy has included the 'soft power' of foreign aid. Soft power refers to foreign aid and donations used to foster international relationships and achieve the donor state's foreign policy goals (Joseph Nye 1990). The PRC has a long, often unacknowledged, history of providing foreign aid to Africa, building upon a common experience to forge what China refers to as mutually beneficial partnerships. In fact, the words 'mutual' and/or 'benefit' were two of the most frequently uttered terms at the 2012 Roundtable on China-Africa Health Cooperation.

The overall amount of Sino-African aid is substantial, particularly over the past twenty years, with close to US$3 billion provided in 2015 (CARI 2018a). In 2009, the PRC pledged a package to the African continent which included: supplying US$3 billion in soft loans and US$2 billion in subsidised credit to trade partners; establishing a US$5 million development fund to encourage Chinese companies to invest in Africa; forgiving a substantial amount of the debt of the poorest countries that had diplomatic relations with the PRC; opening the Chinese market to African products through the removal of customs duties on most goods; and opening economic and trade cooperation areas (Cheng *et al.* 2012). At the 2012 Forum on China-Africa Cooperation, China doubled its 2009 package, pledging US$20 billion to Africa over the next three years (Associated Press 2012). This aid specifically targeted the development of infrastructure that facilitates trade within Africa; builds agricultural technology centres; supports the construction of wells; and trains medical and other professional personnel (Ibid.). 'From 2000 to 2015, the Chinese government, banks and contractors extended US $94.4 billion in loans to African governments and their state-owned enterprises' (CARI 2018b). In 2009, China overtook the United States to become Africa's largest trade partner (OECD Factbook 2012). If this trajectory continues, China is projected to overtake the World Bank as Africa's most important financier (Associated Press 2012). The value of Sino-African trade in 2016 was, however, only US$128 billion; down from a high of US$215 billion in 2014 (CARI 2018c).

Improving the public health of African nations has been a long-term emphasis of China's aid to Africa. Health sector assistance has constituted more than a quarter of China's foreign aid to Africa. Of the US$462 million 2006 Sino-African assistance package, US$126 million was specifically targeted for healthcare (Brautigam, 2008). The PRC has identified four health-related priorities, emphasising (1) the need to develop and promote effective treatments for malaria; (2) the exchange of medical personnel and information; (3) a commitment to disseminate medical teams and equipment to improve medical facilities and train more doctors throughout Africa; and (4) increased technical support, including research exploring the potential of traditional herbs in treating and preventing HIV/AIDS (Youde 2010). Based on the effectiveness of accessible, cost-effective, herbal-based antimalarial treatment throughout the continent, cost-effective herbal therapies for HIV have been in particular demand. In 2009, China pledged additional measures to strengthen Sino-African health cooperation, including plans to train 3,000 practitioners across Africa and a US$73.2 million assistance package, facilitating the construction of 30 hospitals and 30 malaria prevention and treatment centres (Chan, Chen and Xu 2010).

Chinese medical teams (yiliaodui 医疗队)

Chinese Medical Teams (CMTs) have been the chief and persistent force behind the PRC's health aid to Africa. According to Xu, Liu, and Guo (2011: 7), 'China had sent medical teams to 67 countries and regions in Asia, Africa, Latin America, Europe, and Oceania, totalling 21,238 medical professionals who distributed approximately 200 million treatments, of which Africa received the majority'. Since the first medical team arrived in Algeria in 1963 until 2010 'more than 15,000 Chinese medical personnel have served in forty-seven different African states and treated at least 180 million patients' (Youde 2010).

Unlike Western development aid, such as from USAID and DiFID (UK Department for International Development), Chinese foreign aid has been decentralised to the provincial level. Typically, an African country is paired with a Chinese province and aid is then arranged at the provincial level. Informants explained that recipient country representatives would request certain specialists on a medical team, with which the province would comply to the extent of practitioner availability. In addition to biomedical specialists, almost all CMTs included an acupuncturist and a Chinese herbalist. One CMT informant who led the team in Madagascar explained: 'They all request different teams, but they usually want high tech professionals. They want heart and lung surgeons'. However, such requests were usually not feasible for China to provide for more than a short visit. CMTs have been primarily deployed to district and regional level hospitals and are usually not to be found at primary or secondary level rural healthcare facilities.

CMT members commit to two-year periods and are usually replaced by another team at the same hospital upon the completion of their voluntary period. In this respect, CMTs are a highly unusual form of health aid compared to aid projects from most countries, which normally last for a defined period and then end abruptly with little to no follow-up. In contrast, CMTs have served in many of the same hospitals for over 50 years. Of the four hospitals that have had CMTs since 1976 in Madagascar, all four still have CMTs to date.

CMTs target rural and medically under-served and understaffed contexts with limited healthcare access and few specialists. They have been influential in the areas where they have practised. Patients interviewed in Madagascar identified the low cost of treatment and the care, kindness, and professionalism of Chinese medical teams as a reason some would travel as far as 750 km to be treated at a hospital with CMTs. As in China, acupuncture and Chinese herbal medicines were used as both adjunctive and primary therapies. The pharmacies of the CMT hospitals visited carried both Chinese and Western medicines, although herbal medicines were sold as pills rather than as raw herbs.

The unusual longevity of Chinese medical teams over generations in one area coupled with the State sanctioning of Chinese medical practices has played a significant role in popularising acupuncture and Chinese herbal medicines throughout African nations. Yet, other factors have also influenced the adoption of Chinese medical practices at both State and local levels across Africa.

The influence of the World Health Organization on the transplant of Chinese medical practices

The World Health Organization has been influential in promoting the use of 'traditional medicines', in general, and the globalisation of Chinese medical practices in particular. After the

PRC rejoined the WHO in 1973, their purported success with both the barefoot doctors programme and the integration of Chinese medical practices into China's formal healthcare system was adopted as the model by which to address health disparity and achieve universal healthcare coverage (or 'health for all') in many low and middle-income countries (Siddiqi 1995). At the behest of Russian WHO delegates, and possibly influenced by Chinese nationals in positions of power at the WHO at the time, China's model of healthcare integration became an integral component of the Declaration of Alma Ata; a document that became central to the planning of primary healthcare systems and to issues of healthcare access and equity (Ibid.). Ironically, at the same time Mao Zedong's model to address rural healthcare inequity was being promoted globally, it, along with the health brigade units that comprised rural primary healthcare in Maoist China, was being dismantled after Mao's death, almost exactly two years before the 1978 Declaration of Alma Ata, in September 1976, and prior to the collapse of the rural cooperative medical system in the early 1980s (Chapter 44 in this volume; Duckett 2012, 2011).

Subsequently, the WHO opened so-called TRM or 'Traditional Medicine Units' at WHO headquarters in Geneva and at their regional offices. The role of the TRM was to oversee the recommended 'integration' of non-biomedical practices and practitioners into biomedical healthcare systems. The main TRM unit in Geneva has promoted acupuncture and Chinese herbal medicines almost exclusively, while ignoring other equally legitimate and systematised popular ethnomedical practices, such as Āyurveda. For example, although numerous practice and teaching guidelines, textbooks, and standardised nomenclature guides for acupuncture have been published by the WHO, there has been only one publication committed to training in Āyurveda as of the time of this writing (WHO 2010a). And, though 10 of the 25 (or 40%) WHO Collaborating Centres for research in traditional medicine are located in China, two are in India, one of which conducts research on Āyurveda (WHO 2010b). While TCM has become a global commodity that offers education and licensure nearly worldwide, Āyurveda has no formal licensure beyond South Asia and parts of South East Asia.

Several informants have attributed the prioritisation of Chinese medical practices to China's historical role as the primary funder of the Traditional Medicine (TRM) Unit at WHO headquarters, as well as to the aforementioned presence of Chinese nationals in positions of power at WHO,[3] including the past directors of the Traditional Medicine Unit at WHO headquarters. The WHO's promotion of Chinese medical practices is evident not only in their publications of numerous acupuncture guidelines (which extends to the development of a standardised nomenclature), but also in the promotion of particular Chinese herbs, such as their endorsement of the effectiveness of *qinghao* 青蒿 (artemesia annua) and its derived active ingredients for the treatment of malaria. The WHO's prioritisation of Chinese medical practices is also clearly demonstrated by the inclusion of a chapter about TCM in the eleventh version of the WHO's (ICD) International Statistical Classification of Diseases and Related Health Problems (WHO 2019).

The WHO's promotion of Chinese medical practices has clearly influenced the perceived efficacy, acceptance, and adoption of Chinese practices at a multinational level. Chinese medical practices are often prioritised for integration into healthcare systems to the detriment of healthcare practices that are actually local, which ironically often leads to the marginalisation of local practices (Kadetz 2011). For example, according to one representative of the research division of the Ministry of Health of Madagascar:

> The WHO has suggested that to improve the primary healthcare,
> Malagasy must integrate traditional medicine in the primary healthcare,
> and so we introduced acupuncture into our traditional medicine.

Furthermore, by serving as the model for achieving universal primary healthcare in the Declaration of Alma Ata, the Chinese model – with its elements of prioritising rural healthcare access, community health workers, and the integration of local healthcare practices and practitioners of Chinese medicine – has come to symbolise primary healthcare.

The political identification of Chinese medical practices with primary healthcare

From the post-Second World War period, China framed its relations with newly independent African countries through a particular political lens that may have also influenced a local political representation of Chinese medicines, as being accessible to the people, egalitarian, and somewhat revolutionary (Chapters 44 and 45 in this volume; Zhan 2009: 31–61). In interviews with administrators in the Ministry of Health of Madagascar, Chinese medical practices were symbolically linked with primary healthcare.

Mao Zedong's promotion of traditional Chinese medicine as a symbol of socialist healthcare access and equity was represented as antithetical to the capitalistic business of biomedicine (Taylor 2005). Thereby, Chinese medical practices can still be represented as a product of revolution and as an antidote to Western imperialism and multinational corporate greed. Though Chinese medical products are most definitely a profitable traded commodity, they are ultimately more financially accessible to many poorer populations than are most biomedical pharmaceuticals. Furthermore, the PRC's official position in the World Trade Organization on granting access to African countries for generic biopharmaceuticals and trade may also have bolstered a positive perception of Chinese medicine at the state level (described in next section).

Similarly, the PRC understood that building diplomatic relations with post-colonial countries could open the door to communism, particularly as a viable answer to colonial capitalism and the rapid expansion of American imperialism. Sino-African relations in the 1950s and 1960s promoted Maoism as the most appropriate model for international development and anti-colonialism (Zhan 2009; Youde 2010). China also understood that winning African favour could potentially reverse the American/Western European official recognition of the Republic of China (Taiwan) as 'China' and the concomitant dismissal of the PRC on the world stage. China specifically supported the African Liberation Movements of the 1960s (Li 2007).

Unlike the Soviet Union, the Chinese government presented itself to African constituencies as a patron who rejected the imperial mandates of Western powers and understood the unique struggles of 'peasant movements' (Youde 2010). Most importantly, China successfully characterised its relationship with Africa as a strategic partnership of two brother nations, rather than a paternalistic relationship between a superior donor country and its inferior recipient, as was characteristic of the Global North. African governments were to be treated as equal partners and allies in development with one developing country helping another (Chan 2008). In return for China's support, twenty-six African nations supported China in regaining member state status in the United Nations in 1971 and the PRC became officially recognised as 'China' (Huang 2010).

The impact of pharmaceutical trade agreements on Sino-African relations

The PRC took an official position in a World Trade Organization debate to grant African countries access to generic biopharmaceuticals and trade (Kadetz 2013). This position may have bolstered the acceptance, by African nations, of Chinese medical practices, particularly for Chinese herbal medicines. Although China joined the World Trade Organization well after many African countries, it became an important voice in opposition to particular trade policies, such as the Agreement on Trade-Related Aspects of Intellectual Property Rights or TRIPS. TRIPS favours the intellectual property rights of multinational corporations at the expense of low-income countries who, in the case of healthcare, can barely afford to provide essential medical supplies.

Of the 37.9 million people in the world who were living with HIV in 2018, approximately 27 million (68%) were living in Africa, where 62% of all HIV/AIDS deaths occurred in 2018 (UNAIDS 2020). These deaths are not believed to be a consequence of a lack of viable antiretroviral therapies (ARV), but rather due to a lack of access to affordable ARVs. For, as the WHO (2014: 6) identifies: 'Globally, [ARV] programmes averted an estimated 7.6 million (6.9–8.4 million) deaths between 1995 and 2013'. It is estimated that the cost of ARVs would need to be reduced by as much as 95% in order to be accessible to all populations (Chan 2008). However, as a result of the 1994 TRIPS agreement, the cost of ARVs was beyond the income of the average citizen in a low- or even middle-income country.

Under TRIPS, a generic product cannot be registered without the patent holder's agreement for the life of the patent, which can last up to twenty years (Lee, Buse and Fustukian 2002). As a result of the Doha Declaration of the Fourth WTO meeting in 2001, 'developing' countries can override drug patents by issuing compulsory licences to manufacture or import cheaper versions of a pharmaceutical product, if a national health emergency can be justified (Chan 2008). It is specified, however, that such a compulsory licence would need to 'be authorised predominantly for the supply of the domestic market of the Member authorising such use' (African Union 2009: 10). Therefore, countries lacking domestic pharmaceutical industries, which include almost all African nations, do not have a sufficient manufacturing capacity to produce significant quantities of generic pharmaceuticals.

Upon joining the WTO, China strongly sided with low-income countries in criticising TRIPS and its addendum for perpetuating an imbalance in the rights and obligations of developing countries (Chan 2008). China argued that public health rights should take priority over intellectual property rights in government decision-making (Ibid.). Furthermore, China had been active in the African pharmaceutical sector for over three decades, developing pharmaceutical factories in the 1970s that produced drugs for local use in Zanzibar, Mali, Cote d' Ivoire, Kenya, Egypt, and Sudan (Chen and Xu 2012).

Currently, China is the world's largest exporter of active pharmaceutical ingredients (APIs). According to a Chinese informant, low profit margins and regulatory barriers to the production of generic ARVs have resulted in a primary focus of Chinese pharmaceutical companies on the production of APIs, rather than generic ARVs. Informants stated that currently 80% of Africa's ARVs are imported from India, and 80% of the APIs needed for India's ARV production come from China. In recognising the capacity of the Chinese pharmaceutical industry to supply generic ARVs, the government of China could work to incentivise Chinese companies to produce generic ARVs for Africa.

There has also been a marked interest among a number of African governments in collaborating with China on research exploring the use of African herbal medicines to treat HIV/AIDS. Similar to the use of *artemesia* to treat malaria, the goal would be to generate sustainable cost-effective HIV/AIDS treatments that offer a viable alternative to cost-prohibitive ARVs, and to the many issues associated with generic ARV production.

Medical pluralism

Finally, it is important to bear in mind that in actual practice, people often seek more than one kind of healthcare intervention, and usually follow multiple (and sometimes conflicting) understandings of how to maintain their health and manage illness, utilising healthcare in a plural manner. In other words, healthcare can be understood as a grouping of multiple, discrete, bounded local practices offered in tandem with universalised biomedicine (Baer 2004; Hsu 2008). The resulting hodgepodge of practice is an outcome of complex and dynamic historical, socio-cultural, political, and economic forces that may involve: public demand; individual choice; competing medical systems; changing individual and group beliefs; and issues of physical and financial access to available healthcare practices. Hence, such disparate healthcare practices as, for example, diet, prayer, massage, exercise, and meditation will be incorporated with biomedical therapies by the lay public without any concern for the myriad epistemological differences between these practices, or for the different ordered systems of meaning and understanding from whence each practice came.

Medical pluralism is often an outcome of immediate needs, rather than any adherence to rigid epistemologies. Arguably, a healthcare 'system' is, in actuality, created by public demand in a bottom-up manner; via each person's plural use of the practices available to them, rather than through a top-down representation by the State. Therefore, medical pluralism in local African contexts could have easily supported the plural acceptance and use of acupuncture, Chinese herbs, and other Chinese medical practices in combination with local healthcare practices and biomedicine.

The case of the transplant of acupuncture to Madagascar

Context

The former French colony of Madagascar is classified as one of the poorest countries in the world. Madagascar exhibits some of the lowest values for indicators of per capita wealth in Sub-Saharan Africa, with 81.3% of the population living below the poverty line and 35.4% designated as living in severe poverty (UNDP 2013). Per capita gross national income (GNI) in Madagascar of 0.828 is significantly lower than that of Cameroon (2.114), Uganda (1.168) and the average for Sub-Saharan Africa (2.010) (Ibid.). Madagascar experiences a severe double burden of communicable and non-communicable disease, with less than 2.65% of Madagascar's GDP being allocated to healthcare since the mid-1990s.

According to one informant from the Malagasy Ministry of Health (MOH):

> The health system is not well implemented. 80% of Malagasy live in rural areas and 40% live more than 15–20 km from any health facility. There are no laboratories in remote areas. Hospitals are very far away from the average rural person. As a result of accessibility, 80% of mothers and children die within their own communities. In general, only 20% of the operating budget [for healthcare] arrives in the remote areas; 80% is retained in the city.

According to another informant from the MOH, 80% of Malagasy patients will consult and use local healthcare practices before approaching a biomedical physician. Local healers have an association and are regulated by the MOH. Informants reported approximately 10,000 traditional healers in Madagascar, of whom 2,000 are members in their professional association.

Acupuncture in Madagascar

Although Chinese herbs are available in the four CMT hospitals in Madagascar, acupuncture is more frequently used as both a primary and an adjuvant treatment, as well as for anaesthesia throughout the country. The Director of Acupuncture Services at a CMT hospital describes its popularity:

> Patients queue from early morning. In the beginning, in 1975, the [acupuncture] service received 120 patients a day. Now, with the cost of transportation and after the 2009 political crisis [which has reduced State healthcare funding], the number of patients has decreased to 80 per day. Normally, treatment with acupuncture is daily, but in the current [financial] environment, patients come according to their ability. Sometimes, patients rent small apartments near the hospital to avoid commuting long distances during treatment.

Acupuncture was formally introduced to Madagascar in the mid-twentieth century, shortly before CMTs were introduced in Malagasy hospitals. According to a Malagasy acupuncturist; 'in 1960, a Malagasy physician published an article about acupuncture in a review of the Malagasy Academy. He named this technique "*Tevipilo*" from Malagasy "*tevika* = to drill" and "*filo* = needle"'. Acupuncture training has been reserved for Malagasy biomedical physicians. China has provided scholarships for Malagasy doctors to be trained in acupuncture in China. However, according to an MOH informant, 'in 2007, the former President of Madagascar, Marc Ravalomanana, [in office from 2002–2009] suggested that doctors trained in China must pass their knowledge to Malagasy doctors to multiply the availability of acupuncture to the Malagasy'. Subsequently, the MOH, with the support of the WHO, sponsored a national training institute for traditional medicine and acupuncture in Madagascar, staffed by those physicians who received training in China. The current Malagasy training for biomedical physicians lasts two years, with ten physicians trained every year. The first cohort of physicians completed training in 2009.

Acupuncture is not only available in the four CMT-staffed hospitals, but is now accessible wherever trained physicians practise. The growing supply of acupuncture practice in Madagascar is a direct outcome of the growing demand for acupuncture by the Malagasy and the MOH's formal recognition of acupuncture as an effective form of treatment. However, according to the physician who leads the acupuncture training institute:

> We are not yet satisfied, it's time to prepare for shifts. For example, acupuncture is currently not recognised by the faculty of medicine at the University of Antananarivo. We must implement further training for acupuncture specialists.

The cost of Chinese medicines

Given the limited financial resources of the Malagasy, the cost of healthcare is an important factor to consider for healthcare access. The Chinese medical practices – delivered by CMTs as part of Chinese aid packages – were completely free until 1995, when the Ministry of Health of Madagascar insisted that Chinese herbal medicines and biomedicines carry a nominal fee. To ensure that fees would be charged, an import tax was added to Chinese medicines imported to Madagascar. According to informants, the Ministry of Health was mostly concerned about reducing the imbalance between the cost of Malagasy healthcare and what was being provided for free from the CMT hospitals.

Furthermore, regardless of its overall cost-effectiveness, acupuncture was also being provided for free in CMT hospitals, as it was completely funded by the Chinese government. According to the director of acupuncture services of one CMT hospital, 'since January 12th, 2015, this CMT hospital service, known as The Centre of Traditional Chinese Medicine, will be supported by 100,000 Yuan by the PRC'.

However, while my research was concerned with the transplant of Chinese medical practices via official state channels, it needs to be emphasised that Chinese medical practices are also accessible, and in some contexts *only* accessible, via the private practice of Chinese medical practitioners throughout the African continent. Mohan and Tan-Mullins (2009: 589) argue against the conceptualisation of a 'centralized "Beijing consensus" where all actions of Chinese firms and individuals overseas are orchestrated by the Chinese Communist Party (CCP). Rather the proliferation of semi-private and private Chinese firms in Africa points to the need to differentiate between different economic actors and their developmental potentials [...] as many of the migrants in Africa are there independently of any state direction'. Yet, as Hsu (2008) has noted, even when private clinics replace State sponsored CMTs, they are able to build on the social and cultural capital already established by the CMTs.

Hence, practices of Chinese medicines in African contexts may be best portrayed as a dynamic mix of both State-supported and private entrepreneurs in various sectors. Regardless, the CMTs continue to practise, while charging little to nothing for treatments and medicines, particularly compared to what is available locally. Financial access may be one factor in explaining the popularity of Chinese medical practices in many low-income African contexts, but physical access to healthcare is also relevant.

Physical access to healthcare

Similar to the importance of financial access to Chinese medicine, is the issue of physical access to healthcare. Chinese medical teams tend to purposely target rural areas with limited medical access, especially in terms of specialists. This is significant in a context where most patients interviewed travelled an average of 50 kilometres, with some informants travelling 750 kilometres or more, due to the low cost and high quality of care they received. Chinese medical teams are strategically placed and have served the same areas for as long as fifty years (Wang *et al.* 2012; Liu *et al.* 2014). Over such unusually long durations, trust is inevitably fostered.

The importance of building trust

In interviews with Malagasy patients, trust in CMTs was mentioned repeatedly as an important reason for patient satisfaction, and word of mouth was a key factor in the positive appraisal of Chinese medical practices. Although Chinese imports of general products were often identified as cheap and ultimately of inferior quality, Chinese physicians were identified as kind, caring, and concerned about their patients, and not merely seeking financial gain; a characteristic frequently attributed to Malagasy physicians, in addition to poor quality of care. A 42-year-old Malagasy male patient interviewed in a CMT hospital outside of Antananarivo reported:

On 28th of June, I got shot in my leg. I was at the CHU (university hospital centre). I had five bullets in my leg and I had surgery. The surgeon only removed one bullet and he said that the rest will come out alone. I could not believe this. I have already spent much money there. I asked to leave the hospital and I am here [at the CMT hospital]. They [CMT physicians] immediately removed the four bullets left in my leg.

A 65-year-old female Malagasy patient at the same hospital stated:

I'll be blunt. When Chinese doctors treat, they treat thoroughly. Malagasy doctors think first about money. Malagasy doctors ask for money for everything they do. In this [CMT] hospital, patients have a right to choose between Chinese or Malagasy doctors. I chose Chinese doctors, because if I have a problem, Malagasy doctors do not want to take care of me. Chinese doctors do not discriminate. They sympathize with everyone, rich or poor. If there were solely Malagasy doctors in this hospital, the death rate would increase by 75%.

The Malagasy director of acupuncture at one CMT hospital notes:

[CMTs] have a good relationship with Malagasy patients and they do not charge consultation fees. As a result they get many returning patients and referrals.

This trust in Chinese practitioners may have been instrumental in building trust in Chinese practices.

Chinese practices themselves may also have served as a source of trust for many informants. According to one 72-year-old Malagasy male patient:

In the [19]80s, I was a delivery man, and that's how I knew this hospital. I always stay at this hospital whenever I need surgical intervention or care we cannot solve. I choose this hospital first because of cost, and second, because the doctors use both modern and traditional medicine. I do not trust in western medicine if it is used alone.

The adaptability of Chinese medical practices to local cultures

Unlike biomedicine, Chinese medicine can be perceived to address more than just treatment for disembodied diseases, and thereby may be likened to local healthcare practices. By including the patient's experience of illness, as well as the social aspects of sickness, Chinese medical practices may also offer a system of meaning that is more akin to local meanings, than is possible to achieve with biomedicine and its claims to universal application. Thereby, the global aspect of Chinese medicines may be more easily 'glocalised' – whereby the global is adapted to the local – than is possible with biomedicines.

By virtue of shared meanings, an understanding of effectiveness may also be shared. In addition to cost and access, over 90% of informants interviewed cite the fact that Chinese medicine 'works' as a primary reason for its popularity. For example, a 58-year-old Malagasy male patient informant receiving acupuncture for post-stroke rehabilitation revealed:

I had a stroke on January of this year. I could not speak. I could not move my right side. I was in a hospital in Antananarivo for half a month and I was doing kinesiotherapy [sic]. There were no significant results. Now you can hear me. I speak clearly.
I can move my right hand. I am very satisfied. My health is improving day by day.

Similarly, a 64-year-old Malagasy female states:

> The treatment [acupuncture] is miraculous. I had a stroke. I was in coma at the military hospital. When I woke up, I was paralysed and I could not walk…could not stand up. Now, I walk. I'll get out soon.

Interestingly, acupuncture may also hold a particular symbolic capital by virtue of its uniqueness. Zhan (2009: 38) states 'acupuncture was the most captivating feature that distinguished the Chinese medical teams from other teams and aroused interest throughout Africa'. More than one informant had used the term 'magic' to describe acupuncture, mainly due to the fact that their chief complaint was eradicated quickly and for long durations. This was especially shared by patients who were given little to no hope for recovery by biomedical practitioners, such as stroke survivors.

Conclusion: understanding the adoption of Chinese medical practices across Africa as complex processes

This chapter has argued that the transplant of Chinese medical practices to local African contexts results from a complexity of factors that vary from one context to the next. Bilateral, multilateral, and global political economic forces lay at the heart of many of these influential factors. From trade routes to the movement of convicts and labour migrants that remained to form Chinese communities, to China's bilateral aid in the twentieth century and the subsequent acceptance and popularity of Chinese medical teams, to the global endorsement of the use of Chinese medical practices and the linkage of traditional Chinese medicine and primary healthcare by the World Health Organisation. Via these many winding roads, Chinese medical practices have found an adopted home in disparate communities across the African landscape.

Notes

'Please note, portions of this chapter previously appeared in an article published by Frontiers in Human Development. Kadetz, P. (2021) 'About Face: How the People's Republic of China Harnessed Health to Leverage Soft Power on the World Stage', Front. Hum. Dyn, 3:774765. doi: 10.3389/fhumd.2021.774765'

1. In order to avoid portraying the practices of Chinese medical cultures as the outcomes of a fixed and hegemonic system, the term 'Chinese medical practices' will be used throughout this chapter, in place of 'Chinese medicine' and 'traditional Chinese medicine'.
2. The Chinese population in South Africa is estimated to be between 300,000 and 400,000 (Liao and He 2015).
3. For example, Margaret Chan is a Chinese national who served as the Director General of the WHO from 2006 to 2017.

Bibliography

African Union (2009) *Local Pharmaceutical Production in Africa*, 4th Session of the AU Conference of Ministers of Health, Addis Ababa, Ethiopia, 4–8 May 2009. 10.

Armstrong, J. (1997) *The Chinese at the Cape in the DEIC Period, 1652–1795*, Slave Route Project Conference, Cape Town, October 1997.

Associated Press (2012) 'China pledges $20 billion in credit to Africa over the next 3 years', *The Washington Post*, 19 July.

Baer, H. (2004) 'Medical pluralism', in C. Ember and M. Ember (eds) *Encyclopedia of Medical Anthropology*, New York: Kluwer, pp. 109–15.

zu Biesen, C.M. (2010) 'The rise to prominence of Artemisia annua L.–the transformation of a Chinese plant to a global pharmaceutical', *African Sociological Review/Revue Africaine de Sociologie*, 14.2: 24 –46.

Bloomberg News (2017) Chinese businesses Quit Angola after 'Disastrous' currency blow, *Bloomberg*, 20 April, https://www.bloomberg.com/news/articles/2017-04-20/chinese-businesses-quit-angola-after-disastrous-currency-blow, accessed 07/04/21.

Brautigam, D. (2008) *China's African Aid: transatlantic challenges*, Washington, DC: German Marshall Fund of the United States.

CARI: China Africa Research Initiative (2018a) *Data: Chinese foreign aid to Africa*, http://www.sais-cari.org/data-chinese-foreign-aid-to-africa, accessed 27/7/2018.

⸻ (2018b) *Data: Chinese loans to Africa*, http://www.sais-cari.org/data-chinese-loans-and-aid-to-africa, accessed 27/7/2018.

⸻ (2018c) *Data: China-Africa trade*, http://www.sais-cari.org/data-china-africa-trade, accessed 27/7/2018.

Chan, L.H. (2008) *China Engages Global Health Governance: a stakeholder or a system-transformer?* PhD thesis, Griffith University.

Chan, L.H., Chen, L. and Xu, J. (2010) 'China's engagement with global health diplomacy: was SARS a watershed?' *PLoS Medicine*, 7.4: 1–6.

Chen, L. and Xu, J. (2012) The development of China's capacity for supplying generic medicines to Africa, in L. Chen (ed.) *China-Africa Health Collaboration in the Era of Global Health Diplomacy*. Beijing: World Knowledge Publishing House, p. 124.

Cheng, S., Fang, T., and Lien, H.T. (2012) China's international aid policy and its implications for global governance, *Research Center for Chinese Politics and Business Working Paper No. 29*, https://ssrn.com/abstract=2169863 or http://dx.doi.org/10.2139/ssrn.2169863, accessed 2/4/2021.

Cook, S., Lu, J., Tugendhat, H. and Alemu, D. (2016) Chinese migrants in Africa: facts and fictions from the agri-food sector in Ethiopia and Ghana, *World Development*, 81: 61–70.

Duckett, J. (2011) 'Challenging the economic reform paradigm: policy and politics in the early 1980s' collapse of the rural cooperative medical system' *The China Quarterly*, 205: 80–95.

⸻ (2012) *The Chinese State's Retreat from Health: policy and the politics of retrenchment*, London: Routledge.

Freeman, C., and Boynton, X. (2011) *China's Emerging Global Health and Foreign Aid Engagement in Africa*, Washington, DC: Center for Strategic and International Studies.

Hsu, E. (2007a) 'Medicine as business: Chinese medicine in Tanzania', *Communities*, 13.2: 113–24.

⸻ (2007b) 'Chinese medicine in East Africa and its effectiveness', *IIAS Newsletter*, 45: 22.

⸻ (2008) 'Medical pluralism', in K. Heggenhougen and S. Quah (eds) *International Encyclopaedia of Public Health*, vol. 4, Oxford: Elsevier, pp. 316–21.

⸻ (2009) 'Chinese propriety medicines: An "alternative modernity?" The case of the anti-malarial substance artemisinin in East Africa', *Medical Anthropology*, 28.2: 111–40.

⸻ (2012) 'Mobility and connectedness: Chinese medical doctors in Kenya', in H. Dilger (ed.) *Medicine, Mobility, and Power in Global Africa: transnational health and healing*, Bloomington: Indiana University Press, pp. 295–315.

Huang, Y. (2010) 'Pursuing health as foreign policy: the case of China', *Indiana Journal of Global Legal Studies*, 17.1: 105–46.

Jennings, M. (2005) 'Chinese medicine and medical pluralism in Dar es Salaam: globalisation or glocalisation?', *International Relations*, 19.4: 457–73.

Kadetz, P. (2011) 'Assumptions of Global Beneficence: health care disparity, the WHO, and the effects of global integrative health care policy on local levels in the Philippines', *Biosocieties*, 6: 88–105.

⸻ (2013) Unpacking Sino-African health diplomacy: problematizing a hegemonic construction, *St Antony's International Review*, 8.2: 149–172.

Lee, K., Buse, K. and Fustukian, S. (eds) (2002) *Health Policy in a Globalising World*, Cambridge: Cambridge University Press.

Li, A. (2007) 'China and Africa: policy and challenges', *China Security*, 3.3: 69–93.

⸻ (2011) *Chinese Medical Cooperation in Africa: with special emphasis on the medical teams and anti-malaria campaign*, Uppsala: Nordiska Afrikainstitutet.

Liao, W. and He, Q. (2015) 'Tenth world conference of overseas Chinese: annual international symposium on regional academic activities report', *The International Journal of Diasporic Chinese Studies*, 7.2: 85–9.

Liu, P., Guo, Y., Qian, X., Tang, S., Li, Z. and Chen, L. (2014) 'China's distinctive engagement in global health', *The Lancet*, 384.9945: 793–804.

Mohan, G. and Tan-Mullins, M. (2009) 'Chinese migrants in Africa as new agents of development? An analytical framework', *The European Journal of Development Research*, 21.4: 588–605.

Nye, J. (1990) 'Soft power', *Foreign Policy*, 80: 153–71.

Organisation for Economic Co-operation and Development Staff (2012) *OECD Factbook 2011–2012: Economics, Environmental and Social Statistics*, https://www.oecd.org/publications/factbook/oecd-factbook2011-2012.htm, accessed 04/7/221.

Primer, A. (2011) 'The plausibility design, quasi-experiments and real-world research: a case study of antimalarial combination treatment in Tanzania', in P.W. Geissler and C. Molyneu (eds) *Evidence, Ethos and Experiment: the anthropology and history of medical research in Africa*, pp. 197–228.

Reuters (2009) Chinese, Algerians fight in Algiers – witnesses, *Reuters*, 4 August 2009, https://www.reuters.com/article/idINIndia-41535320090804, accessed 07/04/2021.

Siddiqi, J. (1995) *World Health and World Politics: the World Health Organization and the UN system*, London: Hurst & Co.

Smidt, W. G. (2001) 'A Chinese in the Nubian and Abyssinian kingdoms (8th century). The visit of Du Huan to Molin-guo and Laobosa', *Arabian Humanities. Revue internationale d'archéologie et de sciences sociales sur la péninsule Arabique* (International Journal of Archaeology and Social Sciences in the Arabian Peninsula), 9.

Taylor, K. (2005) *Chinese Medicine in Early Communist China, 1945–1963: a medicine of revolution*, London: Routledge.

Tremann, C. (2013) 'Temporary Chinese migration to Madagascar: local perceptions, economic impacts, and human capital flows', *African Review of Economics and Finance*, 5.1: 7–16.

UNAIDS (2020) Global HIV & AIDS statistics — 2020 fact sheet. https://www.unaids.org/en/resources/fact-sheet, accessed 4/04/2021.

UNDP (2013) *Human Development Report. The Rise of the South: human progress in a diverse world*, United Nations Development Programme, http://hdrstats.undp.org/images/explanations/MDG.pdf, accessed 04/04/2021.

US Dept of State (2010) *Background Note: Mauritius*, Washington, DC: U.S. Department of State.

van der Veen, J. (2014) 'The Roman and Islamic spice trade: new archaeological evidence', *Journal of Ethnopharmacology*, 167.C: 54–63.

Wang, K., Gimbel, S., Malik, E., Hassen, S., & Hagopian, A. (2012) 'The experience of Chinese physicians in the national health diplomacy programme deployed to Sudan', *Global Public Health*, 7.2: 196–211.

White, N.J. (2008) 'Qinghaosu (artemisinin): the price of success', *Science*, 320.5874: 330–4.

Willcox, M.L. and Bodeker, G. (2004) 'Traditional herbal medicines for malaria', *Bmj*, 329.7475: 1156–9.

WHO (2010a) *Benchmarks for Training in Ayurveda*, http://apps.who.int/medicinedocs/documents/s17552en/s17552en.pdf, accessed 04/04/2021.

——— (2010b) *Collaborating Centres for Traditional Medicine*, http://www.who.int/medicines/areas/traditional/collabcentres/en/index.html, accessed 30/6/2019.

——— (2014) *Global Update on the Health sector response to HIV*, http://apps.who.int/iris/bitstream/handle/10665/128494/978?sequence=1, accessed 26/7/2018.

——— (2019) *International Statistical Classification of Diseases and Related Health Problems* (11th ed), ICD-11, Geneva: World Health Organization.

Xu, J., Liu, P. and Guo, Y. (2011) 'Health diplomacy in China', *Global Health Governance* 4.2: 1–12.

Yap, M. and Man, D.L. (1996) *Colour, Confusion and Concessions: the history of the Chinese in South Africa*, Hong Kong: Hong Kong University Press.

Youde, J. (2010) 'China's health diplomacy in Africa', *China: An International Journal*, 8.1: 151–63.

Zhang, C. (2010) 'Health diplomacy and soft power development – case of Chinese medical teams to Africa', *Contemporary International Relations*, 3: 49–53.

Zhan, M. (2009) *Other-Worldly: making Chinese medicine through transnational frames*, Durham, NC: Duke University Press.

43
TRANSLATING CHINESE MEDICINE IN THE WEST
Language, culture, and practice

Sonya Pritzker

Translating anything, let alone a medicine with a rich and diverse history of thousands of years, involves a stretch, a striving. It involves taking a step beyond the comfort zone of the familiar. In the West, translation is often conceived of as a bridge between worlds, a 'carrying across' of a crystallised form known as meaning (Bellos 2011: 28–9, 33). Meaning here is often conceived of as in-the-world, separate from language or experience—a constant and stable *thing* that, when packaged in the right linguistic form, can be 'transmitted' across time and culture. This way of conceptualising translation is rife with difficulties. It is fine, of course, when dealing with everyday nouns like table or chair, and can be applied to regular verbs like walk, or even dance. But what happens when we encounter things or actions that don't exist in another setting? In Chinese medicine, there are words—ambiguous nouns—like *qi* 氣 and *xue* 血. These things can be translated as 'energy' or 'blood', as some scholars have chosen to do, but are they exactly the same? There are verbs such as *bu* 補 and *jie* 解 and the actions they are attached to such as *bu xue* 補血 or *jie biao* 解表. We can follow familiar conventions and translate these as 'supplement blood' or 'resolve the exterior', but again we are faced with the fundamental challenge of explaining what it means to do such things in the context of Chinese medicine.

These types of questions have characterised the translation of Chinese medicine from Chinese to English and other Western languages since the first efforts were made to make sense of this diverse tradition across cultures. As is perhaps observable in the examples above, the questions do not merely pertain to linguistic issues. They arise from differences in the way the human body is approached in Chinese medicine and biomedicine. Questions and deep challenges also emerge when trying to embed Chinese medicine into a cultural milieu different from those in which it evolved. Political issues, very much intertwined with issues of power and legitimacy, also present subtle challenges for translators working in this field. While each of these vast networks of connected issues qualifies for a detailed study on its own, in this chapter I provide a broad overview covering of these complex areas—the body, cultural context, and politics. Throughout the chapter, I argue that the common notion of translation-as-transfer is deeply challenged by the ways in which translation in Chinese medicine actually occurs in everyday practice. In my conclusion, I briefly examine alternative ways to conceptualise translation in Chinese medicine, drawing upon several recent advances in Translation Studies and anthropology to rethink the 'living translation' of Chinese medicine as a meaning-making process rather than simply a transfer of meaning (Pritzker 2011, 2012a, 2014).

Different bodies?

If there is one thing that is often assumed to be universal, it is the human body. But as anyone who has studied Chinese medicine knows, there are multiple ways of apprehending the body. Kuriyama (1999: 14) thus writes that 'the body is unfathomable and breeds astonishingly diverse perspectives'. Detailing the differences in the way ancient Chinese and Greek physicians understood the human body, Kuriyama contrasts the Greek emphasis on muscles and tendons with the Chinese emphasis, despite their own detailed surgical study of the same structures (see Li, this volume), on colouring or hue and vitality of the channels and five chief organ networks. Where the Chinese perceived a harmonious network of *qi*, for example, Greek physicians sought a central agent governing the actions of the muscles. Such things are not simply surface differences, or even just different ways of seeing the same thing. 'What we habitually call anatomy is just *one kind* of anatomy', writes Kuriyama (1994: 159).

The notion that there are anatomies other than that accpeted by western bio-sciences challenges some of the core assumptions of contemporary medicine, but remains an important starting point for understanding why translation in Chinese medicine is difficult even at the most fundamental level of the human body. Kuriyama thus further highlights how differences in the way the human body was approached affected language. 'Besides using different words', he says, 'diagnosticians in China and Europe *used words differently*' (Ibid.: 64). In translation, these differences—in ways of conceptualising, seeing, touching, and speaking about the body—pose a significant challenge. Still, over many hundreds of years, scholars have attempted to translate the Chinese medical body into Western languages. They have drawn from a variety of approaches in order to create connections. In early attempts made by various explorers, missionaries, and Jesuits travelling to China, the surface resemblances between the Chinese medical emphasis on *qi* and the Galenic humoral model of the body led to many mistranslations, some of which persist today (Barnes 2005: 55). This approach to translation often hinged on the explicit assumption that there was one right way to see the human body—one chief anatomy. As medicine became increasingly scientific in the eighteenth to twentieth centuries, this assumption increased exponentially. In 1736, for example, 'Du Halde described Chinese knowledge as "depending upon a doubtful System of the Structure of the Human Body"' (Barnes 2005: 87). At this juncture in history, the fact that previous translators had associated the Chinese medical body with Galenic medicine, which was falling out of favour in Europe as anatomical models became more popular, led to an increasingly biased reception of translations from China (Ibid.: 99).

By the late nineteenth and early twentieth centuries, in Europe and the US, the anatomical body of biomedicine had become the *only* body. In China, this created what Karchmer (2004: 75–6) calls the 'anatomy problem', as Chinese physicians struggled to reconcile the anatomical structures emphasised in biomedicine with Chinese medical descriptions of the body—which included organs, such as the *sanjiao* 三焦, that didn't exist in anatomical models, and attributed functions, such as digestion to the spleen or *pi* 脾, to organs that were considered to do other things in biomedicine. 'The most conservative reaction to this [among Chinese writers] was to claim that Chinese bodies were different from European bodies, so that anatomy didn't apply', writes Andrews, 'However, many more medical writers of the time preferred to integrate the new anatomy with their own knowledge' (1996: 245).

Translating Chinese medicine in the west

The integration of Western anatomy into Chinese medicine was not without its difficulties, however. Many factions developed, and the camps were divided on how exactly to relate the two bodies.

In the early twentieth century, advocates for Chinese medicine and those who supported the wholesale importation of the anatomical biomedical body reached a tentative agreement. Very broadly speaking, within this agreement, the anatomical body of biomedicine was seen to pertain mostly or entirely to *structure,* whereas the Chinese medical body described *function*. This division allowed scholars and physicians in China to write about—and translate—concepts that were not included in the anatomical body of biomedicine (like *qi* or the *sanjiao*), while still providing a framework for relating the two within the 'structure/function dichotomy' (Karchmer 2004: 89). Karchmer explains that this dichotomy worked particularly well because it preserved 'essential differences' at the same time as it constructed the two bodies as 'mutually complementary'. Translations of Chinese medicine emerging from this paradigm thus draw upon a kind of same-but-different approach, using a single overarching body as a reference point for depicting the biomedical emphasis on structure and the Chinese medical emphasis on function.

Most translations that can be found today adopt these parameters. Even though it has become common practice to use Western anatomical terms to translate the names for Chinese organs—the *pi* 脾 in Chinese medicine is therefore commonly translated as spleen, the *gan* 肝, liver, etc. (Chapter 1 in this volume)—most translators emphasise that it is not the same spleen or liver or whatever organ as we think about in the biomedical body. Maciocia, author of a series of popular textbooks of Chinese medicine, thus writes that 'when studying the Chinese theory of the Internal Organs, it is best to rid oneself of the Western concept of internal organs entirely' (2005: 97). Maciocia here capitalises the names of organs in Chinese medicine, writing Spleen or Liver, for example, to emphasise the difference, and many have followed this lead. Others use different strategies, calling the organs 'orbs', for example (Porkert 1983), or giving them entirely different names. None of these translations is without controversy, however. Some authors, for example, avoid emphasizing differences in bodies, using lower-case biomedical terms to highlight the fundamental sameness of the bodies—and their disorders—in biomedicine and Chinese medicine. For more information on these debates, see Wiseman (2000) or Pritzker (2014). For our current purposes, it is safe to say that what it often boils down to, very broadly, is a problem of reconciling if there is indeed one body or multiple, different bodies that are all valid. If there are multiple bodies, authors are faced with how to translate the Chinese medical body into a language that emerged to treat the biomedical body. If there is one overarching body, the challenge for translators is how to use language to designate the relationship between terms used in one context with terms used in the other. Confusion often arises because certain translators have a commitment to a stable meaning that can be transferred across languages through the conduit of translated words.

It will perhaps help to get a sense of how Westerners learning Chinese medicine grapple with the translations they are offered in their various texts. These students, whom I have studied intensely over several years, often wrangle with the language of their translations. Especially when they are forced to 'translate' for patients, they struggle with how to explain, for example, spleen versus Spleen, or kidney versus Kidney. For example, one second-year student described her frustration with the limited time interns have to describe the organs to patients during their visit. And yet patients' enquiries about the language interns use requires them to provide explanations, she continued, saying 'you know, you can't explain that to somebody in like three minutes when you're laying them down on the table for an acupuncture treatment' (Pritzker 2014: 170). In the learning process, however, what it often comes down to is students' own embodied experience of *qi*, or of treating *their own* Kidneys/kidneys (2011b). In this sense, the body of learners itself becomes, to borrow from Emad (1998), a site for the cultural translation of Chinese medicine. The problem of one or multiple bodies and the search for coherent meaning is reconciled, in other words, in the bodies of the users of Chinese medicine, in a meaning-making *process* rather than a transfer that, along with Emad, I argue constitutes the actual site of translation. In many ways, the linguistic translation of the Chinese medical body in the contemporary US thus mirrors the way learning occurs in both classical and contemporary Chinese medicine in China, where the transmission of complex ideas occurs through the medium of embodied experience (Farquhar 1994; Sivin 1995; Hsu 1999; Pritzker 2014).

Shifting cultural contexts

Acts of translation are always achieved within a social, cultural, and historical setting. In the contemporary US, Chinese medicine is part of a broad-reaching movement known as complementary and alternative medicine or CAM. This movement, characterised by a holistic, spiritually oriented, natural, and 'person-centred' approach to healthcare (McGuire and Kantor 1988; Goldstein 1999; Greaves 2002; Ho 2007; Harrington 2008; Barcan 2011; Ross 2012), has its roots in a distinct cultural history in the US. This history includes the countercultural movements of environmentalism, feminism, and New Age spirituality (Heelas 1996; Baer 2004; Barcan 2011). It also draws heavily upon a distinctly Western notion of 'person-centred medicine' that has developed in response to a growing dissatisfaction with biomedicine and the growing influence of evidence-based medicine or EBM (Pritzker, Katz and Hui 2013). Within this context, Chinese medicine has become popular as an alternative practice in the contemporary US (Barnes 2003; Zhan 2009; Pritzker 2014; Chapter 40 in this volume).

In China, on the other hand, the cultural setting has been vastly different. Instead of Chinese medicine finding a home in some sort of alternative medicine universe, it has been adopted in concert with biomedicine as a mainstream form of healthcare (chapter 45 in this volume). In China, 'integrative Chinese and Western medicine' (*zhong xiyi jiehe* 中西医结合) is thus practised in hospitals, researched in clinical trials, and discussed in both scholarly and popular formats (Scheid 2002).

Such differences in cultural contexts have, of course, influenced the way translations have been accomplished. In the West, for example, many translators have appealed to the alternative desires of their perceived readership—the many students and practitioners of Chinese medicine who have eschewed mainstream medicine in order to practise a more holistic, spiritually oriented approach. Their language reflects a distinct emphasis on the types of things that this readership cares about, such as 'spirituality' and the interconnectedness of mind and body. The result, several scholars have demonstrated, has been a 'psychologisation', or 'spiritualisation' of Chinese medicine, where the language of Chinese medicine becomes intermeshed with terms borrowed from psychology and New Age spirituality (Barnes 1998; Pritzker 2012c). As Unschuld points out, such a language inaccurately reproduces Chinese medicine, for example by intentionally leaving out battlefield metaphors so common in traditional texts: 'In contrast to reports from the battlefield of modern immunology', he writes, 'the theory of TCM freed of its martial metaphors gives the impression that it can lead patients back to the harmony of the great whole. It offers solace where modern medicine offers only the uncertainty of a murderous battle' (Unschuld 2003: 218–9). Unschuld's observation here reveals two aspects of the cultural adaptation of Chinese medicine in the West as it occurs through the medium of translation. First, essential elements of Chinese medicine are being left out of certain popular translations. Second, Chinese medicine is being translated in opposition to a contemporary form of biomedical practice that did not exist at the time many of the original texts were composed. Transformations such as these lead many scholars to the conclusion that the US is producing its own unique and deeply hybrid form of 'American Chinese medicine' (Hare 1993).

In China, very broadly speaking, many translations have been produced for what is assumed to be an integrative, biomedically oriented audience. Instead of a psychologised or spiritualised language, then, there is more emphasis on a biomedical language that attempts to translate Chinese medicine into terms that make it consonant with an imagined contemporary global medical culture. Here, we find translations that leave out key pieces of traditional Chinese medical thinking as well. Any discussion of spirits, or healing that incorporates a religious component, is often left out (Chapter 27 in this volume). Terminology is often transformed, with contemporary Chinese authors borrowing terms from biomedicine in order to discuss traditional illness categories. Likewise, the *format* of Chinese medicine is also deeply affected by this cultural context, with traditional teachings being reorganised to accommodate a textbook format and institutional instructional technique (Zhan 2009).

Culture is not nearly so simple, however, as those in the field of anthropology can attest to. Within cultures, there are subcultures, and there are individuals who diverge from cultural and even subcultural norms. For this reason, it is important to get a sense of what is happening in specific sites as translations are being created, and consumed, by individuals, especially in a field as complex as Chinese medicine. At this level, the culturally informed nature of translation in Chinese medicine becomes even more complex, as individuals are rarely simple products of their cultural context. Students and practitioners in the US are quite aware of the transformations that have resulted from the various translations they are exposed to, and yet the *way* they consume these translations is often through a filter of what matters most to them at a particular point in time: the kinds of complaints that their patients present with, for example, or their various ideas about what it means to be a good healer in the contemporary US. This statement is equally true for the producers and consumers of translations in China, and in both places, the ways in which translations are consumed tend to shift over time, sometimes even on a daily basis. Such shifts in orientation vis-à-vis different translation strategies point to translation in Chinese medicine as an ongoing *process* or meta-process that encompasses many smaller, everyday processes of decision-making and experience, rather than a simple transfer of meaning.

Power and politics

Translation, many in the field of Translation Studies argue, is always about power—the power of representation (Venuti 1992; Tymoczko and Gentzler 2002). Whoever gets to define the valid body, for example, gets to designate the terminology that is used to describe it. Whoever understands the priorities of the readership will produce translations that are accepted in that cultural context. In Translation Studies, the terms *target-oriented* and *source-oriented* are often used to describe issues like these. In essence, a target-oriented approach to translation works to create a product that appears to have been written in the target language, a product that fits well with the priorities of the target culture. A source-oriented translation, on the other hand, works to preserve the cultural distinctiveness and even *foreignness* of the source text (Nida 2000). A source-oriented approach, Nida explains, 'is designed to permit the reader to identify himself as fully as possible with a person in the source-language context, and to understand as much as he can of the customs, manner of thought, and means of expression' (Nida 2000 [1964]: 129).

One of the greatest controversies in the contemporary translation of Chinese medicine is thus the debate over whether a source-oriented or target-oriented translation strategy should be used. Several scholars argue, for example, that both approaches described in the last section (translating Chinese medicine within a spiritualised/New Age framework and translating Chinese medicine in an integrative, biomedical framework) are essentially target-based. The argument is that, by attempting to fit Chinese medicine into a language and cultural context that differs from the one(s) in which it evolved, translators are committing a deep betrayal of original concepts. In doing so, argues Nigel Wiseman, a prominent linguist and author of several Chinese-English dictionaries of Chinese medicine, they are wielding a linguistic power that overrides traditional culture and pollutes traditional concepts with cultural ideologies from the present (Wiseman 2001). There are also several prominent translators, practitioners, and scholars who argue that a target-oriented approach is appropriate and necessary, however. Consider Professor Xie Zhufan 谢竹藩, one of the foremost authors and researchers in integrative medicine in China, who argues that the use of traditional language in the translation of Chinese medicine places this system of medicine at a disadvantage in the context of mainstream global healthcare. 'In short, the proper use of Western medical terms is necessary', he writes, 'and may facilitate the correct understanding of TCM. Insisting on intentionally keeping TCM terminology apart from Western medical terms in every aspect will make a false impression that TCM is an esoteric system of medicine' (Xie 2003a: 24). In the US, moreover, there are many who argue that, by all means, Chinese medicine *should* be transformed as it moves to the West, and that the language used to translate it will of course participate greatly in that transformation (Beinfield and Korngold 2001).

Underlying all of these arguments is a deeply political, power-laden question of who should have the cultural rights to translate Chinese medicine. Many Chinese translators are thus exasperated with Westerners who argue for a traditional or source-oriented approach to translation, because they feel that they are the true owners of the medicine. On the other hand, many Westerners feel abandoned by contemporary Chinese efforts to biomedicalise Chinese medicine, and see themselves as the true champions of a traditional approach.

The sparks created by this debate have flown in many contexts. They are not simply abstract, philosophical differences. International efforts in creating a standardised terminology for Chinese medicine constitute one arena in which we can witness the political implications of debates over translation. The standardisation of Chinese medicine is a complex affair, one that has been discussed extensively in both China and the West (see Wiseman and Zmiewski 1989; Wiseman 2000; Xie 2002a, 2002b, 2003a, 2003b; Niu 2005, 2006; Felt 2006; Flaws 2006; Li 2008, 2009a, 2009b; Li and Pan 2009a, b and c; Pritzker 2012b, 2014). Briefly, there are several different organisations, including the World Health Organisation, the World Federation of Chinese Medical Societies, and the Korean Institute of Oriental Medicine, who have developed standardised term sets of 40,000+ terms for the translation of Chinese medicine into English. The process by which these standards are agreed upon is deeply political, involving meetings at which experts in various fields discuss and vote on terms. Often, the experts involved are not translators, nor are many of them native English speakers, although on certain occasions Westerners have been invited to participate. As you might imagine, the voting process can be arbitrary, depending on the biases of participants. Not surprisingly, many participants are biased in the direction of biomedical terminology. As of now, however, no single term list has achieved the place of world standard for the translation of Chinese medicine, although the English terms that are used in the 'traditional medicine' section of the *International Classification of Diseases, 11th Edition* arguably come closest to that goal. Even once a standard has been officially instituted, however, a further concern is enforcement. Students, authors, and translators in the US and Europe are deeply resistant to the imposition of a standard English-language terminology by a governing body of any size, in the US or abroad. To these individuals, such enforcement seems like a breach of power that infringes on the freedom of speech, and delimits the flexibility they have in translating Chinese medicine into English. Others resist these international standards on the grounds that they are too biomedical in orientation, and they argue that a more source-oriented set of standards should be developed (Wiseman 2000; Ergil 2001; Ergil and Ergil 2006; Flaws 2006; Unschuld and Tessenow 2011).

From an ethnographic perspective, it is clear that the issues at stake in the politics of translation in Chinese medicine constitute a complex 'field of practice' (Bourdieu 1990; Scheid 2002). Anthropological enquiry reveals that what we are dealing with in these settings is a vast network consisting of multiple layers of moral codes of conduct (what it means to be a good healer, what it means to use language effectively), ideologies of translation (whether translation is even possible, how to create the best translation), and cultural commitments. As I have previously argued (Pritzker 2012b), one possible way to get past some of these political divides is to increase our mutual understanding of some of these deeper issues as they effect individuals in the context of their everyday lives. Here, it helps greatly to adopt the view of translation as a process rather than a unilateral transfer of a meaning that is separate from that process.

Conclusion

By introducing some of the complex challenges facing translators of Chinese medicine from several different vantage points, all deeply intertwined, I have attempted to show the many ways in which translation in Chinese medicine is far more than a simple "transfer" of "meaning." As a process of coming to know the body or bodies, of making sense of Chinese medical concepts within a cultural framework, or of fighting for the rights to set standards, it comes to look more like a meaning-making endeavour that involves multiple interlocutors in an ongoing conversation. This conversation is about the fundamental truths of the human body. It is about reshaping Chinese medicine into forms that are compatible with wildly diverse cultural demands. And it is a heated political debate fraught with conflict and disparities of power. For participants who are learning or practising Chinese medicine, it is an embodied search for the meaning of translated words, a morally situated effort to make sense of Chinese medicine through the lens of personal priorities and desires, and a challenge forcing people to take a stance on language even if they are in the field primarily to contribute to the healing of patients.

For these reasons, I have found it useful to think about translation in Chinese medicine, not as transfer, but as a living, breathing process, or meta-process encompassing multiple ongoing processes, that I call 'living translation' (Pritzker 2011, 2012a, 2014). Inspired by recent developments in Translation Studies and anthropology that encourage an expanded view of translation as an ongoing, dialogic, and embodied series of meaning-making practices (Wadensjö 1998; Montgomery 2000; Schieffelin 2007; Hanks 2010), living translation, as a framework for organising a whole barrage of processes and practices related to translation, points to the ways in which translation in Chinese medicine goes far beyond the transfer of meaning in language. In focussing on the everyday moments in which translation occurs, it emphasises the interactive, embodied, experiential, and practical nature of translation in Chinese medicine, which I have touched upon briefly here. From this perspective, the translation of Chinese medicine is a shifting and ongoing process that instead of being simply interpreted as a transfer, can also be understood as an active *exchange* or *engagement* with multiple bodies, divergent cultures and subcultures, various domains of power, and the notion of meaning itself. For those trying to conduct or simply understand the process, I argue that this broad view of living translation offers readers of multiple kinds of translations an appreciation of the underlying issues involved in creating various texts. This can open new windows of understanding when we encounter vastly different readings of Chinese medicine in our own search for meaning in this complex field.

Bibliography

Andrews, B. (1996) *The Making of Modern Chinese Medicine, 1895–1937*, PhD thesis, University of Cambridge.
——— (2014) *The Making of Modern Chinese Medicine, 1850–1960*, Vancouver; Toronto, ON: UBC Press.
Baer, H. (2004) *Toward an Integrative Medicine*, Walnut Creek, CA: Alta Mira Press.
Barcan, R. (2011) *Complementary and Alternative Medicine: bodies, therapies, senses*, Oxford and New York: Berg.
Barnes, L.L. (1998) 'The psychologizing of Chinese healing practices in the United States', *Culture, Medicine, and Psychiatry*, 22: 413–43.
——— (2003) 'The acupuncture wars: the professionalizing of American acupuncture – a view from Massachusetts', *Medical Anthropology*, 22.3: 261–301.
——— (2005) *Needles, Herbs, Gods, and Ghosts: China, healing, and the west to 1848*, Cambridge, MA; London: Harvard University Press.
Bates, D. (ed.) (1995) *Knowledge and the Scholarly Medical Traditions*, Cambridge: Cambridge University Press.
Beinfield, H. and Korngold, E. (2001) 'Centralism vs pluralism: language, authority, and freedom in Chinese medicine', *Clinical Acupuncture and Oriental Medicine*, 2: 145–54.
Bourdieu, P. (1990) *The Logic of Practice*, Stanford, CA: Stanford University Press.
Bellos, D. (2011) *Is That a Fish in Your Ear? Translation and the meaning of everything*, New York: Faber and Faber, Inc.
Emad, M. (1998) *Feeling the Qi: emergent bodies and disclosive fields in American appropriations of acupuncture*, PhD thesis, Rice University.
Ergil, M. (2001) Considerations for the translation of traditional Chinese medicine into English, https://cewm.med.ucla.edu/wp-content/uploads/CM-Considerations-4.10.14-FINAL.pdf, accessed 29/3/2020.
Ergil, M. and Ergil, K. (2006) 'Issues surrounding the translation of Chinese medical texts into English', *American Acupuncturist*, 37: 24–6.
Farquhar, J. (1994) *Knowing Practice: the clinical encounter of Chinese medicine*, Boulder, CO: Westview Press.
Felt, R. (2006) 'The role of standards in the transmission of Chinese medical information', *American Acupuncturist*, 37: 1820.
Flaws, B. (2006) 'Arguments for the adoption of a standard translational terminology in the study and practice of Chinese medicine', *American Acupuncturist*, 37: 16–27.
Goldstein, M. (1999) *Alternative Health Care: medicine, miracle, or mirage?* Philadelphia: Temple University Press.
Greaves, D. (2002) 'Reflections on a new medical cosmology', *Journal of Medical Ethics*, 28: 81–5.
Hanks, W. (2010) *Converting Words: Maya in the age of the cross*, Berkeley: University of California Press.

Harrington, A. (2008) *The Cure Within: a history of mind-body medicine*, New York: W.W. Norton & Company.
Hare, M.L. (1993) 'The emergence of an urban U.S. Chinese medicine', *Medical Anthropology Quarterly*, 7.1: 30–49.
Heelas, P. (1996) *The New Age Movement*, Malden, MA: Blackwell Publishing.
Ho, E.Y. (2007) '"Have you seen your aura lately?" Examining boundary-work in holistic health pamphlets', *Qualitative Health Research*, 17.1: 26–37.
Hsu, E. (1999) *The Transmission of Chinese Medicine*, Cambridge: Cambridge University Press.
Karchmer, E, (2004) *Orientalizing the Body: postcolonial transformations in Chinese medicine*, PhD thesis, University of North Carolina at Chapel Hill.
Kuriyama, S. (1994) 'The imagination of winds and the development of the Chinese conception of the body', in A. Zito and T.E. Barlow (eds) *Body, Subject, and Power in China*, Chicago, IL; London: University of Chicago Press, pp. 23–41.
—— (1999) *The Expressiveness of the Body and the Divergence of Greek and Chinese Medicine*, New York: Zone Books.
Li Zhaoguo 李照国 (2008) 'WHO xitaiqu yu shijie zhongyi yaoxuehui lianhehui zhongyi mingci shuyu guoji biaozhun yanjiu: zonglun bufen WHO西太区与世界中医药学会联合会中医名词术语国际标准比较研究：总论部分' (Comparative study on WHO western pacific region and World Federation of Chinese Medical Societies international standard terminologies on Chinese medicine: an analysis of the general), *Journal of Chinese Integrated Medicine*, 6.7: 761–5.
—— (2009a) 'Comparative study on WHO Western Pacific Region and World Federation of Chinese Medical Societies international standard terminologies on Chinese medicine: an analysis of the causes of disease (part I)', *Journal of Chinese Integrated Medicine*, 7.3: 284–7.
—— (2009b) 'Comparative study on WHO Western Pacific Region and World Federation of Chinese Medical Societies international standard terminologies on Chinese medicine: an analysis of the causes of disease (part II)', *Journal of Chinese Integrated Medicine*, 7.4: 383–8.
Li Zhaoguo and Pan Shulan 潘淑兰 (2009a) 'Comparative study on WHO Western Pacific Region and World Federation of Chinese Medical Societies international standard terminologies on Chinese medicine: an analysis of body constituents', *Journal of Chinese Integrated Medicine*, 7.1: 79–84.
—— (2009b) 'Comparative study on WHO Western Pacific Region and World Federation of Chinese Medical Societies international standard terminologies on Chinese medicine: an analysis of the five sensory organs', *Journal of Chinese Integrated Medicine*, 7.2: 183–6.
—— (2009c) 'Comparative study on WHO Western Pacific Region and World Federation of Chinese Medical Societies international standard terminologies on Chinese medicine: an analysis of the mechanism of diseases (Part I)', *Journal of Chinese Integrated Medicine*, 7.5: 482–7.
Maciocia, G. (2005) *The Foundations of Chinese Medicine: a comprehensive text for acupuncturists and herbalists*, 2nd ed., London: Elsevier.
Makihara, M. and Schieffelin, B. (eds) (2007) *Consequences of Contact: language ideologies and social transformation in Pacific societies*, New York: Oxford University Press.
McGuire, M.B. and Kantor, D. (1988) *Ritual Healing in Suburban America*, New Brunswick: Rutgers University Press.
Montgomery, S.L. (2000) *Science in Translation: movements of knowledge through cultures and time*, Chicago, IL; London: Chicago University Press.
Nida, E. (2000 [1964]) 'Principles of correspondence', in L. Venuti (ed.) *The Translation Studies Reader*, London; New York: Routledge, pp. 126–40.
Niu Chuanyue 牛喘月 (2005) 'Zhongyi mingci shuyu guoji biaozhunhua gongcheng zhengshi qidong 中医名词术语国际标准化工程正式启动' (Program on international standardization of traditional Chinese medicine nomenclature has been started), *Journal of Chinese Integrated Medicine*, 3.1: 79–82.
Niu Chuanyue (2006) 'Mingyue songjian zhao, qingquan shishang liu: Zailun zhongyi yingyu fanyi ji qi guifanhua wenti 明月松间照，清泉石上流：再论中医英语翻译及其规范化问题' (On some issues concerning the translation and standardization of traditional Chinese medicine), *Journal of Chinese Integrated Medicine*, 4.6: 657–60.
Porkert, M. (1983) *The Essentials of Chinese Diagnostics*, Zurich: Acta Medicinae Sinensis.
Pritzker, S.E. (2011) 'The part of me that wants to grab: embodied experience and living translation in U.S. Chinese medical education', *Ethos*, 39.3: 395–413.
—— (2012a) 'Living translation in U.S. Chinese medicine', *Language in Society*, 41.3: 343–64.

―― (2012b) 'Standardization and its discontents: four snapshots in the life of language in Chinese medicine', in V. Scheid and H. MacPherson (eds) *Integrating East Asian Medicines into Contemporary Healthcare*, London: Elsevier Press, pp. 75–88.

―― (2012c) 'Transforming self, world, medicine: the trend toward spiritualization in U.S. Chinese medicine', in V. Scheid and H. MacPherson (eds) *Integrating East Asian Medicines into Contemporary Healthcare*, London: Elsevier Press, p. 203.

―― (2012d) 'Translating the essence of healing: Inscription and interdiscursivity in U.S. translations of Chinese medicine', *Linguistica Antverpiensia New Series: Themes in Translation Studies*, 11: 151–66.

―― (2014) *Living Translation: language and the search for resonance in Chinese Medicine*, Oxford; New York: Berghahn Books.

Pritzker, S.E., Katz, M. and Hui, K.K. (2013) 'Person-centered medicine at the intersection of East and West', *European Journal of Person-Centered Medicine*, 1.1: 209–15.

Ross, A.I. (2012) *The Anthropology of Alternative Medicine*, London; New York: Berg.

Scheid, V. (2002) *Chinese Medicine in Contemporary China*, Durham, NC; London: Duke University Press.

Scheid, V. and MacPherson, H. (eds) (2012) *Integrating East Asian Medicines into Contemporary Healthcare*, London: Elsevier Press.

Schieffelin, B. (2007) 'Found in translating: reflexive language across time and texts', in M. Makihara and B. Schieffelin (eds) *Consequences of Contact: language ideologies and social transformation in Pacific societies*, New York: Oxford University Press, pp. 140–64.

Sivin, N. (1995) 'Text and experience in classical Chinese medicine', in D. Bates (ed.) *Knowledge and the Scholarly Medical Traditions*, Cambridge: Cambridge University Press, pp. 177–204.

Tymoczko, M. and Gentzler, E. (eds) (2002) *Translation and Power*, Amherst and Boston, MA: University of Michigan Press.

Unschuld, P.U. (ed.) (1989) *Approaches to Traditional Chinese Medical Literature: proceedings of an international symposium on translation methodologies and terminologies*, Dordrecht: Kluwer Academic Publishers.

―― (2003) 'The spread of traditional Chinese medicine in the western world: an attempt at an explanation of a surprising phenomenon', *Studies in the History of Natural Sciences*, 22.3: 215–22.

Unschuld, P.U. and Tessenow, H. (2011) *Huang di nei jing su wen: An annotated translation of huang di's inner classic - basic questions*, 2 volumes: University of California Press.

Venuti, L. (1992) *Rethinking Translation: discourse, subjectivity, ideology*, London: Routledge.

―― (ed.) (2000 [1964]) *The Translation Studies Reader*, London; New York: Routledge.

Wadensjö, C. (1998) *Interpreting as Interaction*, London; New York: Longman.

Wiseman, N. (2000) *Translation of Chinese Medical Terminology: a source-oriented approach*, PhD thesis, Exeter University.

―― (2001) *The Extralinguistic Aspects of English Translation of Chinese Medical Terminology*, http://www.paradigm-pubs.com/paradigm/refs/wiseman/Extralinguistics.pdf, accessed 15/11/2005.

Wiseman, N. and Zmiewski, P. (1989) 'Rectifying the names: suggestions for standardizing Chinese medical terminology', in P.U. Unschuld (ed.) *Approaches to Traditional Chinese Medical Literature: proceedings of an international symposium on translation methodologies and terminologies*, Dordrecht: Kluwer Academic Publishers, pp. 55–66.

Xie Zhufan 谢竹藩 (2002a) 'On standard nomenclature of basic Chinese medical terms (II)', *Chinese Journal of Integrated Medicine*, 8.3: 231–4.

―― (2002b) 'On standard nomenclature of basic Chinese medical terms (IIIA)', *Chinese Journal of Integrated Medicine*, 8.4: 310–1.

―― (2003a) *On the Standard Nomenclature of Traditional Chinese Medicine*, Beijing: Foreign Languages Press.

―― (2003b) 'On standard nomenclature of basic Chinese medical terms (IV)', *Chinese Journal of Integrated Medicine*, 9.2: 148–51.

Zhan, M. (2009) *Other-Wordly: making Chinese medicine through transnational frames*, Durham, NC; London: Duke University Press.

Zito, A. and Barlow, T.E. (eds) (1994) *Body, Subject, and Power in China*, Chicago, IL; London: University of Chicago Press.

PART 7

Negotiating modernity

44
THE DECLARATION OF ALMA ATA

The global adoption of a 'Maoist' model for universal healthcare

Paul Kadetz

Introduction

In May 1977, the World Health Assembly of the World Health Organization (WHO) announced its target of universal healthcare coverage for all by the year 2000, which was codified the following year in the Declaration of Alma Ata. Universal healthcare coverage requires universal access to healthcare resources. Mao Zedong's utilisation of existing resources to achieve extensive rural healthcare coverage – including: the development of the primary level of rural healthcare; the training of community health workers; and the integration of local healthcare practices and practitioners into the biomedical healthcare system – was adopted by the WHO and embedded in the Declaration of Alma Ata as a means to achieve 'health for all'. However, eight months previously, on September 9, 1976, Mao Zedong died and with his death followed the rapid dissolution of rural healthcare coverage in the People's Republic of China (PRC). The Declaration of Alma Ata was adopted at the WHO's World Conference on Primary Healthcare in September 1978. Three months later, Mao's successor, Deng Xiaoping, instituted the first economic reforms of the PRC (*Gaige kaifang* 改革开放) leading to the current economic system of state capitalism accompanied by an ongoing cycle of healthcare policies and reforms that have, to the present, sought to return access to rural healthcare to its former coverage under Mao.

This chapter reviews the complex story of how the PRC's solutions to remedy poor rural healthcare access were adopted by the WHO as a central means to achieve universal healthcare. We will delineate the formation of this model of primary healthcare and critically examine the adoption of this model on the global stage.

The global context for the adoption of a communist Chinese model of healthcare

The World Health Organization was formed as a technical healthcare agency of the United Nations in 1948 (Lee 2009). Vertical disease treatment programming (which typically focuses on the eradication of a communicable disease, such as smallpox, or of a group of communicable diseases) was the dominant paradigm for WHO interventions, regardless of the more horizontal, social medicine, and healthcare resource building orientation of its charter.

Part of the reason for the preferential use of vertical programming at the WHO may be a consequence of the political economic dominance of a US agenda that favoured vertical programming following the departure of the Soviet Union, the PRC, and other communist countries shortly after the WHO began operations in 1949 (Brown 1979). The Washington agenda sought 'modernisation with limited social reform' (Brown *et al.* 2006: 65), which included the global market expansion of American pharmaceutical corporations and a concomitant emphasis on treatment, rather than prevention. Hence, the policy decisions and embedded frameworks of the WHO were, at least in part, a reflection of the agendas of larger international political economic alliances and hierarchies of power. Therefore, the WHO's marked shift in emphasis towards universal primary healthcare and preventative medicine via horizontal programming during the 1970s was novel and noteworthy.

Several factors are believed to have supported the adoption of what came to be known as Comprehensive Primary Healthcare (CPHC) including:

a   a renewed appreciation in the 1960s of the need to strengthen healthcare infrastructure after assessing the failure of several vertical programmes in low-income settings, such as malaria eradication programmes;
b   the WHO's efforts to redress rural healthcare coverage, which date to the 1960s;
c   the influence of the WHO's then director general, Halfden Mahler, and his prioritisation for covering basic health needs;
d   the influence of the Soviet Union and their support for the horizontal promotion of national health services, which developed into primary healthcare services at the community level;
e   the readmission of the PRC as a member state of the World Health Assembly in 1973;
f   the support of several African nations (who had received aid from the PRC) for China's healthcare approach;
g   the world economic recession of the 1970s; and
h   the growing acknowledgement, especially in the West, of the PRC's success in rural health coverage (Litsios 2004; Lee 2009; Huang 2010; Kadetz 2013).

This last factor was particularly instrumental in inspiring the World Health Assembly to launch the 'Health for All by 2000' campaign. According to the former deputy director general of China's Department of Rural Health Management: in attempting to resolve the challenges identified in the 1970s, related to the health-cost burden and an unequal distribution of health resources, the WHO conducted research in nine countries, including four cooperation centres in the PRC (WHO 2008).

Also, during this period, the normative biomedical healthcare model of disease eradication 'was failing to meet the basic needs of populations in low-income countries and became an increasingly untenable model' (Lee 2009: 73). All of these factors – coupled with the marked gains of community-based models (particularly in Latin America and Bangladesh), which reflected significantly improved health outcomes despite resource-poor conditions (particularly in Cuba and Kerala), in addition to the purported success of the community health worker (i.e. barefoot doctor) programme in the PRC – promulgated a paradigm shift in the WHO's approach to healthcare in low-income countries (Ibid.; Chapter 45 in this volume).

This shift from a more vertical to a more horizontal approach, which supported an understanding of health as a human right, was endorsed by all 134 World Health Assembly member states attending the WHO conference at Alma Ata (Brown *et al.* 2006). The outcome of this conference, the Declaration of Alma Ata, was a bold attempt by the WHO to achieve universal healthcare by prioritising the development of primary healthcare resources.

However, the WHO's vision for primary healthcare, which emphasised strong basic healthcare services at the most local community level, was challenged at a post-Alma Ata Rockefeller Foundation-sponsored conference in Bellagio, Italy, for being too expensive, too broad, and too horizontal in implementation to be seriously considered for addressing universal healthcare (Magnussen et al. 2004). In its place, 'a rationally conceived, best data-based, selective attack on the most severe public-health problems' was proposed (Walsh and Warren 1979: 970). In other words, the normative vertical curative approach, addressing the treatment (and, at times, prevention) of particular diseases, was ironically presented as 'the most effective means of improving the health of the greatest number of people' (Ibid.).

This approach, labelled 'Selective Primary Healthcare' (SPHC), was solely concerned with the infant and child health interventions of growth monitoring to fight malnutrition in children, oral rehydration to fight diarrhoeal diseases, breastfeeding, and immunisations (Brown et al. 2006: 67). The Bellagio conference touted SPHC as the best possible option for low-income countries 'until comprehensive primary healthcare can be made available to all' (Walsh and Warren 1979: 970). This approach was supported by then World Bank president, Robert McNamara, and donor agencies including USAID, the Ford Foundation, and the Rockefeller Foundation (Brown et al. 2006). In other words, this decision was heavily orientated towards US policy and political economic considerations. The new director of UNICEF, another American, James Grant, adopted SPHC as the framework through which UNICEF would immediately operate (Ibid.) Thereafter, a partition was placed between WHO's approach of Comprehensive Primary Healthcare and the Selective Primary Healthcare approach of UNICEF, and a debate was born in international health circles, which has continued for the past four decades. Yet, Mao's model to redress rural healthcare access through community health workers and the integration of non-biomedical practices and practitioners into the formal healthcare system was quietly and independently adopted by several countries after Alma Ata. Hence, even though comprehensive primary healthcare was not ultimately globally adopted, Mao's model of primary healthcare – through the integration of non-biomedical practices and practitioners into the biomedical healthcare system – shaped the discourse of healthcare integration, particularly in the WHO. To better understand this model it is necessary to trace its historical development.

Defining healthcare integration

The word 'health' stems from Old English and Germanic words for wholeness. Anthropologist, Mary Douglas, refers to the inherent drive of social groups to make order out of chaos, as 'a unity of experience' (1966: 3). The literature on healthcare integration reflects this idea of a unity of situated rationalities through terms such as 'unification' [Mao Zedong]; 'harmonisation' [ASEAN]; and, of course, 'integration' [WHO] (Kadetz 2012). Thus, the representation of a top-down coherence of order, as opposed to a dynamic, fluid, and grassroots multi-plural order, is depicted by the WHO as necessary and beneficent for effective healthcare.

From extensive field research, I argue that, in actual practice, healthcare integration is neither simply a top-down process solely determined by the state, nor by its biomedical practitioners in positions of authority (Kadetz 2014). Rather, healthcare integration may be best understood as an outcome of the medical pluralism dynamically practised by the lay public.

However, in a context, such as Maoist China, where all healthcare practices and practitioners were controlled by the State, top-down integration may be the most accurate description of how integration was actually carried out.

Ring out the old

The story of healthcare integration in the PRC is, at least in part, an outcome of the intense pressures China experienced from the nineteenth to twentieth centuries to bridge the divide between the old China and the new modern China; between the superstitious and the 'scientific'; and between the authoritative knowledge[1] of the West and local (geographic, political, and historical) understandings. The impact of these pressures on local healthcare practices and practitioners in China reached an impasse in the early Republican period.

The heart of the *Westernisation Movement* (*yangwu yundong* 洋务运动), from the 1860s to 1890s, concerned the adoption of Western knowledge and technology (Ma 1995: 25). Andrews (2014) suggests that the modernisation of the schools of Chinese medicine in particular was an outcome of these pressures to modernise. 'Chinese physicians mobilised western knowledge as a resource to defend themselves and the values they aimed to uphold' (Andrews 1996: 2). Although a distinct shift in the Imperial Court's attitude towards modernity and biomedicine can be identified from the time of the *Self-Strengthening Movement* (from 1860 to 1895), it was not until after the Revolution of 1911 with the collapse of the Qing dynasty, along with the Manchurian Pneumonic Plague (1910–1911), and the subsequent International Plague Conference (hosted by China), that the government embraced biomedicine (Wu 1959; Andrews 2014).[2] A new metropolitan medical elite was quickly established who worked in cities and catered to wealthy patients, and who 'were able to acquire some of the status and authority of the state' (Andrews 1996: 15).

Under Chiang Kai-shek and the Nationalist government, health became equated with national strengthening and the Ministry of Health of China was inaugurated in 1928 and represented the first steps towards a health bureaucracy under the Nationalist regime (Yip 1982). Having embraced Western modernity, the Republican government set out to eradicate the practices of Chinese medicine completely. Starting in 1914, the Ministry of Education proposed to abolish all Chinese medicine practices (Ma 1995).[3] This was followed by the 1929 bill entitled *Abolishing the Old Medicine In Order To Clear the Obstacle for Health and Medicine* that was accepted by the Ministry of Health, and that attempted to ban all publications about Chinese medicines and close all schools of Chinese medicine (Ibid.) The domination of Chinese health organisations by Western-trained physicians 'with their generally unsympathetic attitude toward traditional doctors and their attempts to regulate or even abolish traditional medicine led to a prolonged and often bitter feud between the two groups' (Yip 1982: 1201). In fact, some intellectuals believed that adherence to Chinese medicine would 'lead the country down a dark, autocratic road' (Ma 1995: 205).

However, not all were in agreement with abolishing Chinese medicine. There was a small contingent of scholar-practitioners of Chinese medicine known as the *School of Merging* [Chinese and Western medicine] (*huitong pai* 汇通排), who, since the late nineteenth century to the end of the Republican Era, sought to preserve Chinese medicine by hybridising it with biomedicine (Scheid 2001: 370). They promoted Chinese medicine as 'essential to Chinese culture' (Andrews 1996: 15). This group was part of a larger national movement, which opposed the wholesale adoption of Western culture (Ibid.). Yet, they believed it detrimental to ignore biomedicine and instead sought to validate the impacts of Chinese medicine via biomedicine (Ma 1995). 'Their efforts ranged across a wide spectrum from assimilation of certain western ideas into Chinese medicine to the use of biomedical knowledge to instigate total reform of Chinese Medicine' (Scheid 2001: 370–1). They attempted to standardise the teaching and practice of Chinese medicine through government licencing of schools and practitioners (Ibid.). Yet, they also dismissed many Chinese 'popular medical practices, as ignorant superstition unworthy of scholarly consideration' (Andrews 1996: 16).

The School of Merging called for a *huitong* (会同), or *synthesis* of Chinese medicine and biomedicine. Several of these 'reformers' opened their own schools (Ma 1995). 'The first of these, Liji Medical School, opened in 1885 and offered elementary western medical education courses such as anatomy, psychology, and public health' (Ibid.: 221; Chapter 47 in this volume). The school also offered a teaching hospital and a medical journal (Ibid.). Although several of these modernised Chinese medicine schools opened during this period, none were documented to have survived the 1911 Revolution (Ibid.). Nevertheless, these early attempts at a grassroots integration of Chinese medicine with biomedicine introduced the possibility of 'modernising' the practices of Chinese medicine and marked the conception of a potential future for healthcare integration into the consciousness of several influential reformers; reformers who were later to be subsumed into Mao's integration 'from above'.

Mao's healthcare integration: a means to redress rural healthcare inequities in China

Lucas (1982: 1) argues that the various political upheavals stemming from the 1949 Revolution to the Cultural Revolution did not ultimately alter the basic policies of national medicine first diffused to China in the late 1920s. 'There was little dispute in China that the Ministry of Health, established after the foundation of the People's Republic of China in 1949, was to function entirely on the basis of biomedicine' (Taylor 2005: 6). Mao's original platform for the Chinese Communist Party (CCP) was solely focussed on modernising China. During the formation of the CCP, Mao Zedong wrote his key text, *On New Democracy* (新民主主义论), published in 1940, in which he emphasised: 'We want not only to change a politically oppressed and economically exploited China, but also to change a China which has been ignorant and backward under the rule of the old culture into a China that will be enlightened and progressive' (quoted in Taylor 2001: 344). Four years later, Mao stressed: 'This type of new democratic culture is scientific. It is opposed to all feudal and superstitious ideas; it stands for objective truth and for unity between theory and practice' (Ibid.: 345). In fact, Mao singled out Chinese medicine as a 'hindering factor toward the "new democratic culture," which required "remoulding" to become part of the revolutionary movement' (Ibid.).[4]

There are numerous conflicting reports concerning what Mao thought or said about Chinese medicine depending on when he said it and to whom. However, an interesting anecdotal account is retold by his personal physician, identifying that the conflict may have been within Mao himself. 'Mao attributed China's large population to the efficacy of Chinese medicine. "For thousands of years" he told me, "our people had depended on Chinese medicine. Why were there still people who dismissed Chinese medicine? What I believe is that Chinese and Western medicine should be integrated. Well-trained doctors of Western medicine should learn Chinese medicine, senior doctors of Chinese medicine should study anatomy, physiology, bacteriology, and so on. They should learn how to use modern science to explain the principles of Chinese medicine [...] then a new medical science, based on the integration of Chinese and Western medicine can emerge. That would be a great contribution to the world [...] even though I believe we should promote Chinese medicine,

I personally do not believe in it'" (Zhisui 1994: 83–4). Throughout his biography of Mao, his physician recounts several instances in which Mao personally refuses Chinese medicines, whilst publicly promoting it and politically enforcing its unification with biomedicine.

Regardless of Mao's rhetoric, modernising healthcare in China presented a challenge for the CCP. Of the between 21,000 and 40,000 biomedical physicians practising in China in 1949, the vast majority catered to metropolitan elites; refusing to work in the poor rural areas dominating the Chinese landscape (Sidel 1973: 20). This two-tiered system resulted in a markedly increased inequity of rural healthcare that challenged the purported benefits of communism and resulted in an immediate issue for the new CCP government to address (Ibid.). Mao began to address this disparity. Interestingly, after disparaging Chinese Medicines in *On New Democracy*, Mao also states: 'If we only rely on the new medicine [biomedicine], we will not be able to solve our problems. Of course the new medicine is superior to the old medicine, but if they [the doctors of the new medicine] are not concerned about the sufferings of the people, do not train doctors to serve the people, and do not unite with the thousand old doctors of the old school in order to help them improve, then they will actually be helping the practitioners of witchcraft by callously observing the death of a large number of men. Our task is to *unite* with the old style doctors who can be used, and to help educate and remould them. In order to remould them we must first unite [with them]' (quoted in Taylor 2001: 346).

Hence, Mao's criteria for the development of a 'New Democracy' emphasised that practices, including healthcare practices, must be new (*xin* 新), scientific (*keuxue* 科学), and unified (*tuanjie* 团结) (Taylor 2001: 344). Thereby, the 'old style doctors' (旧医生) are to be modernised and *united* with the 'doctors of the new medicine' (新医), in order to improve human resources that will serve to remedy rural healthcare access; which cannot be addressed through dependence on the urban-elite-practising biomedical physicians. *On New Democracy* foreshadows many of Mao's future concerns and actions regarding rural healthcare.

What ultimately motivated Mao to promote the Chinese medicine he formerly disparaged is up for debate. Hsu (2018: 3) suggests that integrated traditional Chinese medicine (TCM) was the logical outcome of a Marxist dialectic, whereby 'progress results from the escalation between two antitheses, which leads to a great tension followed by a synthesis'. Whereas, Lei (2014) identified that the discourses around TCM represent a middle ground where Chinese scientists and intellectuals might prove a distinctly Chinese modernisation, independent of the West. However, I question if this repackaging and scientisation of the practices of Chinese medicine to suit communist tastes might not have also been employed to display the benefits of the communist system. Mao's situation in the 1950s was similar to the quandary that Castro faced in the 1990s, when what is now known as *Medicina Traditional y Naturale*, was developed in order to fulfil the promise of the Cuban communist constitution for universal healthcare, given the severe reduction of healthcare resources with the dissolution of Cuba's main trading partner, the USSR, and the concomitant increased severity of the US embargo to Cuba (see Chapter 41 in this volume, also Kadetz and Perdomo 2011). Similarly, might integration of Chinese medicine ultimately have been a means to save the CCP from losing face, particularly given the, then, rural majority of the PRC?

Speaking of integration

Mao did not speak of 'integration' at first. During the 1940s, Mao continued to stress the political slogan that 'Chinese and western medicine *should join together*' (*zhongxiyi tuanjie* 中

西医团结) (Taylor 2001: 361). But integration of Chinese medicine and biomedicine was only meant to be a first step towards Mao's final goal of *unification* (Taylor 2005). Mao sought to alter Chinese medicine via biomedicine in order to yield a new unified medicine. With unification, Mao's goal was to achieve a new form of healthcare that was neither traditional medicine nor biomedicine, but a modern Chinese hybrid meant to be greater than the sum of its parts (Ibid.).[5]

However, Mao was ultimately only able to achieve the first step of his goal of unification, i.e. integration. The process of what Mao called integration was meant to 'raise Chinese medicine to a higher' scientific level, comparable to biomedicine. It was therefore a process that was originally intended to be achieved prior to unification (Taylor 2005). From the mid-1950s, Mao specifically called for integration of biomedicine and Chinese medicine under the slogan 'Chinese and Western medicine *should be integrated*' (*zhongxiyi jiehe* 中西医结合) (Taylor 2001: 361). Mao's shift needs to be understood within the particular political context of the breakdown of Sino-Soviet relations; a political event which caused Mao to adopt a kind of practical nationalism that served to justify survival by means of China's own resources. This nationalism is reflected in Mao's statement that 'Chinese medicine is a great treasure house'; inferring that the Chinese cultural genius of the past can be mined for contemporary science.

However, it is not accurate to portray the many practices and schools of Chinese medicine throughout China as a static, hegemonic medical system of 'Chinese medicine'. Unschuld (1985: 5) identifies seven major conceptual systems, from internal sources, or adopted from foreign cultures, that influenced what is referred to as 'Chinese medicine' over a period of approximately 3,500 years and maintains that the changes which occurred in Chinese medicine did so primarily as a result of changes in sociopolitical ideology. Thus, integration in China was the standardisation of the many different schools of Chinese medicine practice along a biomedical framework.

To achieve integration, Mao employed biomedical physicians, primarily from the Ministry of Health, as well as from biomedical schools, such as the Rockefeller-based Peking Union Medical College (PUMC), to comb through the practices of Chinese medicine and remove any superstitious or spiritual elements, whilst maintaining those elements that were more aligned with biomedicine and capable of being standardised (Taylor 2005).[6] In this manner, what was to become the normative representation of healthcare integration by the WHO, could be understood as a hybridisation process of both Chinese deculturation and Eurocentric biomedical acculturation. What was developed by these biomedical physicians is what is now generically referred to (in English) as traditional Chinese medicine or TCM, commonly referred to in Chinese as *zhongyi* (中医)[7] or 'Chinese medicine', as opposed to simply 'medicine' or *yi* (医). Eventually, TCM was incorporated into hospitals and clinics throughout China and employed for any given patient in combination with biomedicine (Hsu 2018).

The confusion often surrounding China's representation of healthcare integration is that the practice of using TCM in hospitals or clinics in conjunction with biomedicine is what is being identified as integration. However, what is said to be integrated, in this instance, is itself an already integrated system of various schools of Chinese medicine practices that was standardised by biomedical physicians. Though TCM is different from the varied practices of its precursors, even the different schools of Chinese medicine were arguably more standardised than most local healing practices worldwide. Hence, it is important to acknowledge that in actuality, few local health practices in the world are as systematised or standardised, or even capable of being as standardised, as integrated TCM or its precursors.

Creating healthcare resources

Healthcare integration in China provided a means of disseminating healthcare resources to rural areas. However, the practice of TCM alone could not reduce health disparity, especially rural health disparity, without training more practitioners. Therefore, in the mid-1960s, Mao began a campaign to train community health workers – or 'barefoot doctors' (*chi jiao yi sheng* 赤脚医生), as they were called, in a combination of biomedicine, public health prevention, and TCM practices, in order to serve as a first line of healthcare practitioners in the rural areas of China, which were lacking medicine or biomedical physicians (Lucas 1982). Barefoot doctors were farmers, who were elected by their peers to train in a commune hospital for a period of three months to two years. Specifically, they were trained in preventative medicine, 'anatomy, physiology, bacteriology, pathology, environmental sanitation, and epidemiology. Some would also learn the delivery and care of pregnant mothers, and all would study acupuncture' and treatment with both traditional Chinese herbs, in addition to 40–50 Western drugs (Ha°klev 2005: 6). Thereby, the barefoot doctor was one of the first fully integrated practitioners. It should be noted, though, as Fang (2012) argues, that in actuality biomedical pharmaceuticals were predominantly disseminated by the barefoot doctors throughout rural China, due to their ease of use for barefoot doctors and the lay public. Thus, according to Fang's extensive oral histories, it is questionable if barefoot doctors, who were meant to serve as agents of integration, were in actuality, more successful in popularising biomedicines throughout rural China.

Was the primary healthcare model attributed to the PRC, actually an American model in Republican China?

Up to this point, we have made a fundamental assumption that this model for rural healthcare coverage began with Mao. However, such assumptions can be challenged if considering the influence of the West, particularly the US, on Chinese healthcare. The work of the Rockefeller Foundation (RF) in the early twentieth century was central to the dissemination of an international health episteme (Rockefeller Foundation 1917). Prior to the 1920s, foreign medical missionaries 'were the only practitioners of Western medicine to extend their reach significantly' into rural areas of China (Andrews 1996: 15). However, the RF may have, in actuality, established the groundwork for what became Mao's comprehensive primary healthcare and Barefoot Doctors programme (Yip 1982).

The RF's John Grant sought 'an organized core of a regionalized system of community healthcare' and envisioned 'health for all' through his China rural programme that predated Alma Ata by more than half a century (Grant 1919; Yip 1982: 1200). Harvard graduate C.C. Ch'en headed the health department of the *Mass Education Movement* in Tingshien in 1929 (Ibid.). As part of this programme: 'Village Health Workers and paramedics received short-term training in first aid and hygiene. These health workers were crucial in providing health information and simple curative and preventive services to their fellow villagers' (Ibid.: 1203). This creation of a network of community health systems, with the expectation for community biomedical healthcare, 'formed an important part of the medical legacy that the Communists inherited' (Ibid.).

In 1931, plans were made for rural health stations that would provide healthcare at the most local population level possible, using minimally trained village health aides (Yip 1982). By 1934, with the help of Grant, Ch'en developed a functioning health district that consisted of a district health centre encompassing administrative offices, a fifty-bed hospital, a laboratory, and classrooms for training, plus seven sub-district health stations that served more than seventy-five rural villages (Ibid.). The Rockefeller Foundation's China Rural Health Programme was initiated in 1935, but ended abruptly in 1937, due to the Japanese invasion of China (Ibid.).

Hence, it is interesting to consider whether the Rockefeller Foundation, and thereby the US, could have been the actual originators of the model for primary healthcare access in rural China that was eventually adopted by the WHO.

Volte face: the PRC in the WHO

An example of the political core of the WHO can be found in the battle for 'which China' would gain UN member state status. China was a member of the WHO from its inception in 1948. In fact, the Chinese delegate at the UN Conference on International Organization in 1945, Dr. T.V. Soong, was identified as 'the first person to suggest the founding of a single international health organisation' (Siddiqi 1995: 110). However, in 1949, with the formation of the People's Republic of China (PRC), China separated from the former Republican government of the Guomingtang, which became the Republic of China (i.e. Taiwan). The UN agencies not knowing which China to recognise as 'China' eventually recognised the Republic of China (ROC) as 'China', with the support of the US and other prominent UN donor Western nations. It was not until 1971, long after the return of full membership to the Soviet Union,[8] that the decision of 1949 was reversed and the PRC replaced the ROC as 'China' in the United Nations, as well as in the World Health Organization in 1973. In light of China's absence from the WHO for nearly the first three decades of its work, it is interesting to consider that the original idea for a single international organisation for health was, at least in part, the idea of a Chinese national.

Problematising the model adopted by the World Health Organization

The adoption of China's model for healthcare integration became an integral component of the Declaration of Alma Ata because of the popular perception of China's success in redressing rural healthcare access, and, possibly, because of Chinese nationals in positions of power at the WHO at the time. However, as discussed, what the WHO has represented as healthcare integration is not the integration of various local Chinese medical practices and biomedicine into the state healthcare system, but rather the integration of what was an already integrated system of TCM with biomedicine. The relevance of the WHO using TCM as a model for healthcare integration with biomedicine, whilst not acknowledging that TCM was itself an already integrated medical system, is that it: (a) falsely projects the idea that any non-biomedical or local healthcare practice can be successfully integrated into a national healthcare system, whilst disregarding the fact that few practices may be as systematised and standardised as TCM; (b) may thereby be facilitating the integration of only those non-biomedical practices and practitioners that have been systematised, standardised, and legitimised by biomedicine; and (c) makes it appear that integration can be achieved by any nation at any point in time, as comprehensively as the PRC appeared to do, regardless of political and socioeconomic contexts.

Basing all healthcare integration on the integration of TCM and biomedicine in Maoist China can pose other important challenges when applied to other countries, or even to China today. First, it is imperative to acknowledge that healthcare functions within a given political economy. The formation of TCM and the Barefoot Doctors programme were conceived and implemented within an egalitarian–authoritarian socioeconomic system, which may not be easily transferable into other socioeconomic systems. Second, although Mao believed that the integration of Chinese medicine could provide an opportunity to redress rural health inequities with the practitioner resources that were readily available, he had to overcome marked opposition from biomedical physicians; Chinese medicine practitioners[9]; the Ministry of Health,[10] and the early CCP; who originally sought to extinguish traditional medical practices (Lucas 1982).[11] Therefore, the rationale for the integration of the various schools of Chinese medicine into TCM could be understood more accurately as a compromise to rectify healthcare disparity within specific political and economic circumstances.

Since healthcare integration in China could not be achieved by complete consensus, coercion and force were employed (Rosenthal 1981). This was accomplished on several fronts: (a) Mao took advantage of the nationalistic fervour that had erupted in China since the Treaty of Versailles (1919) and the *May Fourth Uprising*, which 'catalysed the political awakening of a society that had long seemed inert and dormant'[12] (Meisner 1999: 17). (b) Mao's rhetoric of Chinese medicine, as a 'national treasure' to serve alongside biomedicine was more the product of a Nationalist movement – that included a revival in many of the arts of China – than of any consideration of Chinese medicine's therapeutic value (Taylor 2005). (c) Biomedical physicians were forced to engage in TCM studies. Those who resisted integration were labelled 'bourgeois', criticised, and/or exiled to distant areas (Ibid.). (d) At the onset of the Cultural Revolution, the Ministry of Health was blamed for ignoring rural health. Speaking on the eve of the Cultural Revolution, Mao condemned the Ministry of Health: 'The Ministry of Health serves only 15% of the urban population. It should be renamed the Urban Health Ministry, or the Lords' Health Ministry […] in medical and health work, put the stress on the rural areas!' (Sidel 1973: 28). Eventually, the CCP took over complete administration of healthcare in order to facilitate integration; insisting that practitioners of Chinese medicine be placed in biomedical hospitals and clinics (Rosenthal 1981; Ha°klev 2005). (e) During the Cultural Revolution, the same social pressures that had previously been directed towards biomedical physicians were now directed with even greater ferocity towards those Chinese medicine practitioners who sought to maintain autonomy and ideological separation from biomedicine. Classic Chinese medical texts were burned. Chinese medicine schools were closed. And classic Chinese medicine practitioners became the object of ridicule or physical attack and their practices and pharmacies destroyed (Scheid 2002). The governmental reduction of control and power granted to biomedical physicians and Chinese medicine practitioners, which rendered them politically impotent, offers a sharp contrast to the integration achieved in a bottom-up manner from the political agency of Āyurvedic practitioners in India at approximately the same historical moment (Leslie 1992); a democratic model that was completely overlooked by the WHO.

Finally, Mao's attempt at integration cannot really be considered *complete* healthcare integration. First, as mentioned, Fang (2012) identified the Barefoot Doctors programme was ultimately most successful in educating rural populations to utilise Western pharmaceuticals, rather than any integrated combination. Second, the extent to which non-biomedical practices are integrated in the minds of the public may reveal more about the actual extent of integration. In modern China, TCM has been employed primarily according to what has been perceived as the best interventions for particular ailments or as an adjunctive therapy with biomedicine (Scheid 2002). But, the most important fact that is often overlooked is that not only were many local practitioners *not* integrated into the state medical system, but several types of practitioners were actually prohibited from practising. For example, Ha°klev noted that though herbalists may have been eligible to become barefoot doctors, shamans, and diviners were ineligible (2005: 6). Immediately, such conditionalities – concerning who may or may not be integrated into the healthcare system – problematise the representation of Mao's model of integration as a true or complete integration of a country's plural practices and practitioners into a healthcare system. Rosenthal suggests that Mao's attempt at integration would more accurately be classified as a *selective* form of integration, as was demonstrated in the selective development of TCM, since institutional arrangements left the biomedical physicians 'utilizing combined treatment extensively, but on their own terms' (1981: 610).

And yet, despite these myriad challenges to China's representation of healthcare integration, and despite the fact that rural healthcare was being completely disassembled with the demise of the brigade unit after Mao's death, WHO chose to perpetuate and globalise this model.

Conclusion

The story of how Mao Zedong's attempts to redress rural healthcare inequity were adopted as a global model for universal healthcare access by the WHO, raises more questions than it answers. This chapter has attempted to unpack and critically examine the assumptions embedded in the WHO representation of this model. The success of the transfer of this representation may be specific to a given context. The policy and processes of healthcare integration are not shaped solely by a biomedical agenda, but are very much a product of specific political, economic, philosophical, and cultural forces that may not be replicable from one context to another. Hence, the adoption of a Chinese model for achieving healthcare for all by the WHO was a complex trajectory resulting from international and domestic political economic forces that can be traced from the early work of the Rockefeller Foundation in China to the need for Mao Zedong to demonstrate the value of communism in China through coercive and enforced resolutions to redress rural healthcare inequalities.

Notes

'Please note, portions of this chapter previously appeared in an article published by Frontiers in Human Development. Kadetz, P. (2021) 'About Face: Leveraging China's Soft Power on the World Stage', Front. Hum. Dyn., doi: 10.3389/fhumd.2021.774765'

1 Anthropologist, Bridget Jordan defines authoritative knowledge as the 'one kind of knowledge [that] gains ascendency and legitimacy' even though 'equally legitimate parallel knowledge systems exist' that people may 'move easily between' (1997: 56). A consequence of authoritative knowledge 'is the devaluation, often dismissal, of all other kinds of knowing as backward, ignorant, naive' (Ibid.). Thus, authoritative knowledge can be understood as a kind of Othering of thought.
2 At this conference, the Viceroy of Manchuria, Xi Liang, stated: 'We Chinese have believed in an ancient system of medical practice, which the experience of centuries has found to be serviceable for many ailments, but the lessons taught by this epidemic, which until practically three or four months ago had been unknown in China, have been great, and have compelled several of us to revise our former ideas of this valuable branch of knowledge' (Wu 1959: 48).
3 The Ministry of Education announced 'This Department has decided that all medical schools must provide courses of anatomy, chemistry and other sciences. Without these, one cannot study medicine' (Ma 1995: 215).
4 Interestingly, as discussed in Chapter 41 in this volume, countries have equated Chinese medical practices such as acupuncture with revolutionary movements.
5 It is interesting to note, that though the WHO acknowledges their adoption of Mao's model of integration, any mention of his intended goal of unification towards a new hybrid medical system cannot be located in the WHO literature.

6 According to Zhu Lian, an early proponent of integration in the CCP; 'Reform does not mean that Chinese medicine should drop all its original theories [⋯] instead we need to choose those areas of ancient medicine which are appropriate, drop those areas which are not appropriate, use scientific methods, sort out the experience [...] so as to improve its scientific theory' (quoted in Taylor 2001: 360).
7 Formally, TCM was referred to as *chuantong zhongyi* (传统中医) by Ma Kanwen during the 1970s. However, prior to the introduction of biomedicine from Western missionaries, there was no need to differentiate Chinese medicines from the biomedicine of the West, and medicine in China was simply called *Yiyao* (医药) or medicine. During the Republican Era and the construction of the modern Chinese state, Chinese medicine was known as *Guoyi* (国医), or 'state-sanctioned medicine'; simultaneously associating cultural nationalism with statism (Lei 2014: 110).
8 The Soviet Union was accepted into the UN and the WHO as a member state in 1956 (Brown et al. 2006).
9 For example, many Chinese medicine practitioners were called 'purists' for their rigidity in attempting to ensure the integrity of their system of Chinese medicine and they, like the Ministry of Health, emphasised the importance of maintaining boundaries to guarantee this integrity (Taylor 2005).
10 Despite CCP directives, the Ministry of Health, predominantly composed of graduates from medical colleges, such as the Rockefeller Foundation's Peking Union Medical College, continued to oppose integration of Chinese and biomedicine throughout the 1950s (Farquhar 1994; Lampton 1974; 1977). Mao attacked the Ministry for only being concerned with the urban elite, as well as for their sole focus on curative rather than preventative medicine; for over-dependence on the Soviet healthcare model; and for refusal to integrate TCM and biomedicine (Lampton 1974; Sidel 1973).
11 Though according to Farquhar, the CCP has usually been credited with continual support for Chinese medicine in order to maintain 'a coherent historiography of the party' and its influence on medical history (Farquhar 1994: 13).
12 The May Fourth uprising in 1919 was in reaction to the subsequent transfer of land in Shandong that was formerly ceded to Germany and to Japan without Chinese intervention or return to Chinese sovereignty. The May Fourth uprising also validated that a socialist revolution was possible in China (Meisner 1999).

Bibliography

Andrews, B. (1996) *The Making of Modern Chinese Medicine, 1895–1937*, PhD thesis, University of Cambridge.
——— (2014) *The Making of Modern Chinese Medicine, 1850–1950*, Honolulu: University of Hawai'i Press.
Brown, E.R. (1979) *Rockefeller Medicine Men: medicine and capitalism in America*, Berkeley: University of California Press.
Brown, T., Cueto, M. and Fee, E. (2006) 'The world health organisation and the transition from 'international' to 'global' public health, *American Journal of Public Health*, 96.1: 62–72.
Douglas, M. (1966) *Purity and Danger: an analysis of concepts of pollution and taboo*, Abingdon: Routledge.
Fang, X. (2012) *Barefoot Doctors and Western Medicine in China*, Rochester, NY: University of Rochester Press.
Farquhar, J. (1994) *Knowing Practice: the clinical encounter of Chinese medicine*, Boulder, CO: Westview Press.
Grant, J. (1919) *The Most Efficient Manner in Which the International Health Board May Accomplish Its Fundamental Purpose in China*. Midwifery Education, Midwifery Training School, Beijing (report), folder 372, box 45, series 601, RG1, Rockefeller Foundation Archives, Rockefeller Archive Center.
Ha°klev, S. (2005) *Chinese Barefoot Doctors, a Viable Model Today?* Working Paper, IDSC 11.
Hsu, E. (2018) 'Traditional Chinese medicine: its philosophy, history, and practice', in H. Callan (ed.) *The International Encyclopedia of Anthropology*, Hoboken, NJ: Wiley, pp. 1–10.Huang, Y. (2010) Pursuing health as foreign policy: the case of China, *Indiana Journal of Global Legal Studies*, 17.1: 105–46.
Jordan, B. (1997) 'Authoritative knowledge and its construction', in R. Davis-Floyd and C. Sargent (eds) *Childbirth and Authoritative Knowledge*, Berkeley: University of California Press, pp. 55–79.

Kadetz, P. (2012) *The Representation and Practice of Healthcare Integration: alterity and the construction of healthcare integration in the Philippines*, PhD thesis, University of Oxford.

—— (2013) 'Unpacking Sino-African health diplomacy: problematising a hegemonic construction', *St. Antony's International Review*, 8.2: 149–72.

—— (2014) 'Colonising safety: creating risk through the enforcement of biomedical constructions of safety', *East Asian Science, Technology, and Society*, 8.1: 81–106.

Kadetz, P. and Perdomo, J. (2011) 'Slaves, revolutions, embargoes, and needles: the political economy of Chinese medicine in Cuba', *Asian Medicine: Tradition and Modernity*, 6.1: 95–122.

Lampton, D. (1974) *Health, Conflict and the Chinese Political System*, Ann Arbor: Center for Chinese Studies, the University of Michigan.

—— (1977) *The Politics of Medicine in China*, Boulder, CO: Westview Press.

Lee, K. (2009) *The World Health Organisation (WHO)*, London: Routledge.

Lei, S.H. (2014) *Neither Donkey nor Horse: medicine in the struggle over China's modernity*, Chicago, IL: University of Chicago Press.

Leslie, C. (1992) 'Interpretations of illness: syncretism in modern Ayurveda', in C. Leslie and A. Young (eds) *Paths to Asian Medical Knowledge*, Berkeley: University of California Press, pp. 177–208.

Litsios, S. (2004) 'The Christian medical commission and the development of the World Health Organisation's primary health care approach', *American Journal of Public Health*, 94.11: 1884–93.

Lucas, A. (1982) *Chinese Medical Modernization: comparative policy continuities 1930s–1980s*, New York: Praeger.

Ma, Q. (1995) *The Rockefeller Foundation and Modern Medical Education in China, 1915–1951*, PhD thesis, Case Western Reserve University.

Magnussen, L., Ehiri, J. and Jolly, P. (2004) 'Comprehensive versus selective primary healthcare: Lessons for global health policy', *Health Affairs*, 23.3: 167–76.

Meisner, M. (1999) *Mao's China and After: a history of the People's Republic*, New York: Simon and Schuster.

Rockefeller Foundation (1917) *International Health Board Report of 1917*, Rockefeller Archive Center.

Rosenthal, M. (1981) 'Political process and the integration of traditional and western medicine in the People's Republic of China', *Social Science and Medicine*, 15A: 599–613.

Scheid, V. (2001) 'Shaping Chinese medicine: two cases from contemporary China', in E. Hsu (ed.) *Innovation in Chinese Medicine*, Cambridge: Cambridge University Press, pp. 370–404.

—— (2002) *Chinese Medicine in Contemporary China: plurality and synthesis*, Durham: Duke University Press.

Sidel, V.W. (1973) *Serve the People: observations on medicine in the People's Republic of China*, New York: Josiah Macy, Jr. Foundation.

Siddiqi, J. (1995) *World Health and World Politics: the World Health Organisation and the UN system*, London: Hurst.

Taylor, K. (2001) 'A new scientific and unified medicine: civil war in China and the new acumoxa, 1945–49', in E. Hsu (ed.) *Innovation in Chinese Medicine*, Cambridge: Cambridge University Press, pp. 343–69.

—— (2005) *Chinese Medicine in Early Communist China, 1945–1963: a medicine of revolution*, London: Routledge.

Unschuld, P. (1985) *Medicine in China: a history of ideas*, Berkeley: University of California Press.

Walsh, J.A. and Warren, K.S. (1979) 'Selective primary healthcare: an interim strategy for disease control in developing countries', *The New England Journal of Medicine*, 301: 967–74.

World Health Organisation (2008) 'China's village doctors take great strides', *Bulletin of the World Health Organisation*, 86.12: 909–988, http://www.who.int/bulletin/volumes/86/12/08-021208/en/, accessed 09/10/2019.

Wu, L.T. (1959) *Plague Fighter: the autobiography of a modern Chinese physician*, Cambridge: Heffer and Sons.

Yip, K. (1982) Science, medicine, and public health in twentieth century China health and society in China: public health education for the community, 1912–1937, *Social Science of Medicine*, 16: 1197–1205.

Zhisui, L. (1994) *The Private Life of Chairman Mao*, New York: Random House.

45
COMMUNIST MEDICINE
The emergence of TCM and barefoot doctors, leading to contemporary medical markets

Xiaoping Fang

When the Chinese Communist Party (CCP) overcame the Nationalists and established a new regime in October 1949, it inherited a dire medical situation. Medical resources were deplorably scarce and health indicators were extremely low, as were the administrative blueprint and experimental practices for improving the health of the people (Lucas 1982: 461–89; Yip 2009a: 105). The Communist regime incorporated its political ideology and social mobilisation strategies into its medical and health work in order to overcome resource and personnel constraints (Oksenberg 1974: 375–408; Perry 2007: 15; Thornton 2009: 93). In 1951–52, the government established its four general principles of health work at the National Health Work Meeting, namely 'prevention first; serve workers, peasants, and soldiers; unite Chinese and Western medicine; and combine health work and mass movements' (Wilenski, 1976: 7). These principles underlay the priorities, objectives, and organisational strategies for Communist health work, including strategies regarding Chinese and Western medicine. Throughout the state-building and modernisation processes of the second half of the twentieth century, specific features emerged for Communist medicine, notably the definition of Traditional Chinese Medicine, the promotion of barefoot doctor programmes in rural China, and the rise of national medical markets.

The emergence of traditional Chinese medicine (TCM)

Prior to the nineteenth century, the term 'medicine' (*yi* 医) could refer to any form of medicine from different medical schools and social strata in China. It also encompassed a wide range of practices, including oracular therapy; demonic medicine; religious healing; pragmatic drug therapy; Buddhist medicine and the medicine of systematic correspondence in a broad sense (Unschuld 1985: 5). In 1834, American missionary doctors arrived in Canton, which marked the beginning of Western medicine in China (Barnes 2005: 288–90; Taylor 2005: 79; Yang 2013: 24; Andrews 2014: 53). It was only at this point that the term 'Chinese medicine' (*zhongyi* 中医) first appeared, as a way of distinguishing local practices from those of the missionaries.

The advent of Western medicine in China gave rise to new ideas about 'the presence, nature, and causation of disease; appropriate therapies; and the legitimacy of native, foreign, and foreign-trained healers; the imposition of policing measures in the name of public health; the need for a particular institutional infrastructure; and the intellectual presuppositions themselves of Western medicine' (Cunningham and Andrews 1997: 14). From the early twentieth century onwards, Western medicine started to challenge the legitimacy of Chinese medicine. In 1929, the proposal for 'Abolishing Old-Style Medicine in Order to Clear Away the Obstacles to Medicine and Public Health' was passed by the first National Public Health Conference of the Nationalist government. It became the hallmark event of the legitimacy crisis for Chinese medicine in the first half of the twentieth century.

Throughout the Republican period, reform-minded Chinese medicine practitioners attempted to bring about profound institutional; epistemological and material changes in their field. These reformers strove to launch a national medicine movement, 'scientificize' Chinese medicine, and establish schools and associations (Zhao 1991: 21–37; Farquhar 1994b: 12; Xu 1997: 847–77; Lynteris 2013: 66; Yang 2013: 354; Andrews 2014: 145–84; Lei 2014:101–5). On the one hand, though Chinese medicine doctors were not deprived of their status and their legal right to practise medicine, the state was reluctant to legitimise them, and their position was precarious and continually challenged throughout the Republican period (Croizier 1968: 234). On the other hand, Chinese medicine still enjoyed great practical legitimacy. As a 1935 survey indicated, 1,182 Western-style medicine doctors were practising in Shanghai, while there were 5,477 licenced Chinese medicine physicians, not counting those who were unlicenced (Xu 1997: 847–77). In rural areas, villagers usually resorted to folk healers and professional Chinese medical practitioners (Fang 2012: 20–2).

Ironically, the situation did not change much until the mid-1950s. Though 'the unification of Chinese and Western medicine' was adopted in 1950 as one of the three health work principles at the First National Health Work Meeting with a goal of creating a single 'new medicine', the Communist government gave Chinese medicine very little administrative power within the higher echelons of the party structure (Unschuld 1985: 247; Taylor 2005: 30–1). According to *The Provisional Regulations Governing Doctors of Traditional Chinese Medicine* implemented on May 1, 1951, a Traditional Chinese Medicine doctor was not allowed to prescribe chemically compounded medicines or give injections unless he had received scientific training in medical treatment, and under no circumstances could he induce an abortion (Fang 2012: 45). In the meantime, Chinese medicine was viewed as a 'feudal society's feudal medicine that... needed to be transformed through strict controls on medical practice and reeducation of its practitioners' (Scheid 2002: 69).

According to the new licencing regulations, Chinese medicine doctors had to pass qualification examinations, which required extensive Western medical knowledge. Chinese medicine improvement schools were also established to improve Chinese medicine practitioners' political understanding and scientific techniques and disseminate theoretical and practical knowledge of Western medicine among them as part of the programme of 'Chinese medicine studying Western medicine' (Scheid 2002: 69). Lecturers at these schools were usually doctors of Western medicine who offered a strong, condensed regimen of basic biomedicine, including anatomic physiology, pathology, germs, medical history, and pharmacology (Taylor 2005: 47). Students also studied social sciences, and preventive medicine (infectious medicine and public health), and were encouraged to gradually develop towards preventive medicine and the 'scientificizing' of Chinese medicine (Scheid 2002: 69–70).

However, Chinese medicine gradually acquired legitimacy under the Communist regime. In late 1953, the Ministry of Health was criticised for its policies on Chinese medicine because the licencing and recruitment regulations for Chinese medicine doctors seriously restricted their medical practice. In July 1954, Mao put forward the idea of 'Western medicine studying Chinese medicine' in order to eradicate the boundaries between Chinese and

Western medicine and form a unified Chinese medicine. Soon various training classes of 'Western medicine studying Chinese medicine' were established throughout China. The term 'Traditional Chinese Medicine' (TCM) first appeared in 1955 (Taylor 2005: 84), but it is found only in Western-language literature. No equivalent term is applied in China, where 'Chinese medicine' (*zhongyi* 中医) remains the more proper term (Scheid 2002: 3). Meanwhile, the previous regulations on Chinese medicine doctors were abolished, the China Academy of Traditional Chinese Medicine and Chinese medicine hospitals were founded, and Chinese medicine was integrated into Westernised medical universities, colleges, schools, hospitals, and so on.

In 1956, the 'integration of Chinese and Western medicine' was first proposed by Mao and became the guiding principle from then onwards. During the Cultural Revolution, the integration of Chinese pharmaceuticals and treatment methods with Western diagnostic techniques, treatment, and pharmaceuticals was further promoted and advocated (Lampton 1977: 112; Scheid 2002: 65–88; Taylor 2005: 30–150).

In this way, Chinese medicine, now known as traditional Chinese medicine (hereafter TCM), was transformed from the marginal, sidelined medical practice it had been in the early twentieth century to an essential and high-profile aspect of the national healthcare system. The institutionalisation and standardisation of Chinese medicine in Communist China was completed by 1963, by which time it had begun to be admitted into the primary health care system (Taylor 2005: 12). During the 1960s, TCM was practised in hospitals and clinics and taught in schools. Knowledge of it was systematically recorded in textbooks, and it was divided into categories which parallel those of Western medicine (Scheid 2002: 65; Taylor 2005: 147). By the 1980s, C. C. Chen, who led a rural medical experiment in Ding County, Hebei Province, under the leadership of James Yen in the 1930s and was opposed to TCM, reluctantly admitted that 'each system has its own representation in the central government, as well as its own nationally or provincially administered urban clinics, hospitals, and medical schools… as of 1987, organisational conflict had almost entirely disappeared' (Chen, 1989: 147).

Since the 1970s, TCM has undergone four major tendencies: Westernisation; standardisation; urbanisation and globalisation. The Westernisation process was formally recognised after 1980, when the Ministry of Health listed Chinese medicine, Western medicine, and the integration of the two as the three great pillars of the Chinese medical system. The integration of TCM and Western medicine is usually regarded as a branch within TCM. However, practitioners of both TCM in general and the integration of this with Western medicine usually resort to Western medicine diagnostic techniques and prescribe Western pharmaceuticals.

Second, influenced by the Chinese Herbal Medicine Campaign and Western pharmaceuticals, Chinese patent medicine developed rapidly during the 1970s. To overcome the inconvenience of decoction of Chinese *materia medica*, raw herbal medicines became much more commonly made up into pills, liquids, syrups, and powders than before, and new forms such as granules, instant teas and capsules also came into use. Chinese medicinal products have been further standardised in terms of these extraction and production processes, packaging, dosages, and ingredients.

Third, as barefoot doctors brought new healing styles and a host of Western medicines to rural China from the 1960s onwards, villagers, in turn, developed a preference for Western medicine. Interestingly, as Farquhar found in the early 1990s, urban Chinese 'are these days much more enthusiastic users of Chinese medicine than rural people' (Farquhar 1994a: 476). Her findings were been increasingly verified during the first decade of the twenty-first century.

Last but not least, TCM, represented by acupuncture, massage and herbs, is going beyond China and is now practised in an increasing number of different settings across various continents (Zhan 2009). As Volker Scheid points out (2002: 268–9), 'after a century of struggle against domination by Western medicine, of modernisation and revolution, Chinese medicine now stands at the threshold of emergence as a truly global medicine'. However, this does not mean that Chinese medicine has a dominant position outside China, where it is still an 'alternative medicine'.

The promotion of the barefoot doctor programmes

The establishment of the state medical system in rural areas was highly significant as China had a largely rural population. The rural medical system was proposed and implemented experimentally in the 1930s by the Nanjing-based Nationalist government and the Rural Construction Movement, which was represented by C. C. Chen in Ding County, north China. In both the governmental and non-governmental blueprints and practices, at the lowest level of the organisation were village health workers. The Communist regime basically inherited these practices (Lucas 1982: 479; Chen 1989: 423; Yip 1995: 76–7; Andrews 2014: 108–11). With the beginning of agricultural collectivisation in 1952, villagers were selected to become health workers within mutual aid teams and cooperatives as part of the programme of enhancing agricultural productivity. The selection criteria for these initiatives were the possession of both basic primary educational qualifications and the right political credentials. Selected candidates were required to follow an informal training programme entailing the 'Four Principles of Health Work', which pertained to basic first aid and preventive medical treatment (Fang 2012: 27).

Starting in 1965, the Communist government launched the Socialist Education Campaign targeting the inequality in the distribution of healthcare resources, as well as rural politics and other social issues, including education. Plans were put forward to organise mobile medical service teams for rural areas and to train rural health workers in order to improve the rural medical situation. Under this programme, each production brigade was required to have two 'half-peasant, half-doctors (*bannong banyi* 半农半医)', one of whom was to be a woman who would be in charge of delivering babies. Youths with primary and middle-school education and 'good' family origins; 'correct' political thoughts and 'love for the countryside' were selected after being nominated by the masses; recommended by the association of poor peasants; approved by a party branch or commune and interviewed by a training unit. After receiving training, they returned to their own brigades where they were required to diagnose and treat a number of common diseases using their basic pharmaceutical knowledge, as well as conduct the Patriotic Health Campaigns, while participating in agricultural production (Fang 2012: 29–30). In Shanghai suburban areas, local people usually called these new health workers 'barefoot doctors' (*chijiao yisheng* 赤脚医生) as, in addition to providing villagers with basic healthcare, they also laboured barefoot in the rice paddy fields.

On September 14, 1968, an investigative report entitled 'Fostering a Revolution in Medical Education through the Growth of the Barefoot Doctors' was published in the *People's Daily*, an organ of the Central Committee of the CCP. It described the work of barefoot doctors in Jiangzhen Commune, Chuansha County, Shanghai Municipality. The concept of barefoot doctors was first introduced to the public through newspaper pieces (*Renmin ribao*, 1968a). On December 5, 1968, the same newspaper carried a report with the headline 'Cooperative Medical Service Warmly Welcomed by Poor and Lower-Middle Peasants'. This article introduced the new cooperative medical service of Leyuan Commune, Changyang County, Hubei province (*Renmin ribao*, 1968b).

As one of the 'newly emerged things' that reflected the political ideologies and rural development strategies of the Cultural Revolution, the barefoot doctors were rapidly popularised, and cooperative medical stations were set up in villages nationwide with revolutionary zeal (Sidel 1972). Villagers paid fees to form local 'cooperative medical services' to cover the costs of establishing these medical service stations, which would be presided over by barefoot doctors. When villagers sought treatment at cooperative medical stations, they were given certain services and medicines free of charge. Soon every village had at least one barefoot doctor to provide basic medical care, creating a national network of healthcare services for the very first time. Barefoot doctors formed the lowest level of a three-tiered state medical system that comprised the county, commune, and brigade levels. The health implications of the advent of the barefoot doctors lie in that they carried out the social transformation of rural medicine through the introduction of Western medicine and the marginalisation of TCM in Chinese villages across the spheres of knowledge; pharmaceuticals; healing; institutionalisation and professionalisation.

The barefoot doctor programmes changed the traditional family and apprenticeship-based forms of knowledge transmission in the villages and led to a Western-influenced medical knowledge structure among barefoot doctors themselves and those living in their villages because of the selection criteria, the appearance of unified medical textbooks, and the presence of instructors teaching Western medicine. Through barefoot doctors, Western pharmaceuticals were introduced into Chinese villages on a large scale. Meanwhile, TCM was given official legitimacy due to economic factors, though it was promoted in the name of political discourse and ideology (Fang 2012: 42–93).

Barefoot doctors developed a healing style which was also more oriented towards Western medicine from the start, due to the nature of the knowledge structure, the medical proficiency and the availability of medicines. Their practices included the use of basic modern medical instruments and the prescription of Western medicine as tablets. Villagers also formed comparative medical beliefs about Chinese and Western medicine, such as 'western medicine works quickly, Chinese medicine slowly' and 'western medicine treats (only) symptoms, Chinese medicine treats the root of the disease' and also applied different medicines to different diseases. The interactions of healing styles and medical beliefs completely changed pharmaceutical consumption in Chinese villages, a process in which TCM pharmaceuticals were quickly marginalised (Ibid.: 94–124).

The setting up of medical stations presided over by barefoot doctors not only strengthened the medical community based in each commune, but also completed a mechanism for coordinating a hierarchical medical system for the first time in rural China. Depending on the referral system, barefoot doctors extended and stratified medical encounters in villages, communes and county hospitals. During this process, medical stations and county hospitals grew, while commune clinics – the middle level of the three-tier medical system – experienced a dramatic decline. The pyramidal three-tier medical system evolved towards a dumbbell-shaped structure, with barefoot doctors replacing commune clinics and gaining dominance in the local community (Ibid.: 125–50).

Barefoot doctors gradually developed a group identity from the late 1960s by setting themselves apart from 'competitors' in the local community, including folk healers (legitimate or illegitimate) and other medical practitioners, while forging links with medical station colleagues and barefoot doctor peers. The barefoot doctors' status and respect rose steadily in part as a result of their daily interactions with patients and in part because of the rapid effects of Western medicine. The state contributed to the formation of group identity among barefoot doctors and facilitated their rise in community power relationships over villagers (Ibid.: 151–66).

In 1978, Deng Xiaoping developed the rural socioeconomic reform policies of the household responsibility system following the end of the Cultural Revolution. Their implementation included the dismantling of the people's commune system, which led to the gradual disintegration of the barefoot doctor programme. On January 24, 1985, the health minister Chen Minzhang 陈敏章 announced that the term 'barefoot doctor' would no longer be used in China: 'From now on, those barefoot doctors who reach the proficiency level of secondary technical school (*yishi* 医士) shall be called "village doctors" (*xiangcun yisheng* 乡村医生), while those who cannot reach the *yishi* level shall be called "health workers" (*weishengyuan* 卫生员)' (Chen 1985: 137).

During the changeover, medical examinations and group differentiation re-defined the medical legitimacy of barefoot doctors, and medical proficiency became a key requirement. In this sense, the disintegration had a positive impact on rural health because of the increasing professionalisation of former barefoot doctors. From then on, barefoot doctors effectively became private medical practitioners with their own clinics, though they still undertook public health work in villages. The reforms also resulted in a remarkable continuity in the provision of medical care and public health, even though rural Chinese people faced serious challenges in their efforts to access healthcare services (Rosenthal and Greiner 1982; Fang 2012: 166–76).

The basic structure of the rural medical and health system did not undergo much change from the 1980s until the implementation of the 'integrated management of rural health' medical reforms in 2008–10. County governments established new health service stations and abolished extant village clinics or merged these into health service stations. The regulations specify that these centres and stations must provide 'six-in-one' services to villagers, which encompass prevention; treatment; promotion of health and wellbeing; rehabilitation; health education and family planning advice. Meanwhile, given that the majority of current village doctors began work under the barefoot doctor programme, many will be retiring soon, so county governments started training rural community doctors to fill these forthcoming vacancies in 2009. Senior middle-school students were selected and sent to study clinical medicine for three years at medical college, and will be assigned to health service stations after graduation. As such, in the near future, the barefoot doctors of the Cultural Revolution era will completely disappear from the medical world of China's villages.

The path to the contemporary medical market

The medical market includes doctors and pharmacies. For thousands of years, China's doctors and pharmacies operated separately from one another, with the exception of the 'doctors who sit in the pharmacy (*zuotangyi* 坐堂医)', who used the premises as consulting rooms. Medical practitioners would suggest which pharmacy patient families should buy medicine from, a practice which, to some extent, exerted pressure on the pharmacy (Leung 2002: 354). The owners of these pharmacies usually had some basic medical knowledge and dispensed medicines themselves, since they did not usually hire staff or only had one or two apprentices. The shops were generally small and were usually located inside owners' homes (Zhu 2006: 243). These private Chinese medicine shops had their own medicine supply sources and limited networks, which were confined to certain geographic scopes because of transport issues (Cochran 2006: 4–8). Nonetheless, Chinese consumption of pharmaceuticals (Chinese *materia medica* and patent medicines) was quite limited. The medicine that patients bought was usually *xingjunsan* (行军散 for treatment of heatstroke, diarrhoea, stomach ache and internal heat), *biwendan* (避瘟丹 for treatment of heatstroke, acute gastroenteritis and diarrhoea), and *shayao* (痧药 for treatment of heatstroke) (Qiao 1992: 308–9).

Western medicine shops started appearing in urban and metropolitan areas of China after the late nineteenth century, while transnational pharmaceutical companies moved into the Chinese medical market and competed with newly emerged pharmaceutical companies run by Chinese. Though Chinese people could access Western medicine, especially after the Second World War, supplies were still quite limited and prices were high. For example, a bottle of penicillin was worth the equivalent of 50 kg of rice before 1949 (Shi 1992: 648). As a result, Chinese medicine shops still dominated the Chinese medical market (Fang 2012: 74).

After 1949, the Chinese medical market underwent two significant changes: the integration of doctors and pharmacies at hospitals and the formation of a new state pharmaceutical network based on the pharmaceutical networks and private medicine shops already in existence. The main purposes of these initiatives were to allocate, supply and sell both Western and Chinese pharmaceuticals to customers efficiently and economically, while meeting the growing demands of various medical and health campaigns (Ibid.: 74–7).

The establishment of the state medical system and the pharmaceutical sales system, which were supported by the development of the medical educational system and pharmaceutical industries, contributed to the rise of the Chinese medical market nationwide. In rural areas, the formation of the three-tier medical system made villager patients move from seeking treatment in the broad, mixed sector of folk remedies and healing by exorcism, divination and prayer, to local and regional hospitals. In urban areas, the implementation of free medical services and labour insurance medical services provided the state medical system with a reliable source of patients. These two factors contributed to the rise of a huge medical market in China.

As China is an agricultural country where the population is predominantly rural, the rural medical market is especially significant. County pharmaceutical companies were first set up in the early 1950s, with the aims of managing the wholesale supply of medicines within the county through rural clinic pharmacies; supply and marketing cooperatives and medicine shops, as well as some medicine peddlers. In the mid-1950s, rural pharmacies were incorporated into clinics or rural supply and marketing cooperatives. In this way, a pharmaceutical sales network was gradually established in rural areas.

After 1968, cooperative medical stations were established with the popularisation of barefoot doctors in rural China. Each was presided over by a barefoot doctor with a medical kit. The medical stations and kits extended the pharmaceutical sales network throughout rural China at a rapid pace, and were thus highly significant in the social history of medicine in Chinese villages. Meanwhile, the wholesale pharmaceutical network was further extended to the commune level. Each county's people's disease prevention and treatment hospital (formerly known as the county people's hospital) commissioned commune clinics to serve as medicine wholesalers.

Through these initiatives, prices were radically reduced, which was a crucial factor for villagers. For example, on August 1, 1969, prices for 1,230 kinds of antibiotics, sulphanilamides, fever-reducing medicines, pain-relieving medicines and other medicines were reduced by 37.2%. These products constituted about 72.1% of the total pharmaceuticals available at the time. By 1971, medicine retail prices were only one-fifth of what they had been in 1949 (*Zhongguo yiyao gongsi* 1990: 273). Antibiotics including tetracycline and terramycin became common pharmaceuticals prescribed by doctors and consumed by patients during the 1970s. These medicines were particularly effective for treating common diseases.

Because of the rise of the national medical market, pharmaceutical consumption and medicine expenditure per capita increased steadily throughout the 1970s (Fang 2012: 111–20).

The improvement in basic health indicators was very impressive, and China was promoted as model for developing countries by the World Health Organisation (WHO 1978). From the late 1970s onwards, China's economic reform had a huge impact on the medical system, pharmaceutical sales network and medical market, in both positive and negative ways as the state retreated from medicine and healthcare provision (Duckett 2010). In rural areas, the dismantling of the people's commune system in the early 1980s resulted in the disintegration of cooperative medical service stations presided over by barefoot doctors, which affected public health and medical service to different extents (White 1998; Blumenthal and Hsiao 2005: 1165–9). In urban areas, one consequence of the economic reform, especially the state-owned enterprise reform in the late 1990s, was that workers and employees lost the basic medical welfare they had enjoyed in the socialist era.

Meanwhile, the medical market was becoming increasingly commercialised. On the one hand, because of this commercialisation and marketisation, patients were able to obtain cheaper, more convenient and more effective medical services than before. Alternative medical markets also flourished, including *qigong* 气功, acupuncture, massage and *yangsheng* 养生 (nourishing life) (Farquhar 1996; Hsu 1999; Chen 2003; Farquhar and Zhang 2012). However, the pursuit of profit resulted in over-commercialisation and marketisation, which further led to a further series of problems, such as wide gaps in access to healthcare between rural and urban areas, inter-regional gaps, the over-concentration of medical resources, unaffordable prices for medical attention and a worsening of relationships between doctors and patients.

Among these problems, the overuse of pharmaceuticals and medical technologies is the most serious as the sale of pharmaceuticals as a revenue source for doctors (known as 'supporting doctors by selling medicines', *yiyao yangyi* 以药养医) became the main profit-making mode in the medical market (Farquhar 1996: 244; Fang 2012: 121). Pharmaceuticals account for 45% of China's healthcare expenditure, or 1.6% of its GDP, far above other countries, where pharmaceuticals usually account for one-quarter of total health-related spending (Organisation for Economic Cooperation and Development 2010: 272). This overuse is particularly common among village doctors, who have an incentive to overprescribe, given that they rely on medicine sales for part of their income. For nearly three-quarters of the patients surveyed, the medicine prescribed was an antibiotic, while one-fifth were prescribed two or more drugs (Ibid.: 225). Similarly, this commercialisation and marketisation also resulted in the overuse of medical technologies. The high caesarean section rates at Chinese hospitals is a typical example of this.

In view of these problems, the Chinese government has been making efforts to tackle these shortcomings of the medical market, such as by implementing the National Basic Pharmaceutical Catalogue in the New Rural Cooperative Medical Services and Basic Medical Insurance for Urban Residents to curb the overuse of pharmaceuticals. However, it remains a challenging task for the government, such as 'medical disputes' (*yi'nao* 医闹) over the recent years. Patient families physically attacked and assaulted medical doctors and hospitals due to their mistrust and suspicion towards the medical market and their reluctance to accept the results of medical treatments (Hesketh 2012).

Conclusion

As the key components of the Communist medicine, the emergence of TCM, the promotion of the barefoot doctors programme, and the rise of the national medical market were the landmark events in the social transformation of medicine in China guided and dominated by the state after 1949. During this process, the recognition of the legitimacy of TCM in the 1950s and the launching of the Chinese herbal medicine campaign in the 1960s demonstrated the state's realistic strategies and tactics towards TCM when faced with constraints on resources and personnel. This process also indicated the decisive role of the state in improving health and medicine, including the integration of doctors and medicines in the late 1950s; the implementation of the national barefoot doctor programme after 1968 and the large-scale decrease in pharmaceutical prices from the 1950s onwards.

The social transformation of medicine in China also reveals a dynamic relationship between the state and the local governments. In the post-socialist era, after 1978, the state retreated from health and medicine in terms of investment and allocation of funding, resources and personnel, as well as administrative intervention, which were relegated to local governments (Huang 2004). With the initiation of the economic reform, the Chinese medical market has been greatly commercialised, which is further complicated by the medical technology, services, and management strategies brought by globalisation. The commercialisation of the medical market not only provides Chinese with good, convenient, and effective medical services, but also brings serious challenges to the government, such as the worsening patient-doctor relationships.

Bibliography

Andrews, B. (2014) *The Making of Modern Chinese Medicine, 1850–1960*, Vancouver: UBC Press.
Barnes, L.L. (2005) *Needle, Herbs, Gods and Ghosts: China, healing and the west to 1848*, Cambridge, MA: Harvard University Press.
Blumenthal, D. and Hsiao, W. (2005) 'Privatization and its discontents – the evolving Chinese health care system', *The New England Journal of Medicine*, 353.11: 1165–9.
Chen, C.C. (1989) *Medicine in Rural China: a personal account*, Berkeley: University of California Press.
Chen, M.Z. (1985) 'Chen Minzhang tongzhi zai yijiubawunian quanguo weishengtingjuzhang huiyi shang de zongjie jianghua' 陈敏章同志在一九八五年全国卫生厅局长会议上的总结讲话 (Minister of Health comrade Chen Minzhang's summary speech at the National Health Department director meeting in 1985), January 24, 1985, in H. Ma (ed.) (1992) *Zhongguo gaige quanshu: yiliao weisheng tizhi gaigejuan* 中国改革全书: 医疗卫生体制改革卷 (China Reform: medical and health system reform), Dalian: Dalian chubanshe, pp. 135–8.
Chen, N.N. (2003) *Breathing Spaces: qigong, psychiatry, and healing in China*, New York: Columbia University Press.
Cochran, S. (2006) *Chinese Medicine Men: consumer culture in China and southeast Asia*, Cambridge, MA: Harvard University Press.
Croizier, R. (1968) *Traditional Medicine in Modern China: science, nationalism, and the tensions of cultural change*, Cambridge, MA: Harvard University Press.
Cunningham, A. and Andrews, B. (1997) *Western Medicine as Contested Knowledge*, Manchester: Manchester University Press.
Duckett, Jane. (2010) *The Chinese State's Retreat from Health: policy and the politics of retrenchment*, Abingdon, Oxon: Routledge.
Fang, X.P. (2012) *Barefoot Doctors and Western Medicine in China*, Rochester, NY: University of Rochester Press.
Farquhar, J. (1994a) 'Eating Chinese medicine', *Cultural Anthropology*, 9.4: 471–97.
––––––– (1994b) *Knowing Practice: the clinical encounter of Chinese medicine*, Boulder, CO: Westview Press.
––––––– (1996) 'Market magic: getting rich and getting personal in medicine after Mao', *American Ethnologist*, 23: 239–57.
Farquhar, J. and Zhang, Q.C. (2012) *Ten Thousand Things: nurturing life in contemporary Beijing*, New York: Zone Books.
Hesketh, T., Wu, D., Mao, L. and Ma, N. (2012) 'Violence against doctors in China', *BMJ*, 345.
Hsu, E. (1999) *The Transmission of Chinese Medicine*, Cambridge: Cambridge University Press.

Huang, Y.Z. (2004) 'Bringing the local state back in: the political economy of public health in rural China', *Journal of Contemporary China*, 13: 367–90.
Lampton, D. (1977) *The Politics of Medicine in China: the policy process, 1949–1977*, Boulder: Westview Press.
Lei, H.L. (2014) *Neither Donkey nor Horse: medicine in the struggle over China's modernity*, Chicago, IL; London: University of Chicago Press.
Leung, A.K.C. (2002) 'Mingdai shehui zhong de yiyao' (Medicines in Ming society), *Faguo hanxue* (Sinologie Française), 6: 345–61.
Lucas, A. (1982) *Chinese Medical Modernization: comparative policy continuities, 1930s – 1980s*, New York: Praeger.
Lynteris, C. (2013) *The Spirit of Selfness in Maoist China: socialist medicine and the new man*, Palgrave: Macmillan.
Ma, H. (ed.) (1992) *Zhongguo gaige quanshu: yiliao weisheng tizhi gaigejuan* 中国改革全书: 医疗卫生体制改革卷 (China Reform: medical and health system reform), Dalian: Dalian chubanshe.
Mei, Z. (2009) *Other Worldly: making Chinese medicine through transnational frames*, Durham, NC: Duke University Press.
Oksenberg, M. (1974) 'The Chinese policy process and the public health issue: an arena approach', *Studies in Comparative Communism*, 7.4: 375–408.
Organization for Economic Cooperation and Development (2010) *OECD Economic Survey China 2010*, Volume 2010/6.
Perry, E. (2007) 'Studying Chinese politics: farewell to revolution?', *The China Journal*, 57: 15.
Qiao Qiming 乔启明 (1992) *Zhongguo nongcun shehui jingjixue* 中国农村社会经济学 (Social Economics of Rural China), Shanghai: Shanghai shudian.
Renmin ribao (1968a) 'Cong chijiao yisheng de chengzhang kan yixue jiaoyu geming de fangxiang: shanghaishi de diaocha baogao' 从赤脚医生的成长看医学教育革命的方向: 上海市的调查报告 (Fostering a revolution in medical education through the growth of the barefoot doctors: an investigative report from Shanghai municipality), *Renmin ribao* (The People's Daily), September 14.
—— (1968b) 'Shenshou pinxia zhongnong huanying de hezuo yiliao zhidu' 深受贫下中农欢迎的合作医疗制度 (Cooperative Medical Service Warmly Welcomed by Poor and Lower-Middle Peasants), *Renmin ribao* (The People's Daily), December 5.
Rosenthal, M.M. and Greiner, J.R. (1982) 'The Barefoot Doctor of China: from political creation to professionalization', *Human Organization*, 41: 330–41.
Scheid, V. (2002) *Chinese Medicine in Contemporary China: plurality and synthesis*, Durham, NC: Duke University Press.
Shi Fu 石夫 (ed.) (1992) *Jinhua xianzhi* 金华县志 (Jinhua county gazetteer), Hangzhou: Zhejiang renmin chubanshe.
Sidel, V.W. (1972) 'The barefoot doctors of the People's Republic of China', *New England Journal of Medicine*, 286: 1292–300.
Taylor, K. (2005) *Chinese Medicine in Early Communist China, 1945–1963*, London: RoutledgeCurzon.
Thornton, P. (2009) 'Crisis and governance: SARS and the resilience of the Chinese body politic', *The China Journal*, 61: 23–48.
Unschuld, P. (1985) *Medicine in China: a history of ideas*, Berkeley; London: University of California Press.
White, S.D. (1998) 'From barefoot doctors to village doctor in Tiger Springs Village: a case study of rural health care transformations in socialist China', *Human Organization*, 57: 480–90.
Wilenski, P. (1976) *The Delivery of Health Services in the People's Republic of China*, Ottawa: International Development Research Centre.
World Health Organization (1978) *The Promotion and Development of Traditional Medicine*, Geneva: World Health Organization.
Xu, Xiaoqun (1997). "National essence vs. science: Chinese native physicians' fight for legitimacy, 1912–1937", *Modern Asian Studies*, 31: 847–77.
Yang Nianqun (2013) *Zaizao "bingren": Zhongxiyi chongtuxia de zhengzhi kongjian, 1832–1985* 再造"病人": 中西医冲突下的政治空间, 1832–1985 (Remaking "patients": spatial politics in the conflicts between Chinese and Western medicine, 1832–1985), Beijing: Zhongguo renmin daxue chubanshe.
Yip, K.C. (1995) *Health and National Reconstruction in Nationalist China: the development of modern health services, 1928–1937*, Ann Arbor, MI: Association for Asian Studies.

—— (2009a) 'Disease, society and the state: malaria and health care in mainland China', in K.C. Yip (ed.) *Disease, Colonialism and the State: malaria in modern East Asian history*, Hong Kong: Hong Kong University Press, pp. 103–20.

—— (ed.) (2009b) *Disease, Colonialism and the State: malaria in modern East Asian history*, Hong Kong: Hong Kong University Press.

Zhan, M. (2009) *Other-Worldly: Making Chinese medicine through transnational frames*, Durham: Duke University Press.

Zhao, H.J. (1991) 'Chinese versus western medicine: a history of their relations in the twentieth century', *Chinese Science*, 10: 21–37.

Zhongguo yiyao gongsi 中国医药公司 (China Pharmaceutical Company) (ed.) (1990) *Zhongguo yiyao shangye shigao* 中国医药商业史稿 (The History of Pharmaceutical Commerce in China), Shanghai: Shanghai shehui kexueyuan chubanshe.

Zhu Deming 朱德明 (2006) 'Jindai Hangzhou zhongyaodiantang gouchen' 近代杭州中药店堂钩沉 (The History of Chinese Medicine Shops in Modern Hangzhou), *Zhonghua yishi zazhi*, 36.4: 243–5.

46
CONTESTED MEDICINES IN TWENTIETH-CENTURY CHINA

Nicole Elizabeth Barnes

The apparent ubiquity of Traditional Chinese Medicine (TCM) clinics around the world can mislead one into thinking that Chinese medicine has always occupied a privileged place in Chinese society and is therefore an obvious cultural export. In fact the opposite is true: TCM exists because practitioners of Chinese medicine lost so much cultural prestige that the very survival of their profession came to depend upon the support (and control) of the state. This process occurred slowly over several decades of the late nineteenth and twentieth centuries, in response to both domestic concerns and international pressures. TCM, a 1955 creation of the Communist state, is the culmination of that process (Taylor 2005; Lei 2014), and was discussed in the previous chapter (Chapter 45 in this volume). It is the current manifestation of a cumulative and dynamic body of knowledge that has undergone alternately subtle and profound changes for over two millennia.

This chapter tells the story of dramatic transformations in Chinese medical practice across the twentieth century, the central component of which was the shift in medical legitimacy from Chinese medicine to scientific medicine that originated in the West but became indigenised in China. The primary motivation that spurred the adoption and incorporation of Western medicine was the state's drive to modernise, which peaked at several key moments in the late nineteenth and twentieth centuries under four different governments in response to both internal tensions and external exigencies. The fact that scientific medicine held its appeal in the late Qing dynasty (1644–1911), during the warlord period (1916–27), under Nationalist Party rule (1927–49), and in the People's Republic of China (1949–present) speaks to the centrality of science in modern statehood.

One of the chief factors in the making of a modern medicine is the involvement of the state (Porter 1999). The state has played a dominant role in the construction of a modern medical profession in societies around the world. State action has generally included the creation of a standardised medical school curriculum, administering and revoking licences to practise, and requiring that doctors record infectious disease morbidity and mortality statistics. State interest in regulating the medical profession has tended to arise around the same time that the state begins to enforce health regulations and manage issues of public hygiene, such as drinking water, public toilets, and municipal trash removal. These twin processes of regulating professional medicine and public health are part of the creation of a modern nation-state's political sovereignty, which entails the power to determine the parameters of life and death for all its subjects (Foucault 1995; Mbembe 2003). Thus the history of modern medicine is also the history of the modern state, in China as in many other countries (Rogaski 2004).

Before we enter the twentieth century, several key points need to be clarified. First, when foreign missionaries opened their first medical clinics in China in the early nineteenth century, Chinese medicine was in the middle of a long battle for legitimacy between elite, learned physicians and practitioners of folk medicine. Although the introduction of Western medicine did not instigate this battle, its new presence in the medical marketplace made it an important tool in the ongoing struggle. Second, foreign missionaries did not and could not offer a superior healing system until they had been in China for over a century, by which time many Chinese were also contributing to modern scientific medicine as researchers, scientists, physicians, nurses, and midwives. Third, the impetus to adopt Western medicine came from the Chinese themselves, and it came at a moment when modern Japanese medicine offered a much more compelling model for China than did the medicine of Western missionaries (Andrews 2014; Lei 2014; Barnes 2018). Although the presence of Western medical institutions within China at this point did make an impact, it was not the most important causative factor in the shifts in medical practice.

The limitations of the English language mean that everything from the learned tradition based on ancient medical texts and centuries of practice by family lineages, to folk remedies that incorporate the spirit world and attribute illness to spiteful demons and unsettled ghosts falls under the rubric of 'Chinese medicine'. Much more precision exists in the Chinese language, wherein different terms denote the treatments derived by everyone from the scholar-physician (*ruyi* 儒醫) to the travelling salesman of herbal concoctions and bedside remedies (*lingyi* 鈴醫). Elite physicians who entered the profession either through heredity (having been born into a medical family) or by circumstance (for example, having failed the civil service examinations and wishing to apply years of study to another lucrative endeavour) constantly bemoaned having to compete for patients with itinerant peddlers, temple healers, blind masseurs, and 'granny' midwives. They also chafed at the general lack of respect granted them as specialists in a scholarly tradition whose roots extended back to the *Yellow Emperor's Inner Canon* (*Huangdi neijing* 黃帝內經), reportedly written in the second century BCE. Not surprisingly, then, scholarly physicians expended much energy attempting to explain the superiority of their medical knowledge and treatments vis-à-vis those of illiterate 'quacks'. Yet these so-called 'quacks' treated the majority of the population, for whom the powers of demons and deities explained illness far more compellingly than did abstract theories of *yin* and *yang* and the five agents. And so the battle continued for centuries, with folk medicine commanding the bodies of the masses and elite medicine commanding the minds of the few.

This battle was ongoing when, in 1834, the American missionary Peter Parker (1804–88) opened China's first establishment of missionary medicine in Canton (Guangzhou). Parker named his institution the Ophthalmic Hospital and focussed on cataract surgery, a simple procedure with dramatic effects since it restored clear sight to the nearly blind in a matter of minutes. An ordained Presbyterian minister with degrees in medicine and theology from Yale University, Parker hoped that the sudden change in vision would lead to an equally rapid change of heart in his patients, who might choose to associate the efficacy of the surgical procedure with the loving grace of the Christian God and convert to Christianity. Yet despite the popularity of Parker's clinic, it made a small impact on as vast and populous a country as China, where most people first sought medical assistance from the gods and monks at nearby temples and had no reason to believe that foreign gods might be more powerful than their own. Likewise, when Jennerian vaccination was first introduced to China in the first decade of the nineteenth century, it co-existed alongside the indigenous practice of variolation (developed in China in the sixteenth century) for over a century rather than replacing it (Zhao 1991; Andrews 2014: 42).

Not long after Parker opened his hospital in Canton, British gunboats secured a victory against the Qing court in the first Opium War (1839–42), the concluding treaty of which forced China to cede Hong Kong to the British as a colony and to grant foreign merchants and missionaries free access to five cities along the southern coast, including Canton. After the conclusion of the Second Opium War (1856–60, also known as the Arrow War), the victorious foreign powers won the right to occupy the northern city of Tianjin, dividing it between the Chinese city and the foreign concessions. The Treaties of Tianjin also legalised the opium trade, to the great consternation of the Qing court, whose officials had tried for decades to resist imports of the drug. After these two wars, the Qing government initiated the so-called Self-Strengthening Movement (1861–95), which included thorough reforms of the educational, penal, and diplomatic systems, as well as the creation of arsenals and dockyards where new equipment was built for the army and navy. Other aspects of Western science interested Chinese reformers in this period, but Western medicine offered no apparent advantage over indigenous practices. Although foreign missionaries had gained access to the new treaty port cities and established many more hospitals and clinics in the latter half of the nineteenth century, the majority of their patients came to them either as a last resort, treating their medicine as one tool among many to preserve health, or as an only resort, in the case of very poor patients who patronised missionary establishments because their services were free. In short, medical missionaries treated thousands of people and even managed to convert some to Christianity, but they did not bring about any fundamental change in Chinese medical beliefs or practices. This was partly because their medicine, being quite similar to that of the Chinese, offered little therapeutic advantage, and partly because in their eagerness to attract patients they adapted their practices to Chinese culture so well that their medicine ceased to look novel (Rogaski 2004; Renshaw 2013; Andrews 2014).

This situation slowly began to change at the close of the nineteenth century, which ended with an even more humiliating battle. In the First Sino-Japanese War (1894–95), China fought against Japan over diplomatic control of Korea. While the Japanese did not necessarily possess technological superiority, their victory in this war seemed to many Chinese to serve as proof of Japan's success at modernisation (Elman 2006). After centuries of treating the Japanese as their younger brothers, some Chinese began to see them as teachers and mentors. Looking to Japan for a model of how to apply Western technologies to an Asian society, Chinese reformers gained increasing interest in adopting Western medicine via their Asian neighbour, where from the 1870s onwards Western medicine had served as the foundation of medical institutions of the new Meiji state. A high percentage of the estimated 20,000 Chinese who studied abroad in Japan during this period studied Western medicine (Harrell 1992). Japan's subsequent victory against Russia in the Russo-Japanese War of 1904–05 confirmed this sense that Japan had something to teach China about surviving in a world of competitive empires.

After this sea-change in diplomatic power in East Asia, the twentieth century dawned on a China covered in the blood of Christian converts, foreign missionaries, and rebelling farmers from the parched and barren lands of the north-eastern province of Shandong. A series of severe droughts convinced these farmers that the intrusion of foreign gods had so angered the native gods that they refused to favour the fields with the blessing of rain. By 1900 they had banded together by the tens of thousands and launched what is now called the Boxer Uprising, so named after their 'boxing' style of martial arts which they believed would grant them immunity to bullets. Once the Boxers had killed thousands of Chinese converts and begun attacking missionaries, foreign governments sought their revenge. Eventually, the Qing court asked for help in suppressing this uprising, and the Eight Allied Foreign Armies arrived from Japan, Britain, France, Russia, Austria-Hungary, Germany, Italy, and the United States. Still nurturing anger at the loss of their compatriot missionaries, the eight armies overstayed their welcome in the imperial capital of Beijing and ransacked the Forbidden City and the Qing Emperor's Summer Palace before setting up a more permanent base in the nearby treaty port city of Tianjin (Hevia 2003). Representatives from the eight countries established the Tianjin Provisional Government, which ruled the city's concession territories from 1900 to 1902 and employed its own hygiene police, who exacted harsh and often arbitrary punishments against Chinese who failed to comply with their laws (Rogaski 2004). The association between Western medicine and Western imperial power was now firm and undeniable. Meanwhile, medical missionaries continued to operate hospitals and clinics which they had opened thanks to unequal treaties signed after bloody confrontations, but where they worked hard to associate their medical treatments with the love of a benevolent Christian God. For some Chinese, the first association was more salient, while the latter was more so for others. This often depended on the context in which a person first encountered the foreign medical practice.

One more event affirmed the connection between Western medicine and state power more generally. In 1910, an outbreak of pneumonic plague struck the north-eastern Chinese region known as Manchuria. The homeland of the Manchus who ruled the Qing court, Manchuria abutted Korea and Siberia and included territory that many countries claimed as their own. The failing Qing dynasty, which indeed met its end the following year, lacked the ability to assert unilateral power over the region. Facing an epidemic that killed 60,000 people within a single year, the Qing rulers called in an overseas Chinese doctor from Malaya, Wu Lien-teh (Wu Liande) 伍連德 (1879–1960), who had been among the first Chinese people to study Western medicine abroad (in England, France, and Germany) (Summers 2012). At the time, Dr Wu was Director of the Imperial Army Medical College in Tianjin, and he possessed all the right credentials to use the power of the microscope to confirm the identity of this peculiar plague (which is much more virulent than bubonic plague), discern its transmission route, and devise successful methods of treatment and prevention. While Chinese medicine doctors did successfully treat some individuals, for the most part their treatments proved woefully inadequate, and they failed to understand the importance of enforcing strict quarantine. When contrasted with the experience of Chinese medicine doctors who contracted pneumonic plague from their patients and died, Wu's laboratory work in Manchuria put China on the map as a leader in scientific medical research for the first time. More importantly for our story, this scientific medical authority won political control over the region for the Chinese state, but cost Chinese medicine the privilege of defining health and disease. A new disease category of *chuanran bing* 傳染病—infectious diseases—arose out of this experience and fundamentally changed the way that disease and its prevention were conceptualised in China. Now laboratory scientists had the power to define a disease, and the theories and experience of Chinese medicine doctors ultimately became irrelevant (Lei 2011).

These events—the two Opium Wars, the First Sino-Japanese War, the aftermath of the Boxer Uprising, and the Manchurian Plague Epidemic—were all far more important than was the work of medical missionaries in pushing the Chinese state towards adopting the standards and practices of Western medicine. Moreover, the dramatic changes in Western medicine itself through this period—the adoption of germ theory and the use of science to confirm disease cases and design prevention plans—ensured that it became an attractive option to a state increasingly interested in safeguarding the lives of its people. The revelation that scientific medicine had safeguarded not just Chinese lives but also Chinese sovereignty in Manchuria, coupled with the knowledge that powerful countries like Japan and England had instituted state medicine, solidified the idea that medicine must be a cornerstone of China's modernisation.

At this critical juncture in Chinese history, the long-term battle between elite physicians and folk healers merged with the new dynamics of imperial diplomacy and state building. Each party in this drama had always possessed a certain kind of influence: scholar physicians had the authority of the written word and cultural prestige, while folk healers had the power of popular patronage, which granted them easy access to the majority of patients. Even warlords during the period of political fracture between the fall of the Qing dynasty in 1911 and the stabilisation of the new Republic in 1927 used hygiene and public sanitation measures as a means of advertising their modernity and willingness to break with tradition (Stapleton 2000).

Under the Nationalists, the state took an active interest in medicine but wanted that medicine to be of a particular variety. Most importantly, it had to be effective in fighting contagious disease, and only scientific medicine had proven its mettle at that game. Chinese reformers, too, believed that scientific medicine would strengthen their country and enable it to fend off foreign imperialism. A series of institutions were founded, each one acknowledging only Western scientific medicine as a legitimate healing system: the North Manchurian Plague Prevention Service (1912), the Chinese Medical Association (1915), the Central Epidemic Bureau (1919), the Ministry of Health (1928), the Central Hygienic Laboratory (1929), the National Quarantine Service (1930), and multiple municipal Bureaus of Public Health, many of which were deliberately modelled after that of the colonial Tianjin Provisional Government (Yip 1995).

Alongside these central state institutions, in 1914 the Rockefeller Foundation's China Medical Board established the Peking Union Medical College (PUMC, Xiehe yixueyuan 協和醫學院) in Beijing. Designed to be the 'Johns Hopkins' of China, the PUMC launched the careers of twentieth-century China's leading health professionals and public health officials. In fact, when the Nationalist government founded the new Ministry of Health (later renamed the National Health Administration) in 1928, state officials tapped PUMC faculty member John B. Grant, Director of the Department of Public Health, for his recommendations. Dr Grant named all of his favourite students, each of whom was thereby catapulted into a lifelong career in state medicine (Bullock 2011). Since all of these individuals had been trained in Western scientific medicine in an American system with only English language instruction, they believed Western medicine to be superior to their own, and some used their newfound state power to suppress Chinese medicine.

One such individual was Yu Yan 余巖 (style name Yunxiu 雲岫, 1879–1954). Although not educated at PUMC, Yu had studied Western medicine in Japan at Osaka Medical University, where the link between Western scientific medicine and Japan's new status as an imperial power seemed crystal clear. In 1929, at the inaugural meeting of the Nationalist government's Health Commission, Yu sponsored the 'Proposal to Abolish Old Medicine in order to Remove Obstacles to Medicine and Hygiene'. With the goal of abolishing Chinese medicine altogether, this bill required doctors of Chinese medicine to take courses in Western scientific medicine, and outlawed schools of Chinese medicine, reports on Chinese medicine in medical journals, and all advertisements for Chinese medicine products or practitioners (Croizier 1968; Andrews 2014).

It is important when speaking of medicine not to lose sight of the people who used it. In addition to the international politics explained above, family matters and intimate heartaches shaped the relationship between the two medical systems. Chen Zhiqian 陳志潛 (1903–2000) (also known as C.C. Chen), often revered as the father of public health in China, went to PUMC to study Western scientific medicine after the seventh death from illness in his immediate family (Chen 1989; Barnes 2019). An earlier critic of Chinese medicine, Yu Yue 俞樾 (1821–1907) had written his attack against his country's indigenous medical practices in 1890 after his wife and children died of disease. Forty years later, as Yu Yan was formulating his proposal, this essay was still popular (Zhao 1991).

Because he was not alone in believing Chinese medicine inferior, Yu Yan did not foresee the strength of popular resistance to his proposal. Supporters and practitioners of Chinese medicine organised petitions and parades, established the first national organisation for their profession, and successfully lobbied the government to repeal the motion. In the following year (1930), a few powerful individuals established a new government agency, the Institute of National Medicine (*Guoyiguan* 國醫館, INM). With the dual aim of gaining state support for and scientising Chinese medicine, the INM became the most important organisation for the perpetuation of Chinese medicine, albeit in a highly reformulated version (Lei 2014).

While Chinese medicine had co-existed peacefully alongside Western medicine for nearly a century, the 1929 proposal introduced a new confrontation that threatened Chinese medical theory and practice more than Western medicine had ever done. A new generation of Chinese people preferred scientific medicine and had recruited state power to their cause, forcing the Chinese medical community to respond or face extinction. The battle moved to the level of the state, where the Institute of National Medicine served as the sole representative voice for Chinese medicine. Although local beliefs changed much more slowly, and many practitioners continued to diagnose and treat their patients according to the traditional theories, all who wished to practise legally had to conform to standards dictated by Western scientific medicine, regardless of which medical system they used (Lei 2014).

The 1929 proposal and the reaction that it inspired catapulted Chinese medicine into its modern era, one in which the fundamental tenets of the medicine shifted from indigenous theories and knowledge to the values and practices of modern science. Both Yu Yan and his adversaries had received inspiration from Japan, where vigorous Westernisation in the late nineteenth century had inspired a robust revitalisation movement for Sino-Japanese medicine in the early twentieth century (Andrews 2014). Both movements—the indigenisation of Western medicine and the scientisation of Chinese medicine—were modern phenomena, intimately attached to the aforementioned geopolitical events that pushed the country towards becoming a modern nation-state. In Japan, revitalisation of Sino-Japanese medicine entailed scientific studies on herbal medicines to demonstrate their efficacy in the terms of laboratory science (Andrews 2014: 85–8). This inspired the Chinese to do similar work that continued through the 1950s when it served as a cornerstone of the new TCM created by the Communist state.

China's National Medicine movement introduced a new term, *guoyi* 國醫 or 'national medicine'. Proponents of this movement argued that Chinese medicine should be preserved and maintained as a fountainhead of cultural essence, even as they altered it so much as to make it nearly unrecognisable. Their invention of the phrase 'national medicine' was also an expression of their ultimate goal of tying Chinese medicine to the modern nation-state, as a means of counteracting the privileged position of Western scientific medicine. This agenda appealed to those who wished to modernise their country without losing their own culture, and therefore harked back to the late nineteenth-century mantra of 'Chinese essence, Western utility' (*Zhongti xiyong* 中體西用). Since it underscored the importance of Chinese medicine to the project of strengthening the country and building the modern nation-state, the phrase *guoyi* became very useful rhetoric in the ensuing war.

In July 1937, the Imperial Japanese Army began a full-scale invasion of China that launched an eight-year war known in China as the War of Resistance against Japan (*Kangri zhanzheng* 抗日戰爭, 1937–45). When Japanese soldiers conquered the Nationalist government's capital Nanjing in December 1937 and perpetrated the infamous Rape of Nanking, the entire government fled inland along the Yangtze river, first to Wuhan and later to Chongqing. This latter city in Sichuan served as the nation's capital from October 1938 to the end of the war in August 1945. When the Pacific War began in December 1941 and China's war with Japan became embroiled in the Second World War, Chongqing served as command headquarters for the China-Burma-India theatre of the Pacific theatre of the Second World War.

Meanwhile, the Civil War between the Nationalist army and Communist guerrillas continued despite two United Front alliances, and the Communist base at Yan'an in northwestern China was another centre of anti-Japanese resistance. Throughout this period, when the country was divided into three zones controlled by the Imperial Japanese Army, Chiang Kai-shek's Nationalist government, and the Communists respectively, medicine's political significance was all the more salient. All three parties used medicine and public health as a means of demonstrating their right to control the population under their dominion, and simultaneously to signal their modernity to the rest of the world (Watt 2014). For example, Nationalist health officials worked very hard to clean up Chongqing and make it look like a respectable capital city of a legitimate and powerful nation. They abolished the traditional methods of carrying human waste out of the city to sell to farmers who transformed it into nightsoil fertiliser, and instituted their own transit system (which soon failed). They enforced regular trash pick-up and built rubbish bins throughout the city. They forced beggars off the streets and into asylums where they were required to work for their keep. More strident measures were enforced during the annual epidemic season, when large public gatherings were forbidden, no vendors could sell cold drinks or cut fruits, and the military police could be called in to assist with compulsory vaccinations in a stricken neighbourhood (Barnes 2018).

Since the advocates of 'national medicine' had got started in 1929, throughout the war they were able to use the institutional power of the INM, which moved to Chongqing ahead of the central government, to preserve a place for Chinese medicine in this game of accruing political and symbolic power. The INM sponsored laboratory research on Chinese medicines, regular meetings of research groups, branch societies throughout Nationalist controlled areas, a factory for the production of Chinese medicines, and a Chinese medicine hospital in the wartime capital of Chongqing. The Institute did all of this as an official agency of the Nationalist government, thereby granting the approval of the state to Chinese medicine despite pressure from other agencies to abolish its practice altogether. Such work achieved a dual aim of securing a place for Chinese medicine within the modern state health administration, and cleansing it of its associations with folk medicine and religion. In this way, the institutional success of the INM marked an important stage in the centuries-long battle between elite Chinese medicine and folk Chinese medicine, and firmly tied the former to state power as well as the cultural prestige of modern science. This component of our story illustrates the extent to which Western medicine was a tool in this local battle for medical legitimacy and social power.

The War of Resistance sparked several important changes in Chinese medicine. The national emergency left the Nationalist government and Communist guerrillas begging for increased services and therefore unable to block Chinese medicine practitioners from obtaining government licences and operating legally. In Chongqing they in fact received more licences than did their counterparts in scientific medicine. This meant that less than a decade after health officials had attempted to outlaw the practice of Chinese medicine altogether, it not only remained more ubiquitous than Western medicine but was also operating with the full blessing of the state.

Another impact of the war was geographic. The Nationalist government's move to Sichuan province placed central state institutions in the heartland of Chinese medicine cultivation and marketing. Beginning in the seventeenth century, Chongqing served as the most important port through which products from the East moved West, and vice versa. Herbal, mineral, and animal medicines produced on the Tibetan plateau and in Sichuan and Yunnan provinces regularly moved through Chongqing. Physicians and pharmacists schooled in Chinese medicine also far outnumbered people trained in Western medicine within the city, even after the influx of refugees from the occupied eastern seaboard. The move to the southwest thus gave Chinese medicine a leg-up and helped its supporters to maintain legitimacy.

A third sea-change occurred after the Japanese seizure of Burma and Dutch Indonesia in the spring of 1942. Britain's loss of its colony of Burma forced the closure of the Burma Road, which had been the major land route for bringing all foreign products into China, including medicines. When the Japanese took over Indonesia, the Allied powers lost access to ninety per cent of the world's supply of quinine, the principal prophylactic and treatment drug for malaria. Since malaria was a primary killer among soldiers, the loss of quinine sparked even more research into Chinese medical products that might fully or partially replace imported medicines. Research on *Dichroa febrifuga* (Ch. *Changshan* 常山) and other drugs used to treat malaria was of particular importance in this period (Lei 2014).

Communist authorities in the northwest also preferred scientific medicine and believed that much of folk medicine was 'feudal superstition'. However, they realised that most medical professionals throughout the country were practitioners of Chinese medicine, and during both the War of Resistance and the Civil War they compromised on their principles in order to maximise health services (Soon 2020; Taylor 2005; Watt 2014). The disruption of the Japanese invasion thus provided supporters of Chinese medicine with an opportunity to solidify their relationship with the state and build a modern profession. This last stage in affirming the power of elite medicine over popular medicine culminated in the Communist state's creation, in 1955, of Traditional Chinese Medicine. Despite its name, TCM is anything but traditional. Particularly because it took place in an era of foreign invasion and imperialism, Chinese medicine adopted many new attributes in this final step of becoming modern that signalled a much stronger Chinese state that possessed renewed interest in controlling the populace. Whereas traditional practices kept medical authority in the hands of the doctors themselves, and allowed for the preservation of medical secrets within specific lineages, TCM was a state medical system and required all physicians to attend state-accredited medical schools, study a standardised curriculum with acupuncture terminology that adhered more closely to state institutions (the military) than to ancient medical philosophy, and pass government tests in order to receive a licence to practise. These procedures gave the state much more authority over Chinese medical practice than it had previously had, severely undermined the legitimacy of medical lineages, and fundamentally altered physicians' social networks and means of prestige (Taylor 2005; Scheid 2007).

The global success of TCM in non-Chinese societies from Sydney to Seattle signals the depth of state involvement in its creation. Chinese medicine as practised over thousands of years, in semi-secretive family practices whose elders transmitted knowledge to apprentices, did not allow the state sufficient access to control the population and thereby solidify its political sovereignty. The history of wresting power from Chinese medicine practitioners, then, cannot be divorced from the making of the modern Chinese nation-state. At the same time, because the modern Chinese nation came into being in an era of imperialism, foreign medical standards accrued more political power than did indigenous ones, and after decades of contention Western scientific medicine came to dominate medical practice when its supporters recruited state power to their cause. By the early to mid-twentieth century this medicine's persuasive power relied on more than the politics of imperialism or the activism of Chinese Western-style doctors, as sulpha drugs and later antibiotics proved to be the strongest weapons against contagious diseases, but gunboats and unequal treaties first granted its practitioners access to Chinese bodies. Therefore, domestic concerns and international pressures combined to place a tremendous force on Chinese medicine, and the only way that it could survive into the twentieth century was to transform into a state medicine created in government boardrooms and scientific laboratories. Scientific medicine gained full entrance into Chinese medical practice, both as one strand of the current medical tradition within China, and as the standard against which indigenous medicine would thereafter be measured.

Bibliography

Andrews, B. (2014) *The Making of Modern Chinese Medicine, 1850–1960*, Vancouver: University of British Columbia Press.

Barnes, N.E. (2018) *Intimate Communities: Wartime healthcare and the birth of modern china, 1937–1945*, Berkeley: University of California Press.

——— (2019) 'Serving the people: Chen Zhiqian and the Sichuan provincial health administration, 1939–1945', in D. Luesink, W. Schneider and Zhang Daqing (eds) *China and the Globalization of Biomedicine*, Rochester, NY: University of Rochester Press, pp. 215–231.

Bullock, M.B. (2011) *The Oil Prince's Legacy: Rockefeller philanthropy in China*, Stanford, CA: Stanford University Press.

Chen, C.C. (1989) *Medicine in Rural China: a personal account*, Berkeley: University of California Press.

Croizier, R.C. (1968) *Traditional Medicine in Modern China: science, nationalism and the tensions of cultural change*, Cambridge, MA: Harvard University Press.

Elman, B. (2006) *A Cultural History of Modern Science in China*, Cambridge, MA: Harvard University Press.

Foucault, M. (1995) *Discipline and Punish: the birth of the prison*, 2nd edn., trans. A. Sheridan, New York: Random House Books.

Harrell, P. (1992) *Sowing the Seeds of Change: Chinese students, Japanese teachers, 1895–1905*, Stanford, CA: Stanford University Press.

Hevia, J. (2003) *English Lessons: the pedagogy of imperialism in nineteenth-century China*, Durham, NC: Duke University Press.

Lei, S. Hsiang-lin (2011) 'Sovereignty and the microscope: constituting notifiable infectious diseases and containing the Manchurian plague (1910–11)', in A.K.C. Leung and C. Furth (eds) *Health and Hygiene in Chinese East Asia: policies and publics in the long twentieth century*, Durham, NC: Duke University Press, pp. 73–106.

——— (2014) *Neither Donkey nor Horse: medicine in the struggle over China's modernity*, Chicago, IL: University of Chicago Press.

Leung, A.K.C. and Furth, C. (eds) (2011) *Health and Hygiene in Chinese East Asia: policies and publics in the long twentieth century*, Durham, NC: Duke University Press.

Luesink, D., Schneider, W. and Zhang Daqing (eds) (2014) *China and the Globalization of Biomedicine*, Rochester, NY: University of Rochester Press.

Mbembe, A. (2003) 'Necropolitics', *Public Culture*, 15.1: 11–40.

Porter, D. (1999) *Health, Civilization, and the State: a history of public health from ancient to modern times*, London: Routledge.

Renshaw, M. (2013 (2005)) *Accommodating the Chinese: The American hospital in China, 1880–1920*, Abingdon: Routledge.

Rogaski, R. (2004) *Hygienic Modernity: meanings of health and disease in treaty-port China*, Berkeley: University of California Press.

Scheid, V. (2007) *Currents of Tradition in Chinese Medicine, 1626–2006*, Seattle, WA: Eastland Press.
Soon, W. (2020) *Global Medicine in China: a diasporic history*, Stanford: Stanford University Press.
Stapleton, K. (2000) *Civilizing Chengdu: Chinese urban reform, 1895–1937*, Cambridge, MA: Harvard University Press.
Summers, W.C. (2012) *The Great Manchurian Plague of 1910–1911: the geopolitics of an epidemic disease*, New Haven, CT: Yale University Press.
Taylor, K. (2005) *Chinese Medicine in Early Communist China, 1945–63: a medicine of revolution*, New York: Routledge Press.
Watt, J.R. (2014) *Saving Lives in Wartime China: how medical reformers built modern healthcare systems amid war and epidemics, 1928–1945*, Leiden: Brill.
Yip, Ka-che (1995) *Health and National Reconstruction in Nationalist China: the development of modern health services, 1928–1937*, Ann Arbor, MI: Association for Asian Studies.
Zhan Mei (2009) *Other-Worldly: making Chinese medicine through transnational frames*, Durham, NC: Duke University Press.
Zhao Hongjun (1991) 'Chinese versus western medicine: a history of their relations in the twentieth century', *Chinese Science*, 10: 21–37.

47
PUBLIC HEALTH IN TWENTIETH-CENTURY CHINA

Tina Phillips Johnson

Introduction

Public health focusses on the health of populations rather than on individual treatment of patients, and it designates a particular set of practices and theories about the body, health, and disease. The pathogenic theory of medicine (germ theory) developed by Louis Pasteur and Robert Koch in the mid-1800s maintains that germs are the cause of disease and illness, and that disease and illness can be prevented with proper aseptic, or sterile, methods (Tomes 1997). Public health was at the forefront of medical science in Europe and North America after the First World War, based on the promises of health and hygiene for all. These ideas spread worldwide as colonisers grappled with previously unknown tropical diseases and health challenges of living in new and differing climates (Arnold 1993). In China, medical missionaries since the nineteenth century had come to cure disease and save souls, and they established hospitals and training programmes for medical and paramedical professions like nursing and midwifery (Balme 1921; Liu 1991; Johnson 2009). In the Republican era (1911–37), international government and philanthropic organisations devoted funds and personnel to China, focussing on disease prevention founded on improved nutrition and hygiene widely drawn from the programme developed at Johns Hopkins University School of Public Health in Baltimore (Lucas 1982; Weindling 1997). Chinese officials, modernisers, and intellectuals all weighed in on the importance of public health grounded in biomedical science. These public health efforts shifted in the late 1930s to focus on military medicine and epidemic prevention during the Sino-Japanese War (1937–45) and the Chinese Civil War (1945–49). The founding of the People's Republic in 1949 brought continuations of earlier public health efforts with important differences in structure and method. Reforms after Mao Zedong's death in 1976 included intensified focus on family planning and the decentralisation of medical services. Improving the health of China's vast and diverse population has been a primary challenge throughout the twentieth century and beyond.

Early Republic (1911–27)

The Republican era in China ushered in a proliferation of ideas regarding nation, science, modernity, and economics, together with the publication and dissemination of an amazing array of popular journals and newspapers on public health (Dikötter 2008). Members of the public and the late Qing government alike acknowledged the necessity of public health interventions with a serious outbreak of pneumonic plague in Manchuria in 1910 (Andrews 2014; Lei 2015). Cambridge-educated physician Wu Liande (Wu Lien-teh) 伍連德 (1879–1960) created the North Manchurian Plague Prevention Service that effectively halted the disease by cremating bodies of victims, enacting strict quarantine codes and introducing the public wearing of surgical masks, actions that contradicted traditional Chinese beliefs about behaviour during outbreaks of infectious disease (Lei 2015: 21–44). This event prompted local and national governments to attend to the health of the Chinese people, and in 1916 the Ministry of the Interior began regulating sanitation and epidemic prevention services (Yip 1995: 15–16). However, because of recent humiliations at the hands of foreign powers, many Chinese were ambivalent about what kinds of ideas and inventions to take from the West. Tension between nationalism and modernity arose as China searched for its own identity while aiming to make itself more like developed countries (Esherick, 1999). In the first decades of the twentieth century, warlords built hospitals, local notables and politicians donated money, and the Guomindang, or Nationalist, government passed laws, all to improve the health of China's citizens who would form the foundation of the new nation. Famous modernisers like Liang Qichao 梁啓超 (1873–1929), Chen Duxiu 陳獨秀 (1879–1942), Lu Xun 魯迅 (1881–1936), and Hu Shi 胡適 (1891–1962) heralded science as the saviour of China and promised a linear progression towards modernity and its attendant prosperity and good health (Kwok 1965; Lee 2000). Writers of fiction included tragic scenes of traditional Chinese medical care and looked towards modern biomedicine to heal China's population (Lao 2005; Lu 2009). China's 'sick man of Asia' image resulting from international humiliations could be remedied through public health initiatives like street sweeping, night soil removal, and sanitary house inspections (Dong 2003; Yu Xinzhong 2011; Hanson 2020). The spread of many parasitic and infectious diseases in China, like water-borne cholera, typhoid, and schistosomiasis; and insect-borne malaria, typhus, and bubonic plague; was halted or limited by a clean water supply and a clean environment (Stapleton 2009). Reproductive health provision hit at the very root of China's health problems by treating diseased infants and sickly mothers. It promised to eliminate the 'sick man of Asia' by creating a healthy and robust citizenry beginning at conception (Johnson 2011).

Chinese intellectuals and political leaders reshaped their cities and governments with important public health reforms, establishing hospitals and health initiatives (Stapleton 2000; Shao 2004). The New Policies resulting from the Boxer Rebellion in 1900 included attempts to make China more civilised (*wenming* 文明) by creating a martial citizenry, a modern police force, and urban sanitation programmes (Tsin 1999; Rogaski 2004). In Beijing, poor administration and lack of a strong infrastructure in the 1910s and 1920s created serious public health problems: public urination and littering were rampant, and night soil collectors were unregulated. Beijing municipality incorporated public health administration into the duties of its local police force to regulate public health and sanitation projects including public toilets, street cleaning, ventilation systems, slaughterhouses, and the sewage system (Strand 1989). Guangzhou's progressive Bureau of Health, established in 1912 and run by Western-trained Chinese physicians, established vaccination and sanitation campaigns and set up local dispensaries and public health stations (Yip 1995: 16). As a foreign concession, Shanghai had one of the earliest comprehensive municipal public health systems in China (MacPherson 1987). Foreign missionaries and philanthropic organisations also continued to play a role in public health initiatives. However, despite the efforts of individual city planners, there were few public health departments or government-run institutions countrywide to deal with issues like sanitation, inoculation, and disease prevention. Many public health programmes existed only on paper.

Nanjing decade (1927–37)

Soon after the Guomindang relocated China's capital to Nanjing in 1927, the previously proposed National Board of Health (*weishengbu* 衛生部, renamed the National Health Administration *weishengshu* 衛生署) was founded in 1928. This department, under the newly formed Ministry of the Interior, aimed to suppress communicable diseases, licence physicians and pharmacists, regulate drugs, oversee hospitals, and improve the collection of vital statistics (Yip 1995: 15). The fledgling Ministry of Education administered school hygiene programmes, and the Ministry of Agriculture and Commerce supervised industrial hygiene. The National Midwifery Board oversaw midwifery training and regulation. By this time, plans were in place for a tiered healthcare system with advanced provincial hospitals, mid-tier hospitals and clinics, and lower-level health centres. In the 1930s, Chiang Kai-shek's New Life Movement, a social welfare programme, encouraged short hair, tooth brushing, and physical exercise and hygiene. The movement also enacted bans on opium smoking and public spitting and urination. Public health advocates called for fundamental changes in Chinese living and eating habits, such as eliminating the family bed and promoting the use of individual eating utensils in place of traditional family-style dining (Dirlik 1975; Yen 2005).

Although progressive on paper, in fact Guomindang support of health projects was minimal and the Ministry of Health weak. Chiang Kai-shek had posted unqualified and disinterested officials in charge of the Ministry (Chen and Bunge 1989: 61). Western philanthropic organisations like the League of Nations-Health Organisation (LON-HO) and the Rockefeller Foundation's International Health Division stepped in to fill the need for improved medical care and public health policy by providing consulting support, cash, and personnel. These two organisations formed a partnership in the 1920s to tackle public health problems left in the devastating wake of the First World War (Bashford 2006). The Rockefeller Foundation's Dr John B. Grant (1890–1962) established and headed the premier public health institution in China, Peking Union Medical College's Department of Public Health, which was modelled on the programme developed at Johns Hopkins University School of Public Health in Baltimore, Maryland, USA. Grant developed urban and rural health stations, aided the creation and policymaking of the National Health Administration, and trained public health professionals (Bu 2014). These graduates staffed all levels of municipal and national public health agencies, clinics, and hospitals. One of Grant's students and colleagues, Dr Yang Chongrui (Marion Yang) 楊崇瑞 (1891–1983), helped to reform reproductive health and childbirth through midwifery training programmes, public health visiting nurses, and classes for new and expectant mothers (Johnson 2011).

Rural areas too relied on foreign philanthropy and individual initiatives to reduce poor health. Experimental counties were established to improve the well-being of poor rural inhabitants. They were staffed with health workers whose duties included maternal and child health, vaccinations, and submitting birth and death reports (Johnson 2011: 125–66). The Dingxian 定縣 (Ding County) Mass Education Movement was established in 1929 by Yan Yangchu (James Yen) 晏陽初 (1890–1990) and funded by the Milbank Memorial Fund, individual philanthropists, and the Rockefeller Foundation's China Medical Board and Peking Union Medical College (Hayford 1990). The Central Field Health Station outside Nanjing, run by Harvard-educated Liu Ruiheng (J. Heng Liu) 劉瑞恆 (1890–1961), trained public health personnel, and sponsored public health campaigns, health exhibits, public lectures, pamphlets, posters, lantern slides, and mobile medical units. Other rural initiatives like Jiangningxian 江寧縣 and Qinghe 清河 Experimental District undertook similar work.

Through government regulations and restrictions, the Guomindang attempted – and failed – to eliminate what they deemed to be superstitious and unscientific traditional Chinese medicine (Andrews 2014; Lei 2015). Modern biomedicine often conflicted with the more holistic Chinese paradigm of the body and its relation to the cosmos, society, the mind, health, and disease (Cunningham and Andrews 1997). It also required that patients be attended to by (often young and inexperienced) strangers with unfamiliar dress and even more unusual ideas, methods, and tools. Rather than using interventionist methods to radically change health practices, many public health programmes resorted to only educating the populace and publicising public health methods and local facilities (Johnson 2011: 125–66).

Wartime (1937–49)

The Japanese invasion of China in 1937 circumscribed public health projects that were well established or underway throughout China. Most of the medical schools and health agencies relocated to unoccupied areas in the southwestern and northwestern parts of the country, and many healthcare facilities and personnel were dedicated to the war effort. New health agencies were established in the Guomindang wartime capital of Chongqing, as well as in Sichuan, Kunming, and other remote regions, and several relocated offices and facilities remained after the war. Medical personnel from China and abroad aimed to control epidemics both within the military and among the public in areas devastated by war. They engaged in vaccination and epidemic control campaigns, and established rudimentary urban and rural healthcare systems (Liu 2014). These components remained fundamental to the post-war public health infrastructure. Thus, rather than thwarting the fledgling public health movement in China, the Second Sino-Japanese War spread it southward and westward (Barnes 2018; Brazelton 2019; Watt 2013).

While the Guomindang struggled unsuccessfully to regain control of China at the end of the Japanese War in 1945, the Chinese Communist Party (CCP) implemented intensive public health projects to improve the health of rural inhabitants and popularise support for the CCP (Minden 1979: 310). Health policies in Chinese Communist Party-controlled areas like Yan'an 延安 (in Shaanxi province) and Shaan-Gan-Ning (designating the border regions of Shaanxi, Gansu, and Ningxia provinces) were a crucial part of their success. In these areas, public health programmes incorporated traditional Chinese medicine, especially *materia medica*, combined with modern biomedical techniques like vaccinations and aseptic childbirth (Taylor 2005). The CCP set up short-term training programmes for paramedical personnel, primarily traditional midwives or young boys to establish medical cooperatives in their villages. They provided free midwifery training for both existing traditional and new midwives in 'people's schools', cooperative training programmes administered by the people (Minden 1979: 306–7). In 1944, Mao Zedong asserted that '[w]e must call on the masses to arise in struggle against their own illiteracy, superstitions and unhygienic habits. …This approach is even more necessary in the field of medicine . . . Our task is to unite with all intellectuals, artists and doctors of the old type who can be useful, to help them, convert them and transform them' (Mao 1967: 85–6). Mao also supported the integration of Chinese and biomedicine in Yan'an with the 'scientification of Chinese medicine and popularization of Western medicine' (Taylor 2005: 14).

However, necessity, not only ideology, drove the movement to utilise traditional Chinese medical practitioners, as there were not enough modern physicians to serve this large rural area. Wartime efforts to improve public health may have given the CCP an even greater basis on which to build its medical system, in terms of working with limited resources in a fragmented and unstable situation.

The early People's Republic (1949–76)

After the founding of the People's Republic of China (PRC) in 1949, the Chinese Communist Party made the health of the people a top priority. Article 40 of the 1949 Chinese People's Political Consultative Conference states that 'National physical culture shall be promoted. Public health and medical work shall be expanded and attention shall be paid to the protection of the health of mothers, infants and children' (Hillier and Jewell 1983: 66). Following the Soviet model, the new Ministry of Health, formed in 1954, declared its purpose in the form of three principles: serve the workers, peasants and soldiers; emphasise preventive medicine; and integrate traditional Chinese medicine and biomedicine. The goal of these efforts was to eliminate infectious diseases of poverty. Hygiene and Anti-Epidemic Stations and departments of public health in medical universities focussed on preventive care in the five hygiene areas of labour, radiation, environment, school, and food (Lee 2004).

Beginning with the Patriotic Health Campaign in 1952, the ministry implemented mass mobilisations with various hygiene targets like spitting; mosquitoes, flies, rats, and sparrows (replaced later by bedbugs); and street cleaning (Hillier and Jewell 1983; Core 2014). Anti-schistosomiasis campaigns in southern regions aimed to eradicate the freshwater snails that serve as a vector of the helminthic parasite (Yu 2014; Gross 2016; Zhou 2020). The ministry also established maternal and child health centres and held sanitation and vaccination campaigns against common infectious diseases. These efforts led to marked reductions in maternal and infant mortality as well as in the infectious diseases of kala-azar, malaria, cholera, typhus, plague, and smallpox (Hillier and Jewell 1983: 157–8; Goldstein 1998; Lee 2004; Yip 2009). Magazines like *New Women of China* (*Xin Zhongguo funü* 新中国妇女), the official publication of the All-China Women's Federation, conveyed news and information about campaigns aimed at improving the health of the population, like the annual International Children's Day celebrations and a 1950 anti-tetanus exhibition in Beijing in which a banner was hung exhorting its viewers to 'Struggle to raise the Chinese population' (Goldstein 1998: 158). The Federation sponsored maternal and child health campaigns, and the publication carried stories about modern birth techniques and invectives on the dangers of traditional childbirth.

Like housing and food allotment, healthcare was controlled by one's work unit (*danwei* 單位) in the early PRC. A tiered healthcare structure integrated traditional Chinese medicine and Western biomedicine. During the Great Leap Forward (1959–61) that aimed to push China into modern industrialisation, healthcare was decentralised from state control to the commune level, so that each commune ran its own health centre, which was responsible for medical supplies, mass health campaigns, and individual patient care (Taylor 2005: 116–8). Urban healthcare workers were sent to the countryside to staff these centres and learn from the masses, referring more difficult cases to the higher-level county hospitals. Mobile medical teams travelled to more remote areas for preventive work and patient care. The health system was restructured again in 1965 after Mao accused the Chinese Ministry of Health of ministering to only fifteen per cent of China's population (Mao 1974: 232–3). He was concerned that the focus on cadre privilege, urban care, and high-level medical research meant that only those with money and connections could access quality healthcare, and those less fortunate were left without. Health centres and Rural Cooperative Medical Systems were established nationwide, and training of paramedical personnel began in communes or brigades, while urban healthcare workers were again relocated to rural areas. Localised forms of cooperative medical systems had been established with brigade health auxiliaries as early as 1955. These programmes that combined traditional Chinese medicine with Western pharmaceuticals went nationwide in 1965 (Zhu et al. 1989: 431–2).

In rural China in the 1970s, auxiliary health personnel called Barefoot Doctors covered many aspects of public health including immunisation, sanitation, and maternal and child health, though their responsibilities varied widely depending on local conditions (Fang 2012). In factories and urban areas, their counterparts were Red Worker Doctors, Red Medical Workers, or Red Guard Doctors, who had been given several days or weeks of formal training and continuous on-the-job experience from the local health station doctor (Sidel 1972). Staff midwives, often recruited from among traditional midwives, were retrained in modern childbirth methods in three- to six-month courses. These paramedical workers were responsible for publicising mass health campaigns and were especially important in propagating and recording contraception use, for example, the 'one pregnancy, one live birth; one live birth, one healthy child' movement in the 1950s; and the 'late, spaced, and few' campaign to marry later, space children several years apart, and have fewer children in the 1970s. By the early 1980s, as China's family planning policies became more stringent, Barefoot Doctors were trained to perform vasectomy and female sterilisation (Croll et al. 1985; Greenhalgh and Winckler 2005; White 2009). These medical staff also propagated and advocated new national guidelines on the protection of the health of mothers and infants (Hu and Zhang 1982). As in the Republican era, national policies and local practices continued to diverge considerably in the PRC.

Reform era (1976–2000)

Deng Xiaoping's economic reforms in the 1980s dismantled public health structures and privatised healthcare, creating a market-based medical system. This resulted in considerable inequities in healthcare access and quality in both urban and rural settings, as well as the re-emergence of previously controlled infectious diseases like malaria (Zhu et al. 1989; Yip 2009). The New Rural Cooperative System in the 1990s attempted to address these problems, with expanding individual insurance coverage and subsidised medical care (Xu et al. 2009; Babiarz et al. 2010). Local and international non-governmental organisations and philanthropies like World Vision, the China Medical Board, and the Ford Foundation have also helped to fill social welfare gaps (Saich 2000; Thompson and Lu 2006).

The focus of public health has also changed with China's burgeoning economy. Infectious diseases and ailments associated with malnutrition and poverty have been greatly reduced. However, other infectious diseases like tuberculosis, HIV/AIDS, and sexually transmitted diseases have increased, attributed to factors like migration to urban centres, drug use, the demographic shortage of women, and a rise in commercial sex (Wang et al. 2008; Uretsky 2016). In addition, emerging infectious diseases, including strains of swine and avian flu, tuberculosis, severe acute respiratory syndrome (SARS), and the 2019 coronavirus (COVID-19) are especially problematic in dense urban areas. Furthermore, chronic diseases that are often associated with affluence and an ageing population – cancer, diabetes, obesity-related ailments, and chronic pulmonary and vascular diseases – are of greater concern (Popkin 2008). Environmental degradation and pollution from rapid industrialisation affects quality of air, water, and food supply and is implicated in rising rates of cancer and birth defects (Johnson and Wu 2014). China's Ministry of Health (renamed the National Health and Family Planning Commission in 2013) has expanded its methods of disease surveillance and reporting, and the government continues to struggle with environmental policies (Wang *et al.* 2008).

Conclusion

Characterising public health efforts and their effects over a century is difficult, not only because of China's vast geographic and demographic variations, but also due to the country's enormous political and social upheavals. However, some characteristics of the state of the health of China's population can be measured relatively well, indicating that some efforts to improve health have been successful. China's total mortality rate dropped from 20/1000 in 1949 to 6.43/1000 in 2001 (Lee 2004). The average life expectancy at birth increased from an estimated thirty years in 1900, to thirty-five years in 1949, to 71.4 years in 2000. The infant mortality rate dropped from 200/1000 in 1949 to 28.4/1000 in 2000, while the maternal mortality rate decreased from 150/10,000 in 1949 to 50.2/100,000 in 2000 (Lee 2004). Most of the rise in life expectancy and decrease in mortality rates took place in the first decades of the newly established People's Republic of China between 1950 and 1980, as the government implemented important public health reforms including clean water, sewage treatment, and sanitation efforts; infectious disease eradication initiatives; and development of a basic medical system that included immunisation campaigns and improved prenatal care (Wang 2011). However, in the twenty-first century, China continues to face serious public health challenges with increased urbanisation, environmental pollution, and an ageing population, along with a medical workforce and infrastructure that are not equipped to handle these changes (Babiarz *et al.* 2010; Babiarz *et al.* 2013). Furthermore, although the government has improved health insurance rates, access to affordable quality healthcare is not universal, and significant disparities remain (Xu *et al.* 2009).

Bibliography

Andrews, B. (2014) *The Making of Modern Chinese Medicine, 1850–1950*, Honolulu: University of Hawai'i Press.

Andrews, B. and Bullock, M.B. (eds) (2014) *Medical Transitions in Twentieth-Century China*, Bloomington, Indiana: Indiana University Press.

Arnold, D. (1993) *Colonizing the Body: state medicine and epidemic disease in 19th-century India*, Berkeley: University of California Press.

Babiarz, K.S., Miller, G., Yi, H., Zhang, L. and Rozelle, S. (2010) 'New evidence on the impact of China's new rural cooperative medical scheme and its implications for rural primary healthcare: multivariate difference-in-difference analysis', *British Medical Journal*, 341.7779: 929.

Babiarz, K.S., Yi, H. and Luo, R. (2013) 'Meeting the health-care needs of the rural elderly: the unique role of village doctors', *China & World Economy*, 20.3: 40–60.

Balme, H. (1921) *China and Modern Medicine: a study in medical missionary development*, London: United Council for Missionary Education.

Barnes, N.E. (2018) *Intimate Communities: wartime healthcare and the birth of modern China, 1937–1945*. Oakland: University of California Press.

Bashford, A. (2006) 'Global biopolitics and the history of world health', *History of the Human Sciences*, 19.1: 67–88.

Brazelton, M.A. (2019) *Mass Vaccination: citizens' bodies and state power in modern China*, Studies of the Weatherhead East Asian Institute, Columbia University, New York: Cornell University Press.

Bu, Liping (2014) 'John B. Grant: public health and state medicine', in B. Andrews and M.B. Bullock (eds) *Medical Transitions in Twentieth-Century China*, Bloomington: Indiana University Press, pp. 212–26.

Chen, C.C. and Bunge, F.M. (1989) *Medicine in Rural China: a personal account*, Berkeley: University of California Press.

Core, R. (2014) 'Tuberculosis control in Shanghai: bringing health to the masses, 1928-present', in B. Andrews and M.B. Bullock (eds) *Medical Transitions in Twentieth-Century China*, Bloomington: Indiana University Press, pp. 126–45.

Croll, E., Davin, D. and Kane, P. (eds) (1985) *China's One-Child Family Policy*, New York: St. Martin's Press.

Cunningham, A. and Andrews, B. (eds) (1997) *Western Medicine as Contested Knowledge*, Manchester; New York: Manchester University Press.

Dikötter, F. (ed.) (2008) *The Age of Openness: China before Mao*, Berkeley; Los Angeles: University of California Press.

Dirlik, A. (1975) 'The ideological foundations of the New Life movement: a study in counterrevolution', *Journal of Asian Studies*, 34.4: 945–80.

Dong, M.Y. (2003) *Republican Beijing: the city and its histories*, Berkeley: University of California Press.

Esherick, J. (1999) 'Modernity and nation in the Chinese city', in J. Esherick (ed.) *Remaking the Chinese City: modernity and national identity, 1900–1950*, Honolulu: University of Hawai'i Press.

Fang, X. (2012) *Barefoot Doctors and Western Medicine in China*, Rochester, NY: University of Rochester Press.

Goldstein, J. (1998) 'Scissors, surveys, and psycho-prophylactics: Prenatal health care campaigns and state building in China, 1949–1954', *Journal of Historical Sociology*, 11.2: 153–83.

Greenhalgh, S. and Winckler, E.A. (2005) *Governing China's Population: from Leninist to neoliberal biopolitics*, Stanford, CA: Stanford University Press.

Gross, M. (2016) *Farewell to the God of Plague: Chairman Mao's campaign to deworm China*, Oakland: University of California Press.

Hanson, M. (2020) 'From sick man of Asia to sick uncle Sam', *Current History*, 119.818: 241–4.

Hayford, C.W. (1990) *To the People: James Yen and village China*, New York: Columbia University Press.

Hillier, S.M. and Jewell, J.A. (1983) *Health Care and Traditional Medicine in China, 1800–1982*, London: Routledge & Kegan Paul.

Hu, X.-J. and Zhang, B.-J. (1982) 'Women's health care', *American Journal of Public Health*, 72: 33–5.

Johnson, T.P. (2009) 'Yang Chongrui and the First National Midwifery School: childbirth reform in early twentieth-century China', *Asian Medicine*, 4.2: 280–302.

—— (2011) *Childbirth in Republican China: delivering modernity*, Lanham, MD: Lexington Books.

Johnson, T.P. and Wu, Y.-L. (2014) 'Maternal and child health in nineteenth- to twenty-first-century China', in B. Andrews and M.B. Bullock (eds) *Medical Transitions in Twentieth-Century China*, Bloomington: Indiana University Press, pp. 51–68.

Kwok, D.W. (1965) *Scientism in Chinese Thought, 1900–1950*, New Haven, CT; London: Yale University Press.

Lampton, D.M. (1972) 'Public health and politics in China's past two decades', *Health Services Report*, 87.10: 895–904.

Lao She 老舍 (2005) *Camel Xiangzi*, bilingual edn, trans. Shi Xiaojing, Hong Kong: The Chinese University Press.

Leavitt, J.W. and Numbers, R.L. (eds) *Sickness & Health in America: readings in the history of medicine and public health*, Madison: University of Wisconsin Press.

Lee, L. (2004) 'The current state of public health in China', *Annual Review Public Health*, 25: 327–39.

Lee, L.O. (2000) 'The cultural construction of modernity in urban Shanghai: some preliminary explorations', in W. Yeh (ed.) *Becoming Chinese: passages to modernity and beyond*, Berkeley; Los Angeles; London: University of California Press, pp. 31–61.

Lei, S.H.-L. (2015) *Neither Donkey nor Horse: medicine in the Struggle over China's modernity*, Studies of the Weatherhead East Asian Institute, Chicago, IL: University of Chicago Press.

Leung, A.K.G. and Furth, C. (eds) *Health and Hygiene in Chinese East Asia*, Durham, NC: Duke University Press.

Liu, C. (1991) 'From san gu liu po to "caring scholar": the Chinese nurse in perspective', *International Journal of Nursing Studies*, 28.4: 315–24.

Liu, M.S. (2014) 'Epidemic control and wars in Republican China (1935–1955)', *Extrême-Occident*, 37: 111–40.
Lu Xun (2009) 'Medicine', in J. Lovell (trans.) *The Real Story of Ah-Q and Other Tales of China: the complete fiction of Lu Xun*, London; New York: Penguin Classics, pp. 37–45.
Lucas, A. (1982) *Chinese Medical Modernization: comparative policy continuities, 1930s – 1980s*, New York: Praeger Publishers.
MacPherson, K.L. (1987) *A Wilderness of Marshes: the origins of public health in Shanghai, 1843–1943*, Hong Kong and New York: Oxford University Press.
Mao Zedong 毛澤東 (1967) 'On the united front in cultural work', in *Selected Readings*, Peking: Foreign Language Press, vol. 3, pp. 185–7.
—— (1974) 'Directive on public health, 26 June 1965', in S. Schram (ed.) *Chairman Mao Talks to the People: talks and letters: 1956–1971*, New York: Pantheon Books, pp. 232–3.
Minden, K. (1979) 'The development of early Chinese Communist health policy: health care in the border region, 1936–1949', *American Journal of Chinese Medicine*, 7.4: 299–315.
Popkin, B.M. (2008) 'Will China's nutrition transition overwhelm its health care system and slow economic growth?', *Health Affairs*, 27.4:1064–76.
Rogaski, R. (2004) *Hygienic Modernity: meanings of health and disease in treaty-port China*, 1st edn, Berkeley: University of California Press.
Saich, T. (2000) 'Negotiating the state: the development of social organizations in China', *The China Quarterly*, 161: 124–41.
Shao, Q. (2004) *Culturing Modernity: the Nantong model, 1890–1930,* Stanford, CA: Stanford University Press.
Sidel, V. (1972) 'The barefoot doctors of the People's Republic of China', *New England Journal of Medicine*, 286.24: 1292–300.
Stapleton, D.H. (2009) 'Malaria eradication and the technological model: The Rockefeller Foundation and public health in East Asia', in K. Yip (ed.) *Disease, Colonialism, and the State: malaria in modern East Asian history*, Hong Kong: Hong Kong University Press, pp. 71–84.
Stapleton, K. (2000) *Civilizing Chengdu: Chinese urban reform, 1895–1937*, Cambridge, MA: Harvard University Press.
Strand, D. (1989) *Rickshaw Beijing: city people and politics in the 1920s*, Berkeley: University of California Press.
Taylor, K. (2005) *Chinese Medicine in Early Communist China, 1945–63: a medicine of revolution*, London; New York: RoutledgeCurzon.
Thompson, D. and Lu, X. (2006) *China's Evolving Civil Society: from environment to health*, issue 8, Washington, DC: Woodrow Wilson International Center for Scholars.
Tomes, N. (1997) 'The private side of public health: sanitary science, domestic hygiene, and the germ theory, 1870–1900', in J.W. Leavitt and R.L. Numbers (eds) *Sickness & Health in America: readings in the history of medicine and public health*, Madison: University of Wisconsin Press, pp. 506–28.
Tsin, M. (1999) *Nation, Governance, and Modernity in China: Canton 1900–1927*, Stanford, CA: Stanford University Press.
Uretsky, E. (2016) *Occupational Hazards: sex, business, and HIV in post-Mao China*, Stanford, CA: Stanford University Press.
Wang, F. (2011) 'The future of a demographic overachiever: long-term implications of the demographic transition in China', *Population and Development Review*, 37, Supplement: 173–90.
Wang, L., Wang, Y., Jin, S., Wu, Z., Chin, D.P., Koplan, J.P. and Wilson, M.E. (2008) 'Emergence and control of infectious diseases in China', *The Lancet*, 372.9649: 1598–605.
Watt, J.R. (2013) *Saving Lives in Wartime China: how medical reformers built modern healthcare systems amid war and epidemics, 1928–1945*, Leiden: Brill (China Studies, Book 26).
Weindling, P. (1997) 'Philanthropy and world health: the Rockefeller Foundation and the League of Nations Health Organisation', *Minerva*, 35: 269–81.
White, T. (2009) *China's Longest Campaign: birth planning in the People's Republic*, Ithaca, NY: Cornell University Press.
Xu, K., Saksena, P., Fu, X.Z.H., Lei, H., Chen, N. and Carrin, G. (2009) *Health Care Financing in Rural China: new rural cooperative medical Scheme*, no. 3, Geneva: World Health Organization Department of Health Systems Financing.
Yeh, W. (ed.) (2000) *Becoming Chinese: passages to modernity and beyond*, Berkeley; Los Angeles; London: University of California Press.

Yen, H. (2005) 'Body politics, modernity and national salvation: the modern girl and the New Life movement', *Asian Studies Review*, 29: 165–86.

Yip, K. (1995) *Health and National Reconstruction in Nationalist China: the development of modern health services, 1928–1937*, Ann Arbor, MI: Association for Asian Studies.

⸺ (2009) 'Disease, society, and the state: malaria and health care in mainland China', in K. Yip (ed.) *Disease, Colonialism, and the State: malaria in modern East Asian history*, Hong Kong: Hong Kong University Press, pp. 103–20.

Yu Xinzhong (2011) 'The treatment of night soil and waste in modern China', in A.K.G. Leung and C. Furth (eds) *Health and Hygiene in Chinese East Asia*, Durham, NC: Duke University Press, pp. 51–72.

⸺ (2014) 'Epidemics and public health in twentieth-century China: plague, smallpox, and AIDS', in B. Andrews and M.B. Bullock (eds) *Medical Transitions in Twentieth-Century China*, Bloomington: Indiana University Press, pp. 91–105.

Zhou, X. (2020) *The People's Health: health intervention and delivery in Mao's China, 1949–1983*, Montreal; Kingston; London; Chicago, IL: McGill-Queen's University Press.

Zhu, N.S., Ling, Z.H., Shen, J., Lane, J.M. and Hu, S.L. (1989) 'Factors associated with the decline of the Cooperative Medical System and barefoot doctors in rural China', *Bulletin of the World Health Organization*, 67.4: 431–41.

48
ENCOUNTERS WITH LINNAEUS? MODERNISATION OF PHARMACOPOEIA THROUGH BERNARD READ AND ZHAO YUHUANG UP TO THE PRESENT[1]

Lena Springer

The translation of traditional Chinese *materia medica* knowledge into the context of modern biomedicine and botany has involved complex layers of interlinked chemical, political, social, and linguistic processes. This has taken place in multiple sites, from pharmacies, clinics, and research labs to herbaria, trading centres, and the Chinese hinterland, as well as policy centres and mass-education campaigns. The study of these processes thus involves not only the reading of pre-modern literary Chinese works on *materia medica*, and modern botanical, pharmaceutical, and regulatory literature, but also fieldwork interviewing pharmacists, traders, and manufacturers at urban and rural sites around China. In compiling this chapter, I have drawn on my own fieldwork among suppliers of medicinals in mountainous Western China since 2012, mainly in East Tibet, Sichuan province, in Guizhou province, and along the Qinling mountain range. I performed fieldwork in Kaili City first in 2005 and have returned regularly (Springer forthcoming). My fieldwork in Sichuan began 2012 with a focus on East Tibet (Springer 2015), focussed on Derge County in 2015 and continued among traders continued among traders in Chengdu City as well as the mountains around Neijiang City in 2019. Since 2014, Taibai *materia medica* has extended my perspective from the Southwest towards the Daoist Northwest in China (Springer 2019). Each ethnographic site has fostered my inclusion of textual study into the narratives, life-worlds and material culture of the apothecaries and scientists, part-time producers, and ritual masters I encounter. The history, science, and local literature of *materia medica* are strongly present in contemporary culture of medicinal supply in China.

Making 'Chinese' medicines, like making medicines anywhere, involves an ongoing social-scientific process of selection, testing, and preparation whereby materials and things obtain medical efficacy. Communication about the production and properties of these medicines constitutes a case of produced knowledge and of its migration in the global long twentieth century via localities, print media and domains of emergent disciplines and professions. This chapter summarises how influential scientists and scholars in modern times have chosen what they recognise as 'Chinese medicinals' or 'Chinese drugs' (*zhongyao* 中藥).

Scientists and historians, traders and regulators use two different genres to communicate in Chinese about the various medicinal materials and pharmaceutical products at distinct stages of production and identification: 'materia medica' (bencao 本草) and 'pharmacopoeia' (yaodian 藥典). 'Materia medica' includes crude drugs, pharmaceuticals, and raw materials – but the term also denotes documents about these materials and products, i.e. a genre of textual records archiving them in a corpus of knowledge. 'Pharmacopoeia' now refers to another specific genre: the standard volume of Pharmaceuticals that scientists and a particular nation-state (e.g. the 'British Pharmacopoeia') recognise for pharma-trade. In Chinese, the second character *dian* 典 of *yaodian* 藥典 has legal connotations and can thus mean a 'code' in law, or refer to a reference work, such as an encyclopaedia or, later, dictionary. This indicates that the pharmaceutical content in such a volume was derived from a range of texts across the four classical Chinese historical sources, i.e. classics (*jing* 經), histories (*shi* 史), masters (*zi* 子), and *collectanea* (*ji* 集). The Chinese term pharmacopoeia thus refers explicitly to the orthodoxy of the text, while alluding to its encyclopaedic scale and heterogeneous content.

The authorised 'pharmacopoeia' or 'drug code' (*yaodian* 藥典) in modern China covers the two to five hundred substances that are frequently used as 'Chinese medicine', East Asian oriental medicine, or Traditional Chinese Medicine (TCM). The basic aim of scholarship and regulation then is to provide a standard way to stock one's Chinese 'apothecary cabinet' (*yaogui* 藥櫃) and to tell pharmacists how to control the identity of drugs by reference to a national canon. And yet the relevant scientific discipline of '*materia medica* studies' (*bencao xue* 本草學), and the editorial boards that have updated successive editions of the pharmacopoeia, know and deal with thousands of multi-ethnic, regional and unidentified medicines that fall outside the frame of TCM and the recognised nationwide pharmacopoeia. The never-ending process of identification and selective classification illustrates political and societal changes in the status of medical scholarship and various therapeutic and craft practices throughout present-day China. Thus, Chinese attempts at sorting medicines shed light not only on the transformation of China and on modernisation in Asia, but also on the history of science at large.

The scope of any pharmacopoeia in the modern context of science and of legal trade regulation is national(istic). In modern China, the National Pharmacopoeia Commission (*Guojia yaodian weiyuanhui* 国家药典委员会) produced the *Chinese Pharmacopoeia of the People's Republic of China* (*Zhonghua renmin gongheguo yaodian* 中华人民共和国药典 [commonly abbreviated to *Zhongguo yaodian* 中国药典], 10th edn, 2015), which has become a government-orchestrated standard – legally enforceable and politically binding. China's national pharmacopoeia embraces *materia medica* (*bencao*) on a vast social, temporal, scholarly, and geographical scale of 9.6 million square kilometres, and stretches even further, from the Tibetan highlands to Korea and Japan, from the eastern Chinese coastline to the neighbouring countries to the west, south, north, and beyond. Within the present-day national territory, the *Chinese Pharmacopoeia* covers medicines derived from practices communicated in even more languages than those of the 56 officially recognised nationalities. Besides recording these medicines with names in Chinese characters – a major lingua franca in the historical Chinese empire and in Asian-language world science – the *Pharmacopoeia* illustrates how Latin script, along with botanical nomenclature and taxonomy, has come to dominate scientific discourse in the field. In the pharmacopoeia genre since the late nineteenth century, next to the Chinese characters we find a botanical nomenclature for the mostly plant-based medicines, which follows the Linnaean structure of scientific names in Latin and Greek. Furthermore, this mix of Chinese and Latin characters typically also includes vernacular names in both European and Asian languages. The botanical genus *Ephedra*, for instance, encompasses 57 species of medicinally used plants, while Chinese discourse, in the two characters *ma* 麻 and *huang* 黄, narrows down the range to four of them, i.e. *Ephedra sinica Stapf., Ephedra equisetina Bunge, Ephedra intermedia Schrenk* &

C.A. Mey, and *Ephedra distachya* L. (www.mpns.kew.org, accessed 03/1/2022). Further, pharmaceutical names in Chinese and European languages for the same medicinal indicate the part of the plant that is used, either the stem and root or the 'whole herb' (*quancao* 全草) growing above the ground. For different items in the pharmacopoeia, the relation between Chinese names and scientific names varies, and the scope of authorised medicinal names in Chinese is constantly evolving.

Prior to the English editions of the *Chinese Pharmacopoeia*, more extensive versions were published in mostly Chinese characters. Over its ongoing history, the scientific discourse in Chinese about pharmacopoeia has included the range of studies necessary to identify and classify the medicines scattered throughout China. In this process, pharmacognosists match each regionally specific variety to a continuously updated set of botanically standardised names, as well as pharmaceutical names in standard Chinese pronunciation that is transliterated in Latin letters (e.g. *mahuang*). The set of drugs frequently used and provided in a well-run drugstore for Chinese and other East Asian medicines, be it in China or in the UK, includes the two to five hundred items which China exports via Hong Kong and Taiwan, or directly to international suppliers. Yet, the *Pharmacopoeia* as of 2015 lists over 5,000 entries (5608), including both crude drugs and manufactured ones. Moreover, these thousands of medicines continuously link to a yet wider background of high biological diversity and medical plurality. This biocultural background feeds into the standard set of medicines; responding to availability and substitution, so that the standard is subject to historical change. Experts in the highly specialised field of pharmacognosy in China illustrate their conversations with multiple layers of authority, from the provincial level down to the nitty-gritty of individual local trade licencing. Currently they research and recognise tens of thousands of items.

The *Chinese Pharmacopoeia* was neither a sudden discovery nor is it a fixed list, but it has evolved through many decades of studies. The compilers of this state-standardised reference work, who began their work during the first half of the twentieth century, came from across the scientific community, especially botany and chemistry. Growing up in households of the imperial era, early generations of scholars began their education in the system of classical learning required to participate in the exams leading to literati careers. Later, after the abolition of the Chinese imperial educational system in 1905, most of them studied overseas in Europe, America, and Japan. In a shifting institutional landscape, they organised their collaboration via journals and local associations, through which they developed a format for the Chinese pharmacopoeia, keeping names and entries up to date in successive, and increasingly compendious, editions. At a national and regional level, they began using the new national language – modern standard Chinese[2] – to forge a lingua franca for the field in a multi-linguistic territory. Decades before institutions for TCM were established in the 1950s (Scheid 2002; Taylor 2005), these actors envisioned the political regulation of trade in pharmaceuticals without being in a position to implement it, and their pharmacopoeial studies promised to offer a shared pool of reference, national and global. These processes in the sociocultural history of science went beyond such scientific communities, however, spanning specialised academic disciplines and ethnic knowledge systems, and occurred on many levels of society: bringing together clinicians and philologists, physicians and pharmacists, and even monks and politicians, elite professionals, and lay people.

Overview

The range of pharmacopoeial studies in modern China is outlined below. Regional medicines were included in the dynamic national corpus during phases of increased research. However, open-ended investigation poses a constant challenge to integrating sorting systems of nation-wide *materia medica* knowledge. Historical figures have shaped this evolving field. Their works cover emergent disciplinary fields such as pharmacy and botany but also evidential philology (*kaozheng* 考證) (Zhu *et al* 2005) and historiography, as well as new modern institutions which collect scattered popular knowledge. Finally, a glance at entries in early works by the founding fathers – such as Bernard E. Read in England and Zhao Yuhuang 趙燏黃 who returned from his education in Japan, Na Qi 那琦 in Taiwan, or Xie Zongwan and Wang Xiaotao who conducted surveys all over mainland China – reveals the languages and seeds of increasingly multi-disciplinary studies. Not only the latest edition of the Pharmacopoeia but also previous trade regulations, textual scholarship, and biocultural collections play major roles.

Regional medicines

Tensions between the variety of regional medicines and centralised standards of identification and classification never get completely resolved, whereas new medicines continuously make their way into the authorised selection of pharmacopoeiae. *Ephedra*, or *mahuang* 麻黃 in Chinese, is an example of historic globalisation, i.e. moving, in this case, from imperial China to international European and American scientific discourse. During the twentieth century, chemical extracts of pharmaceutical substances (termed *su* 素 in Chinese) became the main target of laboratory investigation into medicines. Globally the discovery of Ephedrine (the primary active ingredient of *Ephedra*), and the reformulation of effective *Ephedra* as this chemical formula served as a nexus through which scientists in China, Europe, and Japan began to imagine themselves as members of a community. Numerous articles on the subject in scientific journals illustrate the importance that was attached to the translocal and transcultural chemical formulæ, which promised to cut across linguistic and cultural boundaries. However, we have seen above that Chinese *materia medica* studies have already narrowed down the selection of *Ephedra* for medicinal use to four of the 56 known worldwide species. The accuracy of Ephedrine's chemical formula, which circulated within global scientific networks, hardly translated directly to regulatory application in the day-to-day local plant market, where regional and vernacular names needed (and still need) to be used to classify and distinguish the four species, which in that context have already been transformed from raw plants into manufactured plant parts.

Two of the three botanical species of *Ephedra* which are used for the medicinal plant named *mahuang* in Chinese are in fact not native to historical China but have a longer and ongoing history in Central Asian habitats and the Himalayan region (Plants of the World Online). Only one botanical species is native to the region China of earlier historical periods. This historical knowhow about regional shifts in *materia medica* is typical of scattered textual evidence and of oral knowledge and rumours which I have encountered among insiders of supply chains and trends of availability. Similarly, Sichuan pepper and saffron are historic imports of spices via the 'Silk Roads' and both serve medicinal purposes. Sichuan pepper is today regarded as a typical Chinese ingredient and sold to connoisseurs of Chinese cuisine worldwide, yet it was perceived as exotic in historical China. Just as such a precious good, it went through a process of acculturation and authorisation as a constituent part of the Chinese culinary and health-promoting repertoire. This process started as early as in the Han dynasty (Lu and Lo 2015) when other imported plant parts such as pomegranate gained especially high value as exotics (Kong 2017). While this is even one of the major medicinals in Tibetan Sowa Rigpa and its precious pills (Czaja 2015), vice versa, Caterpillar fungus (*dongchong xiacao* 冬蟲夏草) originates from the Tibetan habitat but has been integrated into Chinese *materia medica* since the fifteenth century (Lu 2017). By the Ming dynasty, today's Sichuan pepper had finally become a widely spread part of Sichuan cooking and the Chinese cultural landscape (Dott 2020). As suppliers in East Tibet have explained to me, Saffron is known in Tibetan as coming historically from 'Kashmir', that is India or Persia, to the west of China, even though they are well aware that today it is available from cultivators in Xinjiang province of China. If imported from Iran, a historical origin of materia medica exchange (Laufer 1919), the Tibetan pharmacists assign Saffron the Chinese name 'Hong Kong Saffron' (*gang honghua* 港紅花) to indicate international import through ports – and thus mock the name 'Tibetan red flower' (*zang honghua* 藏紅花), a term dating from historical trade networks, and today used by the less well-informed Chinese traders further East in China (Springer fieldwork in Derge County 2015 and Chengdu City 2019). In historical Tibet and in Han-China, Persia is known as the origin of imported healthy nutrition (Chen 2012, Chapter 32 in this volume). Such regional origin and historical integration manifests in the location words used as prefixes for the names of medicines (*chuan* 川, mostly denoting Sichuan province, for instance). Well before such regional medicines are authorised and listed in the *Chinese Pharmacopoeia*, locational prefixes are used, while suffixes additionally point to the part of the plant to be used (root, rhizome (*gen* 根) or bark, skin (*pi* 皮), for example). The pharmacopoeial nomenclature tolerates a mixture of such additional entries besides the Linnaean scientific names in consideration of the information that it encodes about the origin and preparation of medicines. Currently the prefix *chuan* 川 shows up in eight authorised entries of the national standard.

Fritillaria exemplifies how the addition of prefixes to names of medicines indicates their regional origin. The term *beimu* 貝母 in Chinese includes twelve botanical species, i.e. eleven species of *Fritillaria* plus *Bolbostemma paniculatum* (Maxim.) *Franquet*, while the term *chuan beimu* 川貝母 (Sichuan *beimu*) covers only half that number, i.e. six species (www.mpns.kew.org). The local prefix does not necessarily refer directly to the region of collection or harvesting of the plant (current or historical), but may rather point to the location of those traders who are most familiar with high-quality commodities and who may supply the best available medicines, which need not be of local provenance (Bian 2020). Thus, *chuan beimu* (lit. 'Sichuan' *Fritillaria*) does not grow in the Sichuan basin but thousands of kilometres higher in the Eastern Tibetan highlands of the present-day province. However, the province's capital Chengdu houses a nationally important node in the supply network (Springer 2015). The prefix, by referencing Sichuan, obscures the actual habitat of the medicinal plant, but foregrounds the geographic distribution of trade expertise. Political territories that went under this name in the past include regions that have been outside China or have had several different place names in course of China's long modern and pre-modern history, such as Xikang 西康, 'Western Kham', for the modern Eastern Tibetan region (for a general discussion of historical place names see Hua *et al* 2016).

Global science, local surveys

Historians of botany in East Asia have drawn particular attention to the relation between standardised, state-of-the-art and fragmented practices of scientific and 'folk' pharmacy. Rather than natural history or modern science in a narrow sense, early *materia medica* studies at the beginning of the twentieth century continued to include reference to historical textual archives. These were named 'old *materia medica*' (*jiu bencao* 舊本草) and re-constituted an archive in Chinese characters going back to the first Pharmacopoeia in world history, first commissioned in the seventh century under the Tang dynasty (See Chapter 8 in this volume). Local surveys and fieldwork in modern times, however, increasingly came to refer to new botanical typologies and methods of description. Already in ancient China, as in Greece, studies of medicinal plants predated botanical plant studies. Ancient botany was part of medical studies, i.e. the study of medicinal plants (Métailié 2015; Hardy and Totelin 2016). In the history of science, modern botany changed this field of *materia medica* studies (Métailié 2001) under the impact of Japanese-style botany (1860–80) which, paradoxically, relied extensively on the *materia medica* literature in Chinese characters (Macron 2015; Mayanagi *et al.* 2015; Métailié 2015).

In East Asia, close exchange of findings and ideas could build on *materia medica* texts and comments in Chinese characters, generating a community of knowledge united, and bounded by, Sinographic script. Later, ideas of how to write a modern pharmacopoeia within the framework of emergent nation-states also circulated via scientific publications in European and East Asian languages; leading to editions of national pharmacopoeiae first in Japan 1887, and then China (1930), and Korea (1st edn 1958, 2nd edn 1967, 3rd edn 1976, 11th edn 2014). Korea had played an early role in the transmission of medical and botanical knowledge to Japan (Métailié 2015), yet the exchange of drugs, medical texts, or visiting experts in East Asia did not necessarily lead to the Sinification of pharmacy in these regions, or mutual understanding between them (Trambaiolo 2014). Korean pharmacopoeia studies experimented with a unique local style of nomenclature (see Chapter 34 in this volume; Lee 2015). Due to the political-historical context in East Asia, national(istic) pharmacopoeias exist, for instance, in Taiwan and Hong Kong, too, reflecting regional availability as well as national preferences for format and contents.

Today, the *Chinese Pharmacopoeia* is published in Chinese and in English. Early versions appeared in the 1930s with a small number of entries, years after the first modern pharmacopoeia in Europe or in Japan. The preface of the first edition after the founding of the People's Republic, published in 1953, marked a shift to the new regime by disparaging and distancing this work from these earlier efforts.

China did not, as the modernisation paradigm has it, passively respond to the challenge of scientific technology by imitating an imperialist model. Rather than in the little Linnaean garden in Sweden, the international societies met in various countries including Japan (Schiebinger 2004).

From the Qing dynasty onwards European naturalists, including British, French, and German language authors, conducted surveys of flora with a focus on southwest China instead of the political or cultural centres, but relying on decisive support by local collaborators (Fan 2004; Mueggler 2014). The work of pharmacopoeial scientists in modern China marks a departure from this postcolonial history.

Trade regulation

Three Chinese terms imply divisions in the regulatory and academic *materia medica* field: first medicinal material, second pharmaceutical product, and third the variety of product grades and types.

Chinese medicinal material (*yaocai* 藥材 or medicinal things *yaowu* 藥物) is crude or raw material, yet the workforce providing this crucial resource, and the work that they do, is barely visible behind the discourse of supply chains (Dejouhanet 2014; Pordié and Blaikie 2014; Springer 2015). To discuss the field of knowledge about regional harvest/collecting of *materia medica*, the French term *terroir*, borrowed from high-quality agriculture and cuisine, has become favoured in the field of botany. Under the influence of ethno-botanical and ethno-pharmacological fieldwork and economic theory, that term appears also in recent

English scientific publications on what is in Chinese called authentic medicinal material (*daodi yaocai* 道地藥材) (Zhao *et al.* 2012; Brinkmann 2015; Bian 2020). Geo-authentic *terroir* embraces factors of geography, climate, and seasonal variety of *materia medica* that are crucial during the processes involved in collection and storage. Additionally, processing (*paozhi* 炮製) – part of the standard entries in the *materia medica* and pharmacopoeia genres – needs to take the factors involved in *terroir* into account as it begins with the first steps of manufacture that turn collected or cultivated raw objects into medicinal material.

For the next step of production, expertise in pharmacy is expected. The production and provision of a Chinese 'pharmaceutical product' (*yaopin* 藥品) requires quasi-official personnel to administer the licencing and registration of commodities from local and provincial drugstores up to the national level, wherever the product is to be sold whether directly on the market or recommended by prescription. Besides regulating business between traders and apothecaries, the products thus lie in the dominion of the authorised *Chinese Pharmacopoeia* and the national committee that authors each new edition. The licencing of a product by that committee involves high-ranking politicians, such as the Minister of Health, and thus reflects the layered governance and international voice of Chinese academia. As representatives of science and trade in China, the committee provides a standard to oversee the production and business of the ever-changing scope of Chinese medicinals, from local to national, and other medicinal materials. It is worth noting that, due to the historical and text-centric history of Chinese pharmacy studies, these authors are not just pharmacists or taxonomists but also historians and philologists.

'Grades [and] types' (*pinzhong* 品種) require a third kind of Chinese *materia medica*-expertise: experts test the variety of medicinal products and types, often in response to enquiries from the market regulators, and in light of surveys in the field, of textual studies, and of the modern and the pre-modern research records, which name drugs, plant parts, places, illnesses, and chemicals. A typical question that occupies these experts is whether traders are correct in claiming either that their product is based on historical precedent from ancient literature or that it constitutes a particular local variety of the known authorised pharmacopoeia. Opening and stocking a Chinese pharmacy – be it in any hospital or in an independent local clinic – require compliance with governance by the *China Food and Drug Administration* and *Regulatory Bureau of Chinese Medicine and Pharmacy* (*Zhongyiyao guanli ju* 中醫藥管理局) on the one hand, and additionally with case-by-case judgements of local, provincial, and national product licences. Registration numbers for any such pharmaceutical item in China fall under the two separate categories of either 'national drugs' (*guoyao* 國藥) or 'wellness products' (*baojianpin* 保健品).

Historical Figures Combining Chinese Philology and Scientific Fieldwork

Besides the new modern botanical language, what unites the field is the pharmacopoeial study of individual ingredients instead of formulary (*fangjixue* 方劑學), the composition of prescriptions with several ingredients. 'Crude drugs' (*shengyao* 生藥) are thus the crux of taxonomy while some manufactured drugs (literally 'slices for drinking' *yinpian* 飲片) have come to be included during the twentieth century. After introducing one British, and one Chinese founding father, who have captured some of the pharmacognostic varieties in scientific language while linking it to philological evidence, I will trace records regarding the processing of medicines, and historically recent additions to the pharmacopoeia.

The idea that pharmacopoeial knowledge is hidden in the 'experience' (*jingyan* 經驗) not only of systematised formularies but also individual knowledge holders led to campaigns to collect and systematise knowledge that could be found, using surveys among the population and by recording the practices of the vast majority of unlicenced practitioners and sellers of Chinese medicine (Andrews 2014, Lei 2014). Impelled by the desire for knowledge of medical commodities, as well as by scepticism about earlier surveys of applied knowledge, the search began for a new kind of expert who could sort out *materia medica* records and thus solve a fundamental issue of healthcare in China.

When Bernard E. Read (1930), educated at UK and US institutions and based in China's academia – first at Peking Union Medical College and then at the Henry Lester research institute in Shanghai – finally announced the first *Pharmacopoeia* in 1930 to readers of expert journals, the volume was rather thin and included fewer medicinals than the Ming dynasty model that had been consulted, the *Bencao Gangmu* 本草綱目 (Encyclopaedia of *Materia Medica*, completed 1578) (Nappi 2009, Unschuld 2014). Research at markets rather than in herbariums was one of Read's specialities, but this fieldwork-based approach did not prevent his becoming one of the founding members of a national association for medical historiography (Zhu 2001).

Bernard Read, together with his Chinese co-author J. C. Liu (Liu Ju-ch'iang [Liu Ruqiang] 劉汝強), gave a conference paper underlining 'the importance of botanical identity' for Chinese *materia medica* (Read and Liu 1925).[3] Before expanding on their view, they paraphrased a paper presented at a previous conference held in Japan as having 'advocated a scientific study of Chinese drugs, of their bibliography, pharmacognosy, chemistry, botany, physiological action, and chemical use'. They continued:

> Without the correct botanical identification of a drug, it is impossible to find the scientific literature dealing with the known chemistry of its constituents and their physiological action, as already published by other workers. (⋯) This paper seeks to clarify the identity of two particular drugs Lang-Tang, 茛菪,[4] and Ti-huang,[5] 地黄, and to point out how serious are the consequences if the names of these drugs continue to be misapplied (Ibid.: 987–9).

In 1927, Read and Liu's compilation of plants in China *Flora Sinensis* was a thin volume like a journal, yet this was just the first groundbreaking issue of a future series of reference works. The seed for Read's botanical-chemical and textual research approach is visible already in his 'reference list' (Read 1923). The included lists of literature in this thin volume cite major scientific studies in Japanese and European languages on taxonomy, science, just as on linguistics and historical sources. Accordingly, the structuring units of *materia medica* in modernising China are entries on individual crude drugs in such a 'reference list', as Read had prioritised four columns of content in this genre: names of the medicinal plants in Chinese characters; transliteration; references to new scientific botanical names; and finally to a mix of scientific and historical literature on chemistry and pharmacy. His emphasis on botany and on drug identification through consolidated scientific and historical names is a lasting model for research and regulation.

Read found that his own aims of handling toxicity and ensuring the availably of medicines connected his work directly to that of fellow scientists in China and even to that of Li Shizhen 李時珍 (1518–93), the author of the *Bencao Gangmu* (Chapter 9 in this volume). Read considered himself to be honouring the legacy of Li's approach for rigorous identification through fieldwork, nomenclature and philology. The craze about Li Shizhen during China's modernisation can also be seen through the lens of Joseph Needham's Science and Civilization in China project, via offprints at the East Asian History of Science Library at the

Figure 48.1 Bernard Read and Liu Ju-Ch'iang's Flora Sinensis (1927). Courtesy of Needham Research Institute

Needham Research Institute. Needham's offprints include a stack of sources on Li produced in the context prior to the institutionalisation of Chinese medicine since 1954. In 1958, fellow historian and philologist of medicine in China Wang Chi-min 王吉民 (1889–1972) provided a catalogue from the History of Medicine Museum in Shanghai introducing a 1954 exhibition. This commemorated Li's death 360 years prior and showcased his role in a legacy of pharmacognostic adaptations and continuous scientific developments well into modernising China (Figures 48.1 and 48.2).

Similarly, Zhao Yuhuang 趙燏黄 (1883–1960) published his groundbreaking studies on *materia medica* (Springer forthcoming) in journals rather than as books, and with only three detailed entries plus a list of forty-three more envisioned for future publications. Like Read, Zhao, had also been educated outside China, but in Japan. He was politically influential in the new field of pharmaceutical research and regulation in China, and included applied market-related discussion of production sites and product variety (Zhao 2006 [1932]). He explicitly emphasised that physicians were just broadly responsible for hygiene (*weisheng* 衛生), while pharmacognostic studies required the new expertise in pharmacy. His distinction

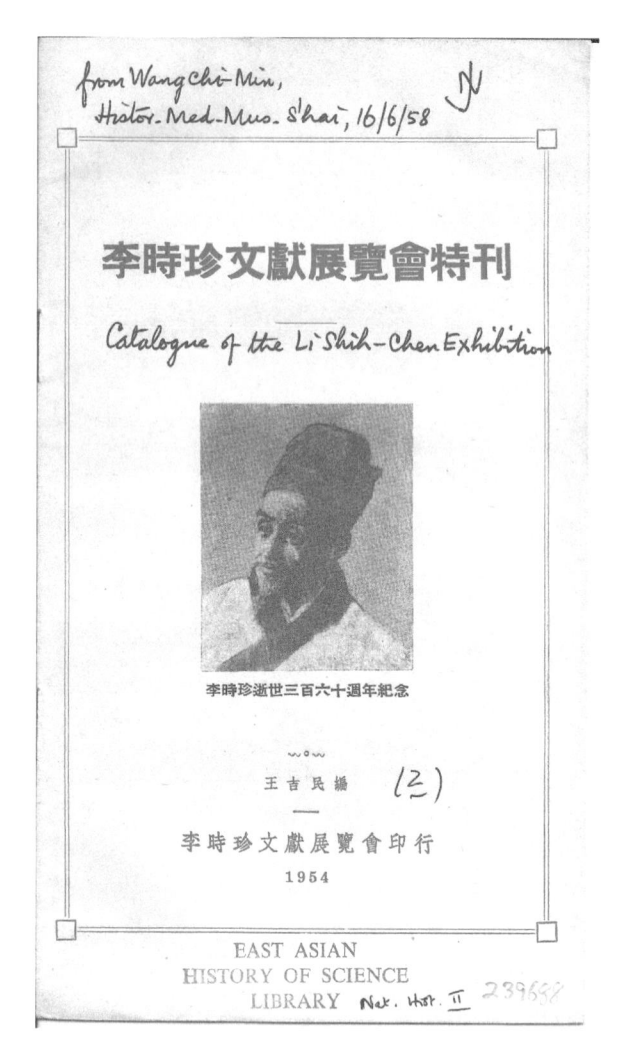

Figure 48.2 'Catalogue of the Li Shih-Chen Exhibition' (1954) History of Medicine Museum, Shanghai. Courtesy of Needham Research Institute

of 'old *materia medica*' and 'new *materia medica*' allowed for the inclusion of ancient texts as well as the newest scientific terminology in German and Latin, photographs, chemical formulæ and plant morphology. While his focus was crude drugs, he also surveyed the historical record and contemporary market at Anguo in Hebei Province (Ibid., 1934, 1936, 2004).

Works

If Read and Zhao were the founding fathers of pharmacopoeia studies in modern China, the official author of the first national pharmacopoeia in 1930 was Meng Mudi 孟目的 (1879–1983), whose career had centred on Nanjing city and academic education. This publication was modelled on scientific taxonomy in Europe, North America and Japan and had the political function of counteracting the 'old' and 'obscure' medical theory of physicians from the past.

Among subsequent editions of the *Chinese Pharmacopoeia*, the prefaces of the recent editions mention the 1953 *Pharmacopoeia* as the first edition after the founding of the People's Republic of China (PRC) in 1949, while the edition of 1963 marked a new development in the field. Experts in the field recognise this edition as the main source for the contents on 'processing' (i.e. *paozhi* 炮製) in entries of Xie Zongwan's *Compilation of Chinese Herbal Medicinals* (*Quanguo zhongcaoyao huibian* 全國中草藥匯編), and as the first integration of Chinese elements into a format that had until then been overly modelled on the global scientific mainstream (Xie 1996).

Japanese-educated Na Qi 那琦 published his *Pents'aology* (*Bencao xue* 本草學) (Na 1974), a term used today in Taiwan and by some graduates of UK institutions, for instance. He was Professor of Pents'aology and Pharmacognosy and Chief of the Chinese Pharmacological Institute at the Chinese Medical College.

In the 1980s, historical and philology-based studies were taken up again in the PRC. An influential personality in combined old- and new-style *materia medica* studies, Wang Xiaotao 王孝洵 (born 1928), published extensively on methods of 'processing' (*paozhi*) (Wang 1981, 1986). Xie Zongwan 謝宗萬 (1924–2004) carried out extensive surveys in the field and archive during the 1950s with a focus on terroir and authenticity of *materia medica* and also on the history of changes in types of single-ingredient pharmaceutical products, and on vernacular names (Xie 1964, 1984, 1996). The task of *materia medica* historians still consisted in outlining the efforts at systematising and sorting the textual records of Chinese medicinals through space and time in Chinese history (Shang *et al.* 1989).

The *Chinese Materia Medica* (*Zhonghua bencao* 1999) is the result of an ambitious and prestigious project that took on momentum in 1988 under the guidance of the newly established (above-mentioned) Regulatory Bureau. This project aimed to cover medicinals known among scattered popular practitioners and was later inspired by the prospects of databases in the digital age. Its latest additions are volumes on Tibetan (2002) and on Mongolian (2004) medicinals. Another similarly ambitious major project in the field that goes back to founding father Zhao Yuhuang has been the illustrated *Complete Books of Materia Medica in China* (Lu *et al.* 1999). In contrast to the current *Chinese Pharmacopoeia*'s relatively handy size of 1809 pages in one volume, both of these publications fill entire shelves with several volumes, complete with very detailed entries on numerous names of medicinals in crude and manufactured versions, as well as their misleading names, fakes, and substitutes.

Entries

In order to write the national pharmacopoeia, surveys of '*materia medica*' were carried out both in the field and in scattered archives and scientific publications. The wide range of materials that this has yielded remains a constant challenge. As noted above, '*materia medica*' comprises not only all the things that are used as medicines in China, but also entries about them in various textual records documenting them: in Han-Chinese and in a range of other languages (Sanskrit, Tibetan, Mongolian, Uighur, Arabic, Japanese, etc., and European languages such as Latin, English, French, and German). This ongoing archival (or textual) project also informs the evaluation and testing of ancient knowledge as well as popular practice. These multilingual records of 'Chinese' *materia medica* serve to position authoritative knowledge about them within a broad field of intellectual, political, epistemological and ethno-economic (Comaroff and Comaroff 2009) controversies about medical cultural relations and Chinese medicine's relevance in society, and about the envisioned future of sociocultural exchange processes.

Socioeconomic and ethnic scope

In addition to combined surveys in the field and archives, societal relations express fundamental change in a conspicuous manner, making it amenable to research.

Throughout Chinese history and to this day, to a striking degree, unlicenced practice feeds into this material repertoire and contributes to setting standards for pharmacological treatments; so that while the Chinese *materia medica* is certainly the product of elite writing and reflection it also constitutes learned responses to more pervasive and popular forms of knowledge. The physicians who prescribe, and often also trade and consume, Chinese medicines share knowledge and terminology with the various collectors, producers, traders, and researchers of these materials and their culture. When suppliers or physicians in China of non-Han ethnicity communicate with their Han-Chinese counterparts about the exchange of medicinals between the two groups, I have observed in the ethnographic field that they speak about 'Han-medicine' (*Hanyi* 漢醫) to denote the portion of *materia medica* that is used and recognised within the orthodox framework of Chinese medicine, and which, by extension, exerts a hegemonic power in relation to local conceptions. Specifically, Han-centred Chinese medicinals continually emerge within the broader scope of the Chinese *materia medica*; thus, they are the product of a centralised authoritative record, as well as changing scientific taxonomy, and of perceived ethnicities.

Assigning new forms of authority in applied medical tradition to living senior masters has added a new layer of ethnicity to the representation of TCM and *materia medica* knowledge. In this context, Jin Shiyuan 金世元 (born 1926) is the only member of the vast, multi-ethnic editorial board of the above-mentioned *Zhonghua bencao* to be appointed 'Great Master of National Medicine' (*guoyi dashi* 國醫大師). He was appointed for his specialisation in *materia medica* and processing and was the first and only knowledge holder of Chinese pharmacy to be authorised as such (Liu 2014: 170–73). The exclusivity of this award silently bestows 'Han' *materia medica* heritage, personified by Jin, the imprimatur of 'National' status, next to TCM physicians, and also the 'scientific' status, since he belongs to the editorial board.

The rich social diversity and even fragmentation of healthcare in China raises the question of how folk medicine – as we encounter it in ethnographic fieldwork or in the non-elite as well as medieval manuscript literature – relates to mainstream medicine (Unschuld and Zheng 2012; Lo and Cullen 2014). Yet China is unique in the extent to which its own 'traditional' medicine – and even rural industries constituting the supply behind it – has been incorporated into modern healthcare and into research institutions of the long twentieth century (Springer forthcoming; Lei 2014). The term 'traditional' was introduced for use in English journals referencing learning medicine since the mid-1950s (Taylor 2005). Even the WHO-orchestrated global science and healthcare policy has, since the late 1970s, claimed to include 'Chinese experience' from communist and earlier modern China (Kadetz 2015), although, as Lei (2014) observes, this inclusion privileged Western scientific forms of knowing. This privileging notwithstanding, the wider non-elite personnel behind 'Chinese medicinals' are restricted neither to figures hailed in heroic 'barefoot doctor' narratives (Fang 2012, Chapter 45 in this volume) nor to ancient healers and practitioners of 'obscure' educational backgrounds. Revolutionaries in China not only developed a concern for an urban science-based healthcare industry, but they also saw holders of *materia medica* knowhow in rural industries as the basis for ensuring this development and for sustaining the availability of new industrial drugs for a future society.

Popular, practical medicine

During the Maoist era, mass science campaigns opened *materia medica* in China to non-elite and folk practices and ethnic minority knowhow, in the attempt to distil the best folk practices and feed them back to the population, together with science-based healthcare. These campaigns propagated an 'alternative modernity' (Hsu 2009) in order to popularise knowledge about medicinal plants across society and on a 'world-revolutionary' scale. In biology 'with Chinese characteristics', genetics, for instance, was politically criticised as overly interested in decadent issues of heritage (Schneider 2003). In that political context, contrary to scholarly refinement, simple prescriptions were celebrated that 'Served the People'. While most Chinese medicinals are prescribed as recipes and are prepared as a mixture with other ingredients, prescriptions of only one medicinal (simples) have a special name in Chinese (*danfang* 單方, or *danweiyao* 單味藥) and are associated with the usage of raw medicinals (*shengyao* 生藥) in folk medicine, especially in southwest China.

That supposedly simpler, less elite, yet biologically rich knowledge is seen as located in southwest and south China, i.e. away from the regions around China's previous capitals in the southeast, northwest, and northeast (on the mainstream region of Chinese medicine see Bretelle-Establet 2010, Hanson 2011, Scheid 2007). This medical geo-history recognises the southwestern province of Sichuan as China's dispensary-province. This view is not without merit: for example, the extraction methods to produce artemisin (*qinghaosu* 青蒿素) that won China's first Nobel Prize in science in 2015 were sourced from southeastern Anhui but reproduced by scientists in Yunnan (the southwesterly province which borders Sichuan).

Individual simple remedies as well as entries of Chinese materia medica are referred to by the technical term *wei* 味, a term which simultaneously refers to their 'sapor', indicating the close connections of pharmacopoeia to refined food culture (Lo and Barrett 2005). As opposed to the historically predominant formulæ, the preference for individual ingredients or very simple prescriptions from any ethnic background, and their integration into the catalogues of the regionally available medicinal plants in developing territories, was underscored under the influence of the WHO in the 1970s, with the publication of the National Essential Medicines List (*Zhongguo jiben yaowu mulu* 国家基本药物目录) (Fang 2012). This period further saw an increased emphasis on individual drugs with the 1973 excavation of four books with medical contents from a tomb at Mawangdui in Changsha dating to 168 BCE. These finds included the 'Formulae for 52 Diseases' (*Wushi'er bing fang* 五十二病方), which was celebrated by scientists at the time as a text recording 396 individual medicinals, i.e. clearly identified ingredients of formulæ. At the same time, in 1972/3, the committee authoring the *Pharmacopoeia* met again after an interval of twenty years, and published the second edition in 1977, shortly after the Great Proletarian Cultural Revolution (1966–1976).

The editorial note of a post-Maoist reference work, the *Compilation of Chinese Herbal Medicinals* (*Quanguo zhongcaoyao huibian* 全國中草藥匯編) (1996: 11–13) mentions the vast extension of authorised entries in the earlier Maoist edition. It had been based on 'summaries since the 1970s of simple tested, handy and healthy pharmaceutical prescriptions from the clinical practice of the broad masses and medical workers' and had included some 'frequently used formulae from Chinese clinical practice'. The preface goes on to outline that for each medicinal it allowed 'five compounds', and 'one to two traditional formulae' as well as 'two to three modern experience-based prescriptions' (author's translations).

A work newly published in this period, which gained authoritative status and then subsequently expanded in the Maoist period, is the *Chinese Dictionary of Pharmacology* (*Zhongguo yaoxue dacidian* 中國藥學大辭典 1979). The editors of the 1982 reissue recall in their preface that after much work had been accomplished from 1958 to 1966, the years from 1972 to 1976 saw a vast extension of the number of entries in the authorised pharmacopoeia. The newly titled *Jiangsu Academy of New Medicine* (*Jiangsu xinyi xueyuan* 江蘇新醫學院), the collaborating *Chinese Medical Hospital* and the *Medical Hospital* in Nanjing republished this book in 1982. Its 5,767 entries on simples (*wei* 味) include 4,773 plants, 740 animals, 82 minerals, as well as 172 additional '*danweiyao*' medicinals. Despite the changing scientific emphases during different political eras of the twentieth century, publications in the *materia medica* field reuse and sometimes cite previous ones.

Publications by authors outside China have continued to contribute to the field. The widely used *Chinese Herbal Medicine: Materia Medica*, for international practitioners of Chinese medicine, was first published in the US (Bensky *et al.* 1986). In 1991, a first translation of the *Chinese Pharmacopoeia* into German was published (Stöger 2021). In China, botanists accomplished the publication of a Chinese plant catalogue in 2004: comprising 125 books it ran to 80 volumes (*Flora Reipublicae Popularis Sinicae* or *Zhongguo zhiwuzhi* 中國植物志). After thirty-five years of preparation, this Chinese catalogue of its flora, was, however, an isolated taxonomy. It had to be translated into the botanical terms and classifications of the global scientific community. A much shorter account in English made *Flora in China* (2015) accessible to a wider audience (www.floraofchina.org). The authors of the latter echo Read's above-mentioned appeal from the 1920s in asserting that 'The *Flora* holds tremendous potential for those wishing to study the medical value of a given species' or 'for searching the relatives of commercially valuable plants that are more resistant to disease or drought'.

Irrespective of political context or changing availability, *materia medica* studies require textual analysis and reading abilities in classical Chinese and familiarity with the often insider jargon and regional knowhow of Chinese medicine practitioners (Unschuld 2012) and/or knowledge of recent developments in folk medicine (Farquhar and Lai 2014) and material histories of toxicity (Akahori 1989; Liu 2015). While the context of Chinese *materia medica* is culturally rich and geographically and biologically vast, expertise in this field mainly calls for old-style philology, as well as familiarity with the newest scientific findings and surveys in the field. The contours of Chinese pharmacy have been shaped over time by economic and natural availability and by multiple bibliographic, epistemological, political, material, and ethnic praxes. It is the task of historians and anthropologists of science and medicine in China to contextualise these variations, in the face of historical as well as ongoing efforts to produce reductive simulacra of this *Wunderkammer* of known and manipulated material objects in the necessarily flexible form of standardised, unified, scientific norms.

Acknowledgements:

This chapter is part of a project that has received funding from the European Research Council (ERC) under the European Union's Horizon 2020 research and innovation programme (Grant agreement No. 856543) and was kindly supported by the Jing Brand Scholarship in Chinese Science and Civilisation at the Needham Research Institute, Cambridge.

Notes

1 Thanks to the Needham Research Institute and John Moffet for the images and kind permission to print them.
2 The new-style *baihua* 白话, as advocated by intellectuals of the May Fourth Movement, notably Hu Shi 胡適 and Chen Duxiu. 陳獨秀. See Elisabeth Kaske (2009) and Chen Pingyuan (1999, ch. 5 *et passim*).
3 Read and Liu (1925) 'The importance of botanical identity', reprint from the Transactions of the 6th Congress of the Far Eastern Association of Tropical Medicine, held in Tokyo.
4 Langdang, also *Tianxianzi* 天仙子; *Hyoscyamus niger L.* (botanical name); *Black Jurinea* or (Common) Henbane (English name).
5 *Rehmannia glutinosa* (Gaertn.) DC. (botanical name).

Bibliography

Akahori, A. (1989) 'Drug taking and immortality', in L. Kohn (ed.) *Taoist Meditation and Longevity Techniques*, Ann Arbor: University of Michigan Center for Chinese Studies, pp. 73–98.
Andrews, B. (2014) *The Making of Modern Chinese Medicine*, Contemporary China Series, Vancouver; Toronto, ON: University of British Columbia.
Bensky, D., Clavey, S., Stöger, E. and Gamble, A. ([1986] 2004) *Chinese Herbal Medicine: Materia Medica*, 3rd edn, Seattle, WA: Eastland Press.
Bian, H. (2020) *Know Your Remedies: pharmacy and culture in early modern China*, New Haven: Princeton University Press.
Bretelle-Establet, F. (2010) 'Is the lower Yangzi river region the only seat of medical knowledge in late imperial China? A glance at the far south region and at its medical documents', in F. Bretelle-Establet (ed.) *Looking at It from Asia: the processes that shaped the sources of history of science*, Dordrecht; New York: Springer, pp. 331–70.
Brinkmann, J. A. (2015) 'Geographical indications for medicinal plants', *World Journal for Traditional Chinese Medicine*, 1.1: 1–8.
Chen Cunren 陳存仁 (1979) *Zhongguo yaoxue dacidian* 中國藥學大辭典 (*Chinese Dictionary of Pharmacology*), Taipei: Shijie shuju.
Chen Ming 陳明 (2012) 'Fachu Bosi: Sanlejiang yuanliukao "法出波斯": "三勒浆"源流考 (It is from Persia: an exploration of the origin of *Sanlejiang*)', *Historical Research*, 335.1: 4–23.
Chen Pingyuan 陳平原 (1999) *Modern Chinese: history and sociolinguistics*, Cambridge: Cambridge University Press.
Comaroff, J. J. and Comaroff, J. (2009) *Ethnicity Inc.*, Chicago: University of Chicago Press.
Cullen, C. (2001) *Yi'an* (Case Statements): The origins of a genre of Chinese medical literature, in E. Hsu (ed.) *Innovation in Chinese Medicine*, Cambridge: Cambridge University Press, pp. 297–323.
Czaja, O. (2015) 'The administration of precious pills: efficacy in historical and ritual contexts', *Asian Medicine*, 10.1-2: 36–89.
Dejouhanet, L. (2014) 'Supply of medicinal raw materials: the Achilles heel of today's manufacturing sector for ayurvedic drugs in Kerala', *Asian Medicine* 9.1.2: 206–35.
Dott, B.R. (2020) *The Chile Pepper in China: a cultural biography*, New York: Columbia University Press.
Fan, F. (2004) *British Naturalists in Qing China: science, empire, and cultural encounter*, Cambridge, MA: Harvard University Press.
Fang, X. (2012) *Barefoot Doctors and Western Medicine in China*, Rochester, NY: University of Rochester Press.
Farquhar, J.B. and Lai, L. (2014) 'Information and its practical order: crafting Zhuang nationality medicine', *East Asian Science, Technology and Society: An International Journal*, 8: 417–37.
Flora of China, *Overview*, http://flora.huh.harvard.edu/china/mss/intro.htm, accessed 03/1/2022.
Flora of China Editorial Committee (2020) *Flora of China*, St. Louis, MO; Cambridge, MA: Missouri Botanical Garden and Harvard University Herbaria.
Guojia yaodian weiyuanhui 国家药典委员会 (National Pharmacopoeia Committee) (eds) (2015) Pharmacopoeia of the People's Republic of China (*Zhonghua Renmin gongheguo yaodian* 中華人民共和國药典), Beijing: Zhongguo yiyao keji chubanshe.
Guojia zhongyiyao guanliju 'Zhonghua bencao' bianwei hui国家中医药管理局《中华本草》编委会 (Editorial Committee of the Chinese Materia Medica) (1999) *Zhonghua bencao* 中華本草 (Chinese Materia Medica), Shanghai: Shanghai kexue jishu chubanshe.
Hanson, M.E. (2011) *Speaking of Epidemics in Chinese Medicine: disease and the geographical imagination in late imperial China*, London: Routledge.

Hanson, M.E. and G. Pomata (2017) 'Medicinal Formulas and Experiential Knowledge in the Seventeenth-Century Epistemic Exchange between China and Europe', *Isis* 108(1).

Hardy, G. and Totelin, L. (2016) *Ancient Botany*, book series sciences of antiquity, London: Routledge.

Hsu, E. (2009) 'Chinese propriety medicines: an alternative modernity? The case of the anti-malarial substance artemisinin in East Africa'', in E. Hsu and G. Stollberg (eds) *Globalizing Chinese Medicine*, special issue of *Medical Anthropology: Cross-Cultural Studies in Health and Illness* 28.2: 111–40.

Hua Linfu, Buell, P.D. and Unschuld, P.U. (eds) (2017) *Dictionary of the Ben cao gang mu*, vol.2, *Geographical and Administrative Designations*, Berkeley: University of California Press.

Institute of the Tibetan Medical Hospital of the Tibetan Autonomous District (西藏自治区藏医院药物研究所) (2002) *Zhonghua bencao, zangyao juan* 中華本草藏藥卷 (Chinese Materia Medica, Volume on Tibetan Medicinals), Shanghai: Shanghai kexue jishu chubanshe.

Jiangsu xinyi xueyuan 江蘇新醫學院 (Jiangsu College of New Medicine) (1986, reprint 2011) *Zhongyao dacidian* 中藥大辭典 (Dictionary of Chinese Medicinals), Shanghai: Shanghai kexue jishu chubanshe.

Kadetz, P. (2015) 'Safety net: the construction of biomedical safety in the global 'traditional medicine' discourse', special issue "efficacy and safety in Tibetan and Chinese medicines: historical and ethnographic perspectives", *Asian Medicine: Tradition and Modernity*, 10.1.2: 121–51.

Kaske, E. (2009) *The Politics of Language in Chinese Education, 1895–1919*, Leiden: Brill.

Kong, X. (2017) 'An annotated translation of *fu* on pomegranate in *Yiwen leiju*', *Early Medival China*, 23: 67–88.

Laufer, B. (1919) *Sino-Iranica: Chinese contributions to the history of civilization in ancient Iran; with special reference to the history of cultivated plants and products* (Field Museum of Natural History, Anthropological series, vol. 15, no. 3), Chicago: Field Museum of Natural History.

Lee, J. (2015) 'Between universalism and regionalism: universal systematics from imperial Japan', *The British Journal for the History of Science*, 1–24.

Lei, S. H.-l. (2014) *Neither Donkey nor Horse*, Chicago, IL: University of Chicago Press.

Liu Changhua 柳长华 (ed.) (2014) *Yidao guan zhu* 医道贯珠 (The Way of Medicine: a string of pearls), Beijing: Zhongguo guji chubanshe.

Liu, Y. (2015) *Toxic Cures: poisons and medicines in medieval China*, PhD thesis, Harvard University.

Lo, V. and Barrett, P. (2005) 'Cooking up fine remedies: on the culinary aesthetic in a six tenth-century Chinese Materia Medica', *Medical History*, 49: 395–422.

Lo, V. and Cullen, C. (2005, 2014) *Medieval Chinese Medicine, The Dunhuang Medical Manuscripts*, London: Routledge.

Lu, D. (2017) 'Transnational travels of the caterpillar fungus in the fifteenth through nineteenth centuries: the transformation of natural knowledge in a global context', *Asian Medicine*, 12: 7–55.

Lu, D. and Lo, V. (2015) 'Scent and synaesthesia: the medical use of spice bags in early China', *Journal of Ethnopharmacology* 167: 38–46.

Lu Jun 鲁军 (ed.) (1999) *Zhongguo bencao quanshu*中国本草全书 (The Complete Books of China's Materia Medica), Beijing: Huaxia chubanshe.

Marcon, F. (2015) *The Knowledge of Nature and the Nature of Knowledge in Early Modern Japan*, Chicago, IL: University of Chicago Press.

Mayanagi, M., Takashi, M. and Vigouroux, M. (2015) 'Yang Shouying and the Kojima Famiily: collection and publication of medical classics', in B. Elman (ed.) *Antiquarianism, Language, and Medical Philology: from early modern to modern Sino-Japanese medical DISCOURSES*, Leiden: Brill, pp. 186–213.

Meng Mudi 孟目的 (1928) *Zhonghua yaodian* 中華藥典 (Chinese pharmacopoeia), 1st ed., *weishengbu bianyin* 衛生部編印 (edited and printed by the Ministry of Health).

Métailié, G. (2001) 'The formation of botanical terminology: a model or a case study?', in M. Lackner, I. Amelung and J. Kurtz (eds) *New Terms for New Ideas: western knowledge & lexical change in late imperial China*, Leiden; Boston, MA; Köln: Brill, pp. 327–38.

——— (2015) *Traditional Botany: an ethnobotanical approach*, Part 4 biology and biological technology, Volume VI, science and civilisation in China series, Cambridge: Cambridge University Press.

Medicinal Plant Name Service (MPNS) Royal Botanic Gardens, Kew, www.mpns.kew.org, accessed 03/1/2022.

Mueggler, E (2014) *Paper Road: archive and experience in the botanical exploration of West China and Tibet*, Berkeley: University of California Press.

Na Qi 那琦 (1974, 1982) *Pents'aology* (*Bencaoxue* 本草學), Taipei: Nantian Shuju.

Nappi, C. (2009) *The Monkey and the Inkpot: Natural history and its transformations in early modern China*, Cambridge, MA: Harvard University Press.

Pordié, L. and Blaikie, C. (2014) 'Knowledge and skill in motion: layers of Tibetan medical education', *Culture, Medicine and Psychiatry*, 38.3: 340–68.

Read, B.E. (1923) *Botanical, Chemical and Pharmacological Reference List to Chinese Materia Medica*, Peking [Beijing]: Bureau of Engraving and Printing.

——— (1936) *Chinese Medicinal Plants from the Pen Ts'ao Kang Mu*, 3rd edn, Peking: Natural History Bulletin.

Read, B.E. and Liu, J.C. (Liu Ju-ch'iang[Liu Ruqiang] 刘汝强) (1925) 'The importance of botanical identity', in Far Eastern Association of Tropical Medicine (ed.) *Transactions of the 6th Biennial Congress Held at Tokyo*, Tokyo: Kyorinsha Medical Publishing Co., pp. 987–99.

——— (1927) *Flora Sinensis, Series A, Volume 1, Plantae Midicinalis Sinensis*, Peking, China, held at the East Asian Library of the Needham Research Institute in Cambridge, United Kingdom.

Scheid, V. (2002) *Chinese medicine in contemporary China: plurality and synthesis*, Durham, NC: Duke University Press.

——— (2007) *Currents of Tradition in Chinese Medicine, 1626–2006*, Seattle: Eastland Press.

Schiebinger, L (2004) *Plants and Empire: colonial bio-prospecting in the Atlantic world*, Cambridge, MA: Harvard University Press.

Schneider, L. (2003) *Biology and Revolution in Twentieth Century China*, Lanham, MD: Rowman & Littlefield.

Shang Zhijun 尚志鈞, Lin Qianliang 林乾良 and Zheng Jinsheng 鄭金生 (1989) *Lidai zhongyao wenxian jinghua* 歷代中藥文獻精華 (Talents from the Archives of Chinese Medicinals throughout the Dynasties), Beijing: Kexue jishu wenxian chubanshe.

Springer, L. (2015) 'Collectors, producers, and circulators of Tibetan and Chinese medicines in Sichuan province', *Asian Medicine*, 10.1–2: 177–220.

——— (2019) 'Taibai Materia Medica: unofficial 'Herb Physicians' in North Western China', *EchoGéo*, 47, http://journals.openedition.org/echogeo/17396, accessed 03/1/2021.

——— (forthcoming) *Arztgeschichten: Zur chinesischen Medizinkultur, 1926–2015* (Physicians' Stories: On Chinese Medical Culture, 1926–2015) (Worlds of East Asia), Berlin: de Gruyter Mouton.

——— (forthcoming) 'The Young Science of Materia Medica Studies in Early Twentieth-century China: botanical and philological surveys by Zhao Yuhuang (1883–1960)'.

Stöger E.A. (2021 [1991]) *Arzneibuch der chinesischen Medizin: Monographien des Arzneibuches der Volksrepublik China 2000 und 2005*, Stuttgart: Deutscher Apotheker Verlag.

Taylor, K. (2005) *Chinese Medicine in Early Communist China, 1945–63: a medicine of revolution*, London: Routledge.

Trambaiolo, D. (2014) 'Diplomatic journeys and medical brush talks: eighteenth-century dialogues between Korean and Japanese medicine', in: O. Gal and Y. Zheng (eds) *Motion and Knowledge in the Changing Early Modern World*, Dordrecht: Springer, pp. 93–113.

Unschuld, P.U. (2014) *Dictionary of the Ben cao gang mu: Chinese historical illness terminology, vol. 1*, Berkeley: University of California Press.

Unschuld, P.U. and Zheng, J. (2012) 'Introductory essay', in *Chinese Traditional Healing: the Berlin collections of manuscript volumes from 16th through the early 20th century*, Leiden: Brill.

Wang Xiaotao 王孝濤 (1981) *Zhongyao yinpian paozhi shuyao* 中药饮片炮制述要 (Essentials of Processing Chinese Medicinal Drug Products), Shanghai: Shanghai kexue jishu chubanshe.

——— (1986a) *Lidai zhongyao paozhi fa huidian, gudai bufen* 歷代中藥炮製法匯典：古代部分 (Compilation of Methods of Processing Chinese Medicinals throughout the Dynasties: the ancient part), Nanchang: Jiangxi kexue jizhu chubanshe.

——— (1986b) *Lidai zhongyao paozhi fa huidian, xiandai bufen* 歷代中藥炮製法匯典：現代部分 (Compilation of Methods of Processing Chinese Medicinals throughout the Dynasties: the modern part), Nanchang: Jiangxi kexue jizhu chubanshe.

Xie Zongwan 謝宗萬 (vol 1: 1964, vol 2: 1984) *Zhongyaocai pinzhong lunshu* 中藥材品種論述 (Discussion of the Grades and Types of Chinese Medicinal Materials), Shanghai: Shanghai kexue jishu chubanshe; (vol. 1: 2nd edn 1990) as *Zhongyao pinzhong lilun yanjiu* 藥品種理論研究.

——— (1996) *Quanguo zhongcaoyao huibian* 全國中草藥匯編 (A Compilation of Chinese Herbal Medicinals), 2nd edn, Beijing: Renmin weisheng chubanshe.

——— (2004) *Zhongyaocai zhengming cidian* 中药材正名词典 (Dictionary of Correct Names of Chinese Medicinal Materials), Beijing: Beijing kexue jishu chubanshe.

Zeitlin, J.Z. (2007) 'The literary fashioning of medical authority: a study of Sun Yikui's case histories', in C. Furth *et al.* (eds) *Thinking with Cases: specialist knowledge in Chinese cultural history*, Honolulu: University of Hawai'i Press, pp. 169–204.

Zhao Yuhuang 赵燏黄 (1932) 'Zhongguo xin bencao tuzhi 中國新本草圖志' (China new materia medica illustrated record), *Guoli zhongyang yanjiuyuan huaxue yanjiusuo*, 3 no 3 and, in 1932, no 6. In (2006) as a book in the series by Wang Zhipu 王致谱 (ed.) Fuzhou: Fujian kexue jishu chubanshe.

——— (2004) Qizhou yaozhi 祁州药志 (Record of Medicinals in Qi Prefecture), Fuzhou: Fujian kexue jishu chubanshe.

Zhao Yuhuang 赵燏黄 and Xu Boyun 徐伯鋆 (1934) *Studies of Crude Drugs* [Pharmacognosy] 生藥學, vol. 1 (vol. 2 by Ye Sanduo 葉三多, 1937), Shanghai: Zhonghua minguo yaoxuehui.

Zhao, Z., Guo, P. and Brand E. (2012) 'The formation of *daodi* medicinal materials', *Journal of Ethnopharmacology*, 140.3: 476–81.

Zhonghua renmin gongheguo guojia weisheng jiankang weiyuanhui 中华人民共和国国家卫生健康委员会 (National Health Commission of the People's Republic of China) 2018, *Guojia jiben yaowu mulu* 国家基本药物目录 -2018 年版 (National Essential Medicines List -2018 edition), http://www.nhc.gov.cn/ewebeditor/uploadfile/2018/10/20181025183346942.pdf, accessed 03/1/2022.

Zhu Jianping 朱建平 (2001) *Zhongguo yiyaoshi yanjiu* 中国医学史研究 (Studies on Chinese Medical History), Beijing: Zhongyi guji chubanshe.

Zhu Jianping 朱建平, Wang Yongyan 王永炎 and Liang Jusheng 梁菊生 (2005) Zhongyaoming kaozheng yu guifan 中药名考证与规范 (*Evidential Research and Standardisation of the Names of Chinese Medicinals*), Beijing: Zhongyi guji chubanshe.

49

YANGSHENG IN THE TWENTY-FIRST CENTURY

Embodiment, belief and collusion

David Dear

Introduction

Yangsheng 养生 literally translates 'nourishing life', 'nurturing life' or 'cultivating life'. The fact that there are multiple English translations of the word gives an immediate indication of the evanescent nature of the concept. This intangible quality is compounded by the breadth of the range of activities than can, or do, claim to have *yangsheng* qualities. They range from widely known forms of callisthenics and meditation to dietetics, sexual regimens, choral singing, calligraphy and even to the keeping of song birds or other pet animals. In its broadest sense *yangsheng* aims to promote the well-being of the individual: health and happiness, and in wider but less focussed terms, the collective societal benefits those goals might entail. *Yangsheng* is and has commonly been associated with Daoism as we will see in the body of this chapter. Equally it can be found as a feature of Chinese Buddhism, or indeed perhaps any religion inasmuch as religious practice cultivates in its acolytes physical disciplines to develop emotional and moral qualities with a spiritual dimension (Salguero 2014: 27–43). *Yangsheng* has further been adopted in a secular twentieth-century socialist lifestyle as a force for public good (Chen 1995; Hsu 1999; Farquhar 2001, 2012).

In its narrowest sense *yangsheng* is a programme of largely self-managed prophylactic healthcare for the avoidance of physical and financial costs of illness. As such, the precepts of *yangsheng* are closely related to Chinese medicine, in both the classical literary and the popularly socially pervasive forms of these two domains; but *yangsheng* concepts are also cross-related to ideas found in religion, philosophy, cosmology and politics (Despeux 1989; Engelhardt 1989, 2000; Stanley-Baker 2006, 2018). They are therefore to be found referenced in many other sections of this volume (Chapters 22, 29 and 30 in this volume).

While much of this chapter is concerned with tradition and the remarkable continuities of daily practices from the distant past, what truly fascinates is how the unique repertoire of practices associated with the term *yangsheng* continually expands and contracts to suit the mood of the time (Campany 2012). To explain this phenomenon, I will propose in this chapter a refinement that I will call 'Collaboration' or 'Collusion' theory. 'Collusion' is a negative term in economic and legal theory which describes a corrupting manipulation of markets and socio-economic conditions among a group of commercial actors (Harrington 2017). In my use of the term, it describes the natural coming together of individual actors and social and political groups to construct a common shared reality. As such it is entirely neutral; it could be 'bad', since it refers directly to selfish forms of self-interest, and it could be 'good', but it is essential to both cognitive processes and the ways in which cognitive processes shape our decisions and activities. It is the way we transcend our solipsistic perceptions to agree on shared meanings; whereby strawberries are 'sweet' and the colour red is to you, what it is to me. It is the construction of necessary and expedient realities that enable a shared social life to exist. This process of collusion, as I shall attempt to illustrate, is key to the birth of twenty-first-century *yangsheng* as *the* independent marker of an authoritative and elite lifestyle.

In the beginning...

There are early references to the term *yangsheng* in *Zhuangzi neipian* 莊子內篇 3 in a chapter that advocates the simple life, but it first appears in the manner in which we now understand the term in the *Yangsheng lu* 養生論 (Discussion on Cultivating Life) of Ji Kang 稽康 in the Three Kingdoms Era (220–265 CE) (Stanley-Baker 2018). But the earliest record of the practices that would now be so recognised are archaeological materials, including manuscripts, excavated from tombs dating from the Western Han dynasty (202 BCE–9 CE) around four hundred years prior to Ji Kang's work. It is these texts from the Mawangdui 馬王堆 (closed 168 BCE) and Zhangjiashan 張家山 (closed 186 BCE) tomb sites that allow us to identify a core of *yangsheng* regimens that pertain to this day: exercise, with a particular emphasis on breath control, dietary and sexual regulation, both in terms of prescriptions and prohibitions (Chapters 2, 6 and 20 in this volume).

The central concept linking these practices and the wider cultural network in which *yangsheng* rests, is *qi* 氣 (historically also written 炁) and its circulation within and without the body. The earliest description of the formal movement of *qi* within the body dates from approximately one hundred and fifty years prior to the Western Han tomb texts and is inscribed on a jade belt pendant of the mid Warring States Era (403–221BCE). It reads:

> Swallow, then it travels; traveling, it extends; extending, it descends; descending, it stabilizes; stabilizing, it solidifies; solidifying, it sprouts; sprouting, it grows; growing, it returns; returning, it is heaven. Heaven – its root is above; earth – its root is below. Follow the pattern and live; go against it and die.
>
> *(Harper 1998: 125–6. The artefact itself is apparently mislaid in the Tianjin Museum)*

Interpretations of this passage variously link it to sexual and alchemical practices (Yang 2018).

The Warring States period was one of political turmoil and rapid intellectual development as well as considerable material innovation. It was a time when scholars sought, and believed possible, universal theories of governance that could be applied to *Tianxia* 天下 'All Under Heaven', to bodies political and corporeal alike. However, mirroring the many disputed philosophies of statecraft in this period (Graham 1989b), the first recordings of the term *yangsheng* already referred to what were contested practices. For Mencius 孟子 (c. 372–289 BCE) it was a question of filial duty of care to one's ancestors and most particularly one's parents; insofar as this included the nurture of the body they gave to you and the duty to procreate (Lau 1970: 123,125,127,130). Writings attributed to Zhuang Zhou 壯周, an approximate contemporary of Mencius, on the one hand, apparently mock the 'huffers and puffers' who exercise and practise sexual regimens in the hope of extending their lives and, on the other hand, appear to endorse breath control exercises as a means of spiritual advancement (Graham 1989a: 84, 1989b: 198; Roth 1999). While the practices of *yangsheng* long predate the terminology as it is now understood, we also find that for most of Chinese history, until very, very recently, these practices were largely an adjunct to a variety of cultures aimed at self-improvement. *Yangsheng* practices are found in subsidiary roles to medicine, martial arts, sociopolitical movements and religious practice in general including the arcane spiritual goals of inner alchemy. So while the concept and its practices have long existed, *yangsheng* was rarely a rubric *per se*, and usually appeared as a descriptive subtitle or explicatory adjunct. This subsidiary position of *yangsheng* is what has changed in the twentieth and twenty-first centuries as, responding to the political climate, it transformed from a nationally sponsored health regimen to a commercial product. This recent transformation is what makes a discrete chapter on the subject necessary.

Yangsheng and *Qi:* thought styles made real

The many approaches to living that *yangsheng* connotes share a kind of formula, a similar rationale and the same prescription: this *yangsheng* culture has created a self-reinforcing network of apparently concordant ideas. This is what Fleck called a *'thought style'* and in the case of *qi* it entails many intersecting 'thought styles' common to specialist activities such as divination, astronomy and medicine, with a common conceptual foundation (see Introduction in this volume, also Fleck 1981: 99). The semantic fields of *qi* link many *yangsheng* practices and their explanations, and as elsewhere in this volume, we need to give an account of qi for the subject in hand. What is *qi*? In the earliest records, it referred to steam, clouds, ghosts and amorphous ever-transforming meteorological phenomena in the external world, it ran in streams and rivers and collected in ponds in the body. Over time the meteorological nature of *qi* became a linguistic abstraction with which one could identify and label bodily experiences, such as sensation emotion, vitality, digestion, heat and cold, which were otherwise vague and unmeasurable (Hsu 2007, 2009). Nevertheless, it is no less real – undoubtedly much more real, as part of the daily 'lived experiences' of those who inhabit the network and its discourse – than the blood that apparently circulates in our bodies. It marks both sensory experiences and the agency of a thing, a substance, or a practice to create those experiences (Lo 1998, 2013: 32–64; Chapter 2 in this volume). Once labelled, the vague and unmeasurable 'experiences' become coterminous with 'things' categorised and placed in relation to one another by their shared qualities. The *qi* of a food, a drug, a movement, a note intoned, can strengthen, warm or calm the body's *qi*. They can be in sympathy, the same kinds of 'thing', eliciting the same kinds of experience, or equally quite different, with the ability to exert a variety of harmonious or antagonistic influences on one another other.

Crucially then, the concept of *qi* and its classifications allow the world to be manipulated. It can be seen as a sort of existential algebra, in that unknown, unmeasurable or sometimes simply subjective phenomena become identified and classified in relation to one and other and indeed in relation to sets of directly observable or palpable physical phenomena (Chapters 1 and 2 in this volume). That the pulse reflects the condition of the inner organs, the spirits of the planets inhabit one's body as spirit animals, that laughter might be associated with summer and the heart – all this depends on the flow of *qi* through channels just as rivers and canals formed the transportation network that unified China. The concept of *qi* allows the building of an epistemic system encompassing all the various elements outlined above, not excluding what many would perceive nowadays as magical phenomena. Ludwig Fleck observes:

> Both thinking and facts are changeable, if only because changes in thinking manifest themselves in changed facts.
>
> *(Fleck 1981: 50)*

Fleck was writing about the contingency of facts on systems of thought which are themselves changeable. I would concur with his observation about the shifting nature of how we manipulate the world, but turn his preceding question – 'Would it not be possible to manage entirely without something fixed?' (Fleck 1981: 50) – on its head: *Would* it be possible to manage at all without something fixed? Fleck himself provided the answer through describing those processes through which thought communities were created, the conversations that located and fixed meaning, the compilations which were canonised, the institutions that endorsed meaning. In the case of *qi*, many of these processes, interactions and institutions are fortunately available to us in the historical record, and form the subject of many chapters of this book (Harper 1998; Sivin 1995; Nylan 2009; Chapters 1 and 2 in this volume). As Shapin has shown us for the history of science, when communities share assumptions about how things work and can be communicated, something does become fixed, and that which is fixed is the beginning of a system. Without any system, systematic action is by definition, impossible. Precisely because of its fluidity, the imagination of *qi* was able to bring sense and order to a world that has had extraordinary longevity. The process of fixing those 'unmeasurables' in a language of *qi* is what allows the *yangsheng* practitioner today to make some order out of ambiguous sensory signals, albeit in a fluid system of facts and judgements. As those involved learn from their own experiences and articulate it through the shared linguistic code, from external concepts and epistemologies transmitted by others over two millennia, the network of 'Collusion' grows.

Qi, often rendered with a capital 'Q' in modern English transliteration, has a very specific meaning in contemporary Chinese medicine, relating to all those signs and signals which the medical practitioner detects in the body of the patient, and the patient experiences as an expression of their health. However, the issues of categorisation and manipulation of things in the world are shared across time. And even though, for example, in different periods the two different *qi* characters (氣 and the less common 炁) may have referred to different aspects of the concept – perhaps *qi* in general and as manifest in the human body – all of those aspects would have been seen as a vitality or life force. The vagueness of *qi* itself is part of its power and a characteristic that has facilitated cross-cultural transmission. Just as some find symbolism and potency in the words and characters of an alien language, the polysemous nature of *qi* allows people to project their own meanings into it.

Yangsheng and Qigong fever

Qigong 氣功, the combination of breathing and projection of *qi* in training the body through repetitive movements, like the gentler martial art, Taijiquan 太極拳 (aka Tai Chi Chuan), now enjoys a widespread popularity outside China and has been adopted by those practising forms of 'mindfulness' – methods of bringing one's attention to experiences occurring in the present moment.

Qigong picks up on the use of *qi* in therapeutic exercise and transforms it into a psycho-medical or meditative practice for health and well-being, physical and spiritual. Though its enthusiasts claim for it an ancient pedigree, it is in fact a twentieth-century invention, which shares elements of its history with Taijiquan (and indeed the broader umbrella category of *yangsheng*), *daoyin* 導引 (lit. 'leading and guiding', the oldest recorded style of therapeutic exercise), Chinese Medicine and *neidan* 內丹 (inner alchemy, cultivating the elixir within the alchemist's person) (Chapters 2, 6 and 30 in this volume). The definitive history of its twentieth-century emanation was written by David Palmer in his 2007 book *Qigong Fever*. The 'fever' in question is the wild upsurge in popularity that *qigong* practice enjoyed in

China in the 1980s and 1990s. This ran in tandem with Deng Xiaoping's 鄧小平 economic and political reforms from 1979 of the period of 'reform and opening up' (*gaige kaifang* 改革開放); in the wake of the years of traumatic repression inflicted by Mao's Cultural Revolution (1966–76).

But *qigong* was first identified as a unique practice in the 1950s, something unique to China, formed from the special resources of Chinese culture, both elite and populist. As Palmer describes it, *qigong* becomes 'a general and autonomous category' post 1949:

> englobing most traditional breathing, meditation, visualization and gymnastic practices to which over the years would be added martial, performance, trance, divination, charismatic healing and talismanic techniques, as well as the *Book of Changes,* the study of paranormal phenomena and UFOs.
>
> *(Palmer 2007: 18)*

A very similar process can be seen taking place in the birth of twenty-first-century *yangsheng*. While *qigong* is not identical to *yangsheng*, the overlap is very substantial. Importantly, they both utilise the same modalities of epistemic syncretism which enable them to expand beyond their original loci of meaning and practice, to encompass much broader ranges.

The story of *qigong* is laced with tales of miraculous cures, amazing special powers, omniscient masters and audacious frauds. Often these came in all and one and the same package. However, there is no doubt that many ordinary practitioners found in it great value both for their physical and emotional well-being.

The public fever for *qigong* came to a chilling halt in April 1999 when 'ten thousand'[1] Falun Gong 法輪功 (Dharma Wheel Practice) practitioners surrounded the Chinese government's Zhongnanhai 中南海 compound in Beijing to protest what they felt were media slanders against their movement. Falun Gong was, and is, a neo-millennialist religious 'cult' mixing teaching of Daoism and Buddhism with *qigong* meditation and exercise practices. It was immensely popular in China in the early 1990s, but its fortunes changed after its founder, Li Hongzhi 李洪志 emigrated to the USA in 1995, and the leadership, in concert with ultra conservative political groups in the USA, started issuing radical denunciations of the Chinese Communist Party (CCP) rule in China. Millennialist religious groups challenging the government and toppling dynastic regimes are another trope of Chinese history that extends back at least two thousand years. That pro-government propagandists should become involved in a war of words with Falun Gong's claims should have come as no surprise to anyone. However, the counter reaction by the Falun Gong followers in Beijing came as a massive shock to the government and resulted in a draconian crackdown on the group and anything of its ilk.

Chameleon Yangsheng: all things to all men?

With the suppression of Falun Gong, the public fever for organised *qigong* was effectively snuffed out. However, the public appetite for what *qigong* meant for the individual could not be so easily dismissed and much of this energy and activity was transmuted into what was now formally identified as *yangsheng*. Interpreted in this light, the practices that were once acceptable under the *qigong* rubric, and had become either outlawed or generally frowned upon, became to some degree acceptable once more.

While her predecessors like Thomas Ots or Nancy N. Chen were, respectively, interested in the inner emotional spaces, or the outer social spaces related to *qigong* practice, anthropologist Judith Farquhar in her books *Appetites* and *Ten Thousand Things* (with Zhang Qicheng) was one of the first to engage with the 'new *yangsheng*' in its own right. *Appetites* contains an engaging series of descriptive essays illustrating the growth of *yangsheng* around the sensual pleasures of food and sex. Published in 2002, however, it prefigures *yangsheng*'s adoption as a marketing slogan for everything from luxury villas to high-end massage parlours (Figure 49.1).

Ten Thousand Things, published a decade later, was based on a collaborative research project with Professor Zhang Qicheng of Beijing's University of Chinese Medicine. The study focusses largely on research among a group of retirees and seniors in west Beijing. While it gives due attention to their embodied qualities of practice, it also has a somewhat 'period feel', reflecting an established way of life from earlier decades. There are a number of reasons for this and they, in themselves, make for valuable insights into the continuously moving research target that is *yangsheng*.

One reason is that the research was mainly conducted in 2002–04. In the first decade of the twenty-first century under the leadership of Secretary-General Hu Jintao 胡錦濤 (1942–), the slogans of the moment were the establishment of the 'harmonious society' and

Figure 49.1 'A decision. A life for living. The Butterfly Allure Gentleman's *Yangsheng* SPA centre'. Poster. Author's own photograph

from 2007, the ambition to build a 'moderately prosperous society' (Hu 2007). Moderation is always hard to achieve in a nation as large and as volatile as China.[2]

Thereafter the urban population of Beijing, and no doubt many other eastern cities, became visibly wealthier almost on a daily basis. This was in part connected to the investment in the city in the run-up to the 2008 Olympics. This investment was not simply restricted to Beijing since the Olympics were also seen as an opportunity to promote the nationwide tourist industry. Following the Olympics, the Chinese government responded swiftly to the global financial crisis of November that year, with a massive release of funds to shore up the internal economy. Two results of this were (a) the boom in car sales, which created gridlock on the streets of many cities and (b) a calculated abandonment of a previously inculcated thriftiness accompanied by a growing bourgeois appetite for luxury goods, demonstrating the wider collusion between the appetites of state and consumer.

The meaning of *yangsheng* also began to shift with the times and the global market for complementary medicine and spa culture. It now became a standalone and distinctive strapline for the promotion of products as diverse as the aforementioned massage parlours, villas, and the more esoteric, not to say wacky and self-indulgent, extremes of Chinese medicine, as well as continuing to be applied to the more conventional forms. Foodstuffs, from commonplace yoghurt sold on street stalls 'made with real milk!', to poached sea cucumber and cordyceps in expensive restaurants (a vital aid to flagging male vigour) all proudly proclaimed their *yangsheng* credentials. And if *yangsheng* was good for you, you should feel good about spending your money and buying it. 'The more expensive, the better', in the minds of purchasers and vendors alike, though perhaps not necessarily for the same reasons.

After Xi Jinping 习近平 took power in 2012, he clearly identified ingrained corruption as the greatest threat to national security and, in equal measure, CCP rule. In the years since then the campaign has gradually ratcheted up in strength, 'trapping tigers and swatting flies', Xi's phrase indicts high- and low-level corrupt officials, as well as targeting political rivals and cracking down on liberal opinion. In these more austere times there has been a general flight from luxury goods and conspicuous consumption among the wealthiest classes. *Yangsheng* has once again shown itself adaptable to the circumstances. With its deep connection to 'traditional' Chinese culture, philosophy and medicine, it became a respectable outlet for the interests of businessmen or retired cadres, when golf (BBC News 2015) or luxury holidays – in non-*yangsheng* locations– might have raised eyebrows.

From the above we can see that *yangsheng* is simultaneously a fashionable commodity and a philosophy of life, or living. The latter is nothing new, but the commercial packaging of a *yangsheng* lifestyle marks a new phase in its history. In the ancient world the way people disported themselves, dressed, ate, even the way they got sick was evidence of their social status (Lo [2007] 2018: 84–86). The *Zunsheng bajian* 遵生八笺 (Eight Treatises on Respecting Life) published by Gao Lian 高濂 in 1592 was a manual for living, aimed at a newly educated and religiously eclectic elite, which combined a mixture of spiritual advice, health tips, guidance for seasonal living and how to acquire and appreciate the finer things of life (Chen 2008: 29–45). It was a bestseller. If that all sounds a bit 'Gwyneth Paltrow', that's just another 'first' for China; the economy of the Ming dynasty was in many regards strikingly modern to our way of thinking.

So with its myriad faces, just what is *yangsheng*? *Yangsheng* is distinct from TCM in that while the latter, like all medical practice since medieval times, has been about the treatment of disease and physical or psychological dysfunction, the focus of *yangsheng* is on the maintenance and promotion of good health. In essence it seeks to produce a bodily equilibrium that obviates the need for medical intervention. The techniques of *yangsheng* involve what can be likened to a form of aspirational Epicureanism, a refining and attenuating of sensory experiences in the pursuit of different goals, in which nothing except excess is necessarily denied. Its key practices, from the very earliest recorded formulations in the Western Han dynasty, are based on meditation and breath control, callisthenic exercises, dietetics and sexology. This remains true to this day and even newly invented or appropriated practices can be judiciously fitted into this framework: it doesn't take a great leap of imagination to connect choral singing with breath control. Practices like raising pet animals (birds, dogs etc.) literally 'nurture life'. But they clearly relate to the equally antique practice of *yangxin* 养心 'nurturing the heart', the heart being more specifically a reference to the residence of the spirit rather than life in general.[3] A key principle that links these practices of *yangsheng* and *yangxin* along with Epicureanism, is the need to avoid extremes of emotion; joy, anger, grief, laugher, etc., and extremes of desire, identified as the causes of disease in aristocratic people from ancient times (Lo 1998).

Patience and selflessness shown in good care for another person or creature are equally good for one's own health, a *yangsheng* reflection of Buddhist principles of merit building which in Buddhism may extend also to the health and well-being of others (Salguero 2013: 344). Western medical studies have also recently asserted the health benefits of pet ownership, albeit completely outside of the *yangsheng* discourse (De la Paz *et al.* n.d.).

The simple pervasiveness of these practices of *yangsheng* and *yangxin*, and the perception of their antiquity, raises the all-important question of how one explains the dynamic between those elements of continuity and transformation in a tradition. To my mind the most theoretically useful in this context is Pierre Bourdieu's work on 'Doxa' and 'Habitus' (Bourdieu 1977: 72–87, 159–97). Doxa accounts for the whole raft of unconsciously absorbed and socially unchallenged influences that form our ways of thought and ways of being. When Doxa becomes explicit it fragments into the Orthodox and the Heterodox; the dialectic struggle between the two is eventually resolved by their subsumation into a reformed Doxa. *Qi* can be described as a doxic phenomenon in East Asian culture. 'Habitus' (a concept that Bourdieu developed from Marcel Mauss, who borrowed from Aristotle's 'Hexis' ἕξις) refers to a form of living practice that creates, reinforces and reproduces the influences, doxic or otherwise, which informed it. In that Aristotle's Hexis related to one's physical constitution and health, Bourdieu's evolved interpretation is remarkably apt in the context of *yangsheng*. Meanwhile a key principle that links the practices of *yangsheng* and *yangxin* (nurturing the heart), along with Epicureanism, is the need to avoid extremes of emotion; joy, identified as causes of disease.

What is important about the habitus of *yangsheng* practice is not simply its validation though antiquity but more that it exists within a network of reinforcing influences (Mauss 1979: 73). Lengthy empirical evidence of efficacy does no harm – 'empirical' in the sense that there are plenty of records of trial and error – but also people only tend to persist, over the long term, with practices that they agree 'work' (a notable exception to this might be found in the use of cinnabar, mercury sulphide, and arsenic, by some elites for at least one thousand years as a medicine to attempt to achieve immortality!) (Needham and Ho 1970). Practices experienced and communicated as successful strengthen belief and therefore the habitus, but habitus is very much an 'à la carte' prescription. No two individuals are exactly the same or experience their practice identically; and there are many practices and prescriptions that were once respected and are now neglected – and still without damage to the reputation of the whole and its founding conceptions. Some 'traditional' medicines still claim to contain cinnabar, mercury and arsenic (though perhaps only in homeopathic quantities); the legend overtops the indisputable scientific evidence of the dangers of heavy metal poisoning.

In twenty-first-century *yangsheng* we have a network of ideas and practices that stretch back in various forms over at least two millennia. To quote Jian Xu in his essay *Body, Discourse and the Cultural Politics of Contemporary Chinese Qigong,* physical practices are vital because 'the body turns the discourse into belief by giving it a material form' (Jian Xu 1999: 985). Many, though not all, of the *qigong* practices of the 1980s and 1990s have come to be considered a respectable part of *yangsheng*. As mentioned above, after the public crackdown on *qigong* cults like the Falun Gong in 1999, many of those who previously would have taught exercises labelled as *qigong* re-branded themselves as *yangsheng* teachers.

Yangsheng's subtlety, its mundane qualities, its ubiquity and utility make it impossible to fully control – though that does not stop the authorities from periodically trying. In 2011, when a number of authors were denounced for the errors in their works, the Beijing authorities published a list of approved *yangsheng* teachers, drawn almost entirely from officially recognised medical institutions. Here the successful collusion is between a modern TCM and the state-sponsored tailoring of tradition to the perceived interests of the nation.

Professor Qu Limin 曲黎敏, Zhang Qicheng's colleague and former student, was one of those whose works were identified as substandard. A philologist by academic specialisation, she had 'seized the moment' to become a popular author and TV lecturer on *yangsheng*. This obvious self-aggrandisement in the search for fame and wealth was evidently considered inconsistent with the proper behaviour of a dignified teacher.

It is striking that a great number of the exercises in *qigong*, *yangsheng* or *neidan* (inner alchemy) practice are, in physical form at least, identical; the defining differential between them is the intention of the practitioner and the way the institutions that sponsor them formulate the practice. Accepted practices included meditating on the *dantian* 丹田 the 'cinnabar field', a centre of power beneath the navel, and on the movement of *qi* along the spine, 'knocking the teeth' (clacking the teeth together to stimulate 'the elixir' beneath the tongue), or standing like a pillar with hands outstretched under trees in the park, which is thought to increase the power of the practice. In the twenty-first-century context, the obvious esoteric alchemical history embedded in these practices sits quietly below the surface.

Nevertheless, the similarity of the exercises across this cultural spectrum adds to the assumption of their authority. The shared elements interlock in mutual support, strengthening belief in the efficacy of practice. In an interview I conducted in 2005 with Professor Zhang Qicheng, he identified two levels of practice: those practices with the simple aim of maintaining physical health and those with a higher intention to progress on a spiritual quest. *Yangsheng* practice in its most basic form acts as entry-level activity for the curious. The apparent simplicity of many of the exercises and practices can also disguise what the practices can achieve for the more deeply committed. While the effect of such commitment and endlessly repeated practice – repeating the same action or movement 'ten thousand times' until finally achieving mastery of it – has consequences both physical and psychological, which become part of the network of influences driving the outcome, the impact occurs through the complex interplay of sociopolitical and cultural collusion described in this chapter.

Yangsheng is both social and highly personal. This also adds to a unique and pervasive sense of efficacy. Both aspects are important in this. The personal results in a tangible sense of increased health and well-being. The social we see in the groups of people meeting to practise together in the parks. Often retirees or middle-aged house wives, meet regularly, daily – weather allowing – to practise their exercises, gossip and swap social news and healthcare tips and anecdotes. Farquhar and Zhang's study identifies a word-of-mouth sharing of information within the practice group as being by some considerable way the main influence on the formation of thinking and practice by group members. In this way they build a more potent shared authority than is possible from information from media outlets like books and television – though these were also judged to have an effect. It does not take a great deal of imagination to appreciate the benefits to the health and satisfaction of the individual in having such a social group to attend and be attached to.

Yangsheng and Guoxue

Guoxue 國學 or 'National Studies' has had a presence in China since the intellectual movements of the early twentieth century. It was probably inspired by the National Studies programme of nineteenth-century Japan (Fogel 2004: 182). It embraced the idea of studying one's mother culture in the manner of an academic discipline of the sort found in the West, as opposed to the traditional Asian methods of study. However, its current incarnation in the People's Republic of China is somewhat ambiguous. It operates as a pseudo- or semi-academic discipline very much focussed on nationalist propaganda, which establishes and justifies a form of 'Chinese exceptionalism' through culture. *Yangsheng*, with its deep reach into popular culture, is one of its most favoured subject areas. Most TV channels broadcast a number of programmes in which 'experts' and audience members trade opinions, very often about food and diet, justified by *yangsheng* theory. Sadly for the programme-makers, many of the more exotic prescriptions and prohibitions associated with *yangsheng* dietetics were recently debunked in a paper from The Beijing University of Chinese Medicine (Lin et al. 2014) – not that that will stop people believing, or *wishing* to believe, in them. *Guoxue*, a blend of collusion between the personal, social and national, became an amplifier for the *yangsheng* marketplace. Zhang Qicheng, Farquhar's collaborator in *Ten Thousand Things* has himself slipped easily from gamekeeper to poacher. Zhang now promotes his own, very successful *yangsheng/guoxue* study courses through his website, Zhang Qicheng *Guoxue* Wang (Zhang Qicheng Guoxue Net): www.zhangqicheng.com.

All about me…?

Against this social and political network of collusion, we can recognise that *yangsheng* is in essence, highly personal. But it is personal in many different ways. It is personal in the sense that it cultivates the individual's relationship with their body. It is personal in that everyone has their own particular views and feelings about what suits their own particular needs. It is personal in that it addresses peoples' sexual and reproductive lives. Though Dikötter (1995. 66–69) describes it as 'relatively new', it is not clear whether he detects this 'newness' in relation to the latest modern formulation of the category, or in relation to the publishing industry of the late imperial era. On the contrary, sex as a matter of both health and pleasure – the two are interlinked in *yangsheng* – seems to have been a constant concern from the Mawangdui tomb literature of the second century BCE to the writings of Mantak Chia, perhaps the strongest promoter of Daoist and *yangsheng* sexual cultivation in the English language. (For a comprehensive list of Chia's works, see Chapters 25 and 26 in this volume.) One aspect of the ancient 'Art of the Bedchamber' (*fangzhongshu* 房中術) – a somewhat misleading catchall phrase that covers a wide variety of sexual interests and purposes historically – that has gained increasing attention since the One Child Policy is the need for an optimal sexual practice by both parties to produce the healthiest and most talented offspring (Wile 1992: 118, 121; Chapter 20 in this volume). Younger people, who very often can afford to take their health for granted, typically start to take an interest in it and seek out some of its regimens when newly wed, and faced with the issue of starting a family. In this context *yangsheng* is the repository for many ancient and modern ideas about ensuring the health of one's progeny, whether as an individual duty to one's ancestors, or from the perspective of a more broad-based eugenic focus on population health (Chapters 25 and 26 in this volume) Typically, the new husband, perhaps even under family pressure, will forego alcohol and tobacco, as well adopting other dietary changes, in order to father the healthiest possible child. Similar strictures will apply to the prospective mother – we can hear an echo of this type in the various nannying pressures put on the behaviour of pregnant women in Western media (see also Raven *et al.* 2007; Holroyd *et al.* 2011: 47–52).

Finally *yangsheng* is deeply personal in that it can encompass the individual's hopes and dreams for 'spiritual' fulfilment. This is where it starts to elide with the practices of *qigong* and the imaginative mysticism of inner alchemy (Chapter 2 in this volume). This is where individuals enter that secondary level in Zhang Qicheng's description. Adepts set out on a quest that may offer the meaning and fulfilment they seek – or their road may lead them into personal trauma and disaster, namely, *Zouhuo rumo* 走火入魔 (obsession or possession by malign spirits) in medieval *neidan*, *piancha* 偏差 or deviant *qigong* in more modern formulation – or psychotic illness, as we might be more inclined to call it. This is why most religious groups that practise these exercises insist that students do so under the guidance of an experienced master. An alternative risk is that a deeply committed belief can pitch an individual or group into conflict with the forces of authority, as in the case of the Falun Gong.

In describing the learning processes involved in *yangsheng* exercises, I have alluded to the way in which the intention and concentration of the practitioner over time are the route to accomplishment. This is very often directed to the simplest physical details of practice without reference to any complex or profound ontology. Practice is repeated, often on a daily basis, sometimes several times per day, leading to an accumulation of thousands and thousands of iterations before 'mastery' is achieved. 'Ten thousand times', and the collusion is complete – embodied, one might say. A good Taiji teacher will always stress the fundamentals of balance, posture and breathing over any more esoteric ideas. Complex ideas or ideals may inspire and motivate but it is only repeated physical activity that leads to embodied change. In this case even something as placid as sitting meditation (*dazuo*打坐) is counted as physical activity.

This raises a further key issue in the nature of *yangsheng* practices; the plasticity of the human brain. Neuroplasticity in the brain is a well-observed phenomenon. It is described thus:

> The brain's ability to reorganize itself by forming new neural connections throughout life. **Neuroplasticity** allows the neurons (nerve cells) in the brain to compensate for injury and disease and to adjust their activities in response to new situations or to changes in their environment.
>
> *Shiel (2017)*

This occurs most obviously in stroke patients where other areas of the brain may restructure or 're-wire' themselves to take over the functions lost in damaged areas. But this 're-wiring' can also occur through study, lifestyle and embodied practice (Münte, Altenmüller and Jäncke 2002). Cognitive neuroplasticity has famously been measured in London taxi drivers who have done 'the knowledge', learning by heart the connective details of roads and routes around the city. The taxi drivers were shown to have a better developed and expanded posterior hippocampus when compared to non-taxi drivers (Maguire, Woollett and Spiers 2006). I would suggest that prolonged and repeated *yangsheng* practice creates 'psycho-plastic' changes to the brain. Essentially, this logic suggests that a repetitive bodily practice or behavioural routines will cause physiological changes in the brain. An example of this type of effect might be seen in the practising of scales on the piano; while there may be some physiological changes in fingers and limbs, the really significant changes giving rise to increased dexterity and ability are surely taking place in the elements of the body that effect coordination, in the nervous system and, most critically, the neural connections of the brain. In fact, I think we can say it goes further. Recent neurological research indicates that repetitive practices lead to physical changes in the brain which then biologise routines and perceptual susceptibilities permanently, though not necessarily irreversibly (Draganski *et al.* 2004). To quote from the abstract of one recent paper in *Nature* on neuroscience:

> The brain is the source of behavior, but in turn it is modified by the behaviors it produces. This dynamic loop between brain structure and brain function is at the root of the neural basis of cognition, learning and plasticity. The concept that brain structure can be modified by experience is not new, but it has proven difficult to address experimentally. Recent developments in structural brain imaging techniques, particularly magnetic resonance imaging (MRI), are now propelling such studies to the forefront of human cognitive neuroscience.
>
> *(Zatorre. et al. 2012: 528)*

Irrespective of whether the changes brought about by *yangsheng* benefit the health of the practitioner objectively, they will in many cases be perceived by the practitioners themselves as beneficial, and as such this will establish a belief in some kind of validity. Thus, in addition to the effects of practice on neural plasticity we have an additional and more immediate response to *yangsheng* practice in the form of the *placebo effect* or, to use Daniel Moerman's less negatively loaded term, a 'meaning response' to the 'felicity' or 'appropriateness' of an intervention.[4]

Some committed to narrow cultures of Evidenced-Based Medicine may belittle placebo as somehow 'unreal' – but clearly it is real – the evidence is there in all those double-blind trials where the placebo drugs' performance to some degree matches that of the substance being tested. Perplexing as this may be, to dismiss it entirely is simply churlish and illogical. Studies have shown on the one hand how response to drug treatment is affected by expectation (Bingel *et al.* 2011) and on the other how inert placebo medicines given openly as such to patients can still produce superior outcomes to the same 'meaning response' interactions without the inclusion of the sugar pill (Kaptchuk *et al.* 2010). Results are complex and inconsistent, and this is often the reason that simplified and comprehensible prescriptions of *yangsheng* have appeal and meaning to ordinary people. Body and 'Mind' are not separate entities any more than 'nature' and 'nurture'. As Gilbert Ryle argues from a philosophical perspective in *The Concept of Mind* (2009), the brain and consciousness are material and physical; the division of the two in early modern European culture is due to a 'category mistake' in Cartesian thinking. And from a purely physiological perspective, were that not the case, how would psychoactive chemicals work? Consciousness without the physical inputs of sensory stimulation would be no consciousness at all. It is a 'two-way street', never bifurcated so simply in medicine in China, so that ailments now deemed physical can be traced to somatised emotional or psychological issues. This point has been made comprehensively by Arthur Kleinman in many publications (see for example Kleinman, 2007).

The suggestion of some, undoubtedly limited, *psycho-plasticity* in combination with so-called 'placebo' responses in controlling the body's behaviour and reactions is far from outlandish; it is worthy of research. We know for example that people can be trained to slow their heart rate, and repeated recent research has shown that people with strong religious belief tend to suffer less in illness and recover more quickly therefrom (Becker *et al.* 2006). The psychological and physical benefits of *Taijiquan* for the mid-age or elderly have been shown in a number of Japanese research papers and indeed the exercises now enjoy a cautious endorsement by the UK's National Health Service (Liu 2005; NHS 2018).

Thus we find a network of ideas and experiences; objective and subjective, societal and individual, social and personal, all of which contribute to conform the practitioners' experience and within which they will seek to conform themselves. 'Culture' is by definition a process of conforming and self-conforming. As such the culture of *yangsheng* and its practices can be considered a form of social, psycho- and physiological collaboration or 'collusion'. Like the ideas of group-think or thought collectives (Denkkollektive) discussed by Fleck (1981), they exist as part of a commonly constructed reality into which all of humankind to a greater or lesser extent must be co-opted in order to sustain a social existence.

However, these repetitive *yangsheng* collaborations or collusive acts are different in that they are not merely abstract social exercises in group-think. As noted by Jian Xu they are physically embodied by their practitioners and this embodying of experience reinforces belief. My very first interviewee on the subject, a university lecturer from Beijing, described how in the late 1980s, after the death of her mother, she underwent an existential crisis and started to seek a non-materialistic meaning for her life through self-taught *qigong*. She would meditate in every spare moment she could find outside her work and busy family life. As she meditated, she would repeat to herself the statement; 'I believe there is a Divine Spirit in the world'. She was quite frank that she did *not*, at that time, believe. Her relentless practice caused strains within her family. One day, after about three years of practice, she suddenly felt she could move her *qi* at will and indeed emit it from her palms. Her longsuffering and sceptical husband returned home and berated her for having failed to do the housework. She demonstrated to him her new found skill. He was dumbfounded but then convinced and quickly became supportive. She subsequently trained to use her powers for *qigong* diagnosis and healing.

Whatever others may make of her anecdote, she was now a true believer and many would seek her out for help with their ailments. Her belief came real and absolute with her perceived embodiment of practice. Her story traverses the bounds of what had become the acceptable medicinal style of *yangsheng* in Communist China and enters into the realms of faith-healing, but the discourse of *qi*, and the network of ideas which are intertwined with it and support it, are what enables this crossover. *Qigong* healing has a recognised, if generally somewhat lowly place in the panoply of Chinese medicine. The *qigong* fever documented by David Palmer saw *qigong*'s social status elevated (in the 1950s), trashed (in the 1960s and 70s) and then hyped to fever pitch (in the 1980s and 90s. This rise, fall and rise again and final 'nemesis' came in response to political diktat and the interpretation of meanings within the context of the social needs of the times, rather than anything much intrinsic to the practices themselves.

From all the above we can see how a gamut of different influences, which form the bedrock of *yangsheng*, coexist. They feed the self-reinforcing network of beliefs and practices; collusive, collaborative, conforming. It is reminiscent of Barbara Ward's (1985) observations of the cultural models held among Hong Kong fishing villages fifty years ago. The villagers were able to recognise three different models of practice, which they used to locate their own identities. First there is their own 'home-made' or immediate model. This is what they actually practise themselves, shaped by circumstance and necessity; it may not fully conform to ideal practice but it is *their* practice, and this is what gives it power and prominence. Second there is the ideal model, the 'believed-in traditional model'. This consists of the practices of those recognised as the educated social and cultural elite; the holders of 'cultural capital', in

Bourdieu's terms. These practices represent a model to which the villagers can aspire, and which, in as far as they are able to conform to it, enhances the authority of their own practice. Finally, there is the 'internal observer model'. This is their view of the heterodox practices of those that they recognise as being related or similar to themselves, which however deviate from the reasoned norms of the first two. The villagers could triangulate between these differing concepts without undue conflict, and indeed the input from each strengthened the rationale for the whole and their particular place in it.

This form of cognitive coherence-making allows for resolution of those potentially dissonant models of identity and actually strengthens the place of the individual or group in the network as a whole. It is at the very heart of 'collusion'. Beyond that, the phenomenon of plasticity in the brain serves to mark the potential for 'collusion' at a biological level. However, simply identifying a practice as 'collusion' is fatuous. In a very real sense, almost everything is a form of collusion, even in the most perverse situations where actors are in direct conflict. The real utility of the concept is to deconstruct and analyse the elements that make up the 'collusion', the better to understand the underlying psychology and motivation of its participants and the external elements that shape their positions.

Conclusion

Yangsheng is a process of empowerment connects individuals through their bodies to a whole network of cultural resources. Repeated physical practice does indeed bring about change in the body, in the brain, in the *xin* 心 – the heart-mind in Chinese culture. It is this sense of empowerment that is its greatest strength. In the first place it makes the individual think and act consciously about their health. Beyond that it might be considered, to some extent, a 'steering wheel' for driving the positive aspects of placebo medicine: the social and psychological factors that prompt an improvement in health status, irrespective of effective medical intervention. And many of its practices are simply the 'common-sense' advice that can be found in any cultural scenario. *Yangsheng* connects a whole lore of popular and populist epidemiology with a cultural network which affirms an individual's place in society and culture at large. The plethora of books, TV shows and popular media, the social connections in the family or in the public parks provide a depth of support that allow these ideas to 'float' in a way that is meaningful to the individual. It provides individuals with a little bit of ownership of a much wider culture that otherwise might just seem confusing and oppressive. Perhaps this is the true meaning and value of *guoxue*.

Yangsheng is currently a buzzword for the marketing of lifestyles and services. Sometimes it is synonymous with discreet luxury pandering to the wealthy, sometimes it is used to dignify rather more lowbrow personal services, but in general it is a signifier of cultural 'distinction' in Pierre Bourdieu's definition of the term.

> Whereas the holders of educationally uncertified cultural capital can always be required to prove themselves, because they are only what they do, merely a by-product of their own cultural production, the holders of titles of cultural production – like the titular members of the aristocracy, whose being is defined by their fidelity to a lineage, an estate, a race, a past, a fatherland or a tradition, is irreducible to any 'doing', to any know-how or function – only have to be what they are because all their practices derive their value from their authors, being an affirmation and perpetuation of the essence by virtue of which they are performed.
>
> (Bourdieu 1984: 23–4)

Yangsheng allows its practitioners to acquire this cultural capital first through practice and then through the sharing in the social networks of knowledge and collusion.

There is another aspect to *yangsheng* which has been implied in much of what has gone before but is worth saying explicitly; *yangsheng* is about pleasure. It is about improving one's life and the enjoyment of it – 'a little bit of what you fancy does you good' – in moderation, of course. Avoidance of illness and the maintenance of good health are the basic premises behind this. Pain and financial expense are both miseries to be avoided by correct practice. There is none of the philosophy of bodily punishment, 'gain through pain', of body loathing, seeing the body as a source of 'wickedness' or indulgence, which lies deeply embedded in much of Western culture. The attitude is much more akin to 'Your body is your only vehicle for this life and you should look after it'. And this gives another clue to the meaning of *yangsheng*: in an Asian society where so much stress is placed on collective 'Confucian' values, obligations to family and society, as opposed to individualistic 'rights', *yangsheng* is a culturally accepted and recognised area for individual expression. It manages this while simultaneously supporting the traditional Confucian norm that to mistreat your body, or indeed to fail to procreate, is to show disrespect – or worse – to your ancestors.

So within this collusive collective set of agreed social norms and acquired individual heuristic philosophy, in Judith Farquhar's words:

> …yangsheng … achieves a kind of lordship. It exerts control over life from within, practicing the civilized arts that give form to life.
>
> *(Farquhar 2005: 323)*

Postscript

The contemporary observations upon which much of this article was based were largely accumulated in Beijing the decade between 2005 and 2015. Since then the changed political weather in the capital and across the nation has had a significant impact on the public expression of lives and livelihoods. In keeping with a rather unimaginative authoritarian puritanism, 'luxuries', while still enjoyed, are not to be flaunted. The atmosphere which sustained many of the examples referred to in this article above has largely evaporated, as indeed have many of the examples themselves. The Yangsheng rubric still exists as a somewhat pompous dignifier to non-controversial practices – and indeed it is still paraded by a few of the bolder chancers and charlatans, as long as they can get away with it![5] The bright economic optimism that prevailed in ordinary people in the years immediately before and after the 2008 Olympics has equally vanished; financial pressures bite hard. In spite of this, the consumerist aspect of Yangsheng still thrives in the cities, particularly through e-commerce, as the stresses of city life invites any form of relief that can be had. Thus Yangsheng has mutated yet again to accommodate the prevailing spirit of the times – it runs, like water in the Daodejing, deep in the soul of Chinese culture.

Highest good is like water.

Because water excels in benefiting the myriad creatures without contending with them *Daodejing* 8, D.C. Lau translation.

Notes

1 'Ten thousand' in Chinese history and culture often simply means a 'huge' number, beyond counting. It is a figure symbolic as much a literal. The salutation to the Emperors was 'Ten thousand years'; in no sense a literal figure, which might imply the Emperor's life was finite!

2 http://www.china.org.cn/english/congress/229611.htm, accessed 3/3/2021.
3 Again the earliest mention of Yangxin can be credited to Mencius. See Lau (1970: 201), Roth (1999) and Rickett (1998).
4 Daniel Moerman is perhaps the best recent commentator on the nature of efficacy and the collusive aspects of the social construction of modern medicine (Moerman 2002; Moerman and Jonas 2002). See also Geertz (1973).
5 Man in China claiming to be Taoist master offering breast implants with 'mind control' https://www.scmp.com/news/people-culture/trending-china/article/3126477/man-china-claiming-be-taoist-master-offering, accessed 3/3/2021.

Bibliography

Pre-modern sources

Beiji qianjin yaofang 備急千金要方 (Recipes for Emergency-Preparedness Worth a Thousand Gold) 652, Sun Simiao 孫思邈 (1955), Beijing: Renmin weisheng chubanshe.

Huangdi neijing (Lingshu) 黃帝內經靈樞, comp. *c*-100BCE–100CE, references to Ming woodblock edition, printed 1995, Beijing: Zhongguo zhongyi yanjiu yuan.

Lüshi Chunqiu 呂氏春秋 (Mr Lü's Spring and Autumn Annals) c. 239 BCE, Lü Buwei 呂不韋, Chen Qiyou 陳奇猷 (ed.) (2002) *Lüshi Chunqiu xin jiaoshi* 呂氏春秋新校釋, Shanghai: Shanghai guji chubanshe.

Mawangdui hanmu boshu 馬王堆漢墓帛書 (The Mawangdui Silk and Bamboo Scrolls), vol. 4, Mawangdui Hanmu boshu zhengli xiaozu (ed.) (1985), Beijing: Wenwu chubanshe.

Mengzi 孟子 attr. Meng Ke 孟軻 (late 4th C BCE), references to *Shisanjing zhushu* and translation by Lau (1970).

Shiji 史記 (Records of the Grand Historian) 94 BCE, Sima Qian 司馬遷 (*c*. 145–90 BCE) (1962), Beijing: Zhonghua shuju.

Shisanjing zhushu 十三經注疏 (Commentaries and Explanations to the Thirteen Classics), Ruan Yuan 阮元 (1764–1849), in Zhu Hualin 朱華臨 et al. (1965) *Chongkan Songben Shisanjing zhushu fujiao kanji* 重刊宋本十三經注疏附校勘記 (Song Commentary on the Thirteen Classics), Taipei: Yiwen yinshuguan.

Xiangqing ouji 閑情偶寄 (Leisurely Reflections on Random Matters) 1670?, Li Yu 李漁 (1610–1680) (2010), Shengyang: Wanjuan chubanshe.

Yashang zhai zunsheng bajian 雅尚齋遵生八箋 (Eight Discourses on Respecting Life from the Studio Where Elegance Is Valued) 1591, Gao Lian 高濂 (1988), Beijing: Shumu wenxian chubanshe.

Yinshu 引書 (The Pulling Book), tomb closed 186 BCE, *Zhangjiashan 247 hao hanmu zhujian zhengli xiaozu* 張家山二四七號漢墓竹簡整理小組 (2001), Beijing: Wenwu chubanshe.

Zhuangzi 莊子 attr. Zhuang Zhou 莊周 (*c*. 4th century BCE) but compiled *c*. late 3rd–early 2nd century BCE). References to translation by Graham (1989).

Zunsheng Bajian 遵生八箋 (Eight Discourses on Respecting Life) 1591, Gao Lian 高濂, in Wang Dachun 王大淳, Li Jiming 李继明, Dai Wenjuan 戴文娟 and Zhao Jiaqiang 赵加强 (eds) (2007) *Zhongyi linchuang bidu congshuse* 中医临床必读丛书 (Chinese Medicine Clinical Essential Reading Series), Beijing; Renmin weisheng chubanshe.

Modern sources

Adamson, P. (ed.) (2018) *Health: a history*, New York: Oxford University Press.
Banton, M. (ed.) ([1965] 2013) *The Relevance of Models for Social Anthropology*, London: Routledge.
BBC News (2015) 'China golf: communist party bans club membership' (22/10/2015): 25th October 2015, BBC News, http://www.bbc.com/news/world-asia-china-34600544, accessed 13/9/2019.
Becker, G., Momm, F., Xander, C., Bartelt, S., Zander-Heinz, A., Budischewski, K., Domin, C., Henke, M., Adamietz, I.A. and Frommhold, H. (2006) 'Religious belief as a coping strategy: an explorative trial in patients irradiated for head-and-neck cancer', *Strahlentherapie und Onkologie*, 182.5: 270–76.
Bingel, U., Wanigasekera, V., Wiech, K., Ni Mhuircheartaigh, R., Lee, M.C., Ploner, M. and Tracey, I. (2011) 'The effect of treatment expectation on drug efficacy: imaging the analgesic benefit of the opioid remifentanil', *Science Translational Medicine*, 3.70: 70ra14–70ra14.

Bourdieu, P. (1977) *An Outline of a Theory of Practice*, trans. R. Nice, Cambridge: Cambridge University Press.
—— (1984) *Distinction: a social critique of the judgment of taste*, trans. R. Nice, Cambridge, MA: Harvard University Press.
Bray, F., Métailié, G. and Dorofeeva-Lichtmann, V. (eds) (2007) *The Warp and the Weft: graphics and text in the production of technical knowledge in China*, Leiden: Brill.
Campany, R.F. (2012) 'Religious repertoires and contestation: a case study based on Buddhist miracle tales', *History of Religions*, 52.2: 99–141.
Certeau, M. de. ([1974] 1984) *The Practice of Everyday Life*, trans. S. Rendall from *L'invention du quotidien*, Berkeley: University of California Press.
Chen Hsiufen 陳秀芬 (2008) 'Nourishing life, cultivation and material culture in the late Ming: some thoughts on *Zhunsheng bajian* 遵生八牋' (Eight Tablets on Respecting Life, 1591), *Asian Medicine*, 4.1: 29–45.
—— (2012) 'Visual representation and oral transmission of *yangsheng* techniques in Ming China', *Asian Medicine*, 7.1: 128–63.
Chen Lai (2009) *Tradition and Modernity: a humanist view*, trans. E. Ryden, Leiden; Boston: Brill.
Chen, N. (1995) *Urban Spaces and Experiences of Qigong in Urban Spaces in Contemporary China*, Washington, DC: Woodrow Wilson Center Press; Cambridge; New York: Cambridge University Press.
—— (2003) 'Healing sects and anti-cult campaigns', *China Quarterly*, 174: 505–20.
Clunas, C. ([1991] 2004) *Superfluous Things: material culture and social status in early modern China*, Honolulu: University of Hawai'i Press.
Csordas, T. (ed.) (1994) *Embodiment and Experience*, Cambridge: Cambridge University Press.
Davidson, R.J. and Lutz, A. (2008) 'Buddha's brain: neuroplasticity and meditation', *IEEE Signal Process Magazine*, 25.1: 176–84.
Despeux, C. (1989) 'Gymnastics: the ancient tradition', in L. Kohn and Y. Sakade (eds) *Taoist Meditation and Longevity Techniques*, Ann Arbor: Center for Chinese Studies University of Michigan, pp. 225–62.
—— (2005) 'Illness prognosis to diagnosis in Tang China', in V. Lo and C. Cullen (eds) *Medieval Chinese Medicine: the Dunhuang medical manuscripts*, London: RoutledgeCurzon, pp. 176–205.
—— (2007) 'Food prohibitions in China', trans. P. Barrett, *The Lantern*, 4.1: 22–32.
Del Rosario, L.R.Q., Yango, A., De la Paz, R.C., Margate, J.C.B. and May, E.R.P (n.d.) 'Pet animals: to own and to love': https://www.uphsl.edu.ph/research/res/ARTS%20_%20SCIENCES/DEL%20ROSARIO, %20Luz%20Remedios/Pet%20animals%20to%20own%20and%20to%20 love.pdf, accessed 13/9/2019.
Dikötter, F. (1992) *The Discourse of Race in Modern China*, London: Hurst & Co.
—— (1995) *Sex, Culture and Modernity in China*, London: Hurst & Co.
Draganski, B., Gaser, C., Busch, V., Schuierer, G., Bogdahn, U. and May, A. (2004) 'Changes in grey matter induced by training', *Nature*, 427 (22 January): 311–12, https://doi.org/10.1038/427311a, accessed 13/9/2019.
Engelhardt, U. (1989) 'Qi for life: longevity in the Tang', in L. Kohn and Y. Sakade (eds) *Taoist Meditation and Longevity Techniques*, Ann Arbor: Center for Chinese Studies University of Michigan, pp. 263–96.
—— (2000) 'Longevity techniques and Chinese medicine', in L. Kohn (ed.) *Daoism Handbook*, Leiden: Brill, pp. 74–108.
Farquhar, J.B. (2001) 'For your reading pleasure: popular health advice and the anthropology of everyday life in 1990s Beijing', *Positions*, 9.1: 105–30.
—— (2002) *Appetites: food and sex in post socialist China*, Durham, NC: Duke University Press.
Farquhar, J.B. and Zhang, Q. (2005) 'Biopolitical Beijing: pleasure, sovereignty, and self-cultivation in China's capital', *Cultural Anthropology*, 20.3: 303–27.
—— (2012) *Ten Thousand Things*, New York: Zone Books.
Fleck, L. (1981) *Genesis and Development of a Scientific Fact*, Chicago, IL: University of Chicago Press.
Fogel, J.A. (ed.) (2004) *The Role of Japan in Liang Qichao's Introduction of Modern Western Civilization to China*, Berkeley: Institute of East Asian Studies, University of California Berkeley, Centre for Chinese Studies.
Furth, C. (1994) 'Rethinking van Gulik: sexuality and reproduction in traditional Chinese medicine', in C.K. Gilmartin, G. Hershatter, L. Rofel and T. White (eds) *Engendering China: women, culture, and the state*, Cambridge, MA: Harvard University Press, pp. 125–46.

Geertz, C. (1973) *The Interpretation of Cultures*, New York: Basic Books.
Gilmartin, Hershatter, G., Rofel, L. and White, T. (eds) (1994) *Engendering China: women, culture, and the state*, Cambridge, MA: Harvard University Press.
Graham, A.C. ([1981] 1989a) *Chuang-Tzu: the inner chapters*, London: George Allen and Unwin.
——— (1989b) *Disputers of the Tao*, La Salle, IL: Open Court.
Harper, D. (1998) *Early Chinese Medical Literature*, London: Kegan Paul International.
——— (2005a) 'Dunhuang Iatromantic Manuscripts', in V. Lo and C. Cullen (eds), *Medieval Chinese Medicine: the Dunhuang medical manuscripts*, London: RoutledgeCurzon, pp. 134–64.
——— (2005b) 'Ancient and medieval Chinese recipes for aphrodisiacs and philters', *Asian Medicine*, 1.1: 91–100.
Harrington Jr, J.E., 'Developing competition law for collusion by autonomous price-setting agents' (August 22, 2017). Available at SSRN: https://ssrn.com/abstract=3037818 or http://dx.doi.org/10.2139/ssrn.3037818
He Jiejun 何介钧 (ed.) (2004) *Changsha Mawangdui er, san hao Han mu* 长沙马王堆二, 三号汉墓 (Tombs 2 and 3 of the Han Dynasty at Mawangdui, Changsha), 2 vols, Beijing: Zhonghua shuju.
Hinrichs, TJ and Barnes, L. (eds) (2013) *Chinese Medicine and Healing: an illustrated history*, Cambridge, MA: Belknap Press of Harvard University Press.
Ho Peng Yoke and Needham, J. (1959) 'Elixir poisoning in mediaeval China', *Janus*, 48: 22–51.
Hudson, W.C. (2007) *Spreading the Dao, Managing Mastership and Performing Salvation: the life and alchemical teachings of Chen Zixu*, PhD thesis, Indiana University.
Holroyd, E., Lopez, V. and Chan, S. (2011) 'Negotiating "Doing the month": an ethnographic study examining the postnatal practices of two generations of Chinese women', *Nursing and Health Sciences*, 13.1: 47–52.
Hsu, E. (1999) *The Transmission of Chinese Medicine*, Cambridge: Cambridge University Press.
——— (2007) 'The biological in the cultural: the five agents and the body ecologic in Chinese medicine', in D. Parkin and S. Ulijaszek (eds) *Holistic Anthropology: emergence and convergence*, New York; Oxford: Berghahn Books, pp. 91–126.
——— (2009) 'Outward form (*xing* 形) and inward *qi* 氣: the "sentimental body" in early Chinese medicine', *Early China*, 32: 103–24.
Kaptchuk, T.J., Friedlander, E., Kelley, J.M., Sanchez, M.N., Kokkotou, E., Singer, J.P., Kowalczykowski, M., Miller, F.G., Kirsch, I. and Lembo, J.L. (2010) 'Placebos without deception: a randomized controlled trial in Irritable Bowel Syndrome', *PLoS ONE* 5(12): e15591, doi:10.1371/journal.pone.0015591, accessed 13/09/2019.
Kleinman, A. (1986) *The Social Origins of Distress and Disease*, New Haven, CT: Yale University Press.
Kleinman, A. and Lee, S. (2007) 'Are somatoform disorders changing with time? The case of neurasthenia in China' *Psychosomatic Medicine* 69(9): 846–9.
Kohn, L. (ed.) (2000) *Daoism Handbook*, Leiden: Brill.
Kohn, L. and Sakade, Y. (eds) (1989) *Taoist Meditation and Longevity Techniques*, Ann Arbor: Center for Chinese Studies University of Michigan.
Kuriyama, S. (1999) *The Expressiveness of the Body and the Divergence of Greek and Chinese Medicine*, New York: Zone Books.
Lally, P. *et al.* (2009) 'How habits are formed: modelling habit formation in the real world', *European Journal of Social Psychology*, 40: 998–1009.
Lau, D.C. (trans.) (1970) *Mencius* (The Book of Mencius), London: Penguin Books.
Li Hongzhi 李洪志 (1998) *Zhuan falun* 轉法輪 (Turning the Dharma Wheel), 2nd edn, Hong Kong: Falun Fofa Publishing Company.
Li Jingming 李敬明 (ed. and trans.) (2008) *Tujie Zunsheng bajian* 图解遵生八笺 (Illustrated Explanation of the Eight Discourses on Respecting Life), Jinan: Shandong meishu chubanshe.
Li Ling 李零 (2006) *Zhongguo fangshu zheng kao* 中国方术正考, Beijing: Zhonghua shuju.
Li Yi 李一 (2010) *Yangsheng you liangfang* 养生有良方, Hangzhou: Zhejiang chubanshe.
Liang, T.T. ([1974] 1977) *Tai Chi Ch'uan for Health and Self-Defence*, New York: Vintage Books.
Lin De 林德, Wang Qiang 王强 and Ding Shuwei 丁树伟 (eds) (2007) *Zhonghua shiwu biandian* 中华食物便典 (Concise Dictionary of Chinese Foodstuffs), Gunagzhou: Guangdong keji chubanshe.
Lin Yin 林殷, Jin Qi 靳琦, Yan Xingli 闫兴丽, Liao Yan 廖艳, Zhang Cong 张聪, Zhang Qingli 张清怡, Zhang Yuping 张玉苹 Zhang Yu 张煜, Yan Ze 严泽, Cui Lian 崔莲, Shu Xiuming 舒秀明 and Fan Ning 范宁 (2014) '*Dangqian ruogan yangsheng redian wenti de bianxi* 当前若干养生热点问题的辨析' (Some Hot Topics in Life Nurturing at Present), *Beijing yiyao daxue xuebao*, 37.9: 581–5.

Liu Lexian 刘乐仙 (2005) 'Love charms among the Dunhuang manuscripts', in V. Lo and C. Cullen (eds), *Medieval Chinese Medicine: the Dunhuang medical manuscripts*, London: RoutledgeCurzon, pp. 165–75.

Liu, Y., Mimura, K., Wang, L. and Ikuta, K. (2005), 'Psychological and physiological effects of 24-style taijiquan', *Neuropsychobiology*; 52. 4: 212–18.

Lo, V. (1998) *The Influence of Yangsheng Culture on Early Chinese Medical Theory*, Phd thesis, SOAS, University of London.

—— (2002) 'Spirit of stone: technical considerations in the treatment of the jade body', *Bulletin of SOAS*, 65.1: 99–128.

—— (2005) 'Pleasure, prohibition and pain: food and medicine in China', in R. Sterckx (ed.), *Of Tripod and Palate*, London: Palgrave Macmillan, pp. 163–86.

—— (2007) 'Imagining practice: sense and sensuality in early Chinese medical illustration', in G. Métailié, V. Dorofeeva-Lichtmann and F. Bray (eds) *The Warp and the Weft: graphics and text in the production of technical knowledge in China*, Leiden: Brill, pp. 385–423.

—— (2013) 'The Han period', in TJ Hinrichs and L. Barnes (eds) *Chinese Medicine and Healing: an illustrated history*, Cambridge, MA: Belknap Press of Harvard University Press, pp. 31–64.

Lo, V. and Cullen C. (eds) (2005) *Medieval Chinese Medicine: the Dunhuang medical manuscripts*, London: RoutledgeCurzon.

Maguire, E.A., Woollett, K. and Spiers, H.J. (2006) 'London taxi drivers and bus drivers: a structural MRI and neuropsychological analysis', *Hippocampus*, 16: 1091–101.

Mauss, M (1973 [1934]) 'Techniques of the body', *Economy and Society*, 2(1): 70–88.

Métailié, G., Dorofeeva-Lichtmann, V. and Bray, F. (eds) (2007) *The Warp and the Weft: graphics and text in the production of technical knowledge in China*, Leiden: Brill.

Moerman, D. (2002) *Meaning, Medicine and the Placebo Effect*, Cambridge: Cambridge University Press.

Moerman, D.E. and Jonas, W.B. (2002) 'Deconstructing the placebo effect and finding the meaning response', *Annals of Internal Medicine*, 136.6 (19 March): 471–6.

Munro, R. (2002) *Dangerous Minds: political psychiatry in China today and its origins in the Mao era*, New York: Human Rights Watch.

Münte, T.F., Altenmüller, E and Jäncke, L. (2002) 'The musician's brain as a model of neuroplasticity', *Nature Reviews Neuroscience*, 3 (June): 473–8.

Needham, J. and Ho Ping-Yü (1970) 'Elixir poisoning in medieval China', in J. Needham, L. Wang, G.Z. Lu and B. He (eds) *Clerks and Craftsmen in China and the West: lectures and addresses on the history of science and technology*, Cambridge: Cambridge University Press, pp. 216–339.

Needham, J., Wang, L., Lu, G.Z. and He, B. (eds) (1970) *Clerks and Craftsmen in China and the West: lectures and addresses on the history of science and technology*, Cambridge: Cambridge University Press.

NHS (2018) *A Guide to Tai Chi*, http://www.nhs.uk/livewell/fitness/pages/taichi.aspx, accessed 13/9/2019.

Ofori, P.K., Biddle S. and Lavallee, D. (2012) 'The role of superstition among professional footballers in Ghana', *Athletic Insight*, 4.2: 115–26.

Ots, T. (1994) 'The silenced body – the expressive Leib', in T. Csordas (ed.) *Embodiment and Experience: the existential ground of culture and self*, Cambridge: Cambridge University Press, pp. 116–36.

Palmer, D. (2007) *Qigong Fever*, New York: Columbia University Press.

Parkin, D. and Ulijaszek, S. (eds) (2007) *Holistic Anthropology: emergence and convergence*, New York; Oxford: Berghahn Books.

Penny, B. (2002a) 'Falun gong, prophesy and apocalypse', *East Asian History*, 23: 149–68.

Qu Limin 曲黎敏 (2008) *Yangsheng shi er shuo* 养生十二说 (Twelve Talks on *Yangsheng*), Beijing: Zhongguo chuban jituan.

—— (2009) *Cong zi dao ren* 从字到人 (From Words to Man), Wuhan: Changjiang wenyi chubanshe.

Raven, J.H., Chen, Q., Tolhurst, R.J. and Garner, P. (2007) 'Traditional beliefs and practices in post-partum period in Fujian province, China: a qualitative study', *BMC Pregnancy and Childbirth*, 7.1: 8.

Rickett, W.A. (1998) *Guanzi: political, economic, and philosophical essays from early China: a study and translation*, Princeton, NJ: Princeton University Press.

Roth, H. (1999) *Original Tao: inward training and the foundations of Taoist mysticism*, New York: Columbia University Press.

Ryle, G. ([1949] 2009) *The Concept of Mind*, London: Routledge.

Salguero, C.P. (2013) 'Fields of merit, harvests of health: some notes on the role of medical karma in the popularization of Buddhism in early medieval China', *Asian Philosophy*, 23.4: 341–9.

Schechter, D. (2001) *Falun Gong's Challenge to China*, New York; Akashic Books.
Shapin, S. (1994) *A Social History of Truth Civility and Science in Seventeenth-Century England*, Chicago, IL: University of Chicago Press.
Shiel, W.C. Jr (2017) 'Medical definition of neuroplasticity', *MedicineNet*, http://www.medicinenet.com/script/main/art.asp?articlekey=40362, accessed 13/9/2019.
Song Shugong 宋书功 (2010) *Gudai fangzhong yangsheng zhenyao* 古代房中养生真要 (The True Essentials of Ancient Bedchamber *Yangsheng*), Beijing: Zhongyi guji chubanshe.
——— (2011) *Fangshi yangsheng tiyao* 房室养生集要 (Essential Collection of Bedroom *Yangsheng*), Haikou: Hainan chubanshe.
Stanley-Baker, M. (2006) *Cultivating Body, Cultivating Self: a critical translation and history of the Tang dynasty Yangxing yanming lu* (Records of Cultivating Nature and Extending Life), MA thesis, Indiana University, Bloomington.
——— (2018) 'Health and philosophy in pre- and early imperial China', in P. Adamson (ed.) *Health: a history*, New York: Oxford University Press, pp. 7–42.
Sterckx, R. (ed.) (2005) *Of Tripod and Palate*, London: Palgrave Macmillan.
Umekawa, S. (2004) *Sex and Immortality: a study of Chinese sexual activities for better-being*, PhD thesis, School of Oriental and African Studies.
van Gulik, R.H. (1951) *Erotic Colour Prints of the Ming*, private edition, Tokyo, Japan.
——— (1974) *Sexual Life in Ancient China: a preliminary survey of Chinese sex and society from ca. 1500 B.C. till 1644 A.D.*, Leiden: E. J. Brill.
Wang Dachun 王大淳, Li Jiming 李继明, Dai Wenjuan 戴文娟 and Zhao Jiaqiang 赵加强 (eds) (2007) *Zhongyi linchuang bidu congshu* 中医临床必读丛书 (Chinese Medicine Clinical Essential Reading Series), Beijing; Renmin weisheng chubanshe.
Ward, B.E. ([1965] 2013) 'Varieties of the conscious model: the fishermen of south China', in M. Banton (ed.) *The Relevance of Models for Social Anthropology*, London: Routledge, pp. 113–37.
Weller, R. (1994) *Resistance, Chaos and Control in China*, Seattle, WA: University of Washington Press.
Wells, M. (2005) *Scholar Boxer*, Berkeley CA: North Atlantic Books.
Wile, D. (1992) *Art of the Bedchamber*, Albany, NY: SUNY.
——— (1996) *Lost Tai-chi Classics from the Late Ch'ing Dynasty*, Albany, NY: SUNY.
Wilkinson, E. (2000) *Chinese History: a manual*, Cambridge; London: Harvard University Asia Center.
Salguero, P. (2014) *Translating Buddhist Medicine in Medieval China*, Philadelphia: University of Pennsylvania Press.
Wilms, S. (trans.) (2007) *Beiji qianjin yaofang* 備急千金要方 (Essential Prescriptions Worth a Thousand in Gold for Every Emergency), volumes 2–4 on Gynecology, Portland, OR: The Chinese Medicine Database.
——— (2010) 'Nurturing life in classical Chinese medicine: Sun Simiao on healing without drugs, transforming bodies and cultivating life', *Journal of Chinese Medicine*, 93.5: 5–13.
China.org.cn (2007) *Hu Jintao's Report at 17th Party Congress*, http://english.china.org.cn/english/congress/229611.htm, accessed 13/9/2019.
Xu, B. (1999) *Disenchanted Democracy: Chinese cultural criticism after 1989*, Ann Arbor: University of Michigan Press.
Xu, J. (1999) 'Body, discourse and the cultural politics of contemporary Chinese qigong', *Journal of Asian Studies*, 58.3–4: 961–91.
Yang, D. (2018) *Prescribing 'Guiding and Pulling': the institutionalisation of therapeutic exercise in Sui China (581–618 CE)*, PhD thesis, University College London.
Zatorre, R.J., Fields, R.D. and Johansen-Berg, H. (2012) 'Plasticity in gray and white: neuroimaging changes in brain structure during learning', *Nature Neuroscience*, 15.4 (18 March): 528–36.
Zhang Wuben 张悟本 (3 vols 2009/12, 2010/15, 2010/17), *Ba chi chulai de bing, chihuiqu* 把吃出来的病吃回去 ('Eating for Better Health' or 'Dietary Recovery/Remedies for Dietary Disease'), Beijing: Renmin ribao chubanshe.
Zhang Xingfa 张兴发 (2006) *Huashuo daojia yangsheng shu* 话说道家养生术 (Talking of Daoist *Yangsheng* Techniques), Jinan: Qilu shushe.
Zheng Jinsheng 郑金生 (2006) 'The vogue for medicine as food in the Song period (960–1279)', trans. P. Barrett, *Asian Medicine*, 2.1: 38–58.

50
LIQUORICE AND CHINESE HERBAL MEDICINE
An epistemological challenge

Anthony Butler

Plant materials have been used for many thousands of years for the relief they can give to humans when illness strikes. For the last century it has been normal, within mainstream medicine and modern pharmacy, to take the medicinal herb and extract the active principle, that is, the substance in the plant material responsible for the herb's therapeutic action. This is accomplished by the use of modern chemical techniques such as chromatography. This substance is then purified, possibly modified chemically, and the product is what is used clinically, rather than the intact plant. One of the advantages of this procedure is that quality control is much easier. As an example, the bark of the cinchona tree (*Cinchona officinalis*) from Peru was used for a number of years as a cure for malaria but the results were unreliable and fraud was widespread (deb Roy 2017). Matters improved greatly when, in the 1820s, the active principle (quinine) was isolated and used in place of the bark. Quinine, but not the bark of the tree, is still used in medicine but only under exceptional circumstances. The Chinese drug artemisinin, extracted from *Artemisia annua*, is a superior drug with fewer adverse side effects. When used quinine is still extracted from the bark of the cinchona tree, grown in plantations in the Far East. Quinine has been synthesised in the laboratory (Woodward and Doering 1944) but the process is very inefficient. Although drugs of herbal origin are also still used, herbalism, a medical system that insists on the use of the whole herb, has been relegated to an alternative or complementary therapy.

However, Chinese herbal medicine, which still uses the intact plant [and often mineral *materia*, which will not be discussed in this chapter], has been practised for at least 2,000 years and still flourishes in Mainland China and in communities of the Chinese diaspora. In a few instances it has influenced biomedicine. For example, the emergence of a new and much needed antimalarial drug (artemisinin, *qinghaosu* 青蒿素 in Chinese) in the 1980s from a Chinese herbal medicine text, the Daoist work, *Recipes to Keep Close at Hand* (*Zhouhou fang* 肘後方), and the award of a Nobel Prize for that work to the Chinese biochemist Tu Youyou 屠呦呦 in 2015, has alerted the biomedical community to the possibility of other innovative drugs from the same source (Chapter 51 in this volume). The careful historical studies of Elisabeth Hsu (2006) have shown the need to adhere fully to the historical accounts of exactly how the plant should be used to optimise its therapeutic value. Subsequent work resulted in isolation of the active principle from the plant. It was then chemically modified to yield a very effective drug (artesunate) in the fight to eliminate malaria (Butler 2019).

However, practitioners of Chinese herbal medicine would claim that isolation of the active principle, and using that alone, is the wrong approach. They retain the practice of prescribing the whole herb or, indeed, complex mixtures of herbs. Modern pharmacy has, for some time, used combinations of synthetic drugs for the treatment of conditions like TB and leprosy. Administering several drugs at the same time seems to enhance their efficacy and prevent drug resistance building up. Maybe, despite the radical differences in epistemological foundation, there is some common ground between the two systems of prescribing. After all, combination therapy is now a very important part of contemporary medicine. Is there anything that can be learnt from Chinese medicine about combining herbs that might be of value to future pharmacy? Is there a common conceptual language or code of practice? In order to pursue this notion we need to know much more about the strategy employed when Chinese physicians compiled their prescriptions. In this article, the benefits of adding liquorice to a prescription will be examined, followed by an analysis of whether its function in Chinese terms can be matched in any way to the processes of modern chemistry.

There are hundreds of remedy collections that survive in printed form from pre-modern China, as well as many *materia medica,* outlining the nature and efficacy of the drugs and lists of simples where single drugs are matched with specific diseases (Chapters 8 and 9 in this volume). In ancient times herbal and mineral medicine was practised in China often in conjunction with spells, incantations and charms but, in the twentieth century the use of herbs and minerals has come to dominate. Some of the earliest known written records of pharmacological remedies were recovered during excavations at the site of Mawangdui 馬王堆 Tomb 3 in 1973, where the son of Lady Dai was buried in 168 BCE (Harper 1998: 15–21; Chapter 3 in this volume). Among many medical manuscripts on bamboo strips and silk scrolls, one silk manuscript discovered there described cures for a number of diseases and disabilities and is dated by scholars to the end of the third century BCE. The remedies are collectively known as *Wushier bingfang* 五十二病方 (Prescriptions for Fifty-two Ailments) and the use of liquorice is mentioned in several of them (Harper 1998: 485).

Nowadays, many Chinese prescriptions consist of a mixture of herbs, generally between four and ten. For example, Bensky and Gamble (1993) recommend a treatment of allergic rashes that combines: forsythia fruit (*lian qiao* 連翹), ephedra stems (*ma huang* 麻黃) red peony root (*chi shao* 赤芍) and liquorice (*gan cao* 甘草). The selection of herbs appears random but, it is claimed, each herb, or group of herbs fulfils a distinct role in the curative process. Three categories were described by Tao Hongjing 陶弘景 (456–526 CE) when he was editing one of the extant editions of the *Shennong bencao jing* 神農本草經 (Divine Husbandman's *materia medica*), which probably dates to the late Han period (25–220 CE).[1] Shennong 神農, The Divine Husbandman, was the legendary founder of Chinese pharmacy. He described the categories as follows (although the last two were elided into one):

The **sovereign herb** (*jun* 君) provides the therapeutic thrust;
The **minister herbs** (*chen* 臣) assist with the therapy;
The **adjunct herbs** (*zuo* 佐) moderate harshness, along with a range of other functions;
The **envoy herbs** (*shi* 使) either guide the sovereign herb to the appropriate organ or have a harmonising effect.

Some prescriptions contain all of these and others a selection. In a survey of Chinese prescriptions it is noticeable that some substances crop up frequently as ingredients, and one of the most common is liquorice.

It is appropriate at this juncture to make a distinction between traditional Chinese drug therapy and Traditional Chinese Medicine (TCM). In recent years the latter acronym has been given to the reinvention of traditional medicine that was approved and promoted by Mao Zedong as a way of bringing better health cheaply to Chinese peasants living far from a hospital. He looked at all the medical practices used in China over the centuries, such as herbalism, acupuncture, *Taiji quan* and moxibustion, and in a speech in the 1950s declared it a 'treasure house' and had a selection made that he felt fitting to healthcare in the modern Communist state (Taylor 2005). That selection became part of the *Barefoot Doctors' Manual* which was given to paramedics as they were sent out into the countryside in an attempt to improve the health of rural communities (Chapter 45 in this volume). Mao never used TCM himself; he preferred Western medicine. In contrast to TCM, traditional Chinese herbalism is redolent of the wisdom, folklore and superstition, accumulated over centuries, concerning the value of certain herbs and minerals, and their combinations in the treatment of disease.

Liquorice and herbal medicine

Liquorice (licorice is the American spelling) is extracted from the root, or more correctly, the stolon of the low growing bushes of the *Glycyrrhiza* family of which *G. glabra* is the most widespread in the Middle East, although in China *G. urelesis* is more common. There are many other varieties which we could cite, but these two are germane to this chapter. The name, given by no less a person than Dioscorides, is derived from the Greek *glukur* (sweet) and *rhiza* (root). In Chinese it is called *gancao* 甘草, meaning 'sweet grass'. *G. glabra* (Figure 50.1) grows naturally in the Mediterranean region, central and southern Russia, Asia Minor and parts of Iran. Presumably its medicinal value was discovered in these regions and that knowledge spread, over many years, westwards to Europe and eastwards to China.

It is said that a specimen of the bush was brought back to Britain by soldiers returning from the Crusades and grown, initially, around Godalming and Pontefract. It flourished particularly well in the soil around Pontefract and was grown in fields or 'garths' and harvested in the autumn. Pontefract became the principal source of liquorice in the UK and there is a

Figure 50.1 Glycyrrhiza glabra Wellcome Collection: Glycyrrhiza glabra (Liquorice or Licorice). Credit: Rowan McOnegal. Attribution-NonCommercial 4.0 International (CC BY-NC 4.0)

liquorice museum there. Because of its Mediterranean origin, liquorice was known locally as 'Spanish' and cultivation flourished until the demand of the confectionery business became so great that cheaper and more plenteous supplies imported from the Near East made it uneconomic to grow it locally.

Although today we associate liquorice with children's sweets, it has been used in medicine far longer than in confectionery. The transition from medicine to sweets appears to date from 1760 when George Dunhill of Pontefract made a cake by adding sugar to liquorice. These were marketed as Pontefract cakes and have the shape of a medicinal lozenge, reflecting the older use of liquoric (Lee 2018: 378–82). For detailed information on all aspects of liquorice, including botany, phytochemistry, pharmacy and the economics of cultivation, a specialist publication in the series 'Springer Briefs in Plant Science' should be consulted (Öztürk 2017). Although there was widespread use of liquorice in European herbal medicine, even greater use of it was made by the Chinese. It was a component of around 50% of all prescriptions and credited with many wondrous cures. To understand why it was a component of so many prescriptions, it is necessary to examine a little of the chemistry of the components of the liquorice bush.

The components of raw liquorice

The crude material, extracted from the stolons with hot water, is a complex mixture of over 400 triterpenes, flavones, isoflavones, chalcone and related compounds. For a full account, the text by Tang and Eisenbrand (1992) should be consulted. The dominant compound is glycyrrhizin (or glycyrrhetinic acid but the first name is preferred), a triterpene and a saponin (soap-like substance). The name is difficult to spell and problematic to pronounce. It is glycyrrhizin that gives liquorice its distinctive sweet taste and some of its curative properties. Glycyrrhizin was first isolated in pure form by the German chemist Paul Karrer (Karrer, Karrer and Chao 1923) in the 1920s and its chemical structure (Figure 50.2) established by Basil Lythgoe in the 1940s (Lythgoe and Trippett 1950). Readers can also consult the data aggregation websites, SuperTCM, developed by the Charite – University Medicine Berlin, and Symmap.org (Wu, Zhang, Yang *et al.* 2018) for an analysis of its chemical components and their molecular targets in the human body.

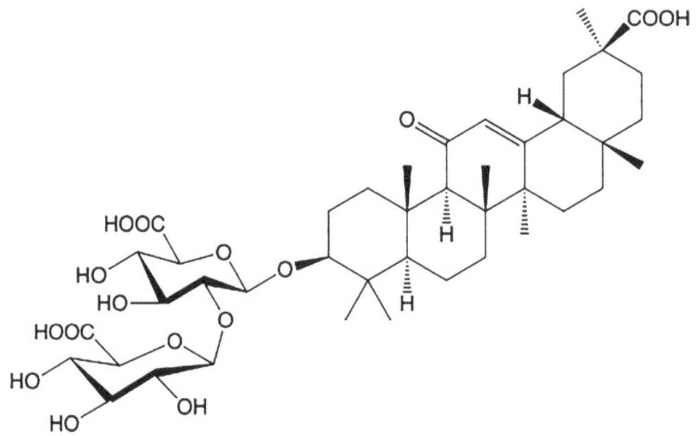

Figure 50.2 Chemical structure of glycyrrhizin. The left part of the molecule is hydrophilic while that on the right is hydrophobic (Drawn by Hazel Nicholson)

The natural colour of liquorice is yellow, due to the presence of flavonoids, and the black colour we normally associate with liquorice sweets is an added vegetable dye. The flavour of liquorice is not due to glycyrrhizin but to anethole, an aromatic, unsaturated ether, found alongside glycyrrhizin. A more plenteous and much used source of this flavour is aniseed oil from *Pimpinella anisum* and Liquorice Allsorts may not contain any liquorice at all. There are at least twenty-five other triterpene components in liquorice and it would be almost certainly incorrect to credit all its curative properties to glycyrrhizin alone. As if over 400 components were not enough, ten new ones were reported in a paper published as recently as 2018 (Schmid, Dawid, Peters and Hofmann 2018). The more sensitive the analytical instrument used, the more components are found.

The sweetness of glycyrrhizin has attracted attention as a low calorie sweetener. It is fifty times sweeter than sucrose. Minor chemical modification of the molecule gives rise to extremely sweet substances (Iijchi and Tamagaki 2005). Without doubt, one of the roles of liquorice in Chinese medicine was that of disguising the unpleasant flavour of herbs with its natural sweetness. In particular this would make the remedy more acceptable to children. Thus, liquorice in a prescription can act as an adjunct or *zuo* herb in Tao Hongjing's rubrics, in that it moderates harshness.

Metabolism

Glycyrrhizin is a molecule of two parts (Figure 50.2). One half is a disaccharide (a sugar) and the other half has a steroid-like structure. The disaccharide is water soluble while the steroid is lipid soluble and this combination gives the molecule special properties, which may be of medicinal value, a matter that will be discussed later. Once ingested, glycyrrhizin may undergo a number of chemical changes. It is first hydrolysed by intestinal bacteria into 18β-glycyrrhetinic acid (Figure 50.3) and a disaccharide. After complete absorption of the former into the blood stream, it is metabolised into 3β-monoglucuronyl-18β-glycyrrhetinic acid in the liver (Kim et al. 1999).

Addition of a glucuronyl group keeps the molecule soluble enough to circulate in the blood stream but it readily releases 18β-glycyrrhetinic acid and it is this molecule that may give glycyrrhizin its biological activity. The Greek letter β merely indicates the stereochemistry of the hydrogen atom at position 18 on the molecule and 18β is the naturally occurring form of the molecule.

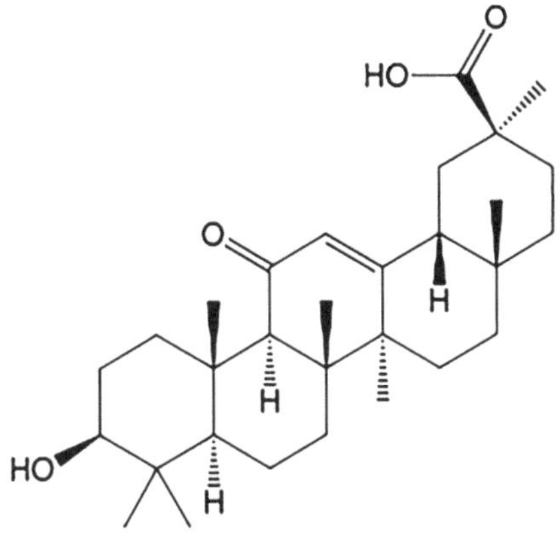

Figure 50.3 Chemical structure of glycyrrhetinic acid, a hydrolytic product of glycyrrhizin metabolism (Drawn by Hazel Nicholson)

Later Chinese remedy literature

Throughout subsequent centuries many compendia of *materia medica* and remedies were assembled but, rather than being *de novo* works, they were largely additions to and revisions of the *Shennong bencao jing*. For example, a number of scholars revised the work sometime between 200 and 250 CE to produce *Lei Gong jizhu Shennong bencao* 雷公集注神農本草 (Shen Nong's *Materia Medica*, Compiled and Annotated by Lei Gong). There were probably other revisions, now lost. The polymath Tao Hongjing, introduced above, made significant revisions to produce *Bencao jing jizhu* 本草經集注 (*Materia Medica* Canon Variorum Edition) and this work, in particular, marked the beginning of a vast *bencao* literature. It included a revision ordered by Emperor Taizu (927–976 CE) to produce *Kaibao chongding bencao* 開寶重訂本草 (*Materia Medica*, Newly Examined and Determined during the Kaibao Period) (Chapters 8 and 9 in this volume). This revision listed 983 simples and was published in 977 CE.

In general, Chinese scholars added to what had been written before, with only a few deletions, and the consequence was that the *bencao* literature became a succession of poorly arranged encyclopaedias filled with repetitions and contradictions (Unschuld 1986). The situation with regard to herbal and mineral medicine in China was completely changed by the work of the scholar Li Shizhen 李時珍 (1518–93 CE), who produced an entirely new work entitled *Bencao gangmu* 本草綱目 (*Materia Medica* Arranged according to Drug Description and Technical Aspects) which, although based on what had been written before, adopted a far more critical approach (Chapters 8, 9 and 21 in this volume). He removed many of the repetitions and discussed the contradictions and did what he could to resolve them. It took him a life-time to complete and was not published until after his death in 1596 CE. It remains, to this day, the definitive account of the herbs and minerals used in Chinese medicine and is rated as one of the great works of Chinese scholarship. Although most of the entries concern herbs, there is some mention of animal and mineral materials used in Chinese medicine. The complete work has been translated into English and annotated by Luo Xiwen (2003).

One of the problems about using the *bencao* literature in a search for new drugs to treat diseases as they are now described and understood is that, in the *bencao* era, diagnoses were made according to very different criteria, such that it is often difficult to identify the disease under discussion. However, there are an increasing number of tools available for those interested but do not read Chinese. Apart from Luo's translation there is a dictionary of the same by Paul Unschuld.[2] Even with these aids, caution must be exercised, especially where illness and substance identification are concerned.

There have been few additions to the *bencao* literature published since the appearance of the encyclopaedic *Bencao gangmu*. In 1666 CE Guo Peilan 郭佩蘭 published *Bencao hui* 本草匯 (Collected Pharmaceutical Knowledge) which, he claims, deals with matters neglected by Li Shizhen (Chapter 48 in this volume). In more recent years the Chinese government has published a pharmacopoeia which gives details of not only modern synthetic drugs used in China but also traditional remedies (*Pharmacopoeia of the People's Republic of China*: 2000). There are also compendia of Chinese remedies written in European languages, or by Chinese doctors practising in the West. One of the most comprehensive is that by Bensky and Gamble (1993).

Throughout all the publications mentioned above liquorice is frequently mentioned as a medicinal herb and also as a component of prescriptions used to treat a wide range of diseases. Whether liquorice is the 'sovereign' drug, providing the therapeutic thrust, or an 'envoy' drug directing the 'sovereign' to the appropriate organ, is not always stated and we must consider both possibilities. How far these prescriptions live up to the claims made for them when tested using modern pharmacological and clinical procedures is one of the concerns of this article.

Testing the therapeutic value of liquorice and glycyrrhizin

Over the last twenty or so years there have been hundreds of reports of investigations into the therapeutic value of both liquorice extract and glycyrrhizin in the treatment of diseases, some of which are mentioned in the *bencao* literature. The reports are of variable quality and some appear in journals with low impact factors. Most are positive, but we must accept these conclusions with caution. The use of liquorice extract is problematic as its composition may vary according to the species, the manner of cultivation and the time of harvest. This makes confirmation or denial of results difficult and so emphasis will be given to studies using purified glycyrrhizin, but if some other component of liquorice is the main chemical agent that drives the efficacy of what we might think of as the sovereign drug then the results of these tests are valueless. Much of the work reported describes *in vitro* studies and all those involved in drug discovery know how frequently successful *in vitro* results disappear when *in vivo* studies are undertaken. Mention has been made already of the metabolism undergone once liquorice is ingested.

Studies, using modern pharmacological methods, have been reviewed by a number of researchers three of whose findings are referenced (Damle 2013: 132–6; Ming and Yin 2013; Mamedov and Egamberdieva 2019). The reviews would have been of greater value had they been more critical of the results. However, Mamedov and Egamberdieva conveniently list the conditions where, according to the *bencao* literature and other studies, liquorice has a beneficial effect. They include antimicrobial, antiviral, anti-inflammatory, anti-ulcer and hepatoprotective activity, as well as dermatological effects. It would be remarkable if they were all correct. We will briefly consider each of these, in turn.

Liquorice is so frequently mentioned in Chinese and European herbal manuals as a treatment for coughing that it has been little studied as its success is well established. However, the value of liquorice in the treatment of other inflammatory conditions has been the subject of much research. This has been reviewed by Yang *et al* (2017) and the data tabulated there are impressive, although in some studies the doses were rather high. A recent paper by Liu *et al* (2018) suggests that glycyrrhizin appears to suppress the production of a number of substances that cause inflammation, such as nitric oxide, PGE2, TFN-α and IL-1β. This applies to both liquorice extract and pure glycyrrhizin, so glycyrrhizin must be the active ingredient, or acting as the sovereign drug by analogy. It would not be unreasonable to suggest that glycyrrhizin is ready for clinical trials as an anti-inflammatory drug. It is interesting to note that a liquorice drink is sometimes given to patients recovering from anaesthesia who have excessive dryness of the mouth (xerostomia) (Kuniyama and Maeda 2019).

Whether Chinese physicians realised that liquorice had antibacterial and antiviral activity is not a question that can be addressed as both concepts (bacteria and viruses) were unknown to them. Instead the infection would have been described in terms of wind invasion and of the Five Agents or the disturbed flow of *qi* (Chapters 1, 4 and 9 in this volume). Yet 300 years before the Five Agents were systematically applied to pharmacological knowledge, liquorice was suggested in the second century BCE remedy manuscript *Wushier bingfang* (Treatment for Fifty-two Ailments) as an aid to wound healing, perhaps because it inhibited bacterial infections. Liquorice was to become a common ingredient in Chinese tonics, a term still used in traditional Chinese medical circles. The *Shennong bencao jing*, which dates to around the first century CE, recommends that it should be consumed regularly to promote long life and prolong sexual activity. In an age when hygiene was little understood, a herb that acted as a mild antibiotic might have given beneficial results. Some modern research (Kowalska and Kalinowska-Lis 2019) suggests that liquorice can be a treatment for certain dermatological conditions such as atopic dermatitis, puritis and acne vulgaris. Its role here could be employing both its antibacterial and anti-inflammatory properties. However, this use of liquorice in the treatment of skin conditions does not appear with any frequency in the *bencao* literature.

Rather surprisingly, clear descriptions of peptic and gastric ulcers are also not common in the *bencao* literature. It could be that they are described in a way that is so different from that of Western medical literature that they are difficult to recognise. Chinese herbal medicine describes its primary target as the spleen, which in traditional medicine is paired with the stomach, and governs digestion. However, liquorice has long been used to ease the discomfort of duodenal and gastric ulcers. In the 1970s Larkworthy showed by an endoscopic study that deglycyrrhizinated liquorice (liquorice from which all the glycyrrhizin had been removed) brought about healing of chronic duodenal ulcers (Larkworthy and Holgate 1975). Clearly glycyrrhizin is not providing the main therapeutic thrust, or acting as the sovereign drug in our analogy. Later (Bennett, Clark-Wibberley, Stamford and Wright 1980) discovered that in the healing of aspirin-induced gastric mucosal damage in rats, a mixture of cimetidine and deglycyrrhizinated liquorice given together was more successful than a dose of either drug alone. This observation gives us a hint of what may be a significant role of glycyrrhizin in herbal medicine. The matter will be discussed in more detail shortly. In the treatment of ulcers, it is possible that glycyrrhizin displays antibacterial properties against *Helicobacter pylori* (Hajaghamohammadi, Zargar, Oveisi et al. 2019).

The use of liquorice in the treatment of liver disease is well authenticated in modern Westernised Asian medicine but in the West is seen as an 'alternative' or 'complementary' therapy. The subject has been summarised at some length by (Li, Sun and Liu 2019). Their conclusion is that the natural ingredients in liquorice relieve liver disease and prevent drug-induced liver injury through targeting a number of therapeutic mechanisms. They claim that further toxicological studies and clinical trials offer an 'alluring prospect' for the use of liquorice in this area of medicine. Time will tell.

An unexpected and recent article on the use of liquorice in dentistry is worthy of special mention. A number of components of liquorice show activity against some common bacteria responsible for dental caries (Sidhu, Shankargouda, Rath et al. 2020). Based on these observations, Hu, He, Eckert et al. (2011) devised a sugar free, orange-flavoured lollipop containing glycyrrizol A, a compound related to glycyrrhizin. Its use over a three week period among preschool children led to a substantial reduction in the number of *Streptococcus mutans* in the mouths of the children. Other infections in the mouth, as well as inflammatory conditions, are said to benefit from treatment with liquorice. In canal root treatment one of the more difficult bacteria to remove completely by irrigation is *Enterococcus faecalis* against which liquorice is claimed to be particularly effective (Badr, Omar and Badria 2011). One of the thrusts of the above article is that, for dental practitioners in emerging nations, local herbal remedies are often just as effective as expensive modern drugs.

The special nature of liquorice

Although liquorice does have therapeutic properties, particularly as a remedy for coughing and other inflammatory conditions, its role is widely described as 'harmonising' the other components of the prescription. Although an attractive term, with a clear meaning when applied to, say, music, this is not so when applied to pharmacy at a molecular level. The liquorice molecule has some unusual properties, as mentioned previously, and these may be a reason for liquorice having been attributed a 'harmonising role' historically. Let us first examine what the *Bencao gangmu* has to say about liquorice. In this instance, there is no difficulty in identifying the plant.

One of the crucial parts of the entry on liquorice in the *Bencao gangmu* is where Li Shizhen quotes from the writings of Tao Hongjing, a quotation that includes the phrase:

此草最為眾藥之主

The translation given in the 2003 English language version of the *Bencao gangmu* by Luo Xinwen, who gives this synopsis:

> *Gancao* (liquorice) *is the principal drug among the drugs*. Most of the classical prescriptions have this drug as an important ingredient… It is also called *guolao* 國老 or 'the principal instructor'… It can bring harmony among all drugs in a prescription and can neutralise the toxins of herbs and *stone drugs*.

The term 'stone drugs' means minerals, which were often used in Chinese prescriptions. The above is just one possible translation and others could legitimately suggest more strongly some property of liquorice that makes its use so widespread.

John Moffett (private comm.) offers the following, including some clauses which Luo left out:

> This herb makes the best master of the multitude of *materia medica*. It is rarely absent from classic formulae, in the same way as incense contains agarwood.

Valerie Pallett (private comm.) translates the whole passage as follows:

> *The herb is one of the most important drugs* and there are very few classical remedies which do not use it, just like agalloch eaglewood among the fragrant herbs. Liquorice is known as the emperor of herbs for, although it is not a sovereign drug, it is revered as such. It can be used to harmonise with other drugs and detoxify.

All three translations emphasise its widespread use and that it can harmonise the drugs in a prescription, an effect that is well-known to practitioners of Chinese medicine. The term 'harmonise' is difficult as it does not correspond to any commonly used term in modern pharmacology. However, it is a term used frequently in Chinese administrative language and, perhaps, in this instance it should be understood as bringing about a co-operative effect between components of the prescription. At least this description suggests that there is something special about liquorice. It may not provide the 'therapeutic thrust' but its presence makes the prescription more potent and much of Chinese pharmacy is concerned with enhancing potency.

Another indication of the value ascribed to liquorice is that the great Tang dynasty physician Sun Simiao in his compendium of prescriptions, charmingly entitled *Qianjin yaofang* 千金要方 (Thousand Ducat Prescriptions), gives details of what should be in a first-aid kit. It includes liquorice.

Liquorice as an aid to drug delivery

The special place given to liquorice may possibly be explained by an examination of the chemistry of the glycyrrhizin molecule. As mentioned previously, one half of the molecule (the sugar half) is hydrophilic, while the other half (the steroid part) is hydrophobic. Molecules of this type are known as amphiphilic. An aqueous environment causes glycyrrhizin molecules to aggregate rather than dissolve, probably in the form of a sphere (although cylindrical aggregates are possible) (Zhang, Wang, Wang and Yu 2006) with the hydrophobic part of the molecules pointing in, and the sugar part on the outside hydrogen bonding to water molecules in the solvent. There is good physical evidence to support this view. A mass spectroscopic study with glycyrrhetinic acid revealed that, in the spectrometer, aggregates of up to eight molecules form readily (Borisenko, Lekar, Vetova et al. 2016).

This result was confirmed by extensive studies using nuclear magnetic resonance spectroscopy (Petrova, Schlotgauer, Kruppa and Leshina 2017). Such aggregates are known as micelles. In very dilute solutions of amphiphilic substances, micelles do not form but at the critical micelle concentration (cmc) aggregation commences and is the dominant state of the molecules (Figure 50.4).

The significance of micelles to drug therapy is that, as a biochemical analogue to the harmonising function as described by Tao Hongjing and taught still today in Chinese medical schools, micelles can enhance the solubility of a drug and thus make it easier to deliver a therapeutic dose (Ahmad, Shah, Siddiq and Kraatz 2014).

For example, glycyrrhizin greatly enhances the solubility and the bioavailability of Atorvastatin (Kong, Zhu, Meleleva et al. 2018) and also shows promise as a carrier of anticancer therapy (Su, Wu, Hu et al. 2017). It does this by incorporating some molecules of the drug into the micelle. Why this happens is not completely clear but it is a well-established phenomenon, which is of great significance for the delivery of drugs *in vivo*. Solubility is a major problem with drugs and it is estimated that 80% of substance undergoing development as drugs have unacceptably low solubility. Hence the importance of drug delivery systems that enhance solubility. Not only does glycyrrhizin increase drug solubility but it may also aid its passage across the cell membrane so that it can enter the cell and reach the target for its therapeutic action (Seylutina et al. 2016). This may mean that, as drug delivery is more efficient, a lower dose can be given and a lower dose means fewer side effects. Experiments using a technique known as atomic force microscopy have shown that glycyrrhizin influences the permeability and elasticity of cell membranes (Seylutina and Polyakov 2019). Thus, the presence of glycyrrhizin constitutes a Drug Delivery System (DDS) and a DDS is just as important as the action of the drug itself in

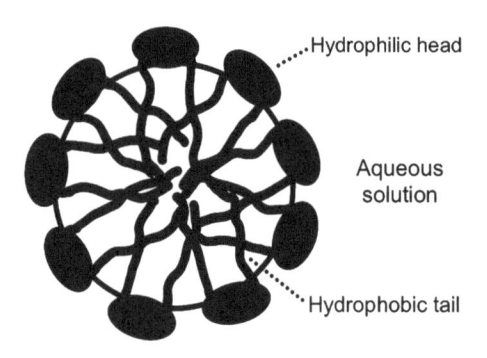

Figure 50.4 Cross section of a micelle. A drug may be contained within the micelle (Drawn by Hazel Nicholson)

providing therapeutic thrust. This function, we might understand as analogous to the *zuo shi* functions of medieval Chinese medicine. The active drug which makes use of glycyrrhizin as a carrier could be from the *jun* herb or from any of the other herbs in the prescription.

Liquorice hazards

The ingestion of liquorice is not without its dangers. There is strong evidence that excessive intake leads to hypertension (van Uum 2005; Penninkilamp, Eslick and Eslick 2017). This effect is known by modern practitioners, who take nosebleeds as an indicative side-effect, and when they occur, will discontinue liquorice. Although the effect is so small that it is not a matter of significance for normal people, those with underlying cardiovascular conditions should take note. Rather more serious for pregnant women is the effect of taking excessive amounts of liquorice on the unborn child. In a carefully executed Scottish-Finnish study (Räikkönen, Pesonen, Heinonen *et al.* 2009) it was found that, because of the similarity of structure between cortisol and glycyrrhizin, mothers who consumed more than 500mg of glycyrrhizin per week gave birth to children with depressed verbal skills, as measured by the Beery Development Test. Although this work should not deter the occasional consumption of a liquorice sweet, pregnant women should be aware of the dangers of an excess.

Conclusion

Liquorice is special among the herbs used in Chinese herbal medicine. Whether or not we accept the rigid hierarchy of herbs in a Chinese prescription, it is not unreasonable to think that each component plays a different role. What is so special about liquorice is that it plays many parts. In some circumstances, as in the case of coughing, it does have a therapeutic role. Undoubtedly its sweetness makes the prescription more palatable, so it is an aid to consumption. Finally, it may, in some instances, provide a drug delivery system for other bioactive components of the prescription. That one herb should provide so much is impressive and valuable but what is even more impressive is that Chinese physicians, with no understanding of modern concepts of drug action, should have recognised that there could be value in adding liquorice to enhance the potency of a prescription.

Amphiphilic compounds like liquorice are used in drug delivery systems in biomedical pharmacy and so there is a sense in which nothing new has been revealed by this examination of the role of liquorice in traditional Chinese herbal medicine. However, it does enhance the reputation of Chinese physicians who developed the prescriptions that sustained the Chinese medical system for centuries. They had no understanding of modern concepts of drug delivery and drug action and yet noticed that the addition of liquorice enhanced the potency of the medicine.

When the Chinese government wanted a new treatment for malaria because there was resistance to chloroquine it recruited a team that included historians. The idea was that the *bencao* literature might offer some clues. But these were not easy to understand as malaria was not described in modern terms and the exact preparation of the drug for use was critically important. Nevertheless, they achieved extraordinary success and proved, despite the prejudices of historians, that one can leap across epistemological domains and map new pathways for scientific discovery. At its best, the prescriptions in the *bencao* literature are the results of clinical trials over thousands of years and with thousands of patients but, to maximise their value to modern pharmacy, they must be read collaboratively by those with historical understanding, not just by pharmacologists.

Note

1 *Shen Nong bencao jing* 神農本草經 (The Divine Farmer's Canon of *Materia Medica*), in Tao Hongjing (ed.) (456–526 CE) *Bencao jing jizhu* 本草經集注 (*Materia Medica* Canon Variorum Edition), Shang Zhijun 尚志鈞 and Shang Yuansheng 尚元勝 (eds) (1994), Beijing: Renmin weisheng chubanshe. For a translation of the *Shennong bencao jing*, see Yang (1998).
2 See Unschuld Dictionary of the *Bencao gangmu*, Volumes 1–3 (Ben Cao Gang Mu Dictionary Project). There are an increasing number of tools available for those interested in the history and contemporary usage of Chinese medicinals, but who do not read Chinese. They must, however, be used with caution, especially where illness terms and substance identification are concerned. Li Shizhen's complete work has been translated into English and annotated by Luo Xiwen (2003).

Bibliography

Pre-modern sources

Bencao jing jizhu 本草經集注 (Materia Medica Canon Variorum Edition) 6th CE, Tao Hongjing 陶弘景 (502–557 CE), in Shang Zhijun 尚志鈞 and Shang Yuansheng 尚元勝 (eds) (1994), Beijing: Renmin weisheng chubanshe.

Modern sources

Ahmad, Z., Shah, A., Siddiq M. and Kraatz, H.B. (2014) 'Polymeric micelles as drug delivery vehicles', *RSC Advances*, 4.33: 17028–35.

Badr, A.E., Omar, N. and Badria, F.A. (2011) 'A laboratory evaluation of the antibacterial and cytotoxic effect of liquorice when used as root canal medicament', *International Endodontic Journal*, 44.1: 51–8.

Bennett, A., Clark-Wibberley, T., Stamford, I.F. and Wright, J.E. (1980) 'Aspirin-induced gastric mucosal damage in rats: cimetidine and deglycyrrhizinated liquorice together give greater protection than low doses of either drug alone', *Journal of Pharmacy and Pharmacology*, 32.2: 151.

Bensky, D. and Gamble, A. (1993) *Chinese Herbal Medicine: materia medica*, Seattle, WA: Eastland Press.

Borisenko, S.N., Lekar, A.V., Vetova, E.V., Filonova, O.V. and Borisenko, N.I. (2016) 'Mass spectrometry study of the self-association of glycyrrhetinic acid molecules', *Russian Journal of Bioorganic Chemistry*, 42.7: 716–20.

Butler, A.R. (2019) 'Artemisinin from Chinese herbal medicine to modern chemotherapy', *Himalaya*, 39.1: 219–28.

Damle, M. (2013) 'Glycyrrhiza glabra (liquorice): A potent medicinal herb', *International Journal Herbal Medicine*, 8.3: 132–6.

Deb Roy, R. (2017) *Malarial Subjects: empire, medicine and nonhumans in British India, 1820–1909*, Cambridge: Cambridge University Press.

Hajaghamohammadi, A.A., Zargar, A., Oveisi, S., Samimia, R. and Reisian, S. (2019) 'To evaluate the effect of adding licorice to the standard treatment regimen of *Helicobacter pylori*', *Brazilian Journal of Infectious Diseases*, 20.6: 534–8.

Harper, D. (1998) *Early Chinese Medical Literature: the Mawangdui medical manuscripts*, London: Kegan Paul.

Hsu, E. (2006) 'Reflections on the "discovery" of the antimalarial *qinghaosu*', *British Journal of Clinical Pharmacology*, 61.6: 666–70.

Hu, C.H., He, J., Eckert, R., Wu, X., Li, L., Tian, Y., Lux, R., Shuffer, J.A., Gelman, F., Mentes, J., Spackman, S., Bauer, J., Anderson, M.H. and Shi, W. (2011) 'Development and evaluation of a safe and effective sugar-free herbal lollipop that kills cavity causing bacteria', *International Journal of Oral Science*, 3.1: 13–20.

Iijchi, S. and Tamagaki, S. (2005) 'Molecular design of sweet tasting compounds based on 3β-amino-3β-deoxy-18β-glycyrrhetinic acid amido functionality eliciting tremendous sweetness', *Chemistry Letters*, 34.3: 356–7.

Karrer, P., Karrer, W. and Chao, J.C. (1923) 'Glucoside VIII. Beitrag zur Kenntnis des Glycyrrhizins', *Helvetica Chimica Acta*, 4.1: 100–12.

Kim, D.H., Lee, S.W. and Han, M.J. (1999) 'Biotransformation of glycyrrhizin to 18-beta-glycyrrhetinic acid-3-O-beta-D-glucuronide by Streptococcus LJ-22, a human intestinal bacterium', *Biological and Pharmaceutical Bulletin*, 22.3: 320–2.

Kong, R., Zhu, X., Meleleva, E.S., Polyakov, N.E., Khvostov, M.V., Baev, D.S., Tolstikova, T.G., Dushikn, A.V. and Su, W. (2018) 'Atorvastatin calcium inclusion complexation with polysaccharide arabinogalactam and saponin disodium glycyrrhizate for increasing solubility and bioavailability', *Drug Delivery and Translational Research*, 8.5: 1200–13.

Kowalska, A. and Kalinowska-Lis, U. (2019) '18β-Glycyrrhetinic acid and its core biological properties and dermatological applications', *International Journal of Cosmetic Science*, 41.4: 325–31.

Kuniyama, A. and Maeda, H. (2019) 'Topical application of licorice for prevention of postoperative sore throat in adults: a systematic review and meta-analyi', *Journal of Clinical Anaesthesia*, 54.5: 25–32.

Larkworthy, W. and Holgate, P.F. (1975) 'Deglycyrrhizinated liquorice in the treatment of chronic duodenal ulcer. A retrospective endoscopic survey of 32 patients', *Practitioner*, 215.1290: 787–92.

Lee, M.R. (2018) 'Liquorice (*Glycyrrhizin glabra*) the journey of sweet root from Mesopotamia to England', *The Journal of the Royal College of Physicians of Edinburgh*, 48: 378–82.

Li, X., Sun, R. and Liu, R. (2019) 'Natural products in licorice for the therapy of liver diseases: Progress and future opportunities', *Pharmacological Research*, 144: 210–26.

Liu, W., Huang, S., Li, Y.L., Li, Y.W., Li, D., Wu, P., Wang, Q., Zheng, X. and Zhang, K. (2018) 'Glycyrrhizic acid from licorice down-regulates inflammatory responses *via* blocking MAPK and P13K/Akt-dependent NF-κB signalling pathways in TPA-induced skin inflammation', *MedChemComm*, 9.9: 1502–10.

Luo Xiwen (trans. and annot.) (2003) *Compendium of Materia Medica by Li Shizhen*, Beijing: Foreign Languages Press.

Lythgoe, B. and Trippett, S. (1950) 'The constitution of the disaccharide of glycyrrhizinic acid', *Journal of the Chemical Society (Resumed)*, 1950: 1983–90.

Mamedov, N.A. and Egamberdieva, D. (2019) 'Phytochemical constituents and pharmacological effects of liquorice: a review', *Plant and Human Health*, 3: 1–21.

Ming, L.J. and Yin, A.C.Y. (2013) 'Therapeutic effects of glycyrrhizic acid', *Natural Product Communications*, 8.3: 415–18.

Öztürk, M., Altay, V., Hakeem, K.R. and Akçiçek, E. (2017) *Liquorice: from botany to phytochemistry*, Cham, Switzerland: Springer.

Penninkilamp, R., Eslick, E.M. and Eslick, G.D. (2017) 'The association between consistent licorice ingestion, hypertension and hypokalaemia: a systematic review and meta-analysis', *Journal of Human Hypertension*, 31.11: 699–707.

Petrova, S.S., Schlotgauer, A.A., Kruppa, A.I. and Leshina, T.V. (2017) 'Self association of glycyrrhizic acid. NMR study', *Zeitschrift für Physikalische Chemie*, 231.4: 839–55.

Pharmacopoeia Commission China (2000) *Pharmacopoeia of the People's Republic of China*, Beijing: Chemical Industry Press.

Räikkönen, K., Pesonen, A., Heinonen, K., Lahti, J., Komsi, N., Eriksson, J.G., Seckl, J.R., Järvenpää, A. and Strandberg, T.E. (2009) 'Maternal licorice consumption and detrimental cognitive and psychiatric outcomes in children', *American Journal of Epidemiology*, 170.9: 1137–46.

Schmid, C., Dawid, C., Peters, V. and Hofmann, T. (2018) 'Saponins from European licorice roots (*Glycyrrhia glaba*)', *Journal of Natural Products*, 81.8: 1734–44.

Selyutina, O.Y., Polyakov, N.E., Korneev, D.V. and Zaitsev, B.N. (2016) 'Influence of glycyrrhizin on permeability and elasticity of cell membrane: perspectives for drug delivery', *Drug Delivery*, 23.3: 848–55.

Selyutina, O.Y. and Polyakov, N.E. (2019) 'Glycyrrhizic acid as a multifunctional drug carrier – From physiochemical properties to biomedical applications: a modern insight on the ancient drug', *International Journal of Pharmaceutics*, 559: 271–9.

Sidhu, P., Shankargouda, S., Rath, A., Ramamurthy, P.H., Fernandes, B. and Singh, K. (2020) 'Therapeutic benefits of licorice in dentistry', *Journal of Ayurveda and Integrative Medicine*, 11.1: 82–8.

Su, X., Wu, L., Hu, M., Dong, W., Xu, M. and Zhang, P. (2017) 'Glycyrrhizic acid: a promising carrier material for anticancer therapy', *Biomedicine and Pharmacotherapy*, 95: 670–8.

SuperTCM, http://bioinf-applied.charite.de/supertcm/home, accessed 18/1/2021.

SymMap: an integrative database of traditional Chinese medicine enhanced by symptom mapping, https://www.symmap.org/, accessed 18/1/2021.

Tang, W. and Eisenbrand, G. (1992) *Chinese Drugs of Plant Origin: chemistry, pharmacology, and use in traditional and modern medicine*, Berlin: Springer-Verlag.

Taylor, K. (2005) *Chinese Medicine in Early Communist China 1945–1963: a medicine of revolution*, London: Routledge.

Unschuld, P.U. (1986) *Medicine in China: a history of pharmaceutics*, Berkeley: University of California Press.

Unschuld, P.U. and Zhang Zhibin 张志斌 (eds) (2015) *Ben Cao Gang Mu Dictionary, Volume 1: Chinese historical illness terminology*, Oakland: University of California Press.

Unschuld, P.U., Hua Linfu 华林甫 and Buell, P.D. (eds) (2017) *Ben Cao Gang Mu Dictionary, Volume 2: geographical and administrative designations*, Oakland: University of California Press.

Unschuld, P.U. and Zheng Jinsheng 郑金生 (eds) (2018) *Ben Cao Gang Mu Dictionary, Volume 3: persons and literary sources*, Oakland: University of California Press.

van Uum, S.H. (2005) 'Liquorice and hypertension', *Netherlands Journal of Medicine*, 63.4: 119–20.

Woodward, R.B., Doering, W.E. (1944) 'The total synthesis of quinine', *Journal of the American Chemical Society*, 67.5: 860–874.

Wu, Y., Zhang, F., Yang, K., Fang, S., Bu, D., Li, H., Sun, L., Hu, H., Gao, K., Wang, W., Zhou, X., Zhao, Y. and Chen, J. (2018) 'SymMap: an integrative database of traditional Chinese medicine enhanced by symptom mapping', *Nucleic Acids Research*, 47.D1: 110–17.

Yang Shouzhong (1998) *The Divine Farmer's Materia Medica: a translation of the Shen Nong Ben Cao Jing*, Portland: Blue Poppy Press.

Yang, R., Yuan, B., Ma, Y., Zhou, S. and Liu, Y. (2017) 'The anti-inflammatory activity of licorice, a widely used Chinese herb', *Pharmaceutical Biology*, 55.1: 5–18.

Zhang, J., Wang, L., Wang, H. and Tu, K. (2006) 'Micellization phenomena of amphiphilic block copolymers based on methoxy poly (ethylene glycol) and either crystalline or amorphous poly (caprolactone-b-lactide)', *Biomacromolecules*, 7.9: 2492–500.

51
DECONTEXTUALISED CHINESE MEDICINES

Their uses as health foods and medicines in the 'global North'

Michael Heinrich, Ka Yui Kum and Ruyu Yao

Many medicines from the Chinese *materia medica* remain highly esteemed by the Chinese government and are used widely by all generations in China, and particularly by the middle-aged and elderly. In the last decades some have also found a specialist niche in healthcare in North America, Europe and other European language speaking regions. There is much controversy as to whether and how to use them with critics frequently focussing on the lack of evidence for either the safe use of single botanical drugs, or for exact formulas. Here we use a case study approach highlighting the different scientific, historical and cultural trajectories through which substances from the Chinese *materia medica* and *dietetica* have entered markets globally, mostly in the last fifty years, and how and why they changed in transit, and what they lost in transmission. We have chosen to highlight the distinctions between herbal medicines like *qinghao* 青蒿 (the aerial part of *Artemisia annua* L.), which is the source of the compound artemisinin, or *yinxingye* 銀杏葉 (the leaves of *Ginkgo biloba* L., Ginkgoaceae, hereafter ginkgo), and substances that have come to be classified as food supplements, like *gouqi* 枸杞 (the fruit of *Lycium barbarum* L., hereafter goji berry) and *renshen* 人參 (*Panax ginseng* C.A.Mey., hereafter ginseng). The uses of these substances outside of China are often more or less decontextualised from their traditional Chinese uses and our analysis will show how different regulatory, economic, medical and sociocultural frameworks have had an impact on the reinterpretation and re-contextualisation of these plants.

The overarching aim of this chapter is to explore the avenues through which substances from the Chinese *materia medica* and *dietetica* have arrived in Europe, Australia/New Zealand and North America and to understand how history and regulatory systems have affected the ways in which the substances have been de-contextualised and re-contextualised. These states and regions are identified in this chapter as 'the global North'. While all of these places may not strictly be in the North, they share a common history of the regulation of medicinal plant use in modern times, with a strong basis in European phytotherapeutic traditions, and each place has benefitted from a wide range of botanical drugs as a result of European colonialist expansion and later liberal trade structures. The introduction of botanical drugs from China into these somewhat similar contexts facilitates the comparative approach adopted in this chapter.

Following the interdisciplinary style of many of the contributions to this Handbook, this chapter uses historical and anthropological approaches to highlight the routes through which certain traditional medicinal plants and fungi from the Chinese *materia medica* and *dietetica* have become successful outside of China and to try to understand why other products have not been approved and put into general use.

First and foremost, the use of the isolated compounds of Chinese medical plants in modern biomedicine inevitably disconnects traditional substances from their local Chinese traditions. At the same time, paradoxically, those that achieve success easily become symbols of the modern and enduring relevance and impact of Chinese medicine. In this context a traditional Chinese medicine becomes TCM, the acronym commonly used to refer to the modernised and standardised twentieth to twenty-first-century forms of Traditional Chinese Medicine. The most iconic example is the derivation of artemisinin from the plant *Artemisia annua* L., and its development into a licensed medicine with full marketing authorisation as a biomedical treatment for malaria. In this process Chinese medicines become decontextualised and take on different roles in what can be called a reductive scientific process. In contrast to those substances which have become biomedical drugs, others such as ginseng are now commonly used food supplements, and have become a familiar part of more alternative approaches to healthcare, with ginseng being used widely for its preventive functions.

In a comparative context, it is essential to understand what framework enables the introduction of a 'new' substance onto new markets globally from a provider country (in this case China) and how this can be achieved. In essence, it is a question of regulation. In other words the transfer of the product and the knowledge that surrounds it is subject to the frameworks used to facilitate or prevent the entry of a specific compound or plant/fungi-based product into a particular market. As a part of the journey of a medicinal plant or fungus derived from Chinese traditions, these species undergo a process of being reinterpreted, based on the regulatory basis in the region where the species or a substance derived from it becomes a commodity. This commodification is based on the socio-economic structure of the receiving country and the medical expectations there. Importantly, in the following analyses we will also highlight examples where such an introduction was not possible, since the use of the medicinal plant or fungus did not fit into these structures.

In general terms, a product newly entering another country may be seen as a food or a medicine. The food-medicine interface has been a topic of debates in the sociocultural sciences for many decades (Etkin and Ross 1982; Etkin 1986; Johns 1990; Prendergast *et al.* 1998), but these studies rarely consider the impact of the receiving state's regulatory framework on the reception of a new botanical arriving from afar. Fundamentally, a medicine[1] or medical substance (in American English a 'drug') is a 'substance or combination of substances that is intended to treat, prevent or diagnose a disease, or to restore, correct or modify physiological functions by exerting a pharmacological, immunological or metabolic action'.[2] Food ingredients (as defined by the regulatory authorities), on the other hand, are 'chemical substances which are used as food additives, food enzymes, flavourings, smoke flavourings and sources of vitamins and minerals added to food'.[3] A specific category within food is supplements, 'foodstuff containing concentrated amounts of nutrients or other substances that are intended to supplement the normal diet'.[4] Thus, coming back to the two examples above, artemisinin as a pure compound derived from a TCM plant is generally regulated as a medicine, while in the case of ginseng it is treated as a food supplement in many countries.

Another common and crucial distinction is between a pure isolated compound and an extract. Clearly, artemisinin, an example of the former, is a single chemical entity used as a medicine, where, in the case of ginseng, the processed botanical drug or an extract is used. In general terms, only very few extracts are licensed as medicines.

Therefore, in the context of this analysis one needs to understand how medicines from China may become an element of biomedical practice or a substance that enjoys popular usage outside China. Key to understanding the transfer of TCM preparations is their role within the new markets globally. There may, for example, be a small niche culture that promotes interest in a substance and the decoction of traditional multi-ingredient recipes, the preparation of which is a highly developed skill generally carried out according to a practitioner's individual training and experience, that is, the substance is compounded by the practitioners who claim specialist knowledge of Chinese medicine. This is, however, outside of the mainstream medical and pharmaceutical practice beyond China and the 'Chinese diaspora', and seen in this chapter as a patient's choice and an alternative to 'normal (biomedical) practice'. The products discussed here have all become widely used in different contexts of healthcare, where they have become highly sought-after commodities.

Nowadays, there are several routes through which Chinese medicinal or nutritional substances can become available outside their country of origin. Practitioners within and outside China may dispense medicines to patients based on the regulations as they exist in their respective countries. Various Chinese medical techniques have become important elements in healthcare practice worldwide; some, like cupping and specialist massage, are elements of alternative practice, while others have, to some degree, entered mainstream medicine (On USA, Phan 2017; see also Chapters 36, 39, 40, 41, 42, 44 and 43 in this volume).[5] Such diverse medical practices cannot be a focus of this chapter, which deals exclusively with botanical foods and medicines and does not even consider the minerals in the Chinese *materia medica*; herbal medicines are normally not available as an element of mainstream medicine/biomedicine.

In technical/regulatory terms, herbal preparations may be treated as some sort of a food substance, such as a health food (like ginseng or goji). However, in other cases, they can be deemed medicines including, for example, as a 'listed' medicine in Australia (a category of medicines with low risks and low levels of claims introduced in 1991) or a traditional herbal medicinal product under a European directive regulating the access of herbal medicines to the European markets – the Traditional Herbal Registration or THR implemented in 2004. In other cases, and for topical[6] preparations, they may, for example, be regulated as cosmetics.

In the following, we will look at a few of these examples and especially their routes into Europe and America and for licensed medicines globally. We will focus on examples which have become widely adopted in these regions providing some case studies on selected cases.

The focus is on some herbal and fungal substances, which have become major commodities worldwide, be it as a health food / supplement or a medicine, and that are available through the various major channels for foods and medicines in Europe, North America and the other regions we identify as the global North. These case studies show the differences in the interpretation inside and outside China, but also highlight the plural pathways to reception, and therefore of the real heterogeneous processes of a pharmacological 'modernity'. This will include essential information on the research on pharmacologically active preparations where it led to their introduction.

Medicines with a full marketing authorisation

Artemisia annua L. is certainly now the most notable example of a modern medicine derived from medical traditions in China (see 50 in this volume, also Hsu 2006). The medical effects on febrile disease of *Artemisia annua* L. had been known since at least 340 CE when it was included in the *Handbook of Prescriptions for Emergency Treatment* (*Zhouhou beiji fang* 肘後備急方) by Ge Hong 葛洪 (284–364 CE) (Stanley-Baker 2021). But it was not until 1967 that Chinese scientists initiated research on new antimalarial drugs using the Chinese *materia medica*. Famously, the entire secret project (Project 523) was initially driven by the Chinese search for new antimalarial agents partly in the context of the Vietnamese War.

After the war ended, the results were shared widely outside China. Key in the *Artemisia annua* L. project had been Tu Youyou 屠呦呦 (1930–). For most of the early development of artemisinin the Cooperative Research Group, and not individual researchers, was named as the author, so it was only much later that Tu was rewarded publicly as the person behind these research and development programmes, and was ultimately awarded the 2015 Nobel prize in Physiology or Medicine for her innovations.

Ten years into the project, Tu was critical in determining the active principle – artemisinin (a sesquiterpene lactone with an unusual endoperoxide ring) (Cooperative Research Group 1977). It showed potent activity against the malarial parasite *Plasmodium falciparum* and especially against chloroquine-resistant malaria. The development was based directly on traditional and local knowledge. Tu Youyou and colleagues had used a range of extraction and processing techniques, and when they had found the extracts not to be (very) active they turned to the historical documents. On the basis of reading Ge Hong's work, 'A modified procedure was designed to reduce the extraction temperature by immersing or distilling *Qinghao* 青蒿 (*A. annua*) using ethyl ether' (Tu 2016). Ge Hong had described 'wringing the juice' (*jiaoqu zhi* 絞取汁) from *A. annua*. This led to a change in the traditional laboratory method, a hot extraction to a cold extraction and ultimately to the isolation of the compound (Figure 51.1).

Artemisinin was then shown to be highly effective in a series of clinical trials in China in a large number of patients. Specifically, the clinical trials showed a clearance of parasitaemia and a reduction of symptoms in patients with malaria, including some with chloroquine-resistant malaria and/or cerebral malaria. After its initial development, it attracted the attention of the World Health Organisation and numerous agencies like the Wellcome Trust (UK) and the National Institute of Health (USA). Later, interestingly, the compound also showed preclinical promise as a potential anticancer agent. Today, derivatives of artemisinin have been developed (ethers, such as *artemether* and *arteether*, and esters, such as *sodium artesunate* and *sodium artenlinate*).

Camptotheca acuminata Decne (*Xishu* 喜樹, tree of joy, Nyssaceae) was first recorded in *Illustrated Catalogue of Plants* (*Zhiwu mingshi tukao* 植物名實圖考 1848) and is not widely

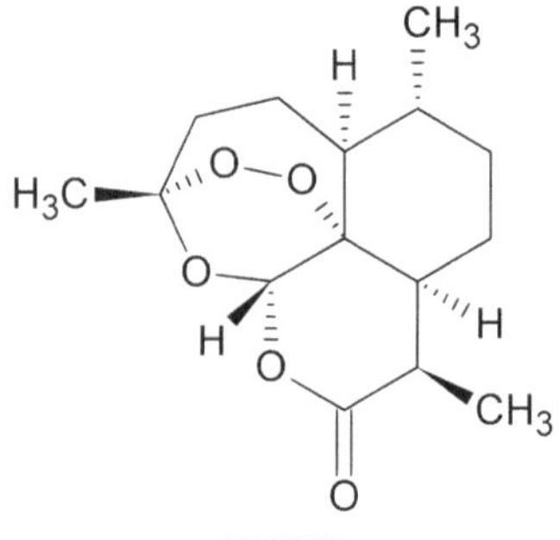

Figure 51.1 Artemisinin from *Artemisia annua*

used in TCM. It was included in 1958 in a screening programme at the Chinese National Cancer Institute (NCI) where it gave positive results. Wood and bark (20kg) were collected for extraction. These extracts were shown to be active in the case of a mouse leukaemia life prolongation assay in which it was unusual to find activity. The fractionation and anticancer testing was a very slow process and finally resulted in the isolation and structural elucidation (in 1966) of *camptothecan*, a highly unsaturated *quinoline alkaloid* with a unique (at the time) structure as an α-hydroxylactone. *C. acuminata* was shown to be extremely active in the life prolongation assay of mice treated with leukaemia cells and in solid tumour inhibition. These activities encouraged the NCI to initiate clinical trials with the water-soluble sodium salt. While the results of some studies conducted in the USA were disappointing, in a clinical trial in China with 1,000 patients the sodium salt showed promising results, for example, against head, neck, gastric, intestinal, and bladder carcinomas.

Several compounds from Chinese medicine are currently under development that may result in clinically approved medications for use in chronic inflammatory conditions. A decoction of *Leigong teng* 雷公藤, *Radix Tripterygii wilfordii* (*Tripterygium wilfordii* Hook.f.; Celastraceae), was first mentioned in the *Supplement to the Systematic Materia Medica* (*Bencao gangmu shiyi* 本草綱目拾遺 1765), the classic pharmacological encyclopaedia produced by Zhao Xuemin 趙學敏 (1719–1805 CE) during the *Qing* dynasty, but seems to be of limited importance in TCM today, due to toxicity problems.

In TCM, it functions as a way of dispelling wind, dehumidification, promoting blood circulation and removing obstruction in channels, reducing the swelling, relieving pain, killing insects and detoxifying.[7] These are common strategies used to treat rheumatic symptoms. Preclinical and clinical development has focussed on potential uses against cancer, chronic nephritis, hepatitis, systemic lupus erythematosus, ankylosing spondylitis (a rare form of arthritis) and a variety of skin conditions. In TCM, a patient who has rheumatism (swelling and pain of the joints or muscles) might well be regarded as having wind, being damp in the body as well as the blood, and her/his *qi* 氣 is described as 'being hindered'. Also, in TCM theory, the kidney is in charge of water and is thus responsible for metabolising human body water. Therefore, a TCM doctor is likely to aim to promote blood circulation and remove obstructions in channels, as well as inducing diuresis to alleviate oedema and to cure nephropathy (Heinrich *et al.* 2009; Heinrich 2013).

These two medications derived from *Artemisia annua* L. and *Tripterygium wilfordii* Hook.f., are the most important ones introduced to the global North in recent decades. Others are also used as licensed medicines, like ephedrine derived from *mahuang* 麻黃 – *Ephedra spp.* (Lee 2011) and its derivatives such as pseudoephedrine. Ephedrine was introduced in the 1920s into European medical markets and had an important place as a cough suppressant (Sneader 1996). Its development was very much driven by phytochemical research in Japan and pharmacological research in China (Ibid.), but it is also found in other *Ephedra* spp., from America and Europe. Its stereoisomer pseudoephedrine is today used more widely. In general, however, all these products have been replaced by compounds like the adrenalin analogues salbutamol and terbutaline both derived from isoprenaline. In contrast to artemisinin, since ephedrine is found in a wide range of Ephedra species distributed over the Northern Hemisphere and these have had a wide range of medical uses (e.g. Moerman 2009), it is not a specific contribution of Chinese medicine to modern medical practice.

In each case, in order to become a licensed medicine, the compound first underwent preclinical development, development of appropriate formulations, then the various stages of clinical development, including a clinical safety assessment, dose finding studies and, lastly, the assessment of its efficacy, normally compared to an existing gold standard (phases 1 to

3 of clinical evaluation), and finally, was subjected to assessment by a competent authority. Once licensed the medicine was also subjected to pharmacovigilance. In essence, for a full clinical development, any substance, whether it is a natural product or not, needs to comply with the existing regulatory framework, which ascertains the licensing for medicines, with an acceptable risk-benefit ratio. By definition, this process cannot take into account the art of traditional Chinese medicine, as it has been known for over 2,000 years, including the skills of decocting multi-ingredient preparations that are personalised to individual patients, and with recipes that change over the course of an illness. Chinese medicines tend to be selected as single ingredient remedies in this regulatory process and become – as with any other modern licensed medicine – a product, which is produced by a standardised process, quality assured and with an acceptable risk-benefit assessment used for a biomedically defined condition/disease.

Meandering between licensed medicine and food

Ginkgo

A special extract obtained from the *leaves* of the maidenhair tree (*yinxingshu* 銀杏樹, *Ginkgo biloba* L., Ginkgoaceae) ginkgo has become one of the most important herbal medicines used in the management of cognitive decline and specific forms of vertigo (dizziness due to circulatory problems in the inner ear – Ménière's disease), and peripheral circulatory problems like 'cold hands' (i.e. Raynaud's syndrome). A special extract is prepared in such a way that the concentration of desired compounds (metabolites) is increased, while undesirable ones are removed. Core is its use in ageing-related disorders to prevent or reduce milder forms of dementia, like the early stages of Alzheimer's disease, and memory deterioration. It is an extremely well-studied medicinal plant, known to enhance cognitive processes, and experimental evidence points to improvements in blood circulation to the brain and other parts of the body. Anti-inflammatory and antioxidant effects are also of pharmacological relevance (Edwards *et al.* 2015; Heinrich 2018). In some countries, it is available as a medicine, while in others it is a food supplement or, in essence, an unregulated product.

So, how did gingko become a part of medical practice in the first in Europe and then the 'global North'? This living fossil[8] survived in China and was mostly grown in monasteries in the mountains, temple gardens or palaces. Without doubt, Ginkgo is an object of veneration and a sacred tree of the East, but the leaves are of no major importance in Chinese medicine and are recorded as effective for cardiovascular diseases. The tree has been seen as a symbol of changelessness, with miraculous powers. In general, it is a symbol of longevity with cultivation by Buddhist monks reported from about 1100 CE. Around 1192 the seeds were taken to Japan (where they were also linked to Buddhism) and Korea. Ginkgo does not seem to have been mentioned in ancient Chinese texts, but in the Song dynasty (eleventh century) it was listed as a species native to Eastern China. It has been depicted in Chinese paintings and appeared in poems.

The successful transmission of Ginkgo seems to be wrapped up with these religious and poetical connotations. Yet, this did not mean that the medicine was processed or used in the original way. In 1691 the German naturalist Engelbert Kaempfer, 'discovered' *G. biloba* trees in Japan. Forty years later (1730) it arrived in Utrecht and thus Europe (Heinrich 2013). East Asian records of medical functions generally relate to the seeds (pseudofruits). Importantly, the leaves are much less frequently used. There are topical preparations including for treating chilblains (frostbites, resulting in swelling, reddening and itching of the skin) and for asthma (as a throat spray). European fascination is linked to its symbolic importance in the context of longevity. Since its European 'discovery', the unusual shape of the leaves has fascinated scientists and poets, including J. W. von Goethe 1749–1832), the famous Romantic poet, statesman, solicitor and natural historian from Frankfurt, who from 1775 onwards worked in Weimar. His poem on ginkgo both crystallises European fascination with the tree and stimulated its spreading fame:

Dieses Baums Blatt, der von Osten	This leaf from a tree in the East,
Meinem Garten anvertraut,	Has been given to my garden.
Giebt geheimen Sinn zu kosten,	It reveals a certain secret,
Wie's den Wissenden erbaut.	Which pleases me and thoughtful people.
Ist es Ein lebendig Wesen	Does it represent One living creature
Das sich in sich selbst getrennt,	Which has divided itself?
Sind es zwey die sich erlesen,	Or are these Two, which have decided,
Dass man sie als eines kennt.	That they should be as One?
Solche Frage zu erwiedern	To reply to such a Question,
Fand ich wohl den rechten Sinn;	I found the right answer:
Fühlst du nicht an meinen Liedern	Do you notice in my songs and verses
Dass ich Eins und doppelt bin?[9]	That I am One and Two?

During the 1960s and 1970s, Dr. Willmar Schwabe Pharmaceuticals (Germany) developed the now famous special extract EGb 761 on the basis of reported usages in traditional Chinese medicine. First, a poorly defined ethanolic extract was developed. The special extract EGB 761 is based on an ethyl acetate extraction and subsequent fractionation and processing. Unfortunately, the research and development that led to this special extract has not been well explained, and it would be desirable to get a better understanding of the company's strategy at the time. With the development of this special extract, a very well-defined therapeutic use based on European phytotherapeutic traditions (rational phytotherapy) has emerged.

The commercial medical preparations contain two major types of pharmacologically active constituents – diterpene lactones, including ginkgolides A, B, and C and bilobalide, as well as flavonoids, most notably biflavone glycosides, such as ginkgetin, isoginkgetin and bilobetin, which also contribute to its activity. Ginkgolic acids are present in the fruit, but normally in only very small amounts in the leaf. Initial research in the mid-1960s identified flavonoid glycosides as active constituents of *G. biloba* leaf extracts. The first patent on the complete extraction and processing was filed in 1971 in Germany and a year later in France, defining the process for obtaining a 'mixture of vasoactive substances'. The intellectual protection for this process was developed within the company and is the foundation for its clinical development into a vasoactive medicine, which was first widely sold in Europe and is now available under a broad range of regulatory schemes in many countries globally.

Is this development derived from Chinese *materia medica*? Based on the existing evidence, this seems unlikely. The symbolic importance in both European and Asian countries was a starting point for research and development, but there seems to be no link into traditional uses in Chinese *materia medica*. It is more plausible that the ideas came from the Romantic European fascination with this 'tree from the East' and its legacy. The symbolic importance and the species' association with longevity (as a temple tree) were certainly a starting point for the initial pharmacological experiments, but J.W. Goethe and others' reinterpretations of the tree were an important interim step and then a 'rational phytomedicine' with a strong evidence base became the objective of the company's research.

Ginseng

Ginseng (*Renshen* 人参, *Panax ginseng* C.A.Mey.) has been used medicinally in China since at least the second century CE, to strengthen the body and increase vitality and virility. In China it is commonly used in a combination decoction. In *Shanghan lun* 傷寒論 (Treatise on Cold Damage) by Zhang Zhongjing 張仲景 (150–219 CE), twenty-one multi-herbal decoctions contain ginseng. Another important medical source for ginseng is the sixteenth-century *Systematic Materia Medica* (*Bencao gangmu* 本草綱目 1596) by Li Shizhen 李時珍, which emphasised that ginseng was used to treat all kinds of symptoms caused by 'deficiency' (Park *et al.* 2012). The unique quasi human shape of the root, like the powerful medicine, poison and hallucinogen, mandrake (*Mandragora officinarum* L.), has always been an important aspect that has stimulated its dissemination and cultural role. The first European mention of ginseng was by Marco Polo in the 1200s, while the earliest evidence of its introduction to Europe is in a document of the East India Company in the late eighteenth century, which suggests it came via Korea and Japan (Loskiel 1794).

While the British East India Company was shipping *Panax ginseng* as Korean ginseng via their employees in Japan, the British and Dutch both competed to source 'similar' species, originally from Japan and Korea, from the Cape of Good Hope region whither they had been transplanted and flourished. There they were known as 'canna' (Foyster 1900). Richard Cock, head of the Hirado branch (Southwestern Japan) of the British East India Company claimed that the processing of the plant in the Cape had a detrimental effect on its quality (Seol 2017).

Knowledge about ginseng was circulated in Europe thanks to the Royal Society in England and Portuguese and Italian Jesuit missionaries, some of whom had lived in China. Alvarus de Semedo, a Portuguese Jesuit priest who lived in China for more than 20 years, described Chinese ginseng and its use in his book *History of the Great and Renowned Monarchy of China*, 1641. Later, Martinus Martini, an Italian missionary, wrote about ginseng in his book *Bellum Tartaricum*, 1654 and was quoted in *La Chine illustree* by Athanase Kircher, 1667 (Appleby 1983). Employees of the Dutch East India Company, such as French orientalist, cartographer, inventor, author, and diplomat Melchisédech Thévenot (1620–92) wrote for the Bulletin of the Royal Society about the tremendous medicinal value of ginseng (Nieuhof 1660; Thévenot 1665). The medicinal claims from these writings caught the attention of the Royal Society (Seol 2017). On 26th June 1679, Dr. Andrew Clench (d. 1692) presented ginseng as:

> a certain root lately brought out of China, called ginseng, of great esteem in China for its virtue in restoring consumptive persons, and those emaciated with long sickness, to their former health and strength. It was valued in China at twice its weight in silver.
>
> *(Birch 1757)*

A fellow of the Society, Dr. Robert Wittie (1613–84), subsequently wrote a letter indicating the degree of interest the Society had in Chinese medical practice (Simpson, Pechey and

Wittie 1680). His successful trials of ginseng tincture for a diverse range of patients were apparently mirrored by Robert Boyle (1627–91), one of the founder fellows of the Royal Society and renowned chemist, who thought it *'a Medicine sent from Heaven to save the Lives of Thousands of Men, Women and Children'*.

While ginseng had been examined and praised for its miraculous effects by academia, it was French Jesuit missionaries who played the most critical role in its globalisation. Pierre Jartoux (1668–1720) gave a detailed description in a letter to the Society of Jesus, including an ethnopharmacology of ginseng in China and Tartary (between the Caspian Sea and the Ural Mountains) a habitat, he believed, to be much like the forests and mountains of Canada (1713).

Brother Joseph-François Lafitau (1681–1746) subsequently used Jartoux's data and searched for the plant in New France (North America) showing the illustration in the letter to the local people. When an Iroquois claimed it was one of their 'common medicines' Lafitau hired local help whereupon they found a similar species near Montreal, later identified as American ginseng – *Panax quinquefolius* L. (Lafitau 1718). Although the native use of American ginseng was different from Asian ginseng, Lafitau apparently adopted some of the new medical indications, which were similar to those of Asian ginseng (Harriman 1973).

In 1718, J.-F. Lafitau published his work in French and presented it to the Royal Academy of Science, Paris, also as a prospective business. French companies not only imported American ginseng but they also re-exported it to China via France in the same year as Lafitau's publication (Hocking 1976). By 1752, c. 7.6 to 13.3 million ginseng plants were re-exported from 'New France' via France, mainly to China (Havens 2020). The international trade in ginseng was significant. In the 1780s, the first trial voyage from New England on the *Empress of China* (under captain John Green) mainly trading American ginseng to Canton resulted in a profit of US$30,727, which was 26% of the cost of the voyage. Although the margin of profit was very high, it was still ten times less valuable than the Chinese ginseng, which was worth its weight in gold (Gibson 1992).

As the ginseng business boomed, its medical claims came under scrutiny. From the sixteenth to nineteenth centuries, many European physicians used ginseng as a tonic, for treating fatigue and as an aphrodisiac, and it was advertised for treating stomach discomfort (Hill 1751; Belchingen and Cope 1786; Devlin 1790; Elliot 1791). Some, however, doubted the clinical effects of ginseng (Swediaur 1784; Wallis 1790), and Rafinesque pointed to the poor quality of American ginseng and lamented the cost of the wild Asiatic variety (Rafinesque 1830).

In the nineteenth century, the study of pharmacology and phytochemistry started to trigger the reform of pharmacopoeias, which had formerly been compiled by local doctors. The British Pharmacopoeia of 1864 attempted to standardise medicine. It aimed to appraise the scientific achievement of the medical profession in the analysis of the effects of opium, atropine, digitalis leaf and other substances (Dunlop and Denston, 1958). Ginseng, with other herbal medicines, was excluded since its chemistry was not well understood at the time, even though it had been commonly used in medical practice and by the public throughout Eurasia long before then (On 2017). It had been included in the 1842 US Pharmacopoeia (USP) but was dropped from 1880 to 2005, when it re-entered the USP 26 (Brinckmann *et al.* 2020).

Lack of official status was no barrier to trade: ginseng-derived products flooded the market with concentrates, medicines, teas, powders, pills, sugar powders, poultices, eye drops, and ointments (On 2017). Only during the Second World War did the ginseng trade pause, but it resumed very quickly once it was over. In 1946, the USA exported almost 186,000 pounds dropping to around 75,000 pounds afterwards as the demand stabilised after the war (Williams 1957). In the 1990s, in *Foreign Trade Statistics*, a *P. ginseng*-derived product was one of the most important herbs listed in Germany (Lange and Schippmann 1997); and in Sweden, ginseng was one of the best-selling herbal drugs (Blumenthal 2003).

Today most ginseng products are sold as a food supplement with no specific therapeutic claim. Such preparations commonly contain a wide range of other food additives, like vitamins and minerals, with the products being advertised as general health supplements without a specific medical claim. Ginseng-derived products now include chewing gum, ice cream, sparkling drinks and toothpaste (Carlson 1986).

After the first clinical trial of *P. ginseng* was published in 1981, the number of studies has increased exponentially, especially post-2000. As of 2020, more than 300 studies concerning ginseng are published annually, and most of them are related to pharmacology and ginsenosides (Xu *et al.* 2017). The raised profile has meant *P. ginseng* was monographed in national pharmacopoeias including those of Sweden, the UK, Germany, France and more (Blumenthal 2003).

The many ginseng medicinal products on the market before 1976 (Committee on Herbal Medicinal Products 2014) enabled a registration under the Traditional Herbal Medicinal Products (THR, see below). In some countries in Europe (Austria, Belgium, Denmark, France, Germany, Ireland, Latvia, Poland, Portugal and Spain) products with a full marketing authorisation (i.e. licensed medicines) are available (Ibid.) with indicated uses for conditions like fatigue, lack of concentration and vitality, help during convalescence and to support or strengthen the immune system. The uses are in the context of claimed adaptogenic effects, that is, helping the body to deal with all types of stress (Panossian *et al.* 2021). All these uses are superficially similar to those in Chinese medicine, but the context of use, its application and explanations are again very different.

The history of ginseng can be contrasted with that of ginkgo, which is a much more recent introduction into medical practice worldwide. Its journey demonstrates the route by which Chinese *materia medica* was introduced into Europe and gained popularity due to business promotion, as well as attention from the academic and medical communities. Clearly, it was not a direct journey from China, but instead, a complex mesh of cultural and economic links furthering its use as a healthy food and general medicine with important contributions based on the use of American ginseng in North America. This decontextualisation of ginseng was driven by a range of stakeholders and their activities: knowledge transfer from the Jesuit missionaries, the pharmacology research focus of the physicians in the eighteenth to nineteenth centuries and the commercial strategy promoting ginseng's use. Thanks to its characteristics, its role as a trade item with wide cultural recognition, ginseng has become one of the best known 'Chinese' medicinal plants.

Traditional herbal medicines

Herbal medicines with a traditional herbal registration (THR) have been available in European countries since 2004. Based on the Traditional Herbal Medicinal Products Directive (THMPD) 2004/24/EC and of the EU Council of 31 March 2004, manufacturers of good quality herbal medicines can register their products as medicinal products. The therapeutic use is not based on clinical evidence, but on traditional use and thus it is possible to make (restricted) therapeutic claims on the packaging and the patient information leaflet (PIL), which has to be approved by the regulator.

Core to the regulatory process is the requirement that a corresponding herbal product (i.e. one derived from the same botanical drug and prepared in a similar way) has been used traditionally for at least 30 years (15 years non-EU and 15 years in the EU, or more than 30 years in the EU). In addition, it is necessary to produce bibliographic data on safety with an expert report and a quality dossier specifying the control mechanisms in place based on the requirements of the regulators. These products can only be used for minor, self-limiting, conditions. While the regulations have ensured good quality medications with detailed information, there have been very few botanical drugs coming from TCM traditions registered under the scheme: two in Holland and one in the UK (at the time of registration a member of the EU and as of 2020 this regulation is transferred into UK national law). The first product was registered in 2012 in Holland (Netherlands Innovation Network 2012) and as at Sept 2020 the traditional medicines registered under the scheme were:

- Dioscoreae nipponicae rhizoma (*Chuanshan long* 穿山龍) from *Dioscorea nipponica* Makino (NL), which is the ingredient of the licensed herbal medicine, *Xinxuekang* 心血康, indicated for the relief of headaches, muscle pains and cramps in neck, back and legs;
- Isatidis radix (*Banlan gen* 板藍根) from *Isatis tinctoria* L. (UK), often known in China under its synonym *Isatis indigotica* Fortune ex Lindl, indicated for cold and flu relief;
- Salviae miltiorrhizae radix et rhizoma (*Danshen* 丹參) (NL) from *Salvia miltiorrhiza* Bunge: indicated for the relief of mild menstrual pains;
- Pueraria lobata radix *Kudzu wortel* (*Gegen* 葛根) (NL) from *Pueraria montana* var. *lobata* (Willd.) Maesen and S.M.Almeida ex Sanjappa & Predeep, indicated for headaches, and neck and shoulder muscle pains;
- Sigesbeckiae orientalis herba (*Xixian cao* 豨薟草) from *Sigesbeckia orientalis* L. (UK), indicated for the relief of backaches, minor sports injuries, rheumatic or muscular pains, and general aches and pains in the muscles and joints.

In addition, a multi-herbal preparation is registered in Germany, which originally came from Japan. The registered medicine is based on a formulation – *Rikkunshito* – defined in the Japanese Pharmacopoeia (JP 2017), and in Europe it is indicated for the relief of mild gastrointestinal disorders, such as loss of appetite, malaise and bloating. It contains defined quantities of eight drugs:

1. rootstock of *Panax ginseng* C.A.Mey. – *Renshen* 人參;
2. rootstock of *Atractylodes lancea* (Thunb.) DC. – *Cangzhu* 蒼朮 (as *Atractylodes japonica*);
3. fruiting body of *Wolfiporia extensa* (Peck) Ginns – *Fuling* 茯苓 (syn. *Poria cocos* F.A.Wolf);
4. rhizome of *Pinellia ternata* (Thunb.). – *Banxia* 半夏 Makino;
5. fruit peel of *Citrus deliciosa* Ten. – *Chenpi* 陳皮 (Unshiu);
6. fruit of *Ziziphus jujuba* Mill. – *Dazao* 大棗;
7. rootstock of *Glycyrrhiza glabra* L. – *Gancao* 甘草 and related spp.;
8. rootstock of *Zingiber officinale* Roscoe. – *Shengjiang* 生薑.

This preparation is, in turn, based on an earlier Chinese one – *Liujunzi tang* 六君子湯 (Six Gentlemen Decoction) – which, however, does not define the quantities of each botanical drug (Kuchta, pers. comm.). Additionally, JP 2017 adds two substances to *Liujunztang*, which is named for the number of ingredients in the traditional recipe, namely, six. There are therefore differences in the preparation itself both in the number of ingredients and their identity. JP2017 uses *Atractylodes Rhizome* or *tractylodes Lancea Rhizome* from *Atractylodes lancea* (Thunb.) DC.), whereas Chinese *Liujunzi tang* uses *Atractylodis macrocephalae rhizoma* (from *Atractylodes macrocephala* Koidz.), called *Baizhu* 白朮.

A widely used Tibetan medicine – Padma 28 – is also available and both registered as a THR and licensed as a medicine. In Switzerland, it is a regulated phytomedicine, in some countries a THR, while most commonly it is sold as a food supplement. It was first marketed in Switzerland – a development driven by links between exiled Tibetans and Swiss alternative medicine practitioners – and later became a THR product in Austria, Lithuania and the UK. In 2020, the company withdrew the product from the UK market due to limited commercial success and the effects of Brexit (Schwabl and Vennos 2015; Schwabl, pers comm 2020).

Importantly, several of these are, while known in China, of very limited if any importance in TCM. Their uses are generally modified and show a limited connection with TCM concepts. THRs are – as spelled out above – for use in minor self-limiting conditions and must have a plausible therapeutic use. Consequently, the use must be acceptable in the context of European therapeutic traditions. For example, in TCM the aerial parts of *Sigesbeckiae orientalis* L. (*Xixian cao*) is important to 'dispel wind-dampness and strengthen the sinews: for wind-heat-damp painful obstructions' (. In Europe, the use is for minor skeletomuscular health problems, since the latter is a plausible therapeutic use within the European regulatory framework.[10]

The above regulatory framework assesses botanical drugs and preparations derived from them on the basis of their traditional use. Yet, that use has to be relevant within the traditions of the European Union including the UK. Other regulatory systems differ fundamentally, but again there are categories which class herbal medicines on the basis of a low risk to the consumer and limited (therapeutic) benefits. The Swiss authorities, for example, use a category called registered medication without therapeutic indication, which cannot be dispensed with a medical claim, limiting the uses of these preparations. In Australia TCM drugs are generally included under 'listed medicines', again a category with limited therapeutic claims (Heinrich *et al.* 2018). Overall, during the last decades, regulatory niches have been found to accommodate traditional medicines where they can be brought on to the market without having to leap all the hurdles involved in becoming a fully licensed medicine.

With the inauguration of a joint Chinese-European Working Group in 2005, TCM drugs have been included in the European Pharmacopoeia (PhEur). A pharmacopoeia provides legally binding standards on the basis of a regulation of the Council of Europe Convention on the Elaboration of a European Pharmacopoeia (1964, Strasbourg, France) with the first edition having been published in 1969. Currently, 39 countries are members, with the standards also being applied in over 120 countries. The aim of including monographs on TCM drugs has been to provide modern quality standards according to the PhEur principles facilitating the use of preparations of these drugs by practitioners for individual patients. The quality standards provide a basis for safe use (Wang and Franz 2015; Fitzgerald *et al.* 2020), but are not a way to obtain an authorisation to market the products. As of Sept. 2020, the Commission for the European Pharmacopoeia has adopted 73 monographs of TCM substances with 32 being under development (R. Bauer, Graz, AT, pers. Com 07/2020; Council of Europe 2021). Currently, many of these botanical drugs are in fact of a very limited importance on the market, since there are no or very few products containing them on the market.

China and the Taiwan Food and Drug Administration (TFDA) of the Ministry of Health and Welfare are observers, as are Japan, Singapore and the Republic of Korea. Quality standards become mandatory on the same date in all states that are parties to the European Pharmacopoeia. Consequently, these monographs are having an impact on the updating of monographs in other countries/regions including the PR China. While these monographs thus contribute to envisioning a modern TCM, there has not been a concomitant increase in the availability of regulated medical products based on TCM.

Food supplements and health foods

Goji berry (fruits of Lycium barbarum *L.)*

Known as goji or 枸杞 (*Gouqi*, in Chinese, the fruit of *Lycium barbarum* L.), goji berry has been used as a traditional medicine and food for over 2000 years in China and other Asian countries (Yao *et al.* 2018a). In some regions, such as Japan and South Korea, the fruit of *L. chinense* Mill. is also used, although these two species have obvious differences in their metabolomic profiles (Yao *et al.* 2018b). While the plant was introduced into the global North in the eighteenth century, it was not a popular health food until the twenty-first century.

A considerable body of evidence demonstrates that the goji plant was introduced from China to Europe ca. 300 years ago. In the 1730s, the third Duke of Argyll imported a *Lycium* plant from China, which was thought to be a tea plant (*Camellia sinensis* (L.) Kuntze). Since then, the species has been known in the UK as the Duke of Argyll's tea tree, and the name is still used on today's plant labels. In *Species Plantarum* (Linne´ 1753), the first taxonomically binding legal record of the plant, it is described as being native to Asia and Africa. Goji was not recorded in later pharmacognosy monographs (Stephenson *et al.* 1834; Pereira *et al.* 1850; Phillips *et al.* 1874). In fact, for a long time the plant was cultivated in Europe as an ornamental tree rather than for food or medicine.

From the seventeenth century CE onwards, when Chinese botanical substances were increasingly introduced into Europe by travellers, and many sections of Chinese medical books were translated into European languages, Chinese medicinal knowledge began to reach a new audience. For example, *Specimen Medicinae Sinicae* (Cleyer 1682) included introductions to several Chinese medical books and Chinese herbals; while it included a list of 298 Chinese botanical drugs, and the root bark, named 'Ti Co Pi' 地骨皮, was cited. Goji berries are included in the Sloane Collection of the Natural History Museum in London, demonstrating that they were introduced into the UK around the 1750s (Zhao *et al.* 2015). As far as we know, goji was introduced to Europe as a medicine not later than 1874, when *La Matière Médicale Chez les Chinois* recorded goji under the names of 'kou-chi-tz', 'go-ci-tsy' and 'kou-ky-tze' (Soubeiran et al. 1874). It was also found in 'Chinese *Materia Medica*' (Stuart and Smith 1911), a later textbook in the New World.

Around the middle of the twentieth century, goji appeared in several European ethnomedical books (Roi 1946), where goji was recorded as a tonic. However, it was not popular as a health food until the twenty-first century. In 2001, several experts (including experienced pharmacognosy researchers) attended *the International Symposium on Lycium and Antiaging Agents* held in Ningxia, China, the main production area of goji. The symposium delivered papers on the frontiers of research into goji in scientific research, agricultural production, and on products derived from it. Subsequent studies on goji went global demonstrating that goji is safe to consume (Adams *et al.* 2006); and there were review articles which gave comprehensive introductions to the public (Potterat and Hamburger 2008; Potterat 2010; Amagase and Farnsworth 2011). At the same time, the agricultural production of goji in China increased several times (Yao *et al.* 2018b). Goji's global acceptance as a 'superfood' has been linked to the identification of its many nutritional compounds, as well as studies on its safety, and the sustainable supply system. In recent years, goji has been cultivated on a large scale in Europe (Italy, Romania, Poland, Lithuania, Turkey) and North America, and several cultivars have been registered in some countries in these regions.

Goji has been adopted in the current European Pharmacopoeia and British Pharmacopoeia. In the USA, goji can be consumed as a food and dietary supplement ingredient; in

Australia, goji is recognised as an active ingredient of 'Listed' medicines; in Canada, goji is classified as a medicinal Natural Health Product; in the European Community goji is allowed to be marketed as a food due to its long use before 1997 (not subject to the Novel Food Regulation (EC) No. 258/97), and could be developed as an active ingredient of a Traditional Herbal Remedy (THR), if a) anyone had a commercial interest that could balance the cost benefit analysis of registration and b) an appropriate indication were to be found (that is, one acceptable under the regulatory framework). Moreover, several types of goji extracts are officially accepted as ingredients for cosmetic products.

While goji can still be found in TCM prescriptions (in a small amount) mostly with wealthier groups of the Global North, its main usage is as a food or food supplement. It is easy to find goji or goji products (such as extracts, juices, cosmetics, etc.) in health food shops or supermarkets in Europe. Those non-Chinese people, who do not have a cultural background in TCM, nor training in the skills of Chinese medicine, have accepted the old oriental botanical drug as a novel food, very likely because of the scientific proofs of phytochemistry, pharmacology, and safety, rather than on the basis of tradition. In China, on the other hand, goji is mostly consumed in traditional ways, either as a medicine or tonic food. It is commonly seen in home recipes; however, 'modern' goji products like the ones found in the global North are also becoming popular with Chinese consumers.

In sum, while in many parts of the world, the uptake and usage of goji is stimulated by the phytochemical and pharmacological evidence, in China traditional TCM theory, and related traditional concepts, promote the consumption of goji. However, there is a convergence developing in the usages between China and beyond. Here again, as with ginseng, we have a journey which transforms goji from a commodity first travelling Westwards and then in a circular flow affecting goji's use at its origins in China. Within a cultural background that incorporates TCM as one form of regulated therapeutics, goji can be formally accommodated as a Chinese medicine. It can be accepted as a (phyto-) pharmaceutical product, as long as its composition, functions, and safety are demystified and adapted to the existing formal regulatory framework.

Reishi (Lingzhi 靈芝)

The basidiomycete white rot fungus, *Ganoderma lucidum*, *lingzhi* 靈芝, is best known in the global North under its Japanese name *Reishi* (as well as *Munnertake*, *Sachitake* and, in Korean, *Youngzhi*). For centuries, this and related species have been used for medicinal purposes, particularly in China, Japan, and Korea. A wide variety of uses have been reported including for migraine, hypertension, arthritis, bronchitis, asthma, anorexia, gastritis, haemorrhoids, diabetes, hypercholesterolaemia, nephritis, dysmenorrhoea, constipation, lupus erythematosus, hepatitis and cardiovascular problems. According to some researchers, it is used for dizziness, insomnia, palpitations, dyspnoea, consumptive cough and asthma. It is practically impossible to establish how widespread the respective uses have been. Whatever the specific use, the cultural importance of this species has been the driving force for developing potential leads from this taxon. Phytochemical research has focussed on bioactive *lingzhi* polysaccharides and triterpenes, especially ganodermic acid. Extracts from *Ganoderma* have been investigated as potential antitumor and antiviral agents and, less so as possible antibacterial agents for antibacterial activity (against Gram-positive bacteria). However, its use as a food remains key to its commercial success.

With their 'journeys' from China to these two examples of food supplements, goji and *reishi*, lost their specific traditional medical functions and their pharmacological activity is of very limited relevance to their regulated use in the global North. They may well retain some original importance to individual consumers, but are not perceived by the medical and other biomedical healthcare professionals as a therapeutic option.

Drugs that 'did not make it'

There are numerous examples of Chinese drugs which are not generally available, like Ephedra species (*mahuang* 麻黄) and *Cordyceps sinensis* ((L.).Fr. – *dongchong xiacao* 冬蟲夏草, a medicinal fungus and medico-culinary food very popular in Tibetan medicine and TCM (Lu 2017)). In the case of Ephedra, while pharmaceutical preparations of the main active metabolite and its derivative are available, the use of preparations derived directly from the plant (dried material and extracts) does not comply with the safety standards of Europe and North America, most notably the associated risks of strokes, seizures, heart attacks and sudden deaths. A key use was and continues to be (for illicit products) as a general stimulant and for weight-loss. Famously it has been linked to the death in 2003 of a baseball player of the Baltimore Orioles, Steve Bechler, who died of a stroke after taking an Ephedra preparation (Harvard Health Publishing 2003). Another famous case of a banned group of medicines is *Aristolichia* spp. like *Guangfangji* 廣防己 (*Aristolochia fangchi* Y.C.Wu ex L.D.Chow & S.M.Hwang) with the entire genus being known for major carcinogenicity, especially of the kidneys (Heinrich *et al.* 2009; Michl *et al.* 2014, 2016).

Cordyceps sinensis (L.).Fr. (currently the accepted name is *Ophiocordyceps sinensis* (Berk.) Sung *et al.* 2007) is a parasite that grows on the larvae of moths (Lepidoptera) of the genera *Hepialus* and *Thitarodes* endemic to alpine habitats (3600–5000m in elevation) on the Tibetan plateau. In China, *C. sinensis* has a long history of medicinal use. It is thought to have been discovered 2,000 years ago with the first formally documented use coming from the *New Compilation of Materia Medica* (*Bencao congxin* 本草叢新 of Wu Yilou 吳儀洛 1757) in the Qing dynasty. Overall, little primary ethnomedical data describing the medical uses of *C. sinensis* exists in the literature. Current ethnomedical reports are limited to its use as a general tonic in China and as an aphrodisiac in Nepal. It first gained worldwide attention when it was revealed that several Chinese runners, who broke world records in 1993, had included this fungus as part of their training programme.

Although there are a wide range of reported uses of *C. sinensis* in the literature, the reports that extracts of this fungus may alter apoptotic homeostasis are most intriguing. That is, based on clinical data, *C. sinensis* may inhibit apoptosis. Since it is not a plant-based substance, it cannot be accepted as a herbal medicine, for example, in the European regulatory framework.

Of course, one could cite a host of other species in the Chinese *materia medica* which are currently of no or only marginal importance market in the global North. However, the examples here highlight what limitations there are for introducing regulated products to a wider audience. All these examples point to the fundamental concept of a risk – benefit assessment. In the absence of contemporary scientific evidence for benefits, and most importantly for risks, it will be most difficult to bring regulated medical products onto a market. With food supplements there would be, on the other hand, very limited claims, but today these too need to be based on a sound body of evidence.

Conclusions

In this chapter we have selected a few examples, to highlight the contemporary significance in Europe of some botanical drugs and other medicines which originated in China. In addition to the use of Chinese medicine by practitioners in the European contexts of alternative medicine (Chapter 39 in this volume), who claim to practise in the traditional manner, the examples here demonstrate a number of pathways by which the use of substances from the Chinese *materia medica* has become fundamentally different in transit. New contexts are most commonly within the category of herbalism and/or food supplements, where there is no immediate link to the original traditional uses.

As a part of their journey from within China to countries outside, these medical substances were adapted to the regulatory framework of the country or region of use. A wide range of factors came to bear on whether they proved feasible for a longer-term sustainable market: the potential commercial benefits, whether the botanical drug and its preparations were intrinsically safe, and styles of evidence germane to phytochemical and pharmacological studies. The traditional uses may have provided hints for the potential benefits, but these were decontextualised, recontexualised and assigned new meanings. This then fed back in a counterflow to China where, for example, people now drink goji juice, or eat goji yogurt. In other words goji has become a global health food. The modern uses are then also reflected back into TCM and how a substance is used and perceived in Asian countries. Their travels also highlight the increasing globalisation of healthcare and medical practice and the identification of globalisation or a globality as a prized attributed of healthcare products. Specifically, artemisinin showcases the innovations that have occurred at each stage, including and beyond the Chinese shift from a closed society, focussed on a continuous revolution in order to serve 'the masses' in a Communist state, to one which has become more and more engaged in the global exchange of goods and (medical) services. Plants like goji and ginseng exemplify these commercial developments over different historical trajectories, and are aligned with many other non-Chinese products often grown not because of their importance in TCM, but specifically as a commercial crop for global markets, for instance, *Hypericum perforatum* – St. John's Wort (Scotti et al. 2018).

Here we have traced the different journeys or histories of these species, where these are known, in order to understand what avenues exist to introduce substances from TCM to the world outside China. There is no single route that exists for developing Chinese medical preparations into herbal medicines to be used elsewhere. Compounds licensed as medicines simply follow globally accepted routes in accordance with national or international regulations. With herbal medicines successful transmission depends on the different ways into a market as outlined above and it is clear that success depends on interpretation at a national or regional (e.g. EU) level. There remains the avenue of unregulated products which, however, restricts the uses and poses risks in terms of long-term supply, quality and safety.

We continuously negotiate what constitutes adequate uses in a complex maze composed of regulators', industries' and consumers' perspectives, but also national interests and aspirations. These different frameworks have resulted in the reinterpretation of these plants and have finally enabled their use in different regions of the global North.

Acknowledgements

We are very grateful to Dr. M Wang (Leiden) for useful information on European THRs as well as to Severin Bühlmann (Zürich) for information on Switzerland, R. Bauer (Graz) for insights into the situation with Chinese *materia medica* (TCM) drugs in the Eur. Phar., K. Kuchta (Göttingen) on Japan, H. Schwabl (Zürich) on Tibetan medicine as well as to Vivienne Lo for her historical challenges to our thinking. Some of the ideas linked to drug research have developed over several years and were discussed in a different context in Heinrich (2013).

Notes

1 Throughout this chapter, the term 'medicine/s' as defined here is used; botanical drug refers to the dried and potentially processed material primarily used in the preparation of (Chinese) medicines.
2 European Medicines Agency, *Medicinal Product*, https://www.ema.europa.eu/en/glossary/medicinal-product, accessed 16/9/2020.
3 European Food Safety Authority, *Food ingredient applications: overview and procedure*, https://www.efsa.europa.eu/en/applications/foodingredients, accessed 16/9/2020.
4 European Commission, *Food supplements*, https://ec.europa.eu/food/safety/labelling_nutrition/supplements_en, accessed 16/9/2020.
5 For acupuncture, see Phan (2017); for the USA, see Chapters 39 and 41 in this volume.
6 A preparation used on the skin, including the *mucosa*, which may be regulated as a cosmetic (and then only acts on the skin's surface and has no pharmacological effect) or penetrate the skin with a pharmacological effect, for example, on pro-inflammatory pathways.
7 Professor Z. Zhongzhen, Hong Kong Baptist University, personal communication, 7/12/2007.
8 A living fossil is a species or other taxon that – in terms of its morphology and genetics – closely resembles extinct organisms (i.e. only known from a fossil record).
9 The second of three known versions written in 1820; cf. https://www.goethe-museum.de/de/gedicht-ginkgo-biloba, accessed 16/9/2020.
10 See Sigesbeckia, *Traditional Use*, http://www.sigesbeckia.com/traditional-use/, accessed 5/10/2020.

Bibliography

Pre-modern sources

Bencao gangmu shiyi 本草綱目拾遺 (Supplement to the Systematic *Materia Medica*) 1765, Zhao Xuemin 趙學敏 (1719–1805 CE) (2007), Beijing: Zhongyi guji chubanshe.

Zhiwu mingshi tukao 植物名實圖考 (Illustrated Catalogue of Plants) 1848, Wu Qijun 吳其濬 (1959), Beijing: Shangwu yinshuguan.

Zhouhou beiji fang 肘後備急方 (Handbook of Prescriptions for Emergency Treatment) Ge Hong 葛洪 (284–364 CE) (1956), Beijing: Renmin weisheng chubanshe.

Modern sources

Adams, M., Wiedenmann, M., Tittel, G. and Bauer, R. (2006) 'HPLC-MS trace analysis of atropine in *Lycium barbarum* berries', *Phytochemical Analysis*, 17.5: 279–83.

Amagase, H. and Farnsworth, N.R. (2011) 'A review of botanical characteristics, phytochemistry, clinical relevance in efficacy and safety of *Lycium barbarum* fruit (Goji)', *Food Research International*, 44.7: 1702–17.

Appleby, J.H. (1983) 'Ginseng and the royal society', *Notes and Records of the Royal Society of London*, 37.2: 121–45.

Beharry, S. and Heinrich, M. (2018) 'Is the hype around the reproductive health claims of maca (*Lepidium meyenii Walp.*) justified?', *Journal of Ethnopharmacology*, 211: 126–70.

Belchingen, C. and Cope, J.A. (1786) *An Essay on the Virtues and Properties of the Ginseng Tea*, [Place of publication not identified]: Gale Ecco.

Birch, T. (1757) *The History of the Royal Society of London for Improving of Natural Knowledge from Its First Rise, in Which the Most Considerable of Those Papers Communicated to the Society, Which Have Hitherto Not Been Published, Are Inserted as a Supplement to the Philosophical Transactions*, Vol. 4, London: A. Millar in the Strand.

Blumenthal, M. (2003) 'Ginseng, Asian', in M. Blumenthal (ed.) *The ABC Clinical Guide to Herbs*, Austin: American Botanical Council, pp. 211–25.

Brinckmann, J., Marles, R., Schiff, P., Oketch-Rabah, H., Tirumalai, G., Giancaspro, G. and Sarmab, N. (2020) 'Quality standards for botanicals – the legacy of USP's 200 years of contributions', *Herbalgram*, 126: 50–65.

Bruton-Seal, J. and Seal, M. (2008) *Hedgerow Medicine: harvest and make your own herbal remedies*, Ludlow: Merlin Unwin Books.

Buenz, E.J., Bauer, B.A., Osmundson, T.W. and Motley, T.J. (2005) 'The traditional Chinese medicine *Cordyceps sinensis* and its effects on apoptotic homeostasis', *Journal of Ethnopharmacology*, 96: 19–29.

Carlson, A.W. (1986) 'Ginseng: America's botanical drug connection to the Orient', *Economic Botany*, 40: 233–49.

Carnes, J., de Larramendi, C.H., Ferrer, A., Huertas, A.J., Lopez-Matas, M.A., Pagan, J.A., Navarro, L.A., Garcia-Abujeta, J.L., Vicario, S. and Pena, M. (2013) 'Recently introduced foods as new allergenic sources: sensitisation to Goji berries (*Lycium barbarum*)', *Food Chemistry*, 137.1–4: 130–5.

Chinese Text Project, https://ctext.org/, accessed 21/8/2020.

Cleyer, A. (1682) *Specimen Medicinae Sinicae sive Opuscula medica ad Mentem Sinensium*, Francofurti: Joannis Petri Zubrodt.

Committee on Herbal Medicinal Products (2014) *Assessment Report on Panax Ginseng C.A. Meyer, Radix*, London: European Medicines Agency.

Cooperative Research Group (1977) 'A novel type of sesquiterpene lactone – qinghaosu, Cooperative Research Group on *qinghaosu*', *Kexue tongbao*, 22: 142.

Council of Europe (2021) *European Pharmacopoeia (Ph. Eur.)* 10th Edition, www.edqm.eu/en/european-pharmacopoeia-ph-eur-10th-edition, accessed 16/9/2020.

DeFeudis, F.V. (1992) *Ginkgo Biloba Extract EGb 761: pharmacological activities and clinical applications*, Paris: Elsevier.

Devlin, M. (1790) *Pillula Salutaris; Or, the Justly Celebrated Dr. Anthony's Irish Pills*, London: Michael Devlin.

Dunlop, D.M. and Denston, T.C. (1958) 'The history and development of the "British Pharmacopoeia"', *British Medical Journal*, 2.5107: 1250–2.

Edwards, S.E., da Costa Rocha, I., Williamson, E.M. and Heinrich, M. (2015) *Phytopharmacy: an evidence-based guide to herbal medicines*, Chichester: John Wiley and Sons.

Elliot, J. (1791) *The Medical Pocket-Book: containing a short but plain account of the symptoms, causes, and methods of cure, of the diseases incident to the human body: including such as require surgical treatment: together with the virtues and doses of medicinal compositions and simples*, London: J. Johnson.

Etkin, N.L. and P.J. Ross (1982) 'Food as medicine and medicine as food: an adaptive framework for the interpretation of plant utilization among the Hausa of Northern Nigeria', *Social Science and Medicine*, 16: 1559–73.

Etkin, N.L. (ed.) (1986) *Plants in Indigenous Medicine and Diet: biobehavioral approaches*, New York: Gordon and Breach Science Publishers.

European Medicines Agency, *Medicinal Product*, https://www.ema.europa.eu/en/glossary/medicinal-product, accessed 16/9/2020.

Fitzgerald, M., Heinrich, M. and Booker, A. (2020) 'Medicinal plant analysis: a historical and regional discussion of emergent complex techniques', *Frontiers in Pharmacology*, 10: 1480.

Foyster, W. (1900) *Letters Received by the East India Company: from its servants in the East, Vol. 5*, London: Sampson Low, Marston & Co.

Gibson, J.R. (1992) *Otter Skins, Boston Ships, and China Goods: the maritime fur trade of the northwest coast, 1785–1841*, Montreal: McGill-Queen's University Press.

Granger, S. (2013) 'French Canada's quiet obsession with China', *Journal of American-East Asian Relations*, 20.2–3: 156–74.

Harriman, S. (1973) *The Book of Ginseng*, New York: Pyramid Books.

Harvard Health Publishing (2003) *The Dangers of the Herb Ephedra*, https://www.health.harvard.edu/staying-healthy/the-dangers-of-the-herb-ephedra, accessed 16/9/2020.

Havens, T.R.H. (2020) *Land of Plants in Motion*, Honolulu: University of Hawai'i Press.

Heinrich, M. (2013) 'Ethnopharmacology and drug discovery', *Comprehensive Natural Products II: Chemistry and Biology, Development and Modification of Bioactivity*, 3: 351–81.

——— (2018) 'Comment on "How similar is similar enough? A sufficient similarity case study with *Ginkgo biloba* extract" by Catlin et al.', *Food and Chemical Toxicology*, 118: 328–39.

Heinrich, M., Chan, J., Wanke, S., Neinhuis, C. and Simmonds, M.S. (2009) 'Local uses of *aristolochia* species and content of aristolochic acid 1 and 2 – a global assessment based on bibliographic sources', *Journal of Ethnopharmacology*, 125: 108–44.

Heinrich, M., Barnes, J., Prieto-Garcia, J.M., Gibbons, S. and Williamson, E.M. (2018) *Fundamentals of Pharmacognosy and Phytotherapy*, Edinburgh: Churchill Livingstone.

Hill, J. (1751) *A History of Materia Medica: containing descriptions of all the substances used in medicine; their origin, their characters... and an account of their virtues, and of the several preparations from them now used in the shops*, London: T. Longman.

Hocking, G.M. (1976) 'A chronology of ginseng', *Quarterly Journal of Crude Drug Research*, 14.4: 168.

Hsu, E. (2006) 'Reflections on the "discovery" of the antimalarial *qinghaosu*', *British Journal of Clinical Pharmacology*, 61.6: 666–70.

Japan, Kōsei Rōdōshō (2017) *The Japanese Pharmacopoeia XVII (JP XVII)*, Tokyo: The Ministry of Health, Labour and Welfare.

Jartoux, F. (1713) 'The description of a tartarian plant, call'd gin-seng; with an account of its virtues. In a letter from Father Jartoux, to the Procurator General of the Missions of India and China. Taken from the tenth volume of letters of the Missionary Jesuits, printed in Paris in octavo, 1713', *Philosophical Transactions of the Royal Society of London (1683–1775)*, 28.337: 237–47.

Johns, T. (1990) *With Bitter Herbs They Shall Eat It: chemical ecology and the origins of human diet and medicine*, Tucson: University of Arizona Press.

Lafitau, J.-F. (1718) *Mémoire présenté à Son Altesse Royale Monseigneur le duc d'Orléans... concernant la précieuse plante du gin seng de Tartarie: Découverte en Canada par le P. Joseph-François Lafitau*, Paris: chez Joseph Mongé, ruë S. Jacques, vis-à-vis le college de Louis le Grand, à Saint Ignace.

Lange, D. and Schippmann, U. (1997) *Trade Survey of Medicinal Plants in Germany: a contribution to international plant species conservation*, Bonn: Bundesamt für Naturschutz (German Federal Agency for Nature Conservation).

Lee, M.R. (2011) 'The history of Ephedra (*ma-huang*)', *Journal of the Royal College of Physicians of Edinburgh*, 41.1: 78–84.

Li, F.S. and Weng, J.K. (2017) 'Demystifying traditional herbal medicine with modern approach', *Nature Plants*, 3: 171–209.

Linné, C. (1753) *Species Plantarum: exhibentes plantas rite cognitas ad genera relatas, cum differentiis specificis, nominibus trivialibus, synonymis selectis, locis natalibus, secundum systema sexuale digestas*, Holmiæ [Stockholm]: Impensis Laurentii Salvii.

Loskiel, G.H. (1794) *History of the Mission of the United Brethren among the Indians in North America*, London: Christian Ignatius La Trobe.

Lu, D. (2017) *Transnational Travels of the Caterpillar Fungus, 1700–1949*, PhD thesis, University College London.

Michl, J., Ingrouille, M.J., Simmonds, M.S.J. and Heinrich, M. (2014) 'Naturally occurring aristolochic acid analogues and their toxicities', *Natural Product Reports*, 31.5: 676–93.

Michl, J., Kite, G.C., Wanke, S., Zierau, O., Vollmer, G., Neinhuis, C., Simmonds, M.S.J. and Heinrich, M. (2016) 'LC-MS- and 1H NMR-based metabolomic analysis and in vitro toxicological assessment of 43 *Aristolochia* species', *Journal of Natural Products*, 79.1: 30–7.

Moerman D.E. (2009) *Native American Medicinal Plants: an ethnobotanical dictionary*, Portland, OR: Timber Press.

Netherlands Innovation Network (2012) *First Traditional Chinese Medicine to Be Sold in the EU*, https://netherlandsinnovation.nl/life-sciences-health/first-traditional-chinese-medicine-to-be-sold-in-the-eu/, accessed 16/9/2020.

Nieuhof (1660) *Ambassade des Hollandois a la Chine, ou Voyage des ambassadeurs de la Compagnie Hollandoise des Indes orientales vers le Grand Chan de Tartarie, maintenant empereur de la Chine. Traduite sur deux manuscrits hollandois*, Paris: S. Mabre-Cramoisy.

On, H. (2017) 'The medicinal usage and restriction of Ginseng in Britain and America, 1660–1900', *Korean Journal of Medical History*, 26.3: 503–44.

Panossian, A., Efferth, T., Shikov, A.N., Pozharitskaya, O.N., Kuchta, K., Iinuma, M., Mukherjee, P.K., Banerjee, S., Heinrich, M., Wu, W., Guo, D.A. and Wagner, H. (2021) 'Evolution of the adaptogenic concept from traditional use to medical systems: pharmacology of stress- and age-related diseases', *Medicinal Research Reviews*, 41.1: 630–703.

Park, H.J., Kim, D.H., Park, S.J., Kim, J.M. and Ryu, J.H. (2012) 'Ginseng in traditional herbal prescriptions', *Journal of Ginseng Research*, 36.3: 225–41.

Pereira, J., Crosby, T.B.S., Rees, G.O. and Taylor, A.S., King's College, London, St. Thomas's Hospital, Medical School, London (1850) *The Elements of Materia Medica and Therapeutics*, 3rd ed., London: Longman, Brown, Green and Longmans.

Phan, T. (2017) *Chinese American Medicine*, PhD thesis, University College London.

Phillips, C.D.F. and Baxter, E.B., Royal College of Surgeons of England and King's College London (1874) *Materia Medica and Therapeutics Vegetable Kingdom*, London: J. & A. Churchill.
Potterat, O. (2010) 'Goji (*Lycium barbarum* and *L. chinense*): phytochemistry, pharmacology and safety in the perspective of traditional uses and recent popularity', *Planta Medica*, 76.1: 7–19.
Potterat, O. and Hamburger, M. (2008) 'Goji juice: a novel miraculous cure for longevity and well-being? A review of composition, pharmacology, health-related claims and benefits', *Schweizerische Zeitschrift für Ganzheitsmedizin*, 20.7/8: 399–405.
Prendergast, H.D.V., Etkin, N.L., Harris, D.R. and Houghton, P.J. (eds) (1998) *Plants for Food and Medicine: proceedings of the joint conference of the society for economic botany and the international society for ethnopharmacology, London, 1–6 July 1996*, Kew: Royal Botanic Gardens.
Qian, D., Zhao, Y., Yang, G. and Huang, L. (2017) 'Systematic review of chemical constituents in the genus *Lycium* (Solanaceae)', *Molecules*, 22.6: 911.
Rafinesque, C.S. (1830) *Medical Flora; or, Manual of the Medical Botany of the United States of North America: containing a selection of above 100 figures and descriptions of medical plants, with their names, qualities, properties, history, & c.: and notes or remarks on nearly 500 equivalent substitutes*, Philadelphia: S. C. Atkinson.
Roi, J. (1946) *Atlas des plantes médicinales chinoises*, Paris: P. Lechevalier.
Schwabl, H. and Vennos, C. (2015) 'From medical tradition to traditional medicine: a Tibetan formula in the European framework', *Journal of Ethnopharmacology*, 167: 108–14.
Sung, G.H., Hywel-Jones, N.L., Sung, J.M., Luangsa-ard, J.J., Shrestha, B., Spatafora, J.W. (2007) 'Phylogenetic classification of *Cordyceps* and the clavicipitaceous fungi', *Studies in Mycology*, 57: 5–59.
Scotti, F., Löbel, K., Booker, A. and Heinrich, M. (2018) 'St. John's Wort (*Hypericum perforatum*) products – How variable is the primary material?', *Frontiers in Plant Sciences*, 9: 1973.
Seol, H. (2017) 'Ginseng utilization and limitations in Western medicine, 1660–1900 –Focusing on the UK and the US', *Korean Journal of Medical History*, 26.3: 503–44.
Simpson, W., Pechey, J. and Wittie, R. (1680) *Some Observations Made Upon the Root Called Nean Or Ninsing Imported from the East-Indies: shewing its wonderful virtue in curing consumptions, ptissicks, shortness of breath, distillation of rhume, and restoring nature after it hath been impaired by languishing distempers and long fits of sickness*, London: Printed for the author.
Sneader, W. (1996) *Drug Prototypes and Their Exploitation*, Chichester: Wiley.
Soubeiran, J.L., de Thiersant, C.-P.D. and Gubler, A. (1874) *La Matière médicale chez les Chinois*, Paris: G. Masson.
Stanley-Baker, M. (2021) 'Ge xianweng zhouhou beiji fang 葛仙翁肘後備急方 JY146 (Transcendent Ge's Emergency Recipes to Keep Close at Hand)', in Lai Chi-Tim 黎志添 and E. Valussi (eds) *Daozang jiyao – tiyao* 道藏輯要・提要 (Explanatory Catalogue of Essentials of the Daoist Canon), Hong Kong: Chinese University of Hong Kong Press.
Sul, H. (2017a) 'The medicinal usage and restriction of ginseng in Britain and America, 1660–1900', *Uisahak*, 26.3: 503–44.
——— (2017b) 'The perception of ginseng in early modern England and its significance in the globalising trade', *East Asian Journal of British History*, 6: 1–39.
Stephenson, J., Burnett, G.T. and Churchill, J.M., King's College London, St. Thomas's Hospital, and Webb Street School of Anatomy and Medicine (1834) *Medical Botany; or, Illustrations and Descriptions of the Medicinal Plants of the London, Edinburgh, and Dublin Pharmacopoeias*, London: J. Churchill.
Stuart, G.A. and Smith, F.P. (1911) *Chinese Materia Medica, Pt. 1, Vegetable Kingdom*, Shanghai: Presbyterian Mission Press.
Swediaur, F. (1784) *Practical Observations on the More Obstinate and Inveterate Venereal Complaints*, London: J. Johnson, Edinburgh: C. Elliot.
Thevenot, M. (1665) 'Relations de divers voyages curieux', *Paris Philosophical Transactions*, 1: 249.
Tu Youyou (2016) 'Artemisinin – a gift from traditional Chinese medicine to the world (Nobel lecture)', *Angewandte Chemie International Edition*, 55.35: 10210–26.
Wallis, G., Bath Medical Library and University of Bristol (1790) *Annual Oration, Delivered March 8th, 1790, Before the Medical Society, Bolt Court, Fleet Street, London*, London: G.G.J. and J. Robinson.
Wang, M. and Franz, G. (2015) 'The role of the European Pharmacopoeia (Ph Eur) in quality control of traditional Chinese herbal medicine in European member states', *World Journal of Traditional Chinese Medicine*, 1: 5–15.
Williams, L.O. (1957) 'Ginseng', *Economic Botany*, 11.4: 344–8.

Xu, W., Choi, H.K. and Huang, L. (2017) 'State of *panax ginseng* research: a global analysis', *Molecules*, 22.9: 1518.

Yao, R., Heinrich, M. and Weckerle, C.S. (2018a) 'The genus *Lycium* as food and medicine: a botanical, ethnobotanical and historical review', *Journal of Ethnopharmacology*, 212: 50–66.

Yao, R., Heinrich, M., Zou, Y., Reich, E., Zhang, X., Chen, Y. and Weckerle, C.S. (2018b) 'Quality variation of goji (fruits of *Lycium* spp.) in China: a comparative morphological and metabolomic analysis', *Frontiers in Pharmacology*, 9: 151.

Zhao, Z., Zhao, K. and Brand, E. (2015) 'Identification of ancient Chinese medicinal specimens preserved at Natural History Museum in London', *China Journal of Chinese Materia Medica*, 40.24: 4923–7.

INDEX

Note: **Bold** page numbers refer to tables; *italic* page numbers refer to figures and page numbers followed by "n" denote endnotes.

18β-glycyrrhetinic acid 711

Abercromby, D. 559
abortion 284, 291, 373, 376, 639
acu-moxa (or acupuncture and moxibustion) 42, 134, 189, 190, 192, 193, 201, 220, 284n1, 285, 287, 293, **313;** diagrams 190–195, 202–203; points 63, 189, 190–195, 201–203
acupuncture: acu-moxa point diagrams 190–195; acupuncture bronze statues and diagrams 192, *193;* from ancient to modern 203–204; birth of acupuncture societies 567; Chamfrault, translation of ancient medical texts 569; channel diagrams 194–199; collateral branches of regular channels 201; from diagrams to theories 202–203; differentiated conditions of practice 564–565; 'divergent' channels 201; duality between 'traditionalists' and 'modernists' 570–571; Eight Extra Channels (*Qijing bamai* 奇經八脈) 199, *200,* 201, 435, 558; evolution of practice in France 565; failure of merger 569–570; figure of acupuncture specialist 566–567; fourteen channels 192, 194, *194;* institutionalisation of knowledge and practices 567; Italian acupuncture departs from 'French traditionalist acupuncture' 572–573; knowledge in progress 564–565; migration of (*see* migration of acupuncture); Mingtang *triptychs* (Tang dynasty) 190, *191;* 'muscle region of the regular channels' 201; practice of acupuncture in Italy 572; practices 3, 4, 8; research questions and reflection 202–204; *Société d'Acupuncture* 568; *Société*

Française d'Acupuncture 567–568; without *qi* theory 36; *see also Maishu* 脈書, 1st record of needling; *Zhenjiu jiayi jing* 針灸甲乙經 (Systematic Classic of Acumoxa)
acupuncturists 285, **529,** 574n6, 578–583, 587, 589; Borsarello, J. 571; Bossy, J. 571
Adams, W.R. 592
adjunct herbs (*zuo* 佐) 708
African contexts: Chinese Medical Teams (CMTs) 602, 607–608; Chinese medicine 599–600; impact of pharmaceutical trade agreements on Sino-African relations 604–605; influence of World Health Organization 602–604; medical pluralism 605–606; People's Republic of China's diplomacy 601–602; political identification of Chinese medical practices with primary health care 604; trade and historical diaspora of Chinese 600–601
Ahern, E.M. 314
Albala, K. 321
alchemy 5, 23, 67, 137, 328, 332, 334, 408, 428, 434, 446, 451–452, 479; non-sexual 449; *nüdan* 女丹 (female inner alchemy) 373, 449–452; sexual 384–385, 446–448; time in (*see* time in Chinese alchemy); *see also neidan* 內丹
Allan, S. 76
Alma Ata Declaration, 1978 8, 594, 603–604, 625–637
almanacs 65, 80, 86n33, 93, 96, 283, 294n5, 305, 310, 312, 314–316; *see also rishu* 日書
Ambroise Paré 208
American: acupuncturists in Latin America 587, 589; American Chinese medicine

617; becoming 'legal' 578–580; becoming 'traditional Chinese medicine (TCM)' 577–578; Chinese herbal prescriptions 576–577; Chinese medicine becoming 'alternative' 580–583; Complementary and Alternative Medicine 580, 596n1; counter-culture a site for Daoist medicine 5; hegemony and imperialism 585, 604; imperial force on China 6; influence on Chinese child-birthing ideas 374–375; model as foundation of PRC healthcare 632–635; pharmaceutical corporations 626; pharmaceutical law and Chinese herbs 721–736; psychotherapy and Chinese medicine 38; recognition of Taiwan 604; research on acupuncture 571–573; search for 'pure' Eastern tradition 38; sexology 383, 385, 390, 393, 395; tradition of medical history 248
Amiot, J.J. 560–562
amphiphilic 716–717
amulets *see* talismans
Ānāpānasati 11
aniseed 711
anatomy 94, 168–170, 174, 189, 203, 206–214, 218, 289, 374, 381, 497, 560, 574n13, 614–615, 629, 632, 635n3
Anatomy Trains: Myofascial Meridians for Manual and Movement Therapists (Myers) 203
Andrès, G. 572
Andrews, B. 186, 273, 614, 628
anger (*nu* 怒) 236
animal magnetism 37
anmo 按摩 (therapeutic exercises/massage) 111, 117, 202, 288; *see also daoyin* 導引; *tuina* 推拿
Annals of the Three Kingdoms (*Sanguo zhi* 三國志) 445
Annotation and Explanation of the Discussion of Cold Damage see Zhujie shanghan lun 注解傷寒論
anomalous qi (*zaqi* 雜氣) 226, 247
Anti-Spiritual Pollution Campaign (*Qingchu jingshen wuran yundong* 清除精神污染運動) 393
Arcane Essentials from the Imperial Library see Waitai miyao 外臺秘要
archaeological evidence *see* manuscripts and artefacts
aristocrats 60, 66, 109, 110, 112, 113–114, 394, 694, 700
Arrow War (or the Second Opium War) 651
artemisinin 9, 221, 226, 600, 721, 724–725, 736
artesunate 707, 724
arts of the bedchamber *see fangzhong shu* 房中術
Association Française d'Acupuncture (AFA) 570
Association Scientifique des Médecins Acupuncteurs de France (ASMAF) 569

astro-calendrical practices 23, 81; branches of Shang calendar 95; stems and branches 95; *see also* numerology; *shushu* 數術

Bai Yuchan 白玉蟾 440
Bangyak hap-pyeon 方藥合編 (Compilation of Formulas and Medicinals) 508
Barefoot Doctor programmes 8, 603, 626 633–634, 641–643, 646
Barlow, J. 579
Barnes, L. 283, 578, 582
Barthes, R. 391
Bashiyi nanjing 八十一難經 (Eighty-one difficulties) 163; *see also* Nanjing 難經
Basic Questions see Huangdi neijing 黃帝內經; *Suwen* 素問
bedchamber techniques *see fangzhongshu* 房中術
Beijing: *Beijing Airusheng Zhongyi dianhai* 北京愛如生中醫典海 313, **313**; Chinese medicine 3
Beiji qianjin yaofang 備急千金要方 (Essential Formulae Worth a Thousand Gold in Emergencies) 111, 115, 125, 138, 223, 233, 305, 312, 314, 332, 339, 341, 345, 362, 370
bencao 本草 (basic herbs or fundamental simples) 51, 133–134; advanced literature 136; 'Catalogue of the Li Shih-Chen Exhibition' 678; 'Chinese medicinals' or 'Chinese drugs' 669; Chinese philology and scientific fieldwork 675–678; collections 133–134; collections, drug therapy 133, 139–140; entries 679; and food 328–336; as a genre 669–685; global science, local surveys 673–674; Liu Ju-Ch'iang's 劉汝強 Flora Sinensis 677; *materia medica* knowledge 669–670, materia medica studies (*bencao xue* 本草學) 669, 679; and modern science 8–10; 'old *materia medica*' (*jiu bencao* 舊本草) 673; overview of styles of knowing 671–672; pharmacopoeia or drug code (*yaodian* 藥典) 670–671; popular, practical medicine 681–682; regional medicine 672–673; repertoires 7; socioeconomic and ethnic scope 680; standardisation of 138; trade regulation 674–675; works 678–679
Bencao 本草 (*Materia Medica*) editions: *Bản thảo thực vật toàn yếu* 本草食物纂要 (Comprehensive Collection of *Materia Medica* and Food) 499; *Beikao shiwu bencao gangmu* 備考食物本草綱目 (*Materia Medica* Organised by Headings with Individual Listings for Foodstuffs for Consideration) 334; *Bencao congxin* 本草叢新 (New Compilation of *Materia Medica*) 735; *Bencao gangmu* 本草綱目 (Systematic *Materia Medica*, also known as Categorical and Itemised Pharmacopoeia, Compendium of *Materia*

Index

Medica, Encyclopaedia of *Materia Medica*, *Materia Medica* Arranged according to Drug Description and Technical Aspects, Comprehensive *Materia Medica*, Classified *Materia Medica*, *Materia Medica* Organised by Headings, and *Materia Medica* Arranged According to Monographs and Technical Criteria) 140, 151, 282, 283, 286, 288, 304, 329, 333–334, 343, 456–458, 484, 517, 676, 712, 715, 728; *Bencao hui* 本草匯 (Collected Pharmaceutical Knowledge) 712; *Bencao jing* 本草經 (*Materia medica*) see *Shennong bencao jing* 神農本草經 *in this entry*; *Bencao jing jizhu* 本草經集注 ([Divine Husbandman's] *Materia Medica*, with Collected Annotations, also known as Materia Medica Canon Variorum Edition) 137, **141**, 330, *712*; *Bencao pin hui jingyao* 本草品匯精要 (Collected Essentials of *Materia Medica* Species) 484; *Bencao shiyi* 本草拾遺 also known as *Bencao gangmu shiyi* 本草綱目拾遺 (Supplement to the Pharmacopoeia, also known as Supplement to the Systematic *Materia Medica*, also known as Gleanings of *Materia Medica*) 294n11, 477, 725; *Bencao tujing* 本草圖經 **126**, 140, **141**, **331**, 482; *Bencao xue* 本草學 (Pents'aology) 679; *Bencao yanyi* 本草衍義 (Further Discussion on *Materia Medica*, also known as Dilatations on Materia Medica) 140, **141**, 482; *Chongxiu zhenghe bencao* 重修正和本草, also known as *Chongxiu zhenghe jingshi zhenglei beiyong bencao* 重修正和經史證類備用本草 (*Materia Medica* Proven and Classified Throughout History for Use from the Zhenghe 正和 Period Revised) 330–**331**; *Dayuan bencao* 大元本草 (Pharmacopoeia of the Great Yuan) 483; *Haiyao bencao* 海藥本草 (Overseas *Materia Medica*) 329, 421, 479, 490n1; *Honzōkōmoku keimō* 本草綱目啓蒙 (Illumination of the Principles and Varieties of *Materia Medica*) 517; *Honzō wamyō* 本草和名 (Japanese Names for the *Materia Medica*) 516; *Hu bencao* 胡本草 (Western Pharmacopoeia) 477; *Jiayou bencao* 嘉祐本草, also known as *Jiayou buzhu Shennong bencao* 嘉祐補注神農本草 (Annotated Shennong *Materia Medica* from the Jiayou 嘉祐 (1056–1063) period) **126**, 127, 139, **141**, **331**; *Jingshi zhenglei beiji bencao* 經史證類備急本草 (*Materia Medica* for Emergencies Proven and Classified through History) *see Zhenglei bencao* 證類本草 in *Bencao* 本草 *editions*; *Kaibao bencao* 開寶本草 (*Materia Medica* from the Kaibao Period 開寶 (968–976) Period) 139, **141**, **331**, 482, 712; *Lei Gong jizhu Shennong bencao* 雷公集注神農本草 (Shen Nong's *Materia Medica*, Compiled and Annotated by Lei Gong) 712; *Shennong bencao jing* 神農本草經 (Divine Farmer's *Materia Medica*, also known as Divine Husbandman's Materia Medica, also known as *Materia Medica* Canon of Shennong) 66, 127, 135, **141**, 310–311, 328–330, **331**, 333, 405, 485, 708, 712, 713; *Shiliao bencao* 食療本草 (Materia Dietetica) 332, 479; *Shixing bencao* 食性本草 (*Materia Medica* of Food Qualities) 332; *Taiqing zhu bencao* 太清諸本草 (Various *Materia Medica* of the Taiqing 太清 Era) 333; *Tangye bencao* 湯液本草 (Tang *Materia Medica*) 140, 149, 332; *Xinxiubencao* 新修本草, also known as *Tang bencao* 唐本草 (Newly Revised *Materia Medica* or) 137, **141,** 329, **331**, 477, 480, 515; *Yamato honzō* 大和本草 (Japanese *Materia Medica*) 517; *Zhenghe bencao* 正和本草, also known as *Zhenghe xinxiu jingshi zhenglei beiyong bencao* 正和新修經史證類備用本草 **331;** *Zhenglei bencao* 證類本草 also known as 140, 141, 330, **331,** 482

Benedict, C. 265, 266, 267, 271
Bennett, A. 714
Bensky, D. 708
Berlin collection: folk healers' manuscripts 284; itinerant healers' manuscripts 284; magico-religious healers 284; non-medical authors 284; pharmacists 284; practitioners of martial arts 284; regular physicians 284
Bertin, H. 560
Bian Que (or Bianque) 扁鵲 58, 63, 64, 66, 120, 164–166, 307, 401, 466, 467
Binhu maixue 瀕湖脈學 151
Black Death in Europe 255
Blake, William 385
Blas Roca Calderio 593
blood (*xue*) 血 17, 21n6, 41, 152, 208, 235, 290, 309, 315, 341, 371, 451, 558, 613
bloodletting 7, 288
boils disorder 270
Bondt, Jacob de: *De Medicina Indorum* 553
bone disorders 218, 223
bone setting and massage 288–290
Book of Songs see Shijing 詩經
The Book of the Dark Woman see Xuannü jing 玄女經
Book of The Generation of the Fetus see Taichan shu 胎產書
The Book of the Plain Woman see Sunü jing 素女經
botany 142n1, 151, 553, 669, 671–674, 676, 710
Boxer Rebellion 660; *see also* Boxer Uprising
Boxer Uprising 7, 652
Boyle, R. 729
Boym, M. 553, 554, 557; *Clavis medica ad Chinarum Dotrinam de pulsibus* 557–558
branch diseases (*biaobing* 標病) 149

breath practices: breathing meditations 3; breathing to lighten spirit 66; 418; *see also qi*
Bréquigny, L.O. 560
bubonic plague 266–267, 270
Buddhism 113–115, 138, 292, 329, 370, 402, 418, 421, 439, 444, 460, 514, 516, 687, 691, 694, 726; *Ānāpānasati* 115; introduction of 5; *Śamatha* 115; *Vipaśyanā* 115
Buddhist medicine: absorption meditations 418; 'apocryphal sutras' 420; breathing techniques 418; concentration exercises 418; exoteric practices 418; Great Elements 417; list of Chinese texts 421–422; *Lotus Sutra* 419; Mahāyāna Buddhism 418; ophthalmological surgical procedures 419; pilgrim Yijing 419; *Sutra of Golden Light* 419–420; *Sutra of the Master of Medicines Buddha* 420; Tantric or Esoteric Buddhism 418; Three Humours or Three Defects 417; translation strategies 420–421; *Vimalakīrti Sutra* 419
Buque zhouhou baiyi fang 補闕肘後百一方 (One Hundred and One Supplementary Formulas to Keep up One's Sleeve) 137

Cadden, J. 356
caesarian delivery 377
Cảnh Nhạc 497; *see also* Zhang Jiebin 張介賓
Cao Shuji 曹树基 267
Carnot, P. 565
Carpenter, E. 385
Case Record Guide to Clinical Presentation see Linzheng zhinan yi'an 臨証指南醫案
castration 7
Castro, F. 587, 589, 630
Categorical and Itemised Pharmacopoeia (*Bencao gangmu* 本草綱目) *see Bencao* 本草 editions
cauterisation in magico-religious context *see* moxa-cautery
Central Epidemic Bureau 653
Central Hygienic Laboratory 653
Chamfrault, A. 569, 570; translation of ancient medical texts 569
changshan (*dichroa febrifuga*) 常山 656
channels and networks *see jingluo* 經絡 30, 33, 149
channels: diagrams of ten channels 195–196, *197*; diagrams of twelve regular channels 196, *198*; illustration of channels in Lifesaving Book 196, 198–199, *199*; medieval medicine 98–100, *99*, **102**; *Qi* in 40
chaotic (*luan* 亂) 34
Chao Yuanfang 巢元方 112, 116, 222, 223, 370
Chavannes, E. 62
Chen, C. C. 640, 641
Chen Duxiu 陳獨秀 36, 660
Chen Minzhang 陳敏章 643
Chen Qide 陳其德 245, 246, 256, 272
Chen Shaowei 陳少衛 434

Chen Shiduo 陳士鐸 237
Chen Shigong 陳實功 206, 208
Chen Shiliang 陳士良 332
Chen Xiuyuan 陳修園 209–211
Chen Yan 陳言 224, 236
Chen Yingning 陳攖寧 385
Chen Yun-ju 陳韻如 51, 218
Chen Zaiqi 陳藏器 477
Chen Zhiqian 陳志潛 654
Chen Ziming 陳自明 224, 363, 371, 372
Cheng Congzhou 程從周 172, 186
Cheng Wuji 成無己 147
childbirth practices 291
China origin of plague hypothesis 255
China Sex Museum 393–394
Chinese Civil War 7, 659
Chinese Communist Party (CCP) 638, 662
Chinese Dictionary of Pharmacology (*Zhongguo yaoxue dacidian* 中國藥學大辭典) 681
Chinese disease monograph 266
Chinese Medical Association 653
Chinese medicine (CM): end of 9–10; Ming Dynasty 151–152; Qing dynasty 152–155; Song Dynasty 146–151; *Treasured Mirror* 509
Chinese mentalities 3
Chinese traditional medicine and diet: basic correspondences 322, **323**; calendrical schedule 322; culinary technology 320; earliest evidence 321–322; 'five fruit and vegetables a day' 326; medieval synthesis 323–324; old foodways for new appetites 324–326; technologies of *qi* (the stuff of life) 320; *wei* 味 (flavours) 320–321
cholera epidemics 249, 254
Chongzhen 崇禎 268
Chow, R. 391
Chu silk manuscript 85n25; *see also* silk manuscript; manuscripts and artefacts
Chunqiu fanlu 春秋繁露 20, 30, 80, 83, 93, 96–97, 105n8–9, 105n19
Chunyu Yi 淳于意 19, 20, 58, 64, 122, 164, 182, 183, 187, 220, 231, 307, 369
Chu Văn An 朱文安 499
Cibot, Pierre-Martial 225
Cinggis-Qan (or Genghis Khan) 275
circulating *qi* 35
Classic of the Plain Girl see Sunü jing 素女經
Classified Collection of Medical Prescriptions (*Uibang yuchi* 醫方類聚) 505
Classified Materia Medica (*Bencao gangmu* 本草綱目) *see Bencao* 本草 editions
Classified Materia Medica (*Zhenglei bencao* 證類本草) *see Bencao* 本草 editions
Clavis medica ad Chinarum Dotrinam de pulsibus (Boym) 557–558
Clench, A. 728
cold damage *see shanghan* 傷寒

Index

Cold War commentators 395
Collected Essentials of Materia Medica *Species (Bencao pin hui jingyao* 本草品匯精要) *see* Bencao 本草 editions
Collected Notes on Medical Recipes (*Yifang jichao* 醫方集抄) 282
Collected Pharmaceutical Knowledge (*Bencao hui* 本草匯) *see Bencao* 本草 editions
communist medicine: barefoot doctor programmes 641–643; contemporary medical market 643–645; emergence of TCM 638–641
Compendium of Life-Saving Recipes Ready-made with Local Korean Herbs see Hyangyak jesaeng jipseongbang 鄉藥濟生集成方
Compendium of Materia Medica (*Bencao gangmu* 本草綱目) *see Bencao* 本草 editions
Compilation of Chinese Herbal Medicinals see Quanguo zhongcaoyao huibian 全國中草藥匯編
Compilation of Formulas and Medicinals see Bangyak hap-pyeon 方藥合編
Complete Collection of the Four Treasuries see Siku quanshu 四庫全書
The Complete Works of Jingyue see Jingyue quanshu 景岳全書
Comprehensive Collection of Materia Medica and Food (*Bản thảo thực vật toàn yếu* 本草食物纂要) *see Bencao* 本草 editions
Comprehensive Compendium of Good Formulas for Women see Furen daquan liang fang 婦人大全良方
Comprehensive Compendium of Worthy Prescriptions for Women see Furen daquan liang fang 婦人大全良方
Comprehensive Materia Medica (*Bencao gangmu* 本草綱目) *see Bencao* 本草 editions
Comprehensive Record of Sagely Beneficence see Shengji zonglu 聖濟總錄
contagion in Chinese paediatrics 274
contaminationist perspective 247, 253
contested medicines: adoption of germ theory 652–653; Arrow War 651; Boxer Uprising 652; China's National Medicine movement 654; Civil War 655; fall of Qing dynasty 652–653; First Sino-Japanese War 651; folk medicine, 'feudal superstition' 656; foreign missionaries 650; geographic 656; Institute of National Medicine (INM) 654, 655; missionary medicine in Canton 650–651; 'national medicine'655; 1929 proposal and reaction 654; Opium War 651; Russo-Japanese War 651; Self-Strengthening Movement 651; TCM clinics 649; in twentieth-century China 649–657; War of Resistance against Japan 655–656; Western medicine and state power 650, 652
control (*zhi* 治) 34
Cook, C.A. 74

Cooter, R. 250
cord and hook design 95, *96*
cordyceps, uses of 10
Core Prescriptions of Medicine see Ishinpō
coronavirus (COVID-19) 664
coughing 32, 97, 309, 713–717
Couplet, P. 553, 554
crude drug (*shengyao* 生藥) 133
Cuba: embrace of modernity 587–588; indentured servitude, Chinese in Cuba 586; *Medicina Tradicional y Natural* (MTN) 587–588, 596n1; migration of acupuncture to 586; mother of invention 588–589
Cullen, C. 185, 283
Cultural Revolution 8, 385, 391–392, 539
Cunningham, A. 250
cupping 135
cupping frenzy 581
Cure of Myriad Diseases (*Wanbing huichun* 萬病回春) 517

Daguan bencao 大觀本草 also known as *Jingshi zhenglei Daguan bencao* 經史證類大觀本草 *see Zhenglei bencao* 證類本草 in *Bencao* 本草 editions
Dai Sigong 戴思恭 233–235
Dai Tianzhang 戴天章 173
Danlun juezhi xinjian 丹論訣旨心鑑 (*Insights on the Purport of the Alchemical Treatises*) 433
Daode jing 道德經 (Book of the Way and Its Virtue) 428, 429, 431
Daoism 114; or Buddhist healing practices 3; creation of European and American countercultures 5; 'merging the pneumas' 339; 'naturalism' 385; ritual practices 5; Shangqing practitioners 114; Way of the Celestial Masters 114; Way of the Great Peace 114; *xian* techniques and *yangsheng* practices 114
Daoism and medicine 5; 'Arts of the Bedchamber' (*fangzhong shu* 房中術) 408; blended practices 408–410; bureaucratic imagination 407; Celestial Master and Great Peace 403; composite ritual solutions in early Daoist Church 407; delaying death 405; diet (*fushi* 服食) 408–409; disease transmission 407; 'fasting from grains' (*bigu* 辟穀) *409*, 409–410; 'guiding and pulling' (*daoyin* 導引) 408; as historical term 410–411; illness theodicy 407–408; long life 405; *materia medica* tradition 402; plagues 403; preventing disease 404–406; problem of communicable illness 407; problem of evil 407; 'spiritual' health practices 401; state, epidemics and individual destiny: the complex problem of Ming 402–403; talismans and incantations (*fuzhou* 符咒) 410, *411*, *462*; transcending mortality 405

Index

Daoist priest 135, 370
Daoist sexual practices: choosing crucible 448; female principle 444; non-sexual alchemy 449; *nüdan* and female physiology 451–452; Qing dynasty: non-sexual practices and *nüdan* 449; sexual alchemy in Ming period 446–448; sexual practices in early period 445–446; wayward women 449–451
daoyin 導引 35, 55–57, 109–117, 408
Dasheng bian 達生編 (Treatise on Easy Childbirth) 373
Daston, L. 175
Da Vinci 168
Daybooks 54, 74, 93, 95, 105n7, 306; *see also* almanacs; *rishu* 日書
Dean, K. 266, 497
Declaration of Alma Ata: abolishing Chinese medicine 628; adoption of communist Chinese model of healthcare 625–627; defining healthcare integration 627–628; PRC in WHO 633; School of Merging 628–629; Self-Strengthening Movement 628; Westernisation Movement 628
decontextualised Chinese medicines: *Artemisia annua* L. 723–725, 724; Chinese *materia medica* and *dietetica* 721–722; drugs 735; food-medicine interface 722; food supplements and health foods 733–735; with full marketing authorisation 723–726; ginkgo 726–728; ginseng 728–730; goji berry 733–734; herbal preparations 723; meandering between licensed medicine and food 726–732; medical plants 722; or nutritional substances 723; *reishi* (*lingzhi* 靈芝) 734–735; traditional herbal medicines 730–732; transfer of TCM preparations 723
defensive *qi* (*wei qi* 衛氣) *see qi* 氣
deities or gods 4–5, 82, 222, 292, 418–419
De la Füye, R. 566, 567, 568, 570, 572
De Medicina Indorum (Bondt) 553
Deng Xiaoping 鄧小平 392, 393, 625, 643, 664, 691
dentistry 714
Desfosses, P. 565
Despeux, C. 169, 273, 283, 311, 313, 342, 434
deviant *qi* doctrine 33, 253
diabetes 218, 222, 320, 526, 664, 734
diagnosis: facial 4; pulse (*see* pulse diagnostics); tongue (*see* tongue diagnosis); visual (*see* visual diagnosis)
diagrams 189, 211, 212, 290, 292, 369, 484, 485; acupuncture illustration 189–204
Diagrams for Preserving Perfection (*Cunzhen tu* 存真圖) 211
dian 癲 232–233; and *kuang* 狂 232; and *xian* 癇 232–233; *yin-yang* dualism 232
diaspora of Chinese 316, 586, 600–601, 707

diet 28, 35, 110, 112, 224, 251, 304, 307, 320–326, 403, 404, 405, 408–409, 410, 418, 520, 606, 696
dietary 65, 74, 149, 155, 220, 223, 251, 284n1, 291, 305, 310, 311, 313, 316, 320–326, 328–335, 406, 478, 479, 481, 483, 599, 688, 697, 733
dietetics 291–292, 293, 311, 499, 564, 687, 694, 696
differentiated flesh (*fenrou* 分肉) 207
Dilatations on Materia Medica (*Bencao yanyi* 本草衍義) *see Bencao* 本草 editions
Ding Chengzong 丁成宗 462
Ding Ganren 丁甘仁 579
Discourse on Cold Damage and Miscellaneous Diseases 155n2
Discourse on Warm Disease (*Wenre lun* 溫熱論) 153
Discussion of Cold Damage see Shanghan lun 傷寒論
Discussion of the Utmost Method Under the Sky (*Tianxia zhidao tan* 天下至道談) 338
disease 217–226, 248–251; acute febrile epidemics 225–226; bone disorders 218, 223; cold damage (*shanghan* 傷寒) 221, 223; cure of disease 221–222; emergence of learned practitioners 219; histories of 218, 248–251; holism and personalised medicine 217; *huoluan* 霍亂 (acute vomiting and diarrhoea) 218, 223, 462; *see* madness; medical manuscripts organising prescriptions 220–221; medical revolution 222; 'night blindness' 223; 'nutrient deficiency diseases' 223; physicians of Han dynasty 220; poxes 225; pre-modern disease categories 222; recording of symptoms 223; regulation of blood and *qi* 224–225; ritual techniques 219–220; silk manuscript 219; Southern Song dynasty 224; in Sui and Tang dynasties 223; transferability of 219; Western medicine 226; wind (*feng* 風) 222–223
The Diseases of China: their causes, conditions, and prevalence contrasted with those of Europe (Dudgeon) 251
disease transmission, human-to-human 406–407
divination: in Daoism 5; magico-religious practices 293; numerology 73–74
Divine Farmer (or Divine Husbandman) *see Shennong* 神農
Divine Farmer's Materia Medica (*Shennong bencao jing* 神農本草經) *see Bencao* 本草 editions
Divine Husbandman's Materia Medica (*Shennong bencao jing* 神農本草經) *see Bencao* 本草 editions
Doctor Sex: bedchamber techniques 384; case of 382–384; 'correct path of sex' 383; modernisation of intimacy 382–384; 'Third Kind of Water' 383–384

Dodoens, R. 517
Dominique P. 174
Dong Zhongshu 董仲舒 20, 85n28, 93
Dongui suse bowon 東醫壽世保元 (Longevity and Life Preservation in Eastern Medicine) 508
Dongxuan zi 洞玄子 (Master Dongxuan) 347, 339, 385
Douglas, M. 304, 590, 627
Dropsy, Dialysis, Transplant: a short history of failing kidneys (Peitzman) 250
drug delivery 716–717
drugs, categorisation (Tao Hongjing 陶弘景): 'fruits' 136; 'grains' 136; 'herbs and trees' 136; 'insects and animals' 136; 'jade and stones' 136; 'miscellaneous' 137; 'vegetables' 136
drug taking 66
drug therapy 133; crude drug (*shengyao* 生藥) 'fresh drug' 133; Imperial Pharmacy 140; *materia medica* collections 133, 139–140, **141, 142**; printing and disseminating earlier literature 139; standardising 138–141
Du You 杜佑 476
Duan Chengshi 段成式 479
Duden, B. 356
Dudgeon, J. *The Diseases of China: their causes, conditions, and prevalence contrasted with those of Europe* 251
Duke Jing 219
Dumoutier, G. 497
Dunhuang 敦煌 62, 138, 283, 287, 295n26, 306, 332, 421, 479–481
Dương Bá Bành 497, 498

e'rou 惡肉 (muscular flesh) 213
East Asian medicine 120, 129, 505–507
Eastern Han 13, 34, 65, 66, 81–84, 115, 135, 146, 207, 221, 359, 402, 405–406, 410, 475
ecological determinism, perspective of 254
Edo period 221, 519
Eight Extra Channels (*Qijing bamai* 奇經八脈) 199, *200*, 201
eight-principle differentiation (*bagang* 八綱) 149
Eisenberg, David M. 580
Eisenbrand, G. 710
Eison 叡尊 516
ejaculation 337, **339**; 340, 359, 383, 384, 390; female 345–346; male 342–345; 'short-term satisfaction' 343
elementary studies 384
Ellis, H. 383
Emergency Prescriptions Made with Local Korean Herbs (*Hyangyak gugeupbang* 鄉藥救急方) 503–504
emotional disorders 235–237; *dian* 癲, *kuang* 狂 and *xian* 癇 236–237; five major human emotions (anger, joy, pensiveness, sorrow and fear) 236; 'flower *dian*' (*huadian* 花癲) 237; grief and fright 236; 'Seven Emotions' 236; 'States of Mind' 236, 239; Women's 'flower *dian*' (love madness) 237
Emperor Enyū 円融 338
Emperor Huizong 徽宗 169
Emperor Jiaqing 嘉慶 493
Emperor Kangxi 康熙 174, 225
Emperor Qianlong 乾隆 461
Emperor Renzong 仁宗 139
Emperor Shunzhi 順治 269
Emperor Taizong 太宗 (Song) 139
Emperor Taizong 太宗 (Tang) 478
Emperor Taizu 太祖 712
Emperor Wanli 萬曆 192, 199, 235, 268, 270
Emperor Xi Zong 僖宗 479
Emperor Xuan Zong 玄宗 480
Emperor Yu 禹 81
Emperor Zhengtong 正統 202
Empress Kōmyō 光明 515
Encyclopaedia of Materia Medica (*Bencao gangmu* 本草綱目) *see Bencao* 本草 editions
Engelhardt, U. 291
envoy herbs (*shi* 使) 708
epidemics 247–248, 269–271; acute febrile 153–154, 225–226; apotropaic talismans and objects 461, 466; Central Epidemic Bureau (Nationalist period) 653; collapse of the Han dynasty 402–403; and concept of ming 402–403; as diagnostic lens 245–257; duanwu festival 458; during Sino-Japanese war 659, 662–663; government responses 147, 418, 506–507, 515, 653, 655, 660, 662–663; Korean approaches to 506–507; late imperial epidemiology 263–276; Manchurian plague and reassessment of value of Chinese medicine 635n2, 652; and multiple applications of herbs 457; new texts in Japan 515; numerology 79, 102, 104; obesity 320; and personalised medicine 220–221; public hygiene measures 655; and religion 402–403; sha rashes 295 n. 19; Spanish colonisation and hybridisation of Euro-Mayan medicine 591; and theodicy 407; and tongue diagnosis 165, 167, 173–174; Viet texts 499; *zhangyi* 瘴疫 outbreak after campaign into Yue 494; *see also shanghan* 傷寒; smallpox; warm disease (*wenbing* 溫病 or *wenyi* 瘟疫)
epistemology 39, 223, 255, 527, 551
Epler, D.C. 266
erotic heritage 394
Essays Written during the Rain (*Vũ trung tuy bút* 雨中隨筆) 496
Essences *see jing* 精
Essential Prescriptions For Every Emergency Worth a Thousand Pieces of Gold see Beiji qianjin yaofang 備急千金要方

Index

Essential Prescriptions Worth a Thousand Gold, for Urgent Need see Beiji qianjin yaofang 備急千金要方
Essential qi 33; *see also jing* 精
Essential Readings from the [Orthodox] Medical Lineage (*Yizong bidu* 醫宗必讀) 153
Essentials of the Golden Cabinet see Jingui yaolue 金匱要略
Essentials of the Jade Chamber see Yufang zhiyao 玉房指要
Essentials of the Plain Questions see Somun daeyo 素問大要
Ethnographic Plague: Configuring Disease on the Chinese-Russian Frontier (Lynteris) 276
Europe: acupoints and pulse points *557*; compilation of travelling physician 553–556; contribution of Jesuits in eighteenth century 559–561; *De Indiciis morborum ex Linguae coloribus & affectionibus* 555–556; *De Pulsibus Libros quatuor e Sinico translatos, De eplanatione pulsuum regulae* 554; difficult confrontation between two medical cultures 561–562; European scholars inspired by Chinese medicine 559; *Excerpta Literis eruditi Europaei in China* 555; first scholarly writings on Chinese medicine 552–553; *Fragmentum Operis Medici ab erudito Europaeo conscripti* 554–555; pulse diagram *556*; *Schemata ad meliorem praecedentium Intelligentiam* 555; scholarly treatise on Chinese sphygmology 557–559; 'shanghan pulse' 560; tales and testimonials from travellers 551–552; *Tractatus de Pulsibus ab erudito Europaeo Collectus* 554
European healers 10
European medicine 2, 170, 517, 520, in China 374
European physician (or physicians) 167, 170, 226, 552, 560, 561
Evans, H. 391, 393; *Histories of Post-Morten Contagion* 276
evidence-based medicine (EBM) 616
evidential scholarship movement 152–153
excavated manuscripts *see* manuscripts and artefacts
exhaustion disorder 273
exorcistic 'sword methods' 287
exoteric practices 418
Exposition of Cultivating the True by the Great Immortal of the Purple Gold Splendour (*Zijin guangyao daxian xiuzhen yanyi* 紫金光耀大仙修真演義) 342
external injuries 134–135
external medicine 206, 208, 210–211, 214
external ulcers'(*waiyang* 外瘍) 209

facial diagnosis 4
family planning, introduction of 393
famine and war 245–246
Fang Renyuan 方仁淵 213
Fang, X. 632
fangji 方技 (Remedies and Arts, Recipes and Methods, Recipes and Techniques, or Techniques and Recipes) 56, **57**, 62, 73, 75, 117n2, 405, 412n10
fangzhong shu 房中術; in Japanese *bōnaijutsu* 57, 62, 66, 338–339, 340, 360, 362, 370, 384–385, 395, 405, 408, 438, 696
Farquhar, J. 186, 640, 692, 695, 701
Farr, W. 264, 265
fasting from grains (*bigu* 辟穀) *409*, 409–410
fear (*kong* 恐) 236
Fédération des Acupuncteurs pour leur Formation Médicale Continue (FAFORMEC) 574n14
Fei Boxiong 費伯雄 236
female ejaculation and female essence 345–346
feminist scholarship 365
Feng Zhaozhang 馮兆張 or 馮氏 496, 497, 499
Festivals *Duanwu* 端午 457–458, 469; and herbs 457–460; *Qingming* 清明 459–460; Spring Festival, or Chinese New Year 460
fetal education 371, 375
fire needling technique 202
first Opium War 651
First Sino-Japanese War 651, 652
First World War 659
Fisher, C.T. 265, 267
Five Agents 五行 (Wood, Fire, Earth, Metal, and Water) 1, 3, 18–19, 95–96; *Basic Questions*, examples 16–17; Beijing Airusheng Zhongyi dianhai 313, **313**; *Beiji qianjin yaofang* 備急千金要方 312–314; category 'must not eat' 314; complexities of Qin and Han period prohibitions and taboos 311; creation of taboos around women 314; early divinatory practices 81; 'eat liver in spring' 310; *Essential Prescriptions* 310; excavated early Han manuscripts, examples 13–16; fifth, Earth 92, 93; first, Water 92, 93; five flavours (pungent, salty, bitter, sweet and sour) 309; fourth, Metal 92, 93; Han ruling house 308; *Huangdi* 黃帝 (Yellow Emperor) corpus 309; *Huangdi neijing suwen* 黃帝內經素問 (Yellow Emperor's Inner Canon: Basic Questions) 309; to *mai* 脈 309; Mawangdui 馬王堆 tomb 312; medical prohibitions 311; Mongol period *Yinshan zhengyao* 飲膳正要 (Principles of Correct Diet) 313; mutual conquest or controlling or conquering cycle 46n24, 92, 93; mutual generation or generating cycle 46n24, 92; and Phenomena in *Suwen* 519; preliminary survey 311–312, **312**; seasonal homoeopathy 309–310; second, Fire 92, 93; *Shennong bencao jing* 神農本草經 310–311; in *Suwen* 素問 **19**; on *Suwen* 309; third, Wood

Index

92, 93; *xiangke* 相剋 conquest cycle 308–310; *xiangsheng* 相生, cycle of generation 308–310; to Yin-Yang Channels in *Suwen* 19

five sapors (*wuwei* 五味: pungent, sweet, salty, sour, and bitter) 134

Fleck, L. 689, 690, 699

Floyer, J. 559

foetal breathing (*taixi* 胎息）35

Folio on Damp Heat Diseases (*Shirebing pian* 濕熱病篇) 154

folk medicine: abortion 291; authors and purposes 283–284; bloodletting and minor surgery 288; bone setting and massage 288–290; 'bottom up' approach 294; childbirth practices 291; dietetics 291–292; 'feudal superstition' 656; lamp wick cauterisation 288; magico-religious practices 292–293; in manuscripts 285; marginalised practices 287; pharmacotherapy 286–287; piercing and cauterisation in magico-religious context 287; and practitioners 283–285; success of TCM 656–657; therapies 286–293

food and dietary medicine (*shiliao* 食療): *Bencao gangmu* 329, 333–334, 335n5; *Chongxiu zhenghe jingshi zhenglei beiyong bencao* 重修正和經史證類備用本草 330–331; Galenic system 329; *Huangdi neijing suwen* 黃帝內經素問 330; *Jiu huang bencao* 救荒本草 334; *Jujia biyong shilei* 居家必用事類 335; *materia dietetica* 329, 332–334; *materia medica* books 328–332, 334; *Shennong bencao jing* 神農本草經 328–330; Song Dynasty official herbals **331**; *Xinxiu bencao* 新修本草 329; *Yinshan zhengyao* 飲膳正要 332, 333, 334, 478

food and sex 4

food supplements and health foods: goji berry 733–734, 736; *reishi* (*lingzhi* 靈芝) 734–735

Former Shu 前蜀 dynasty 479

formulary (*fangji xue* 方劑學) 675

Formulary for Curing all Disorders (*Zhi baibing fang* 治百病方) 135

Formulary of the Pharmacy Service for Benefiting the People in an Era of Great Peace (*Taiping huimin hejiju fang* 太平惠民和劑局方) 147

Foucault, M. 382, 390; *History of Sexuality Volume 1* 389

four gates (*simen* 四門) 97, 98

Four lands of the Shang 94–95

Framing Animals as Epidemic Villains (Lynteris) 276

France: evolution of practice of acupuncture 565; Italian acupuncture departs from 'French traditionalist acupuncture' 572–573

Francis Hsu 253, 254, 264

Francisco de Herrera Maldonado 552

free-handed method 343, 344

fresh drug 133

Fryer, J. 484

Fukane Sukehito 深根輔仁 516

functionalist anthropology 253–255

Furen daquan liang fang 婦人大全良方 (Comprehensive Compendium of Good Formulas for Women) 224, 363, 371, 372

Furth, C. 150, 274, 356, 359

Further Discussion on Materia Medica (*Bencao yanyi* 本草衍義) *see Bencao* 本草 editions

Gai Jianmin 蓋建民 411

Galen 168, 329, 591

Galison, P. 175

Gamble, A. 708, 712

Gao Ruona 高若訥 127

Gao Wu 高武 194

Garcìa, M. 588

Ge Hong 葛洪 137, 221, 225, 333, 405, 446, 461, 724

gender 356; 'category of analysis' 356; sex-gender binary and sexual difference 356

General Discussion of the Disease of Cold Damage see Shanghan zongbing lun 傷寒總病論

ginkgo 10, 726–728

ginseng 10, 728–730

Gleanings of Materia Medica (*Bencao shiyi* 本草拾遺) *see Bencao* 本草 editions

the global North 10

glycyrrhizin xvi, 710–711

Goethe, J.W. 727

goji berry 10, 733–734

Golden Mirror of the Medical Lineage (*Yizong jinjian* 醫宗金鑒) 153, 289

Gong Tingxian 龔廷賢 496, 497, 517

Gong Xin 龔信 233

gongyang 供養 (Giving offerings; Sk. *pūjā*) 418

Gotō Konzan 後藤艮山 518

Grant, J. 363, 632

Grant, John B. 653, 661

Great Essentials on the Golden Elixir by Master Shang Yang (*Shangyangzi jindan dayao* 上陽子金丹大要) 343

Great Leap Forward (1958–61) 8

Gros, F. 495

Guan Rongtang 管榮棠 211

Guanyin 觀音 (The Bodhisattva Sound-Perceiver; Sk. Avalokiteśvara) 418, 460

Guatemala: Chinese medicine and Mayan medicine 591–593; Guatemala Acupuncture and Medical Aid Project (GUAMAP) 589; influence of Spanish colonisation on Mayan medical order 590–591; migration of acupuncture 589; *Popul Vuh* and *Chilam Balam* 590; pre-colonial Mayan health practices 589–590

guiding and pulling *see daoyin*

Index

Gujin tushu jicheng 古今圖書集成 (Synthesis of Past and Present Books and Illustrations) 236
guoyi (national medicine) 國醫 577, 636n7; *Guoyiguan* 國醫館 (Institute of National Medicine, INM) 654; *guoyi dashi* 國醫大師 (Great Master of National Medicine) 680
gynecology: 'bedchamber techniques' 370; caesarian delivery 377; 'clear menstrual blockage' 373; emergence of 'separate prescriptions for women' from Han to Tang dynasties 369–370; European medicine in China 374; 'female inner alchemy' 373; fetal education 375; hospital birth policy 377; imperial period 369; midwifery reform in Nanjing decade 375; neo-Confucian gynecology in Song dynasty 371–372; and obstetrics in period of economic reforms 376–377; 'oracle bones' 369–370; People's Republic of China 375–376; popularising classical gynecology in Qing dynasty 373; rethinking blood and gender in Jin, Yuan and Ming dynasties 372–373; scholarship 369; state-building in biomedical age 374–375; technology of gender 368; women's diseases and childbirth 369–370

haemorrhoid surgery 7
Haklev, S. 634
Han (early) medical manuscripts: medicine of systematic correspondence in Han China: a new style of thought 19–20; *Yin-yang* and five agents, examples 13–16, 21n1, 97
Han Anguo 韓安國 62
Hanaoka Seishū 華岡青洲 519
Han dynasty 16, 25–27, 30, 32–33, 45n7, 73, 77, 95, 110, 114, 115, 122, 136, 142n4, 163, 181, 182, 196, 207, 219–221, 247, 338, 369, 385, 394, 401–410, 445, 493, 494, 672, 688, 694; Huang-Lao movement 123; importance of *Inner Classic* 122–123; sexual pracice 338; tomb, text recording prescriptions 57–59, 62, 135, 187n1; use of governmental model 123
Han Dynasty Medical Bamboo Strips from Wuwei (*Wuwei Handai yijian* 武威漢代醫簡) 135
Hanfei zi 韓非子 164–165
Hansen's disease or outdated leprosy 273
Harper, D. 27, 65, 93, 283, 292, 360, 445
Hart, R. 251
Hartwell, R. 253
Hawkins, J.P. 592
He Bingyuan 和炳元 187
He yinyang 合陰陽 345–346
Headland, F.W. 484
healers 10, 38, 237, 256, 283, 284–286, 294n10, 307, 314, 364, 370, 372, 373, 418, 419, 456, 460, 461, 464, 466–468, 515, 577, 606, 638, 639, 642, 650, 653, 680

Heart wind 233–235
hedonism 394
hemerology: calendrical systems 306; celestial circulation 307; concept of *renshen* 人神 306; *ganzhi* 干支 cycle 305; *Hama jing* 蛤蟆經 (Toad Canon) 306; Jiudian (JD) and Fangmatan (FMTA) 306; Taiyi system 306
Heo Jun 許浚 507, 509
herbs: adjunct herbs 708; application of 135; and astrology 457–460; 'celestial herb' 458; 'Compendium of *Materia Medica*' 456–457; 'Continent of Spirits' 457; envoy herbs 708; and incantations 466–469; 'magical' or 'superstitious' 456; minister herbs 708; sovereign herb 708; and talismans 460–466
hernia repair 7
Hershatter, G. 391
heterogenous *qi see zaqi* 雜氣
Hinrichs, TJ 283
historicalist-conceptualist approach 248–251, 253, 256; on end-of-Ming epidemics 272–273; scholarship, on history of Chinese medicine 1997 onwards 273–275
Histories of Post-Mortem Contagion (Lynteris and Evans) 27
History of Sexuality Volume 1 (Foucault) 389
History of the Ming dynasty (*Mingshi* 明史) 234
History of the Song (*Songshi* 宋史) 445
HIV/AIDS crisis 581, 664
Ho, E.Y. 39, 256
Hoàng Đôn Hoà 黄敦和 499
Hobson, B. 374
Hōjō Masako 北条政子 516
holism 217
Hong Taiji 皇太極 269
Honig, E. 392
Honzō wamyō 本草和名 (Japanese Names for the *Materia Medica*) 516
hoodlums (refugees) 246
hospital birth policy 377
Houtian jing 後天精 ('postnatal essence') 156n21
Hsu, E. 37, 253, 608, 630, 707; *Innovation in Chinese Medicine* 274
Hu, C.H. 714
Hu Jintao 胡錦濤 692
Hu Shi 胡適 660
Hu Sihui 忽思慧 324, 332, 478, 483
Hua Shou 滑壽 194, 553
Hua Tuo 華陀 207, 221, 534
Huainanzi 淮南子 85n11
Huang Di (or Huangdi) 黃帝 41, 53, 404; *see also* the Yellow Emperor; founding culture hero 3; legendary patron of scholarly medicine 3
Huang Longxiang 黃龍祥 207
Huangdi mingtang jing 黃帝明堂經 190, 207

752

Huangdi neijing 黃帝內經 (Inner Canon of the Yellow Emperor or Yellow Emperor's Inner Classic) 13, 30, 31, 42, 51, 53, 60, 74, 75, 91, 94, 98, 99, 100, 104, 116, 120–129, 146, 148, 150, 152–153, 164, 165, 182, 195, 196n3, 240n4, 207, 231–232, 288, 357, 401, 404–405, 555, 650; archaeological discovery 121; canonicity as historical processes 129; categories of *qi* in medicine of 95; corpus, formation of 121–122; Han dynasty 122–123; *Lingshu* recensions 74, 306; *Lingshu* 靈樞 (Numinous Pivot) 40–42, **43–44**, 53, 60, 91, 100, 101, 106n28, 106n32, 116, **121**, 122, **126**, 128, 187n2, 190, 195, 240n2, 240n3, 241n42, 308, 515, 555; period of division, Sui, and Tang 123–125; Song dynasty 125–129; Song editors of 128; structure of 121–122; *Suwen* 素問 (Basic Questions, also known as Simple Questions and Inquiry into the Fundamentals) 6, 10 & 13–14, 15–19, **19**, 20, 38–39, 41–42, **43–44**, 44 & 46–49, 106n33–35, 74–75, 83–84, 85n13 &16, 86n39, 91–97, 101, 103–105, 105n3–4, 116, **121**, 123–125, **126**, 127–128, 130n2, 164, 187n2, 207, 214n2, 221, 222, 241n31, 309–310, 330, 369, 401, 404, 515, 562n4; titles of 120–121; *Yin-yang* and Five agents, examples 16–17; *Yin-Yang* of the Body **17**; see also *Suwen* 素問; *Taisu* 太素 recensions; *Zhenjiu jiayi jing* 針灸甲乙經

Huangfu Mi 皇甫謐 124, 130n3 see also *Zhenjiu jiayi jing* 針灸甲乙經

Huangsu fang 黃素方 (Methods of the Yellow Fundamental) 125

huanjing bunao 還精補腦 (returning the semen/essence to the brain) 344, 359, 384, 438

Huard, P. 497

Hudson, C. 449, 453

Hui Lin 慧琳 478

Human Extinction and the Pandemic Imaginary (Lynteris) 276

huoluan 霍亂 (acute vomiting and diarrhoea) 218, 223, 462

Hyangyak jesaeng jipseongbang 鄉藥濟生集成方 (Compendium of Life-Saving Recipes Ready-made with Local Korean Herbs) 504

Hymes, R. 275

hypertension 65, 529, 593, 717, 734

Ibn al-Nadim 481

Ibn Butlan 332, 333

Illumination of the Principles and Varieties of Materia Medica (*Honzōkōmoku keimō* 本草綱目啓蒙) see *Bencao* 本草 editions

Illustrated Classic of Materia Medica (*Bencao tujing* 本草圖經) see *Bencao* 本草 editions

Illustrated Manual 190, *193*, 202

immortality 35, 54, 56, 65–67, 114, 322

imperial (late) epidemiology: *anomalous qi* (*zaqi* 雜氣) 247; Black Death in Europe 255; 'boils disorder' 270; 'bubonic plague' 266–267, 270; Chen's human greed 248; 'China origin of plague' hypothesis 255; Chinese disease monograph 266; cholera epidemics 249, 254; Cold Damage 248, 266; conceptual and material methods 263; contagion in Chinese paediatrics 274; contaminationist perspective 247, 253; 'deviant *qi* doctrine' 253; disease 248; emic and etic viewpoints 249; end-of-Ming epidemics as plague 267–273; exhaustion disorder 273; famine and war 245–246; functionalist anthropology and natural-realist demography 253–255; genetics of aDNA 275; Hansen's disease or outdated leprosy 273; historicalist-conceptualist approach 248–251, 253, 256, 265–266, 272–275; historiography in 1980s–1997 264–267; 'hoodlums' (refugees) 246; human and bacterial DNA 263–264; *longue-durée* history of Chinese medical governance 265; naturalist-realist method 248–255, 268–271; 'numbing wind' (*mafeng* 痲瘋) 273; perspective of ecological determinism 254; plague 250, 252, 254, 256; political-history perspective 265; polytomy 275; 'rat epidemics' 274; retrospective epidemiology in China 251scientific transformations of plague in China 275–276; 'sexually transmitted diseases in modern China' 266; 'Sheep's Wool heat epidemic' 271; Shen's excessive taxation 248; smallpox epidemic 225, 269–271,; tuberculosis 273; two approaches to history of disease 248–251; warm-disease epidemics 247–248, 248; *Yangzi bing* 252; *Yersinia pestis* 264, 275

Imperial Grace Formulary of the Great Peace and Prosperity Reign Period see *Taiping shenghuifang* 太平聖惠方

imported to China 480, 656

Imura Kôzen 井村哮全 253

incantations (*fuzhou* 符咒) 410, *411*, *462*; *Chuanya* 串雅 467–468, 469; folk healers 467–468; in folk medical manuscripts 467; and herbs 466–469; 'incantations with characters' 467; 'incantations with drugs' 467; *jinshu* 禁書 or *jinfa* 禁法 466; *Mingyi lei'an* 名醫類案 466; recipe collection 'Sea of Medicine' 469; spirit-possession 468–469

inflammation 592–593, 713

Indian Kāmasūtra 337

The Indian Massage 111

Indian medicine: impact of 5; medical doctrines 421

Indian *mudrās* 293

injaculation and sexual energy 340

Inner Canon, Yellow Emperor's see *Huangdi neijing* 黃帝內經

Inner Classic, Yellow Emperor's see *Huangdi neijing* 黃帝內經
Inquiry into the Fundamental (*Suwen* 素問) see *Huangdi neijing* 黃帝內經; *Suwen* 素問
Insights on the Purport of the Alchemical Treatises see *Danlun juezhi xinjian* 丹論訣旨心鑑
Institute of National Medicine (INM) 654, 655
internal medicine (*neike* 內科) 206, 209–210
internal sores (*neichuang* 內瘡) 209
internal ulcers (*neiyang* 內瘍) 209
International Conference on Population and Development (ICPD) 394
Ishinpō 醫心方 195, 338–340, 343–344, 346–347, 385, 514
Islamic healers 10
Italian acupuncture: departs from 'French traditionalist acupuncture' 572–573; practice of acupuncture in 572

Jābir ibn Hayyān 484
Japan: 'bronze doll' or 'channels doll' 516; Bureau of Medicine 513, 515; Chinese-style medicine 513–517; decline and revival of Chinese-style medicine 519–521; development of autonomous medical traditions 517–519; introduction of Buddhism from Baekje 514; *Kanpō* (or *Kampō*) 漢方 513, 514, 518, 520–521; legal codes 514–515; materia medica compendium (*Honzō wamyō* 本草和名) 516; Neo-Confucian style 514
Japanese Materia Medica (*Yamato honzō* 大和本草) see *Bencao* 本草 editions
Japanese Names for the Materia Medica (*Honzō wamyō* 本草和名) see *Bencao* 本草 editions; *Honzō wamyō* 本草和名
Jartoux, F. 729
Jesuit missionaries: arrival of 174; foetus and placenta 174, *175*
Ji Dan 姬旦 219
Ji Kang 嵇康 688
ji 肌 (muscly) part of flesh 207
Jia Sixie 賈思勰 311, 317
Jiajing 嘉靖 265
Jiang Guan 江瓘 186
Jiang Shan 姜姍 34, 35, 39
Jiang Yinsu 江應宿 186
Jiangsu Academy of New Medicine (*Jiangsu xinyi xueyuan* 江蘇新醫學院) 682
Jiaozheng yishu ju 校正醫書局 (Bureau for the Revision of Medical Texts) 126, 147, 170–171
Jiayou bencao 嘉祐本草 (*Jiayou* Materia Medica) **126**, 127, 139, **141**, **331**,
Jiayou Materia Medica (*Jiayou bencao* 嘉祐本草) see *Bencao* 本草 editions
Jin dynasty 137, 269, 372
Jin Weng 金翁 516

Jin-Yuan dynasty: Li Gao 李杲 149–150, 372, 516; Liu Wansu 劉完素 149; Zhang Congzheng 張從正 149; Zhu Zhenheng 朱震亨 150–151
Jing Fang 京房 432
Jing 精 17, 33, 35, 92, 152, 156n21, 222, 341, 342–345, 358–359, 384, 428, 429, 435, 438, 445, 447, 451
jingmai 經脈 33, 220
jingluo 經絡 30, 33, 55, 149
Jingyue quanshu 景岳全書 (The Complete Works of Jingyue) 363
Jingzheng huoshi 驚症火式 289
jinrou 筋肉 (tendinous flesh) 207
jiong 腘 (bulky flesh) 207
jirou 肌肉 (muscular flesh) 206–207
Joachim Bouvet 174
Johann Storch 362
Johns Hopkins University School of Public Health 7
Johnson, V.E. 347
Jolan Chang (or Chang Chung-lan 章鐘蘭) 390
Joseon dynasty 504–506
joy (*xi* 喜) 236
juemai jiejin 訣脈結筋 206
Jurchen Jin dynasty 372
Juyan 居延 60–62

Kahn, Ida 康愛德 374
Kaibao Materia Medica (*Kaibao bencao* 開寶本草) see *Bencao* 本草 editions
Kaibara Ekiken 貝原益軒 517, 520
Kajiwara Shōzen 梶原性全 516
Kamakura period 196, 516
Kang Youwei 康有為 384
Kangri zhanzheng (War of Resistance against Japan) 抗日戰爭 655
kaozheng 考證 (evidential philology, evidential research or evidentiary verification) 152, 509, 672
Kaptchuk, T.J. 573; *Web That Has No Weaver* 36
Karchmer, E. 614, 615
Karrer, P. 710
Katz, P.R. 266
Keegan, D.J. 74, 121, 122
Keightley, D.N. 72
Kelley, L.C. 498
Kerrie MacPherson 264
Kespi, J-M. 570, 572
King Jing 靖王 60
King Pacorus II 475
King Sejong 504
King Wu Ding 武丁 218, 219
King Zhao Mei 趙眜 60, 66, 67, 219
King Zheng of Qin 秦王政 96; see also Qinshi huangdi 秦始皇帝
the kingly way (*wangdao* 王道) 212

754

Index

Kircher, A. 728
Kitab Taqwim as-sihha (Almanac of Health) 332
Kleinman, A. 698
Koch, R. 659
Kohn L. 342
Korean medicine (KM) 537; *Annals of the Joseon Dynasty* records 506; cold damage or warm diseases 509–510; 'Four Constitutional Medicine' or 'Sasang Constitutional Medicine' 509; medical innovations 506–507; period of compilation of East Asian medicine 505–507; period of exchanges between Asian traditional and Western medicine 510–511; period of independent development 507–510; period of local Korean herbal medicine 503–504
Kou Zongshi 寇宗奭 140, 148
Kristeva, J. 391
Kuang 狂: damage to *yangming* 陽明 (Yang Brightness) vessel 231; *dian* 癲 232–233; *yin-yang* dualism 231
Kublai Khan 483
Kuriyama, S. 165, 266, 288, 614

La Route de la Sole (Mazahéri) 475
Lafitau, J.-F. 729
Lai Jizhi 來集之 462
Laignel-Lavastine, Maxime 565
lamp wick cauterisation 288; *see also moxa-cautery*
lancing abscesses 7
languages 26, 39, 170, 369, 390, 670, 679
Laoguanshan 老官山, also known as Tianhui 天回 tomb 3 *see* manuscripts and artefacts
Laqueur, T. 356, 357
Laufer, B. 475
Lawrence, D.H. 385
Lê Hữu Trác 494, 496–499
League of Nations-Health Organisation (LON-HO) 7, 661
Lee Gyujun 李圭晙 508
Lee Jema 508
Lee, M. 579, 581
Lei, S.H.-l. 630, 680
Leriche, René 565
Leung, A.K.C. 265, 273, 274, 356, 361
Li Cang 利蒼 56
Li Chan 李梴 208, 496, 497
Li Daochun 李道純 429, 440
Li Dongyuan 李東垣 *see* Li Gao
Li Gao 李杲 149, 150, 208, 372; 516
Li Hongzhi 李洪志 691
Li Jianmin 李建民 169, 359
Li Miyi 李密醫 480
Li Shizhen 李時珍 140, 150, 151, 283, 288, 304, 333, 334, 343, 484, 517, 676, 712, 715, 728
Li Xiyue 李西月 447
Li Yi 李一 53

Li Yuechi 李月池 151
Li Zicheng 李自成 270
li 理 (principle) 35
Liang Qichao 梁啓超 384, 660
licorice *see* liquorice
Lifesaving Book that Categorises the Presentations of Cold Damage see Shanghan leizheng huoren shu 傷寒類證活人書
Lingshu 靈樞 (Numinous Pivot) *see Huangdi neijing* 黃帝內經
lingyi 鈴醫 (traveling medicine peddler) 151, 650
Linzheng yide fang 臨證一得方 (Comprehensive Recipes for Clinical Syndromes) 213
Linzheng zhinan yi'an 臨証指南醫案 (Case Record Guide to Clinical Presentation) 153, 173
Liqi 癘氣 (pestilential *qi*) 154, 247
liquorice: as aid to drug delivery 716, 716–717; *Artemisia annua* 707; cinchona tree *(Cinchona officinalis)* 707; components of raw liquorice 710, 710–711; future pharmacy 708; glycyrrhizin 716, 716–717; hazards 717; and herbal medicine 709, 709–710; later Chinese remedy literature 712–713; metabolism 711, 711; micelles 716, 716–717; mixture of herbs 708; pharmacological remedies 708; special nature of liquorice 714–715; testing therapeutic value of liquorice and glycyrrhizin 713–714
Liu Dalin 劉達臨 393, 394
Liu He 劉賀 61
Liu Huayang 柳華陽 441, 449
Liu Ruiheng 劉瑞恆 661
Liu Wansu 劉完素 149, 150, 152, 153, 234
Liu Wuyuan 劉悟元 447
Liu Yiming 劉一明 441
Liu Zenggui 劉增貴 311
Liu, J.C. (Liu Ju-ch'iang[Liu Ruqiang] 刘汝强) 676
Liuren astrolabe 82, *82*
Liye 里耶, well 754–755, 134
Lo, V. 30, 53, 283, 287, 736
localisation 5, 565, 573
locus classicus 177n1
Loeper, M. 565
London 3, 275, 374
Longevity and Life Preservation in Eastern Medicine see Dongui suse bowon 東醫壽世保元
longue-durée: and formation of institutions and traditions 2–3; history of Chinese medical governance 265
Lou Juzhong 婁居中 332
Lou Ying 樓英 149, 150, 152, 153, 234, 235
Lü Buwei 呂不韋 20, 80, 96
Lu Gwei-Djen 魯桂珍 553
Lu Xixing 陸西星 447
Lu Xun 魯迅 36, 660

Lucas, A. 629
Luo Xiwen 羅希文 712
Luo Zhenyu 羅振玉 62
Lüshi Chunqiu 呂氏春秋 20, 29, 46n23, 75, 80–81, 85n10, 86n31, 86n31–n33, 92, 96, 105n5, 307, 458
Lynteris, C.: *Ethnographic Plague: Configuring Disease on the Chinese-Russian Frontier* 276; *Framing Animals as Epidemic Villains* 276; *Histories of Post-Morten Contagion* 276; *Human Extinction and the Pandemic Imaginary* 276
Lythgoe, B. 710

Ma Kanwen 馬堪溫 8, 636
Ma Peizhi 馬培之 213
Ma Yuan 馬援 494
MacPherson, K.L. 264
macrocosm 83, 84, 92–94, 102, 357, 360, 440; *see also microcosm*
Madagascar: acupuncture in 606–607; adaptability of Chinese medical practices to local cultures 609–610; context 606; cost of Chinese medicines 607–608; importance of building trust 608–609; Malagasy Ministry of Health (MOH) 606; physical access to healthcare 608
madness, pre-modern: classification of disease 231–233; emotional disorders 235–237; heart disorders 235; *kuang* 狂 230–233; medical treatments 237–238; phlegm-fire 234–235; wind malady 233–234
magico-religious practices: amulets 292; calendrics, divination, and demonology 293; pharmaceutical substances 286; ritual movements 293; rituals 292; spells and incantations 293
mai 脈 14, 33, 34, 55, **57**, 58, 60, 98, 105, 151, 152, 165, 167, 170, 182, 195, 196, 199–200, 201, 206, 220, 231, 307, 309, 342, 359, 435, 485, 515, 554, 558, 559
Maijing 脈經 (Pulse Classic) 34, **126**, 165, 515, 554
mailuo 脈絡 30; *see also jingluo* 經絡; *jingmai* 經脈
Maishu 脈書 (Vessel or Channel Book) 14, 30–32, 55, 58, 60, 106n23, 308
malaria 1, 10, 221, 226, 248, 250, 276, 461, 494, 600–606, 626, 656, 660, 663–664, 707, 717, 722–724
male essence and male ejaculation 342–345; *see also huanjing bunao* 還精補腦
Mancheng 滿城 60
Mangin, A. 497
Manson, P. 252, 255
Mantak Chia 390, 452, 696
manuscripts and artefacts: archaeological record 67; characteristics 65; *Ci shu* 刺數 (Principles of Piercing) 52; classical acupuncture theory 65–66; cultural communication 64; Day Books 65; foods and drugs 66; geographic distribution of manuscripts 63; illnesses and remedies 66; manuscripts and artefacts 53–63; Mawangdui manuscripts 64; medical manuscripts, importance of 52–53; numerological cultures 64; oral tradition, medical practitioners 53; physical grouping of manuscripts 63; pre-existing histories of origins of Chinese philosophy 63–64; remedy and recipe literature 66; scientisation and modernisation 65; *Shiji* 史記, 'Records of the Historian' 67; talismans 463; *tuina* 推拿 manuscripts 289
manuscripts and artefacts, sites: Dunhuang 6, 62, 138, 283, 287, 295n26, 306, 332, 421, 479, 480, 481; Hantanpo 旱灘坡 unspecified tomb 63; Juyan 居延 60–62; Laoguanshan 老官山 tomb 15, 30, 57–59, *59*, 57–59, *59*, 60, 63, 66, 99; Liye 里耶, well 154–155, 134; Mancheng 滿城 60; Marquis of Haihun's 海昏 tomb, tomb 1 61–62, 66; Mawangdui 馬王堆, tomb 3 14–16, 20, 25, 30, 54, *56*, 56–57, **57**, 58, 60, 62–64, 66, 73, 98–99, 105n22, 106n23, 109, 121–122, 134, 219, 231, 283, 291, 312, 322, 325, 338, 344, 360, 365n6, 369, 384, 394, 404, 405, 408, 445, 681, 688, 708; Nanyue 南越 Kingdom 60, 67; Shuangbaoshan 雙包山 30, 58, 60, *61*, 99, *100*, 106n26; Shuanggudui 雙古堆, tomb no. 1 56, *82*; Shuihudi 睡虎地, tomb 11 54, 64, 65, 73, 182; Western Han medical manuscripts on bamboo slips 59; Zhangjiashan 張家山, tomb 247 13–14, 15, 20, 21n3, 30, 55, 58, 60, 66, 98, 99, 106n23, 110, 181, 688; Zhoujiatai 周家臺, tomb 30 54, 55
Mao Duishan 毛對山 468
Mao Zedong 毛澤東 8, 538, 541, 587, 593, 603, 604, 625, 629, 635, 659, 662, 709; creating healthcare resources 632; healthcare integration 629–630; primary healthcare model, PRC 632–633; redress rural healthcare inequities in China 629–630, 635; speaking of integration 630–631; traditional medicine 8
marginalised practices 287
Marquis of Haihun's tomb 海昏侯墓, tomb 1 *see* manuscripts and artefacts
martial arts 8, 64, 117, 284, 290, 564, 565
Martini, Martinus 728
Martiny, Marcel 565, 567, 570
al-Marwarruzi, Abu Yaḥyā 483
Mass Education Movement 661

Index

mass printing 125
massage 3, 109, 110, 111, 115, 116, 117n2, 135, 149, 182, 202, 204n3, 284, 285, 288–289, 295n20, 418, 451, 452, 460, 513, 518, 520, 521, 564, 565, 606, 641, 645, 692, 693, 723; *see also ammo* 按摩; *tuina* 推拿
The Massage Technique of Laozi 111
Master Dong Xuan see Dongxuan zi 洞玄子
Masters, W.H. 347
matching stems, monthly cycle 431–432, **432**
material culture 5, 135, 669
Materia Dietetica (*Shiliao bencao* 食療本草) *see Bencao* 本草 editions
materia medica *see Bencao* 本草
Materia Medica Arranged according to Drug Description and Technical Aspects (*Bencao gangmu* 本草綱目) *see Bencao* 本草 editions
Materia Medica Arranged According to Monographs and Technical Criteria (*Bencao gangmu* 本草綱目) *see Bencao* 本草 editions
Materia Medica Canon of Shennong (*Shennong bencao jing* 神農本草經) *see Bencao* 本草 editions
Materia Medica Canon Variorum Edition 712 (*Bencao jing jizhu* 本草經集注) *see Bencao* 本草 editions
Materia Medica for Decoctions (*Tangye bencao* 湯液本草) *see Bencao* 本草 editions
Materia Medica for Emergencies Proven and Classified through History (*Jingshi zhenglei beiji bencao* 經史證類備急本草) *see Zhenglei bencao* in *Bencao* 本草 editions
Materia Medica Newly Revised in Detail during the Kaibao 開寶 (968–976) *Period* (*Kaibao xin xiangding bencao* 開寶新詳定本草) *see Kaibao bencao* 開寶本草 in *Bencao* 本草 editions
Materia Medica of Food Qualities (*Shixing bencao* 食性本草) *see Bencao* 本草 editions
Materia Medica of the Kaibao Reign (*Kaibao bencao* 開寶本草) *see Bencao* 本草 editions
Materia Medica Organised by Headings (*Bencao gangmu* 本草綱目) *see Bencao* 本草 editions
Materia Medica Organised by Headings with Individual Listings for Foodstuffs for Consideration (*Beikao shiwu bencao gangmu* 備考食物本草綱目) *see Bencao* 本草 editions
Materia Medica Proven and Classified through History from the Daguan 大觀 *period* (*1107–1110*) (*Jingshi zhenglei Daguan bencao* 經史證類大觀本草) *see Zhenglei bencao* in *Bencao* 本草 editions
Materia Medica Proven and Classified Throughout History for Use from the Zhenghe 正和 *Period Revised* (*Chongxiu zhenghe bencao* 重修正和本草) *see Bencao* 本草 editions

[*Divine Husbandman's*] *Materia Medica, with Collected Annotations* (*Bencao jing jizhu* 本草經集注) *see Bencao* 本草 editions
Matsuo Bashō 松尾芭蕉 519
Mauss, M. 694
Mawangdui 馬王堆: in Hunan province 14, 121, 134, 394; tomb 3, 56, 109, 219, 312, 338, 708; *see also* manuscripts and artefacts
Mayan medicine 591–593; influence of Spanish colonisation 590–591; pre-colonial Mayan health practices 589–590
May Fourth New Culture Movement 383
Mazahéri, A.: *La Route de la Sole* 475
McHale, Shawn F. 498, 499
McNeill, W.H. 254–256, 267
medical governance 265, 526
medicalisation 32, 362
medical offices and officers 111, 333
medical prohibitions: behaviours of 'other' people 304–305; blood contamination 315; Chinese diet 304; concepts of emic and etic 304; diversity of prohibitions 303; Five Agents (Wood, Fire, Earth, Metal, and Water) 308–314; hemerology 305–306; nutritional prohibitions 304; Placenta Spirit (*Taishen* 胎神) 316; *Qimin Yaoshu* 齊民要術 317; Shang calendar 316; therapeutic strategies and prohibitions 303; time-sensitive, content of almanacs 315–316; in twentieth century 314–317; *Xuanyuan beiji yixue zhuyou shisanke* 軒轅碑記醫學祝由十三科 314–315; Yin, Yang and seasonal restraint 307–308
medical text, emergence of 171
Medicine Buddha 5
medicine for women *see* women's medicine
medicine/s 23, 721–736, 737n1; notion of 2
medieval medicine: Channels 98–100, *99*, **102**; five and six in 97; organs 100–101, *102*
Meiji period 381, 519; *see also* Meiji state
Meiji Restoration (1868) 7
Meiji state 651
Meiji tennō 明治天皇 519
Meng Shen 孟詵 332
Meng Xi 孟喜 432
Mengzi (or Mencius) 孟子 28, 29, 245, 381, 688, 702n3
Mentzel, C. 554
Menuret de Chambaud, J.-J. 559, 560
merging *qi* 35; *see qi* 氣
Methods of the Yellow Fundamental see Huangsu fang 黃素方
Mey, C.A. 671
Miasma (*zhangqi* 瘴 or *zhangqi* 瘴氣) 226, 494
micelle 716, 724
Michuan dieda bofang 秘傳跌打秘方 (Secretly Transmitted Recipes [Treating Injuries from]

757

Knocks and Falls, [Strung Together like] Cymbals, MS 8111) 290
microcosm 34, 84, 92–94, 247, 360, 434, 440, 457; *see also* macrocosm
migration of acupuncture: to Cuba 586–589; to Guatemala 589–593; in Philippines 593–596; through *Imperium Hispanicum* 586–596
mijiao 密教 (Esoteric or Tantric Buddhism) 418
mind-body medicine 37
Ming dynasty: case records 185–186; consolidation of medical knowledge 151–152; Li Shizhen 李時珍 151–152; sexual alchemy in 446–448; Zhang Jiebin 張介賓 152
mingmen 命門 152, 225, 555
Mingtang 81
Mingtang diagrams 190, *191*, 192, 194, 195, 199, 202–203, 207, 515
Mingyi bielu 名醫別錄 (Separate Records of Famous Physicians) 137
Minh Mạng 495, 499
minister herbs (*chen* 臣) 708
Ministry of Health 653
minor and superficial surgery 7
minority medicine: academic development 542–544; China Medical Association of Minorities 541; *Collections of Collation of Minority Medicine by the SATCM* 542; 'Demonstrative Research on Key Techniques of Minority Medicine Development' 543; Eight-Character Guiding Principles 540; Han people 537; institutionalisation of 539–542; 'Minority Medicine Literature Collation and Appropriate Technology Selection and Propagation' 543; *minzu yi* 民族医 (doctors of minority medicine) 537–538; open-door policy 540; oral traditions 545; *Pharmacopeia of the People's Republic of China* 544; policies and projects issued by government of China **538**; Research Office for Minority Medicine 541; State Administration Departments **539**; 'Suggestions for Promoting Minority Medicine' 541; tenth five-year state plan 541; 'The Twelfth Five-year Plan on the Informatisation of Chinese Medicine' 542
Min Yide 閔一得 441, 449
Miraculous Drugs from the South (Nam dược thần hiệu) 498
Misono Isai 御園意斎 518
Mitchell, P. 250
modernisation 7–9, 65, 177, 382, 510–511, 514, 525, 527, 531, 533, 535, 539, 587, 626, 628, 630, 638, 641, 651, 653, 669–670, 674–676
Moerman D.E. 698
Mohamed, Mashayihei 馬沙亦 黑馬哈麻 483
Mohan, G. 608
Mongolian medicine 537
Monnais, L. 498
monthly ordinances 81

movement of the channel or pulse (*dongmai* 動脈) 34
moxa-cautery or moxibustion 灸 36, 54, 58, 63, 116, 121, 135, 149, 182, 189, 194, *197*, 238, 239, 287, 306, *313*, 402, 411, 470n5, 485, 497, 504–507, 513, 520, 521, 578, 709; *see also* acu-moxa
Mu Zhaosi 穆昭嗣 480

Nagata Tokuhon 永田徳本 518
Nagoya Gen'i 名古屋玄医 518
Nakayama Tadanao 中山忠直 520
Nanjing 南京 decade (1927–37): public health 661–662
Nanjing 難經 (Classic of Difficult Issues) 101, 106n29–31, 130n1, 163, 201, 209, 232, 240n9, 359
Nanyue Kingdom (Southern Yue) 南越 60
Nappi, C. 151
Naquin, S. 265, 266
National Certification Commission for Acupuncture and Oriental Medicine (NCCAOM) 580
Nationalist Party rule 649
National Quarantine Service 653
National Sex Civilisation Survey (Zhongguo xing wenming diaocha baogao) 393
native plants to China 672, 726
naturalist-realist method 248–255, 268–271; of end-of-Ming epidemics as only plague 268–271; of end-of-Ming epidemics as plague 267–268
Needham, J. 2, 253, 322, 334, 385, 389, 390, 553, 676, 677
neidan 內丹 (internal or inner alchemy) 152, 427–428, 429, 431, 441n2; 'chariot' metaphor 437, *437*; *Chart of the Inner Warp* 436, 437; greater celestial circuit 438–439; lesser celestial circuit (*dumai* 督脈 and *renmai* 任脈 vessels) 435–436, 441n3; time in 434–441, 447, 690, 695, 697
Neijing 內經 (Inner Canon) *see Huangdi neijing* 黃帝內經
neike 內科 (internal medicine) 206, 209
Neo-Confucianism 148
New Compilation of Materia Medica (*Bencao congxin* 本草叢新) *see Bencao* 本草 editions
New Culture 383
New Edition of Universal Relief (*Jaejung sinpyeon* 濟眾新編) 508
Newly Revised Materia Medica (*Xinxiubencao* 新修本草) *see Bencao* 本草 editions
New Significance of the History of Disease in China (Fan) 264
Nguyễn Thị Dương 496, 499
Nguyễn Trần Huân 497
Nguyen, V.N. 569, 570, 572
Niboyet, J.E.H. 571

Index

Nieuhof, J. 552, 559
night blindness 223
Nine Fascicles (*Jiujuan* 九卷) **121**, 123
Nixon, R. 577
Ni Zan 倪瓚 334, 335
nong 膿, management and elimination of 206
nongzhi 膿脂 (pus and oil-fat) 212
non-reproductive sexual behaviour 341
non-same-sex partners 341
non-sexual alchemy 449; *see also* alchemy
Northern and Southern Dynasties 136
Northern Song dynasty 112, 126, 147, 148, 171, 192, 195, 196, 274, 332
Northern Wei dynasty 111, 136
North Manchurian Plague Prevention Service 653
numbing wind (*mafeng* 痲瘋) 233, 273, 274
numerology: capitals, use of 84n1, 105n2; and Chinese medicine 91–92; cultural elements 72; and divination 73–74; Eastern Han, number six 82–84; five and six in medieval medicine 97–101; four and five in classical medicine 94–97; *Huangdi neijing Suwen* 74–75; microcosm and macrocosm 92–94; numbers and imperial ritual 80–82; *Numinous Tablets of the Secret Instructions on the Yin Elixir Orally Transmitted by the Perfected Person of Wangwu* (*Wangwu zhenren koushou yindan mijue lingpian* 王屋真人口授陰丹祕訣靈篇) 341; Qin, number five 80–82; Shang, number four 76–78; short history in Chinese thought 75–76; *shushu* 數術 category 73–74; of Warring States, number two 78–80; Western Han, number five 80–82; *Wuyun Liuqi* 五運六氣 102–105; *Yiwenzhi* 藝文志 (Treatise on Literature) category of 'Recipes and Methods' 73
Nurhaci 努爾哈赤 269
nutrient deficiency diseases 223
nutrition 320–322, 324–326, 328, 659, 673, 723
nutritional prohibitions 304
nutritive or camp *qi see Qi* 氣; *ying qi* 營氣

objectivity, rise of 175–176
Obringer, F. 273
Ohsawa, G. 520
One Child Policy 393
One Hundred and One Supplementary Formulas to Keep up One's Sleeve see Buque zhouhou baiyi fang 補闕肘後百一方
Ono Ranzan 小野蘭山 517
Open Door Policy 393
Opium War 651
original *qi see Qi* 氣; *yuan qi* 元氣
Ots, T. 37, 39, 691
Ou Xifan's Diagrams of the Five Organs see Ou Xifan wuzang tu 歐希範五藏圖
Ou Xifan wuzang tu 歐希範五藏圖 (Ou Xifan's Diagrams of the Five Organs) 211

Overseas *Materia Medica* (*Haiyao bencao* 海藥本草) *see Bencao* 本草 editions

Packard, R. 249–251; *The Making of a Tropical Disease: a short history of malaria* 250
Palmer, D. 37, 690, 691, 699
palpation 4, 31, 163, 177n1, 209, 518, 552, 554, 558, 560, 590
Pan Guangdan 潘光旦 383
Pan Mingde 潘明德 211
Pan Shizheng 潘師正 114
Pan Suiming 潘綏銘 394, 395
Pandemic Disease in the Medieval World: Rethinking the Black Death (Green) 275–276
Pang Anshi 龐安時 104, 147
paozhi 炮製 (processing [of medicinals]) 484, 675, 679
Parker, P. 650, 651
Pasteur, L. 659
pathogen 85n17, 117n1, 233, 247
pathogenesis 19, 42
pathology 99, 102, 139–141, 171, 207–208, 234, 239, 593
patient information leaflet (PIL) 730
Pei Du 裴度 222
Peitzman, S.: *Dropsy, Dialysis, Transplant: a short history of failing kidneys* 250
Peking Union Medical College (PUMC) 653
pensiveness (*si* 思) 236
Pents'aology (*Bencao xue* 本草學) *see Bencao* 本草 editions
People's Republic of China (PRC) 585, 649, 663
Persian medicine: Chinese Institutions and Markets 482–484; importing knowledge through translations and tales 477–479; lasting influence 484–485; local texts and uses of western drugs 479–482; *materia medica*, translated terms for 485–489; Sassanid Empire of Persia 476–477; travellers, traders and texts 475–477
personalised medicine 217
Phạm Đình Hổ 496, 497
Phan, T. 582
pharmacology, pre-standardised: formative era, the earliest texts 134–136; second stage, expansion and systematisation of knowledge 136–138; third stage, standardising drug therapy and integrating with canonical doctrines 138–141
Pharmacopoeia of the Great Yuan (*Dayuan bencao* 大元本草) *see Bencao* 本草 editions
pharmacopoeia or drug code *see bencao*
pharmacotherapy 286–287; external and simple recipes 286; fakes and tricks 286; pharmaceutical substances 286
phenomenology 29, 37, 45n2, 67
Philippines: acupuncture in 593–596; Communist Party of the Philippines (CPP)

594; Community-based Health Programs (CBHP) 594, 595; 'Morong 43' 595; New People's Army (NPA) 594–595; *Partido Komunista ng Pilipinas* (PKP) 594
phoenixes soaring 345
Phùng Thị 馮氏 *see* Feng Zhaozhuang 馮兆張
pi 皮 (skin) 206
piercing in magico-religious context 287
Pimpaneau, J. 390
pinzhong 品種 (grades [and] types [of medicinal products]; variety) 675
Piwei lun 脾胃論 150
plague 250, 252, 254, 256; in China since 2011, scientific transformations 275–276; naturalist-realist critique of end-of-Ming epidemics as only 268–271; *see also* epidemics
Plague and the City (Engelmann, Henderson, and Lynteris) 276
Plague Prevention and Politics in Manchuria, 1910–1931 (Nathan) 253
political-history perspective 265
polytomy 275
poxes 225
practices of spirits and religious orthodoxies 4–5
pregnancy 196, 223, 291, 311–314, 316, 324–325
pre-modern disease categories 222
pre-natal (*xiantian* 先天) *qi* 33, 35; *see also* Qi 氣
Prescriptions for Fifty-Two Ailments see Wushier bingfang 五十二病方
Prescriptions of the Public Pharmacy of the Era of Great Peace and of the Bureau of Medicines see Taiping huimin hejiju fang 太平惠民和劑局方
Prescriptions Worth a Thousand Gold Pieces for Urgent Situations see Beiji qianjin yaofang
Pritzker, S. 39, 581, 582, 615
prohibitions: medical (*see* medical prohibitions); nutritional 304
proven remedies or efficacious formulæ 134
public health: early People's Republic (1949–76) 663–664; early republic (1911–27) 659–661; germ theory 659; Nanjing decade (1927–37) 661–662; reform era (1976–2000) 664–665; in twentieth-century China 659; wartime (1937–49) 662–663
Pulling Book see Yinshu 引書
pulse (*mai* 脈) 34, 55, **57**, **126**, 152, 156n17, 182–184, 209–210, 213–214, 221, 232, 237, 241n40, 309, 485, 507, 508, 551–561, 591, 689; pulse taking 163–177; *see also mai* 脈
pulse diagnostics: conflict with European conceptions of body 167–170; cumbersome mastery 165–166; *cunkou* 寸口 (inch opening) 165, 209; European medical theory 168–169, *169*; floating pulse *vs.* flooding pulse 166; *hanxie* 寒邪 (cold evil) 166; limits of pulse diagnosis in acute medical scenarios 166–167;

mai 170; pulse taking *168, 169*; *rexie* 熱邪 (heat evil) 166; socioeconomic influences 167; tongue qualities 166–167; twenty-four diagnostic states of pulse 166; vocabulary for 209

pus: 'pus and blood' 208; 'swelling blood' 208

qi 氣 3; abbreviation for variety of different types of *qi* 40; absorbing during sex practice (*shi shenqi* 食神氣) 341; categories in *Inner Canon*, by Jiang Shan 39–43; chaotic (*luan* 亂) 34, 404; characteristic or tendency 41; counterpart of blood (*xie* 血) 41; in Chinese medicine 208; control (*zhi* 治) 34; cultivation of 8; deviant *qi* (*xieqi* 邪氣) 33, 41, 117n1, *hanxie* 寒邪 (cold evil) and *rexie* 熱邪 (heat evil) 166; defensive or guard *qi* (*wei qi* 衛氣) 33; of drugs (*siqi* 四氣: hot, warm, cold, and cool) 134; formless elements of body, as opposed to physical form 40; four seasons (*siqi* 四氣) 105; functioning of any body parts 40; in early medical writing 30–32; emerging concepts 26–27; epistemological itch 37–39; essence *qi* and spirit (*jing qi shen* 精氣神) 427–428, 451; essential *qi* (*jingqi* 精氣) 14, 33; and five agents, political cosmos 29–30 (*see also wuxing* 五行); grammar 24; graphs with fire component 25, *25*; graphs with grain component 25, *25*; of Heaven and Earth 83–84; imaging and sounding *24*, 24–26; in imperial medicine 33–35; internal sensations in body 41; intestinal gas 42–43; as medium for self-cultivation 27–29; modern forms 35–37; movement of the channel or pulse (*dongmai* 動脈) 34; nutritive, supply or camp *qi* (*yingqi* 營氣) 33, 40; original *qi* (*yuan qi* 元氣) 33, 212, 446, 451; overall term for various *qi* that flowed through body 41; pathogenesis and pathomechanism 42; pre-natal (*xiantian* 先天) *qi* 33, 152, 428; *qi* in network of channels (*jingmai* 經脈, *jingluo* 經絡) 33, 40; *qi*-talk 39; rebellious (*ni* 逆) 34, 123, 404, 427; response to needling 42; and *spirit* (*shen* 神) 33; substances in nature 42; true *qi* (*zhenqi* 真氣) 33, 438; typologies organised by number **43–44**; upright or orthodox *qi* (*zhengqi* 正氣) 33

Qianjin fang 千金方 or *Qianjin yaofang* 千金要方 *see Beiji qianjin yaofang* 備急千金要方
Qianyi 錢乙 148
Qian Yunzhi 錢允治 334
Qi Bo (or Qibo) 岐伯 18, 40, 41
qie 切 (palpation of the pulse) 163
qigong 氣功 5, 23, 36–38, 117, 452, 564, 645, 690–691, 695, 697, 699
qigong re 氣功熱 (*qigong* fever) 37, 690–691
Qijing bamai kao 奇經八脈考 152

760

Index

Qin dynasty: earliest textual record of prescriptions dates to 134; numbers and imperial ritual 80–82
Qing Code (Daqing lüli 大清律例) 234
Qing dynasty: case records 186–187; evidential scholarship movement 152–153; force of doctrine 153; non-sexual practices and *nüdan* 女丹 449; synthesis 153–155; Warm Disease 153–154; Wu Tang 吳塘154–155; Xue Xue 薛雪 154; Ye Gui 葉桂 153
Qinshi Huangdi 秦始皇帝 55, 66; *see also* King Zheng of Qin 秦王政
Qiu Jun 丘濬 192
Quanguo zhongcaoyao huibian 全國中草藥匯編 (Compilation of Chinese Herbal Medicinals) 679, 681
quinine 7, 226, 656, 707
Qu Limin 曲黎敏 695

Rashīd al–Dīn Faẓl Allāh Hamadānī 2, 167, 485; translations of medical works 2
rat epidemics 274
Ravalomanana, M. 607
Rawski, E.S. 265, 266
al-Rāzī, Muḥammad ibn Zakariya 481, 483, 484
Read, Bernard E. 伊博恩 304, 672, 669, 676–678
Reagan, Ronald 579
rebellious (*ni* 逆) 34
received texts 20, 58, 74, 91, 182, 421
Recipes for Fifty-two Disorders see Wushier bingfang 五十二病方
Recipes for Nurturing Life see Yangsheng fang 養生方
recipe traditions 3
Reconstructing Treasured Mirror of Eastern Medicine see Uigam jungma 醫鑑重磨
Records on Nourishing the Disposition and Prolonging the Mandate of Life see Yangxing yanming lu 養性延命錄
Red Emperor or Divine Husbandman *see* Shennong 神農 (Divine Farmer)
Regulations for Medical Doctors see Uisa gyujeuk
Regulations for Medical Students see Uisaeng gyuchik
reishi (lingzhi 靈芝) 734–735
religious healing 4, 287, 494, 638
remedy collections 3, 63, 66, 708
Republic of China (ROC) 531, 633
The Re-revised Materia Medica *from the Kaibao Period* (*Kaibao chongding bencao* 開寶重定本草) *see Kaibao bencao* 開寶本草 in Bencao 本草 editions
retrograde ejaculation 344
Rhapsody on the Great Pleasure in the Mutual Joys of Heaven and Earth, Yin and Yang see Tiandi yinyang jiaohuan dale fu 天地陰陽交歡大樂賦
Richardson, C. 579

rishu 日書 54, 64–65, 74, 93; *see also* daybooks
rituals 292
ritual techniques 219–220
Rocher, Émile 251, 252, 255
Rockefeller Foundation (RF) 632, 661
Roh Moo-Hyun 511
root diseases (*benbing* 本病) 149
Rosenberg, C.E. 249, 250
Royle, J.F. 484
Rubin, G. 356
Rumphius, Georg Eberhard 554
Russo-Japanese War 651
ruyi 儒醫 (scholar-physicians) 148, 650; *see also* scholar-physicians
Ryle, G. 698

Sābūr ibn Sahli 沙卜而撒哈里 483
Sadava, D. 576
Saillant, C.J. 560–562
Sallet, A. 497
Śamatha 115
Samuel Baron 497
San Francisco 3, 582
sanjiao 三焦 101
Sanyin jiyi bingzheng fanglun 三因極一病證方論 (Treatise on Three Categories of Pathogenic Factors) 224
SARS epidemic, 2002–2003 247–248
Schatz, Jean 572
Scheid, V. 39, 187, 641
Schiebinger, L.L. 356
Schipper, K.M. 264, 266, 390
scholar-physicians 148–149, 151, 153, 156n13, 284–285, 466, 468–469, 650
Schonebaum, A. 282
scientific transformations of plague in China 275–276
Scott, J.W. 356, 357, 365
Second Opium War 651
second Sino-Japanese War (1937–45) 7
Secret Decisions in the Jade Chamber see Yufang mijue 玉房祕訣
Secrets of the Jade Chamber see Yufang mijue 玉房祕訣
self-cultivation, practices of 66
self-strengthening movement 628, 651
self strengthening reform movement 7
Separate Records of Famous Physicians see Mingyi bielu 名醫別錄
severe acute respiratory syndrome (SARS) 664
sex and modernity: case of 'doctor sex' 382–384; Communist China and Cultural Revolution 390–393; discourse of *Xing*: modern terms for modern ideas 381–382; Eastern *ars erotica* 389; fate of tradition 384–385; 'Healing Tao' 390; heritage and happiness 393–395; sexual alchemy in modern China 384–385;

spirituality of postmodern age 389–390; Western *scientia sexualis* 389
Sex Histories (*Xingshi* 性史) 382
sexing Chinese medical body: feminist historiographies of medicine for women (*see* women's medicine); gendering history of Chinese medicine 360–365; *Yellow Emperor's* body 357–360
sexual alchemy 384–385; in Ming period 446–448
sexual blindness 392
sexual body techniques: combinations thereof with visualisations 343; ejaculation, 'short-term satisfaction' 343; female ejaculation and female essence 345–346; free-handed method 343, 344; general aspects of sexual scenario 340–342; 'injaculation' and 'sexual energy' 340; male essence and male ejaculation 342–345; methods to avoid emission and ejaculation 343; non-reproductive sexual behaviour 341; non-same-sex partners 341; performing sexual encounter 346–348; 'sexual vampirism' 342; textual sources to sexual knowledge culture 337–340; urethral pressure method 343, 344; *vajrolī mudrā* (lightning seal) 342
sexual cultivation *see fangzhong shu* 房中術
sexual encounter: 'five sounds' or vocalisations 347–348; foreplay, or 'playful ways' 346; linking passage in *Tianxia zhidao tan* 天下至道談 347; onset phase, 'approaching the matter' or 'approaching the ride' 346–347; performing 346–348
Sexual Knowledge see Xing de zhishi 性的知識
sexually transmitted diseases in modern China 266
sexual practices in early period 445–446; *Huangshu* 445–446; *Huangting jing* (Scripture of the Yellow Court) 445; Shangqing tradition 446
sexual revolution 393
sexual vampirism 342
Shadow of the Double Plum Tree Anthology see Shuangmei jing'an congshu 雙梅景闇叢書
Shang dynasty 19, 181, 218, 369, 401
Shang king 76, 78, 94, 305
Shang Yang 商鞅 343
Shang Zhijun 尚志鈞 480
Shang: Four Emblematic animals 77, 77; four lands 76–77; Han Dynasty 77; sexagenary cycle **76**, 76–78; ten Heavenly Stems 76; twelve Earthly Branches 76; Xishuipo 西水坡 tomb (M45) in Henan 77–78, 78
shanghan 傷寒 (cold damage) 4, 171, 209, 221, 223, 248, 266, 560

Shanghan leizheng huoren shu 傷寒類證活人書 (Lifesaving Book that Categorises the Presentations of Cold Damage) 147; *see also shanghan* 傷寒; *Shanghan lun* 傷寒論
Shanghan lun 傷寒論 (Treatise on Cold Damage) 123, 135, 146, 211, 509, 518, 554–555, 728
Shanghan zongbing lun 傷寒總病論 104, 147; *see also shanghan* 傷寒, *Shanghan lun* 傷寒論
Shangqing 上清 114, 344, 409, 446
Sheep's Wool heat epidemic 271
Shen Nong's Materia Medica, *Compiled and Annotated by Lei Gong* (*Lei Gong jizhu Shennong bencao* 雷公集注神農本草) *see Bencao* 本草 editions
Shen 神 (spirit or spirits) 246, 249, 256, 272; early 1640s epidemics, Zhejiang province 246; excessive taxation 248
Shengji zonglu 聖濟總錄 (Comprehensive Record of Sagely Beneficence) 140, 142, 224
Shennong bencao 神農本草 (*Shennong bencao jing* 神農本草經) *see Bencao* 本草 editions
Shennong 神農 (Divine Farmer) 51, 311, 321, 333, 409, 464, 466, 485, 708, 712, 713; legendary patron of earliest *materia medica* 3
Shi Cangyong 石藏用 192, 195; *The Study of Contagious Diseases in China* 253
shi 式 or *shipan* 式盤 81–82
Shiji 史記 (Record of the Grand Historian) 206
Shijing 詩經 (Book of Songs) 13
shiliao 食療 *see* food and dietary medicine
Shishan yi'an 石山醫案 (The Stone Mountain Medical Case Histories) 363
shōgun Yoshimune 八代將軍吉宗 517
Shōmu tennō 聖武天皇 515
Short Essay on Classical Formulae (*Jing fang xiaopin* 經方小品) 125
Shozen Kajihara 梶原性全 196fi
Shuangbaoshan 雙包山 *see* manuscripts and artefacts
Shuanggudui 雙古堆, tomb no. 1 *see* manuscripts and artefacts
Shuangmei jing'an congshu 雙梅景闇叢書 (Shadow of the Double Plum Tree Anthology) (Ye Dehui 葉德輝) 384, 385
Shuihudi 睡虎地 tomb 11; *see also* manuscripts and artefacts
Shujing 書經 'Great Plan' 92, 93
Shushu 數術 5, 73–75, 81
sickness and healing 3–4
sida 四大 (The Four Great Elements; Sk. *mahābhūta*) 417
Sigismond III 553
Siku quanshu 四庫全書 153, 497
silk manuscript 78, 85n25, 109, 219, 708; *see also* manuscripts and artefacts
Silk Road 115, 138, 476

Sima Chengzhen 司馬承禎 114
Sima Qian 司馬遷 63, 96, 163, 182, 220, 308, 404
simple questions *see Huangdi neijing* 黃帝內經
Singapore: actions by religious organisations 534; Chinese medicine 524–525; Medical Plan 526; Medical Registration Ordinance of 1905 525; Poisons Act 526; Republic of China (ROC) 531; Republic of Singapore 526; research institutes 533; roles of professional associations 528, **529–530**, 531–534; Singapore Chinese Physicians' Association (SCPA) 526, 531–532; state laws and policies 525–528; TCM associations 524, 527, *527*, **529–530**; training system 528, 533
Sino-British Opium War 374
Sinographic medicine: diversity of interlinked traditions 5–6
Sino-Iranica (Laufer) 475
Sino-Japanese War 7, 659
Sivin, N. 122, 265, 292
Six Dynasties period 136, 138
Six *qi* 24, 27, 44, 75, 83–84, 91, 94, 101–104, 149, 222
sixty hexagrams, daily cycle 431
Skinner, G. W. 265, 267
smallpox 225, 269–271
Smith, H.A. 276
social norms 3
Società Italiana di Agopuntura (SIA) 572
Società Italiana di Ricerca in Agopuntura ed Auricoloterapia (SIRAA) 572
Société d'Acupuncture (SA) 567, 568
Société Française d'Acupuncture (SFA) 567–568
Société Internationale d'Acupuncture (SIA) 568
Song dynasty: Anglophone scholarship 146–147; flowering of Cold Damage 147; four great masters of Jin-Yuan 149–151; governmental oversight 147; Jin and Yuan dynasties 148; mass printing 125; medical texts 126, **126**; record keeping in 183–184; scholarly medicine 148; systematisation in diagnostics and pharmacology 148–149; Taisu 139; technology of woodblock printing 125; *Treatise on Cold Damage* 127
Song Yingxing 宋應星 35
sorrow (*you* 憂) 236
Soulié de Morant, G. 520, 566–569, 571
Southeast Asia 138, 421, 476
sovereign herb 708
spells 64, 219, 291, 293
Spivak, G.C. 391
Standard for Diagnosis and Treatment see Zhengzhi zhunsheng 證治準繩
state institutions 514, 653, 656
statutes and regulations 111, 570

Stein, M.A. 62
Stone, Mary 374
The Stone Mountain Medical Case Histories see Shishan yi'an 石山醫案
Sugiyama Wa'ichi 杉山和一 518
Sui dynasty 109–119, 123, 136, 222–223, 385, 408, 419, 421, 479, 514
Sun Ruzhong 孫汝忠 447
Sun Simiao 孫思邈 111, 115, 125, 138, 164, 190, 221, 223, 305, 311, 312, 314, 323, 324, 332, 333, 339, 362, 363, 370, 371, 477, 514, 715
Sun Yat-sen 7
Sun Yikui 孫一奎 172, 185, 187, 233, 236
Sunü jing 素女經 (The Book of the Plain Woman) 339, 346, 385
Supplement to the Pharmacopoeia (*Bencao shiyi* 本草拾遺) *see Bencao* 本草 editions
Supplement to the Systematic Materia Medica (*Bencao shiyi* 本草拾遺) *see Bencao* 本草 editions
supply *qi see ying qi* 營氣
surgery: Cold Damage or febrile diseases 209; haemorrhoid 7; 'internal sores' (*neichuang* 內瘡) 209; minor 288; Pus 208; sweating 208
suturing wounds 7
Suwen 素問 74–75, 123–124; Heaven and Earth 79–80, 83; movement and human bodily fluids, relationship between 207; numerology and divination 73–74; recension, commentaries and evolution of 75; *Suwen* 2 92, 105n4, 106n48; *Suwen* 3 94; *Suwen* 4 92, 94; *Suwen* 5 **19**, 92, 103, 105n3, 105n11, 105n18; *Suwen* 6 94; *Suwen* 8 94, 105n12; *Suwen* 9 94, 106n46; *Suwen* 22 **19**; *Suwen* 25 **19**; *Suwen* 66 103, 106n38, 106n39; *Suwen* 67 103, 106n38; *Suwen* 68 106n39; *Suwen* 69 103, 106n44; *Suwen* 71 106n47; *Suwen* 74 106n49; totality of ideal human lifespan 92; Wang Bing 王冰 75, 85n16; *Wuyun liuqi* 五運六氣 tradition 75–76
sweating 208
Synopsis of the Medical World see Uilim chalyo 醫林撮要
Synthesis of Past and Present Books and Illustrations see Gujin tushu jicheng 古今圖書集成
Systematic Classic of Acumoxa see Zhenjiu jiayi jing 針灸甲乙經
Systematic Differentiation of Warm Disease see Wenbing tiaobian 溫病條辨
Systematic Materia Medica (*Bencao gangmu* 本草綱目) *see Bencao* 本草 editions
systematisation 95, 97, 136–138, 141, 148–153

taboos 3; *see also* medical prohibitions
Tabori, P. 391
Tacuinum Sanitatis (The *Taqwim* of Health) 333

Taichan shu 胎產書 (Book of The Generation of the Fetus) 369
taiji 117, 580, 697, 709
Taipei 3
Taiping huimin hejiju fang 太平惠民和劑局方 (Prescriptions of the Public Pharmacy of the Era of Great Peace and of the Bureau of Medicines) 147, 224
Taiping shenghuifang 太平聖惠方 (Imperial Grace Formulary of the Great Peace and Prosperity Reign Period) 224
Taishō Tripitaka 424n1
Taisu 太素 recensions **121**, 124, **126**, 139, 187n2, 187n3, 240n10, 240n12, 241n42; *see also* *Huangdi neijing* 黃帝內經
Taiwan Food and Drug Administration (TFDA) 732
Taiyi 太乙 (Grand Unity) 305–306
Taiyi ju 太醫局 (Imperial Medical Service) 171
Taiyin yangming lun 太陰陽明論 (On the theory of Taiyin and Yangming) 207
Taiyishu 太醫署 (Imperial Medical Office) 111
Taizong 太宗 (Tang) 478
Taizong 太宗 (Song) 139
Takashi Hōyoku 高志鳳翼 518
Takeda Shōkei 竹田昌慶 516
Talismans amulets 287, 291–293; carving of talismans 466; 'Divine Tree that Heals Disease' 465, *465*; elaborate talismans 461, *463*; healing talismans 461; Heavenly Physician 463, *464*; manuscripts 463; nineteenth-century medicinal jar 463, *464*; physical support 465; secret characters 461–462, *462*; Talisman and Incantation (*fuzhou* 符咒) 410, *411*; thirteen categories of illness 462–463
Tamba Yasuyori 丹波康賴 138, 195, 385, 516
Tan Mingxun 譚銘勳 392
Tan-Mullins, M. 608
Tang bencao 唐本草 (Tang Dynasty *Materia Medica*) *see Bencao* 本草 editions; *Xinxiu bencao* 新修本草)
Tang dynasty 136
Tang Materia Medica (*Tang bencao* 唐本草) *see Bencao* 本草 editions; *Xinxiu bencao* 新修本草
Tang Shenwei 唐慎微 330, 482
Tang Zonghai 唐宗海 35, 170
Tanksuqnama 167, 177n.
Tānksūqnāmeh 485
Tansuqnama-i ilkhan dar funun-I'ulum-i khatayi (The Treasure Book of the Ilkhan on Chinese Science and Techniques) 2, 167 *See also* Tanksuqnama; Tānksūqnāmeh
Tao Hongjing 陶弘景 114, 135–137, 225, 329–330, 708, 711, 712, 715, 716
Taoist Ritual and Popular Cults of South-East China (Dean) 266

The Tao of Love and Sex: The Ancient Chinese Way to Ecstasy (Chang) 390
Tao Zongyi 陶宗儀 483
Tashiro Sanki 田代三喜 516
TCM emergence of in 1950s 8, 577, 656; and Alma Ata 594, 630–635; and Barefoot Doctors 638–645; and Daoism 221, and gynaecology 368–369; and decontextualised Chinese medicine 721–736; drugs and regulation 732; and drug therapy 709; enduring relevance of 217; and Kim Taylor 565; in Korea 511; the limits of 9; and Mayan medicine 591–593; in modern China 72; and modernisation and standardisation 723–725; and official regulation 545; and pharmacopoeia 670–671; pre-evidence based medicine 456, 459; and *Qi* 36; in Singapore 524–535; and social status 649; and translation 616–618; and *yangsheng* 693
techniques of meditation 66
Temkin, O. 266
Temple of Heaven (*Tiantan* 天壇) 81
Temple, W. 559
Ten Interviews or Ten Questions (*Shiwen* 十問) 57, 338, 341, 343
Ten Rhyne, W. 553
therapeutic exercise 3; *anmo* 111; aristocrats 113–114; Buddhists 115; Daoists 114; *daoyin* exercises 109–110, 112–113; medical teaching staff at Imperial Medical Academy 111, **111**; physicians 115–117; process of unification and centralisation 111–112; State medical education 111; Sui dynasty 110–111; *yangsheng* practices 110, 112, *112*; *see also daoyin* 導引
therapies 286–293
Thevenot, M. 728
Thompson, C.M. 495, 498
Three Kingdoms Era 688
three-word-principle 294n9
Tiandi yinyang jiaohuan dale fu 天地陰陽交歡大樂賦 (Rhapsody on the Great Pleasure in the Mutual Joys of Heaven and Earth, Yin and Yang) 340
Tianhui 天回 *see* Laoguanshan 老官山
Tibetan medicine 537, 732
time in Chinese Alchemy: application to alchemy 429; *Daode jing* 道德經 sequence 428; emblematic time cycles 431–433; 'matching stems,' monthly cycle 431–432, **432**; *neidan* 內丹 or Internal Alchemy 427–428, 434–439; post-celestial to pre-celestial 428–429; sixty hexagrams, daily cycle 431; time as metaphor 439–441; 'Twelve-stage Ebb and Flow': Yearly Cycle 432–433, **433**; *waidan* or External Alchemy 427–428, 433–434; *Yijing* 易經 sequence 428–429, 430
Tokyo 3
tonics 325, 713, 729, 733–735

tongue diagnosis: *Guang wenyi lun* 廣瘟疫論 (Expanded Treatise on Febrile Epidemics) 173; *Linzheng zhinan yi'an* 臨証指南醫案 (Medical Case Records as a Guide to Clinical Practice) 173; *Mingyi lei'an* 名醫類案 (Classified Case Records) 173; rise of 165; *Shishan yi'an* 石山醫案 (Stone Mountain Medical Case Records) 172–173; *Touding qingshen san* 透頂清神散 171, *172*; white tongue 173–174; *zhen wu tang* 真武湯 (True Warrior Decoction) 173
tongue drawings 171
toxic mire (*duxian* 毒陷) 210
trade regulation 674–675
Traditional Chinese Medicine (TCM) 8, 565; clinics 649; 'drug' – *yao* 133; emergence of 638–641; evidence-based 456; growth of 368; limits of 9; success of 656–657
Traditional Herbal Medicinal Products Directive (THMPD) 730
traditional herbal registration (THR) 730
Trần Nghĩa 495
translated modernity 383
translation: different bodies 614–616; evidence-based medicine (EBM) 616; power and politics 617–619; shifting cultural contexts 616–617; in West 613–614
traumatology 209, 283–285, 290
Treasured Mirror of Eastern Medicine (*Dongui bogam* 東醫寶鑑) 505–508, 517
Treatise on Cold Damage see *Shanghan lun* 傷寒論
Treatise on Cold Damage and Miscellaneous Disorders (*Shanghan zabing lun* 傷寒雜病論) see *Shanghan lun* 傷寒論
Treatise on Easy Childbirth see *Dasheng bian* 達生編
Treatise on Febrile Epidemics see *Wenyi lun* 瘟疫論
Treatise on the Origins and Manifestations of the Myriad Diseases see *Zhubing yuanhou lun* 諸病源候論
Treatise on Three Categories of Pathogenic Factors see *Sanyin jiyi bingzheng fanglun* 三因極一病證方論
Treaty of Tianjin 6
Treaty of Versailles 383, 634
True *qi* (or True breath, *zhenqi* 真氣) 33, 43, 124, 438; *see also qi* 氣
tuberculosis 273, 664
Tuệ Tĩnh 497–499
tuina 推拿 202, 204n3, 288, 295n20, 564; *tuina* manuscripts 289; *see also anmo* 按摩
Tujing bencao 圖經本草 (*Bencao tujing* 本草圖經) see *Bencao* 本草 editions
Tu Youyou 屠呦呦 461, 707, 724
Twelve-stage Ebb and Flow: Yearly Cycle 432–433, **433**

two-dimensional diagrams or *Mingtang* diagrams 202
two-sex model of human bodies 359

Uigam jungma 醫鑑重磨 (Reconstructing Treasured Mirror of Eastern Medicine) 508
Uilim chalyo 醫林撮要 (Synopsis of the Medical World) 504, 506
Uisaeng gyuchik 醫生規則 (Regulations for Medical Students) 510
Uisa gyujeuk 醫士規則 (Regulations for Medical Doctors) 510
ulcers 185, 209–213, 463, 468, 714
Uniting Yin and Yang (*He yinyang* 合陰陽) 57, 338, 343, 345, 348
Unschuld, P.U. 122, 151, 282, 284, 616, 631, 712
upright, proper or orthodox *qi* (*zhengqi* 正氣) 33, 42; *see also qi* 氣
urethral pressure method 343, 344
US Food and Drug Administration 580
Uyghur medicine 537

vajrolī mudrā (lightning seal) 342
van Gulik, R.H. 389, 390; *Sexual Life in Ancient China* 389
Various Materia Medica of the Taiqing 太清 Era see *Bencao* 本草 editions; *Taiqing zhu bencao* 太清諸本草
Verran, H. 37
Vessel Book see *Maishu* 脈書
vessels: damage to *yangming* (Yang Brightness) vessel 231; lesser celestial circuit (*dumai* 督脈 and *renmai* 任脈 vessels) 435–436, 441n3
Vietnam in pre-modern period: defined 493; famous doctors 498–499; further research 499; historiography 497–498; medicine in 494; 'northern medicine' 493; northern (and southern) visions of south 494–495; Red River delta 493; traces of medical history 495–497
Vipaśyanā 115
visual diagnosis see *wang* 望
Visual Representations of the Third Plague Pandemic project 276
Vossius, I. 559

waidan 外丹 (external alchemy) 427–428, 431, 433–434, 436, 437; time in 433–434
waike 外科 (the curriculum of external medicine) see surgery
Waitai miyao 外台秘要 126, 127, 138, 142, 223, 477, 514
Wakeman, F., Jr. 269
Wang Bing 王冰 16, 75, 84, 85n16, 91, 97, 101, 103, 104, 124, 309; *Suwen* 85n16
Wang Chong 王充 308
Wang Guowei 王國維 62

Wang Haogu 王好古 140, 149, 150, 332, 334
Wang Ji 汪機 172, 363
Wang Kentang 王肯堂 166, 233–234, 236, 239
Wang Qixian 王奇賢 63
Wang Shixiong 王士雄 154, 211
Wang Shuhe 王叔和 165, 166, 485, 554, 558
Wang Shuo 王碩 224
Wang Tao 王濤 138, 190, 223, 514
Wang Wenbin 王文彬 392
Wang Xiaotao 王孝淘 672, 679
Wang Xuequan 王學權 212
Wang Yanchang 王燕昌 213
Wang Ying 王穎 334
Wang Yirong 王懿榮 213
Wang Yuanzhi 王遠知 114, 117
Wang Zuosu 王作肅 198
wang 望 (visual inspection, visual diagnosis, patient complexion) 163, 164, 170–171
Ward, B. 699
warlord period 649
warm disease (*wenbing* 溫病) 153–154, 254, 256, 272, 506; epidemics 247–248
the way of the tyrant (*badao* 霸道) 212
wayward women 449–451
Web That Has No Weaver (Kaptchuk) 36
Wei Zhixiu 魏之琇 186
wen 問 (asking, propensity for particular flavours) 163
wen 聞 (auditory, the timbre of the voice) 163
Wenbing tiaobian 溫病條辨 (Systematic Differentiation of Warm Disease) 154
Wendi of the Sui dynasty 隋文帝 55, 110–113, 115–117
Wenyi lun 瘟疫論 (Treatise on Febrile Epidemics) 246
Western-educated intellectuals 395
Western Jin 61, 113
Western Liang dynasty 116
Western medicine 7, 226; advent of 638–639
Western Pharmacopoeia (*Hu bencao* 胡本草) *see Bencao* 本草 editions
Wilson, A. 248–249, 267
wind (*feng*) 風 19; absent-mindedness (*shi xin feng* 失心風 234; black sand fright wind (*wusha jingfeng* 烏沙驚風) 288; caused by fire 173; changing orthography 234, 247, 266, 273–274, 286, 288, 314, 407, 417, 470n5, 509, 561, 590, 593, 713, 725, 732; fright-wind (*jingfeng* 驚風) 213, 233, 236, 286; in the *Huangdi neijing* 黃帝內經 40, 42, 46n27; invasion and depletion 220; leprosy: numbing wind (*mafeng* 痲瘋) or great wind (*dafeng* 大風) 233, 273–74; as madness: *feng* 瘋 230–242 (*see also* madness); one of the six pathogens 26, 33, 83–84, **94**, 101, **102**, 103, 111; prevented by *daoyin* 導引 114; as a primary disease category 222–224; shift away from single-pathogen diseases to *qi* transformation 148; wind malady: *dian* 癲, *kuang* 狂 and *xian* 癇 230–234, 239; wind stroke (*zhongfeng* 中風) 233

Wiseman, N. 615, 618
Wittie, R. 728
womanhood 361–362, 374–376
women's 'flower *dian*' 花癲 (love madness) 231, 237
women's medicine: Blood and *qi* 362; cacophony of voices 360; construction of sexed body 361–362; deconstructing female difference 362–365; feminist historiographies of *fuke* 婦科 361–364, 366n7; *Golden Mirror* 363; 'infinitive' body 364; medicalisation of menstruation 362; Ming-Qing medicine 363; post-Han to Song dynasty 361–362; post-Song to Qing 362–365; womb in Chinese medicine 363, 366n9
woodblock printing, technology of 125
World Health Organization (WHO) 524; adoption of China's model for healthcare integration 633; Barefoot Doctors programme 633–634; integration of TCM 633–635; May Fourth Uprising 634; PRC in 633; Treaty of Versailles 634
World Health Organization 524, 599, 602–604, 610, 618, 625–626, 633, 645
Worthington, A. 176
Wotton, W. 559
wounds 7, 134, 207–211, 213, 223, 288
Wu Hai 吳海 210
Wu Jingda 吳景達 116
Wu Kun 吳崑 236
Wu Lien-teh (or Wu Liande) 伍連德 252, 254, 255, 256, 652, 660
Wu Tang 吳塘 154–155, 238
Wu Youxing 吳有性 153, 226, 246, 253, 254, 256, 272, 273
Wu Zetian 武則天 170, 515
Wu, Y.-L *see* Yi-Li Wu
Wuliangguang fo 無量光佛 (Buddha of Infinite Life; Sk. Amitābha) 418
Wushier bing fang 五十二病方 (Fifty-two Remedies; also known as Prescriptions for Fifty-Two Ailments and Recipes for Fifty-two Disorders) 134–135, 232, 322
Wuyun liuqi 五運六氣: Cold Damage theory 104; Five Movements and Six *qi* theory 91, 102–103; practical applications of 104–105

xian 仙 techniques 114, 117n1
Xianbei 鮮卑 113
xiangke 相剋 or *xiangsheng* 相勝 (mutual conquest) cycle 81, 92, 103, 308–310
xiangsheng 相生 (mutual generation) cycle 92, 93, 103, 308–310

Index

Xie Guan 謝觀 183
xieqi 邪氣 (deviant/evil *qi*) *see* Qi 氣
xiesui 邪祟 469
Xie Zhufan 謝竹藩 618
Xie Zongwan 謝宗萬 672, 679
Xi Jinping 習近平 585, 693
Xing de zhishi 性的知識 (Sexual Knowledge) 392–393
Xingqi ming 行氣銘 27
Xiongnu 匈奴 113
Xu Cheng 許澄 116
Xu Chunfu 徐春甫 233
Xu Dachun 徐大椿 150, 210, 469
Xu Guozhen 許國禎 481
Xu Shen 許慎 13, 24, 26
Xu Shiheng 徐士衡 462
Xu Shuwei 許叔微 150, 184
Xu Sibo 徐嗣伯 233
Xu Wei 徐渭 235
Xu Yinzong 許胤宗 116
Xu Zhicang 許智藏 116
Xuannü jing 玄女經 (The Book of the Dark Woman) 339
Xue Daoguang 薛道光 439
Xue Ji 薛己 372
Xue Xue 薛雪 154

Yan Fu 嚴復 36, 38
Yan Zhitui 顏之推 114
Yang Chongrui 楊崇瑞 375, 661
Yang Zijian 楊子建 371
Yangdi 煬帝 of the Sui dynasty 109–117
yangsheng 養生 (nourishing life or cultivation of life) 28, 32, 110, **112**, 113–115, **116**, 117n5, 117n8, 206, 211, 338, 370, 390, 404, 438, 446, 479, 645; *Chameleon Yangsheng* 691–696; 'collaboration' or 'collusion' theory 687, 699–700; Evidenced-Based Medicine 698; and *Guoxue* 國學 696; learning processes 697; marketing of lifestyles and services 700; 'nature' and 'nurture' 698; neuroplasticity 697; neuroscience 698; One Child Policy 696; pleasure 701; postscript 701; practices 114, 116, **116**; psycho-plasticity 698–699; and *qi* 689–690; *and qigong fever* 690–691; translations 687; in twenty-first century; Warring States Era 688; Western Han dynasty 688
Yangsheng fang 養生方 (Recipes for Nurturing Life) **57**, 112, 338
Yangsheng yaoji 養生要集 (Compendium of Essentials for Nourishing Life) 110, 113
Yangxing yanming lu 養性延命錄 (Records on Nourishing the Disposition and Prolonging the Mandate of Life) 339
Yangzi bing 癢子病 252
Yao Kecheng 姚可成 334

Yaodian 藥典 (The [Chinese] Pharmacopoeia) 544, 670
Yaoshifo 藥師佛 (The Master of Medicines Buddha; Sk. Bhaiṣajyaguru) 418
Yasuo Yuasa 湯浅泰雄 37, 39
Ye Dehui 葉德輝 339, 384, 385, 394, 395; *Shadow of the Double Plum Tree Anthology* 384, 385
Ye Gui 葉桂 150, 153–155, 186, 187; *see also* Ye Tianshi 葉天士
Ye Tianshi 葉天士 173, 186
Yellow Emperor 2, 3, 18, 29, 51, 53, 62, 92, 123, 124, 128, 321, 363, 404
Yellow Emperor's body: abundance of 'Kidney *qi*' 357–358; body, rethinking sexual difference 357–360; difference between male and female bodies 358–359; *locus classicus* of Chinese medicine 360; relationship of Blood and Essence 358, *359*; sexual maturation of boys and girls 357; 'two-sex model' of human bodies 359; *see also Huangdi neijing*
Yellow Emperor's Inner Classic (*Huangdi neijing*) *see Huangdi neijing* 黃帝內經
Yellow Emperor's Inner Classic, Plain Questions (*Huangdi neijing: suwen*) *see Huangdi neijing* 黃帝內經
Yellow Emperor's Systematic Classic of Acumoxa in Three Parts (*Huangdi sanbu zhenjiu jiayi jing* 黃帝三部針灸甲乙經) *see Zhenjiu jiayi jing* 針灸甲乙經
Yersin, Alexandre É.J. 251
Yersinia pestis 264, 275
Yi Eng 一雲 390
Yi Jing 義淨 419–420, 478
Yi-Li Wu 283, 363, 364
yi'an 醫案 (case records) 181–183; compilation of patient notes 182; Encyclopaedias of 186; medical knowledge and diagnostic practices 182–183; Ming dynasty case records 185–186; physician Xu Shuwei 184–185; Qing dynasty case records 186–187; record keeping in the Song 183–184; *Shuihudi* 睡虎地 tomb 11 182; *Zhangjiashan* 張家山 tomb 181–182
Yijing 易經 (Book of Changes) 80, 84, 85n20, 428–429, *430*, 431
yin 陰 and *yang* 陽 17–18; *Basic Questions*, examples 16–17; of body in section four of *Basic Questions* **17**; to body physiology 4; 'complementary opposition' 322; correspondence in Mawangdui manuscripts and *Cheng* **16**; correspondences between imperial bureaucratic structures and twelve viscera **18**; excavated early Han manuscripts, examples 13–16; Five Circuits 103; four-fold system 80, *80*; Fuxi 伏羲 and Nüwa 女媧 79, *79*; Heaven and Earth 79–80, 91–92; medicine of systematic correspondence

767

in Han China 19–20; principles of 1, 3; relationship of Blood and Essence 358, *359*; sexual maturation of boys and girls 357–358

ying qi (or *yingqi*) 營氣 33, 40, 760

Yinqueshan 銀雀山 tomb 79

Yinshu 引書 14, 30–32, 55, 106n23, 110

Yizong jinjian 醫宗金鑑 (Golden Mirror of Medical Learning) 211, 212

Yoshimasu Tōdō 吉益東洞 518

Yu Chang 喻昌 153, 208

Yu Fan 虞翻 432

Yu Yan 余巖 653, 654

Yu Yue 俞樾 654

Yuan dynasty 171, 256, 372, 481–482

yuan qi 元氣 (primordial or original *qi*) 33, 212, 446, 451; *see also qi* 氣

Yüfang mijue 玉房秘訣 (Secret [Decisions] in the Jade Chamber) **339**, 339–340, 342–343

Yüfang zhiyao 玉房指要 339, 344, 385

Yuzuan yizong jinjian 御纂醫宗金鑒 (Imperially Commissioned Golden Mirror of Medicine) 363

zang fu (or *zangfu*) 臟腑 36, 149, 192, 196, 201, 203, 225, 558

Zang fu zhizhang tu shu 藏府指掌圖書 (Illustrated Guide to the *Zangfu* Organs) 212

zaqi 雜氣 226, 247

Zen monk Mubun 無分 518

Zeng Zao 曾慥 446

Zhang Boduan 張伯端 439, 440

Zhang Congzheng 張從正 149, 150, 238, 467

Zhang Daoling 張道林 406

Zhang Ding 張鼎 332

Zhang Ji 張機 127, 135, 146, 221, 224, 232, 266, 310, 370, 518, 519, 728; *see also* Zhang Zhongjing 張仲景

Zhang Jiebin 張介賓 152, 194, 236, 363, 496, 497; *see also* Zhang Jingyue 張景岳

Zhang Jingsheng 張競生 382–384

Zhang Jingyue 張景岳 152

Zhang Lei 張耒 63

Zhang Lu 張璐 225

Zhang Qicheng 張其成 692, 695–697

Zhang Sanfeng 張三丰 447

Zhang Yuansu 張元素 149, 150

Zhang Zhan 張湛 110, 113

Zhang Zhongjing 張仲景 127, 135, 310, 509, 518–519, 519, 728

Zhangjiashan 張家山, tomb 247 *see* manuscripts and artefacts

zhanzhuang 站樁 38

Zhao Hu 趙胡 60

Zhao Mei 趙眛 60, 66, 67

Zhao Yuanyi 趙元益 484

Zhao Yuhuang 趙燏黃 669, 672, 677, 679

Zhao Zhiyi 趙志一 392

Zheng Jinsheng 郑金生 282, 291

Zhenghe shengji zonglu 政和聖濟總錄 *see Shengji zonglu* 聖濟總錄

Zhenghe 政和 Reign Period 140

Zhengzhi zhunsheng 證治準繩 (Standard for Diagnosis and Treatment) 236

Zhenjiu jiayi jing 針灸甲乙經 (Systematic Classic of Acumoxa) **121**, 124, **126**, 126–127; *Huangdi jiayi jing* 黃帝甲乙經 (Numbered Classic of the Yellow Emperor) 515

Zhiyi 智顗 115, 420

Zhong-Lü 鍾呂 439, 440

Zhongjing-[style] external medicine 211

Zhongti xiyong (Chinese essence, Western utility) 中體西用 7, 654

zhou 咒 (incantation, spell; Sk. *dhāraṇī*) 418; *see also* incantations (*fuzhou* 符咒)

Zhou Dingwang 周定王 484

Zhou Jianren 周建人 383

Zhou Zuliang 周祖亮 6

Zhoujiatai 周家臺 *see* manuscripts and artefacts

Zhu Danxi 朱丹溪 *see* Zhu Zhenheng 朱震亨

Zhu Feiyuan 朱費元 212, 213

Zhu Gong 朱肱 147, 196

Zhu Lian 朱璉 (36

Zhu Siben 朱思本 600

Zhu Xi 朱熹 35, 150, 151, 517

zhuyou 祝由 238, 284n1, 292, 295n23, 315, 401, 410, 412n9; *see also* incantation

Zhu Zhenheng 朱震亨 149–152, 224, 234, 235, 372, 516

Zhubing yuanhou lun 諸病源候論 (Treatise on the Origins and Symptoms of Medical Disorders, 610) 109, 222, 223, 370, 514

Zhujie shanghan lun 注解傷寒論 (Annotation and Explanation of the Discussion of Cold Damage) 147

Zidanku 子彈庫 silk manuscript 78

zigong 子宮 (uterus, womb) 211–212, 366

Zou Yan 騶衍 29, 67, 81

Zuo Maodi 左懋第 269